Up and Running with AutoCAD 2022

Up and Running with AutoCAD 2022

2D and 3D Drawing, Design and Modeling

Elliot J. Gindis
President - Vertical Technologies, Los Angeles, CA, United States

Robert C. Kaebisch
Construction Sciences - Architecture/Structural, Gateway Technical College
Sturtevant, WI, United States

ELSEVIER
ACADEMIC PRESS
An imprint of Elsevier

Academic Press is an imprint of Elsevier
125 London Wall, London EC2Y 5AS, United Kingdom
525 B Street, Suite 1650, San Diego, CA 92101, United States
50 Hampshire Street, 5th Floor, Cambridge, MA 02139, United States
The Boulevard, Langford Lane, Kidlington, Oxford OX5 1GB, United Kingdom

Library of Congress Cataloging-in-Publication Data
A catalog record for this book is available from the Library of Congress

British Library Cataloguing-in-Publication Data
A catalogue record for this book is available from the British Library

ISBN: 978-0-323-89923-9

For information on all Academic Press publications visit our website at https://www.elsevier.com/books-and-journals

Publisher: Katey Birtcher
Acquisitions Editor: Steve Merken
Editorial Project Manager: Michelle Fisher
Publishing Services Manager: Shereen Jameel
Production Project Manager: Rukmani Krishnan
Cover Designer: Alan Studholme

Typeset by TNQ Technologies

Contents

Level 1
Answers to review questions

Level 2
Chapters 11–20

11. Advanced linework

12. Advanced layers

Spotlight on: electrical engineering

13. Advanced dimensions

14. Options, Shortcuts, CUI, Design Center, and Express Tools

Spotlight on: drafting, CAD management, teaching, and consulting

About the authors

ELLIOT GINDIS

Elliot J. Gindis started out using AutoCAD professionally at a New York City area civil engineering company in September 1996, moving on to consulting work shortly afterward. He has since drafted in a wide variety of fields ranging from all aspects of architecture and building design to electrical, mechanical, civil, structural, aerospace, and rail design. These assignments, including lengthy stays with IBM and Siemens Transportation Systems, totaled over 50 companies to date.

In 1999, Elliot began teaching part-time at the Pratt Institute of Design, followed by positions at Netcom Information Systems, RoboTECH CAD Solutions, and more recently at the New York Institute of Technology. In 2003, Elliot formed Vertical Technologies Education and Training https://www.verticaltechtraining.com/, an AutoCAD training firm that continues to train corporate clients nationwide in using and optimizing AutoCAD.

Elliot holds a bachelor's degree in aerospace engineering from Embry-Riddle Aeronautical University and a master's degree in engineering management from Mercer University. As of 2016, he resides in the Los Angeles area and is a flight test engineer for the United States Air Force. He also continues to be involved with AutoCAD education and CAD consulting. *Up and Running with AutoCAD 2022* carefully incorporates lessons learned from over 20 years of teaching and industry work. His 2012 textbook has also been translated and is available in Spanish from Anaya Multimedia. He can be reached at Elliot.Gindis@gmail.com or via his website: www.verticaltechtraining.com.

Previous textbooks by Elliot Gindis

English

- *Operational AutoCAD 2008.* New York: Netcom Inc. (out of print)
- *Up and Running with AutoCAD 2009* [e-book only]. Waltham, MA: Elsevier, Inc.
- *Up and Running with AutoCAD 2010.* Waltham, MA: Elsevier, Inc.
- *Up and Running with AutoCAD 2011: 2D Drawing and Modeling.* Waltham, MA: Elsevier, Inc.
- *Up and Running with AutoCAD 2011: 2D and 3D Drawing and Modeling.* Waltham, MA: Elsevier, Inc.
- *Up and Running with AutoCAD 2012: 2D Drawing and Modeling.* Waltham, MA: Elsevier, Inc.
- *Up and Running with AutoCAD 2012: 2D and 3D Drawing and Modeling.* Waltham, MA: Elsevier, Inc.

- *Up and Running with AutoCAD 2013: 2D Drawing and Modeling.* Waltham, MA: Elsevier, Inc.
- *Up and Running with AutoCAD 2013: 2D and 3D Drawing and Modeling.* Waltham, MA: Elsevier, Inc.
- *Up and Running with AutoCAD 2014: 2D and 3D Drawing and Modeling.* Waltham, MA: Elsevier, Inc.
- *Up and Running with AutoCAD 2015: 2D and 3D Drawing and Modeling.* Waltham, MA: Elsevier, Inc.
- *Up and Running with AutoCAD 2016: 2D and 3D Drawing and Modeling.* Waltham, MA: Elsevier, Inc.
- *Up and Running with AutoCAD 2017: 2D and 3D Drawing and Modeling.* Waltham, MA: Elsevier, Inc.
- *Up and Running with AutoCAD 2018: 2D Drafting and Design.* Waltham, MA: Elsevier, Inc.
- *Up and Running with AutoCAD 2019: 2D Drafting and Design.* Waltham, MA: Elsevier, Inc.
- *Up and Running with AutoCAD 2020: 2D Drafting and Design.* Waltham, MA: Elsevier, Inc.
- *Up and Running with AutoCAD 2021: 2D and 3D Drawing, Design and Modeling.* Waltham, MA: Elsevier, Inc.

Spanish

- *AutoCAD 2012 - Dibujar y Modelar en 2D y 3D.* Madrid, Spain: Anaya Multimedia.

ROBERT KAEBISCH

Robert C. Kaebisch started using AutoCAD in 1989 with R10 and continued using each subsequent release through college and throughout his career. Most of his drafting work has involved architecture, interior design, mechanical, electrical, plumbing, and other construction-related industries, from schematic designs through full construction documents.

After starting his architectural career in the mid-1990s, and spending several years in the field, Robert also worked as a part-time adjunct instructor in the School of Design and Drafting at IIT in 2001. In 2004, he stepped away from teaching to focus on architecture and obtained his architect's license in 2005. In 2008 he returned to teaching and became a full-time instructor in Gateway Technical College's Architectural-Structural Engineering Technician program. He continues to teach and be involved in architectural design and consulting, as well as AutoCAD, Revit, and other software training.

Robert holds a bachelor's degree in architecture from the University of Wisconsin–Milwaukee's School of Architecture and Urban Planning, and a certificate in Project Management from the University of Wisconsin-Stout. He maintains his architect's license, technical college teaching license in WI, and national certifications in both Autodesk AutoCAD and Autodesk Revit. He currently lives in the Milwaukee area and can be reached at Robert.Kaebisch@gmail.com.

Previous textbooks by Robert Kaebisch

- *Up and Running with AutoCAD 2018: 2D Drafting and Design.* Waltham, MA: Elsevier, Inc.
- *Up and Running with AutoCAD 2019: 2D Drafting and Design.* Waltham, MA: Elsevier, Inc.
- *Up and Running with AutoCAD 2020: 2D Drafting and Design.* Waltham, MA: Elsevier, Inc.
- *Up and Running with AutoCAD 2021: 2D and 3D Drawing, Design and Modeling.* Waltham, MA: Elsevier, Inc.

Preface

What is Autocad?

AutoCAD is a drafting and design software package developed and marketed by Autodesk, Inc. As of 2021, it has been around for approximately 39 years—several lifetimes in the software industry. It has grown from modest beginnings to an industry standard, often imitated, sometimes exceeded, but never equaled. The basic premise of its design is simple and is the main reason for AutoCAD's success. Anything you can think of, you can draw quickly and easily. For many years, AutoCAD remained a superb 2D electronic drafting board, replacing the pencil and paper for an entire generation of technical professionals. In recent releases, its 3D capabilities finally matured, and AutoCAD is now also considered an excellent 3D visualization tool, especially for architecture and interior design.

The software has a rather steep learning curve to become an expert but a surprisingly easy one to just get started. Most important, it is well worth learning. This is truly global software that has been adopted by millions of architects, designers, and engineers worldwide. Over the years, Autodesk expanded this reach by introducing add-on packages that customize AutoCAD for industry-specific tasks, such as electrical, civil, and mechanical engineering. However, underneath all these add-ons is still plain AutoCAD. This software remains hugely popular. Learn it well, as it is still one of the best skills you can add to your resume.

About this book

This book is not like most on the market. While many authors certainly view their text as unique and novel in its approach, we rarely reviewed one that was clear to a beginner student and distilled AutoCAD concepts down to basic, easy to understand explanations. The problem may be that many of the available books are written by either industry technical experts or teachers but rarely by someone who is actively both. One really needs to interact with the industry and the students, in equal measure, to bridge the gap between reality and the classroom.

After years of AutoCAD design work in the daytime and teaching nights and weekends, we set out to create a set of classroom notes that outlined, in an easy to understand manner, exactly how AutoCAD is used and applied, not theoretical musings or clinical descriptions of the commands. These notes eventually were expanded into this book that you now hold. The rationale was simple: An employer needs this person to be up and running as soon as possible to do a job. How do we make this happen?

Teaching methods

My teaching approach has its roots in a certain philosophy I developed while attending engineering school many years ago. While there, I had sometimes been frustrated with the complex presentation of what in retrospect amounted to rather simple topics. My favorite quote was, "Most ideas in engineering are not that hard to understand but often become so upon explanation." The moral of that quote was that concepts can usually be distilled to their essence and explained in an easy and straightforward manner. That is the job of a teacher: Not to blow away students with technical expertise but to use experience and top-level knowledge to sort out what is important and what is secondary and to explain the essentials in plain language.

Such is the approach to this AutoCAD book. I want everything here to be highly practical and easy to understand. There are few descriptions of procedures or commands that are rarely used in practice. If we talk about it, you will likely need it. The first thing you must learn is how to draw a line. You see this command on the first few pages of Chapter 1, not buried somewhere in later sections. It is essential to present the "core" of AutoCAD, essential knowledge common to just about any drafting situation, all of it meant to get you up and running quickly. This stripped-down approach proved effective in the classroom and was carefully incorporated into this text.

Text organization

This book comes in three parts: Level 1, Level 2, and Level 3:

Level 1 (Chapters 1–10) is meant to give you a wide breadth of knowledge on many topics, a sort of "mile wide" approach. These 10 chapters comprise, in my experience, the complete essential knowledge set of an intermediate user. You then can work on, if not necessarily set up and manage, moderate to complex drawings. If your CAD requirements are modest or if you are not required to draft full time, then this is where you stop.

Level 2 (Chapters 11–20) is meant for advanced users who are CAD managers, full-time AutoCAD draftspersons, architects, or self-employed and must do everything themselves. The goal here is depth, as many features not deemed critically important in Level 1 are revisited to explore additional advanced options. Also introduced are advanced topics necessary to set up and manage complex drawings.

Level 3 (Chapters 21–30) is all about 3D... Solid knowledge of the previous two levels is highly recommended before starting these chapters. The 3D material covers all aspects of AutoCAD solid modeling including lights and rendering.

Throughout this book, the following methods are used to present material:

- Explain the new concept or command and why it is important.
- Cover the command step by step (if needed), with your input and AutoCAD responses shown so you can follow and learn them.
- Give you a chance to apply just-learned knowledge to a real-life exercise, drawing, or model.
- Test yourself with end-of-chapter quizzes and drawing exercises that ask questions about the essential knowledge.

You will not see an extensive array of distracting "learning aids" in this text. You will, however, see some common features throughout, such as.

Commands: These are presented in almost all cases in the form of a command matrix, such as the one shown here for a line. You can choose any of the methods for entering the command.	*Keyboard:* Type in **line** and press Enter *Ribbon:* **Home tab→Line**
Tips and tricks: These are seen mostly in the first few chapters and one is shown here. They are very specific, deliberate suggestions to smooth out the learning experience. Do take note.	**TIP 1:** The Esc (escape) key in the upper left-hand corner of your keyboard is your new best friend. It gets you out of just about any trouble you get yourself into. If something does not look right, just press the Esc key and repeat the command. Mine was worn out learning AutoCAD, so expect to use it often.
Step-by-step instructions: These are featured whenever practical and show you exactly how to execute the command, such as the example with line here. What you type in and what AutoCAD says are in the software's default font: Courier New. The rest of the steps are in the standard print font.	**Step 1.** Begin the line command via any of the previous methods. • AutoCAD says: Specify first point: **Step 2.** Using the mouse, left-click anywhere on the screen. • AutoCAD says: Specify next point or [Undo]: **Step 3.** Move the mouse elsewhere on the screen and left-click again. You can repeat step 2 as many times as you wish. When you are done, click Enter or Esc.
Learning objectives and time for completion: Each chapter begins with this, which builds a "road map" for you to follow while progressing through the chapter, as well as sets expectations of what you will learn if you put in the time to go through the chapter. The time for completion is based on classroom teaching experience but is only an estimate. If you are learning AutoCAD in school, your instructor may choose to cover part of a chapter or more than one at a time.	In this chapter, we introduce AutoCAD and discuss the following: • Introduction and the basic commands • The Create Objects commands • The Edit and Modify Objects commands • The View Objects commands, etc. By the end of the chapter, you will ... Estimated time for completion of chapter: 3 hours.
Summary, review questions, exercises: Each chapter concludes with these. Be sure to not skip these pages and to review everything you learned.	**Summary** **Review questions** **Exercises**

What your goal should be

Just learning commands is not enough; you need to see the big picture and truly understand AutoCAD and how it functions for it to become effortless and transparent. The focus after all is on your design. AutoCAD is just one of the tools to realize it.

A good analogy is ice hockey. Professional hockey players do not think about skating; to them, it is second nature. They are focused on strategy, scoring a goal, and getting by the defenders. This mentality should be yours as well. You must become proficient through study and practice, to the point where you are working with AutoCAD, not struggling against it. Using the software becomes "transparent" and you focus only on the design, not the tools and techniques, to truly perform the best architecture or engineering work of which you are capable.

If you are in an instructor-led class, take good notes. If you are self-studying from this text, pay very close attention to every topic; nothing here is unimportant. Do not skip or cut corners. Do complete every drawing assignment. Most important, you must practice, daily if possible, as there is no substitute for sitting down and using the software. Not everyone can learn on the job while getting paid; companies want ready-made experts and do not want to wait. If that is the case, you must practice on your own in the evening or on weekends. Just taking a class or reading this book alone is not enough.

It may seem like a big mountain to climb right now, but it is completely doable. Once on top, you will find that AutoCAD is not the frustrating program it may have seemed in the early days but an intuitive software package that, with proficiency of use, becomes a natural extension of your mind when working on a new design. That, in the end, is the mark of successful software; it helps you do your job easier and faster. Feel free to contact myself or my coauthor with questions or comments at Elliot.Gindis@gmail.com and Robert.Kaebisch@gmail.com. Also, be sure to visit www.verticaltechtraining.com, where you will find other resources such as exercises, videos, articles, and a discussion board to get your AutoCAD questions answered.

Instructors using this book can access and download LMS-compatible chapter review questions, as well as figures from the book, by registering at our Elsevier Educate website: https://inspectioncopy.elsevier.com/book/details/9780323899239.

Acknowledgments

A textbook of this magnitude is rarely a product of only one person's effort. I thank all the early and ongoing reviewers of this text and Chris Ramirez of Vertical Technologies Consulting for research and ideas when most needed as well as using the text in his classroom. A big thank you also to Karen Miletsky (retired) formerly of Pratt Institute of Design, Russell and Titu Sarder at Netcom Information Technology, and everyone at the New York Institute of Technology, RoboTECH CAD Solutions, Future Media Concepts, and other premier training centers, colleges, and universities for their past and present support.

Extensive gratitude also goes to Aravindakshan Joseph P. Hayton, Todd Green, Stephen R. Merken, Peter Jardim, Jeff Freeland, Kathleen Chaney, Gnomi Schrift Gouldin, Kiruthika Govindaraju, Maria Ines Cruz, Grace Lander, and the rest of the team at Elsevier for believing in the project and for their invaluable support in getting the book out to market. Thank you also to Denis Cadu of Autodesk for all the support at the Autodesk Developers Network.

A big thank you to my coauthor, Robert Kaebisch of Gateway Technical College, for joining the team and assisting in bringing the 2018 edition to the next level and continuing his excellent work for the 2019, 2020, 2021 and now the 2022 book.

Finally, I thank my friends and family, especially my parents, Boris and Tatyana Gindis, for their patience and encouragement as I undertake the annual effort of bringing this textbook to life. This book is dedicated to the hundreds of students who have passed through my classrooms and made teaching the enjoyable adventure it has become.

Elliot J. Gindis
May 2021

First and foremost, I must thank Elliot Gindis for asking me to join him on this journey in 2017 and trusting me to help with the evolution of his creation over the years. Thank you to the teachers and professors who gave me the freedom to explore technical drafting, AutoCAD, architecture and all their associated technologies: Mr. Peter Wilson, Mr. Paul Barry, and Mr. Rick Jules.

Thank you to the team at Elsevier, especially Stephen R. Merken and Grace Lander, for promoting the project and helping us getting it published this year. Thank you also to Denis Cadu and the team at Autodesk for all the help from the Autodesk Developers Network.

I could not do the work I do without the support of the Construction Science faculty: Ray Koukari, Jr., Pat Hoppe, and Steve Whitmoyer. Another big thank you to the Gateway Technical College administration: Dr. Bryan D. Albrecht, Zina Haywood, Matt Janisin, and Dr. John Thibodeau for helping me grow as an instructor.

Finally, I must thank my children, Alex and Emma, my mom Sharon Kaebisch, and my amazing wife, Dr. Sandra Van Den Heuvel for their continued support. This journey is proving to be a great learning experience, making me a better architect, instructor, and learner.

Robert C. Kaebisch
May 2021

Level 1

Chapters 1–10

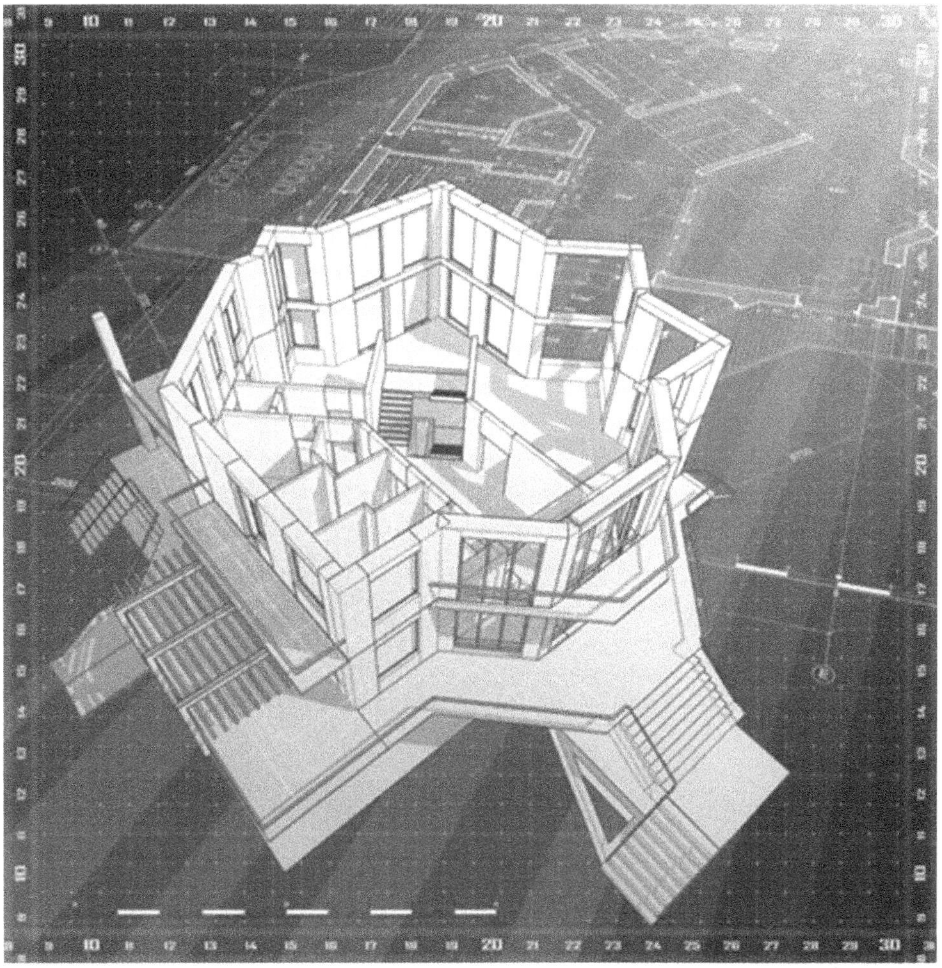

Level 1 is the very beginning of your studies. No prior knowledge of AutoCAD is assumed, only basic familiarity with computers and some technical aptitude. You are at a slight advantage if you have hand drafted before, as many AutoCAD techniques flow from the old paper and pencil days, a fact alluded to later in the chapters.

We begin Chapter 1 by outlining the basic commands under Create Objects and Modify Objects followed by an introduction to the AutoCAD environment. We then introduce basic accuracy tools of Ortho and OSNAP.

Chapter 2 continues the basics by adding units and various data entry tools. These first two chapters are critically important, as success here ensures you will understand the rest and be able to function in the AutoCAD environment.

Chapter 3 continues to layers, then each succeeding chapter continues to deal with one or more major topics per chapter: text and mtext in Chapter 4, hatching in Chapter 5, and dimensioning in Chapter 6. In these six chapters, you are asked to not only practice what you learned but apply the knowledge to a basic architectural floor plan. Chapter 7 introduces blocks and Chapter 8 arrays. At this point, you are asked to draw another project, this time a mechanical device. Level 1 concludes with basic printing and output in Chapter 9 and finally advanced printing and output (Layouts/Paper Space) in Chapter 10.

Be sure to dedicate as much time as possible to practicing what you learn; there really is no substitute.

Chapter 1

AutoCAD fundamentals—Part I

Learning objectives

In this chapter, we introduce AutoCAD and discuss the following:

- Introduction and the basic commands
- The Create Objects commands
- The Edit/Modify Objects commands
- The View Objects commands
- The AutoCAD environment
- Interacting with AutoCAD
- Practicing the Create Objects commands
- Practicing the Edit/Modify Objects commands
- Selection methods—Window, Crossing, and Lasso
- Accuracy in drafting—Ortho (F8)
- Accuracy in drafting—OSNAPs (F3)

Up and Running with AutoCAD 2022. https://doi.org/10.1016/B978-0-323-89923-9.00001-9

By the end of this chapter, you will learn the essential basics of creating, modifying, and viewing objects; the AutoCAD environment; and accuracy in the form of straight lines and precise alignment of geometric objects via OSNAP points.

Estimated time for completion of this chapter: 3 hours.

1.1 Introduction and basic commands

AutoCAD is a very complex program. If you are taking a class or reading this textbook, this is something you probably already know. The commands available to you, along with their submenus and various options, number in the thousands. So, how do you get a handle on them and begin using the software? Well, you must realize two important facts.

First, you must understand that, on a typical workday, *95% of your AutoCAD drafting time is spent using only 5% of the available commands, over and over again.* So, getting started is easy; you need to learn only a handful of key commands; and as you progress and build confidence, you can add depth to your knowledge by learning new ones.

Second, you must understand that even the most complex drawing is essentially *made up of only a few fundamental objects that appear over and over again* in various combinations on the screen. Once you learn how to create and edit them, you can draw surprisingly quickly. Understanding these facts is the key to learning the software. We are going to strip away the perceived complexities of AutoCAD and reduce it to its essential core. Let us go ahead now and develop the list of the basic commands.

For a moment, view AutoCAD as a fancy electronic hand-drafting board. In the old days of pencil, eraser, and T-square, what was the simplest thing that you could draft on a blank sheet of paper? That, of course, is a line. Let us make a list with the following header, "Create Objects," and below it add "line."

So, what other geometric objects can we draw? Think of basic building blocks, those that cannot be broken down any further. A *circle* qualifies and so does an *arc*. Because it is so common and useful, throw in a *rectangle* as well (even though you should note that it is a compound polyline object, or four lines acting as one). Here is the final list of fundamental objects that we have just come up with:

Create Objects

- Line
- Circle
- Arc
- Rectangle

As surprising as it may sound, these four objects, in large quantities, make up most of a typical design, so already you have the basic tools. We will create these on the AutoCAD screen in a bit. For now, let us keep going and get the rest of the list down.

Now that you have the objects, what can you do with them? You can *erase* them, which is probably the most obvious. You can also *move* them around your screen and, in a similar manner, *copy* them. The objects can *rotate*, and you can also *scale* them up or down in size. With lines, if they are too long, you can *trim* them, and if they are too short, you can *extend* them. *Offset* is a sort of precise copy and one of the most useful commands in AutoCAD. *Mirror* is used, as the name implies, to make a mirror-image copy of an object. Finally, *fillet* is used to put a curve on two intersecting lines, among other things. We will learn a few more useful commands a bit later, but for now, under the header "Edit/Modify Objects," list the commands just mentioned:

Edit/Modify Objects

- Erase
- Move
- Copy
- Rotate
- Scale
- Trim
- Extend
- Offset
- Mirror
- Fillet

Once again, as surprising as it may sound, this short list represents almost the entire set of basic Edit/Modify Objects commands you need once you begin to draft. Start memorizing them.

To finish up, let us add several View Objects commands. With AutoCAD, unlike paper hand drafting, you do not always see your whole design in front of you. You may need to *zoom* in for a close-up or out to see the big picture. You may also need to *pan* around to view other parts of the drawing. With a wheeled mouse, virtually universal on computers these days, it is very easy to do both, as we soon see. To this list we add the *regen* command. It stands for *regenerate*, and it simply refreshes your screen, something you may find useful later. So here is the list for View Objects:

View Objects

- Zoom
- Pan
- Regen

So, this is it for now, just 17 commands making up the basic set. Here is what you need to do:

1. As mentioned before, memorize them so you always know what you have available.
2. Understand the basic idea, if not the details, behind each command. This should be easy to do because (except for maybe *offset* and *fillet*) the commands are intuitive and not cryptic in any way; *erase* means erase, whether it is AutoCAD, a marker on a whiteboard, or a pencil line.

We are ready now to start AutoCAD, discuss how to interact with the program, and try out all the commands.

1.2 The AutoCAD environment

It is assumed that your computer, whether at home, school, work, or training class, is loaded with AutoCAD. It is also assumed that AutoCAD starts up just fine (via the AutoCAD desktop icon or Start menu) and everything is configured "as-is" with the default setup. If not, ask your instructor for help, as there are just too many variables on any PC or laptop, and it is beyond the scope of this book to cover these unique situations.

Startup AutoCAD and if all is well, you should see the Start tab welcome screen depicted in Fig. 1.1. This is a change from older versions of AutoCAD, which defaulted to a blank drawing area when first opened. We do not spend too much

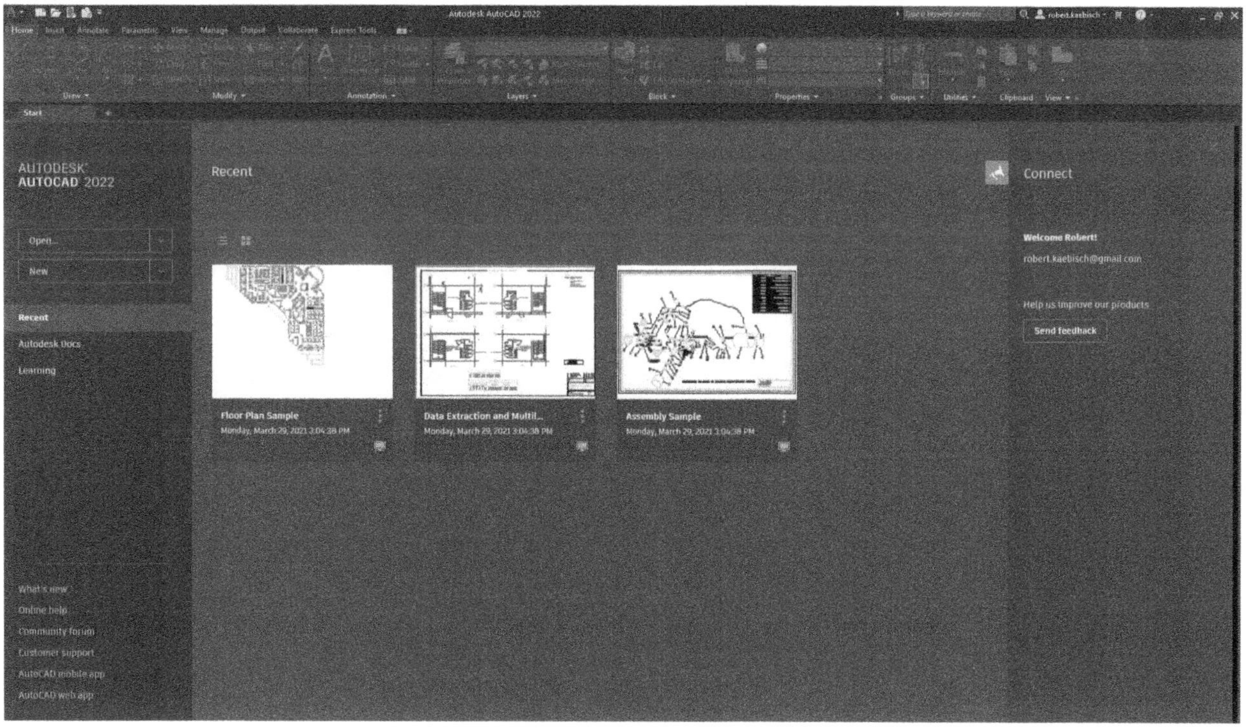

FIGURE 1.1 AutoCAD—Start tab.

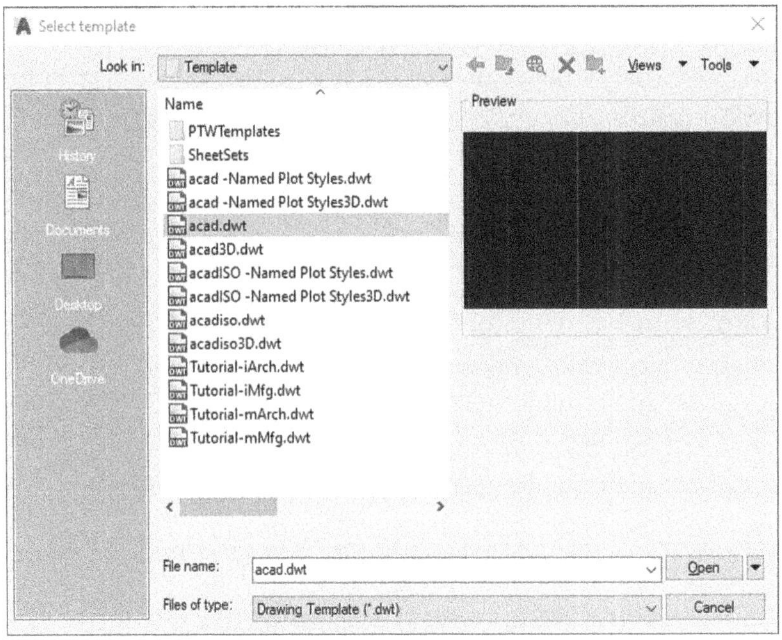

FIGURE 1.2 Select template—acad.dwt.

time here but note the following. This tab contains the *Get Started* column (with the huge Start Drawing button), the *Recent Documents* column (with some sample files or recent files for quick access), and *Notifications/Connect* column (where you can get product updates and give feedback to Autodesk). Take a quick tour of these if you wish, and when you are ready to open a new blank file, press the "sheet of paper with the folded corner" icon at the very top left of the screen near the Big A (application menu); the universal icon for a new file. A Select template dialog box opens, already set to open an acad.dwt file (Fig. 1.2). Press the Open button at the lower right, and a blank file opens (Fig. 1.3).

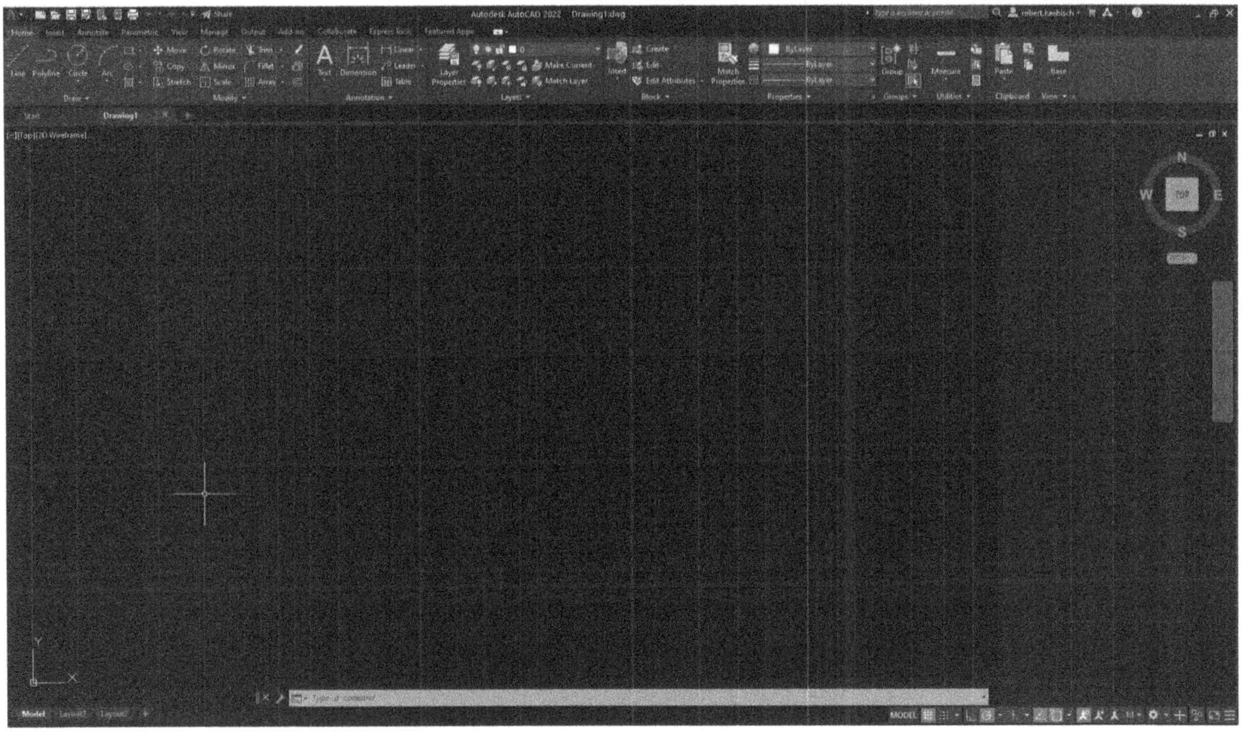

FIGURE 1.3 AutoCAD (Running on Windows 10).

This is your basic "out-of-the-box" AutoCAD environment for the Drafting and Annotation workspace. From both the learning and teaching points of view, this screen layout (and the layout of the last few releases) is an improvement over older versions of AutoCAD, and Autodesk has done an admirable job in continuing to keep things clean and (relatively) simple.

The Ribbon replaced traditional Toolbars and Cascading menu bars as the AutoCAD default setup more than 10 years ago. After these many years, it is a well-established way of interacting with AutoCAD, and we discuss it in detail. Other screen layouts or workspaces are available to you; however, they are related to 3D work and we discuss them in the Level III, 3D sections Chapters 21-30. The Drafting and Annotation workspace you are in now is all you need to get started.

Note: To bring back some classic AutoCAD elements, you can go to the Quick Access Toolbar, Customize Quick Access Toolbar-view options down arrow at the top left of the screen. Choose Show Menu Bar from the drop-down list. After the Menu Bar appears above the Ribbon, go to Tools→Toolbars→AutoCAD in the cascading menu system to display any of the many, many toolbars. This system of accessing command will not be covered in Chapters1-20 as it has been replaced by the Ribbon and by simply typing in commands. The Menu Bar, Cascading Menus and Toolbars are helpful for the Level III, 3D sections in Chapters 21-30 and is discussed there as well.

For clarity and ease of learning, we need to change the look of a few items on the screen. First, we lighten everything. By default, AutoCAD opens with a dark theme because it is easier on the eyes to work long hours with a dark screen. We switch to the light theme and change the work area color to white. This not only saves ink for this publication but makes it a lot easier for you to see and understand the graphics on the printed page. We then turn the command line from its single-line configuration to a multiline one, so you can follow command sequences more easily. Finally, we make a few other minor tweaks to the drawing and construction aids at the bottom right, such as turning off the Grid (which you can do right away by pressing F7). You need not do all of this; a lot of the steps are based on personal taste of how you want AutoCAD to look. However, it is necessary for the textbook and will make learning AutoCAD a bit easier for you initially as well. You may want to keep things dark or go back to dark later down the road however.

Let us now take a tour of the features you see on the screen. Along the way you will see descriptions of how to incorporate the previously mentioned changes. Your screen will then look like Fig. 1.4. Study all the screen elements carefully, so you know exactly what you are looking at as we progress further.

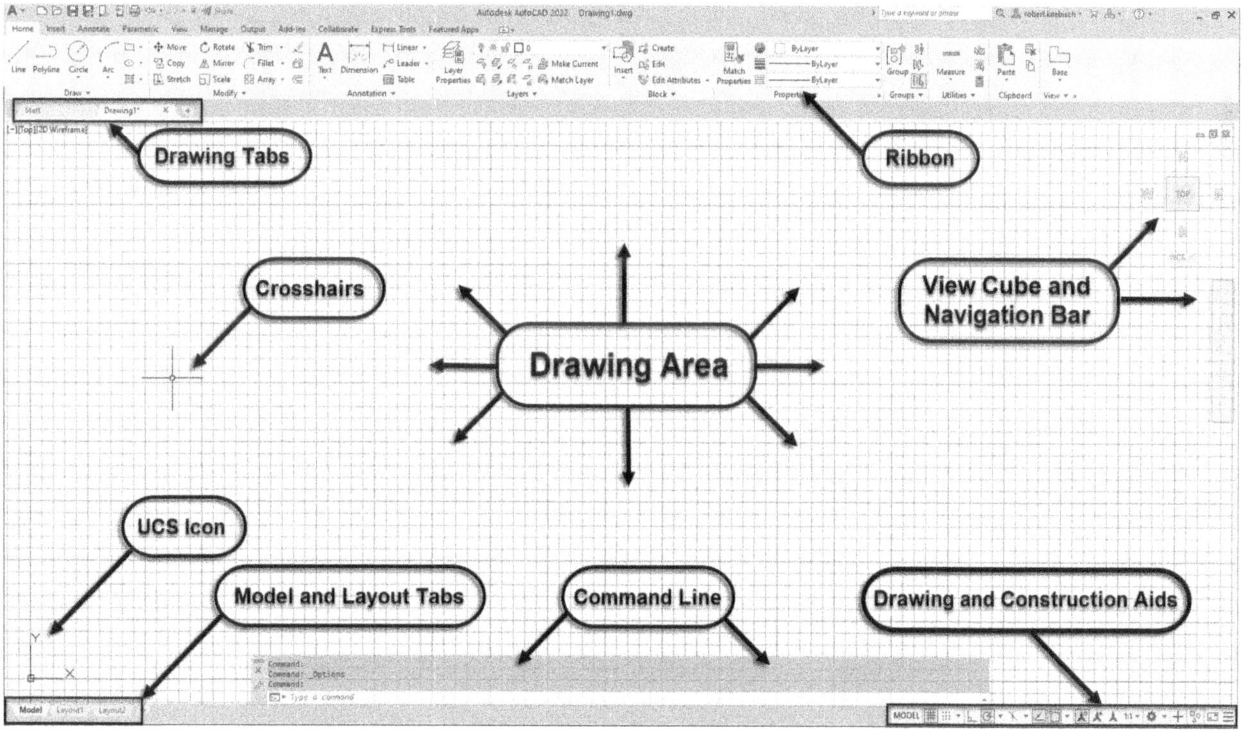

FIGURE 1.4 AutoCAD screen elements.

- **Drawing Area:** The drawing area takes up most of the screen and is colored a very dark gray in the default version of the environment. This is where you work and your design appears. Some users change it to all-black to further ease eyestrain, as less light radiates toward you. In this textbook, however (as in most textbooks), the color always is white to conserve ink and for clarity on a printed page. If you wish to change the color of the drawing area, you need to right-click into Options …, choose the Display tab, then Colors …. Finally, change the color from the drop-down menu on the right. We cover all this in more detail in Chapter 14, so you should ask your instructor for assistance in the meantime, if necessary.
- **Drawing Tabs:** One new change that was added back in the 2015 release is that the drawing area now has "tabs," visible at the upper left, which allow you to cycle through multiple drawings. As described earlier, AutoCAD defaults to the Start tab. You can close it while you are working on another drawing, but it will return if no other drawings are open.
- **Command Line(s):** Right below the drawing area are the command lines or, by default, just one floating line. It can be turned off and on via Ctrl+9. This is where the commands may be entered and where AutoCAD tells you what it needs to continue. Always keep an eye on what appears here, as this is one of the main ways that AutoCAD communicates with you. Although we discuss the "heads-up" on-screen interaction method in Chapter 2, the command line remains very relevant and often used. It is colored gray and white by default but can be changed (in a manner like the drawing area) to all white if desired. We leave the color as default but dock it at the bottom and stretch it upward to include three lines of history for clarity. These are all minor details, so you can leave your command line completely "as is" if you wish.
- **UCS Icon:** This is a basic X-Y-Z (Z is not visible) grid symbol. It will be important later in advanced studies and 3D. It can be turned off via an icon on the Ribbon's View tab, but generally it is kept where it is by most users. The significance of this icon is great, and we study it in detail later. For now, just observe that the Y axis is "up" and the X axis is "across."
- **Model and Layout Tabs:** These Model/Layout1/Layout2 tabs, like those used in Microsoft's Excel, indicate which drawing space you are in and are important in Chapter 10, when we cover Paper Space. Although you can click on them to see what happens, be sure to return to the Model tab to continue further.
- **View Cube and Navigation Bar:** The View Cube is used mostly for navigation of the model space when working in 3D, though it can be useful for "spinning" your current plan view as well. The Navigation Bar contains other navigation tools like the Navigation Wheel, Pan, Zoom, Orbit and Show Motion. As powerful and tempting as these tools may seem to explore, it is easiest to navigate 2D space using the wheel on your mouse which we discuss later in this chapter.
- **Crosshairs:** Crosshairs are simply the mouse cursor and move around along with the movements of your mouse. They can be full size and span the entire screen or a small (flyspeck) size. You can change the size of the crosshairs if you wish, and full screen is recommended in some cases. For now, we leave it as is. The crosshairs have been empowered with a new intelligence a few releases ago. They now show, via a small icon, what you are doing while executing certain commands.
- **Drawing and Construction Aids:** These various settings assist you in drafting and modeling, and we introduce them as necessary. They were also redesigned some time ago. More important, however, some of them now contain selectable menus (look for the little black triangles), where you can set or choose even more options. This was one of the more significant interface changes, a few years back, and gives you more power and convenience at your fingertips. The full menu of what is available can be seen by clicking the tab with the three short horizontal lines (Menu customization button or "hamburger menu") all the way on the right of the construction aids. If an option in this menu is checked, the corresponding button is visible in the horizontal menu. For all the drawing aids, remember, if they are gray, they are *off*; if they are blue, they are *on*.
  - *NOTE:* **Be sure to turn all of them to "off" for now. We activate them as needed throughout the chapters.**
- **Ribbon:** It comprises a large panel that stretches across the screen and is populated with context-sensitive tabs. Each tab, in turn, contains many panels. The Ribbon can be turned off and on but this is not suggested except for those legacy users who wish to use Toolbars and/or Cascading menus. We have more to say about the Ribbon in the next section.

Review the above bulleted list carefully; it contains a lot of useful "get to know AutoCAD" information. Once again, you do not have to change your environment exactly as suggested, but be sure to understand how to do it if necessary.

1.3 Interacting with AutoCAD

OK, so you have the basic commands in hand and ideally a good understanding of what you are looking at on the AutoCAD screen. We are ready to try out the basic commands and eventually draft something. So, how do we interact with AutoCAD and tell it what we want to do? Two primary ways are shown throughout the text and below. There are also four outdated methods, Cascading Menus, Toolbars, the Tablet and the Screen Side Menu (dating back to the very early days of AutoCAD), but we do not cover them.

Method 1: *Type in* the commands on the command line (AutoCAD v1.0–current).
Method 2: Use the *Ribbon* tabs, panels, and **Tools** (AutoCAD 2009–current).

Details of each method, including the pros and cons, follow. Most commands are presented in those two ways, and you can experiment with each method to determine what you prefer. Eventually, you will settle on one way of interacting with AutoCAD or a hybrid of the two.

Method 1. Type in the commands on the command line

This was the original method of interacting with AutoCAD and, to this day, remains the most foolproof way to enter a command: good old-fashioned typing. AutoCAD is unique among leading CAD software in that it has retained this method while almost everyone else moved to graphic icons, toolbars, and ribbons. If you hate typing, this will probably not be your preferred choice.

However, do not discount keyboard entry entirely; AutoCAD has kept it for a reason. When the commands are abbreviated to one or two letters (Line = L, Arc = A, etc.), input can be incredibly fast. Just watch a professional typist for proof of the speed with which one can enter data via a keyboard. The commands are literally at your fingertips. The disadvantage is of course that you must type.

To use this method, simply start typing in the desired command, as seen in Fig. 1.5, and press Enter. There is no need to move your cursor to the command line to start typing. The command initiates after you press Enter (or spacebar) and you can proceed to the next steps or inputs.

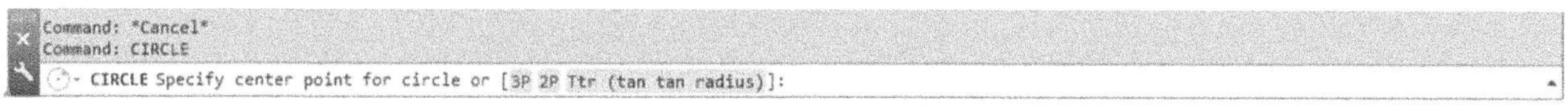

FIGURE 1.5 AutoCAD command line (CIRCLE typed in).

Several shortcuts are built into AutoCAD (try using just the first letter or two of a command), and we learn how to make our own shortcuts in advanced chapters. This keyboard shortcut method is still preferred by many new and "legacy" users—a kind way to say they have been using AutoCAD forever. Note that AutoCAD has an AutoComplete (as well as an AutoCorrect) feature which calls up a possible list of commands based on just the first letter or two, as seen in Fig. 1.6 and Fig. 1.7. This can be useful as you start out your learning the software. If you are wondering if a command exists, just start typing. It even works if you misspell a command.

FIGURE 1.6 AutoCAD suggestions when typing in LINE at command line with drawing aids turned off.

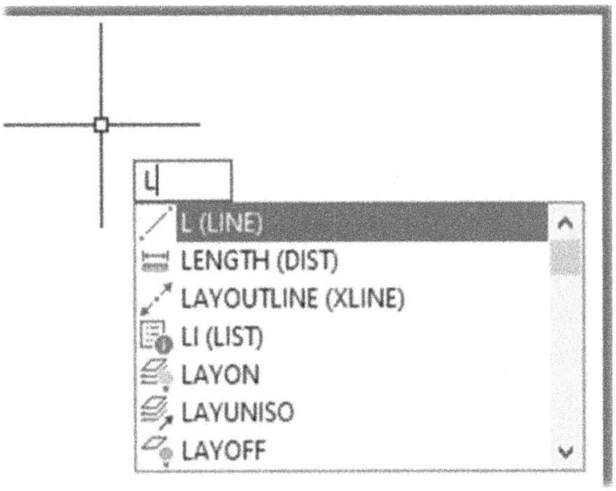

FIGURE 1.7 AutoCAD suggestions when typing in L for LINE with Dynamic Input figure drawing aid turned on.

For a right-handed user, try to keep your left hand on the keyboard and your right hand on the mouse, by using keyboard shortcuts and the spacebar as the enter key, your command entry will go very quickly.

Method 2. Use the Ribbon tabs, Panels, and Tools

This is the most recently introduced and common method for new users of interacting with AutoCAD The Ribbon has several tabs, which are broken up into Panels that contain specific command buttons and sometimes additional tools in drop down menus as well (see Fig. 1.8).

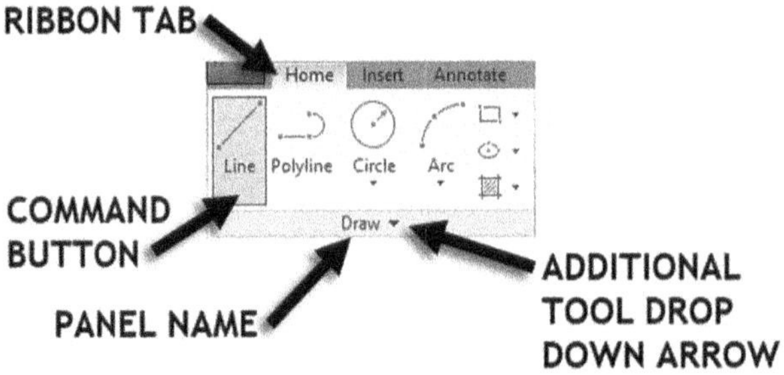

FIGURE 1.8 AutoCAD Ribbon details.

It follows the trend in software user interface design, such as that used by Microsoft Office (seen in Fig. 1.9 with Word). Notice how tools have been displaced by "tabbed" categories (Home, Insert, Page Layout, etc.), where information is grouped together by a common theme. So, it goes with the new AutoCAD. The Ribbon was introduced with AutoCAD 2009, and it is here to stay. The entire ribbon is shown in Fig. 1.10.

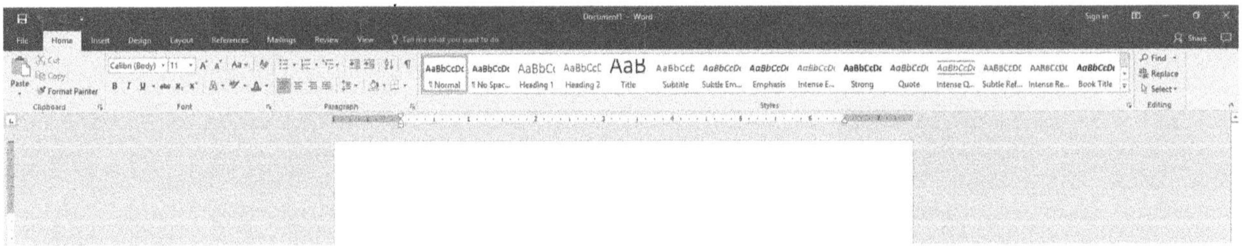

FIGURE 1.9 MS Word® Ribbon.

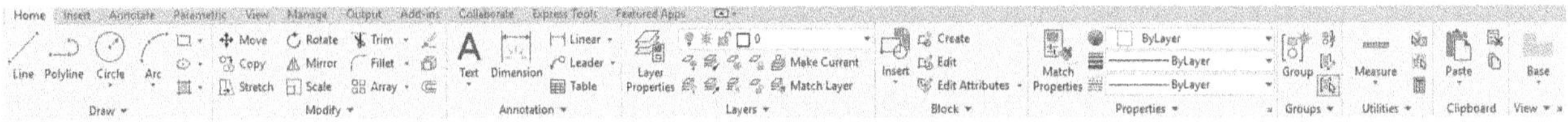

FIGURE 1.10 AutoCAD Ribbon.

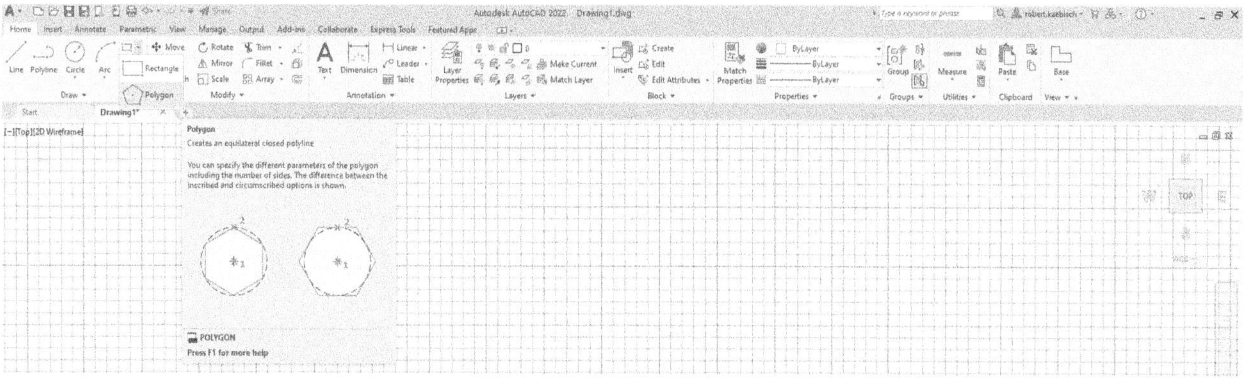

FIGURE 1.11 Polygon command and tool tip.

Notice what we have here. A collection of tabs, indicating a subject category, is found at the top (Home, Insert, Annotate, etc.), and each tab reveals an extensive set of tool Panels (Draw, Modify, Annotation, etc.). At the bottom of the Ribbon, additional options can be found by using the small drop arrows near the Panel titles. In this manner, the tool have been rearranged in what is, in principle, a more logical and space-saving manner.

Additionally, tool tips appear if you place your mouse over any tool for more than a second. Another second yields an even more detailed tool tip. In Fig. 1.11, you see the Home tab selected, followed by additional options via the drop arrow, and finally the mouse placed over the polygon command. A few moments of waiting reveal the full tool tip; pressing the icon activates the command.

Familiarize yourself with the Ribbon by exploring it. It presents some layout advantages. Contextual, command specific ribbon tabs and specialized Ribbon configurations can be displayed via other workspaces. The Ribbon is the best way for new users to AutoCAD's user interface and represents a jump forward in user/software interaction and has proven to be the best way for new users to learn AutoCAD.

Before we try out our basic commands, let us begin a list of tips. These are a mix of good ideas, essential habits, and time-saving tricks passed along to you occasionally (mostly in the first few chapters). Make a note of them as they are important. Here is the first, most urgently needed one.

TIP 1

The Esc (Escape) key in the upper left-hand corner of your keyboard is your new best friend while learning AutoCAD. It gets you out of just about any trouble you get yourself into. If something does not look right, just press the Esc key and repeat the command. Mine was worn out learning AutoCAD, so expect to use it often. NOTE: Esc should not be a substitute for ending commands "cleanly" with enter, spacebar or right click menus.

1.4 Practicing the create objects commands

Let us try the commands now, one by one. Although it may not be all that exciting, it is extremely important that you memorize the sequence of prompts for each command, as these are really the ABCs of AutoCAD.

Remember: If a command is not working right or you see little blue squares (they are called *grips*, and we cover them very soon), just press Esc to get back to the Command: status line, at which point you can try it again. Two methods of command entry are presented: typing, and the Ribbon. Alternate each method until you decide which one you prefer. It is perfectly OK to use a hybrid of methods.

Line

Keyboard: Type in **line** and press Enter
Ribbon: **Home tab→Line** 

Step 1. Begin the line command via any of the preceding methods.
- AutoCAD says: `Specify first point:`

Step 2. Using the mouse, left-click anywhere on the screen.
- AutoCAD says: `Specify next point or [Undo]:`

Step 3. Move the mouse elsewhere on the screen and left-click again. You can repeat Step 2 as many times as you wish. When you are done, click Enter or Spacebar or Right-click and select Enter from the pop-up menu. You should have a bunch of lines on your screen, either separate or connected, as shown in Fig. 1.12.

You need not worry about several things at this point. The first is accuracy; we introduce a tremendous amount of accuracy later in the learning process. The second is the options available in the brackets at the command line, such as `[Undo]`; we cover those as necessary. What is important is that you understand how you got those lines and how to do it again. In a similar manner, we move on to the other commands.

Circle

Keyboard: Type in **circle** and press Enter
Ribbon: **Home tab→Circle** 

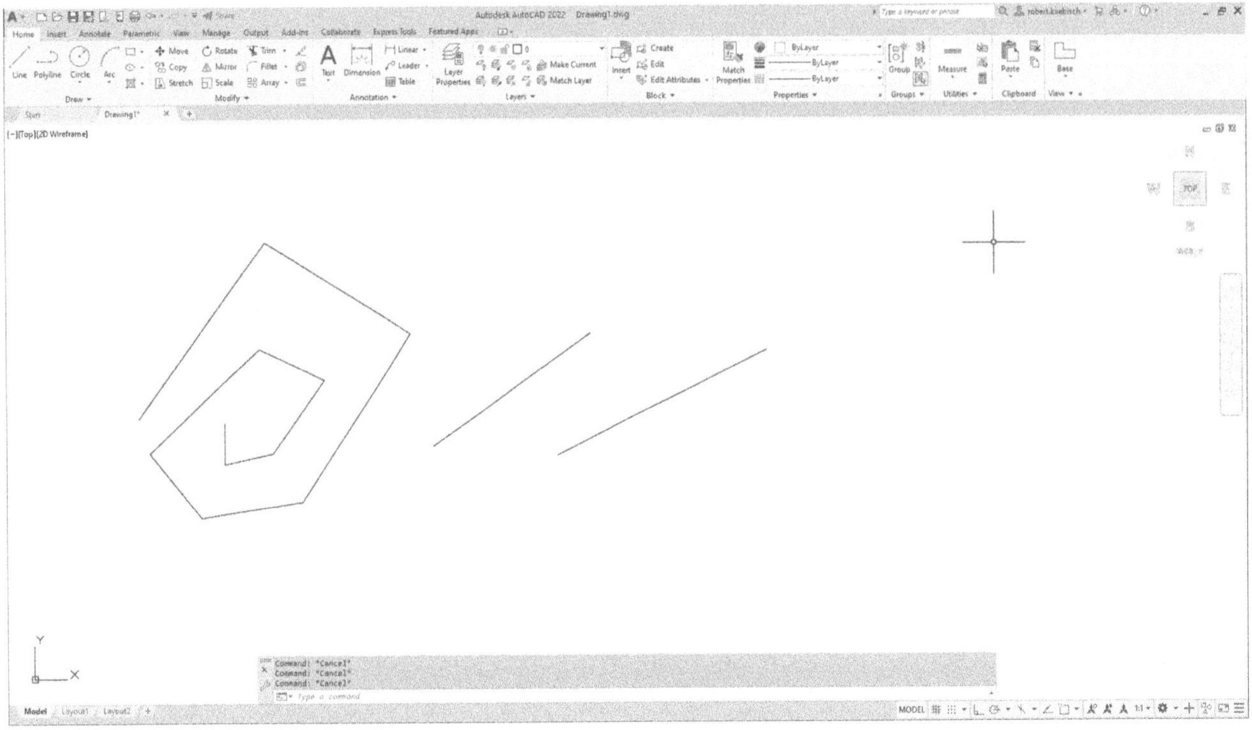

FIGURE 1.12 Lines drawn.

Step 1. Begin the circle command via any of the preceding methods.

- AutoCAD says: `Specify center point for circle or [3P/2P/Ttr (tan tan radius)]:`

Step 2. Using the mouse, left-click anywhere on the screen and move the mouse out away from that point.

- AutoCAD says: `Specify radius of circle or [Diameter] <1.9801>:`

Notice the circle that forms; it varies in size with the movement of your mouse. The value in the <> brackets will of course be different in your case or may not even be there at all if this is the first circle you just created.

Step 3. Left-click to finish the circle command. Repeat Steps 1 through 3 several times, and your screen should look like Fig. 1.13.

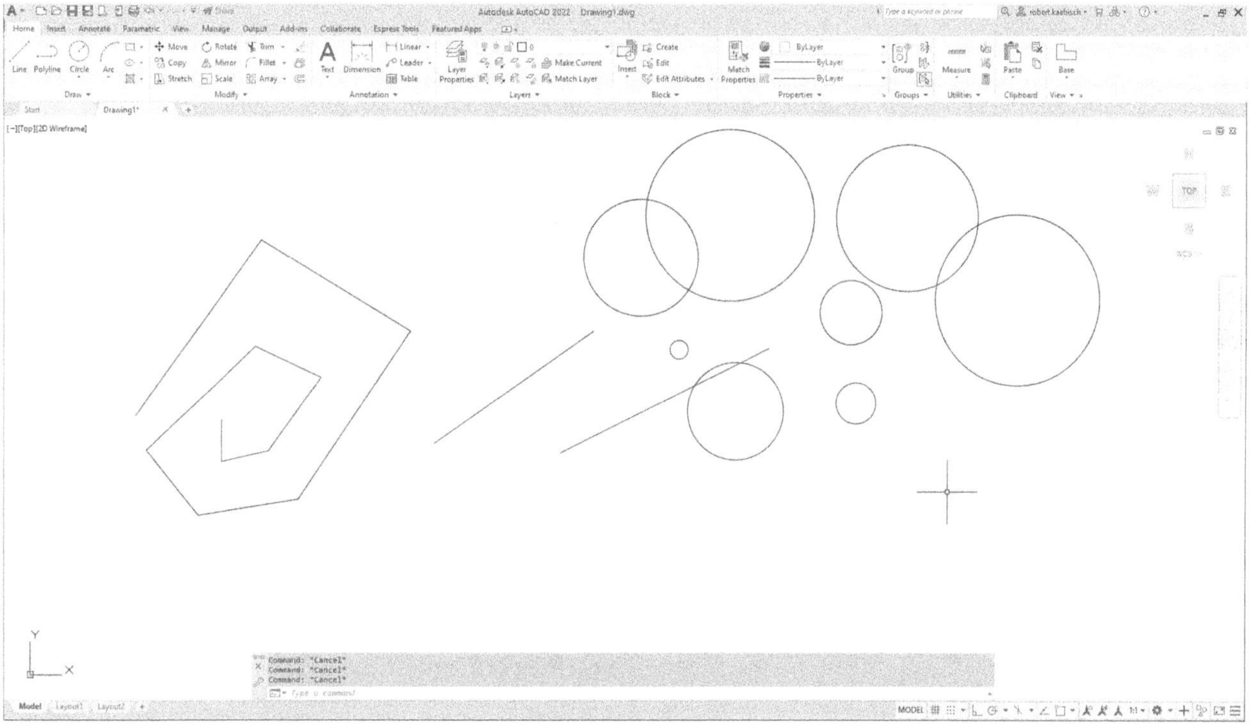

FIGURE 1.13 Circles drawn.

The method just used to create the circle was called *Center*, *Radius*, and you could specify an exact radius size if you wish, by just typing in a value after the first click (try it). As you may imagine, there are other ways to create circles—six ways to be precise, as seen with the Ribbon (Figs. 1.14 and 1.15).

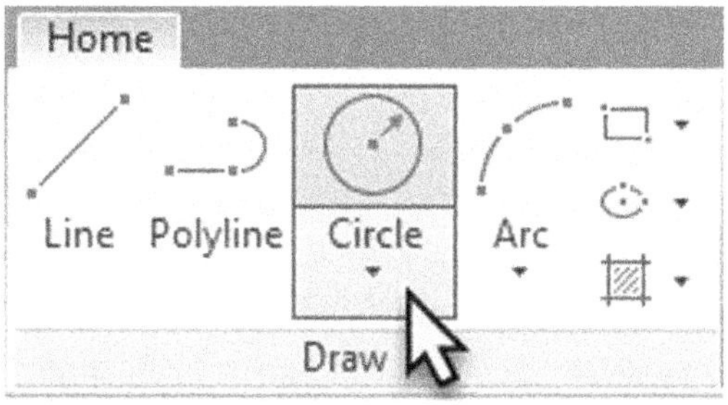

FIGURE 1.14 Additional circle options location.

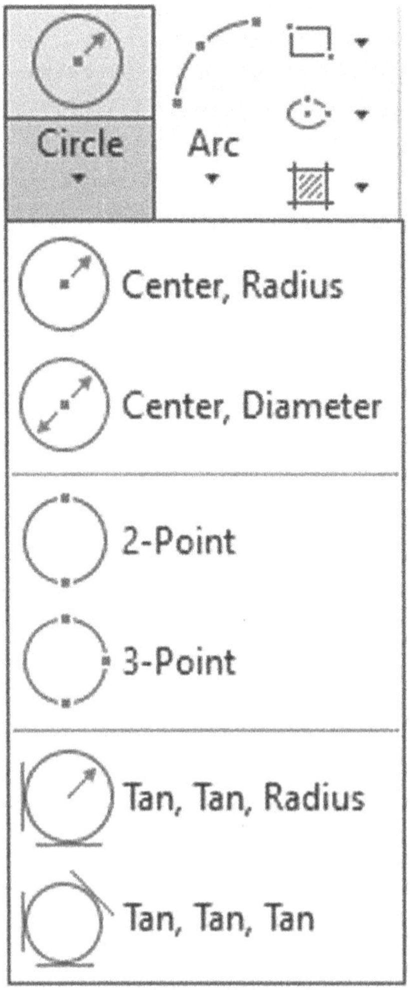

FIGURE 1.15 Additional circle options expanded menu.

Special attention should be paid to the Center, Diameter option. Often students are asked to create a circle of a certain diameter, and they inadvertently use radius. This is less of a problem when using the Ribbon, but you need to watch out if typing, as you will need to press `d` for `[Diameter]` before entering a value; otherwise, guess what it will be. We focus much more on these bracketed options as the course progresses. *You get a chance to practice the other circle options as part of the end of chapter exercises.*

Arc

Keyboard: Type in **arc** and press Enter
Ribbon: **Home** tab→**Arc**

Step 1. Begin the arc command via any of the preceding methods.

- AutoCAD says: `Specify start point of arc or [Center]:`

Step 2. Left-click with the mouse anywhere on the screen. This is the first of three points necessary for the arc.

- AutoCAD says: `Specify second point of arc or [Center/End]:`

Step 3. Click somewhere else on the screen to place the second point.

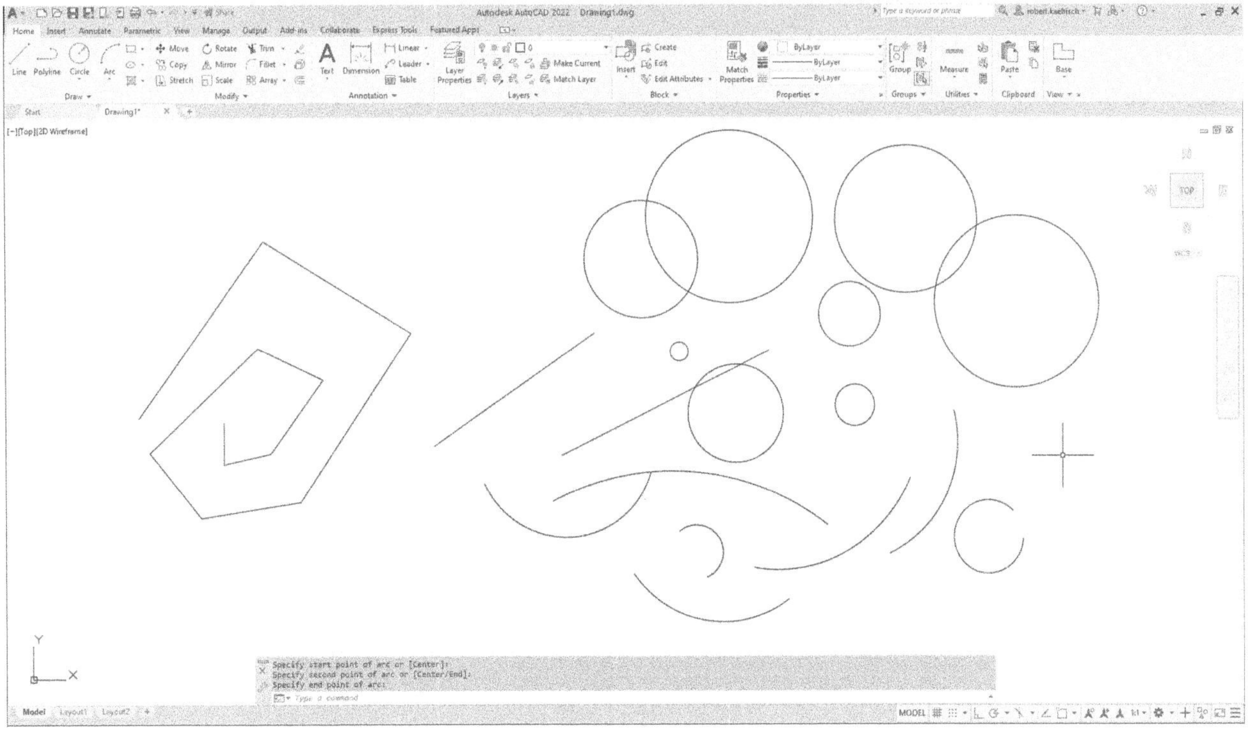

FIGURE 1.16 Arcs drawn.

- AutoCAD says: `Specify end point of arc:`

Step 4. Left-click a third (final) time, somewhere else on the screen, to finish the arc.

Practice this sequence several more times to fully understand the way AutoCAD places the arc. Your screen should look something like Fig. 1.16.

The method just used to create the arcs is called *3 Point* and, without invoking other options, is a rather arbitrary (eyeball) method of creating them, which is just fine for some applications. Just as with circles, there are, of course, many other ways to create arcs—11 ways to be precise, as seen with the Ribbon (Figs. 1.17 and 1.18).

Not all these options are used, and some you will probably never need, but it is worth going over them to know what is available. *You get a chance to practice the other arc options as part of the end of chapter exercises.*

Rectangle

Keyboard: Type in **rectangle** and press Enter
Ribbon: **Home** tab→**Rectangle**

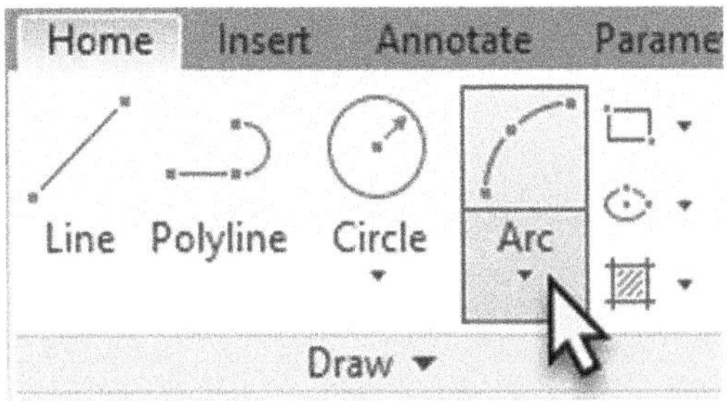

FIGURE 1.17 Additional arc options location.

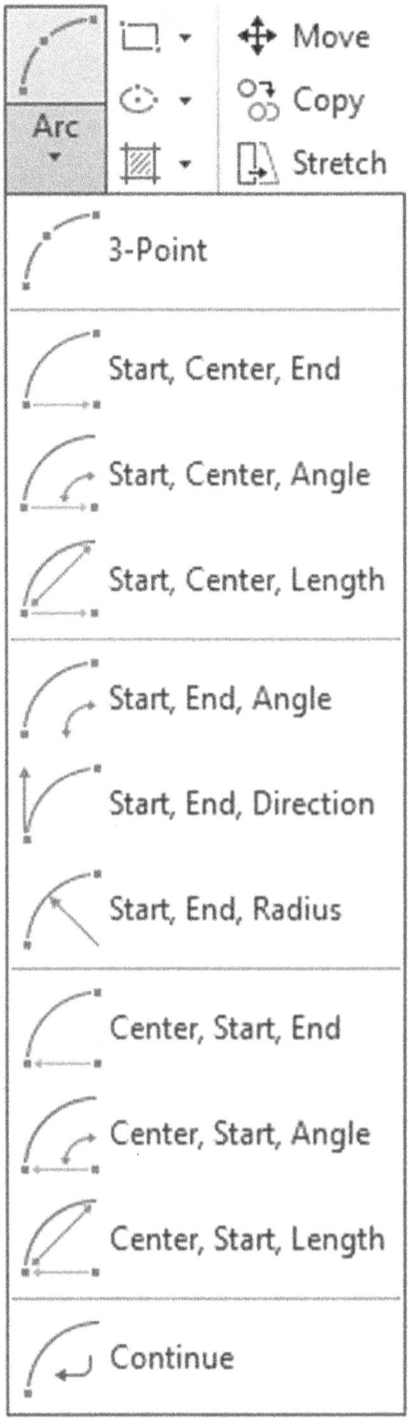

FIGURE 1.18 Additional arc options expanded menu.

Step 1. Begin the rectangle command via any of the preceding methods.

- AutoCAD says: `Specify first corner point or [Chamfer/Elevation/Fillet/Thickness/Width]:`

Step 2. Left-click, and move the mouse diagonally somewhere else on the screen.

- AutoCAD says: `Specify other corner point or [Area/Dimensions/Rotation]:`

Step 3. Left-click one more time to finish the command. Your screen should look more or less like Fig. 1.19.

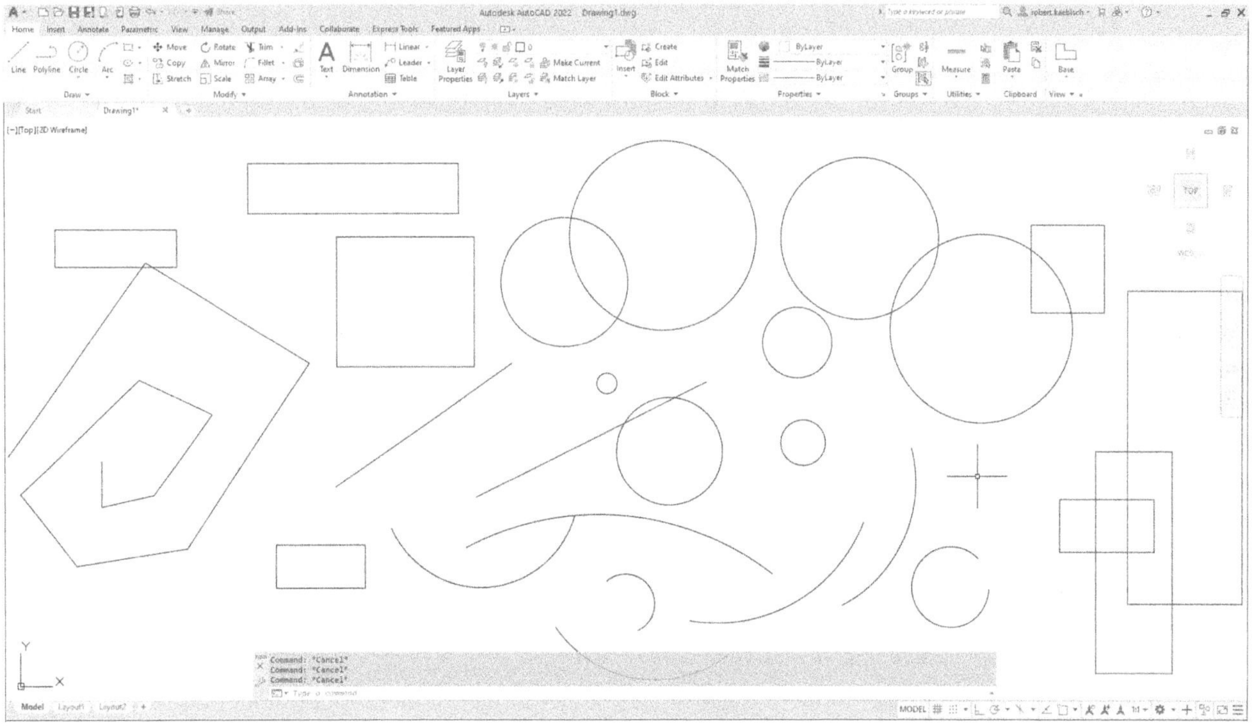

FIGURE 1.19 Rectangles drawn.

There are of course more precise ways to draw a rectangle. In Step 3, you can press `d` for `Dimensions` and follow the prompts to assign length, width, and a corner point (where you want it) to your rectangle. Try it out, but as with all basic shapes drawn so far, do not worry too much about sizing and accuracy, just be sure to understand and memorize the sequences for the basic command—that is what is important for now.

So, now you have created the four basic shapes used in AutoCAD drawings (and ended up with a lot of junk). Let us erase all of them and get back to a blank screen. To do this, we need to introduce Tip 2, which is quite easy to just type. Notice the use of the `e` abbreviation; we will see more of this as we progress.

TIP 2

To quickly erase everything on the screen, type in `e` for `erase`, press Enter, and type in `all`. Then press Enter twice. After the first Enter, all the objects are grayed out, indicating they were selected successfully for erasing. The second Enter finishes the command by deleting everything. Needless to say, this command sequence is useful only in the early stages of learning AutoCAD, when you do not intend to save what you are drawing. You may want to conveniently forget this particular tip when you begin to work on a company project. For now, it is very useful, but we will learn far more selective erase methods later.

So, let us put all of this to good use. Create the drawing shown in Fig. 1.20. Use the rectangle, line, circle, and arc commands. Do not get caught up trying to make everything line up perfectly, just do a quick sketch. Then, you can tell everyone that you learned how to draw a house in AutoCAD within the first half hour. Well, sort of.

At this point it is necessary to mention the zoom and pan commands. These are very easy if you have a mouse with a center wheel (as almost all computers do these days).

1.5 View objects

Zoom

Put your finger on the mouse wheel and turn it back and forth without pressing down. The house you just created gets smaller (roll back) or larger (roll forward) on the screen. Remember though, the object remains the same size; just your view is zooming in and out.

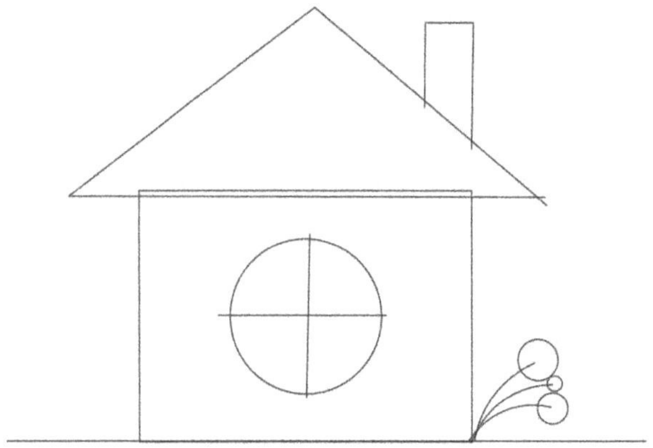

FIGURE 1.20 Simple house sketch.

Pan

Put your finger on the mouse wheel, but this time press and hold. A hand symbol appears. Now move the mouse around while keeping the wheel pressed. You can pan around your drawing. Remember again, the drawing is not moving; your view is moving around it.

Regen

This is the easiest command you will learn. Just type in regen (or *re*) and press Enter. The screen refreshes. One use for this is if you are panning over and AutoCAD does not let you go further, as if you hit a wall. Simply regen and proceed.

While the mouse method is easiest, there are other ways to zoom and pan. One alternative is using the Navigation Bar which has the Zoom (magnifying glass) and the Pan (hand) icons for this purpose, as seen in Fig. 1.21. Explore the icons, their drop-down menu options (small triangles) and the Navigation Bar display options (small circle at the bottom right) on your own.

FIGURE 1.21 Navigation bar.

One zoom option is critically important: Zoom to Extents. If you managed to pan your little house right off the screen and cannot find it, this comes in handy, and a full tip on Zoom to Extents is presented next.

TIP 3

Type in z for `zoom`, press Enter, type in the letter e, and then press Enter again. This is called Zoom to Extents and makes AutoCAD display everything you sketched, filling up the available screen space. If you have a middle-button wheel on your mouse, then double-click it for the same effect. This technique is very important to fit everything on your screen.

What if you did something that you did not want to do and wished you could go back a step or two? Well, like most programs, AutoCAD has an undo command, which brings us to the fourth tip.

TIP 4

To perform the undo command in AutoCAD, just type in u and press Enter. Do not type the entire command—if you do that, more options appear, which we do not really need at this point, so a simple u suffices. You can do this as many times as you want to (even to the beginning of the drawing session). You can also use the familiar *Undo* and *Redo* arrows seen at the very top of the screen in the Quick Access Toolbar. AutoCAD also has an *oops* command. This command just restores the last thing you erased, even if it was a while ago, without undoing any other steps. Keep it in mind; it can be useful as well. To try it, erase something, then type in `oops` and press Enter.

Now that you have drawn your house, erase it and let us move on. We now must go through all the Edit/Modify Objects commands, one by one. Each of them is critical to creating even the most basic drawings, so go through each carefully, paying close attention to the steps involved. Along the way, the occasional tip is added as the need arises.

Just to remind you: The Esc key gets you out of any trouble you may be in and returns you to the basic command line. Be sure to perform each sequence several times to really memorize it. As you learn the first command, erase, you also are introduced to the *picking objects* selection method, which is similar throughout the rest of the commands. Also, keep a close eye on the command line, as it is where you and AutoCAD communicate for now, and it *is* a two-way conversation.

Finally, before you get started, here is another useful tip that students usually appreciate early on to save some time and effort.

TIP 5

Hate to keep having to type, pull a menu, or click on an icon each time you practice the same command? Well, no problem. Simply press the spacebar or Enter, and this will repeat the last command you used, saving you a few minutes per day for that coffee break.

1.6 Practicing the Edit/Modify objects commands

Erase

Keyboard: Type in **erase** and press Enter
Ribbon: **Home** tab→**Erase**

Step 1. Draw a random-sized rectangle on your screen. Begin the erase command via any of the preceding methods.

- AutoCAD says: `Select objects:`

Step 2. Select the rectangle by moving the mouse to position the cursor over the rectangle. When you are "on" the object, a red X appears over the crosshairs and the object becomes faded. Left-click to select the object.

- AutoCAD says: Select objects: 1 found. Select object: You saw this before when we erased everything in Tip 2. It means the object was selected. AutoCAD continues to ask you: `Select objects:` Watch out for this step. AutoCAD always asks this, in case you want to select more objects, so get used to it. You are done, however, so …

Step 3. Press Enter, right-click, or Spacebar and the rectangle disappears. No figure is provided for this command as it is very straightforward and completely static.

Practice this several times, using the undo command to bring the rectangle back. You can of course also draw a few more rectangles and practice selecting more than one object, as this is the whole point of being asked to select again. As you click on each one, it becomes faded. You can do this until you run out of objects to select; there is no limit. To deselect any objects, hold down the Shift key and click on them. They are removed from the selection set and will not be erased when you end the command with a right-click, Enter or Spacebar.

Move

Keyboard: Type in **move** and press Enter
Ribbon: **Home** tab→**Move**

Step 1. Draw a random-sized circle and begin the move command via any of the preceding methods.
- AutoCAD says: `Select objects:`

Step 2. Select the circle by positioning the mouse directly over it (not the empty space) and left-clicking once.
- AutoCAD says: `Select objects: 1 found`. The selected circle becomes highlighted.
- AutoCAD then asks you again: `Select objects:`

Step 3. Unless you have more than one circle to move, you are done, so press Enter, right-click or Spacebar to stop selecting objects.
- AutoCAD says: `Specify base point or [Displacement] <Displacement>:`

Step 4. Left-click anywhere on or near the circle to "pick it up"; this is where you are moving it *from.*
- AutoCAD says: `Specify second point or < use first point as displacement>:`

Step 5. Move the mouse somewhere else on the screen and left-click to place the circle in the new location. This is where you are moving it *to.*

Notice how the original circle fades in color slightly, and a rubber banding line follows your mouse until you indicate the new location for the circle. Fig. 1.22 shows the move command in progress. Undo what you just did and practice it again.

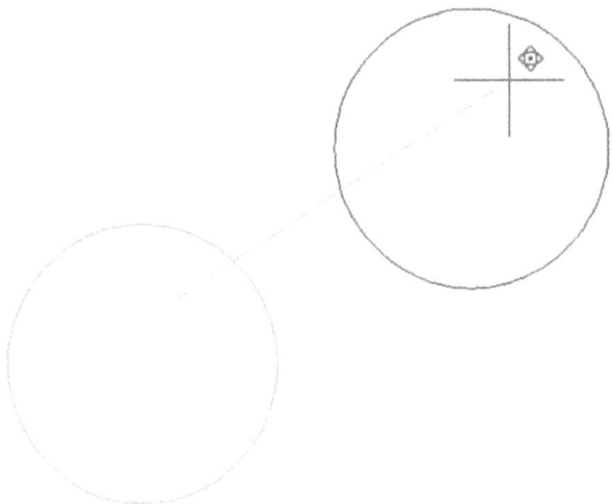

FIGURE 1.22 Move command in progress.

Copy

Keyboard: Type in **copy** and press Enter
Ribbon: **Home** tab→**Copy**

Step 1. Leave the circle you just used on the screen. Begin the copy command via any of the preceding methods.
- AutoCAD says: Select objects:

Step 2. Select the circle by positioning the mouse over it and left-clicking once.
- AutoCAD says: Select objects: 1 found. The circle will become highlighted when selected.
- AutoCAD then asks you again: Select objects:

Step 3. Unless you have more than one circlet to copy, you are done, so press Enter.
- AutoCAD says: Specify base point or [Displacement/mOde]<Displacement>:

Step 4. Left-click anywhere on or near the circle to "pick it up"; this is where you are copying it *from.*
- AutoCAD says: Specify second point or [Array] <use first point as displacement>:

Step 5. Move the mouse somewhere else on the screen and left-click to copy the circle to its new location. This is where you are copying it *to.* Notice that a lightened copy of it remains in its original location until you complete the command.

Step 6. You make one copy OR as as you want, pressing Enter to finish before the additional copies are placed,
- AutoCAD says: Specify second point or [Array/Exit/Undo] <Exit>:

Fig. 1.23 shows the copy command in progress. The highlighted, original circle remains there as a "shadow," along with the rubber banding line attached to your mouse, until you finish the copy command. Undo what you just did and practice it again.

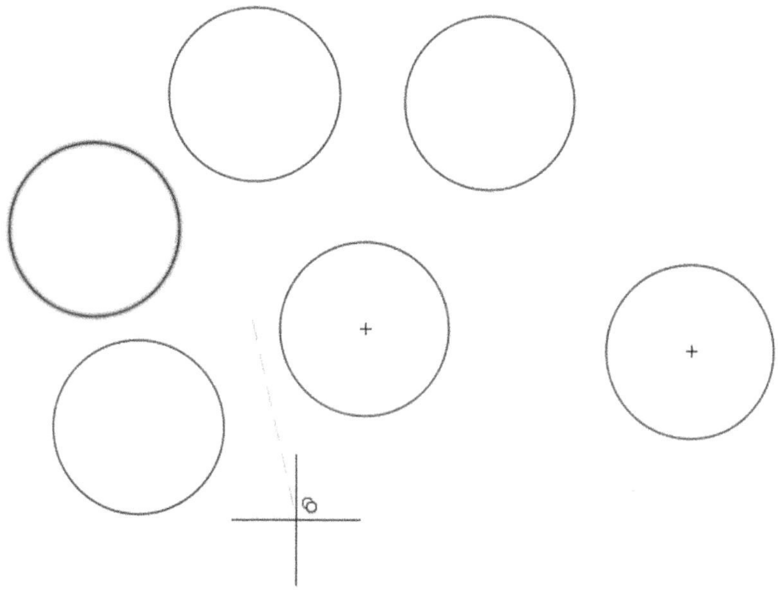

FIGURE 1.23 Copy command in progress.

Rotate

Keyboard: Type in **rotate** and press Enter
Ribbon: **Home** tab→**Rotate**

Step 1. For this command erase the circle and draw a rectangle. Begin the rotate command via any of the preceding methods.

- AutoCAD says: `Current positive angle in UCS: ANGDIR = counterclockwise ANGBASE = 0`, and on the next line: `Select objects:`

Step 2. Select the rectangle, remembering to press Enter. So far, it is quite similar to erase, move, and copy.

- AutoCAD says: `Specify base point:` This means select the pivot point of the object's rotation (the point about which it rotates). If you do use a circle for your object, try to stay away from the center point, as your rotation efforts may be less than obvious.

Step 3. Click anywhere on or near the selected object.

- AutoCAD says: `Specify rotation angle or [Copy/Reference] <0>:`

Step 4. Move the mouse around in a wide circle and the rectangle rotates. Notice how the motion gets smoother as you move the mouse farther away. You can click anywhere for a random rotation angle or you can type in a specific numerical degree value.

Fig. 1.24 shows the rotate command in progress. The dotted rectangle remains there as a "shadow," along with the rubber banding line attached to your mouse, until you rotate it into the new locations. Undo what you just did and practice it again.

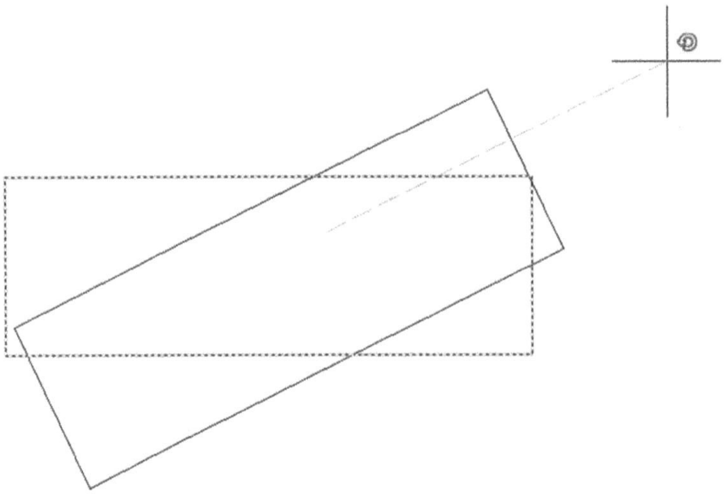

FIGURE 1.24 Rotate command in progress.

Scale

Keyboard: Type in **scale** and press Enter
Ribbon: **Home tab→Scale**

Step 1. We reuse the previous rectangle to do scale. Begin the scale command via any of the preceding methods.

- AutoCAD says: `Select objects:`

Step 2. Select the rectangle as before, remembering to press Enter again after the first selection.

- AutoCAD says: `Specify base point:` This means select the point from which the scaling of the object (up or down) is to occur.

Step 3. Click somewhere on or near the rectangle or directly in the middle of it.

- AutoCAD says: `Specify scale factor or [Copy/Reference].`

Step 4. Move the mouse around the screen. The rectangle gets bigger or smaller. You can randomly scale it or enter a numerical value. For example, if you want it twice as big, enter `2`; half size will be `.5`.

Undo what you just did and practice it again. Fig. 1.25 shows the scale command in progress. The dotted rectangle remains there as a "shadow," along with the rubber banding line attached to your mouse, until you finish the scale command.

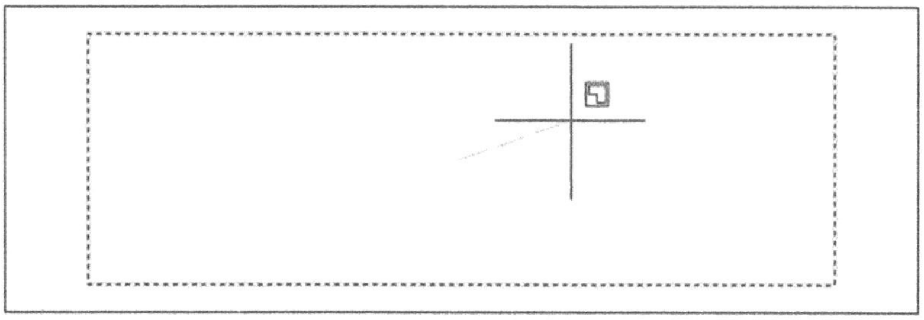

FIGURE 1.25 Scale command in progress.

Trim

Keyboard: Type in **trim** and press Enter
Ribbon: **Home tab→Trim**

Trim was all new for AutoCAD 2021, the first change since its inception. In all previous versions, you had to first select cutting edges or boundaries to Trim to and then select the object parts you wanted to trim off to those cutting edges or boundaries. Now the default Quick Mode selects all potential boundaries to trim to by default. So, after you start the TRIM command, simply select the objects to be trimmed.

The new, and default, Quick Mode is much better and easily understood by a new user. If you have an older version of AutoCAD or want to restore the previous Trim command behavior for practice, start the Trim command and change the Mode option to Standard. Stop the Trim command with Enter and then restart the Trim command to complete the switch to Standard mode. Start the Trim command again, then first select cutting edges, stop selecting cutting edges with Enter, then chose what part of an object you want to Trim off, and finally stop selecting things to trim with Enter. Let's keep the Mode on the new default, Quick Mode for now.

Restart the Trim command, picking line segments to remove to practice as you wish. You can even keep trimming until an object is completely removed, meaning Trim can be used to Erase an object completely. After starting the command again for more practice, try to single click a point in open space to start a fence line or click and drag from a point in open space to create a freehand (rope-like) selection line, making it easier to Trim many objects in less clicks (See Fig. 1.26)

To practice this command, you first need to draw a series of random intersecting lines. Once this is done, go ahead and perform the sequence that follows.

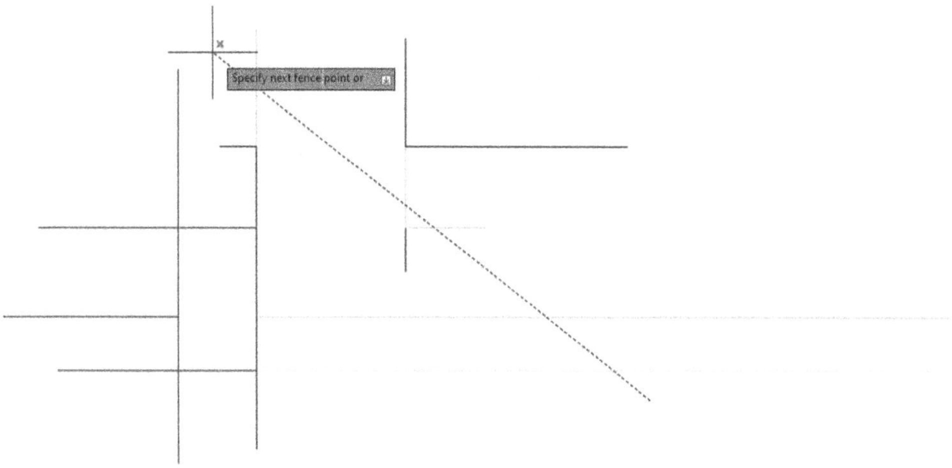

FIGURE 1.26 Trim command using a Fence in progress.

Step 1. Begin the trim command via any of the preceding methods.

- AutoCAD says:

  ```
  Current settings: Projection UCS, Edge None, Mode=Quick
  Select object to trim or shift-select to extend or
  [cuTting edges/Crossing/mOde/Project/eRase]:
  ```

Step 2. Simply select one line to trim.

- AutoCAD says: `Select objects or <select all>: 1 found`

Step 3. Press Enter.

Extend

Keyboard: Type in **extend** and press Enter
Ribbon: **Home tab→Extend**

Extend of course is also all new for AutoCAD 2021 as well. In all previous versions, you had to first select boundary edges to extend your lines to and then select the objects you wanted to extend. Now the default "Quick" Mode selects any and all potential boundaries to extend to by default. So, after you start the Extend command, simply select the objects to be extended. (See Fig. 1.27)

You can change the command to behave like previous versions as mentioned in the Trim command section above, but it is recommended to keep the default "Quick" mode set.

To practice this command, you first need to draw a series of random intersecting lines. once this is done, go ahead and perform the sequence that follows.

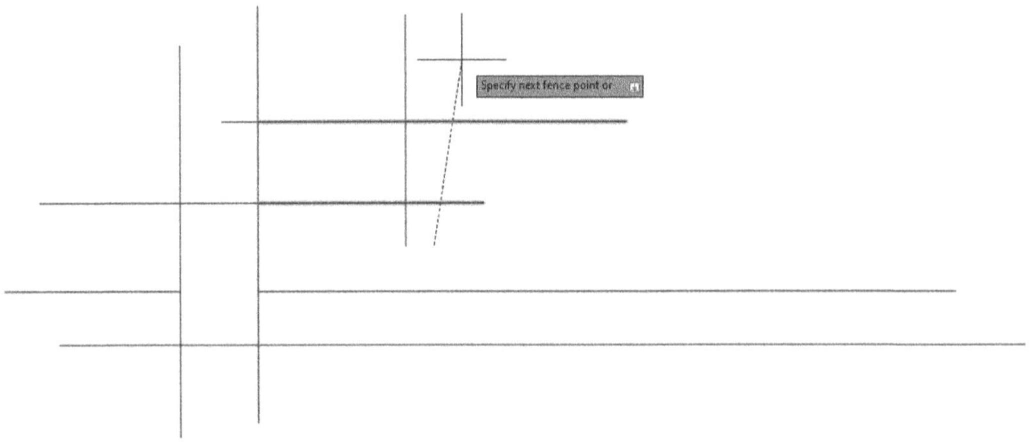

FIGURE 1.27 Extend command using a Fence.

Step 1. Begin the extend command via any of the preceding methods.

- AutoCAD says:

  ```
  Current settings: Projection UCS, Edge None, Mode=Quick
  Select object to extend or shift-select to trim or [Boundary edges/Crossing/mOde/Project/Edge/
  Undo]:
  ```

Step 2. Select one line to extend. If the line highlights and shows a preview of the extension, go ahead and click it. If it shows a No symbol, that means there is no available object to extend the selected object to. For now, just pick something else.

- AutoCAD says: `Select object to extend or shift-select to trim or [Boundary edges/Crossing/mOde/Project/Undo]:`

Step 3. Press Enter to end the command.

You can repeat the command as many times as you want, practicing with single object, fences and freehand selections. Practice Extend and Trim several times by pressing U (for undo) then Enter and repeating the steps.

While using Trim or Extend you can hold the Shift key to switch to the other command. If you keep practicing Extend, make a few selections to extend and then hold down Shift key to switch to the Trim command while inside of Extend. Let go of Shift to continue extending again until you are done. Of course this works in reverse when inside of the Trim command.

Offset

Keyboard: Type in **offset** and press Enter
Ribbon: **Home tab→Offset**

This command is one of the more important ones you will learn. Its power lies in its simplicity. This command creates new lines by the directional parallel offset concept, where if you have a line on your screen, you can create a new one that is a certain distance away but parallel to the original or in the case of circles, lines and rectangles, concentric to the original. To illustrate it, first draw a random vertical or horizontal line anywhere on your screen.

Step 1. Begin the offset command via any of the preceding methods.
- AutoCAD says: Specify offset distance or [Through/Erase/Layer] <Through>:

Step 2. Enter an offset value by drawing an imaginary, relatively short, line on screen to specify a distance. (Note: we will begin distance entry in Chapter 2).
- AutoCAD says: Select object to offset or [Exit/Undo] <Exit>:

Step 3. Pick the line by left-clicking on it; the line becomes highlighted (shown dashed below for clarity).
- AutoCAD says: Specify point on side to offset or [Exit/Multiple/Undo] <Exit>:

Step 4. Pick a *direction* for the line to go, which is to one of the two sides of the original line. It does not matter exactly how far away from the original line you click, as the distance has already been specified. You can keep doing this over and over by selecting the new line then the direction (see Fig. 1.28). If you want to change the offset distance, you must repeat Steps 1 and 2.

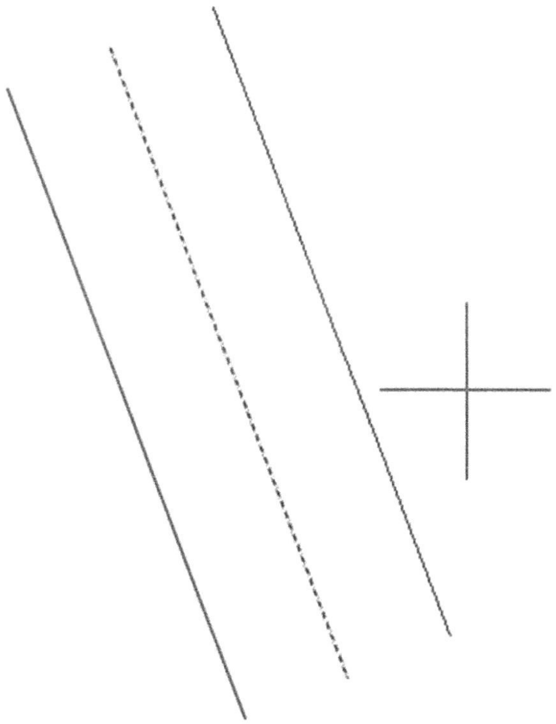

FIGURE 1.28 Offset command in progress.

The significance of the offset command may not be readily apparent. Many students, especially ones who may have done some hand drafting, come into learning AutoCAD with the preconceived notion that it is just a computerized version of pencil and paper. While this is true in some cases, in others, AutoCAD represents a radical departure. When you begin a new drawing on paper, you literally draw line after line. With AutoCAD, surprisingly little line drawing occurs. A few are drawn, and then the offset command is employed throughout. Learn this command well, you will need it.

Mirror

Keyboard: Type in **mirror** and press Enter
Ribbon: **Home tab→Mirror**

This command, much as the name implies, creates a mirror copy of an object over some plane. It is surprisingly useful because often just copying and rotating is not enough or takes too many steps. To learn this command, you must think of a real-life mirror, with you standing in front of it. The original object is you, the plane is the door with the mirror attached (which can swing on a hinge), and the new object is your mirrored reflection. To practice this command, draw a triangle by joining three lines together, as seen in Fig. 1.29.

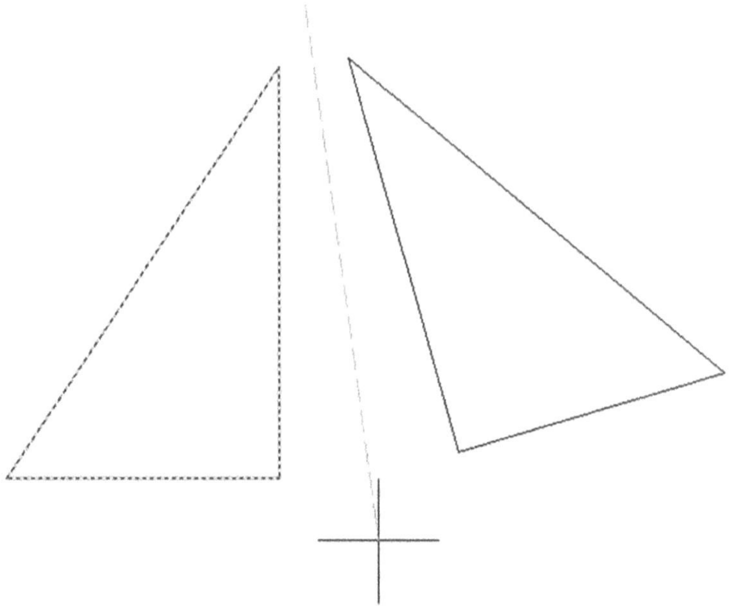

FIGURE 1.29 Mirror command in progress.

Step 1. Begin the mirror command via any of the preceding methods.
- AutoCAD says: `Select objects:`

Step 2. Select all three lines of the triangle and press Enter.
- AutoCAD says: `Specify first point of mirror line:`

Step 3. Click anywhere near the triangle and move the mouse around. You notice the new object appears to be anchored by one point. This is the swinging mirror reflection.
- AutoCAD says: `Specify second point of mirror line:`

Step 4. Click again near the object (make the second click follow an imaginary straight line down). You notice the new object disappear. Do not panic; rather check what the command line says.
- AutoCAD says: `Delete source objects? [Yes/No] <N>:`

Step 5. The response in Step 4 just asked you if you want to keep the original object. If you do (the most common response), then just press Enter and you are done. Fig. 1.29 illustrates Step 3.

Fillet

Keyboard: Type in **fillet** and press Enter
Ribbon: **Home** tab→ **fillet**

Fillets are all around us. This is not just an AutoCAD concept but rather an engineering description for a rounded edge on a corner of an object or at an intersection of two objects. These edges are added to avoid sharp corners for safety or other technical reasons (such as a weld). By definition, a fillet needs to have a radius greater than zero; but in AutoCAD, a fillet can have a radius of zero, allowing for some useful editing, as we see soon. To illustrate this command, first draw two perpendicular lines that intersect, as seen in the first box of Fig. 1.30.

Step 1. Begin the fillet command via any of the preceding methods.
- AutoCAD says:`Current settings: Mode = TRIM, Radius = 0.0000`
  `Select first object or [Undo/Polyline/Radius/Trim/Multiple]:`

Step 2. Put a radius on the fillet. Type in `r` for *radius* and press Enter.
- AutoCAD says:`Specify fillet radius <0.0000>:`

Step 3. Specify a radius by drawing an imaginary, relatively short, line on screen to specify a distance. (Note: we will begin distance entry in Chapter 2).
- AutoCAD says:`Select first object or [Undo/Polyline/Radius/Trim/Multiple]:`

Step 4. Select the first (horizontal) line, somewhere near the intersection. It becomes faded, and a proposed fillet appears. This is just a preview for you to get an idea of the result.
- AutoCAD says:`Select second object or shift-select to apply corner or [Radius]:`

Step 5. As your mouse moves toward the second (vertical) line, you continue to see a preview of the resulting fillet (if the fillet is valid). Go ahead and click the second line somewhere near the intersection; a fillet of the radius you specified is added, the unneeded lines disappear, and everything becomes solid.

These are the steps involved in adding a fillet with a radius. You can also do a fillet with a radius of zero, so repeat Steps 1 through 5, entering 0 for the radius value. This is very useful to speed up trimming and extending. Intersecting lines can be filleted to terminate at one point, and if they are some distance apart, they can be filleted to meet. All this is possible with a radius of zero. In these cases, the preview feature shows where the lines intersect. Try out all the variations. Fig. 1.30 summarizes everything described so far. The 1 and 2 indicate a suggested order of clicking the lines for the desired effect.

Well, this is it for now. We covered the basic Draw and Modify commands. These commands form the essentials of what we call *basic CAD theory*. All 2D computer-aided design programs need to be able to perform these functions to work, although other software may refer to them differently, for example using *displace* instead of *move*, but the capability still must be there. Be sure to memorize these commands and practice them until they become second nature. They are truly the ABCs of basic computer drafting, their importance cannot be overstated, and you need to master them to move on.

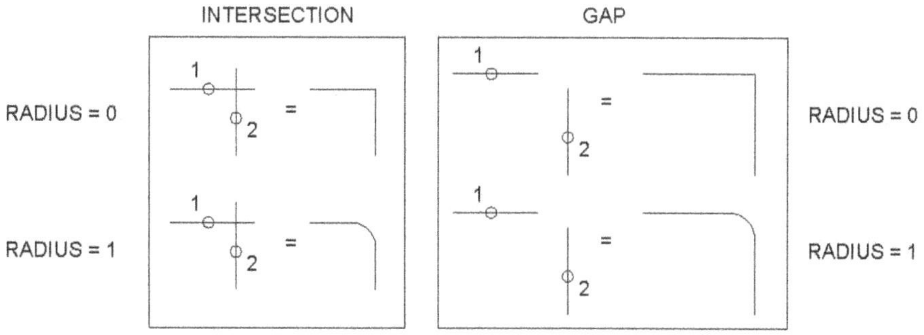

FIGURE 1.30 Fillet command summary.

1.7 Selection methods

Before we talk about the accuracy tools available to you, we need to discuss more advanced types of selection methods. The classic approaches are called *Window* and *Crossing*, and as of AutoCAD 2015 release, you also have the new *Lasso* tool.

You may have already seen the Window, Crossing, and Lasso tools by accident: When you single-click randomly on the screen without selecting any objects and move your mouse to the right or left, a rectangular shape forms. It is either dashed (with a green fill) or solid (with a blue fill) and goes away once you click the mouse button again or press Esc. Alternatively, if you single-click randomly and keep *holding down* the left mouse button while moving the mouse you will get the Lasso. These all happen to be powerful selection tools. They are defined as follows:

- **Crossing:** A crossing appears when you single-click and move to the *left* of an imaginary vertical line (up or down) and is a *dashed* green shaded rectangle. *Any object it touches or is inside of it gets selected.*
- **Window:** A window appears when you single click and move to the *right* of an imaginary vertical line (up or down) and is a *solid* blue-shaded rectangle. *Any object falling completely inside it is selected.*

Fig. 1.31 shows what both the Crossing and the Window look like. It is not shown in the figure, but note that, depending on settings, your Crossing and Window may be filled with a color—green for crossing, blue for window. Practice selecting objects in conjunction with the erase command. Draw several random objects and start up the erase command. Then, instead of picking each one individually (or doing Erase, All), select them with either the Crossing or Window and make a note of how each functions. This is a very important and useful selection method.

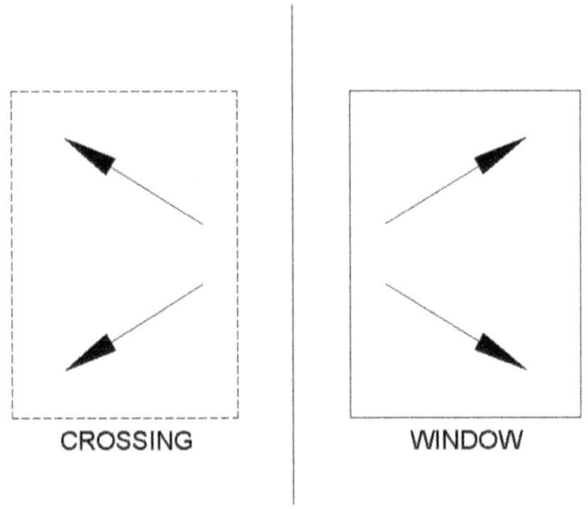

FIGURE 1.31 Crossing and Window.

Finally, let us discuss the new Lasso function. This is a technique borrowed from Photoshop and similar software, and it frees you from any geometric constraints whatsoever, allowing you to "free-hand" choose the selection area by "roping" around (or through) it. You may end up liking this new method the most. It can be applied both as a Window Lasso and a Crossing Lasso.

To try out the *Window Lasso*,

Step 1. Draw a set of circles, as seen in Fig. 1.32.
Step 2. Begin the erase command via any method you prefer.
Step 3. Do not type anything in, rather *press and hold down* the left mouse button, and drag it to the *right* (thus making it a *Window* Lasso) around several of the circles, as seen in Fig. 1.32.
Step 4. Continue to smoothly drag the mouse *around* the circles you wish to select, and right-click when done.

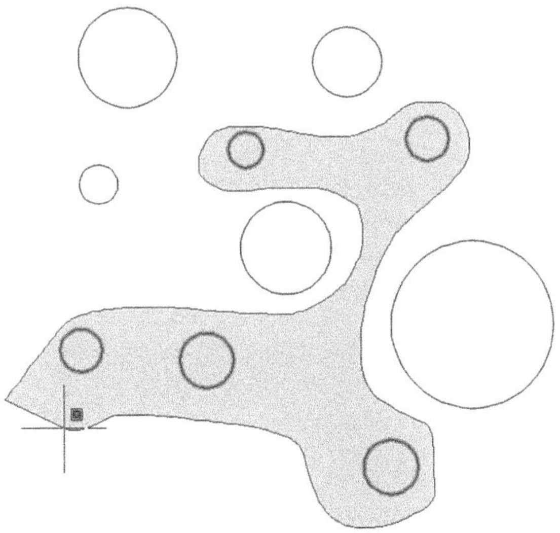

FIGURE 1.32 Window Lasso.

To try out the *Crossing Lasso*,

Step 1. Draw another set of circles, as seen in Fig. 1.33, or just undo your previous steps.
Step 2. Begin the erase command via any method you prefer.
Step 3. Do not type anything in, rather press *and hold down* the left mouse button and drag it to the *left* (thus making it a *Crossing* Lasso) around several of the circles, as seen in Fig. 1.33.
Step 4. Continue to smoothly drag the mouse *through* the circles you wish to select, and right-click when done.

Note how, in both cases, this tool makes the selection process very smooth and free form, as opposed to being restricted to straight lines or rectangles. This may well end up being the primary approach to selecting objects in subsequent releases of AutoCAD, with all the other methods as a secondary backup.

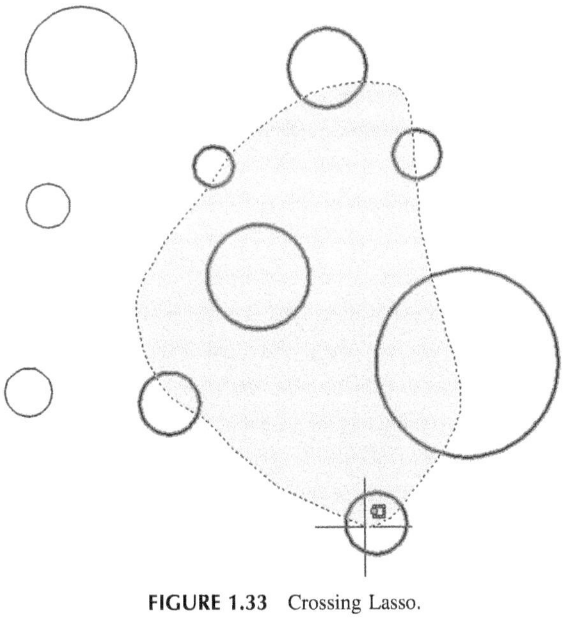

FIGURE 1.33 Crossing Lasso.

1.8 Drawing accuracy—part 1

Ortho (F8)

So far, when you created lines, they were not perfectly straight. Rather, you may have lined them up by eye. This of course is not accurate enough for most designs, so we need a better method. Ortho (which stands for *orthographic*) is a new concept we now introduce that allows you to draw perfectly straight vertical or horizontal lines. It is not so much a command as a condition you impose on your lines prior to starting to draw them. To turn on the Ortho feature, just press the F8 key at the top of your keyboard or click to depress the ORTHOMODE button at the bottom right of the screen in the drawing aids (looks like an "L" angle). Then, begin drawing lines. You see your mouse cursor constrained to only vertical and horizontal motion. Ortho is usually not something you just set and forget about, as your design may call for a varied mix of straight and angled lines, so do not be shy about turning the feature on and off as often as necessary, sometimes right in the middle of drawing a line.

1.9 Drawing accuracy—part 2

OSNAPs

While drawing straight lines, we generally like those lines to connect to each other in a very precise way. So far, we have been eyeballing their position relative to each other, but that is not accurate and is more reminiscent of sketching, not serious drafting. Lines (or for that matter all fundamental objects in AutoCAD) connect to each other in very distinct and specific locations (two lines are joined end to end, for example). Let us develop these points for a collection of objects.

ENDpoint

ENDpoints are the two locations at the ends of any line. Another object can be precisely attached to an endpoint. The symbol for it is a square, as illustrated in Fig. 1.34.

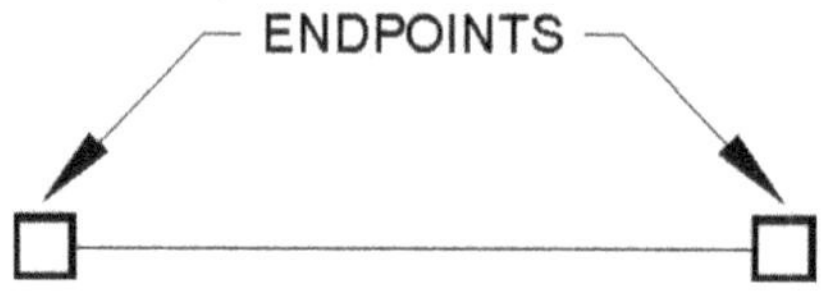

FIGURE 1.34 Endpoints.

MIDpoint

MIDpoint is the middle of any line. Another object can be precisely attached to a midpoint. The symbol for it is a triangle, as illustrated in Fig. 1.35.

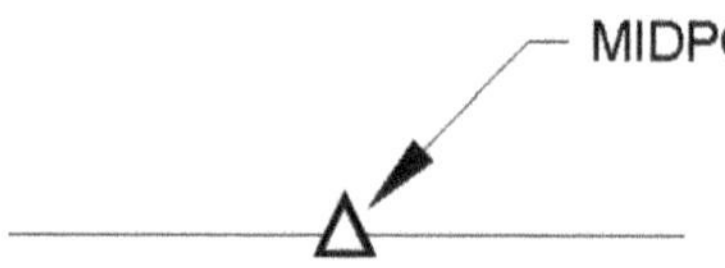

FIGURE 1.35 Midpoint.

CENter

CENter is the center of any circle. Another object can be precisely attached to a center point. The symbol for it is a circle, as illustrated in Fig. 1.36.

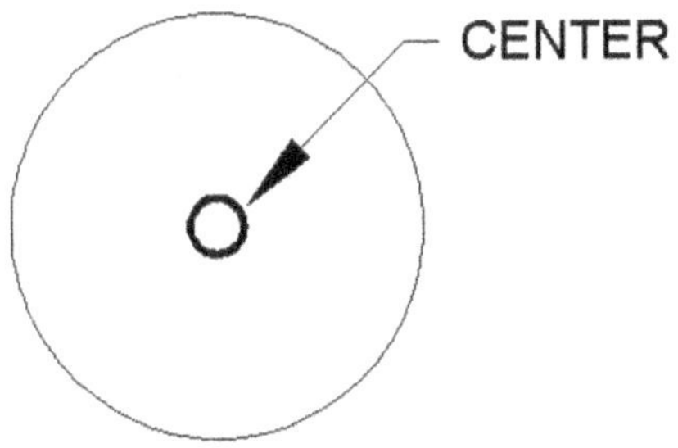

FIGURE 1.36 Center.

QUADrant

QUADrants are the four points at the North (90 degrees), West (180 degrees), South (270 degrees), and East (0 degrees, 360 degrees) quadrants of any circle. Another object can be precisely attached to a quadrant point. The symbol for it is a diamond, as illustrated in Fig. 1.37.

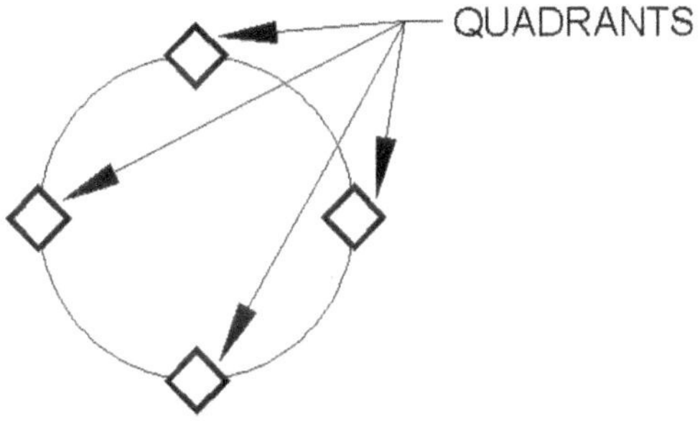

FIGURE 1.37 Quadrants.

INTersection

INTersection is a point located at the intersection of any two objects. You may begin a line or attach any object to that point. The symbol for it is an X, as illustrated in Fig. 1.38.

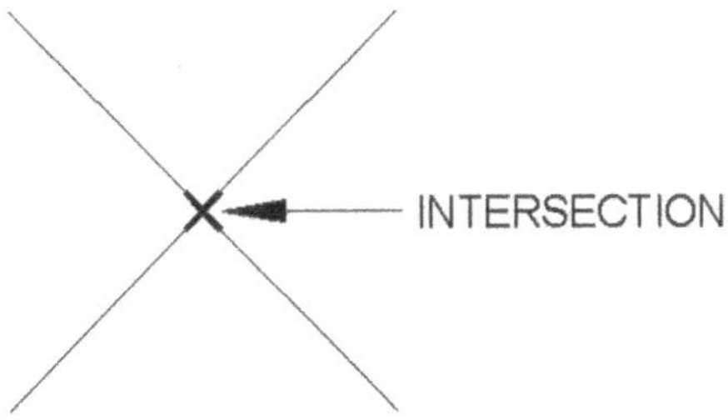

FIGURE 1.38 Quadrants.

PERPendicular

PERPendicular is an OSNAP point that allows you to begin or end a line perpendicular to another line. The symbol for it is the standard mathematical symbol for a perpendicular (90 degrees) angle, as illustrated in Fig. 1.39.

These are the six fundamental OSNAP points we mention for now. There are others of course, such as Geometric Center, Node, Extension, Insertion, Tangent, Nearest, Apparent Intersection, and Parallel (a total of 14), but those are used far less often. We touch on a few of them, such as Tangent, Nearest, Node, and Geometric Center (covered in upcoming chapters), but the remaining ones are not discussed.

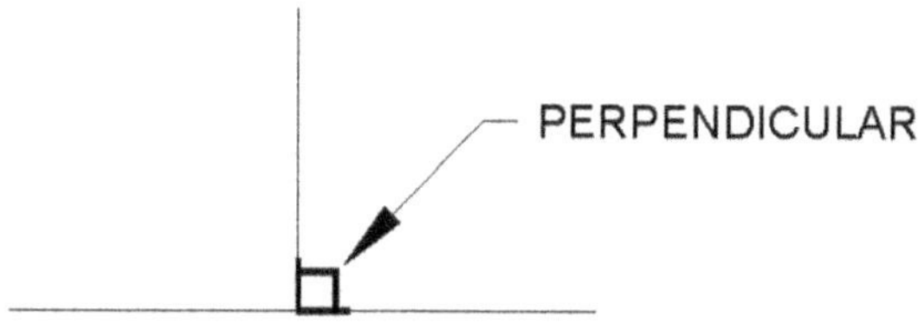

FIGURE 1.39 Perpendicular.

So, how do you use the OSNAPs that we have developed so far? There are two methods. The first is to type in the OSNAP point as you need it (only the capitalized part needs to be typed in: END, MID, etc.). For example, draw a line. Then, begin drawing another line, but before clicking to place it, type in `end` and press Enter. AutoCAD adds the word of, at which point you need to move the cursor to roughly the end of the first line. The snap marker for endpoint (a square) appears, along with a tool tip stating what it is. Click to place the line; now your new line is precisely attached to the first line.

In the same manner, try out all the other snap points mentioned (draw a few circles to practice QUAD and CEN). In all cases, begin by typing in line, then one of the snap points, followed by clicking Enter. Next, bring the mouse cursor to the approximate location of that point and click.

1.10 OSNAP drafting settings

As you may imagine, there is a faster way to use OSNAP. The idea is to have these settings always running in the background, so you need not type them in anymore. To set them up, you need to type in osnap or click on the black triangle at the OSNAP icon in the drawing aids area in the lower right of the screen. The Drafting Setting dialog box appears (Fig. 1.40). First, make sure the Object Snap tab is selected, then select the Clear All button. Finally, check off the six settings mentioned previously (Endpoint, Midpoint, Center, Quadrant, Intersection, Perpendicular), and press OK.

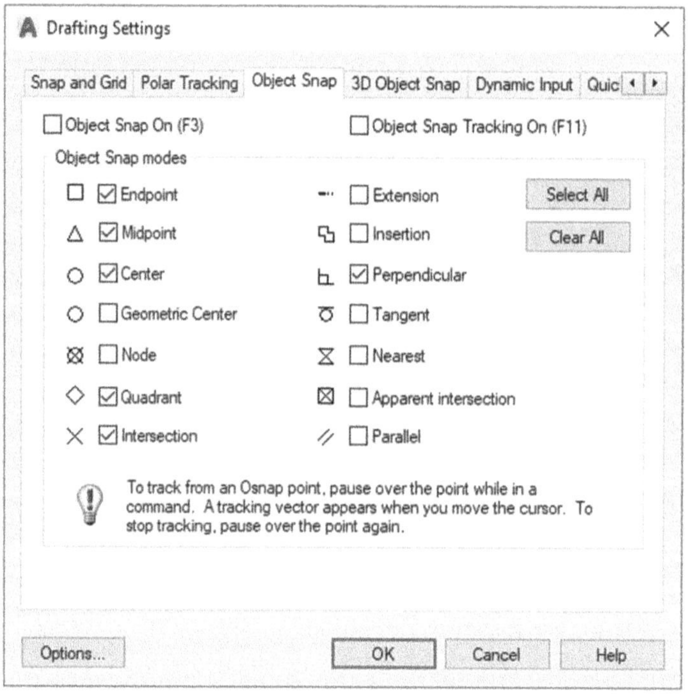

FIGURE 1.40 Drafting Settings dialog box showing the suggested Running OSNAPS.

So now you have set the appropriate settings, but you still need to turn them on. For that, we need to mention another F key. If you press F3 or press the OSNAP button at the bottom right of the screen in the drawing aids (looks like a square), the OSNAP settings become active. Go ahead and do this, then begin the same line drawing exercise; see how much easier it is now as the snap settings appear on their own. This is the preferred way to draw, with them running in the background. Feel free to press F3 anytime you do not need them temporarily. Running OSNAPs are typically something that you want all the time but turning on more than the suggested six is not recommended as it tends to overwhelm new users.

Summary

You should understand and know how to use the following concepts and commands before moving on to Chapter 2:

- AutoCAD drawing environment
  - Drawing Area, Command Line, Ribbon, Navigation Bar, Etc
- Create Objects
  - Line
  - Circle
  - Arc
  - Rectangle
- Edit/Modify Objects
  - Erase
  - Move
  - Copy
  - Rotate
  - Scale
  - Trim
  - Extend
  - Offset
  - Mirror
  - Fillet
- View Objects
  - Zoom
  - Pan
  - Regen
- Selection methods
  - Crossing
  - Window
  - Lasso
- Ortho (F8)
- OSNAP (F3)
  - ENDpoint
  - MIDpoint
  - CENter
  - QUADrant
  - INTersection
  - PERPendicular

Additionally, you should understand and know how to use the following tips introduced in this chapter:

TIP 1: The Esc (Escape) key in the upper left-hand corner of your keyboard is your new best friend while learning AutoCAD. It gets you out of just about any trouble you get yourself into. If something does not look right, just press the Esc key and repeat the command. Mine was worn out learning AutoCAD, so expect to use it often Esc should not be a substitute for ending commands "cleanly" with enter, spacebar or right click meus.
TIP 2: To quickly erase everything on the screen, type in `e` for `erase`, press Enter, and type in `all`. Then press Enter twice. After the first Enter, all the objects are grayed out, indicating they were selected successfully for erasing.

The second Enter finishes the command by deleting everything. Needless to say, this command sequence is useful only in the early stages of learning AutoCAD, when you do not intend to save what you are drawing. You may want to conveniently forget this particular tip when you begin to work on a company project. For now, it is very useful, but we will learn far more selective erase methods later.

TIP 3: Type in `z` for `zoom`, press Enter, type in the letter `e`, and then press Enter again. This is called Zoom to Extents and makes AutoCAD display everything you sketched, filling up the available screen space. If you have a middle-button wheel on your mouse, then double-click it for the same effect. This technique is very important to fit everything on your screen.

TIP 4: To perform the undo command in AutoCAD, just type in `u` and press Enter. Do not type the entire command—**if you do that, more options appear, which we do not really need at this point, so a simple** `u` suffices. You can do this as many times as you want to (even to the beginning of the drawing session). You can also use the familiar **Undo** and **Redo** arrows seen at the very top of the screen in the Quick Access Toolbar. AutoCAD also has an **oops** command. This command just restores the last thing you erased, even if it was a while ago, without undoing any other steps. Keep it in mind; it can be useful as well. To try it, erase something, then type in **oops** and press Enter.

TIP 5: Hate to keep having to type, pull a menu, or click an icon each time you practice the same command? Well, no problem. Simply press the spacebar or Enter, and this repeats the last command you used.

Review questions

Answer the following questions based on what you learned in this chapter. You should have the basic commands memorized, so try to answer most of the questions without looking up anything:

1. List the four Create Objects commands covered.
2. List the 10 Edit/Modify Objects commands covered.
3. List the three View Objects commands covered.
4. What is a workspace?
5. What two main and two outdated methods are used to issue a command in AutoCAD?
6. What is the difference between a selection Window, Crossing, and Lasso?
7. What is the name of the command that forces lines to be drawn perfectly straight horizontally and vertically? Which F key activates it?
8. What six main OSNAP points are covered? Draw the snap point's symbol next to your answer.
9. What command brings up the OSNAP dialog box?
10. Which F key turns the OSNAP feature on and off?
11. How do you terminate a command in progress in AutoCAD?
12. What sequence of commands quickly clears your screen?
13. What sequence of commands allows you to view your entire drawing?
14. How do you undo a command?
15. How do you repeat the last command you just used?

Exercises

1. Draw the following sets of lines on the left and practice the 2-Point, 3-Point, and Tan, Tan, Tan circle options. You may use the Ribbon, or typing to start the commands. In the first two cases, use ENDpoint OSNAPs to connect the circle precisely to the drawn lines. In the third case, the TANgent OSNAP will be set for you automatically when using the Tan, Tan, Tan circle creation method. The result you should get is shown on the right. (Difficulty level: Easy; Time to completion: <5 minutes.)

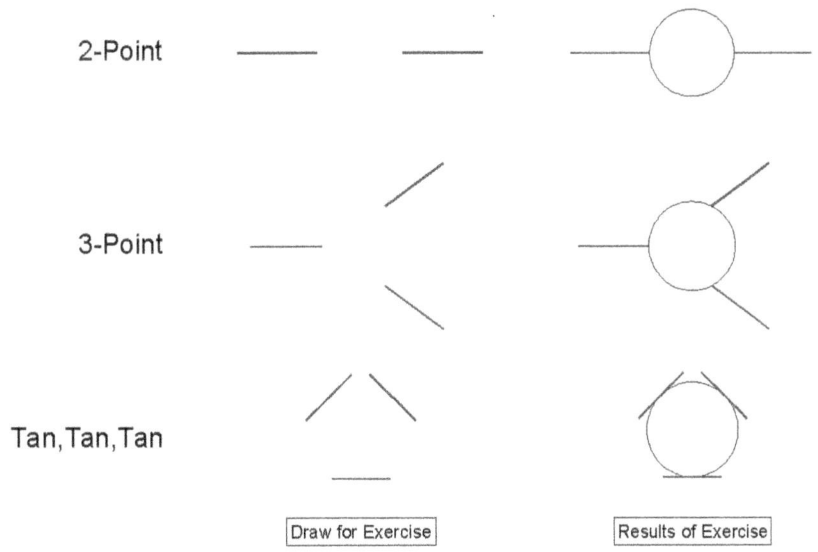

2. Draw an arc using the Start, Center, and End option (a commonly used one) as follows. First, draw the three lines shown on the left, making use of the mirror command as needed, and then draw the arc itself. Be sure to select from the bottom up, as shown, and use the ENDpoint OSNAPs. The result you should get is shown on the right. Try the same exercise by picking from the top down, noticing the difference. (Difficulty level: Easy; Time to completion: <5 minutes.)

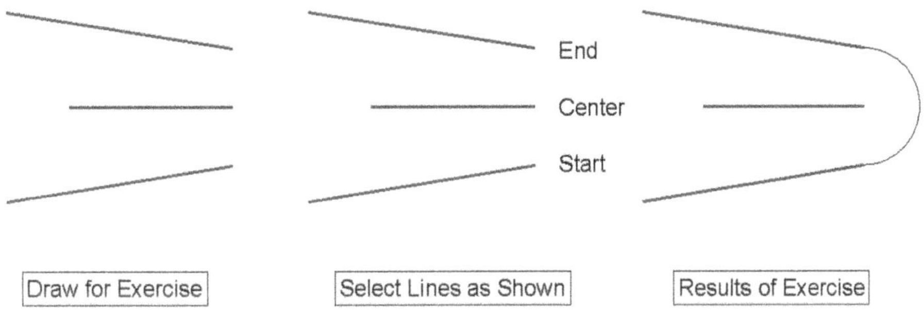

3. Draw a simple door and door swing using the Center-Start-End Arc method. This requires some precise mouse clicking and we introduce an alternate method in Chapter 3, but for now this is a good exercise to try. First draw, copy, rotate, and position the lines as seen in Steps 1 through 4, and then add the arc as shown in Step 5 (use Ortho and OSNAPs throughout). You may have to try it a few times to get the sequence correct. Finally, erase the bottom line to complete the door as seen in Step 6. (Difficulty level: Easy; Time to completion: <5 minutes.)

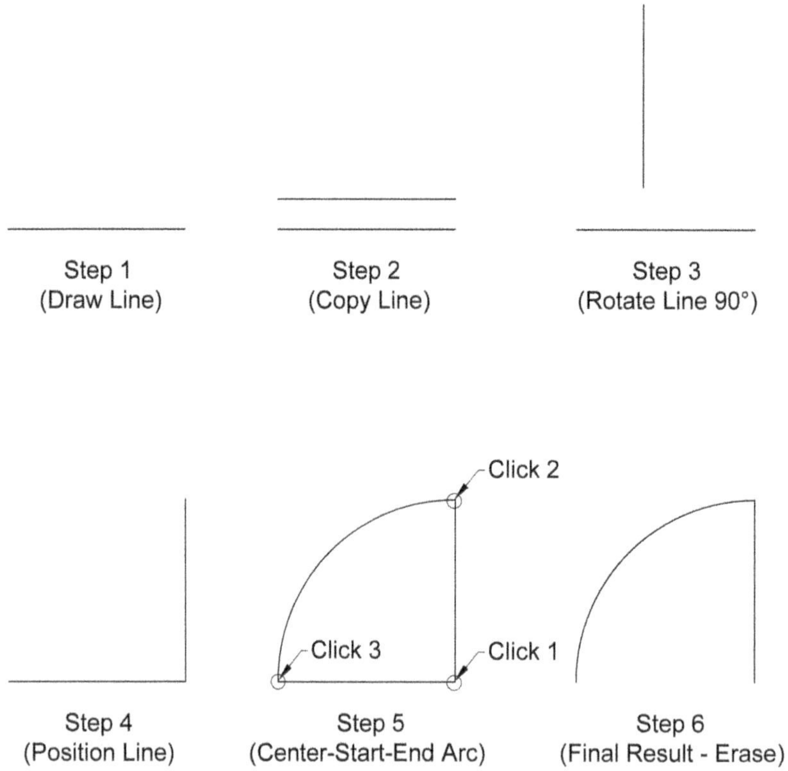

4. Draw the following sets of arbitrary electrical connection shapes. You need only the basic Create and Modify commands, including rectangle, circle, line, offset, fillet, and mirror, as well as ENDpoint, QUADrant, MIDpoint, PERPendicular, OSNAPs, and Ortho. Specific sizes are not important at this point, but make your drawing look similar, and use accuracy in connecting all the pieces. (Difficulty level: Easy; Time to completion: <10 minutes.)

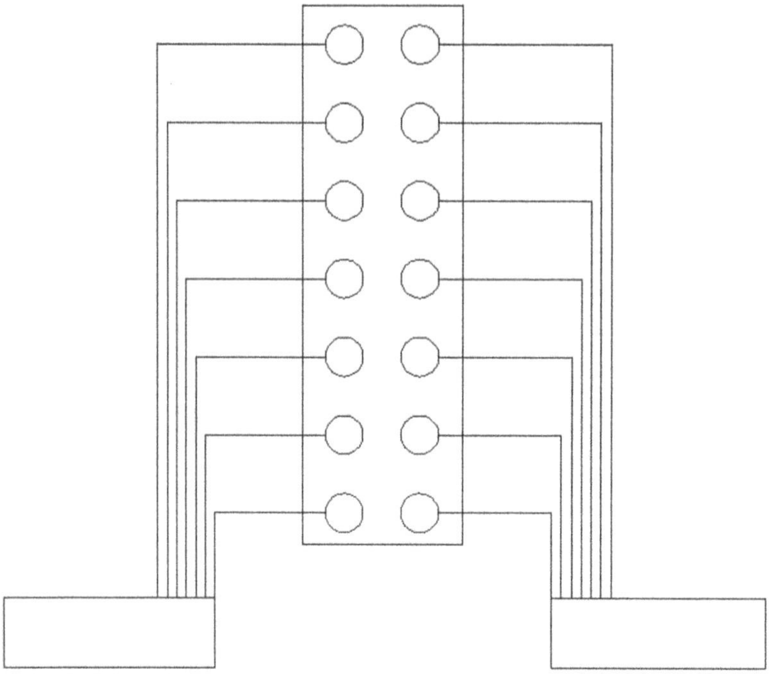

5. Draw the following house elevation using what you learned. You need only the basic Create and Modify commands. The overall sizing can be approximated; however, use Ortho (no crooked lines) and make sure all lines connect using OSNAPs. (Difficulty level: Intermediate; Time to completion: <15 minutes.)

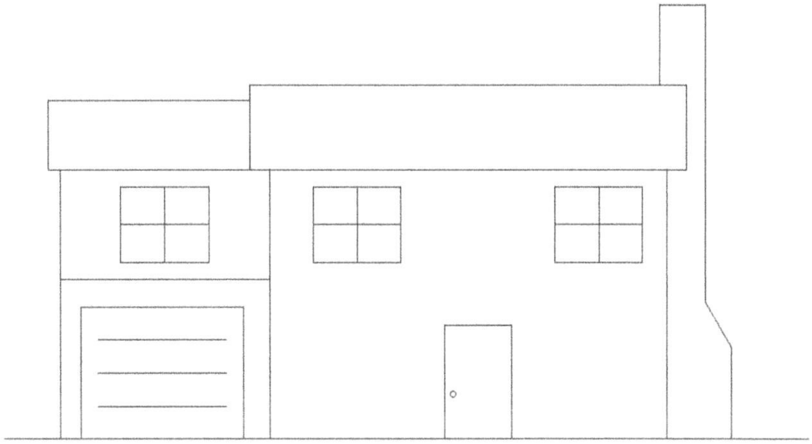

6. Draw the following sets of random shapes. You need only the basic Create and Modify commands, but you use almost every one of them, including multiple cutting edges trim. The overall sizing is completely arbitrary, and your design may not look exactly like this one, but be sure to draw all the elements you see here and use OSNAP precision throughout. (Difficulty level: Intermediate; Time to completion: <15 minutes.)

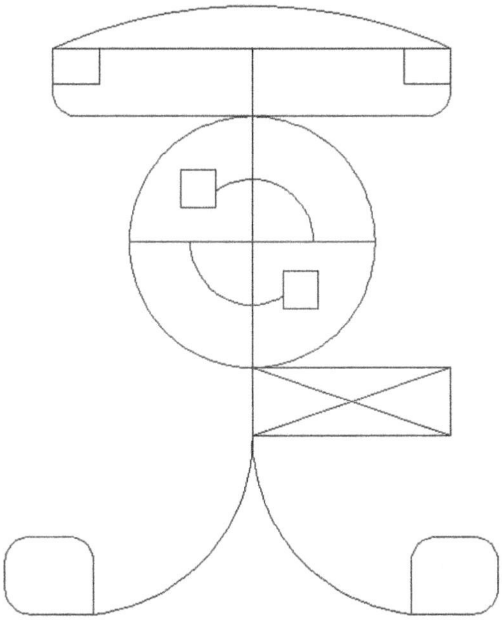

7. Draw the following side view of a car using what you learned. You need only the basic Create and Modify commands. The overall sizing (including multiple fillets needed on most corners) can be approximated; however, use accuracy and make sure all lines connect using OSNAPs. (Difficulty level: Intermediate; Time to completion: <20 minutes.)

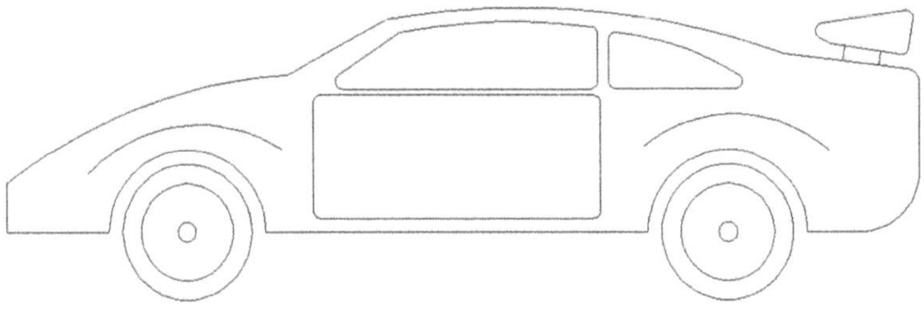

8. Draw the following elevation of a building using what you learned. You need only the basic Create and Modify commands. The overall sizing (windows, door, etc.) can be approximated; however, use accuracy and make sure all lines connect using OSNAPs. There is extensive use of mirroring to save time. (Difficulty level: Intermediate; Time to completion: <20 minutes.)

9. Logic diagrams are a familiar sight in electrical engineering; they are made up of components such as AND, OR, Not OR (NOR), Not AND (NAND), and others, as well as various inputs and outputs (I/O). They are used to construct signal flow diagrams by representing voltage flows as True (+V) and False (0V). Truth tables (you may have seen them in junior high math classes) are derived from logic diagrams. These schematics lend themselves easily to AutoCAD. Draw the following diagram using basic Create/Modify tools. Extensive use is made of the copy command. (Difficulty level: Intermediate; Time to completion: <20 minutes.)

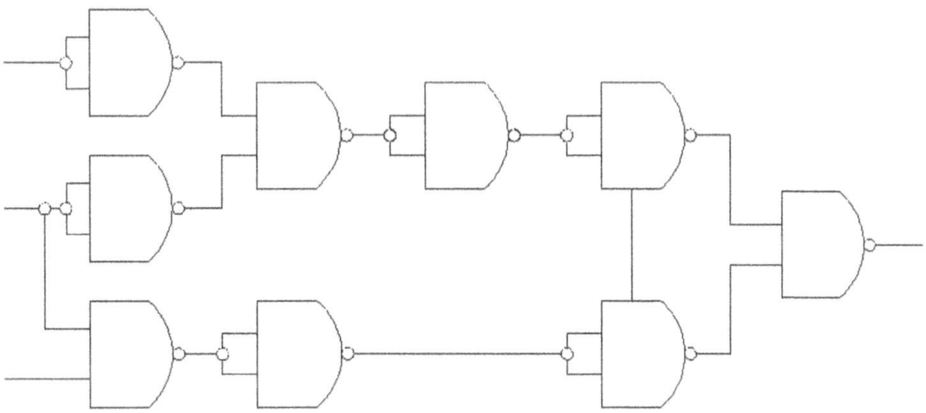

10. Draw the following set of pliers using just basic Draw/Modify commands. Extensive use is made of the copy and mirror commands, along with trim. The rest is just lines, arcs, and a few circles. (Difficulty level: Intermediate; Time to completion: <30 minutes.)

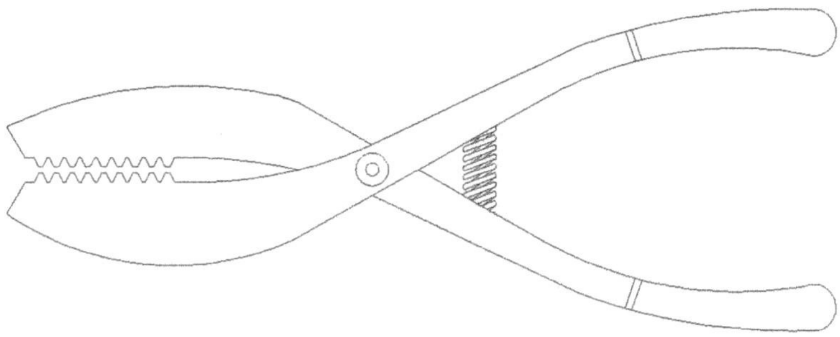

Chapter 2

AutoCAD fundamentals—Part II

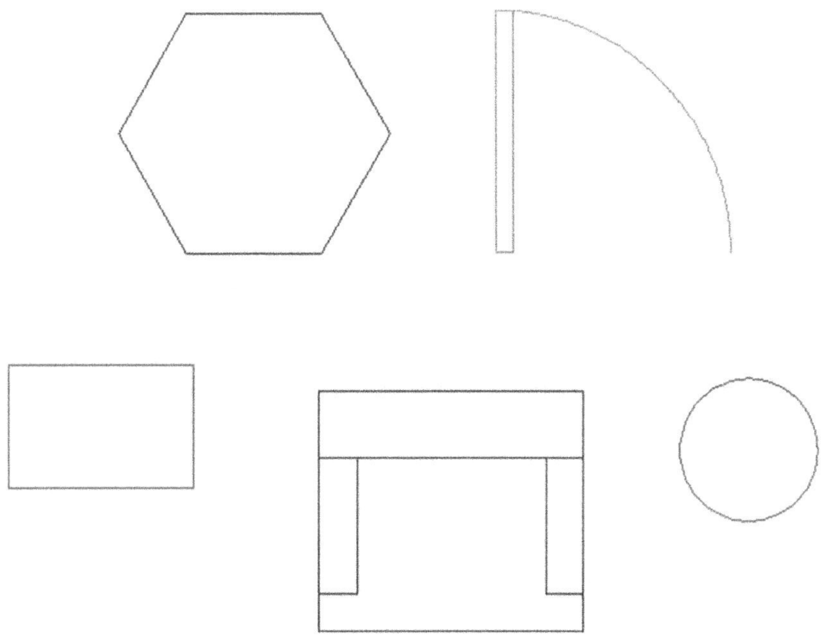

Learning objectives

You learned a great deal in Chapter 1, AutoCAD Fundamentals—Part I, in fact the majority of what you need to know to get started on a design. This chapter serves as a "cleanup" and concludes the basic introduction to AutoCAD, as there are just a few more topics to get to. Here, we discuss the following:

- Using grips
- Setting units
- Using Snap and Grid
- Understanding the Cartesian coordinate system
- Geometric data entry
- Inquiry commands
  - Area
  - Distance
  - List
  - ID
- Explode command
- Polygon command
- Ellipse command
- Chamfer command
- Templates
- Setting limits

Up and Running with AutoCAD 2022. https://doi.org/10.1016/B978-0-323-89923-9.00002-0

- Save
- Help files
- TANgent OSNAP
- Time

By the end of this chapter, you will have learned all the necessary basics to begin drafting your first project, including setting proper units, analyzing what you have done, and saving your work. After an introduction to layers in Chapter 3, you will begin a realistic architectural design. Estimated time for completion of this chapter: 2 hours.

2.1 GRIPS

We begin this chapter with grips, a natural progression from learning the OSNAP points. These are relatively advanced editing tools, and chances are you have already made them appear by accident by clicking an object without selecting a command first. That is OK; just press Esc and they go away. But, what are these mysterious blue squares? Take a close look; after learning OSNAPs, these should look familiar.

Grips are control points, locations on objects where you can modify that object's size, shape, or location. They are quite like the "handles" that you find in Photoshop or Illustrator if you are familiar with those applications.

To try out the concept, first draw some lines, arcs, circles, and rectangles. Then, activate grips by selecting all of them (as seen in Fig. 2.1). The objects become highlighted and the grips appear. When your mouse hovers over one of them, the grip turns pink. If you click on it, it becomes red, meaning it is active (or hot). With the grip activated, move the mouse around. The object changes in one of several ways. A line or circle can be moved around if the center grip is selected. A circle can be scaled up or down in size if the outer quadrants are selected. Arcs can be made larger or their length increased. Other grips, such as the endpoints of a line or corners of a rectangle, change (or stretch) the shape of that object if activated and moved. Try them all out.

There is of course much more to grips. Formally referred to as *multifunctional grips*, they perform other construction tasks via an available vertex menu that appears if you hold your mouse over them. We explore this in depth in a later chapter.

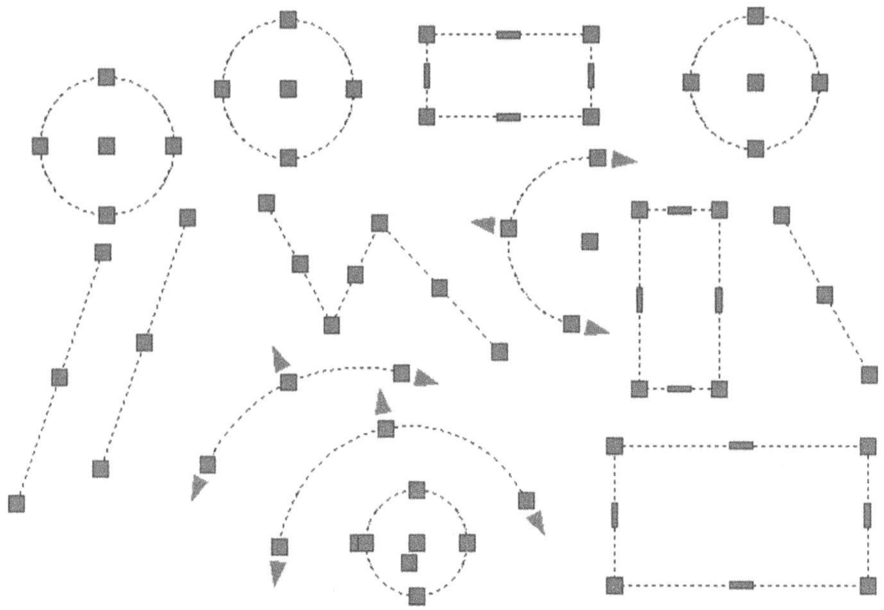

FIGURE 2.1 Grips on circles, rectangles, arcs, and lines.

2.2 Units and scale

We have not yet talked about any drawing units, such as feet, inches, or centimeters. When you practiced the offset command, you entered a value, such as 1 or 2, but what did it mean? Well, as you inadvertently discovered, the default unit system in AutoCAD, until you change it, is a simple decimal system of the type we use in everyday life. However, it is unitless, meaning that the 1 or 2 you entered could have been anything—feet, inches, yards, meters, miles, or millimeters. Think about it: It is an easy and powerful idea, but it is up to you to set the frame of reference and stick to it. So, if you are designing a city and 1 is equal to a mile, then 2 is 2 miles and 0.5 is half a mile. AutoCAD does not care; it gives you a system and lets you adapt it to any situation or design. Then, you just draw in real-life units or, as we say in CAD, 1-to-1 units. This is the Golden Rule of AutoCAD. *Always draw everything to real-life size, full size, 1:1. Do not randomly draw or scale objects up or down.*

A decimal system of units is OK for many engineering applications, but what about architecture? Architects prefer their distances broken down to feet and inches because it is always easier to understand and visualize a distance of 22′-6″ than 270″. To switch to architectural units, type in `units` then Enter and the dialog box in Fig. 2.2 appears.

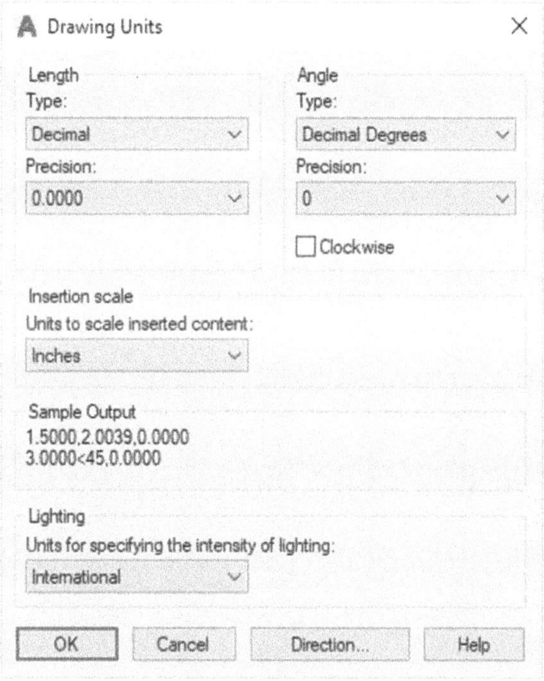

FIGURE 2.2 Drawing Units dialog box.

In the upper left, Length area, simply select Architectural from the Type: drop-down menu and ignore the rest of the options in the dialog box; there is no need to change anything else at this point.

Now, as you draw and enter in units, they are recognized as feet or inches. To enter 4 feet (such as when using the offset command), type in the number with an apostrophe sign, 4′. To enter a distance in whole inches, say 5 inches, there is no need to enter the inch sign (quotes or "), just type in the number 5. AutoCAD understands that an absence of a foot marker means the number is in inches. To mix feet and inches, type in 5′6, which is of course 5′-6″; there is no need for a hyphen as the ‘ (apostrophe) acts as a unit separator between the feet and inches. If you need a fraction, such as 5 feet, four and 1/4 inches, then type in 5′4.25. You can enter actual fractions as well with a dash between whole inches and fractions of an inch as well; simply type 5’4-1/4. Students sometimes make errors with this type of data entry, so it may be best to convert all fractions to decimals. Therefore, 18 feet 3 and 13/16 could be 18′3.8125 or 18′3-13/16.

2.3 Snap and Grid

These two concepts go hand in hand. The idea here is to set up a framework pattern on the screen that restricts the movement of the crosshairs to predetermined intervals. That way, drawing simple shapes is easy because you know that each jump of the mouse on the screen is equal to some preset unit. Snap is the feature that sets that interval, but you need the grid to make it visible. One without the other does not make sense, and usually they are set to equal values. Snap and the visible Grid are not used often in the industry by AutoCAD veterans, but it is a great learning tool, and we do a drawing at the end of this chapter using them. It also helps knowing about this feature, as it can be inadvertently activated by pressing the Snap or Grid buttons at the bottom of the screen. Lately, having a Grid on has also been the default setting in the acad.dwt template file, so many students just leave it on.

To set Snap

Type in `snap` and press Enter.

- AutoCAD says: `Specify snap spacing or [ON/OFF/Aspect/Legacy/Style/Type]<0'-0 1/2">:`

  Type in `1`, setting the snap to 1 inch.

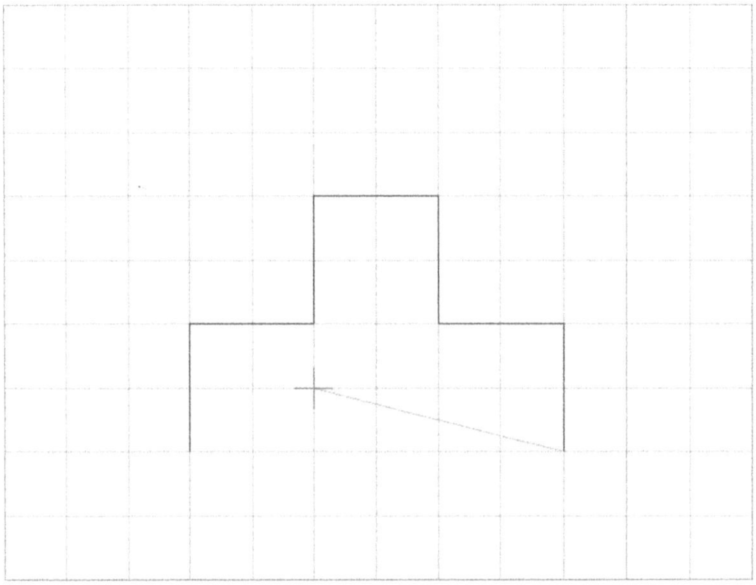

FIGURE 2.3 Figures drawn with Snap and Grid.

To set Grid

Type in `grid` and press Enter.

- AutoCAD says: `Specify grid spacing(X) or [ON/OFF/Snap/Major/aDaptive/Limits/Follow/Aspect]<0'-0 1/2">:`

  Type in `1`, setting the grid to 1 inch.

Snap and Grid are now set, although in some cases you may have to zoom to extents (Tip 3) to see everything. In older versions of AutoCAD, the grid was hard-to-see dots. In AutoCAD 2012 newer versions, it became reminiscent of graph paper, as seen in Fig. 2.3. Try drawing a line; you notice right away that the cursor "jumps" according to the intervals you set. F9 turns Snap on and off, while F7 takes care of Grid, and you can always toggle back and forth using the Drawing and Construction Aids buttons at the bottom of the screen just to the right of the MODEL icon. Grid looks like a grid, and snap looks like a bunch of dots.

Fig. 2.3 is a screenshot of Snap and Grid in use while drawing a simple shape. You can, if desired, change the color of the Grid's major and minor lines via the right-click Options…, Display tab, Colors… button. The entire Options dialog box is described later in the text, but you should explore this on your own as well.

2.4 Cartesian coordinate system

AutoCAD, and really all CAD programs, rely on the concepts of axes and planes to orient their drawings or models in space. In 2D drafting, you are always drawing somewhere on AutoCAD's X−Y plane, which is made up of the intersections of the X and Y axes. When you switch to 3D, the Z axis is made visible to allow drawing into the third dimension. The X, Y, and Z axes are part of the Cartesian coordinate system, and we need to briefly review this bit of basic math and illustrate why it is helpful to know it. Fig. 2.4 illustrates the 2D coordinate system, made up of the X and Y axes and numbered as shown.

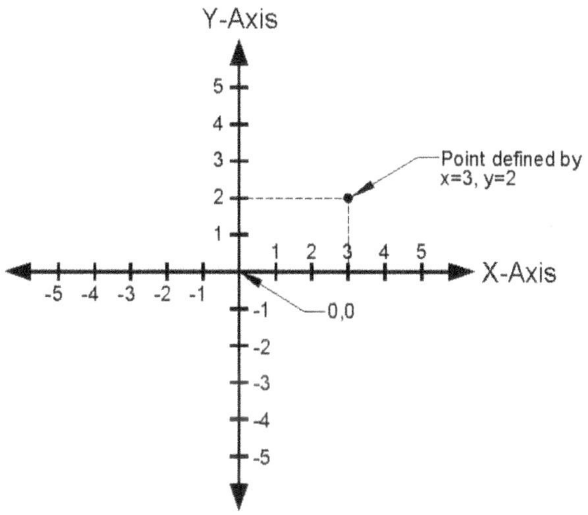

FIGURE 2.4 The Cartesian coordinate system.

This system is important not only because you may want to know where in space your drawing is, but so you can create objects of a given size by entering their X and Y values. Moreover, we refer to this system often for a variety of other construction needs and distance entries, including starting and/or inserting drawings precisely at the point 0, 0, so review and familiarize yourself with it again, as it is likely you may not have seen this in a while. Next, we introduce the various distance entry tools. As you go through Section 2.5, knowing the Cartesian coordinate system helps you understand what is going on.

2.5 Geometric data entry

The ability to accept sizing information when creating basic objects is the "bread and butter" of any drafting or modeling software. Geometric data entry refers to the various techniques needed to convey to AutoCAD your design intent, which may involve everything from the length and angle of a line to the diameter of a circle, or the size of a rectangle.

AutoCAD has two precise methods to enter distances, angles, and anything else you may need. These techniques are the older "manual" method, where you enter data via the keyboard and it appears on your command line, and the newer "dynamic" method, where you still use the keyboard, but the data appears on your screen via what is known as a "heads-up" display. This type of display can of course be used with all commands, not just for data and sizing entry. Most students probably gravitate toward the newer method, and indeed it offers many advantages, not the least of which is faster data input. We consider this the primary method in this text but present the manual method as an optional alternative. Experience has shown that some students dislike the additional on-screen clutter that Dynamic Input adds, finding the manual method less distracting. Ultimately, it is up to you to decide what to use.

Dynamic Input

Erase anything that may be on your screen and turn off all drawing aids at the bottom of the screen, including Snap, Grid, and anything else you may have on. Then, go to the drawing/construction aids menu via the Customization button (three horizontal bar "hamburger" menu at the very bottom right) to turn on the Dynamic input button visibility. Now turn on the Dynamic Input, DYNMODE drawing aid, as seen in Fig. 2.5.

FIGURE 2.5 Dynamic Input turned on.

This type of input allows you to specify distances and angles directly on the screen. Begin drawing a line and you see an example of Dynamic Input in action. An arbitrary example is shown in Fig. 2.6. Your values of course may be different, but the basic display should be the same.

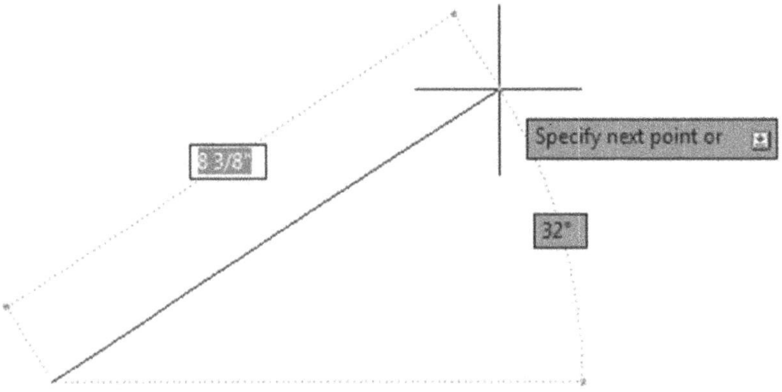

FIGURE 2.6 Drawing a line with Dynamic Input.

Notice what AutoCAD is allowing you to do. The 8 3/8″ value in Fig. 2.6 is surrounded by a box, meaning it is editable. All you have to do is enter a desired value, so enter a small value, such as 5, but do not press Enter yet. What if you wanted to change the angle of the line? Simply press the Tab key, tab over to the 32 degrees value, and enter something else, perhaps a 45. Finally, press Enter. Your line 5 units long at 45 degrees is done. The line command continues of course, and you can repeat this as much as needed. Very quick and efficient.

This works with Ortho on as well. In that case, you enter only distance values, no angles, as Ortho forces the lines to all be 90 degrees. Be sure to "lead" the mouse (and in turn the lines being created) by slightly moving the mouse in the direction you want to go.

To use Dynamic Input to draw a rectangle (5" × 3" in this example), make sure DYNMODE is on, begin the command, and click anywhere to start the rectangle. You see fields in which to enter the values for the height and width (Fig. 2.7). As before, enter a new value in the first text field (5) and Tab over to the next field, entering the other value (3). As you press Enter, the 5 × 3″ rectangle is created. Note of course that the first value is the width and the second value is the height.

The usefulness of Dynamic Input extends out to just about any shape with which you may work. Fig. 2.8 shows DYNMODE with an arc (left) and a circle (right). As before, you can specify on-screen data needed to create that particular shape.

You may notice also that some of the "heads-up" menus show a downward pointing arrow. You can use the down arrow on your keyboard to drop down the menu and reveal other choices, which of course mimic what is available in the parentheses on the command line. Fig. 2.9 is one example with the offset command. You can cycle through these options to select what you need.

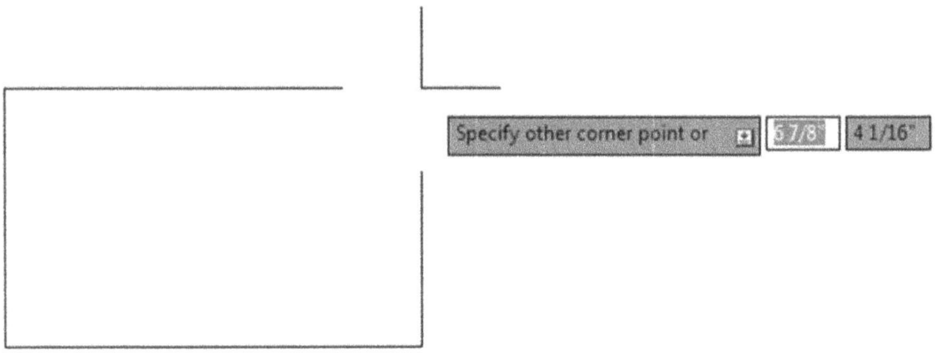

FIGURE 2.7 Drawing a rectangle with Dynamic Input.

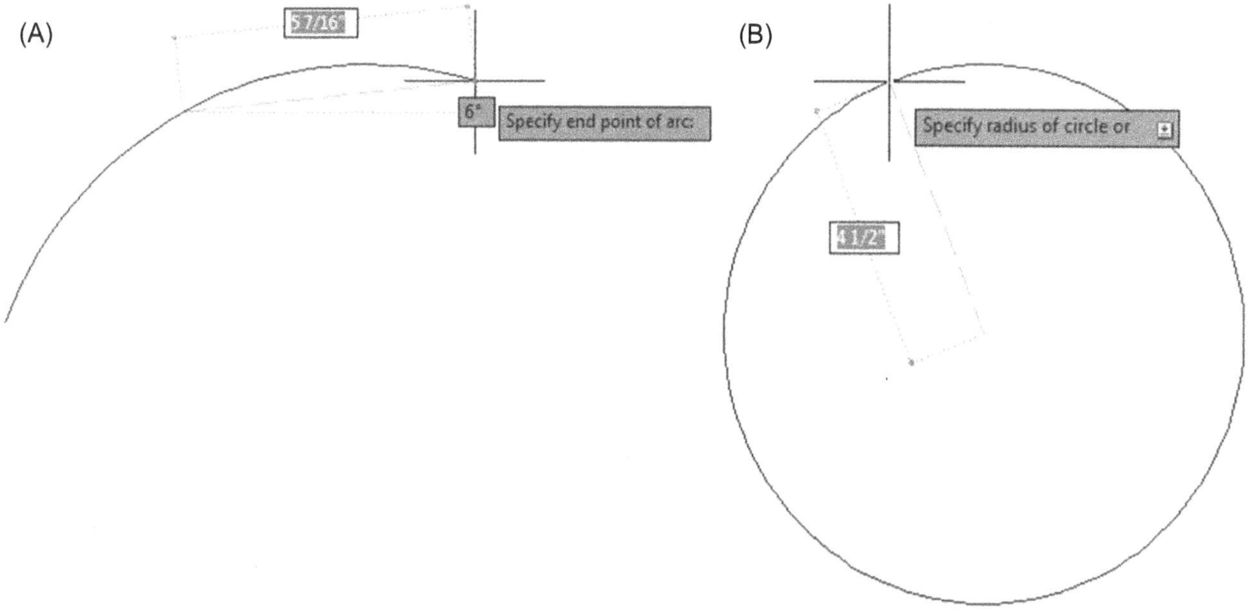

FIGURE 2.8 Drawing an arc (left) and a circle (right) with Dynamic Input.

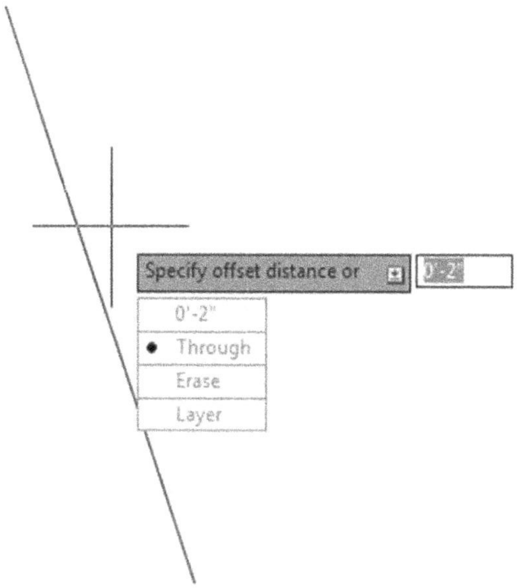

FIGURE 2.9 Dynamic Input menu, offset command.

Manual Input

Using the manual (command line) method, also called Direct Distance Entry, you can achieve the same thing that can be done with the DYNMODE input. Let us run through a few scenarios and, along the way, introduce Absolute and Relative Entry. The simplest case just involves lines. Turn off the DYNMODE button and begin the line command, clicking where you want to start. Now, all you have to do is point the mouse in the direction you want the line to go (moving the mouse a bit in that direction), type a length value on the keyboard, press Enter, and the line appears. You can continue this "stitching" indefinitely. Note that we have not yet explored how to draw the lines at various precise angles, except for 90 degrees; for that, just turn on Ortho. If using Ortho, be sure to once again "lead" the linework as before, by moving your mouse in the direction you want to go, so AutoCAD understands your intent.

Whether done manually or via DYNMODE, this point-by-point stitching, called *Direct Distance Entry*, is a useful technique. One example of its use occurs after you do a basic room or building perimeter survey. This may just be a wall-by-wall sequential field measurement of a space. When you return to the office you can quickly enter the wall lengths into AutoCAD one at a time via this technique. Do take a few minutes to practice it.

Direct Distance Entry can be done manually one other way: *Absolute Coordinate Entry*. While not a common technique for day-to-day drafting, it really gets you to practice and understand the Cartesian coordinate system. What you do is enter X and Y values (in pairs) one at a time. To practice this, begin the line command, then, instead of clicking anymore, type in the following values, pressing Enter after each one:

`0,0 0,2 2,2 2,4 4,4 4,2 6,2 6,0` then `c` for `close`

These are the X and Y values for the lines to follow; and if you typed correctly, then the shape seen in Fig. 2.10 should emerge on your screen one segment at a time.

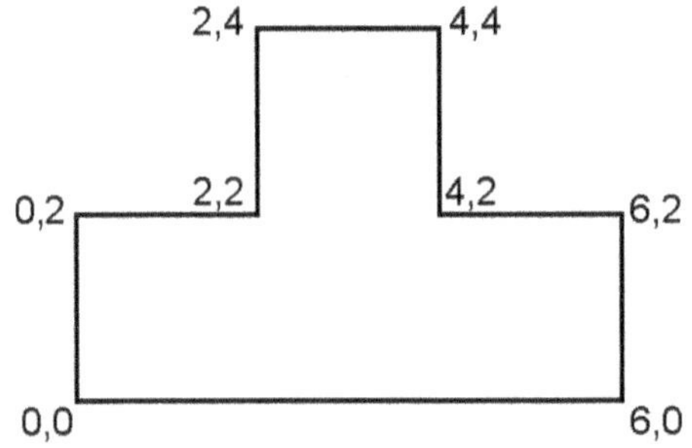

FIGURE 2.10 Shape created via Direct Distance Entry.

Another variation on Direct Distance Entry is called *Relative Data Entry*. This is useful for quickly and easily creating shapes, such as rectangles, anywhere on the screen. It also allows you to enter in various angles for lines besides the Ortho-induced 90 degrees. To do either you need to know two new symbols, @ and <.

The @ symbol frees you from the absolute 0,0 and can be interpreted as meaning "relative to." The last point in the case of a rectangle, this means you can create it relative to wherever you are in space by simply starting the rectangle command and specifying an X and a Y value preceded by the @ symbol. So, for a 5 × 3″ rectangle,

1. Start the rectangle command.
2. Click somewhere on your screen to anchor the first point.
3. Type in: @5,3.

The rectangle then appears.

Now that you know about the @ symbol and what it does, you can also draw lines at any angle, from any point, by employing the "less than" symbol (<). This is called Relative Distance Angle Entry. Any numerical value after this "alligator teeth" symbol is interpreted as an angle. The format is as follows: @Distance<Angle. Therefore, to create a line that is 10 units long at 45 degrees,

1. Start the line command.
2. Click somewhere on the screen to anchor the first point.
3. Type in: @10 < 45.

The line (10 units long, at 45 degrees) appears.

As we conclude Section 2.5, remember that, when it comes to creating a rectangle, the dynamic and basic manual methods are not the only ones available. A brief mention was made in Chapter 1, when the rectangle was first introduced, that you can also access length and width data as part of the command line suboptions, such as pressing d for Dimensions. You can vary that by specifying a total area than either the length or width. It may be worth it to briefly review all this again to have the full set of options available to you. You need to create quite a few rectangles as you begin drafting. They are the basis of furniture, mechanical components, and even entire floor plans, as we soon see. You should be well versed in all the ways to create them.

2.6 Inquiry commands

Our next topic in AutoCAD fundamentals is the Inquiry commands set. These commands are there to give you information about the objects in front of you on the screen as well as other vital data, like angles, distances, and areas. As such, it is important for you to learn them early on; the information these commands provide is especially valuable to a beginner student and will continue to be essential as you get more experienced. First seen in AutoCAD 2020 is Quick Measure. The other inquiry commands were also upgraded just a few releases ago, and they currently comprise the following:

- *Quick gives you the dimensions, distances, and angles within a 2D drawing of objects near the cursor.*
- *Area* gives you the area of an object or any enclosed space
- *Distance* gives you the distance between any two points
- *List* gives you essential information on any object or entity
- *ID* gives you the coordinates of a point
- *Radius* gives you the radius of a circle or arc
- *Angle* gives you the angle between two lines
- *Volume* gives you the volume of a 3D object
- *Mass properties* lists various mass properties of a 3D solid

We cover the first five inquiry commands in detail (quick, area, distance, list, and ID) and include a brief mention of radius and angle. Volume and mass properties are not discussed here, as those are for 3D work. As before, most commands can be accessed through the usual ways of typing and the Ribbon.

FIGURE 2.11 Ribbon Home tab, Utilities panel.

Quick

Keyboard: **MEASUREGEOM**
Ribbon: **Home tab→Measure, Quick**

With the new Quick option, measuring has become much faster and more intuitive. With this option, you can quickly review the dimensions, distances, and angles within *and* between 2D drawing objects just by moving your cursor near objects. Ninety-degree angles will also be shown with squares in the corners where appropriate. It is best to zoom in to complex areas to avoid quick dimension clutter and to improve computer performance. *Note that the only way to get to this new command is through the Ribbon. It is not available through a keyboard entry* (Figs. 2.12 and 2.13).

Area

Keyboard: Type in **area** and press Enter
Ribbon: **Home tab→Measure, Area**

Area is a very important and useful Inquiry command. It is used often by architects to verify the area of a room or an entire floor plan. Area calculations are performed in two ways: either on an object or, more commonly, point by point (such

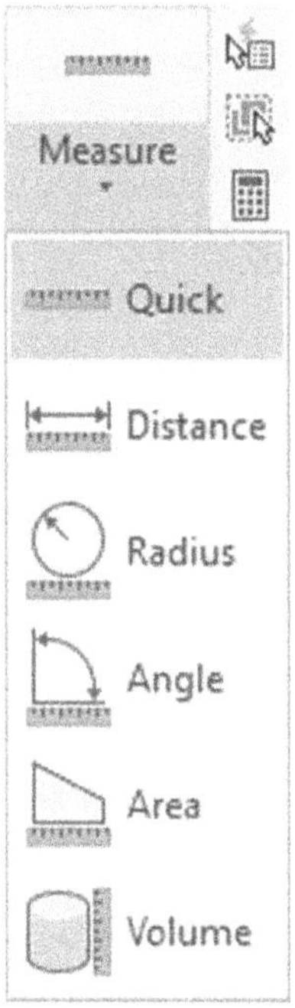

FIGURE 2.12 Quick Measure ribbon access.

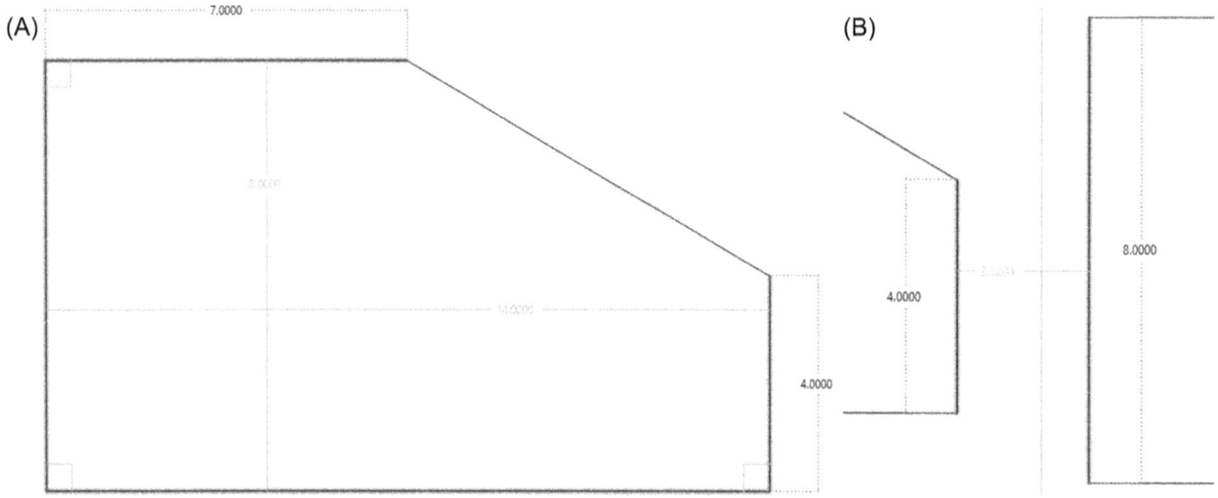

FIGURE 2.13 Quick Measure inside geometry and outside geometry measurements.

as the area of a room). Draw a rectangle of any size. We go over both methods, although with a rectangle you would really use only the object method, saving the point-by-point approach for a more complex floor plan.

Point by point

This is used when measuring an area defined by individual lines (not an object).

Step 1. Make sure your units are set to Architectural and you have a rectangle on your screen.

Step 2. Begin the area command via any of the preceding methods.

- AutoCAD says: `Specify first corner point or [Object/Add area/Subtract area] <Object>:`

Step 3. Activate your OSNAPs (F3) and click your way around the corners of the rectangle until the green shading covers the entire area to be measured (four clicks total).

- AutoCAD says:

Specify next point or [Arc/Length/Undo]:
Specify next point or [Arc/Length/Undo]:
Specify next point or [Arc/Length/Undo/Total] <Total>:
Specify next point or [Arc/Length/Undo/Total] <Total>:

Step 4. When done, press Enter. You get a total for area and perimeter. Mine said:
Area = 67.9680, Perimeter = 34.2044

Yours of course will likely say something different.

Object

This is used when you have an actual *object* like a rectangle or circle (not a visually "closed" shape, defined by individual lines) to measure.

Step 1. Begin the area command via any of the preceding methods.
- AutoCAD says: Specify first corner point or [Object/Add area/Subtract area/eXit] <Object>:

Step 2. Press the letter o for Object.
- AutoCAD says: Select objects:

Step 3. Select the square. You should of course get the same area and perimeter measurements.

Distance

Keyboard: Type in **dist** and press Enter
Ribbon: **Home tab→Measure, Distance**

The distance command gets you the distance between any two points. Although we practice on the corners of a rectangle, the command works on any two points, as selected by a mouse click, even an empty space.

Step 1. Make sure your units are set to Architectural, a rectangle is on your screen, and your OSNAPs are activated.

Step 2. Begin the distance command via any of the preceding methods. (Be sure to read the note after Step 4.)
- AutoCAD says: Specify first point:

Step 3. Select the lower left corner of the rectangle.
- AutoCAD says: Specify second point or [Multiple points]:

Step 4. Select the upper right corner across the rectangle. AutoCAD then gives you some information. Mine said:
Distance = 10.9152, Angle in XY Plane = 23, Angle from XY Plane = 0
Delta X = 10.0440, Delta Y = 4.2731, Delta Z = 0.0000

Yours will likely say something different since you are measuring a different shape.

Note: There is an important difference in the visual output if you use the keyboard or the Ribbon with DYNMODE on or off. With DYNMODE on, using the Ribbon you get a graphical representation of the distance and associated geometry, as seen in Fig. 2.14. With DYNMODE on using the keyboard, you only get the one distance you chose. With DYNMODE off using the Ribbon or keyboard all distance reporting will be in the command line history. (press F2 to bring up the command line history if needed.) See Fig 2.14.

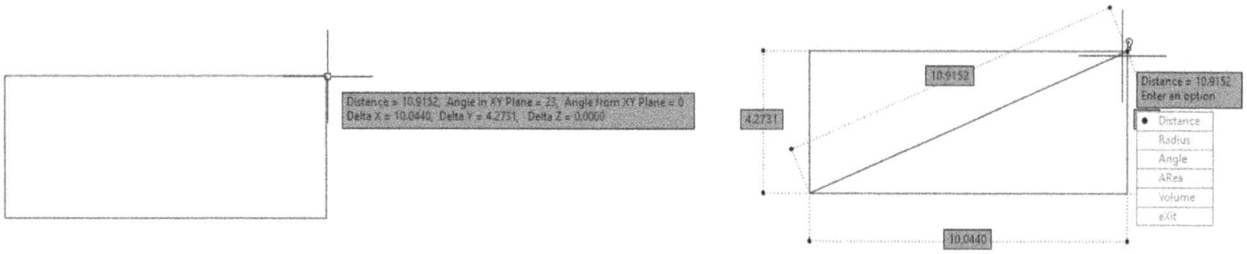

FIGURE 2.14 Distance command DYNMODE on, Keyboard vs. Ribbon results.

List

Keyboard: Type in **list** and press Enter
Ribbon: **Home tab→Properties** List

The list command gives you information about not just the geometric properties of the object but also its CAD properties, such as what layer it is on, the subject of the next chapter. It is an extremely useful command that clues you in on what you are looking at (whether a beginner or an expert, you will sometimes ask yourself this question).

Step 1. Make sure you have something on your screen to investigate. Draw a line, rectangle, or circle.

Step 2. Begin the list command via any of the preceding methods.

- AutoCAD says: `Select objects:`

Step 3. Select your object (it becomes highlighted) and press Enter. The command line history pops up above the command line. Look it over and close it with F2 (or just click in the command line) when done. What mine shows is a list about an arbitrary rectangle which is shown in Fig. 2.15.

```
Command:
Command:
Command: _list
Select objects: 1 found
Select objects:
                  LWPOLYLINE  Layer: "0"
                            Space: Model space
                   Handle = 216
           Closed
    Constant width    0.0000
              area   42.9192
         perimeter   28.6342
          at point  X=  18.0289  Y=    9.8461  Z=    0.0000
          at point  X=  28.0729  Y=    9.8461  Z=    0.0000
          at point  X=  28.0729  Y=    5.5730  Z=    0.0000
          at point  X=  18.0289  Y=    5.5730  Z=    0.0000
```

FIGURE 2.15 List command AutoCAD text window.

ID

Keyboard: Type in **id** and press Enter
Ribbon: **Home tab→Utilities, ID Point**

This command is the simplest of all. It tells you the coordinates of a point in space, such as the intersection of two lines. This is useful for inserting blocks (see Chapter 7) and just plain old figuring out the general location of an object, something that may be needed for Xrefs (Chapter 17).

Step 1. Make sure you have something on your screen to ID. Create a line or a rectangle.

Step 2. Begin the ID command via any of the preceding methods.

- AutoCAD says: `Specify point:`

Step 3. Using OSNAPs, select a point, such as the corner of a rectangle or end of a line. AutoCAD gives you the X, Y, and Z coordinates. Here is what I got; you of course will likely get different values:

`X = 11.1504 Y = 6.7753 Z = 0.0000`

Radius and Angle

You should now have a good feel for the Inquiry toolset, so go ahead and practice the remaining two commands on your own. Radius and angle cannot be typed in; you must use the Ribbon.

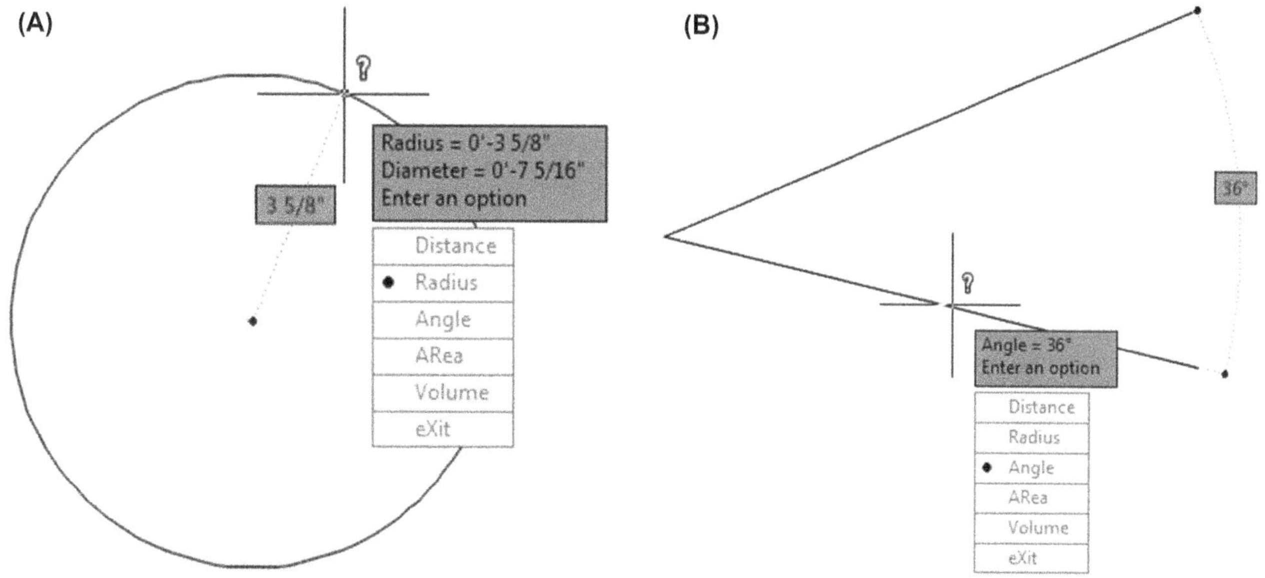

FIGURE 2.16 Radius (left) and Angle (right) measurements.

To practice radius, draw an arbitrary circle, then activate the command and select it. A radius is displayed, and you can just press Esc to get out. The result is shown in Fig. 2.16 (left). To get an angle reading, create two "alligator teeth" lines, activate the command, and select both lines. The angle is displayed, and again, you can press Esc to exit. The result is shown in Fig. 2.16 (right). Note that these commands are "dynamic" in nature, so DYN activates when you use them.

This concludes the section on Inquiry commands. We are now in the home stretch and will explore some commands and concepts that do not quite fit in neatly anywhere else but will prove to be very useful as you start a realistic project in the next chapter and, of course, everyday AutoCAD use.

2.7 Additional drafting commands

We are going to take a moment to explore four additional drafting commands: explode, polygon, ellipse, and chamfer. They are not mentioned in Chapter 1 because they are not part of the essential set of drafting commands. The need for lines and rectangles in a typical drawing is far greater than the need for ellipses or polygons. However, in the few cases you do need one of those shapes, there really is no substitute. We start off with the unusual explode command.

Explode

Keyboard: Type in **explode** and press Enter
Ribbon: **Home tab**

Explode is used to make complex objects simpler. That means a rectangle becomes four separate lines, and other objects we have yet to study (blocks, text, dimensions, hatch, etc.) are simplified into their respective components. This command is mentioned in almost every chapter, as virtually all AutoCAD entities can be exploded—although many should never be! To try it out, we again need a rectangle of any size. Make sure you are using the actual rectangle command, not just four connected lines. Those are already in their simplest form and explode does not work on them.

Step 1. Begin the explode command via any of the preceding methods.
- AutoCAD says: `Select objects:`

Step 2. Pick the rectangle and it becomes highlighted.
- AutoCAD says: `1 found`

Step 3. Press Enter and the rectangle is exploded.

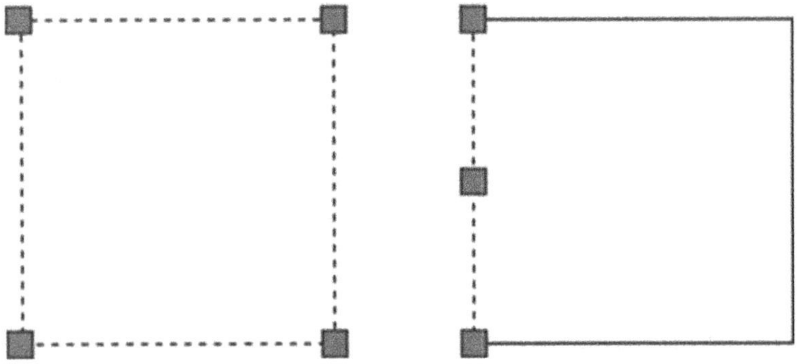

FIGURE 2.17 Unexploded rectangle (left) and exploded rectangle (right).

Click any of its lines and you notice that each line is independent, as seen in Fig. 2.17. Keep in mind that you cannot explode circles, arcs, and lines (previously mentioned), as those are already the simplest objects possible. There is a way to reverse explode, in addition to an immediate Undo. You can use the Join command or Pedit (discussed in chapter: Advanced Linework), but that applies only to simple compound objects, such as a rectangle. If you explode a hatch pattern discussed in Chapter 5, (it is possible, but do not do it) and save your work before using *undo*, that is it…, no going back. This is why you see appropriate warnings when we get to those chapters.

Polygon

Keyboard: Type in **polygon** and press Enter
Ribbon: **Home tab→Draw**

The polygon command is a useful tool that we need in later chapters for a project. A polygon, strictly speaking, is any shape with three or more sides to it. Of course, in those cases when it has three sides (triangle), you can draw it with existing line techniques. Same thing with four sides (square or rectangle). However, what about five sides or more, such as a pentagon or hexagon? Here, a polygon command is needed, as it is a real chore to draw these shapes using the line command—just try figuring out all the correct angles.

Drawing polygons, however, presents a complication you do not have with simpler shapes. With a rectangle, you specify the height and width and that is it, but how do you do that with, say, a pentagon? Fig. 2.18 is a picture of one. What is the height? What is the width? It is not an easily measured shape.

The solution that drafters came up with a long time ago (many of AutoCAD's techniques are just adaptations of pencil-and-paper board-drafting methods used for decades) is to enclose the shape in a circle of a known radius or diameter. Of course, you can also enclose the circle inside the polygon.

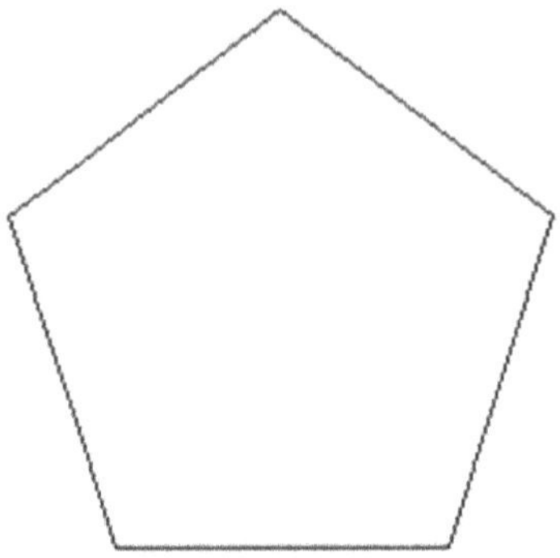

FIGURE 2.18 Basic pentagon.

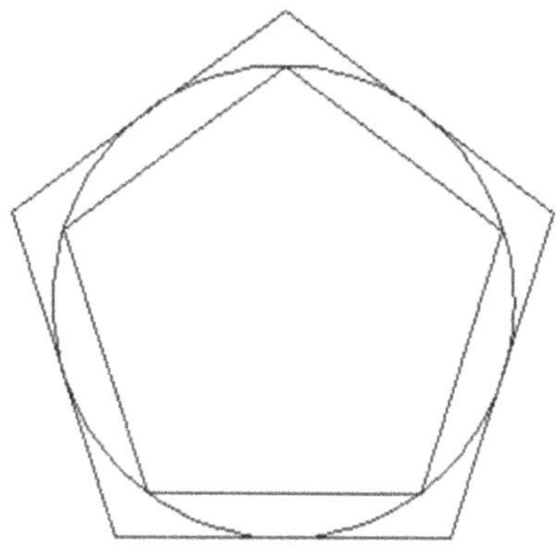

FIGURE 2.19 Inscribed and circumscribed pentagons.

With either approach, it is much easier to specify a size. You just say, "I need a pentagon that fits inside (or outside) a circle of 10 inches in diameter." This is where the terms *inscribed* and *circumscribed* come from. Fig. 2.19 is a diagram of two pentagons, one inscribed in a circle and one circumscribed about a circle. Note the tangencies. In one case, the points of the polygon touch the circle; in the other case, its sides touch it.

Let us go through the steps in making a polygon inscribed in a circle. Begin by drawing a circle with an arbitrary radius of 5. Note that we do not need this step to create a polygon, but for your first try, it is a great help in visualizing what you are doing. Then proceed as shown next.

Step 1. Begin the polygon command via any of the preceding methods.
- AutoCAD says: `Enter number of sides <4>:`

Step 2. Enter the number 5 for five sides (making it a pentagon), and press Enter.
- AutoCAD says: `Specify center of polygon or [Edge]:`

Step 3. With your OSNAPs on, click to select the center of the circle.
- AutoCAD says: `Enter an option [Inscribed in circle/Circumscribed about circle] <I>:`

Step 4. Now that you know what these terms mean, go ahead and press Enter to select `I` for `Inscribed in circle` (the default value). You can just as easily select c as well.
- AutoCAD says: `Specify radius of circle:`

Step 5. Enter the value `5`. A screenshot of the result is shown in Fig. 2.20.

You can now erase the circle, since it was just a guide. In the future you do not need it; just click to place the polygon anywhere on the screen. Go ahead and practice this command, but this time select a different number of sides (six or more) and use the Circumscribed option. This new command leads us into the first challenge exercise.

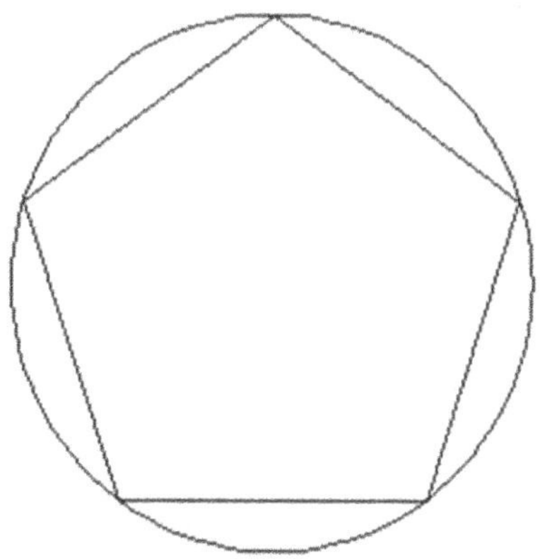

FIGURE 2.20 Results after Step 5.

Drawing challenge—star

Try this exercise based on what you learned. Draw a perfect five-sided star, such as the one shown in Fig. 2.21. Think about everything we talked about up to this point. Only four commands are needed to create the star. Take a few minutes to try to draw it. If you cannot get it to look right, the answer is at the end of this chapter, just before the summary.

FIGURE 2.21 Drawing challenge—star.

Ellipse

Keyboard: Type in **ellipse** and press Enter
Ribbon: **Home tab→ Draw**

The ellipse command is easy to use and is needed when we draw furniture in Chapter 4, Text, Mtext, Editing, and Style, so it is included here. The command can be executed in one of the two ways: Center or Axis, End. You can also create a partial ellipse by defining just a section of it.

Step 1. Begin the ellipse command via any of the preceding methods.
- AutoCAD says: `Specify axis endpoint of ellipse or [Arc/Center]:`

Step 2. Click anywhere on the screen.
- AutoCAD says: `Specify other endpoint of axis:`

Step 3. With Ortho on (not required but good to have), click again somewhere off to the left or right of the first point. Then, move the mouse back toward the center as the ellipse is formed, as seen in Fig. 2.22.
- AutoCAD says: `Specify distance to other axis or [Rotation]:`

Step 4. Click anywhere again to complete the ellipse. You can also specify a rotation angle along the major axis (the line between clicks 1 and 2). This would be as if the ellipse were rotating away (or toward) you. For a variation on this, try the Arc submenu of the ellipse, which is essentially similar in execution until the final step, when you trace how much of the ellipse you want to see, as shown in Fig. 2.23.

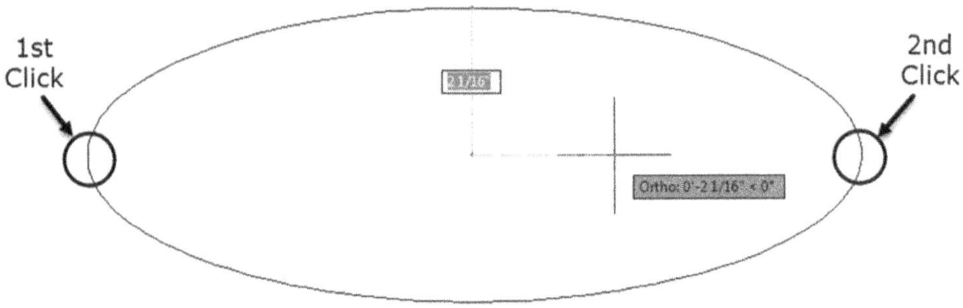

FIGURE 2.22 Ellipse after Step 3.

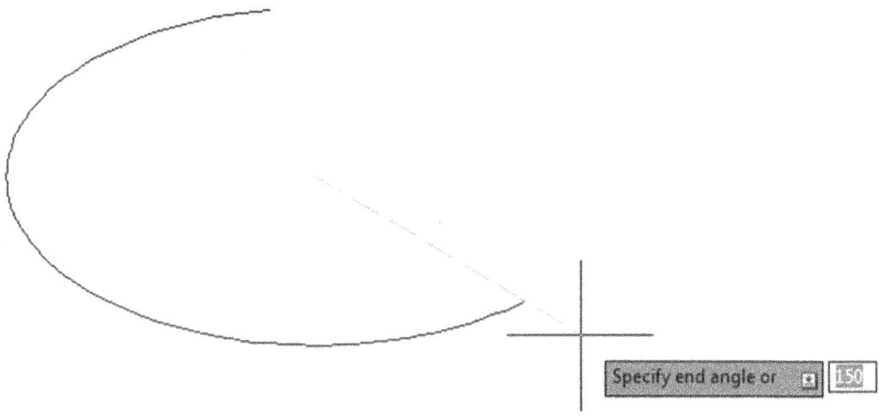

FIGURE 2.23 Ellipse, arc variation.

Chamfer

Keyboard: Type in **chamfer** and press Enter
Ribbon: **Home tab→ Draw**

You may have already tried the chamfer command, as it is combined with Fillet command drop down in Ribbon. While normally not used as often as a fillet, it is still very much worth investigating. A chamfer is a beveled edge connecting two surfaces. Think of it as a fillet that is angled instead of round. Chamfers may have a 45-degree corner, but this is not required; AutoCAD allows for asymmetrical chamfers. Just like Fillet, a Chamfer with both distances set to 0 will make corners

To try out the command, draw two perpendicular lines, like the fillet exercise in Chapter 1 (reproduced in Fig. 2.24), and follow these steps.

Step 1. Begin the chamfer command via any of the preceding methods.

- AutoCAD says:`(TRIM mode) Current chamfer Dist1 = 0.0000, Dist2 = 0.0000`
`Select first line or [Undo/Polyline/Distance/Angle/Trim/mEthod/ Multiple]:`

Step 2. No distance has been entered yet, and any attempt at chamfer results in a chamfer of zero. Press `d` for `Distance`.

- AutoCAD says: `Specify first chamfer distance <0.0000>:`

Step 3. Enter a small value, such as `1`.

- AutoCAD says: `Specify second chamfer distance <1.0000>:`

Step 4. Here the `1` is not needed again, so just press Enter.

- AutoCAD says: `Select first line or [Undo/Polyline/Distance/Angle/Trim/mEthod/Multiple]:`

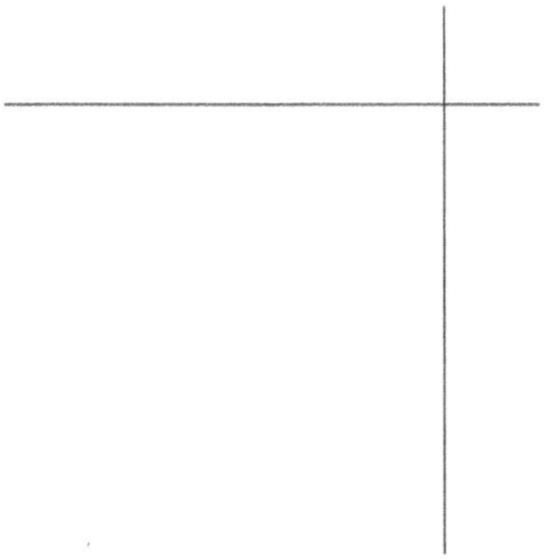

FIGURE 2.24 Chamfer setup.

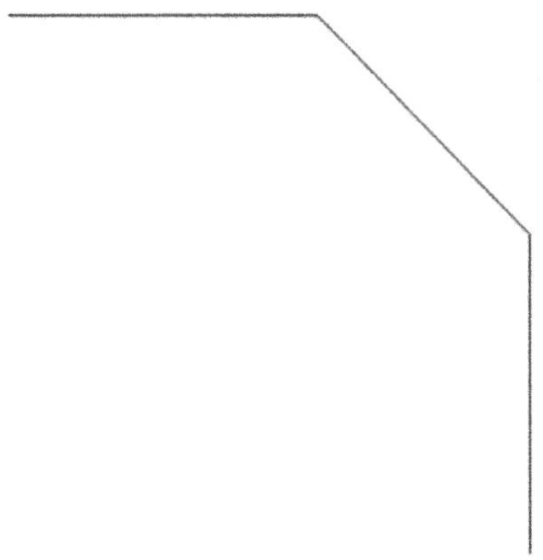

FIGURE 2.25 Chamfer executed.

Step 5. Select the first line.

- AutoCAD says: `Select second line or shift-select to apply corner or[Distance/Angle/Method]:`

Step 6. As you move the mouse toward the second line, you again see a preview of the completed command (similar to fillet). Select the second line and the chamfer is created as seen in Fig. 2.25. Zoom in if needed to see it.

If the chamfer fails, it is possible that you selected values too large for the available lines. Another way to create a chamfer is using the length of one line segment together with an angle, using the Angle option. Experiment with both.

Templates

Templates are something of a mystery for a beginner AutoCAD student. What exactly are they? While templates can get complicated in what they contain, they are just files with some information embedded in them, with all the proper settings but no actual design work present. So it is a blank page with some intelligence built in and everything set up and ready for you to draw. These templates can be saved under the *.dwt* extension, and you can certainly make your own.

In this case, using the Application Menu (Big red A) New button or the Quick Access Toolbar New icon, you default to the screen in Fig. 2.26. You should have some déjà vu with this. It is the same as Fig. 1.2, and it is how you opened a blank

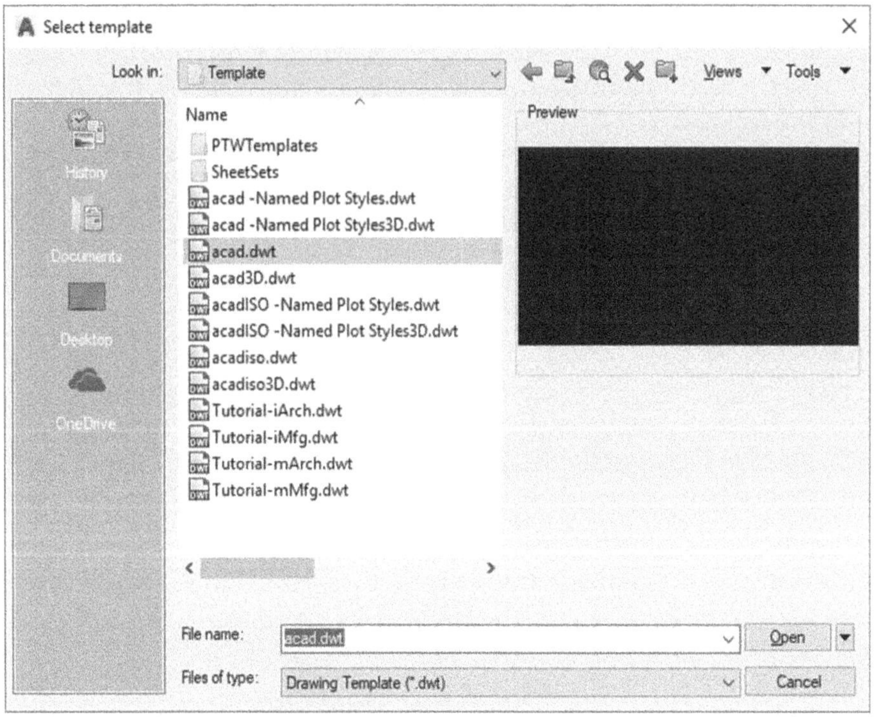

FIGURE 2.26 An acad.dwt template.

file when AutoCAD was first introduced in Chapter 1. Open the highlighted acad.dwt and you have a blank file, and that is it. Except now you know a bit more about what it is. Note the other preset templates available. In Chapter 10, we come back here to grab an Architectural title block.

Limits

This is yet another topic that causes confusion. Limits are not technically necessary because we have Paper Space Layouts, but many textbooks persist in having you set them up. All they are is an indication of the size to which your drawing area is preset in model space. If they are turned on, you could be limited in how much of the limitless model space you could draw in. Remember, however, that model space is limitless, so all Limits refer to when they are turned off, is the area you can work on without a regen. If you draw a shape that is way beyond the Limits, you simply regenerate the screen via a Zoom, Extents (see Tip 3 in Chapter 1 for a reminder) and the Limits become irrelevant. If you would like to set them anyway, just type in `limits`, press Enter, and set the lower left corner to 0,0, then the upper right corner to perhaps 100′, 50′, or anything else you wish. (Talk to an AutoCAD veteran about the days before Paper Space layouts for a long winded discussion on this as well.)

Save

Needless to say, you have to save your work often. While it is pretty rare these days, AutoCAD (like any software) can lock up or crash. If not, Windows will, there will be a power failure, you may spill coffee on your laptop, or an asteroid can hit you, so it is just a good habit to get into. Starting in the next chapter, you need to start saving your work. To do so, you can type in save and press Enter or press the graphic of a disk at the top of the screen. Another way is Application Menu (Big red A) at the upper left of the screen (Fig. 2.27) and choose Save or Save As. You then get the usual Save Drawing As dialog box from which you should know how to proceed. Once your file has been named something other than Drawing 1.dwg, just use the Save command, there is no need to keep renaming the file.

Help files

AutoCAD, like pretty much every software application, has extensive Help files, and indeed that is one way (though not recommended) to attempt to learn the software. A better application of Help is to explore it once you know how to use AutoCAD and need assistance with an aspect of a command. Tool tips that appear with every command can also be considered as part of the Help files.

AutoCAD's Help files have come a long way in recent releases. The original method to access them was generally to press F1 (same as for other software), and the installed files would come up. Early on, they became context sensitive and showed only relevant results based on what you were doing. You could also just search for what you wanted and access additional information, including examples. Autodesk even offered a printed Help files manual for those who preferred paper.

These files, however, were part of AutoCAD, could not be updated by Autodesk between releases, and offered very little interactive content and access to other tools. All that began to change as Internet access became virtually guaranteed as part of working on a computer. The monumental shift with Help files in recent years was to move them completely online and link them with additional tools and functionality. Today, online content is accessed if the Internet is available, and built-in files are not even offered with AutoCAD (though you can download them, as we discuss in a bit). As far as paper help manuals? I am not aware of them even being offered anymore, but you can probably print those downloads if you want to.

Help files today are part of a larger, interactive suite of online tools, applications, and resources, which we discuss shortly. To access just the regular Help files, you can still just press F1 (or the question mark at the very top right of your screen, Fig 2.30). You can do this with no commands active (resulting in Fig. 2.28) or in "context-sensitive" mode, while in the middle of a command (such as shown in Fig. 2.29 with "line").

Here, you can search for any other commands and find related tasks, references, and concepts. Note also how, on the left-hand side of the Help files, you can filter what you see based on your level of interest and responsibility. You can now access pretty much anything that has to do with that command, including programming tools, related tutorials, and even relevant blog entries. It is an interesting, if overwhelming, approach and signifies a major shift in Help file philosophy. You are now getting content-based, Google-like content to cover every angle, not just a cookie-cutter text file to explain the command.

As mentioned, there is far more to the Help files than just command explanations, and the extras can be found under the Help drop down menu, as seen in Fig. 2.30.

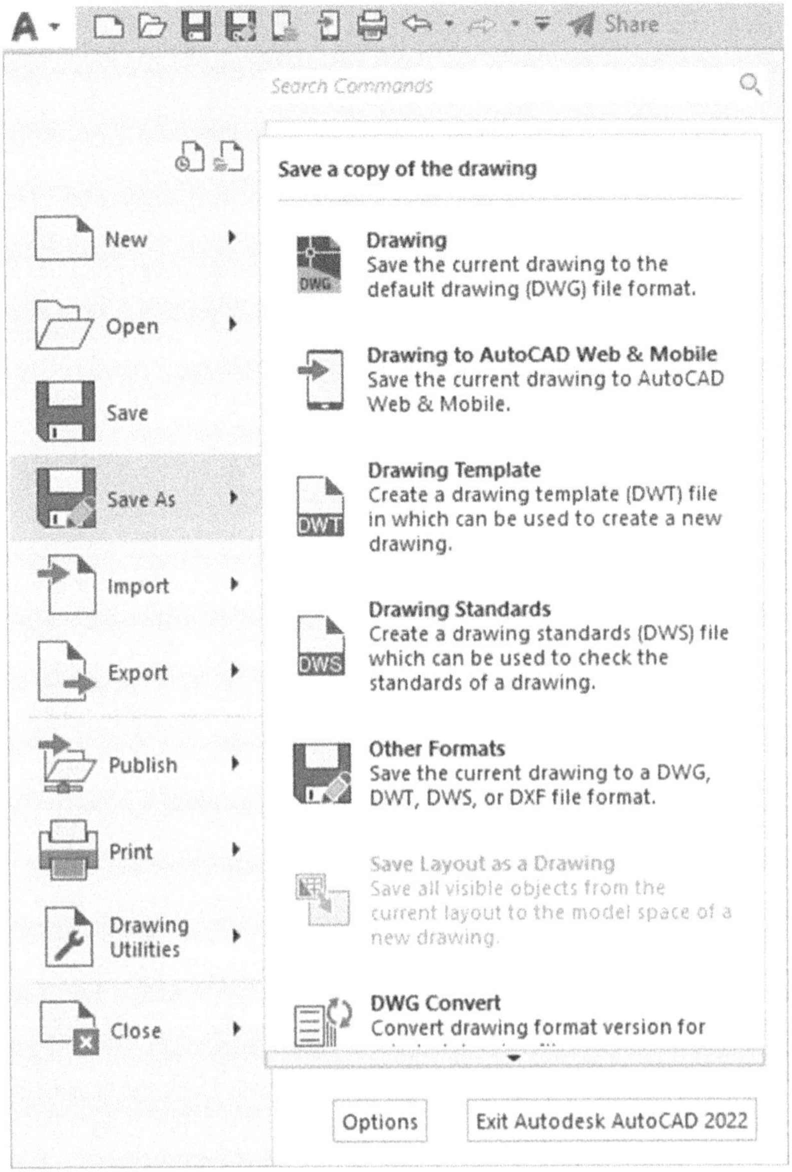

FIGURE 2.27 AutoCAD Application Menu with Save As selected.

You already have called up the regular command Help files via F1 (or the question mark icon), so no need to select the first item from the top, but what about the rest of these choices? Let us go over a few of them. You can access AutoCAD's Download Offline Help choice from Fig. 2.30, it gives you the chance to download AutoCAD Help documentation—all 203 MB of it as of this writing. Just choose your language and download away, ideally with a fast connection. When it all downloads, you get a self-extracting *.exe file, which needs to be installed. If you anticipate working offline a lot and generally want local access to the files, this may be a good idea. They are not regularly updated or interactive but do the job when offline.

When you click Send Feedback you are taken to a website where you can fill out a form with your comments, thoughts and suggestions for AutoCAD directly to Autodesk.

Selecting the Download Language Packs choice, takes you to a web site where you can get additional language options for your current version of AutoCAD. If you choose Desktop Analytics… you will have the choice to turn on or off Autodesk's collection of data on how you use AutoCAD.

Finally, the About Autodesk AutoCAD… choice takes you to all the information about your currently installed AutoCAD software, including the version number. More importantly there is a button here to manage your license if you need to do so.

Another helpful resource is AutoCAD Support Knowledge Base. You can search for this on the internet, or as of this printing, go there directly at: https://knowledge.autodesk.com/support/autocad. You get the Autodesk web page called

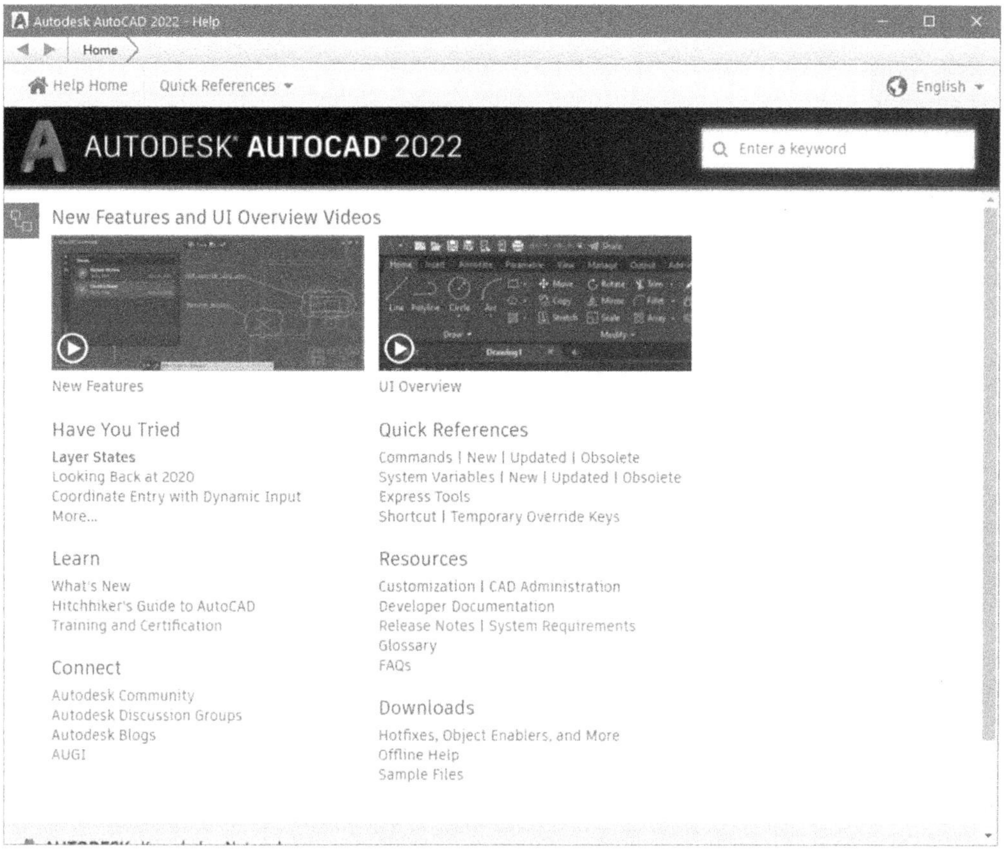

FIGURE 2.28 AutoCAD Help files (general).

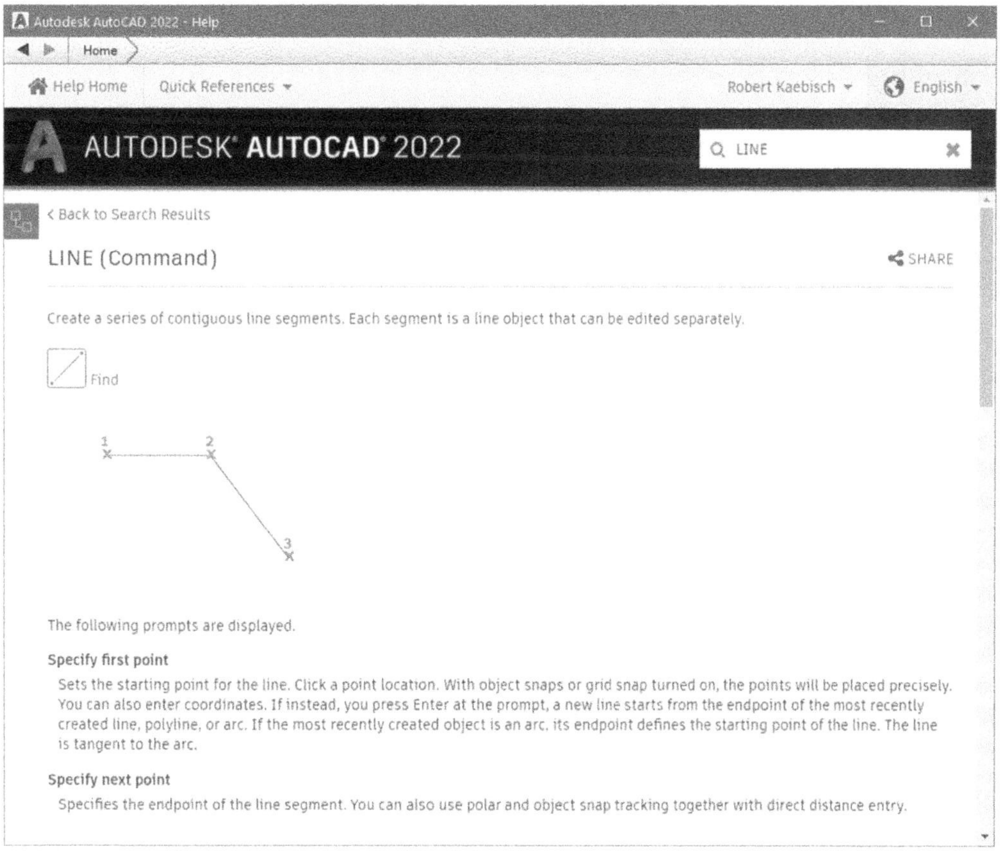

FIGURE 2.29 AutoCAD 2020 Help files (LINE Command).

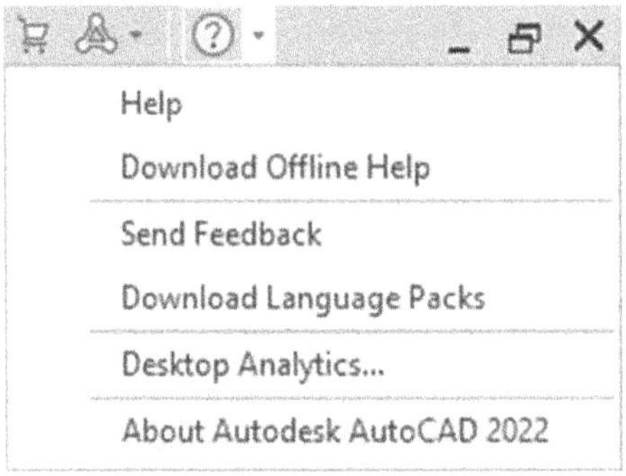

FIGURE 2.30 AutoCAD Help menu.

Knowledge Network, seen here in Fig. 2.31. Note that you may see a slightly different design, as web pages can certainly be updated after this textbook is released. Here, you find some troubleshooting guidance as well as Service Pack downloads and a vast array of instructional data.

As we wind down our tour, a new feature of the Help files, first introduced with AutoCAD 2015, deserves mention. Look at Fig. 2.29 again. Do you see the word *Find* just to the right of the line icon? If you click on this while the Ribbon is active, it finds the relevant command for you and points a bobbing red arrow at it. The arrow can be seen all the way to the top left in the screenshot in Fig. 2.32 (minus the bobbing up and down, of course). This clever assistant works with any command contained in the Ribbon and if you have the Ribbon displayed. Starting with AutoCAD 2016 and continuing with this release, it now also works with application menus and the Status bar.

Finally, we have the Autodesk Exchange (Fig. 2.33). It is not part of the Help files per se, but this is a good time to introduce it. You can access it at any time via a small shopping cart toward the upper right of the screen, to the right of the Autodesk 360 Sign In. The Exchange is also geared toward developers and "power users," with various tools and apps to customize, extend, or otherwise enhance AutoCAD. Although, as a beginning user, you probably will not need all this additional content, make a note of it for future reference. The Exchange's web address, for independent access to it, is https://apps.autodesk.com/ACD/en/Home/Index.

As you can see, a tremendous amount of useful information is found in all these references. You will need to occasionally explore these files for answers, so get comfortable with them at the outset.

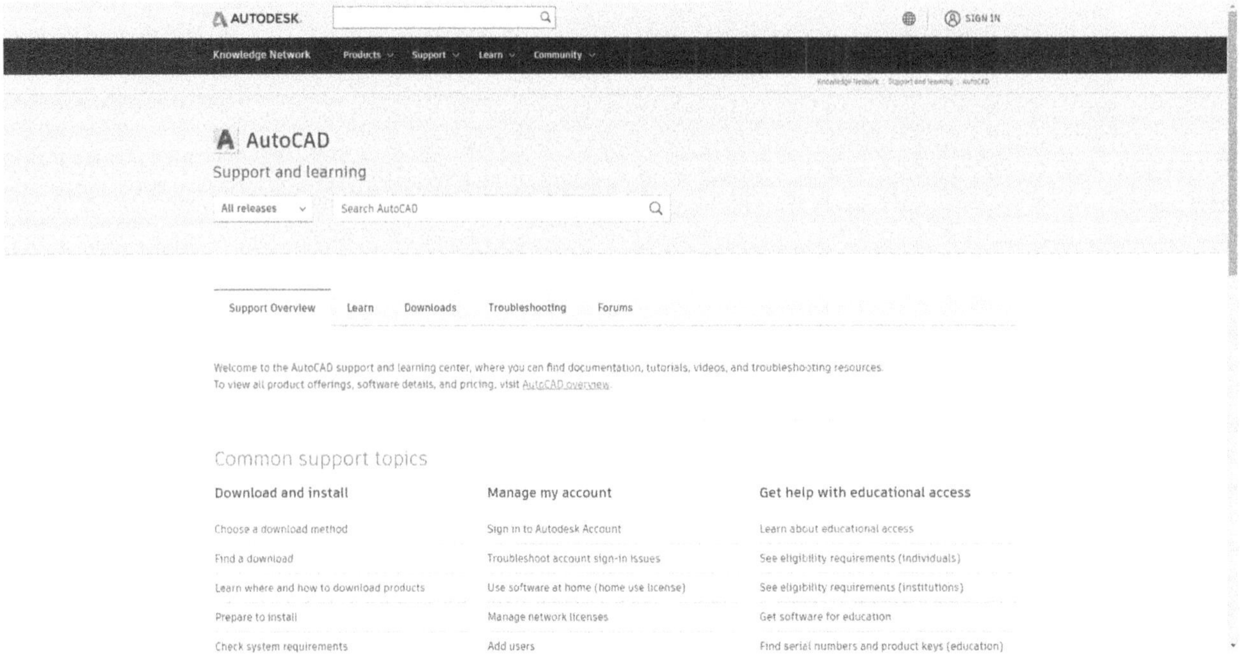

FIGURE 2.31 Autodesk's Knowledge Network web page—Overview tab.

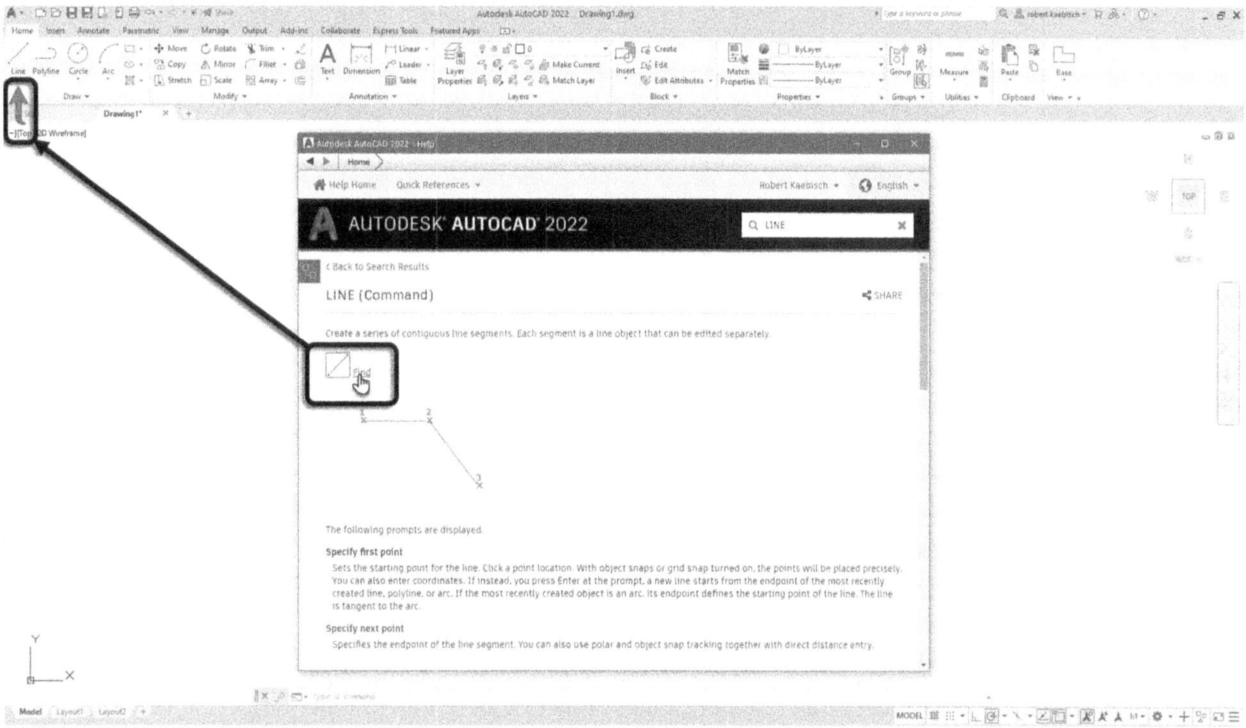

FIGURE 2.32 Find command tool.

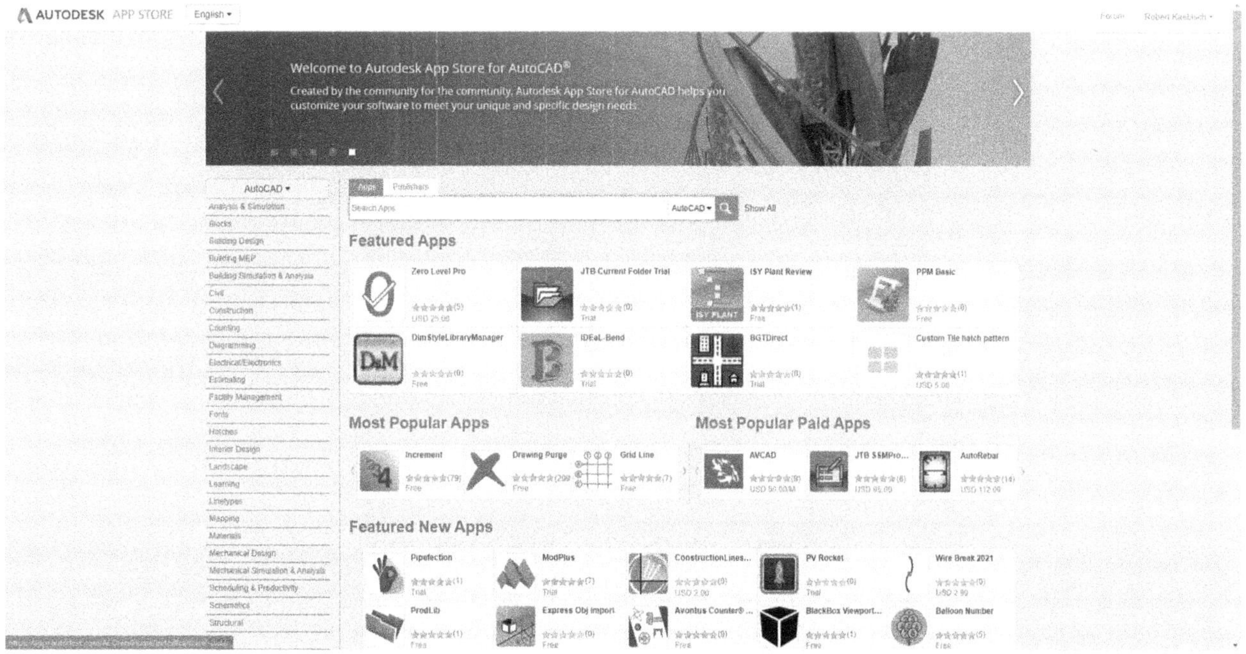

FIGURE 2.33 Autodesk Exchange App Store in Google Chrome browser.

TANgent OSNAP

The Tangent OSNAP point is necessary to draw part of Exercise 7 and is one of the four additional OSNAP points mentioned throughout the upcoming chapters. Tangent works by allowing you to create a line that maintains perfect tangency to a circle or an arc from any other point. Because this OSNAP is used quite rarely it is not recommended to set it up as a running Osnap. Just right-click for a list of temporary Osnaps to choose from (see Fig 2.34) or type in *tan* when you

FIGURE 2.34 Right-click temporary OSNAP menu.

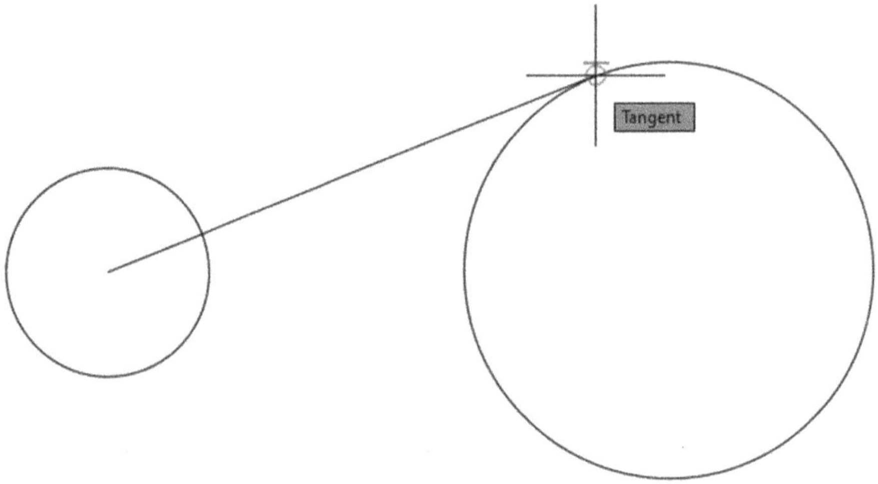

FIGURE 2.35 Tangent OSNAP.

need to create the tangency; you see the symbol shown in Fig. 2.35 with a line drawn from a Center Osnap to a point Tangent on a circle. Once started, create another tangency point on another circle or arc using the right-click menu or by typing Tan agin.

Time

This interesting command allows you to keep track of various time factors involved with a drawing, such as when it was created, last updated, or the total time spent on it. This could be useful for billing purposes when doing contract drafting or good when you are the boss and want to see how long an employee spent on a file. For those students just starting out, or if you are not anyone's boss yet, keep this command in mind, someone may be watching you! To use it with a drawing open, just type in time and press Enter. You will see the command line history pop up above the command line with useful time data.

Review everything presented so far. In Chapter 3, we introduce the layers concept and put all the preceding together to begin drawing a floor plan. Remember that learning AutoCAD is cumulative, and you need everything we have covered so far, no more and no less, to successfully deal with the upcoming material, so make sure you understand it all before moving on.

Drawing the star

As promised, the technique of drawing the star is as follows. First, create a pentagon. Then, noticing that it is a five-point geometric shape, draw lines from endpoint to endpoint, connecting those vertices. Finally, erase the pentagon and use trim to erase the inner lines, as shown in the progression in Fig. 2.36.

The reason students do this simple exercise is to underscore a very important point. AutoCAD is used across a vast array of industries, each with its own symbols and unique objects. It would be impossible for the software to include all of them in its database. Instead, AutoCAD gives you basic tools and leaves it up to your creativity and skill to create the shape you need. So, when at a loss as to how to draw something you imagined in your mind, stop and think about the basics that you learned and how you can put them together into something new.

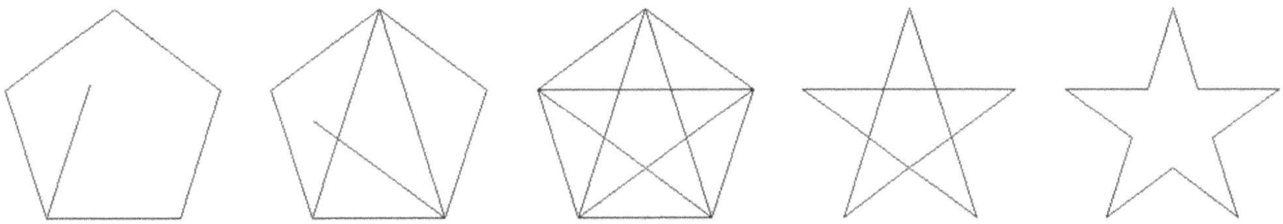

FIGURE 2.36 Drawing challenge.

Summary

You should understand and know how to use the following concepts and commands before moving on to Chapter 3:

- Using grips
  - Activating them, using Esc to get out
- Setting units
  - Architectural and Engineering (Decimal) units
- Using Snap and Grid
  - Setting both, turning them on and off
- Understanding the Cartesian coordinate system
  - X and Y axis ordered pair (X,Y)
- Distance Entry tools
  - Direct Distance Entry
  - Relative Distance Entry
  - Dynamic Distance Entry
- Inquiry commands
  - Area
  - Dist
  - List
  - ID
  - Radius and angle

- Explode command
- Polygon command
  - Circumscribed and inscribed
- Ellipse command
- Chamfer command
- Templates
  - acad.dwt
- Setting limits
- Save
- Help files (F1)
- Tangent OSNAP
- Time

Review questions

Answer the following questions based on what you learned in this chapter. You should have the basic commands memorized, so try to answer most of the questions without looking up anything:

1. How do you activate grips on AutoCAD's geometry?
2. How do you make grips disappear?
3. What are grips used for?
4. What command brings up the Drawing Units dialog box?
5. The snap command is always used with what other command?
6. Correctly draw a Cartesian coordinate system grid. Label and number each axis appropriately, with values ranging from −5 to +5.
7. On the preceding Cartesian coordinate system grid, locate and label the following points: (3,2), (4,4), (−2,5), (−5,2), (−3,−1), (−2,−4), (1,−5), (2,−2).
8. What are the two main types of geometric data entry?
9. Describe how Dynamic Input works when creating a line at an angle.
10. List the six inquiry commands covered.
11. What command simplifies a rectangle into four separate lines?
12. What is the difference between an inscribed and a circumscribed circle in the polygon command?
13. What is a chamfer as opposed to a fillet?
14. What template should you pick if you want a new file to open?
15. What is the idea behind limits?
16. Why should you save often?
17. What is the graphic symbol for save?
18. How do you get to the Help files in AutoCAD?

Exercises

1. Open a new file, set units to Architectural, and set both your Snap and Grid to 1″. Then, draw the following shape. Keep in mind the spacing between each dot is now 1″. You would be wise to use the mirror command to complete this faster. The linework in this example has been darkened slightly for clarity. (Difficulty level: Easy; Time to completion: <5 minutes.)

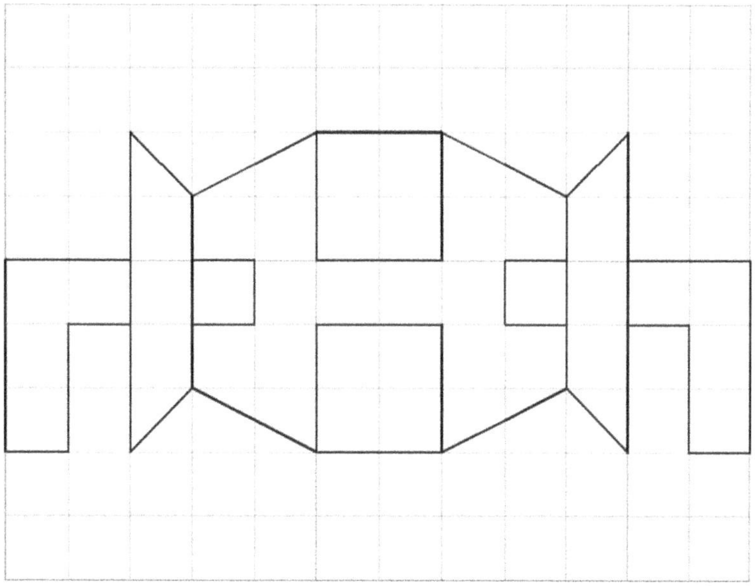

2. Open a new file and set units to Architectural. Turn off ORTHO, GRID and grid SNAP. Then, draw the following figure using the DYNMODE drawing aid. When done, use the area command to find its area. Redraw the shape via the manual method. (Difficulty level: Easy; Time to completion: <5 minutes.)

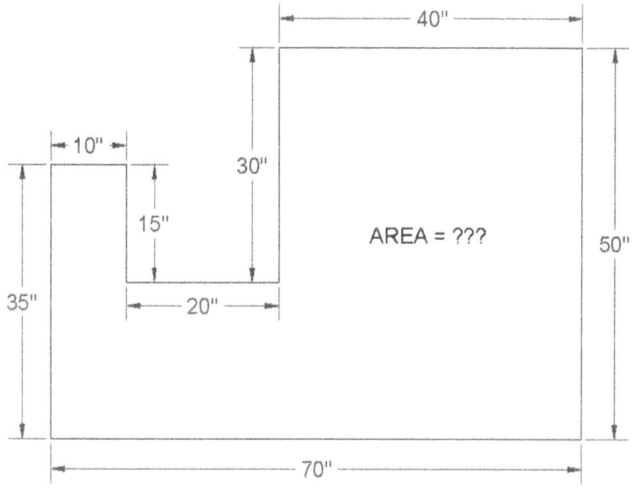

3. Open a new file and set units to Architectural. Turn off ORTHO, GRID and grid SNAP. Draw the following shapes using the manual Relative Distance Entry (using the @ sign) method. Repeat the exercise using the DYNMODE method. Repeat again using the rectangle command's length and width suboptions. Finally, use the area command (Object option) to find the area and perimeter of each shape. The sizes are 24″ × 36″, 17.5″ × 3″, 12″ × 12″, and 3″ × 40″. Remember that the first number is always the X, the second is the Y, as shown in the exercise figure. (Difficulty level: Easy; Time to completion: <10 minutes.)

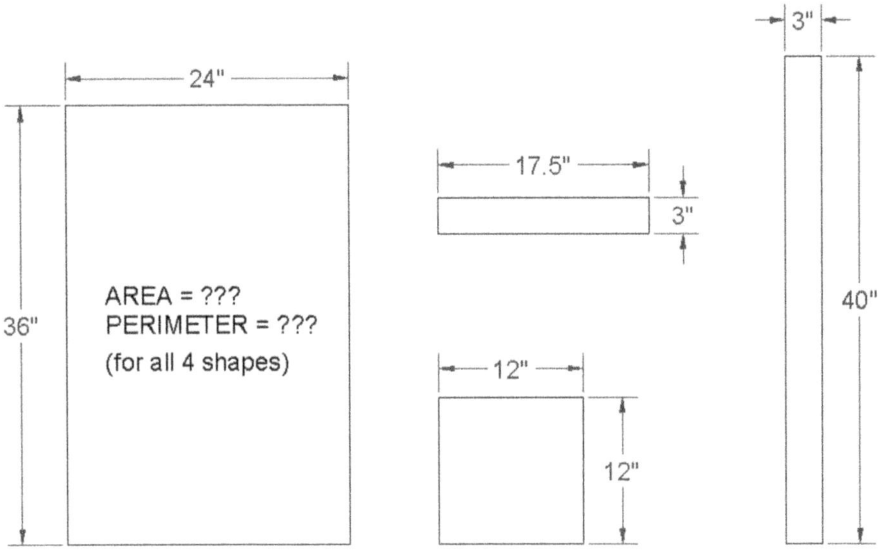

4. Open a new file and set units to Architectural. Draw the following figures using the basic circle, line, offset, trim, and fillet commands. Take note of how the front and side views are aligned horizontally to each other. Create the front view first, using circles and then project lines from the quadrants of the circles to create the side view. The dimensions should be used but do not need to be duplicated in the drawing. (Difficulty level: Easy; Time to completion: 5–7 minutes.)

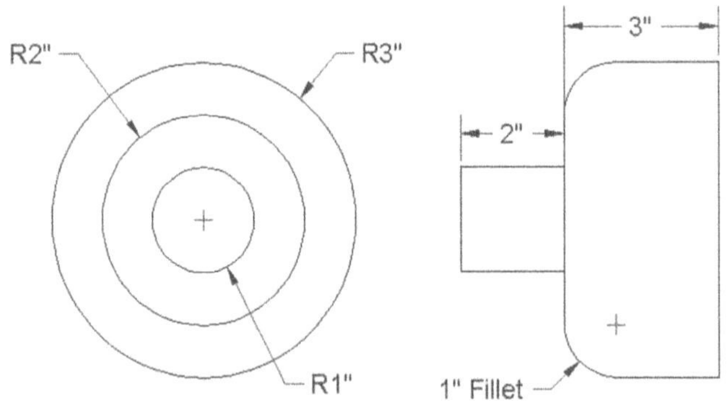

5. Open a new file and set units to Architectural. Draw the following bookshelf. Depending on your exact approach, you may need the rectangle, offset, explode, trim, rotate, copy, and line commands. The size and rotation of the books can be anything you want. The dimensions should be used but do not need to be duplicated in the drawing. (Difficulty level: Intermediate; Time to completion: <10 minutes.)

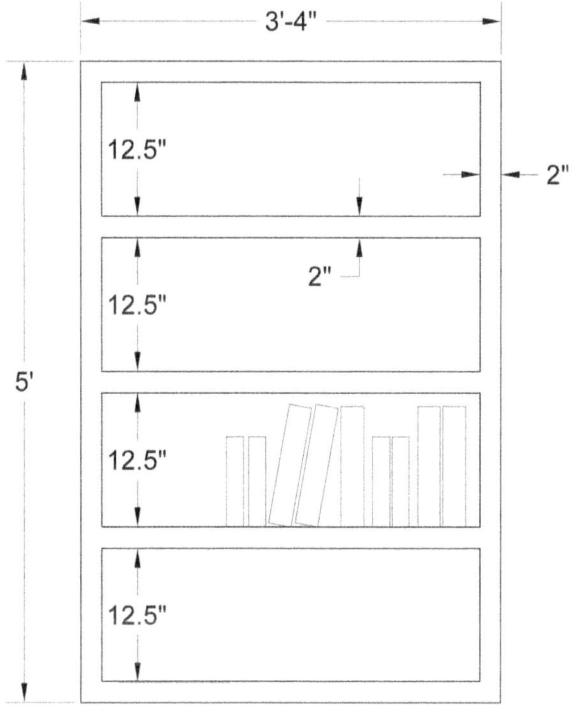

6. Electrical engineering schematics, because they generally comprise basic 2D linework, lend themselves easily to AutoCAD, and we have a few such layouts as exercises in this textbook. Open a new file and set units to Decimal. Draw the following basic capacitor, resistor, and diode schematic. The dimensions should be used to get you started but do not need to be duplicated in the drawing. All sizing is otherwise approximated and Grid and Snap are helpful here. (Difficulty level: Easy; Time to completion: <10 minutes.)

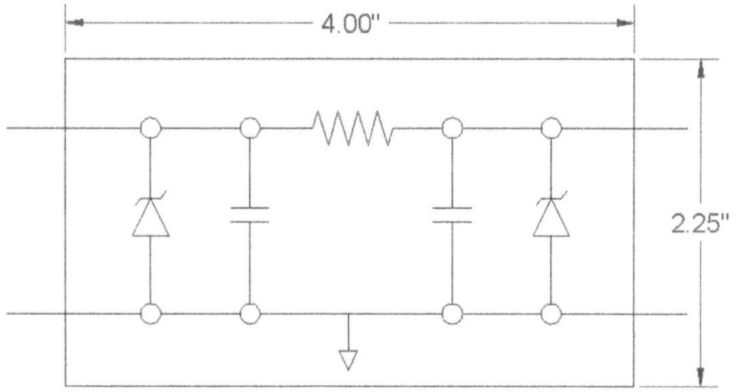

7. The remainder of the exercises in this chapter is mechanical in nature. Open a new file and set units to Decimal. Draw the following mechanical gasket. Notice how the top and front views are aligned vertically to each other. The dimensions and centerlines should be used but do not need to be duplicated in the drawing; we have not yet covered how to create them. You need to use the Tangent OSNAP among others. (Difficulty level: Intermediate; Time to completion: 10 minutes.)

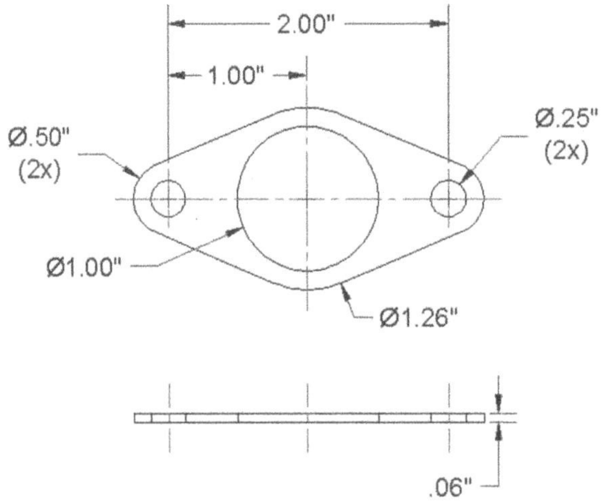

8. Open a new file and leave the units as Decimal. Draw the following simplified bolt diagram. You have all the information needed but may have to carefully project some parts of the side view. Take note of how the front and side view are aligned horizontally to each other. You may need the polygon, circle, line, trim, chamfer, and other commands. (Difficulty level: Intermediate; Time to completion: 10 minutes.)

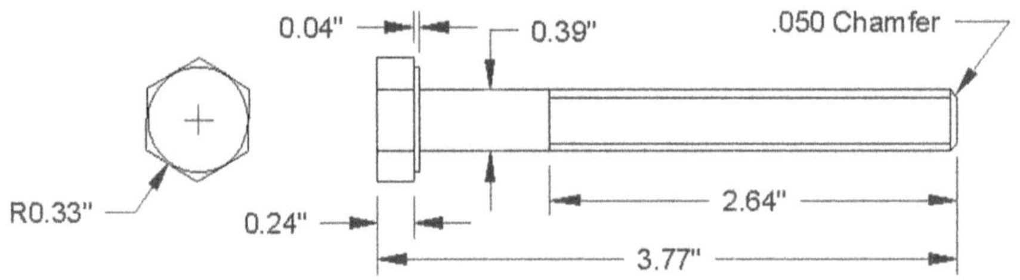

9. Open a new file and leave the units as Decimal. Draw the following mechanical bracket. You have all the information needed except for the bolts—you can improvise their sizing and position. You may need the rectangle, circle, line, trim, fillet, polygon, and other commands. (Difficulty level: Intermediate; Time to completion: 20 minutes.)

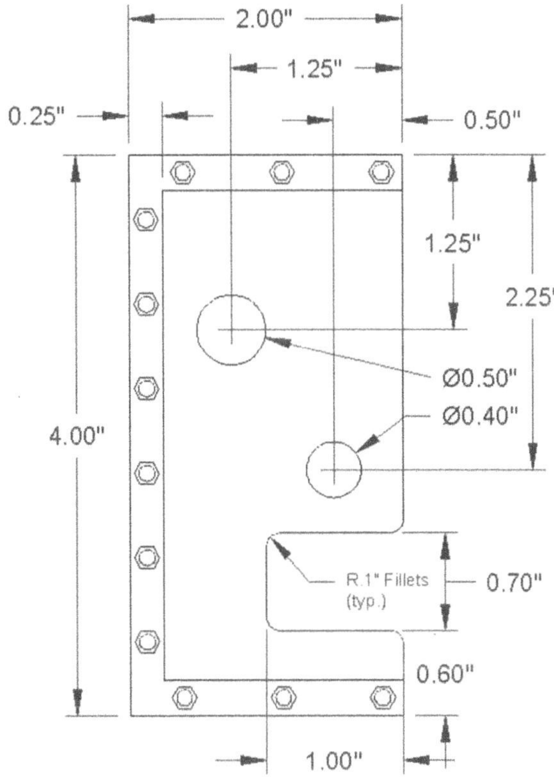

10. Open a new file and leave the units as Decimal. Draw the following mechanical bracket. You may need to use the rectangle, circle, fillet, offset, and other commands. The middle slot is made up of two circles, positioned as shown. Then, two lines are drawn using the Tangent OSNAP, with a trim to complete the slot. (Difficulty level: Intermediate; Time to completion: 30 minutes.)

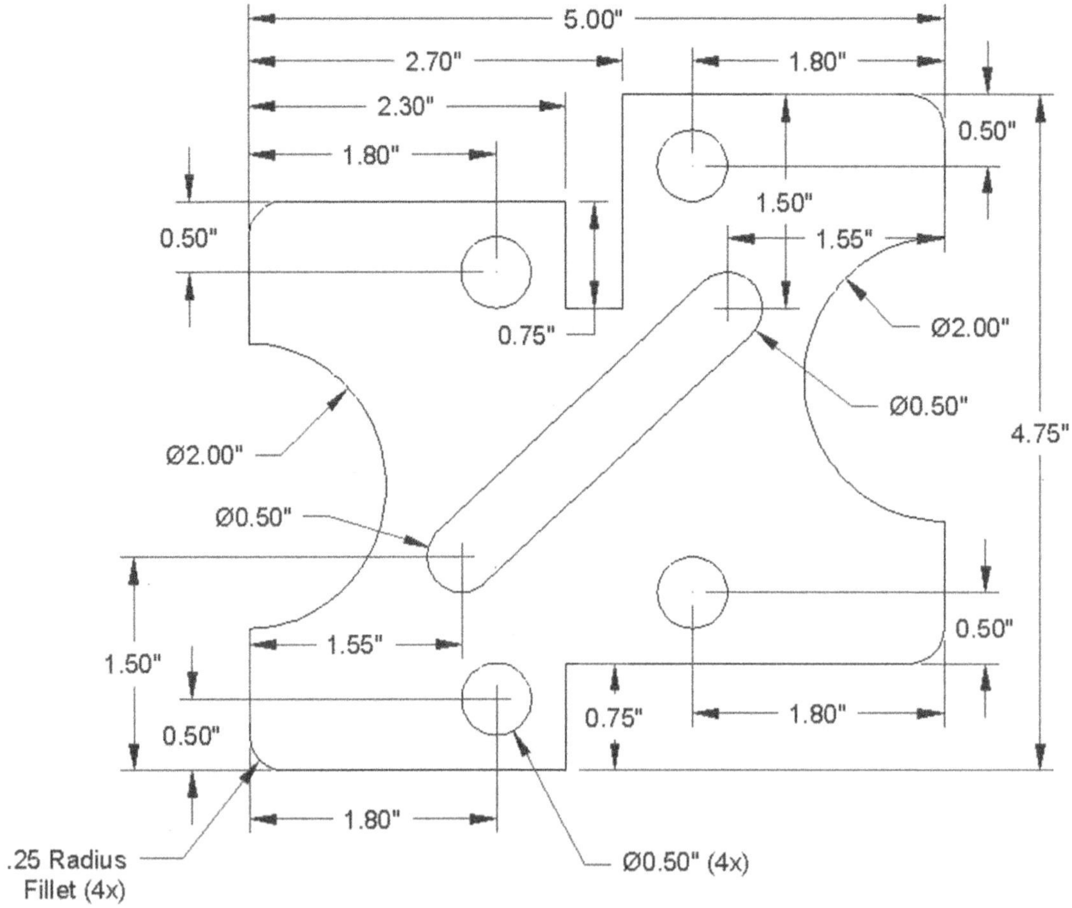
5.00"
2.70"
1.80"
2.30"
1.80"
0.50"
0.50"
1.50"
1.55"
0.75"
Ø2.00"
Ø0.50"
4.75"
Ø2.00"
Ø0.50"
1.50"
1.55"
0.50"
0.75"
1.80"
0.50"
1.80"
.25 Radius
Fillet (4x)
Ø0.50" (4x)

Spotlight on: architecture

AutoCAD is the undisputed CAD software of choice among most architects, and indeed the software finds its widest application in architecture and related fields. Therefore, it makes sense to feature architecture in our first Spotlight On feature.

Architecture is the art and science of designing and constructing buildings and other physical structures for human shelter or use. It is a far-reaching field that encompasses not only the construction and layout of structures but also some civil engineering, landscape design, interior design, and quite a bit of project management.

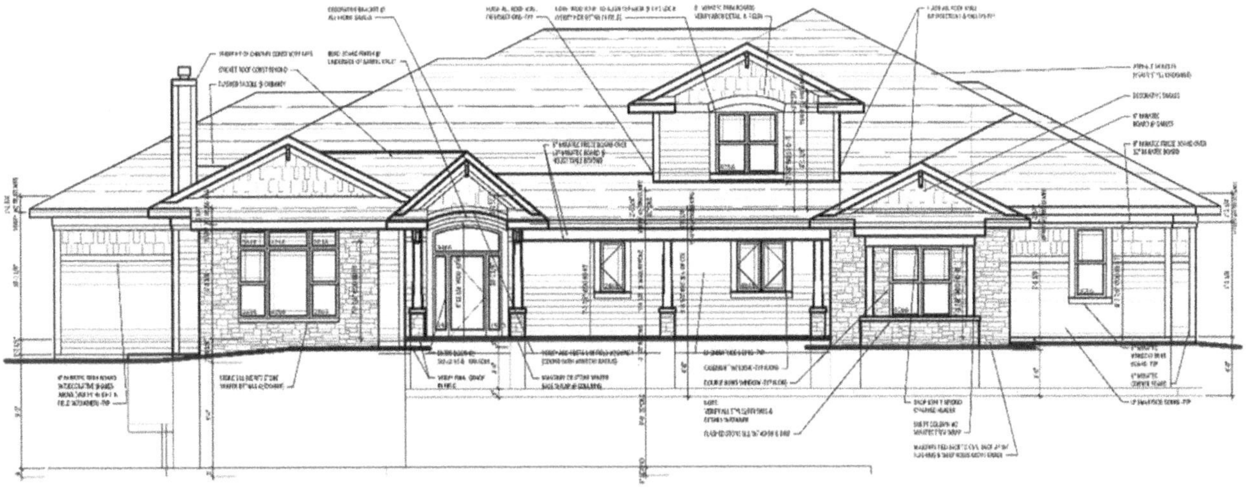

FIGURE 1 *Image Source: Robert C. Kaebisch.*

Architects are often described as a cross between engineers and artists. They not only need extensive knowledge of statics, strength of materials, construction methods, and codes but also artistic flair and well-developed aesthetics. Although architects often work with specialists to complete projects, many are well versed in electrical, plumbing, HVAC, landscape, and lighting design. Add to that the necessary ability to manage projects and deal with clients as well as financial and legal matters, and one can see that architecture is a demanding and multifaceted profession.

In the United States, requirements to become a licensed architect vary by state. Generally, a student needs to complete a 4- to 5-year Bachelor of Architecture program, followed by a master's in architecture (if desired). The student then must gain a specified number of work experience hours, often while earning a relatively low salary, before being eligible to sit for the Architect Registration Examination (ARE), to be licensed in the state of his or her residence. While earnings for partners and senior architects can be substantial, it does take quite a while to get established and build a reputation and client base.

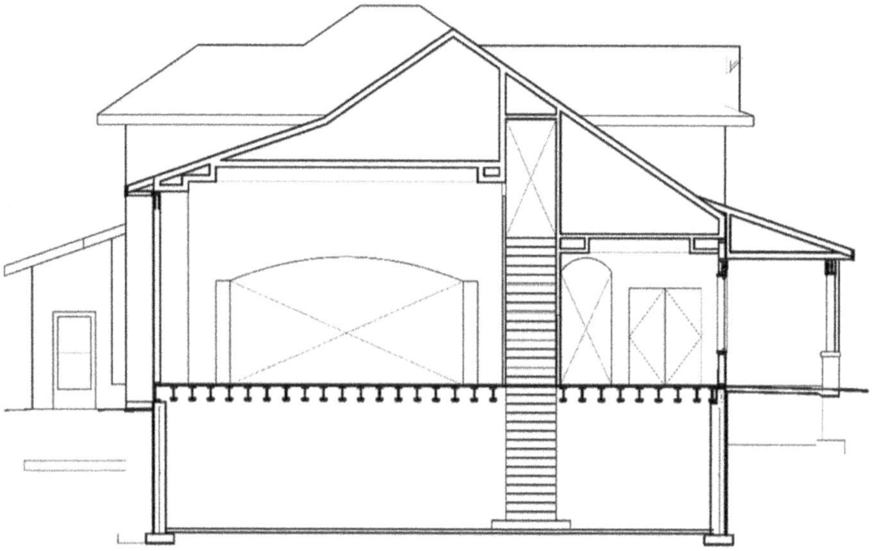

FIGURE 2 *Image Source: Robert C. Kaebisch.*

Virtually all architects today have switched to computer-aided drafting, and a group of CAD drafters is a common sight in most offices. Drafting is time-consuming, so most senior architects focus on bringing in new business and creating top-level concepts and designs, leaving the details and drafting to project designers and CAD drafters (sometimes the same person). This works out well, as project designers then get an opportunity to learn by doing actual, useful work while "learning the ropes" of the business. Acquiring cutting-edge AutoCAD skills is, therefore, a great way to get your foot in the door of the profession.

Around 2015, the architecture profession finally started to rebound somewhat from the decline of 2008–13. This decline was tied to the housing market and the recession of the previous 7 years. Because of lack of demand in new construction and, in turn, new design work, architects have seen many of their clients cancel projects or postpone nonessential upgrade work. With an upturn in the economy and the rebound of the housing and real estate market, this should all continue to improve. Surveys done in 2012 and 2013 already indicated a designer shortfall for 2014 onward due to many architects leaving the profession, and as a result, there seemed to be enough new ones entering to steadily increase the overall amount of newly registered architects in those same years; a good sign indeed.

Despite the downturn, compensation for employed architects has remained steady or risen. The median annual wage for a licensed architect is $76,930 per year. The average salary for a graduate with a Bachelor of Architecture degree is about $63,000, and for a graduate with a Master of Architecture degree, it was $74,000. Professional architects with 20 or more years of experience earn an average income of over $110,000.

So how do architects use AutoCAD, and what can you expect? A typical medium-sized project, such as a residential home, features a set of plans that may include

- T1 Title Sheet
- SP-1 Site Plan: The property surrounding the structure.
- A1.1 Demo Plan: What needs to be demolished to make room for the new design, if applicable.
- A1.2 Foundation Plan: The foundation, basement, and/or finished basement plan with mechanical systems.
- A1.3–A1.4 Floor Plan(s): The actual design, which may have multiple floors.
- A1.5 Roof: Shows construction of roof beams, rafters, braces, joists, and other elements.
- A2.1–A2.2 Exterior Elevations: Shows exterior views, heights, and materials
- A3.1–A3.2 Building sections, wall sections, and details: "cuts" through the house to show details of construction.
- A4.1 Interior Elevations and Details: Shows cabinets and appliances and other interior elevation elements.
- A4.2 Bathroom Elevations: Shows bath, shower, sink, and other features.
- A5.1 Schedules: Lists of doors, windows, appliances and other items, including sizing, quantity, etc.
- E1.1 Electrical Plan: The outlets, switches, and panels for each floor level (may include switching, ceiling, and wall lighting).

- E1.2 Electrical Lighting Plan (RCP-reflected ceiling plan): Switching, wiring schematic, and ceiling/wall lighting.
- GN-1 General Notes: Describes the overall job and any special instructions to the contractor (could be on other sheets).
- P1.1 Plumbing Riser Diagram: Describes the flow of pipes and plumbing features.
- S1.1 Framing Plan: Describes the construction of the house frame.
- S2.1 Structural Details: Additional details of design; may run multiple pages.

Extensive use of AIA layering, or an in-house system based on it, is employed, as seen in Figs. 3–5 show some architectural details. You should also expect to see extensive use of advanced Level 1 and Level 2 concepts, such as

- *Xrefs* (see chapter: External References (Xrefs)): The main plan is externally referenced to the electrical and RCP plans.
- *Paper Space* (see chapter: Advanced Output—Paper Space): Each of the plans is a separate layout tab.
- *Attributes* (see chapter: Attributes): Found in the title block, door and window schedules, and other elements.

Architectural drafting is a demanding and skill-intensive endeavor. Many jobs are started with measurements of existing structures if any exist. If the project calls for a brand-new building, not a renovation, then only a survey and site plan is needed and the design can get started right away. For commercial buildings, typically, a building core is laid out (which then becomes the Xref), and the architect then works on designing the internal details of the building. When the overall design is set, other systems are added, such as electrical, plumbing, HVAC, security, fire protection, mechanical, and much more. Corporate and commercial jobs are typically more involved than residential ones, and some of the previously mentioned subsystems are applicable mostly to those. Finally, extensive construction notes, details, and schedules need to be developed. The final document set may feature 20–30 sheets and sometimes much more.

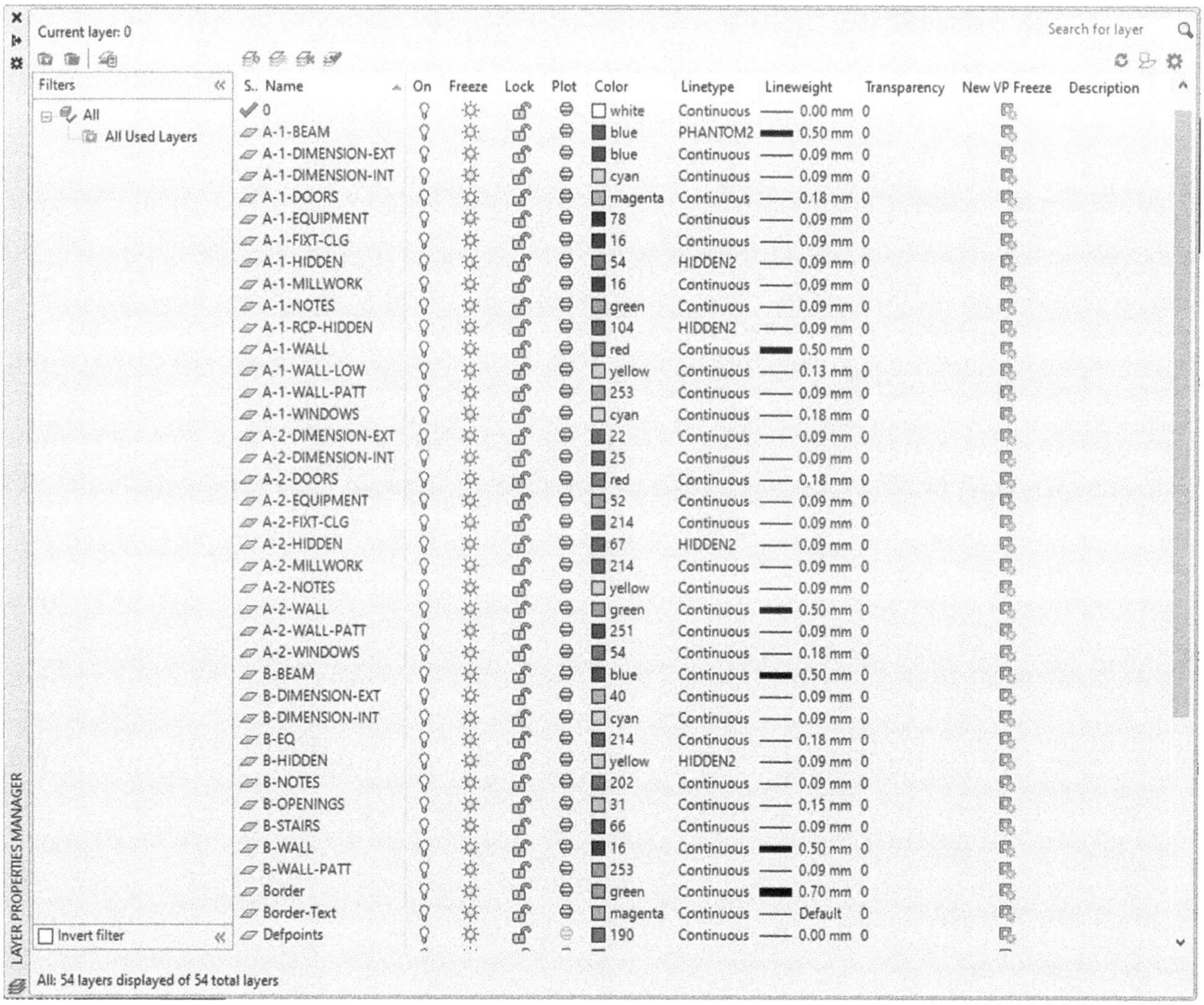

Name	Color	Linetype	Lineweight	Transparency
0	white	Continuous	0.00 mm	0
A-1-BEAM	blue	PHANTOM2	0.50 mm	0
A-1-DIMENSION-EXT	blue	Continuous	0.09 mm	0
A-1-DIMENSION-INT	cyan	Continuous	0.09 mm	0
A-1-DOORS	magenta	Continuous	0.18 mm	0
A-1-EQUIPMENT	78	Continuous	0.09 mm	0
A-1-FIXT-CLG	16	Continuous	0.09 mm	0
A-1-HIDDEN	54	HIDDEN2	0.09 mm	0
A-1-MILLWORK	16	Continuous	0.09 mm	0
A-1-NOTES	green	Continuous	0.09 mm	0
A-1-RCP-HIDDEN	104	HIDDEN2	0.09 mm	0
A-1-WALL	red	Continuous	0.50 mm	0
A-1-WALL-LOW	yellow	Continuous	0.13 mm	0
A-1-WALL-PATT	253	Continuous	0.09 mm	0
A-1-WINDOWS	cyan	Continuous	0.18 mm	0
A-2-DIMENSION-EXT	22	Continuous	0.09 mm	0
A-2-DIMENSION-INT	25	Continuous	0.09 mm	0
A-2-DOORS	red	Continuous	0.18 mm	0
A-2-EQUIPMENT	52	Continuous	0.09 mm	0
A-2-FIXT-CLG	214	Continuous	0.09 mm	0
A-2-HIDDEN	67	HIDDEN2	0.09 mm	0
A-2-MILLWORK	214	Continuous	0.09 mm	0
A-2-NOTES	yellow	Continuous	0.09 mm	0
A-2-WALL	green	Continuous	0.50 mm	0
A-2-WALL-PATT	251	Continuous	0.09 mm	0
A-2-WINDOWS	54	Continuous	0.18 mm	0
B-BEAM	blue	Continuous	0.50 mm	0
B-DIMENSION-EXT	40	Continuous	0.09 mm	0
B-DIMENSION-INT	cyan	Continuous	0.09 mm	0
B-EQ	214	Continuous	0.18 mm	0
B-HIDDEN	yellow	HIDDEN2	0.09 mm	0
B-NOTES	202	Continuous	0.09 mm	0
B-OPENINGS	31	Continuous	0.15 mm	0
B-STAIRS	66	Continuous	0.09 mm	0
B-WALL	16	Continuous	0.50 mm	0
B-WALL-PATT	253	Continuous	0.09 mm	0
Border	green	Continuous	0.70 mm	0
Border-Text	magenta	Continuous	Default	0
Defpoints	190	Continuous	0.00 mm	0

FIGURE 3 Typical architectural project layers.

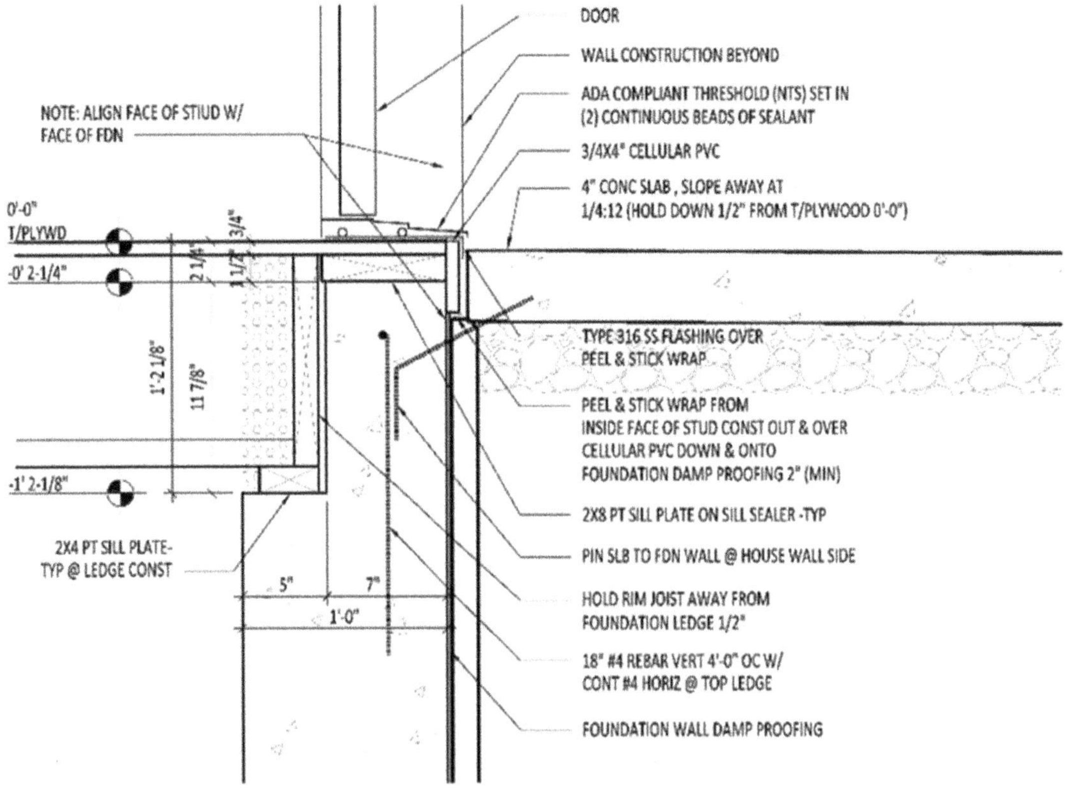

FIGURE 4 Architectural details. *Image source: Robert C. Kaebisch.*

Several factors make architectural drafting challenging:

- A complex building design may use virtually all of AutoCAD's tools and concepts; you need to be well versed in all of them. This means mastering every topic in Levels 1 and 2 of this book.
- There is a need for extreme accuracy, as the original design plans of the building structure are the anchor for the rest of the design. Mistakes made early on come back to haunt you.
- There is rarely a luxury of excess time. Architects bid on a job and must work within a set budget. A slow draftsperson who is not proficient (or efficient) takes too long to create the drawings and diminishes profits. Working quickly and accurately is a must. Rarely do you learn the basics and are you allowed to mature "on the job" anymore; it is just too expensive for a small architecture office.

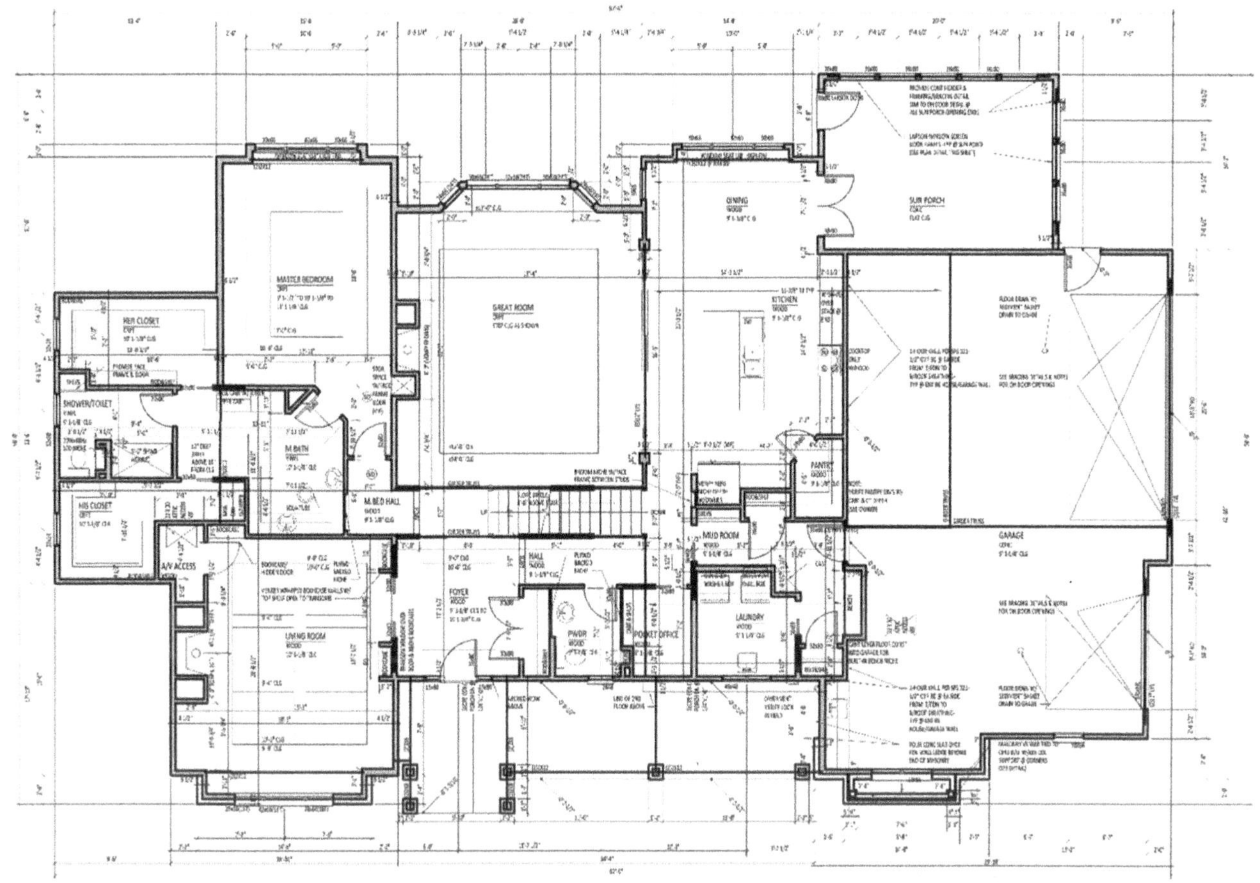

FIGURE 5 Residential architectural floor plan. *Image source: Robert C. Kaebisch.*

Despite of these challenges, or maybe because of them, architectural drafting is enjoyable and rewarding, especially if you are a new graduate and picking up knowledge in your profession along the way. Even if you just do nondesign drafting, the pay is good (salaries vary from $35,000 to $55,000 and up), and you acquire some frighteningly good AutoCAD skills very quickly in this fast-paced environment. Most of today's AutoCAD experts (including the authors) can trace their roots to using AutoCAD in architecture; it is the cutting edge of computer-aided drafting.

Chapter 3

Layers, colors, linetypes, and properties

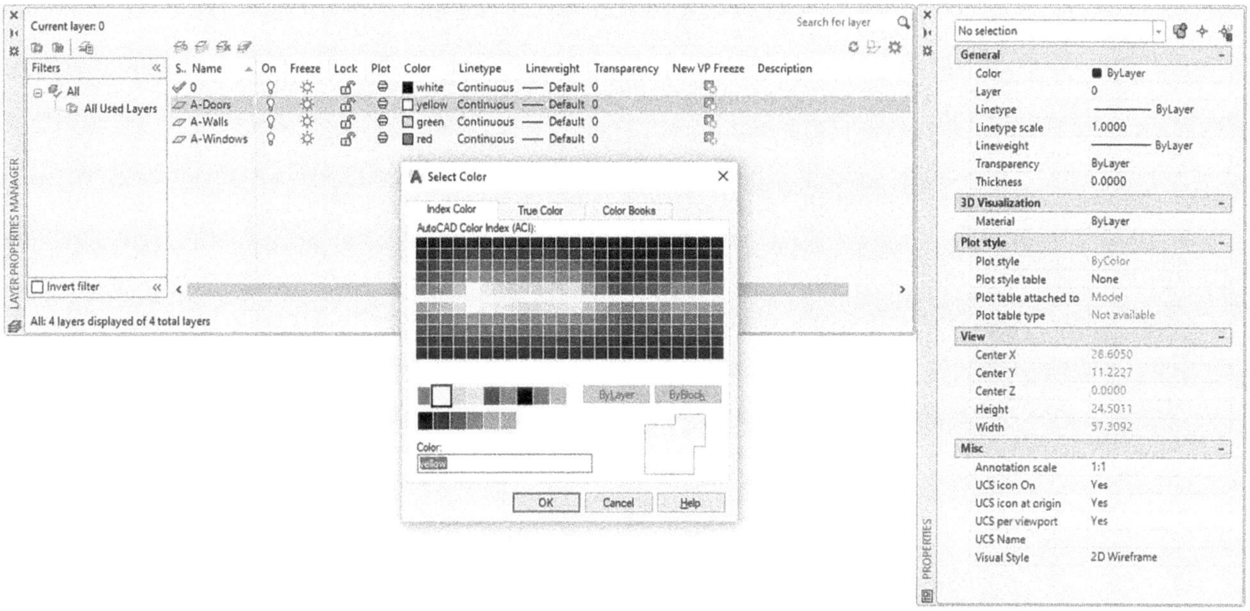

Learning objectives

Layers are an essential concept and need to be introduced before a serious drawing is attempted. In this chapter, we introduce layers and discuss the following:

- What are layers?
- Why use them?
- Creating and deleting layers
- Making a layer current
- Assigning layer colors
- Index color
- True Color
- Color Books
- Layer Freeze/Thaw
- Layer On/Off
- Layer Lock/Unlock
- Loading and setting linetypes
- Properties palette
- Match Properties
- Layer Manager palette

Here, you will learn all the remaining tools to begin creating a floor plan, which we start at the end of this chapter.

Estimated time for completion of this chapter: 3 hours (lesson and project).

Up and Running with AutoCAD 2022. https://doi.org/10.1016/B978-0-323-89923-9.00003-2

3.1 Introduction to layers

Layers is our first major topic after the previous chapters of basics. It is a very important concept in AutoCAD, and no designer or draftsperson would ever consider drawing anything more complex than a few lines without using them.

What are layers?

"Layers" is a concept that allows you to group drawn geometry in distinct and separate categories according to similar features or a common theme. This allows you to exercise control over your drawing by, among other things, applying properties to the layers, such as assigning colors and linetypes. You can also manipulate each individual layer, making it visible or invisible for clarity, as well as being able to lock them to prevent editing.

For example, all the walls of a floor plan will exist on the "Walls" layer, all the doors on the "Doors" layer, windows on the "Windows" layer, and so on. You can think of layers as transparency sheets in an anatomy textbook, with each sheet showing a unique and distinct category of data (skeleton, muscle, circulatory system, etc.), but when viewed all at once they show the complete picture of a person. Similarly, in AutoCAD, there can be any number of layers, with all of them holding their respective data and combining into the complete design.

Why use them?

As just stated, the key word here is *control.* If all your geometry were lumped together on one layer, as some beginners do, then whatever you do to one set of objects, you do to all; certainly not what you want. It is very important that you learn about layers early in your AutoCAD education and apply them right away to your first real drawing.

Let us look at the Layers Properties Manager. This is AutoCAD's Layers dialog box, where everything important related to layers happens. Open a new file and look for the Ribbon icon that calls up the Layers Property Manager. Once you call up the Layers Property Manager, you will see Fig. 3.1.

Keyboard: Type in **layer** or **la** and press Enter
Ribbon: **Home** tab→**Layers**

Examining the dialog box from left to right, notice the filters taking up some room on the left. We do not talk about filters until Chapter 12, Advanced Layers, and you can minimize that whole area by clicking the ≪ arrows. To the right of the word *Filters*, notice the main category headers, such as "Status," "Name," "On," and "Freeze." Below the categories, you find the only existing layer thus far, 0, and its properties. Notice that it is white with a "continuous" linetype. Above the categories are some action buttons for creating and deleting layers, which we cover next.

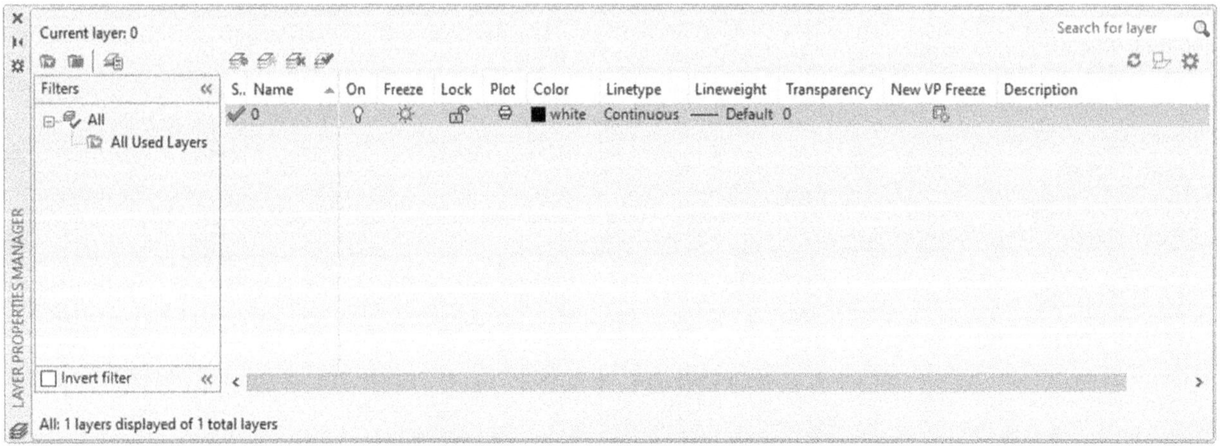

FIGURE 3.1 AutoCAD Layer Properties Manager.

Creating and deleting layers

Assuming you are starting from a blank file (using the acad.dwt to start), each new layer initially needs to be typed according to what you (or the company you work for) require.

Press the New Layer icon at the top of the layer box (two icons to the left of the red X), which looks like a stack of paper with a sunburst next to it. You can also hold down the Alt key and press N. Either way, once you do that, a new band saying Layer 1 appears. Just type in your new layer's name followed by pressing Enter, and a new layer is created. Here are some helpful tips:

- If you notice that you made a spelling mistake after you already pressed Enter, you can just click back into the layer name once or twice until the cursor enters Edit mode and retype a new or revised name.
- Do not attempt to change the 0 layer's name. It is the default layer and cannot be deleted or renamed.
- If you want to delete a layer (perhaps an accidentally made one), simply highlight the layer and click the red X icon, or press Alt + D. If the layer contains hasn't been used, is not current and contains no geometry, it will be deleted.

Following this procedure, create the following three layers:

- A-Doors
- A-Walls
- A-Windows

What you should see is shown in Fig. 3.2. The letter A in front of each layer simply means it will be for Architectural objects.

Making a layer current

The current layer is the active layer, on which you can draw at any given moment. You are always on some layer (even if you are not immediately sure which one exactly). So, logically, if you are about to draw a wall, you should be on the A-Walls layer and so on. To specify what layer to have as current, either click on the green checkmark button, to the right of the red X, or double-click on the layer name itself. At the very top left of the Layers dialog box, it says Current layer: followed by the name of the layer. Check there if you are unsure what the current layer is. For now, go ahead and make A-Walls the current layer.

Assigning layer colors

Assigning colors to layers is very important and should be done as you create them or soon after. The reasons to have color are twofold. On the most basic level, a colorful drawing is more pleasant to look at on the screen, but more important, it is easier to interpret what objects are on which layer and work with them accordingly.

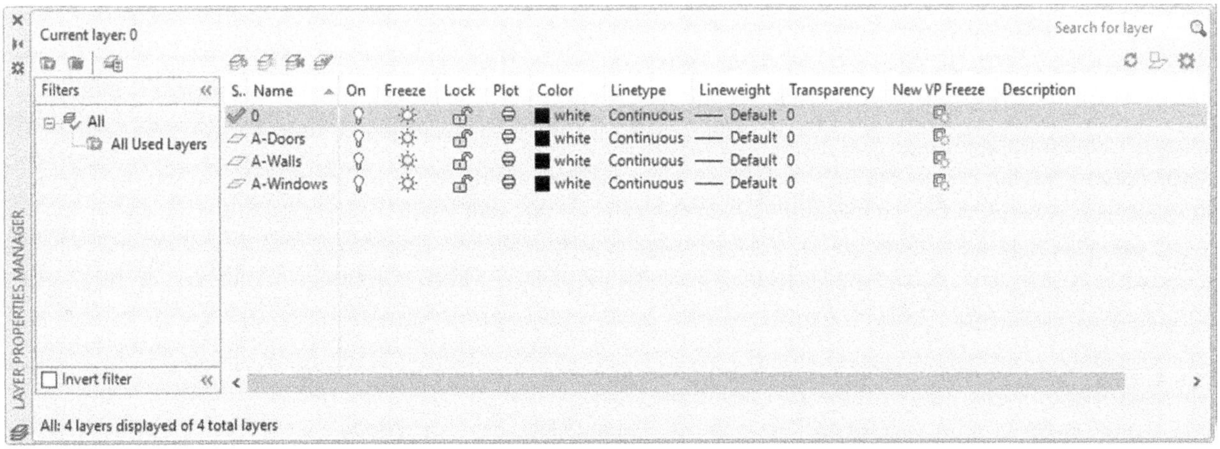

FIGURE 3.2 Layer Properties Manager with three new layers.

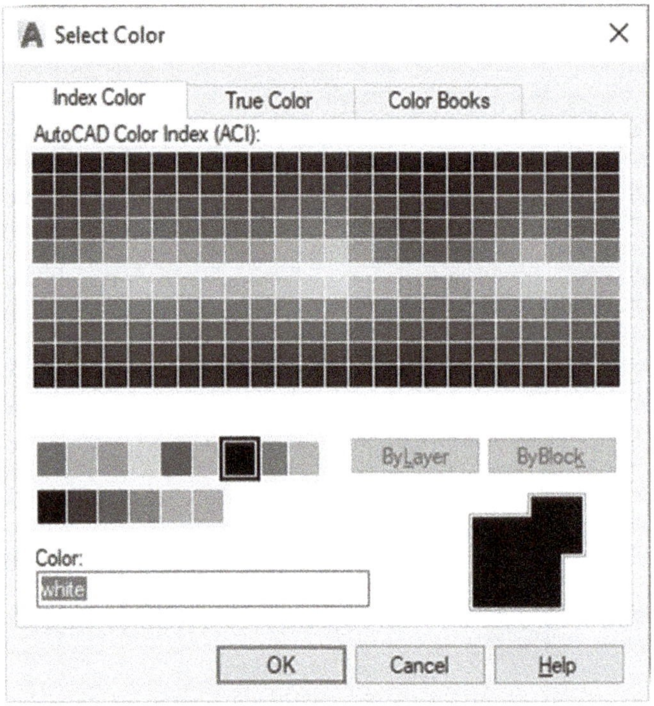

FIGURE 3.3 Select Color, Index Color.

If, e.g., your walls are green and you keep this in mind, then your eye can quickly see and assess the design. Likewise, yellow doors are easy to spot. This *color separation* greatly aids in drafting. Imagine a scenario where your building floor plan is made up of all white lines; you would be unable to tell one line from another. As that drawing grew more complex, all the lines would melt into one monochrome mess. Finally, colors are assigned pen thicknesses (an advanced topic) and contribute to more professional-looking output. We address this in Level 2.

To assign colors, let us turn our attention back to the Layers dialog box. Highlight, by clicking once, layer A-Walls. Now, follow the blue ribbon across until you see a box with the word *white* next to it. It will be under the Color column. Click on it, and the Select Color dialog box appears, as seen in Fig. 3.3.

Index Color

Let us take a tour of this Index Color tab. The two large 24 × 5 stripes of colors you immediately see at the top make up the AutoCAD Color Index, and it contains 240 colors. Do not be too quick in selecting one of those right away. Take a look below; there are two smaller color stripes. The rainbow-like stripes 1 through nine are the ones in which we are most interested. If you take a close look, they are the *primary* colors and the 240 colors above them are just different shades of the primaries. You should always use up those nine colors first—they are very distinct from each other—and only then move on to the others. Also, this makes setting pen weights easier later down the road, as you need not scroll through dozens of colors to find the ones you used.

The colors from left to right are Red, Yellow, Green, Cyan (not Aqua), Blue, Magenta (not Maroon), then White and two shades of Gray, 8 and 9. The color stripe below that contains just additional shades of gray, and we do not need them for now. So, finally, pick a color for the A-Walls layer (green is what many architects use for walls), click on it, then press Enter; you are taken back to the Layers dialog box to assign more colors to more layers. Yellow is recommended for A-Doors and red for A-Windows.

True Color

There is more to colors, of course. Go back to the Select Color dialog box. Now, click on the second tab, called True Color (Fig. 3.4). Here is where some experience in graphic design may come in handy.

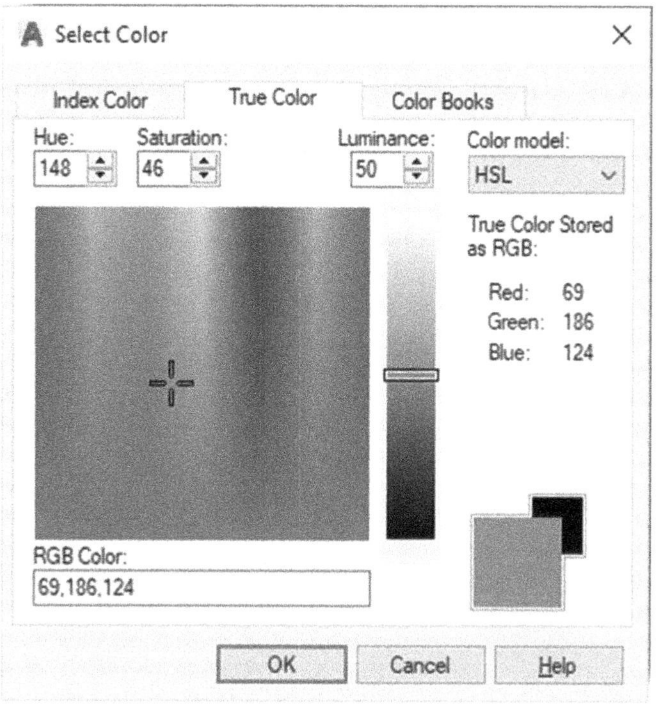

FIGURE 3.4 Select Color, HSL.

The main color window you are looking at now contains over 16 million colors and should be familiar to you if you use Photoshop, Illustrator, Paint Shop, or any of those types of design programs. You can now move your mouse around, selecting colors based on the HSL method. HSL stands for Hue (the actual colors, moving across), Saturation (moving up or down), and Luminance (brighter or darker, moving the slider up or down next to the main window). You may also use the RGB (red, green, blue) model of color selection by using the drop-down selector at the upper right (Fig. 3.5).

Whichever color you settle on is shown by both color models. Designers rarely use these colors for 2D work (where you want objects to be bright and visible), but do pick these more subtle colors for realistic-looking 3D models.

Color Books

Here is where things get even fancier. Color Books are Pantone colors. If you are an architect or an interior designer, you already know what these are. If you are shaking your head and have never heard of Pantone, do not worry about it, chances are you will not need to use them. Just to try, however, scroll through the palettes and select a color. Fig. 3.6 is a screenshot of the Color Books tab.

Layer Freeze/Thaw and On/Off

Here, we come to a very important part of the Layers dialog box. Freeze and Thaw are opposites of the same command, one that makes a layer's objects appear or disappear from view. The command freezes or thaws the layer for the entire drawing. Pick a layer, highlight it, follow the band across to the Freeze column and underneath you will find a yellow sun icon. If you click on it, that layer "freezes" and the icon becomes a blue snowflake. Clicking again undoes the action. So, what does freeze mean? It makes the layer (or more precisely all the objects on that layer) disappear, but remember, they are not deleted only invisible. You will use this feature routinely in drawings to control what is displayed. The On/Off feature represented by the light bulb (yellow to blue and back) is essentially the same as Freeze/Thaw, with some subtle (and minor) details. Objects on a Layer set to Off can still be drawn on (a warning dialog box will appear) and even erased with the *Erase; All* command sequence. This is strange but true. You can use either method, but let us stay with Freeze/Thaw just to keep things consistent.

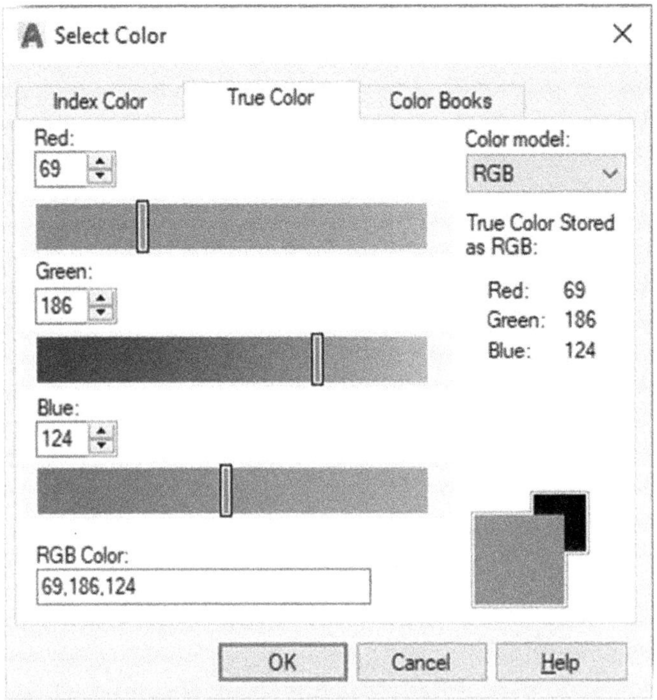

FIGURE 3.5 Select Color, RGB.

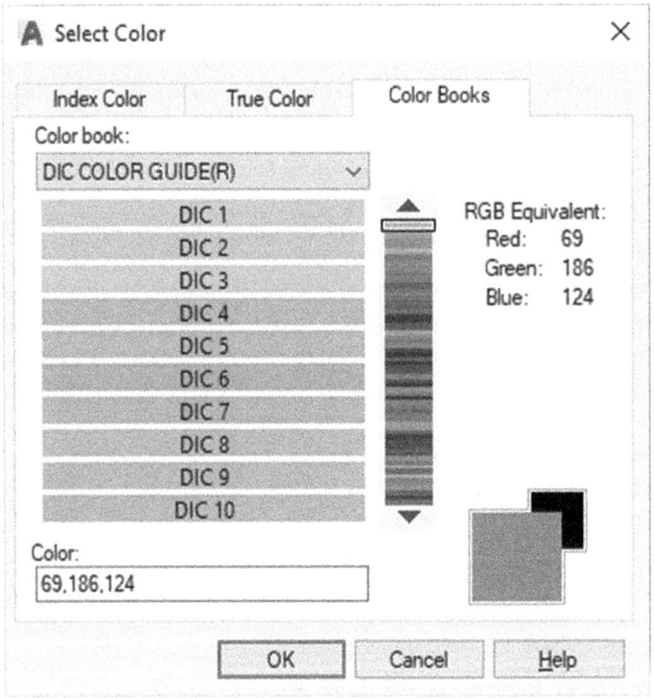

FIGURE 3.6 Select Color, Color Books.

Layer Lock/Unlock

This layer command, represented by the padlock, does exactly that; it locks the layer so you cannot edit or erase it. The layer remains visible but untouchable. The unlocked layer's icon is blue and looks "unlocked," whereas a locked layer's padlock turns yellow and changes to a "locked" position. To let you know which layers are locked, the layers fade somewhat, so you can tell them apart from unlocked geometry. You can control the level of this fading via a slider on the Ribbon. Go to the Home tab and drop down the arrow in the Layers panel corner. You see a slider that says Locked layer fading. Use it to make the locked layers darker or lighter to your liking.

Experiment with everything mentioned thus far to get a feel for layer controls. When done, make sure everything is reset (not locked or frozen) and click on the X in the upper corner of the layer box to close it.

3.2 Introduction to linetypes

Linetypes are the different lines that come with AutoCAD. As a designer, architect, or engineer, you may need a variety of lines to convey different ideas in your design. Just as in hand drafting, where you create Dashed, Hidden, Phantom, and other line types to show cabinets, demolition work, or hidden geometry in part design, in AutoCAD you load the lines and use them as necessary.

The idea here is to load them all in the beginning, then assign the linetypes to either a layer or sometimes just a few items. You can also load them as needed, but it is recommended to get this out of the way so you have them at your fingertips.

The procedure to do so is as follows:

Step 1. Type in `linetype` and press Enter. The Linetype Manager (Fig. 3.7) appears. If it is full of linetypes, it means they have been loaded earlier. If not, and the Linetype Manager is nearly empty as shown, go on to Step 2.
Step 2. Press Load … The Load Linetypes box comes up (shown expanded to its entirely in Fig. 3.8).
Step 3. Scroll down as needed and select the BATTING linetype.
Step 4. Then scroll to the bottom of the list, hold down SHIFT and select the ZIGZAG linetype to quickly select all of the linetypes from BATTING all the way to ZIGZAG in a few clicks.
Step 5. Press OK and the dialog box disappears. All the linetypes are loaded now, so press OK again in the Linetype Manager, and you are done.

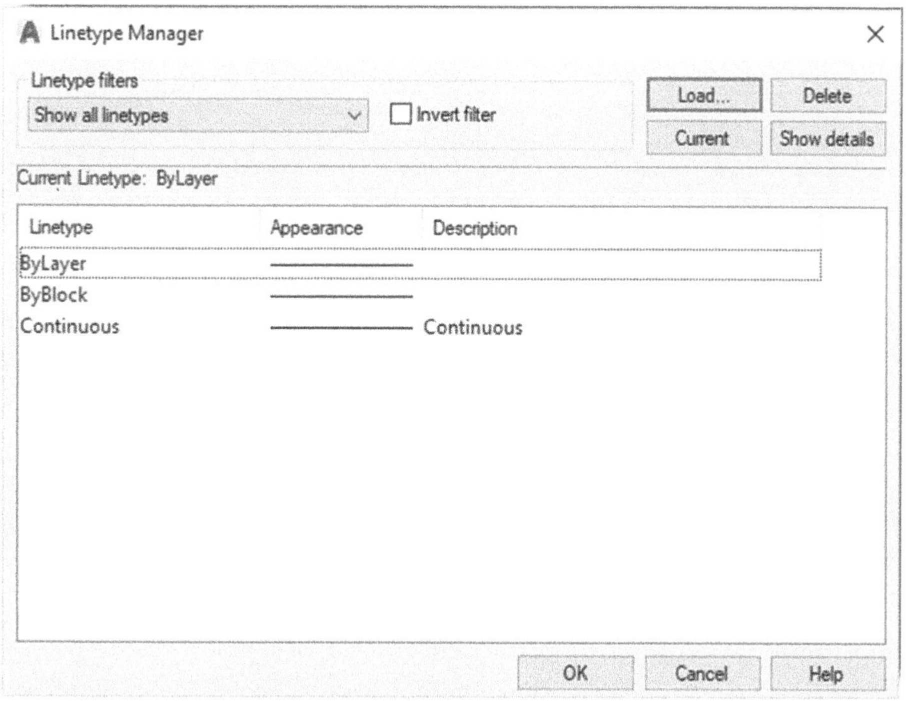

FIGURE 3.7 Linetype Manager.

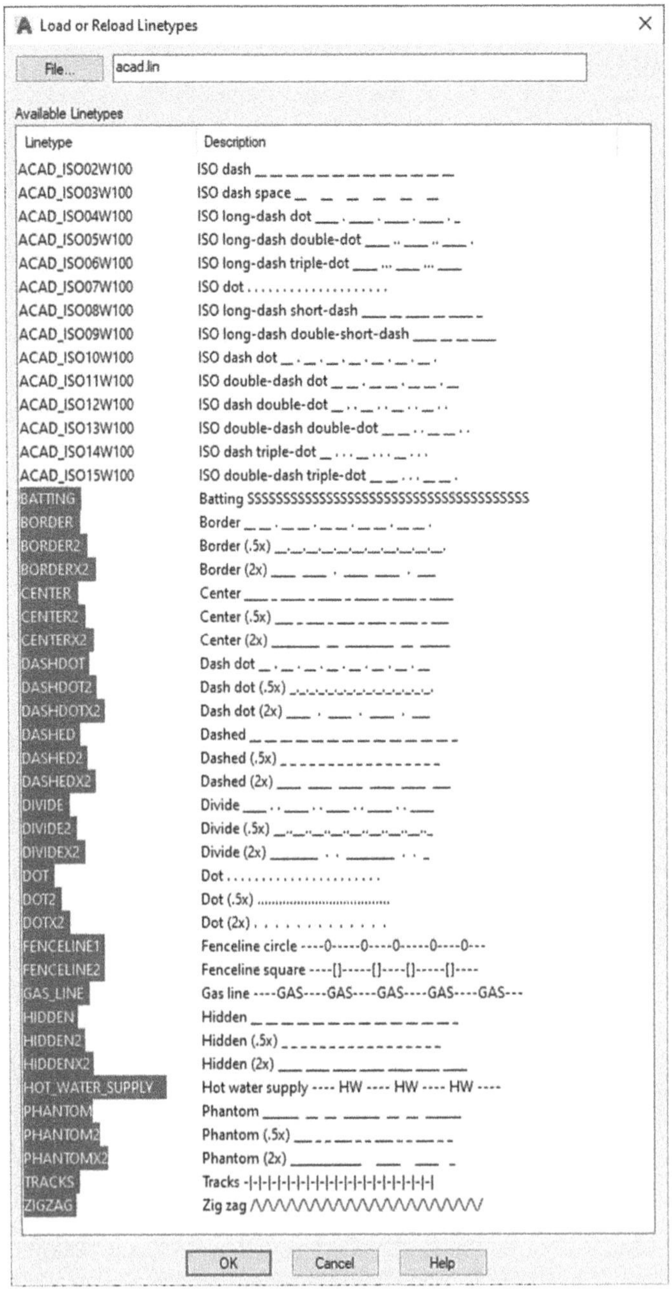

FIGURE 3.8 Load (or Reload) Linetypes.

We can now use the loaded linetypes and assign a linetype to a layer. Go back to the Layers dialog box:

Step 1. Create a new layer called Hidden and assign a linetype to it (the "Hidden" linetype, naturally).
Step 2. Make sure the Hidden layer is highlighted with the blue bar and move your mouse to the right until it matches up with the header category Linetype. It should say Continuous right now.
Step 3. Click on the word Continuous, and the Select Linetype (Fig. 3.9) dialog box comes up with all your linetypes preloaded. If we did not load them earlier, the box would be practically empty and we would have to do the loading at this point.
Step 4. So now, all we must do is scroll up and down the list to find the appropriate linetype (Hidden in this case), select it, and press OK.

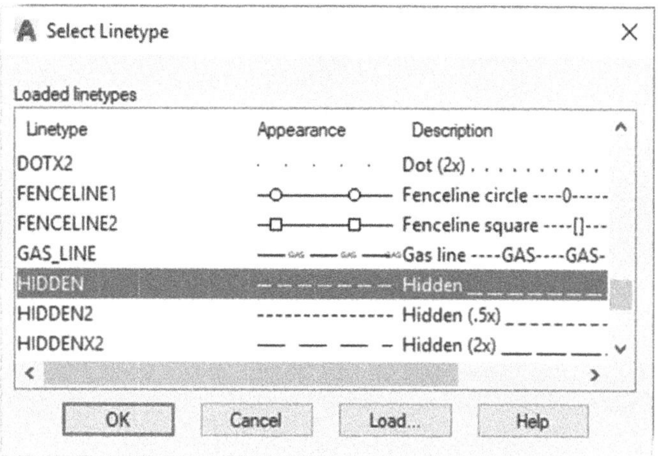

FIGURE 3.9 Select Linetype.

Make a note of the important linetypes for future reference: Center, Dashed, Divide, Hidden, and Phantom are some of the key ones. You did not load we do not suggest using the linetypes named ACAD_ISO… for the simple reason that their names are not descriptive enough to recognize at a glance. Also, make a note of why there are three types of the same linetype (such as DASHED, DASHED2, and DASHEDX2). By checking the appearance carefully, you see that the scale is different. Finally, let us get back to the Layers dialog box. If you set your Hidden layer to Current, then everything you draw is a dashed line. Go ahead and try it out.

We conclude with one final important note on linetypes. If you cannot see the linetype, and it appears as a solid line when it should be something else, chances are the scale of the dashes or dot spacing is too small to see. Type in ltscale (linetype scale) and enter a larger number (12, 24, 48…) followed by a Regen (re [Enter]). This is mentioned again as a tip in Chapter 4, Text, Mtext, Editing, and Style.

3.3 Introduction to properties

Sometimes, you need to change objects from one layer to another (perhaps because of an error in originally placing it on the incorrect layer or a later decision to change to another layer). Also, you may need to change the color, linetype, or other properties of objects. AutoCAD has a collection of tools to change the properties of objects, with the three most common methods being the Properties (and QuickProperties) palette, the Match Properties tool, and the Ribbon's Layer panel. We explore each in order.

Properties palette

Keyboard: Type in **ddchprop** or **properties** and press Enter
Ribbon: **Home** tab→**Properties**→**panel right corner arrow**

Once you do any of these (also: Ctrl+1 or Right-click pop up menu), the Properties box shown in Fig. 3.10 appears.

Look at the palette in detail from top to bottom. There are many categories (General, View, etc.) and many features that can be modified in those categories (Color, Layer, Linetype, etc.). Also, there are two columns, a gray, one on the left and a mix of gray (inactive) and white (active) fields on the right. To use this palette, first draw several lines on your screen, each of which is on a different layer. Then, follow these steps:

Step 1. First, make sure you have selected one or more items to change. They become highlighted. If you selected two items, only those properties common to both that can be changed are available.
Step 2. In the Properties box, find the category you want to be changed, color, for example.
Step 3. Click on the right-hand column directly across from that property.
Step 4. An arrow drops down. Click on it and view the choices within that category.

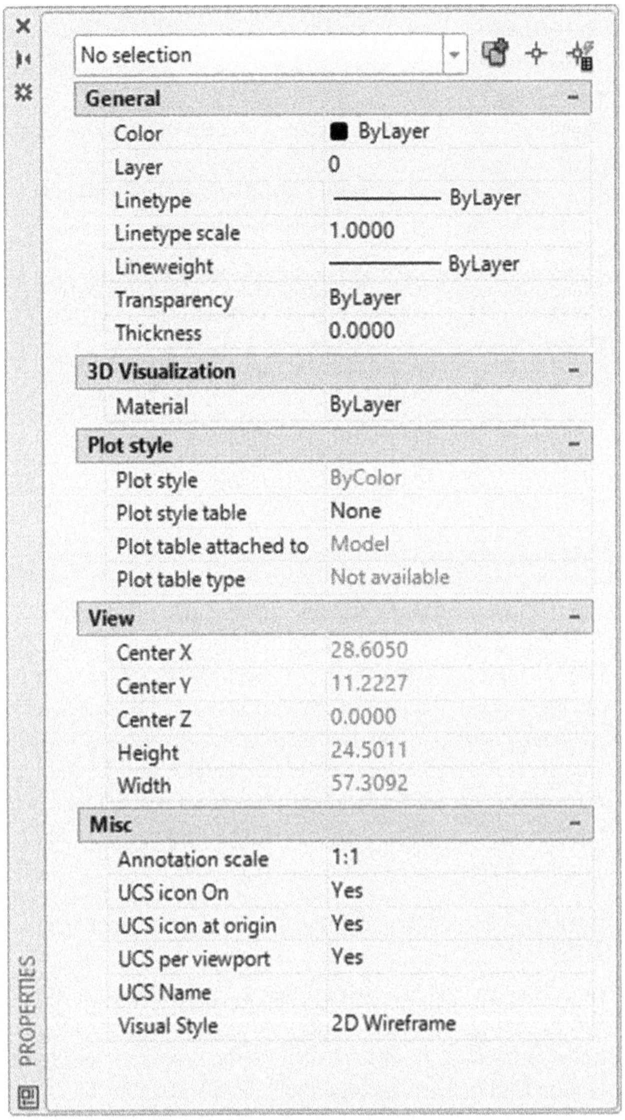

FIGURE 3.10 Properties palette.

Step 5. Select the different property that you want the new objects to have, such as a new color.
Step 6. Close the Properties box (the X in the upper corner) and press Esc once or twice; the objects have that property.

The Properties palette is a quick, straightforward way to change many properties but there is another way, shown next. There is also a QuickProperties variation to the full Properties palette just described. This feature is accessible by just double-clicking on any of the objects in the drawing. The QuickProperties palette appears, as seen in Fig. 3.11. Notice a

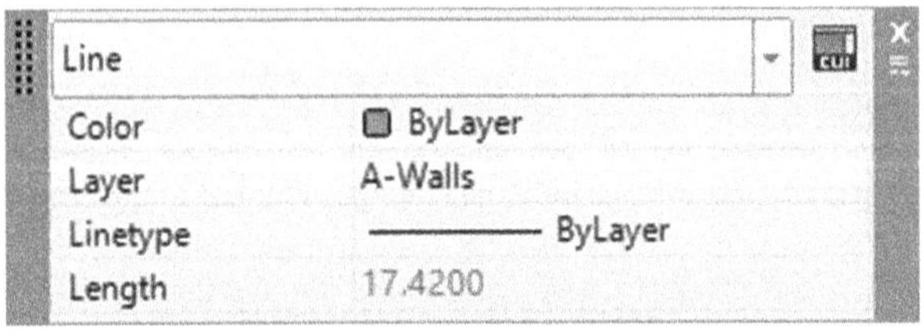

FIGURE 3.11 QuickProperties palette.

major difference here. The QuickProperties palette is much smaller and shows only the essential properties, such as Color, Layer, and Linetype. It omits various geometry data, which is probably okay because you are unlikely to use them. Notice also that the QuickProperties palette is *context sensitive*, meaning that it varies slightly, depending on what you have selected, line, arc, and so forth. The process of using it, however, is the same as just described for the full Properties palette. Go ahead and try it out, then close it by just pressing Esc.

Match Properties

Keyboard: Type in **matchprop** and press Enter
Ribbon: **Home** tab→**Clipboard**

This is a fantastic way to very quickly match up the properties of one object with the properties of another. The procedure is simple. Start the command via the preceding methods, select the first object whose properties you like, then click on another object that you want to have those properties.

Match Properties matches almost anything: layers, colors, linetypes, linetype scales, text sizes, hatch patterns, and many more factors. Finally, we have the Ribbon's Layers panel, which has its own method of layer changes and deserves its own separate discussion.

Layers and the Ribbon

The Ribbon Layers panel dropdown is shown in Fig. 3.12 with the three layers added.

To change an object's layer using this dropdown, simply select whatever object you would like to change to another layer first (what you selected becomes highlighted and its Grips are shown), then drop down the Layer menu in the panel and pick the new layer name on which layer you want the highlighted geometry to move to, and it moves to that layer. You may have to press Esc to deselect the objects afterward.

This Layer panel is also useful for quickly freezing, turning off, and locking layers with their associated control icons. It is also a great visual reference to determine what layer is current, which of course is the layer being shown when the dropdown is collapsed. The Layer panel has many useful icons, although the only one we have covered so far was the big one on the left that brings up the Layer Properties Manager. These icons provide quick, visual tools to control a layer or layers state by selecting objects rather than sorting through the layer list.

This is it for basic layers and associated topics. You should know the following:

- What layers are and why they are needed.
- How to open the Layers dialog box.
- How to enter a new layer, delete it, and change its name.
- How to make a layer current.
- How to assign a color to a layer and all the color options.
- How to Freeze/Thaw, turn Off/On, and Lock/Unlock the layers.
- How to select and set linetypes.

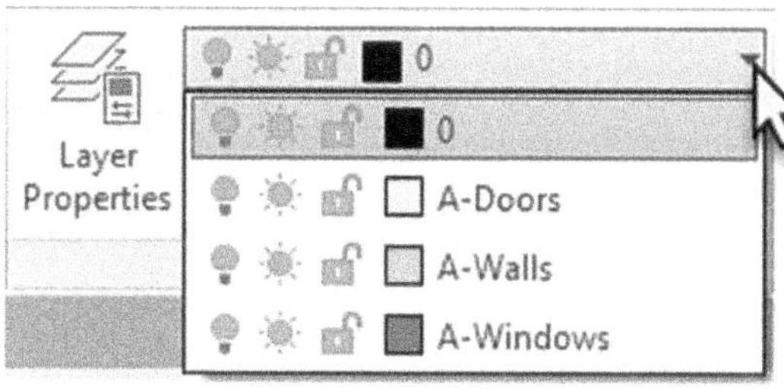

FIGURE 3.12 Ribbon Layers panel dropdown.

- How to use the Properties and QuickProperties palettes.
- How to use Match Properties.
- How to use the Ribbon's Layers panel.

We did not yet cover a few items, such as the previously mentioned Filters and Layer States, Plot, Lineweights, and a few other features of the Layers dialog box. We get to them in due time in advanced chapters. For now, it is time to put everything to good use and draw a full apartment floor plan. Review everything thus far and begin the assignment in Section 3.4.

3.4 In-class drawing project: floor plan layout

This is your first major drawing assignment. You need to put together everything you have learned so far and apply it to a realistic drawing situation. Although this floor plan is very simple compared to major architectural or engineering projects created in the industry, it still features components of a professional drawing, just fewer of them, so you do not get overwhelmed right away. Follow all instructions carefully and proceed slowly. Speed comes later when all the commands become second nature. Our goal is to draw the floor plan shown in Fig. 3.13.

This is a basic one-bedroom apartment. The kitchen is on the upper left, bedroom on the upper right, bathroom on the lower right, living room right in the middle, and entrance foyer in the lower left corner. There are two closets, six windows, and six doors (the kitchen has no door). The dimensions are only for you to look at and use, not draw in yet, as we do not cover them until Chapter 6. The numbers inside the rectangles are the door opening sizes. All walls are 5″ thick.

Outlined next is a complete discussion on how to start drawing the floor plan, intended for those students studying AutoCAD by themselves without an instructor. If you are taking a class, your instructor will assist you but should

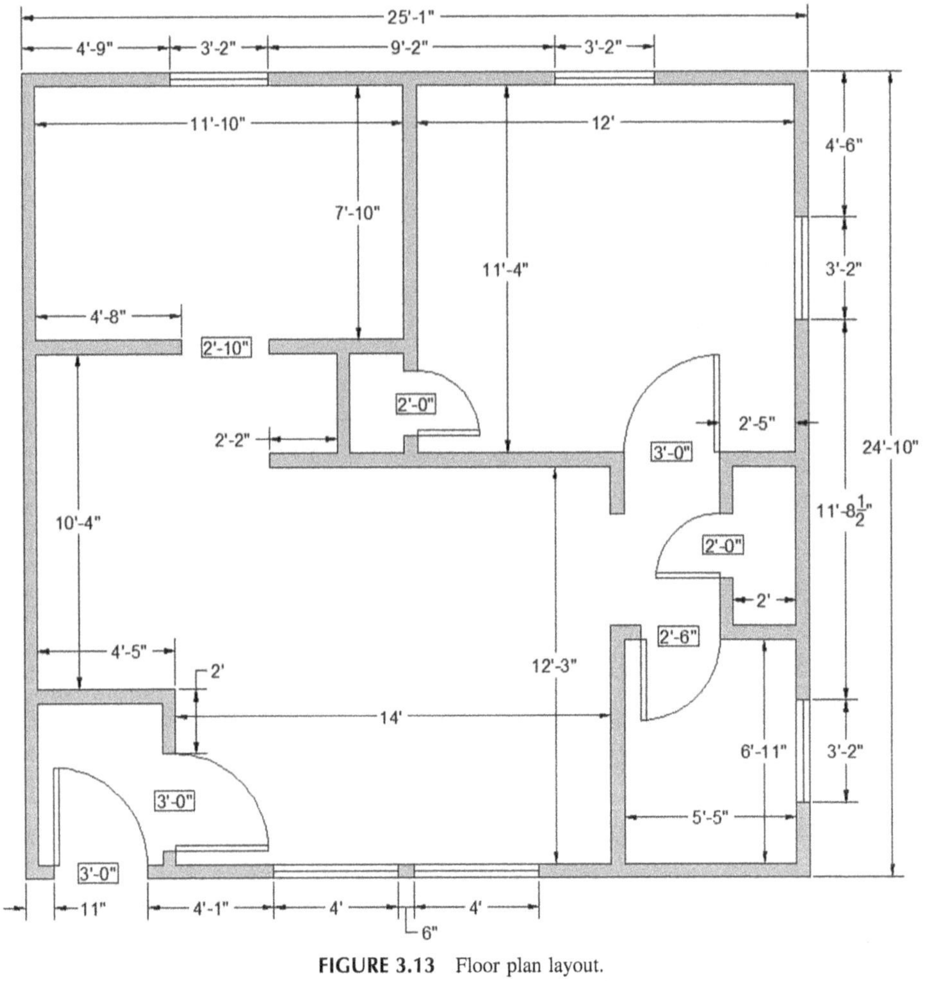

FIGURE 3.13 Floor plan layout.

essentially follow the same format. Once you learn all the steps involved and build confidence, aim to draw it again in less than 60, then 30, and finally less than 10 minutes, starting from a blank screen. This may sound unreasonable now, but that is the required industry speed, and you will be able to get to that quite soon with some practice. An outline of steps with additional information (if needed) follows.

Basic file preparation

1. Open a brand-new AutoCAD file using the acad.dwt template. Save it as YourName_AptProject1.dwg. We start from the very beginning for practice purposes.
2. Set your units to Architectural using the units command.
3. Set up layers: A-Walls (green), A-Windows (red), and A-Doors (yellow).
4. Load all your linetypes (we need them later). If you were to save and close this file right now and use it later for another project, then it would be called a *template*, as first explained in Chapter 2.

Starting the floor plan

Now that the basic file is set up, how would you start this design from the very beginning? The exterior walls would be the first step, and there are a few ways to go about it. We could certainly draw line by line, but that is not too efficient. Think back to your set of tools learned thus far. The best approach is to draw a rectangle of given dimensions (25′-1″ × 24′-10″) and offset it 5″ to the inside. This first rectangle represents the outer walls and the offset one represents the inner walls.

5. Draw a rectangle that is 25′-1″ wide and 24′-10″ tall. Be careful and use the correct method of manual Relative Distance Entry or Dynamic Input, your choice. You can also just use the rectangle command, indicating the dimensions as allowed by one of the suboptions. If you do it manually, you start the Rectangle command, press Enter, left-click anywhere in the screen, then type in `@25'1,24'10` exactly as written. If you use DYN, start the rectangle and enter the values on screen, using the Tab key to hop between the first and second values. If you use the Rectangle Dimension option, start the rectangle, press d, and enter its length and width when prompted. Once again, make sure the units are Architectural or you will immediately run into problems.
6. Offset the rectangle 5″ to the inside to create the inner walls (offset command).
7. Explode both wall rectangles. This is needed to break apart the rectangles for use as the basis for the interior walls as well as doorways and windows. You should now have a total of eight individual lines, as seen in Fig. 3.14.

Drawing the inner wall geometry

As mentioned in Chapter 1, one of the biggest misconceptions beginners have about drafting in AutoCAD is that they will spend lots of time drawing line after line. That expectation is natural; after all, in hand drafting, that is exactly what you would do—pick up the pencil and T-square and draw lines connecting them in intricate ways to create a design. Not so in AutoCAD. You still draw lines here and there, but a lot of the time you use the offset command to create new geometry, such as in this case. For example, we use the inner main walls as a basis for the room walls and offset as needed.

8. Offset the inner left vertical wall 11′-10″ to the right to create part of the kitchen wall. Then, right away, offset 5″ for the thickness of the wall. Repeat with the inner top horizontal wall (offset 7′-10″) to get the lower part of the kitchen. Offset 5″ below that line for thickness. Fig. 3.15 illustrates this. Note that, along the way, you will run into fractional lengths to offset (11′-8½″ for example). Do not forget that to enter fractions you just simply type in a decimal equivalent (11′8.5 in this case).
9. Fillet the two sets of walls to create a sharp corner. This is where the fillet command's advantage really shows. Follow it carefully and select the two inner wall lines first, then the outer.
10. Create a 2′-10″ opening for the kitchen door. Once again use the offset command. Offset the inner wall 4′-8″ to the right as shown on the plan. Then, offset again 2′-10″ to frame the opening. Next trim it out by selecting all the lines using a crossing, press Enter, and click on the lines you do not need. You use this technique to create the windows in the same way.
11. Trim out the extra line where each inner wall meets the main wall. Fig. 3.16 shows you what you should see so far (be sure to ask your instructor if something is not working out).

FIGURE 3.14 Floor plan, Step 7.

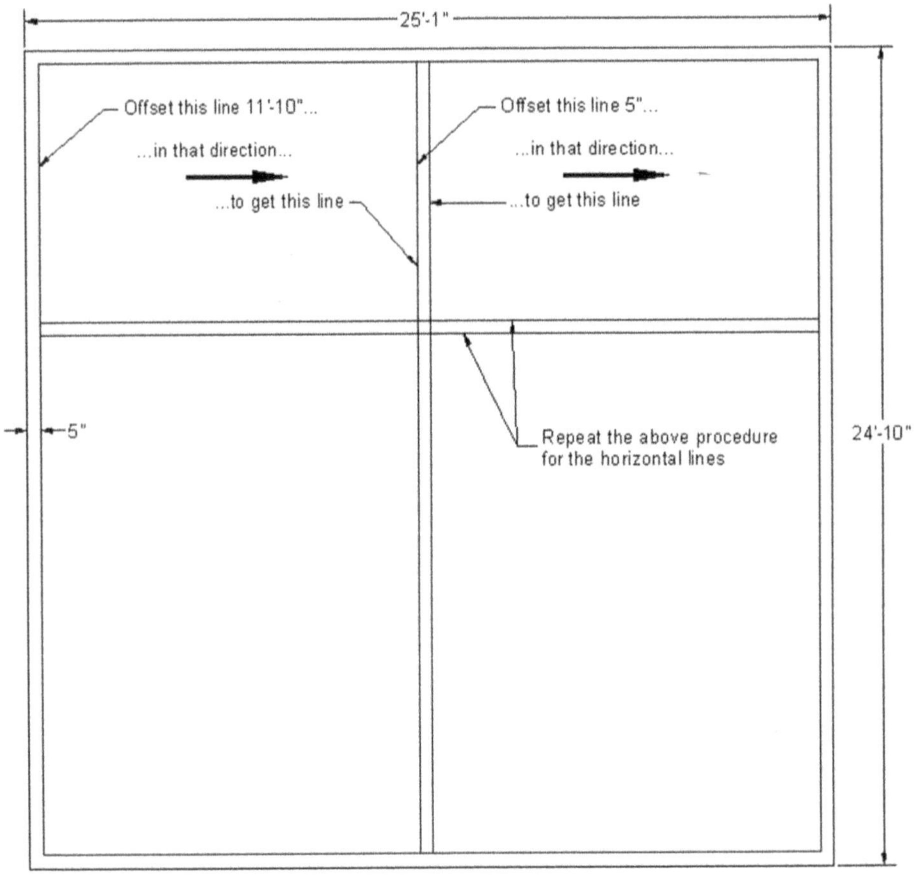

FIGURE 3.15 Floor plan, Step 8.

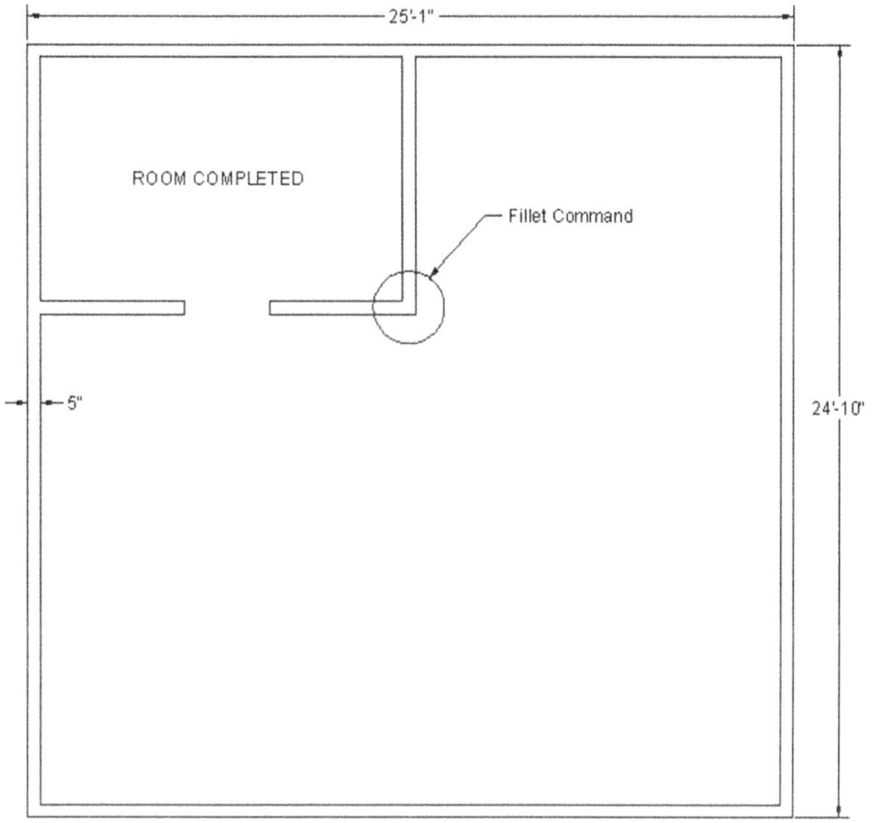

FIGURE 3.16 Floor plan, Step 11.

If you understood the previous steps, then you should have a good idea of what is needed to draw the rest of the floor plan. It really is mostly similar. You of course must get creative on occasion, but just stop and think of the basic tools you have available first. If it seems overly complicated, you are doing it wrong. After the kitchen, go on to do the bathroom in the lower right corner, then the bedroom and entrance foyer.

Drawing the doors and windows

Create the openings for the doors and windows early in the process but do the actual doors and windows last. Be sure to use the correct layers. Both a door and a window are shown in Fig. 3.17.

Here are some additional tips to help you along.

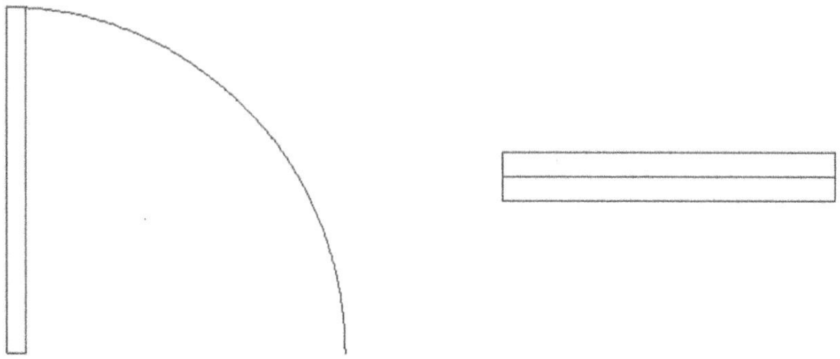

FIGURE 3.17 Door and Window symbols.

Doors

Once you have an opening in a wall of a certain size, adding the door involves creating a similarly sized rectangle, oriented in the proper direction. Then, you need to add the door swing to complete the door. You already know how to create rectangles in one of several ways from Chapter 1. You can even use what you did in Exercise 3 of Chapter 1. However, if you want the doors to look more authentic, you need to give them a thickness. We use a thickness of 2″ for all doors in this floor plan. Use DYN or any preferred method to create the door rectangle. A sample result is shown in Fig. 3.18.

To create the door swing, you can carefully position a three-point arc (as in the previously mentioned in a Chapter 1 exercise) or, as a more accurate alternative, create a circle based at the hinge point and with a radius equal to the door width, as seen in Fig. 3.19. Then, just trim the unneeded three quarters of the circle by using the top of the door and the left wall as cutting edges.

Windows

To create windows, draw a rectangle of the proper size to fit into the opening and add a line through it. Then, position it into the window opening as seen in the four-step process of Fig. 3.20. These are of course very simplified windows; expect to see much more complex contraptions if you do architectural drafting. Finally, it bears repeating, for both the doors and the windows, be sure to be on the correct layer for each.

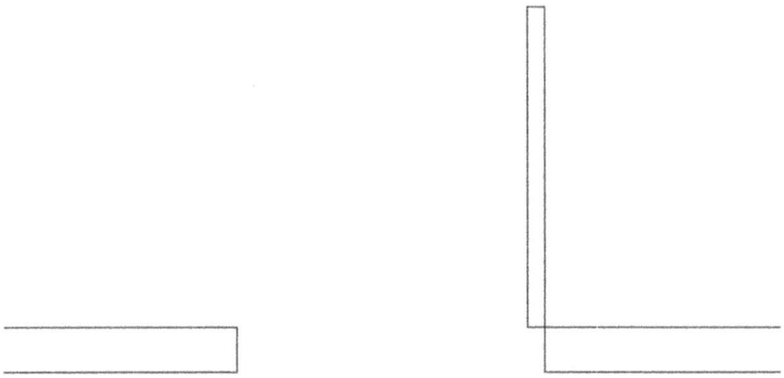

FIGURE 3.18 Creating the door symbol.

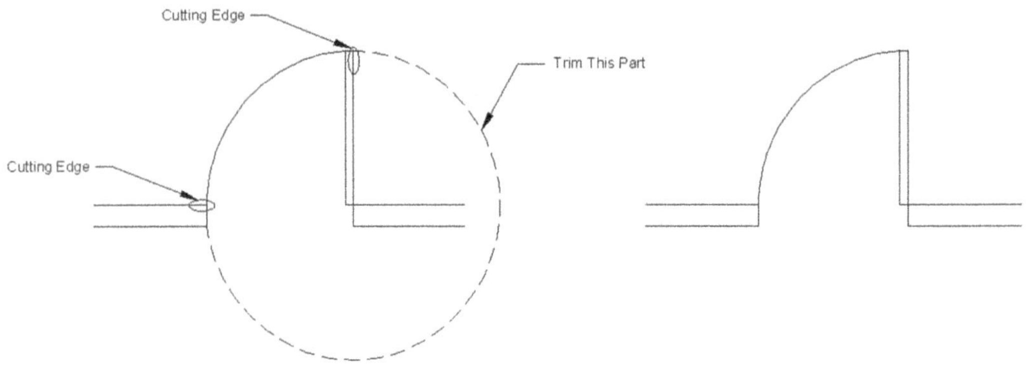

BEFORE AFTER

FIGURE 3.19 Creating the door swing.

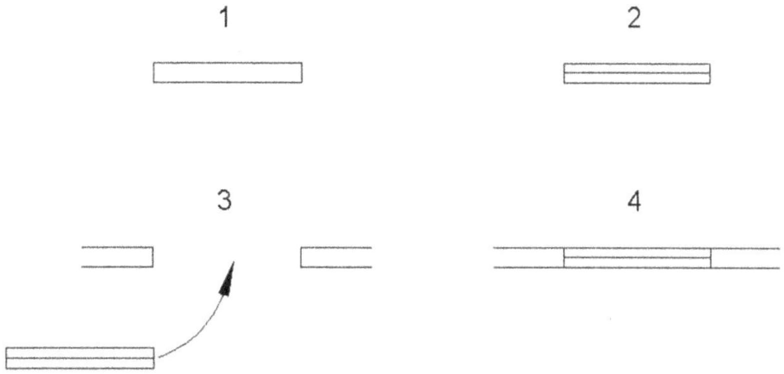

FIGURE 3.20 Creating a simple window.

Summary

You should understand and know how to use the following concepts and commands before moving on to Chapter 4:

- Layers
  - Basic concept and purpose
  - Creating and deleting
  - Current layer
- Layer colors
- Setting and changing colors
  - Index color
  - True color
  - Color Books
- Layer Freeze/Thaw
- Layer On/Off
- Layer Lock/Unlock
- Linetypes
  - Loading and selecting
- Properties palette
- Match Properties
- Layers panel

Review questions

Answer the following based on what you learned in this chapter:

1. What is a layer?
2. What is the main reason for using layers in AutoCAD?
3. What typed command brings up the Layers dialog box?
4. List two ways to create a new layer.
5. How do you delete a layer?
6. How do you make a layer current?
7. What three Color tabs are available?
8. What does freezing a layer do?
9. What does locking a layer do?
10. What are some of the linetypes mentioned in the text?
11. What are three ways to bring up the Properties palette?
12. What two other ways to change properties or layers are discussed?

Exercises

1. **Step 1:** In a new file based on the acad.dwt, load all linetypes, next open the Layers dialog box, and set up the following layers, with index colors, and linetypes.
   - A-Walls-Demo, Gray, Hidden;
   - A-Walls-New, Green, Continuous;
   - A-Windows-New, Red, Continuous;
   - A-Doors-Demo, Gray, Hidden;
   - A-Doors-New, Yellow, Continuous;
   - A-Shelves, Blue, Hidden;
   - A-Furniture, Magenta, Continuous
   - A-Text, Cyan, Continuous
   - A-Dims, Cyan, Continuous
   - A-Carpeting, Gray, Continuous
   - E-Outlets-Duplex, Red, Continuous
   - E-Switches, Red, Continuous
   - E-Switches-3 way, Red, Continuous

   Step 2: Then, Lock all E-layers, make A-Walls-New current, and freeze A-Text. (Difficulty level: Easy; Time to completion: <5minutes.)
2. In a new file, set units as Decimal; load all linetypes; set up layers M-Part, Hidden, and Center; and assign them proper linetypes and colors of your choosing. Then, draw the following mechanical object, including all hidden and centerlines. Align the front and side views as shown to project the sizes of different geometry from one view to another as needed until both views are complete. NOTE: (CL stands for centerline) All text and dimensions are for your reference only to create the geometry accurately; we learn how to create text in Chapter 4 and dimensions in Chapter 6, and you do not need to create them here. (Difficulty level: Easy; Time to completion: 10–15 minutes.)

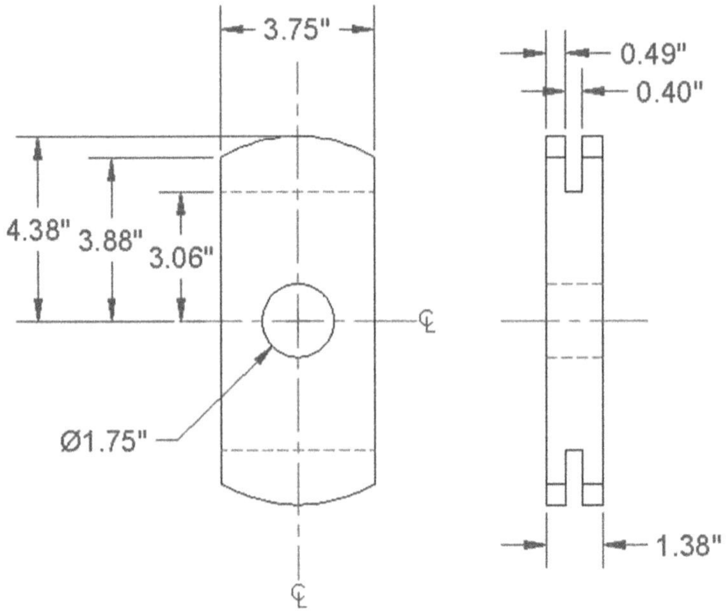

3. In a new file, set units as Decimal; load all linetypes; set up layers M-Part and Hidden, and assign them proper linetypes and colors of your choosing. Then, draw the following mechanical object, including all hidden lines. Align the top and front views vertically as shown to project the sizes of different geometry from one view to another as needed until both views are complete. NOTE: All text and dimensions are for your reference only to create the geometry accurately; we learn how to create text in Chapter 4 and dimensions in Chapter 6, and you do not need to create them here. (Difficulty level: Easy; Time to completion: 10–15 minutes.)

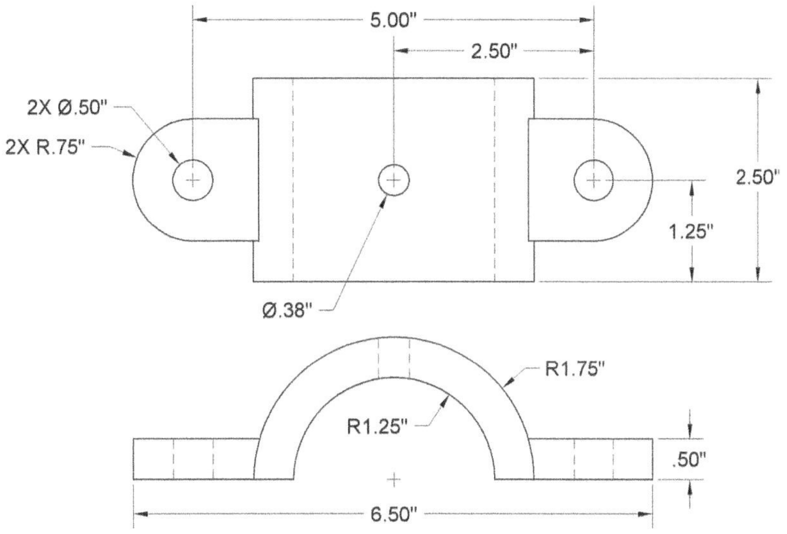

4. In a new file, set units as Decimal; load all linetypes; set up layers M-Part, Hidden, and Center; and assign them proper linetypes and colors of your choosing. Then, draw the following mechanical object, including all hidden and centerlines. NOTE: All text and dimensions are for your reference only to create the geometry accurately; we learn how to create text in Chapter 4 and dimensions in Chapter 6, and you do not need to create them here. (Difficulty level: Easy; Time to completion: 15–20 minutes.)

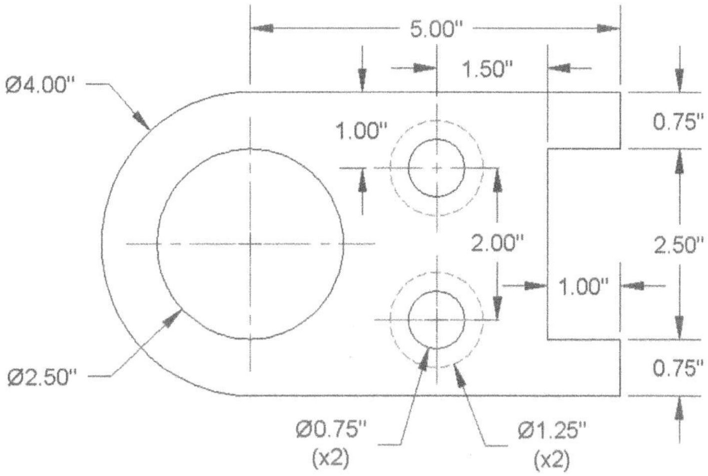

5. In a new file, set units as Decimal; load all linetypes; set up layers M-Part, Hidden, and Center; and assign them proper linetypes and colors of your choosing. Then, draw the following mechanical object, including all hidden and centerlines. NOTE: All text and dimensions are for your reference only to create the geometry accurately; we learn how to create text in Chapter 4 and dimensions in Chapter 6, and you do not need to create them here. (Difficulty level: Easy; Time to completion: 15–20 minutes.)

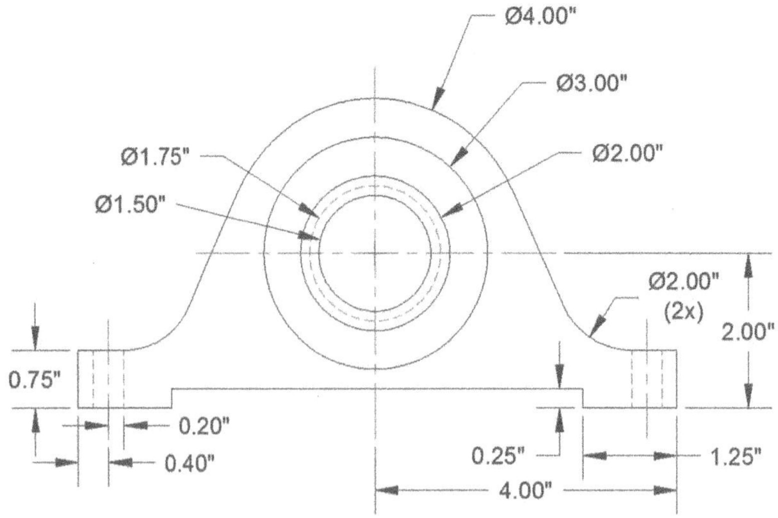

6. In a new file, set units as Decimal; load all linetypes; set up layers M-Part, Hidden, and Center; and assign them proper linetypes and colors of your choosing. Then, draw the following mechanical object, including all hidden and centerlines. Align the front and side views as shown. NOTE: All text and dimensions are for your reference only to create the geometry accurately; we learn how to create text in Chapter 4 and dimensions in Chapter 6, and you do not need to create them here. (Difficulty level: Intermediate; Time to completion: 20–25 minutes.)

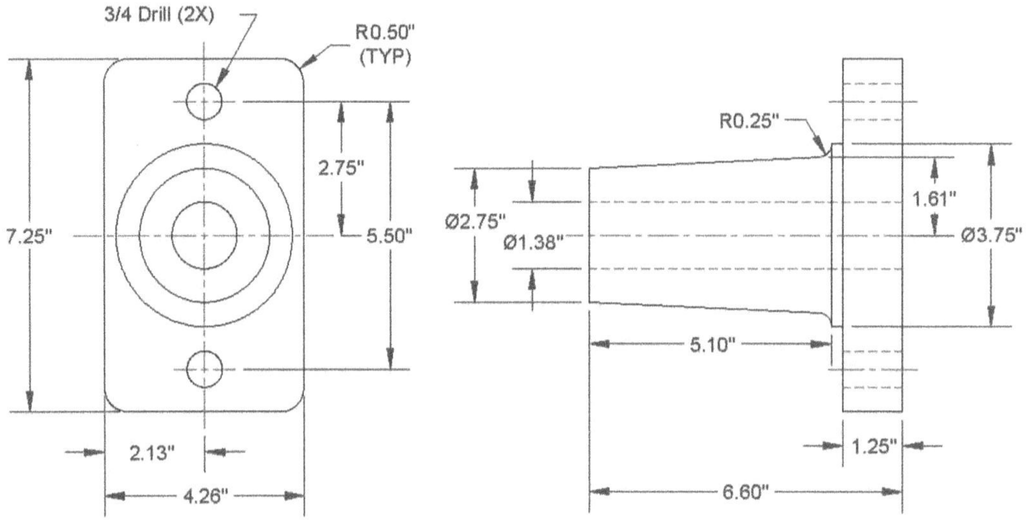

7. Often, in design work, you need to draw standard parts because no CAD files are available. Such is the case with the following electrical termination blocks, breakers, and thermostat equipment. You are provided minimal dimensions, and your job is not to design the part but simply to document its appearance, so an engineer can insert it into a larger modular design. In a new file, leave units as Decimal and set up layer E-Part, assigning it any color you wish. Then, draw the electrical hardware, sizing everything relative to the overall provided dimensions and estimating most of the internal geometry. This is an excellent exercise in basic shape drafting. Using Grid and grid Snap settings will help you quickly draw accurately here. Be sure to use the copy command wisely to avoid significant duplication of drawing effort. (Difficulty level: Intermediate; Time to completion: 45 minutes.)

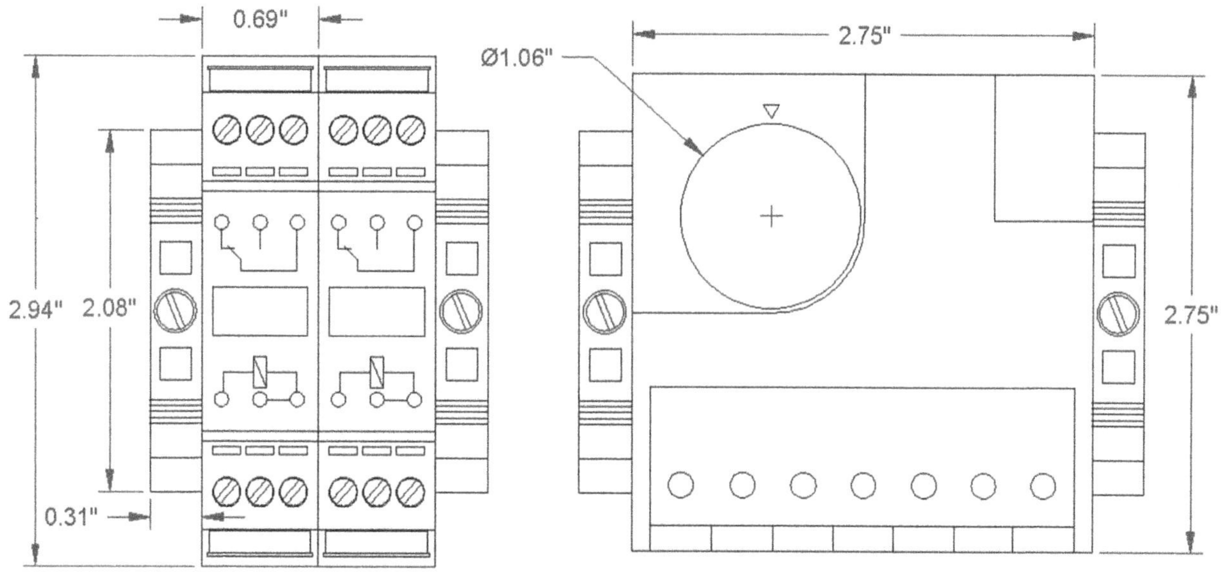

8. The next three exercises will all be Architectural in nature. In a new file, set up Architectural units, the correct layers (only A-Walls and A-Doors are needed), and draw the following floor plan. NOTE: Dimensions, text, and the solid hatch pattern are for clarity and construction purposes only. We learn them in upcoming chapters. (Difficulty level: Intermediate; Time to completion: 45–60 minutes.)

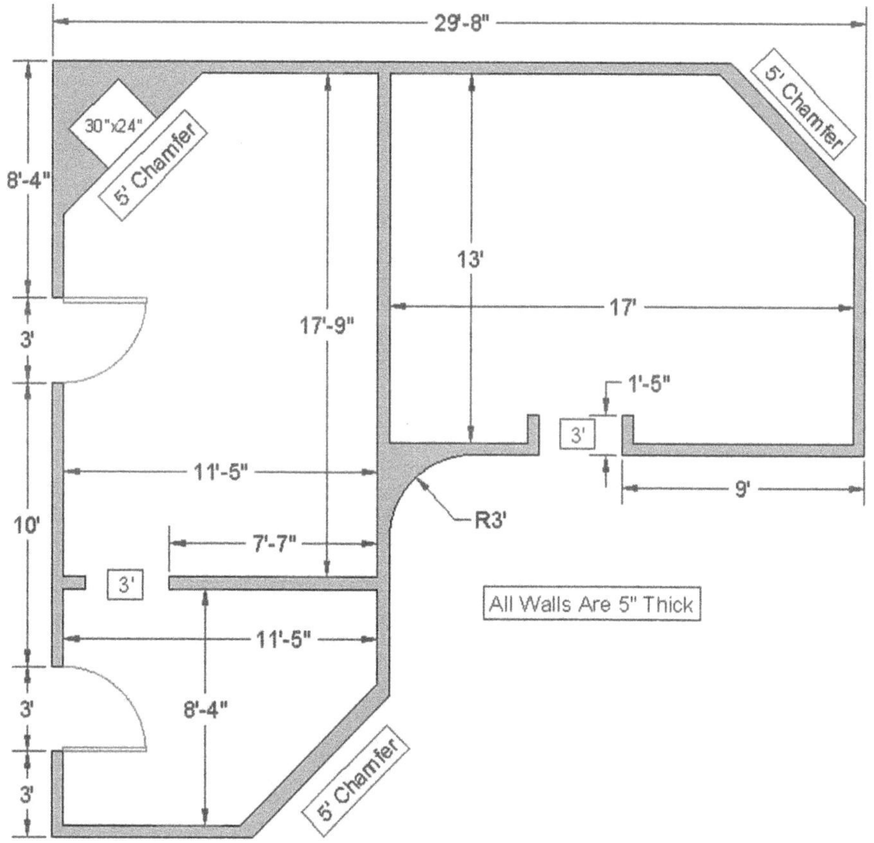

9. In a new file, set up Architectural units, the correct layers (A-Walls, A-Doors, and A-Windows), and draw the following floor plan. NOTE: Dimensions, text, and the solid hatch pattern are for clarity and construction purposes only; we learn them in upcoming chapters. (Difficulty level: Intermediate; Time to completion: 60–90 minutes.)

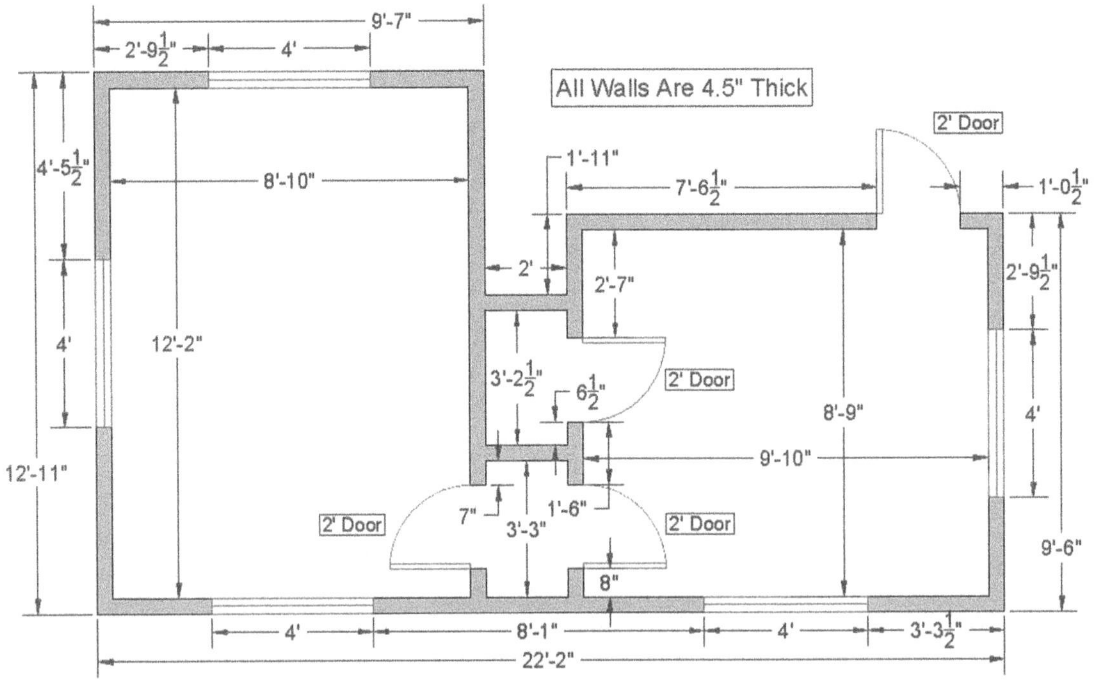

10. In a new file, set up Architectural units, the correct layers (A-Walls, A-Doors, A-Windows, and A-Furniture), and draw the following floor plan including the furniture. NOTE: Dimensions, text, and the solid hatch pattern are for clarity and construction purposes only; we learn them in upcoming chapters. Those dimensions not expressly given (such as with the furniture, kitchen, and bathroom elements and other minor details of the floor plan itself) can be assumed and free-drawn to an appropriate size. (Difficulty level: Intermediate; Time to completion: 60–90 minutes.)

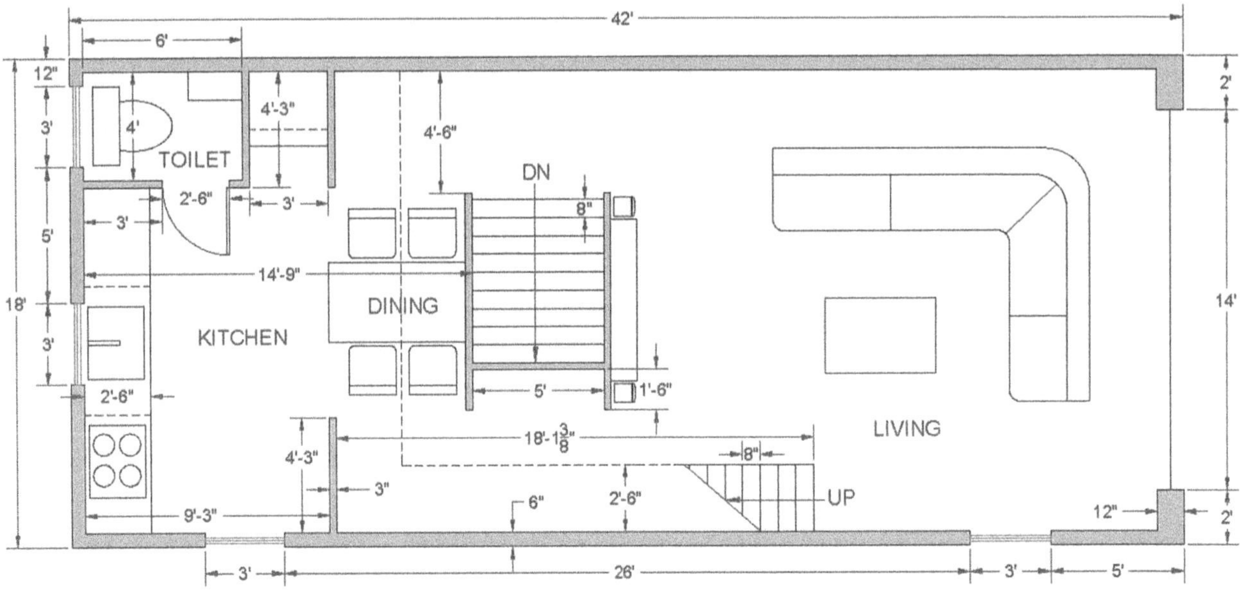

Chapter 4

Text, mtext, editing, and style

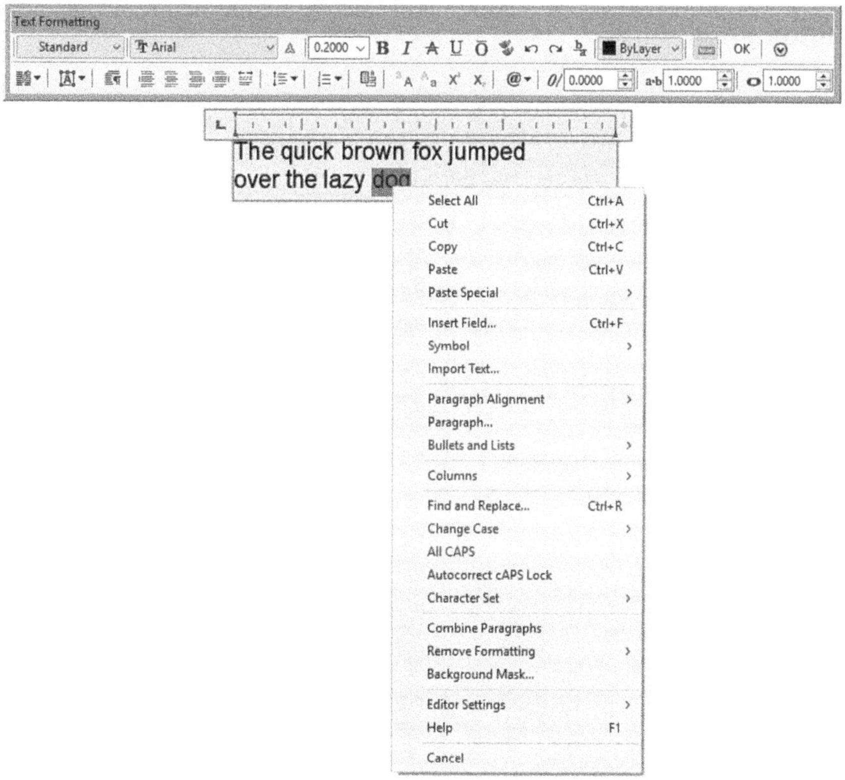

Learning Objectives

Text allows you to communicate beyond what can be described in lines on a drawing. In this chapter, we introduce text, mtext, style, and editing and discuss the following topics:

- Text
- Properties and applications of text
- Editing all types of text
- Mtext
- Properties and applications of mtext
- Mtext formatting
- Mtext symbols
- Style
- Spell check
- Nearest OSNAP

At the end of this chapter, you will be able to annotate your floor plan as well as add additional features, such as furniture and stairs.

Estimated time for completion of the chapter: 2 hours (lesson and project).

Up and Running with AutoCAD 2022. https://doi.org/10.1016/B978-0-323-89923-9.00004-4

4.1 Introduction to text and mtext

Adding text to your AutoCAD drawing is the next logical step once you learn how to create and edit basic designs. After all, the purpose of most drawings is to describe how to build something or show what a design looks like. Text goes a long way in assisting in this and, along with dimensions, is usually the next item to be added to a new design.

Text in AutoCAD comes in two versions: regular text (single-line text) and mtext (multiline text). The two share some overlaps and similarities, and we look at how to create and edit both types. We then look at how to choose and set fonts and conclude the chapter by adding text to the previously designed floor plan.

4.2 Text (single line)

This is your basic text creation command, and it creates a field anywhere you click on the screen, into which you can type whatever text you need. It does have a carriage return that goes to another line on pressing Enter, so you need to press Enter twice to get out of the command field. While an unlimited number of lines of text can be typed, typically regular text is used for only a single line, as multiple lines of text are not joined together in paragraph form and cannot be formatted to any significant extent. Let us summarize:

- Single Line Text is used mostly when only one or a few lines are needed.
- The text cannot be formatted nor any special effects added, beyond underlining and a few other minor effects.
- Multiple lines of text are not in paragraph form; rather they remain as individual lines.

Open a new file, find the Single Line Text toolbar command (Fig. 4.1), and let us create a few sample lines.

Keyboard: Type in **text** and press Enter
Ribbon: **Home** tab→**Single Line Text**→ A

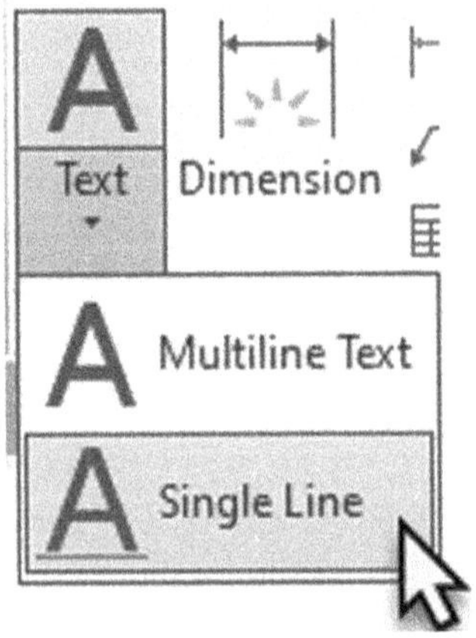

FIGURE 4.1 Text toolbar.

Step 1. Begin the text command via any of the previous methods.

- AutoCAD says:

```
Current text style: "Standard" Text height: 0.2000
Annotative: No Justify: Left
Specify start point of text or [Justify/Style]:
```

Step 2. Left-click anywhere on the screen.

- AutoCAD says: `Specify height <0.2000>:`

Step 3. Enter a new height if desired, say 1.0, and press Enter.

- AutoCAD says: `Specify rotation angle of text <0>:`

You can rotate the text, but there really is no need to, so just type in `0` for now, and press Enter.

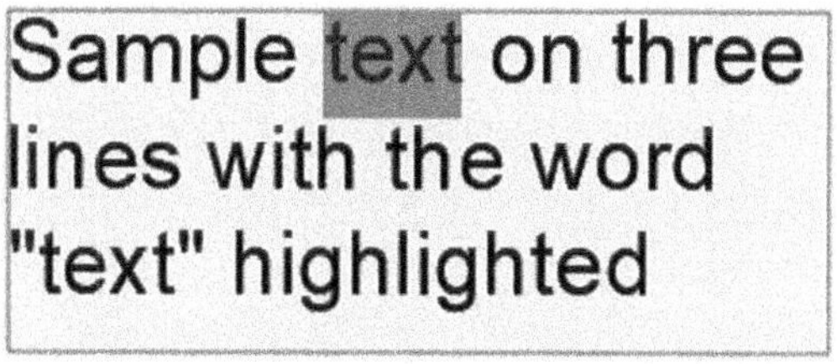

FIGURE 4.2 Basic text.

Step 4. A text field opens up with a blinking cursor. Go ahead and type something, pressing Enter once to go to the next line or twice to finish. Fig. 4.2 shows what the basic text looks like hanging out in the middle of a blank screen (just before you press Enter one last time). If the text is too large to see completely or too small to be even readable, just zoom in or out to fit it nicely on the screen. You can zoom and pan even while you are typing.

The text is still rather simple; we have not yet addressed font, sizing, or any effects. Notice that, once you complete the command, the text is on three separate lines. You can edit, move, copy, or erase any of the lines without affecting the others. So how would you edit this text?

Editing Text

Editing, or changing, what the text says is very simple and follows the same procedure for both text and mtext, so we go over it right away for both cases. The easiest way to edit the text is to *double-click* on it, which opens both the text and mtext fields.

It is worth noting that an older typed-in command also exists, called ddedit. If you type that in, press Enter, and click on the text, you can edit it. This may come in handy with older versions of AutoCAD you may run acros.

Whichever method you use, when you are done editing, just press Enter again. Do not press Esc, as that cancels whatever editing you just did.

Here is a trick with regular text that you will not hear about much anymore. If you want to underline it, type in %%u just before the text string you want underlined. This old code dates back to DOS days, and it will automatically and immediately add an underline to your basic text. You can also experiment with %%c, %%p, and %%d. See what those three do. None of this is needed with mtext, as discussed next.

4.3 Mtext

Mtext is short for *multiline* text. In other words, this is the command you use when you anticipate typing in a paragraph as opposed to one or two lines and, more important when you require some advanced formatting and effects. As an added bonus, it also has a spell checker built in. Many designers use mtext for all their text needs, just in case formatting needs to be applied to even just one word. The only downsides are that it takes a few seconds longer to set up, and the features make it a more complex command. The idea here is to define an area where your text will go. Once you do that, the new paragraph fits into that area. Let us summarize the mtext features and try the command:

- Used primarily for writing paragraphs
- Has extensive formatting and editing features
- Has an extensive symbol library and spell check
- Can accept extensive text importing

A few releases ago the mtext command was enhanced even further and has inherited more features from a dedicated word-processing program, like MS Word, including auto bulleting, better auto stacking of fractions, and other useful tricks.

Keyboard: Type in **mtext** and press Enter
Ribbon: **Home** tab→**Multiline Text**→ A

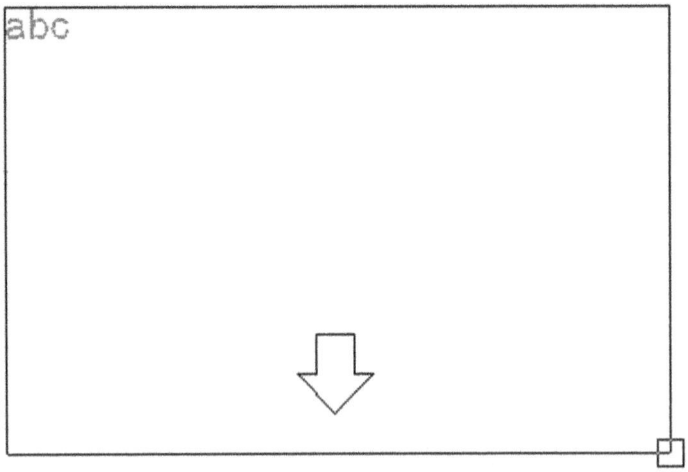

FIGURE 4.3 Mtext field.

Step 1. Begin the mtext command via any of the previous methods.

- AutoCAD says: `Current text style: "Standard" Text height: 0.2000 Annotative: No`
  `Specify first corner`

Step 2. Left-click anywhere on the screen, noticing the abc next to the crosshairs. A rectangle with an arrow appears, as shown in Fig. 4.3. Continue to move your mouse down and across to the right, making the rectangle bigger. This is your text field; make it as large or small as you need it to be to hold all the text. As you do this,

- AutoCAD says: `Specify opposite corner or [Height/Justify/`
  `Line spacing/Rotation/Style/Width/Columns]:`

Step 3. Click again when you have defined the field.

You should see the text field where you can type in your text with a toolbar just above it (but not attached to it), as seen in Fig. 4.4, (NOTE: If you do not see the Text Formatting toolbar, it is okay as all the tools are also duplicated in the contextual tab of the Ribbon.) Type in multiple lines of text using Enter to go to the next line or paragraph. To finish up, press OK in the upper right of the toolbar, click the green check mark Close Text Editor button at the right end of the Ribbon, or just click anywhere outside the text field, Do not use Esc as you may lose your mtext work. We cover mtext's extensive formatting tools in the next section.

Formatting mtext

First of all, notice something: If you click on mtext, it remains a paragraph, a big difference from a few lines of regular text. To edit mtext, as already mentioned, double-click on it, or type in ddedit.

Now, on to formatting. Mtext has a lot of additional features, and once you input the text, you can modify it significantly. Some intent was made here to mimic MS Word's text editing abilities, and although AutoCAD cannot really come close to a dedicated word-processing program, the available tools are still quite extensive. We take a look at the non-Ribbon toolset first and cover the tools again, this time using the Ribbon's Text Editor (in much less detail, as the tools essentially are the same). Looking at the Text Formatting toolbar from left to right, as shown in Fig. 4.6, we have the following groups:

Group 1. This is where you set *name*, *font*, and *size*, although this is generally done using *style*, to be discussed later. You can also make the text annotative, to be covered in Level 2.

Group 2. These are the standard *Bold*, *Italic*, *Strikeout*, *Underline*, *Overline*, and *Undo/Redo* buttons. This is also where you set *color*, although that is usually done by layer. The Ruler icon turns off the ruler grid; the OK button closes the field; and the down arrow brings up additional options, to be covered soon.

Group 3. Here you find a tool for making *columns* and some of the associated settings, *paragraph justification* tools, as well as tools to change *line spacing*, *numbering*, and *fields*.

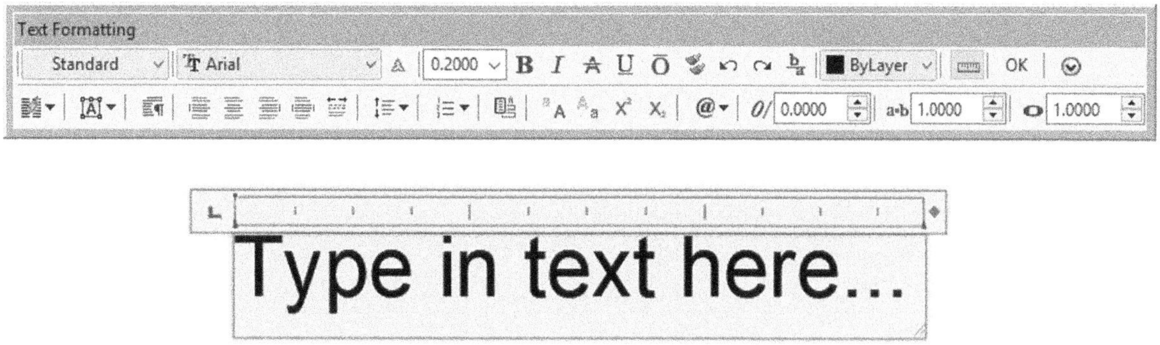

FIGURE 4.4 Text formatting field and toolbar.

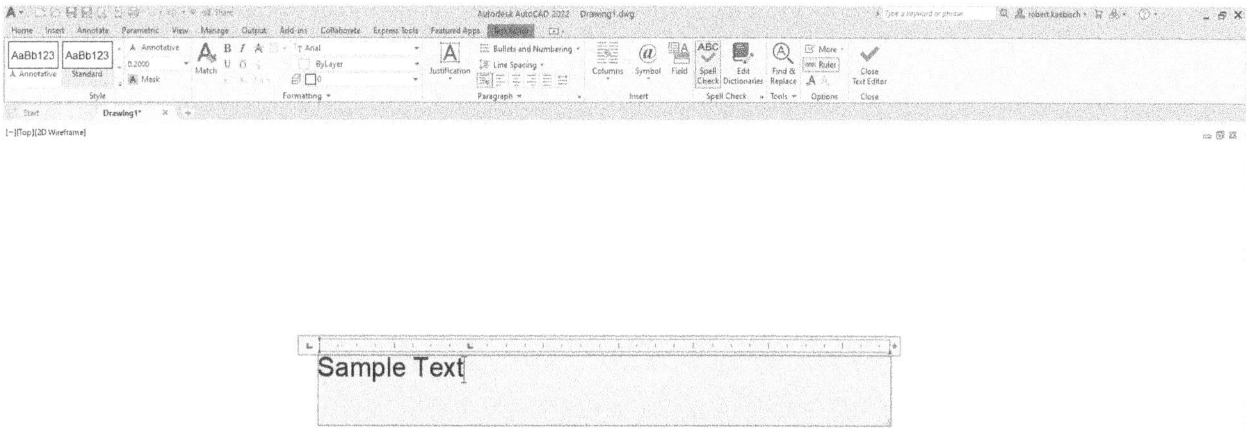

FIGURE 4.5 Text formatting via the Ribbon's Text Editor.

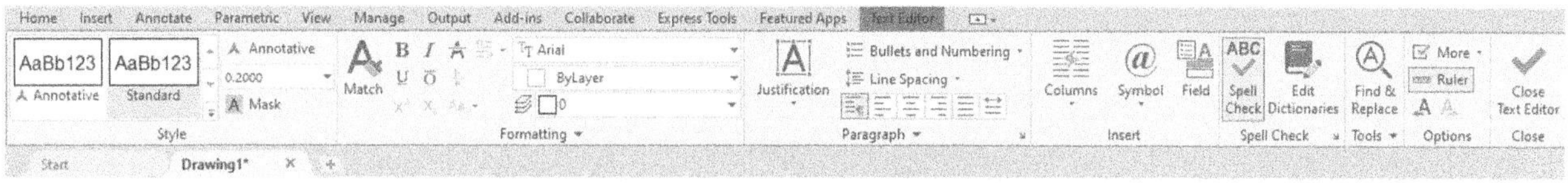

FIGURE 4.6 Text Editor Ribbon Panels.

Group 4. Here you find tools to change from *upper to lower case* (and vice versa), *superscript* and *subscript* (introduced in AutoCAD 2015), *oblique angle*, *tracking*, and *width factor tools.* The @ character is used to bring up additional symbols (to be discussed shortly).

You may be familiar with most of these if you ever typed a document in MS Word. Just as in Word, you need to highlight the text to which you want the changes to apply. Experiment with the buttons to see what they all do and try to duplicate what you see in Fig. 4.7. Notice also how Arial is the default font in AutoCAD.

There is more to the mtext command than just the contextual Text Editor tab of the Ribbon detailed in Fig. 4.6, although the Text Editor tab and toolbar contain most of the often-used commands. Two additional menus are accessed by either right-clicking while inside the mtext box or pressing the @ Symbol icon in Group 4. We cover this @ Symbol menu first; it introduces a vast array of available symbols.

Fig. 4.8 shows what this menu looks like. Examine it closely; it has interesting and useful options. In architecture and engineering, one typically encounters many industry- or trade-specific symbols. AutoCAD provides a significant database of them for your use. Browse through and try a few by clicking and making them appear in your text field.

In case none of those symbols is what you are looking for, there are more. At the bottom of the menu, click on Other …. A *character map* appears, as shown in Fig. 4.9.

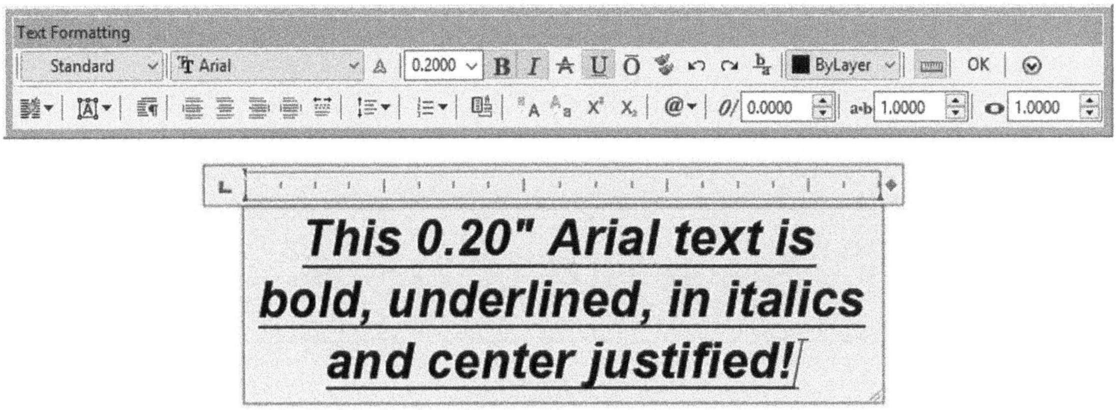

FIGURE 4.7 Mtext sample.

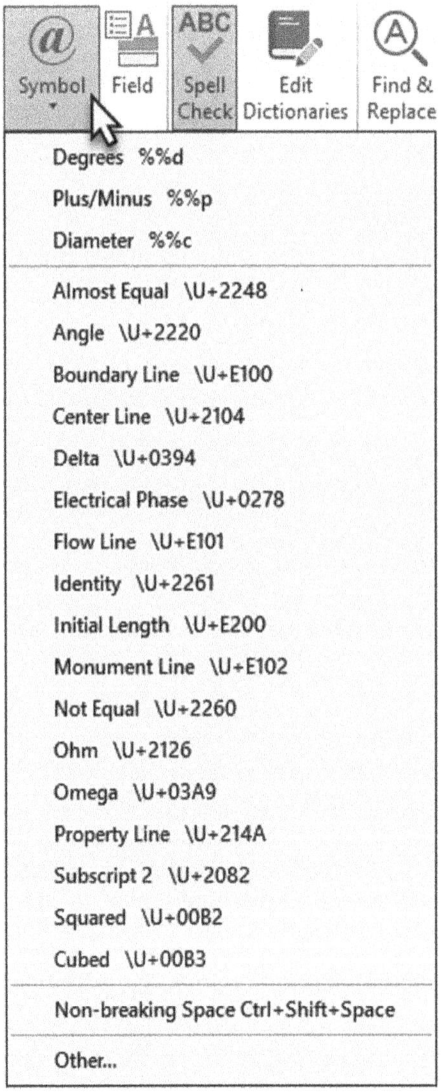

FIGURE 4.8 Mtext @ Symbols.

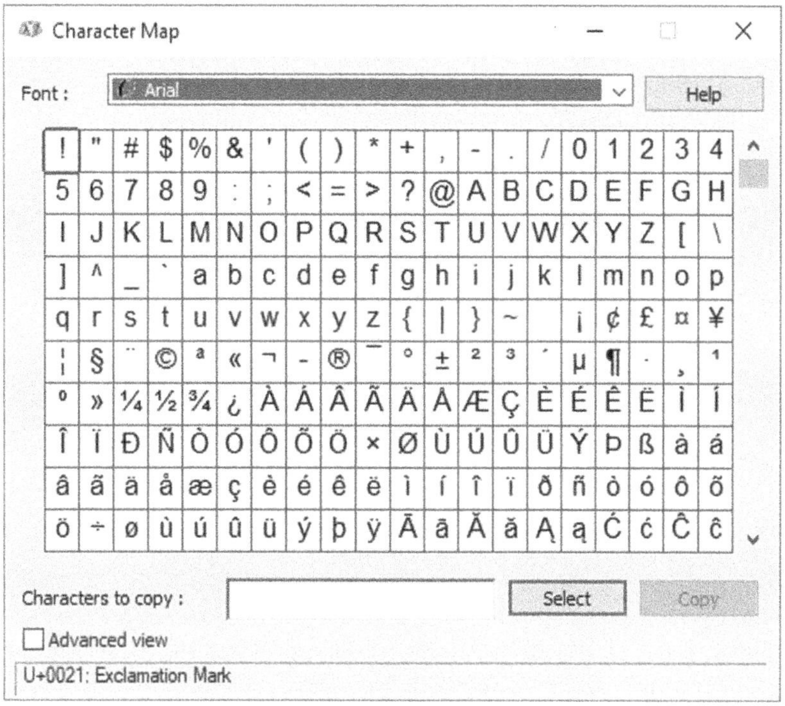

FIGURE 4.9 Character map.

It is like the character map used by MS Word and functions the same way. This is basically a database of available letters (in English and several other languages) as well as symbols and characters. The choices change somewhat from font to font. To use this tool, simply click on a symbol you want (it will temporarily increase in size so you can examine it closer) and press the Select button. Then, minimize the Character Map box (or close it if you do not need it anymore) and right-click to paste the symbol into your mtext field.

Character maps are generally for symbols not found in the main list of the drop-down table (Fig. 4.8). If you see the symbol you want there first, use it. To introduce the final menu, right-click in the mtext field and you see the menu in Fig. 4.10.

Some of these menu choices duplicate what is already seen in the main mtext menu bar, and you can access all the symbols through here as well. Aside from some Cut and Paste choices, this menu just presents an alternate way of making option selections. As a final note, be aware that pressing Esc will exit you out of the Mtext window, and you need to save if you want to keep what you did. AutoCAD, of course, prompts you to do this.

Toolbar text editing

As mentioned before, the Toolbar has essentially the same formatting tools as in the Ribbon but presented in a presented manner. Fig. 4.11 is the broken Text Editor, broken down into 4 categories for easier learning. The Ribbon and Toolbar can be displayed and used at the same time.

The Ribbon's Text Editor categories are detailed next.

- *Style*: Annotation and size can be changed here.
- *Formatting*: Bold, italics, underline, overline, font, color, background mask (color), oblique, spacing, and width factor can be changed here.
- *Paragraph*: Justification, line spacing, bullets, and numbering can be changed here.
- *Insert*: Columns, symbols, and fields can be changed here.
- *Spell Check*: You can run the spell check and set dictionaries and settings here.
- *Tools*: Find and Replace, Import Text, and AutoCAPS are found here.
- *Options*: Here, you can make the Editor toolbar come back (AutoCAD never really gets rid of anything), add the ruler grid, and use Undo/Redo. You can also change the background of the Editor to opaque.
- *Close*: Closes the Editor, similar to the OK button.

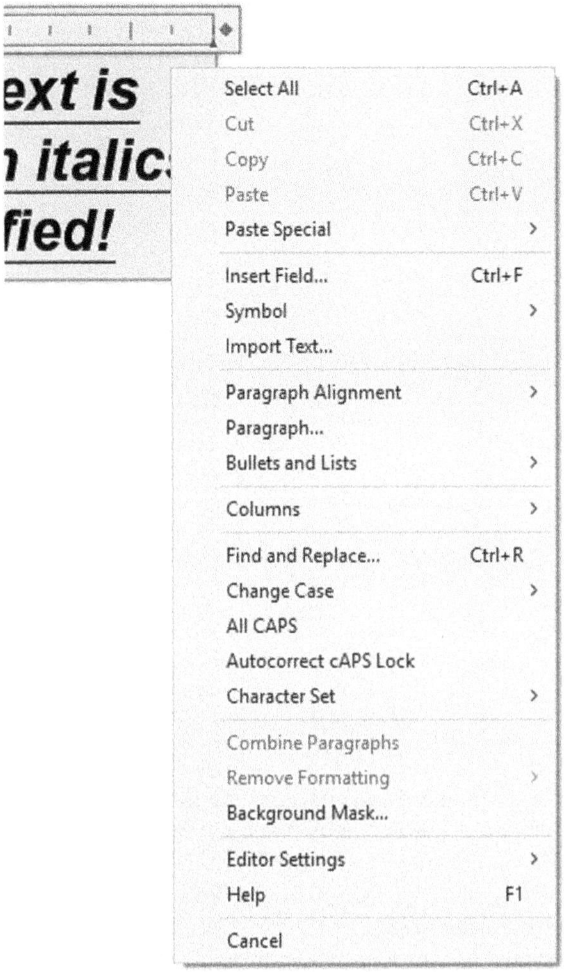

FIGURE 4.10 Mtext right-click menu.

FIGURE 4.11 Toolbar Text Editor.

Finally, before we move on to discussing style, a new feature needs to be mentioned that was just added recently. You can now put your mtext into a box frame, which was one of the top requests from past AutoCAD users. To do this, call up the Properties palette (first seen in Fig. 3.10). Then click your mtext one time to select it, which will make the Properties palette display four categories. Scroll to the bottom of the Text category and find Text frame on the left. Across from it will be a Yes/No drop-down menu. Select Yes and you are done. Your mtext will have an automatic frame around it, a very useful feature.

4.4 Style

The idea behind the style command is very straightforward. Pick a font, give it a name and a size, and use it throughout your drawing. Drawings typically use only one font throughout the main design and perhaps another, fancier one in the title

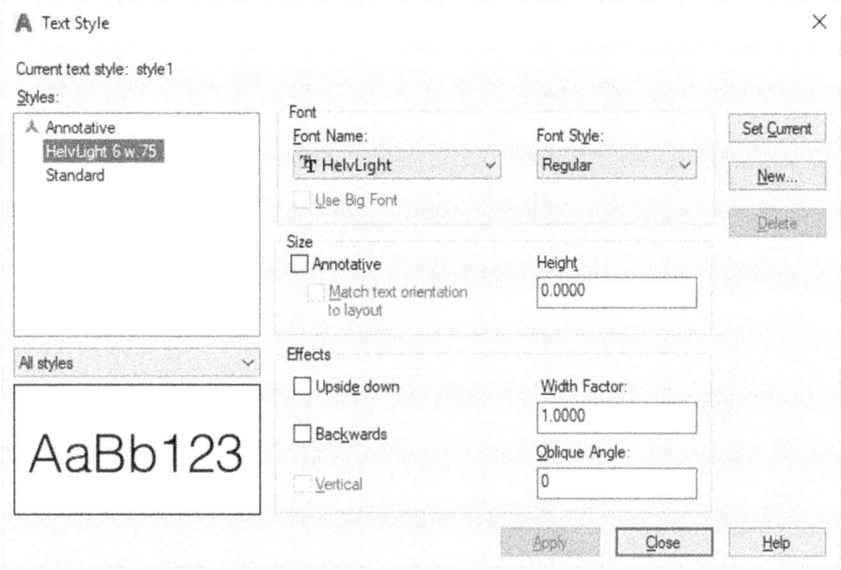

FIGURE 4.12 Text Style dialog box.

block area for logos and other designations, so it makes sense to set one style and stick to it. You already may have changed the font while learning the mtext command, but it was only for *that* instance. You need to make sure that font is set globally, meaning for the entire drawing.

The default font is Arial (the default size is 0.2), so you may want to stay with this popular font and not change it. If you need to, however, here is the procedure to create a Helvetica Light text style using the HelvLight font with a fixed height of 6 and a width factor of .75 *(Note: If your computer does not have this font installed, choose one of your liking):*

Keyboard: Type in **style** and press Enter
Ribbon: **Annotate** tab→**Text**→**arrow**

Step 1. Begin the style command via any of the preceding methods. Whichever method you use, the Text Style dialog box of Fig. 4.12 appears. Looking at this dialog box, let us say Arial is not what we want here; we would like to set a Helvetica font that is 6″ high, and slightly narrower instead. Note, if HelvLight is not available on your computer, choose any Font you would like to try and name your Style appropriately.
Step 2. Press the New… leave the name as *style1* for now
Step 3. Pick HelvLight from the Font Name: drop-down menu (or any font you choose to try or if HelvLight is not available).
Step 4. Highlight the 0.0000 in the Height field and type in 6 (no need for the inch symbol).
Step 5. Highlight the 1.0000 in the Width Factor field and type 0.7500.
Step 6. Right-click on *style1* and select Rename from the pop-up menu to rename the Style to HelvLight 6 w.75 (or name according to your font, height, and width choice).
Step 7. Press Apply and Close.

All text that you type from now on will be HelvLight, 6″ font. To create another font just repeat these steps. To size your font up and down, you can also create new font styles, but in this case, it may be more practical to just use the scale command. (Note: If you leave the font height set to 0, the font can be used at any height when prompted using the text or mtext commands.) The additional options in the style dialog box are not used that often but review them just in case. A list follows. All effects can be previewed in real time in the Preview box in the lower right.

- *Upside down*: Flips the text upside down.
- *Backwards*: Flips the text backwards.

FIGURE 4.13 Misspelled word underlined in red.

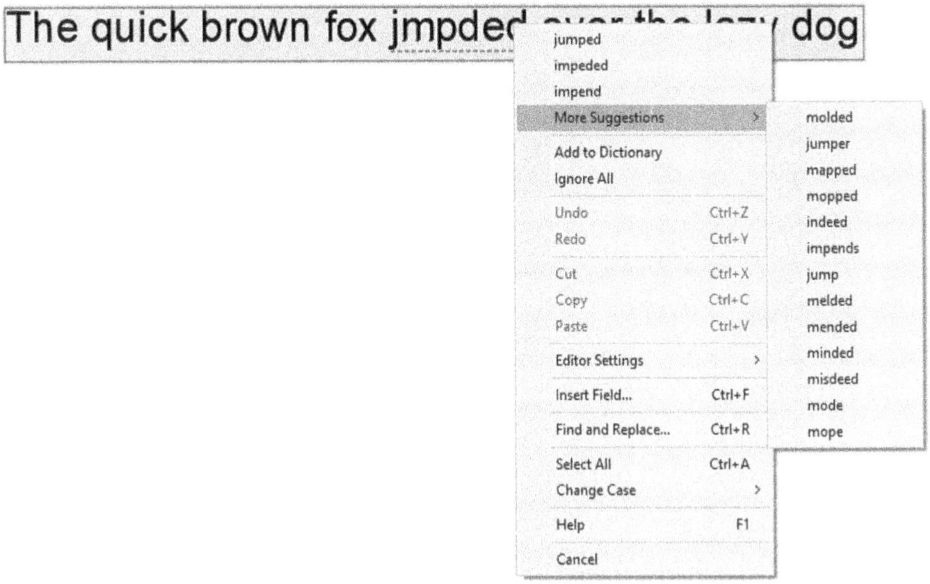

FIGURE 4.14 In-place spell check.

- *Vertical*: Stacks the text vertically.
- *Width Factor*: Widens the text if (>1), narrows text if (<1).
- *Oblique Angle*: Leans the text to the right if positive (+) and to the left if negative (−).
- *Annotative*: An advanced topic for Chapter 10.

4.5 Spell check

This section briefly describes AutoCAD's in-place, stand-alone spell-checking abilities. Yes, AutoCAD does have spell check, and why not? For the price the software costs, it had better. It is very easy to use, so let us try it out on the misspelled phrase created via regular Text, as shown in Fig. 4.13.

The first thing you notice is that the misspelled word (jumped) is underlined by a dotted line. If you right-click on the word while in editing mode, you get the in-place spell check with suggestions (Fig. 4.14). Simply select the word you want to correct, and AutoCAD obliges.

Now, what if you already have a significant amount of text (perhaps from an import) and need to spell check it? Then, you can use the stand-alone utility, as described next.

Keyboard: Type in **spell** and press Enter
Ribbon: **Annotate** tab→**Check Spelling**

Using any of these methods, the Check Spelling dialog box appears, as seen in Fig. 4.15.

Using the drop-down menu of the Check Spelling dialog box, you can select specific paragraphs or lines of text or just press Start to initiate spell checker for all the text in the drawing. In any case, Check Spelling appears, as seen in Fig. 4.16.

Much as with the MS Word spell checker, you can run through some options, such as ignoring the word, changing it, or adding it to the dictionary so it is not flagged again (good for names and acronyms). Even a variety of languages are available. When done, spell check tells you, and you can close it.

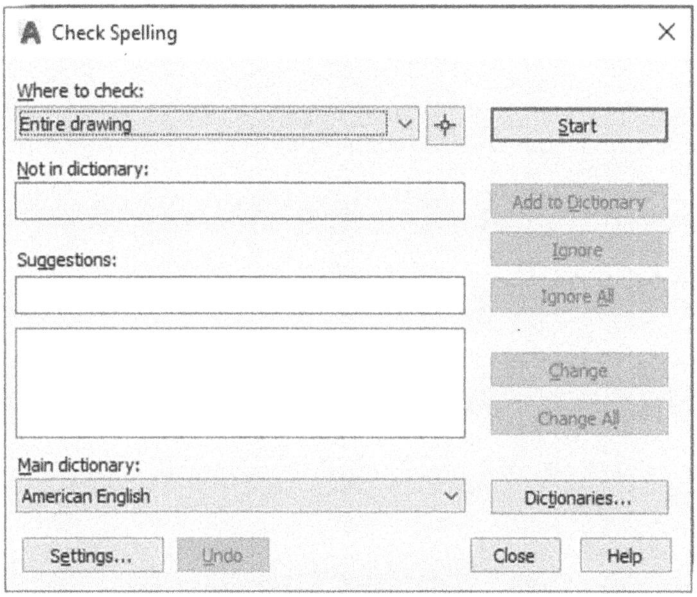

FIGURE 4.15 Check Spelling.

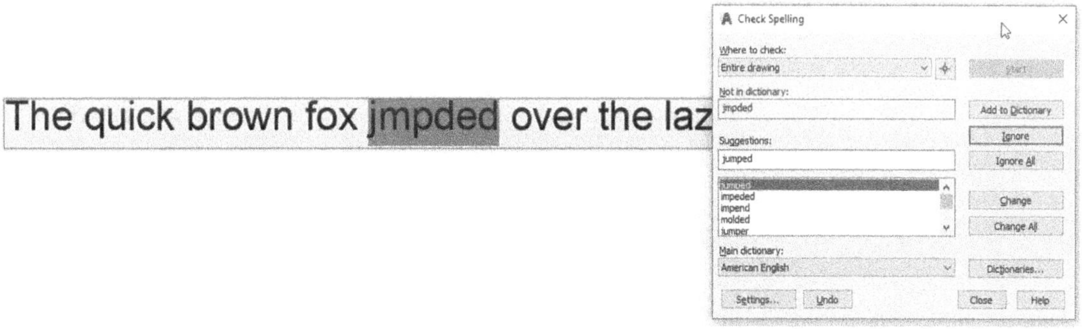

FIGURE 4.16 Check Spelling initiated.

4.6 In-Class drawing project: adding text and furniture to floor plan layout

Let us now apply what you learned to our floor plan. Shown in Fig. 4.17 is the same floor plan you worked on in Chapter 3, with the addition of text in the rooms, closet shelving (the dashed lines), new furniture, and appliances.

Here are two general tips before you get started:

- Make your own layers and colors for each set of data, being sure the names are logical. For example, call kitchen appliances something along the lines of A-Appliances and so on.
- None of the furniture is drawn to scale, so approximate all the sizes. However, draft everything carefully, connecting all lines with OSNAP and using Ortho for straight lines. Although you, in theory, make blocks out of this furniture, we have not covered that topic yet, so this is just a basic drafting exercise. Most shapes are based on rectangles, with some fillets, arcs, and circles. All were done in the simplest manner possible, and you should be able to create them with basic techniques learned in previous chapters. Some hints are listed next.

Some additional specific drawing tips:

- All kitchen furniture on the left side of the kitchen is formed using basic rectangles, drawn to no particular size. At the top left is a refrigerator with a microwave on top. Next to it is a countertop and a sink. Next to that is the range top with four burners. Fig. 4.18 is a suggested procedure for creating the range top. In Chapter 6 we cover some new ways to create center marks, just introduced in AutoCAD 2017, but here is the classic approach.

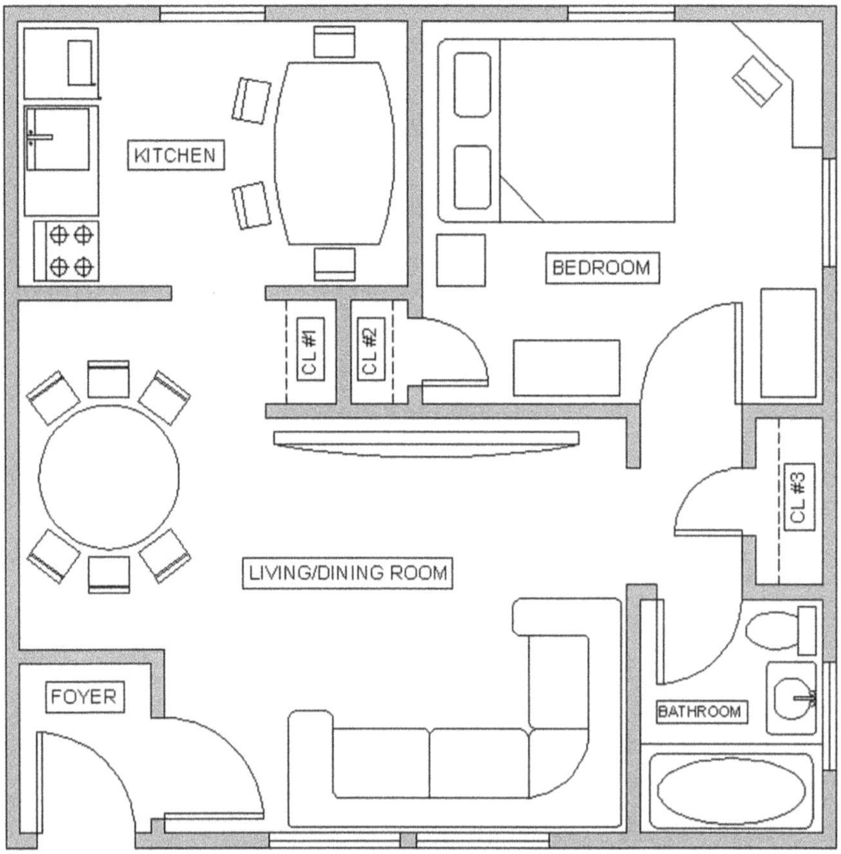

FIGURE 4.17 Floor Plan layout.

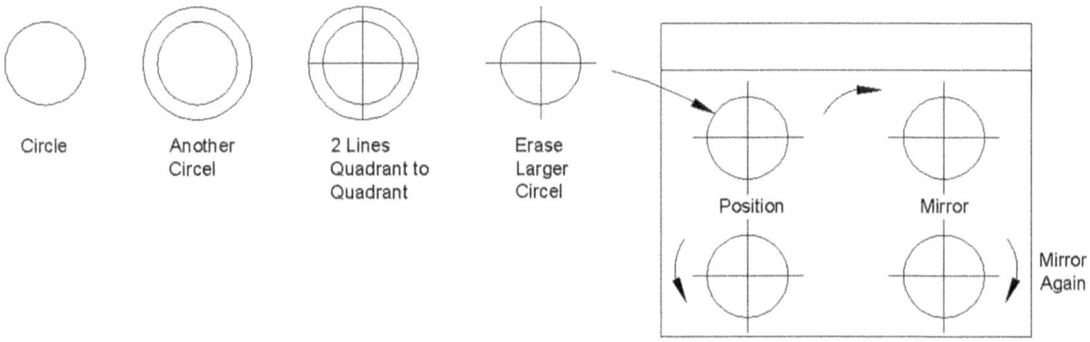

FIGURE 4.18 Constructing the range top.

- To create the kitchen table, you could use a rectangle with a small straight guideline sticking out from the midpoint of the left side. That could be the anchor for the second point of the arc. Then, a mirror followed by an "explode" and "erase" finishes it off, as seen in Fig. 4.19.
- Draw rectangles around each of the room text fields; we need this later for hatch.
- The bathtub and toilet are done using the basic ellipse command, introduced in Chapter 2.
- The corner desk in the bedroom, the pulled-back bed cover, and all closet shelves are constructed using lines and a new OSNAP point, the second (of three) that was not initially introduced in Chapter 1, a description follows.

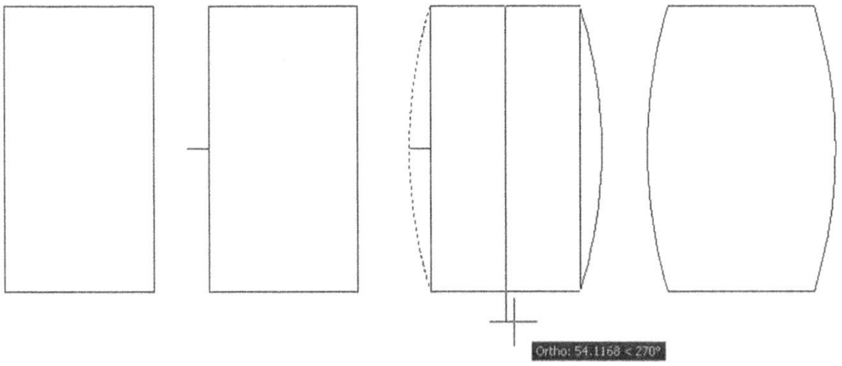

FIGURE 4.19 Constructing the table.

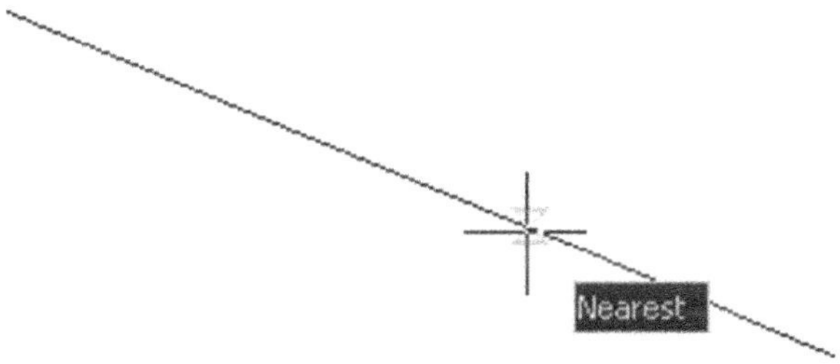

FIGURE 4.20 Nearest OSNAP.

NEArest OSNAP

You may occasionally need a new OSNAP point called *NEArest.* It begins or ends shapes (usually new lines) along random spots of other geometry (usually other lines). In a sense, NEArest is the exact opposite of the precise end, mid, and other points you learned. NEArest simply shadows a line or any object and allows you to begin a new object anywhere along that perimeter. We do not generally leave NEArest running along with the other OSNAP points, as it makes using them more difficult; rather, just type in nea and press Enter or use the toolbar as needed. An hourglass figure appears; just click on wherever you want to start a line. Fig. 4.20 is a screenshot of the NEArest OSNAP in action.

Finally, we have two more tips. Tip 6 has to do with ltscale, something briefly mentioned in Chapter 3. Tip 7 shows you how to get rid of the Universal Coordinate System (UCS) icon, something many students ask about.

TIP 6

If you draw your hidden lines for the closet shelving and they look solid not dashed, then the problem may be incorrect linetype scaling, a common error to watch out for. Type in ltscale, press Enter, and input a new value (a larger one, as your floor plan is bigger than the default original space of 8.5″ × 11″). Do this a few times until you get it to look just right (Try 12, 24, 48, etc.).

TIP 7

Want to get rid of that pesky UCS icon hanging out in the drawing area? It is somewhere near the lower left-hand corner of your screen, looks like a big L, and has a Y and an X on its tips. To make it go away, type in ucsicon, press Enter, then type in N for Noorigin and it will remain in the lower left corner of the screen regardless of where 0,0 is or go with OFF to remove it completely. Of course, if it does not bother you and you like it floating around, then ignore this whole tip.

Summary

You should understand and know how to use the following concepts and commands before moving on to Chapter 5:

- Text
  - Creating basic text
- Mtext
  - Creating basic mtext
  - Mtext formatting options
    - Bold
    - Underline
    - Italics
    - Justification
    - Symbols
    - Other
- Editing text
  - Double-click
  - ddedit
- Style
  - Font name
  - Font type
  - Font size
  - Width factor
  - Oblique angle
- Spell check
- Nearest OSNAP

Review questions

Answer the following based on what you learned in this chapter:

1. What are the two types of text in AutoCAD? State some properties of each and describe when you would likely use each type.
2. What is the easiest method to edit text in AutoCAD? What is another, typed method?
3. What feature allows you to add extensive symbols to your mtext?
4. What are the three essential features to input when using Style?
5. What is the width factor?
6. What is an oblique angle?
7. How do you initiate spell check?
8. What new OSNAP is introduced in this chapter?
9. How do you get rid of the UCS icon?
10. What can you do if you cannot see the dashes in a Hidden linetype?

Exercises

1. Creating tables of data, such as a Bill of Materials, is an important part of many designs. AutoCAD has a table creation tool (similar to MS Word), which we cover in Level 2, but for this exercise, you need to create one from scratch with basic linework and the offset command, populating it with text data, as shown. Open a blank file and create a new style called RomanS 0.2—choosing font: RomanS, height: 0.2, and width factor: 0.80. Then, create the table shown, using the suggested sizing. Finally, fill in the data, using single line text for everything except the SIZE DWG column. That is best suited for Mtext with a height of 0.1. Position and size the text and mtext accurately using Osnaps and justification grips as needed. (Difficulty level: Easy; Time to completion: 20 minutes.)

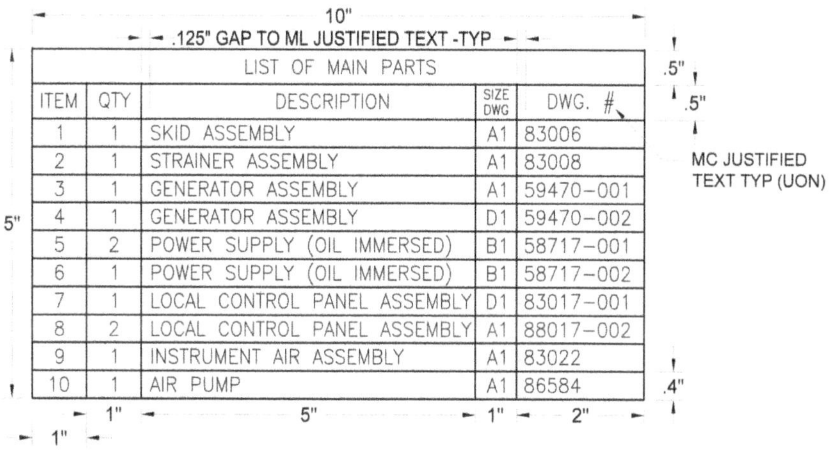

LIST OF MAIN PARTS				
ITEM	QTY	DESCRIPTION	SIZE DWG	DWG. #
1	1	SKID ASSEMBLY	A1	83006
2	1	STRAINER ASSEMBLY	A1	83008
3	1	GENERATOR ASSEMBLY	A1	59470–001
4	1	GENERATOR ASSEMBLY	D1	59470–002
5	2	POWER SUPPLY (OIL IMMERSED)	B1	58717–001
6	1	POWER SUPPLY (OIL IMMERSED)	B1	58717–002
7	1	LOCAL CONTROL PANEL ASSEMBLY	D1	83017–001
8	2	LOCAL CONTROL PANEL ASSEMBLY	A1	88017–002
9	1	INSTRUMENT AIR ASSEMBLY	A1	83022
10	1	AIR PUMP	A1	86584

2. This next exercise is similar to the previous one. Using a combination of text and mtext, create the following table of tooling part numbers based on the starter dimensions given, including the text height. Exact text positioning is not specified and can be approximated. Use text styles similar to the previous example but leave the width factor as 1.0 and adjust the height to the value shown. Create the fractions via mtext or by hand. (Difficulty level: Easy; Time to completion: 20 minutes.)

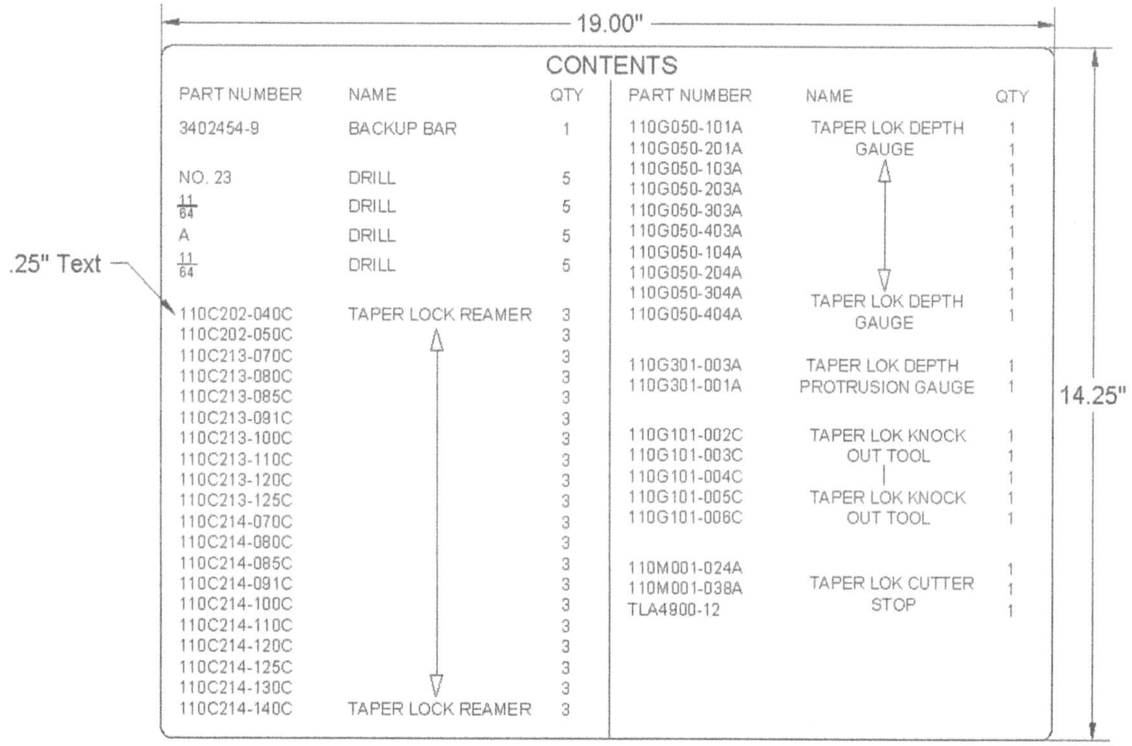

3. As seen with Exercise 7 of Chapter 3, Layers, Colors, Linetypes, and Properties (electrical pieces), you may be asked to draft something based on a picture and use arbitrary, not strict, design dimensions. You need to be fast and efficient while maintaining strict accuracy. A good grasp of the basics is essential, and you must know exactly how you will draft something in your mind, just moments before your fingers do it via mouse and keyboard. Such is the nature of this next exercise—to build up this accuracy and speed. Draw the following set of P&ID (piping and instrumentation diagram) symbols. They are not to scale and are approximated, but you have all the necessary tools at your disposal. Before drafting each shape, recite to yourself the approach you will use, and then execute it. Finally, add the text next to

the respective symbols. Use the Arial font of appropriate size. (Difficulty level: Intermediate; Time to completion: 35–45 minutes.)

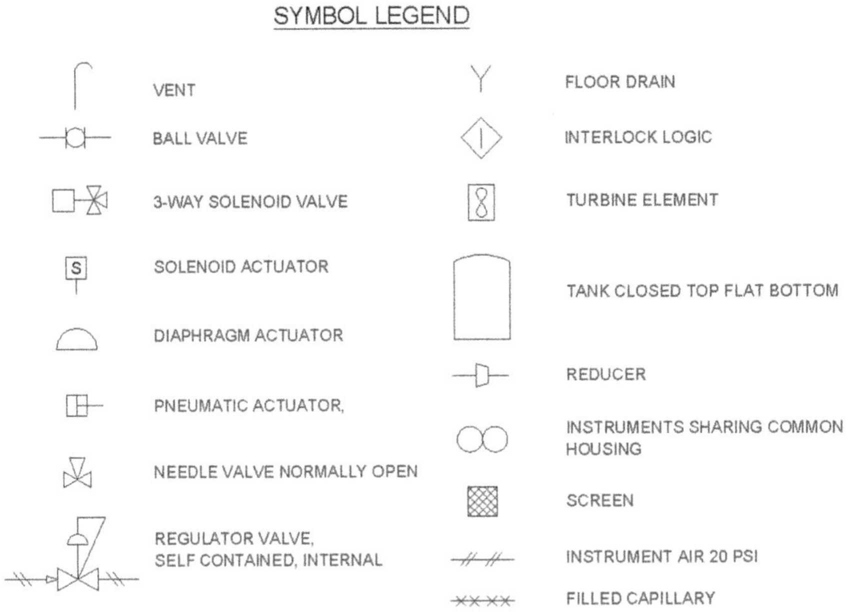

4. This exercise shows a simple P&ID schematic using some additional industry symbols. Draft it based on the reference dimensions given, with all internal pieces and text sizing estimated in relation to those. Use accuracy, connecting all lines with OSNAPs. (Difficulty level: Easy/Intermediate; Time to completion: 20–30 minutes.)

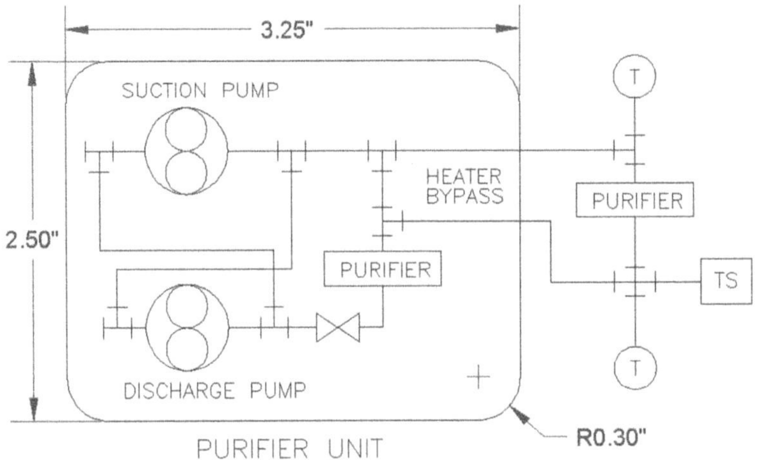

5. While we are on the subject of P&IDs, let us create a simplified design of a pipe. Open a new file and set up Architectural units, creating proper layers, colors, and linetypes (Hidden and Center), as shown. Then, draw the following pipe design. Be careful in how you approach the hidden lines, noting that they are on the same plane as the solid lines. This is commonly seen on mechanical drawings. (Difficulty level: Intermediate; Time to completion: 20–30 minutes.)

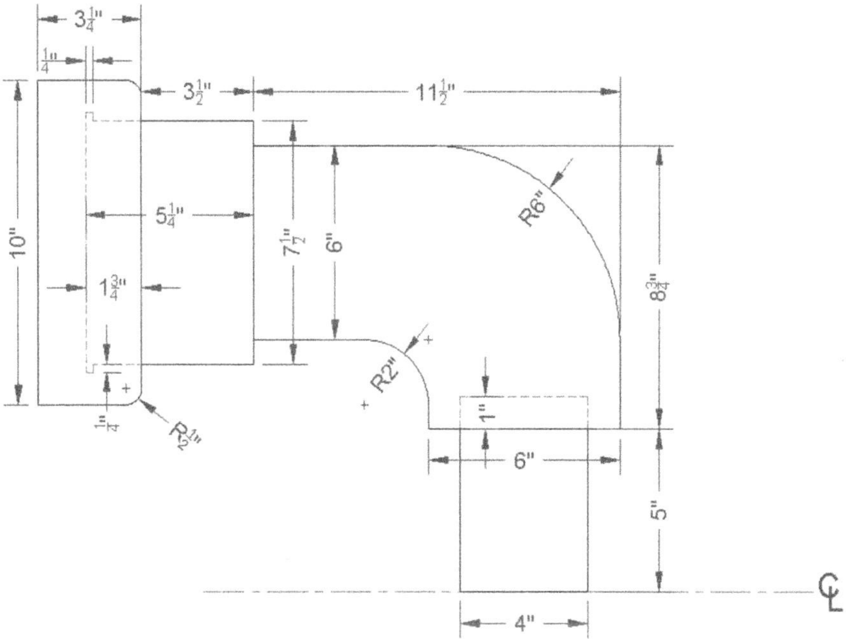

6. For this drawing, we return to basic bracket design, although this one is a bit more intricate. Set up the appropriate layers and linetypes and create the following mechanical part. (Difficulty level: Intermediate; Time to completion: 20–30 minutes.)

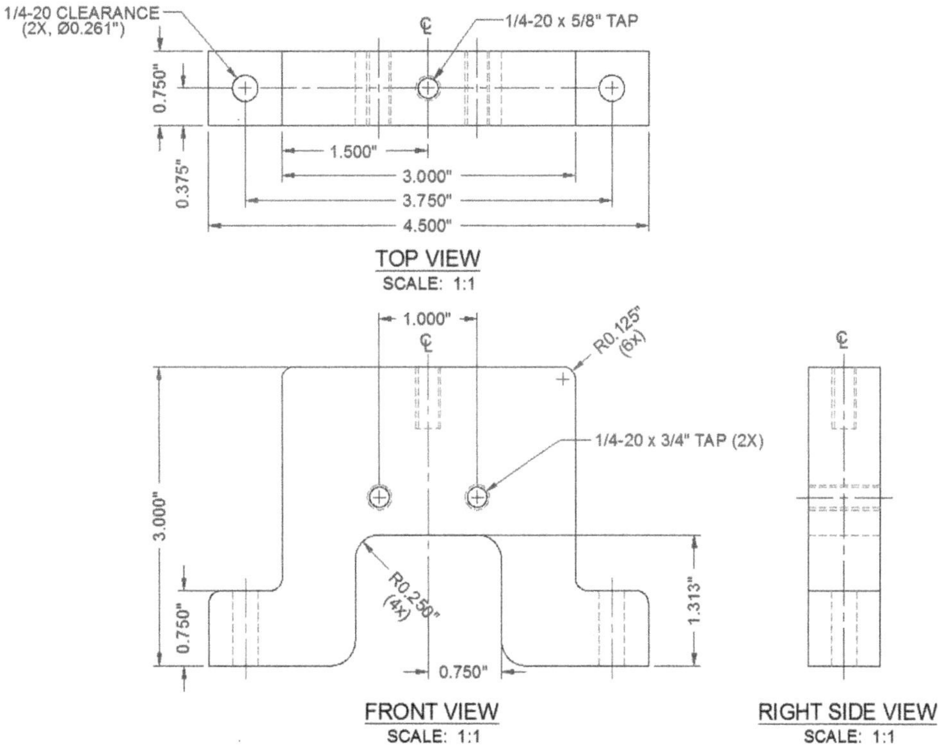

7. This exercise involves a floor plan. Like previous activities, open a new file, set up A-Walls and A-Doors layers and colors, and leave your units as Architectural. Then, complete the following basic floor plan. All walls are 6″ thick. All the doors are 3′ wide and have no thickness; you may give them thickness if you like. (Difficulty level: Intermediate; Time to completion: 45–60 minutes.)

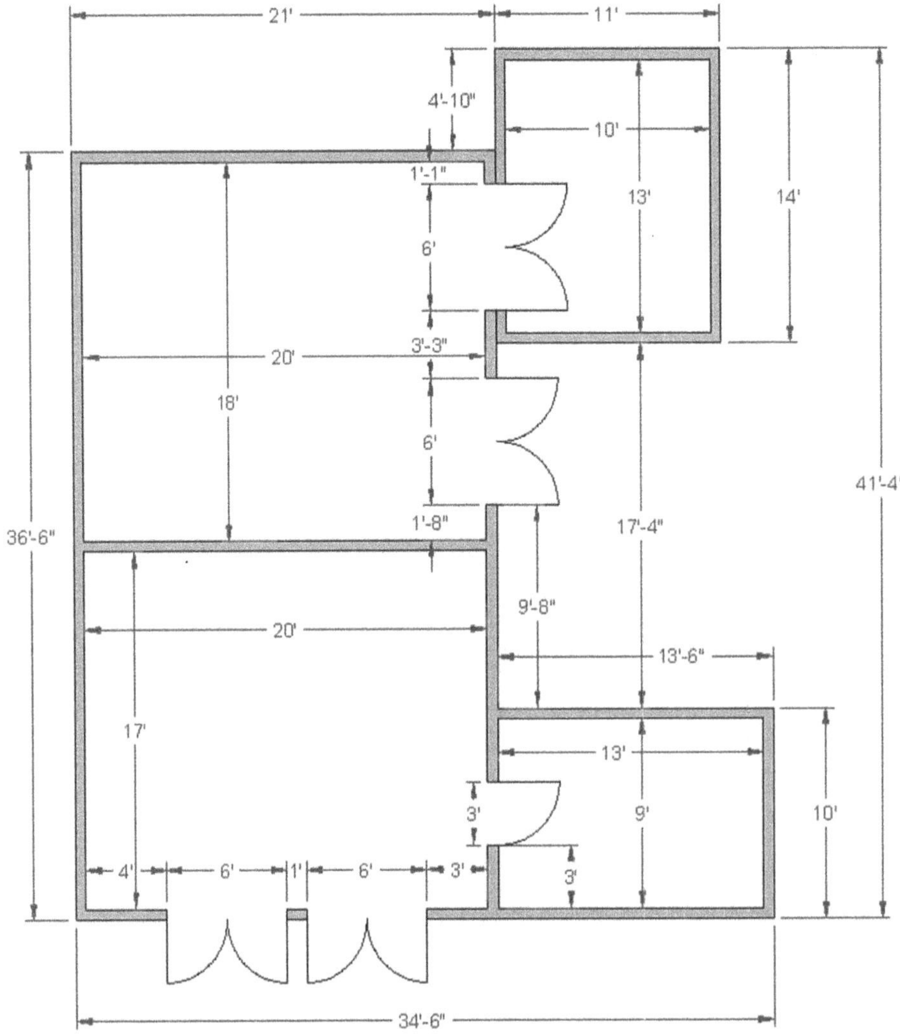

8. This exercise is a similar, if slightly more involved, floor plan. As before, open a new file; set up A-Walls, A-Door, A-Door-Swing and A-Furniture layers colors; and linetypes; setting your units to Architectural. Then, complete the following floor plan layout. All exterior walls are 7.5″ thick with interior a 5″ thick. All the doors are either 3′ or 4′ wide and have a 2″ thickness. The furniture sizing can be approximated. Dimensions and shading are for clarity only. (Difficulty level: Intermediate; Time to completion: 45–60 minutes.)

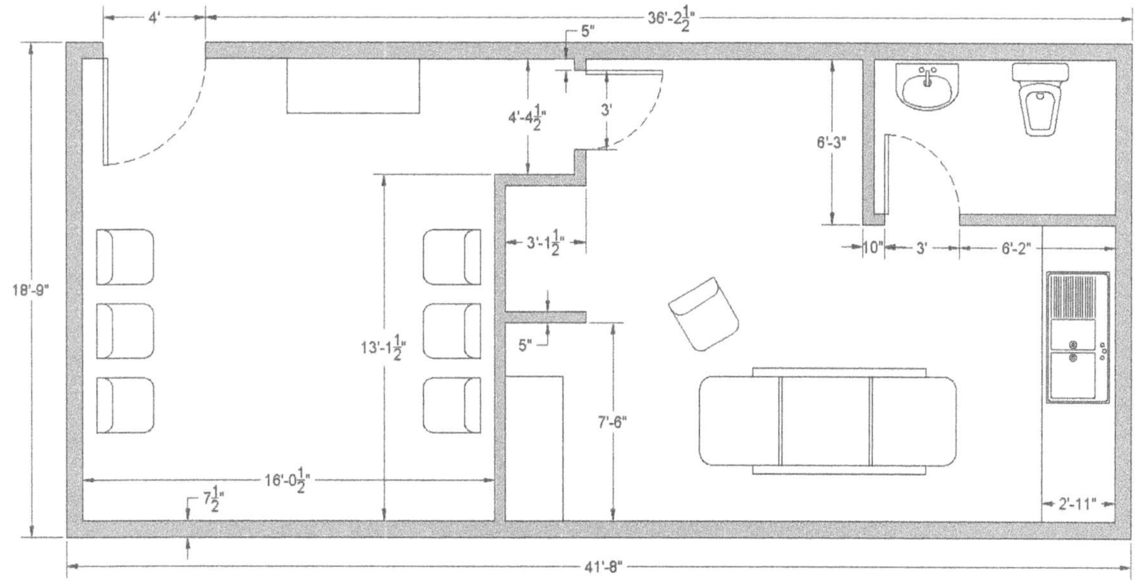

9. Open a new file and set up layers of your choosing. Then, create the following electrical power strip according to the critical dimensions shown. Where no dimensions are given, assume an appropriate value. Add text where shown. Dimensions and some hatching are shown just for clarity and for construction purposes. (Difficulty level: Easy; Time to completion: 20–30 minutes.)

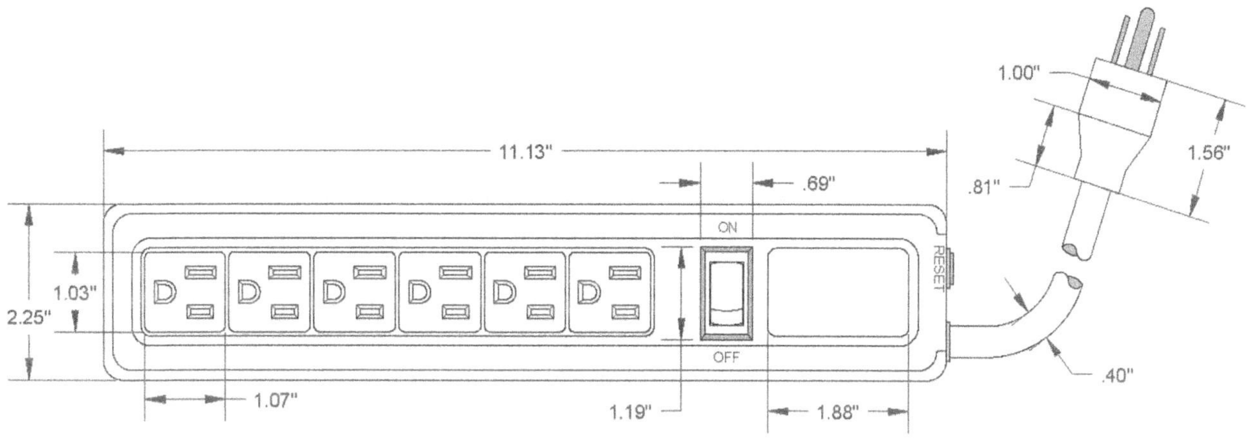

10. This exercise is another mechanical bracket. In a fashion similar to Exercise 6, it is projected in top, front, and side views, which is very typical for mechanical design drawings. Open a new file, set up your units as Decimal, and draw the shapes as shown. Line up the three views to project the sizing of pieces from one view to another. The dimensions and shading are for your use and clarity; we cover both shortly. (Difficulty level: Intermediate/Advanced; Time to completion: 60 minutes.)

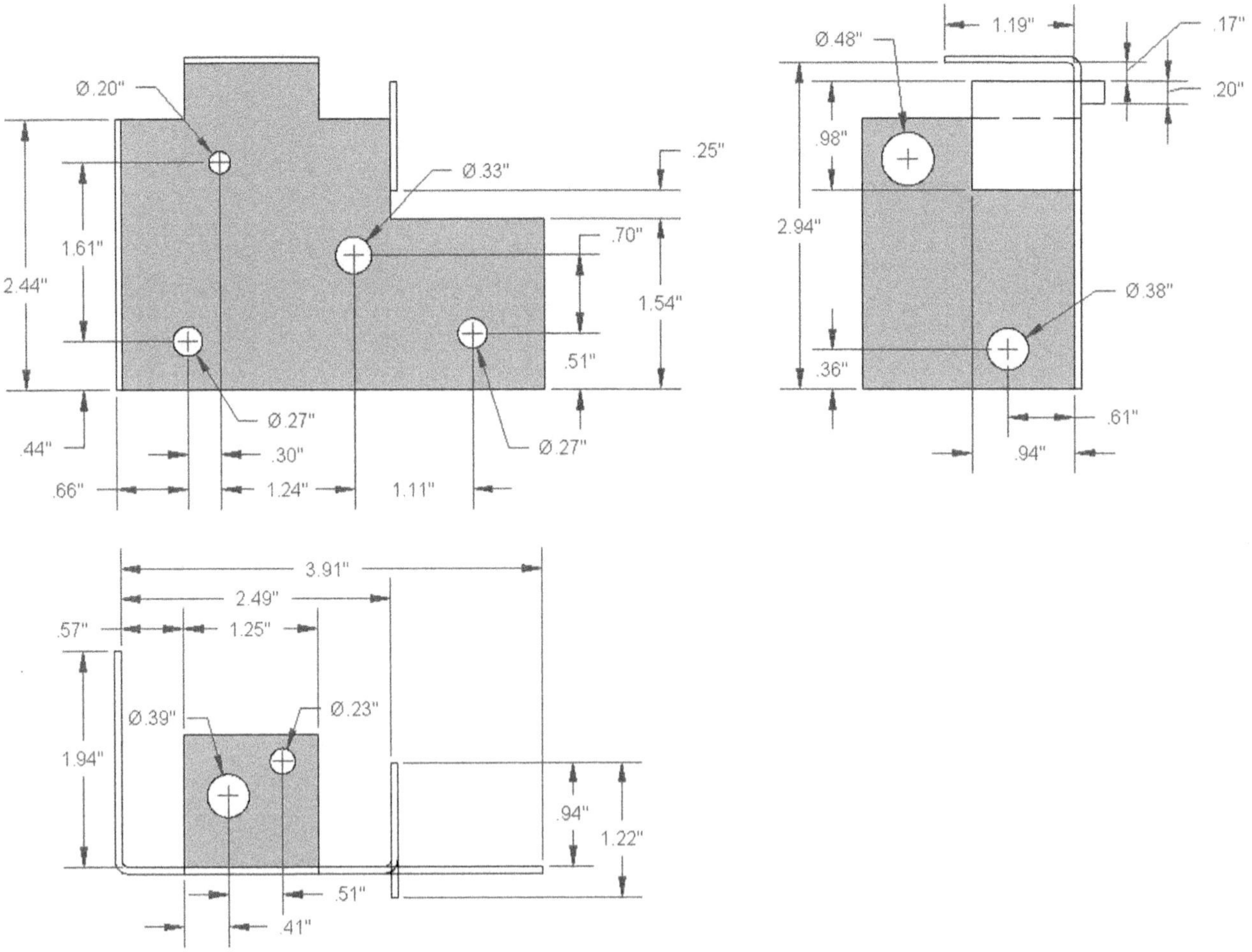
Ø.20"
Ø.33"
.25"
.70"
1.61"
2.44"
1.54"
.51"
Ø.27"
.44"
.30"
Ø.27"
.66"
1.24"
1.11"
1.19"
.17"
Ø.48"
.20"
.98"
2.94"
Ø.38"
.36"
.61"
.94"
3.91"
2.49"
.57"
1.25"
1.94"
Ø.39"
Ø.23"
.94"
1.22"
.51"
.41"

Chapter 5

Hatch patterns

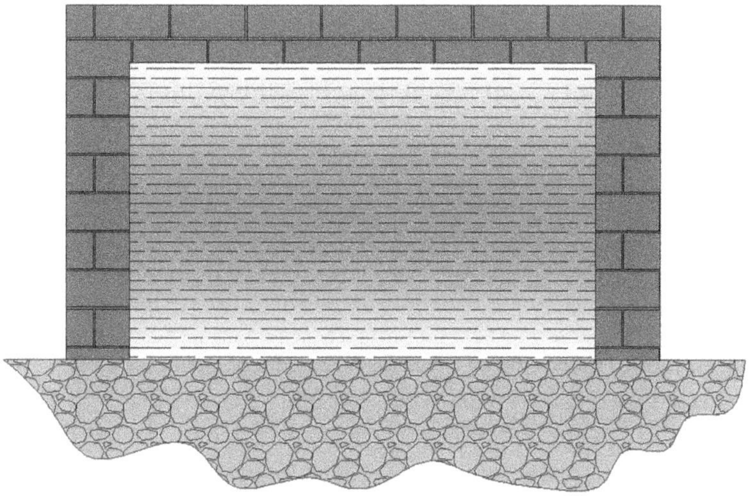

Learning Objectives

Hatch patterns allow you to indicate a variety of surfaces, from wood to concrete, on your design. In architecture and engineering, they are also used to indicate materials of surfaces being cut in cross-sections. In this chapter, we introduce boundary Hatch and discuss the following:

- Hatch fundamentals
- Picking patterns
- Picking area or objects to Hatch
- Adjusting scale and orientation
- Working with Hatch patterns
- Gap tolerance
- Gradients and solid fills
- SuperHatch

At the end of this chapter, you will be able to add Hatch patterns to your floor plan.
Estimated time for completion of this chapter: 2 hours (lesson and project).

Up and Running with AutoCAD 2022. https://doi.org/10.1016/B978-0-323-89923-9.00005-6

5.1 Introduction to Hatch

This is a chapter on the concept of boundary Hatch or just *Hatch* for short. This is simply a tool for creating patterns and adding fills to your design. These are not merely decorative; rather, they serve an important function of indicating types of materials, ground covering, floor coverings, and cross sectional materials. As such, this topic is quite important and is the next logical concept to master in AutoCAD.

Hatch patterns are not an AutoCAD invention. In the days of hand drafting, there was also a need to visually indicate what sort of material one was looking at. Architects and engineers created easy to understand repeating patterns that somewhat, if not closely, resembled the actual material or at least gave you a very good idea of what it was. Just a few of the uses of Hatch patterns follow.

Architecture

- *Brick*: An important and popular pattern to render the exterior and (sometimes) interior of buildings. AutoCAD has numerous brick patterns available.
- *Herringbone, parquet*: Important for flooring designations.
- *Honeycomb*: Insulation designation.

Civil engineering

- *Concrete, sand, clay, earth, gravel*: All used in designating surfaces in civil and site plan design.

Mechanical engineering

- *Various ANSI diagonal patterns*: Used to designate the visible inside of an object cut in cross section.

With AutoCAD, creating these patterns is easy, as they are all predrawn and saved in a library, which is called up anytime you use the command. Often a designer can purchase or download additional patterns if the basic ones are inadequate. You can even make your own and save them for future use, although this is a tedious process and only briefly covered in Appendix D.

5.2 Hatch procedures

At the most basic level, AutoCAD's Hatch command requires only four steps:

Step 1. Pick the Hatch pattern you want to use.
Step 2. Indicate where you want the pattern to go.
Step 3. Fine-tune the pattern by adjusting scale and angle (if necessary).
Step 4. Watch the preview of the pattern and accept it if it is OK.

Memorize the four steps and their sequence; this will help you when looking at the Hatch dialog box.

Before we get into the details, it goes without saying that you first need something to Hatch. The command will not work on a blank screen or a bunch of unconnected lines. You need a closed area (although later on, we violate that rule), and the easiest way to do this is to form a circle or rectangle. Open a new file and draw a square, sized 10 × 10. Follow the steps carefully as described next. All advanced Hatch functions flow from these basics.

Step 1. Pick the Hatch pattern you want to use

Keyboard: Type in **hatch** and press Enter
Ribbon: **Home** tab→**Hatch**

Start the Hatch command via the Draw Panel of the Home Tab of the Ribbon or simply type Hatch. The Hatch Creation contextual tab of the ribbon (Fig. 5.1).

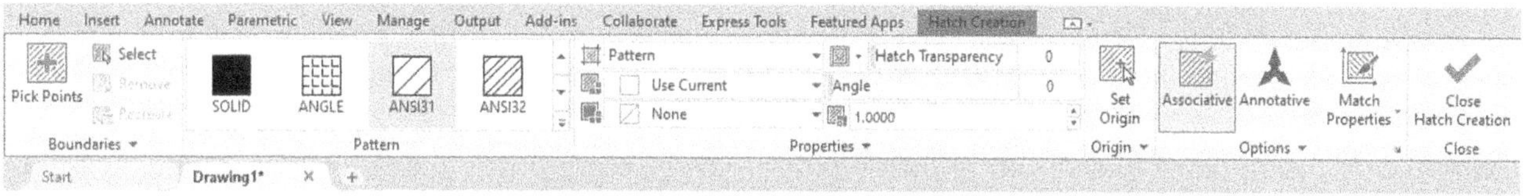

FIGURE 5.1 Hatch Creation contextual Ribbon tab.

Now select a pattern. The choices are listed in Pattern panel of the Hatch Creation tab. Use the scroll bars at the right of the patterns to see a few at a time (Fig. 5.2).

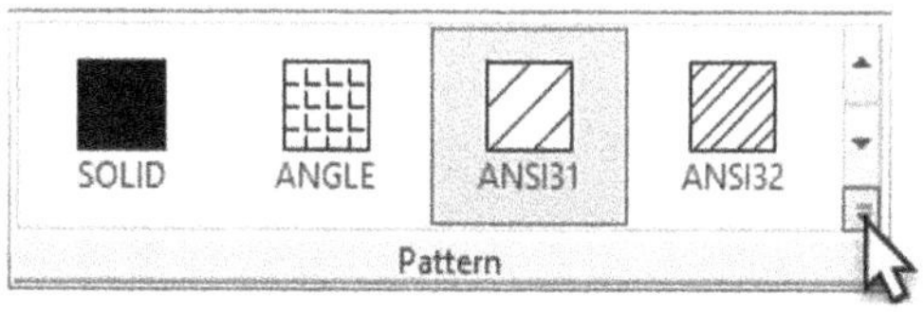

FIGURE 5.2 Pattern panel of the Hatch Creation contextual tab of the Ribbon.

Look through the many patterns: ANSI, ISO, User defined, Gradients, and Solid to name a few. We have a use for some of the ANSI patterns, not so much for the ISO ones, and User is to go if for simple straight lines at a given spacing in single or double (cross hatch patterns). You can expand the list with the down arrow below the scroll arrows and even stretch out the icon preview box to show all the patterns available in the Other Predefined tab. Go ahead and select he down arrow and expand the pattern box by stretching its lower right corner; you will see what is shown in Fig. 5.3.

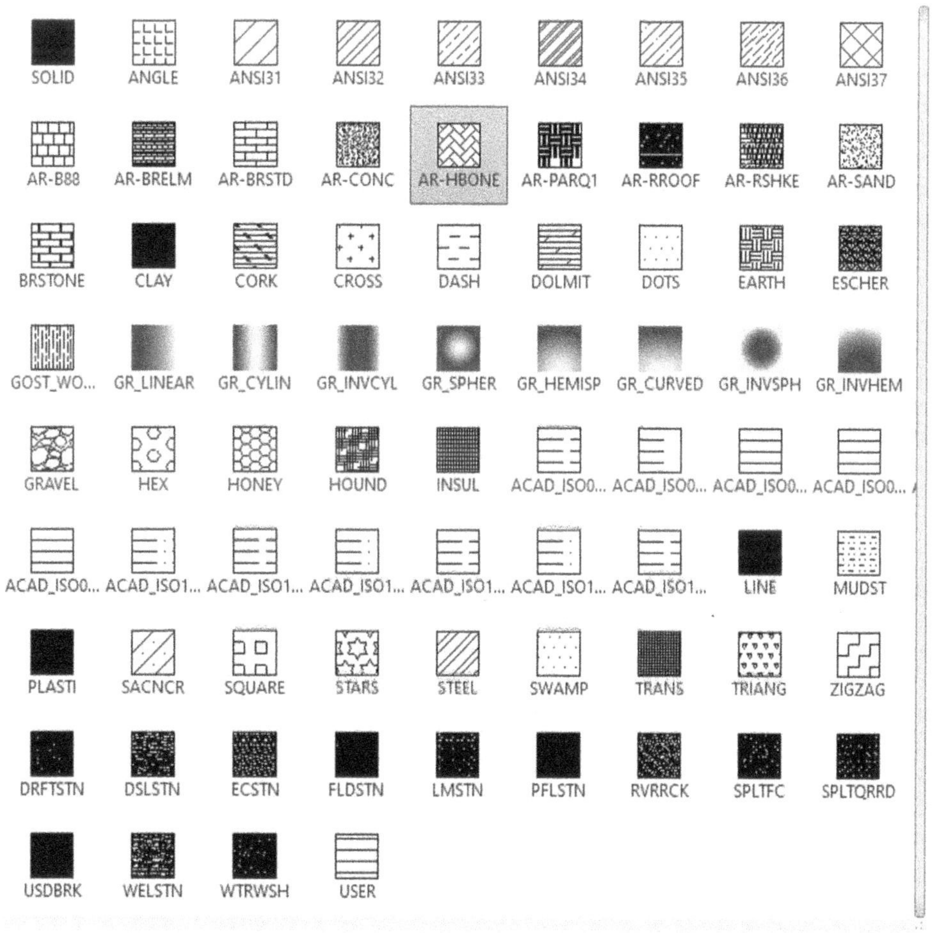

FIGURE 5.3 Hatch Pattern palette fully expanded with AR-HBONE selected.

FIGURE 5.4 Steps 1 and 2 completed.

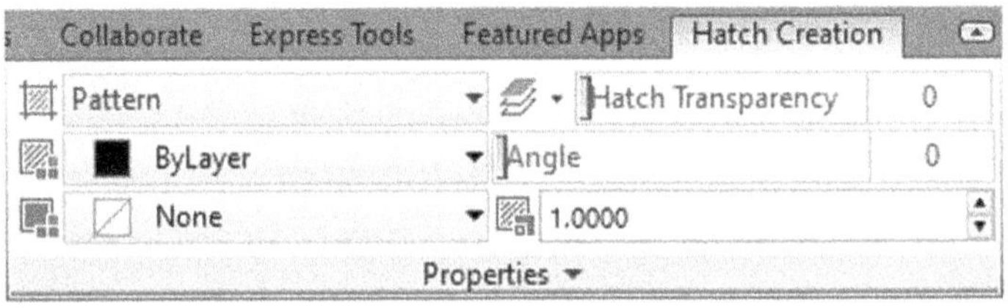

FIGURE 5.5 Properties panel of the Hatch Creation contextual tab of the Ribbon.

Scroll up and down the patterns. You will notice some of the ones mentioned at the start of this chapter. Go ahead and pick one that you like, preferably one that is distinct and stands out; AR-HBONE is a good choice and is used in this example. When you click on it, the pattern is highlighted blue. Go ahead and click on OK. We are done with Step 1, and the new pattern choice is reflected in the Swatch area.

Step 2. Indicate where you want the pattern to go

You can indicate where to put the pattern in two ways, either by *directly picking* an object that will contain the pattern or by *picking a point inside* that object. Which method to use is easy to determine. If the object is actually a single entity making a closed shape such as your rectangle or circle, then the shape can be picked directly. If not and the closed "object" is just a collection of connected lines defining an area, then the best way is to pick a point in the middle of that area, and indeed this is the more common situation. The pattern then behaves like a bucket of paint spilled in the middle of the room. It flows outward evenly and stops only when it hits a wall. Once again, no holes or gaps are allowed yet; that is addressed later.

In the Boundaries panel of the Creation tab of the Ribbon (far left), notice the two choices just discussed—Pick Points and Select objects. Choose.

Pick Points

Then hover your cursor somewhere *inside* of an object, or closed area. You should see a a preview of your Hatch pattern. Click to accept it.

Successfully picking a boundary can sometimes be a tricky business. The existence of gaps, no matter how small, can be a source of occasional frustration to designers; and until relatively recently, it was difficult to tell where those gaps were hiding in order to fix them. As of AutoCAD 2010, red circles now appear where the gaps are, aiding you in locating them, but this all can be avoided in the first place by not doing sloppy drafting (i.e., not connecting lines together properly). It is a sign of advanced AutoCAD skills when a complex area Hatches right the first time.

In cases where boundary selection is not successful, complex areas can often be broken down to smaller ones by using lines to divide the area into smaller pieces and the nonproblematic portions Hatched first, although this is a last resort. A few releases ago, AutoCAD added the gap tolerance command, which ignores gaps up to a preset limit. This is a great tool, which we cover shortly, along with the entire gap issue in general, but it is somewhat of a Band-Aid and still does not address the underlying sloppy drafting but merely allows you to get away with it.

Remember, you get only the level of accuracy that you put in. It is essential to master the fundamentals of drafting early on. The Hatch command is an "early warning" to students. If they are having problems using it smoothly, they need to go back and refine their basic drafting (linework and accuracy) skills.

Step 3. Fine-Tune the pattern by adjusting scale and angle (if necessary)

You are almost done and, at this point, could probably just press the big green check mark for Close Hatch Creation (or Enter, Spacebar, or Right-click) and finish the Hatch pattern creation; however, you may want to make adjustments.

Here, you would go ahead and right-click or press Enter. Sometimes, however, the pattern is too big or too small. You can usually tell it is too big by either not being able to see it or seeing a small part of it. If it is too small, the problem is worse because you may not be able to finish the pattern at all. In those cases, nothing can be seen and AutoCAD tells you the following:

For learning purposes, only let us practice changing the scale as the final step in our Hatch creation. In the Properties panel, change the scale to a smaller number, such as 0.2500, by typing it in (see Fig. 5.6). In some cases where the scale is way off, you may need to do this several times until the Hatch pattern is just right. In the same manner, you may adjust the angle of the entire pattern if desired.

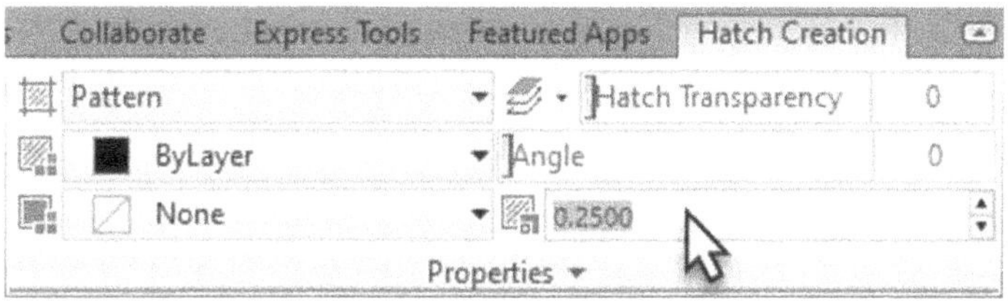

FIGURE 5.6 Scale changed to 0.2500.

Step 4. Watch the Preview of the pattern and accept it if it is OK

Finally, if you are done press the big green check mark for Close Hatch Creation (or Enter, Spacebar, or Right-click). If you followed the previous steps and used a rectangle, the AR-HBONE pattern, and the scale suggested, then you should get results as shown in Fig. 5.7 or something reasonably close.

FIGURE 5.7 Final Hatch pattern.

5.3 Working with Hatch patterns

In this section, we mention some of the additional tools and concepts that pertain to Hatch patterns. First , the most basic question at this point is How do you edit a Hatch pattern after you have created it? It is very easy: To edit a Hatch pattern, simply single-click on it. Make sure you get one of the lines of the pattern, not just empty space. A contextual tab of the Ribbon appears called Hatch Editor (Fig. 5.8), but it is essentially the same as the Hatch Creation contextual Ribbon tab. Here, you can change whatever you need to. just like when the hatch was created.

Exploding Hatch patterns

This may be one of the worst things you can do. Never explode Hatch patterns; depending on the size, they turn into hundreds of pieces that cannot be put back together again in the same editable form. All the reasons designers give for exploding Hatch patterns (making them fit, trimming them, moving the containment border, etc.) can be addressed with additional Hatch tools.

Hatch pattern layers and colors

Hatch patterns should be on a generic A-Hatch layer or whatever layer name best describes what the pattern represents (A-Carpet, M-Cross-Sec, C-Gravel, etc.). Choose any color you want for the Hatch patterns, though you should not go with anything too bright, as it may overwhelm your eyes while working on the design. Shades of gray are common for cross sections in mechanical design, as are various shades of rusty orange for brick. As a side note, Hatch patterns can be easily erased by the usual erasing methods. The patterns can also be moved, copied, rotated, and mirrored, although there is rarely a reason to do so.

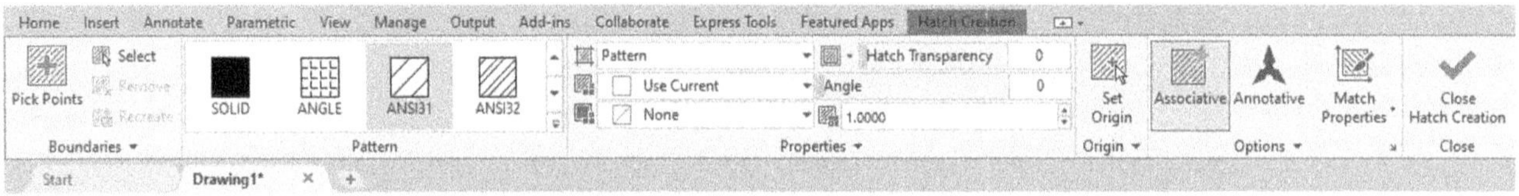

FIGURE 5.8 Hatch Editor contextual tab of the Ribbon.

As of AutoCAD 2011, you can assign a color directly to a Hatch pattern regardless of what layer and associated color, it is on (Bylayer is a better option here). You can also assign a solid background color. To do this when first creating the Hatch, you can select both the Hatch color and the background color (Fig. 5.9) from two drop-down menus in the Properties Panel of the Hatch Creation contextual Ribbon tab. One sample result of assigning a Hatch and background color is shown in Fig. 5.10.

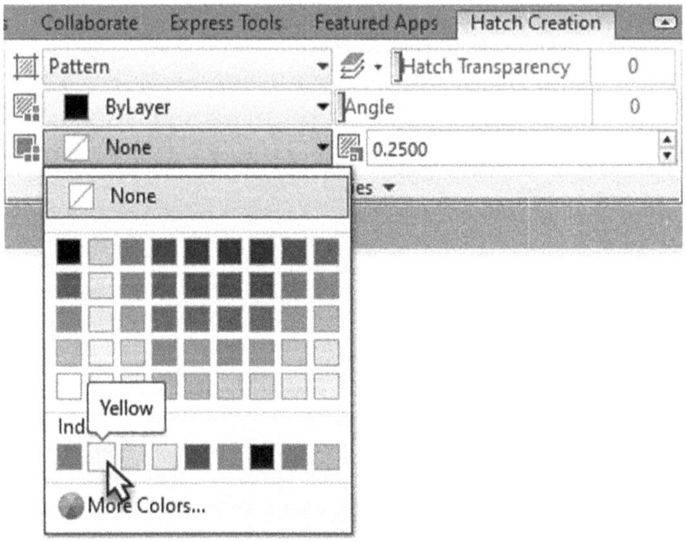

FIGURE 5.9 Hatch Lines (ByLayer) and Background (Yellow) color selection.

FIGURE 5.10 Hatch with a background color assigned.

Advanced Hatch topics

Here, we summarize some additional advanced options available for working with Hatches, including a more detailed discussion concerning the gap tolerance feature. You should explore these on your own as well; this is an important part of learning AutoCAD and greatly enhances your confidence with the software.

Origin

Set Origin: This allows you to begin Hatch patterns from a precise location (origin) as opposed to a random fill-in based on the 0,0 of the drawing. AutoCAD chooses the origin on the pattern itself, and you get to choose to what point on the object that origin is aligned. Try it out.

Options

- *Associative*: Keep this checked. This simply means that, if you move the border and it stays closed and intact, the Hatch moves and stretches with it, very useful in case of future design updates
- *Annotative*: This will not be covered until Level 2.
- *Match Properties*: This is a handy tool when need to remember all the settings of an existing hatch in your drawing that you are trying to match.
- *Gap Tolerance*: (Options drop down menu) See below for detailed description.
- *Create separate Hatches*: (Options drop down menu) If there are multiple separate areas to Hatch, this option keeps them separate; sometimes useful, sometimes not.
- *Island Detection*: (Options drop down menu).
- This gives AutoCAD guidance on how to interpret boundaries inside boundaries (e.g., a chair in the middle of your floor). It simply tells AutoCAD to ignore them or, if not, how to deal with them. Try drawing another rectangle inside your first one and running through the options represented by the pentagon shapes.
- *Draw order options*: (Options drop down menu) This is exactly what it sounds like; it gives you options on how you want to stack the Hatch and its boundary. Check out the options in the drop-down menu.

Gap Tolerance

This is basic Hatching aid that was mentioned earlier. Create a new rectangle with values close to what you did earlier, then create a gap in one of the sides, such as what is seen in Fig. 5.11. To do this, you can use a trim command or the new break command (which we officially introduce in Chapter Advanced Design and File Management Tools). Then, if you try to fill this shape with a Hatch pattern, you get what you see in Fig. 5.11. Note that the red circles may or may not be present. If they are, they can be removed via the regen command.

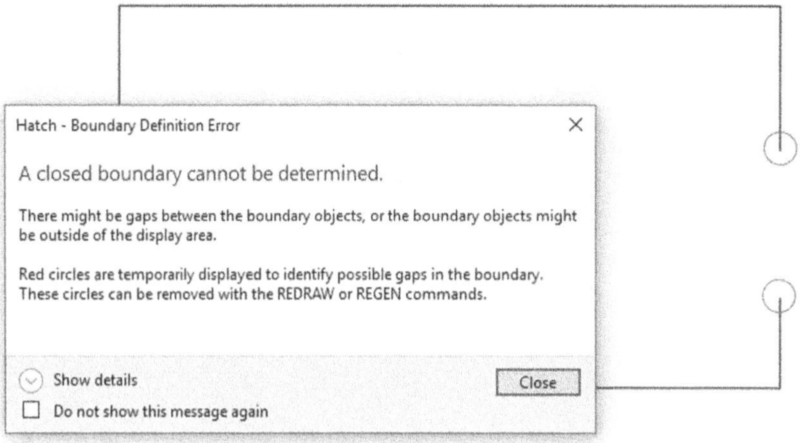

FIGURE 5.11 Boundary not closed alert.

Close the alert and press Esc to continue. Then, returning to the main dialog box, set a value for the gap tolerance (you can use 10 for starters) and re-Hatch the area. You get another warning, which can be eliminated from future occurrence by checking off *Always perform my current choice*, as seen in Fig. 5.12.

Select *Continue Hatching this area*, and AutoCAD temporarily closes the gap to continue the Hatch command. After successful completion, the gap shows again but the Hatch remains, as seen in Fig. 5.13. Some designers have used the command to ignore door openings when Hatching carpeting in floor plans. You can try this as well on your floor plan later.

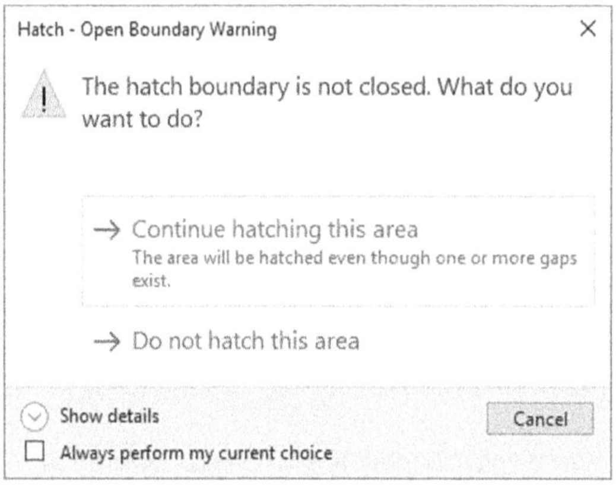

FIGURE 5.12 Open boundary warning.

FIGURE 5.13 Hatch with gap tolerance.

5.4 Gradient and solid fill

Create and enclosed boundary and start the Hatch command again. Select a Gradient from the Pattern panel or in the upper left of the Properties paenl of the Hatch Creation contextual Ribbon tab. The left side of the box changes, while the right-hand side stays essentially the same, as shown in Fig. 5.14.

The gradient option is just a solid fill with some flair. You can choose the desired pattern, in one or two colors, and adjust the Shade, Tint, Orientation, and Angle of the gradient. The rest of the steps are similar: Pick point or Select object, check your results, and done.

Shown in Fig. 5.15 is a possible use for Gradient in rendering a brick wall. A brick pattern is created and a brick-colored fill tops it. Note that you may need to use Options panel to position the brick above the gradient fill (or the gradient fill behind the brick, depending on what you hatch first). Try it.

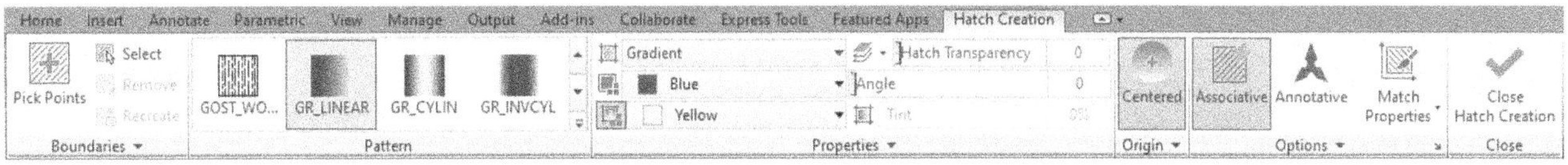

FIGURE 5.14 Gradient pattern selected.

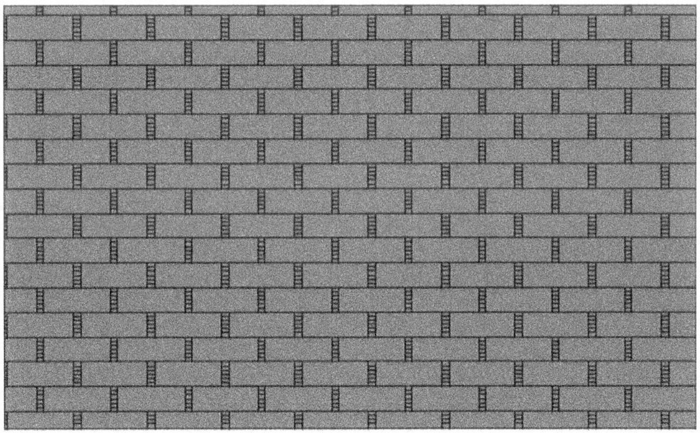

FIGURE 5.15 Brick wall created with Hatch and Gradient.

Solid fill

This fill is the solid black box, top left of the Other Predefined tab Pattern panel selections and deserves special mention as our final topic. This solid fill is like a gradient but with no fancy color mixing, just one color. It is very handy as a fill for walls, some types of cross sections, or any location where you want a dark area.

It becomes even more useful when paired up with a lineweight feature called *screening*, which is discussed in detail in Chapter 19, Advanced Output and Pen Settings. It has to do with being able to fade out (or screen) the intensity of any color, typically down to 30% and is perfect for indicating "area of work" on key plans, some types of carpeting, and even pavement on civil engineering drawings.

A way to mimic this now without screening is to change the solid fill's color to Gray 9; the soft gray haze then seen is distinct and visible, yet not overwhelming. If you are wondering how the walls were filled in on the apartment drawing, this is how.

5.5 SuperHatch

SuperHatch is an interesting variation of the Hatch command. The basic idea is that instead of using a premade pattern, you select a source image of your own, and AutoCAD duplicates whatever you chose into a pattern. Over several years the command evolved to allow you to choose nonimage items as well such as Blocks and Xrefs, but the most common application is still patterning images.

When using this command, be careful with the sizing of the original image. For example, if you use a photograph that is about 1 MB in size (a commonly encountered size for a high-resolution photo), then if your SuperHatch effort creates 1000 of these—well, you can see how the drawing file size will balloon. It may also take a while to SuperHatch a large area with large source images, and it is a great way to bring even a powerful workstation to a crawl.

With that in mind though, let us look at the command. Create a rectangular sample area to SuperHatch inside of and prepare a good candidate for the source image. In this textbook example, we will pattern a photo of some flowers, but obviously, you would pick something that serves a purpose in the context of your design. You generally use a SuperHatch only if needed to fulfill a certain Hatch need that is not addressed by more conventional patterns.

Step 1. Type in `superhatch` and press Enter. You will see the dialog box shown in Fig. 5.16.

Step 2. Click on Image… and browse around to find your chosen source image. Double click it and you will see the dialog box in Fig. 5.17.

Step 3. Click OK and position the image (using ENDpoint) at the lower left of the rectangle. AutoCAD will say: `Base image size: Width: 1, Height: 0.662338, Inches. Specify scale factor <1>:` Your sizing may, of course, be

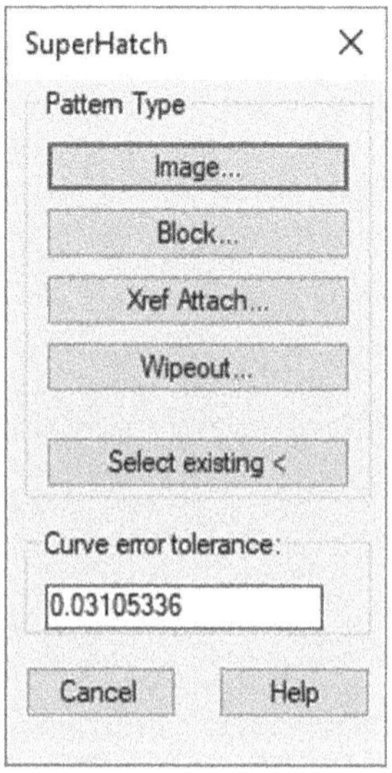

FIGURE 5.16 SuperHatch.

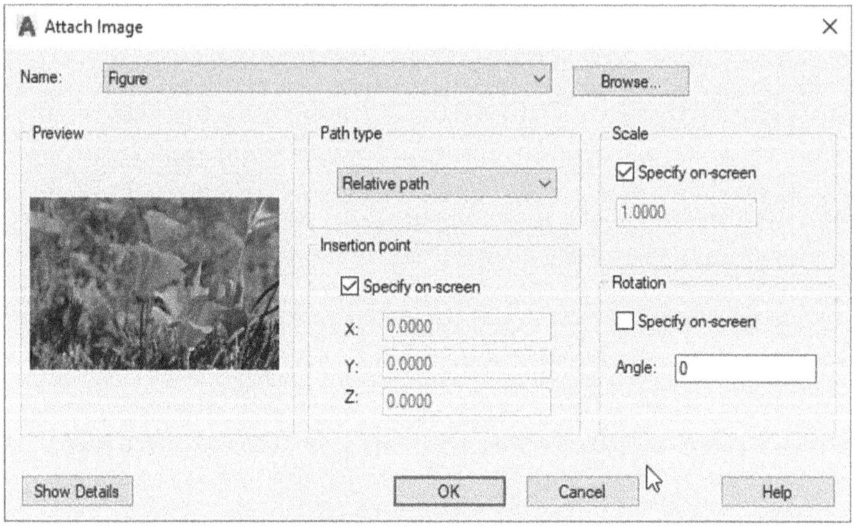

FIGURE 5.17 Attach Image.

FIGURE 5.18 Scale Factor.

different. Go ahead and adjust the scale factor by either typing in a value or just dragging the mouse out. Make the image relatively small compared to the overall area (Fig. 5.18).

Step 4. Click when done and one copy of the image will appear where you placed it (Figure 5.19).

If you like where it is, press Enter, or select Yes and AutoCAD will say: `Selecting visible objects for boundary detection... Done.`

Step 5. AutoCAD will say: `Specify an option [Advanced options] <Internal point>:` Click anywhere in the rectangular area and press Enter. AutoCAD will say: `Preparing Hatch objects for display... Done`. The SuperHatch should then form a pattern (Fig. 5.20).

FIGURE 5.19 Source image positioning.

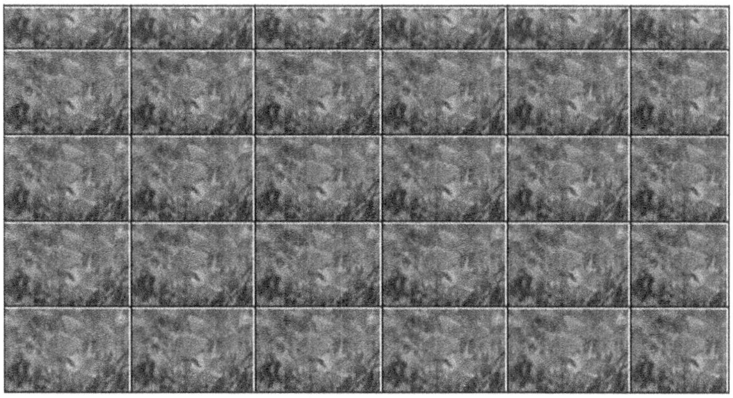

FIGURE 5.20 Final output.

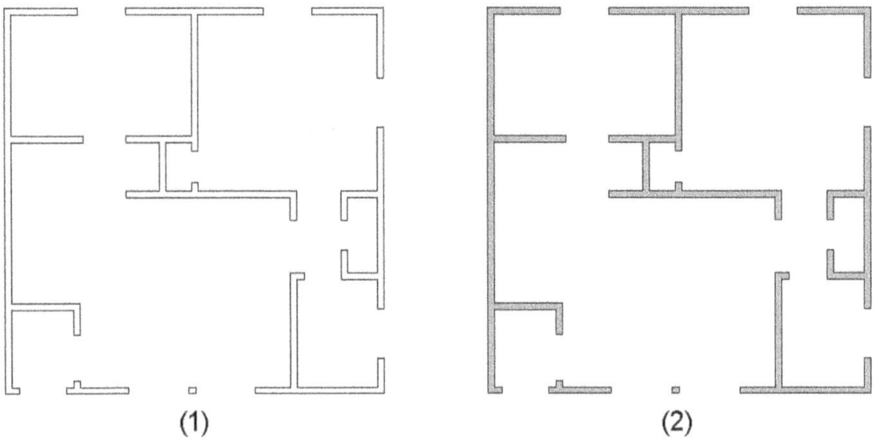

FIGURE 5.21 Wall Hatching before (1) and after (2).

5.6 In-Class drawing project: adding Hatch to floor plan layout

We continue to work on the floor plan and add Hatch to the rooms and walls. Some tips follow:

- To add Hatch to rooms, freeze the A-Doors layer and instead draw *temporary* lines closing off the door entrances. After Hatching, remove the lines. You may also be able to use Gap Tolerance set to at least 3 feet, but that is not recommended.
- Make sure the Hatch patterns are on the proper layers, whatever they may be (Carpeting or just Hatch_1, Hatch_2; you decide). You do not have to pick the Hatch patterns shown, but make sure they are reasonable and at the proper scale.

To fill in the apartment walls, freeze everything except the A-Walls and the Hatch layer. Then, use the solid fill, color Gray 9, as described earlier. If you created everything properly, there should be no gaps or breaks in the wall islands. Check to make sure the windows are drawn correctly, meaning the wall has to have an actual closed gap into which the window is inserted. Fig. 5.21 shows what everything should look like prior to wall Hatching (1) and right after (2). One possible approach to Hatching the apartment is shown in Fig. 5.22.

One possible approach to Hatching the apartment is shown in Fig. 5.22. (Note: You may have to add lines to close off rooms or represent material changes in floor hatches.)

Summary

You should understand and know how to use the following concepts and commands before moving on to Chapter 6:

- Hatch
  - Picking a pattern
  - Selecting area or object
  - Scale and orientation adjustments
  - Preview
- Editing a Hatch
  - Double-click
- Additional options
  - Islands
  - Inherit properties
  - Associative
  - Draw order
  - Gap tolerance
- Gradients and fills

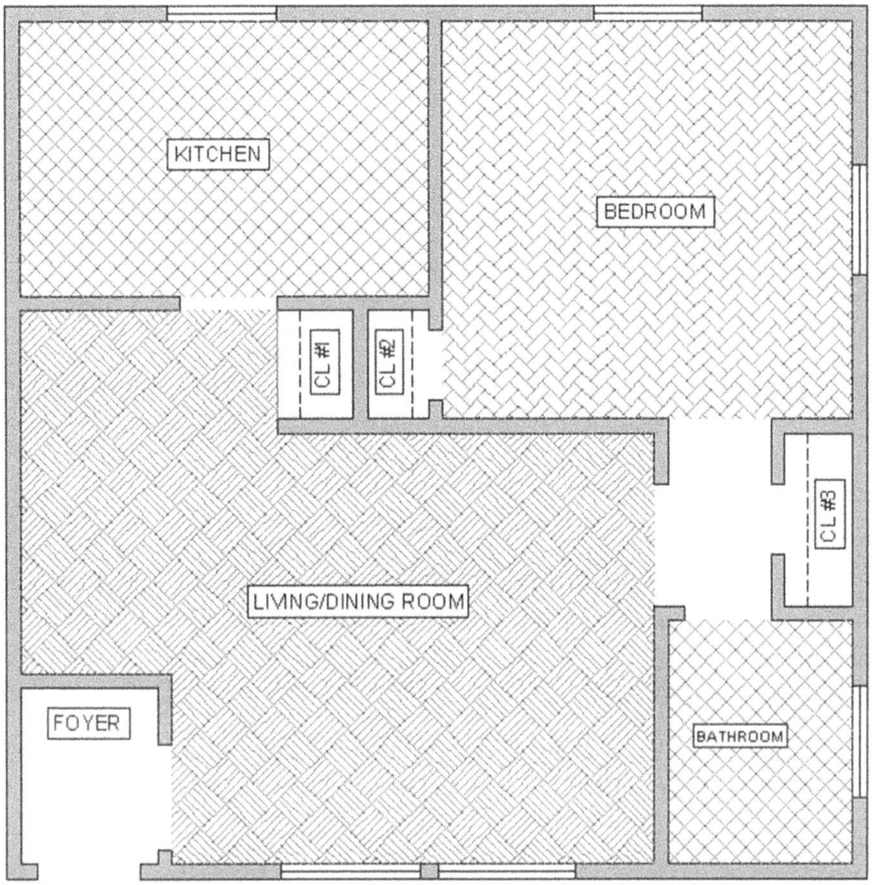

FIGURE 5.22 Floor plan with wall and floor Hatching.

Review questions

Answer the following based on what you learned in this chapter:

1. List the four steps in creating a basic Hatch pattern.
2. What two methods are used to select where to put the Hatch pattern?
3. What commonly occurring problem can prevent a Hatch being created?
4. What is the name of the tool recently added to AutoCAD to address this problem?
5. How do you edit an existing Hatch pattern?
6. What is the effect of exploding Hatch patterns? Is it recommended?
7. What especially useful pattern to fill in walls is mentioned in the chapter?

Exercises

1. In a new file, create the following 10″ rectangle, offset 2″ to get the smaller ones, and fill them all in with the shown Hatch patterns. Be sure to create a layer for the rectangles and a separate layer for the Hatch patterns. Do try to duplicate what is shown, including Hatch patterns, scale, and angles as closely as possible. (Difficulty level: Easy; Time to completion: <5 minutes.)

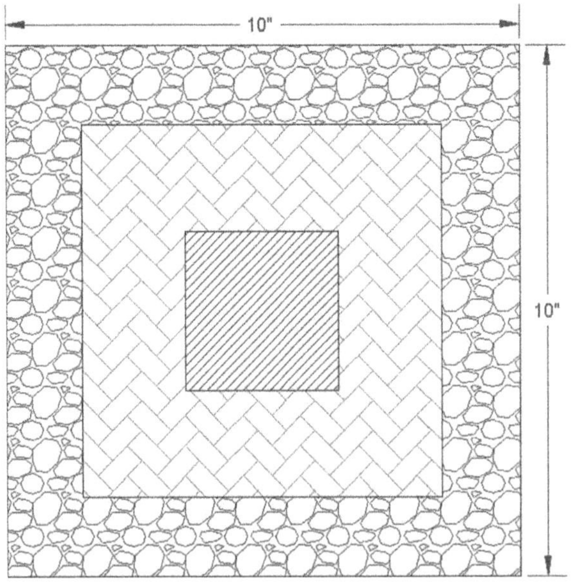

2. In a new file, create the following shapes and Hatch patterns. Some overall sizing is shown, but you have to improvise the rest via the offset and line commands. Be sure to create a layer for background linework and a separate layer for the Hatch patterns. Do try to duplicate what is shown, including Hatch patterns, scale, and angles as closely as possible. (Difficulty level: Easy; Time to completion: <5 minutes.)

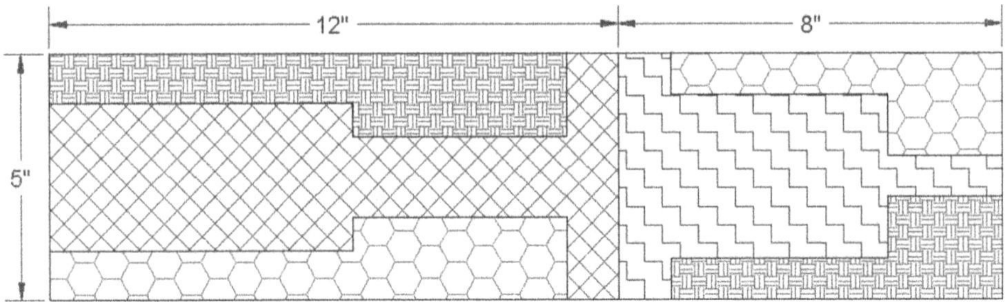

3. Open up (or redraw) Exercise 5 from Chapter 1 and add the following Hatch patterns. Be sure to create a layer for the background linework and a separate layer for the Hatch patterns. Do try to duplicate what is shown, including Hatch patterns, scale, and angles as closely as possible. (Difficulty level: Intermediate; Time to completion: 10 minutes.)

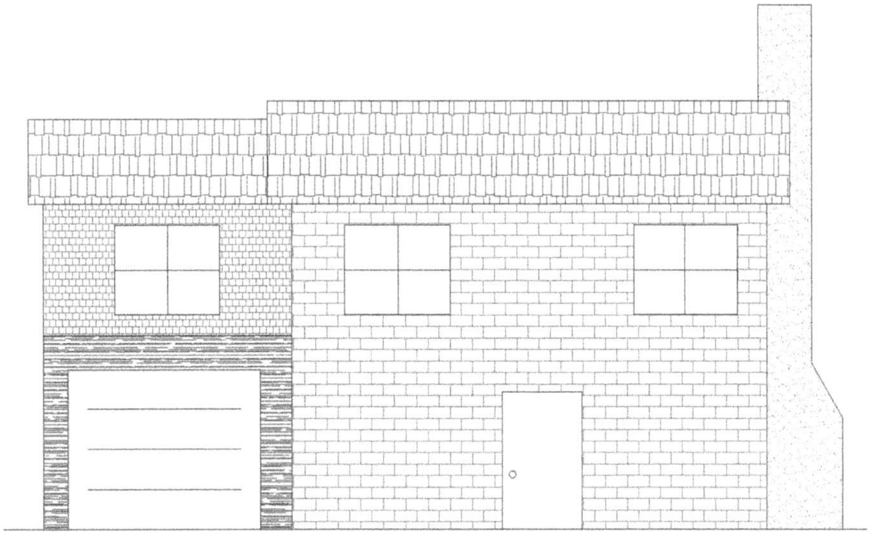

4. Open a new file and set units to Architectural. Create the following brick wall design, a slightly modified version of what is seen on the cover page of this chapter. Note that some of the backgrounds are gradients. Sizing and scale is up to you, but do try to duplicate what you see. (Difficulty level: Easy; Time to completion: <10 minutes.)

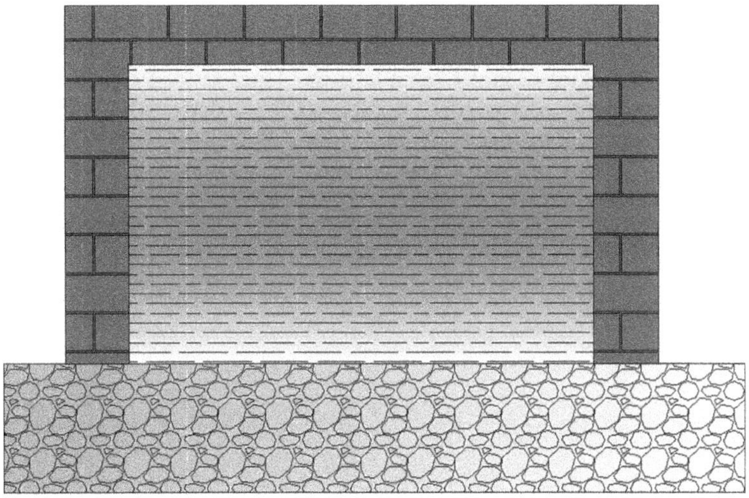

5. Open a new file, create the appropriate layers, and set units to Architectural. Create the following stair layout and Hatch patterns. Any sizes and dimensions that are not given can be estimated, as can the scale of the patterns. (Difficulty level: Intermediate; Time to completion: 20–30 minutes.)

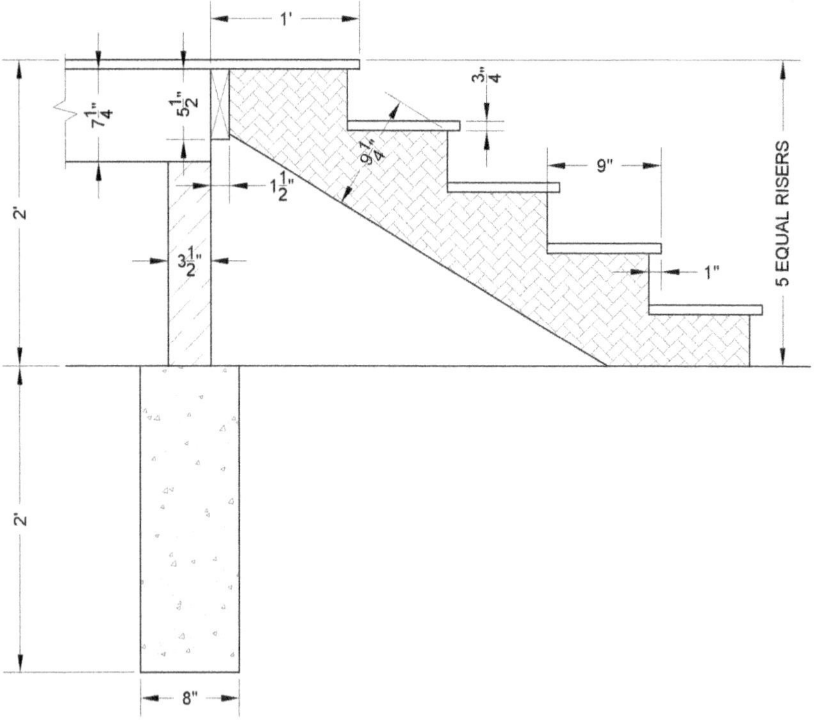

6. Open a new file and set units to Architectural. Create the following layout of a fictional set of cross-sectioned cylinders. Dimensions are provided, but sizing is of secondary importance to creating the Hatch and gradient patterns. Notice how gradients are used to depict curvature inside the pipe cross sections, a good trick to add realism to 2D drawings. (Difficulty level: Intermediate; Time to completion: 20 minutes.)

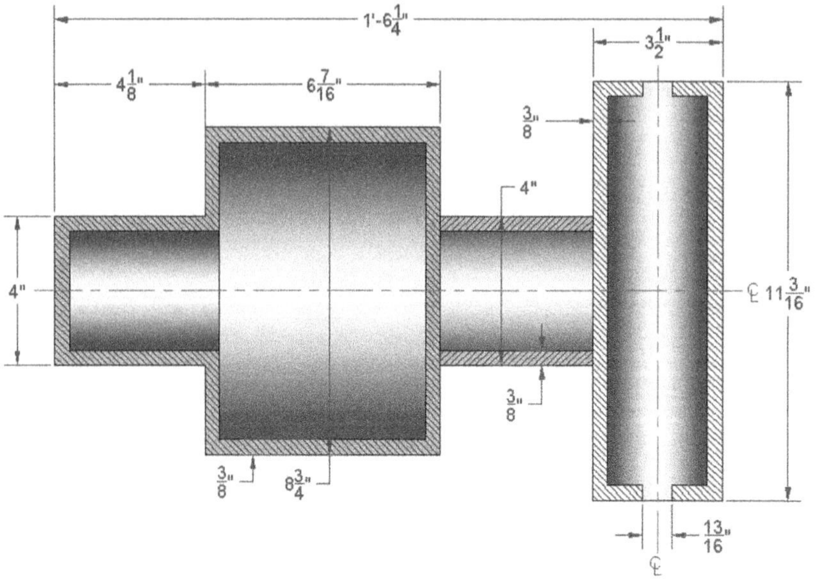

7. Create the following basic road signs using the starter dimensions given. All other sizing is to be approximated off of them. Add solid Hatch fill as shown. (Difficulty level: Easy; Time to completion: 15 minutes.)

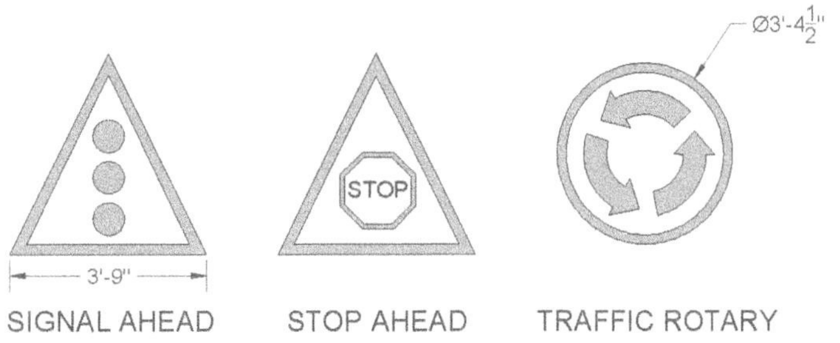

8. This exercise features some additional (slightly more complex) road signs. Create them using the starter dimensions given. All other sizing is to be approximated off of them. Add solid Hatch fill as shown. (Difficulty level: Easy; Time to completion: 25 minutes.)

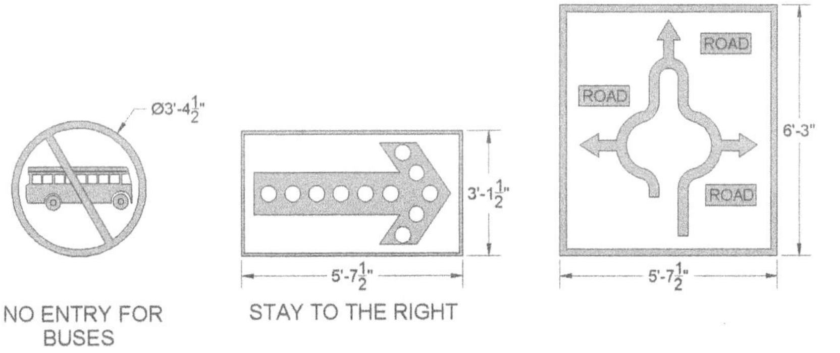

9. Create the following architectural detail. Many (but not all) of the dimensions are given. Use them as a basis to size the rest. Then, approximate the Hatch patterns as shown. The squiggly vertical one is not a Hatch pattern but a linetype, called *Batting*, often used to show insulation. Adjust its ltscale until it looks like what is shown. The rest of the Hatch patterns should be familiar. (Difficulty level: Easy; Time to completion: 25–35 minutes.)

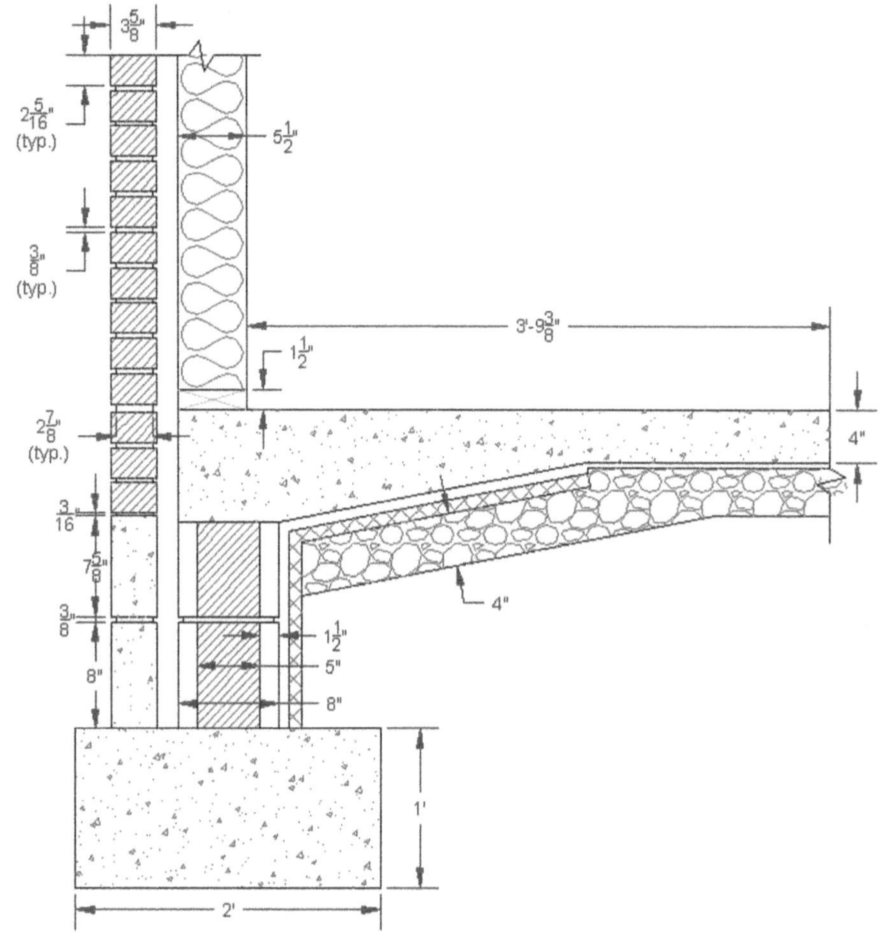

10. For the final exercise of this chapter, set up your file with the appropriate layers and text settings and draft the following detail of the roof and support beams. (NOTE: We will add the dimensions and leaders in the next chapter.) All necessary dimensions are found in either the dimensions or the notes. Any dimensions not given can be assumed to be standard construction sizes. Remember to draw actual sizes for dimensional lumber: 2 × 4s (1-1/2″ × 3-1/2″), 2 × 8s (1-1/2″ × 7-1/4″) or 2 × 10s (1-1/2″ × 9-1/4″). Then, approximate the Hatch patterns as shown. The 4:12 above the roof indicates its slope (rise:run) and can be drawn with *X, Y* distance entry (*X* of 12, *Y* of 4) or by calculating the angle via some basic trigonometry. (Difficulty level: Moderate; Time to completion: 30–40 minutes.)

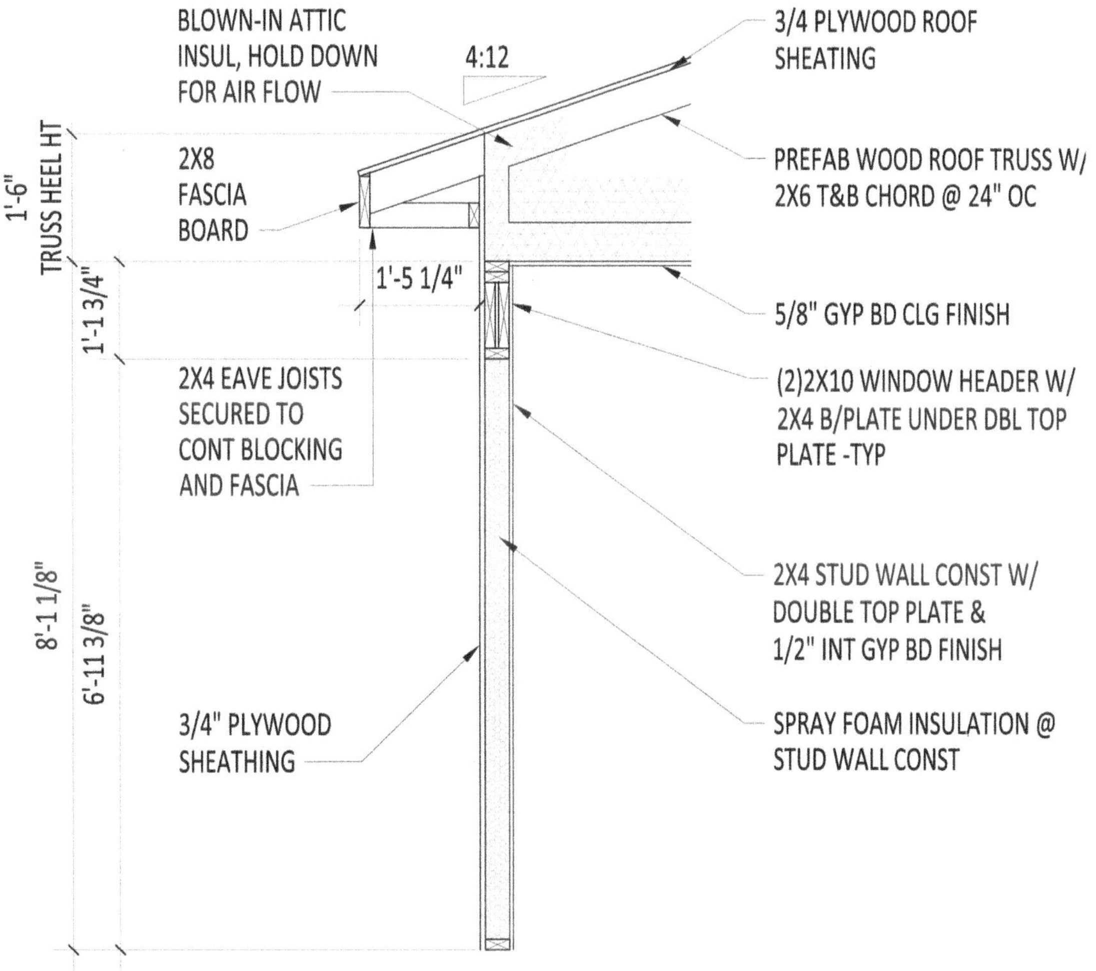

Spotlight on: mechanical engineering

Mechanical engineering is the science of designing and manufacturing mechanical systems or devices. It is one of the oldest and most widespread engineering specialties and is often the profession one has in mind when thinking of a "typical" engineer. Mechanical engineering is also by far the most diverse of the engineering specialties, and "mechies" can be found working in fields as wide-ranging as aerospace, automotive, naval, rail, power, infrastructure, robotics, manufacturing, consumer products, and much more. It is truly a profession that keeps our society moving.

FIGURE 1 *Image Source: Chrysler.*

Mechanical engineers need to master a wide variety of engineering sciences, structures, materials, vibrations, machine design, control systems, electrical engineering, and more. They also need to be well versed in business practices, cost analysis, and risk assessment, as design and production of just about anything are closely tied to these. Mechanical engineers have been the driving force behind the development of high-end software tools, such as finite element analysis (FEA) for structural work.

FIGURE 2 *Image Source: Mercruiser.*

Education for mechanical engineers starts out like education for the other engineering disciplines. In the United States, all engineers typically have to attend a 4-year ABET-accredited school for their entry-level degree, a Bachelor of Science. While there, all students go through a somewhat similar program in their first 2 years, regardless of future specialization. Classes taken include extensive math, physics, and some chemistry courses, followed by statics, dynamics, mechanics, thermodynamics, fluid dynamics, and material science. In their final 2 years, engineers specialize by taking courses relevant to their chosen field. For mechanical engineering, this includes classes on structures, controls, vibration analysis, thermodynamics, and machine design, among others.

On graduation, mechanical engineers can immediately enter the workforce or go on to graduate school. Although not required, some engineers choose to pursue a Professional Engineer (P.E.) license. The process first involves passing a Fundamentals of Engineering (F.E.) exam, followed by several years of work experience under a registered P.E., and finally sitting for the P.E. exam itself. Mechanical engineering is also an excellent undergraduate degree for entry into law, medicine, or business school. Just some of the careers you can pursue with an engineering bachelor's followed by a JD, MD, or an MBA include patent attorney, biomedical engineer, and manager or CEO of an engineering firm. If you can handle the educational requirements, the mechanical engineering undergraduate degree may truly be one of the most useful ones in existence.

Mechanical engineers can generally expect starting salaries (with a bachelor's degree) in the $62,000/year range, which is in the middle of the pack among engineering specialties, behind computer, petroleum, and chemical engineering. This, of course, depends highly on market demand and location. A master's degree is highly desirable and required for many management spots. The median salary is around $80,000, with the top 10% making over $120,000. The job outlook for mechanical engineers is excellent, as the profession is so diversified that there is always a strong demand in some of the sectors, even if others are in a downturn.

So, how do mechanical engineers use AutoCAD and what can you expect? Industry-wide, AutoCAD enjoys a significant amount of use, even as its share shrinks as companies switch to low-cost 3D solutions such as SolidWorks and Inventor. In truly high-end design (aerospace, automotive, and naval, for example), the dominant software applications continue to be CATIA, NX, and Pro/Engineer (Creo); and AutoCAD does not compete with these. However, this still leaves many applications where AutoCAD is appropriate, such as small part design, schematics, and overall layout of systems. Many companies also do not manufacture but rather assemble modular pieces into new products; here, too, AutoCAD is the appropriate low-cost solution. Let us take a look at a few examples.

The acronym P&ID stands for piping and instrumentation diagrams, a major mechanical engineering field, commonly referred to as *process control*. Engineers use the principles of fluid flow, material science, and hydrodynamics to design piping and valves that regulate fluid flow. This can be anything from a chemical plant to an electrochlorination system to waste treatment. Usually, systems are assembled out of standard industry parts, so AutoCAD is the appropriate tool to put them together into a design. Fig. 3 shows a typical piping valve profile diagram.

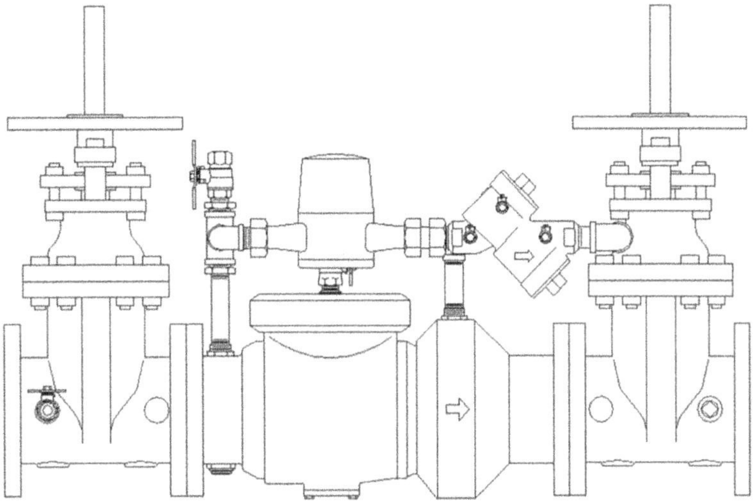

FIGURE 3 Piping valves. *Image Source: Elliot Gindis.*

Small device assemblies are another example of mechanical design. If the parts are standard and no manufacturing is necessary, only assembly drawings, then AutoCAD can be used to depict the ideas. Fig. 4 shows an AutoCAD drawing of an instrument tuning peg assembly.

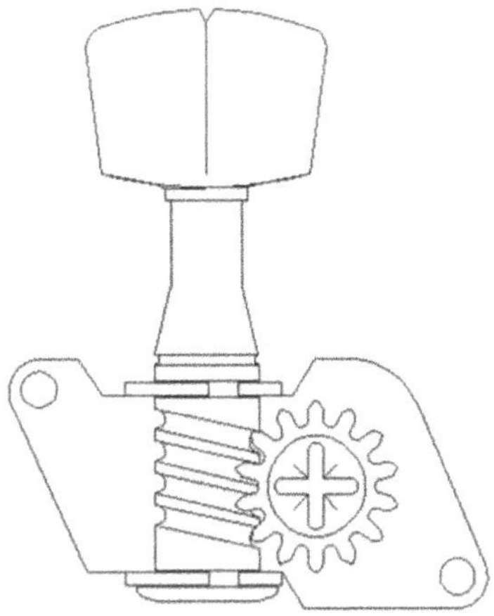

FIGURE 4 Tuning peg assembly. *Image Source: Elliot Gindis.*

Chapter 6

Dimensions

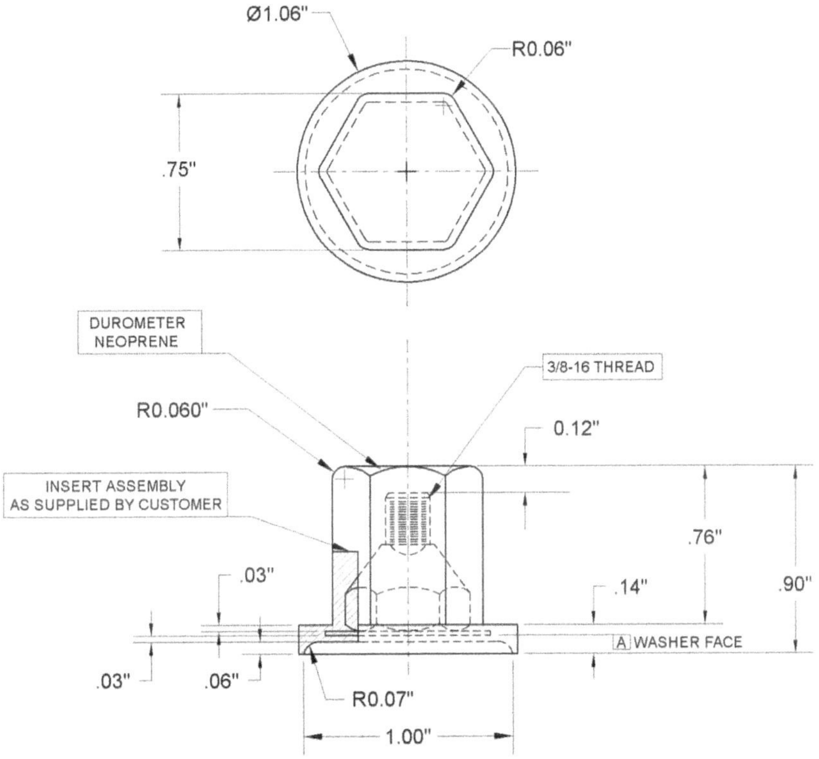

Learning objectives

In this chapter, we introduce the first parts of AutoCAD's extensive dimensioning capabilities and discuss the following:

- Linear (horizontal and vertical) dimensions
- Aligned dimensions
- Diameter and radius dimensions
- Angular dimensions
- Continuous and baseline dimensions
- Leader and multileader
- Arc length dimensions
- Jogged and jogged linear dimensions
- Ordinate dimensions
- ddedit
- ddim
- Dimension units
- Dimension font
- Dimension arrowheads
- Dimension overall size
- Center marks and center lines

At the end of this chapter, you will be able to add dimensions to your floor plan.
Estimated time for completion of this chapter: 3 hours (lesson and project).

Up and Running with AutoCAD 2022. https://doi.org/10.1016/B978-0-323-89923-9.00006-8

6.1 Introduction to dimensions

Dimensioning in AutoCAD is a major topic, one so extensive that it is split into two separate discussions. In this chapter, we address the basics of what dimensions are, which ones are available, and how to use and edit them. In Chapter 13, Advanced Dimensions, we look at extensive customization and additional options and explore some relatively new additions to the dimensioning family: parametric dimensioning and the concept of constraints and dimension-driven design.

So, what are dimensions? They are simply visible measurements for purposes of conveying information to the audience that will be looking at your design. It is how you describe the size of the design and where it is in relation to everything else. Dimensions can be natural, meaning they display the actual value, or forced, when they display an altered value, such as when you dimension an object with a break line.

6.2 Types of dimensions

The first step in learning dimensions is to know what is available, so you can use the appropriate type in any situation. The primary and secondary dimensions available to you are shown next. Commit them to memory, as you may need most of them on a complex project. AutoCAD features some rather fast methods for entering basic dimensions. We need to go over these but also mention other ways for those using slightly older releases.

Primary dimensions are the dimensions used most often by the typical architectural or engineering designer:

- Linear (horizontal and vertical)
- Aligned
- Diameter
- Radius
- Angular
- Continuous
- Baseline
- Leader

Secondary dimensions are used less often or are quite specialized but still need to be learned:

- Arc length
- Jogged
- Jogged linear
- Ordinate

Let us first take the primary list then the secondary and discuss the dimensions one by one to understand how each is used. In each case, you need to draw the appropriate shape to practice with, then add the new dimension(s) by following the step-by-step instructions.

The preferred methods for dimension commands entry in AutoCAD are the Ribbon and command-line typing. As it happens, both approaches lead to the same revised DIM command, which gives you a brand-new "one-touch" method of dimension creation and placement. For users of older versions of AutoCAD (that is, 2015 and older), the closest you have for this is the Quick Dimension (or QDIM). You can also rely on the "legacy" linear methods of selecting OSNAP points for the new dimensions. The number of different methods can be a bit overwhelming, but you will quickly gravitate toward one over another. All methods are detailed exhaustively with the first linear dimension, and less so for the other ones, as the approach is quite similar.

We can find most dimension tools on the Home tab of the Ribbon in the Annotate panel with all of the basic and more advanced tools on the Annotate Tab of the Ribbon in the Dimensions Panel (Fig. 6.1).

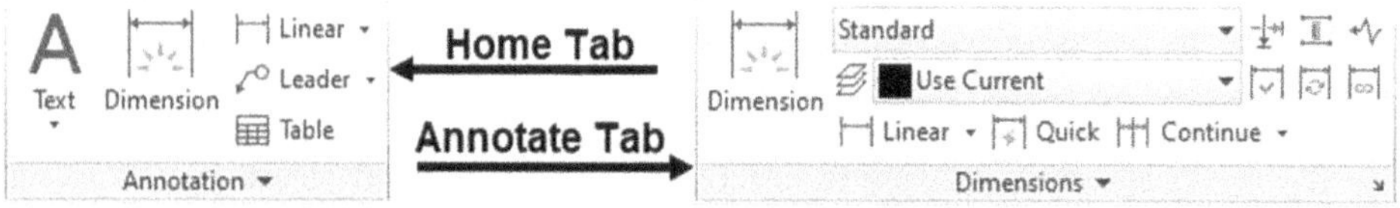

FIGURE 6.1 Ribbon locations for Dimension tools.

Note an important point if typing dimension commands: You need to type in `dim` and press Enter. This gets you into the dimension command line submenu, and you can operate by making new dimensions (and related features) but not with any other commands. Simply press Esc to get out of this.

Linear dimensions

These are any dimensions that are strictly horizontal or vertical (Fig. 6.2). We describe a general approach next, applicable to either the horizontal or vertical dimension, so be sure to practice both. Before you begin, create a generic square or rectangle as seen in Fig. 6.2.

Keyboard: Type in **dim** or **qdim** or **dimlinear** and press Enter
Ribbon: **Annotate** tab→**Dimensions** panel→ or or

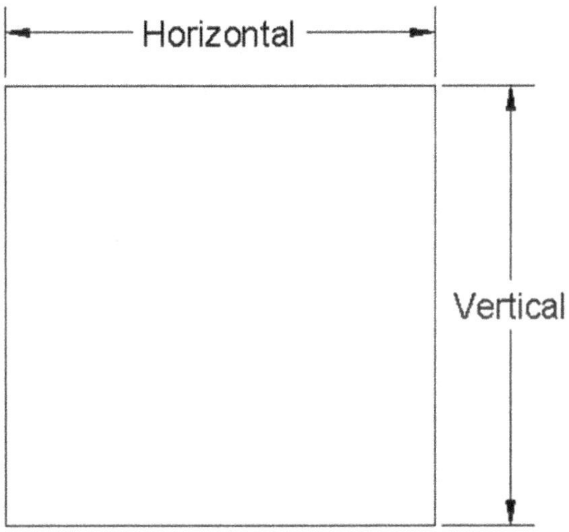

FIGURE 6.2 Linear (horizontal and vertical) dimensions.

For Ribbon input

Step 1. Press the Ribbon icon (with the sunburst pattern), or type in `dim` and press Enter.

- AutoCAD says: `Select objects or specify first extension line origin or [Angular/Baseline/Continue/Ordinate/aliGn/Distribute/Layer/Undo]:`

Step 2. Hover the mouse over any vertical or horizontal line making up the square shape. It becomes highlighted and a horizontal or vertical dimension appears in "preview" mode; shifting around as you move the mouse.

- AutoCAD says: `Select line to specify extension lines origin:`

Step 3. Click once to select the line to dimension, moving the dimension moving the dimension to its appropriate location, just to the outside of the square.

- AutoCAD says: `Specify dimension line location or second line for angle [Mtext/Text/text aNgle/Undo]:`

Step 4. Click one last time to fix the dimension in place.

- AutoCAD says: `Select extension line origin as baseline or [Continue]:`

You are done at this point and can move your mouse to another line that you may wish to dimension. This "one-touch" method of entry was introduced 3 years ago with AutoCAD 2016, somewhat speeding up dimensioning, and it continues with this version of AutoCAD.

For Typing, (Quick Dim)

Step 1. Type in `qdim` and press Enter.
- AutoCAD says: `Associative dimension priority = Endpoint`
  `Select geometry to dimension:`

Step 2. Pick any of the lines of your square, with the entire square becoming highlighted.
- AutoCAD informs you: `1 found.`

Go ahead and press Enter.

Step 3. Drag the dimension out to its appropriate location, just to the outside of the square. You will not yet see the actual dimension, only the extension lines, as you move the mouse up and down.
- AutoCAD says: `Specify dimension line position, or [Continuous/Staggered/Baseline/Ordinate/Radius/Diameter/datumPoint/Edit/seTtings] <Continuous>:`

Step 4. Click one last time to fix the dimension in place. You are done at this point and can repeat the command for another location.

For Typing or Ribbon Input (Linear Dim, "Legacy" Method)

Step 1. Type in `dimlinear` and press Enter, or press the appropriate Ribbon icon (with a basic horizontal dimension).
- AutoCAD says: `Specify first extension line origin or <select object>:`

Step 2. With the ENDpoint OSNAP select one of the corners of the square.
- AutoCAD says: `Specify second extension line origin:`

Step 3. With the ENDpoint OSNAP select any other nearby corner of the square—vertically or horizontally. A dimension appears; go ahead and drag it to its appropriate location.
- AutoCAD says: `Specify dimension line location or [Mtext/Text/Angle/Horizontal/Vertical/Rotated]:`

Step 4. Click one last time to fix the dimension in place. The dimension appears and the command line shows the same value you already saw on the screen (e.g., Dimension text = 2.8422).

To summarize, as you just saw, there are essentially three ways to create virtually any dimension. The first is via the relatively new "one-touch" method. This can be done only via the Ribbon. This method allows you to begin the dimensioning command and "touch" any geometry you want to dimension, with the appropriate dimension appearing automatically. Second is the qdim approach. It has been around for a while and can be accomplished via all four methods of input. It is a bit more cumbersome than Dim, but not by much, giving you the dimensions of some, but not all, geometry selected. It is mostly limited to linear and radius/diameter dimensions. The final version is the much older, but still effective, Linear command. It can also be executed via all four input methods. It requires two points, not a line, to create a linear dimension and takes a bit more time to do.

For users of AutoCAD 2015 and older releases, you can also type in Dim:, press Enter, and type in hor for a horizontal or vert for a vertical dimension, and then proceed to place it point by point following the previous Linear description. This method of command input is not available from AutoCAD 2016 on, but that does not matter much, as the more advanced Dim command can handle point-by-point dimension entry just as easily as the "one-touch" approach if you happen to pick corners instead of lines.

In general, linear dimensions are by far the most common in drawings and are also the basis for the continuous and baseline dimensions, to be covered later. Be sure to run through both the horizontal and vertical linear dimensions, with all methods shown, before settling on one. Next is the aligned dimension, which gives the true distance of a slanted surface. We describe it using the brand-new one-touch approach, but the command matrix shows other options.

Aligned dimension

This dimension measures a slanted line or object, as seen in Fig. 6.3. To practice it, you first need to create a small- to medium-sized square and chop off any corner, as also seen in Fig. 6.3. Only the first Ribbon approach is described in detail, but experiment with the other approaches (listed in the command matrix) if you prefer them.

Keyboard: Type in **dim** or **dimaligned** and press Enter
Ribbon: **Annotate** tab→**Dimensions panel**→  or 

Step 1. Press the first Ribbon icon (with the sunburst pattern).
- AutoCAD says: `Select objects or specify first extension line origin or [Angular/Baseline/Continue/Ordinate/aliGn/Distribute/Layer/Undo]:`

Step 2. Hover your mouse over the slanted line of the square shape. It becomes highlighted and an aligned dimension appears in preview mode, shifting around as you move the mouse.
- AutoCAD says: `Select line to specify extension lines origin:`

Step 3. Click once and drag the dimension out to its appropriate location, just outside the square.
- AutoCAD says: `Specify dimension line location or second line for angle [Mtext/Text/text aNgle/Undo]:`

Step 4. Click one last time to fix the dimension in place.
- AutoCAD says: `Select extension line origin as baseline or [Continue]:`

Note that after selecting the line, the type of dimension you get, Aligned or Linear, depends on the relationship of the cursor/mouse location to the object. Be sure you are getting the type of dimension you want before placing the dimension object. You are done at this point and can move your mouse to another line that you may wish to dimension.

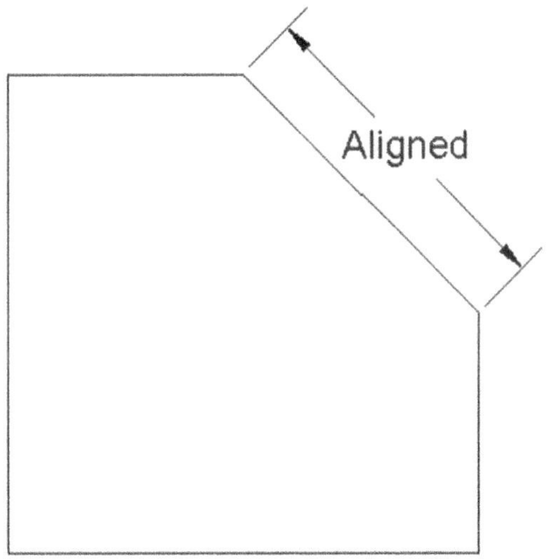

FIGURE 6.3 Aligned dimensions.

The next two dimensions, diameter and radius, are essentially similar, although we cover them separately for clarity. Note the primary differences: the symbol indicating diameter (Ø) and the symbol for radius (R) prior to the value. Although you can set this up, by default there is no line crossing the circle halfway for radius or all the way across for diameter, so you must read the values carefully and look for the appropriate symbol to know what you are looking at.

Also, note the location of the values. They are positioned at roughly 10, 2, 4, and 8 o'clock relative to the circle. Do not just put them anywhere; hand-drafting rules still apply in CAD, and designers have typically left them in these positions for clarity, consistency, and style.

Diameter dimension

This dimension measures the diameter of a circle or an arc, as seen in Fig. 6.4. To practice it, you first need to create a small- to medium-sized circle, as also seen in Fig. 6.4. Only the first Ribbon approach is described in detail, but experiment with the other approaches (listed in the command matrix) if you prefer them.

Keyboard: Type in **dim** or **qdim** or **dimdiameter** and press Enter
Ribbon: Annotate tab→**Dimensions panel**→ [icon] or [icon] or [icon]

Step 1. Press the first Ribbon icon (with the sunburst pattern).

- AutoCAD says: `Select objects or specify first extension line origin or [Angular/Baseline/Continue/Ordinate/aliGn/Distribute/Layer/Undo]:`

Step 2. Hover your mouse over any part of the circle. It becomes highlighted and a diameter dimension (along with a center mark) appears in preview mode, shifting around as you move the mouse.

- AutoCAD says: `Select circle to specify diameter or [Radius/Jogged/Centermark/Angular]:`

Step 3. Position your mouse somewhere to the upper right of the circle (rarely inside, usually outside) at 2 o'clock or at any of the other accepted positions.

- AutoCAD says: `Specify diameter dimension location or [Radius/Mtext/Text/text aNgle/Undo]:`

Step 4. Click one last time to fix the dimension in place. You are done at this point and can move your mouse to anything else that you may wish to dimension. You have a value with a diameter symbol (e.g., ø2.25).

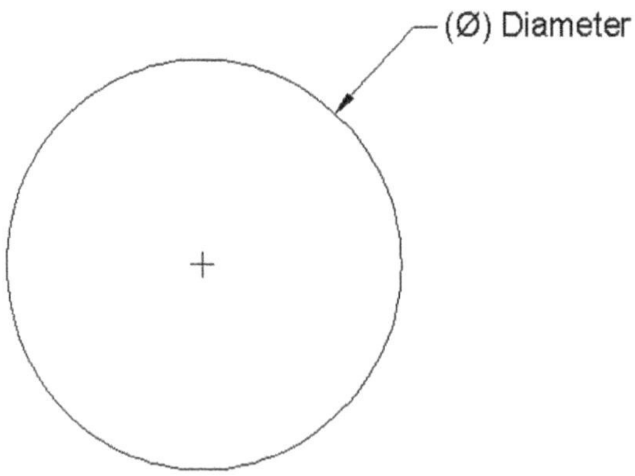

FIGURE 6.4 Diameter dimension.

Radius dimension

This is a dimension that measures the radius of a circle or an arc, as seen in Fig. 6.5. To practice it, you first need to create a small- to medium-sized circle (or retain the one you already have), as also seen in Fig. 6.5. Only the first Ribbon approach is described in detail, but experiment with the other approaches (listed in the command matrix) if you prefer them.

Keyboard: Type in **dim** or **qdim** or **dimradius** and press Enter
Ribbon: Annotate tab→Dimensions panel→ or or

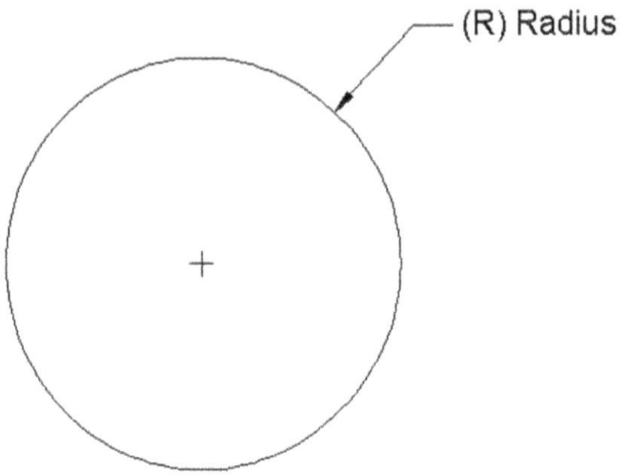

FIGURE 6.5 Radius dimension.

Step 1. Press the first Ribbon icon (with the sunburst pattern).

- AutoCAD says: Select objects or specify first extension line origin or [Angular/Baseline/Continue/Ordinate/aliGn/Distribute/Layer/Undo]:

Step 2. Hover your mouse over any part of the circle. It becomes highlighted and a radius dimension (along with a center mark) appears in preview mode, shifting around as you move the mouse.
Note an important point: If AutoCAD shows the diameter dimension (left over from the last attempt) instead of the expected radius, then simply press r [Enter] for Radius, as seen in the command line and drop down menu near your cursor (when DYNMODE is on), and AutoCAD switches to a radius instantly. The same goes for the other way around, if you are expecting a diameter and get a radius, press d when you get to this point.

- AutoCAD says: Select circle to specify diameter or [Radius/Jogged/Center mark/Angular]:

Step 3. Position your mouse somewhere to the upper right of the circle (rarely inside, usually outside) at 2 o'clock or at any of the other accepted positions.

- AutoCAD says: Specify diameter dimension location or [Radius/Mtext/Text/text aNgle/Undo]:

Step 4. Click one last time to fix the dimension in place. You are done at this point and can move your mouse to anything else that you may wish to dimension. You have a value with an R (e.g., R1.25).

Angular dimension

This dimension measures the angle between two lines or objects, as seen in Fig. 6.6. To practice it, you first need to create two lines at some random angle to each other (they need not be connected at their ends), as also seen in Fig. 6.6. For this dimension command, using the Ribbon's Home tab, Annotate panel or the Annotate tab, Dimensions panel icon is easier than the one-touch newer method, so we detail that one, but experiment with the other approach for practice.

Keyboard: Type in **dim** or **dimangular** and press Enter
Ribbon: **Annotate** tab→**Dimensions panel**→ or

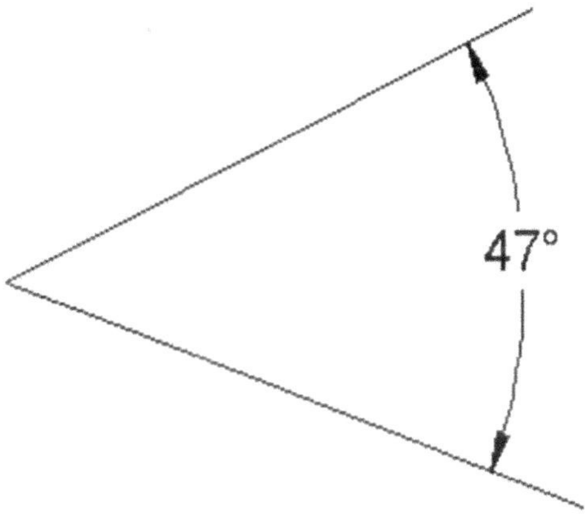

FIGURE 6.6 Angular dimensions.

Step 1. Begin the angular dimension command via the Ribbon's Angular icon.

- AutoCAD says: _dimangular
Select arc, circle, line, or<specify vertex>:

Step 2. Pick the first line (either one).

- AutoCAD says: Select second line:

Step 3. Pick the second line, and an angular dimension appears, attached to every move of your mouse.

- AutoCAD says: Specify dimension arc line location or [Mtext/Text/Angle/Quadrant]:

Step 4. Drag the mouse out and move it around, selecting the best position for the new dimension and click when you find a spot you like. Be careful, as the supplement of the degree value shows if you move behind the lines, which of course may be desirable in some cases. On that final click, AutoCAD gives you a value on the command line that matches what is on the screen (e.g., Dimension text = 47).

Note that, if you do use the new one-touch approach, the procedure is more or less the same as in all the previous dimensions, but AutoCAD does not recognize that you are asking for an angle value right away when you touch the first line and attempts to give you an aligned or linear reading. Only when you touch the second line does the software "get it" and gives an angular dimension. So, as stated earlier, in this case, the new approach is a bit less visual.

Angular completes the basic set of new fundamental dimensions. Next, we have continuous and baseline, which are not new, as we will see, but rather an extension (or automation) of the linear dimension learned earlier. We then conclude with a leader, not so much a true dimension but a useful member of that family.

Continuous dimensions

These are continuous *strings* of dimensions, as seen in Fig. 6.7.

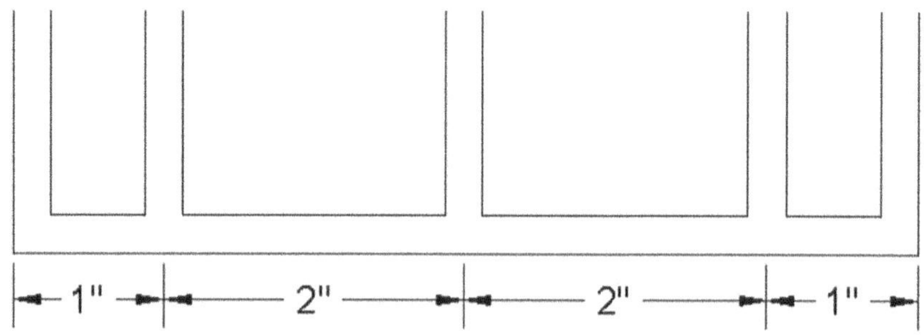

FIGURE 6.7 Continuous dimensions.

A short explanation is in order, as continuous and baseline dimensions sometimes give new students initial problems. Continuous dimensions are nothing more than a string of familiar horizontal or vertical ones. The idea here is to create one of those two types of linear dimensions, and then start up the continuous dimension where you just left off and let AutoCAD quickly fill them in as you pick contact points. It is really the same as using linear dimensions repeatedly but faster and more accurate as it is partially automated.

The easiest way to practice continuous dimensions is to draw a set of squares attached to each other. These represent a simplified view of building and room walls. Draw one random-sized square and copy it, endpoint (lower left corner) to endpoint (lower right corner), until you have what is shown in Fig. 6.8.

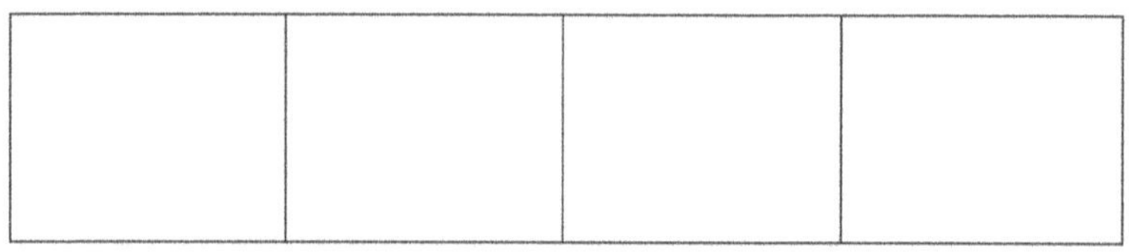

FIGURE 6.8 Setup to practice continuous dimensions.

The next step is to draw one horizontal linear dimension, as you already have done (it could have easily been a vertical line, too, if the rectangles were stacked vertically). Locate it about where you would like the entire string to go, as seen in Fig. 6.9.

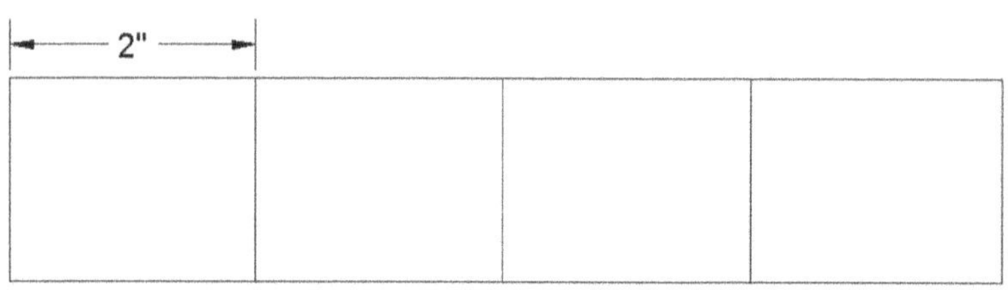

FIGURE 6.9 First step to create continuous dimensions.

This is the essential first step in creating both continuous and baseline dimensions. Note that you *can* create an initial linear dimension then do something else for a while. AutoCAD remembers (at least in most cases) the last place you left off and starts the new string from there. This changes a bit once you create another similar dimension somewhere else. Then, AutoCAD does not know to which one you are referring, so you must slightly modify your approach and pick the dimension that will be the start of the continuous string. In the instructions to follow, we assume you began to create the string right after making an initial linear dimension.

Keyboard: Type in **dim** or **dimcontinuous** and press Enter
Ribbon: **Annotate** tab→**Dimensions** panel→

Step 1. Once you have the squares and one linear dimension, as seen in Fig. 6.9, start up the continuous command via any of the preceding methods.

- AutoCAD says: `Specify a second extension line origin or [Undo/Select]<Select>:`

Step 2. A new dimension appears. Pick the *next* point (corner) along the string of rectangles and click on it (always using OSNAP points, no eyeballing).

- AutoCAD says: `Dimension text = 2.0`

Your value, of course, may be different.

Step 3. You can continue this until you run out of objects to dimension; at that point, just press Esc. Your result is shown in Fig. 6.10. Notice how all the values are neatly spaced in a row.

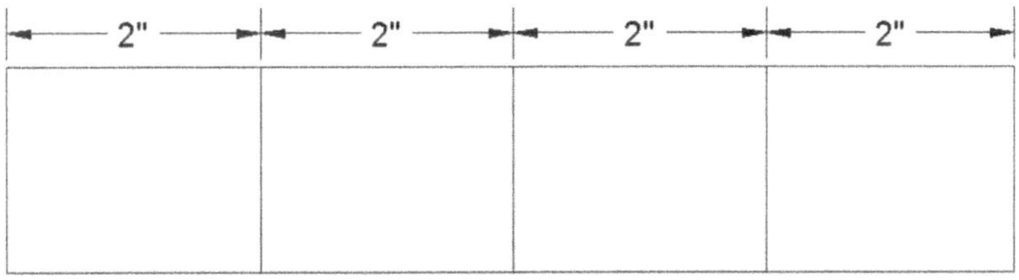

FIGURE 6.10 Continuous dimensions result.

Occasionally (and for a number of reasons), AutoCAD may not automatically know what the source for the continuous string may have been and may ask you to `Select continued dimension:`. Just pick the linear dimension you wish to serve as the "source" and continue with Step 2 as before.

Baseline dimensions

These are continuous *stacks* of dimensions, as seen in Fig. 6.11.

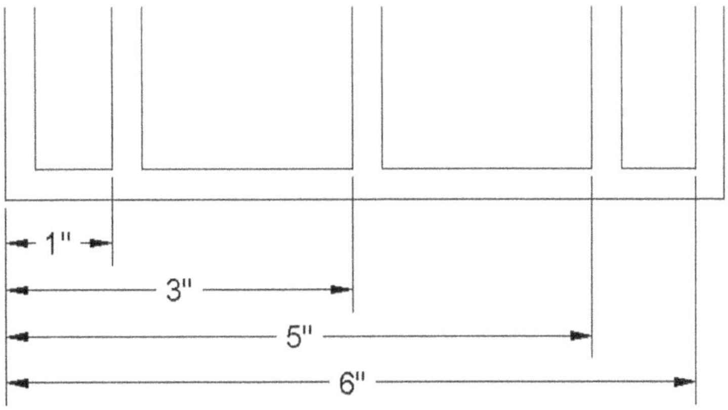

FIGURE 6.11 Baseline dimensions.

The baseline dimension, as mentioned before, is very similar in principle to the continuous dimension. The goal is to make a neat stack of evenly spaced dimensions that all start at one point (the base in baseline). To begin, erase the previous continuous dimensions (leaving the squares) and once again draw one linear (horizontal) dimension, as seen before in Fig. 6.9.

Keyboard: Type in **dim** or **dimbaseline** and press Enter
Ribbon: **Annotate** tab→**Dimensions** panel→

Step 1. Once you have the squares and one linear dimension, start up the baseline command via any of the preceding methods.

- AutoCAD says: `Specify a second extension line origin or [Undo/Select]<Select>:`

Step 2. A new dimension appears. Pick the *next* point (corner) along the string of rectangles and click on it (always using OSNAP points, no eyeballing).

- AutoCAD says: `Dimension text = 4.0`

Your value, of course, may be different.

Step 3. You can continue this until you run out of objects to dimension; at that point, just press Esc. Your result is shown in Fig. 6.12.

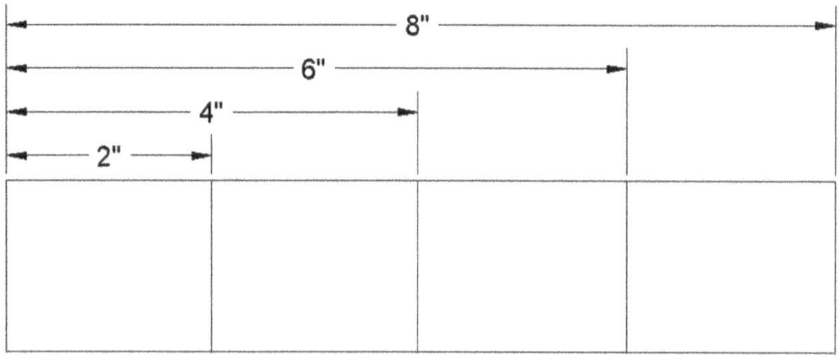

FIGURE 6.12 Baseline dimensions result.

Occasionally, just like with the continuous dimension, AutoCAD may not automatically know what the source for the baseline string may have been and ask you to Select base dimension:. Just pick the linear dimension you wish to serve as the "source" and continue with Step 2 as before.

Leader and multileader

This is an *arrow and label* combination pointing at an object to further identify or describe it (Fig. 6.13).

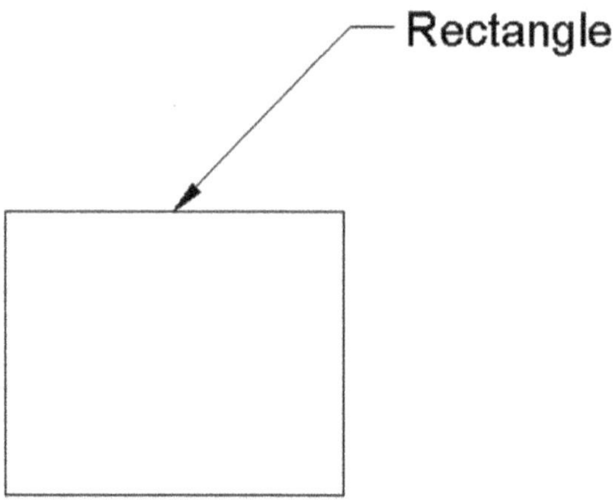

FIGURE 6.13 Leader.

While a leader is not a true dimension, it is still a very common and necessary member of the dimension family. The values entered by the user for the leader's content can be numerical or text comments from the designer, concerning the part or object to which it is pointing.

Leaders are so important that AutoCAD has given them a major rework in recent releases. We go over the two main options in increasing order of complexity, starting with the basic leader command, followed by the more feature-rich Multileader, with its Add, Remove, Align, and Collect options.

Leader

Step 1. Draw a small box on your screen, Making sure Ortho is off, type in the command *leader* and press Enter (there is no Ribbon equivilant).

- AutoCAD says: `Specify leader start point:`

Step 2. Click with OSNAP precision on your shape, perhaps a midpoint on the top side, as seen in Fig. 6.13.

- AutoCAD says: `Specify next point:`

Step 3. Move your mouse at a 45 degrees angle away from the first point and click again at a reasonable distance away. You create the *leader line and arrowhead* seen in Fig. 6.14.

- AutoCAD says: `Specify next point or [Annotation/Format/Undo] <Annotation>:`

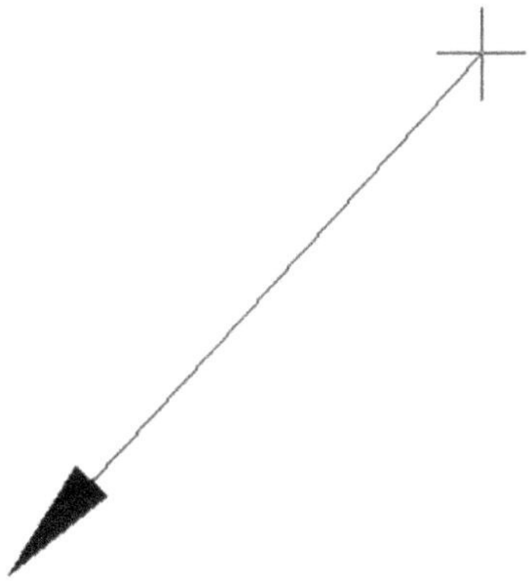

FIGURE 6.14 Leader, Step 3.

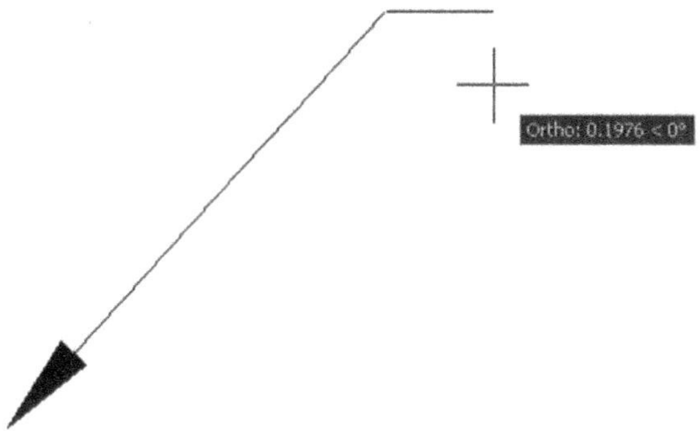

FIGURE 6.15 Leader, Step 4.

Step 4. Turn Ortho back on and draw a short line to the right, clicking when done; this is your *horizontal landing*, as seen in Fig. 6.15.

- AutoCAD says: `Specify next point or [Annotation/Format/Undo] <Annotation>:`

Step 5. You are done, so press Enter.

- AutoCAD says: `Enter first line of annotation text or <options>:`

Type something in, pressing Enter if you wish to do another line, and Enter again if you are done, as seen in Fig. 6.16.

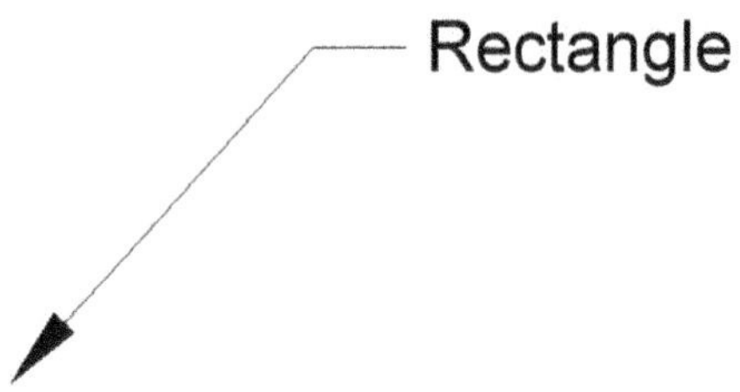

FIGURE 6.16 Leader, Step 5.

Multileader

This method is preferred and is meant to have much more flexibility and usefulness (and inadvertently some complexity) to the leader basic command. To use the mleader, look for the command on the Home tab of the Ribbon in the Annotate panel with basic and advanced tools on the Annotate Tab of the Ribbon in the Leaders panel, shown in Fig. 6.17.

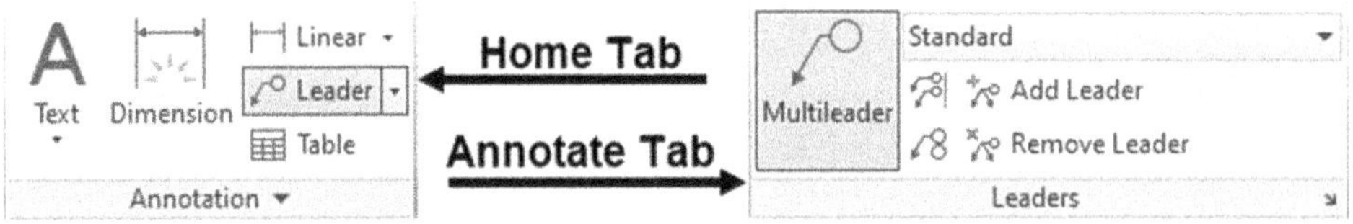

FIGURE 6.17 Ribbon location for Multileader tools.

Keyboard: Type in **mleader**
Ribbon: **Annotate tab→Leaders panel→**

With Ortho off, try the mleader command from the Ribbon or by simply typing mleader. Select a point on an object for your arrow and then click a point away from the object for your leader landing and text. It is very automated when compared to the old leader command.

There is something new to the Multileader that you did not have with the basic leader. You can set up a Multileader style, so all leaders have the exact look you want. In our sample case, we give our Multileader a detail bubble in which to add text. Start the mleader style tool by typing mleaderstyle or using the Ribbon's Home tab Annotation drop down menu (mleader with a paint brush) or the Annotate tab Leader panel arrow and the dialog box shown in Fig. 6.18 appears.

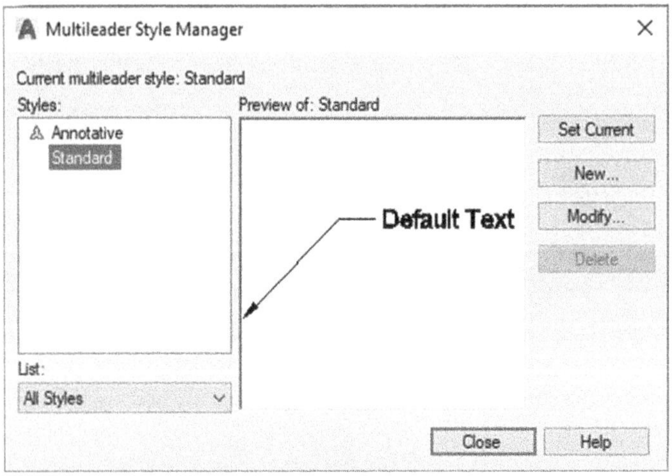

FIGURE 6.18 Multileader Style Manager.

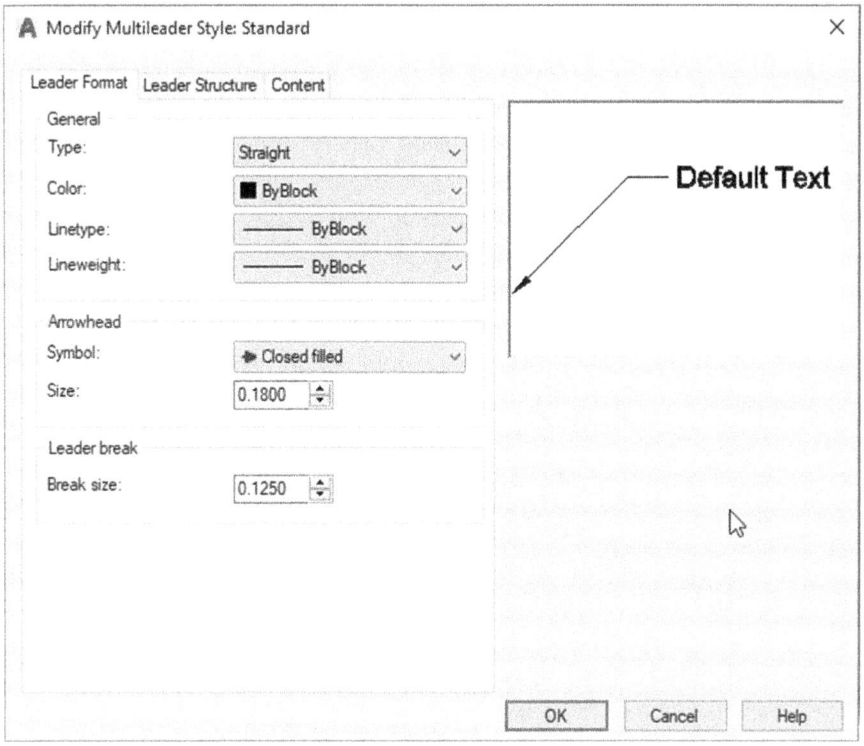

FIGURE 6.19 Modify Multileader Style.

Press New … and give the Multileader a name, such as Sample Style, and press Continue. You are taken to the Modify Multileader Style dialog box (Fig. 6.19). Examine the three tabs, Leader Format, Leader Structure, and Content, carefully. Most of the options are reasonably self-explanatory, and you will see some of them again later. We will only make one simple change. Under the Content tab, select Multileader type: as Block and Source block: as Detail Callout. Finally, press OK, Set Current, and Close. Now, let us try out our new Multileader style.

Step 1. Start the Multileader command via any of the preceding methods.

- AutoCAD says: Specify leader arrowhead location or [leader Landing first/Content first/Options] <Options>:

Step 2. Click anywhere and extend your mouse (Ortho off) at about a 45 degrees to the right.

- AutoCAD says: Specify leader landing location:

Step 3. Click again to place the Multileader and bubble. You see something like Fig. 6.20.

- AutoCAD says: Enter view number <VIEWNUMBER>:

Step 4. Enter in some value.

- AutoCAD says: Enter sheet number<SHEETNUMBER>:

Step 5. Enter in another value.

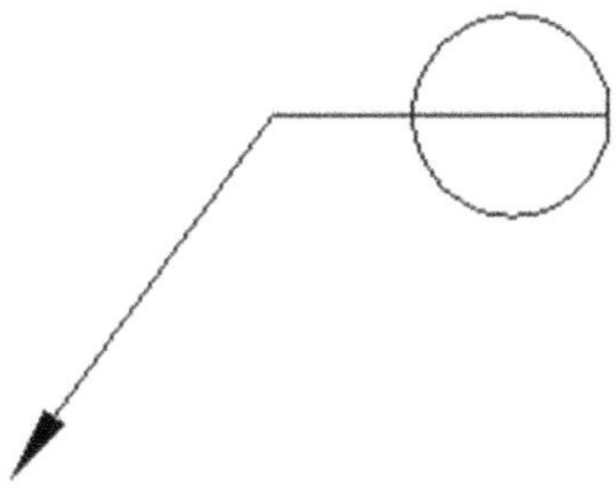

FIGURE 6.20 Multileader.

Once you do this, you see the result (Fig. 6.21), the appearance of which can, of course, be fine-tuned. Now that you have one leader on the screen, it is easy to experiment with other interesting options. Explore the mleader tools in the Leaders panel of the Annotate tab of the Ribbon, starting first with Add Leader (Fig. 6.22) and get rid of it via Remove Leader. Then, after adding it back in, create a few new leaders and line them up using Align Multileaders (Fig. 6.23), which aligns the leaders with one that you pick as the primary. Finally, Collect Multileaders (Fig. 6.24) combines them into one string.

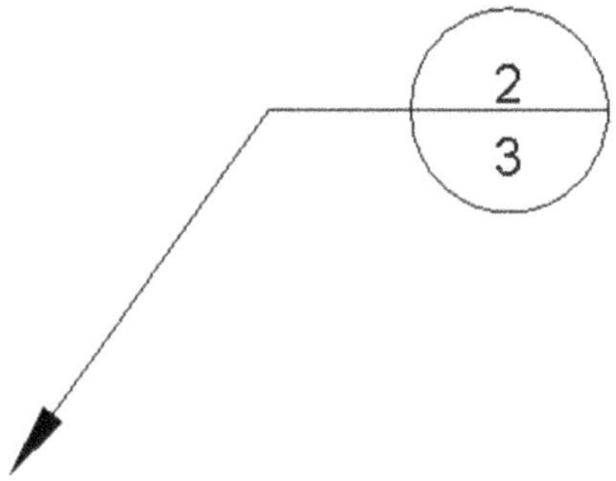

FIGURE 6.21 Multileader with block.

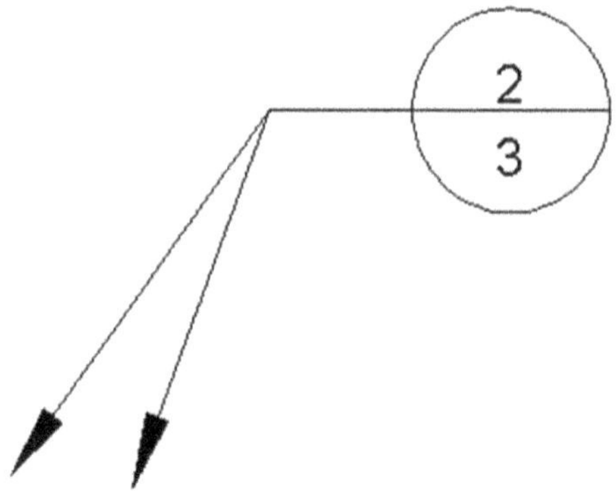

FIGURE 6.22 Add leader.

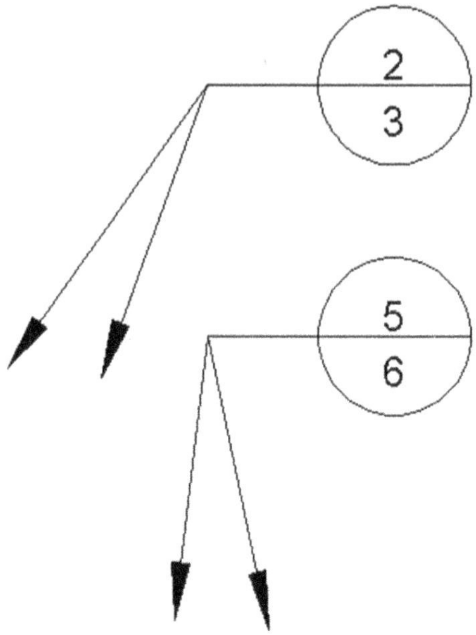

FIGURE 6.23 Align Multileaders.

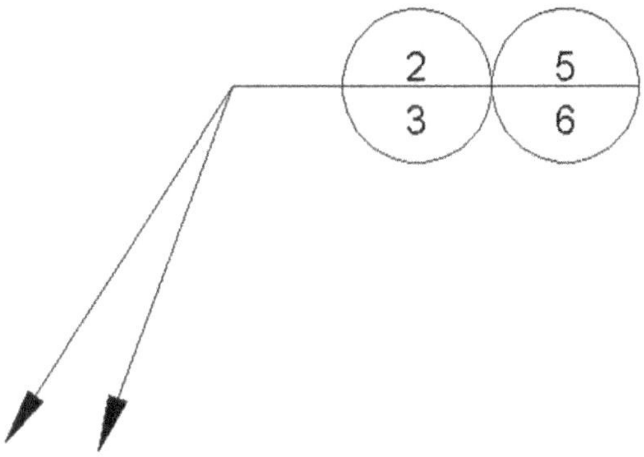

FIGURE 6.24 Collect Multileaders.

Secondary dimensions

It was mentioned that some dimensions are secondary, meaning they are not used as often. We briefly discuss them here, and you may want to also go over them in detail on your own, especially if you feel one or more of them may be extremely useful to your particular field. They are

- Arc length
- Jogged
- Jogged linear
- Ordinate

Arc length, as you may have guessed, measures the length of an arc. Simply select the Ribbon icon from the Annotate or Dimension panels already discussed and click on the arc you want to measure, as seen in Fig. 6.25. You can place the dimension value outside or inside the arc itself.

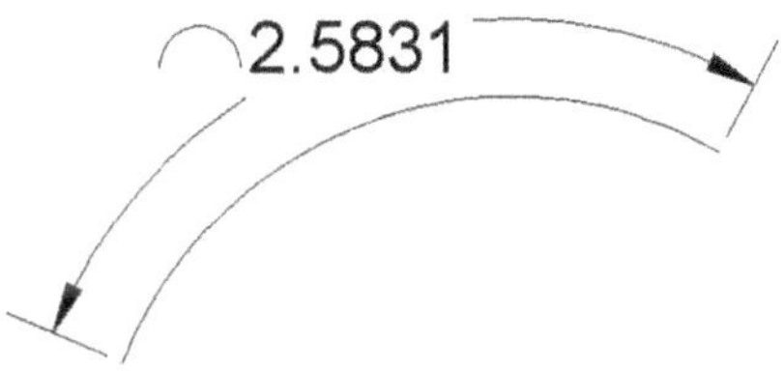

FIGURE 6.25 Arc length dimension.

Jogged dimensions are useful when the radius or diameter dimension's center is off the sheet of paper and cannot be shown directly, so a "break line" of sorts is used. You simply select the arc or circle then the center location override, and a jogged dimension appears (Fig. 6.26).

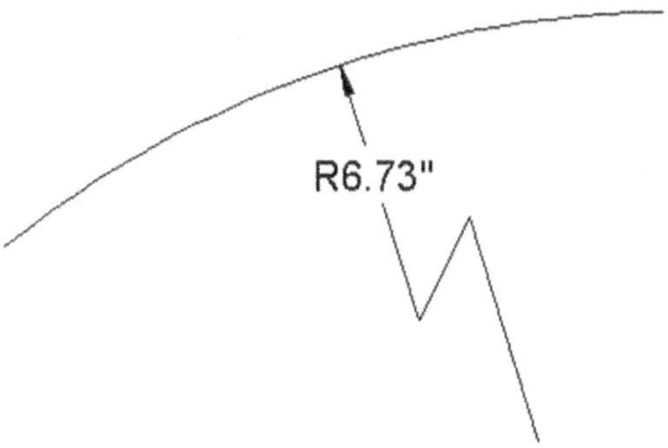

FIGURE 6.26 Jogged dimension.

Jogged linear dimensions are useful for representing break lines in a linear format. Jogged linear is not a dimension type as much as a modification to a linear (horizontal, vertical, or aligned) dimension. To try it out, create a horizontal dimension, as you have done numerous times before, and press the Jog linear icon on Ribbon Annotate or Dimensions panel. Then, click the dimension line of the linear dimension you wish to modify and one more click to position the jog. If you do not select a location for it, AutoCAD will do it for you. The result is seen in Fig. 6.27.

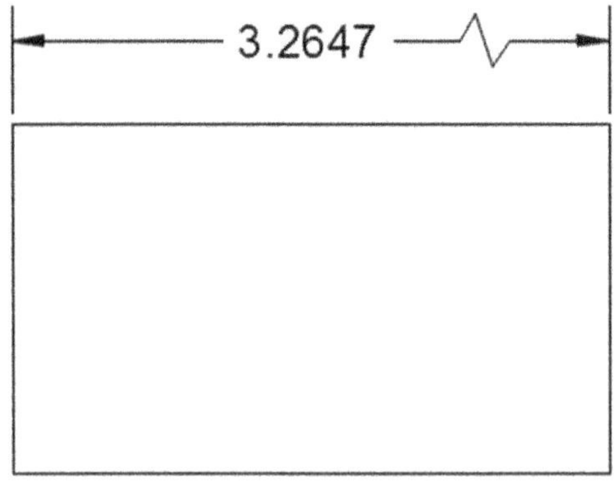

FIGURE 6.27 Jogged linear dimension.

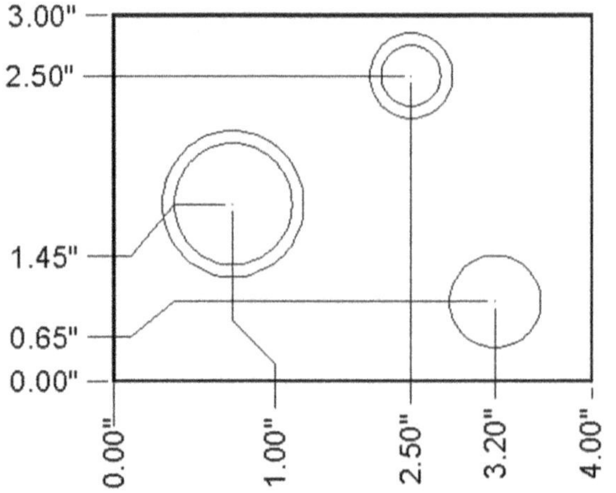

FIGURE 6.28 Ordinate dimension.

Ordinate dimensions (Fig. 6.28) measure horizontal or vertical distances from a datum point (0,0 in this case). They are used in manufacturing to prevent accumulation of errors that can occur using continuous measurements. Draw the shapes shown in Fig. 6.28, positioning the lower left corner of the rectangle at 0,0, and begin the ordinate dimension. Then, click major points along the way in either horizontal or vertical directions to get precise X or Y values at that location from the origin.

There are other dimension options we have not yet explored. We come back to them in Level 2. For now, we must move on and learn how to do editing and customizing.

6.3 Editing dimensions

You should now be quite familiar with the types of dimensions available and how to apply them to basic shapes. The next step is to be able to edit them or change their values, if needed. This is a very simple and short topic, but before we start, note something very important. You should not be changing dimension values manually, but rather seek to find out why your dimension value is wrong. Perhaps you did not attach it to the geometry correctly (did you use Osnaps?) or the geometry itself is flawed (poor drafting?). Try to get to the root of the selection points or drawn objects problem, and change dimensions manually only as a last resort!

You already learned that you can edit text and mtext by double-clicking on it. The text in the dimensions is essentially editable mtext, so go ahead and double-click any part of the dimension to edit the text. Recall also, in Chapter 4, Text, Mtext, Editing, and Style, we covered ddedit and mentioned it was also useful for editing purposes. You can use that command as well, and indeed with AutoCAD versions prior to 2013, you needed to use ddedit, as double-clicking on the dimension brings up the Properties palette. This "issue" was resolved for later releases.

Just like with Mtext, you will get a contextual Text Editor tab in the Ribbon and a toolbar for you to edit your dimension or mleader text as shown in Fig. 6.29. You can just edit the value and press OK or the big green check mark or click in open space away from the dimension you are editing.

Here is a good tip. If you want to reset your overwritten dimension value back to its natural value and you forgot what that was, just type in <>. These two "alligator teeth" brackets restore the natural value. Type them in instead of the forced value and click on OK. Try it yourself.

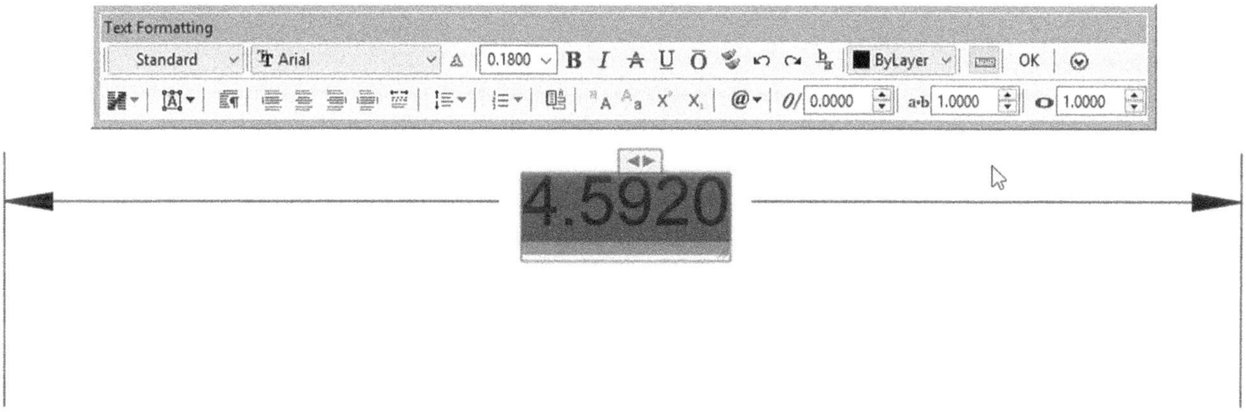

FIGURE 6.29 Mtext edit of dimensions.

6.4 Customizing dimensions

We have one more topic to cover on our way to proficiency in basic dimensioning. It is customization; and as mentioned at the start of this chapter, it is an extensive subject, necessitating it being split into the fundamentals here and advanced customization in Chapter 13. The goal here is to learn what experience has shown to be the most important four customization tools. This allows you to be productive in most of the drafting situations you are likely to encounter as a beginner. Later, you will add to this knowledge by learning tools used by CAD administrators and senior designers.

So, what customization is necessary at this level? We need to learn how to

- Change the *units* of the dimensions.
- Change the *font* of the dimensions.
- Change the *arrowheads* of the dimensions.
- Change the *fit* (size) of the dimensions.

To be able to change any of these, we need to introduce a new tool, the Dimension Style Manager dialog box, or *dimstyle* for short. There, you find a button to bring up the Modify Dimension Style dialog box. It is a "one-stop-shopping" dialog box for dimension modification that you will eventually need to learn backward and forward.

Dimstyle

Keyboard: Type in **dimstyle** or **ddim**
Ribbon: **Annotate** tab→**Dimensions** panel→**arrow** ↘

Using any of the preceding methods, bring up the Dimension Style Manager, as seen in Fig. 6.30.

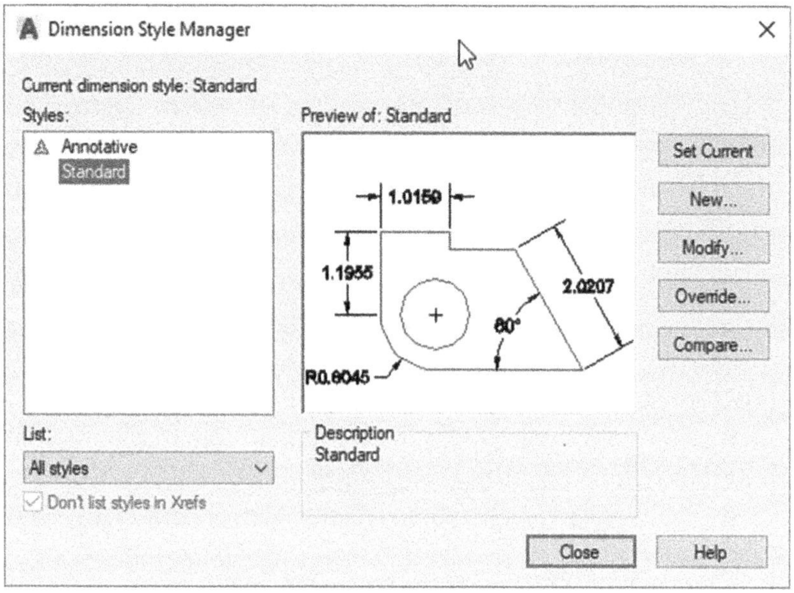

FIGURE 6.30 Dimension Style Manager.

The Dimension Style Manager dialog box has some options for creating a new dimension style or just modifying an existing one. For us at this point, it really does not matter which we pick, so just press Modify ….

You then are taken to the Modify Dimension Style dialog box, shown in Fig. 6.31. Note that you may not necessarily see the Lines tab as the open tab, as seen in Fig. 6.31. If you do not, just click on it to have the same thing on your screen as what is in the text.

This rather imposing dialog box contains tools to significantly modify any dimension. We, in fact, do so in Chapter 13, Advanced Dimensions, as we explore nearly every feature. For now, we just want to focus on learning how to modify the handful of items shown in the bulleted list at the start of Section 6.4.

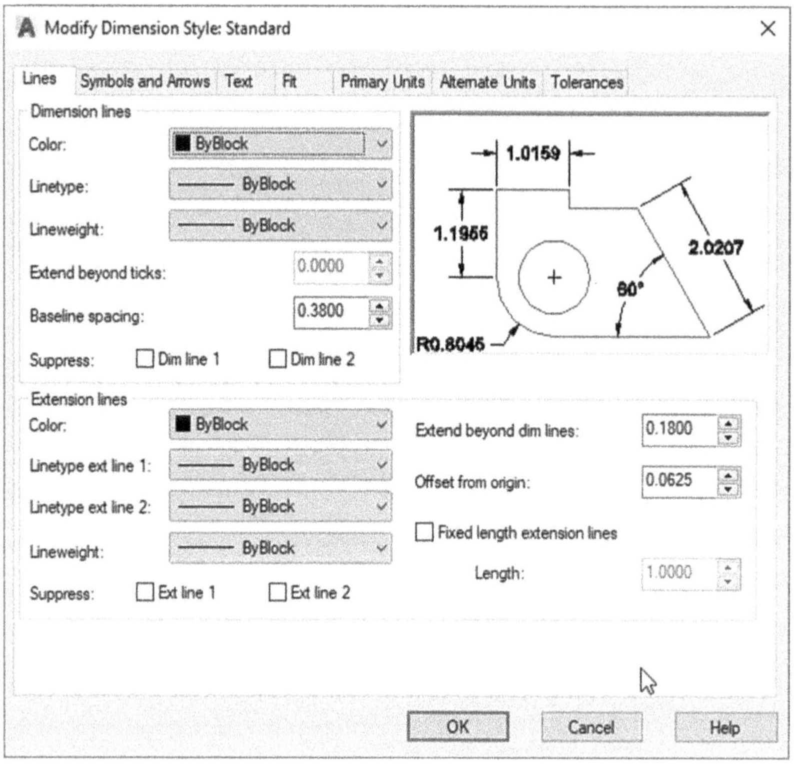

FIGURE 6.31 Modify Dimension Styles (Lines tab).

Step 1. Change the units of the dimensions

Pick the Primary Units tab and simply select the drop-down menu at the very top left, where it says Unit format:. The default value is Decimal; change it to Architectural and select the appropriate Precision in the menu just below it (the default value is fine). That is all we need from that tab for now. Notice how the preview window on the upper right reflects your choice of units, as seen in Fig. 6.32.

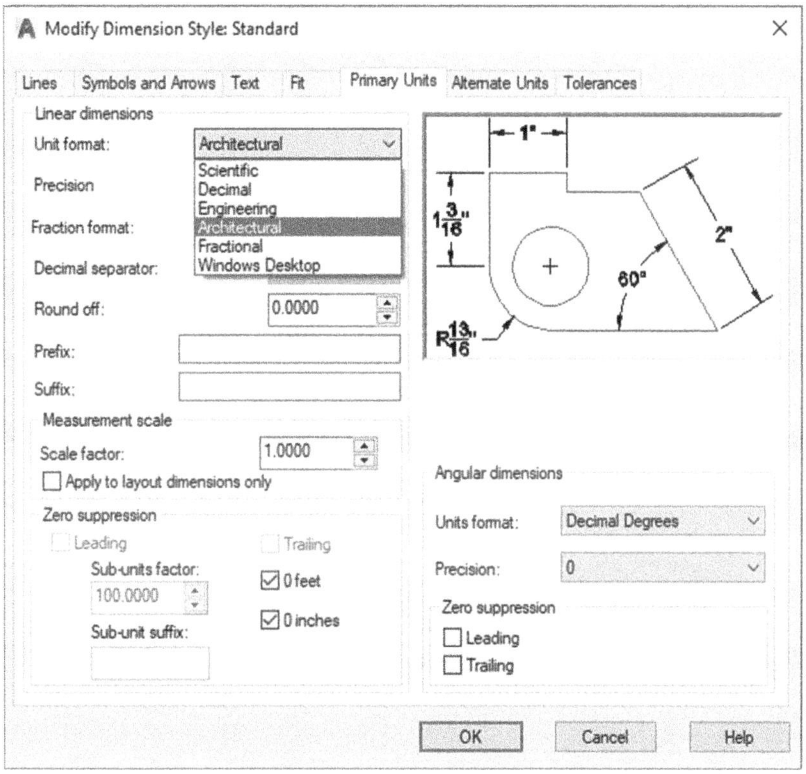

FIGURE 6.32 Selecting Architectural units.

Step 2. Change the font of the dimensions

To do this you need to click on the Text tab. You then see Text style: on the upper left. If you just opened a new AutoCAD file to practice dimensions, you probably do not have a font set and get only Standard as your choice if you click the down arrow. Fortunately, Standard is now a more attractive Arial font, not the stick-figure Simplex it used to be, so changing fonts is optional but still useful to know. For example, the HelvLight font is an alternative discussed in Chapter 4.

Something else to keep in mind is that, in a real working drawing, you would likely set your text style before you work on any dimensions (an important point, take note). Then, you would just select the font from the list. The idea here is to match up your regular font used in text and mtext with the font used in dimensioning, another important point. In general, you should stick to one font in a given drawing and just vary the size as needed. The title block is exempt from this, as that may have special fonts, logos, and the like.

AutoCAD, of course, anticipates that you may not have set a font style when you first set up dimensions and allows you to do this from the Text tab by bringing up the Style box when you click on the button just to the right of the Text style menu (with the three dots, …). Go ahead and set a different font if you wish, HelvLight perhaps with a Text Height of 0.25, and the new setting appears, as seen in Fig. 6.33.

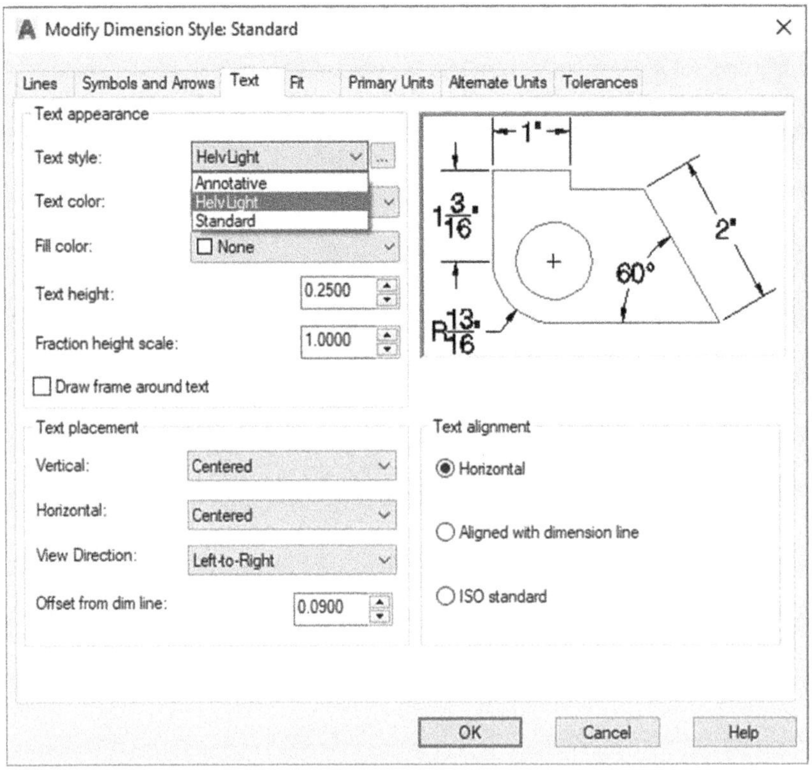

FIGURE 6.33 Setting a new font.

Step 3. Change the arrowheads of the dimensions

The default for all dimensioning is the standard arrowhead. You can easily change that to any other type, including the Architectural tick, popular with architects. You can even create a custom arrowhead (not something we try to do here).

Click over to the Symbols and Arrows tab. In the upper left corner under the Arrowheads section, you see First:. That is the first arrowhead type, and if you change it, AutoCAD assumes you want the second one to be the same and changes it as well. Simply click the down arrow and select the Architectural tick. Leave everything else the same. The result is shown in Fig. 6.34.

Step 4. Change the fit (size) of the dimensions

This function is quite important. Located in the Fit tab, just underneath the preview window, it features only a few lines and is shown in Fig. 6.35.

The idea here is to increase the size of the dimensions proportionately, so everything scales up evenly and smoothly in step with the size and scale of your overall drawing. This concept, however, opens up a number of questions. Chief among them: What value do you enter there and why? We are not ready to discuss this yet, but we do in Chapter 10, Advanced Output—Paper Space (on paper space). So, for now, do not enter in anything but be aware of how to do it when needed. When you get to dimensioning the floor plan in the next section, enter 15 in that space.

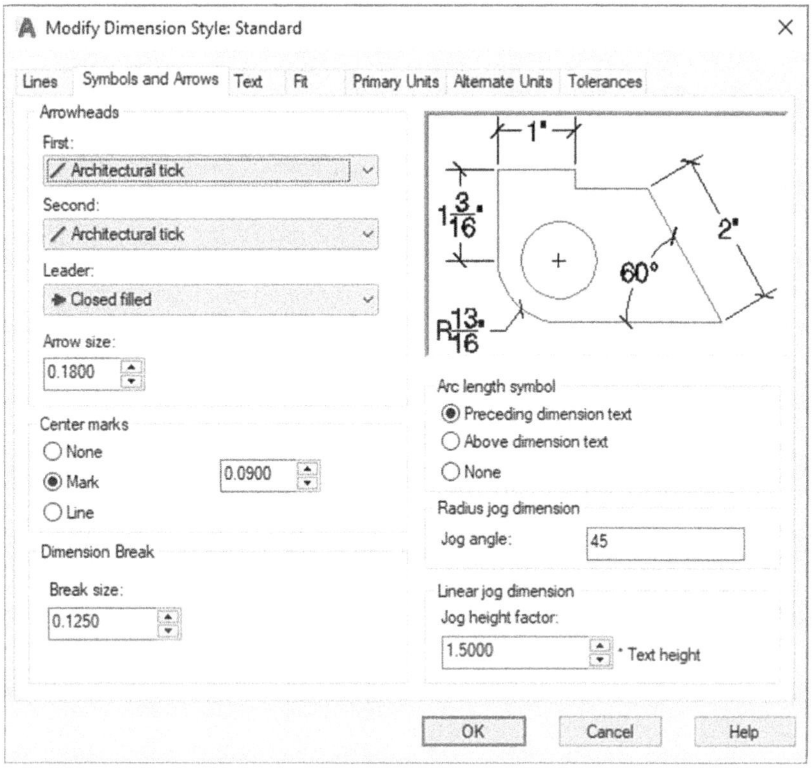

FIGURE 6.34 Changing the arrowheads.

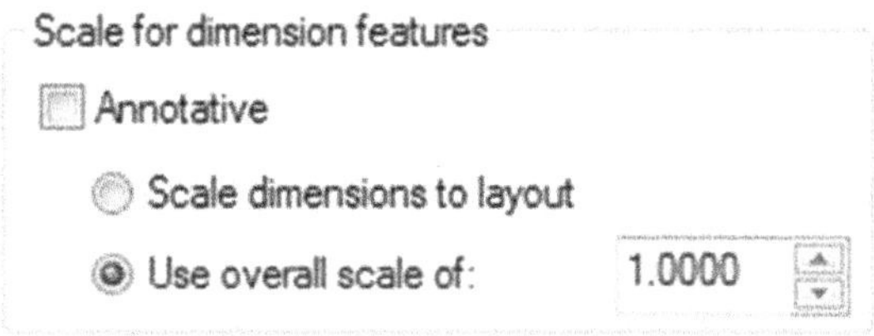

FIGURE 6.35 Scale for dimension features.

This is it for now; just click on OK and AutoCAD returns you to the Dimension Style Manager, where you can just press Close. Try creating a dimension and see what it looks like now. Fig. 6.36 shows a set of horizontal dimensions before the changes we just went through and another one afterward. The difference is obvious. For now, this is all you need to make professional-looking dimensions. Review and memorize everything you learned.

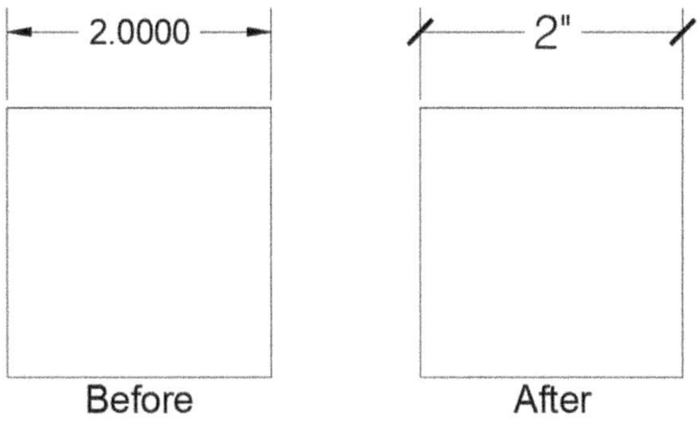

FIGURE 6.36 Scale for size of dimension.

6.5 Center marks and centerlines

This feature was introduced in AutoCAD 2017 and is a very welcome addition. For years creating center marks and especially centerlines was a somewhat tedious process that involved a few steps. A simple example of creating circle centerlines is shown in Fig. 4.18 back in Chapter 4. Good drafting practice requires these centerlines to be even and properly centered, with use of the correct linetype. A typical example of two circular and one rectangular part with centerlines is shown in Fig. 6.37.

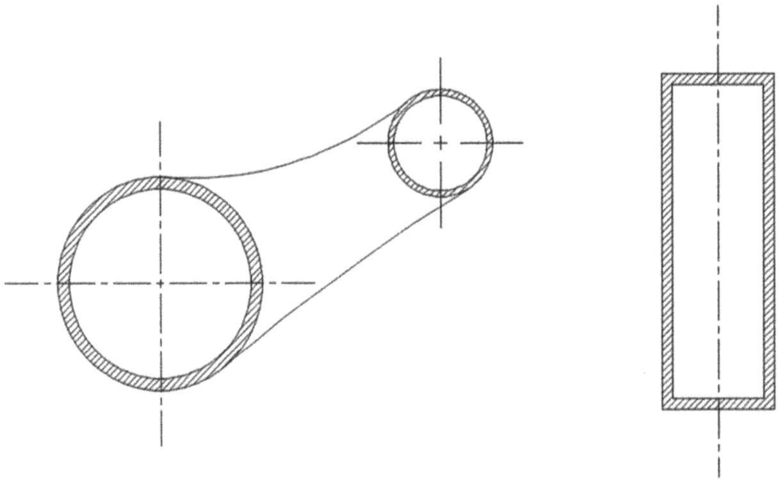

FIGURE 6.37 Centerlines in use.

These center marks and centerlines can now be created quickly and easily with just a few clicks—though what you will get will be default centerlines that use the Center2 linetype, and of a fixed size relative to the part you are annotating. If you need something different, outside those parameters, including an actual center mark, things do get slightly more involved. The command is still a big time-saver and speeds up the process.

To try this out, first create a few circles, then for some variety, a rectangle and a pair of parallel lines, as seen in Fig. 6.38. Next, locate the new buttons on the Ribbon, under the Annotate tab, Centerline panel, as seen in Fig. 6.39.

Go ahead and click the Center Mark button. AutoCAD says: Select circle or arc to add centermark: at which point, just click each of the circles. In the same manner select the Centerline button and select the parallel lines (both the rectangle and free drawn). The result is shown in Fig. 6.40.

These centerlines are editable via grips, and of course, additional options are available to customize them further. In Fig. 6.41 a double-click of the centerline brought up the options palette, and two of the extensions have been changed to a new length and color.

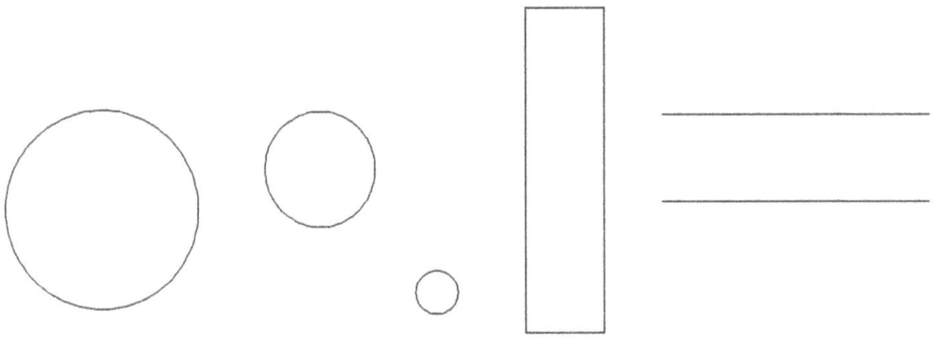

FIGURE 6.38 Basic shapes for center mark and center lines.

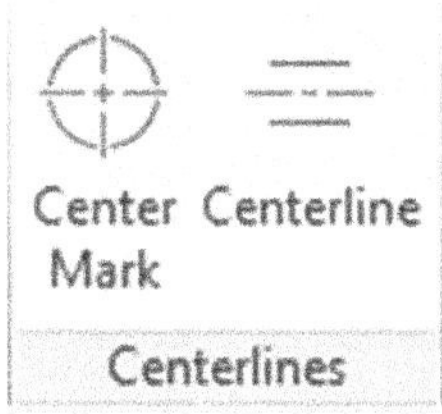

FIGURE 6.39 Centerlines panel.

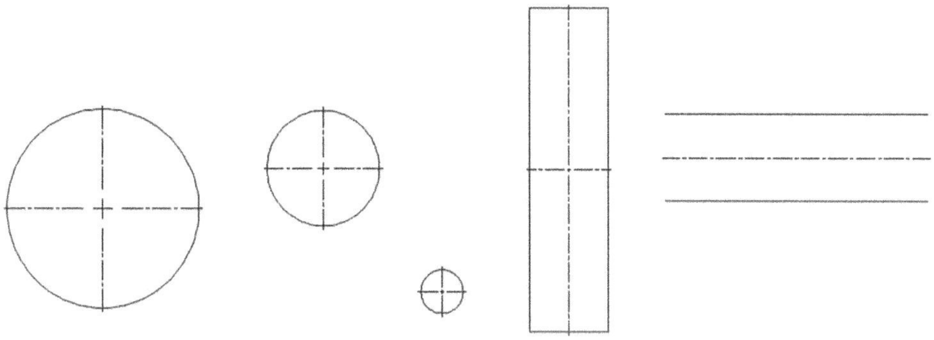

FIGURE 6.40 Completed center lines.

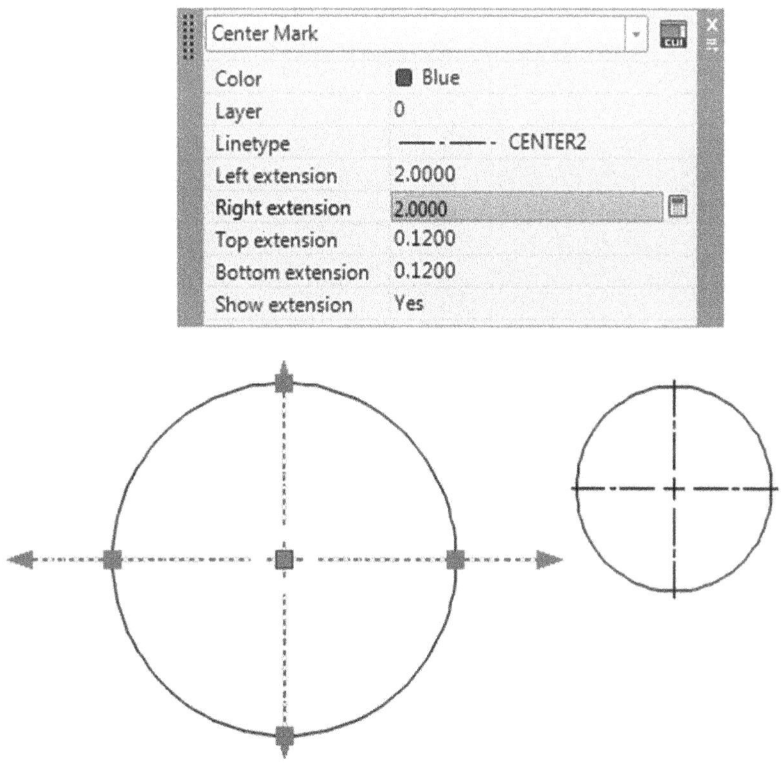

FIGURE 6.41 Manual modification of centerlines.

Some additional commands to control center mark and center line appearance include

- CENTEREXE: This controls the length of extension line overshoot (the amount of line outside of the shape). Determining what looks right may require some trial and error. Setting a value here will then apply it to all future centerlines. The ones already created remain the same until manually updated.

- CENTERMARKEXE: This command creates true center marks, with no extension lines (basically a small "plus" sign inside the circle). It is of the On/Off variety; either it is a center mark or a centerline. Setting a value here applies it to all future marks. The ones already created remain the same until manually updated.
- CENTERLTYPE, CENTERLAYER, and CENTERLTSCALE: These specify the linetype, layer, and linetype scale, respectively. Enter the desired type, layer, and scale when promoted. The settings apply globally to all future center marks or centerlines. The first two items can also be entered locally on a case-by-case basis via the properties palette seen in Fig. 6.41.

There are a few more of these commands, and the full listing can be found in the Help files, but these are the essential ones needed to customize this useful new feature of AutoCAD. Some of these commands hint at what are called *system variables*. We mention them again in Chapter 15, Advanced Design and File Management Tools.

6.6 In-Class drawing project: adding dimensions to floor plan layout

Let us now apply what we learned to the floor plan. Open the file and freeze all the layers except the walls and windows. If you really want to get fancy, put the wall solid hatch on its own (visible) layer and freeze the regular floor hatch layer so the carpeting and floors do not show. Next, create a new layer for the dimensions, A-Dim. Finally, set up the dimensions: Use Architectural units and Arial 6″ font, leave the arrowheads as they are, and change the scale under the Fit tab to 15. Dimension the floor plan any way you want; what is shown in Fig. 6.42 is a guide but not the only way to do it. Note that the outer dimensions are continuous.

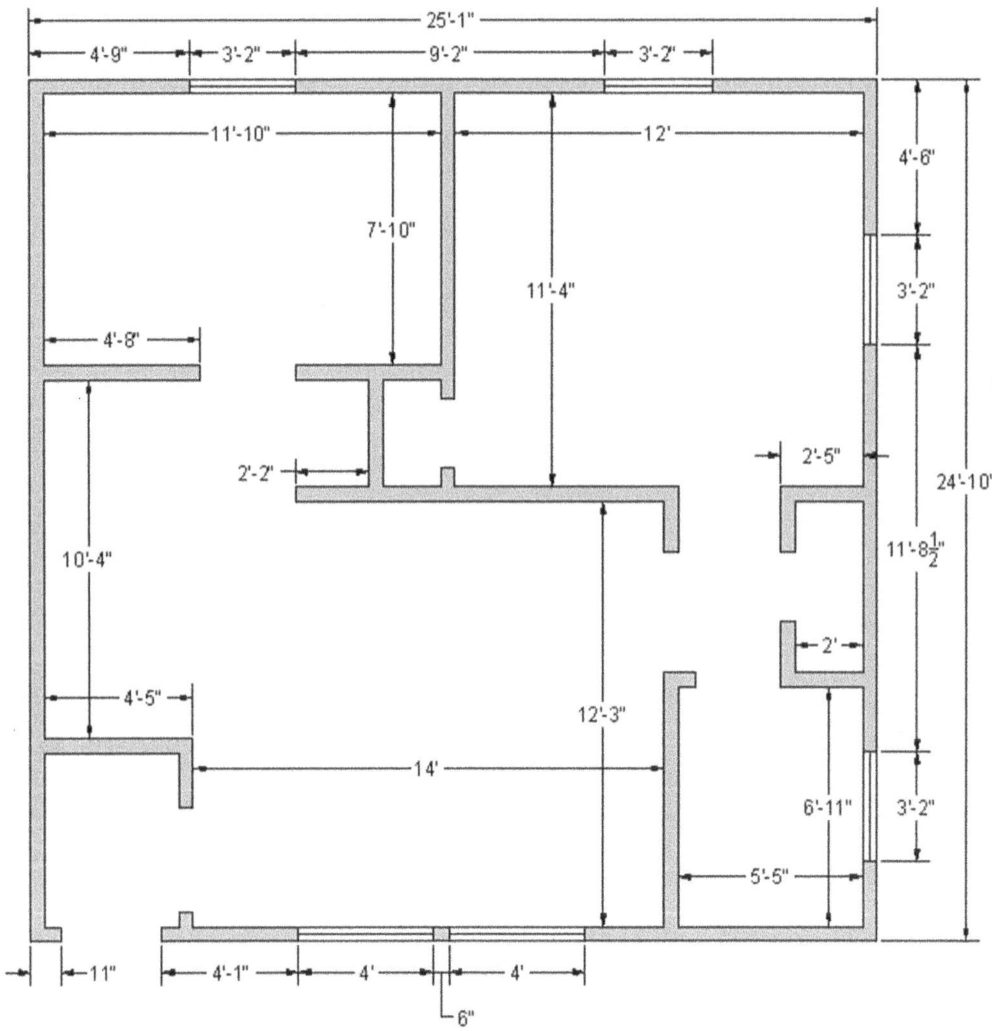

FIGURE 6.42 Adding dimensions to floor plan.

Summary

You should understand and know how to use the following concepts and commands before moving on to Chapter 7:

- Dimensions
  - Linear (horizontal and vertical)
  - Aligned
  - Diameter
  - Radius
  - Angular
  - Continuous
  - Baseline
- Leader and Multileader
- Arc length
- Jogged
- Jogged linear
- Ordinate
- Editing dimension values
  - ddedit
  - The effect of <>
- ddim command
- Dimension units
- Dimension font
- Dimension arrowheads
- Dimension overall size
- Center marks and centerlines

Review questions

Answer the following based on what you learned in this chapter:

1. List the 11 types of dimensions discussed.
2. What command is best for editing dimension values?
3. What is the difference between natural and forced dimensions?
4. How do you restore a natural dimension value when you have forgotten what it was?
5. List the four items that needed to be customized in our dimensions.
6. What command brings up the Dimension Style Manager?
7. What type of arrowhead do architects usually prefer?

Exercises

1. Retrieve Exercises 2, 3, 4, 5, and 6 from Chapter 3, Layers, Colors, Linetypes, and Properties, then set up and add all shown dimensions. (Difficulty level: Easy; Time to completion: 10 minutes/exercise.)
2. Retrieve Exercises 5, 6, 7, 8, and 9 from Chapter 4, Text, Mtext, Editing, and Style, then set up and add all shown dimensions. (Difficulty level: Easy; Time to completion: 10–15 minutes/exercise.)
3. Retrieve Exercises 5, 6, 9, and 10 from Chapter 5, Hatch Patterns, then set up and add all shown dimensions. (Difficulty level: Easy; Time to completion: 10–15 minutes/exercise.)
4. In a new file, set up the appropriate layer, colors, and linetypes. Then, draw and fully dimension the following mechanical part. Select Decimal as the dimstyle, with a precision of 0.00 and inch symbols (") for a suffix. (Difficulty level: Easy/Moderate; Time to completion: 35–45 minutes.)

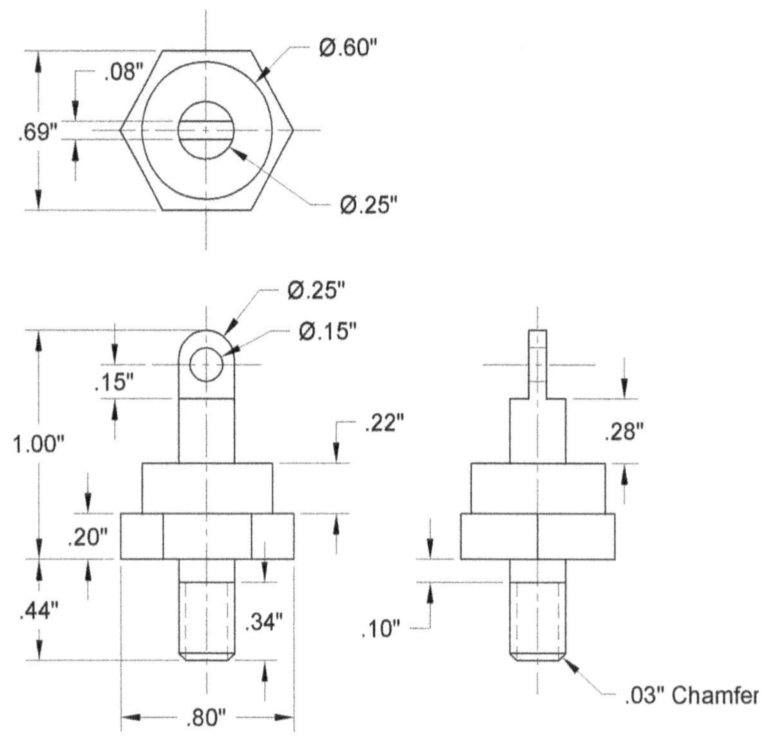

5. In a new file, set up the appropriate layer, colors, and linetypes. Then, draw and fully dimension the following mechanical part. Select Decimal as the dimstyle, with a precision of 0.00 and inch symbols (") for a suffix. Make careful use of mirror to minimize redundant work. All sizing is based on the smallest two circles and their position from an arbitrary centerline. (Difficulty level: Moderate; Time to completion: 35–45 minutes.)

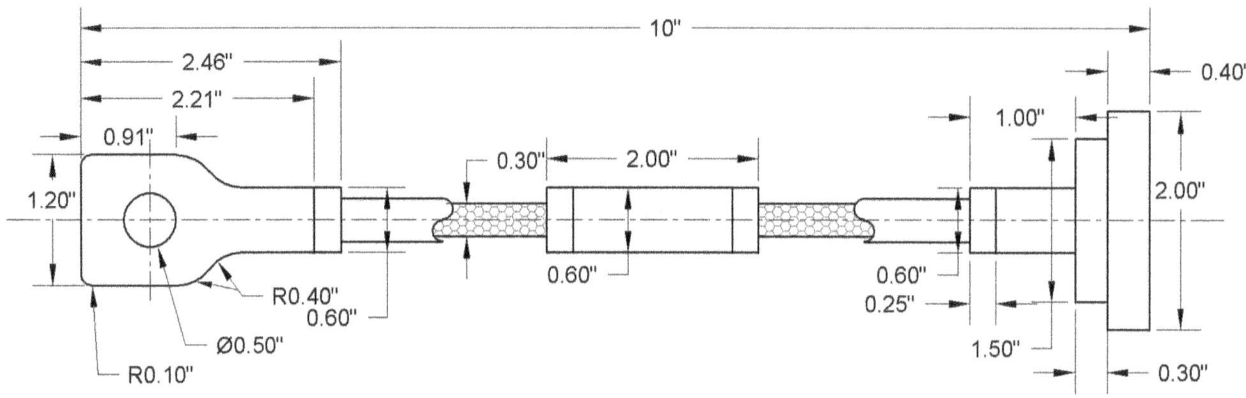

6. In a new file, set up the appropriate layer, colors, and linetypes. Then, draw and fully dimension the following mechanical part. Select Decimal as the dimstyle, with a precision of 0.00 and inch symbols (") for a suffix. (Difficulty level: Moderate; Time to completion: 35–45 minutes.)

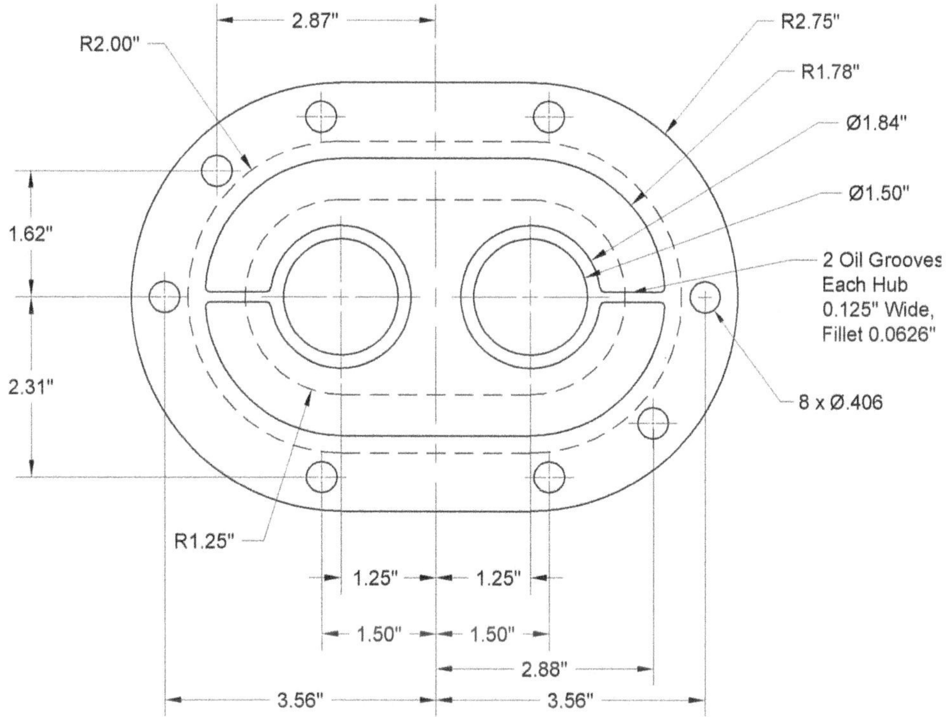

7. In a new file, set up the appropriate layer, colors, and linetypes. Then, draw and fully dimension the following mechanical part. Select Decimal as the dimstyle, with a precision of 0.00 and inch symbols (") for a suffix. (Difficulty level: Moderate; Time to completion: 35–45 minutes.)

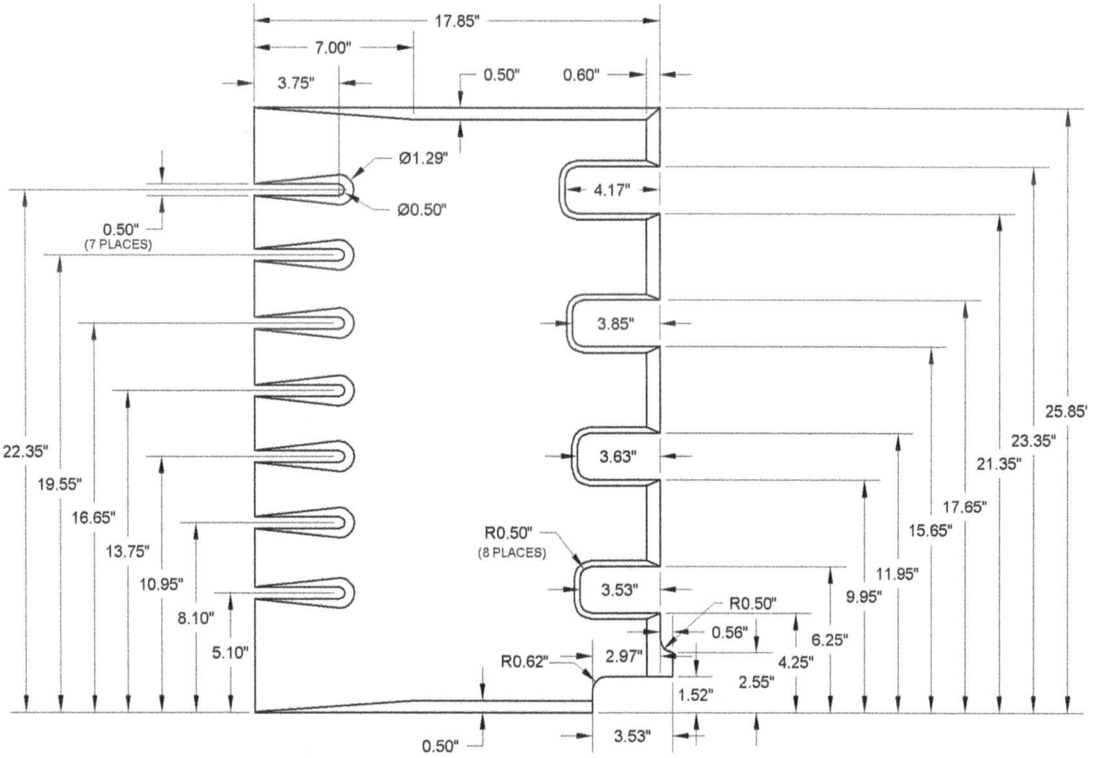

8. In a new file, set up the appropriate layer, colors, and linetypes. Then, draw and fully dimension the slightly simplified version of the bolt assembly seen on the opening of this chapter. Select Decimal as the dimstyle, with a precision of 0.00 and inch symbols (") for a suffix. Note that no dimensions are available for some internal parts of the bolt; you have to improvise. (Difficulty level: Moderate/Advanced; Time to completion: 60 minutes.)

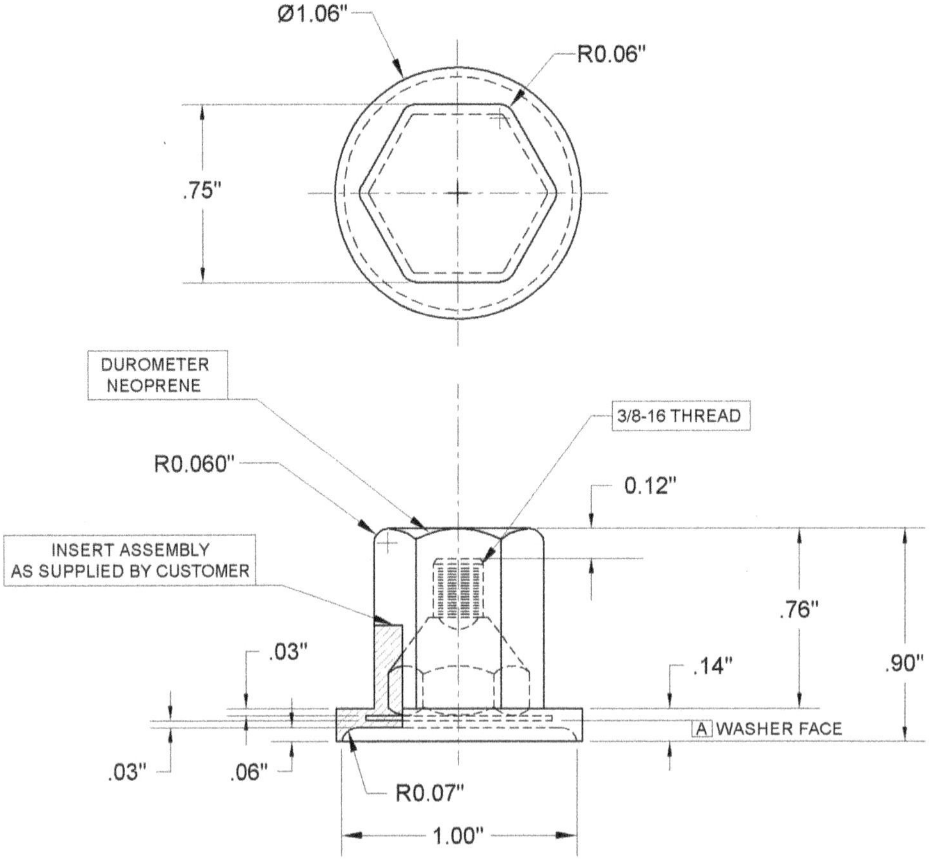

9. Open CH 5-Ex10 and continue drafting the foundation on to your existing wall section geometry (or start from a new file, setting up the layers as needed). Then set up the appropriate architectural dim style and mleader styles for a 1/2"=1'-0" printed drawing (1:24 scale) adding all of the dimensions and notes as shown. (Difficulty level: Moderate/Advanced; Time to completion: 30–60 minutes.)

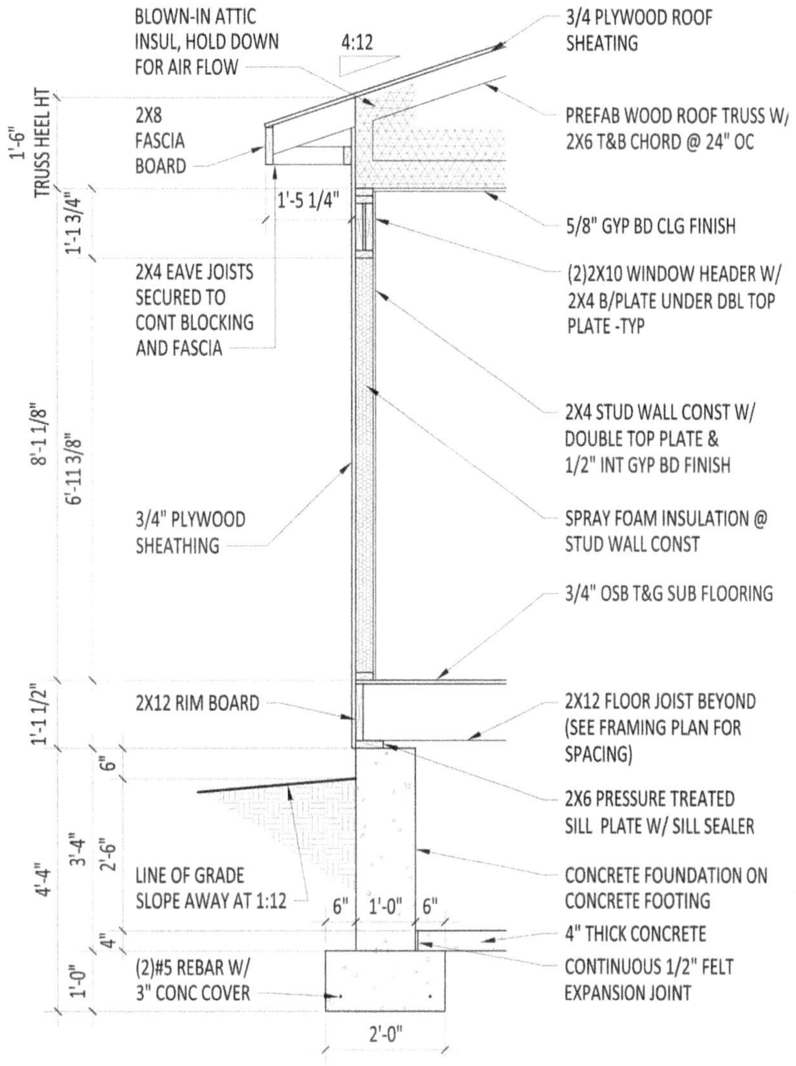

10. The final exercise of this chapter is a basic floor plan to keep your drawing skills sharp. It also features some electrical symbols, which you should get used to seeing, as they are standard in any electrical layout. Create one copy of each, and duplicate them throughout. We cover blocks for this purpose in the following chapter. The numbers inside the rectangles are the doorway dimensions. A few secondary dimensions are not given due to space limitations; you need to make a best estimate. Use proper layers and add text and dimensions as shown. (Difficulty level: Moderate/Advanced; Time to completion: 90 minutes.)

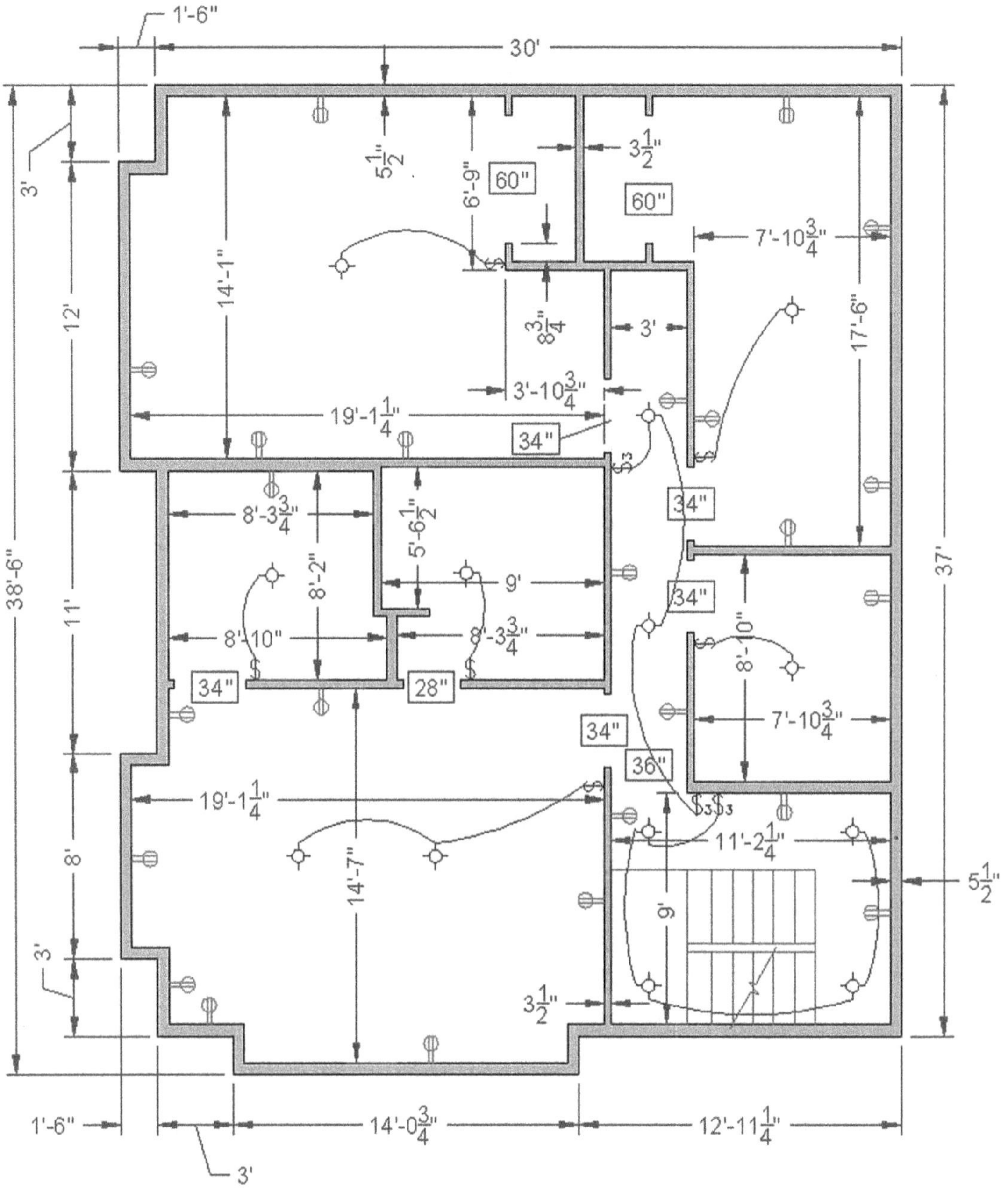

Chapter 7

Blocks, Wblocks, dynamic blocks, groups, and purge

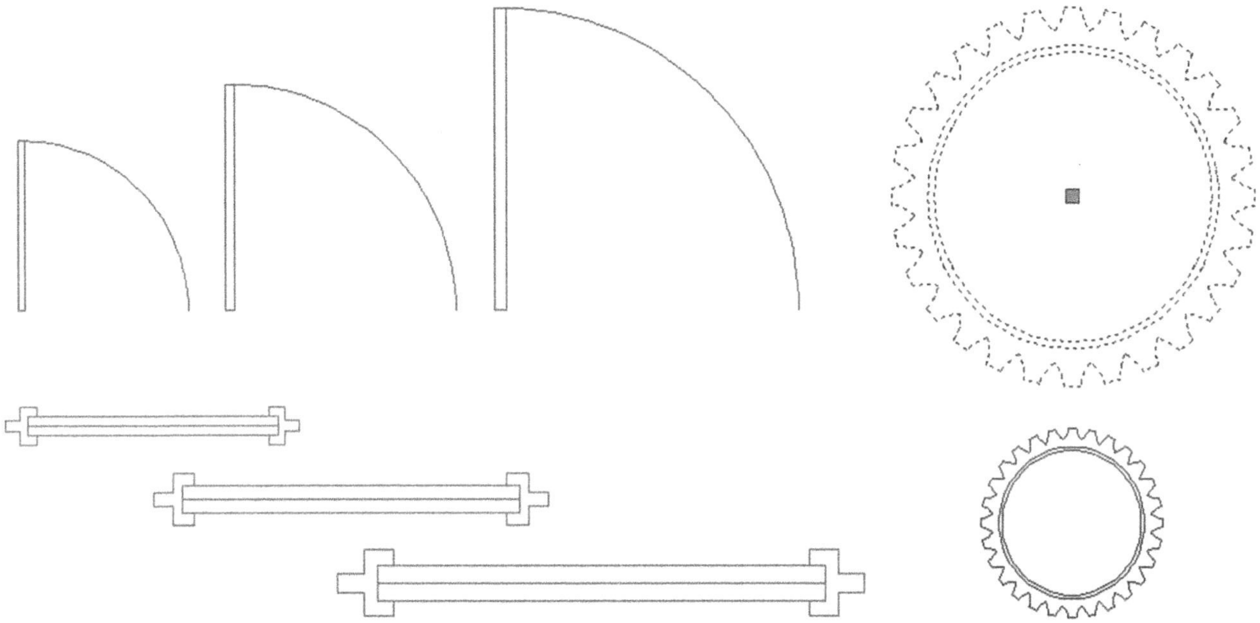

Learning objectives

In this chapter, we introduce the concept of blocks and discuss the following:

- Creating and working with blocks
- Retain, convert, and delete options
- Redefining blocks
- Inserting blocks
- Purge
- Creating and working with wblocks
- Creating and working with dynamic blocks
- Creating and working with groups

At the end of this chapter, you will be able to create blocks, wblocks, dynamic blocks, and groups as well as insert symbol libraries and purge your drawing.

Estimated time for completion of this chapter: 1–2 hours.

Up and Running with AutoCAD 2022. https://doi.org/10.1016/B978-0-323-89923-9.00007-X

7.1 Introduction to blocks

This chapter is about some very useful AutoCAD concepts that greatly simplify your workload. Blocks, wblocks, dynamic blocks, and groups are all members of the same family. They exist around the idea that objects can and should be grouped together, if possible, to allow for easier handling and storage for future reuse.

Let us formally define a block. It is a collection of objects that are grouped and held together under some identifying name. These objects can then be copied, moved, erased, and just about anything else, all as one unit. Think of blocks as electronic "glue" that binds the objects they contain to each other.

There are many benefits to this. If you have an office table with eight chairs around it, and you need to move them all around many times while optimizing a room layout, it is a lot easier to click on a block once than attempt to select all nine pieces using a Window or a Crossing or one at a time. Imagine an engineering example involving a simple gear with all its intricately designed teeth. Dozens if not hundreds of lines and arcs may make up the gear. It would be not just wise but mandatory to form them all into a single block.

There is, however, another, more compelling, reason for blocks. That is the concept of *reuse* and *symbol libraries*. Blocks can be saved and reused for future drawings that may need the same exact object. That way you never have to draw the same thing twice, an AutoCAD Golden Rule. A collection of these objects cataloged in some orderly fashion becomes a symbol library, an indispensable part of just about any designer's toolset. Likely, candidates for these libraries include doors, windows, furniture, bolts, standard parts, fasteners, and just about anything that can be used in more than one design or drawing.

Blocks versus the wblock command

Ideally, you are sold on the importance of blocks as we go through the specifics. There is confusion in industry about a block versus a wblock. The differences between blocks (objects) and the wblock (a command) are confusing, but they are also alike in many ways. The key difference is that blocks are saved *internal* to the file you are working on and wblock command creates an *internal to the current file block and external .dwg file*. As such, the main use of blocks is for grouping objects together to reuse in the current AutoCAD drawing. While they certainly can be moved between files via the clipboard Copy/Paste tool, this is not really what they are intended for. Over time you want to have a large file library of *.dwg* files that when inserted become blocks.

The wblock command is intended for creating symbol libraries "on the go." Wblocks are saved in the current drawing *and* to an external file, as an independent, stand-alone AutoCAD drawing, which you can open and work on if needed. The wblock command is ideal for creating a block to be used in your current drawing and saved to your library of .dwg files that you insert into other drawings as blocks.

An alternative approach to the wblock command would be to start a new drawing, create the objects you want to use as a symbol in other drawings, and save that file as a .dwg file without creating a block at all. Then go back to your original, main project file, and use the insert command and browse to your .dwg file to insert it. It will become a block upon insertion into the main project file. This is outlined in greater detail below.

Creating a block

To create a block, you first need something from which to make a block. Once you draw something, only three things need to be done:

1. Select the object(s).

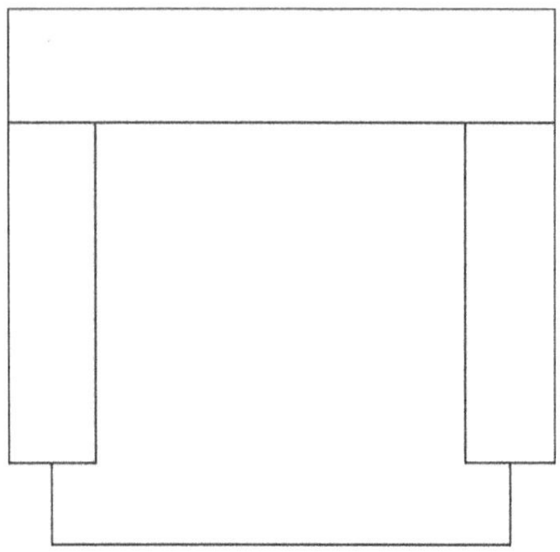

FIGURE 7.1 Basic chair symbol.

2. Give the block a name.
3. Select a base point.

Let us try this out. Draw a chair, as shown in Fig. 7.1. It is made up of three rectangles and three lines. Use accuracy and drawing tools as needed.

We need to now group these lines all together into one object.

Keyboard: Type in **block** and press Enter
Ribbon: **Insert** tab→**Create Block**

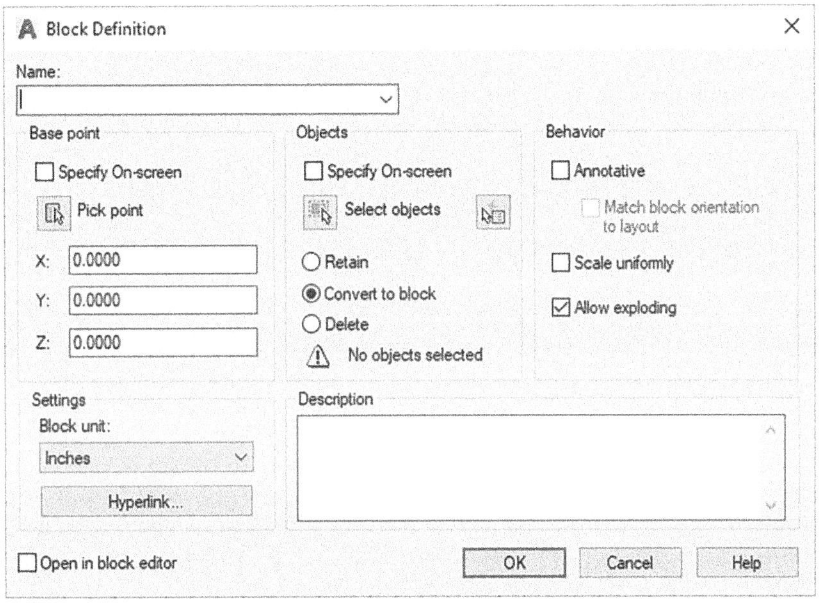

FIGURE 7.2 Block Definition dialog box.

Start the create block command via any of the preceding methods. The Block Definition dialog box appears, as seen in Fig. 7.2.

Step 1. Press the Select objects button in the middle column. Select all of the geometry of the chair and press Enter.
Step 2. Enter a name for this block in the top left field under Name.
Step 3. Pick a base point with the Pick point button at the upper left. Select a corner or midpoint somewhere on the object.

The significance of this becomes more apparent when we insert the object back into the file.

There are a few other items in the dialog box under Settings, Behavior, and Description, of course, but they are not usually needed (although feel free to enter a description). Annotative is covered in advanced chapters. The one setting of importance, however, is the Retain, Convert to block, and Delete choices under the Objects column near the select objects button. They are explained next and appear with almost similar wording in wblock as well.

Retain

This means that the block is created and stored in memory, but the original object (the chair you are looking at) remains in separate pieces, useful if you wish to make several blocks of the chair, one after the other, each slightly different yet based on the same original.

Convert to block

This option is what you are likely to use the most. It creates the block and stores it in memory and makes a block out of the original object. Use it when you are sure one block is all you need to make out of this one object and you have a need for it right away.

Delete

Here, the block is created and stored in memory, but the original is deleted from the drawing. This is useful when you are making a block for future use but have no need for it now.

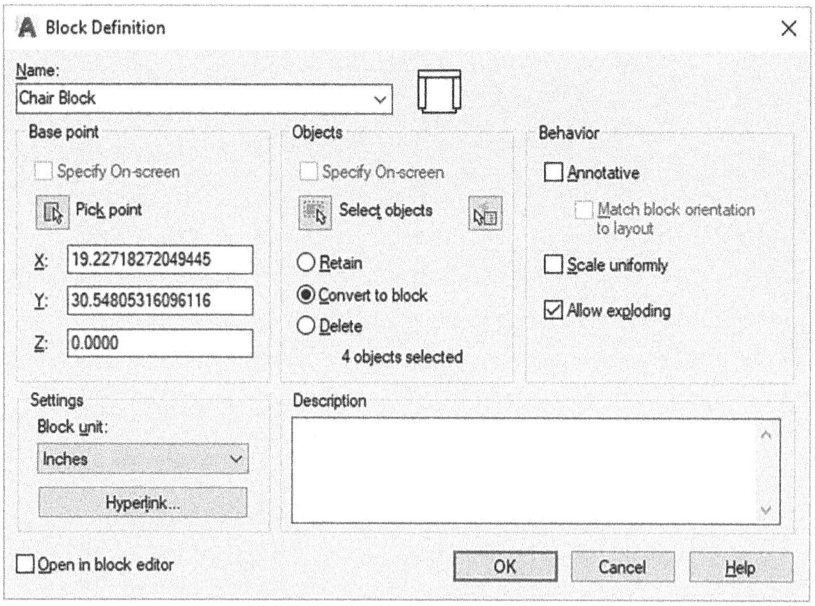

FIGURE 7.3 Completed block definition.

The common theme with all three of these choices is that, whichever one you choose, the block is created and saved to the memory of the current drawing for later use. The only real choice is what to do with the original. Finally, you are ready to click on OK. The Block Definition dialog box should look like Fig. 7.3. Notice the small preview window in the upper middle of the dialog box.

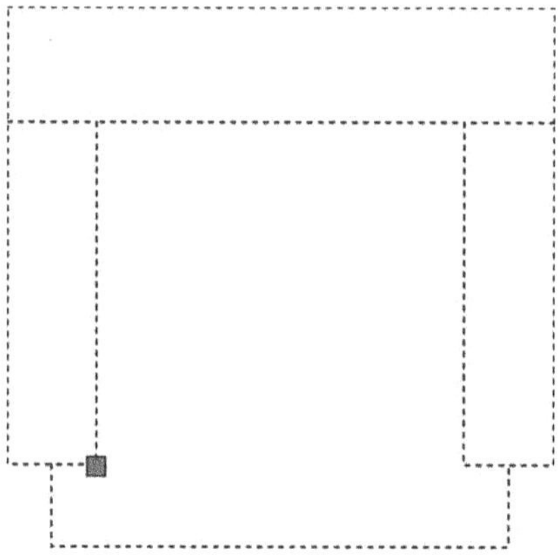

FIGURE 7.4 Chair as a block.

The chair is now a block, which can be proven by clicking on it once with the mouse. Notice that only one blue grip point appears (wherever you picked the base point; see Fig. 7.4). If you now erase it, the entire chair will disappear. Go ahead and erase it; we need to do this to introduce two new commands before we discuss wblocks.

7.2 Insert

As mentioned before, the process of making the block adds it into the memory of the file you have open. Then, you can insert it back into the drawing when needed.

Keyboard: Type in **insert** and press Enter
Ribbon: **Insert** tab→**Insert**

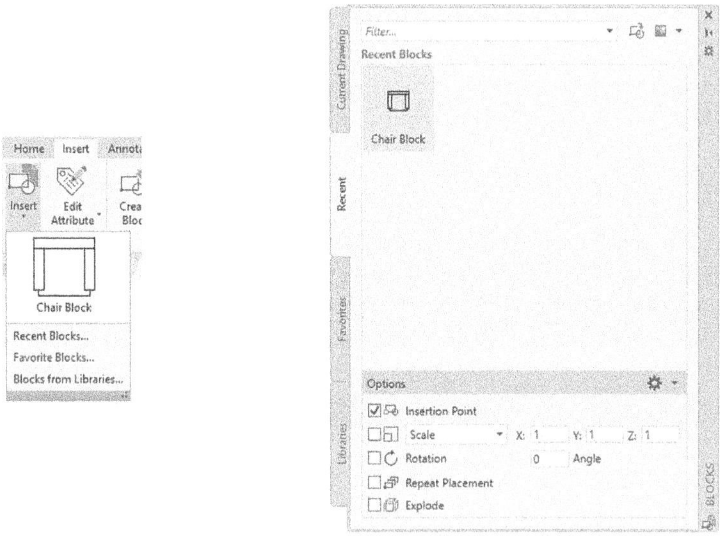

FIGURE 7.5 Insert Command from Insert Ribbon Tab or Directly to Palette by Typing Command.

Start up the command via any of the preceding methods. If you start with the button in the Ribbon you can select a block to insert from the box that appear below the tool button. If you select Blocks from Libraries… you can browse for other drawings to insert into this one. If you select Recent Blocks… (or start the command from the command line by typing *insert* or simply *i*) the Insert Palette appears, as seen in Fig. 7.5, which and was new for AutoCAD 2020. Notice the chair is cued up and ready to go, as it is the only block created so far.

Look over the various options near the bottom, such as Insertion point, Scale, Rotation, Angle, and Explode, but at this point, you need not do anything extra. Go ahead and single click on the Chair Icon then move your cursor into the drawing area; the chair appears attached at the insertion point to the mouse. The point of attachment was specified when you made the block. (Consider this point carefully when making blocks in the future to make using your block easier.) Click anywhere on the screen and the block becomes part of the drawing. This process can be repeated as often as necessary for any blocks saved in this drawing. Once inserted, a block can be modified (copied, rotated, moved, scaled, etc.) as needed. Note that you can always specify a precise point of insertion for the block by typing in the coordinates (e.g., 0,0,0).

There is something important to remember concerning redefining blocks. If a block is saved under a certain name and another block is created and saved under the same name, the first block is redefined, a sort of overwriting process. Sometimes, this is bad, as you are destroying the old block and replacing all previously inserted blocks with the new definition. Sometimes, however, this is desirable, such as when you are looking to update the appearance of the block. When there are many blocks in the drawing under the same name, all are updated in this manner for better or worse.

There are some new features and enhancements to block insertion since AutoCAD 2020. Notice all the side tabs in the Insert Palette. The Current Drawing tab will display only the blocks available in the current drawing. The Recent tab will display a list of blocks you have inserted recently from the current drawing *and* other drawings you have recently been using. The other drawing tab will allow you to browse for other drawings (look for the ellipsis, three dots near the top of the palette) to insert that drawing in its entirety as a block (look for the asterisk near the name) or pick out individual blocks from with that drawing. These are powerful tools that make adding content to your drawings easier and organizing a block library easier.

7.3 Purge

Between block creation and the wblock concepts is a good time to take an intermission of sorts and discuss a command called *purge*. Purge is a very useful tool to "clean up" your drawing of unwanted and often unseen geometry and data. It is especially useful when it comes to unused blocks lurking in the background memory of your drawing, hence its inclusion here.

Conceptually, here is the idea. A complex AutoCAD drawing often grows in file size and becomes weighed down, like an iceberg floating in the ocean. Much like the mountain of ice, mostly hidden under the surface, what you see on screen may be only a small fraction of what is present in the background data of the file.

The reason why becomes apparent when you think about the drafting process for a moment. If you create 50 layers for anticipated future use but end up needing only 40 of them, the other 10 are uselessly hanging out in your layer dialog box. Same thing with unused fonts or linetypes (remember, loading *all* of them but using only a few). Layers, linetypes, dimensions styles, and text styles really take up little space, but with blocks and wblocks, it is a whole different story.

Blocks are miniature drawings all to themselves, and while many are just simple doors or window symbols, others can be quite large, such as entire furniture sets or appliances in an architectural layout. Worse yet, these blocks rarely come alone; they bring friends, lots of them. Complex multistory sprawling building layouts feature not dozens but *hundreds* of blocks. They all take up room in a file, ballooning it up to a rather huge size.

If all these blocks are needed, then fine; this is a necessary evil. But if not—say, a type of chair was deleted from the specs—they need to be removed. Erasing them all is easy enough, but are you done? Are they all gone for good? Unfortunately, no. They may not be visible, but they are still there weighing down your file, waiting to be used as part of that fie, slowing down computer performance and just about everything else. This is where purge comes into play.

Purge gets rid of items that are not actually used in the drawing file and permanently deletes them. You can use the command to selectively purge items or purge everything all at once. The three main points to remember are the following:

- Purge does not get rid of objects that are visible and present in the drawing—it is not an erase command.
- Purge is permanent (unless you immediately undo it), so be careful of what you purge. If you anticipate eventually needing something, do not purge it out.
- Purge is a good practice at the end of a project, when it is ready for archiving.

Let us give the command a try.

Keyboard: Type in **purge** and press Enter
***Ribbon:* Manage Tab→Cleanup Panel→Purge**

First, erase the chair from your screen. Then, start up the purge command via one of the methods shown in the command matrix, and the dialog box in Fig. 7.6 appears.

As you can see from looking at the list of items in the dialog box, quite a few items are eligible for purging. Some were already mentioned, like Layers and Linetypes, while others you may not have heard of, like Materials and Shapes. Either

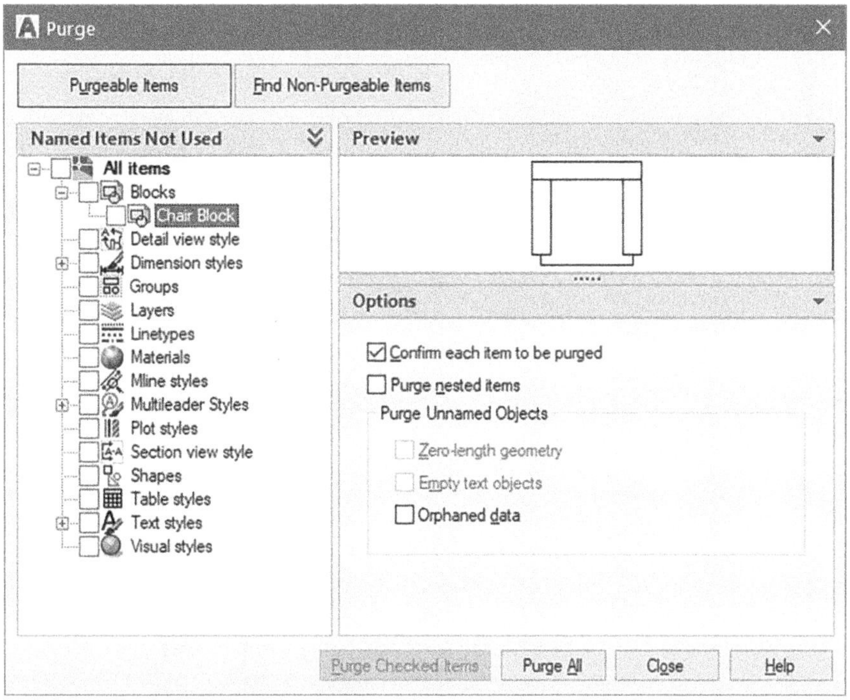

FIGURE 7.6 Purge dialog box.

way, what is important is whether the item has a plus sign next to it. If it does, you can click on the plus sign and expand the folder to see what specific items are in those general categories. In our case, you see the chair block under the Blocks category. You may or may not have a few other plus signs. If you do, then whatever may be there will just get purged along with the chair block.

The actual process of purging is easy. You can drill down into each line and select each item one at a time (usually done only when you want to selectively purge) or you can click on Purge All at the bottom. If purging all, be sure the Purge nested items box is checked and the one above it is not, as seen in Fig. 7.6. If you need to confirm each item to be purged, you may not be ready to purge. When all the items are purged, there are no more plus signs and the buttons gray out as well. Press Close after you are done and that is it.

You can also try the Find Non-Purgeable Items button in the Purge dialog box. There are times when you are looking to remove a layer or other item from the drawing and it is difficult to find it due to nested blocks (blocks within other blocks) dimensions styles, or layering. This is a way to peek behind the curtain to see where the items you are trying to get rid of are hiding.

Experienced users purge often keep the drawing cleaned up as they go along, keeping little useless junk in the file unless it is necessary for use later in their drafting or design process. Purging is certainly something that could be done prior to saving the file for the day or emailing it, and definitely before archiving it for storage and record-keeping. Be careful though, and do not get rid of needed items. If in doubt, do not purge the items.

7.4 Wblocks

Pretty much everything said about blocks applies to the wblock command as well, and so this section is a bit of a review. The central idea to keep in mind is that a wblock is a command to create a block internal to the current drawing and write an external *.dwg* file for later use in other drawings. As such, you need to tell AutoCAD where to put the *.dwg* as you make it. Let us try this out. Clear your screen and make sure all items from the previous exercise are purged. Now go ahead and draw another furniture object, perhaps a sofa this time.

You can start up this command via the Ribbon's Insert tab or just by typing in `wblock` and pressing Enter. Once you do that, the dialog box in Fig. 7.7 appears. For the screenshot in this figure, a specific path has not been selected, but you can do that by clicking on the button with the three dots, … to the right of File name and path, and browsing around until you find where you want the file to go. Be sure to also change the name of the file at the end of the string, so it does not just say

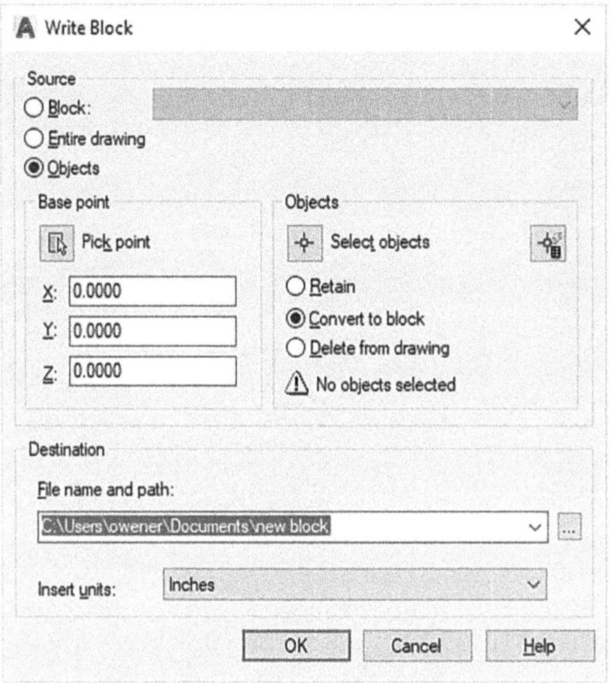

FIGURE 7.7 Write Block dialog box.

"new block." Then, select the sofa and pick one of the three options to Retain, Convert, or Delete from drawing the original. When done, press on OK and the block is created in the current file and a *sofa.dwg* file is written where you indicated in the path.

It is important to review again the three options you have for dealing with the original drawing objects selected for the wblock command. The last of these three is worded slightly differently. Here they all are again:

Retain

This means that the new block is created and stored in memory of the current file and a *.dwg* file is written to the specified location. The original objects (the sofa or chair you are working with) remain in separate pieces. This is useful if you wish to make several blocks of the sofa, one after the other, each slightly different yet based on the same original.

Convert to block

This option is what you are likely to use the most. It creates a block of the currently selected objects for use in the current drawing and writes a *.dwg* file to the specified location.

Delete from drawing

Here, the block is created and stored in memory, but the original is deleted from the drawing. This is useful when you are making a block for future use but have no need for it now.

Inserting blocks from other files or entire *.dwg files*

Start a new drawing and use the Insert command to insert your sofa into a new drawing. You may see the sofa in the Recent tab on the Blocks Palette. Ignore that for now and go to the Other Drawing tab. Using the Browse button (look for the ellipsis, three dots near the top of the palette) locate the location where you saved your sofa.dwg. When you find it, double-click the icon or single click it and then click Open in the dialog box. An icon is displayed with your *sofa.dwg below it. The asterisk before the file name * means that you are inserting an entire drawing file. That is fine in this case as the objects will become a block when inserted. Single click the icon and move your cursor into the drawing to place your sofa block.

Any *.dwg* file can be inserted this way, from simple symbols to entire complex drawings. Generally, a *.dwg* file containing a single object like a chair, sofa, sink, arrow, etc., that will be used as a block should not have a block in it. The

.dwg file becomes a block when it is inserted. If a *.dwg* file contains a block of the same name within that file (*sofa.dwg* has a block within it called *sofa*), AutoCAD will give you an error when trying to insert that file.

7.5 Dynamic blocks

Dynamic blocks were introduced back in AutoCAD 2006. Like many new features, they are not a radically new concept. Rather, they are an evolution of the regular block and wblock. The idea here is to be able to modify the block in response to different design conditions. A simple example is of a door. If you insert a block of a 3′ door, opened at 45 degrees, and realize that you need a 3′-6″ door, shown in a closed position, you can change that in place without redefining the block or inserting another one. You can even go a step further and, using the visibility option, change the block to something else; in other words, not just its orientation and size, but what it is completely, akin to having many blocks in one.

So what we are essentially doing here is adding a level of intelligence and automation to the standard block and wblock concept. We use this for scaling, stretching, mirroring, and otherwise modifying the block's sizing and orientation. We also use this intelligence to add menus to a block, and it allows you to select many different versions of the block as dictated by your design needs. AutoCAD provides options for almost every conceivable design intent, so let us take a look at this from the basics on up by first considering very simple parameter changes.

Actions and parameters

To create a generic dynamic block, you need to modify a regular block by adding parameters and actions. Three things make up a dynamic block: the *geometry* itself, at least one *parameter*, and at least one *action* associated with that parameter. A parameter defines the features of the geometry, and the action defines the modification you would like to act on that parameter. The result is a block that has one or more special grip points, each defined with a certain modifying property (e.g., stretch, rotate, scale). The block can then be modified in place as needed. For a simple example, recreate or retrieve our existing chair block, and follow the procedure outlined next.

Keyboard: Type in **bedit** and press Enter
Ribbon: **Insert** tab→**Block Editor**

Regardless of which method you use to start the command (you can also just double-click on the block itself), you see the dialog box in Fig. 7.8. Select the block you want to work with, which in this case is the only one available.

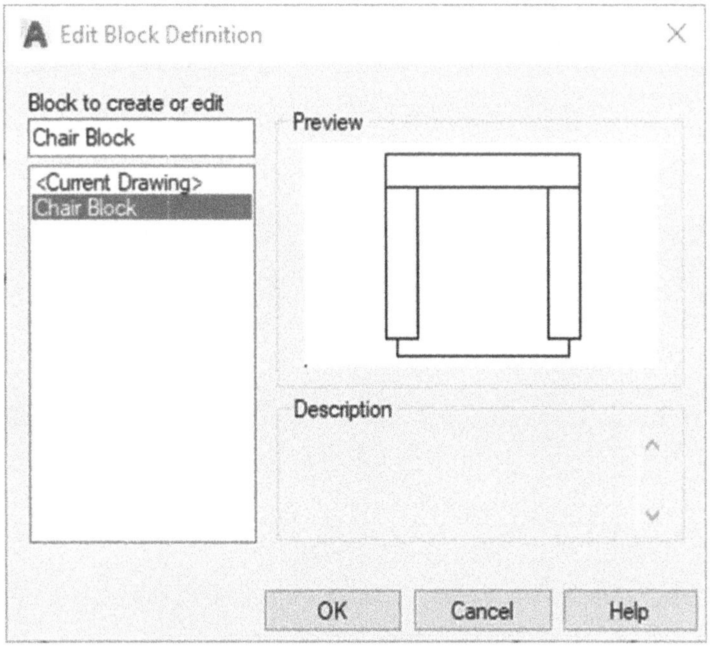

FIGURE 7.8 Edit Block Definition.

After you select Chair Block, you are taken to the main editing screen, called a *Block Editor* (recognizable, unless changed via Options, by its medium gray background), shown in Fig. 7.9. This is where the settings are applied.

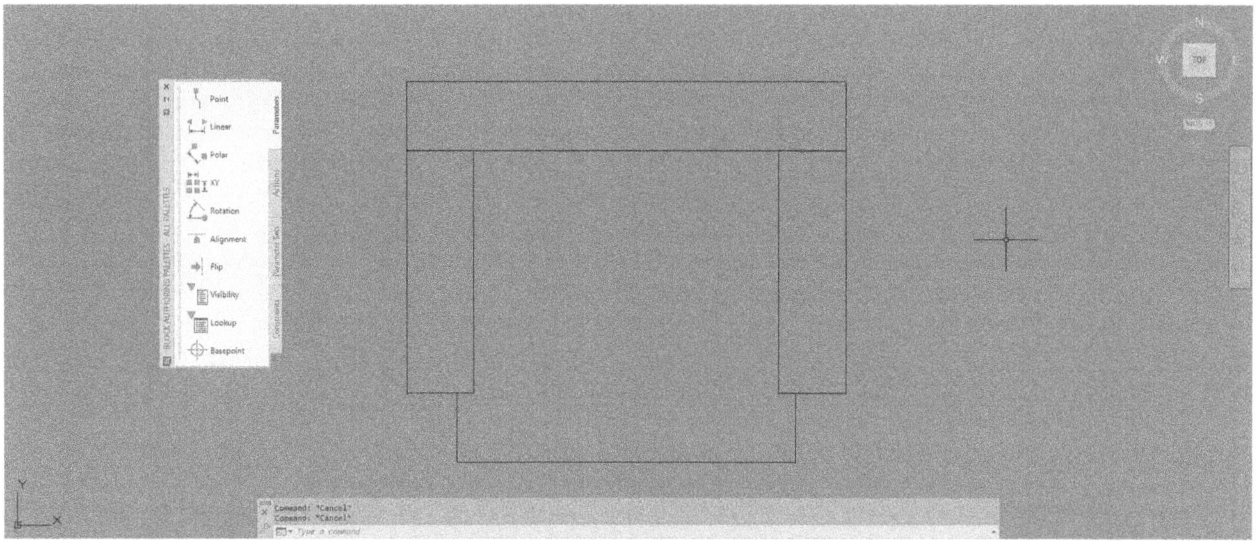

FIGURE 7.9 Dynamic Block Editor.

A Block Authoring palette also appears. Sets of parameters and actions are available from this palette, and the block can be set up to have any of those. When done, press Close Block Editor (top right of screen) to return to the regular drawing space. So what can we do here? Well, let us add two actions and parameters to the chair that give it some scaling and rotational abilities.

Step 1. Open the Parameter tab on the palette and pick the Linear parameter. That, after all, is the way scale works, by linearly scaling up an object. First, pick the upper left corner start point, then pick the upper right corner endpoint of the chair (like adding a horizontal dimension). The result is shown in Fig. 7.10. Note that a yellow exclamation point is seen as a reminder that no action has yet been added to this new parameter.
Step 2. While you are there, add a rotation by running through the required Base Point (lower left), Radius (any value), and Angle (30 degrees) prompts. The result is shown in Fig. 7.11.
Step 3. Now, you can add actions associated with those parameters. Pick the Actions tab and pick Scale. Then, select the Parameter and the Objects (the chair), as prompted. Repeat the same steps for the rotation. The result is shown in Fig. 7.12. Notice the new (faded) scale and rotate symbols, and the disappearance of the exclamation point.

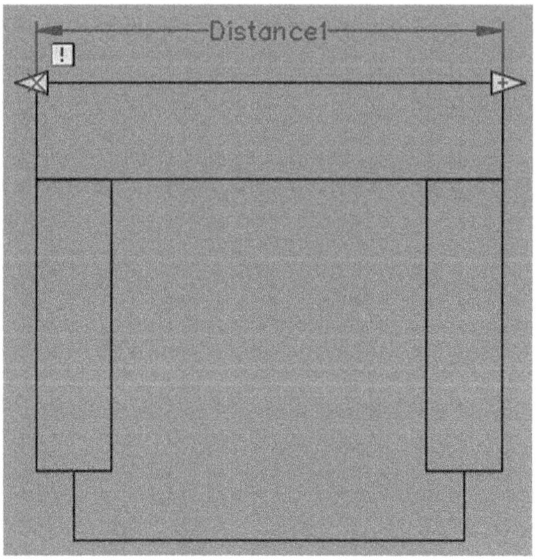

FIGURE 7.10 Dynamic Block, Step 1.

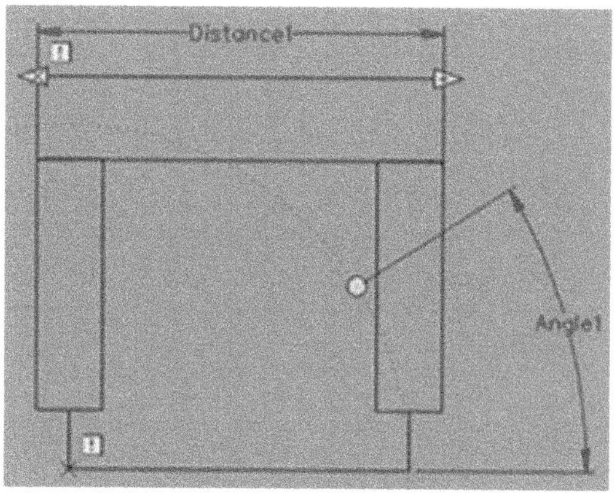

FIGURE 7.11 Dynamic Block, Step 2.

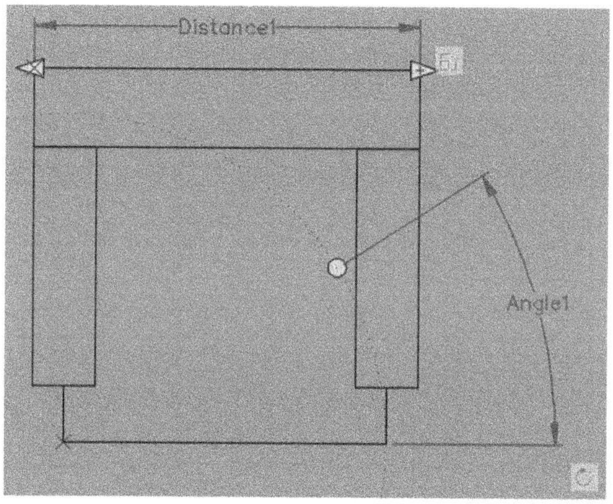

FIGURE 7.12 Dynamic Block, Step 3.

Close the Block Editor (button at upper right of the Ribbon) and you are prompted to save your work, as seen in Fig. 7.13.

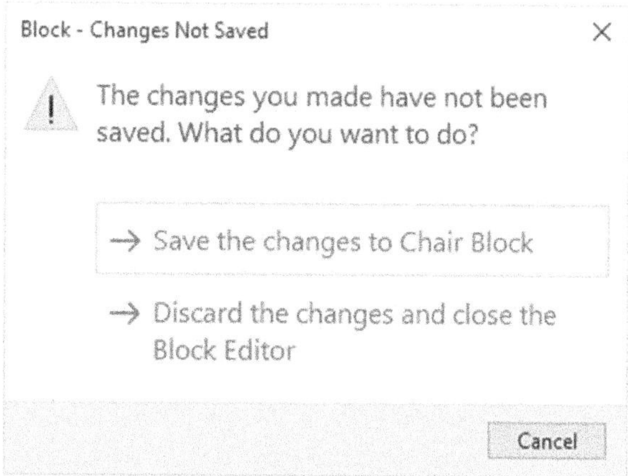

FIGURE 7.13 Save changes prompt.

Finally, you are back where you started. Click on the new block and notice the dynamic grips, as seen in Fig. 7.14. You can use these to change the size and rotation of the block, as seen in Fig. 7.15 with scaling.

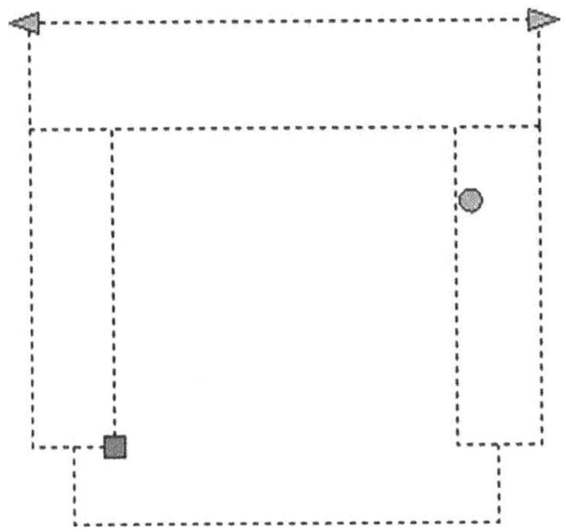

FIGURE 7.14 Dynamic Block grips.

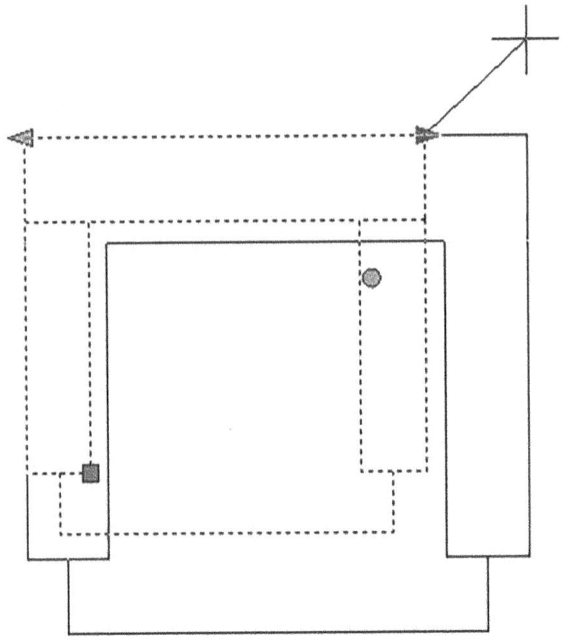

FIGURE 7.15 Dynamic Block modification.

This procedure features the absolute basics of creating a dynamic block. Let us try this again, with a door this time, and discuss a few more features. Create the door shown in Fig. 7.16, and call the block Door_36. Be sure to add in the door swing, the hatching, and the door jams. The dimensions are only for your reference.

Go into dynamic block edit mode and let us add the Flip action. This allows you to reverse the opening of the door after it has been inserted into place.

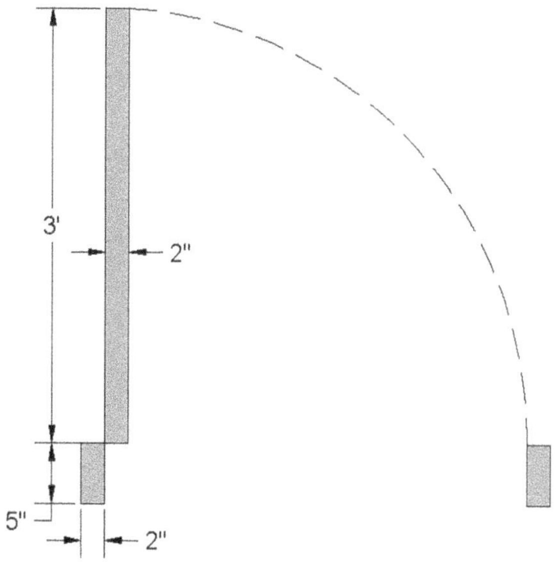

FIGURE 7.16 36″ Door.

Step 1. Add a horizontal reference line across the bottom from the left side of the door to the far right end of the door swing.

Step 2. Add the Flip parameter using the midpoint of the reference line and tracing upward using Ortho. Position the Flip state1 somewhere near the line. You can rename this text string to Flip Left-Right or something similarly descriptive if you wish, using the change properties palette.

Step 3. Add the flip action by selecting the icon, then the parameter, then all the objects. Fig. 7.17 shows what you should have on your screen. You can go ahead and erase the horizontal guide line at this point.

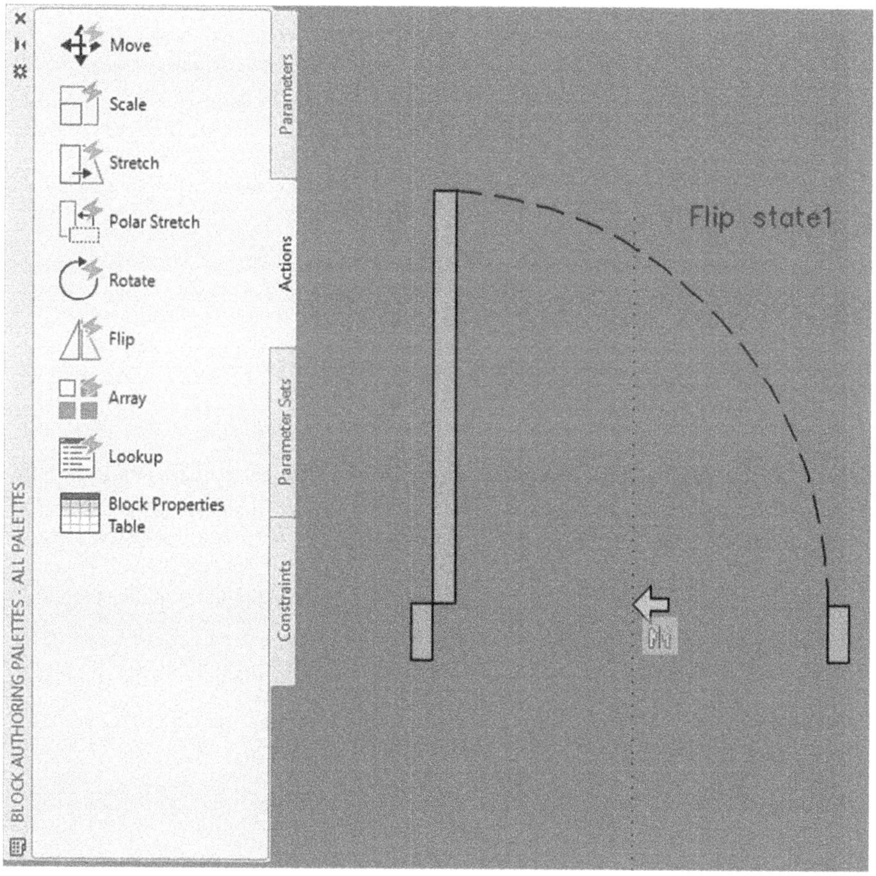

FIGURE 7.17 Flip parameter and action.

Step 4. As a final step, press the Test Block button at the upper left of the Ribbon. This takes you into an intermediate area (still in block edit mode) where you can test out what you have done. In our case here, click on the door and you see a blue arrow pointing to the left at the bottom. Click on that and the door's swing flips around (Fig. 7.18).

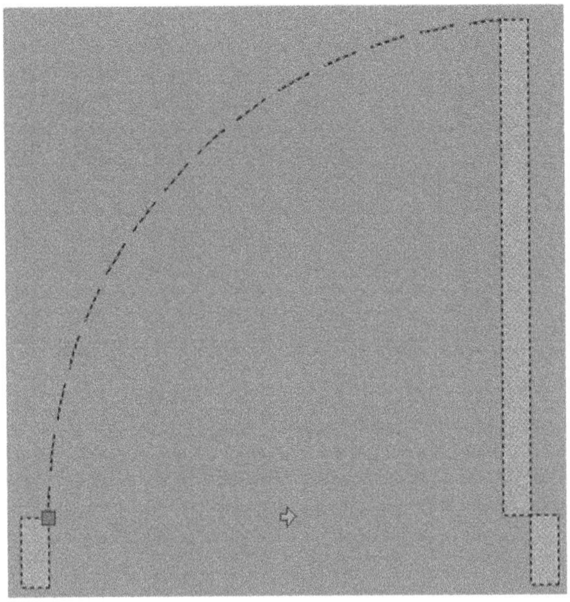

FIGURE 7.18 Flip executed.

You can also set up the flip so the door opens "downward" and not "inward," as it does now, of course. Your reference point would then be the midpoints of the two door jamb blocks. Try that as an exercise.

Another useful feature you can try is called *Value Set*. Add a linear parameter and action to the door (exactly like Distance 1 in Fig. 7.10). Then, while Distance 1 is selected, go to the Properties palette, Value Set category, and for the Dist type on the left side, select List on the right. Just below that for Dist value list, enter a few incremental values greater than 36″ (the current door size), such as 48, 60, and 72″, into the Add Distance Value dialog box. After you complete this step and go to test the door by clicking on the scale grip, you see a series of tick marks appear and you can increase the sizing only incrementally, as opposed to arbitrarily. There is quite a bit more to dynamic blocks, of course, and you can combine many features to create new ones.

Visibility

The final item to cover with dynamic blocks is a very useful feature called *Visibility*. This option is a favorite among quite a few designers because it allows you to combine virtually unlimited variations of a block into one, then simply select what you want via a drop-down menu. It can take a while to set up the numerous blocks and set the visibility states, but the payoff in the end is significant.

Here is the basic idea. You make a block by combining a number of other blocks (or even nonblock geometry by itself) by putting them one on top of the other, usually with a common reference point. You then create named visibility states—with each state representing a version of the block you wish to see when you click that name. You then set all but the relevant block as invisible. The process gets repeated until you process all of the geometry in such a manner. The result is one large block that contains numerous sets of geometry, but only one set is visible at a time. To pick which is the visible one, you select from a drop-down menu, which appears when you click on the block. There, you see all of your visibility state names, as you click each one the relevant geometry appears and all others disappear. Let us give this a try with a very simple example.

Step 1. Create five shapes and make a block out of each—a square, a circle, a hexagon, a triangle, and an ellipse—as seen in Fig. 7.19.

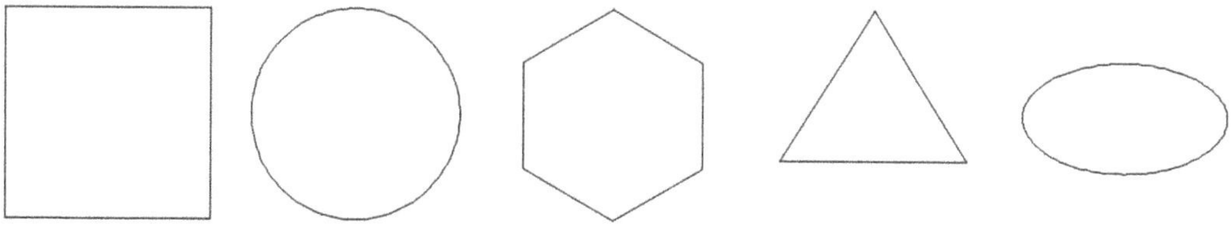

FIGURE 7.19 Basic block shapes.

Step 2. Put all the shapes on top of each other, centering them at the top. Make a block out of all of them calling it Main_Block. Then go into the dynamic block editing mode via bedit or just double-clicking on Main_Block. You should see what is shown in Fig. 7.20.

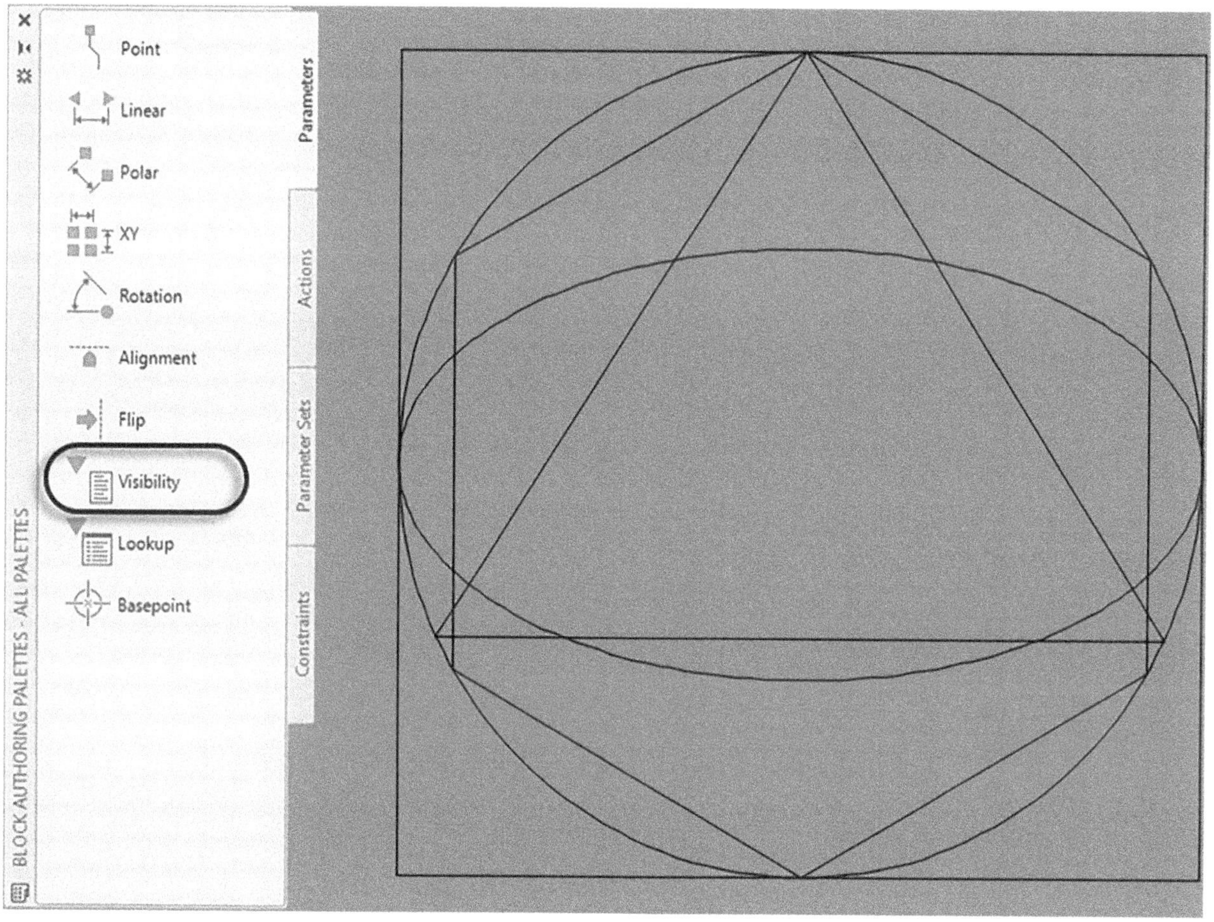

FIGURE 7.20 Combined shapes in bedit.

Step 3. Now add the visibility parameter (highlighted with a black rectangle in Fig. 7.21) by clicking on the Parameters tab. AutoCAD asks where to locate it. Select any convenient spot next to the shapes, as seen in Fig. 7.21.

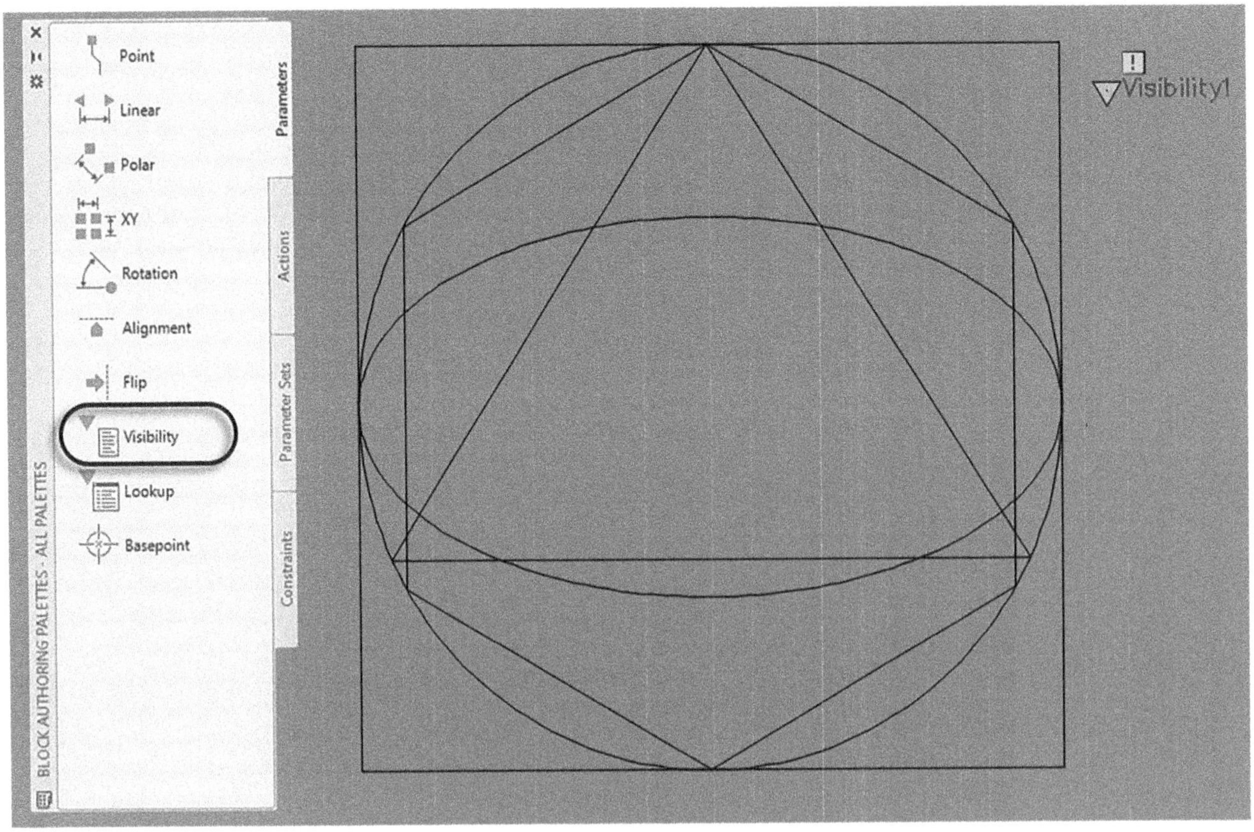

FIGURE 7.21 Visibility parameter added.

Step 4. Select the Visibility States button in the Ribbon's Block Editor tab, Visibility panel (Fig. 7.22). The Visibility States dialog box appears. Rename the existing state to Square and add the other four via the New... button (leaving any options as default in the dialog box). The result is seen in Fig. 7.23. Press OK.

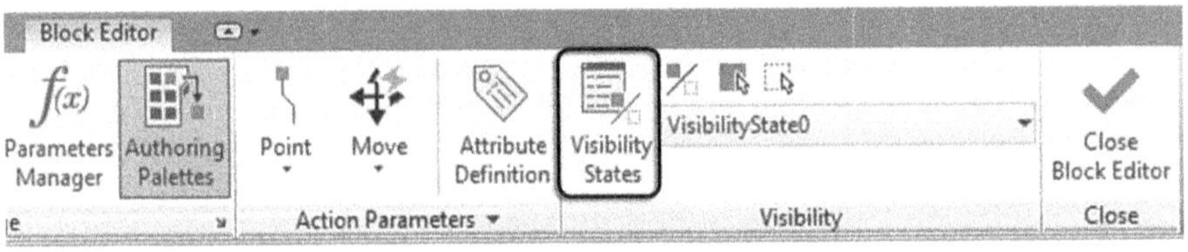

FIGURE 7.22 Visibility States.

Step 5. You now need to actually set the invisibility parameter of each visibility state. Go up to the Ribbon and find the drop-down menu just to the right of the Visibility States button (it can be seen in Fig. 7.22). Drop it down and

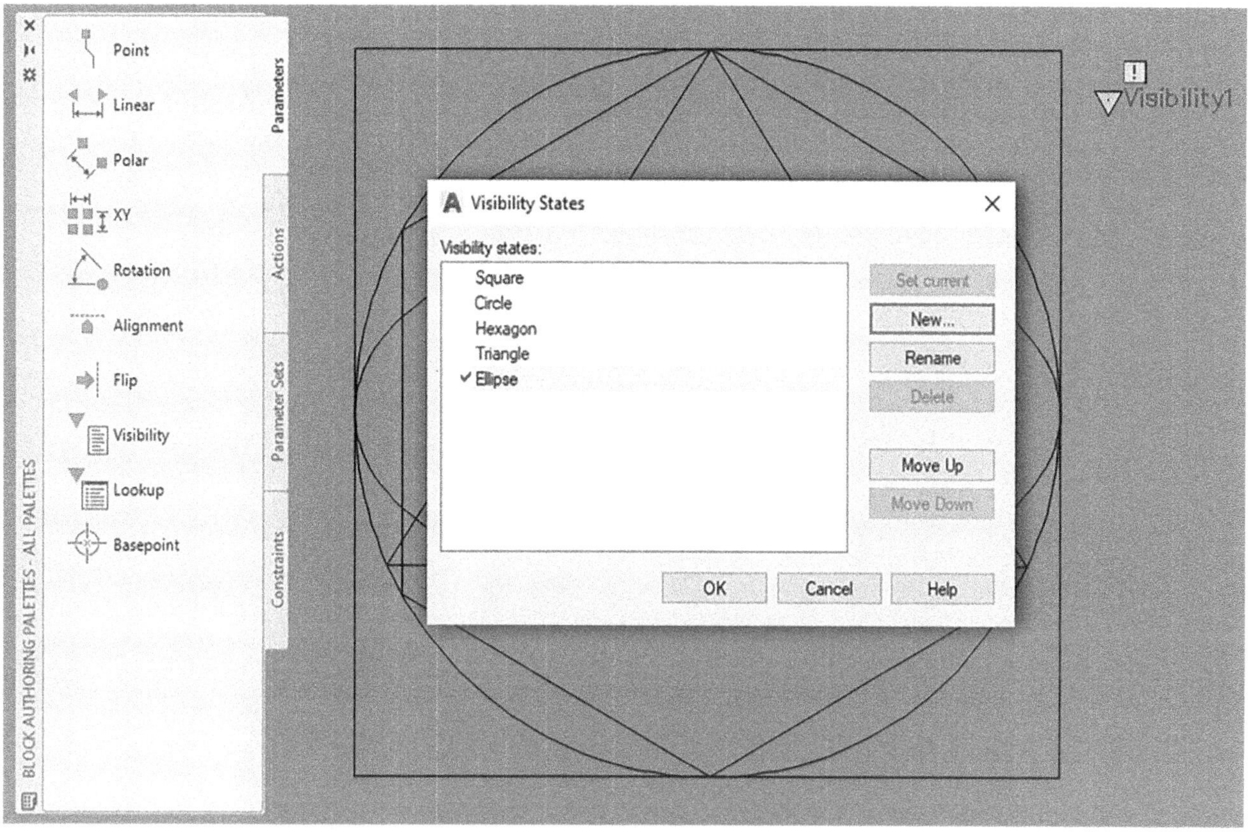

FIGURE 7.23 Visibility States completed.

double-click on the Square state. It is now the current visibility state, and it is here that you need to make all the other shapes invisible. To do that, press the button just above the drop-down menu called *Make Invisible* (it looks like a faded square with an arrow and can also be seen in Fig. 7.22). AutoCAD says Select objects to hide:. Go ahead and pick *all* of the shapes *except* for the square. The selected shapes all fade to a gray color, as seen in Fig. 7.24.

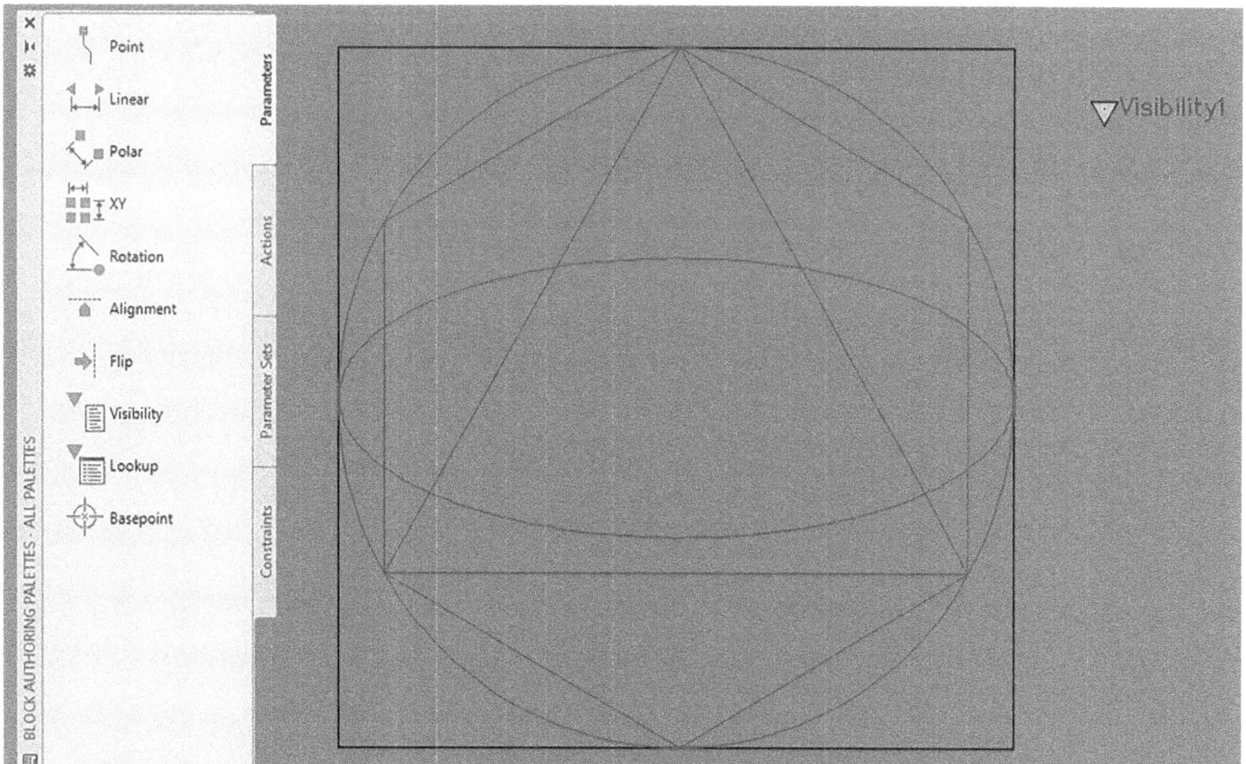

FIGURE 7.24 Square visibility state.

Step 6. Repeat this process for the other shapes by going again to the drop-down menu and double-clicking on the next shape on the list (the circle). All the shapes reappear, and now you have to use the Make Invisible button to select *all* of the shapes *except* the circle. This is why it was remarked that it may take a while if you have several dozen or more blocks embedded in one. The effort is well worth it, however.
Step 7. When you completed setting all of the visibility states, test them out via the Test Block button all the way on the left of the Ribbon's Block Editor tab. A new tab opens; go ahead and click on the block. Fig. 7.25 shows the first visibility state (square) with the drop-down menu revealing the rest of the choices. Click through each one to test each visibility state. When done, exit the test tab and save the dynamic block. It is now ready for use.

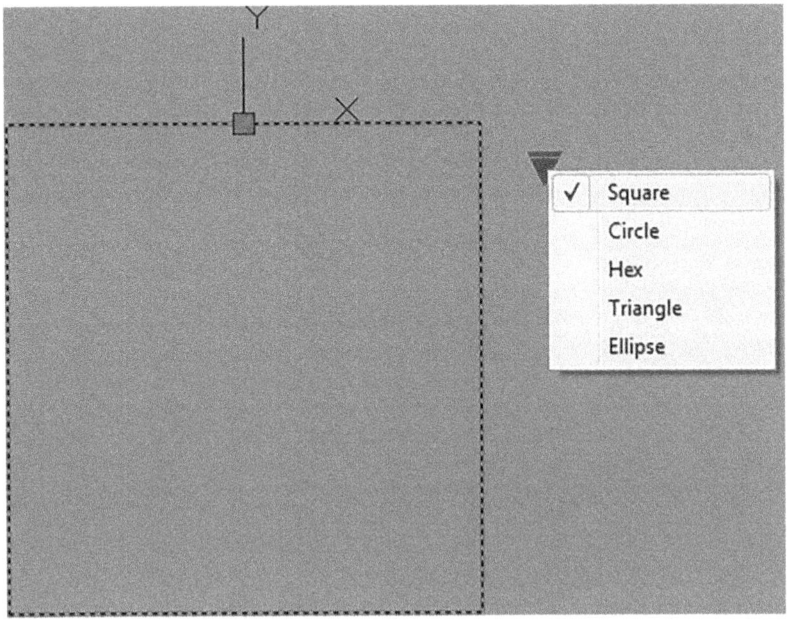

FIGURE 7.25 Visibility state menu.

This of course was a very basic example, but the process for more complex, realistic parts is identical. Keep them as blocks, stack them one on top of the other, create and set visibility states, save, and use. Here is a real-life example from the world of electrical (power) engineering. Fig. 7.26 is a top view of an actual relay switch commonly found at power substations.

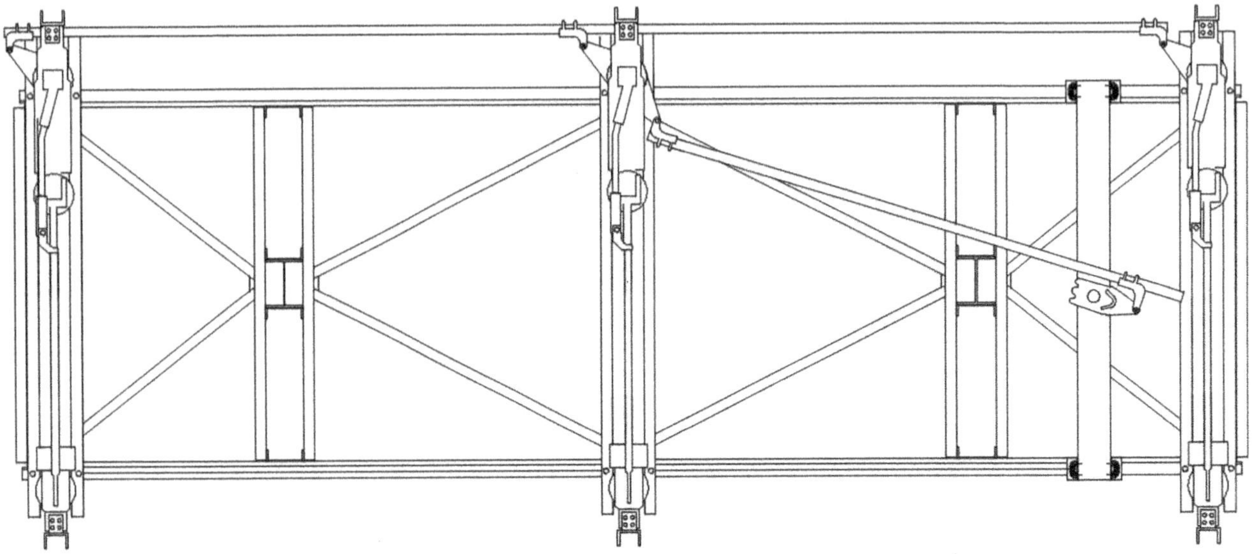

FIGURE 7.26 Relay switch.

As it happened in this case, a number of switches were at the disposal of the electrical engineer, all of them essentially the same in their basic function but very different in how they looked. It made sense to create one main block of them and just select what was needed for a particular job. That way the designer need not keep track of numerous blocks in a library. Such was the approach to drafting the switch in Fig. 7.26, as seen in the next image (Fig. 7.27). Notice the menu of available visibility state alternatives, with "Disconnect – Vertical Break SS20220 (B-42)" as the currently visible one.

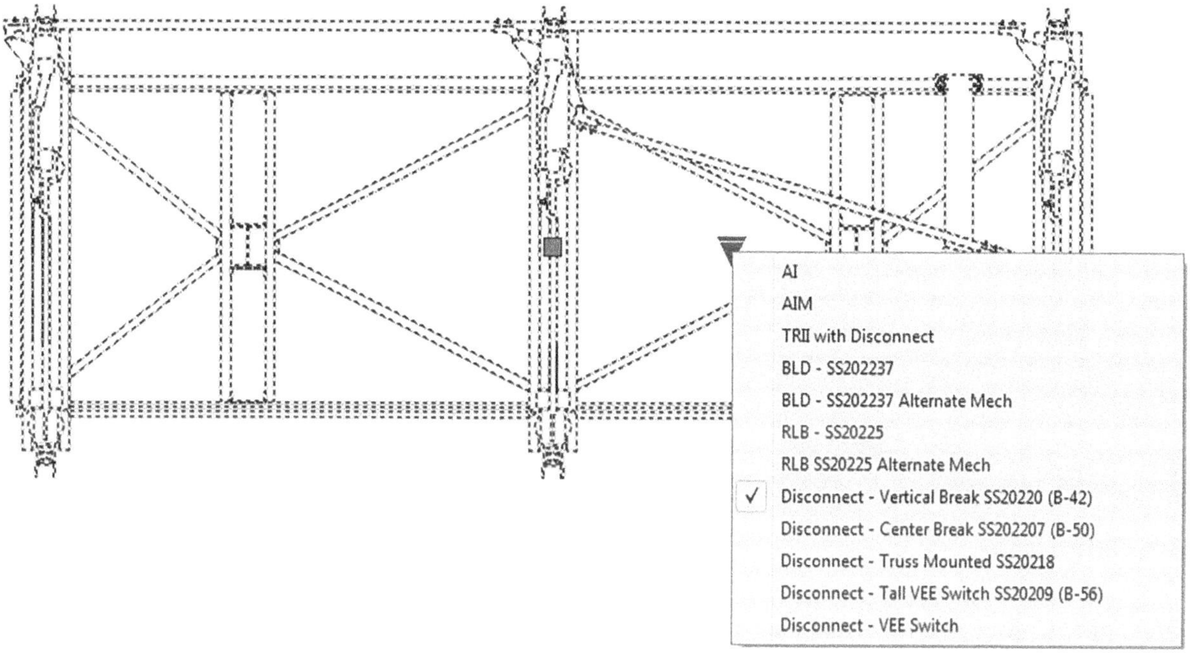

FIGURE 7.27 Available visibility states.

Figs. 7.28 and 7.29 show two other versions of the same switch. As you can see, they are complex blocks and quite different from one another but all sourced from the same file. Be sure to practice and get well versed in setting up these types of dynamic blocks, visibility states are very useful in actual design work!

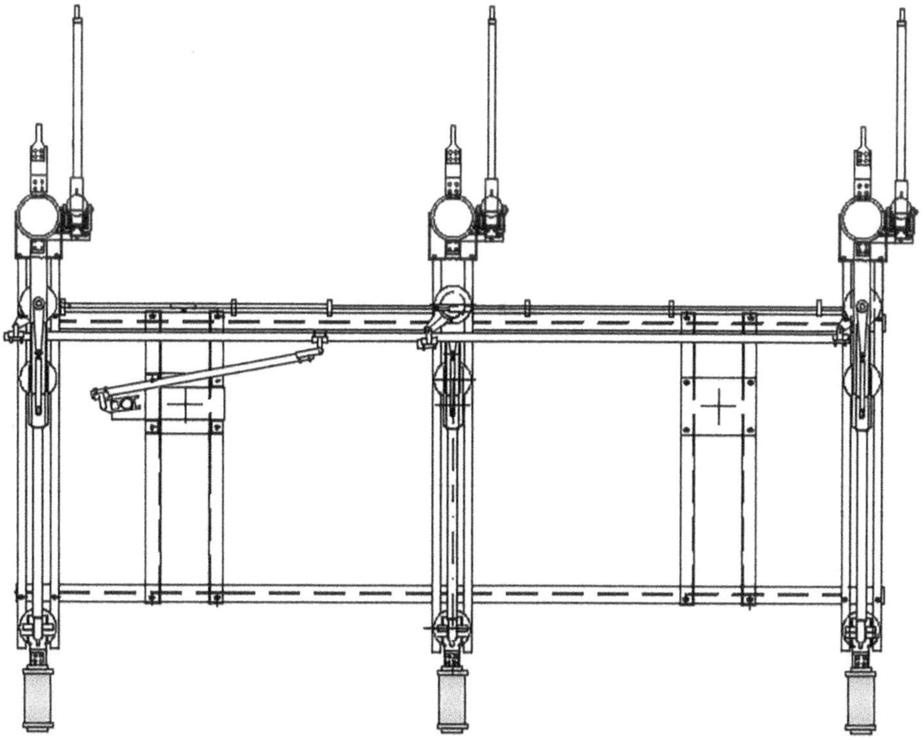

FIGURE 7.28 Alternate visibility state 1.

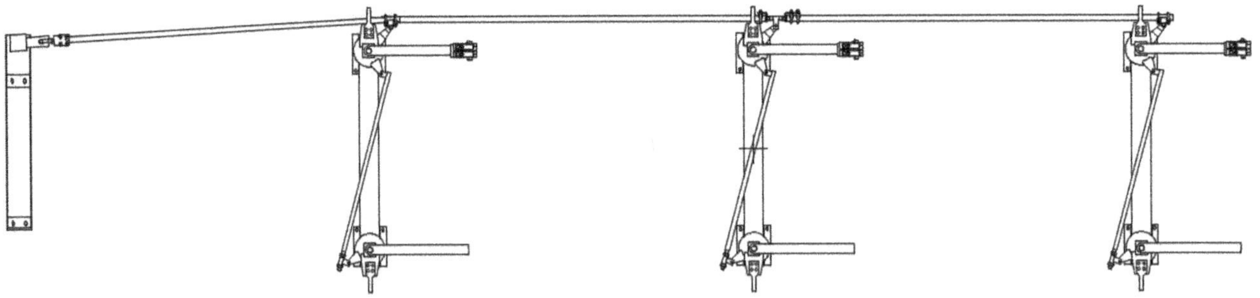

FIGURE 7.29 Alternate visibility state 2.

So, how do dynamic blocks fit into the big picture? Are they used as often as they could be? The results are mixed. Clearly, they are a big step forward for blocks and the built-in intelligence is very useful. However, one potential problem for long-time users is that, if they already have an extensive library of nondynamic blocks in use, there may not be any real incentive to rework them into dynamic blocks. This is an effort that sees more payoff if done from the beginning of block and block library creation. Some experienced users may not want to upgrade their blocks and long-standing libraries if they have been in use and working fine for all these years. It is worth the effort, however, as dynamic blocks, by allowing multiple variations on a single block, reduce the sizes and complexities of these libraries.

The bottom line for you as a student, however, is that it is a good skill to pick up, so try to explore further this interesting topic on your own. Chances are that, in the future, these will be the main types of blocks you will run into out there in the industry.

7.6 Groups

A group is a useful, but often overlooked, command. It is, perhaps, the easiest way to collect objects together for easy modifications like copy, rotating and moving. It presents some advantages (but also some disadvantages) over the familiar block, as shown next.

Like a block, a group can

- be given a name,
- bind a collection of objects together,
- be moved around, copied, and rotated.

Advantages of a group are these:

- A group can have objects added and subtracted from it.
- A group can have individual objects (and their properties) edited.

Disadvantages of a group are these:

- A group cannot be transferred between drawings (for "local" use only).
- An update to one group does not update other groups with the same name.

You can use the command line or Ribbon (Fig. 7.30) to create and manage the groups, but the classic Object Grouping dialog box, which has been around forever, is still available as one of the options and is discussed later.

FIGURE 7.30 Group Panel on Home tab of Ribbon.

To try this command, first draw a random collection of objects, as seen in Fig. 7.31, then follow the steps shown.

Keyboard: Type in **group** and press Enter
Ribbon: **Home** tab→**Groups** Panel→**Group**

Step 1. Begin the group command via any of the preceding methods.

- AutoCAD says: `Select objects or [Name/Description]:`

Step 2. Press `n` for `Name`.

- AutoCAD says: `Enter a Group Name or [?]:`

Step 3. Type in a descriptive name: Sample_Group in this case. Note that spaces in the name are not allowed.

- AutoCAD says: `Select objects or [Name/Description]:`

Step 4. Select all the objects.

- AutoCAD says: `Group "SAMPLE_GROUP" has been created.`

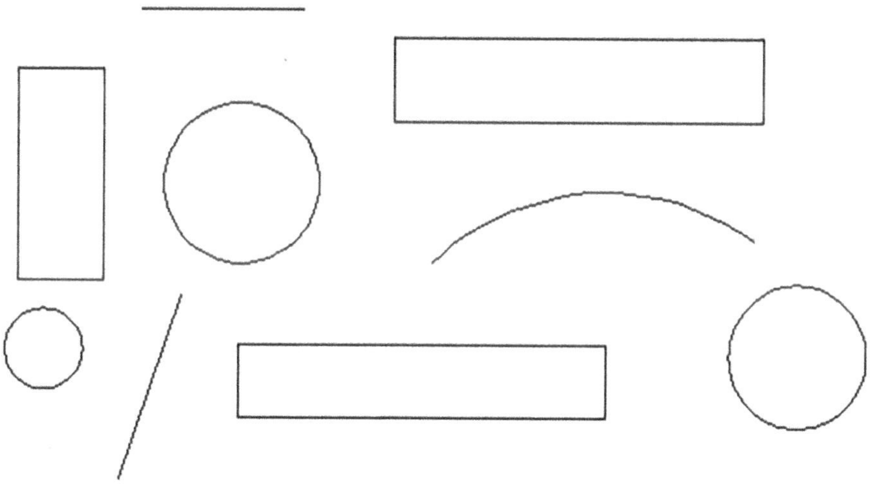

FIGURE 7.31 Random objects for grouping.

The group is now completed, and you can see results by clicking on any object (Fig. 7.32). Note how there is a rectangle around the objects and a single grip in the middle. You can now manipulate all the objects together as one unit.

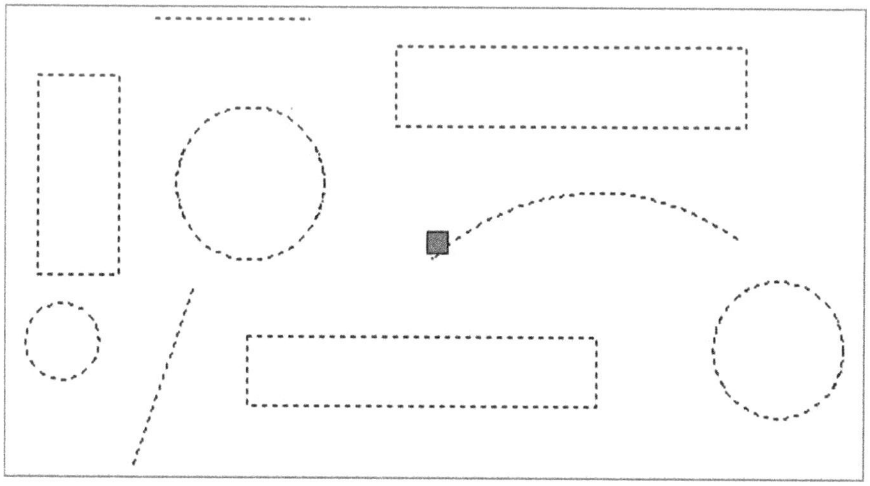

FIGURE 7.32 Completed group.

Let us take a look now at some of the features of a group and some additional functionality. One property you can take advantage of right away, and the main reason to use a group, is the ability to alter the properties of the objects in the group (or the shapes of the objects themselves). Double-click on any of the shapes, and a palette called *QuickProperties* appears, as seen in Fig. 7.33.

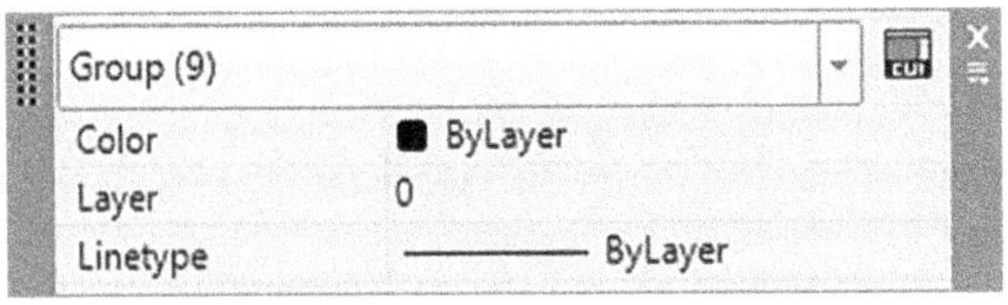

FIGURE 7.33 Group QuickProperties.

Drop down the top menu and you see a breakdown of the group, object by object. There are three rectangles (polyline), two lines, one arc, and three circles, for a total of nine objects. Select Circle (3) in the menu, and the palette shows only the properties of the circles. Go ahead and change the color to red and observe the result (Fig. 7.34).

You can also work with individual objects in the group by clicking on the lowest right icon on the Group panel of the Home tab of the Ribbon, called *Group Selection On/Off*. You can do the same by typing in pickstyle and entering 0 when prompted. Pickstyle is a system variable, and more is said about system variables in Chapter 15, Advanced Design and File Management Tools. For now, go ahead and modify any of the shapes, and when done, press the same icon or type in the same system variable (entering 1 this time) to get back to the grouping. Fig. 7.35 shows the editing in progress.

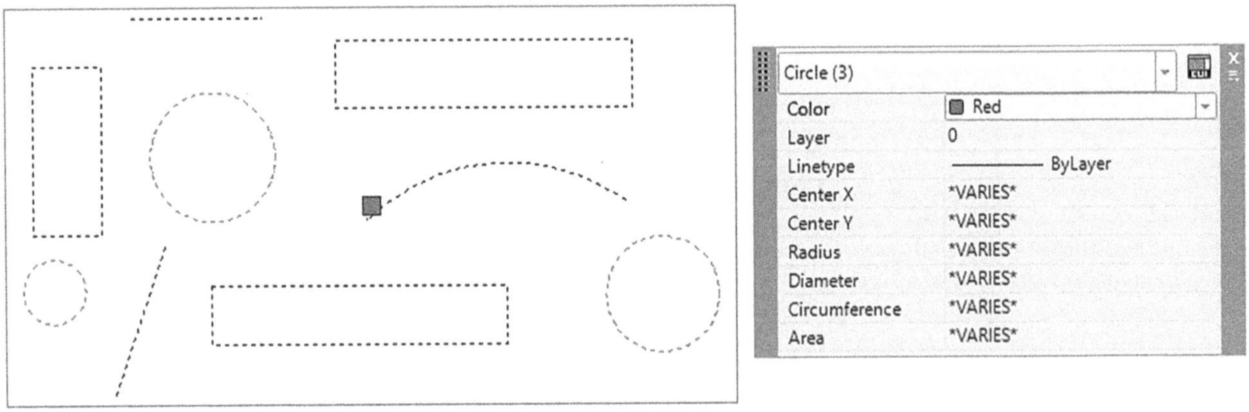

FIGURE 7.34 Color change to circles in group.

Look at the other upper right icon on the Group panel, called *Ungroup*. As you would expect, it removes the grouping from the objects. Note that exploding the group does *not* delete the group itself but rather just explodes the objects inside of it—only the rectangles in this case.

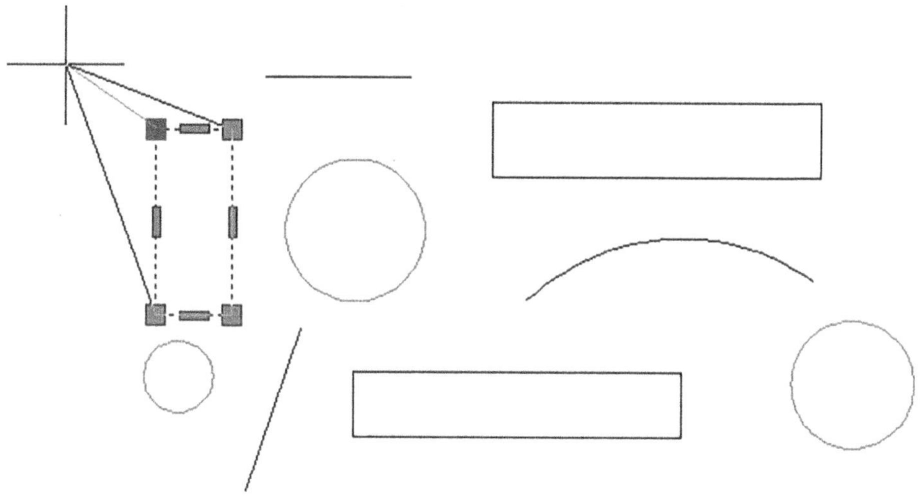

FIGURE 7.35 Group object editing.

The next icon to the right (with the ±), called *Group Edit*, allows you to add an object to or subtract an object from the group. Go ahead and try this on your own by deleting a rectangle via the Remove objects option and adding a new object (a polygon in this example) via the Add objects option. An example of the result is shown in Fig. 7.36.

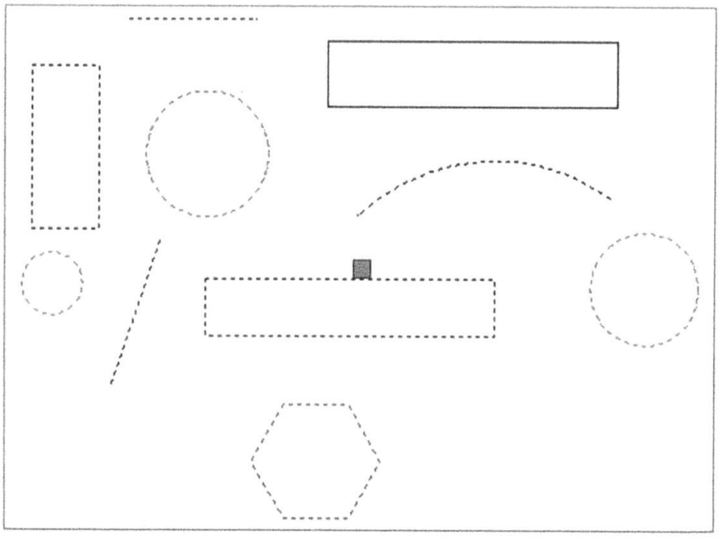

FIGURE 7.36 Object added (polygon) and another subtracted (rectangle).

Finally, it is necessary to mention the legacy Object Grouping dialog box (Fig. 7.37). It serves two purposes here. One is to provide you an alternate way of managing your groups. The second is to familiarize you with the process if you end up using an older version of AutoCAD, prior to the 2012 release, where this was the only way to create and manage groups.

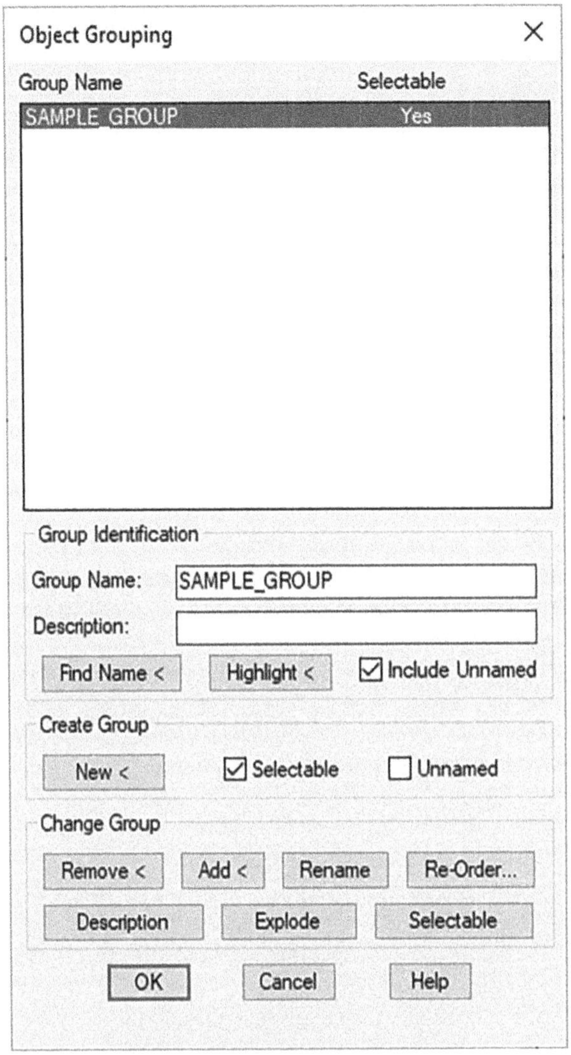

FIGURE 7.37 Object Grouping dialog box.

The Object Grouping dialog box can be called up by typing in `classicgroup` and pressing Enter. Note that, in this example, SAMPLE_GROUP is already created and is highlighted in the dialog box to reveal a multitude of options that should generally be familiar to you, based on the preceding discussion.

From this dialog box, you can create new groups, remove and add objects, and execute many other functions. Take some time to go over this dialog box on your own to complete your tour of the group command.

Summary

You should understand and know how to use the following concepts and commands before moving on to Chapter 8.

- Block
  - Select objects
  - Name the block
  - Select a base point

- Wblock
  - Select objects
  - Select path and enter block name
  - Select a base point
- Options
  - Retain
  - Convert to block
  - Delete
- Insert command
- Purge command
- Dynamic blocks
  - Parameters
  - Actions
  - Visibility
- Groups

Review questions

Answer the following based on what you learned in this chapter:

1. What are blocks and why are they useful?
2. What is the difference between blocks and the wblock command?
3. What are the essential steps in making a block? The wblock command?
4. How do you bring a block or wblock back into the drawing?
5. Explain the purpose of purge.
6. Explain the basic idea behind dynamic blocks. What steps are involved? What is visibility?
7. What are the advantages and disadvantages of using groups?

Exercises

1. In a new file, create the following—architectural table, chair, keyboard, phone, and computer screen—and make a block out of the individual pieces as well as the overall "assembly." Nested blocks, such as these, are quite common in AutoCAD design. Insert the block back into the file. Only general dimensions are given, so you must improvise for the individual pieces. Also, naming the blocks is up to you, just make them descriptive. (Difficulty level: Easy; Time to completion: 15–20 minutes.)

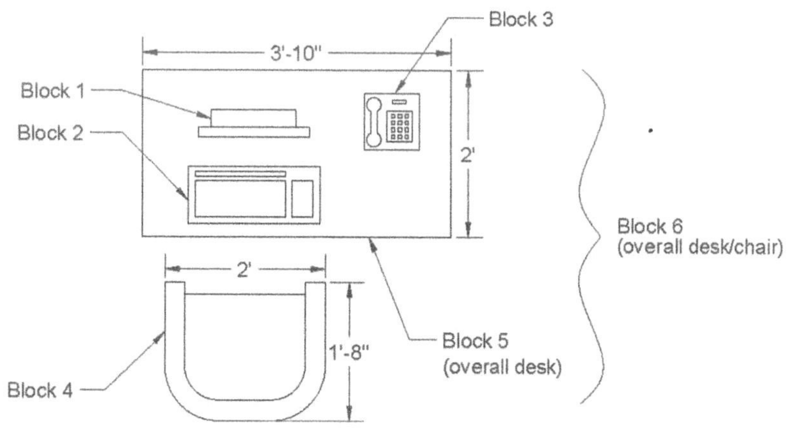

2. In a new file, create the following set of bathroom toilet and sink fixtures, doors, and other items. The overall dimensions are provided, but you need to create the details on your own. Be sure to use the appropriate layers. Create basic blocks out of the fixtures and copy as needed. (Difficulty level: Easy/Moderate; Time to completion: 30–45 minutes.)

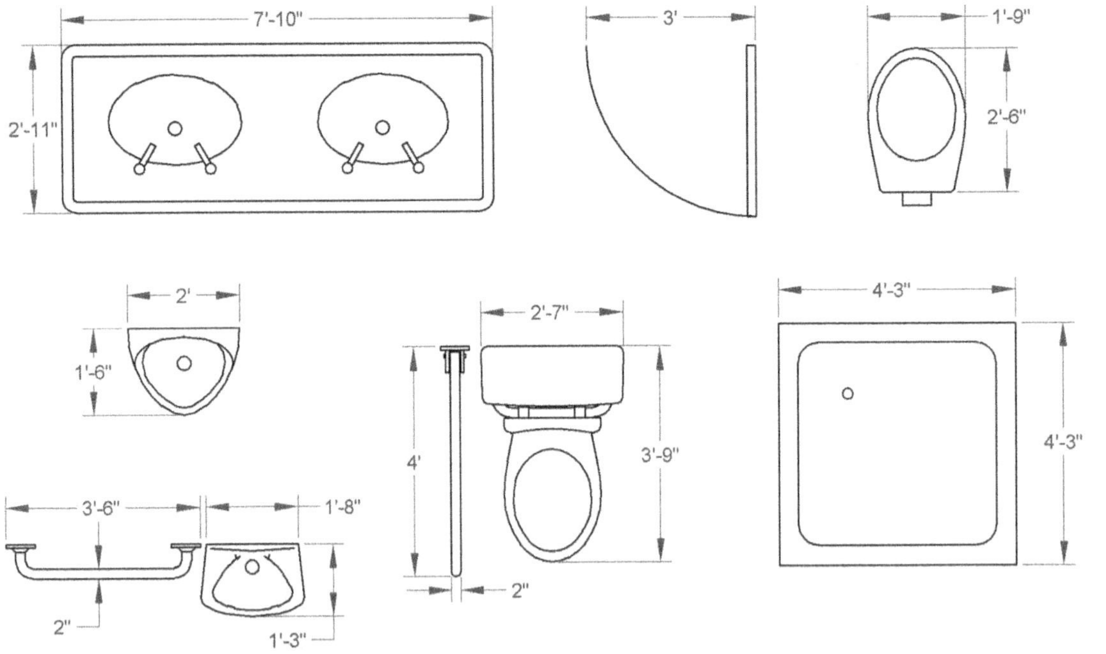

3. In a new file, create the floor plan depicted next. Insert the blocks from Exercise 2 into the floor plan as shown, adding the wall hatching as a last step. (Difficulty level: Moderate; Time to completion: 20–30 minutes.)

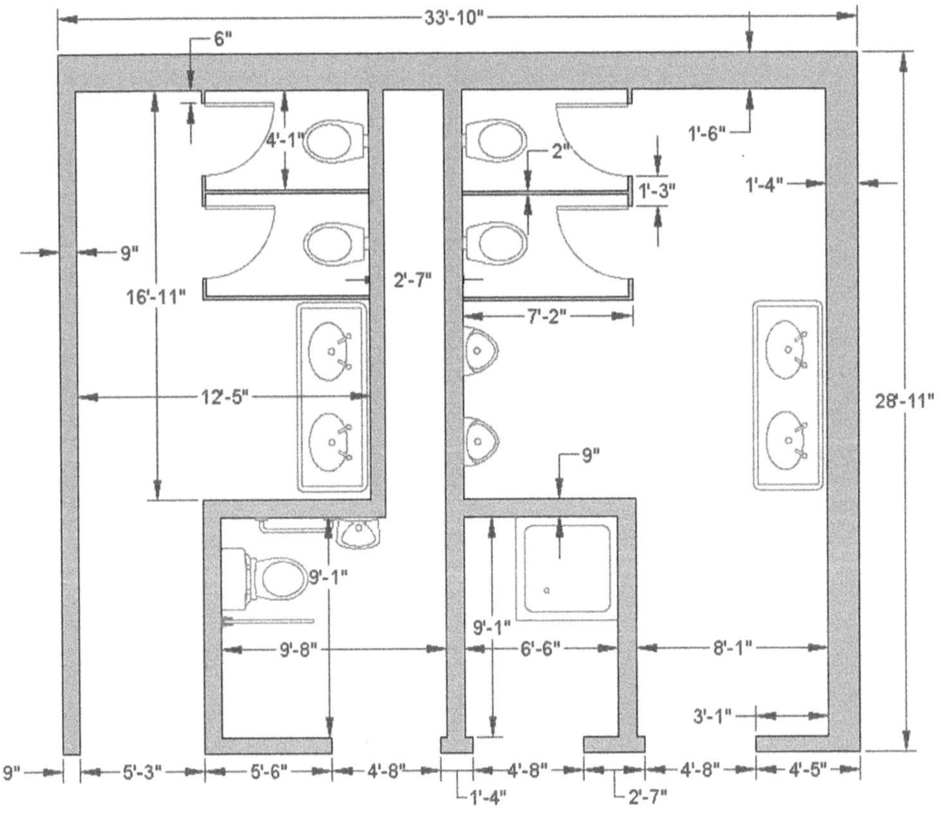

4. In a new file, create the following set of doors and make a wblock out of each one (Door_24in, Door_30in, Door_36in). Drop them all into a premade folder called Doors. Insert each wblock back into the file. When all three doors are inserted, explode them all and purge the file. Then, create dynamic blocks out of all three. Add the linear parameter followed by the scale action. (Difficulty level: Easy; Time to completion: 5–10 minutes per block set.)

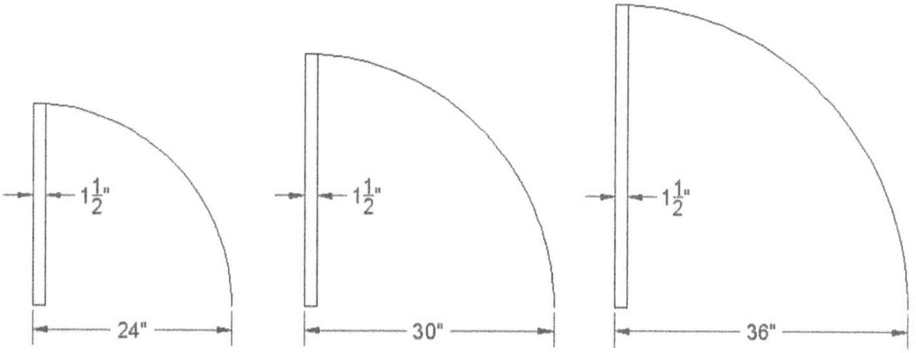

5. Create the following dynamic block double door. The basic dimensions are 2″ × 24″, with the door handle sized off that. Add in shading, arbitrary-sized door jams, and the hidden line door swing. Create a variety of parameters and actions, including Flip, Rotate, Scale, and others as you see fit. (Difficulty level: Easy; Time to completion: 15–20 minutes.)

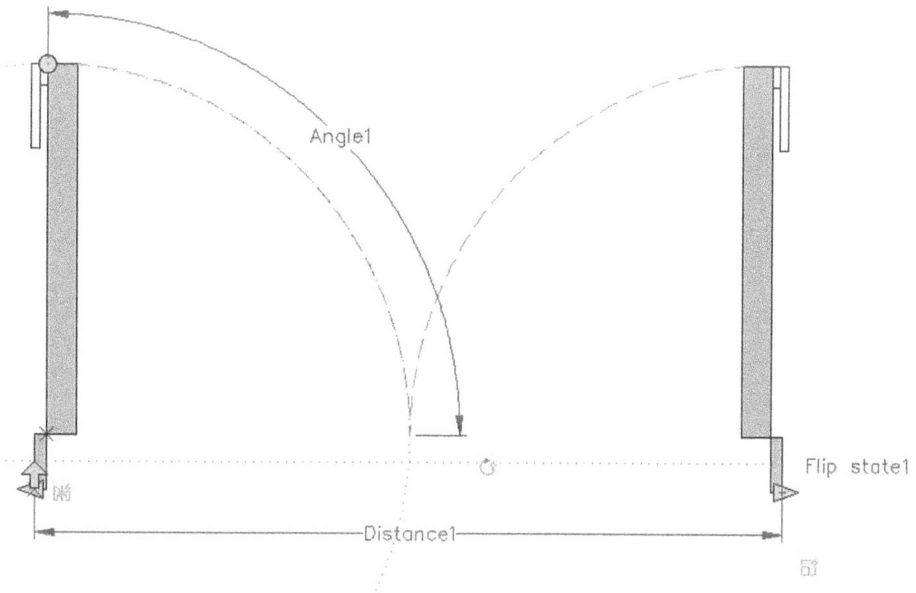

6. Electrical symbols are all usually blocks, and electrical designers/draftspersons have an extensive library at their fingertips. You have already drawn a few electrical symbols in the previous chapter; now let us turn our attention to a few more. Draw the following symbol library and create a block out of each symbol. While you are drawing these symbols, try to memorize them as well. You will see them again in almost any architecture-related field. (Difficulty level: Moderate; Time to completion: 30–45 minutes.)

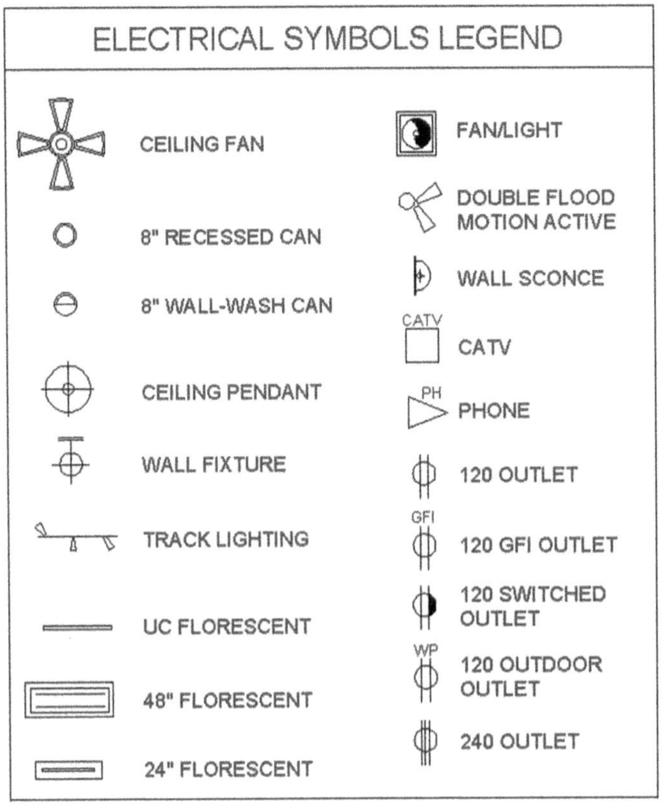

7. This exercise is another example of electrical schematics. Draft all the linework and add text as shown. Like most schematics, the drawing is not to any scale, but accuracy must still be maintained as far as connecting lines and other shapes, as well as text placement. Create blocks whenever practical and make use of the copy and mirror commands to expedite the drafting. (Difficulty level: Moderate; Time to completion: 30–45 minutes.)

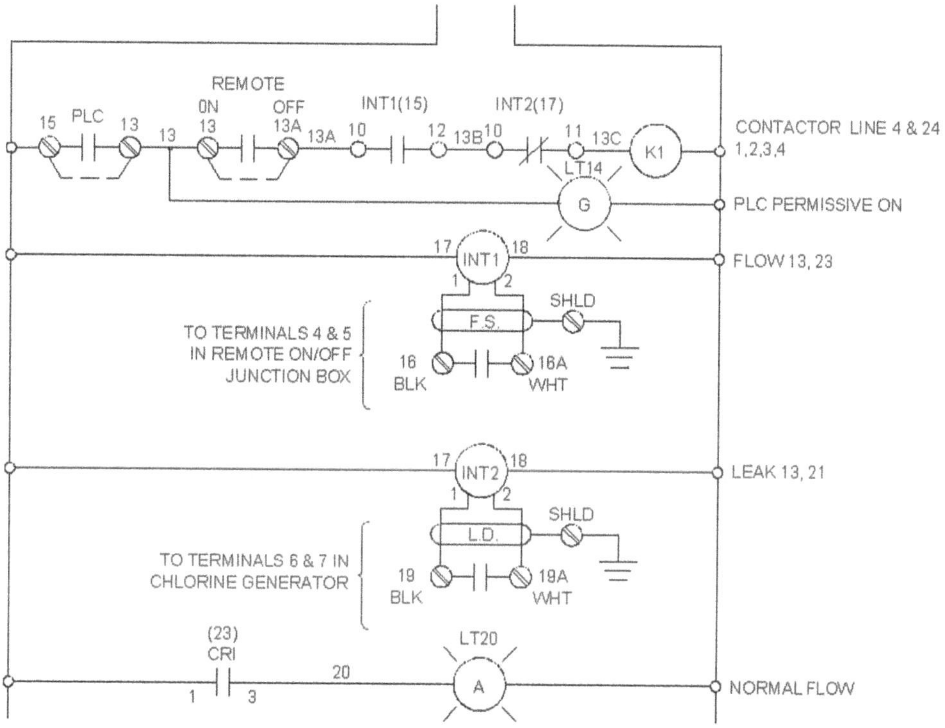

8. For this exercise, we go back to mechanical drafting while incorporating elements of what we just covered. Set your units to Mechanical; set up all appropriate layers, fonts, linetypes, and mechanical dimensions (with 0.000 accuracy); and create the following mechanical fixture, including the two section views. The symbols used for sections are typical in the industry. Be sure to make a dynamic block out of them with the Flip action, so they can easily be reversed to point in the other direction if needed. For additional practice, add in all hatching and dimensions/text. (Difficulty level: Moderate; Time to completion: 45–60 minutes.)

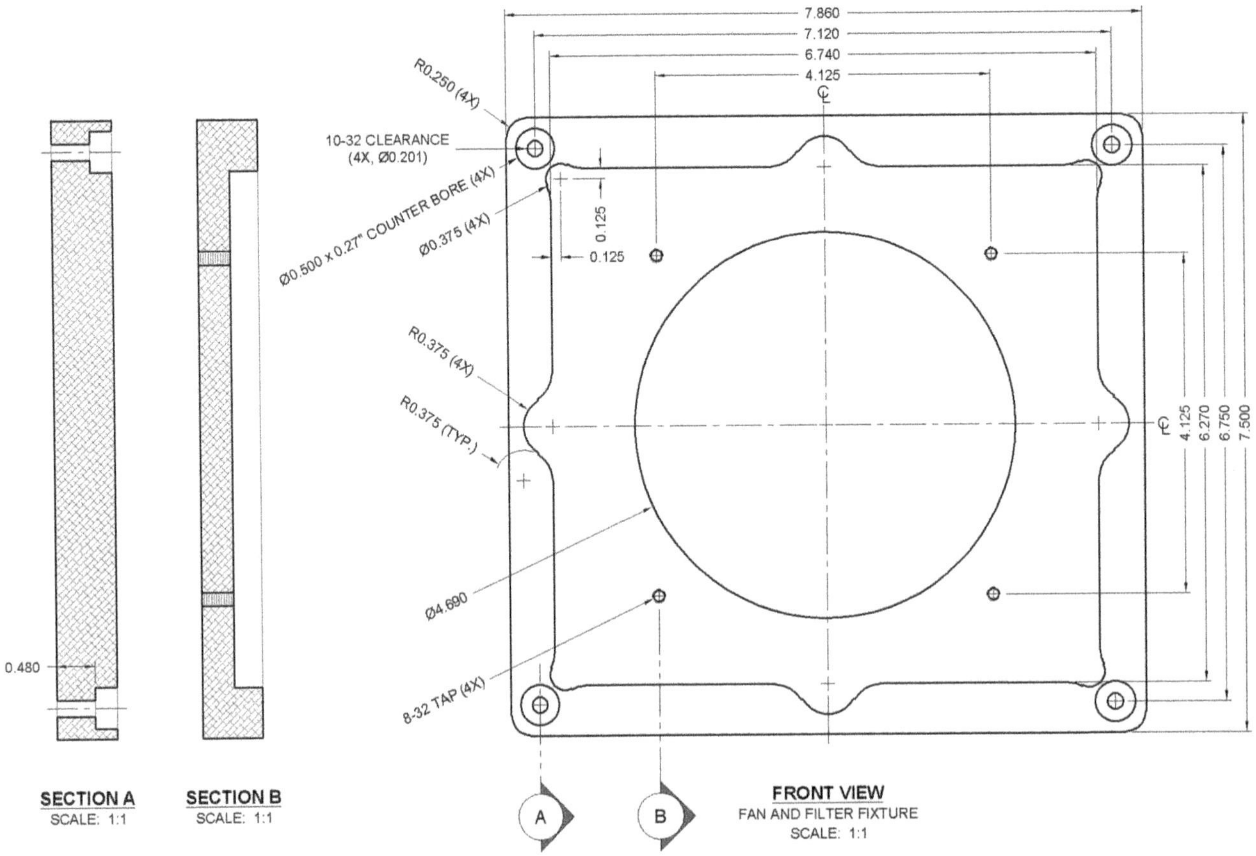

9. Valves are an important part of P&IDs. In a new file, set up all appropriate layers, fonts, linetypes, and mechanical dimensions (with 0.00 accuracy and a " suffix), then, draw the valve shown. Be sure to use the mirror command wisely to avoid duplication of drawing effort. The hex nuts are not specifically detailed out; you must improvise. Make a block out of each hex nut with the Flip action added. (Difficulty level: Intermediate/Advanced; Time to completion: 45–60 minutes.)

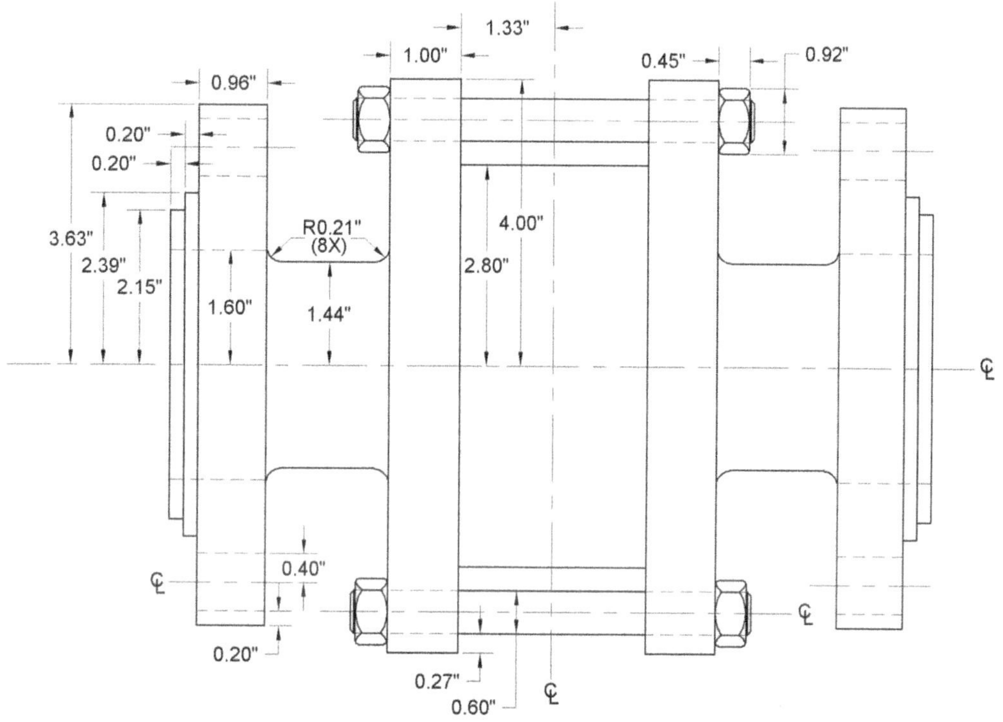

10. The final exercise of this chapter focuses on the basic drafting of an architectural detail. Set up the proper layers and draft what you see, based on the given primary dimensions. Where none are shown, you must do a reasonable estimate. Note how lines and arcs were used to create borders for some of the hatches, then deleted. This is a common technique that we try again with plines and splines in Chapter 11, Advanced Linework to achieve a smoother effect. (Difficulty level: Moderate; Time to completion: 60–90 minutes.)

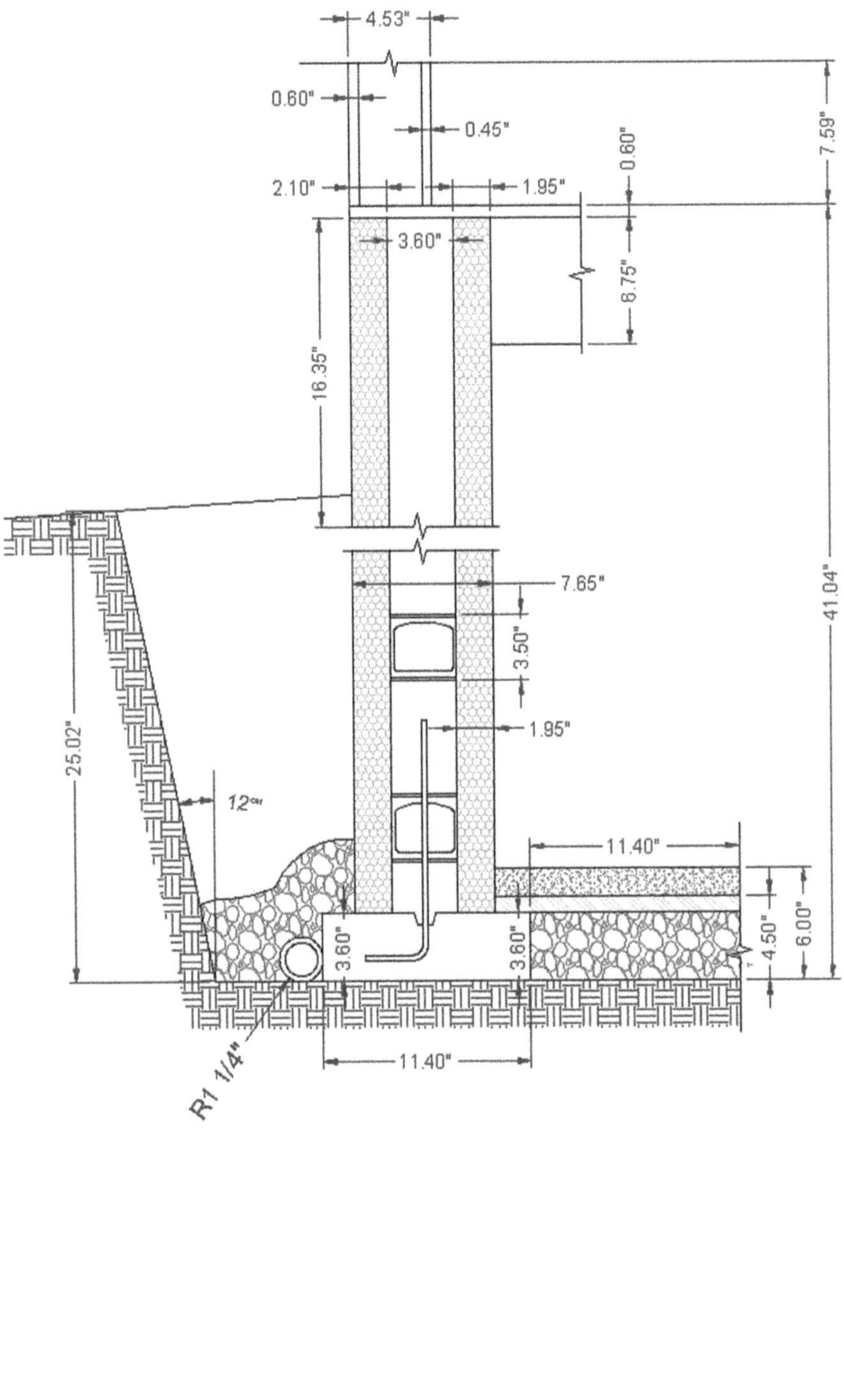

Chapter 8

Polar, rectangular, and path arrays

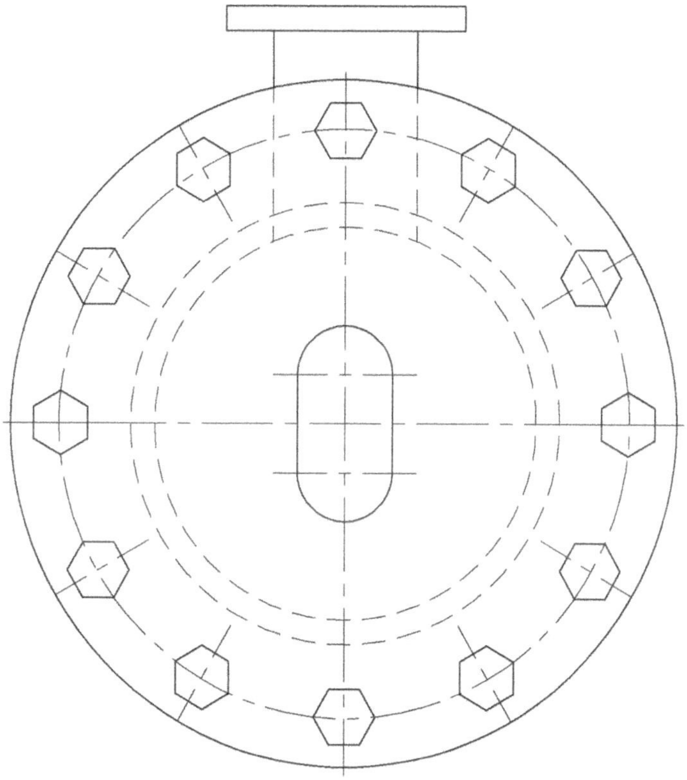

Learning objectives

In this chapter, we introduce the concept of an array and discuss the following:

- Creating a polar array
- Object, center, quantity, and degrees
- Additional operations with polar arrays
- The legacy polar array
- Creating a rectangular array
- Object, rows and columns, and distances
- Additional operations with rectangular arrays
- Legacy rectangular array
- Creating a path array
- Additional operations with path arrays

At the end of this chapter you will be able to create polar, rectangular, and path arrays. You then start a new in-class mechanical project.

Estimated time for completion of this chapter (lesson and project): 3 hours.

Up and Running with AutoCAD 2022. https://doi.org/10.1016/B978-0-323-89923-9.00008-1

Arrays are very useful tools in AutoCAD to create patterns of objects. These patterns can be of a circular, linear, or path type, hence the polar array, rectangular array, and path array discussions that follow. In a strict sense, these tools are not something entirely new but merely automate what you can do (albeit tediously) by commands learned in Chapter 1, such as Move, Copy and Rotate, and they are huge time-savers.

8.1 Polar array

A polar array is a collection of objects around some common point arranged in a circle (or part thereof). An example in architecture is a group of chairs arranged around a circular table. The idea is to draw the table, then one chair, and after positioning the chair exactly where it will be in relation to the table, use a polar array to copy it around the table to create, let us say, 10 of them, as seen in Fig. 8.1 on the left. The array command copies and rotates them into position and spaces them out evenly, a task that would have taken a while to do one chair at a time.

In mechanical engineering, an example may be the teeth of a sprocket gear. You draw the circle representing the wheel of the gear; put in one tooth properly drawn, detailed, and centered on top; then array to get the full gear. Fig. 8.1 shows illustrations of what was just discussed. There are, of course, many other examples.

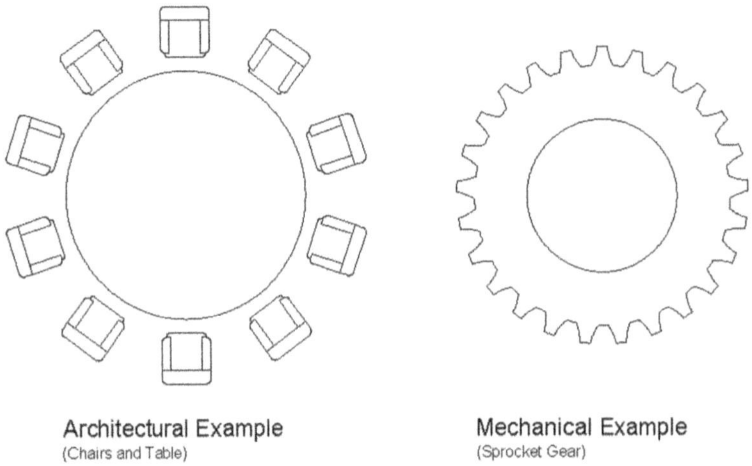

FIGURE 8.1 Polar array examples.

Steps in creating a polar array

First, we need to create something to array. So, go ahead and draw a circle of any size to represent the table just discussed, then create a convincing-looking chair. You may want to make a block out of it after you are done for convenience.

Next, position the chair at the top of the table, as seen in Fig. 8.2. Be sure to line it up perfectly on the vertical axis, or else the array will be skewed off center; and move it back a bit from the table with Ortho on. What does AutoCAD need to know to do the array? The following four items are critical to a polar array, although note that only the first two items are explicitly asked for by AutoCAD as you create the array:

- It needs to know *what object(s)* to array (that would be the chair, not the table).
- It needs to know *the center point* of the array (center of the table).
- It needs to know *how many objects* to create (let us stick to 10 or so).
- It needs to know if the pattern goes all the way around (i.e., is it 360 degrees or less).

It is useful to memorize these steps and realize what you need to do before even starting up the array command; that way you have a clear idea of what to do and what data to enter as you get going. Once you have the table and chair drawn, follow the steps shown to create the polar array.

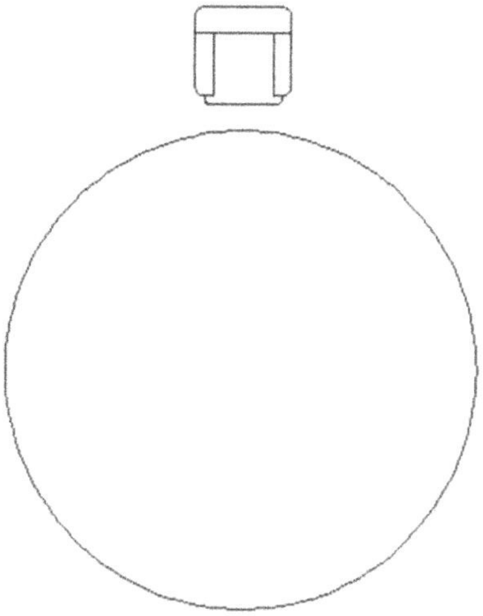

FIGURE 8.2 Chair and table, initial setup.

Keyboard: Type in **arraypolar** and press Enter
Ribbon: **Home tab→Modify→Polar Array**

Step 1. Start up the polar array command via any of the preceding methods.

- AutoCAD says: `Select objects:`

Step 2. Select the chair block (or the chair geometry if not using a block)

- AutoCAD says: 1`found`

Step 3. Press Enter.

- AutoCAD says: Type = `Polar Associative = Yes`

`Specify center point of array or [Base point/Axis of rotation]:`

Step 4. You successfully selected the *object to array*. You now need to specify the *center point of the array*. Simply select the center of the large circle (the table) using the CENter OSNAP. AutoCAD creates a set of chairs around the table.

- AutoCAD says: `Enter elect grip to edit array or [ASsociative/Base point/Items/Angle between/Fill angle/ROWs/Levels/ROTate items/eXit]<eXit>:`

Step 5. Note a few things at this point. The array is automatically associative (look for that button to be active in the Ribbon). This means all the pieces are linked and not separate, a very desirable new property of arrays. Also note the details shown in the Ribbon. Here, you can modify the number of items (chairs), and the angle to fill under the Items tab. Make sure you have 10 chairs for this exercise.

Step 6. Your array is essentially done, so just press Enter to accept. If you wish, experiment with some of the variables such as Fill and Between. Also note how the chairs are rotated to face the table via the Rotate Items button that was automatically set (under Properties tab). You can, of course, turn that feature off. The completed polar array is shown in Fig. 8.3.

FIGURE 8.3 Polar array completed.

Additional operations with polar array

Your new array behaves like a group (assuming Associative was setduring creation). Click on one of the chairs with your mouse and it becomes selected, with several varieties of grips visible. If you hover your mouse over the chair grip, you see a menu of additional options. Finally, a new Ribbon Array tab appears. All of this is seen in Fig. 8.4. Let us take a closer look at the Ribbon and the Grip menu.

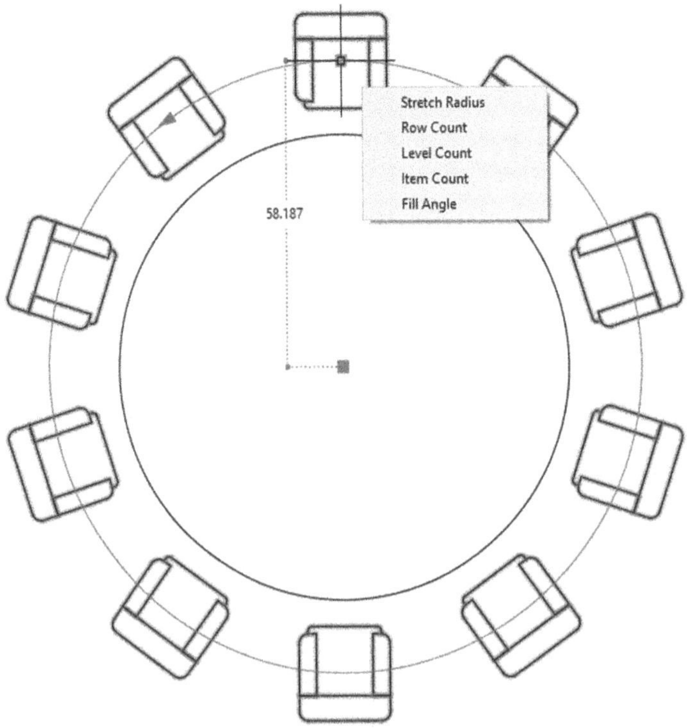

FIGURE 8.4 Polar array editing.

The Polar Array Ribbon has three areas of interest: the Items, Properties, and the Options/Close category, as seen in Fig. 8.5. The Items category we already mentioned, and it is relatively self-explanatory. Top to bottom, the first value (10) is the number of items in the array, and you can change it for an instant update. Below that is the angle between all the objects (36 degrees), and finally, below that, the total angle filled (360 degrees, or a full circle). This should make sense as $36 \times 10 = 360$.

The Properties category's first button allows you to select a different base point for the rotation, useful if you did not quite get the first one right. The second button rotates the objects to face the rotation point as opposed to them all facing in one direction. The last button changes the direction of the array, from clockwise to counterclockwise, or vice versa. The last two buttons may be activated already on your sample table/chair array.

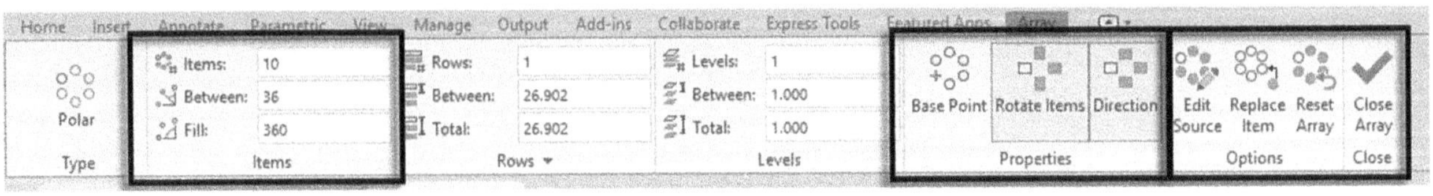

FIGURE 8.5 Polar array tab: Items, Properties and Options/Close.

The Options category's tools are a bit more involved. They are therefore described in more detail next:

- *Edit Source*: This option allows you to edit and modify the source object. These changes are then reflected in the overall array. If you forget which object the original source was, it does not matter; you can pick any of the chairs and work on it. Execute the command, say "Yes" if you see the Array editing State dialog box, select the object, modify it, and when done, type in arrayclose, or click on Save Changes on the Ribbon at the far right. Fig. 8.6 shows Edit Source in progress.

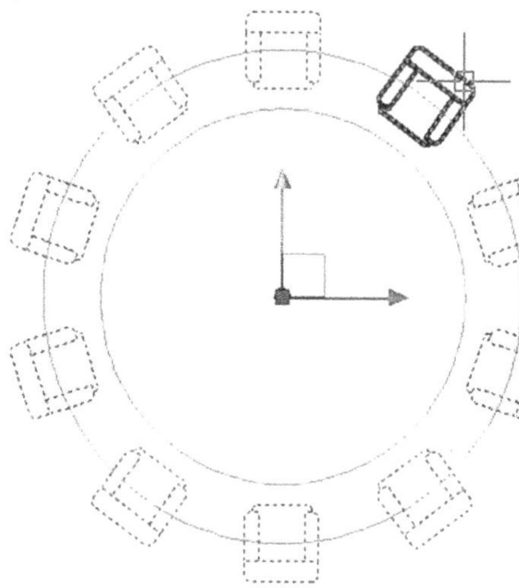

FIGURE 8.6 Edit Source.

- *Replace Item*: If editing the source is too time-consuming, then just replace it with another design via this option. AutoCAD asks you to select the replacement object and a base point. Use the Key Point option to pick a point on the replacement object. Finally, select an item (or several items) in the array to replace. Fig. 8.7 shows Replace Item in action as squares replace the chair symbol.

FIGURE 8.7 Replace Item in progress.

- *Reset Array*: This is an easy one. Autodesk figured you will mess up a few arrays while learning this tool (speaking from experience, of course) and included a Reset button. Its function is apparent immediately upon use, though it is limited to just restoring erased items and some overrides.

Next, we have the menu seen when the mouse is held over the chair grip. Run through all the options as you read the descriptions. It contains the following tools:

- *Stretch Radius*: This option simply expands the radius of the overall array. The effect is the same as if you just activate the grip and move the mouse. Stretch Radius in action is seen in Fig. 8.8.
- *Row Count*: This option adds another set of rows of items in front or behind the main row, as seen in Fig. 8.9.
- *Level Count*: This option adds 3D levels, and its discussion is not included in this 2D-only chapter.
- *Item Count*: This option adds or deletes items (like the Ribbon function), or it can just be used to tell you how many items there are in the array, also a useful feature.
- *Fill Angle*: This option (also via the triangle grip) changes the included angle, currently set to 360 degrees. An example is shown in Fig. 8.10, where the fill angle has been rolled back to 270 degrees.

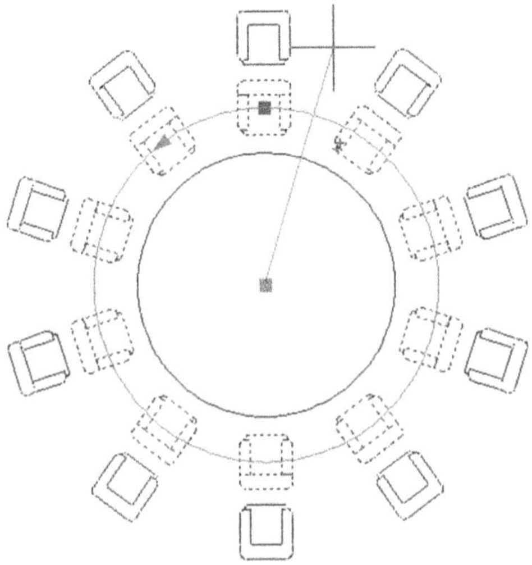

FIGURE 8.8 Stretch Radius in progress.

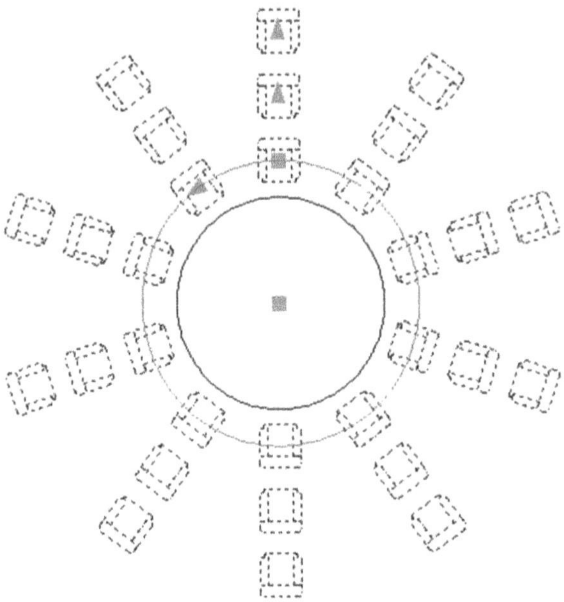

FIGURE 8.9 Row Count in progress.

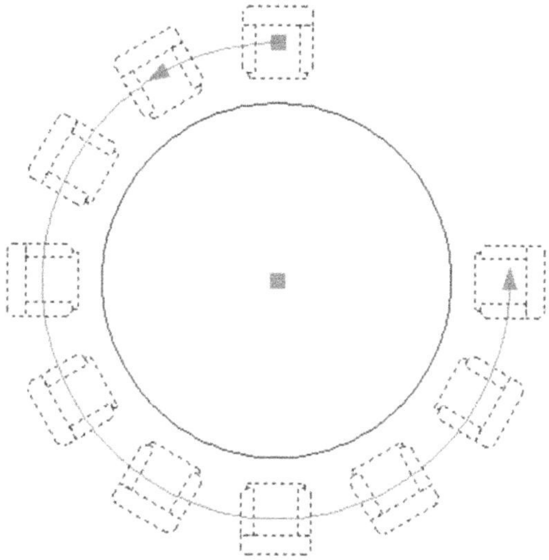

FIGURE 8.10 Fill Angle in progress.

Go over all these features until you are comfortable working with them. Be aware that you also can explode the array, which makes it un-associative, where each piece is separate.

8.2 Rectangular array

The idea behind a rectangular array is not fundamentally different from a polar one. We are looking to replicate objects, but this time in rows (horizontal) and columns (vertical). In 3D, you can also do levels, which we do not address here. An example, as shown in Fig. 8.12, may be columns of a warehouse. If they are spread out evenly, then this is the perfect tool to create them. You simply draw one I-beam column, make a block out of it, and array it up and across the appropriate area.

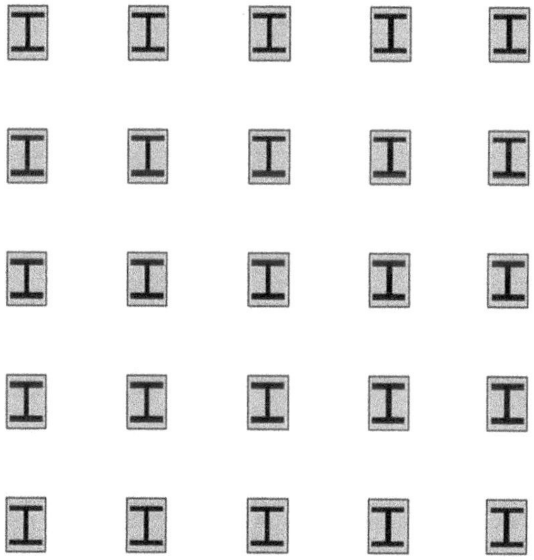

FIGURE 8.11 Rectangular array example.

Steps in creating a rectangular array

The first thing to do is to draw a convincing-looking I-beam. Because we need actual distances, you cannot just randomly draw any object; use known sizing this time around. With one of the columns in Fig. 8.11 as a model, create a 20″ wide by 26″ tall rectangle. Inside of it, put an I-beam drawn with basic linework and a solid hatch fill, and shade the background a light gray (Fig. 8.12). Finally, make a block out of it. That sets up the column for an array.

Now, what does AutoCAD need to know to create a rectangular array? The following three items are critical to a rectangular array; although, note that only the first item is explicitly asked for by AutoCAD as you create the array:

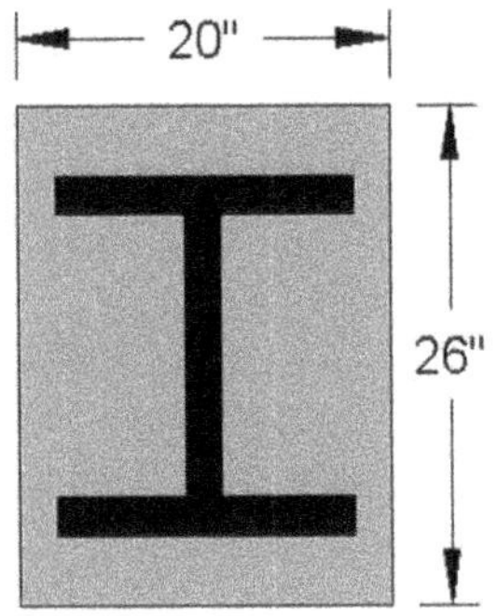

FIGURE 8.12 I-beam for column.

- It needs to know *what object(s)* to array (the I-beam column).
- It needs to know *how many rows* and *columns* to create (rows = across, columns = up and down).
- It needs to know the *distance between the rows and the columns* (from centerline to centerline).

Keyboard: Type in **arrayrec** and press Enter
Ribbon: **Home tab→Modify→Rectangular Array**

Step 1. Start up the array command via any of the preceding methods (it is recommended to have the Ribbon turned on).
- AutoCAD says: `Select objects:`

Step 2. Select the I-beam column block.
- AutoCAD says: 1`found`

Step 3. Press Enter. A small array of items is created, with grips visible.
- AutoCAD says: `Type = Rectangular Associative = Yes`
  `Select grip to edit array or [ASsociative/Base point/COUnt/Spacing/COLumns/Rows/Levels/eXit]<eXit>:`

Step 4. You have now successfully created the basic rectangular array. You still need to define the exact number of rows and columns, however, as well as the distance between them, as this current array is rather arbitrary. You can do this either via command line, dynamically (using grips), or via the Ribbon. Fig. 8.14 shows how you can pull the items diagonally in a dynamic fashion using the center grip. You can also do this via the outer arrow grips. This dynamic method is not appropriate for setting precise distances between the columns, which are random values for now, but it does make creating the pattern very easy.

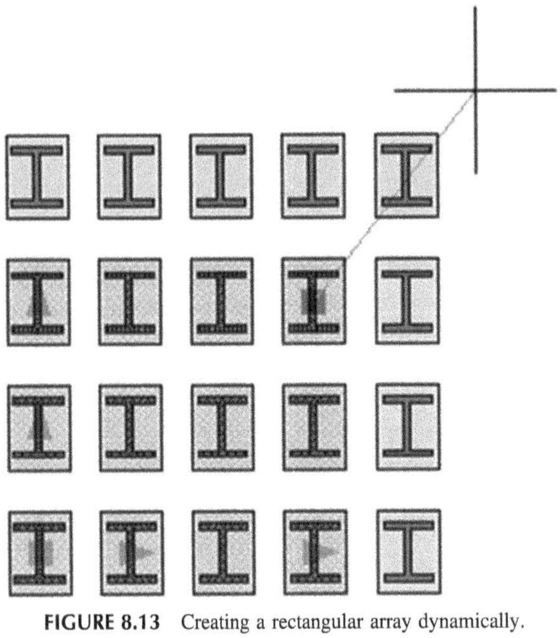

FIGURE 8.13 Creating a rectangular array dynamically.

Step 5. If you have the Ribbon up, then you just enter values. We want to create a 5 × 5 array, so start by entering those values into the Ribbon's Columns: and Rows: fields. For distances between the columns and rows use the value 72.
Step 6. If using the command line, you just must follow the prompts for Columns, Rows, and Spacing. Regardless of which you start with (Rows or Columns), AutoCAD asks you for the spacing as needed. Press Enter when done. Your rectangular array should look like Fig. 8.13.

Additional operations with rectangular array

After a lengthy discussion on additional operations with the polar array, you will find these options with the rectangular array to be quite familiar. Click on the new array of I-beam column, and notice a set of grips, a menu, and an Array tab in the Ribbon, all like the earlier polar array (Fig. 8.14).

The Rectangular Array Ribbon also has three areas of interest: the Columns, Rows, and Option/Closes categories, as seen in Fig. 8.15. The Columns category allows you to adjust the number of columns, the distance between them, and the total distance. The Rows category is the exact same thing but for rows. Finally, in the Options category, we already discussed the functionality of Edit Source, Replace Item, and Reset Array. They are all the same as with the polar array.

Next, we have the menu seen when the mouse is held over the top right I-beam column grip. The two choices are Row and Column Count and Total Row and Column Spacing. Their functions should be relatively easy to see by just trying them out. If you activate the first choice and move the mouse, the number of rows and columns increases, but the spacing remains the same. If you activate the second choice, the spacing increases but not the number of rows and columns.

Finally, try out the other grips on the array, specifically the ones in the lower left corner. You find that the square grip moves the entire array and the inner triangles increase the spacing. The outer triangle grips (toward the edges of the array), however, are used to increase the count. Experiment with all the options.

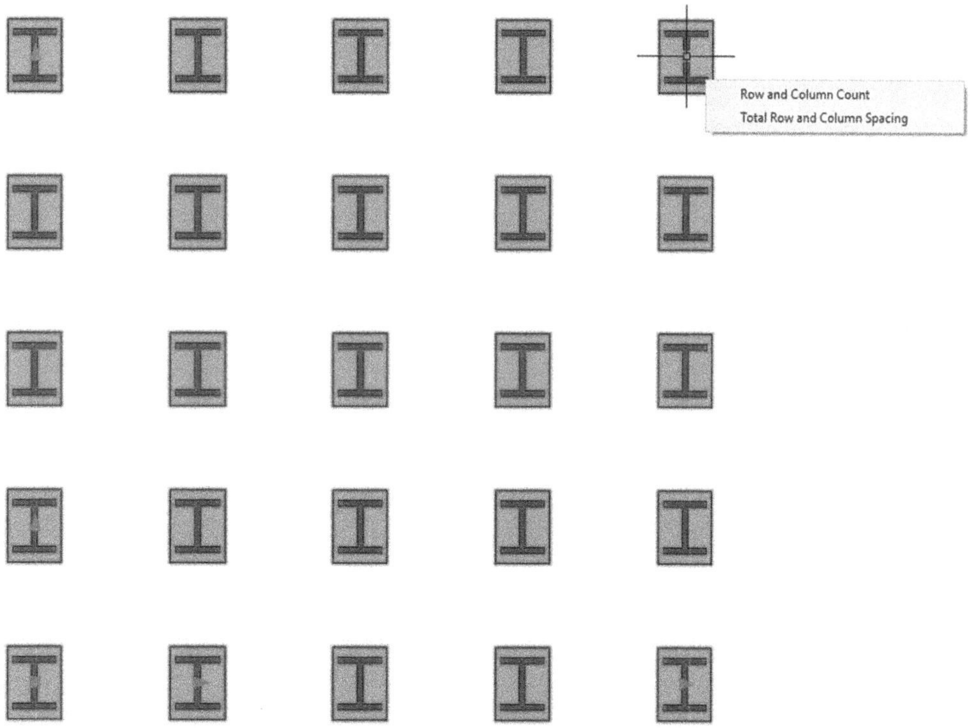

FIGURE 8.14 Rectangular array editing.

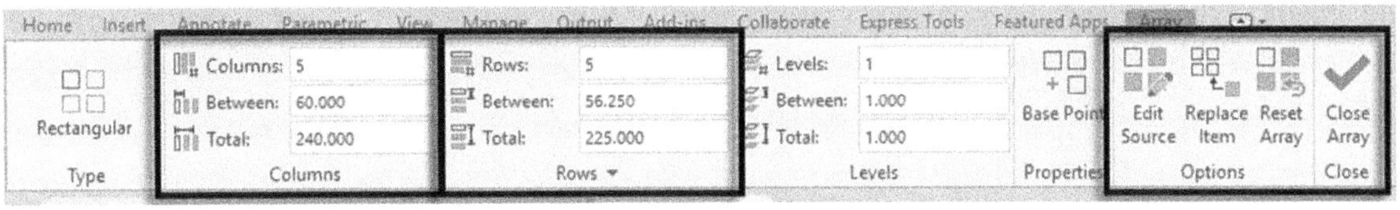

FIGURE 8.15 Rectangular array tab: Columns, Rows, and Options/Close.

8.3 Path array

The path array is the (relatively) new kid on the block. Requested by users, this feature was finally added in AutoCAD 2012. The idea is simple, and it is surprising that this feature was not added earlier. A path array creates a pattern along any *predetermined pathway*, of any shape, as opposed to a strictly circular or linear one. This path can be an arc, a pline, or a spline. Because plines and splines have not yet been discussed (a Chapter 11, Advanced Linework topic), we focus on arcs as the pathways. A path array can also be created in 3D, but we do not get into that right now. An example of a path array in progress is shown in Fig. 8.16.

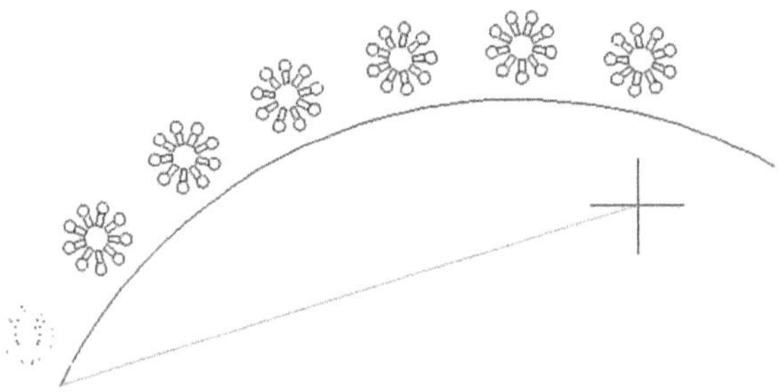

FIGURE 8.16 Path array in progress.

Steps in creating a path array

Boiled down to its essence, the path array just needs an *object to array* and a *path to follow*. Much like the previous two arrays, the objects can be spread out dynamically with mouse movements (as seen being done in Fig. 8.19) or with an input of a specific value. All other options flow from this basic approach. Let us give it a try. Create an arc and some random object to array; a rectangle will do, as seen in Fig. 8.17.

FIGURE 8.17 Setup for a path array.

So, what does AutoCAD need to know to do the array?

- It needs to know *what object(s)* to array (the rectangle).
- It needs to know *the path* (the arc).
- It needs to know *how many objects* to create (let us stick to 10 or so).

Keyboard: Type in **arraypath** and press Enter
Ribbon: Home tab→Modify→Path Array

Step 1. Start up the array command via any of the preceding methods.
- AutoCAD says: `Select objects:`

Step 2. Select the rectangle.
- AutoCAD says: 1 `found`

Step 3. Press Enter.
- AutoCAD says: `Type = Path Associative = Yes`
  `Select path curve:`

Step 4. You successfully selected the *object to array*. Now, select the *path* by clicking on the arc. Several copies of the rectangle appear along the arc.
- AutoCAD says: `Select grip to edit array or [ASsociative/Method/Base point/Tangent direction/Items/Rows/Levels/Align items/Z direction/eXit]<eXit>:`

Step 5. You have essentially created the basic path array, but you can do a tremendous amount of modification to this basic layout. The easiest way to modify it is via the Ribbon. Under the Items: tab you have the most basic of the "mods," such as total number of items and the distance between them. Also note how you can alter the path array via dynamic movement of the grips. Be sure to keep the array associative, as with the previous examples. The completed array is seen in Fig. 8.18.

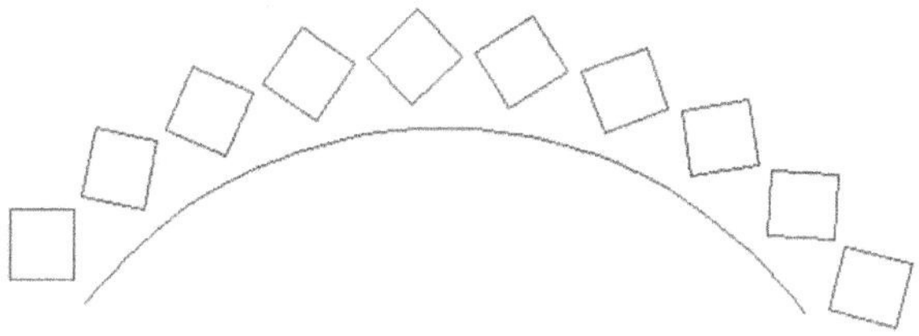

FIGURE 8.18 Path array completed.

Additional operations with path array

Just as with the previous two arrays, the path array has its own Ribbon tab, which is revealed by selecting one of the elements in the array, as well as its own menu, revealed as you hover the mouse over the grip, as seen in Fig. 8.19.

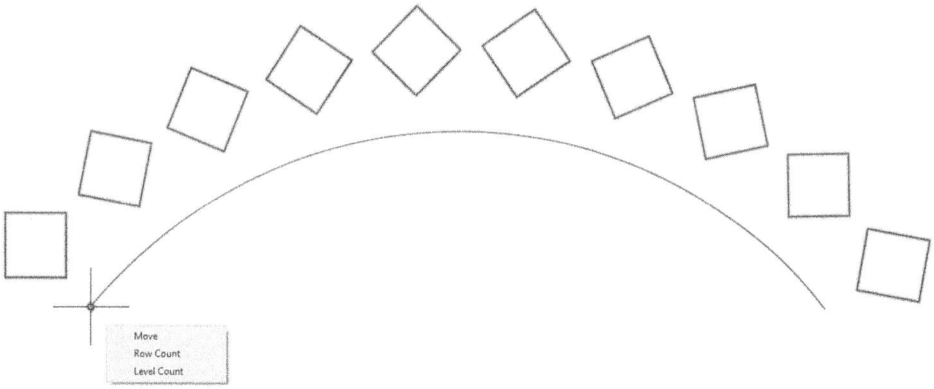

FIGURE 8.19 Path array editing.

The Items category panel is of some interest, and at this point should be somewhat familiar. In the Items panel of the Ribbon, you can vary the *Items* number when the *Divide* option is current in the Properties panel. Alternately, you can vary the *Between* distance when the Measure option is current in the properties panel. AutoCAD does the math for you in the Total category (Fig. 8.20). Under the Properties category, experiment with redefining the base point—this shifts your pattern around. Also check out Align Items—this makes all your elements face the same direction versus angling with the curve. Edit Source, Replace Item, and Reset Array all have the same meanings as with the previous polar and rectangular arrays. Spend a few moments understanding the cause and effect of these options.

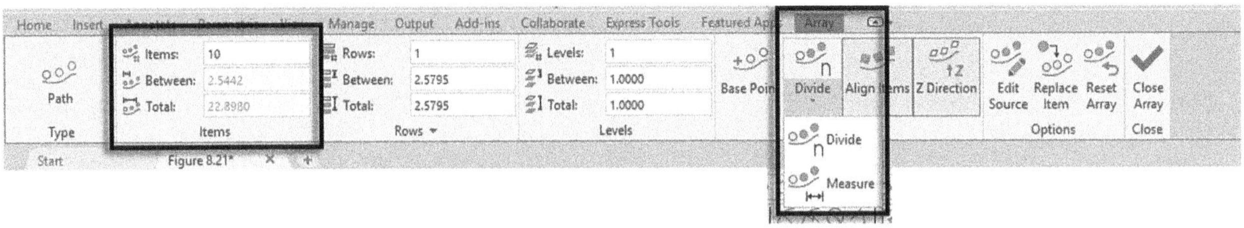

FIGURE 8.20 Path array Ribbon editing—Divide/Items and Measure/Between Properties options.

Finally, let us look at the grip menu (Fig. 8.19). It also should be more familiar after your experiences with the previous two arrays. Move moves the array away from its originating path and triggers a warning along the way (Fig. 8.21). You may have seen the warning with earlier arrays as well, if you tried to move them around. Simply press Continue and carry on.

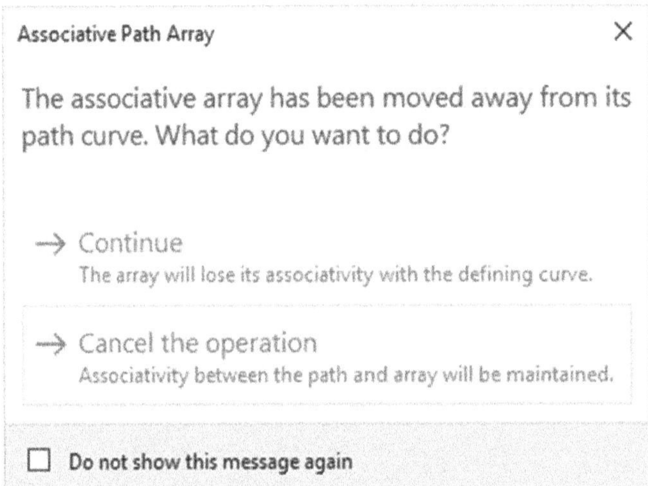

FIGURE 8.21 Associative Path array warning.

Row Count creates additional rows of the pattern, as seen in Fig. 8.22. This option also repeats what is available with the other arrays. Level Count is not covered here, as it is a 3D function.

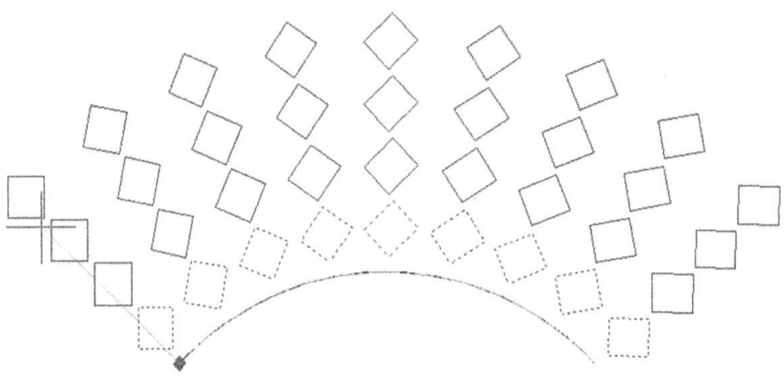

FIGURE 8.22 Row count in progress.

8.4 In-class drawing project: mechanical device

Let us now put what you learned to use by drawing a mechanical gadget (a suspension strut of sorts), created years ago for a class. Although what it is exactly is still debatable, the gadget proved to be an excellent drawing example and features quite a bit of what we covered in this and preceding chapters as well as a few minor nuances of AutoCAD drafting that, once learned, raise your skill levels considerably. The trick is to make your drawing look exactly like the example. If you approximate and cut corners, you miss out on some of the important concepts that give a more professional look to your drawings.

A full-page view of the completed device is found toward the end of the chapter, see Fig. 8.34. Take a good look at it and decide on a strategy to draw it. Over the course of the next few pages, you will see a step-by-step process to do it. Follow the steps carefully, and do not worry about the thick border, i.e., a polyline, to be covered in Level 2. The title block is otherwise just a simple collection of lines and text.

Step 1. For starters, open a new file, give it a name (Save As...), and set up your layers. We need, at a minimum,
M-Part, Color: Green (for most of the parts of the device).
M-Text, Color: Cyan.
M-Dims, Color: Cyan (for the dimensions).
M-Hidden, Color: Yellow, Linetype: Hidden (for all hidden lines).
M-Center, Color: Red, Linetype: Center (for the main centerlines).
M-Hatch, Color: 9 (a gray for cross-section cutaways).

You may also want additional layers for the spring and the title block. You can decide on the exact layers and their colors; the preceding is just a guide. Also, the images in this chapter remain black and white, for clarity on a white page.

Step 2. Set up all the necessary items for smooth drafting, such as the Text Style (Arial, .20), the Dim Style, and Units (either Architectural or Mechanical is fine) and get rid of the UCS icon if you do not like it.

Step 3. To begin the drawing itself, we need to start with the hex bolts. Go ahead and draw one of them according to the information in Fig. 8.23. Be very careful, all sizes are diameters not radiuses; you must tell AutoCAD this by pressing d for Diameter while making a circle. This has tripped up many students in the past. Also, do not forget to make layer M-Part current. If the bolt is anything but green in color, you missed this step. The dimensions are only for your reference for now; there is no need to draw them.

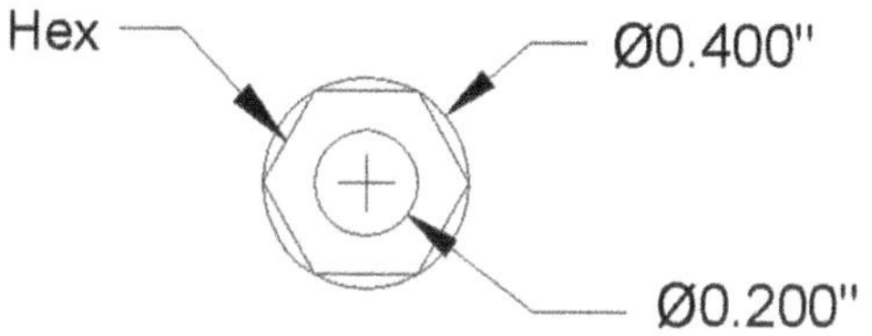

FIGURE 8.23 Bolt layout.

Step 4. Now make a block out of the bolt, calling it Bolt—Top View.

Step 5. Complete the rest of the top view by drawing the remaining circles and positioning the bolt at the top, as shown in Fig. 8.24. Use accuracy (OSNAPs), although the bolt can be any distance from the top of the circle. Be careful to use diameter, as indicated.

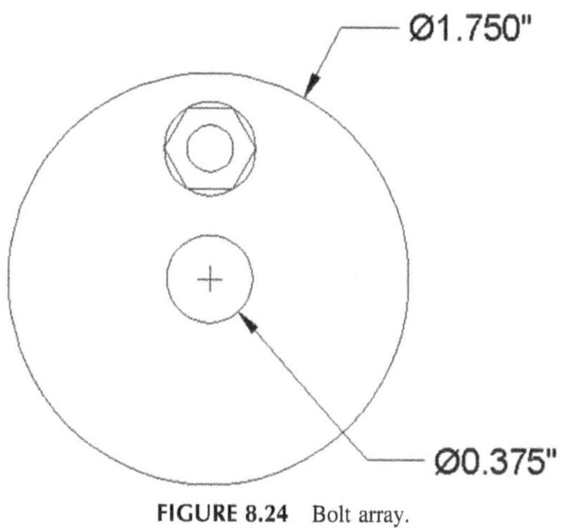

FIGURE 8.24 Bolt array.

Step 6. Array the bolt around the top view four times, as shown in Fig. 8.25.

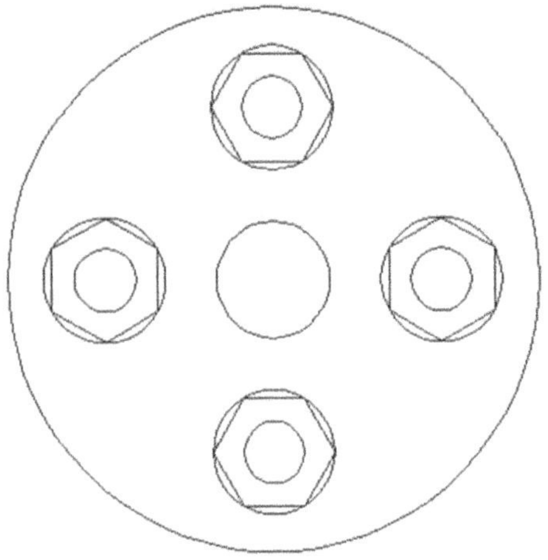

FIGURE 8.25 Bolt array completed.

Step 7. You now must project the main body of the part based on this top view. The best way is by drawing two horizontal guidelines, of any length, starting from the top and bottom quadrants of the big circle; then, draw an arbitrary vertical line to mark the beginning of the part on the right, as shown in Fig. 8.26.

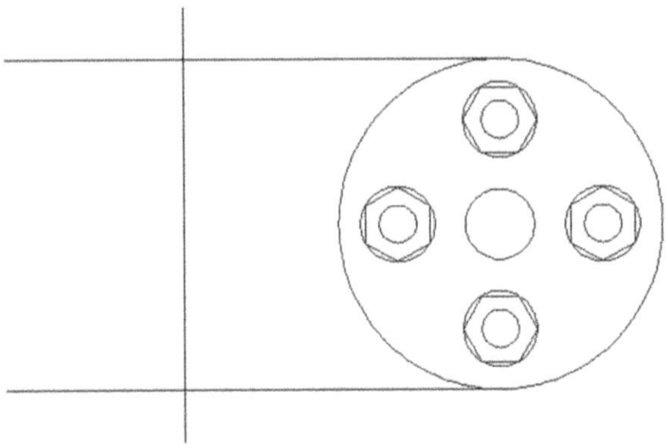

FIGURE 8.26 Projection lines.

Step 8. Trim or fillet the main vertical guideline and continue offsetting, based on the given geometry, to "sketch out" the basic outline of the shape, as seen in Fig. 8.27. Do not yet add dimensions; they are for construction only.

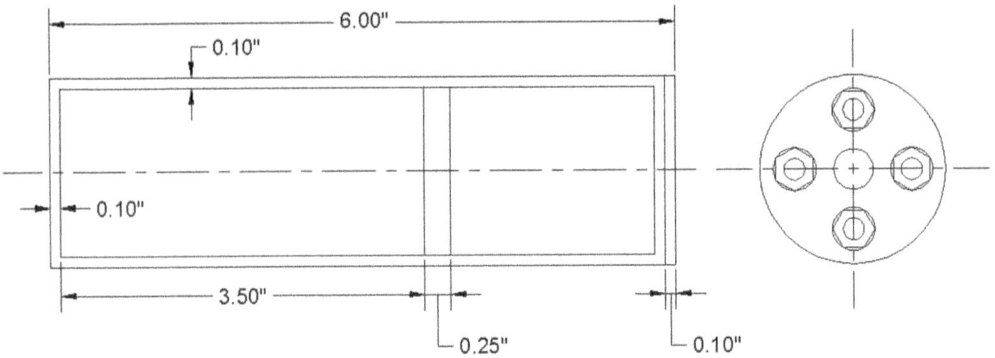

FIGURE 8.27 Body of part.

Step 9. Continue adding the basic geometry to the side view, as well as hidden and centerlines, as shown in Fig. 8.30. Set LTSCALE to .4.

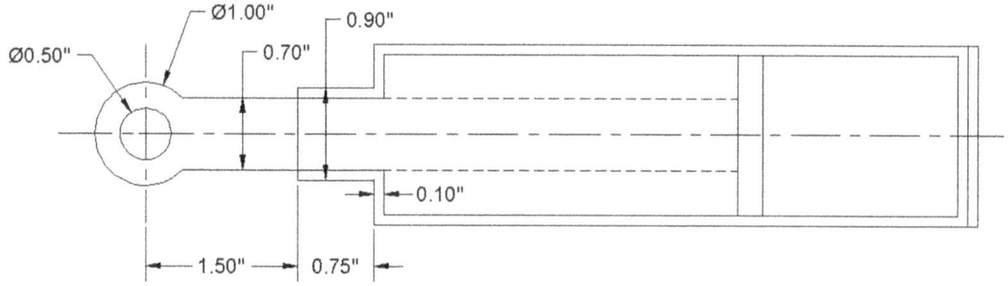

FIGURE 8.28 Creating side view.

Step 10. Add the bolt projections, as seen in Fig. 8.29 (the depth into the side view can be approximated). Finally, add hatch patterns, using the Angle option to reverse the hatch directions, which is a common technique in mechanical design to differentiate the parts from each other in a cross section.

Step 11. Now, how do you do the spring? Well, this is where the Circle/TTR comes in. Although we have not practiced this version of the circle command, it is very simple to do. Start it up and select the TTR option. Then, pick the vertical and the horizontal lines where the circle needs to go (seen in Fig. 8.30) and specify a radius (0.10″). One way of proceeding is to draw a straight arbitrary line down from the center of that circle, rotate the line 15 degrees about that center point, then offset 0.1″ in each direction, erase the original line, extend the two new ones to the bottom (if necessary), and repeat the process, as seen in Fig. 8.30. Once a set is done, you can mirror the rest and trim carefully to create the impression of a coiled spring.

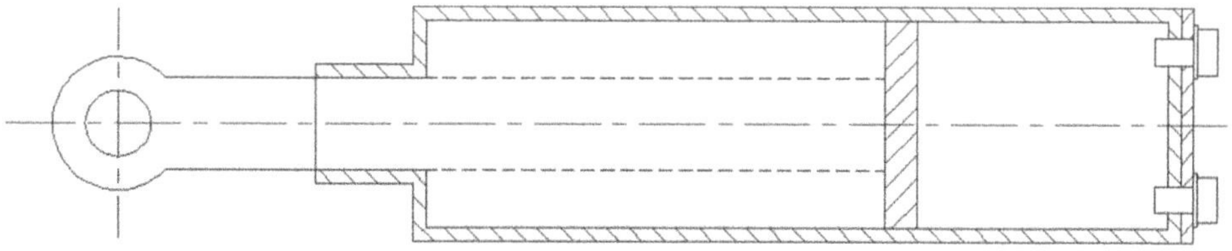

FIGURE 8.29 Hatching side view.

Step 12. Finally, add dimensions (Fig. 8.31) and a title block (Fig. 8.32) by use of rectangles. The title block and its contents are typical of a mechanical engineering drawing.

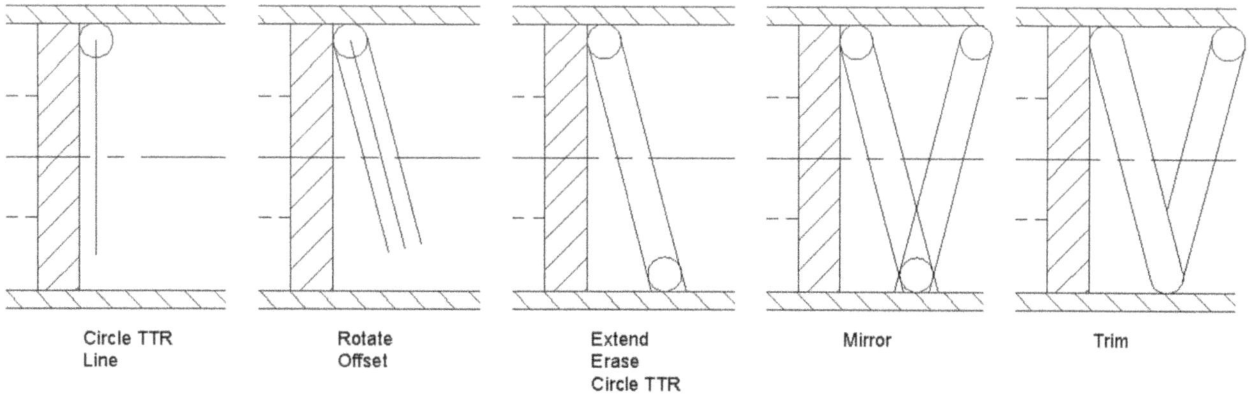

FIGURE 8.30 Drawing the spring.

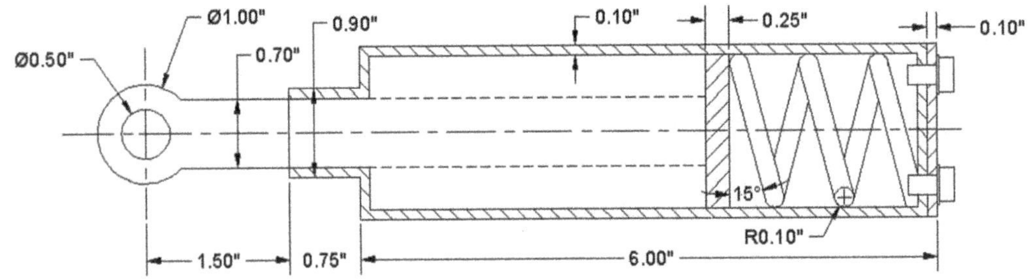

FIGURE 8.31 Completed side view.

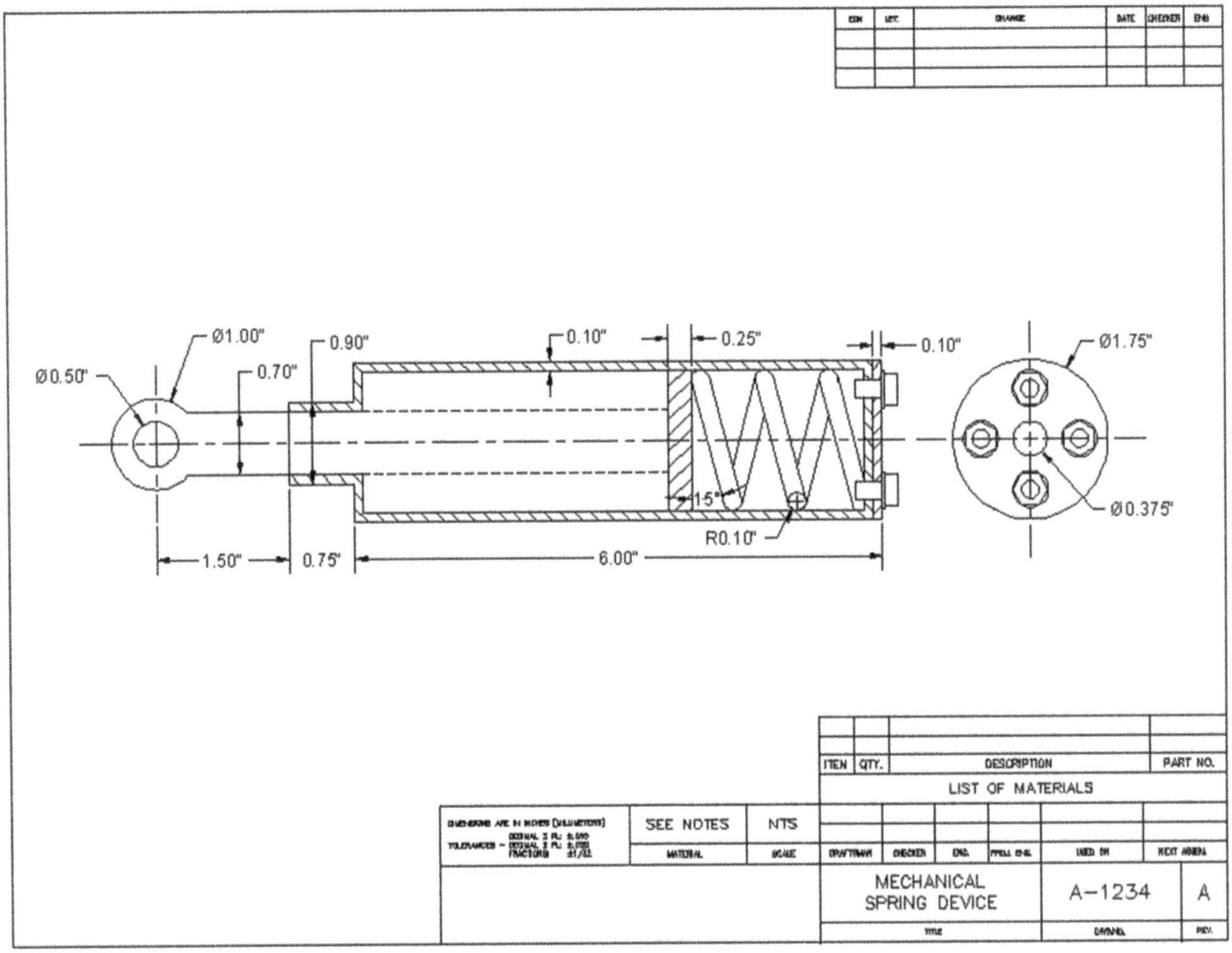

FIGURE 8.32 Mechanical project—final.

Summary

You should understand and know how to use the following concepts and commands before moving on to Chapter 9:

- Polar array
  - Select object
  - Select center point of array
  - Number of objects to create
  - Angle to fill
  - Additional operations and options

- Rectangular array
  - Select object
  - Number of rows and columns
  - Distance between rows and columns
  - Additional operations and options
- Path array
  - Select object
  - Select path
  - Additional operations and options

Review questions

Answer the following based on what you learned in this chapter:

1. List the three types of arrays available to you.
2. What steps are needed to create a polar array?
3. To array chairs around a table, where is the center point of the array?
4. What steps are needed to create a rectangular array?
5. What steps are needed to create a path array?

Exercises

1. In a new file, draw the following five row by five column rectangular array. The chairs are 20″ by 20″ and the offset between them is 50". The array is also at a 45 degrees angle. (Difficulty level: Easy; Time to completion: <10 minutes.)

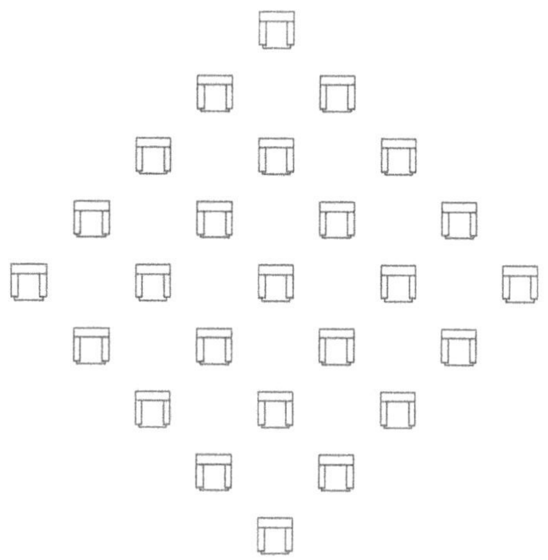

2. In a new file, draw the following seven rows by six column rectangular array, including the centerline grid and the grid lettering and numbering. Offset the column grids and I-beams as shown. (Difficulty level: Easy; Time to completion: 10–15 minutes.)

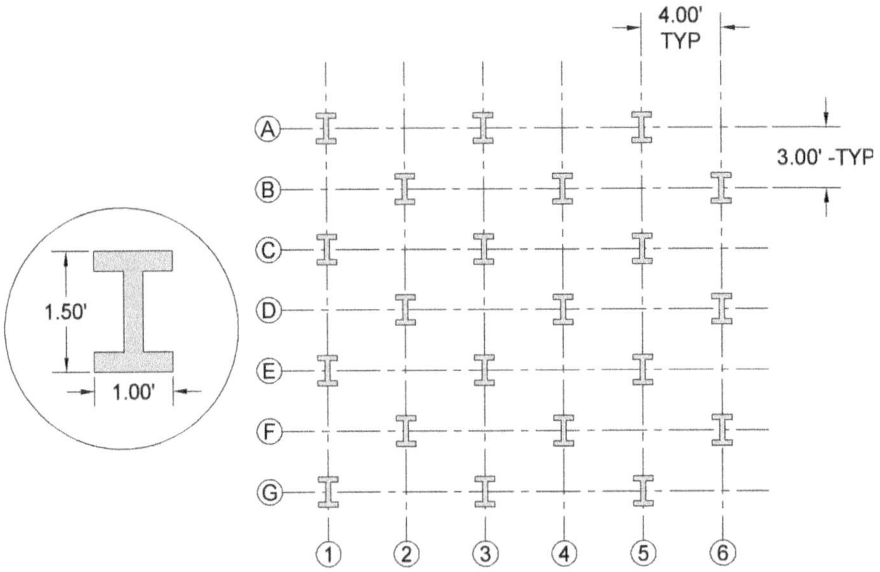

3. In a new file, draw the following set of path arrays with multiple rows. The two paths are simple arcs, and the symbols used are scaled up and shown next to the patterns. (Difficulty level: Easy; Time to completion: <15 minutes.)

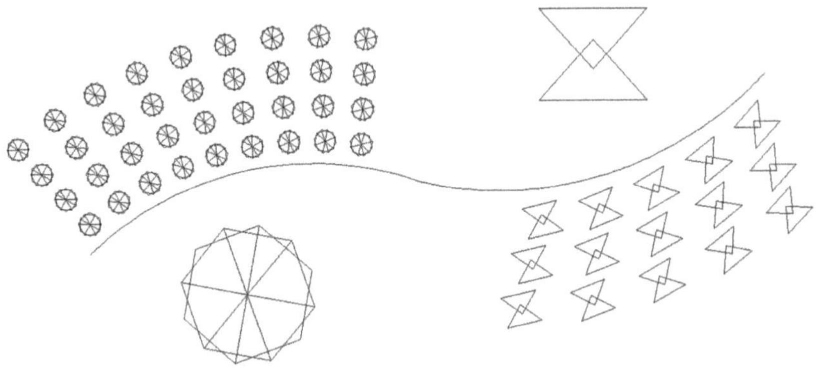

4. In a new file, draw the following set of path arrays with multiple rows. The three paths are simple arcs, and the symbols used are basic top view sketches of a flower pot. (Difficulty level: Easy; Time to completion: <15 minutes.)

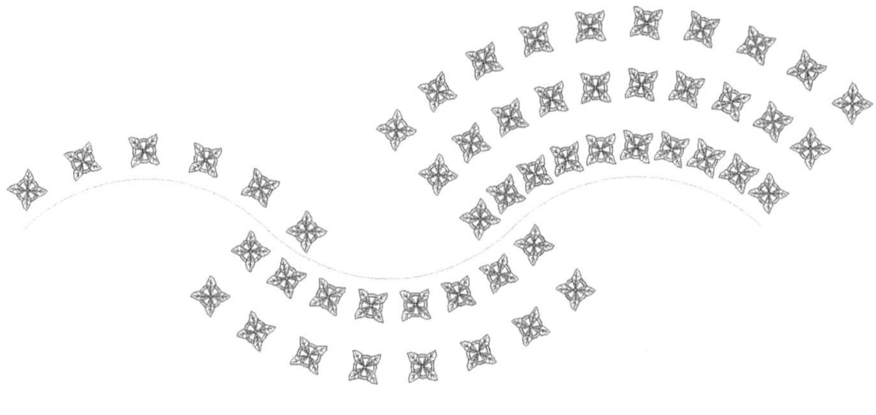

5. In a new file, draw the following *polar* array, made up of a table, chairs, and computers. No sizing is given and all elements can be approximated. Place each element on the appropriate layer and make a block out of it prior to using it in the array. (Difficulty level: Moderate; Time to completion: <20 minutes.)

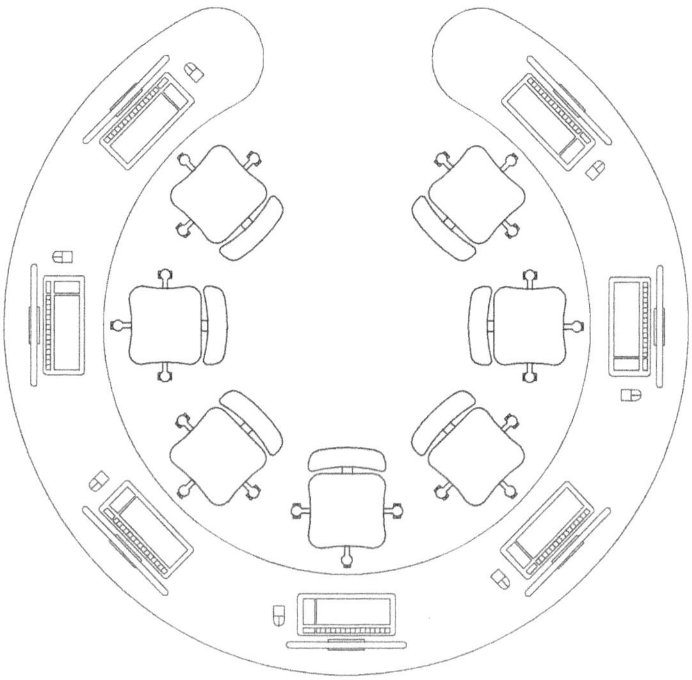

6. In a new file, draw this slightly altered version of the image found at the opening of this chapter, shown here with all the necessary dimensions. For practice, add the dimensioning as well. (Difficulty level: Easy; Time to completion: 15–20 minutes.)

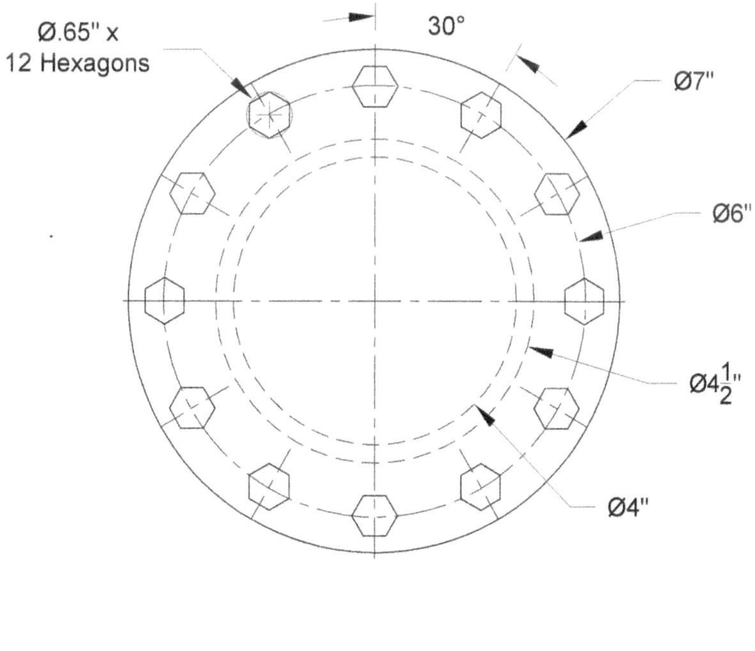

7. In a new file, draw the following gear array. Begin with circles of the indicated diameter. Then, focus on creating the first tooth, as seen in the detail. Make sure it is centered and aligned. Finally, array the tooth (24 teeth are used in the example) and clean up the design with trimming. The result follows. (Difficulty level: Moderate; Time to completion: <20 minutes.)

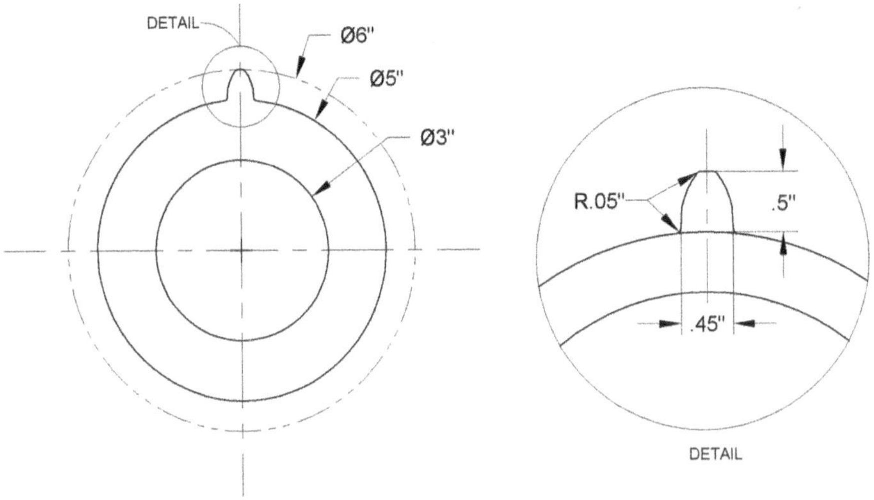

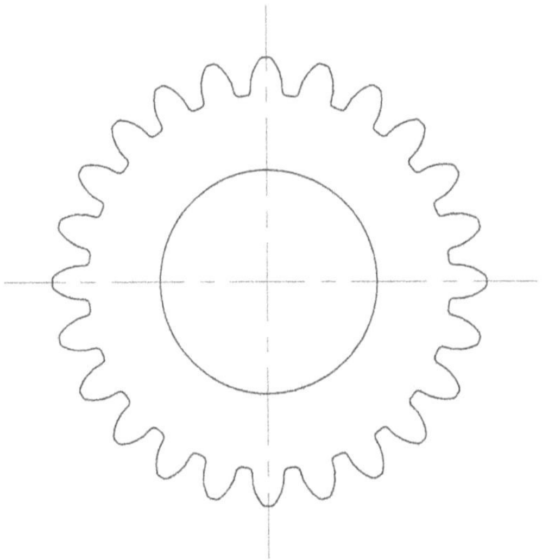

8. In a new file, draw the following pulley and its section view using polar array. All necessary diameters and widths are given. Load the center and phantom linetypes and set units and dimensions to Decimal (accuracy 0.00, with a " suffix). (Difficulty level: Moderate; Time to completion: 30–45 minutes.)

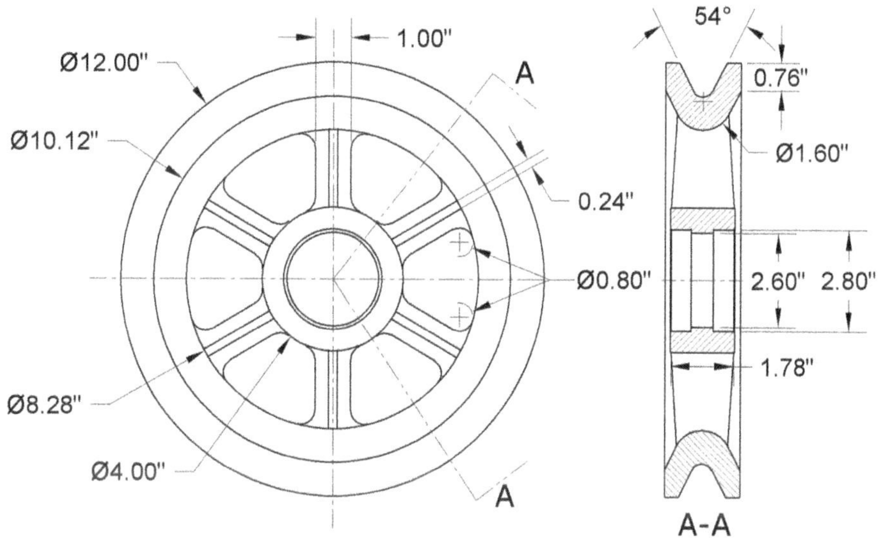

9. In a new file, draw the following fan and its side view using polar array. All necessary diameters and widths are given. Load the center linetype and set units and dimensions to Decimal (accuracy 0.00, with a " suffix). (Difficulty level: Moderate; Time to completion: 30–45 minutes.)

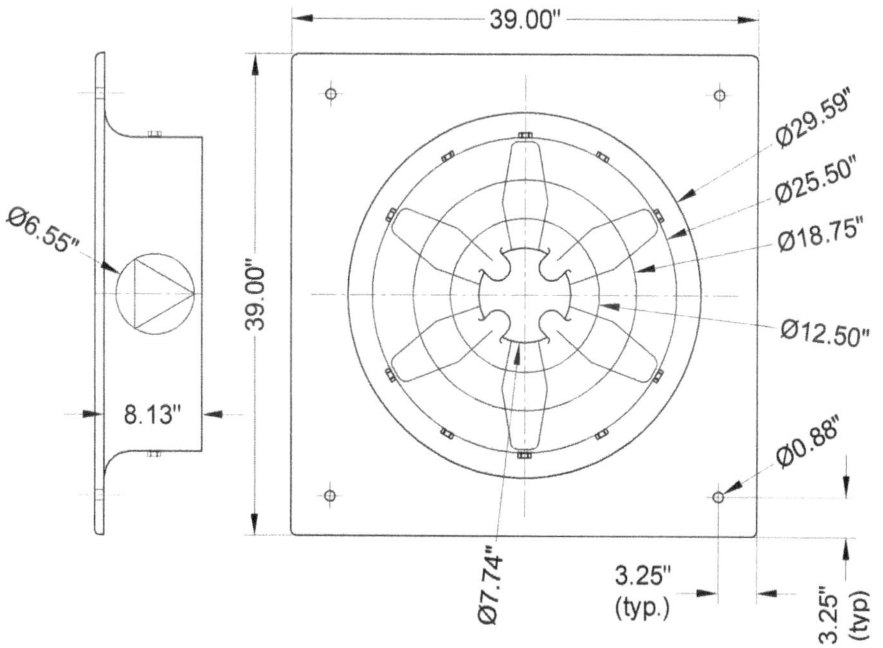

10. In a new file, draw the following mechanical device utilizing several polar arrays. All necessary diameters are given. Load the Center2 and Hidden2 linetype and set units and dimensions to Decimal (accuracy 0.00, with a " suffix). The details are nothing more than scaled up copies of the circled areas; be sure to draft them as well. (Difficulty level: Moderate/Advanced; Time to completion: 60–90 minutes.)

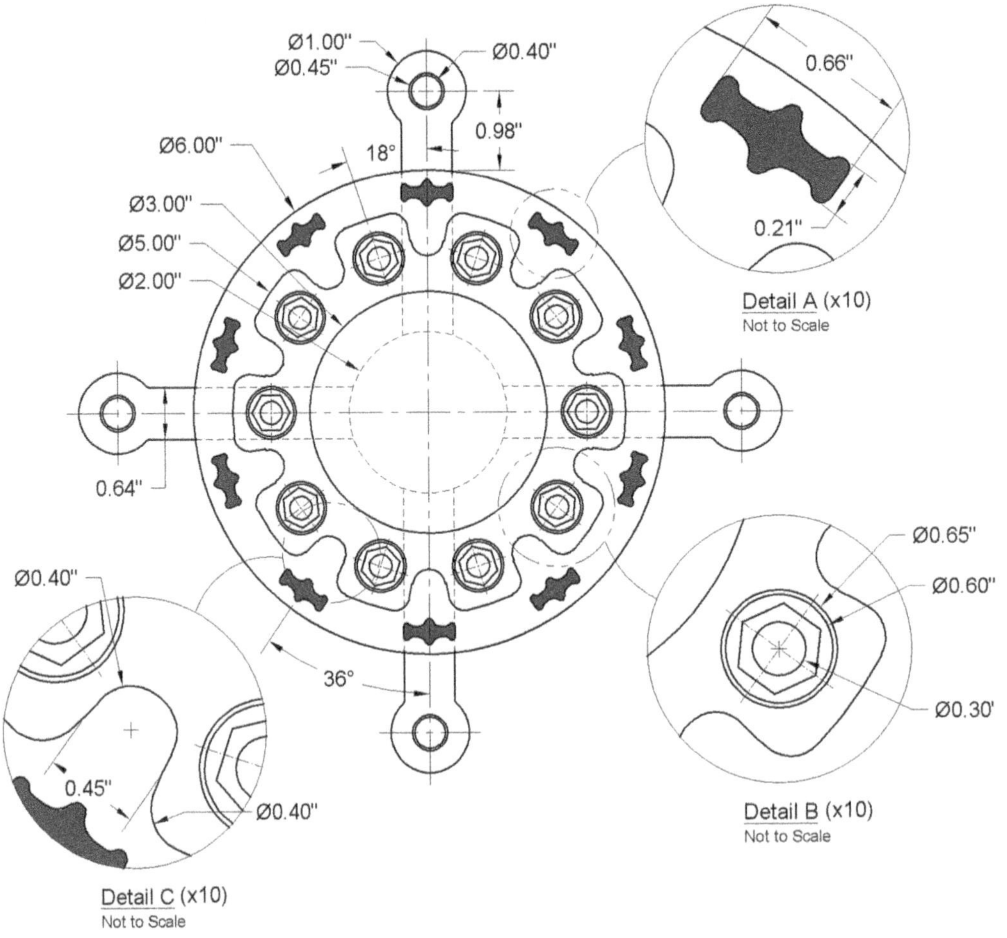

Spotlight on: interior design

Interior design is a major branch of architectural design that concerns itself with space planning and design. The goal is to create a personal, work, or public space that is both functional and aesthetically attractive by the creative use of colors, lighting, and materials. Interior design is not to be confused with interior decorating. Interior designers are often well versed in the principles of architecture, engineering, and materials and must generally hold at least a 4-year degree. They can be called on to create challenging spaces that are tailored to certain ecological or handicapped-accessible needs. It is a creative profession, but one that still requires analytical skills.

FIGURE 1 *Image Source: www.coldcoldlake.com.*

Education for interior designers (in the United States) involves a 2-year associate or 4-year baccalaureate degree in interior design. Master's degrees (MS, MA, MFA, and recently the MID) in interior design are also available, although this advanced degree is not as common. Many professionals pursue advanced degrees in related subjects, such as industrial design, fine art, or education. Doctoral programs in interior design are increasing in number at various institutions of higher education. It is worth noting that some of the more notable interior designers held no formal education, though this is the exception, not the rule.

Following formal training, graduates usually enter a 1- to 3-year apprenticeship to gain experience before taking a national licensing exam or joining a professional association. The National Council for Interior Design Qualification (NCIDQ) administers the licensing exam. To be eligible to take the exam, applicants must have at least 6 years of combined education and experience in interior design, of which at least 2 years constitute postsecondary education in design. Once candidates have passed the qualifying exam, they are granted the title of Certified, Registered, or Licensed Interior Designer, depending on the state. Some states require continuing education units to maintain the license.

Interior designer earnings vary, based on employer, number of years of experience, and the reputation of the individual. For residential projects, self-employed interior designers usually earn a per-hour fee plus a percentage of the total cost of furniture, lighting, artwork, and other design elements. For commercial projects, they may charge per-hour fees or a flat fee for the whole project. The median annual pay for interior designers is about $48,000, with the top 10% earning over $75,000.

So, how do interior designers use AutoCAD and what can you expect? Very often, as far as AutoCAD goes, interior design is tied in with the general architectural design of a building. Therefore, interior designers work side by side with architects on their CAD files. The layering convention often follows AIA standards. You find layers such as A-Furniture, A-Carpeting, and the like. Designers also work with blocks, attributes, and tables and make extensive use of custom hatch patterns and a wide array of colors (such as Pantone).

Interior designers often work in 3D, as the whole idea is often to present and sell the design concept, such as the rendering shown in Fig. 2. In these cases, they must master 3D AutoCAD as well as additional rendering software. With 3D work, AutoCAD is not the only choice, and one can find Rhino, Form Z, and other applications running side by side with AutoCAD or even bypassing AutoCAD completely.

FIGURE 2 *Image Source: www.revolutionp.com.*

Shown in Fig. 3 is a 2D plan view of a residential interior. It makes use of solid shading and custom hatch patterns to give the client a good idea of the outcome of the design effort.

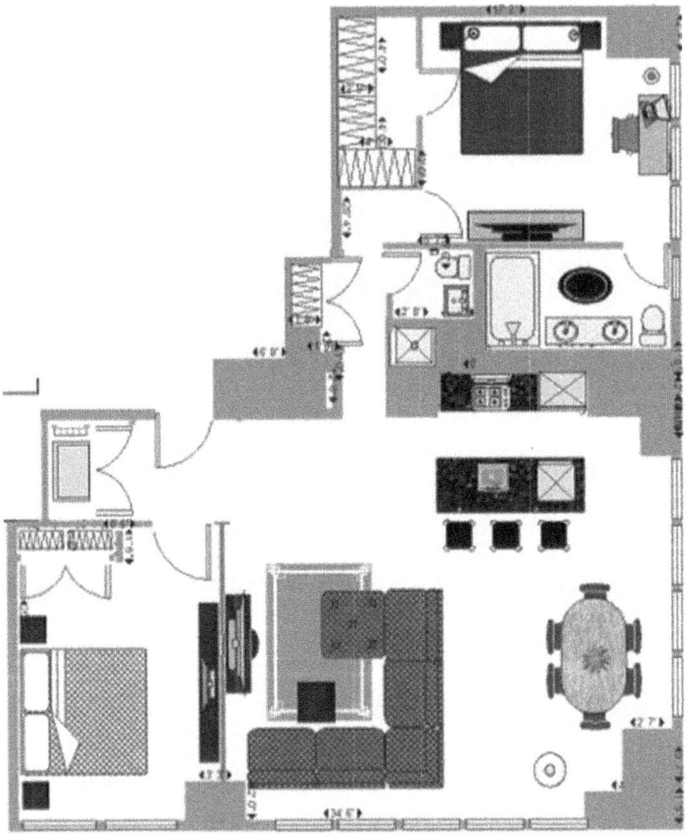

FIGURE 3 2D plan view of a residential interior. *Image source: www.homein.design.com.*

Shown in Fig. 4 is a 3D photorealistic rendering of another design. Although AutoCAD can be used to easily create the underlying architecture, additional rendering tools are usually used for this level of sophistication and realism.

FIGURE 4 3D photorealistic rendering of a design. *Image source: www.spazio3d.com.*

Chapter 9

Basic printing and output

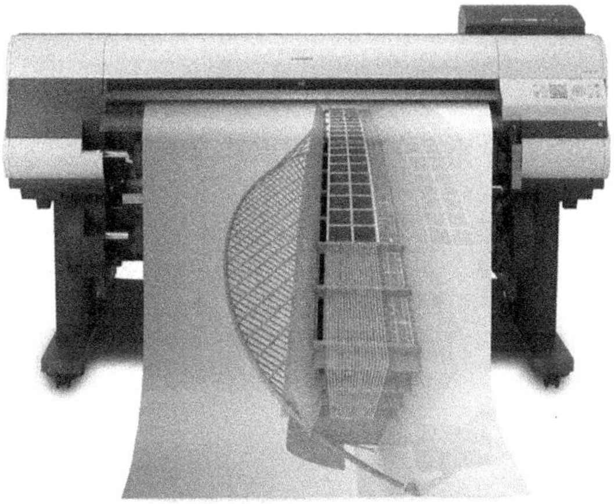

Learning objectives

In this chapter, we introduce the Plot dialog box and the Page Setup Manager and go through all associated options to produce a print or a plot, including the following:

- What *printer* or *plotter* to use.
- What *paper size* to use.
- What *area* to plot.
- At what *scale* to plot.
- What *pen settings* to use.
- What *orientation* to use.
- What *offset* (if any) to use
- The Page Setup Manager.

Upon the completion of this chapter, you will be able to recognize and set up all the features needed to print or plot any type of drawing from any AutoCAD workstation.

Estimated time for completion of this chapter: 1 hour.

Up and Running with AutoCAD 2022. https://doi.org/10.1016/B978-0-323-89923-9.00009-3

9.1 Introduction to printing and plotting

Designers and their clients usually want something tangible to look at and use. Because of this, most design works in AutoCAD must be created with the goal of displaying it on a piece of paper, not just on screen. Indeed, the fundamental immediate "product" of most design work is paper output. This output is then used to manufacture, fabricate, build, or assemble the design. It also may be needed for pre- and postdesign purposes, such as sales or archiving. These needs influence the evolution of a drawing from the outset, and a skilled AutoCAD designer selects a paper size, output scales and lineweights early in the process so the inevitable output proceeds smoothly.

Professional looking output in AutoCAD is no accident and comes with its own set of challenges. Although you may be able to walk up to an expertly set up drawing and just select File→Plot→OK, such smooth results come from only a thorough understanding of what is involved. More important, if the requirements call for a different output of the same drawing (e.g., for an informal check of a small section of the drawing vs. a final full-size submission to the client), you must know what to change and by how much.

AutoCAD allows for a wide array of output, each method with its own tricks and details. Output spans this chapter as well as part of Chapters 10 and 19. Our goal right now is only to familiarize you with the essentials and what AutoCAD needs to know to print or plot a drawing.

Take note, as you go through this chapter, that although some specifics, such as type of printer or plotter, may be unique to your office, home, or school, most of what you need to know stays constant from situation to situation and from one release of AutoCAD to another (even if the locations of the buttons and menus change). Therefore, it is important that you memorize and understand the essential underlying concepts, not just "which buttons to push."

9.2 The essentials

You need to give AutoCAD the following information when setting up a drawing for printing or plotting (See Fig. 9.1):

- What *printer* or *plotter* to use.
- What *paper size* to use.

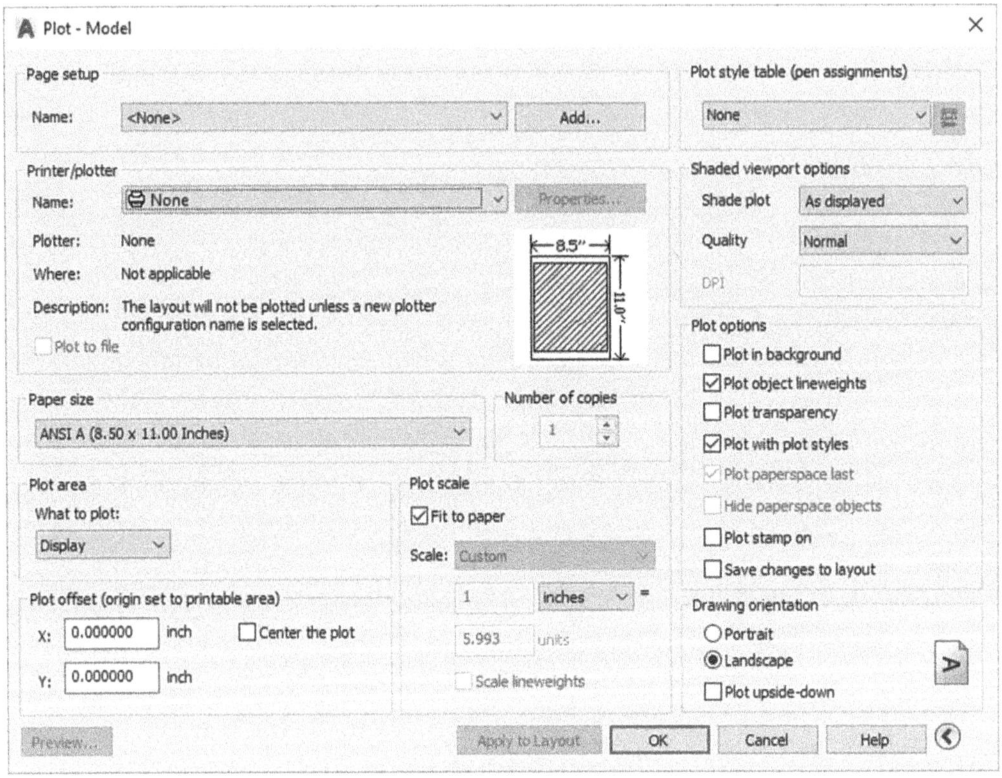

FIGURE 9.1 Plot dialog box.

- What *area* to plot.
- At what *scale* to plot.
- What *pen settings* to use.
- What *orientation* to use.
- What *offset* (if any) to use.

What printer or plotter to use

This is where you select what output device to use. In most architecture offices, engineering companies, and schools, you have a plotter (color or black and white) for C-, D-, and E-sized drawings (paper sizes are discussed next), and a LaserJet printer (color or black and white) for A- and B-sized drawings. Your typical AutoCAD station usually is networked to access both, as requirements may change daily. You need to select which device is getting the drawing. (AutoCAD does not care—it will print a floor plan of a sports stadium on a postcard if you let it.) This choice, of course, is tied in with the scale and the ultimate destination of the printed drawing (Is it a check plot or for a the project manager or a full set for construction?).

What paper size to use

You would think that paper sizing is a cut-and-dried, straightforward matter (somewhere a printing company manager is laughing), but this is certainly not the case. Paper comes in a bewildering array of international sizes and standards in both inches and metric (millimeter) sizing. Some common standards include ANSI, ISO, JIS, and Arch. For the purposes of this discussion, we need to simplify matters and present all paper as existing in the following basic ANSI sizes as viewed in Landscape mode, meaning the first (larger) value is the horizontal size:

- ANSI A size: 11″ × 8.5″ sheet of paper (letter).
- ANSI B size: 17″ × 11″ sheet of paper (ledger).
- ANSI C size: 22″ × 17″ sheet of paper.
- ANSI D size: 34″ × 22″ sheet of paper.
- ANSI E size: 44″ × 34″ sheet of paper.

These are the maximum paper dimensions. Some standards list essentially the same sizes but with the border accounted for and removed, so an ANSI D-size sheet is 34″ × 22″ and so on. Note that legal (14″ × 8.5″) size paper, while certainly can be used, is not common in engineering or architecture.

As a side note, we can now define precisely the difference between printing and plotting.

- *Printing* usually refers to the output of drawings on A- or B-sized paper, which is easily done on most office desktop size printers or office copy machines. Most home printers can do only A size, but this, of course, varies with the size (and cost) of the printer or office copy machines.
- *Plotting* usually refers to the output of drawings on C-, D-, or E-sized paper, which must be done on plotters, as a desktop printer generally cannot accept paper of this size.

Memorize the paper sizes presented here, as it is pretty much the industry standard as far as AutoCAD design is concerned. You will hear these sizes asked for and referred to, come printing time; and if there is any deviation from these sizes, rest assured it will not be anything radical, as printers and plotters can accept only so much flexibility, at least among the models that small companies and schools can afford. Architectural (Arch) sizes are similar and follow a similar letter naming convention.

What area to plot

What "area to plot" can be taken quite literally, as in, "What do you want to see on the printed paper?" The essential choices really boil down to two. Do you want to see the entire drawing (usually) or parts thereof (sometimes)? Your choices in that regard include the following.

Extents

This option prints everything visible in the AutoCAD file. However, this means everything visible to AutoCAD (and not necessarily to you right now on screen), so if you have a floor plan with some stray lines that you did not notice located some distance away, AutoCAD shrinks the important floor plan to make room for those unimportant stray lines. This is the same as typing in Zoom, then Extents (from Chapter 1). So, it is a very good idea to zoom to extents prior to printing. Remember this acronym: What you see is what you get (WYSIWYG). If you can see it on the screen after zooming to the extents, it will print unless prior arrangements are made (freezing a layer erasing the "stray" objects, etc.). For a carefully drawn layout, Extents is the one good choice when you want to plot the entire drawing.

Window

As you may have guessed, this is for those situations where you want to see just a piece of the drawing. You simply select that choice, click on the Window button that appears to the right of it, and draw a selection window around what you want to see.

Display and limits

The remaining two choices are of little use in most situations. Display plots only what is seen on screen in the current view of the Model Space (or the current Paper Space view in a Layout tab, something we have not covered yet). One can generally use a Window plot selection in those situations, but Display is fine as well if you want only the current on screen visible objects printed. Limits plots the entire drawing from 0,0 to the preset Limits, something most users do not set anyway.

At what scale to plot

When plotting from model space without title blocks and borders surrounding your drawings, you generally want your design to fit on the desired output paper size to some known scale. The last thing you want is for someone to get ahold of your design drawing with few or no dimensions, assume the scale and start using it to build something. Use the Scale: option to select from typical mechanical and architectural scales until your drawing fits on the chosen sheet size. There are of course industry scale standards to follow, but choosing an actual known, standard scale from the list is more desirable than scaling to fit.

But wait… what if you do not care about scale and just want a print to look over? Here, the pendulum swings in the other direction and things become very easy. Just check off the Fit to Paper option box above the Scale: list and the drawing fits on whatever paper you are using, with no regard to scale whatsoever. While this technique is not for final prints of to-scale architectural layouts, it is used very often for non-scaled work, such as electrical wiring schematics, network design, conceptual layouts, and of course check plots.

Of all options in Plot, scaling of drawings causes the most confusion to a beginner AutoCAD user, for the simple reason that scale is closely tied in with a very important topic: Paper Space. The essential idea here is that while, yes, you can assign a scale from the Plot dialog box and print directly from Model Space, this is not usually done for most drawings, except simple and small mechanical pieces, such as bolts and brackets that would fit on a letter size sheet at 1:1 scale. What is done instead is that a sheet with a title block is set up in Paper Space and the design (in model space) is displayed and scaled through viewports (much more on this in Chapter 10), then the Layout/Paper Space/Sheet is printed at 1:1 (one-to-one) and a properly set up drawing(s) seen through a viewport prints to scale.

What pen settings to use

This is a topic that has an entire chapter all to itself (see Chapter 19), and we just do a brief overview here. Proper pen settings are a major factor in giving your drawings a professional look. The concept is simple, usually addressed in hand drafting, yet often overlooked in the computerized world of AutoCAD. It deals with the fact that linework in a drawing is not all the same size width or thickness. Primary features, such as walls on a floor plan, should appear darker (thicker) on a print, while secondary features, furniture, for example, should appear slightly lighter (thinner).

For those students old enough to remember the hand drafting days, this was the reason for the existence of 2H, 4H, and other pencil lead hardness grades. When hand drafting ink on mylar you would select different ink pen widths for different features. This is the reason for the existence of the Pen Style Table or pen assignments in AutoCAD.

With AutoCAD, different colors represent the different ink pen widths, and a strict standard needs to be followed as to what drawing features have what colors assigned to them. In this manner, drawings come out with different thicknesses of linework, that have enough differentiation to be noticed by the eye; and as a result, the drawings are much more readable and professional looking. These settings are created, customized, adjusted, and saved under the ctb extension, for all to use according to company standards set by management.

For now, select monochrome.ctb as your ctb file; this will set all pens to print as black regardless of the color of the layer chosen. The full set of ctb and pen setting customization tools are presented in Chapter 19.

What orientation to use

This is simply a question of whether you want the drawing to be oriented Portrait or Landscape; in other words, do you want the longer side of the sheet of paper oriented up/down or across? This should be familiar to just about every computer user who has printed anything from a word processor or other software.

One twist here is that you can check off a box to plot upside down. A useful application for this option appears when you are printing many sets of similar-looking (but different) drawings. If you want to know which drawing is coming out of the plotter without waiting until it plots completely, send it in backward. That way, the title block (with the identifying drawing number) comes out first and the question is instantly resolved.

What offset to use

This refers to whether you need to shift your drawing around to account for a border and stapling area on the left (or elsewhere). If you check off Center the plot, it will be automatically centered on the paper, and in most cases, this is what you would do when printing from Model Space.

Miscellaneous

There are several other features in Plot. For example, you can plot to a file (*.plt) and take it with you to another computer. This is useful for large volume drawing duplication at stores that have no AutoCAD software, only printers, such as FedEx or other local print shops. You can also turn on a plot stamp so information such as date and time of the printing can appear in the border. Some plotting features are unique to 3D and solid modeling. We explore some of these topics as well in Chapter 19. Finally, there is a Preview window and several copies to print selection, the functions of those being self-explanatory.

You can also select a printer that will generate PDF files (AutoCAD PDF of various qualities or DWG to PDF, pc3) or a graphic format (Publish to Web PNG.pc3 or Publish to web jpg pc3). Rather than paper coming out of a printer, you will be asked where you want to save your PDF, JPG or PNG file for viewing electronically. We will return to the PDF topic again in Chapter 16.

9.3 The plot dialog box

Let us now take a close look at the plot dialog box and identify the locations of all the features described earlier. Open the floor plan from the earlier chapters and bring up Plot via any of the following methods.

Keyboard: Type in **plot** and press Enter. Also **Ctrl-P**
Ribbon: **Output** tab→**Plot**

The Plot dialog box appears, as seen in Fig. 9.2. It should open to full size right away. If for some reason it is not full size, press the small arrow at the bottom right to expand it fully to show all the features we discuss previously in Fig. 9.3.

Go through all the settings, referencing the explanations in the previous discussion, until you completely understand what each one does. Set the final settings as shown next.

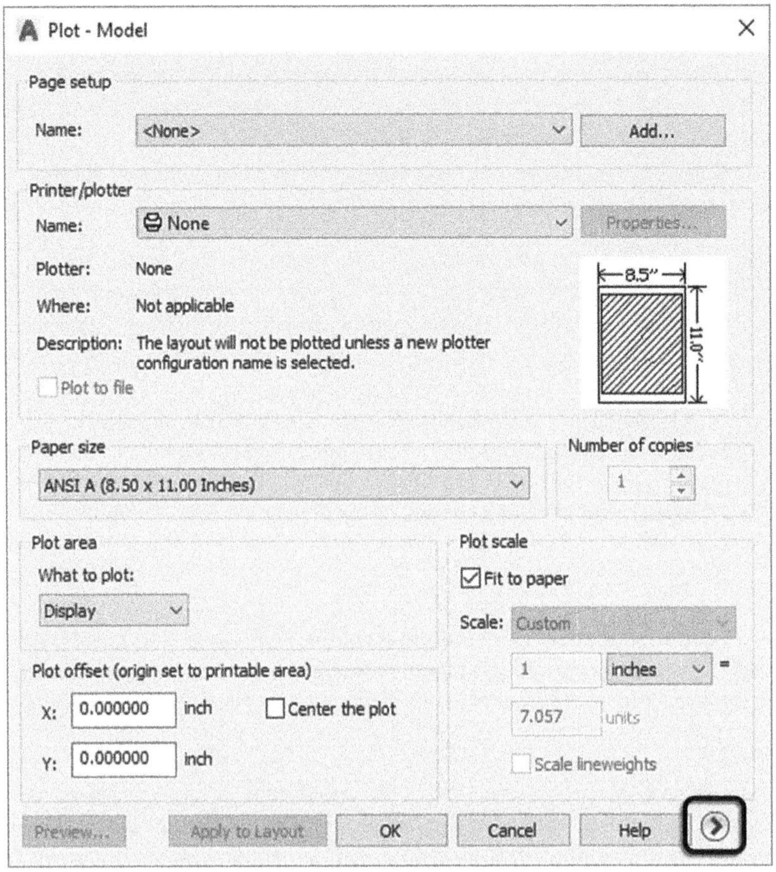

FIGURE 9.2 Plot dialog box collapsed, with expand arrow.

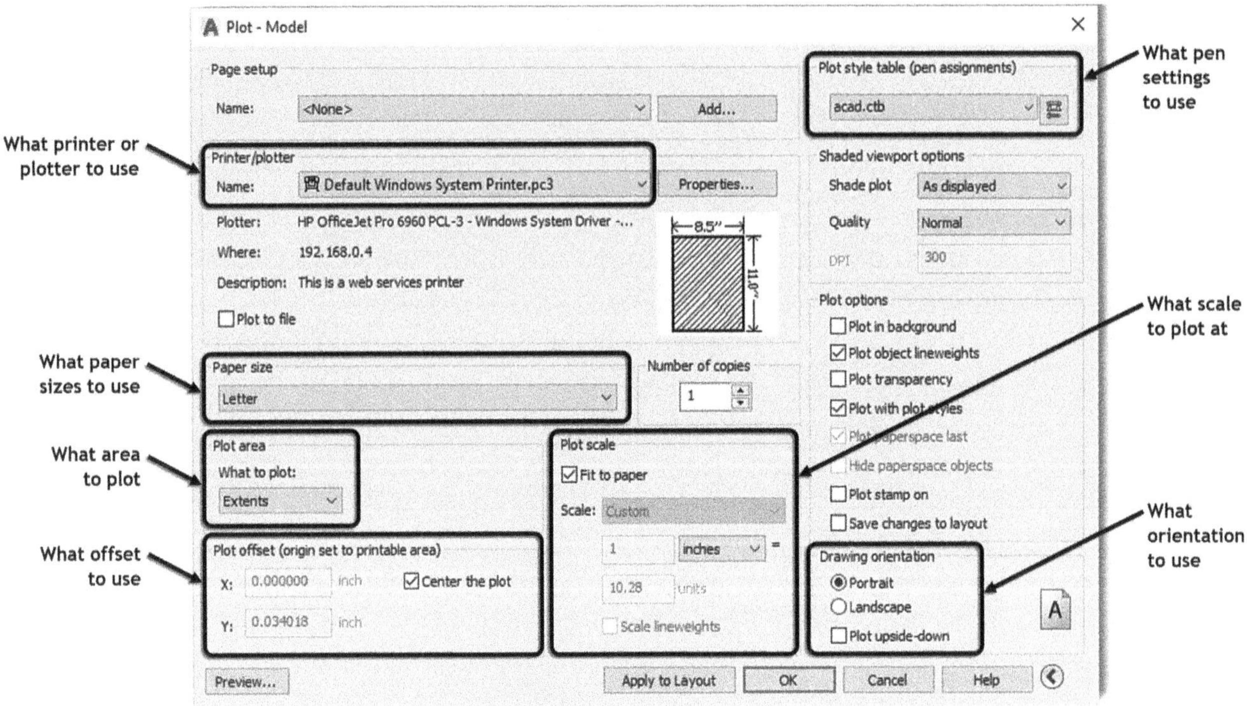

FIGURE 9.3 Plot features.

- *What printer or plotter to use*: Set whatever printer is available at your location (it likely differs from what you see here). If you have a plotter, ignore it for now and use a printer instead, as we are using A-size paper for this exercise.
- *What paper size to use*: Select from the drop-down list. We want any ANSI A (8.50 × 11.00) or Letter size is fine.
- *What area to plot*: Select Extents.
- *At what scale to plot*: Check *Fit to paper* (for this exercise).
- *What pen settings to use*: Select monochrome.ctb and click on Yes if prompted to use this for all layouts.
- *What orientation to use*: Select Landscape for this example.
- *What offset (if any) to use*: Check *Center the plot*.

Do *not* press OK yet. The Plot box should look like (or close to) Fig. 9.3. We need to first discuss an important feature of printing: the preview.

Preview

Much like the situation with creating an array or hatch, a Preview button is available. With printing and plotting drawings, it is crucially important to preview the drawing before sending it to the printer or plotter. Paper and ink cost money, and everyone regardless of skill level makes mistakes in setting up printing. This really should become a habit, as errors can be costly, especially on large print jobs. On an industry-wide scale, this simple Preview button saves tons of paper and barrels of ink. Not only printing setup errors but even linework drawing errors can be caught during the preview process.

Go ahead and press Preview… (lower left corner), and the Preview screen in Fig. 9.4 appears. You can zoom and pan around using standard mouse techniques, and when you are ready to send the drawing to the printer (or return to Plot for adjustments), right-click and select Exit or Plot, respectively.

Go ahead and plot. If everything was set up correctly and your computer is connected to the printer, the floor plan should print pretty much as expected and seen in the preview of Fig. 9.4.

If something does not look right anywhere in the plotting process, review the complete set of steps, one by one, returning to the Preview screen to check up on the effects of the changes.

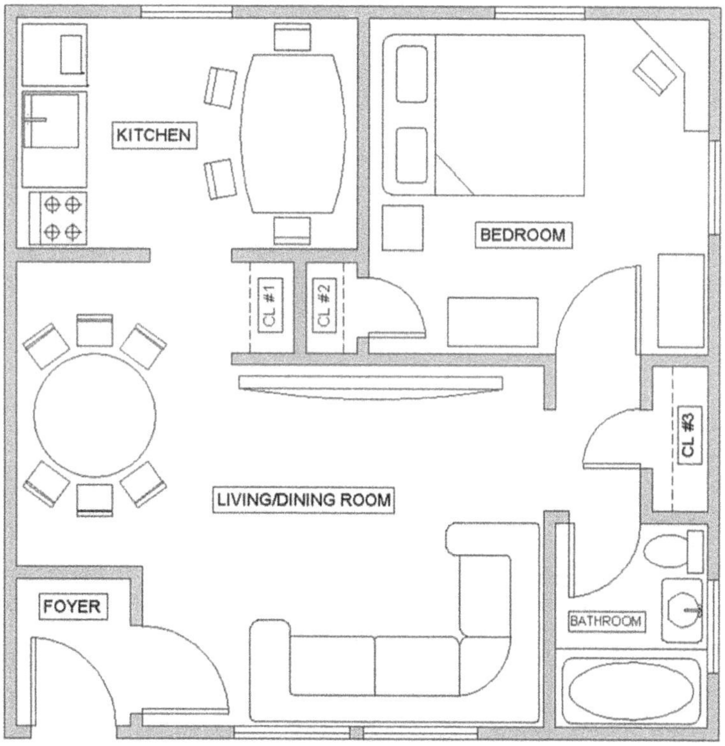

FIGURE 9.4 Plot preview.

9.4 Page setup manager

There is one last topic to learn in this chapter. The settings on the print job you just sent are not saved in memory. Test this by typing in plot again. Unless you happened to click the Apply to Layout button by chance, the settings have not been saved and you must reset them all again. Another technique to "save" your settings involves a feature of plotting called *Page Setup Manager.*

The idea here is as follows. You can enter all the data for a print job in a dialog box that is, similar-to the Print dialog box and save it under a descriptive name. for future use Then, save the drawing with these settings, and when you do the actual plot, the name you saved it under appears under Page setup (upper left corner). No more entering settings one item after another; just select the Page setup (it may already be cued up), preview, and print.

This has two advantages. The first is obvious: It speeds things up a lot by keeping all the drawing's print settings ready for use. The second may not be as obvious. Because you can save as many setups as you want, you can have one drawing set up for multiple printers or multiple settings.

For example, your office may need the final output sent in color to a D-sized plotter at a certain scale, but in-house checks are done on 17″ × 11″ sheets, in black and white and no scale. You can easily set up two distinct setups and name them Client_Layout and Check_Layout. It is not unusual to see up to half a dozen setups on one drawing, each with its own purpose: presentation, archiving, optimized for color, check plots, and so forth.

A question usually arises when this topic is presented. Does this need to be done every time a new drawing is created from a blank file? The answer is yes, but it is something that needs to be done only once and saves you much time down the road. Obviously, using templates negates the need to do this at all, as all settings carry over.

The Page Setup Manager (Fig. 9.5.) is found using the Application Menu (Big A): Print→Page Setup… or the Output tab, Plot Panel, Page Setup Manager tool on the Ribbon. or if you right-click on the current Model or Layout Tab you can get there as well.

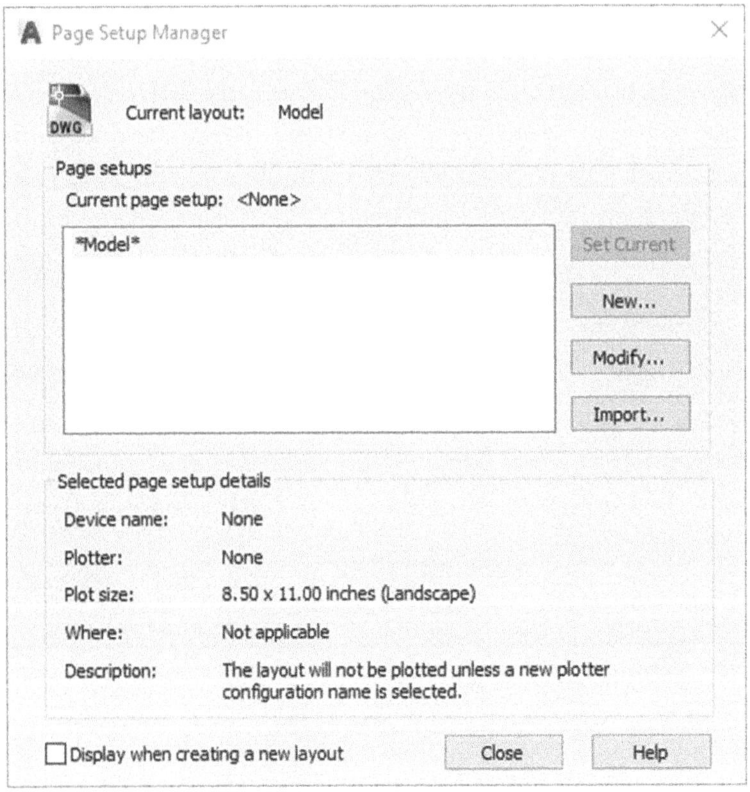

FIGURE 9.5 Page Setup Manager.

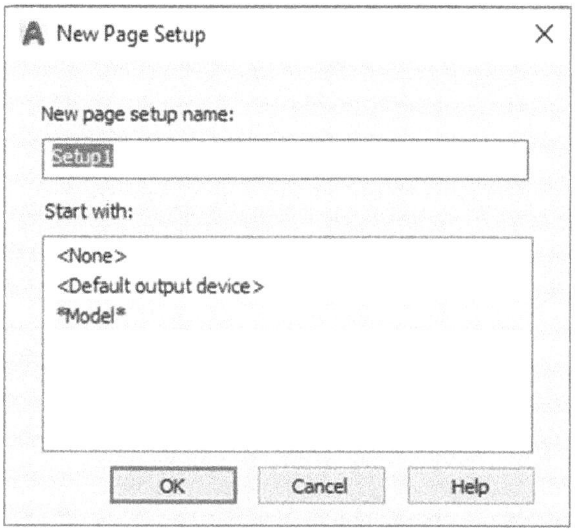

FIGURE 9.6 New Page Setup.

Press the New… button, and the New Page Setup box appears, as shown in Fig. 9.6. Enter the name of the new setup and press OK.

You are then taken to the now familiar-looking Page Setup dialog box. Upon completion of the settings, a preview (yes, still needed), and a press of the OK button, you are redirected to the Page Setup Manager, ready for another setup if needed. If not, press Close.

To see the benefit of this, go to Plot and select the name of the setup from the Page setup in the upper left corner; all your settings appear. Plot the drawing and save it. You never have to enter settings again (unless something changes, like the arrival of a new printer). So, finally, after learning all this new material, you can now reduce printing to a pressing of the Preview and the OK buttons—our goal all along.

As stated before, there is still a lot we have not covered, and we revisit this topic again in Chapters 10 and 19. Printing is something you usually receive assistance with at a new job, training center, or consulting gig. However, the most you should expect to be told is what printer is commonly used (up to 10 connected to one networked computer is not unheard of) and what is the standard ctb to use. The rest is up to you, and with the knowledge you gained in this chapter, you should be proficient in all printing matters just short of the level of AutoCAD management and administration.

Summary

You should understand and know how to use the following concepts and commands before moving on to Chapter 10.

- What printer or plotter to use
- Differences between plotting and printing
- What paper size to use
  - ANSI A size
  - ANSI B size
  - ANSI C size
  - ANSI D size
- What area to plot
  - Extents
  - Window
  - Display
  - Limits
- At what scale to plot
  - 1:1
  - Fit to paper
  - Other industry standard scales

- What pen settings to use
- ctb files
- What orientation to use
  - Portrait or Landscape
- What offset (if any) to use
  - Center the plot
  - X or Y offset
- Preview
- Page Setup Manager

Review questions

Answer the following based on what you learned in this chapter:

1. List the seven essential sets of information needed for printing or plotting.
2. List the five paper sizes and their corresponding measurements in inches.
3. What is the difference between printing and plotting?
4. List the four plot area selection options.
5. What scale setting ignores scale completely?
6. What is the extension of the pen settings file?
7. Explain the importance of Preview before printing and plotting.
8. Why is the Page Setup Manager useful?

Exercise

1. Open the mechanical device drawing from Chapter 8, Polar, Rectangular, and Path Arrays, and practice printing with a variety of scales and other settings. (Difficulty level: Easy; Time to completion: <10 minutes.)
   - Select printer (specific to your school, home, or office or print to a built-in AutoCAD PDF printer).
   - Select paper size; stay with A size.
   - Select Extents for the area; observe the results in Preview.
   - Select Window for the area; select an area and observe the results in Preview.
   - Try a variety of scales such as 1:1, 1:2, and so on; observe the results in Preview.
   - Pick the monochrome.ctb file.
   - Try both Landscape and Portrait orientation.
   - Center the plot.

The following figures show the mechanical device in an Extents Preview (first part) and Windows Preview (second part).

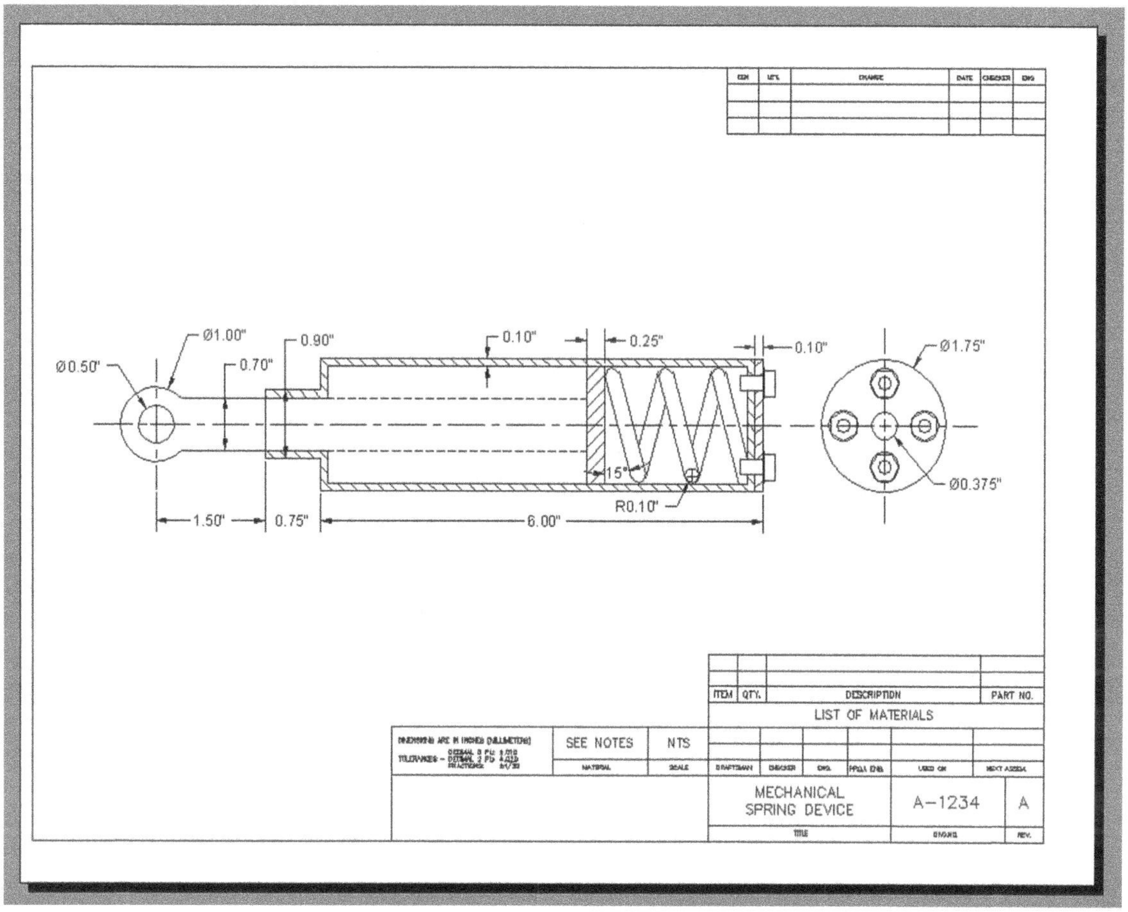
Ø1.00"
Ø0.50"
0.90"
0.70"
0.10"
0.25"
0.10"
Ø1.75"
15°
R0.10"
Ø0.375"
1.50"
0.75"
6.00"
ITEM
QTY.
DESCRIPTION
PART NO.
LIST OF MATERIALS
SEE NOTES
NTS
MECHANICAL SPRING DEVICE
A–1234
A

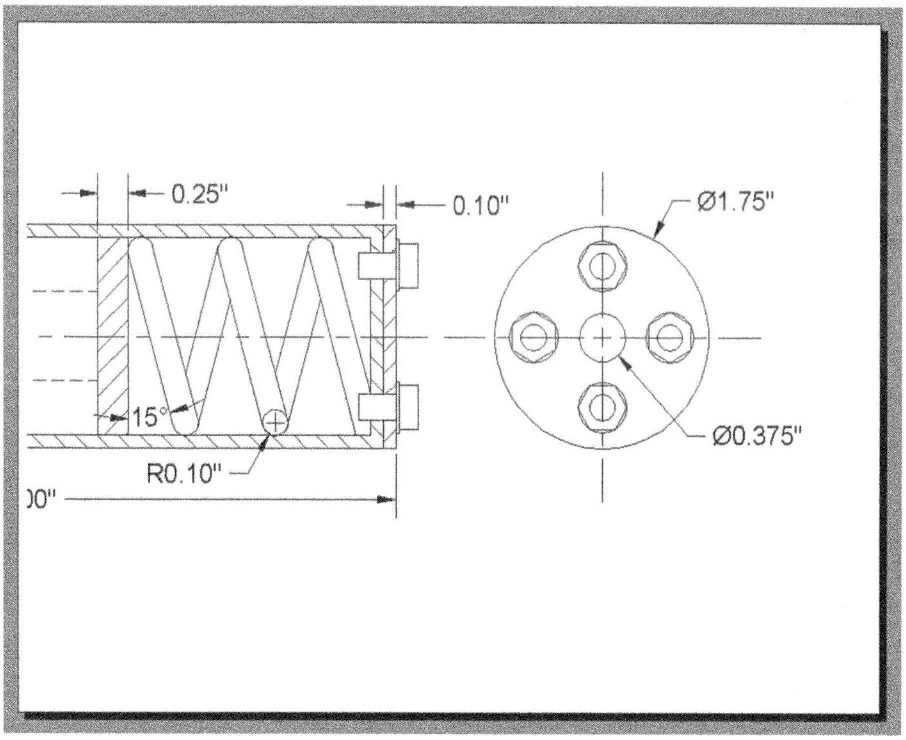
0.25"
0.10"
Ø1.75"
15°
R0.10"
Ø0.375"

Chapter 10

Advanced output—Paper Space

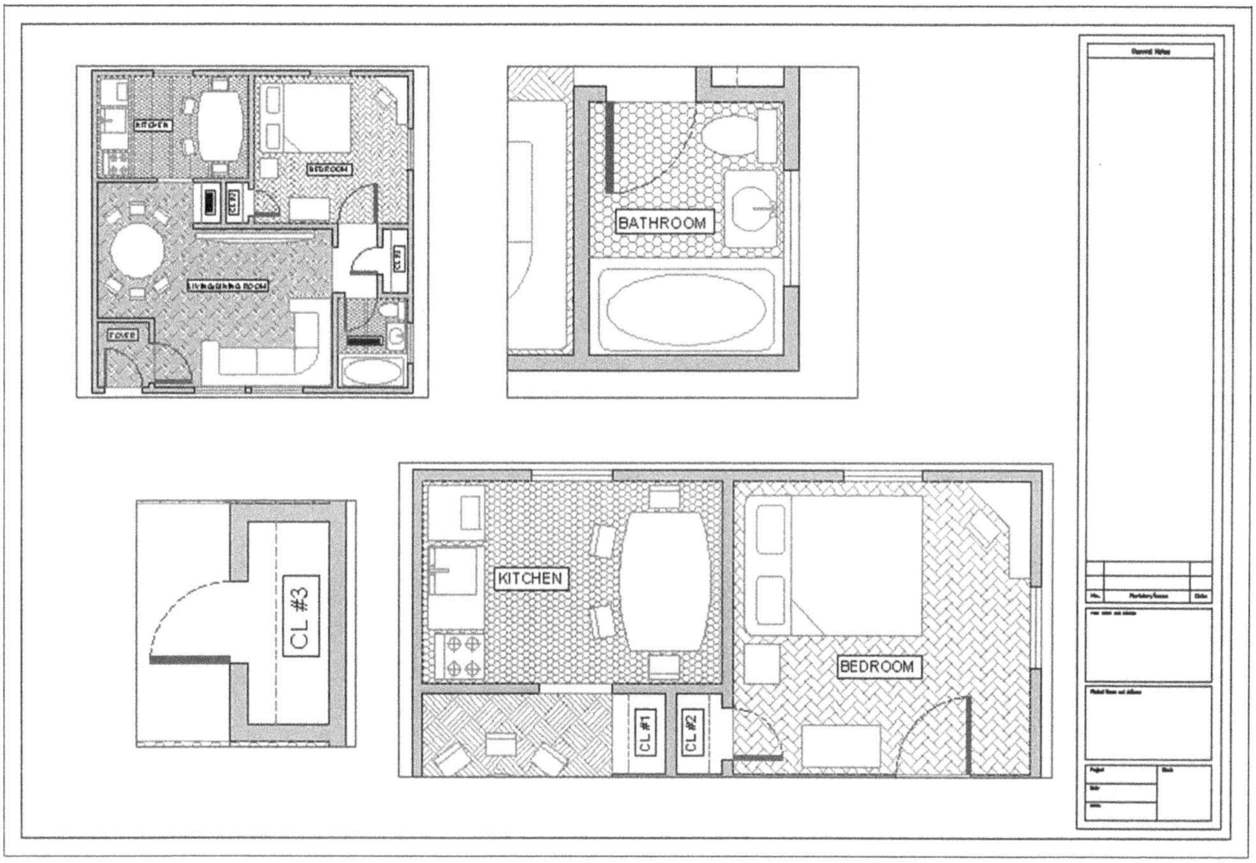

Learning objectives

In this chapter, we introduce and thoroughly cover the concept of Paper Space. We specifically introduce

- The primary reasons for using Paper Space
- Entering Paper Space and setting up title blocks (Layouts)
- Viewports and their properties
- Viewport scale assignments
- Controlling what is visible in viewports
- Setting up text and dim styles based on the viewport scale
- The new Annotation feature that simplifies the setup of text and dims

By the end of this chapter, you will be well versed in applying this critical concept to your design work. Estimated time for completion of this chapter: 3 hours.

Up and Running with AutoCAD 2022. https://doi.org/10.1016/B978-0-323-89923-9.00010-X

10.1 Introduction to Paper Space

Paper Space is an extensive and critically important topic in AutoCAD. Learning it is a rite of passage for all AutoCAD designers, and proper setup and use of Paper Space is a mark of an expert user. The topic is not necessarily difficult but can be somewhat tedious, requiring care and attention from the designer who really must "dot the Is and cross the Ts" to assure everything is done right. It is also a topic that needs to be understood well in theory, first and foremost. The actual "button pushing" process to set things up is much easier when you understand what Paper Space is and why it is used. The result of learning all this, however, is a professional drawing set, i.e., accurate and easy to read, understand, and print. This is a necessity and a goal well worth striving for. We explore Paper Space from all angles, one step at a time. Read and understand each section before moving on to the next, as the knowledge is cumulative.

What is Paper Space?

When you create a design of a building with pencil and paper, this design needs to be drawn to a certain scale, as opposed to full size, for the simple reason that it would not fit on a piece of paper if you did not do this. This is the reason why architects carry with them an architectural scale, a portion of which is shown in Fig. 10.1.

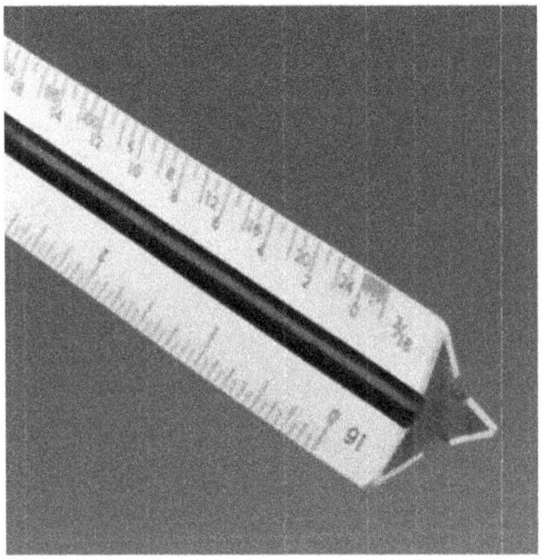

FIGURE 10.1 Architectural scale.

Then, they can simply pick the appropriate scale for the size of their intended design (1/8″ = 1′-0″ is common), and now, because 8 feet are represented by 1 inch, the building floor plan can fit nicely on a large sheet of paper.

In AutoCAD, however, you do not draw "to scale." Because you now have an infinite-size canvas on which to create your design, everything is drawn one to one (1:1) or "life-size," meaning that, if you were to jump into your design through the computer monitor, you would find yourself in proportion to (and would be able to freely walk around inside) your building.

This is one of the Golden Rules of AutoCAD, and indeed in all computer-aided design, and marks a significant departure in philosophy and method from a hand-drafting approach. In fact, attempting to draw "to scale" results in complications and potential for error when creating, and later changing, the design and is simply not done in practice now that we have computers.

Drawing one to one (1:1), however, creates a whole new set of problems. How do you output a design that may be dozens or hundreds of feet across, onto a paper that is generally only 3 or 4 feet wide? Furthermore, this design must be framed by a title block or title page that itself must be drawn one to one (1:1), so as to not violate the Golden Rule. The solution is elegant, simple in principle, and has stood the test of time.

The idea is as follows: Draw the design, full-size (1:1) in one area of AutoCAD, the *Model Space*, and then draw the border with title block, also full size (1:1) in another somewhat unconnected and unrelated area, called *Paper Space*. Finally, connect the two together via a tool called a *Viewport*, which is simply a "window" or a hole looking from

Paper Space onto Model Space. This viewport is then assigned an appropriate scale, and if everything is drawn and printed full-size, one to one (1:1) from Paper Space, then the design comes out at a scale on the full-size sheet and the problem is resolved.

This, is Paper Space: Nothing more than a technique to separate the two full-size worlds and then reconcile them in a meaningful way to get around the tricky problem of a finite-size piece of paper holding a full-size, and usually much larger, design.

Be sure you understand the previous few paragraphs completely. Everything outlined from here on is easier to understand once you have a good idea of why we are doing and what we are doing.

10.2 Paper Space concepts

The devil, as they say, is in the details, and the study of Paper Space involves a number of separate concepts, each of which is an important part of the complete picture and needs to be looked at closely:

- *Layouts*: Entering Paper Space and setting up title blocks
- *Viewports*: Viewports and their properties
- *Scaling*: Viewport scale assignments
- *Layers*: Controlling what is visible in viewports
- *Text and dims*: Setting up text and dim styles based on viewport scale
- *Annotation*: A feature that simplifies the setup of text and dims
- *Summary*: Putting it all together

These topics are listed roughly in the order a designer would go through them as he or she sets up and does the project. As an example drawing, we continue using the apartment floor plan that you have been working on since the start of Level 1. It happens to fit nicely on an ARCH D-size (36″ × 24″) sheet of paper at 1/2”=1’−0” scale to illustrate the concepts. If you want to use another design, perhaps one drawn in one of the end-of-chapter exercises or one your instructor assigned to you, then use that. If you stick with the apartment floor plan, then turn on and thaw all the layers. It may be messy, but we fix that later. Let us begin with the first topic, Layouts (aka Paper Space).

Layouts

The first step involves switching over to Paper Space and familiarizing yourself with the Layout tabs. They are found at the bottom of your drawing area and resemble Microsoft Excel tabs. In the default acad.dwt, you will find Model, followed by Layout1, Layout2, and a plus for creating more tabs, as in Fig. 10.2.

Go ahead and switch into Paper Space by clicking on Layout1. You notice a different environment, as shown in Fig. 10.3. For starters, the UCS icon is a triangle and the Layout1 tab is highlighted. Your background color is white, and finally you have a variety of background features (rectangles, shadows, viewports, and others). You now know how to enter and exit Paper Space.

Nothing is inherently special about Paper Space. Remember, it is just a technique in AutoCAD to create two separate, full-size worlds. You can draw in Paper Space as if it is your regular drawing area with no adverse effects, but of course, that would defeat the whole purpose of what we are doing.

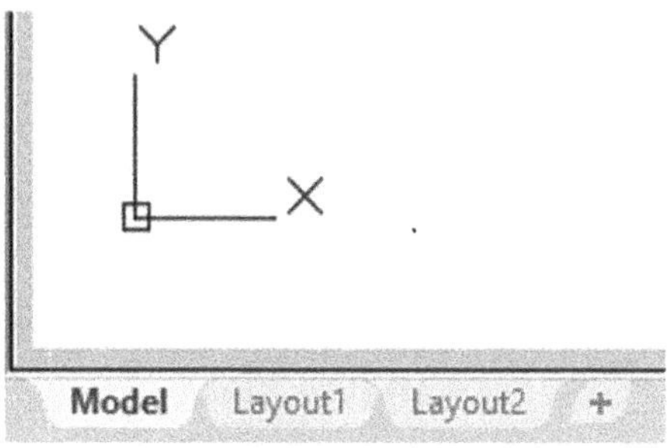

FIGURE 10.2 Layout tabs.

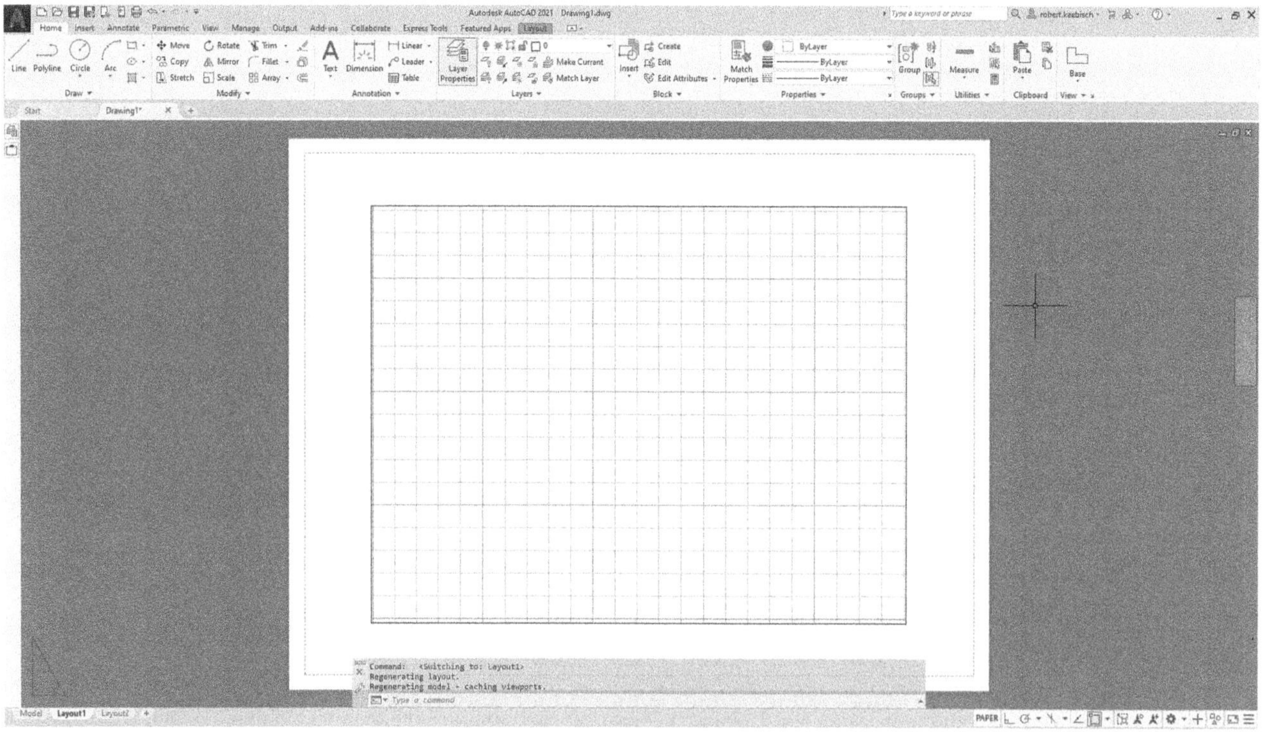

FIGURE 10.3 AutoCAD Paper Space.

You have the option to display or hide the Paper Space shadow, printable area, and other features. Some experienced users prefer their Paper Space to look as close to Model Space as possible. It is ultimately up to you if you want to do this, but most new users keep the image of the paper in tact to know when they switch between spaces.

If you choose to do this, you need to right-click in an open area and select Options... from the popup menu (or go the the Application Menu and select Options), then select the Display tab and take a look at the Layout elements (lower left). Uncheck all but the very first and last choice, as seen in Fig. 10.4.

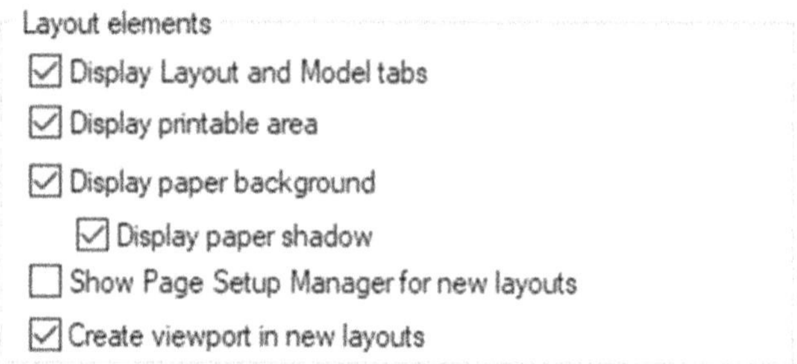

FIGURE 10.4 Paper Space display options.

If you choose to go do this, the Paper Space will appear as shown in Fig. 10.5: a blank area with the triangular shaped UCS and a floating viewport displayed. Again, for new users, this is not recommended as it can be disorienting.

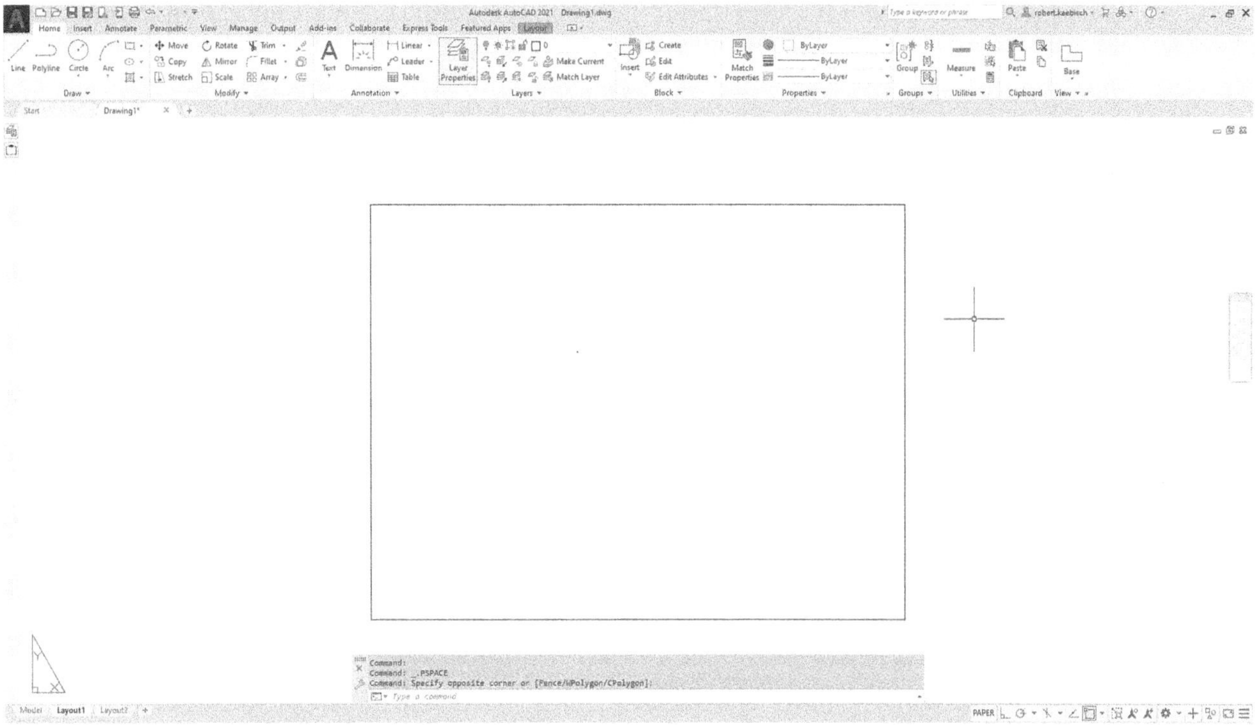

FIGURE 10.5 Paper Space, with printable area, background, and shadow turned off.

So, what are these layout tabs in real life? They are virtual sheets of paper, ready to print when the job complete. The design (done in Model Space) is "presented" in Paper Space with each layout being one sheet of the total drawing set, i.e., the Architectural, Mechanical, Electrical, Details, and so on There can be one, and only one, Model Space but as many Layout sheets as needed for your drawing set.

If you are thinking that we need to learn how to make title blocks at this point and perhaps create more layouts, then you are correct. We lend a title block from a built-in AutoCAD template. This may or may not be the way you will do it at your job. If you are in an architecture office, there may be a predrawn title block that you can just retrieve and paste in or have it built in to their template, but for our learning purposes, we create a new one.

Click the Layout1 tab then *right-click* to bring up the menu shown in Fig. 10.6.

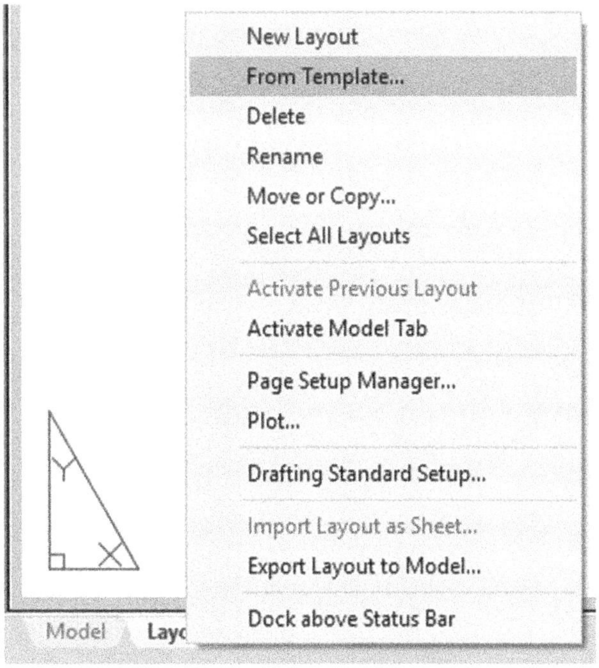

FIGURE 10.6 Layout tab 1, right-click menu.

We work with this menu for the next few pages. Let us, first of all, try creating a title block from an existing template by clicking on the *From template …* choice. You get the dialog box seen in Fig. 10.7. Pick the Tutorial-iArch.dwt template and click on Open.

Another small dialog box, Insert Layout(s), appears, as shown in Fig. 10.8. It is simply showing you the list of available Layouts in the selected template file. The only one available here is layout is called D-size Layout. Select it and click on OK.

Notice what happened: You have a new tab, called *D-Size Layout*, positioned right after Layout1 and Layout2. Click on it and you see a D-sized sheet of paper with a title block, It is a reasonable representation, minus the fancy logos, of what a typical architectural office may use. Inside the title block is a blue rectangle through which you may or may not see your building; this is the viewport, and we cover that topic next. For now, go ahead and delete the viewport, as we make our own soon.

What you have now is shown in Fig. 10.9.

The next step is to modify the layout tabs by renaming them to something more descriptive and removing the unnecessary ones. Right-click on the new D-Size layout tab to get the menu to appear again. Select the Rename choice and

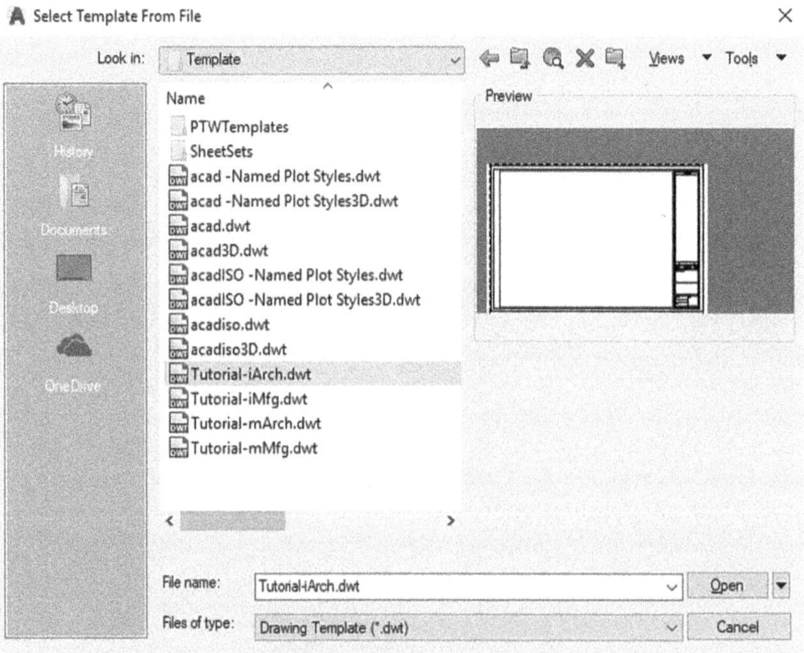

FIGURE 10.7 Select template from file.

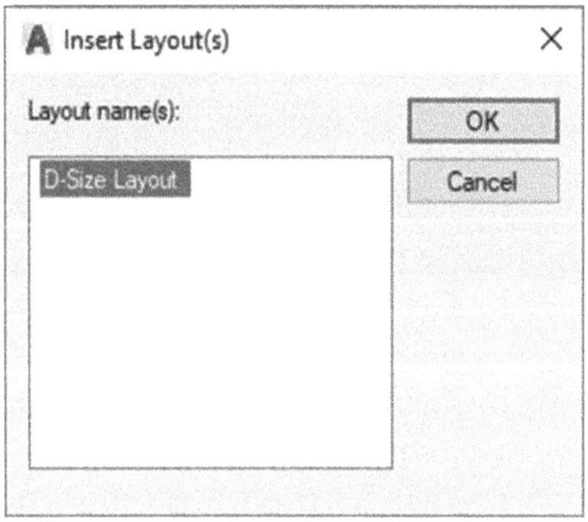

FIGURE 10.8 Insert Layout(s).

FIGURE 10.9 Tutorial-iArch.dwt D-size Layout.

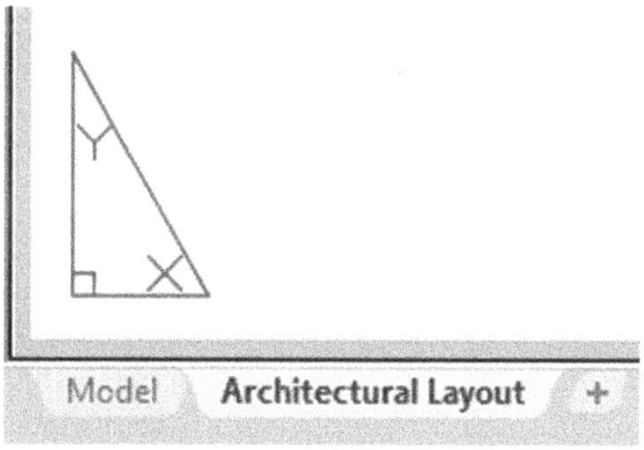

FIGURE 10.10 Model and Architectural Layout tabs.

the tab's name becomes editable; type in Architectural Layout. Now, get rid of the Layout1 and Layout2 tabs, as they are empty and not needed. Begin by clicking on Layout1, then right-click to get the menu up again. Select Delete and press OK when the warning box pops up. Repeat the procedure for Layout2. After these steps, we have only two tabs remaining, Model and Architectural Layout, as seen in Fig. 10.10.

Take a moment to look over the title block in the Architectural Layout tab. As a reminder, delete the original viewport (the blue rectangle), as we will make our own from scratch. Hover your cursor over the lines of the title block. Notice how the whole thing highlights? Double click it and the Enhanced Attribute Editor will appear. Modify the attribute values of the FNAME-File Name: to Plan Setm, the USER -Author: to your name, and the REVDATE, -Revision date: to the current date by selecting each row and then editing the values in the lower Value: area of the dialog box. This is optional, but also serves as a preview of Chapter 18.

Now, we need to create more of these sheets with title blocks on them, for the Mechanical and Electrical layouts. Go back to the menu (right-click on the Architectural Layout tab) and select the Move or Copy … choice. The dialog box in Fig. 10.11 appears. Note that you can also just click the plus sign found to the right of the tabs, create a generic layout, and then rename it. The disadvantage here is that the layout is blank; your title block is not carried over to this new tab.

If you use the Move or Copy approach, be sure to do two things. First of all, check off the Create a copy box at the bottom left and click on the (move to end) choice, which simply places the new title sheet next in line after the existing Architectural Layout and may be important later when it comes to the printing order. Finally, click on OK and observe what happens with tabs. A new one appears, saying Architectural Layout (2), and it is a perfect copy of the previous existing tab. Go ahead and rename it Mechanical Layout. Repeat the whole procedure to create a third tab, called Electrical Layout, and finally, a fourth tab, called Landscaping Layout. The result is in Fig. 10.12.

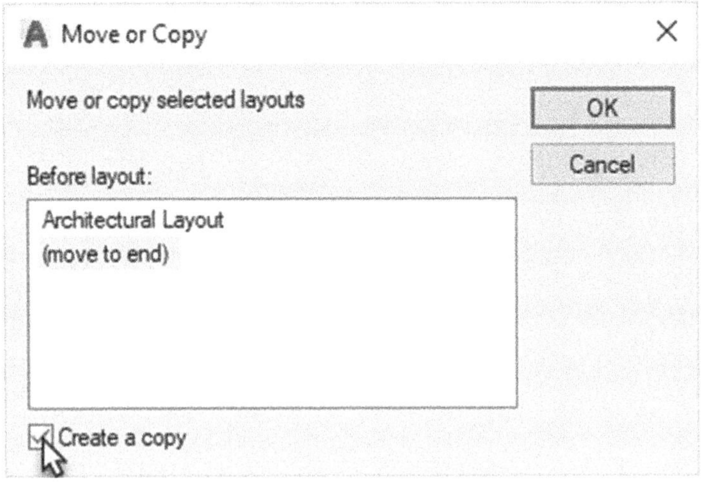

FIGURE 10.11 Move or Copy.

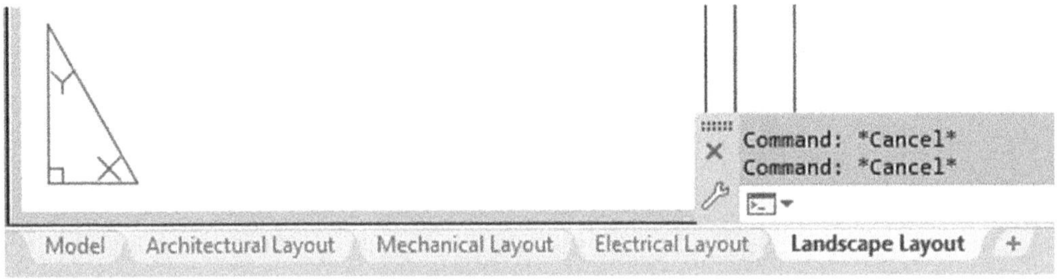

FIGURE 10.12 All tabs.

Review this entire section; it is an important first step to learning Paper Space and is a common procedure needed to set up a project that has multiple pages, or add pages as your design grows in scope and complexity. It should be clear as to what each sheet (tab) represents by its name. Next, we work on the sheets to set up viewports, so the design can be viewed at the proper scale.

Viewports

Viewports, as mentioned in the introduction, are simply windows from the Paper Space world looking into the Model Space world. It is as if we ripped a hole in the sheet of Paper Space and observed what was behind there. We assign scales to these viewports later and, in this manner, reconcile the different scales or sizes of Paper Space and Model Space. For now, we need to first learn how to create and modify the viewports and go over some of their properties. Your new layouts already have a premade viewport (the blue rectangle), but as mentioned earlier, you should have now erased it; we will learn to create a new one from scratch.

Just a reminder: To go through the following steps, make sure you have an accurate, full-size design in model space to look at, which you already should have, as suggested earlier in this chapter. For example in this text, the apartment plan from the previous chapters continues to be used. It is slightly modified to include both furniture and floor hatching. You are welcome to duplicate the same thing. If all is well, then let us proceed to learn about the viewport creation process.

Click on the Architectural Layout tab, entering Paper Space, type in Mview, and press Enter. This is the keyboard command for creating new viewports. Alternatively, a contextual tab of the Ribbon appears any time you are in Paper Space called *Layout.* Go to Layout tab → Layout Viewports panel, → Rectangular tool. Note that *you can create more than one viewport at a time, but we do only one for now.*

- AutoCAD says: `Specify corner of viewport or [ON/OFF/Fit/Shadeplot/Lock/Object/Polygonal/Restore/LAyer/2/3/4]<Fit>`

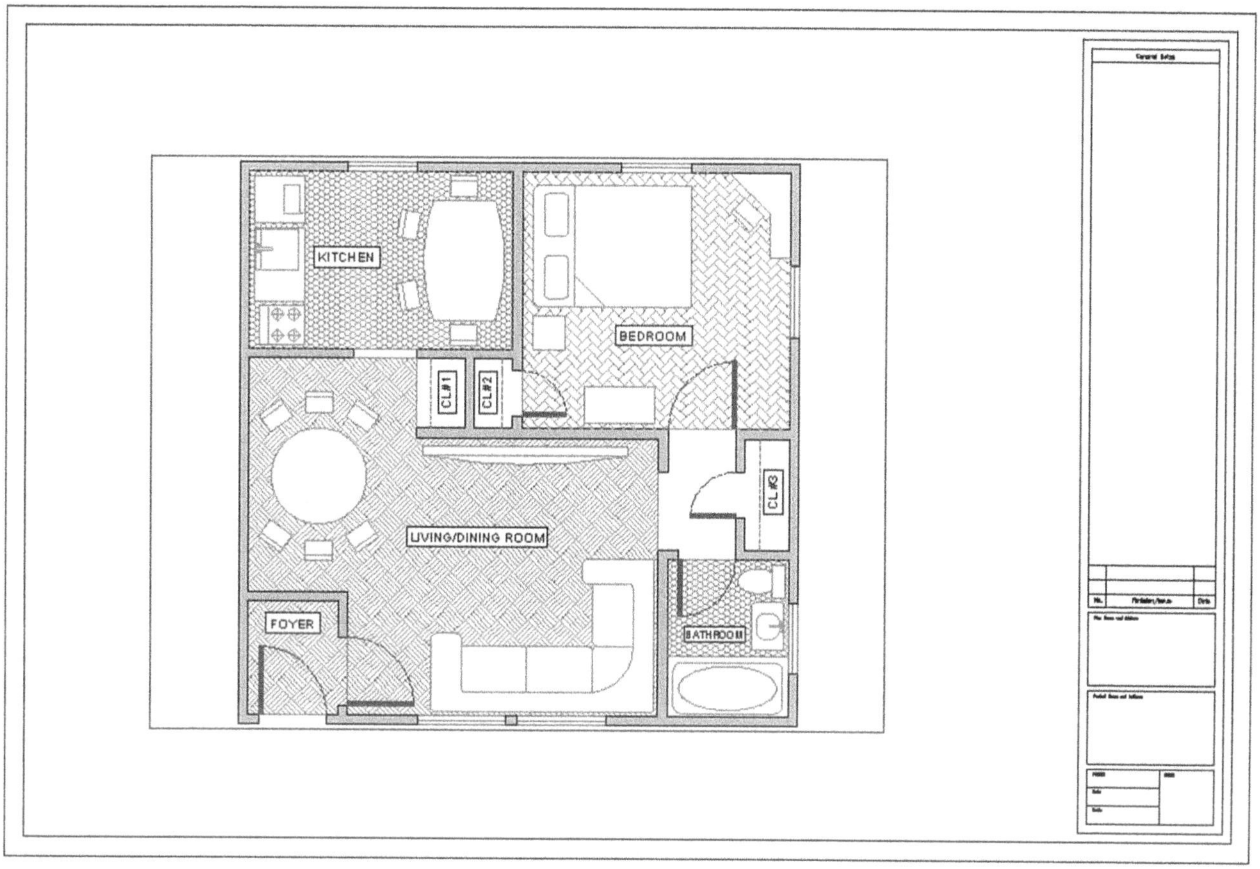

FIGURE 10.13 New viewport.

Then simply click near the upper left corner of the sheet and move the mouse across your screen, effectively drawing a large rectangle from one side of the sheet to the other (avoiding the actual title block in the lower right corner). AutoCAD says: Specify opposite corner: as you are doing this and Regenerating model as you complete the viewport. You should now be able to see the design inside the "rectangle" (actually the viewport) you created, as shown in Fig. 10.13.

You can create more viewports in the other layouts if you like for practice, but we focus on this one, in the Architectural Layout.

Floating Model Space

Double-click inside the viewport. Its border becomes thicker and you notice your mouse is "inside" the viewport, as shown in Fig. 10.14. Note also how the UCS icon changes to the Model Space version. The advantage of having full-width crosshairs (set at 100) is evident here, as the viewport "clips" the ends of the crosshairs and it becomes clear that you are "in" the viewport this mode. Pan and zoom while inside the viewport and you notice the design shifts around while the Paper Space sheet stays still. This Floating Model Space mode is sort of a "no man's land" between the Model and Paper Spaces and is critical in scaling the view of the design.

For now, double-click anywhere outside of the viewport to get back to Paper Space. Be careful not to double-click on the viewport itself. That takes you to the vpmax mode (maximum/minimum viewport size), which can be used for editing in Floating Model Space but is not needed here. If you do accidentally double-click on it, press the button shown in Fig. 10.15 to get back (bottom right of the screen in the Drawing and Construction Aids).

Practice jumping in and out of the viewport several times; this is very important for the rest of this chapter. Note also the big picture discussed so far. You are in Paper Space and created a viewport to look onto the design in Model Space. When you are inside of a viewport, you are technically in Model Space but still in Paper Space as well. You need this for scaling purposes only. Design work on either the floor plan or the title sheet or block is done in their respective spaces.

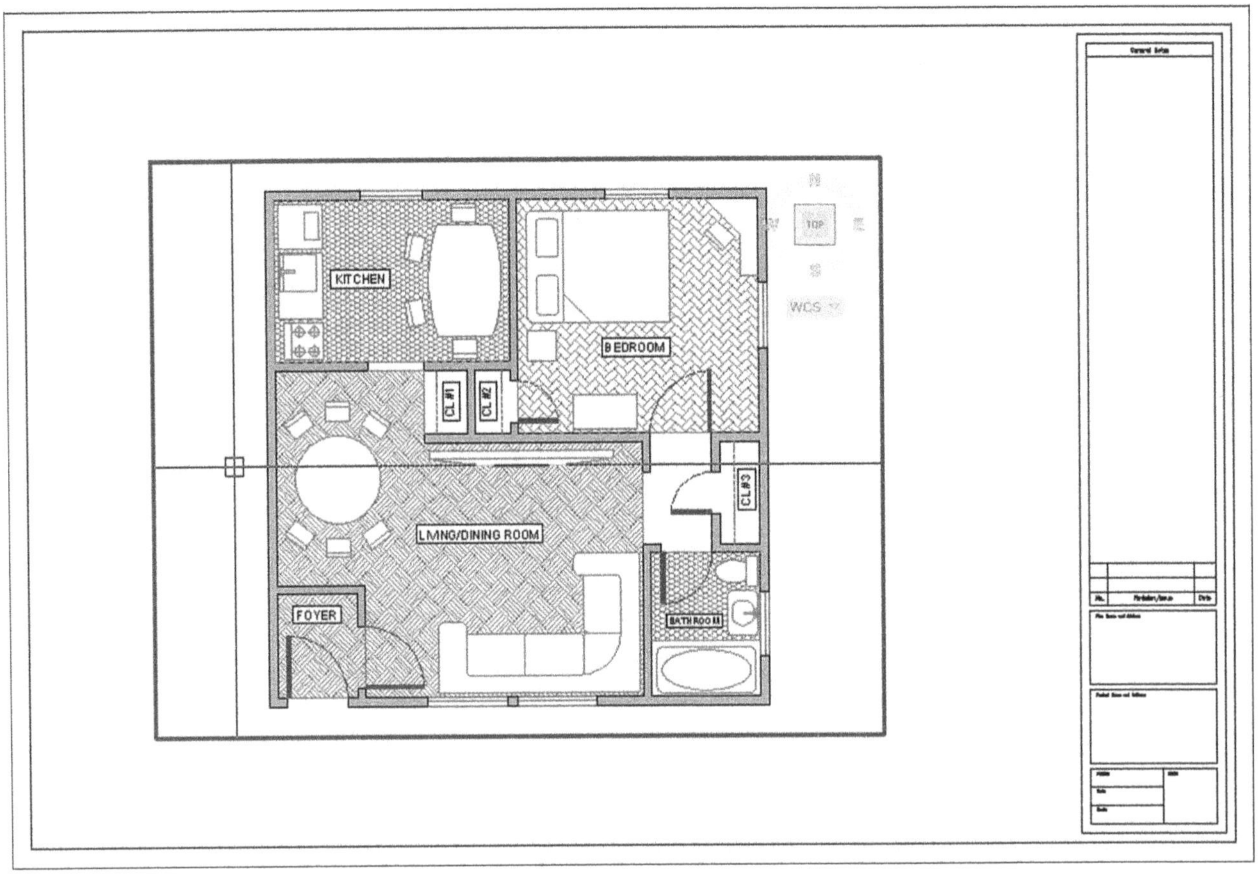

FIGURE 10.14 Floating model space.

FIGURE 10.15 Minimize Viewport tool.

Next, we need to learn how to modify the viewport(s). Note that they are not static displays but rather dynamic tools you can and should mold and shape them to your needs, as outlined in the next section.

Modifying the Viewport

First, make sure you are in Paper Space, and then click once on the viewport object (not inside of it but on the lines of the actual viewport rectangle). The viewport becomes highlighted and grips appear. Using them, you can resize the viewport to any size desired by clicking on any corner grip, activating it (it turns red), and adjusting the viewport's size and shape. In general, you can:

- Adjust the viewport(s) to any shape.
- Move the viewport(s) anywhere you like.
- Erase the viewport(s).
- Copy the viewport(s) as many times as needed.

Take a few minutes to go through all these features. Make your viewport smaller, and copy it three times. Move all four around the screen and then erase one. Finally, create a new viewport with Mview or from the Layout Ribbon tab. In the end, try to have what is shown in Fig. 10.16 and make sure you understand viewport manipulation, the mview command,

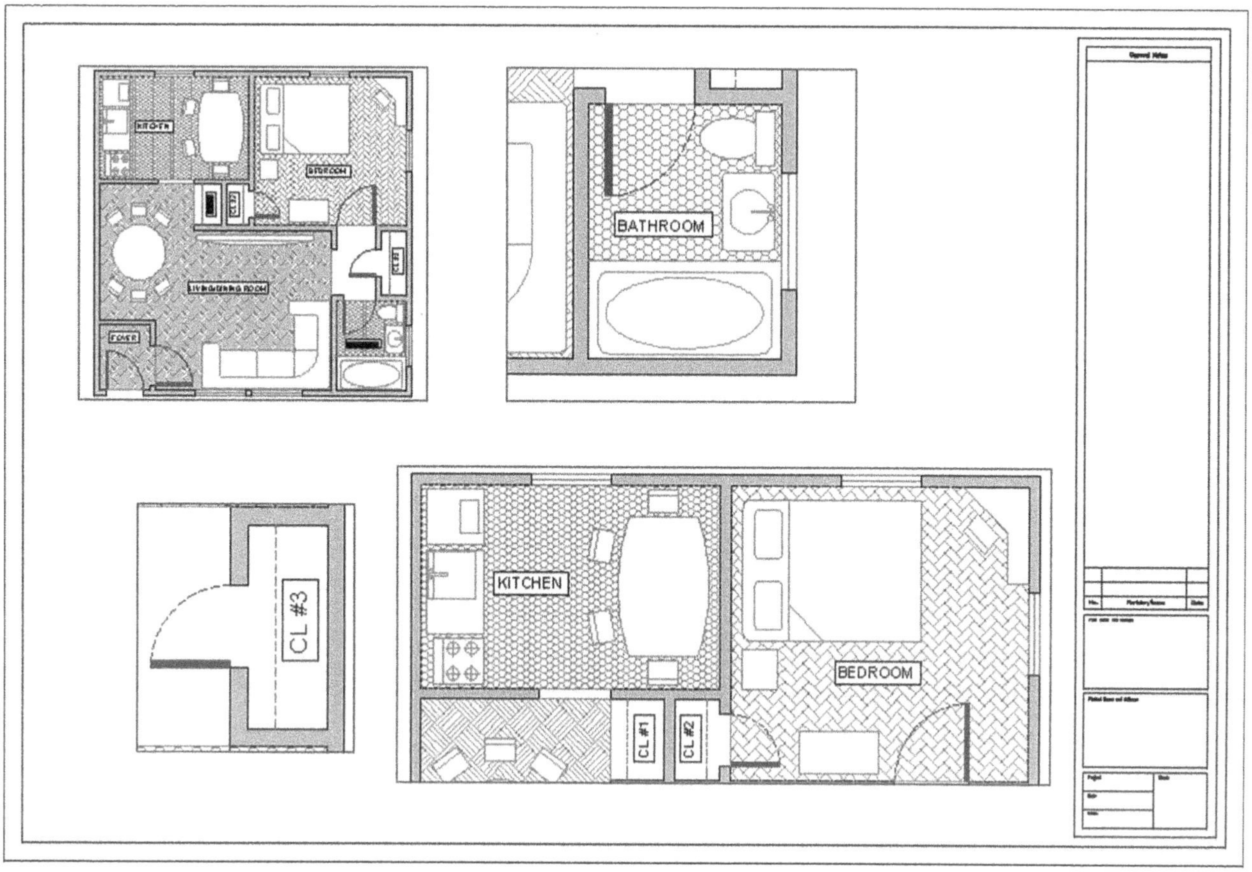

FIGURE 10.16 Four random viewports.

and getting in and out of Floating Model Space. When you are inside the viewport, do not forget to zoom in and out on the entire drawing or select parts of it and also zoom to extents. We focus on a precise scale later.

Polygonal Viewport

The shape of the viewport need not be rectangular or square; it can be any shape necessary by using the Polygonal option. To practice this, click over to the Mechanical Layout and create a new viewport using the mview command; however, before clicking to draw the actual viewport, select p for Polygonal in the lengthy menu.

- AutoCAD says: `Specify start point:` and after the first click `Specify next point or [Arc/Length/Undo]:`

Click your way around in some random fashion (do not overlap the clicks) and press c for Close to close the viewport. What you get after creating a starburst sort of pattern is shown in Fig. 10.17. You can see the design inside the oddly shaped viewport.

Of course, this is not really what this tool is for. Rather it is to make viewports that weave their way around possible revision blocks or notes on the drawing, thereby freeing you from being forced to use a rectangle and wasting space. An example is shown in Fig. 10.18, and this is often the case in real-life designs, making polygonal viewports very important for creating an efficient layout.

There are a few other important items to cover with viewports, such as locking them and making them invisible, but we introduce these features a bit later, as the need arises. What we need to look at next is the "core" of Paper Space: scaling the viewports and their contents.

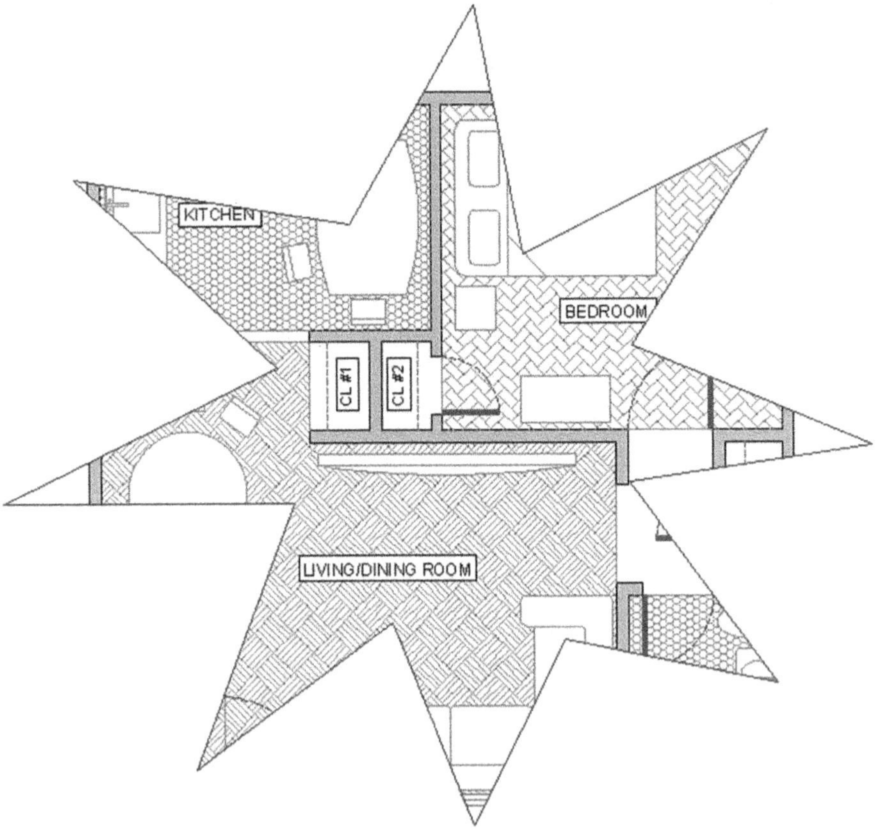

FIGURE 10.17 Polygonal viewport 1.

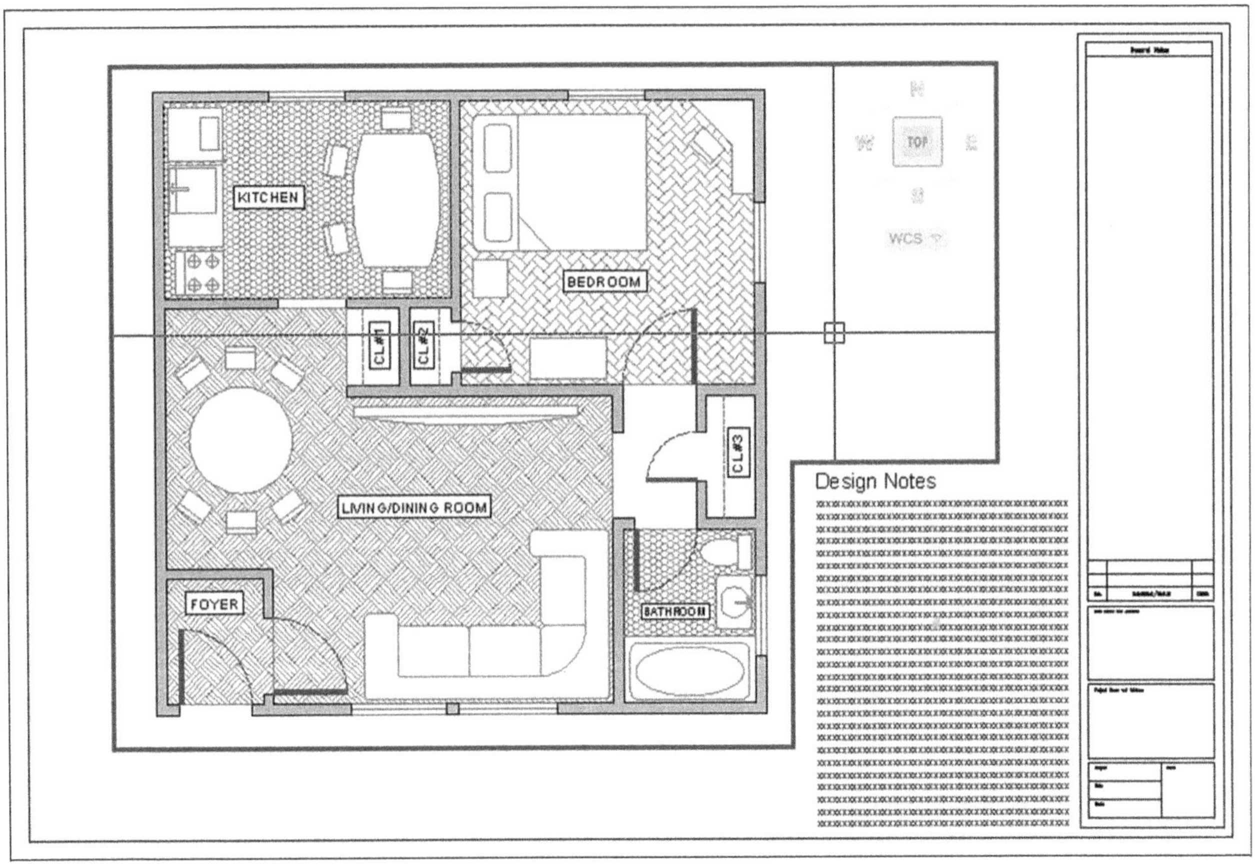

FIGURE 10.18 Polygonal viewport 2.

Scaling

Here, we finally put everything together, because really what we have done so far was just the setup. We have our design in Model Space and our title blocks on sheets in Paper Space, connected by one or more viewports. Now, we need to give these viewports a scale, which was our goal all along.

To make things simple, let us just have one viewport in our Architectural Layout; so delete any others you may have, including polygonal ones, and leave one large viewport similar to Fig. 10.13.

So, how do we scale a viewport? Well, you may have already guessed it has something to do with zooming. After all, as the design is zoomed in and out (while in Floating Model Space), the scale continuously changes and is always at some value. So scaling the design properly is just a matter of knowing how much to zoom it in the viewport. The big question of course is, How much? Another way to put it is, What scale factor do we apply?

There are two ways to assign a scale to any viewport. The first one is a technique used for just about every AutoCAD release up until the last few. It is a bit more involved but really gets you to understand what is going on when using the other method. Then, we cover the other, easier method.

Method 1. Double-click inside the viewport, Pan and zoom as required to "fit" your entire design in the viewport. You do not need to be precise here. Then type in zoom, and press Enter.

- AutoCAD says: `Specify corner of window, enter a scale factor (nX or nXP), or [All/Center/Dynamic/Extents/Previous/Scale/Window/Object]<real time>:`

Notice the `enter a scale factor (nX or nXP)` choice. This is what we are interested in right now, specifically nXP.

What nXP means is that you need to enter a ratio relative to Paper Space (which is the 1:1 "anchor," so to speak) and the design scales to that ratio. So, what is it? Well, for 1/4″= 1′–0″, the n is 1/48 (followed by the XP). So type in 1/48xp and press Enter. The design zooms to precisely ¼″ = 1′–0″ scale, as shown in Fig. 10.19.

Now, this seems to be too small, does not it? No problem, just type in zoom and press Enter again. Then, type in 1/24xp and press Enter again. The design will zoom to 1/2″= 1′–0″, as shown in Fig. 10.20. which seems to be a better size for this design and title sheet.

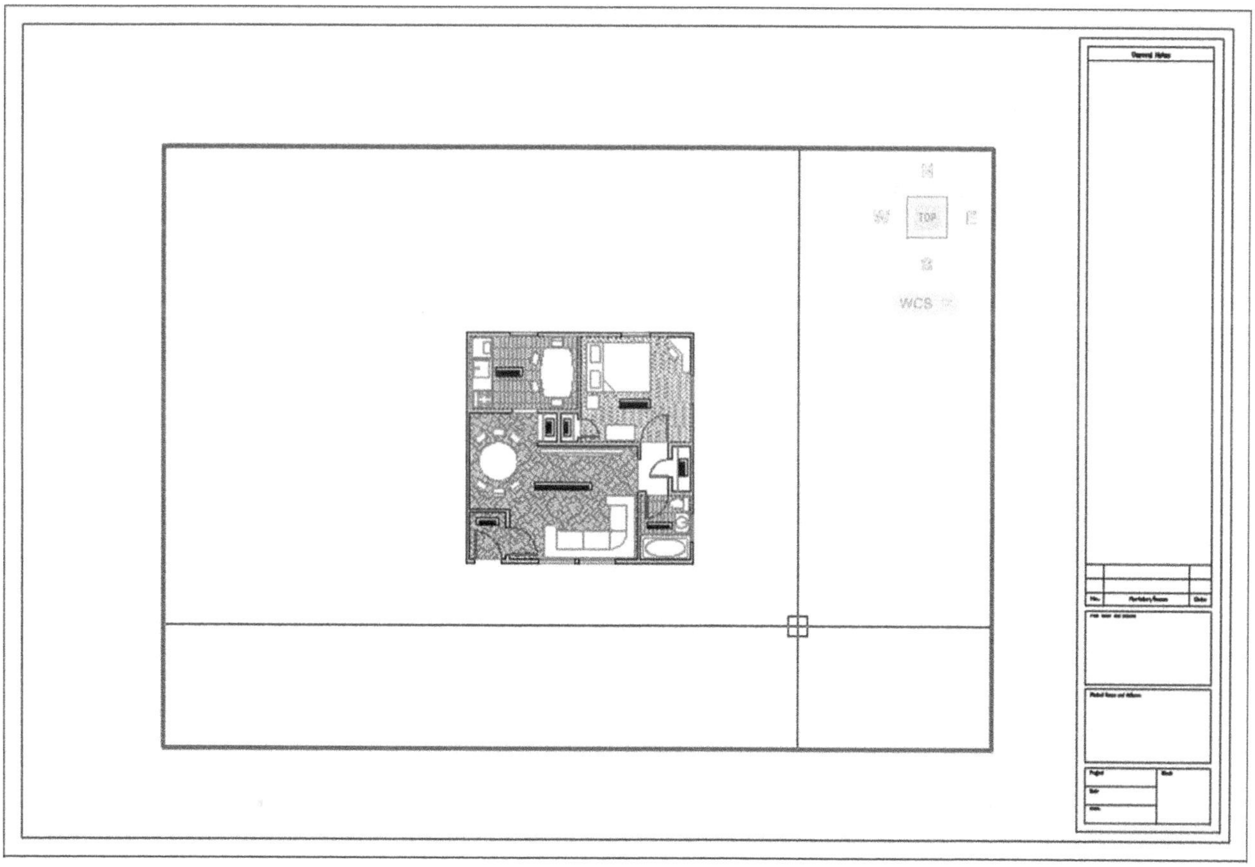

FIGURE 10.19 Design zoomed to 1/4"= 1′-0″ scale (1/48xp).

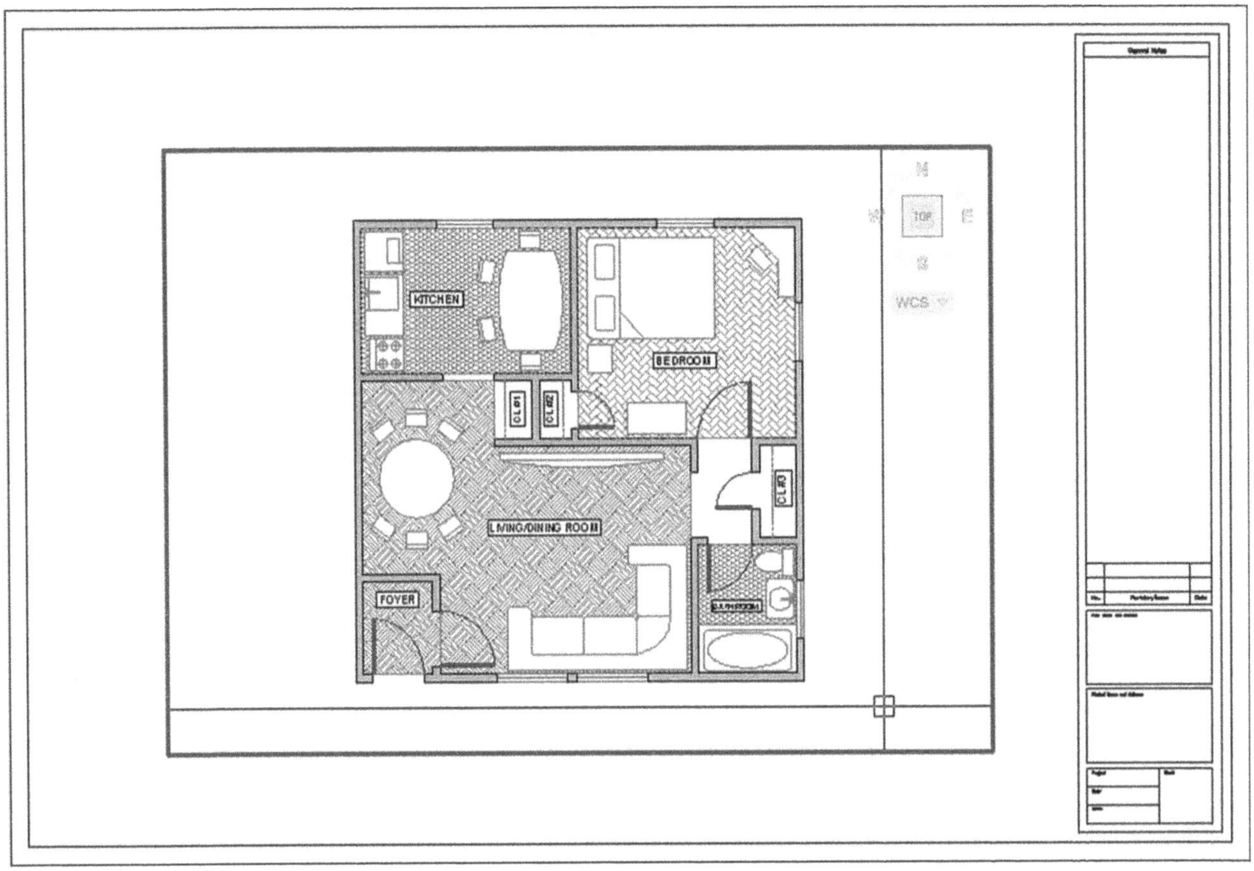

FIGURE 10.20 Design zoomed to 1/2"= 1′-0″ scale (1/24xp).

So, where did 1/48xp and 1/24xp come from? They are just ratios developed from the fact that there are 12 inches in 1 foot and the scale is 1/4 or 1/2. Therefore, for 1/4, it is 1/48 (1/4 inch scale × 1/12 inches in a foot = 1/48 relative to the 1:1 Paper Space). For 1/2, it is 1/24 (1/2 inch scale × 1/12 inches in a foot = 1/24 relative to the 1:1 Paper Space) and so on. There are additional scales such as 1/16″, 1/8″, 3/8″, 5/8″, and others, all of which can be derived in a similar manner. Now that you understand basic viewport scaling, let us apply this method in a somewhat easier way.

Method 2. Double-click back to Paper Space and single click the Viewport border, you will see triangular scale grip, shown in Fig. 10.21. (the design in Floating Model Space has been removed for clarity).

Click the triangular scale grip near the center of the viewport and select the desired scale from the list as shown in Fig. 10.22. The scale you choose is applied to the viewport and on the border of the Viewport automatically adjusts to maintain the clipping boundaries of the view, as shown in Fig. 10.22.

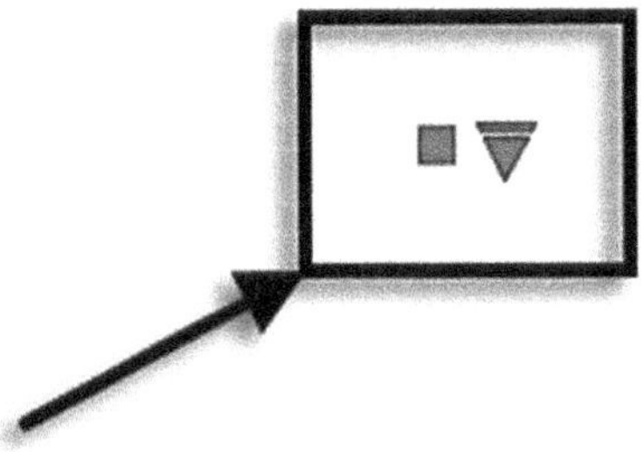

FIGURE 10.21 Viewport Grip.

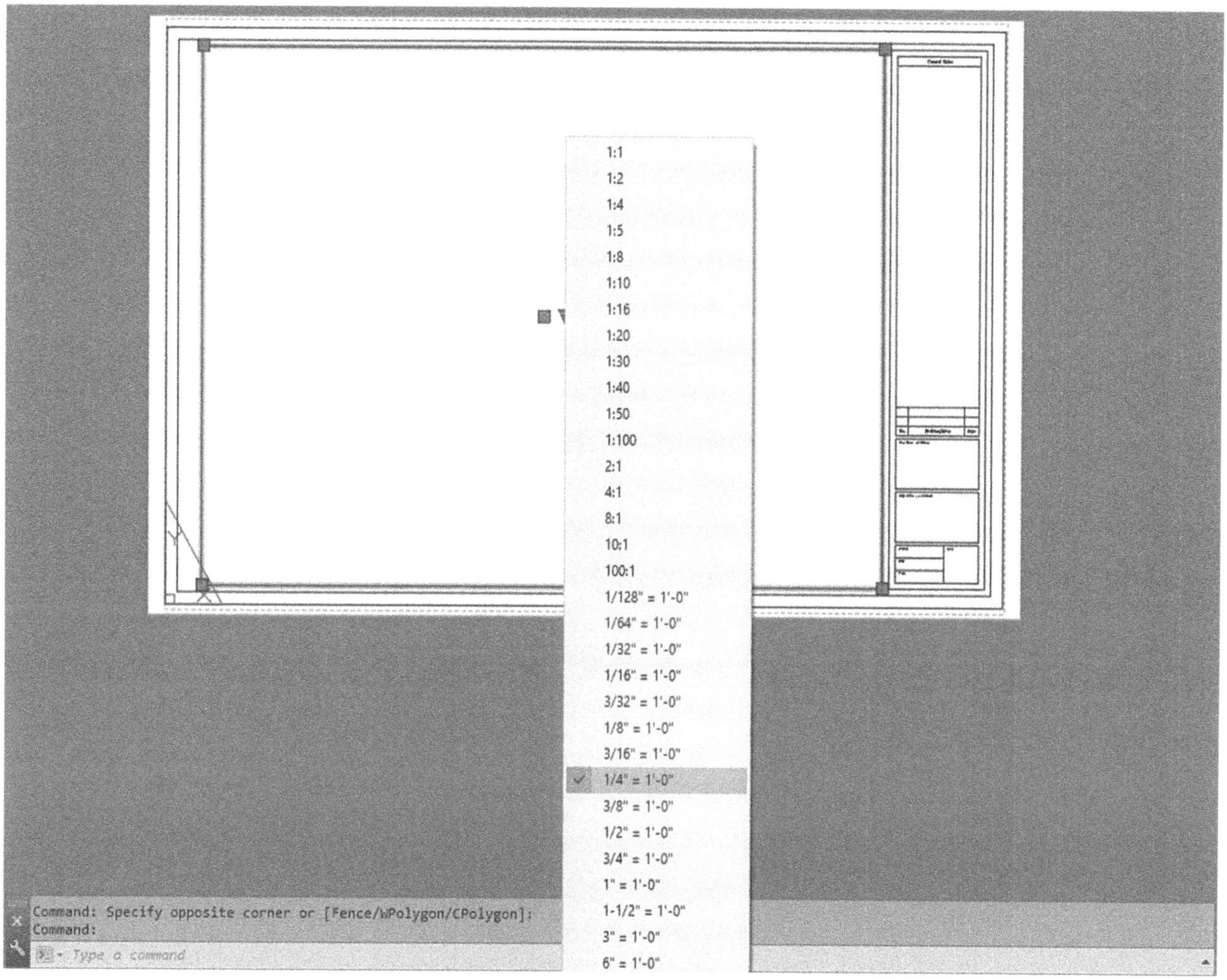

FIGURE 10.22 Viewport scaling via grip.

Layers

We have done a lot already in setting up of the drawing in Paper Space. We now turn our attention to several important details that are either new or change significantly when dealing with Paper Space.

One of those is layers. While the fundamental layer concepts do not change, a new ability is added to the Layers dialog box. It turns out that, when viewports are present, you can freeze and thaw layers in specific viewports independent of each other viewport and the main drawing back in model space. This is very significant, as you want to be able to display different items as you go from viewport to viewport.

Perhaps the most important example of that is when you jump between the Architectural, Mechanical, and Electrical Layouts. Each tab is a new viewport, yet all views of the design originate from one actual copy of that design. In each layout, you need to see only what is relevant to that layout, and it would be a major problem if you could turn layers on and off only in the main Model Space drawing. Then, each tab would display the exact same thing.

Fortunately, this is not the case, and in each viewport, you can turn layers on and off as needed to get the design points across. Therefore, the Architectural Layout has only architectural layers and so on. It is the same idea with multiple viewports on one layout (such as the case with detail sheets). Each viewport is truly a world unto itself. To see an example of this, let us create four viewports in the Architectural Layout, as shown in Fig. 10.23. (The scales assigned in the example are random in this case).

Double-click into the top left viewport and bring up the Layer Properties Manager. The dialog box with some additional columns appears, as shown in Fig. 10.24. All relevant columns have been expanded for maximum visibility of the contents. Look for the VP Freeze column (circled); this is the column in which we are interested right now. It only appears when you access this dialog box while in a floating viewport.

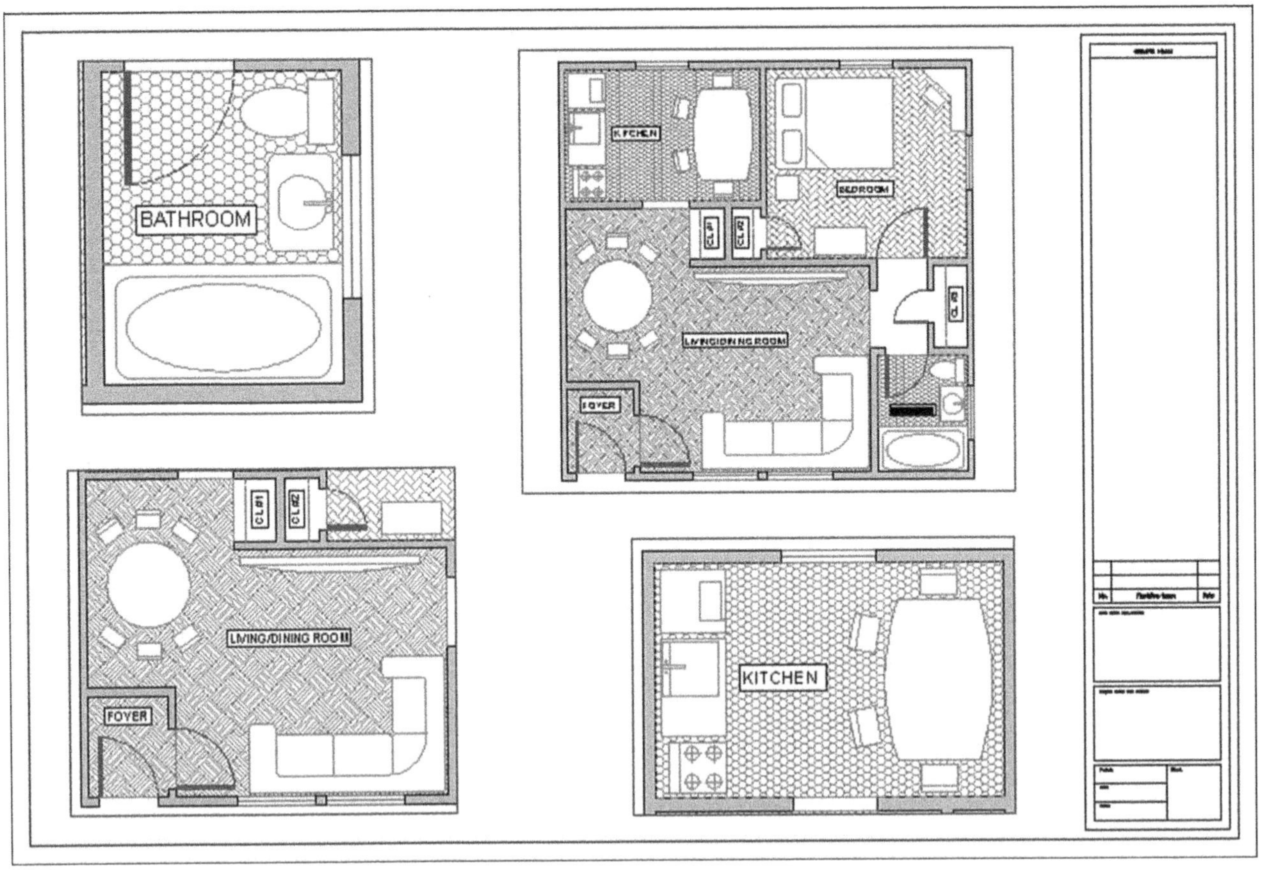

FIGURE 10.23 Four viewports for layers demo.

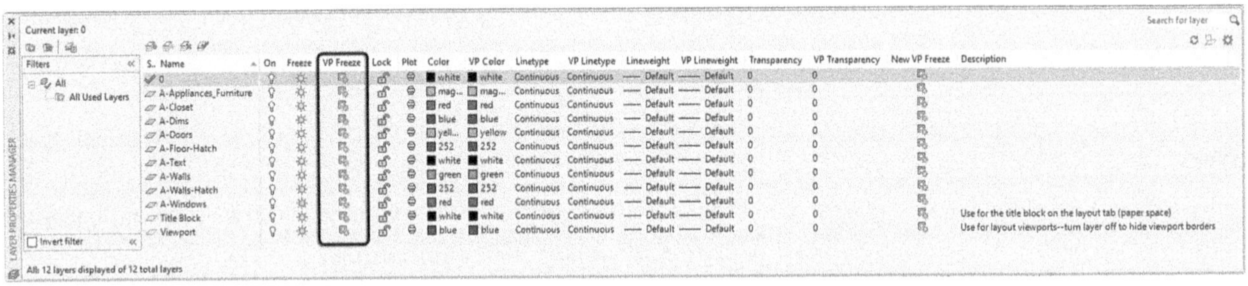

FIGURE 10.24 Layer Properties Manager with VP freeze column.

In that column, you can freeze all the unneeded layers for that layout without affecting either the main layout in Model Space or the other viewports. Go ahead and freeze the A-Floor-Hatch layer. The result is shown in Fig. 10.25. Note how the neighboring viewports are not affected by what you did in the top left one. Now click over to the top right one and freeze the A-Doors layer only. Click over to the bottom left viewport and freeze the A-Floor-Hatch, and the A-Appliances_Furniture layers. Finally, in the bottom right viewport, freeze everything except the walls and windows. Note the distinctly different layer settings in each viewport; the displayed views are completely independent.

Lock VP

We now cover two other notable features of viewports. The first is the ability to lock the viewport, which simply means you cannot change the viewport's scale and all zooming attempts fail (the entire Paper Space drawing zooms instead). This can be useful to prevent accidental zooming while inside the viewport and should be set after the design is laid out and set up. To do this, bring up the Properties box. Then, click on the viewport once and select (under Misc.) the Display locked option, setting it to Yes, as shown in Fig. 10.26.

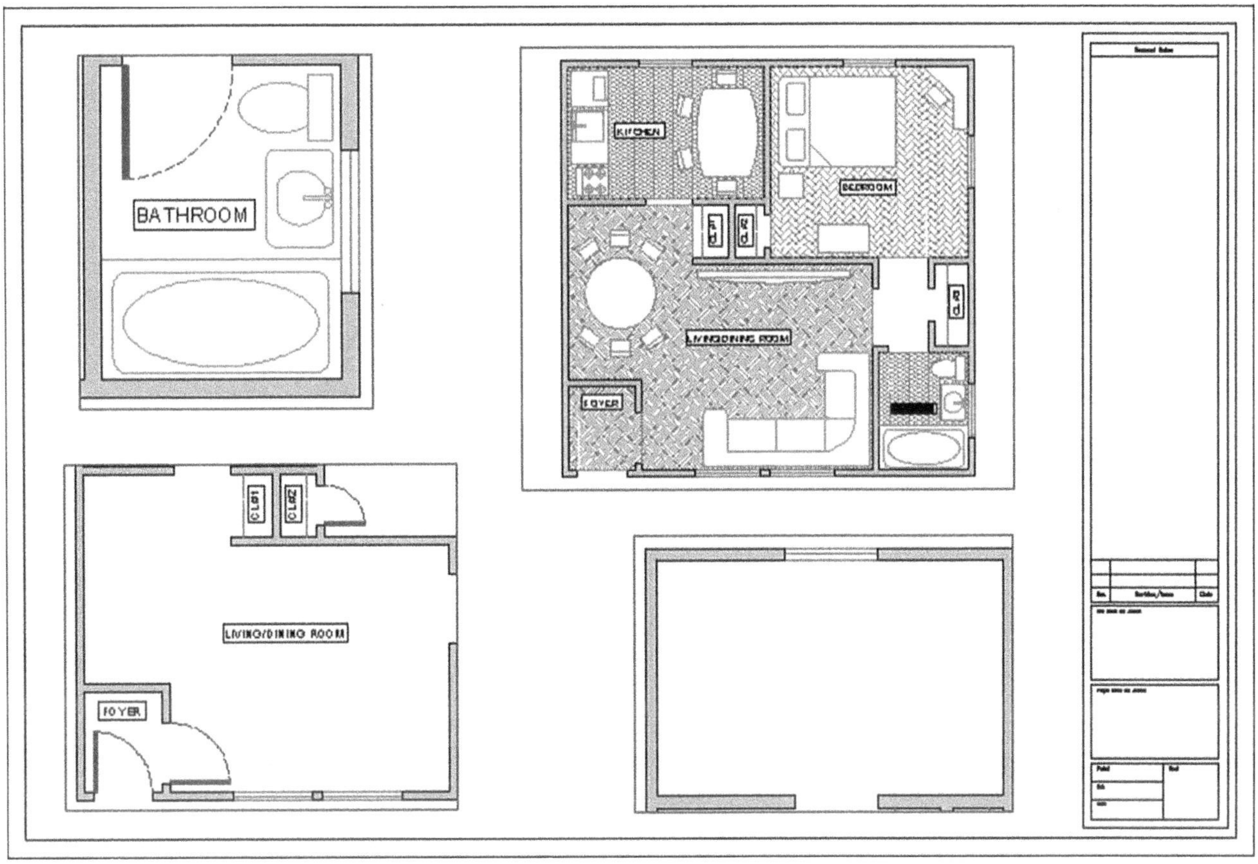

FIGURE 10.25 VP Frozen layers.

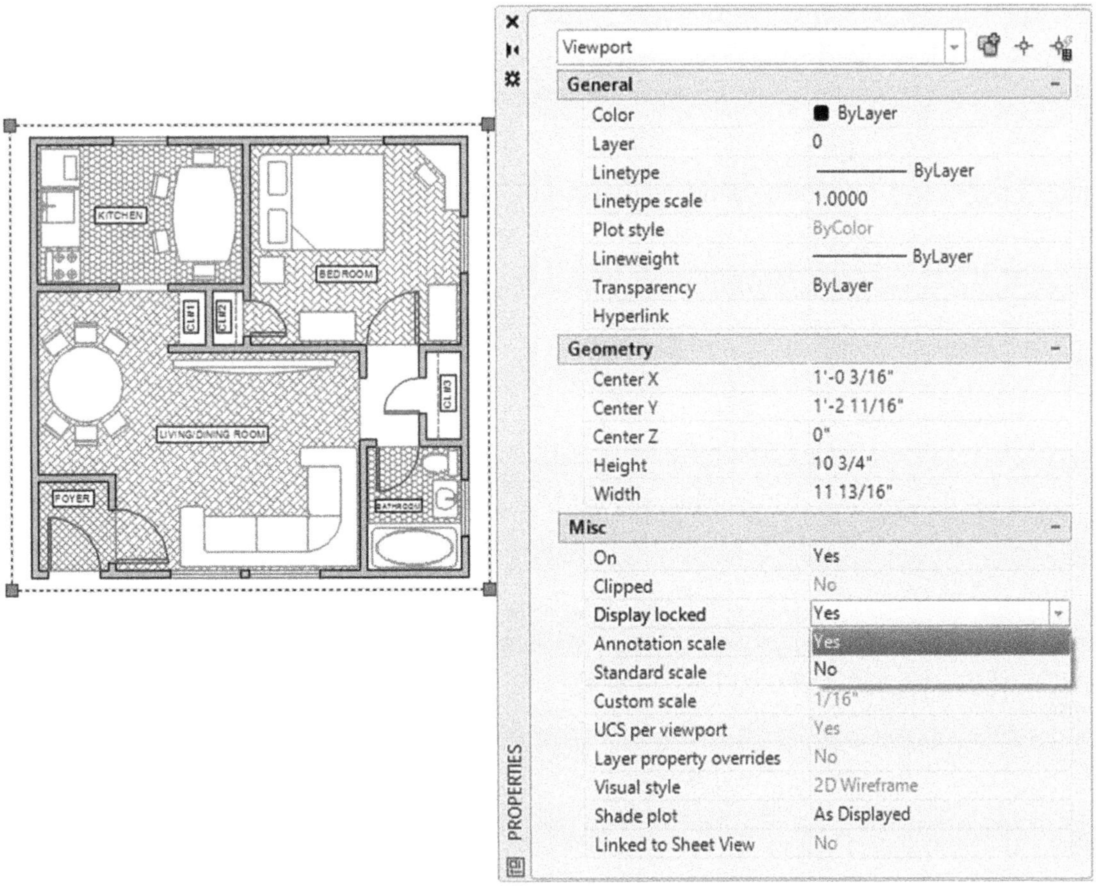

FIGURE 10.26 Locking the viewport.

Hiding the VP

Generally, you do not want a big fat rectangle surrounding each view on your final plots, which will go to the client. This is especially true of the main layouts that feature just one large viewport. The visible viewport rectangle can be mistaken for part of the border and is unnecessary. Even with detail sheets, each of which may display many details, the trend is to just separate them adequately but not have any rectangles visible. You have the option, outlined below, to either "freeze" the viewport layer or simply prevent the viewport layer from printing but remain visible.

First, we need to put the viewports on an appropriate layer if they are not there already. Designers generally create a viewport or_VP layer for them. Notice a trick here. The VP layer has an underscore in front of it. This forces the layer (or any layer that has that underscore) to rise to the top of the layer list, above the 0 layer. This makes it easy to find the title block and sheet or the VP layers, if you do not overdo it and stick too many up there with the underscore. Go ahead and change the color of the layer to gray and freeze it. Then, finally, change all the viewports to that frozen_VP layer. If you did everything correctly, then the viewport boxes disappear, as shown in Fig. 10.27.

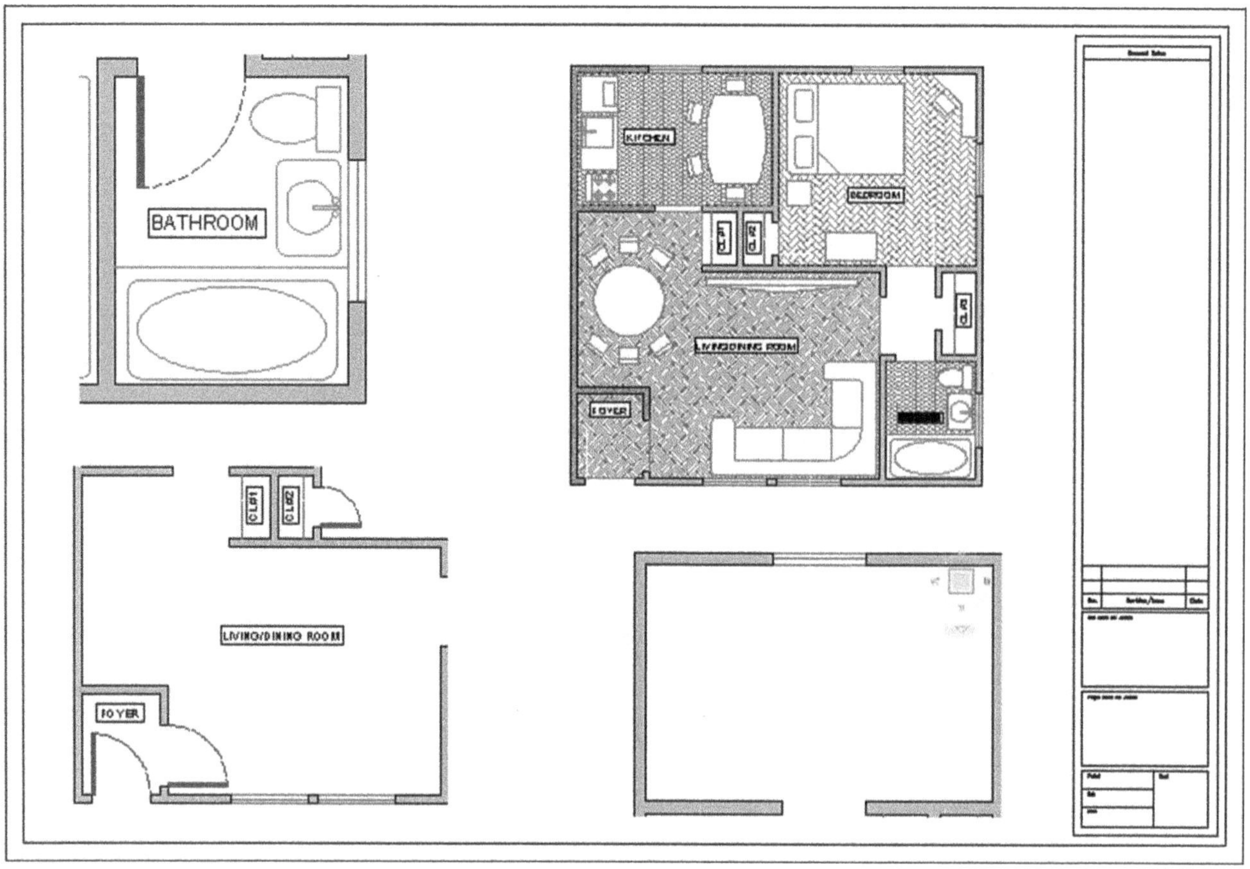

FIGURE 10.27 Viewports frozen.

There are other alternatives, of course, for those who want to see the viewports but not print them. You can set the _VP layer to not print in the Layer Property Manager by clicking the printer icon. You can also put the Viewports on the Defpoints layer. This layer does not ever print (it is the layer AutoCAD creates for the definition points of dimensions).

We are getting to the end of Paper Space; only two major topics are left: text and dims and the Annotation tools.

Text and dims

This topic is probably the most confusing one in Paper Space and really requires the AutoCAD designer to not only understand everything thoroughly but also work carefully and deliberately. In practice, much of this is done only once and saved as part of a template, but for educational purposes, we assume you are being tasked with the text and dims setup from scratch and must know all the steps.

Our goal here is quite straightforward: We want all the text and dimensions to be the same size when printed, regardless of in what scale viewport they appear. This uniformity is a basic requirement of a professional drawing. Architects and engineers specify something to the effect of: "I want the text height to be 1/4″ on all drawings, both in layouts and in all details," and it is up to the AutoCAD designer to ensure that this happens. With hand drafting, this was not an issue; the text and dims were drawn all at once using the same size test regardless of the drawing scale.

With AutoCAD, due to our use of Model Space and Paper Space, things are not so simple. The basic fact remains that all text and dimensions have a certain fixed measurable size when drawn in Model Space and appear as different sizes when viewed through the variously scaled viewports, just as the design itself does. Some designers try to simplify everything by dimensioning the design on the actual sheet in Paper Space to achieve uniformity (AutoCAD lets you do this), but this is not advised. The dimensions and all text must remain with the design in Model Space. The reasons for this are many, not the least of which is that any movement of the design geometry in Model Space misaligns the corresponding annotation done in Paper Space.

So the text and dims need to stay where they are supposed to be, and we need to understand how to properly size text and dimensions based on the anticipated scale of the viewport in which they appear. This all just boils down to a game of "match-up." The basic theory is as follows. You already know that viewport scale is a function of zooming. Therefore, when you have uniform text in Model Space and view the same set of text through the 1/4″= 1′−0″ versus the 1/2″= 1′−0″ scale viewport in Paper Space, the text in the 1/2″ viewport appears two times as large, as the viewport is a larger scale. To counter this effect, you need to make sure that the text that appears in that 1/2;″ viewport is drawn to be half as big, so it appears uniform with the 1/4″ text when viewed through a Viewport (just as an example).

Here is another way to put it: imagine that we are going to print a floor plan at 1/4" = 1′-0". The viewport scale for that plan would be set to 1/4" = 1′-0" (or 1/48xp). Think of this as zooming a camera 48x away from the floor plan. If we want the labels for the rooms to print out through a Viewport at 1/8″ high, we would have to make the text 48X bigger so when we zoom the viewport "camera" 48X away the text appears to be 1/8″ tall when printed; 48 x 1/8″= 48 ÷ 8″ or 6". 6″ high text in model space yields 1/8″ high text printed text in a 1/4″= 1′-0″ Viewports.

The trick is to be consistent and careful so that the wrong size text does not appear with the wrong viewport scale. So, how is this done in practice? How do you know what size text and dims are appropriate for a particular scale? We have tables for this, such as the one shown in Fig. 10.28.

VIEWPORT SCALE	DRAWING TEXT HEIGHT			SCALE FACTOR	LT SCALE
	1/8	3/16	1/4		
Architectural					
1 1/2" = 1'-0"	1	1.5	2	8	3
1" = 1'-0"	1.5	2.25	3	12	4.5
3/4" = 1'-0"	2	3	4	16	6
1/2" = 1'-0"	3	4.5	6	24	9
3/8" = 1'-0"	4	6	8	32	12
1/4" = 1'-0"	6	9	12	48	18
1/8" = 1'-0"	12	18	24	96	36
1/16" = 1'-0"	24	36	48	192	72
1/32" = 1'-0"	48	72	96	384	144
Engineering					
1" = 10'	15	22.5	30	120	45
1" = 20'	30	45	60	240	90
1" = 30'	45	67.5	90	360	135
1" = 40'	60	90	120	480	180
1" = 50'	75	112.5	150	600	225
1" = 60'	90	135	180	720	270

FIGURE 10.28 Text and scale chart.

To use this chart, you must know what scale viewports you are using. going to use. Architects and Engineers have been using industry standard scaled drawings in their sets for millennium. Often a sheet features only one large plan of a known scale 1/8″= 1′–0″ is commonly used for commercial architecture plans; therefore, if the architect desires all text output to be 1/8″ in height (also a common size), then by reading the chart from left to right until we meet the $^{3}/_{16}$″ column, we arrive at a height of 12 inches, which is to be used for all text in that drawing in Model Space. For a 1/4″= 1′–0″ scale detail view of the same plan, the text size is now 6 inches. We are zoomed in 2x closer so the text is half as big to print at the same size from a Layout.

With details, which may come in all the scales listed, the procedure is no different; you just must be very careful in knowing ahead of time what scale viewport you are using, to use the appropriate text height. All this comes with experience, and after much practice, you will be able to quickly assess the necessary scale of a given detail to ensure that all aspects of it are visible and clear to whoever is reading the plans. This is the only factor in determining the appropriate text size.

The chart shows not only the text height but also the dimension sizes under the Scale Factor column. Here, you retrieve the appropriate size (24 in the 1/2″= 1′–0″ example) and enter that value in the Use overall scale of: text field under the Fit tab of the DDIM (Dimension Style) dialog box. Also shown is the appropriate scale for linetypes (9 in the case of ½″ = 1′–0″). You need this to ensure you can see a hidden line's spaces and dashes at.

Most of the information contained in Fig. 10.28 needs to be entered in AutoCAD in the form of text styles and dim styles. This is the job of the CAD manager on initially setting up an architectural or engineering office's CAD standards and templates. Yes, it is as tedious as it sounds; the chart is abbreviated and does not contain every scale used, so the work is substantial. Fortunately, it needs to be done only once, and a template is formed for all users to start their drawings with from there on out. Enforcing this usually turns out to be the hardest part of all, as the "system" is only as good as the people who follow it. A user simply switches their text style or height, dimensions style, and Mleader style to the appropriately scaled choice. It is also imperative that the different scales of text, dimensions and Mleaders be placed on separate layers so that they can be frozen through viewports of the "wrong" sizes You must be careful to use the proper style (Arial_.25, for example) and be sure to set that style as current while working on the relevant design or detail. The payoff comes later in the process when you view the design in Paper Space. Uniformity and professionalism are the final goals. Go ahead and practice these concepts.

Step 1. Go to the Style dialog box and create three new text styles, all Arial fonts. They are set up for a final text height of 1/4″ and are viewed through three viewports of the following scales:
1/2″= 1′-0" (therefore, size: 6″)
1/4″= 1′-0" (therefore, size: 12″)
1/8″ = 1′-0" (therefore, size: 24″)

Step 2. Go to the DDIM (Dimension Styles) dialog box and create three new dim styles. All must use the appropriate text from the previous procedure (therefore, the 1/2″= 1′-0″ dimension style uses the Arial_.5 style of 6″ height, etc.). Then, set the fit as shown on the chart and reproduced here:
1/2″= 1′-0" (therefore, Use overall scale of: 24″)
1/4″= 1′-0" (therefore, Use overall scale of: 48″)
1/8″= 1′-0" (therefore, Use overall scale of: 96″)

Step 3. One by one, in Model Space, set each text and dimension style as current and create one set of text and one set of random dimensions next to each other, as shown in Fig. 10.29.

This may take you a few minutes to set up and execute. Be sure to work carefully, checking and rechecking all your settings. As expected, the three scales go up in size twice for each jump from 1/2″ to 1/4″ to 1/8″. We now want to see them uniform in size in Paper Space.

Go into Paper Space and create three viewports next to each other. Use the tools learned previously to set the first one to the 1/2″= 1′–0″, the second to 1/4″= 1′–0″, and the third to 1/8″ = 1′–0″. Pan around until the appropriate text and dims combination is in the respective viewport (you may have to go to Model Space and move them apart slightly). The result is shown in Fig. 10.30.

FIGURE 10.29 1/2″, 1/4″, and 1/8″ text and dimensions.

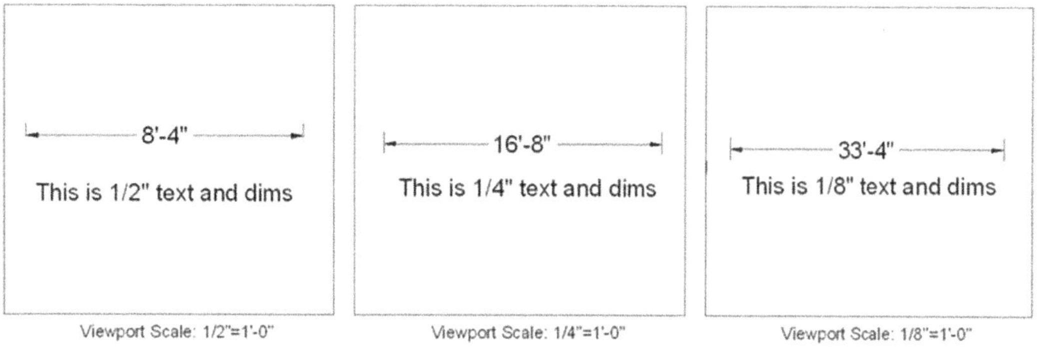

FIGURE 10.30 Paper Space text and dimensions uniformity.

Notice the complete uniformity of the text and dims in Paper Space, as opposed to how they looked in Model Space in Fig. 10.29. This is the essence of what we are doing here. If you can set up these three viewports and all the text and dims correctly, you can do any drawing. Be sure to review and learn everything thus far presented. Ask your instructor for assistance if needed.

Annotation and Annotative Objects

The final Paper Space feature we will discuss first appeared in AutoCAD 2008. It is called *Annotation*, and *Annotative Objects*. Its purpose is to present an alternative approach to the manual, style creation and switching methods of the previous section. The Annotative Scaling approach is different in that it allows you to display many scales for one text style, so the proper scales are visible in the proper viewports with no further input from the user. Several items can be set to *Annotative*: text styles, dimensions, blocks, and hatches. We explore annotative text first , as outlined in the following procedure, and then present a more detailed example, with dimensions.

Open a brand-new file using ACAD.dwt, type in style, and press Enter. The familiar text Style dialog box comes up. Create a new style, calling it *Arial_Anno*, font Arial. Check off the *Annotative* box under Size and leave the Paper Text Height as 0. The new text has a triangular symbol next to it, as seen in Fig. 10.31. This symbol represents then end view of an architectural, mechanical or civil engineers ruler called a scale. Be sure to set your new style as Current.

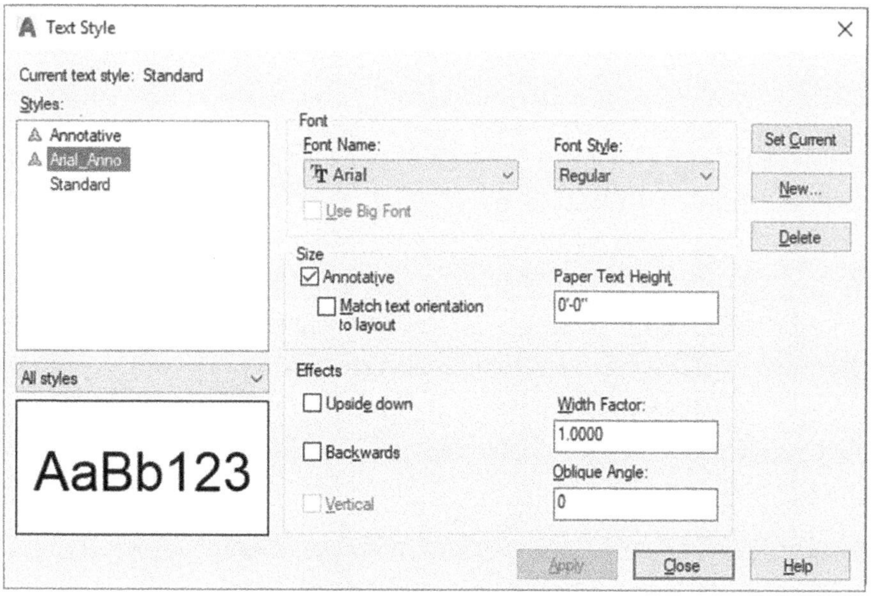

FIGURE 10.31 Annotative style.

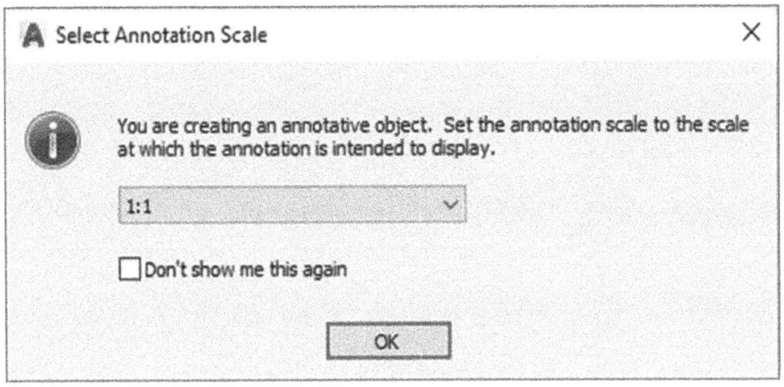

FIGURE 10.32 Select Annotation Scale.

Now switch to Model Space and create a line of text. The first time you do this, a dialog box, such as shown in Fig. 10.32, appears. Set the scale at which you anticipate this text to show up (such as inside a 1/4″ = 1′–0″ viewport, as done here).Then press OK and finish typing the text.

Now, switch to Paper Space and create a viewport and set it to the 1/4″ scale. Expand or contract it to fit nicely around the text, as shown in Fig. 10.33.

Copy that viewport to the right and change its scale to 1/8″. The text briefly shrinks to a smaller size, then balloons back up to the same size you set (¼″), as shown in Fig. 10.34. Note that this only works with the AutoScale (annotative "lightning bolt," see Fig. 10.35) button turned on. If AutoScale is off, the text will disappear but only if the Annotation Visibility (annotative "light bulb," see Fig. 10.35) button is also off. If Annotation Visibility is on, the viewport will show smaller text as we zoomed farther away.

This is the essence of annotation. The Annotation Scale is adjusted at the bottom right of the taskbar, as shown in Fig. 10.35, so when new text is being created in Model Space, you need to set its anticipated viewport scale before creating text or other annotative objects.

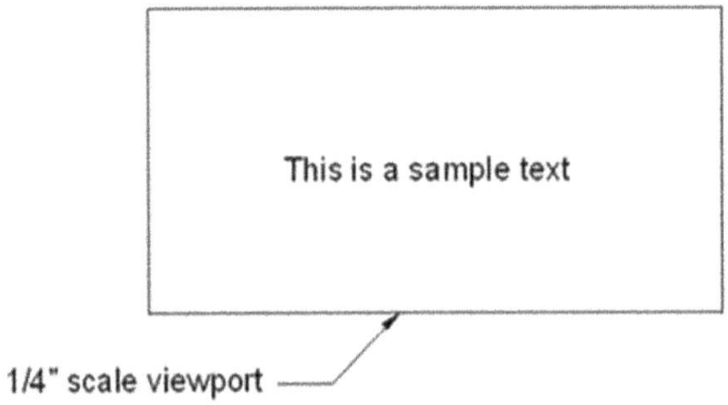

FIGURE 10.33 Annotation text in sample viewport.

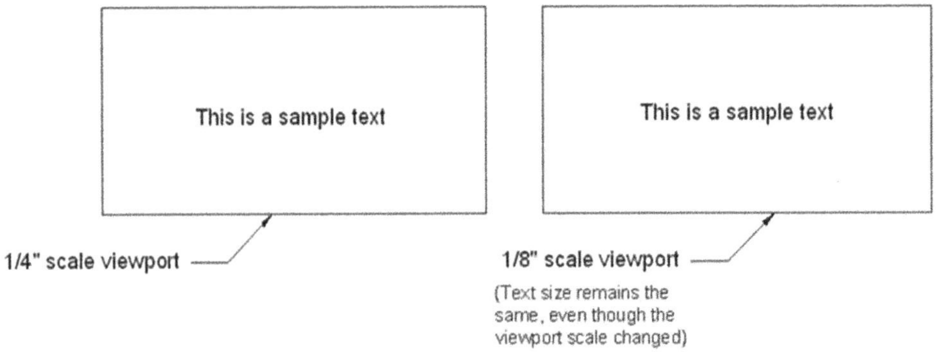

FIGURE 10.34 Annotation text in sample viewports.

FIGURE 10.35 Annotation scale icons.

You can also set multiple annotation scales to the same text if you anticipate it being visible in two or more viewports set to different scales. In model space, click on the text and a "double" of the text pops up (Fig. 10.36). This double is not a true copy but a twin that will contain the same words no matter which scale is displayed. With AutoScale turned on, go ahead and set the new annotation scale from the pop-up list in Fig. 10.35 and move the text to a new position using its grip (usually right on top of where the original was).

Everything done here with annotative text can be done with dimensions, so let us go ahead and do one more example, this time with dimensions.

Step 1. Open Exercise 5 from Chapter 3. If you do not have it ready, go ahead and recreate it, leaving just the object (no dimensions or text) on your screen. It should look like what you see in Fig. 10.37.

Step 2. Set the *Annotative* dimension style current in the Ribbon's Home tab, Annotation panel drop-down as seen in Figs. 10.38.

Step 3. Go to the Layout1 tab as your active view. In Layout1, erase the existing viewport that shows a view of your model space if it exists. You should now have a blank piece of paper, ready for new viewports as seen in Fig. 10.39. Now turn the Annotation Visibility and AutoScale off in the status bar.

FIGURE 10.36 Annotation text copy.

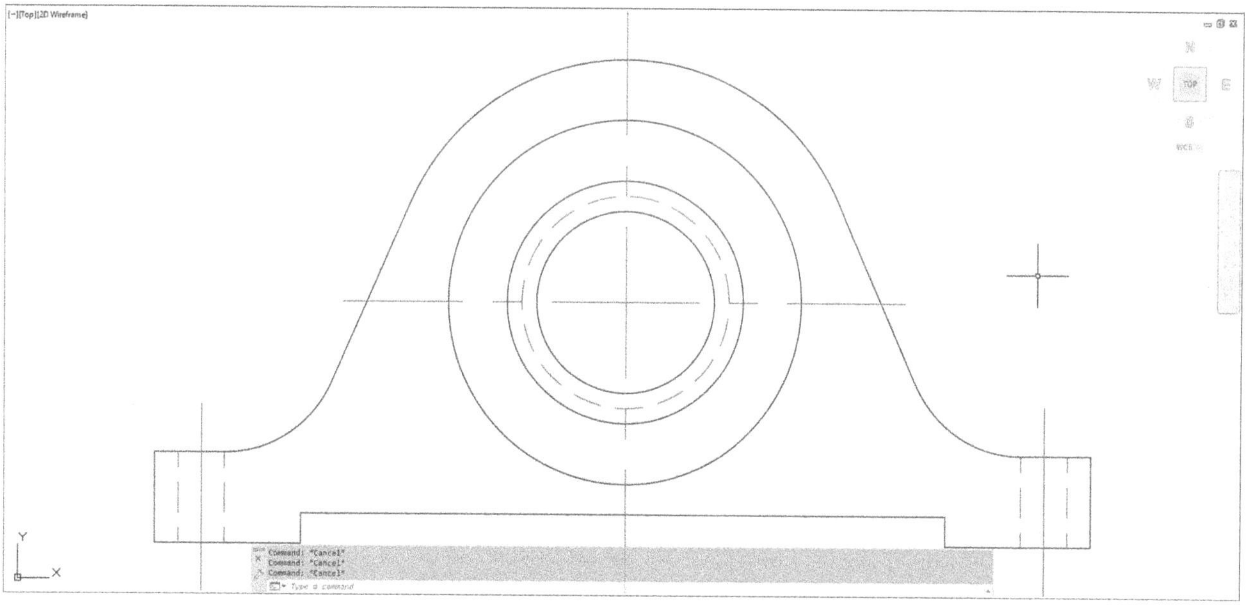

FIGURE 10.37 Exercise 5, Chapter 3.

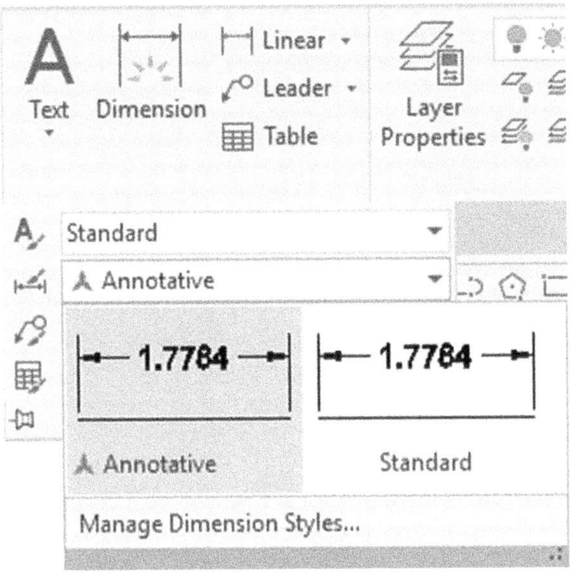

FIGURE 10.38 Set Annotative dimension style.

FIGURE 10.39 Blank Paper Space sheet.

Step 4. On the Layout tab of the Ribbon (a contextual tab that only appears when in a Layout/Paper Space), click the Viewports Rectangular button, and press "4" Enter for the create four viewports at once option, and then create a rectangular viewport as inside of the printable area of the given sheet seen in Fig. 10.40. The model will appear equally zoomed to extents in all four viewports.

Step 5. Double-click within the top-left viewport to activate it. Then in the Options/Status bar (bottom right), find the viewport scale setting (it will show a decimal number) and click it. A lengthy selection menu will pop up. Set the scale to 1:2 (Fig. 10.41) and lock the viewport scale with the Viewport Lock button.

Step 6. Click in the upper right viewport to activate it. Pan and zoom manually to focus on the lower left base of the part. Then, following the same procedure as Step 5, set the scale to 2:1 and lock the viewport scale.

Step 7. Click in the lower right viewport to activate it. Pan and zoom manually to focus on the lower right base of the part. Again, following the same steps, set the scale to 2:1, and lock the viewport scale.

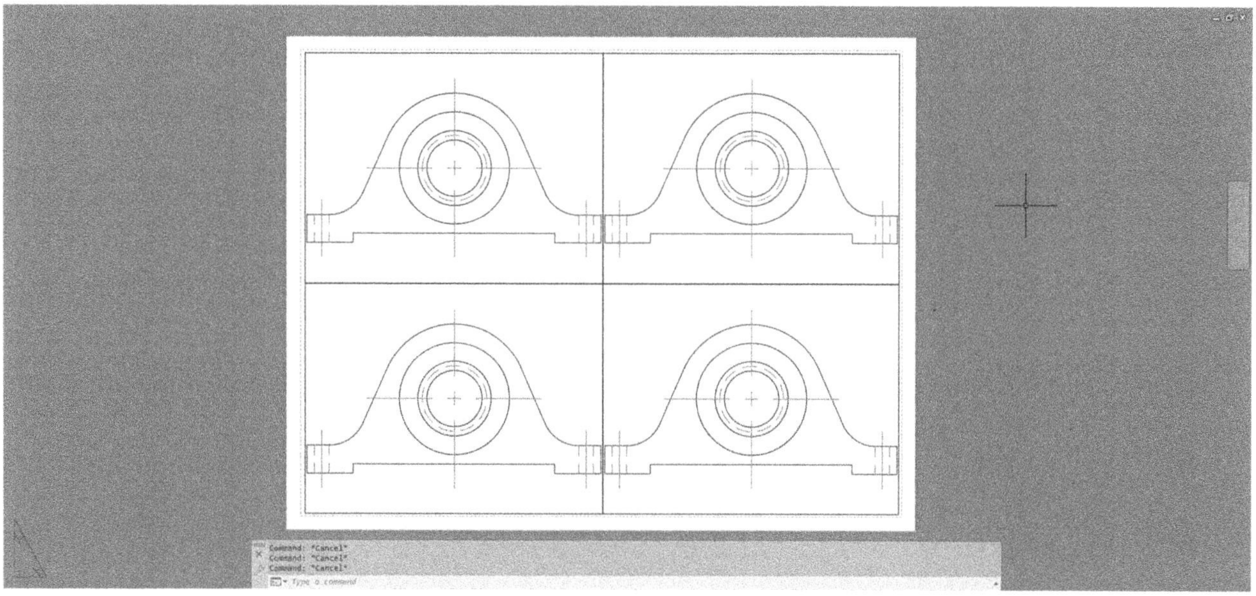

FIGURE 10.40 Four Viewports in Paper Space.

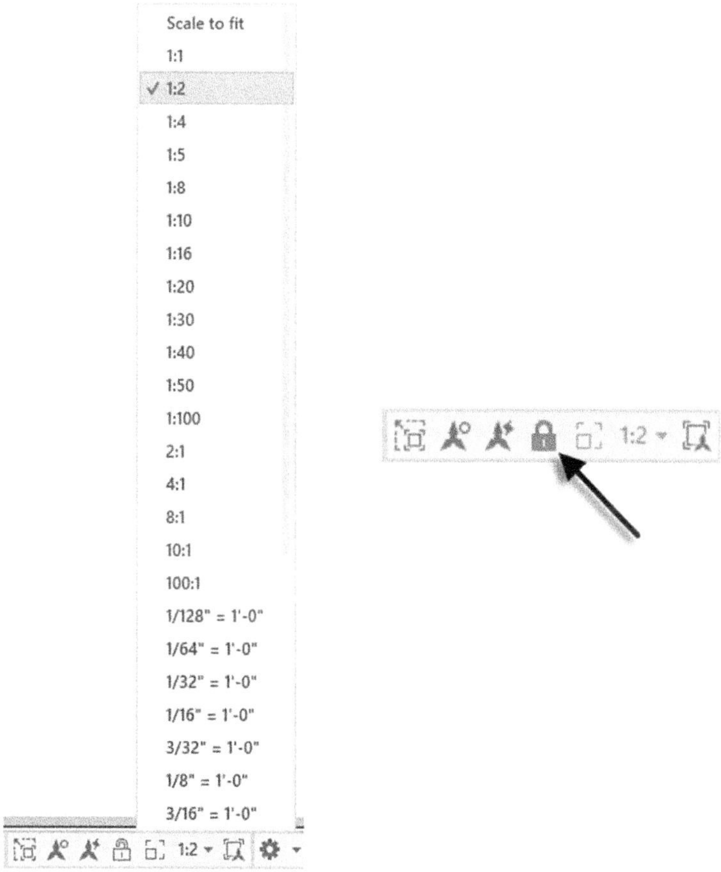

FIGURE 10.41 Set Scale and Viewport Lock.

Step 8. Finally, click in the lower left viewport to activate it. Pan and zoom manually to focus on the right quadrant of the center circles. Again, following the same procedure, set the scale to 4:1 and lock the viewport scale. Your result is shown in Fig. 10.42.

Step 9. Double check that the Dimension Style is set to Annotative and make sure both Annotation Visibility and Auto-Scale buttons are set to off (gray) as shown in Fig. 10.43 before we start adding dimensions through the Viewports.

Step 10. Add dimensions as follows by double clicking into each viewport to activate it (result is shown in Fig. 10.44):

- Using Linear Dimensions add an overall height and width, as shown, to the *upper left viewport.*

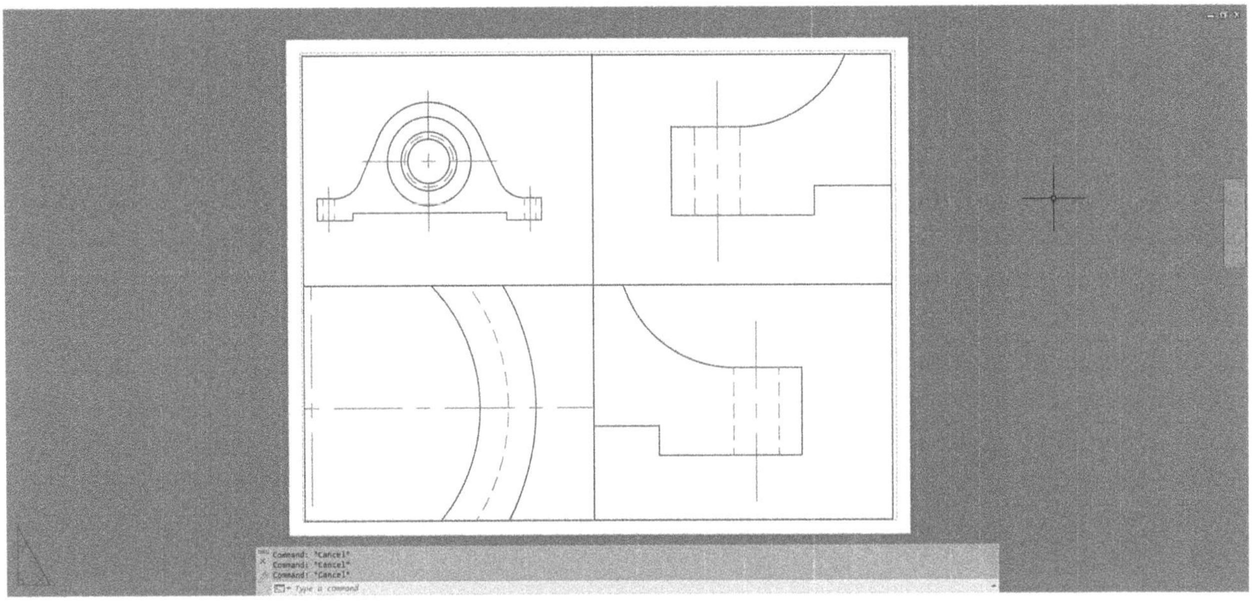

FIGURE 10.42 Final Viewport setup.

FIGURE 10.43 Annotative settings.

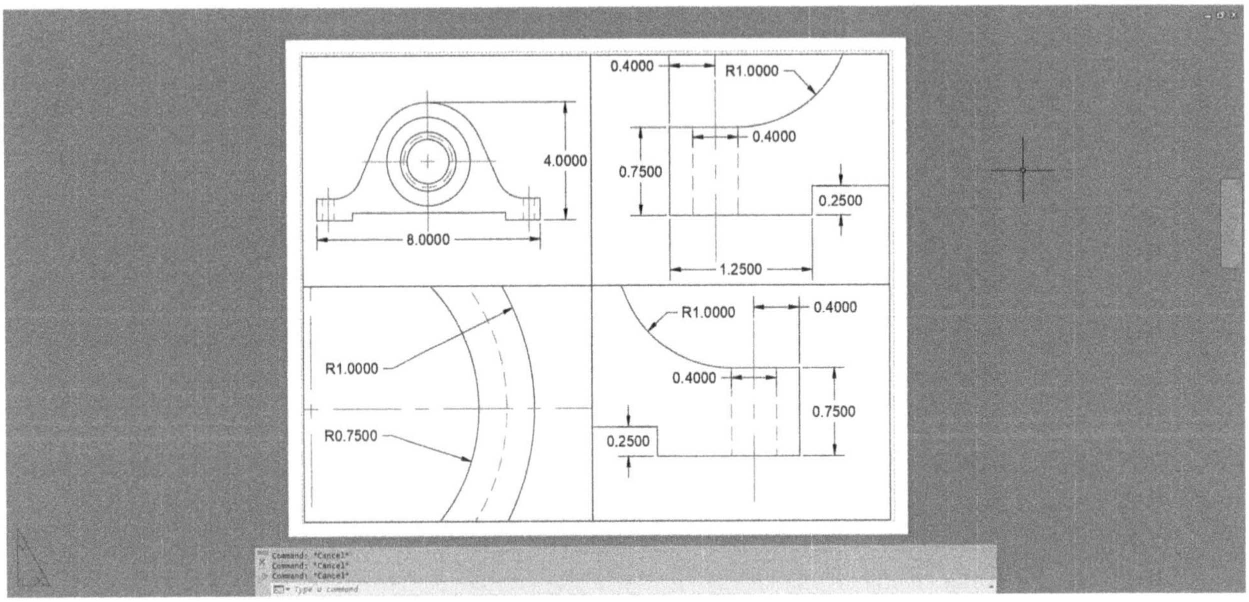

FIGURE 10.44 Final dimensioning.

- Using Linear Dimension and Radius Dimension, add dimensions as shown to the *upper right viewport.*
- Using Linear Dimension and Radius Dimension, add dimensions as shown to the *lower right viewport.*
- Using Diameter Dimensions, add dimensions as shown to the *lower left viewport.*

When done, double-click outside of the boundary of the final viewport you are working on to get back onto Paper Space.

So, let us do a quick analysis of what has just happened as it may not be obvious yet. All your dimension features (text height arrow size, extension line gaps, etc.) are all the same in each viewport even though the scale of each viewport is different in most of them. AutoCAD has essentially automated what would have been a tedious dimension setup process. This of course closely mirrors what you would have in hand drafting: no matter what scale of drawing you were producing, the annotation (text and dimensions) would be the same for all details.

To see how things "really are," click back into Model Space and turn on the Annotation Visibility button (blue is on as shown in Fig. 10.45). The Annotative Scale should remain at 1:1 for now.

You will see your full-size model with all the dimensions as "actual size" based on the viewport scale they were created in as seen in Fig. 10.46. Only the dimension features were scaled up or down according to the scale of the viewport (not the dimension distances or sizes). The closer the viewport was zoomed in to the object the smaller the dimension features became to account for that. The farther out the viewport was zoomed the bigger the dimension features became to account for that.

Now let us reverse that process and add Annotative Dimensions directly in Model Space. This would be a more typical procedure in industry because you would likely be familiar with the scale of the viewport you were going to show your drawing in on the Layout/Paper Space. You would simply draw your objects full size as usual and then before adding any annotation, set your Annotative Scale for the anticipated viewport scale. Let us try that on the current drawing.

Step 11. Set the current Annotative Scale in Model Space to 1:2 and turn off Annotation Visibility (Fig. 10.47)

FIGURE 10.45 Annotation Visibility on in model space.

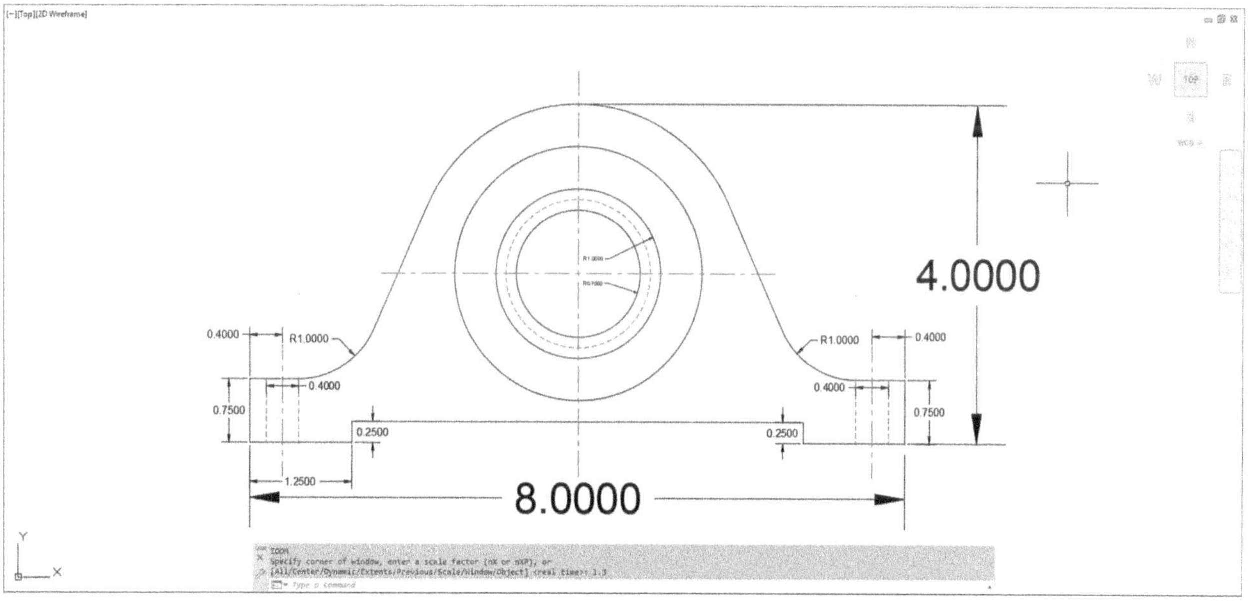

FIGURE 10.46 Object with annotative dimensions in Model Space.

FIGURE 10.47 Annotation Visibility.

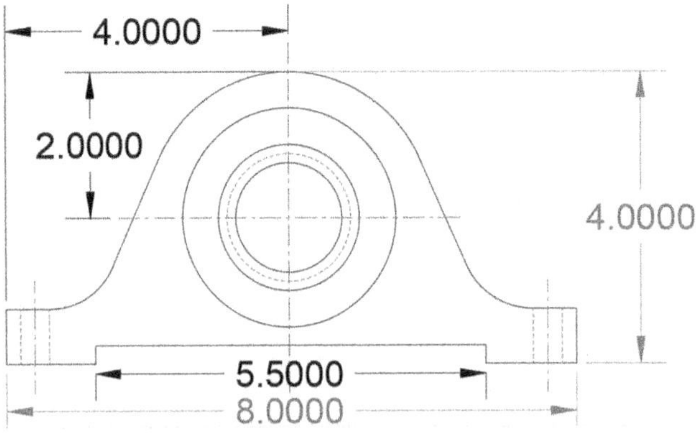

FIGURE 10.48 Additional dimensions added in at 1:2 scale.

All dimensions should disappear except those created at the 1:2 scale. Now go ahead and add in additional dimensions as shown in Fig. 10.48. Note the new dimensions are shown bold and the existing dimensions are faded for clarity only.

Now change your Annotative Scale in Model Space to 2:1 and add some more dimensions as shown in Fig. 10.49. Note the new dimensions are shown bold and the existing dimensions are faded for clarity only.

Finally change the Annotative Scale in Model Space to 4:1 and add additional dimension as shown in Fig. 10.50. Note the new dimensions are shown bold and the existing dimensions are faded for clarity only.

Now go back to the Layout/Paper Space. You will notice that the only the dimensions created at a scale that matches the viewport zooming are shown through that viewport (Fig. 10.51).

Now in Layout/Paper Space turn on the Annotation Visibility button as seen in Fig. 10.52.

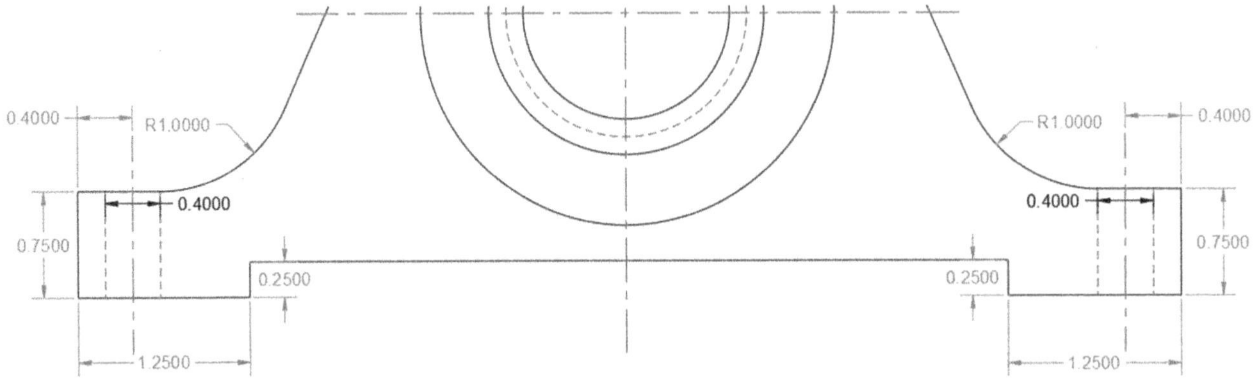

FIGURE 10.49 Additional dimensions added in at 2:1 scale.

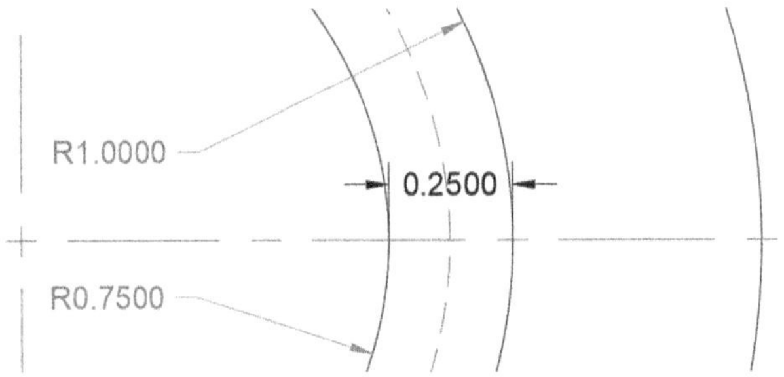

FIGURE 10.50 Final dimension added in at 4:1 scale.

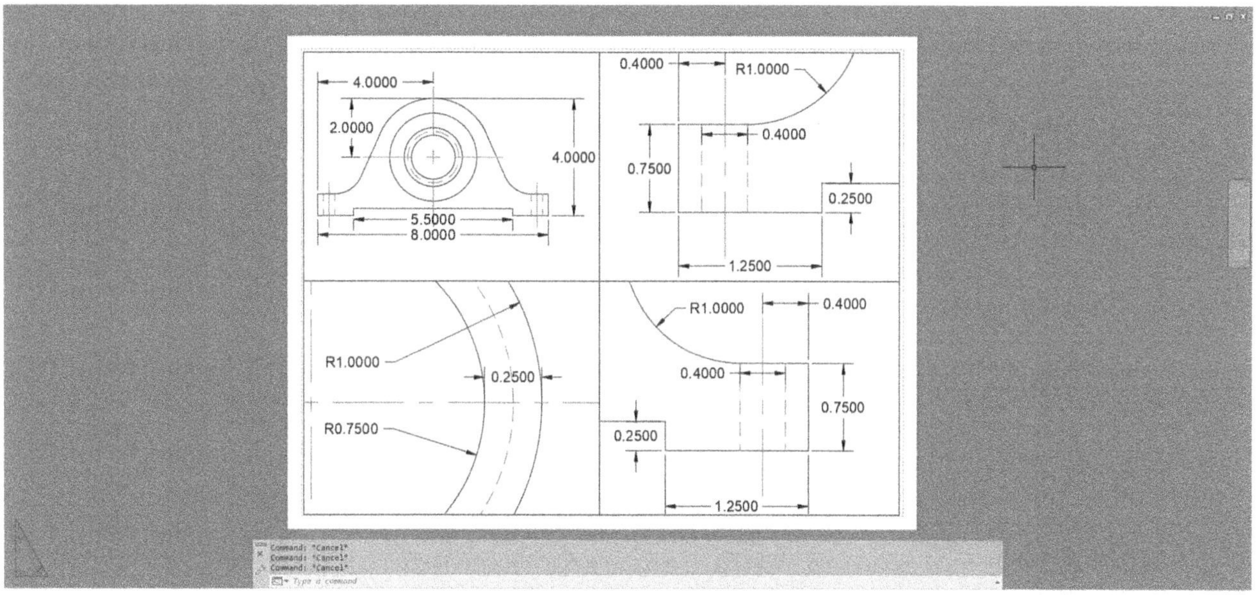

FIGURE 10.51 Final set of dimensions added in.

FIGURE 10.52 Annotation Visibility turned on in Layout/Paper Space.

Now all dimensions at all annotative scales are shown in all viewports (Fig. 10.53). Not a desirable view, but it shows the drastic difference in the size of the dimension features required to show dimensions at different scales accurately as we saw in model space.

Finally, turn off the Annotation Visibility button again to hide the "incorrect" dimensions and only display the dimensions that match the scale of the viewports they are displayed in.

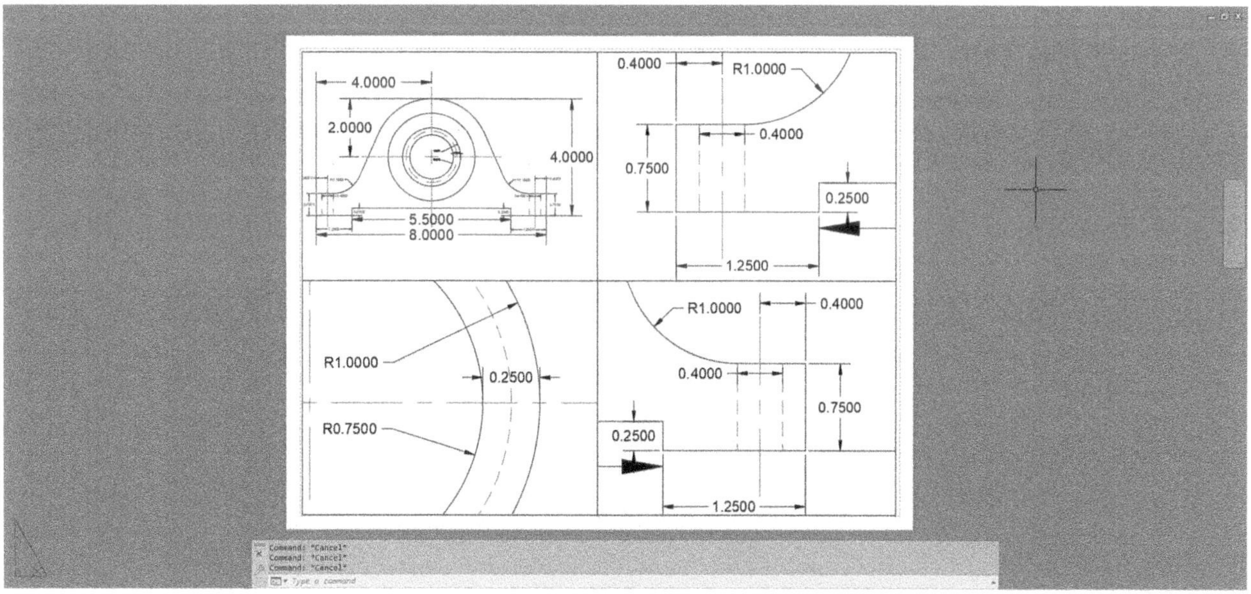

FIGURE 10.53 Annotation Visibility on in Layout/Paper Space.

So, what is the real benefit here? The first is that you could nearly get by with one annotative text style, one annotative dimension style, and one annotative Mleader style for a given project. You also could nearly get by with only one layer for text, one layer for dimensions, and one layer for leaders because only the annotative objects that show through a viewport are the ones that match that viewport's scale.

When using annotative styles for text, dimensions and Mleaders in model space always draw your objects full size, 1:1 as usual. Then set the annotative scale to your desired printed scale/viewport scale before adding any dimensions, text, leaders, etc. Changing the annotative scale in model space does not affect the drawing of your full size, 1:1 objects. If you are not sure of the annotation scale you will need for your final print, go ahead and make a viewport in your Layout/Paper Space sheet and set the scale desired. You can then go back to model space and set an annotation to match before adding your annotation or double-click in the viewport and start adding annotations.

If you choose to use nonannotative styles, you would have to create a separate style for each scale you would want to use as you did earlier in this chapter. Then you would switch between styles and set a proper unique layer current before adding annotation at different scales. This would allow you to see the "properly" sized dimensions and Viewport Freeze the layer for the "improperly" sized dimensions. In the end, annotative scaling of annotation created with annotative styles set properly can allow for fewer text styles, dimension styles, Mleader styles, and fewer layers to manage.

As discussed earlier, the AutoScale (Annotative Symbol with Lightning Bolt) button can be turned on (blue) to automatically add annotation scales to all annotative objects when the annotation scale changes. This means that any time you change the annotative scale, all annotative objects will be shown at additional scales including the one they were created in initially. This can seem like a quick fix if you have not set your annotative scale correctly before adding text, dimensions, and Mleaders, but there are some dangers here. If trying to view an object, or parts of it at different scales through several viewports, you may start seeing these additional scaled annotations in viewports that you did not intend to see. If you need annotative objects to show in several viewports at different scales, it is a much safer choice to select those objects and use the Quick Properties or Properties palette to add or remove annotative scales from objects.

Summary

We go over a lot of information in this Paper Space chapter. Ideally, you have absorbed all of it and are able to use this valuable and important tool. Many use it improperly or not at all, and being an expert in Paper Space sets you apart from the crowd and adds a true level of professionalism to your designs. This summary serves as a capstone to this chapter and presents the big picture once again, now that the details have been mastered.

Do not think of Paper Space as a rigid or fixed system, where buttons are pushed and results pop out. While an element of this is present, the proper way of looking at it is as a flexible and loose collection of tools that allows you to shape and mold the presentation of your design in a way, i.e., efficient, clear, professional, and of course, pleasing to the eye. There is no *one* way to do things. Rather there is a constant tug of war between (sometimes conflicting) requirements.

For example, you must consider

- *The design*: This is the first and foremost concern. Your design takes priority over everything else, so focus on that in Model Space, creating your architectural layout or engineering device, as per client need.
- *The size of the available plotter and paper*: After the design is completed, the output is considered. What do you have to work with? Most companies have a plotter capable of 36″ × 24″ plots. Some have the larger 48″ × 36″ plotters. Know your paper options. The larger plotters take E-sized paper, the smaller only up to D-size. You cannot print on what you do not have unless you outsource to a print shop or print to PDF.
- *The scale needed to clearly see all facets of the design*: This is the next important step. Look at how your design will look on the D- or E-size sheets. Can you see everything? Are any parts of the design not clearly visible? Generally, a 25′×25′ building (such as the one used as an example in this chapter) fits easily on a D-size sheet at easily 1′–0″ scale. If it is much bigger, then you can either switch to an E-size sheet (if possible) or to another, smaller scale. A smaller scale, makes some details harder to see, when viewing the entire plan. In that case additional Viewports can be added to stay show more views zoomed into room or details at larger scales.

As you can see, many variables come into play in setting up Paper Space properly; and with experience, all this will be in your head at the start of a project and you will know instinctively what paper, scale, and viewports to go with, based on the size and expected overall shape of the design.

At the completion of this chapter, you have graduated from a beginner to an intermediate level of AutoCAD knowledge. If you have understood everything up to now and have not skipped any exercises or reading, then great job! If all you need AutoCAD for is basic drafting or just knowing enough to oversee other full-time AutoCAD designers, then you may stop here, you have what you need. If your work requires expert level knowledge, you need to go on to Level 2, Chapters 11–20. See you there!

You should understand and know how to use the following concepts and commands before moving on to Level 2:

- Paper Space
- Layouts
- Viewports
- Scaling a design inside a viewport
- Working with layers in individual viewports
- Scaling text and dims in viewports
- The Annotative concept

Review questions

Answer the following based on what you learned in this chapter:

1. What is Paper Space?
2. What are layouts?
3. How do you create, delete, rename, and copy layouts?
4. What are viewports?
5. How do you create, scale, and shape viewports?
6. How do you scale a design inside a viewport?
7. How do you turn layers on and off in individual viewports?
8. How do you scale text and dims in viewports?
9. Describe the Annotation concept.

Exercises

1. Set up a total of five title blocks in Paper Space using the Tutorial-iMfg.dwt template file. Name the tabs as seen next and delete any extra tabs. (Difficulty level: Easy; Time to completion: <10 minutes.)
   - Front View
   - Side View
   - Section A-A;
   - Section B-B;
   - Section C-C
2. Bring up the floor plan you did in Chapter 4, Text, Mtext, Editing, and Style, Exercise 7 (or redraw it). Then, go into Paper Space and create an Architectural Layout tab, followed by several viewports, as in the top figure that follows. Assign each of the viewports a scale as shown in the bottom figure. Freeze the dims layer in the two detail viewports. Note that the floor plan has been rotated 90 degrees clockwise for this exercise. (Difficulty level: Easy; Time to completion: 10–15 minutes if the floor plan is completed, 55–75 if not.)

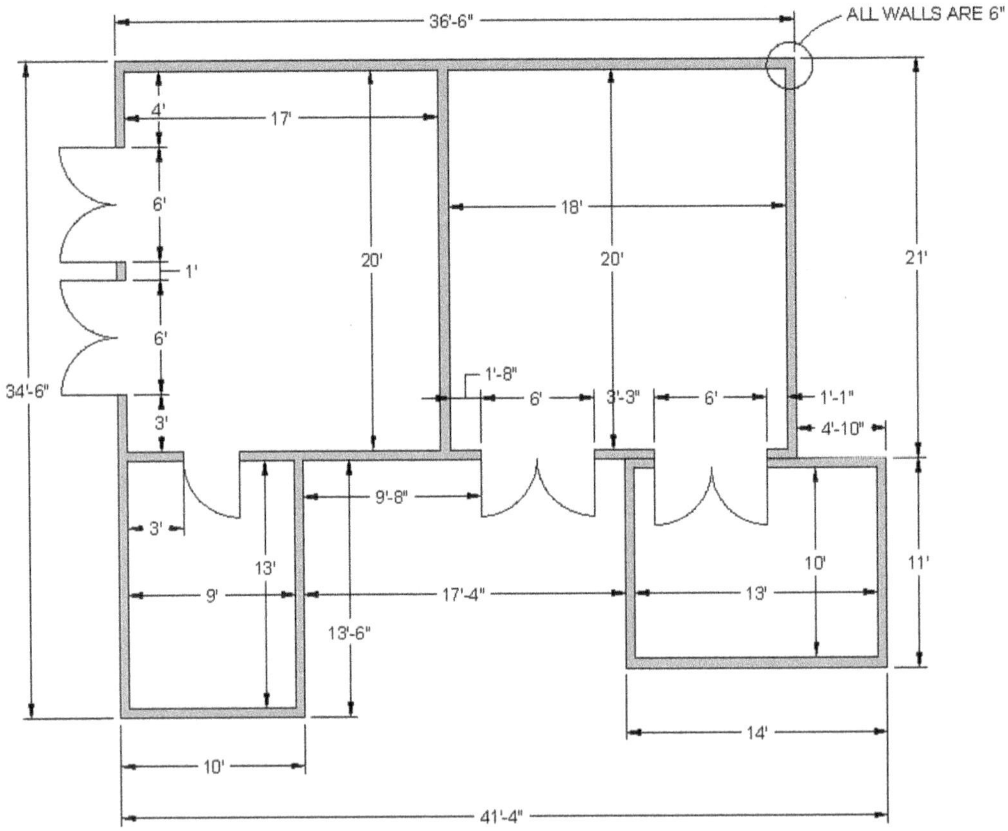
36'-6"
ALL WALLS ARE 6"
4'
17'
18'
6'
1'
20'
20'
21'
6'
34'-6"
1'-8"
6'
3'-3"
6'
1'-1"
4'-10"
3'
9'-8"
3'
13'
10'
11'
9'
17'-4"
13'
13'-6"
14'
10'
41'-4"

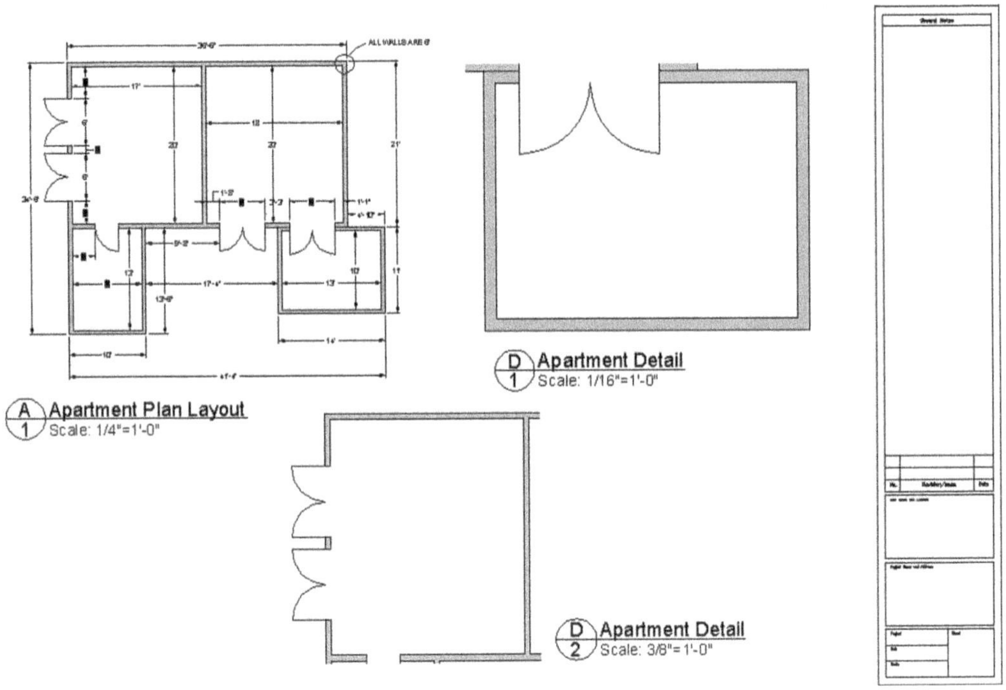
D Apartment Detail
1 Scale: 1/16"=1'-0"
A Apartment Plan Layout
1 Scale: 1/4"=1'-0"
D Apartment Detail
2 Scale: 3/8"=1'-0"

Level 1

Answers to review questions

Chapter 1: Review questions

1. List the four Create Objects commands covered.
   Line, circle, arc, and rectangle.
2. List the 10 Edit/Modify Objects commands covered.
   Erase, move, copy, rotate, scale, trim, extend, offset, mirror, and fillet.
3. List the three View Objects commands covered.
   Zoom, pan, and regen.
4. What is a workspace?
   A workspace is a customized screen layout.
5. What four main methods are used to *issue a command* in AutoCAD?
   Command line, toolbars, cascading menus, and the Ribbon.
6. What is the difference between a selection Window, Crossing, and Lasso?
   A Window is a solid rectangle produced by moving the mouse to the right of an imaginary line and selects only what is completely inside of it. A Crossing is a dashed rectangle produced by moving the mouse to the left of an imaginary line and selects anything it touches. A Lasso can behave like either a Window or Crossing but is a free-form shape, not limited to a rectangle/square. It is invoked via a click-and-drag technique.
7. What is the name of the command that forces lines to be drawn perfectly straight, horizontally, and vertically? Which F key activates it?
   Ortho, F8.
8. What six main OSNAP points are covered? Draw the snap point's symbol next to your answer.
   Endpoint, Midpoint, Intersection, Center, Quadrant, and Perpendicular. See the text for the symbols.
9. What command brings up the OSNAP dialog box?
   OSNAP.
10. Which F key turns the OSNAP feature on and off?
    F3.
11. How do you terminate a command-in-progress in AutoCAD?
    Esc key.
12. What sequence of commands quickly clears your screen?
    Erase → All, Enter, Enter.
13. What sequence of commands allows you to view your entire drawing?
    Zoom → E. Also double-click center mouse wheel (if equipped).
14. How do you undo a command?
    Type in u for undo *or use the Undo arrow in the Standard toolbar.*
15. How do you repeat the last command you just used?
    Press the space bar or Enter.

Chapter 2: Review questions

1. How do you activate grips on AutoCAD's geometry?
   Click on any of the objects.
2. How do you make grips disappear?
   Press the Esc key.
3. What are grips used for?
   To modify the shape or position of an object.
4. What command brings up the Drawing Units dialog box?
   Units.
5. The snap command is always used with what other command?
   Grid.
6. Correctly draw a Cartesian coordinate system grid. Label and number each axis appropriately, with values ranging from −5 to +5.
   See the text, Fig. 2.4.
7. On the preceding Cartesian coordinate system grid, locate and label the following points: (3,2) (4,4) (−2,5) (−5,2) (−3,−1) (−2,−4) (1,−5) (2,−2).
   Student activity.
8. What are the two main types of geometric data entry?
   Dynamic Input and Manual Input.
9. Describe how Dynamic Input works when creating a line at an angle.
   With DYN activated, begin drawing a line, entering a distance value in the first highlighted box. Then, tab over to the second highlighted box, entering an angle value.
10. List the six Inquiry commands covered.
    Area, distance, list, ID, radius, and angle.
11. What command simplifies a rectangle into four separate lines?
    Explode.
12. What is the difference between an inscribed and a circumscribed circle in the polygon command?
    Inscribed means the polygon appears inside a given circle, with vertices tangent. Circumscribed means the polygon appears outside a given circle, with the sides tangent.
13. What is a chamfer as opposed to a fillet?
    A chamfer creates an angled edge, whereas a fillet creates a smooth circular edge or none at all.
14. What template should you pick if you want a new file to open?
    acad.dwt.
15. What is the idea behind limits?
    Limits define the outer boundaries of your work area.
16. Why should you save often?
    Because all software and computers are subject to crashes and malfunctions.
17. What is the graphic symbol for save?
    A computer floppy disk.
18. How do you get to the Help files in AutoCAD?
    By pressing F1 at any time.

Chapter 3: Review questions

1. What is a layer?
   Layers is a concept that allows you to group drawn geometry in distinct and separate categories according to similar features or a common theme.
2. What is the main reason for using layers in AutoCAD?
   To maintain control over your drawn geometry.
3. What typed command brings up the Layers dialog box?
   Layer or just `la`*.*
4. List two ways to create a new layer.
   New Layer icon or, once a new layer exists, just press Enter while selecting any layer.
5. How do you delete a layer?

The red X.

6. How do you make a layer current?
   Double-click on it or select it and press the green check mark symbol.
7. What three available Color tabs are available?
   Index Color, True Color, and Color Books.
8. What does freezing a layer do?
   It makes all objects on that layer invisible.
9. What does locking a layer do?
   It makes all objects on that layer uneditable.
10. What are some of the linetypes mentioned in the text?
    Hidden, Center, Dashed.
11. What are three ways to bring up the Properties toolbar?
    Double-click on any object, press Ctrl+1, or type in `ch`.
12. What two other ways to change properties or layers are discussed?
    Match Properties and Layers toolbar.

Chapter 4: Review questions

1. What are the two types of text in AutoCAD? State some properties of each, and describe when you would likely use each type.
   Text: single line, limited formatting. Mtext: paragraph form, advanced formatting.
2. What is the easiest method to edit text in AutoCAD? What was another typed method?
   Double-click on the text. `Ddedit`.
3. What feature allows you to add extensive symbols to your mtext?
   Character Map.
4. What are the three essential features to input when using Style?
   Font name, font, and size.
5. What is the width factor?
   How wide a font is; 1 is standard width.
6. What is an oblique angle?
   The "tilt" of the font.
7. How do you initiate spell check?
   Right-click on the underlined misspelled word, or type in spell.
8. What new OSNAP is introduced in this chapter?
   Nearest.
9. How do you get rid of the UCS icon?
   Type in `ucsicon`, *then* `off`.
10. What can you do if you cannot see the dashes in a Hidden linetype?
    Use ltscale.

Chapter 5: Review questions

1. List the four steps in creating a basic hatch pattern.
   Pick hatch pattern, indicate where you want it to go, fine-tune the pattern via scale and angle, and preview pattern.
2. What two methods are used to select where to put the hatch pattern?
   Directly picking an object or picking a point inside that object.
3. What commonly occurring problem can prevent a hatch being created?
   A gap.
4. What is the name of the tool recently added to AutoCAD to address this problem?
   Gap tolerance.
5. How do you edit an existing hatch pattern?
   Double-click on it.
6. What is the effect of exploding hatch patterns? Is it recommended?
   The hatch breaks up into individual components. No.
7. What especially useful pattern to fill in walls is mentioned in the chapter?
   Solid Fill.

Chapter 6: Review questions

1. List the 11 types of dimensions discussed.
   Linear, Aligned, Diameter, Radius, Angular, Continuous, Baseline, Leader, Arc length, Ordinate, and Jogged.
2. What command is best for editing dimension values?
   Double-click.
3. What is the difference between natural and forced dimensions?
   Natural dimensions show the actual value, while forced dimensions are added manually and may be false values.
4. How do you restore a natural dimension value when you forgot what it was?
   Type in <>.
5. List the four items that needed to be customized in our dimensions.
   Units, font, arrowheads, and fit.
6. What command brings up the Dimension Style Manager?
   Dimstyle or ddim.
7. What type of arrowhead do architects usually prefer?
   Architectural tick.

Chapter 7: Review questions

1. What are blocks and why are they useful?
   A block is a collection of objects grouped together under a common name. They are useful to keep multiple objects together and for reuse via symbol libraries.
2. What is the difference between blocks and wblocks?
   A block is internal to a file, and a wblock is external.
3. What are the essential steps in making a block? A wblock?
   For a block, you need to specify a name, select objects, and pick a base point. For a wblock, you also need to specify a path.
4. How do you bring the block or wblock back into the drawing?
   Via the insert command.
5. Explain the purpose of purge.
   Purge rids the drawing of unseen elements present in the file, such as blocks, layers, or fonts.
6. Explain the basic idea behind dynamic blocks. What steps are involved? What is visibility?
   Dynamic blocks are intelligent blocks that can be modified without being redefined. To create a dynamic block, you add parameters and actions. Visibility allows you to combine many blocks together and control, which is visible at any one time via visibility states and a drop-down menu contained in the main block.
7. What are the advantages and disadvantages of using groups?
   Advantages of a group: A group can have objects added and subtracted from it and can have individual objects (and their properties) edited. Disadvantages of a group: A group cannot be transferred between drawings (for "local" use only), and an update to one group does not update other groups with the same name.

Chapter 8: Review questions

1. List the three types of arrays available to you.
   Polar, rectangular, and path arrays.
2. What steps are needed to create a polar array?
   You need to specify what object(s) to array, what the center point of the array is, how many objects to create, and whether you want the pattern to fill out a full circle or part thereof.
3. To array chairs around a table, where is the center point of the array?
   The center of the table.
4. What steps are needed to create a rectangular array?
   You need to specify what object(s) to array, how many rows and columns, and the distance between those rows and columns.
5. What steps are needed to create a path array?
   You need to specify what object(s) to array, the path, and how many objects to create.

Chapter 9: Review questions

1. List the seven essential sets of information needed for printing or plotting.
   What printer or plotter to use, what paper size to use, what area to plot, at what scale to plot, what pen settings to use, what orientation to use, and what offset (if any) to use.
2. List the five paper sizes and their corresponding measurements in inches.
   A size: 11 × 8.5″ sheet of paper (letter), B size: 17 × 11″ sheet of paper (ledger), C size: 22 × 17″ sheet of paper, D size: 36 × 24″ sheet of paper, and E size: 48 × 36″ sheet of paper.
3. What is the difference between printing and plotting?
   Printing usually refers to the output of drawings on A- or B-sized paper; and plotting usually refers to the output of drawings on C-, D-, or E-sized paper.
4. List the four plot area selection options.
   Extents, Window, Display, and Limits.
5. What scale setting ignores scale completely?
   Fit to paper.
6. What is the extension of the pen settings file?
   ctb.
7. Explain the importance of Preview before printing and plotting.
   Mistakes can be spotted before a print is sent to the printer or plotter, thereby saving paper.
8. Why is the Page Setup Manager useful?
   Because it allows you to set up and save print and plot settings for future use.

Chapter 10: Review questions

1. What is Paper Space?
   Paper Space is an AutoCAD concept that allows you to draw your design in full size (1:1) units and still print it to scale using viewports.
2. What are layouts?
   Layouts are Paper Space pages that hold the title block and viewport.
3. How do you create, delete, rename, and copy layouts?
   All these tools are accessed by right-clicking on the Layout tabs.
4. What are viewports?
   A viewport is an AutoCAD concept that allows you to link together Model and Paper Space.
5. How do you create, scale, and shape viewports?
   You can create a viewport via the mview command. Scaling and shaping it are accomplished via the viewport grips.
6. How do you scale a design inside a viewport?
   You can use the icons on the Viewport toolbar or via keyboard entry using scale, then 1/xp.
7. How do you turn layers on and off in individual viewports?
   By clicking inside a viewport, calling up the Layer Manager, and freezing the layers under the VP Freeze column.
8. How do you scale text and dims in viewports?
   By using the appropriate charts and matching the text and dim sizes with the correct scale.
9. Describe the Annotation concept.
   Annotation gives text, dims, and other AutoCAD features the ability to scale to viewports automatically based on the assigned scaling of those viewports.

Level 2

Chapters 11–20

Now that you have completed Level 1, beginner to intermediate, you should have the basic skills necessary to work on a drawing of moderate complexity with a certain amount of speed and accuracy. Level 2 is the second part of your journey in becoming proficient in AutoCAD. The goal at the end of this level is for you to learn most of the remaining practical information about AutoCAD in two dimensions. This will allow you to not only work on very complex drawings but also set up and manage them. Then, with some additional work experience, you will have the skills of a CAD manager or administrator, a necessity in many engineering or architectural firms, even if your job is not just AutoCAD.

Starting with the next chapter, this textbook does not feature any of the basic introductions concerning AutoCAD and what it is. It is assumed you have already read through all the introductory information and are familiar with all the materials in Chapters 1–10. There is minimal explanation or review of earlier material, so if you need to refresh anything from Level 1, do so before proceeding. With this level, we dive into the new material right away and do not look back, as there is much more to cover.

Level 2 consists of several parts. Chapters 11–13 are all about advanced features of basic concepts: lines, layers, and dimensions. We add significant depth to these Level 1 topics. Chapters 14–16 cover a broad set of topics that introduce various advanced features, such as interacting with other software, the Design Center, Express tools, the CUI, and many shorter topics. This clears the way for some remaining "big ideas" in AutoCAD: external references, attributes, and advanced output and pen settings, which make up Chapters 17–19. These are very important topics, especially external references, as they are critical to many architectural designs. We complete Level 2 with isometric design, which is a very useful tool and a great lead-in to 3D if you go on to learn that aspect of AutoCAD.

The pace of the course picks up somewhat as well, and more responsibility is placed on you to follow up and do your own research into additional AutoCAD functions. We also introduce a drawing project that spans the entire level from Chapter 11 to Chapter 20. It is a more in-depth floor plan that, while still greatly simplified when compared to full construction documents, attempts to replicate a real-life architectural layout. There also are fewer generic drawings to do, in favor of more topic-specific exercises at the end of each chapter. One constant remains: You still need to practice to ensure your success.

Chapter 11

Advanced linework

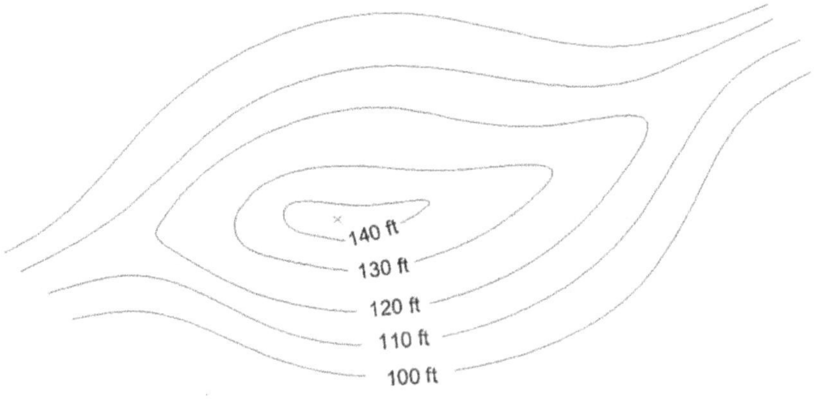

Learning objectives

In this chapter, we revisit the line concept and expand on it by discussing the following:

- Polyline, pedit, and Boundary
- Xline
- Ray
- Spline
- Mline, mlstyle, and mledit
- Sketch

By the end of this chapter, you will learn how to use and apply these new types of lines and be able to incorporate them into future design work.

Estimated time for completion of this chapter: 2–3 hours.

Up and Running with AutoCAD 2022. https://doi.org/10.1016/B978-0-323-89923-9.00011-1

11.1 Introduction to advanced linework

We begin Level 2 in the same way we started Level 1, with the simple line command. At the time, this was enough to start with, and no mention was made of the many other types of lines AutoCAD has. Classroom experience had shown that introducing all the other types of lines at an early stage was a sure way to scare off brand-new students, and though you may have explored them on your own. Now that you are more advanced, we can come back and look at other line types in depth.

As it turns out, while the line command is useful for most drawing situations, it is inadequate or inconvenient for a small but important number of other applications. Here, we look at other variants of the line including the *Pline* (polyline), the *Xline* (construction line), its close relative the Ray, a special case of the polyline called a *Spline*, the *Mline* (multiple line), and the most unusual one of all, the *Sketch* command.

11.2 Pline (polyline)

A pline is a regular line with some unique properties:

Property 1. A pline's segments are joined together (many lines acting like one object).
Property 2. A pline can have thickness.
Property 3. A pline can have straight and/or curved segments.
Property 4. A pline can be open or closed.

These three properties add tremendous importance and versatility to the single line command, which creates only individual segments that maintain a constant zero thickness. Fig. 11.1 provides an illustration of the first property. The top "W" is drawn with four line segments. As you can see, they are all separate, with only the left segment clicked on, and the grips reflect that. The "W" below it is drawn using the pline command and is all one unit, as seen by clicking it once. Notice the slightly compressed center grips.

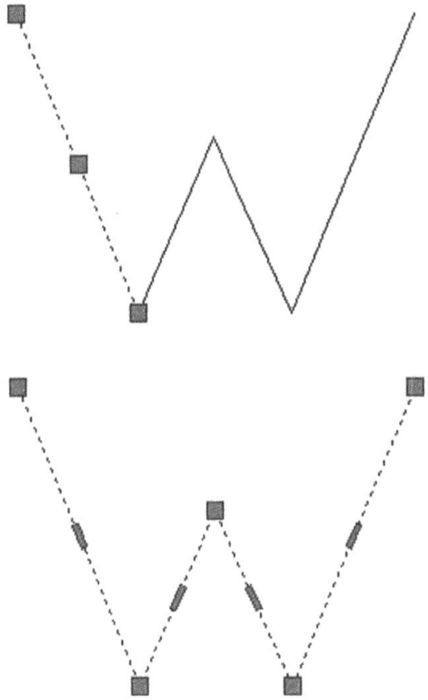

FIGURE 11.1 Line versus pline command (property 1).

One example of applying this feature may include drawing lengthy runs of cabling or piping. Then, if the run needs to be deleted, only one click is necessary to remove all the segments. Let us try to draw a pline.

Keyboard: Type in **pline** and press Enter.
Ribbon: **Home** tab→**Polyline**

Step 1. Start the pline command via any of the preceding methods.

- AutoCAD says: `Specify start point:`

Step 2. Click anywhere on the screen.

- AutoCAD says: `Current line-width is 0.0000 Specify next point or [Arc/Halfwidth/Length/Undo/Width]:`

Step 3. Ignoring the lengthy menus for now, click around until you get the same W shape seen earlier.

- AutoCAD says: `Specify next point or [Arc/Close/Halfwidth/Length/Undo/Width]:`

When done, press Enter to end the command. Then, click on your new pline to see that it is all one unit not separate lines.

Pedit

What if you wanted to apply thickness to the pline (Property 2) or create a pline out of a bunch of individual lines with common endpoints? You can do some of this while creating the command through the options seen in the command line (Fig. 11.2). You can switch to Arc segments and back to line segment and change the width desired for each segment on the fly. More on this below. After the pline is created, you need pline's companion command, called *pedit* or *polyline edit*. Let us try this out on the pline you created and give it thickness.

Keyboard: Type in **pedit** and press Enter.
Ribbon: **Home** tab→**Modify** panel→**Edit Polyline**

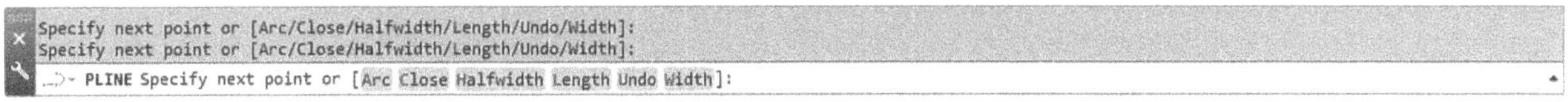

FIGURE 11.2 Polyline creation options in Command Line.

Step 1. Start the pedit command via any of the preceding methods.

- AutoCAD says: `Select polyline or [Multiple]:`

Step 2. Click on any of the W polyline segments.

- AutoCAD says: `Enter an option [Close/Join/Width/Edit vertex/Fit/Spline/Decurve/Ltype gen/Reverse/Undo]:`

Step 3. Press `w` for `Width`, and press Enter.

- AutoCAD says: `Specify new width for all segments:`

Type in a small value, say, 0.25, and press Enter. The pline adopts the new width for all segments, as shown in Fig. 11.3. You can then press Esc to completely exit out of the pedit command. Now, try making plines out of lines. Draw the W once again out of ordinary lines, then follow the procedure shown next:

FIGURE 11.3 Pline with thickness (property 2).

Step 1. Start the pedit command via any of the previous methods.

- AutoCAD says: `Select polyline or [Multiple]:`

Step 2. Click on any of the W line segments.

- AutoCAD says: `Do you want to turn it into one? <Y>`

Step 3. Press Enter to accept "yes."

- AutoCAD says: `Enter an option [Close/Join/Width/Edit vertex/Fit/Spline/Decurve/Ltype gen/Reverse/Undo]:`

Step 4. Select `j` for `Join`, and select all the lines of the W, pressing Enter when done. The collection of separate lines is now a polyline and you may press w for Width as before if you wish. Press Esc to completely exit out of the pedit command.

Another notable option in the pedit command is spline, and if you press s while in the pedit command, your W will look like Fig. 11.4.

FIGURE 11.4 Pline with the spline option.

To undo this, you can press `d` for `Decurve`, yet another pedit option. Spline is an important case of a polyline. We look at spline as a separate command later in this chapter.

Exploding a pline

Plines can be easily exploded using the explode command. The effect is twofold:

- The pline loses its thickness.
- The pline is broken into individual lines.

Putting it back together requires the pedit command, so explode and pedit are opposites of each other. Be careful though, and do not explode plines unless necessary, as there may be many individual lines and arcs to account for once exploded.

Additional pline options

A few more options are worth mentioning with the pline command. First, you can start out drawing a pline with a preset width (or half width). Here is the command sequence for that, presented in a shortened "actual" format, as seen from the command line by the user:

Command: pline (via any of the previously shown methods)

```
Specify start point:
Current line-width is 0.0000
Specify next point or [Arc/Halfwidth/Length/Undo/Width]:w
Specify starting width<0.0000>:.25
Specify ending width<0.2500>:.25
Specify next point or [Arc/Halfwidth/Length/Undo/Width]:
Specify next point or [Arc/Close/Halfwidth/Length/Undo/Width]:
```

As you can see, the W option was chosen and a width of 0.25 was entered for both start and end. If those lengths are not similar, you get something akin to Fig. 11.5. In that example, a zero start width was followed by a 0.75 end width to create the triangular first segment of the pline. You can change these widths "on the fly" to create some interesting shapes.

FIGURE 11.5 Pline with varying width.

Another option is L for Length. This is like Direct Distance Entry from Level 1. Finally, there is the A for the Arc option. This creates arcs on the fly and is very useful for drawing organic-looking shapes for everything from landscape design to furniture to cabling and wiring, as shown in Fig. 11.6. You can also alternate from line segments to arc segments and back within one pline.

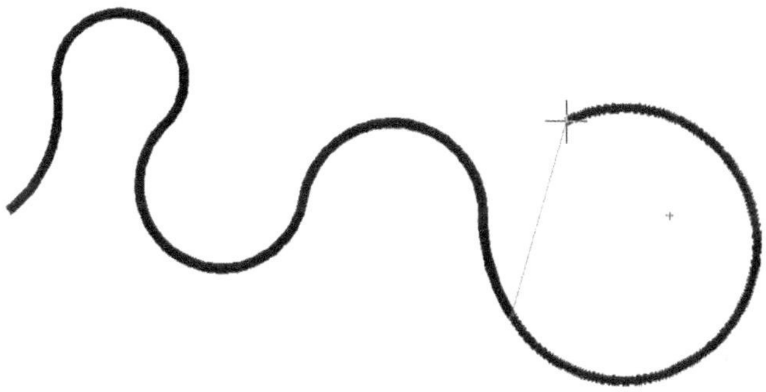

FIGURE 11.6 Pline of 0.2 width with arc option.

Finally, several releases ago, AutoCAD introduced some additional tools to work with plines (including unexploded rectangles, which are plines by definition). These tools come in the form of a new menu, which appears when you hover the mouse over a grip belonging to a pline or a rectangle (Fig. 11.7).

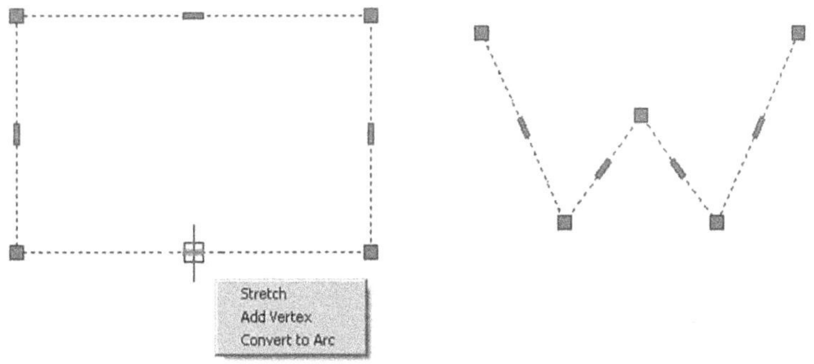

FIGURE 11.7 Grip menu.

As you can see, you can stretch, add a vertex, or even convert the line to an arc, allowing some new design possibilities and even saving some time.

Stretch

This option, also covered as a separate command in Chapter 15, Advanced Design and File Management Tools, allows for stretching that line and anything attached to it in one of several ways, with one possibility seen in Fig. 11.8.

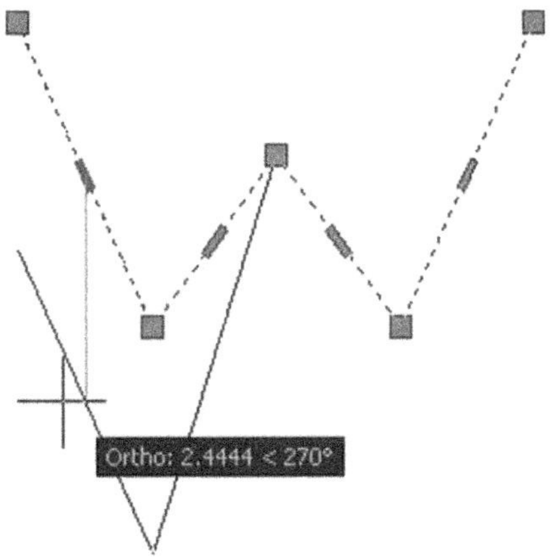

FIGURE 11.8 Stretch option (with Ortho on).

Add Vertex

This option allows you to "break" the line with the addition of another control point (grip), as seen in Fig. 11.9. You can then create two angled lines out of one straight line.

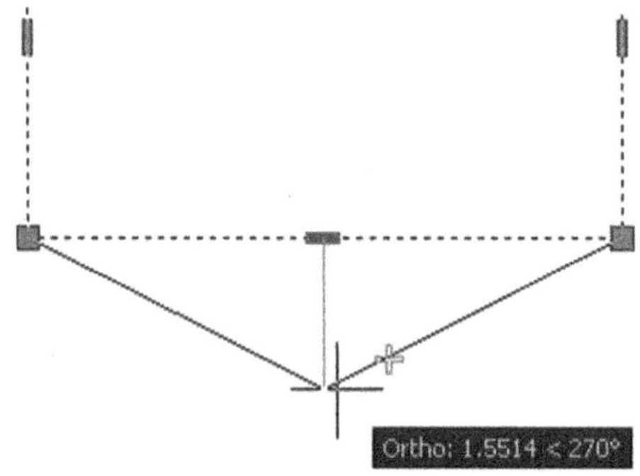

FIGURE 11.9 Add vertex option (with Ortho on).

Convert to arc

This somewhat similar option also adds a vertex but one that allows an arc to be formed, as seen in Fig. 11.10.

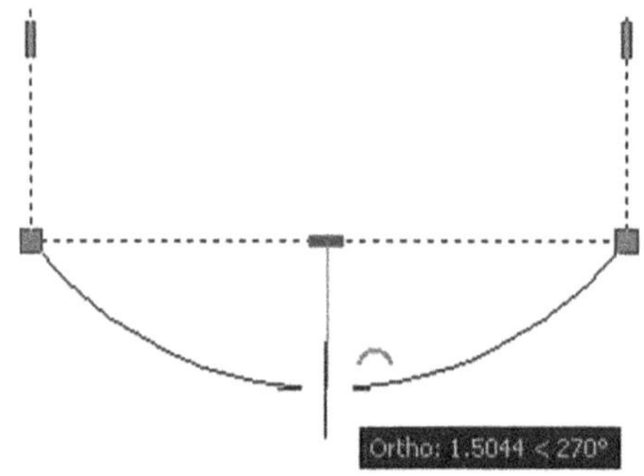

FIGURE 11.10 Convert to arc option (with Ortho on).

Boundary

The boundary command is part of the pline family and is a bit of a "hidden gem" from a long time ago, when it was called *bpoly*. It is no longer listed under that name in the Help files and has been permanently renamed *boundary*, but bpoly still works if typed in on the command line (Autodesk never seems to permanently get rid of any old commands).

The premise behind boundary is very simple. If you have a collection of regular lines forming a closed area, then you can use this command to just click inside that area and form a pline on top of all the regular lines. You need not select anything because the bpoly command works like the hatch command's *Add: Pick points option*, in the sense that the command "flows" outward, like a bucket of spilled paint, until it hits the edges of the area. Every element of the edge that it touches contributes to the new pline.

Keep in mind, the command does not touch any of the original linework; it merely forms the new pline on top of it. You can then use this new pline border for area calculations, hatching, or any other purpose. If needed, you can delete the pline on completion of any subsequent tasks. To try this out, clear your screen and draw a simple set of ordinary lines to form a closed shape, as seen in Fig. 11.11. These lines can represent the outline of a room in an architectural plan or the property lines on a civil engineering site plan.

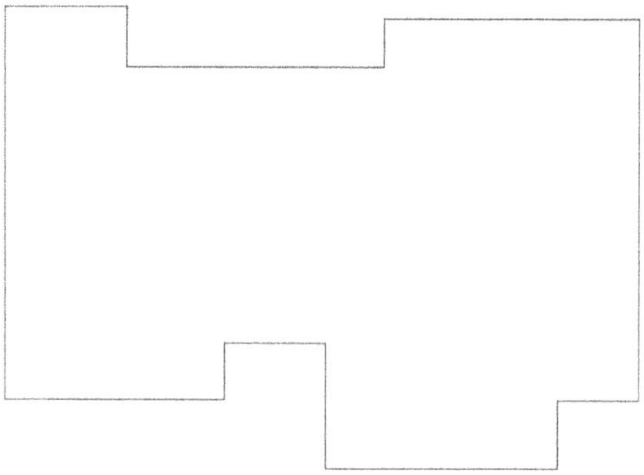

FIGURE 11.11 Set of lines for bpoly.

Start the command via any of the following:

Keyboard: Type in **boundary** or **bpoly** and press Enter.
Ribbon: **Home** tab→**Draw**→**Boundary...**

The dialog box shown in Fig. 11.12 appears.

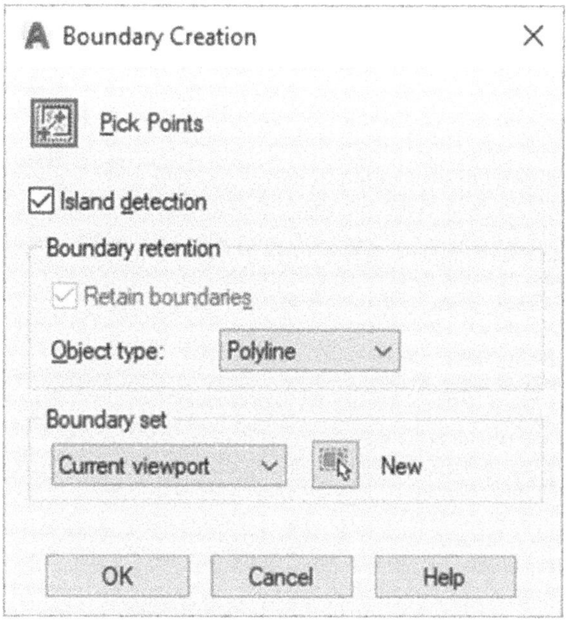

FIGURE 11.12 Bpoly boundary creation.

Note that the command features a very "hatchlike" *Island detection* option, which you can uncheck if you wish to ignore any elements inside the boundary, but in most cases, the only thing you need to do is just click on the Pick Points button.

- AutoCAD says: `Pick internal point:`

Go ahead and click anywhere on the inside of the shape. AutoCAD runs a short analysis, as seen here:

```
Selecting everything ...
Selecting everything visible ...
Analyzing the selected data ...
Analyzing internal islands ...
```

Press Enter.

- AutoCAD says: `BOUNDARY created 1 polyline`

At this point you are done. Click on the newly formed boundary and note how it is a separate entity. In Fig. 11.13, this new pline boundary has been moved slightly down and to the right, and one of the lines underneath has also been clicked to show that the original structure remains as a collection of ordinary lines.

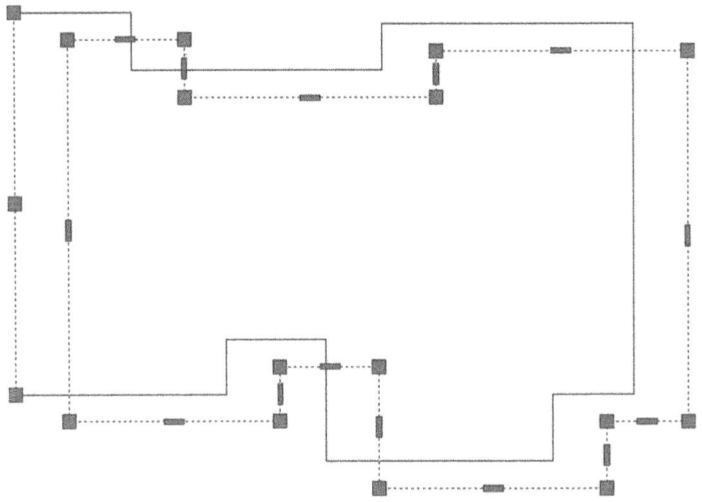

FIGURE 11.13 Bpoly created and moved aside.

11.3 Xline (construction line)

From the relatively involved pline, we swing over to the very simple xline. This is a continuous line that has no beginning or end. The zoom command has no effect on it. The construction part of the name refers to the days of hand drafting, when you would draw, in light pencil, a guideline that stretched from one end of the paper to the other and was used as a reference starting point for a string of elevations (or anything else) to be drawn one after another.

To use it, make sure Ortho and OSNAP are not activated, and do the following:

Keyboard: Type in **xline** and press Enter.
Ribbon: **Home** tab→**Draw**→**Construction Line**

Step 1. Start the xline command via any of the previous methods.

- AutoCAD says: `Specify a point or [Hor/Ver/Ang/Bisect/Offset]:`

Step 2. Click anywhere and an xline appears.

- AutoCAD says: `Specify through point:`

Step 3. Click randomly anywhere along the screen and multiple xlines appear. Fig. 11.14 shows one result of clicking in a circular manner.

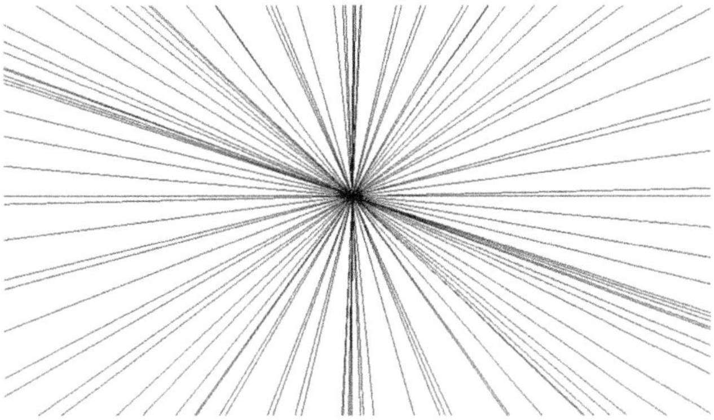

FIGURE 11.14 Random xlines.

Fig. 11.14 is perhaps not the most useful of creations, but landscape designers may find some use for the result of placing a circle at the intersection of those lines and trimming. Then, via some creative scaling, rotating, and copying, you get the reasonable representation of trees or shrubs in Fig. 11.15.

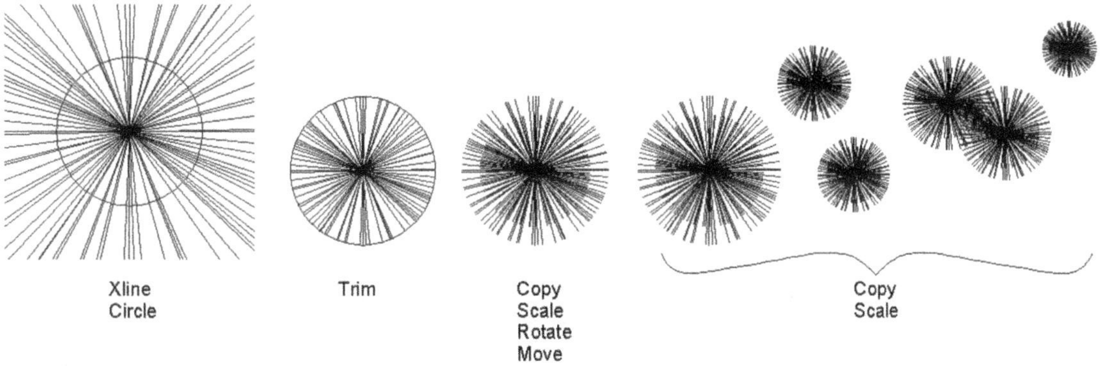

FIGURE 11.15 An application of xline.

As seen in the menu that follows the xlines, you can also create perfectly horizontal, vertical, or angular xlines. The same can be accomplished by having Ortho on, although the options allow you to create multiple xlines more easily. Clear your screen, start xline again, and when you see the following menu, try out the various options (results are seen in Fig. 11.16): `Specify a point or [Hor/Ver/Ang/Bisect/Offset]:`

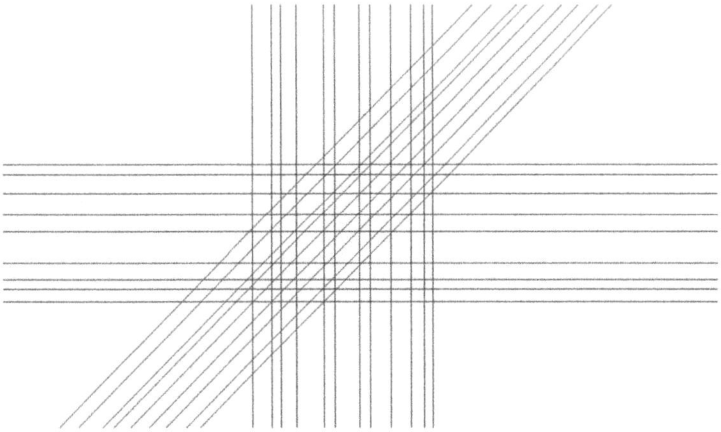

FIGURE 11.16 Hor, Ver, and Ang (set to 45 degrees) options in xline.

11.4 Ray

A ray is simply an xline but infinite in only one direction. It is easy to use and quite like xline, except there are no options for horizontal, vertical, or angular; to get a perfectly straight ray, you just turn on Ortho.

Keyboard: Type in **ray** and press Enter.
Ribbon: **Home** tab→**Draw**→**Ray**

Step 1. Start the ray command via any of the preceding methods.

- AutoCAD says: `Specify start point:`

Step 2. Click anywhere and a ray is created (Fig. 11.17).

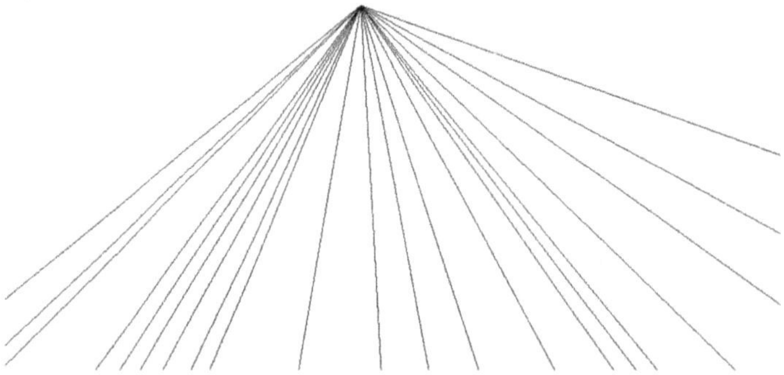

FIGURE 11.17 Ray command.

11.5 Spline

Spline is a unique and valuable command. Its origins are in the mathematical field of numerical analysis, and it is called a *piecewise polynomial parametric curve* (or a NURBS curve, nonuniform rational B-spline). While knowing the details is not necessary for using this command, what is important is that a spline is ideal for approximating complex shapes through curve fitting. What you do is create a path made of point clicks, and AutoCAD interpolates a "best-fit" curve through those points.

The result is a smooth, organic shape useful for a great number of applications, from contour gradient lines in civil engineering to landscape design to electrical wiring and many other shapes and objects. The term *spline*, by the way, comes from the flexible spline devices used by shipbuilders and draftspersons to draw smooth shapes in pre-computer days. To draw a spline, do the following:

Keyboard: Type in **spline** and press Enter.
Ribbon: **Home** tab→**Draw**→**Spline Fit**

Step 1. Start up the spline command via any of the preceding methods.

- AutoCAD says: Current settings: `Method = Fit Knots = Chord`

`Specify first point or [Method/Knots/Object]:`

Step 2. Click anywhere and a spline appears.

- AutoCAD says: `Enter next point or [start Tangency/toLerance]:`

Step 3. Click on another point somewhere on the screen, in an up and down wave pattern, as seen in Fig. 11.18.

- AutoCAD says: `Enter next point or [end Tangency/toLerance/Undo/Close]:`

As early as this step, you can select `c` for `Close` to close the spline tool. In this case, however, continue with the spline curve to add more points.

Step 4. Click on a few more points in a similar wave pattern. AutoCAD keeps saying the same thing as in Step 3. Continue clicking several more times (refer again to Fig. 11.18). When done, simply press Enter.

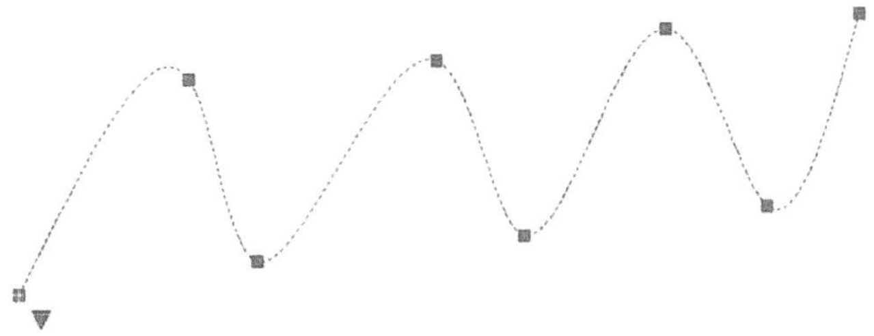

FIGURE 11.18 Spline command, grips enabled.

This is your basic "out-of-the-box" spline. Notice how, as you click on points, AutoCAD fits a smooth curve that best approximates the path. Each grip in Fig. 11.18 represents a click of the mouse, a control point. You can manipulate the positions of these grip points in the standard manner, with the spline following suit. This makes the spline a very flexible tool, adjustable to the exact demands of the designer. Try activating the grips and moving them around.

The command was redesigned somewhat back in AutoCAD 2011, and many of its advanced features are hidden away in submenus, which is a good thing, as most users need only the basic spline and not extensive options for tangencies and other details.

That is not to say they are not there if needed. Click on the down arrow on the bottom left of the spline (grips enabled), and select Show Control Vertices, which is another way to design the spline, useful for 3D work. Go over the various other options on your own with the Help files; there are too many to go into here but learning those helps you understand this very intriguing command.

Through the examination of spline, you should now have a good idea of how to create and control it. The points of the spline do not have to be random. You can set up a framework for the shape and use OSNAPs to precisely place where you want the spline to go, as shown in Fig. 11.19. What you see here are four lines, representing a framework. Then, using OSNAP points, a spline is added in click by click, from an endpoint to a midpoint to another midpoint, then to the final endpoint, at the bottom left.

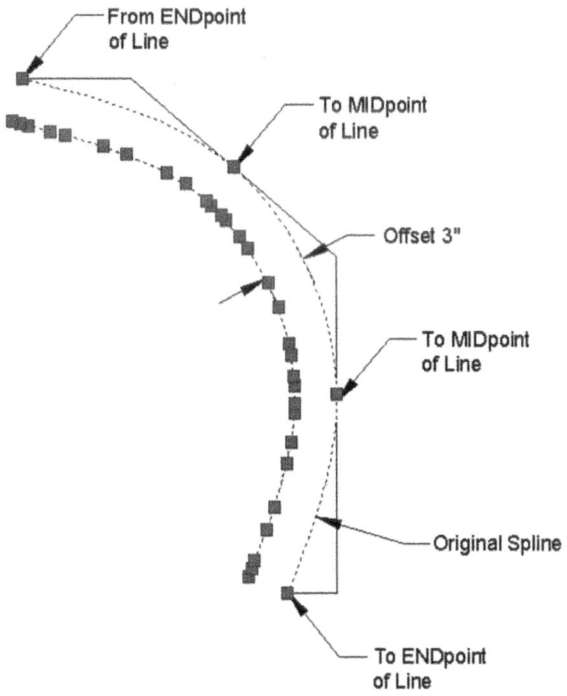

FIGURE 11.19 Spline with framework, offset, and grips.

A spline can then be offset using the standard offset command, although be careful in areas where the spline takes a sharp turn. If the offset distance is too great and the spline cannot make a smooth turn after the offset, it breaks down into separate lines. The offset spline also has many more grip and control points than the original spline. A 3″ spline offset, with all grips enabled, can be seen in Fig. 11.19.

Examples of spline use are numerous. Already mentioned were contour lines (this chapter's cover image), electrical wiring, piping, network cables, landscape design (golf courses, curved driveways, and walkways), interior design (Jacuzzis, swimming pools, furniture), and much more can be constructed with this tool.

11.6 Mline (multiline)

A multiline is a very useful tool that does pretty much what you would expect from the name: It is used to draw multiple lines all at once. This, in theory, should save you time when drawing everything from architectural walls to multilane highways. As we see soon, the basic mline is easy to draw but not too useful. The trick is to set it up for useful applications using additional tools, as outlined in the next sections. Try out the basic mline to get a feel for it, as shown next.

Keyboard: Type in **mline** and press Enter.
Ribbon: none

Step 1. Start up the mline command via any of the preceding methods (no Ribbon button is available).

- AutoCAD says: `Current settings: Justification = Top, Scale = 1.00, Style = STANDARD Specify start point or [Justification/Scale/STyle]:`

Step 2. A generic mline appears. Click again.

- AutoCAD says: `Specify next point or [Undo]:`

Step 3. Click somewhere else again.

- AutoCAD says: `Specify next point or [Close/Undo]:`

You can continue in this manner until finished. The default distance between the lines is 1.0, and of course, if the mline is too small or too large, you can zoom in or out. Fig. 11.20 shows what you should see after properly executing the previous steps.

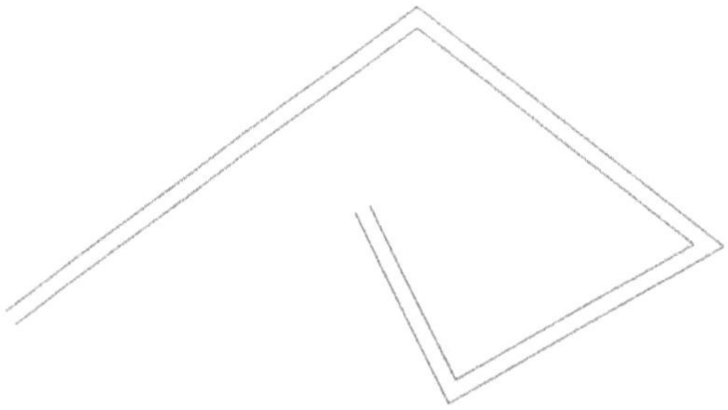

FIGURE 11.20 Basic mline.

Modifying the mline

While simple, the preceding was not too useful. Specifically, you may want to change one or more of the following properties:

- Number of lines making up the mline.
- Distances between those lines.
- Linetypes making up the lines.
- Colors of the lines.
- Fills, end caps, and other details.

All these and more can be input and modified, thereby making the mline more useful and sophisticated. The key command for setting up mlines is mlstyle.

Mlstyle (multiline style)

The multiline style command opens a dialog box that allows you to set all the preceding parameters. Type in `mlstyle` and press Enter (there is no Ribbon alternative). The dialog box shown in Fig. 11.21 appears.

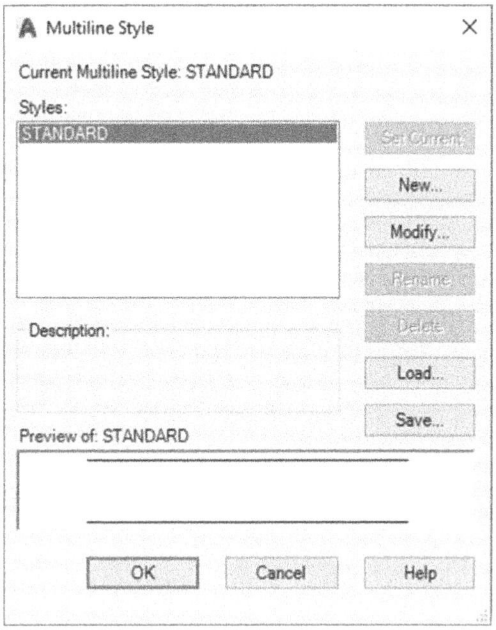

FIGURE 11.21 Multiline Style dialog box.

Let us create a new style. Press New … and you see a smaller Create New Multiline Style box. Type in a name for your mline (no spaces allowed in the name; use underscores if needed), and press Continue. You then see the main dialog box shown in Fig. 11.22. The name created for this example was a generic MY_NEW_MLINE.

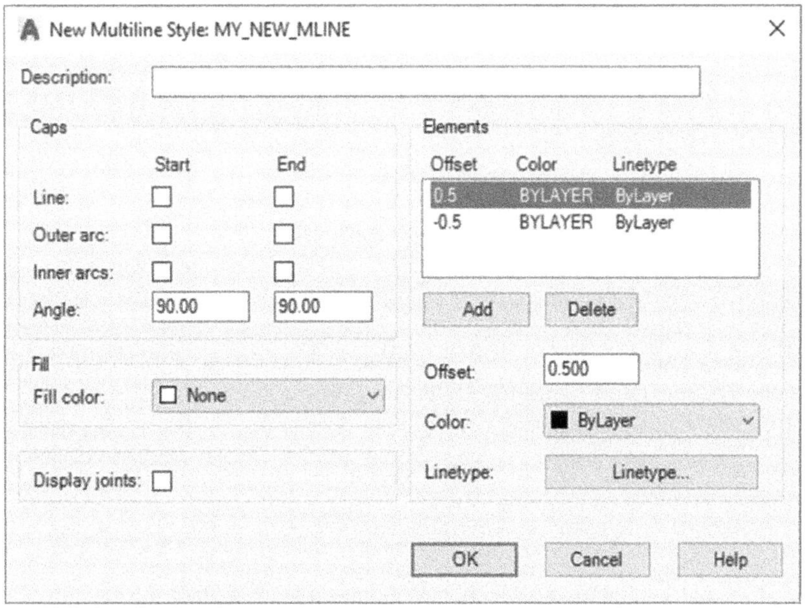

FIGURE 11.22 New Multiline Style dialog box.

We can now go through the essential changes to the mline. Our goal is to create a typical wall used in construction (see Fig. 11.23), often shown in cross section in architectural elevations. For simplicity, it is 6″ wide and features a 1″ drywall panel on either side, with insulation in between, and room for a wooden stud (not shown here).

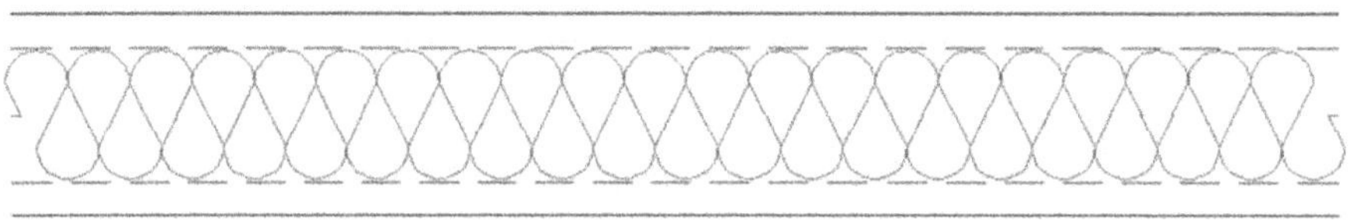

FIGURE 11.23 Architectural wall with insulation.

This mline, once set up, can be used to draw building walls faster and easier than offsetting and changing properties one line at a time. The process to do this follows.

Step 1. You first need to determine the *number of lines* making up the mline, and the distances between those lines. Both can be set inside the Elements area (right-hand side) of the New Multiline Style dialog box using the Add button and the Offset field. The idea is to add more line elements (via the Add button) and enter an offset distance for each of these line elements (via the Offset field). This offset distance spaces the lines at the desired interval to create the wall. Before you do all this, however, you need to decide what the origin of the offset is. This is your starting point for the offsets. The two choices are positioning zero all the way on the leftmost line element (then, all values are positive) or in the middle element (then, half the values are negative). With the first choice, the outer walls are at 0 and 6, and with the second choice, they are at −3 and +3, as shown graphically in Fig. 11.24.

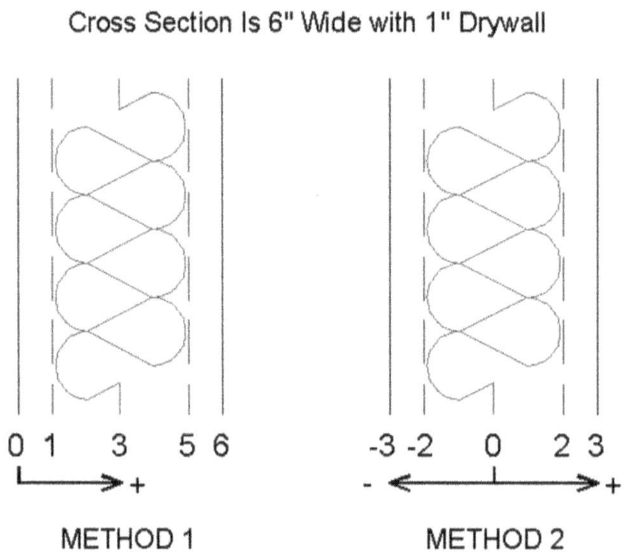

FIGURE 11.24 Offset origins.

For this example, choose Method 1, and after adding additional lines, for a total of six, set the walls to 0, 1, 3, 5, and 6 units (there are no feet or inches in these settings; you use decimal units, representing inches, throughout).

Step 2. Next you need to determine the linetypes and colors making up the lines of the overall mline.

Both can be set in the Elements area using the Color: drop-down menu and the Linetype ... button, as shown in Fig. 11.25. Simply highlight the line element you wish to change and select a new color or linetype. Remember that you may not have the linetypes loaded yet, so you may need to do this (refer to Chapter 3 for a reminder).

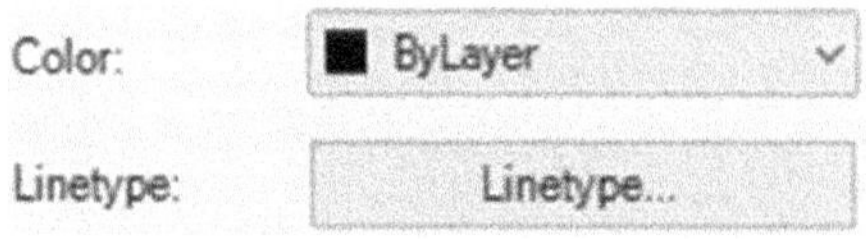

FIGURE 11.25 Color and Linetype settings.

In Figs. 11.23 and 11.24, the outer wall color was selected to be blue and the inner walls red with a hidden linetype. The centerline is a special linetype, often used for insulation, called BATTING. Feel free to choose your own colors and linetypes; those shown are just for illustration, but they are a good starting point. Fig. 11.26 shows the Elements area as you go through Steps 1 and 2.

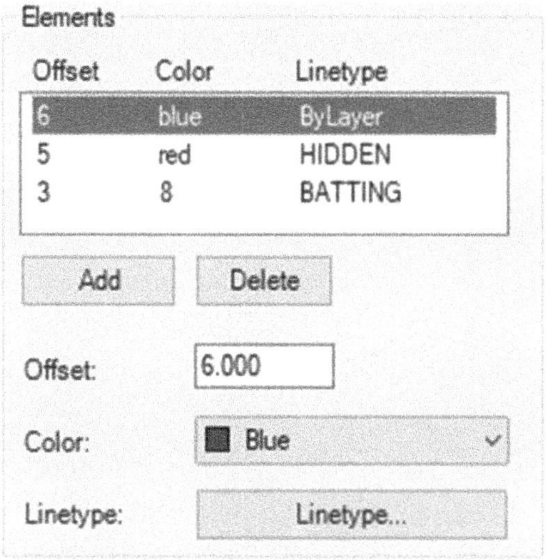

FIGURE 11.26 Elements settings.

Step 3. As the final step, you can add in fills and start/end caps.
End caps, as the name implies, are various ways to finish the start and end of your mlines. You can add regular lines or two types of arcs. Often mlines are joined to each other, so these may not be necessary. You can also add color fills to the mlines, although this is rarely done in practice. All of these can be done via the check-off boxes and drop-down menu on the left-hand side of the New Multiline Style dialog box, as seen in Fig. 11.27. You can also display the joints

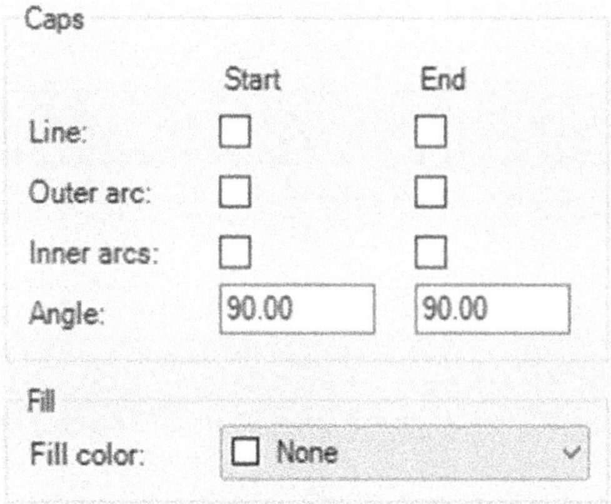

FIGURE 11.27 Caps and fill.

of a mline via a checked box, but that is also rarely called for. This example has no caps or fills, but feel free to experiment on your own.
After your new mline has been created, go ahead and click on OK; you are taken back to the dialog box of Fig. 11.21, with your new mline previewing in the window at the bottom. The other choice should be Standard, the default mline you created by just typing in mline and saw in Fig. 11.20. Set your new mline as the current one (Set Current button) and click on OK. Now, go ahead and type in mline and draw part of a rectangle, as seen in Fig. 11.28.

FIGURE 11.28 The new mline.

The linetypes may not look right in this example. This is due to a low linetype scale being set. If you recall from Chapter 3, Layers, Colors, Linetypes, and Properties, the way to fix this is to type in ltscale and enter a new, more appropriate value. The value used is 4.8. The result is shown in Fig. 11.29.

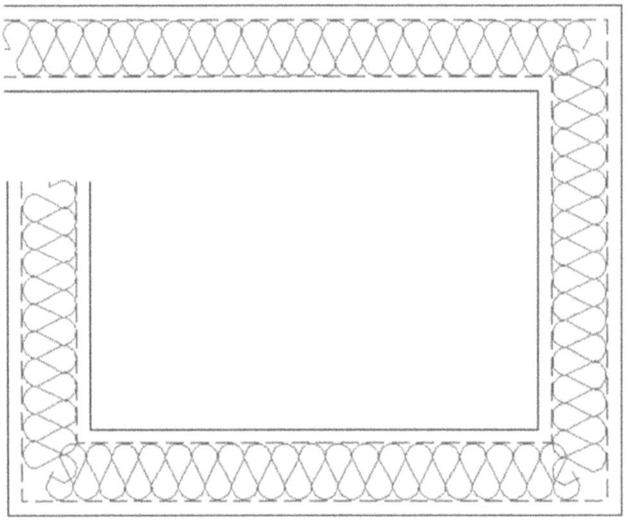

FIGURE 11.29 The new mline with a ltscale of 4.8.

Mledit (Multiline Edit)

Multiline Edit is a dialog box that deals with the practical applications of mlines, such as how to trim intersecting mlines and how to deal with joints, cuts, and vertexes. To try out some of the features, draw two intersecting mlines, as shown in Fig. 11.30.

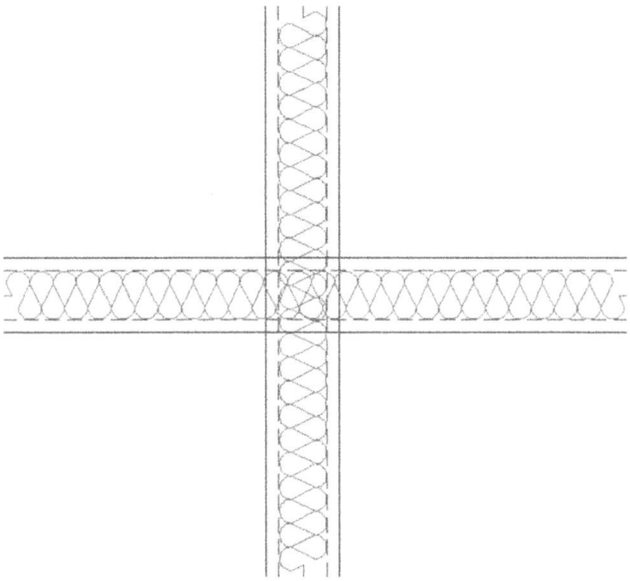

FIGURE 11.30 The new mline with a ltscale of 4.8.

Then, type in mledit, and press Enter the dialog box shown in Fig. 11.31 appears.

Select Closed Cross in the upper left. The dialog box disappears, and you are asked to select the first mline, click on it, then the second mline, and click on that also. After the second click, Fig. 11.32 appears. As you can see, the mline's intersection has been modified.

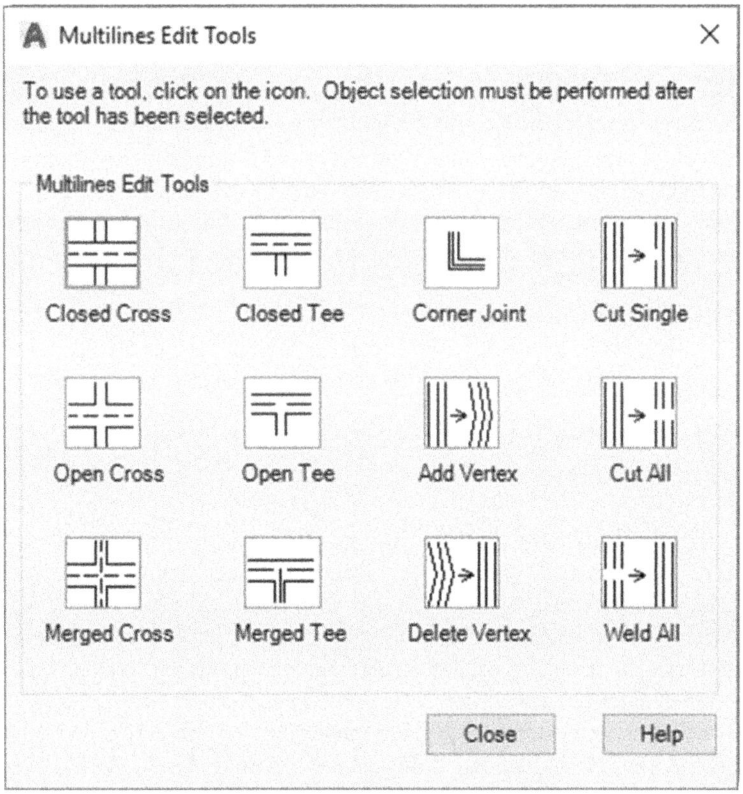

FIGURE 11.31 Multiline edit tools.

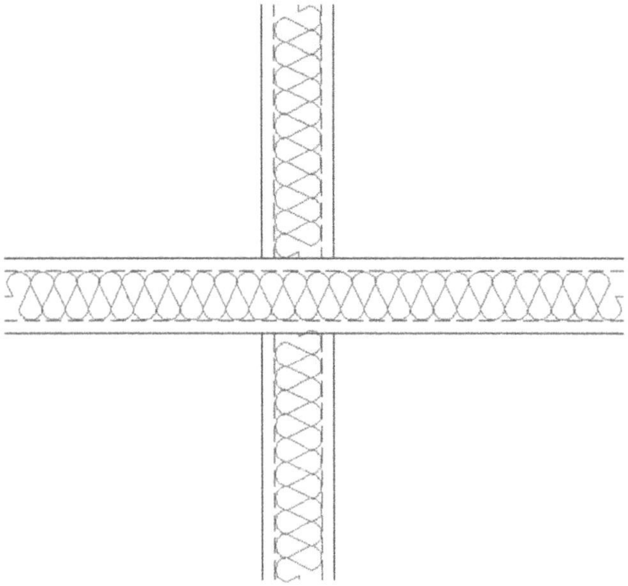

FIGURE 11.32 Multilines edited.

In the same manner, try out all the features of mledit at your own pace; they are all relatively straightforward and self-explanatory. A Cut breaks the line; a vertex adds a "kink" in the mline; Close, Open, and Merge deal with intersections; and so on.

Other Mline properties

- *Grips*: All mlines have grip points at either the top or bottom of the overall line (this is set using Justification when first creating the mline). Feel free to use these grips to stretch the line around as needed.
- *Exploding the Mline*: Mlines can be exploded, which separates them into individual component lines. This may not necessarily be a bad thing. An architectural floor plan may be quite complex, and you may not be successful in getting a mline to do exactly what you want all the time. If the mline is exploded, then you can use the fillet, trim, and extend commands to coax it into the shape you want. Much time has already been saved by using a mline in the first place, so exploding it may be the right solution to finish off the floor plan linework.
- *Trim and Extend*: Mlines can be trimmed and extended without exploding them (although filleting is not possible). However, an option that pops up is this: `Enter the mline junction option [Closed/Open/Merged]<Open>:` The Closed, Open, and Merged options have to do with the intersections, so you need to pick one option to continue the trim or extend. Go through them all as practice.
- *Modification of Mlines*: Once created, a new mline style cannot be modified once you start using it (but can be beforehand). If you wish to change a mline after you have a few drawn, you need to create a new style that incorporates your desired changes.

11.7 Sketch

This is perhaps one of the most unusual tools in AutoCAD. It allows you to draw objects in freehand. This may seem counterproductive and unnecessary. After all, you spent so much time learning precision and accuracy; but sketch finds a wide array of applications, as we soon see.

To truly sketch in freehand on a computer you need bitmap graphics (such as those found in Photoshop). Then, the maximum resolution is determined by the pixels. In almost all cases, the pixels are so small that it truly seems like you are drawing smooth, continuous freehand lines and shapes. You would have to zoom in close to see what is called *pixilation* or the breakdown of smooth lines into the little squares (pixels) that make up the drawn geometry.

This is all fine and would be useful for AutoCAD if not for one problem: AutoCAD is a "vector program," meaning it does not use pixels. Instead, all the geometry is represented by mathematical formulas, which completely describe the shape, size, and location of everything seen on the screen. This approach is necessary for many reasons, not the least of which is to ensure the integrity of the geometry when zoomed in close. So then, how would you sketch in freehand? The solution is simple: Instead of pixels, use very short lines, which look continuous if they are small enough and you are not too close to them. This is exactly how sketch works. The command uses small lines, with their size determined by something called a *record increment*, which we discuss in a moment.

The sketch command has another twist to it. A few releases ago AutoCAD allowed you to turn sketched linework into a spline or a polyline. This can occasionally be useful if you are drawing simple freehand shapes and want maximum smoothness, although the pline option is not recommended, as it adds a tremendous amount of control points to the linework. We discuss mostly the "classic" sketch command here. First, we need to set the record increment to a very small value (use 0.01). Let us try this out.

Step 1. Type in `sketch` (there is no Ribbon for the classic sketch command) and press Enter.
- AutoCAD says: `Type = Lines Increment = 0.1000 Tolerance = 0.5000 Specify sketch or [Type/Increment/toLerance]:`

Step 2. First, make sure `Type = Lines`, as that is what we want to practice first. If not, press `t` for `Type` and select `l` for `Lines`. Then, press `i` for `Increment`, and press Enter.
- AutoCAD says: `Specify sketch increment <0.1000>:`

Step 3. Type in 0.01 and press Enter.
- AutoCAD says: `Specify sketch or [Type/Increment/toLerance]:`

Step 4. Nothing further needs to be set, so press Enter.
- AutoCAD says: `Specify sketch:`

Step 5. Move the mouse around to begin drawing.
- AutoCAD says (and continues to say): `Specify sketch:`

You will see what is essentially a smooth line, green in color, and completely freehand generated. To "lift the pen," you need to left-click once; then click again to "put it down." You can also click and hold the left mouse button down when you want to draw, which is more realistic for some. To "set" the drawing, press Enter; the lines turn white (or some other color depending on the current layer). Be careful not to press Esc until you set the lines, as everything disappears. Fig. 11.33 shows what you can get after a few minutes of doodling.

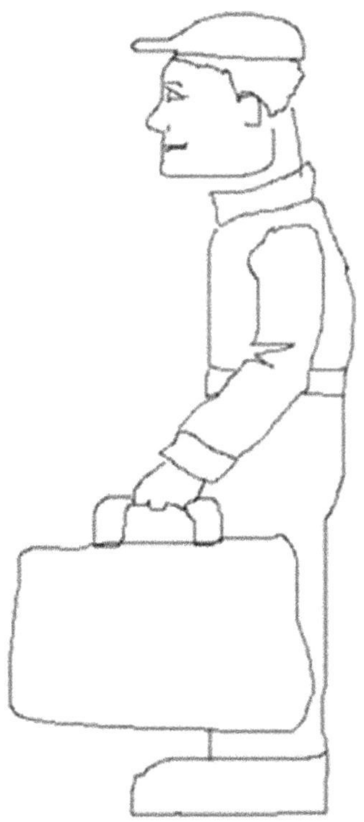

FIGURE 11.33 Basic sketch.

As mentioned before, you can also turn the sketched lines into a spline by selecting the Spline option. You will still sketch freehand at first, but on pressing Enter, the lines smooth out into the familiar spline shapes. The process is the same with the Pline option. You should spend a few minutes experimenting with both. You can also access the Spline option directly via the Ribbon's Surface tab (look for the spiral on the Curves panel), but for that you need to change the workspace to 3D Modeling.

Applications of sketch

I hope whatever you tried to draw was better than the odd-looking little man with a briefcase. It is not that easy to draw something well with a mouse; occasional erasing and fixing up is needed to get a good sketch. Sometimes, a graphic design tablet and a penlike stylus may be used. Zoom in close on whatever you created and you notice the figure is made up of the small lines we talked about. In the close-up of the hand and briefcase handle in Fig. 11.34, you can clearly see the individual line segments. Be careful making a dense drawing. All these lines increase the size of the file very quickly.

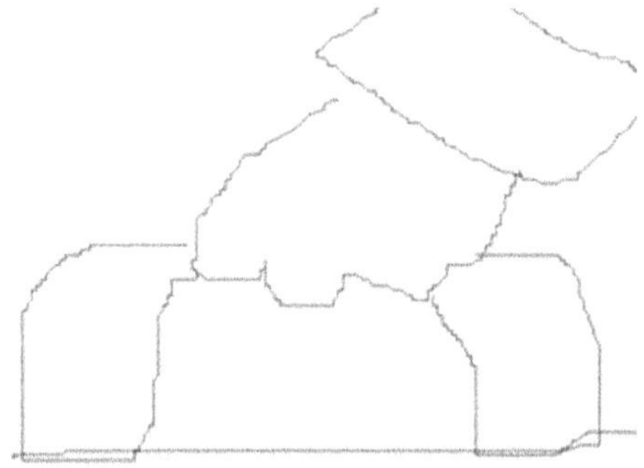

FIGURE 11.34 Line increments visible in the sketch.

Sketch has many applications, which are further expanded via the spline variation of the command. Some examples include:

- Architectural elevations featuring sketched people for comparative sizing.
- Sketched people for traffic sign design and civil engineering studies.
- Trees, shrubs, and other landscape design objects.
- Realistic drawings of small cables and wiring (via spline).
- Anytime an organic, nonrigid shape is needed.

There are numerous websites that can provide free blocks, which are usually uploaded and shared by other users. Just a few examples of these sites include cad-blocks.net, cadforum.cz, cadblocksfree.com, cad-block.com, and many others. Many of these have blocks that feature the sketch command in creation of vegetation for landscaping or people for traffic and civil engineering as well as interior design and architecture layouts. Fig. 11.35 shows some examples pulled from these sources.

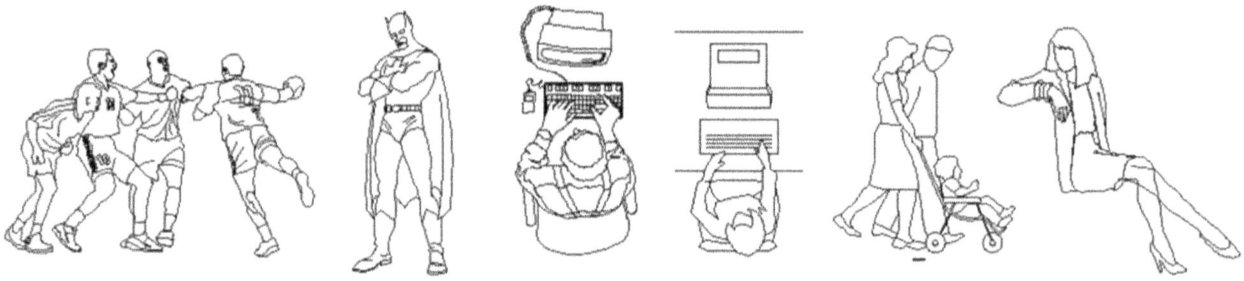

FIGURE 11.35 Creative use of the sketch command.

GEOmetric center OSNAP

This brand-new OSNAP point, introduced a few years back with AutoCAD 2016, was requested by past AutoCAD users and really should have been included a long time ago. It allows you to snap from (or to) the geometric center of objects such as triangles, multisided polygons, and various nonsymmetrical shapes. In technical terms, you are snapping from (or to) the mathematical "centroid" of that shape. This can be quite useful; in the past, you needed to draw guidelines and find that center yourself. The only rule is that this shape must be a closed polyline; hence, we delayed discussing this OSNAP point until the pline, pedit, and spline commands were introduced.

To try out this OSNAP point, first draw a small collection of shapes, as shown in Fig. 11.36. In all cases, make sure they are closed polylines or splines. You need to do this with the first and third shapes if they are made of basic lines or lines and arcs. The second one (octagon created with the polygon command) is already closed, and the fourth shape is a spline, also closed.

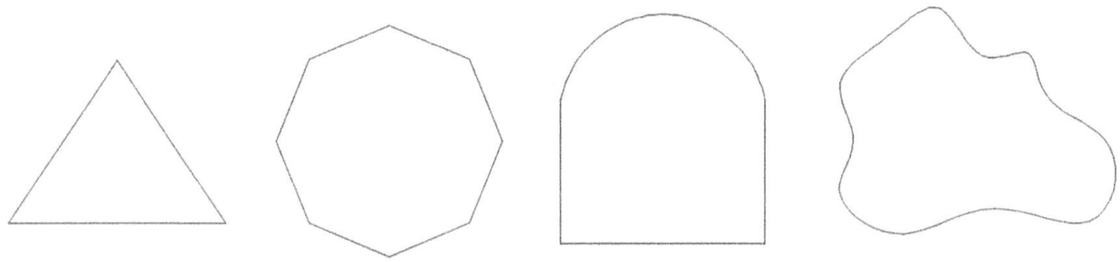

FIGURE 11.36 Shapes for the geometric center OSNAP.

Now open the Drafting Settings dialog box and set the Geometric Center OSNAP. Its symbol is a circle, and it is located just below the Center OSNAP. For learning purposes, turn off all the other ones. Now, begin a line command, hovering the mouse in the center of each shape. AutoCAD recognizes these as closed shapes and places a special symbol at the centroid (as seen in Fig. 11.37). Click to begin a line; that is all there is to it. Fig. 11.38 shows all the shapes with a line beginning from the geometric center of each.

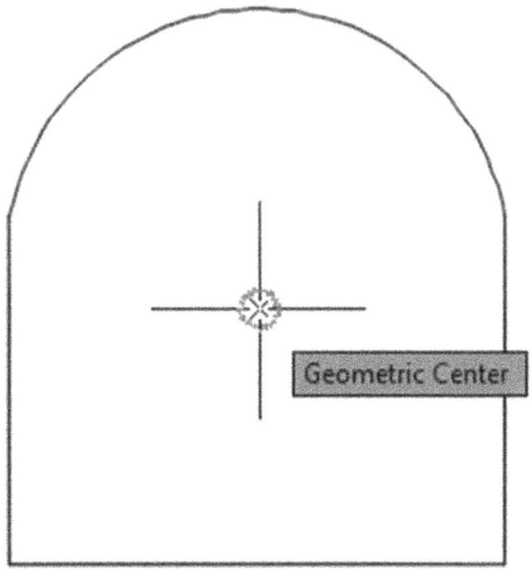

FIGURE 11.37 Geometric center recognized.

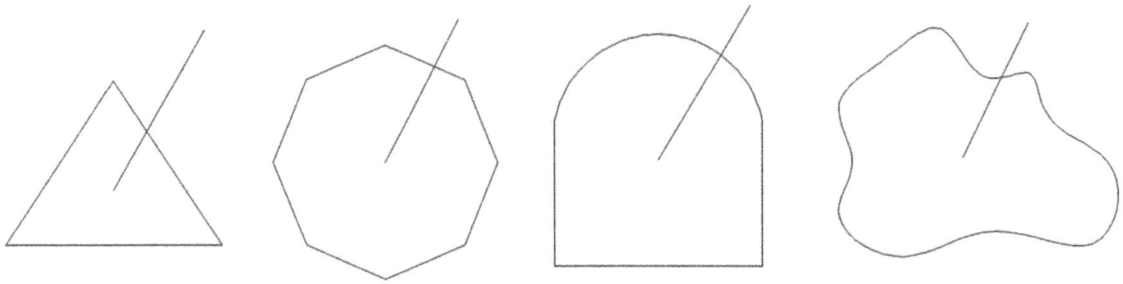

FIGURE 11.38 Lines starting from geometric centers of each shape.

11.8 Level 2 drawing project (1 of 10): architectural floor plan

Level 2 features a drawing project that is designed chapter by chapter, beginning with this one and completing in Chapter 20. It is an architectural layout, like what you did in Level 1, although a bit more involved. The project enables you to use all the cumulative knowledge you acquired up until now and includes features you learn from Level 2 chapters.

The project features many of the basic elements of a "real-world" design and is a step closer to the professional world. You will see a moderately complex floor plan and electrical, HVAC/mechanical, and landscaping plans. You then set up Paper Space and printing with lineweights and other details. A real house design also is likely to feature numerous details, riser diagrams, and sections of doors, windows, and walls. We include some of these elements here and there to increase realism.

Follow all the steps closely, but feel free to embellish the design and get creative. This is excellent practice for the real thing, and if you get through the project smoothly, you will feel right at home with a real design.

Step 1. Open a new file and save it as House_Plan.dwg.

Step 2. Set up units as Architectural and create the following initial layers. The drawing is monochrome in this textbook for clarity, but you need to follow CAD standards as required in the industry (color suggestions shown). For now, you need:

- A-Deck (Blue).
- A-Deck-Hatch (Gray_9).
- A-Dims (Cyan).
- A-Walls-Exterior (Green).

Step 3. Draw the floor plan shown in Fig. 11.39.

Step 4. The dimensions are not just for your reference anymore; you also need to draw them in. The floor plan will eventually be printed at 1/4″= 1′-0″, so the scale of the dimension features should reflect that. The dimensions feature size shown here are for clarity only and are much too big for an actual print. They also serve as checks on your design. Use layers with proper colors, linetypes, and lineweights. Interior walls, doors, and windows are added in Chapter 12. It should take you about 30 minutes or less to set up and complete this design.

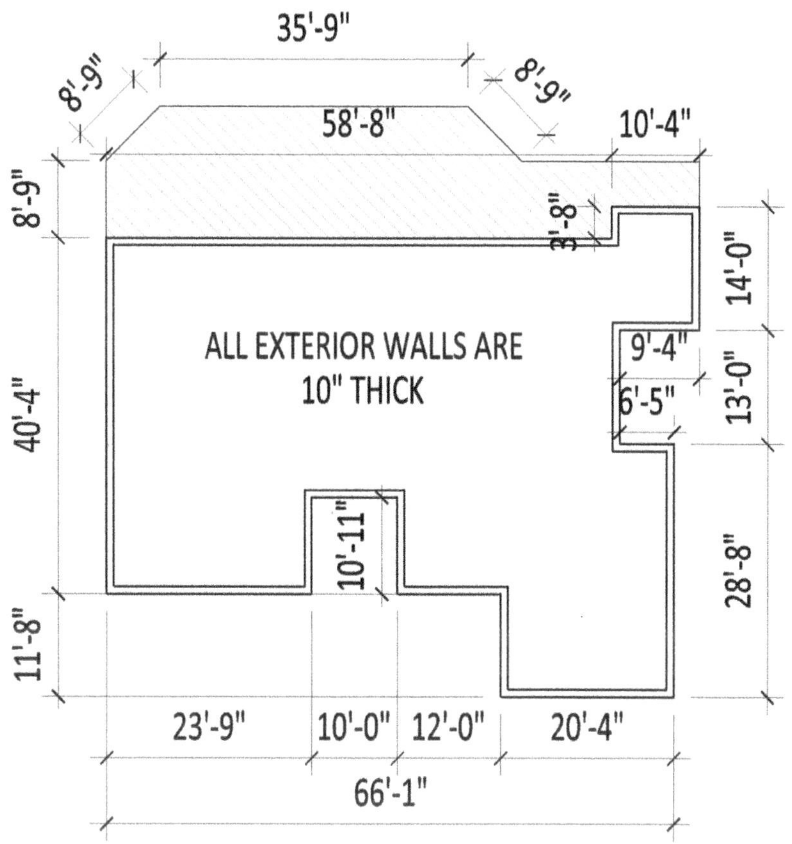

FIGURE 11.39 Level 2 project—overall layout.

Summary

You should understand and know how to use the following concepts and commands before moving on to Chapter 12*:*

- Pline (polyline)
  - Creating a pline
  - Pedit
  - Setting constant width (thickness)
  - Setting variable width (thickness)
  - Changing lines into plines
  - Adding spline curves and decurving
  - Other pline options (Length, Arc, Offset)
  - Exploding a pline
  - Bpoly command
  - Xline (construction line)
  - Creating a random xline
  - Creating a horizontal, vertical, or angular xline
- Ray
  - Creating a random ray
  - Creating a horizontal or vertical ray (with Ortho)
- Spline
  - Creating a random spline
  - Offsetting the spline
  - Grips and modifying the spline
- Mline (multiline)

- Creating a generic mline
- Mlstyle
- Mledit
- Other mline properties
- Sketch
  - How sketch works
  - Setting a record increment
  - Applications of sketch
- Geometric center

Review questions

Answer the following based on what you learned in this chapter:

1. What are the two major properties a pline has and a line does not?
2. What is the key command to add thickness to a pline?
3. What is the key command to create a pline out of regular lines?
4. What option gives the pline curvature? What option undoes the curvature?
5. What two major things happen when you explode a pline?
6. What other options for pline were mentioned?
7. What is the bpoly command?
8. Define an xline and specify a unique property it has.
9. How is a ray the same and how is it different from an xline?
10. Describe a spline and list its advantages.
11. List some uses for a spline in a drawing situation.
12. What is a mline? List some uses.
13. What are some of the items likely to be changed using mlstyle?
14. What are some of the uses of mledit?
15. Describe the sketch command. What is a record increment? What size is best?
16. Describe the basic function of a Geometric Center OSNAP point.

Exercises

1. In a new file, set your units to Architectural and do the following pline exercise. Step A: Draw a pline with 10 segments, each 2″ long and at right angles to each other as shown. Step B: Give the pline a thickness of 0.2″. Step C: Spline the pline. Step D: Explode the spline. (Difficulty level: Easy; Time to completion: <5 minutes)

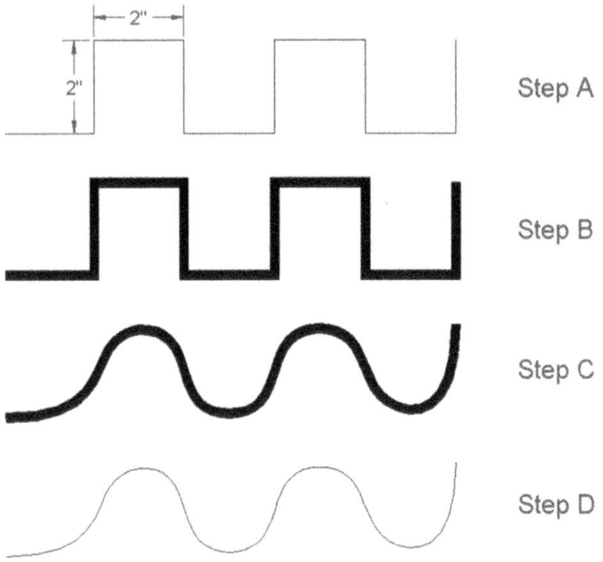

2. In a new file, set your units to Architectural and do the following pline exercise. Begin at the bottom and proceed to the right, following the given pline length and width information. (Difficulty level: Easy; Time to completion: <5 minutes)

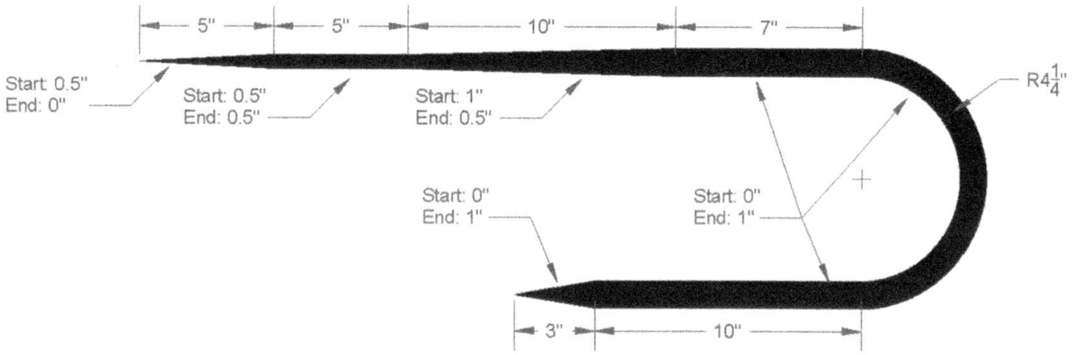

3. Using xline, circle, trim, copy, and rotate, draw a tree symbol as seen in Fig. 11.15, reproduced here. Repeat the same procedure using the ray command. (Difficulty level: Easy; Time to completion: <5 minutes)

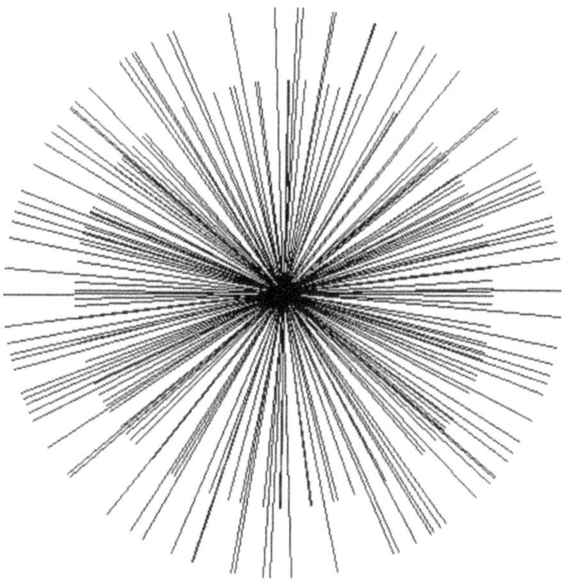

4. Using the spline and text commands, draw contour lines as seen in this chapter's cover image and reproduced here. To create the "gaps" in the splines where the height values are positioned, draw two long vertical lines, and trim between them. We cover another method to do this in Chapter 15, Advanced Design and File Management Tools. (Difficulty level: Easy; Time to completion: 5–7 minutes)

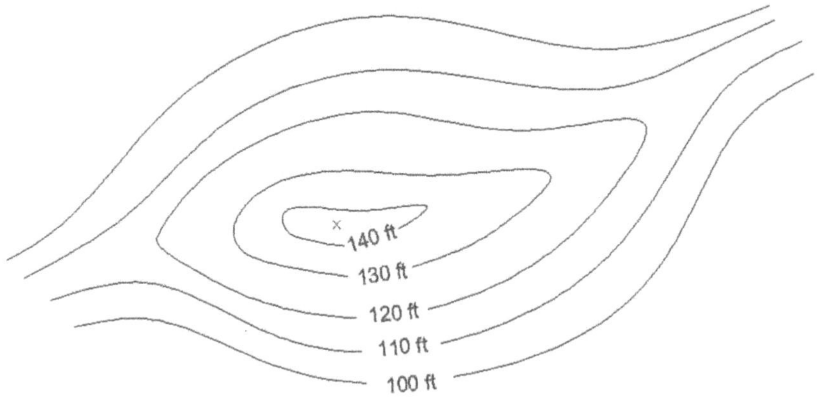

5. Use the mline command to create the following wall cross section. Be sure to replicate the linetypes shown, including the central BATTING linetype. (Difficulty level: Easy; Time to completion: 5–7 minutes)

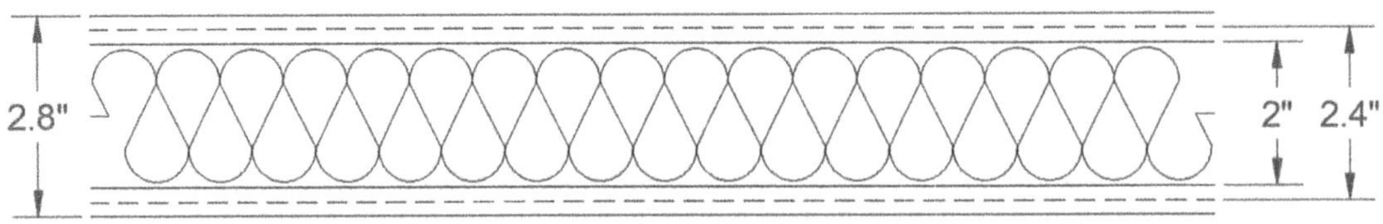

6. Using the mline, line, and hatch commands, draw a representation of a divided highway, as shown. You must convert feet and inches to just inches, such as 4′–10″ = 58″, to set up the mline offsets. (Difficulty level: Intermediate; Time to completion: 15–20 minutes)

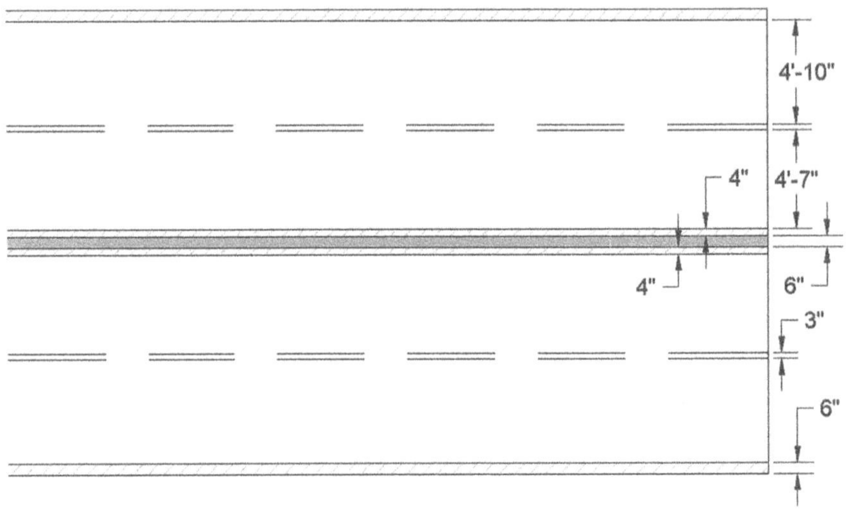

7. Using the pline or spline command, create the following cross-sectional contour of an airplane wing leading edge. Then, add in the additional hardware detail, text, and shading, as shown. (Difficulty level: Intermediate; Time to completion: 30−45 minutes)

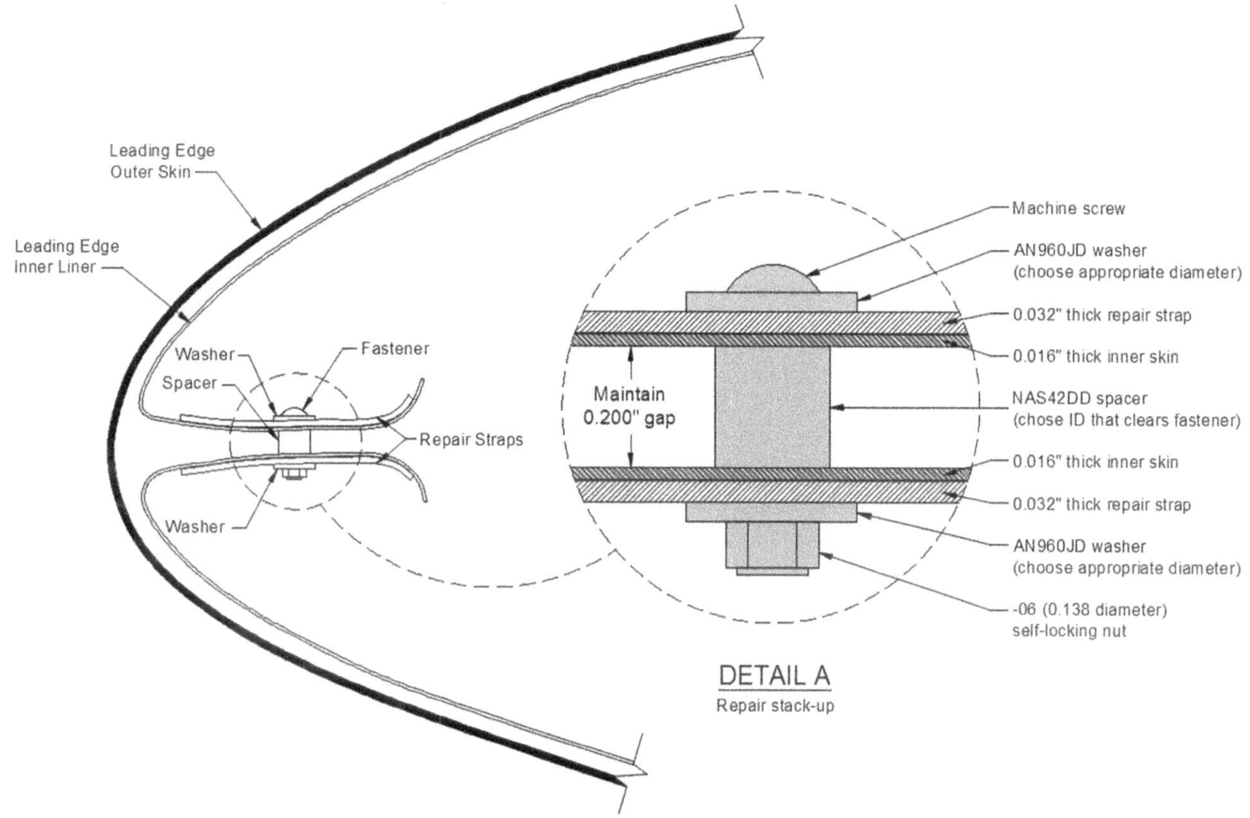

Chapter 12

Advanced layers

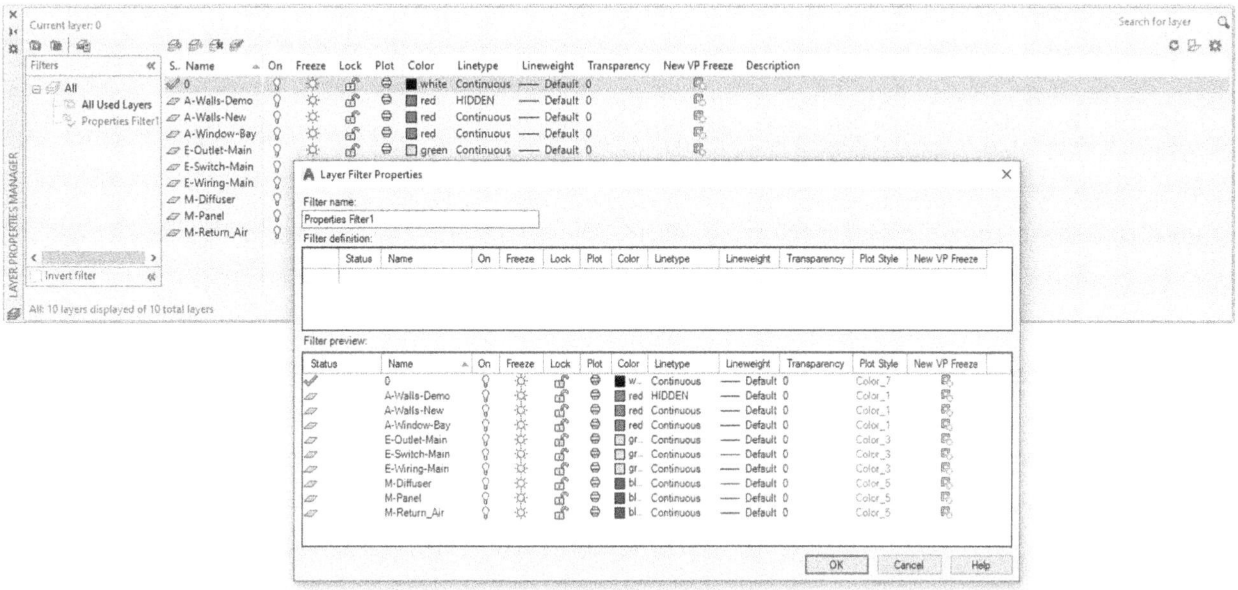

Learning Objectives

In this chapter, we introduce advanced layers and discuss the following:

- The need for automation
- Script files
- Layer State Manager (LSM)
- Layer filters
- Layer Freeze, Isolate, and Walk.

By the end of this chapter, you will learn how to use and apply these new types of layer tools and be able to incorporate them into future design work where many layers are present in the file.

Estimated time for completion of this chapter: 1–2 hours.

Up and Running with AutoCAD 2022. https://doi.org/10.1016/B978-0-323-89923-9.00012-3

12.1 Introduction to advanced layers

After the basic commands in Level 1, we introduced the layer concept. This was just a basic primer, showing you how to create them and set a variety of items, such as color and linetype. Also, several methods of control were covered, such as freezing and locking. Here, we look at additional powerful features that allow for unprecedented control over your layers. They are script files, the LSM, and layer filtering.

So, why do you need advanced layer control tools? Well, imagine the following scenario. You work at a large architecture or engineering consulting firm that specializes in complete commercial building and site design. Among your staff are a number of

- Architects and interior designers to lay out the overall exterior and interior of the building.
- Civil engineers to assist in structural design and site plans.
- Mechanical engineers to design the HVAC and mechanical systems.
- Fire protection and security engineers to design the electrical, sprinkler, and security systems.
- Landscape designers to integrate landscape design into the site plan.
- Additional design professionals as needed for specialized subsystems.

All these people contribute to the complete design in their own way. Generally, all their work is based on (or ends up in) one main building design file. This common file contains not only the architectural layout of the building but also the data from all these other disciplines. Each discipline has its own set of layers, usually following the AIA convention or another in-house layer naming method. Therefore, architects have their own set of layers starting with an A- (e.g., A-Demo-Wall, A-Bay-Window). Civil engineers have their own (C-Struct-Column, etc.), electrical engineers their own (E-Light-Switch), and so on.

You may already see where this is going as far as layers are concerned. As the project progresses, the layers grow in number. Toward the end, when everyone has added his or her respective design, the file may contain as many as 500 layers (not common, but not entirely unusual either!). There needs to be a way to impose some order and control over them, so only the needed layers are visible to each discipline. This can certainly be done by manually freezing and thawing layers but is very time-consuming. If you guessed that some sort of simple automation is available to you for this purpose, then you are correct.

This is advanced layer management, a process where the layers, once set, are saved under a descriptive name, such as Arch_Layout or Elect_Layout, and recalled when necessary by simply clicking on the name. AutoCAD then freezes and thaws the layers automatically, exactly the way you initially told it to. This is what script files, the LSM, and layer filtering do. These tools allow sorting, classifying, and filtering of layers according to a variety of properties and are indispensable when working on a large project.

12.2 Script files

A script, in most general terms, is simply a text file with one command on each line. Its purpose is to automate a sequence of commands when executed. It is perhaps the simplest type of programming you can do. All you need is a text editor (such as Notepad). You next list the commands you want the script to go through, one per line. Then you save it under a name and a.scr extension (e.g., MyScript.scr), close the file, and run it from AutoCAD's command line by typing in script and pressing Enter (alternatively via the Ribbon's Manage tab→Run Script). The Script dialog box appears, the script is double-clicked on (executed), it does its →thing; and if you got the command sequence correct, a short task is automatically performed.

Scripts are generally not used as much now for purposes of layer control as they once were, but we briefly cover them because knowing how they perform gives you insight into the other remaining tools. It is also a good skill to have overall. You can automate a variety of AutoCAD routines. It may be worth your time to read up on further script file details in the Help files, as we do not go too much in depth here.

We present a very simple example of a script file. It freezes all layers after first thawing them and setting 0 as the current layer. As scripts do not work with dialog boxes, everything must be command-line driven, such as this version of the layer command (generally any commands that are preceded by a hyphen [-] go to the command line, directly bypassing any dialog boxes). After the -layer is a set of answers to prompts: t means thaw, * means all, s means set, and f means freeze. This exactly answers everything AutoCAD asks for if you were to do this by hand at the command line. The last two keystrokes (→ →) are Enter and Enter.

```
-layer
t
*
s
```

```
0
f
*
→ →
```

This reads as Layer, Thaw, All, Set, 0, Freeze, All, Enter, Enter. Try this out with an AutoCAD file that has a bunch of layers. In a real application, you of course must manually type in the layers deemed not relevant to the state (which therefore need to be frozen) in place of the last * and before the two Enters. The steps prior to that (t, *, s, 0) are necessary in case a layer that needs to be frozen is set to current (you cannot freeze the current layer), so the 0 layer is set as current.

This was a lengthy process and is the main reason why script files gradually fell out of favor, replaced by the LSM for layer control, as shown next. Keep in mind, however, that the LSM works in pretty much the same way, except you select layers and do everything graphically through the Layers dialog box. Now that you have read through the underlying theory, we can rapidly go through the functions of the LSM.

12.3 Layer State Manager

As described previously, the LSM is essentially a graphical version of script files. As you should now have a basic understanding of what you need to do, let us just set up a collection of layers and go through the mechanics of how to set up several layer states. While our example is simple—we set up only three layer states, an Arch_Layout, Mech_Layout, and Elect_Layout—the theory can be easily extended to many layers and states. Open a new AutoCAD file and set up the following layers:

A-Walls-Demo
A-Walls-New
A-Window-Bay
E-Outlet-Main
E-Switch-Main
E-Wiring-Main
M-Diffuser
M-Panel
M-Return_Air

Assign color to these layers as shown in Fig. 12.1. What we want to do now is create a layer state called Arch_Layout. This means only the A- layers are needed, with the electrical and mechanical layers frozen. So, go ahead and freeze the six layers beginning with E- and M-, as shown in Fig. 12.1.

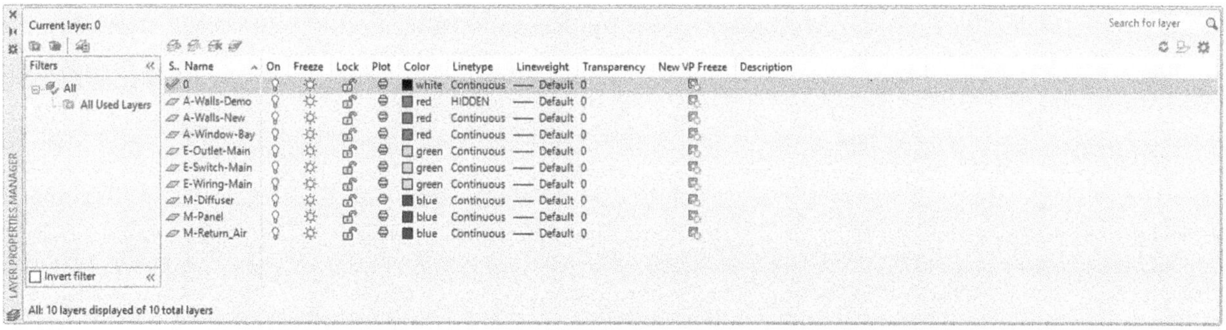

FIGURE 12.1 Layer setup for Layer State Manager exercise.

Now that the layers have been frozen, we need to save this "state" under the appropriate name. Click on the LSM button in the upper left of the Layers dialog box, as seen in Fig. 12.2. The LSM then appears (Fig. 12.3). It can be expanded (if it is not already) by pressing the arrow in the lower right corner.

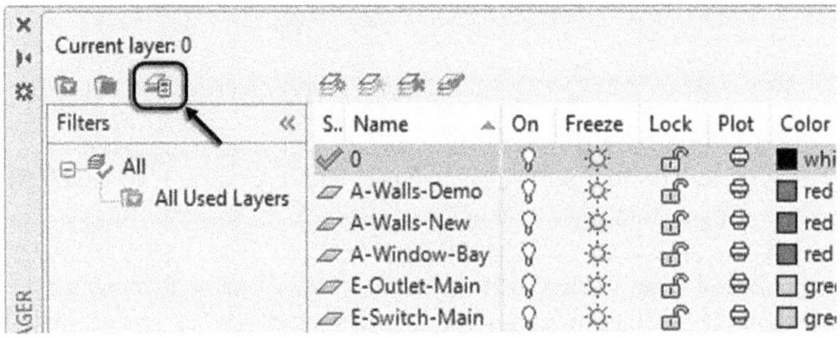

FIGURE 12.2 Layer State Manager icon.

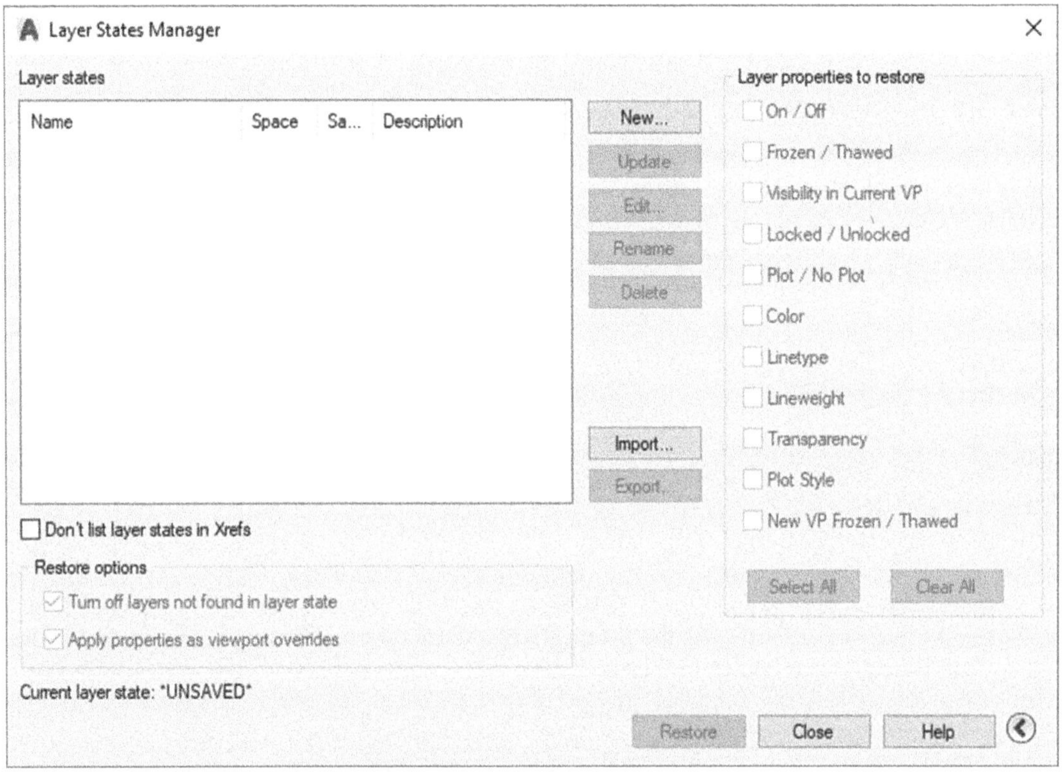

FIGURE 12.3 Layer State Manager (LSM).

We now create a new layer state by clicking on New … and filling in the name and description (if desired), as shown in Fig. 12.4, then clicking on OK. Once the layer state is entered, it appears in the LSM, as shown in Fig. 12.5, and you can press Close to exit out of the LSM dialog box.

FIGURE 12.4 New Layer State to Save.

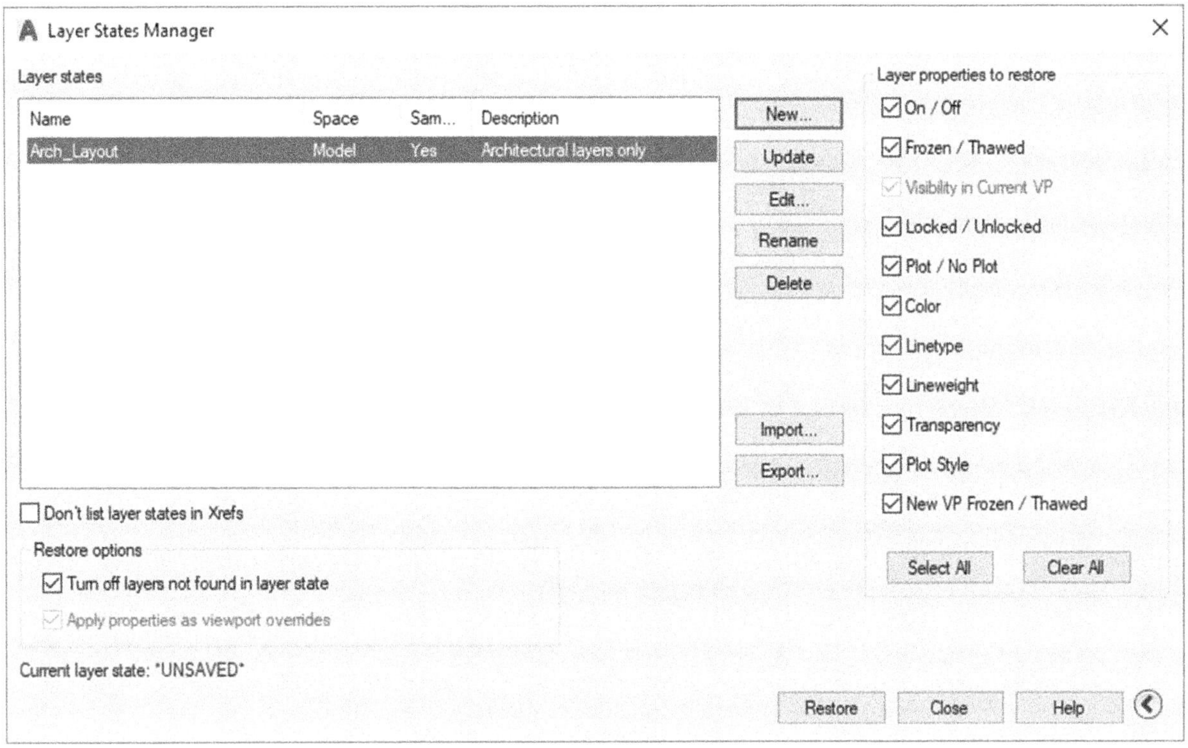

FIGURE 12.5 LSM with new layer state.

In the same manner, go back to the Layers dialog box and create the other two states: Elect_Layout (only E-layers visible) by turning off the unneeded layers and activating the LSM and Mech_Layout (M-layers visible). The result is shown in Fig. 12.6.

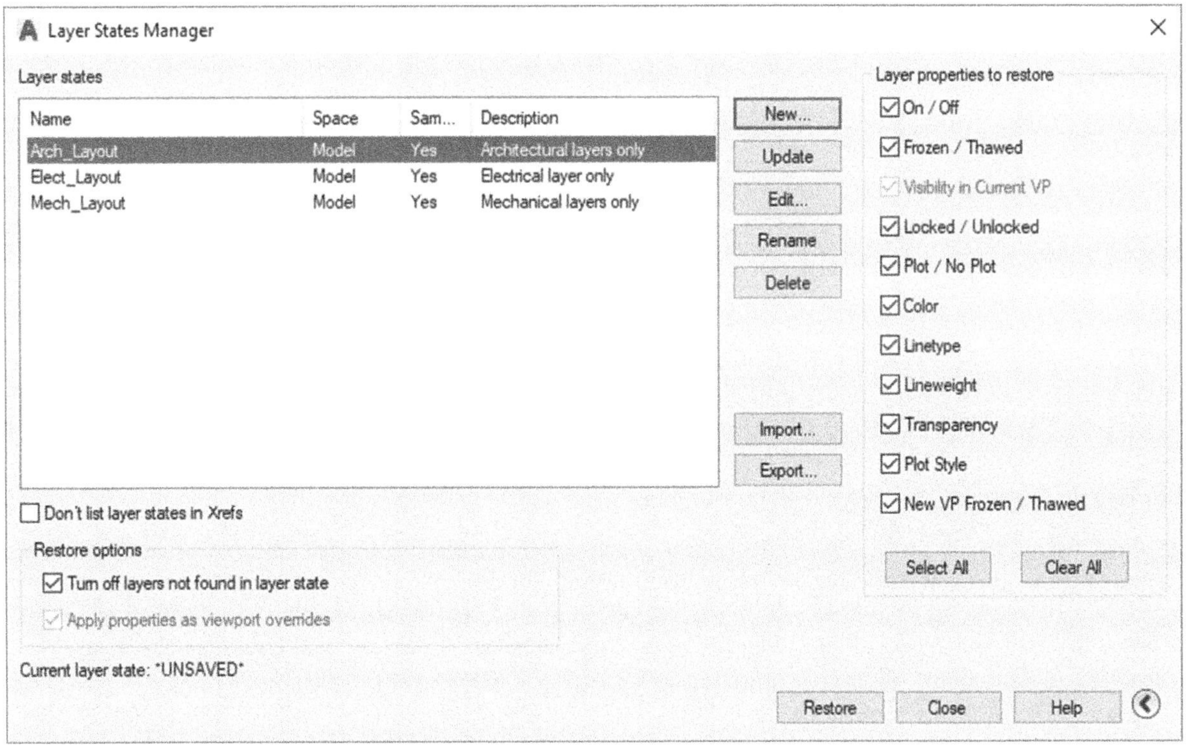

FIGURE 12.6 LSM with remaining layer states.

To activate these states when needed, you simply go to the Layers dialog box, click on the LSM icon, and when it appears, double-click on the state name or highlight it with one click and press Restore. Instantly, the correct layers are frozen or thawed (you can see this happen in real time in the Layers dialog box as you click on one state or another).

The purposes of the buttons found to the right of the main LSM window are as follows:

- *New*: Displays the New Layer State to Save dialog box, which we already used.
- *Save*: Saves the selected named layer state.
- *Edit*: Displays the Edit Layer State dialog box, where you can modify the layer state.
- *Rename*: Allows editing of the layer state name.
- *Delete*: Removes the selected layer state.
- *Import*: Layer states (*.las extension) can be imported from another file by means of this option, although this is not done often in practice.
- *Export*: Layer states (*.las extension) can be exported to another file by means of this option, although this is also rarely done in practice.

The additional options available in the LSM dialog box can be made visible or hidden if you click on the small arrow in the extreme bottom right, as previously mentioned (Fig. 12.7).

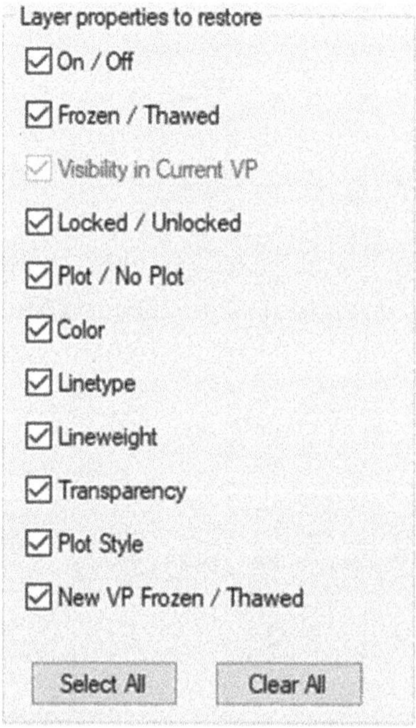

FIGURE 12.7 Additional LSM options.

Layer properties to restore are simply a listing of every property a layer typically has, and you can check off different properties if you want them restored. If all are checked, as is more typical in practice, all the layer's properties are affected.

12.4 Layer filtering

Layer filtering is a tool that allows you to filter out (or selectively choose) layers that fall under a certain description. You can create filters of your own choosing and assign them names. For example, if for some reason you want only the red layers to show or ones assigned a hidden linetype, this can be done. Layer names can also be filtered. If you want only those layers that start with an A- to show, this can be arranged, although in this case you are essentially doing what was described previously with the LSM.

The key to layer filtering is to set up filter definitions. These can include full names or partial names (using the wildcard * character) and property definitions (color, linetype, etc.). The entire left side of the Layers dialog box is dedicated to filters.

When used properly, these are very effective tools. We highlight some of the more important features essential to using them in this chapter. You are encouraged to explore them further on your own as well.

Two types of filters are available as icons on the top left of the Layers dialog box, as shown in Fig. 12.8.

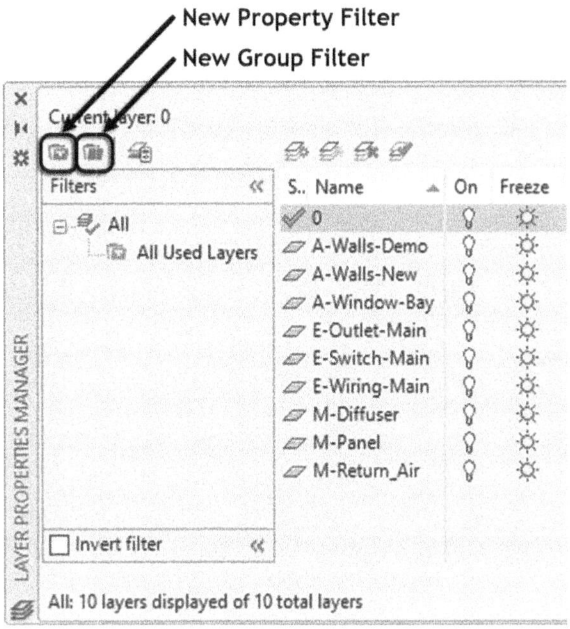

FIGURE 12.8 Layer filters.

The New Property Filter is what you use to set up the definitions. If you press that icon, the box in Fig. 12.9 appears, with the previously entered layers visible.

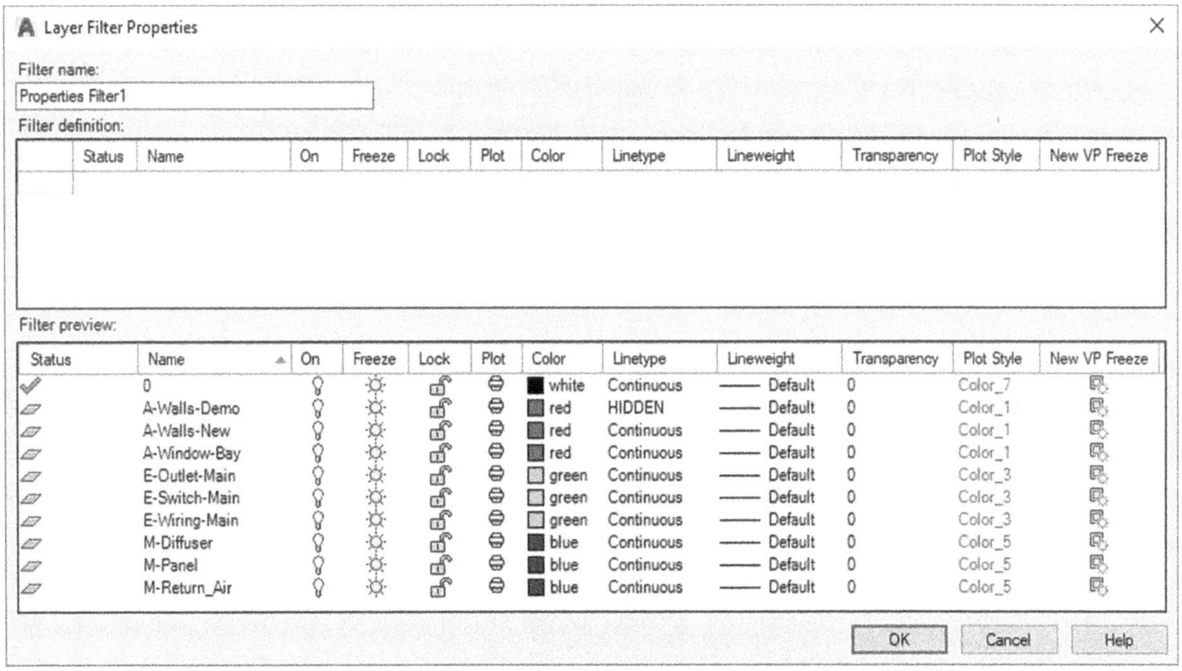

FIGURE 12.9 Layer Filter Properties.

The Layer Group Filter is a general filter that includes all the layers put into the property filter when you define it, regardless of their names or properties. Selected layers can then be added from the layer list by dragging them into the filter. This type of filtering is generally not used but can be employed to make a new filter based on the layers of another filter. To begin, click on the icon, create a name, and click and drag layer names from the right side of the Layers box into the group filter.

Remember an important point with filters: By themselves, they do little except sort the layers. It is up to you to then do something useful with this, such as freeze or lock them as needed. Therefore, think of filtering as simply a way to get the layers in some sort of order for further action on them.

Here, you first enter the name of your filter in the text field in the upper left. Then, it is just a matter of careful selection of the properties or names (or both) of the layers you want to see. You may use the wild character * in conjunction with typed parts of the layer name to indicate to the filter to include anything before or after the specific letters. Any property can also be used as a filter; color and linetype are common ones.

Go through the filter at your own pace and experiment with the settings. Shown in Fig. 12.10, as a simple example, are the layers that are green and the letter W somewhere in the name.

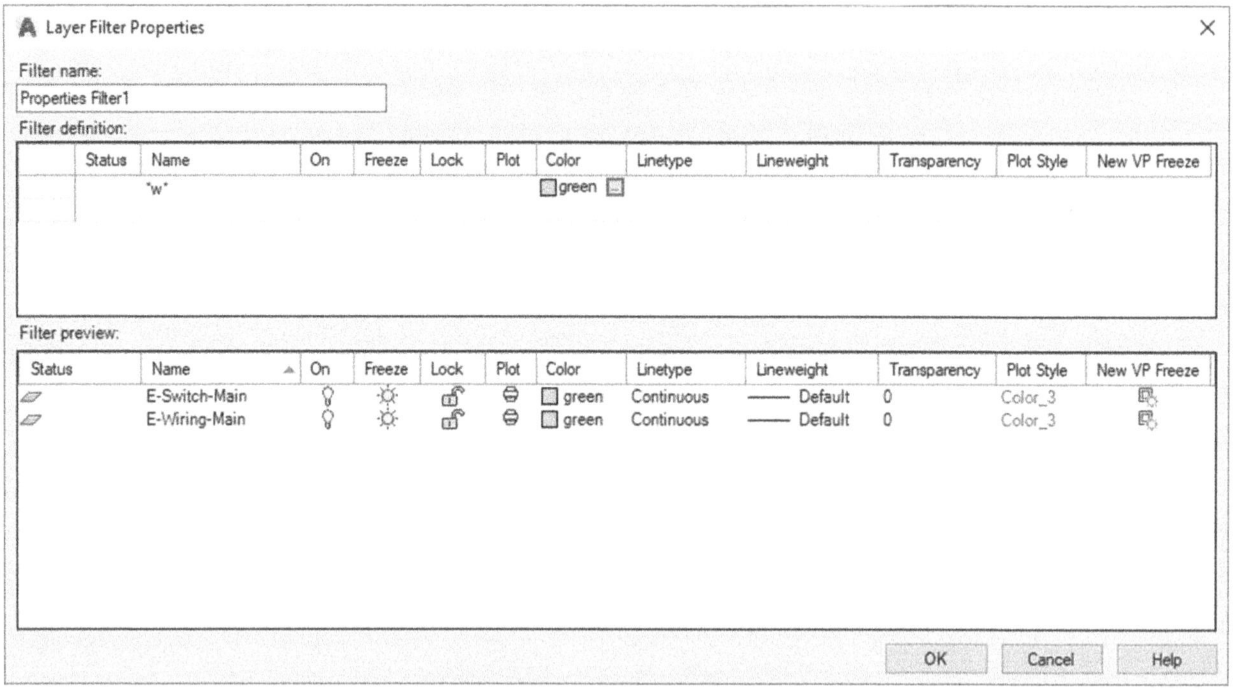

FIGURE 12.10 Example of layer filtering.

12.5 Layer Freeze, Isolate, and Walk

These very useful layer manipulation commands were once part of the Express Tools (to be covered in detail in Chapter 14), but their importance has been acknowledged by an upgrade to permanent status within regular AutoCAD menus. Although there are a bunch of them (many are redundant), we focus on the primary three that are important:

- Layer Freeze (command line: `layfrz`)
- Layer Isolate (command line: `layiso`)
- Layer Walk (command line: `laywalk`)

In addition to typing, these tools are accessible via the drop-down menus under Format→Layer tools, as seen in Fig. 12.11.

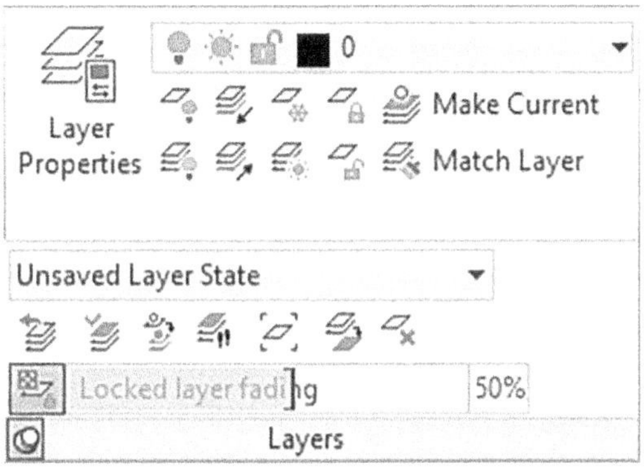

FIGURE 12.11 Layer tools in Home tab, Layers panel of the Ribbon.

Layer Freeze

In a complex drawing, layers may number in the dozens or even hundreds. Any tool that can simplify dealing with them is a welcome addition. Layer Freeze allows you to freeze a layer merely by selecting an object that is on that layer. This bypasses the tedious process of finding out what layer an object is on (if not known), then freezing it manually via the layer properties manager or the Ribbon layer list dropdown. Simply type in `layfrz` and press Enter or select the tool from from the Layer panel of the Home tab of the Ribbon.

- AutoCAD says: `Select an object on the layer to be frozen or [Settings/Undo]:`

Pick an object that represents the layer you want frozen, and all the objects on that layer disappear because the layer is frozen. This can be repeated as often as needed by selecting more objects and is a good way to clear up unneeded objects without losing them from a busy, dense drawing. (Note: A quick way to thaw all layers, is to use *laythw*, located directly below layfrz in the Layer panel of the Home tab of the Ribbon.)

Layer Isolate

Layer Isolate allows you to isolate a layer (or many layers) merely by selecting an object on it. The command turns off (not freezes in this case) all the other layers not selected, leaving the selected one(s) on. This bypasses the tedious process of doing it manually. Simply type in `layiso` and press Enter or select the tool from the Layer panel of the Home tab of the Ribbon.

- AutoCAD says: `Current setting: Hide layers, Viewports=Vpfreeze`
  `Select objects on the layer(s) to be isolated or [Settings]:`

Pick an object that represents the layer or layers you want isolated, press Enter, and all the others disappear because their layer has been turned off. It is another good way to clear a dense drawing without losing objects. (Note: A quick way to get back to where you were before isolating layers, is to use *layuniso*, located directly below layfrz in the Layer panel of the Home tab of the *Ribbon*.).

Layer Walk

This is perhaps the most interesting of all the layer controls and a good way to shuffle through an entire drawing layer by layer looking for stray or mis-layered objects Layer Walk presents a list of layers and isolates one of them. Then, you use the arrow keys on your keyboard to select your way up or down the list. AutoCAD turns layers on or off in succession, allowing you to inspect each layer, one at a time (walk through them). To try this, open any file that has several layers. Then, type in `laywalk` and press Enter or select the tool from the Layer panel dropdown of the Home tab of the Ribbon.

A dialog box appears, listing all the layers visible from layer 0 until the last one, as shown on a sample file (Level 2 project) in Fig. 12.12. Click on layer 0; it shows only objects on that layer and all the rest disappear from view. Use the arrow keys to cycle down through each layer and notice how appear and disappear as their turn comes up in the listing.

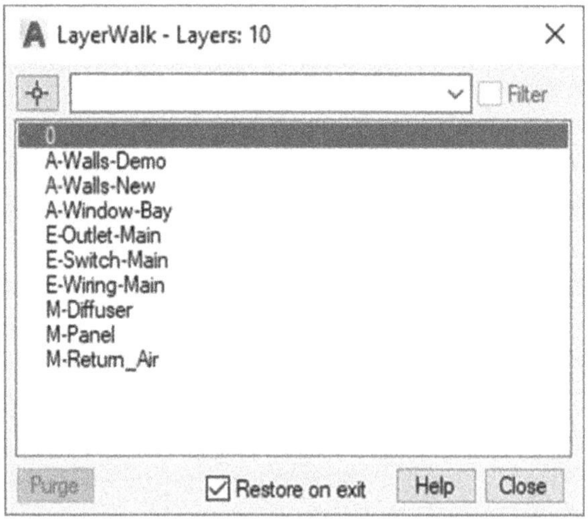

FIGURE 12.12 Layer walk.

This is a great tool to review all the layering on an unfamiliar drawing as well as one of your own to ensure accuracy. Sometimes AutoCAD designers let their layering accuracy slip as the project gains complexity, and by the end, layers are sloppy, with objects residing where they are not supposed to be.

After inspecting all the layers, you can stop on any one of them for editing; if the Restore on exit selection box is checked, then all the layers come back on when you press Close.

Most of the other layer commands listed in Fig. 12.11 are redundant, as you can already perform these functions via the Layer Properties Manager, so we do not discuss them further, but review them at your own pace.

12.6 Level 2 drawing project (2 of 10): architectural floor plan

For Part 2 of the drawing project, you continue to add to the walls of the floor plan, with a lot of new work on the interior, as well as adding doors, windows dimensions, and text. We can now also hatch the exterior and interior walls.

Step 1. As always, before starting to draft something new, create the appropriate layers, as follows:

- A-Doors (Yellow).
- A-Text (Cyan).
- A-Walls-Hatch (Gray_9).
- A-Walls-Interior (Green).
- A-Windows (Red).
- A-Roof (Magenta)
- A-Roof-Wall Below (Blue, Dashed)
- A-Roof-Text (Green)

Step 2. Add in the exterior windows and all doors. Be sure to follow the layering guidelines carefully. Make all windows and doors blocks for convenience. The exterior dimensions are shown in Fig. 12.13. Note that the deck and deck hatching is frozen, as it is not relevant to the new material you are adding.

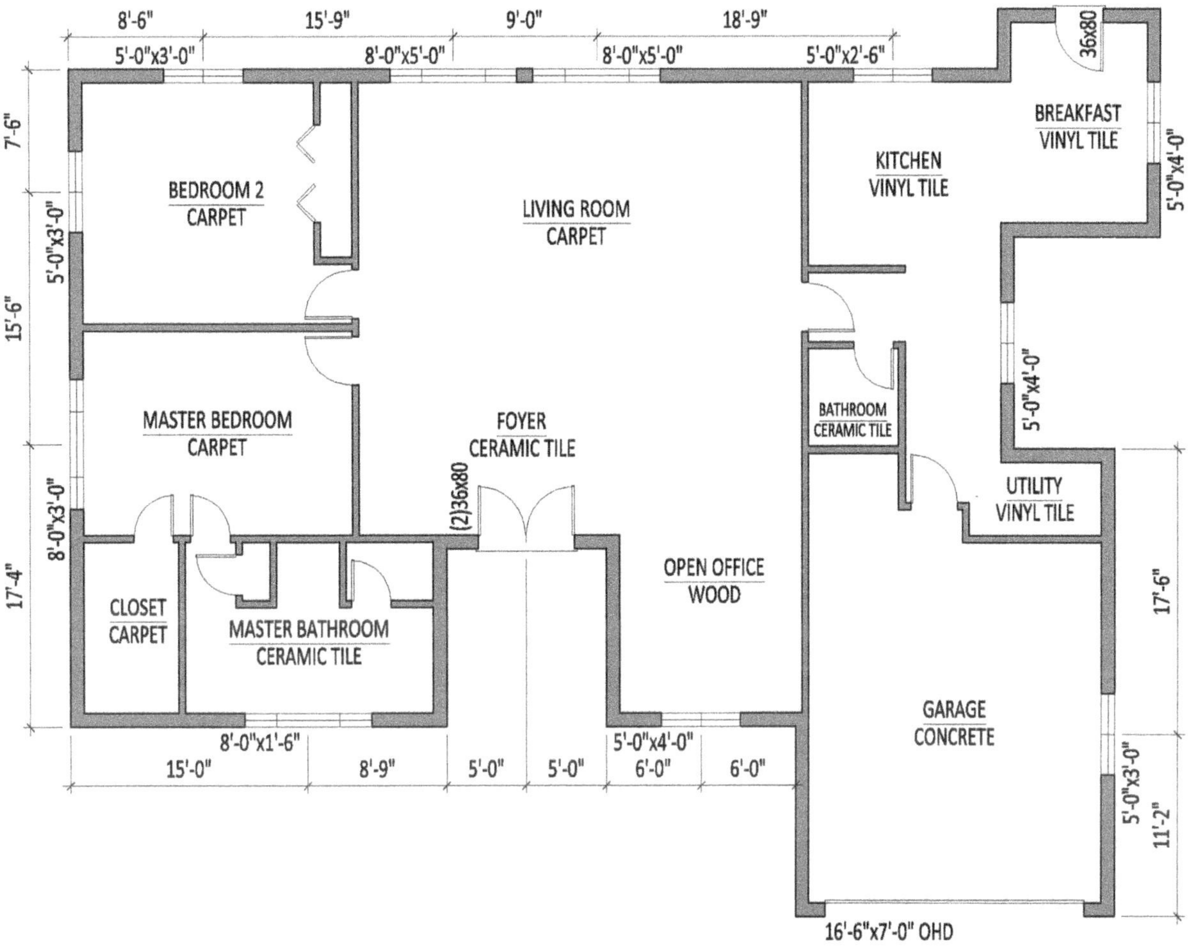

FIGURE 12.13 Exterior windows and doors.

Step 3. Add in the interior walls and doors as shown in Fig. 12.14. All interior walls are 5″ thick. Also add in the actual dimensions themselves for ¼″ = 1′-0″ final printing as shown in Figs. 12.15 and 12.16. As mentioned before, they are very useful as a check of your drafting accuracy.

Step 4. Add in the hatch (using the light gray color 9). This should be the very last step because you need to erase the hatch if you find a mistake in the layout.

Step 5. Finally, outline the house footprint with a polyline on an A-Roof-Wall Below layer. Turn off all other layers except the layers for the roof. Create new roof plan lines based on the A-Roof-Wall Below lines adding more layers as needed. Complete the roof plan and notes as shown. Your roof plan will be drawn directly on top of the floor plan with the layer visibility being controlled with freeze/thaw in model space and through viewport freeze/thaw later.

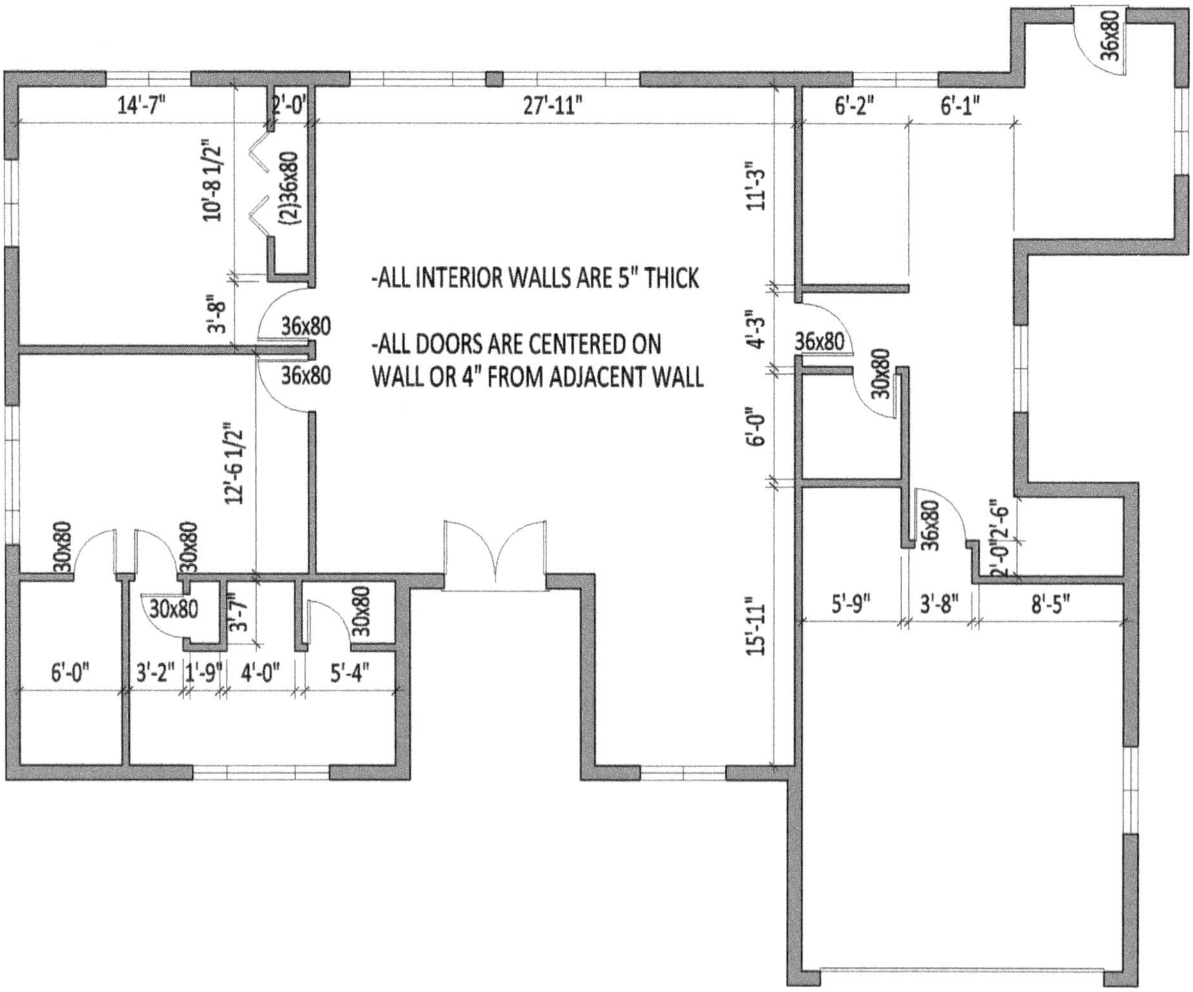

FIGURE 12.14 Interior walls and doors.

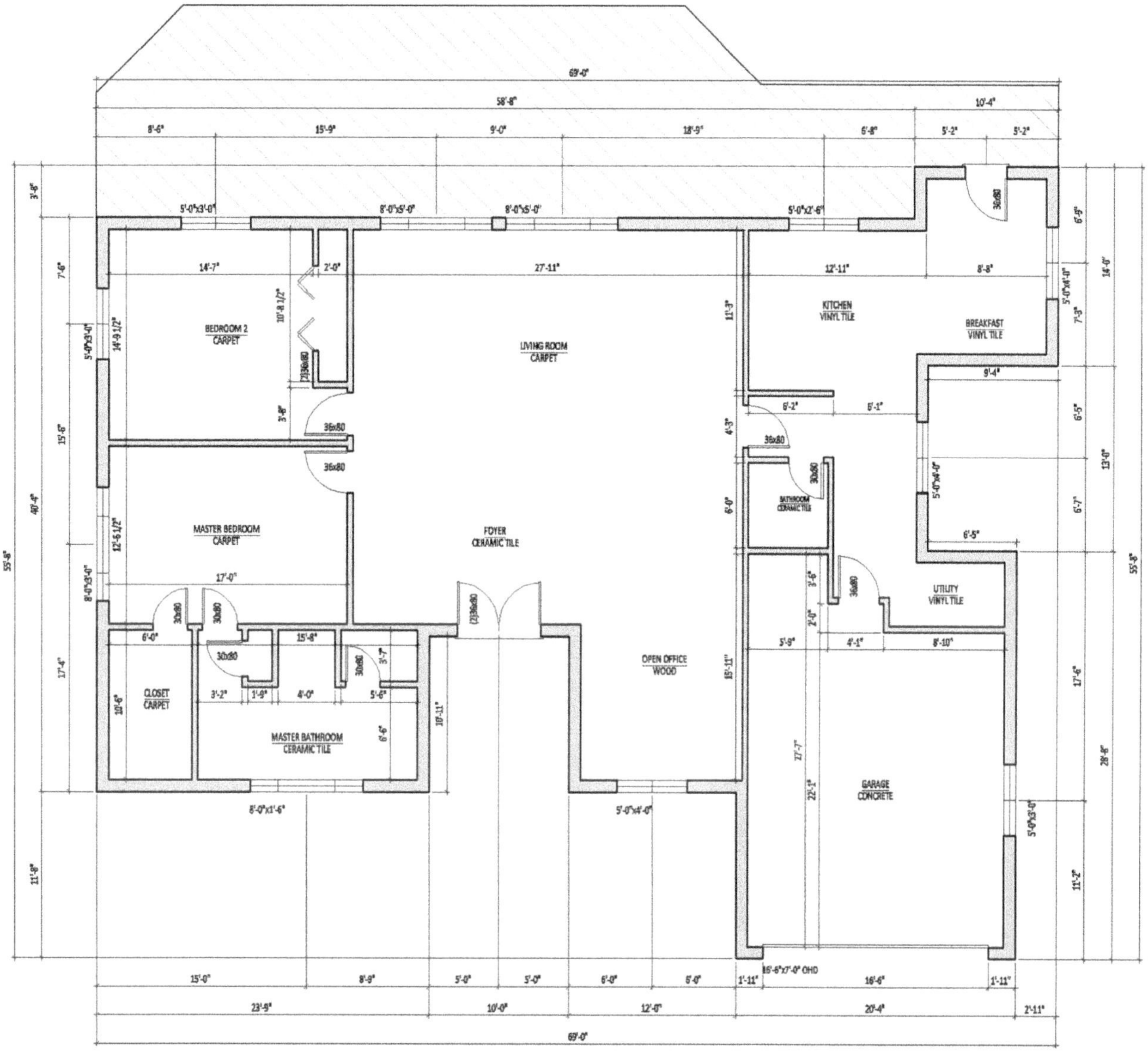

FIGURE 12.15 Floor plan with all notes and dimensions for ¼″ = 1′-0″ printing.

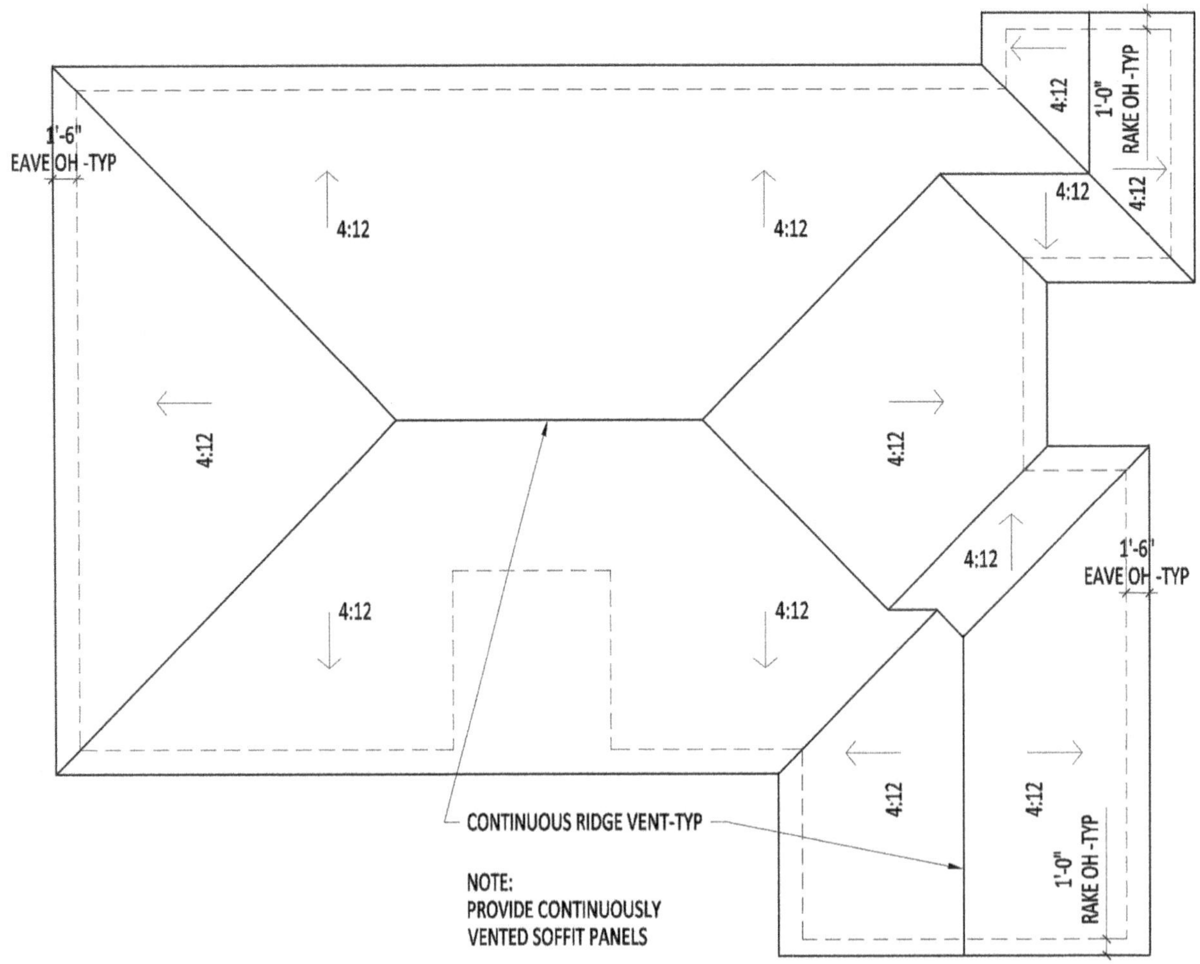

FIGURE 12.16 Roof Plan.

Summary

You should understand and know how to use the following concepts and commands before moving on to Chapter 13:

- Script files
  - Basic concept
  - Writing a simple script file
- LSM
  - New
  - Save
  - Edit
  - Rename
  - Delete
  - Import
  - Export
  - Restore
  - Layer properties
- Layer filters
  - New Property filter
  - Layer Group filter
- Layfrz
- Layiso
- Laywalk

Review questions

Answer the following based on what you learned in this chapter:

1. Describe what is meant by *layer management.* Why is it needed?
2. Describe the basic premise of a script.
3. What can a script file do for layer management?
4. Describe what the LSM does.
5. What are the fundamental steps in using the LSM?
6. What is layer filtering? What are the two types?
7. Describe layfrz, layiso, laywalk.

Exercises

1. Create six layers (and colors) as shown next. Then, write a script file to thaw all the layers; set layer 0 as current. Finally, freeze all and thaw just the even numbered layers. (Difficulty level: Easy; Time to completion: 10 minutes.)
   - A-Walls_1 (Green)
   - A-Walls_2 (Red)
   - A-Walls_3 (Green)
   - A-Walls_4 (Red)
   - A-Walls_5 (Green)
   - A-Walls_6 (Red)
2. Using the same six layers, set up layer states Odd_Layers and Even_Layers and run them. (Difficulty level: Easy; Time to completion: <5 minutes.)
3. Using the same six layers, create filters to sort and capture only the green layers. (Difficulty level: Easy; Time to completion: <5 minutes.)
4. To maintain your drafting skills, draw the following two sets of mechanical valves. Not all dimensions are shown, so improvise when needed. The upper and lower valve assemblies have many features in common. Use the detail images to assist in drafting, but there is no need to draw them separately. (Difficulty level: Intermediate; Time to completion: 40–50 minutes.)

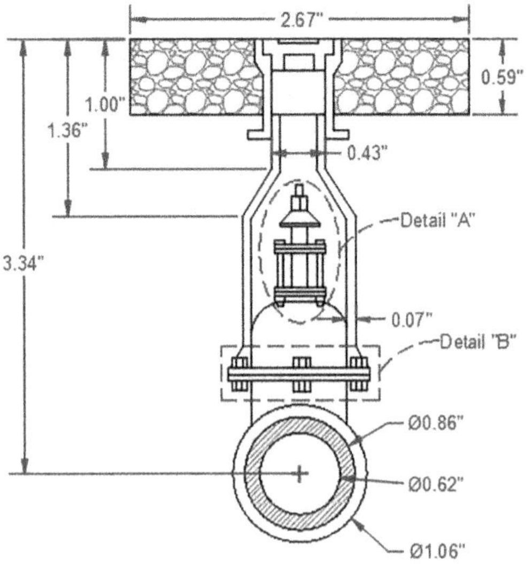

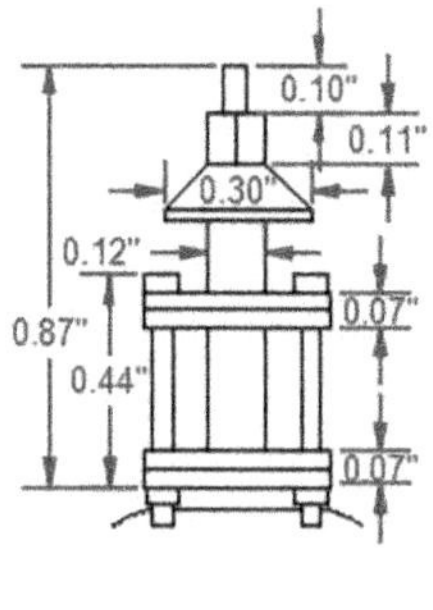

Detail "A"

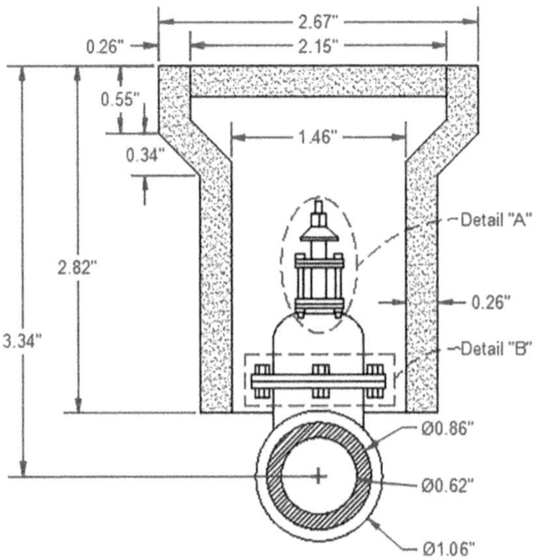

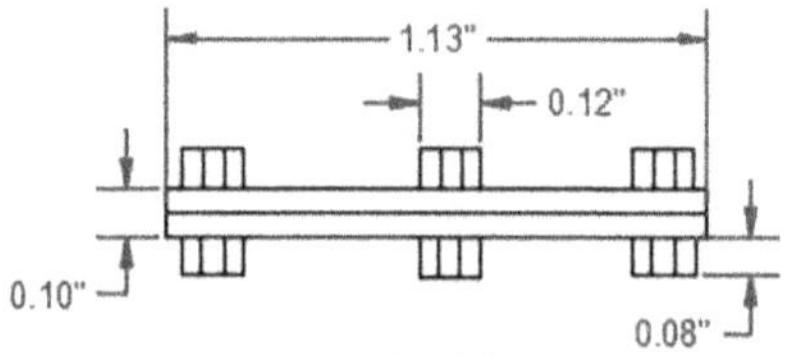

Detail "B"

Spotlight on: electrical engineering

Electrical engineering is the branch of engineering that deals with electricity and its applications. This can be on a large scale, such as power generation and transmission or small scales, such as the design of chips and consumer electronics, sometimes referred to as *electronics engineering*. The profession appeared in the mid-1800s, as major advancements in the understanding of electrical phenomena were put forward by Ohm, Tesla, Faraday, Maxwell, and others. Today, electrical engineering is a wide-ranging profession vital to our society. Few human-made devices are without any electrical components, and electrical engineering figures prominently in virtually every product or device you see before you. Electrical and electronic engineers held over 300,000 jobs in, making this the second largest branch of the US engineering community (behind software engineering).

Just some of electrical engineering's specialties include power, controls, electronics, microelectronics, signal processing, telecommunication, and computer engineering. Electrical engineers need a mastery of a wide variety of engineering sciences, such as basic and advanced circuit theory, electromagnetics, embedded systems, signal processing, controls, solid state physics, and computer science and programming, among others.

Education for electrical engineers starts out like that for the other engineering disciplines. In the United States, all engineers typically must attend a 4-year ABET-accredited school for their entry-level degree, a Bachelor of Science. While there, all students go through a somewhat similar program in their first 2 years, regardless of future specialization. Classes taken include extensive math, physics, and some chemistry courses, followed by statics, dynamics, mechanics, thermodynamics, fluid dynamics, and material science. In their final 2 years, engineers specialize by taking courses relevant to their chosen field. For electrical engineering, this includes courses in most or all of the sciences just listed.

On graduation, electrical engineers can immediately enter the workforce or go on to graduate school. Though not required, some engineers choose to pursue a Professional Engineer (P.E.) license. The process first involves passing a Fundamentals of Engineering (F.E.) exam, followed by several years of work experience under a registered P.E., and finally sitting for the P.E. exam itself.

Currently, electrical engineers receive starting salaries averaging $63,000 with a bachelor's degree and $73,000 with a master's degree. The median pay is around $88,000. This, of course, depends highly on market demand and location. A master's degree is highly desirable and required for many management spots. The job outlook for electrical engineers was projected as good, with a 6% growth rate per year for the next 5 years. The profession is extremely diversified; and by selecting a specialization carefully, an engineer can join one of many growing fields, such as controls, nanoengineering, robotics, and other such career paths.

So how do electrical engineers use AutoCAD and what can you expect? Industry-wide, AutoCAD enjoys a significant amount of use for the simple reason that most electrical work is 2D in nature and plays into AutoCAD's strength. AutoCAD layering is far simpler in electrical engineering than in architecture. There is no AIA standard to follow, only an internal company standard. As always, the layer names should be clear as to what they contain. The overall standard for much of electrical engineering work is provided by the IEEE, which may include CAD standards as well.

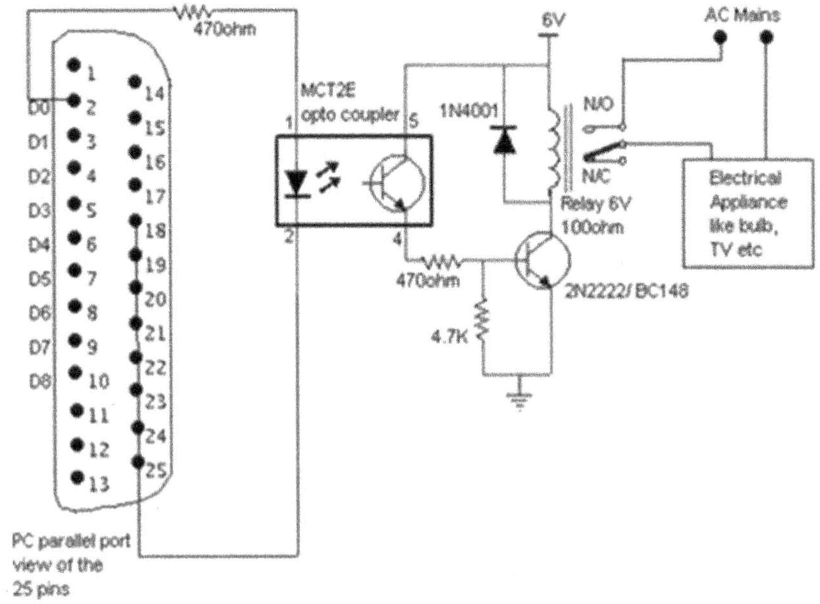

FIGURE 1 Electrical schematic 1. *Image Source: Elliot Gindis.*

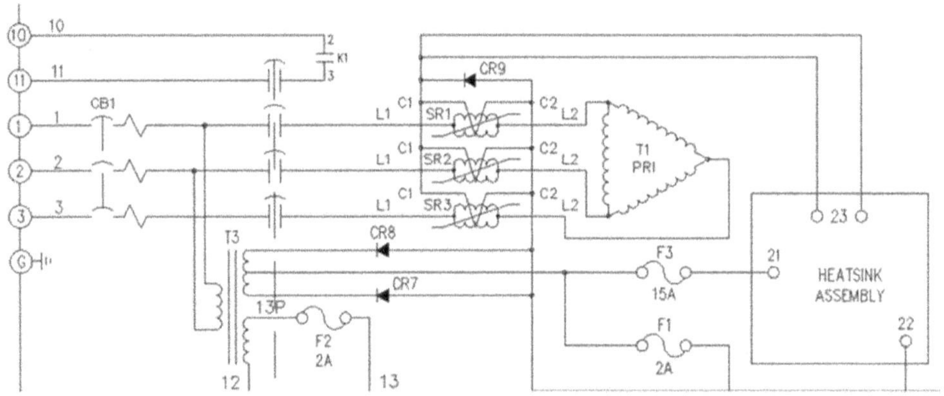

FIGURE 2 Electrical schematic 2. *Image Source: Elliot Gindis.*

In electrical engineering, you will not likely need advanced concepts such as Paper Space and Xrefs, but often you need to build attributes that hold information on connectors and wires. You also need to have a library of standard electrical symbols handy, so very little needs to be drawn from scratch every time. The key in electrical engineering drafting is accuracy (lines straight and connected), as the actual CAD work is quite simple. A related field, network engineering, is similar. Here, you may use plines or splines that indicate cables going into routers and switches. Shown in Figs. 1 and 2 are examples of electrical schematics. Notice the extensive use of symbols for capacitors, diodes, transformers, and switches.

Fig. 3 is an example of some racked equipment in a network/electrical engineering drawing. There is nothing complex in this image, just careful drafting, and a lot of copying. You will see this example again when wipeout is discussed. Fig. 4 shows some power transmission equipment from a substation.

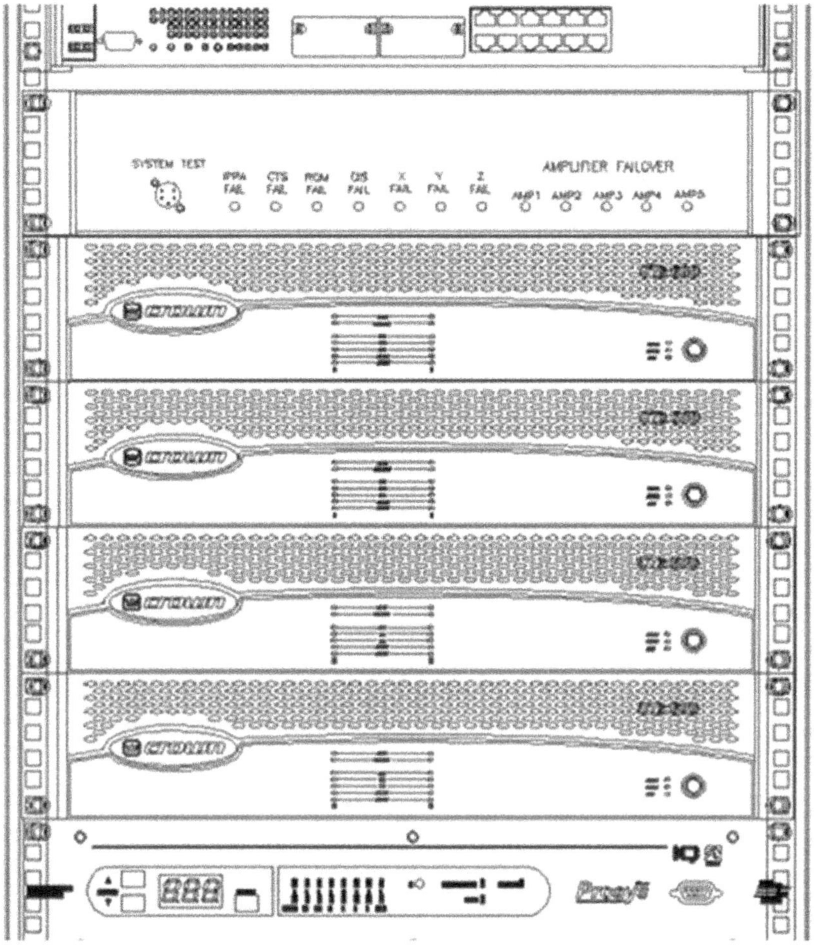

FIGURE 3 Racked equipment. *Image Source: Elliot Gindis.*

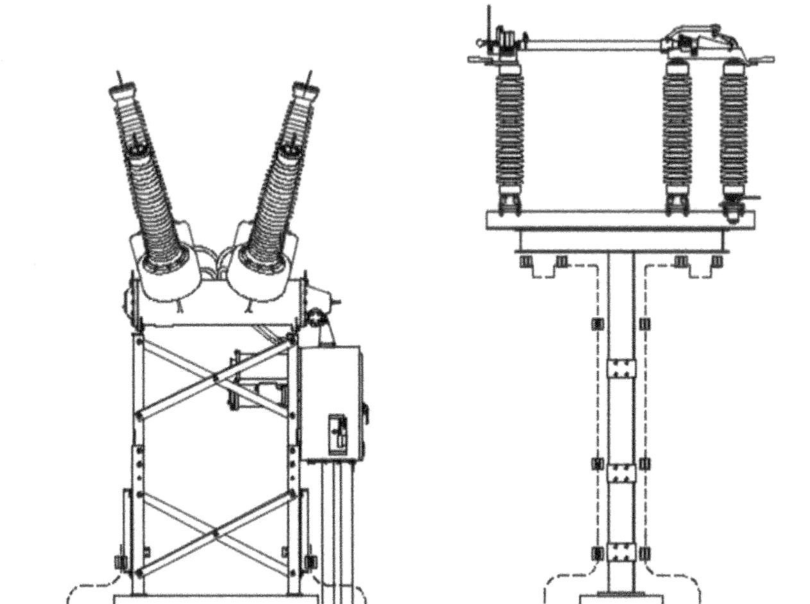

FIGURE 4 Substation power equipment. *Image Source: Elliot Gindis.*

Chapter 13

Advanced dimensions

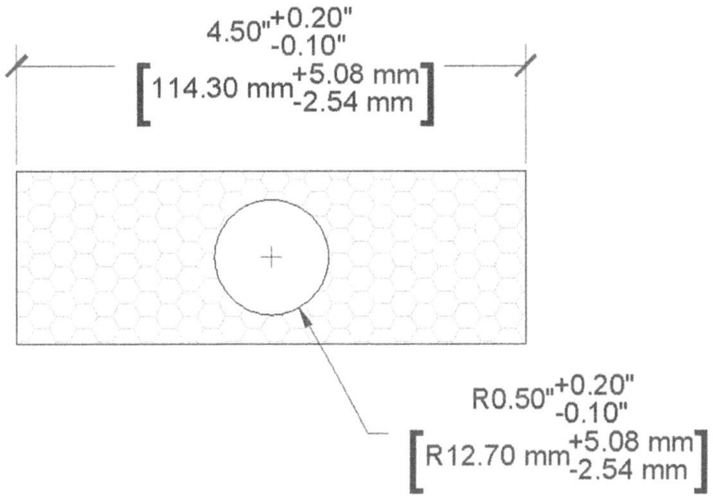

Learning objectives

In this chapter, we introduce advanced dimensions and discuss the following:

- Lines tab
- Symbols and Arrows tab
- Text tab
- Fit tab
- Primary Units tab
- Alternate Units tab
- Tolerances tab
- Geometric constraints
- Dimensional constraints

By the end of this chapter, you will learn additional, advanced features of the Dimension Style Manager and be capable of setting up complex dimensions. We also discuss two advanced dimensioning features called *geometric* and *dimensional constraints.*

Estimated time for completion of this chapter: 2–3 hours.

13.1 Introduction to advanced dimensions

This chapter is meant to complete your knowledge of AutoCAD's dimensioning features. Specifically, we look at the New Dimension Style (NDS) dialog box and Parametrics. In Level 1, we focused primarily on defining the types of available dimensions and how to properly dimension geometry. Little was mentioned of this dialog box except for the four essential

Up and Running with AutoCAD 2022. https://doi.org/10.1016/B978-0-323-89923-9.00013-5

features deemed most important: Arrowheads, Units, Fit, and Text Style (not needed as much anymore, as the better-looking Arial is the default font). In this chapter, we go through the entire NDS dialog box and discuss other options and features, some more than others, according to their usefulness.

The reason you need this knowledge is because AutoCAD allows for a tremendous amount of power, flexibility, and variation with its dimensioning. As a beginner user, you may not come to appreciate what is available because much of it may already be "set up" for you. As a CAD manager and advanced user (for whom Level 2 is geared), you may be charged with doing this "setting up" and need to know what to do. We conclude this chapter with a feature introduced back in AutoCAD 2010 called *Parametric Dimensions and Constraints*. These are routinely found in advanced 3D software but are relatively new for AutoCAD.

13.2 Dimension Style Manager

To start off, let us take a closer look at the Dimension Style Manager. You saw it in Chapter 6, so as a reminder, here is how we got it up on our screen.

Keyboard: Type in **dimstyle** or **ddim**
Ribbon: **Annotate** tab→**Dimensions**→**arrow** ↘

Create a new style, by pressing the New... button, and name it Sample Style (Fig. 13.1). Click on Continue when done. You then see the Create NDS dialog box. We now take a much closer look at the seven tabs: Lines, Symbols and Arrows, Text, Fit, Primary Units, Alternate Units, and Tolerances.

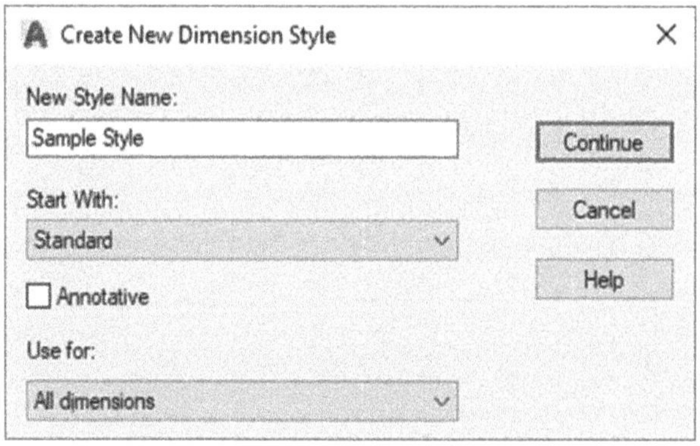

FIGURE 13.1 Create New Dimension Style.

Keep in mind as we go through these that our goal is not to cover every possible button and menu in extreme detail. Many of the settings are rarely used, and to describe them all completely would take dozens of pages and bewilder even the most dedicated AutoCAD student.

A much better approach is to give a basic overview of each tab and focus on only the features that are of utmost importance and widely used. You will see those features identified and described in more detail, with emphasis on how they fit in with the design process and how they benefit you as an AutoCAD user. If you wish to learn more, use this information as a starting point and consult the Help files.

Let us look at the first tab, all the way on the left of the NDS dialog box, called *Lines* (Fig. 13.2). If it is not already visible, click on it to bring it up. Throughout the process, carefully observe the Preview window at the upper right to see the results of changing or adjusting whatever we discuss.

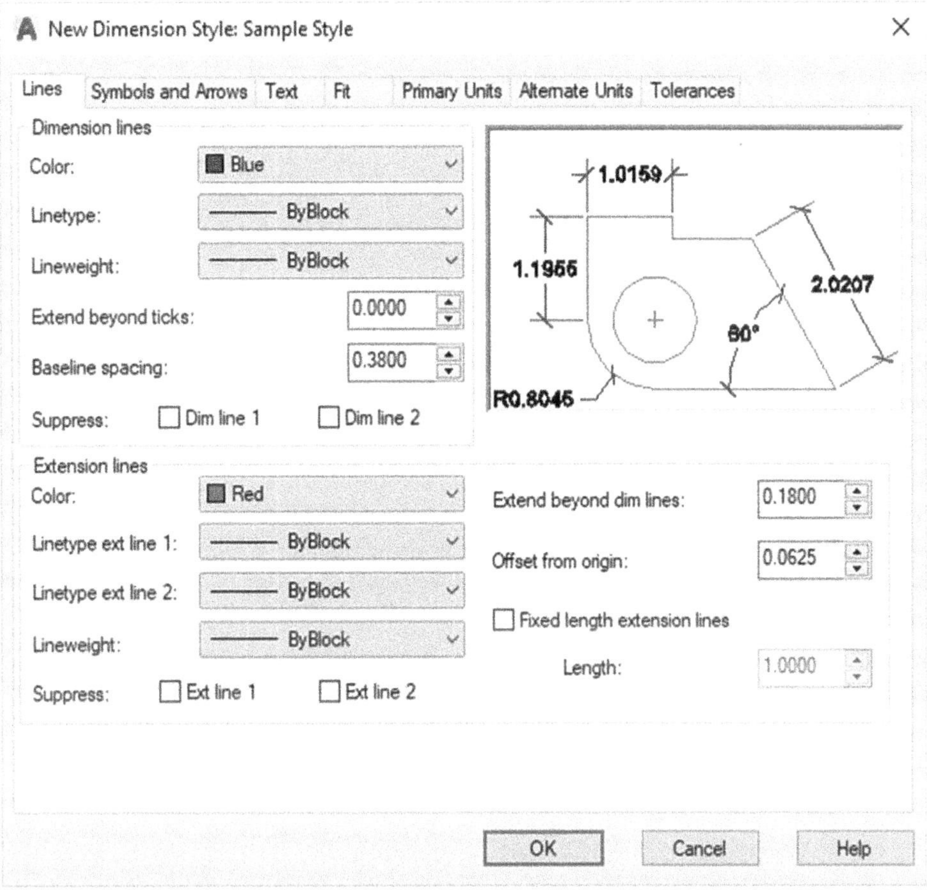

FIGURE 13.2 Lines tab.

Lines tab

This first tab is one that most users leave "as is" and generally do not modify. Here, we can make mostly cosmetic changes to two items specifically: Dimension lines (upper left) and Extension lines (bottom). It is, of course, instructive to know what they are. Dimension lines extend from the numerical value and contain arrowheads or another symbol. The color Blue is selected for them in Fig. 13.2. Extension lines extend from the part you are dimensioning and "frame" the rest of the dimension. The color Red is selected for them in Fig. 13.2.

The Lines tab is all about working with these two features. For both the dimension and extension lines, you can change their color, linetype, and lineweight. You simply click on the down arrow and change the property. You can also suppress either one or both extension or dimension lines, which make them disappear. This has some value if, e.g., a dimension is sitting between two walls and the extension lines are not visible anyway. You can then suppress them. This, however, would be a "local" thing, applied to just that one dimension via the Properties dialog box. There is little use in suppressing either extension or dimension lines for the entire drawing. Next, a few values are presented:

- Extend beyond ticks: (grayed out).
- Baseline spacing.
- Extend beyond dim lines.
- Offset from origin.

All these values refer to the various spaces between dimension elements or length of elements. They can all be fine-tuned one at a time. However, it is very important to understand that you do not want to do that here in this tab. If you wish to change the size of a dimension, you need to do this by increasing the overall scale, not one element or space at a time. We discuss this in detail later, when we cover the Fit tab. If you try to adjust the values one at a time, you end up with a lopsided dimension, so leave them all as default.

Finally, we have a check box for fixing the length of an extension line (and assigning a permanent value). This has the effect of making the extension line a specified length regardless of how far the dimension line is placed away from the objects being dimensioned. It is rarely used, as it takes away some flexibility in dimension placement. Just for practice, go ahead and change the colors of the dimension lines to blue and the extension lines to red, as seen in Fig. 13.2.

Symbols and Arrows tab

The most important feature found under this tab is the ability to change the arrowheads of your dimensions (upper left of tab). You already did this briefly in Chapter 6: changed your arrowheads to tick marks. Go ahead and do this again, making a note of the other available choices. You can make the second arrowhead different from the first, of course, but that is rarely done. The leader is generally left as an arrow since its main purpose is to point at something. Just below the leader is a field to change arrow sizing. This also falls under the "do-not-do-it-here" category, as discussed in the previous tab. Arrowheads need to be sized according to the overall scale, not individually. We discuss this under the Fit tab.

Moving below the Arrowheads, we have the Center marks. This has to do with measurements of a circle and what you want to see in the center of one when a diameter or radius dimension is added. Mark or Line is typically selected; let us go with Line. Notice how the Preview window reflects both this and the arrowhead changes.

Below Center marks and to the right, we have some relatively minor options. The Arc length symbol concerns the placement of the symbol, and the Radius jog dimension has to do with the angle of the jog "wiggle." Leave them, as well as the Linear jog dimension, as default all around. Do the same thing for the Dimension break field. It sizes up or down according to overall scale settings, so leave it as is for now. Your Dimension Style dialog box should look like Fig. 13.3.

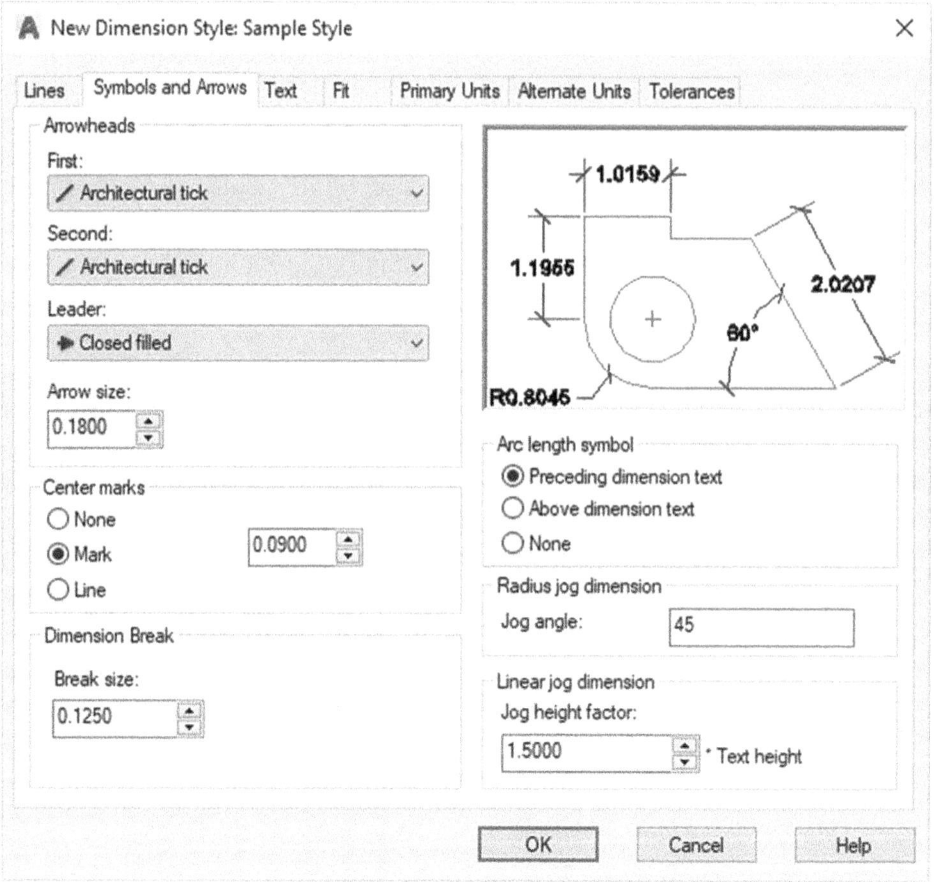

FIGURE 13.3 Symbols and Arrows tab.

Text tab

This tab is all about what dimension text looks like (appearance) and where it goes (placement and alignment). Look at the Text appearance category at the upper left. You already practiced changing the text under Text style. The goal was to make the dimension text the same as the rest of the text in the drawing: same font and usually the same size. In , when this was done, the text you selected on the floor plan was Arial 6″.

In this case, create another font from scratch by pressing the button to the right of the Text style: field (more options button with the three dots). Create a nonannotative Text Style called *HelveticaLight 0.5*, setting the Font Name to HelvLight and Size, Height to 0.5000 (use another font if HelvLight is not available on your computer). This technique is first outlined in Chapter 4, so review that chapter if you have any problems doing this.

Moving down the Text appearance category, we find

- *Text color*: Usually not set separately and remains "By Layer." Here, for practice, we set it to the color Magenta, as seen in Fig. 13.4.
- *Fill color*: Creates a color highlight behind the text. Try it out for practice, but in this example, no fill is set permanently. To add a frame around the text, there is a check box for this further on down, which we check off. Often a fill and a box are used together.
- *Text height (grayed out)*: This is usually not adjusted here. Use the previously mentioned Style dialog box.
- *Fraction height scale (grayed out)*: This applies to the size of fractions. The field can be set only when architectural units are being used. We look at this setting in detail soon.

These settings take care of pretty much every text appearance issue. Next, we look at how the text is placed in the drawing.

The Text placement category (bottom left) allows you to position the dimension value in every conceivable way near the rest of the dimension. You can place the values above, below, or centered on the dimension line. You can also move the values around the extension line, lining them up. Experiment with all the possible settings, observing the results in the preview window to the upper right. The default setting for both Vertical and Horizontal is Centered, as seen in Fig. 13.4. View Direction: flips the values around, and Offset from dim line: is best left alone and is adjusted as part of the overall scale settings.

With Text alignment (bottom right), the common default is Horizontal because people generally do not want to tilt their heads to the side every time they check a dimension on a printout. Do try the other two options (aligned and ISO) to see what they do, though. Your Dimension Style dialog box should look like Fig. 13.4.

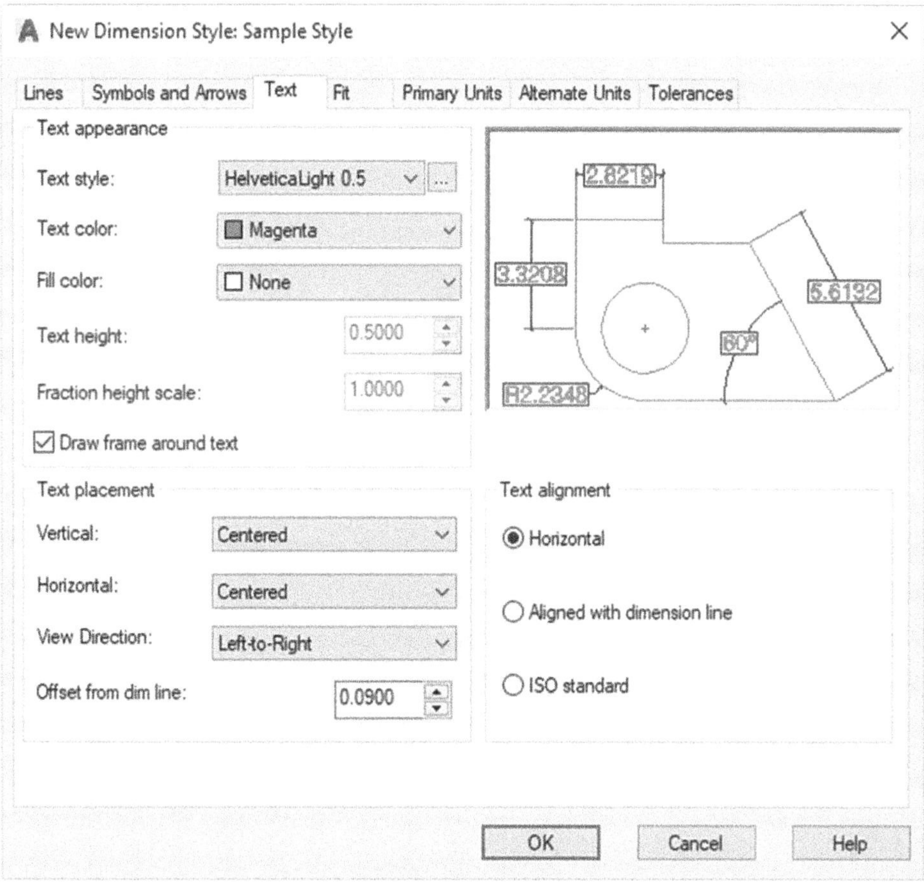

FIGURE 13.4 Text tab.

Fit tab

The contents of this tab can be a bit confusing at first; after all, didn't we just focus on text placement under the Text tab? Yet, here is another set of categories to fine-tune text placement. You can think of the Fit tab as a continuation of the Text tab with additional Fit options, Text placement, and Fine tuning categories.

Sometimes when you dimension small areas, the extension lines are so close together that the dimension value just does not fit in between them. The Fit options category, at the upper left, simply asks: "If the dimension doesn't fit, what do you want to move?" The best answer here is to select the first option, *Either text or arrows (best fit)*, and let AutoCAD decide.

You can manually force other choices as seen in the listing (arrows, text, etc.), of course, but then this choice is applied to all cases uniformly, which may not always be correct. You can even suppress arrows if they do not fit between the extension lines, but that is rarely a good idea and makes for some odd-looking dimensions.

The Text placement category (bottom left) is just a minor setting and should be left in the default first choice. At the very bottom right, we have the Fine tuning category. These two boxes are not all that important, as you rarely want to place text completely manually (adds extra workload) and asking AutoCAD to always draw a dim line between extension lines makes no sense, as the numerical value is then "crossed out."

The most important category to discuss here is the *Scale for dimension features* area just above the Fine tuning category. You already saw the all-important Use overall scale of: field in Chapter 6 (you entered the value 15). Then, in Chapter 10 there was a full discussion of what values go in that field and why. Recall that this value boosts (or lowers) the overall sizing of the dimension, and there is no need to adjust all the individual components that we mentioned in the previous tabs. For now, just enter the value 5.0000.

Your Dimension Style dialog box should look like Fig. 13.5.

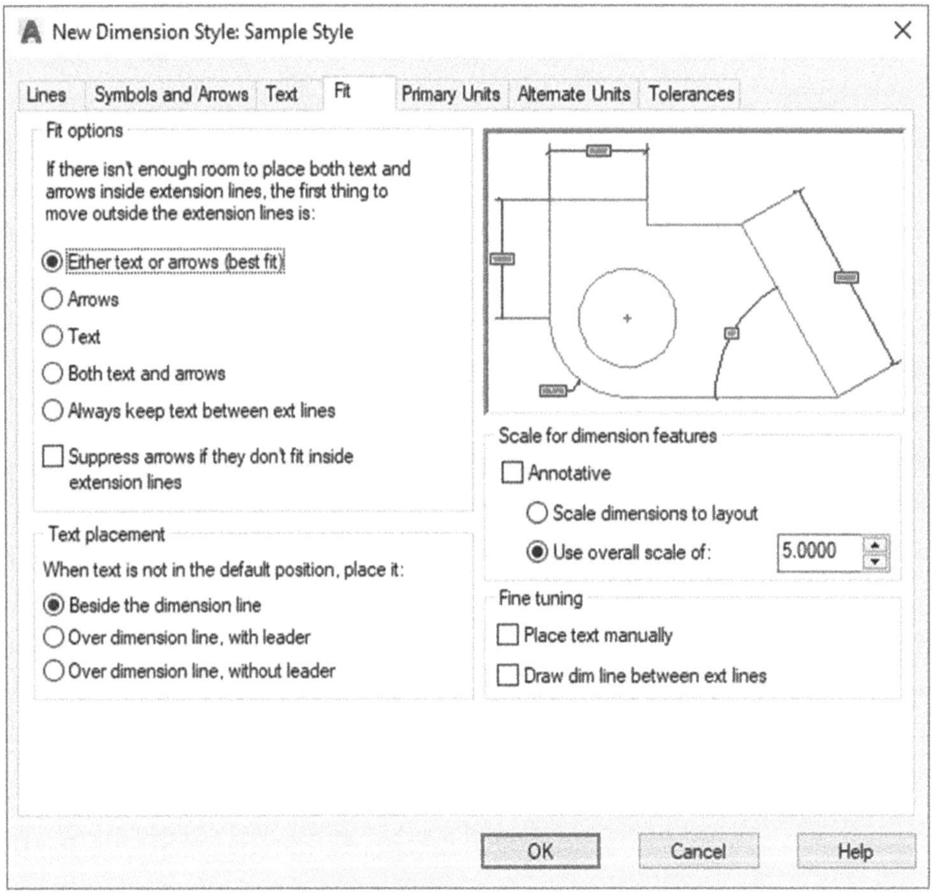

FIGURE 13.5 Fit tab.

Primary Units tab

This tab features quite a lot of useful settings. In the Linear dimensions category (upper left), you already practiced changing the Unit format: from Decimal to Architectural in Chapter 6. Generally, the dimension units are the same as the drawing units, so if you draw in architectural feet and inches, you would want to dimension in similar units. For this exercise, we leave the units in Decimal mode. In both cases, the Precision setting just below the units can be set to whatever value is appropriate. The range is 0 to 0.00,000,000 for Decimal and 0′−0″ to 0′−1/256″ for Architectural. Make a note also of the other units that can be selected, although they are rarely used.

Just below Precision you find the Fraction format:, Decimal separator:, and Round off: fields. The first two should be relatively clear in their function. Fraction format gives you a few choices (Horizontal, Diagonal, and Not Stacked) in how you want the fractions to look if using architectural units. This field is grayed out if using decimal dimensions.

The Decimal separator gives you a few choices (Period, Comma, and Space) in how you want the decimal values to look (European engineers do not use a comma when writing 1,000, for example). This field, of course, is grayed out if using architectural dimensions. You do not want to use Round off: at all. This field, of course, is grayed out if using architectural dimensions. Some other features further down the column are outlined next.

- *Prefix/Suffix*: Very useful features allowing you to add text, symbols, or values before and after the dimension values. We add the inch symbol in the suffix as an example.
- *Measurement scale*: Scales dimension values; not necessary in common usage.
- *Zero suppression*: Another very useful feature allowing you to suppress zero values either before or after the main significant digits when using decimal units. For example, a 0.625 value becomes just .625 and a 1.250 becomes 1.25. This is used in conjunction with the Precision setting to "clean up" dimension values, removing extra zeros. With the Architectural units setting, the idea is the same but now 0 feet and 0 inches are suppressed.
- *Angular dimensions*: This is the final category (bottom right). Here, you can select other units for angles such as Gradians, Radians, and Degrees/Minutes/Seconds (useful for surveyors). For most uses, however, Decimal Degrees are enough and the Precision: field below that allows for setting 0 to 0.00,000,000. Angular dimensions can also have zero suppression option for leading and trailing zeros.

Run through all the options available and, when done, set your Dimension Style dialog box to what is shown in Fig. 13.6.

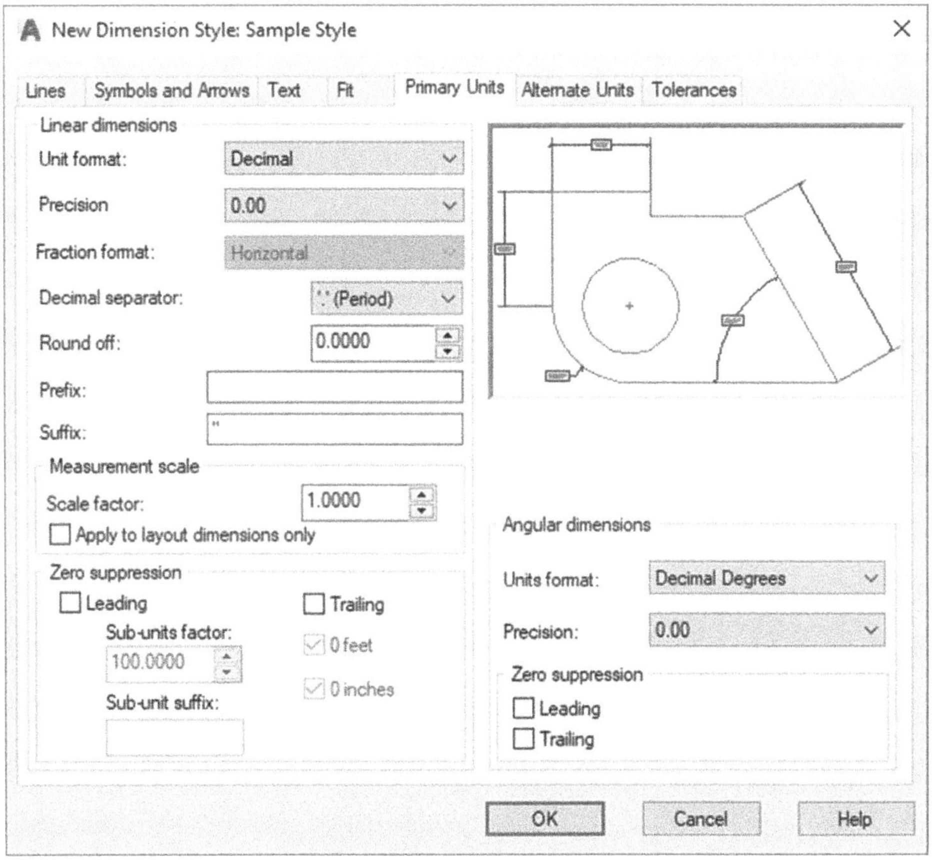

FIGURE 13.6 Primary Units tab.

Alternate Units tab

This is a tab we did not explore at all in Level 1. Alternate units are a secondary set of dimensions that reside inside parentheses attached to the main set of units. Their purpose is to present the main units in an alternate form, such as metric versus English (just one example).

The first thing you must do, as you click over to the Alternate Units tab, is to turn on the alternate units if they are all grayed out. Check off Display alternate units at the upper left, and the entire tab activates and the fields become editable. Most of our discussion centers on the Alternate units category (upper left). The Zero suppression and Placement categories should be familiar by now.

Under this category, we have:

- Unit format:
- Precision:
- Multiplier for alt units:
- Round distances to:
- Prefix:
- Suffix:

Most of these should also be familiar to you, such as Unit format and Precision. These values generally match the main set of units. Skipping Multiplier, we have Round (never a good idea to use) and finally the previously explained Prefix: and Suffix: fields.

Understanding alternate units then boils down to fully grasping the Multiplier for alt units: concept. It is a powerful and simple idea. Instead of presenting a collection of "canned" multipliers, AutoCAD allows you to set your own. So, if the main unit is 1 mile, to express this in feet (as an alternate unit), type in 5280 as a multiplier. You are in no way limited as to what sort of value you can enter (inches to miles or millimeters to yards is completely acceptable, if not necessarily practical), but of course you must know exactly what the multiplier is or else, as they say, "garbage in, garbage out." Therefore, an incorrect multiplier gets you false alternate unit readings.

Explore all the features of Alternate units and try entering several multipliers. For example, the number of yards equal to 1 mile (1760). Note also the default multiplier: 25.4. That is how many millimeters there are in an inch; this is useful for the most common application of alternate units, which is presenting metric units next to English ones.

In our case, let us go with the number of centimeters in an inch (2.54) and enter that as the alternate unit. Also add cm in the suffix. Your Dimension Style dialog box should look like Fig. 13.7.

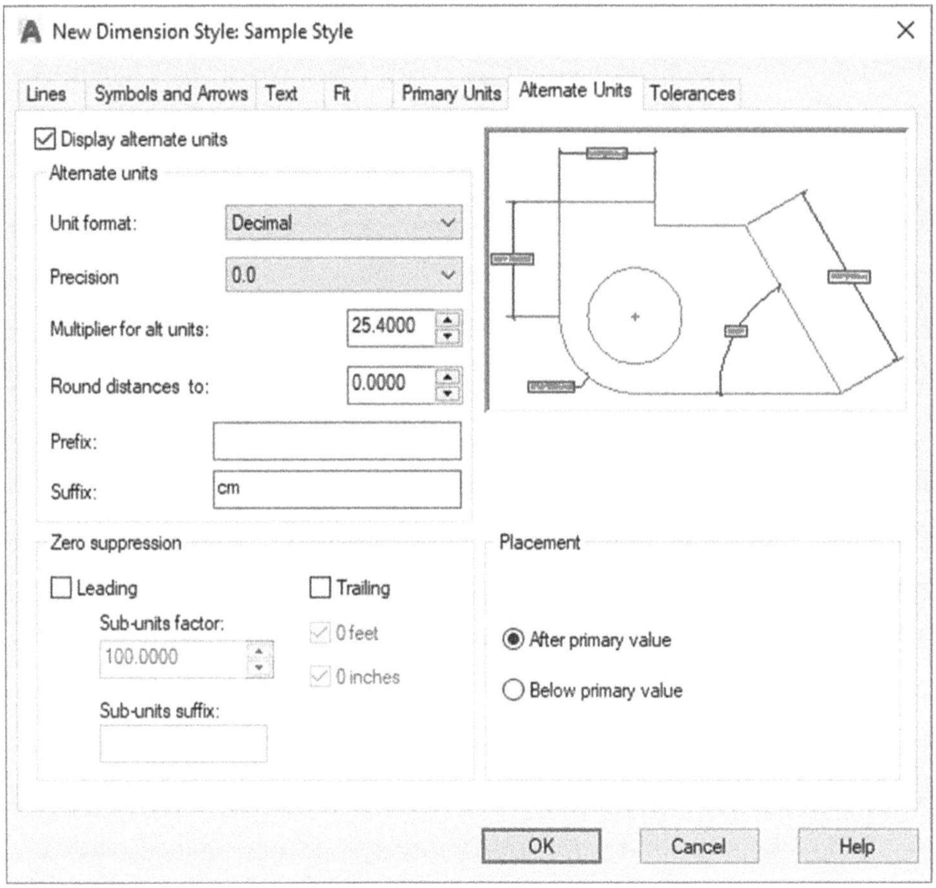

FIGURE 13.7 Alternate Units tab.

Tolerances tab

This final tab also has not been examined yet. Tolerances are generally defined as deviations from an ideal stated value, as used in manufacturing specifications of engineering devices. Tolerances are relevant mostly to engineering and manufacturing and less so to architecture, but knowledge of this tab is useful for all AutoCAD students.

The underlying need for specifying tolerances hinges on the fact that, in manufacturing, closer tolerances may be costlier to achieve, while larger tolerances, although cheaper, may adversely affect performance of a part. Tolerances are a way of expressing just how much a part may be off the mark and still be acceptable, based on engineering need and cost analysis. Aerospace, automotive, and biomedical engineering tolerances are closer than most consumer products, for example.

Tolerances are a familiar topic to mechanical engineers and machinists. To them, no further explanation is needed, and AutoCAD's tolerance choices are familiar: Symmetric, Deviation, Limits, and Basic. Other options under this tab, such as Zero suppression and Placement, should look familiar to all. A short description of each type of tolerance is presented next.

- *Symmetric tolerances*: These indicate the overall deviation, which is symmetric for upper and lower values; therefore, you set only one (the upper value).
- *Deviation tolerances*: These indicate nonsymmetrical deviations in either direction with a stacked ±.
- *Limits tolerances*: These are similar in theory and indicate the acceptable limit in one or both sets of values, but no ± is shown.
- *Basic tolerances*: These are the same as none, the first choice.

Tolerances are an interesting topic to explore further, but because a large percentage of this textbook's readers will not need them in their current or future work (as related to architecture), we do not go any further in depth. If you wish to learn more, there is a tremendous amount of information online. For now, feel free to experiment and set your tolerance to Deviation, as seen in Fig. 13.8.

So, what did we end up with after all these settings? Well, in the interest of trying out as much as possible, we did not really create a very realistic looking dimension. Let us look at it. Click on OK in the Dimension Style box; it will disappear and take you back to the Dimension Style Manager. Then, press Set Current in the next box, followed by Close. Now draw a 10″ × 10″ rectangle and dimension it using the basic Horizontal dimension. It should look similar or close to Fig. 13.9.

Let us try another sample dimension on a hypothetical section of roadway. Draw something like Fig. 13.10, with the two vertical lines 3 units apart. Run through all the tabs again, using the following data:

- Architectural ticks.
- Red-colored extension lines.
- Blue-colored dimension lines.
- Arial font, .25″, color: Red.
- Fit: 1.000.
- Primary units: Decimal. Precision = 0.
- Suffix added on main units → mile(s) and alternate units → feet.
- Alternate units, use 5280 as the multiplier.
- No tolerances.

You should get a result like Fig. 13.10. Note how simple it is to use alternate units if you know the multiplier.

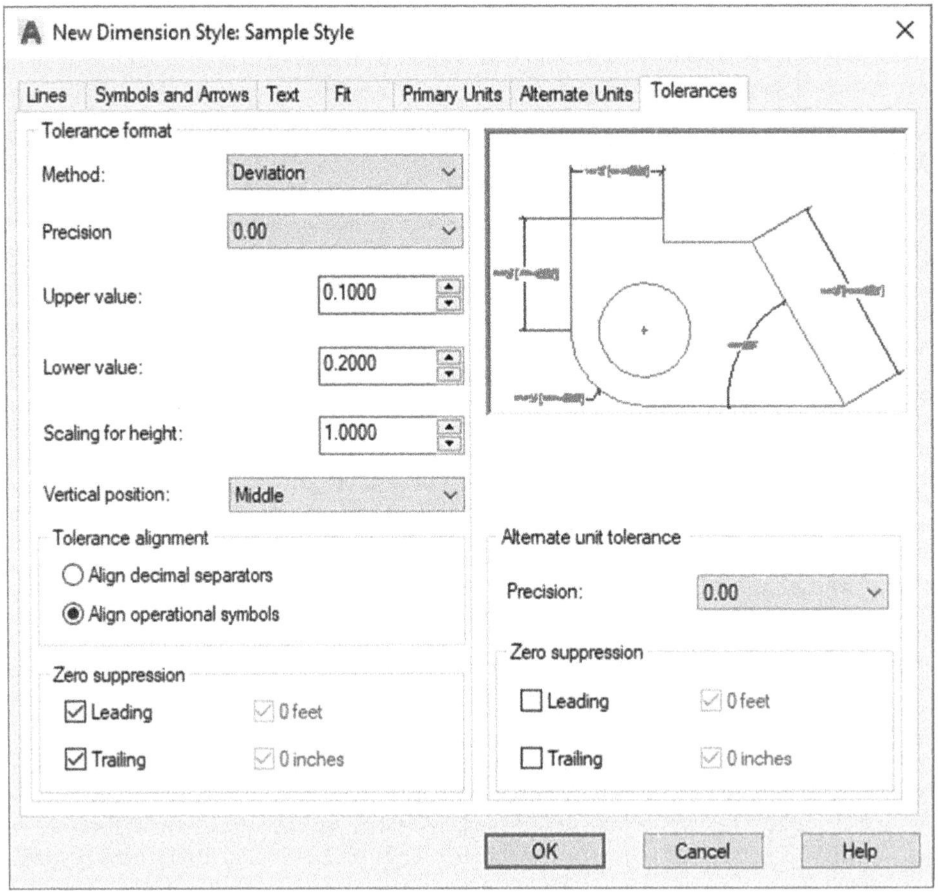

FIGURE 13.8 Tolerances tab.

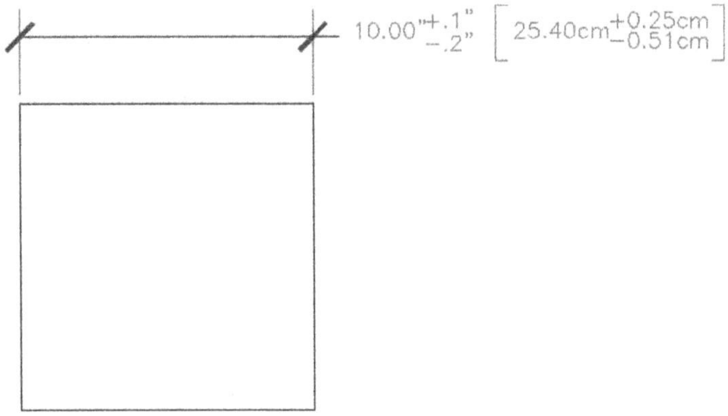

FIGURE 13.9 The new dimension style.

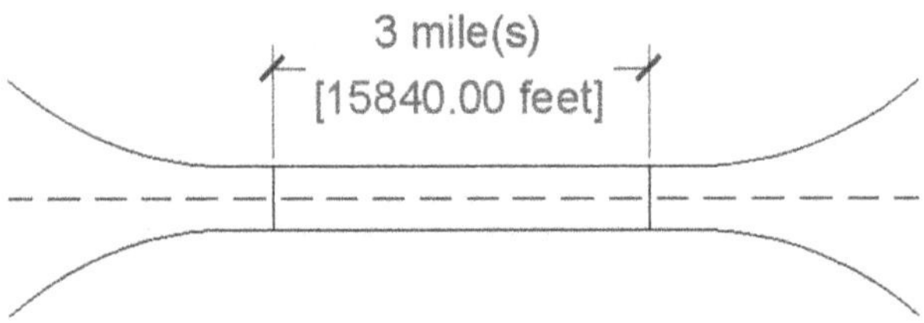

FIGURE 13.10 Sample dimension.

13.3 Introduction to constraints

AutoCAD was always a top-notch 2D drafting software application but one with a bit of an inferiority complex. Not regarding any other 2D competitors but rather toward 3D solid modeling and design software. These sophisticated products were developed in their own world, apart from AutoCAD, via development firms that never competed against or had anything to do with Autodesk or architecture. These programs included CATIA, NX, Pro-Engineer (now called *Creo*), and SolidWorks among others. Their developers invented quite a few revolutionary concepts and methods that transformed the way engineering design is done. Autodesk eventually jumped on board with Inventor and later Revit, both of which are excellent 3D products (one for engineers and the other for architects).

That left the AutoCAD development team looking at how its software could be enhanced by incorporating some of these sophisticated 3D tools without fundamentally altering AutoCAD's identity or formula for success. If you go on to study AutoCAD 3D, you will notice how many "borrowed" concepts found their way into the software as AutoCAD reached ever further into the high-end 3D world.

Some of these concepts also found their way into 2D, such as the ones we are about to discuss, namely, Parametrics, which is really two topics, geometric constraints and dimensional constraints; the latter is also referred to as *dimension-driven design.* If you are an engineer or an engineering school student who used the previously mentioned 3D solid modeling software, you quickly recognize these concepts. AutoCAD literally took a page out of their playbook. Let us discuss each concept separately.

13.4 Geometric constraints

The fundamental idea behind geometric constraints, regardless of what software we deal with, is to restrict (constrain) the possible movement of drawn geometry and force that geometry into certain positions. Ortho is a horizontal and vertical constraint on lines. So, if you want two parallel or perpendicular lines, a constraint can be placed on them that forces them to always stay parallel or perpendicular as well as to be that way in the first place. Expanding to more general terms, geometric constraints set "relationships" between geometries, such as parallel, perpendicular, tangent, and coincident.

This overall concept is quite important in setting what is referred to as *design intent.* If a drilled hole in a bracket absolutely must be concentric to the bracket's corner fillet to fulfill the design intent, then that is the main driving factor in its design (*concentric* means the circles and arcs share the same center point). Other aspects of the design, such as perhaps notches and other geometry on the bracket, however, may be less critical. The constraints then "hold" the critical geometry, while you design the rest around it.

This is a very simple example; with a bracket, you can probably keep track of the critical relationships without help. Geometric constraints really earn their keep with more complex designs, where many constraints are needed to preserve the design intent. Without them, it would be easy to forget one or two important relationships and discover later down the road that the part you are working on does not function properly because a tangency somewhere was not preserved.

Almost since the beginning of high-end 3D software, geometric constraints were an important part of the design approach, but they were not included in AutoCAD until release 2010; and AutoCAD designers needed to keep track, and check over, their geometric relationships manually. With these new tools, AutoCAD's mechanical engineering design capabilities have been greatly enhanced.

Types of geometric constraints

Listed next are most of the constraints available to you. We do not go through every single one but cover only the more commonly used ones, leaving the rest for you to explore. The best way to set constraints is via the Geometric panel on the Parametric tab of the Ribbon (see Fig. 13.11) or the Ribbon's Parametric tab.

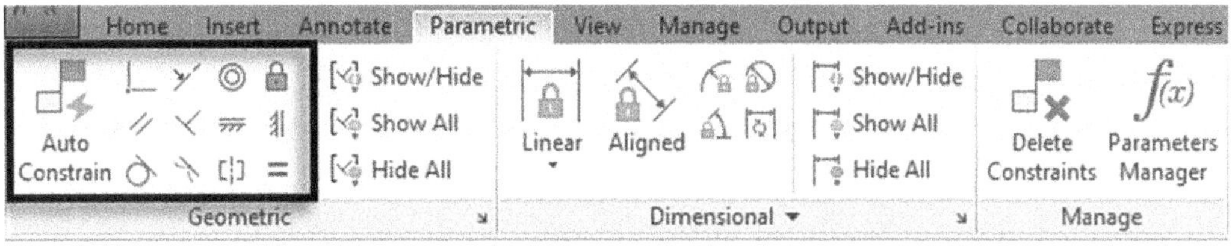

FIGURE 13.11 Geometric constraints Ribbon panel.

The critical constraints we discuss are the following:

- *Perpendicular*: Constrains two lines or polylines to perpendicular 90 degrees angles to each other.
- *Parallel*: Constrains two lines to the same angle.
- *Horizontal*: Constrains two lines to lie parallel to the X axis.
- *Vertical*: Constrains two lines to lie parallel to the Y axis.
- *Concentric*: Constrains circles, arcs, or ellipses to maintaining the same center point.
- *Tangent*: Constrains a tangency between curves and lines.
- *Coincident*: Constrains two points together (to be discussed along with dimensional constraints).

Adding geometric constraints

Adding geometric constraints involves simply picking the constraint you wish to add and selecting the first object (which serves as the reference point) followed by the second object. The constraint then appears next to the objects, reminding you it is set. The constraint tool also forces the objects into that constraint, so if you draw two lines that are not quite parallel, once you set the constraint, they become parallel right away.

Having read this, you may be tempted to completely do away with Ortho and even OSNAP to create accurate geometry, allowing constraints to "fix" drawn shapes. This is not generally recommended, of course, as sloppy drafting catches up to you sooner or later (though we do try this in the next example). While geometric constraints are very useful new tools, use them when appropriate, without forgetting or throwing away basic accuracy habits.

Let us try this out with the first constraint on the previous list: Perpendicular.

Keyboard: none
Ribbon: **Parametric** tab→**Perpendicular**

Step 1. Draw two lines that are roughly perpendicular, as seen in Fig. 13.12 (left).

Step 2. Select the perpendicular geometric constraint via any of the preceding methods.

- AutoCAD says: `_GcPerpendicular`
  `Select first object:`

Step 3. Select the first line (the vertical one).

- AutoCAD says: `Select second object:`

Step 4. Select the second line.

Immediately after the selection of the second line, you see it snaps to perpendicular and two sets of geometrical constraints are added next to each line, as seen in Fig. 13.12 (right). Note that, if you were to select the bottom line first, then the vertical one would snap to it, not the other way around. Selection order matters.

Let us try one more, the concentric one. We leave the rest as an in-class exercise.

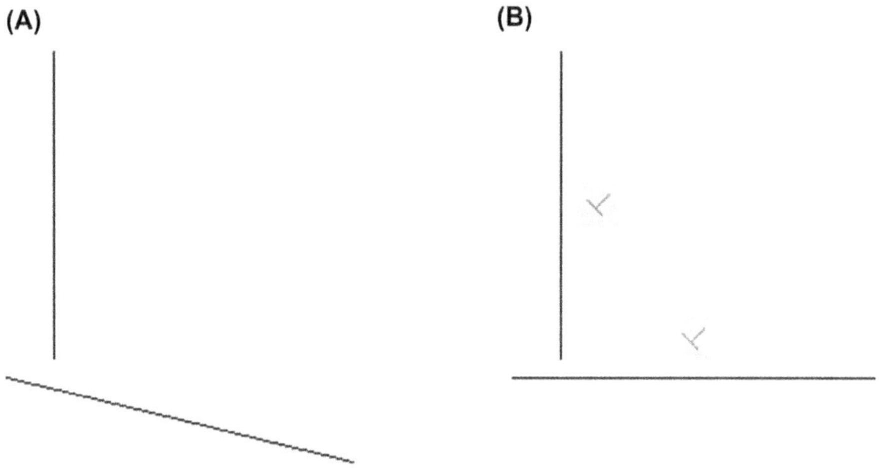

FIGURE 13.12 Perpendicular geometric constraints.

Keyboard: none

Ribbon: **Parametric** tab→**Concentric**

Step 1. Draw a circle, then an arc above it, as seen in Fig. 13.13 (left).

Step 2. Select the concentric geometric constraint via any of the preceding methods.

- AutoCAD says: `_GcConcentric`
  `Select first object:`

Step 3. Select the circle.

- AutoCAD says: `Select second object:`

Step 4. Select the arc.

Immediately after the selection of the arc, you see one or both snap to a concentric setting (their centers are the same), as seen in Fig. 13.13 (right). Note that, if you were to select the arc first, the circle would adjust to it, not the other way around. Once again, selection order matters.

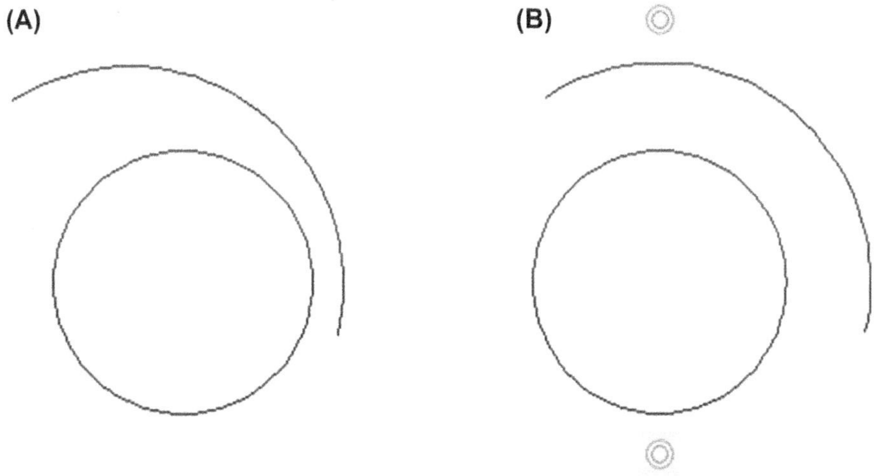

FIGURE 13.13 Geocentric geometric constraints.

Try out the remaining geometric constraints on your own as an in-class exercise.

Hiding, showing, and deleting geometric constraints

There are two ways to access these additional options to work with constraints. You can either right-click on one of the constraints to reveal a small menu, as seen in Fig. 13.14, or find the same options on the Ribbon, as seen in Fig. 13.15, along with the rest of the constraints.

The Constraint Settings dialog box, which can be accessed via the small drop-down arrow at the bottom right of the Geometric tab, is just for global settings, for which geometric constraints to show. Leave all the choices selected and the transparency at 50%, as seen in Fig. 13.16. Finally, Fig. 13.17 shows all seven of the constraints mentioned, as you should see them on your screen after completing the in-class exercise.

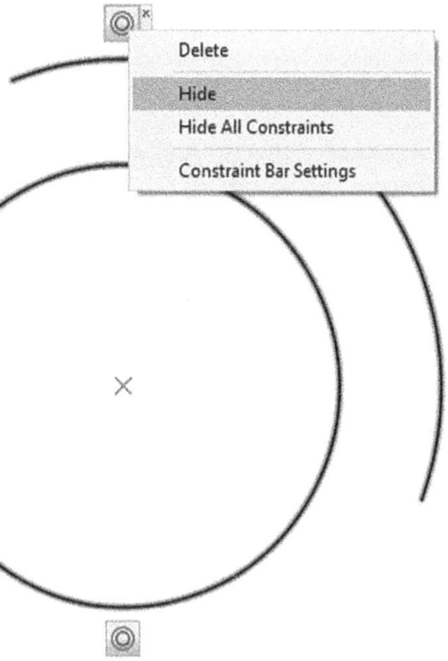

FIGURE 13.14 Geometric constraints options, right-click menu.

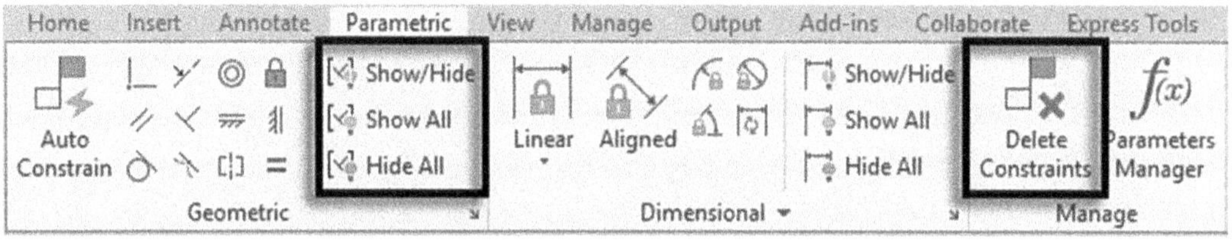

FIGURE 13.15 Geometric constraints options, ribbon.

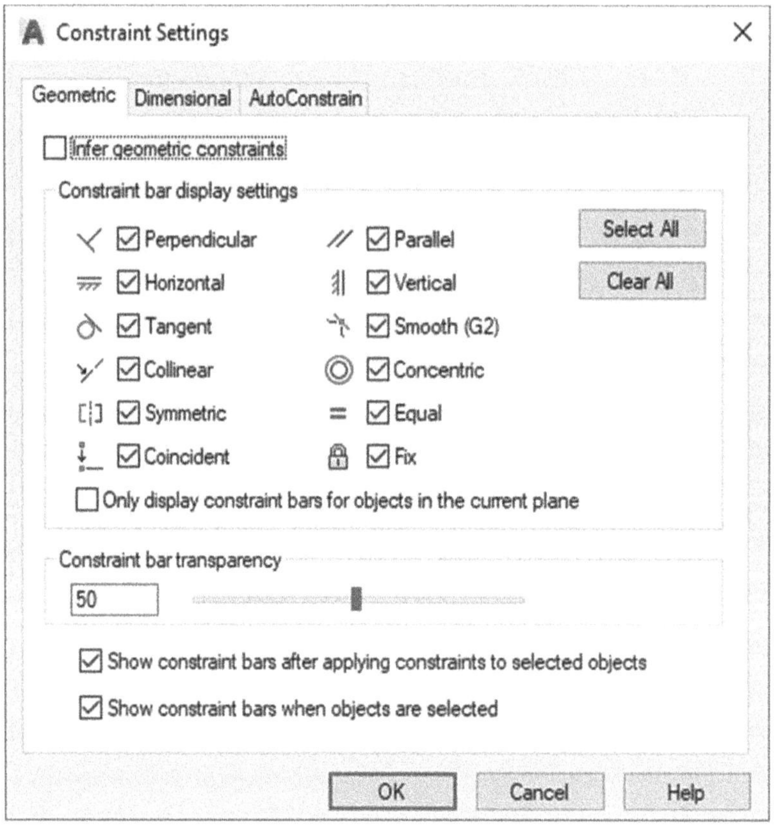

FIGURE 13.16 Constraint settings.

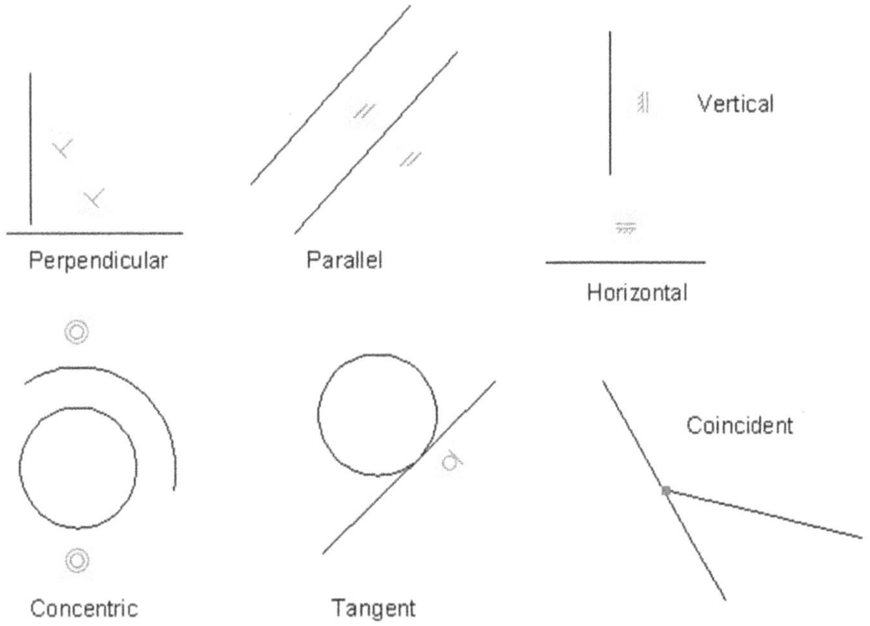

FIGURE 13.17 Geometric constraints.

13.5 Dimensional constraints

Since we already discussed the basic idea of constraints, this introductory section is shorter. The idea here is to move from constraining pieces of geometry (and their positions relative to each other) to constraining the actual dimensions of the drawn design. So, if two circles, representing drilled holes, absolutely need to be 3″ apart, then constraining that dimension ensures it does not change regardless of what other design work goes on around them.

Dimensional constraints are also a major part of 3D parametric design and can be easily transferred to 2D work, as was done here in AutoCAD. Further, because the geometry must obey the constrained dimension, you can change the geometry by simply changing the dimension value, which is quite a departure from the normal order of business with AutoCAD. This ability to change geometry by updating dimensions is called *dimension-driven design*. With high-end 3D solid modeling software, this is really the only way to change the size of an object (usually called a *feature*). AutoCAD lacks this ability in 3D but, as just mentioned, can do this with 2D shapes.

At this point, you may start to see why all the preceding is referred to as *parametric design*. We are designing by changing or assigning parameters to our geometry. The geometry itself is almost secondary to the relationships and data that drive its existence and creation. This type of philosophy and approach is one of the several steps critical to assuring that what we have is a valid engineering design, not just a pretty picture. The data behind the design must make sense.

Working with dimensional constraints

Learning the basics of dimensional constraints is straightforward. You can either add them in right away to your design or convert existing regular dimensions to constraints. Dimensional constraints can be accessed via the Dimensionsla panel of the Parametric tab of the Ribbon (Fig. 13.18),and their settings can be changed via the additional options arrow at the lower right of the Dimensional panel (Fig. 13.19).

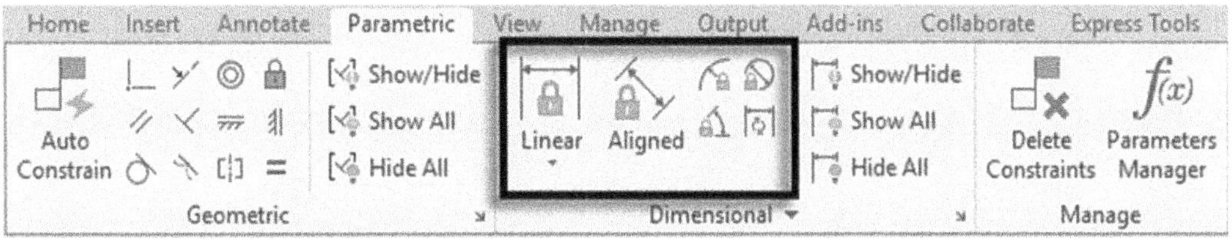

FIGURE 13.18 Dimensional constraints Ribbon panel.

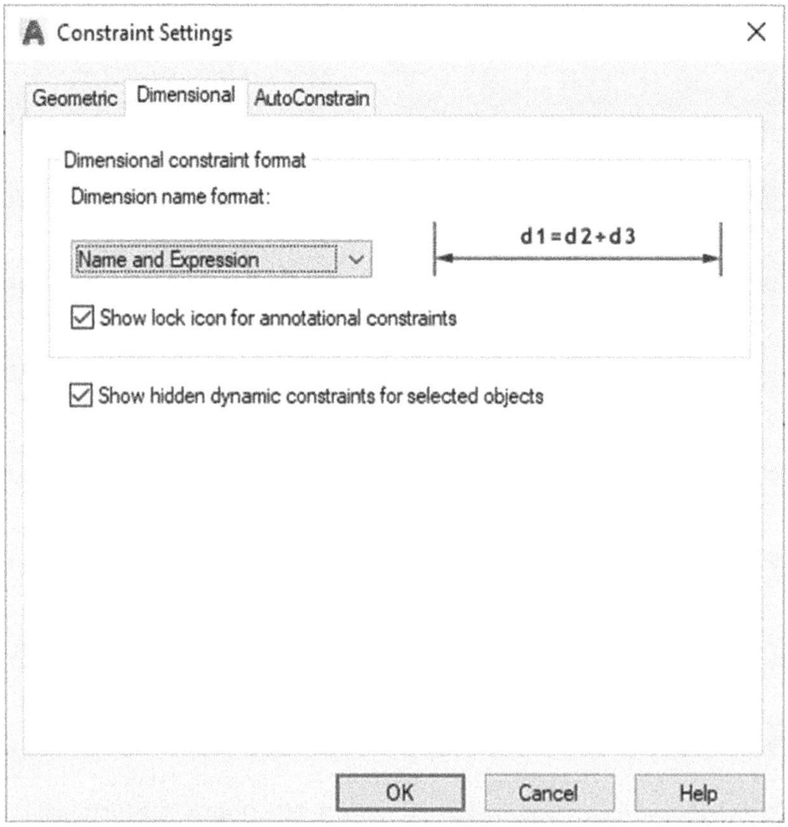

FIGURE 13.19 Dimensional constraints Settings dialog box.

The available dimensional constraints mirror the regular dimensions closely:

- Horizontal
- Vertical
- Aligned
- Radius
- Diameter
- Angle

To practice them, draw the bracket seen in Fig. 13.20, but do not dimension it; use the shown values only to create the geometry.

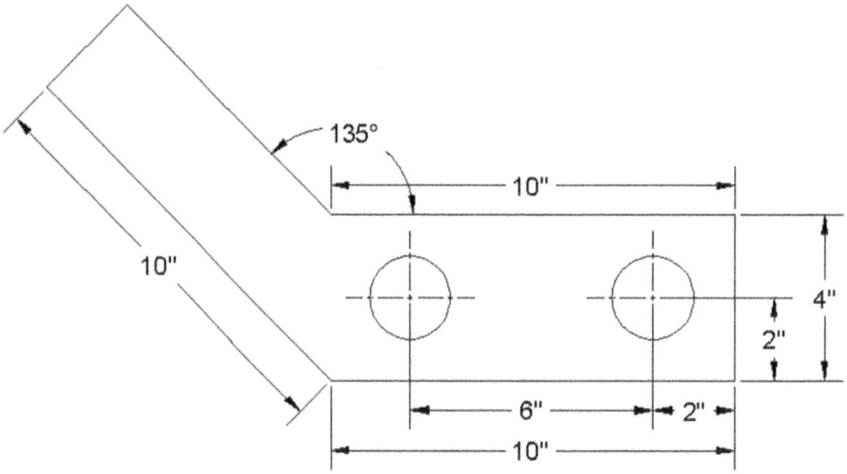

FIGURE 13.20 Bracket.

Keyboard: none
Ribbon: **Parametric** tab→**Linear**

Step 1. Start up the dimensional constraint via any of the preceding methods.
- AutoCAD says: `Specify first constraint point or [Object] <Object>:`

Step 2. Pick one end of the bottom part of the bracket (a red circle with an X appears).
- AutoCAD says: `Specify second constraint point:`

Step 3. Pick the other end of the bottom part of the bracket (a red circle with an X appears again).
- AutoCAD says: `Specify dimension line location:`

Step 4. Locate the dimensional constraint some distance away from the part, like a regular dimension. AutoCAD displays the value (d = 10.0000) with a graphic of a lock, as seen in Fig. 13.21. Press Enter.

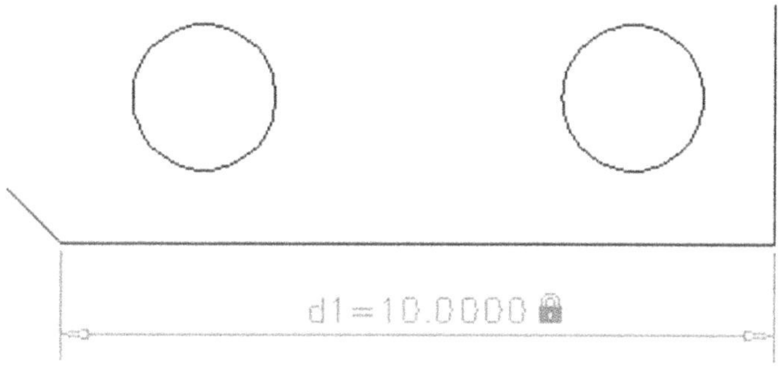

FIGURE 13.21 Dimensional constraint.

In a similar manner, you can add additional dimensional constraints to the rest of the bracket or convert existing dimensions (if you had some) via the Convert button in the Ribbon's Parametric tab. Note the central idea behind this tool. The dimensional values are "locked in," as represented by the lock graphic and do not budge regardless of how other surrounding geometry may change.

13.6 Dimension-driven design

Let us put the preceding information together into a concise approach to design, as well as add in some comments about how to work smoothly with dimensional constraints. First, you can change the value of the dimensional constraint by just double-clicking on it and typing in something else (Fig. 13.22).

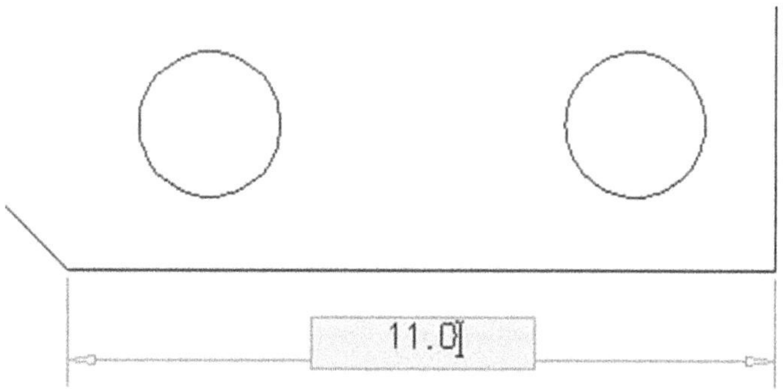

FIGURE 13.22 Altering linear dimensional constraint.

The value forces the line to move to the right, leaving a gap, which is not quite what you intended. Try the same with a circle and diameter constraint (Fig. 13.23).

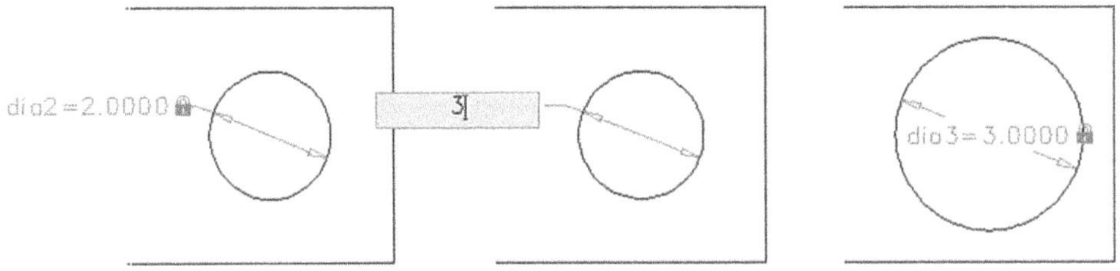

FIGURE 13.23 Altering diameter dimensional constraint.

This leads to much better results; the circle has changed its size. There is a very important point to understand here to take full advantage of these tools. With linework and angles, dimensional constraints need to work together with geometric constraints. Used just by themselves, dimensional constraints change only the dimension value and drag the lines along for a ride; but if you add a coincident geometric constraint, the lines are "glued" together so you can make entire sections of the design properly constrained. A coincident constraint asks you to select one of the lines, then the second, and create a point where they intersect. Try this out on our bracket's bottom pieces, and try changing the bottom line's size again, leading to quite a different effect.

A lot more can be written about constraints in general; it is a broad and important topic. You are encouraged to explore further and practice setting up dimensional constraints on critical parts of a design. It will take some getting used to and applying dimensional constraints intelligently can be awkward at first. The key is to apply these constraints only to the critical geometry and not constrain every single line and circle on the screen. Fortunately, AutoCAD prevents you from doing this and issues a warning that you are over constraining or there are conflicts, like high-end solid modeling applications.

13.7 Level 2 drawing project (3 of 10): architectural floor plan

For Part 3 of the drawing project, you complete the architectural/interior design by adding furniture and appliances to the house.

Step 1. You again need new layers:
- A-Appliances (Magenta).
- A-Furniture (Red).
- A-Cabinets (Yellow).
- A-Cabinets-Overhead (Yellow, Hidden)

Step 2. Dimensions for the furniture and appliances are not specifically given, but draw in reasonable-size designs, as seen in the layout. Feel free to add more than is shown. The full plan is in Fig. 13.24.

Step 3. Add in text describing the furniture and appliances. Put it on the text layer. It should take you about an hour to an hour and a half to add in the new features. Use copy often!

Summary

You should understand and know how to use the following concepts and commands before moving on to Chapter 14*:*

- Creating a new dim style (ddim)
  - Lines tab
  - No permanent settings, but know functions of all
  - Symbols and Arrows tab

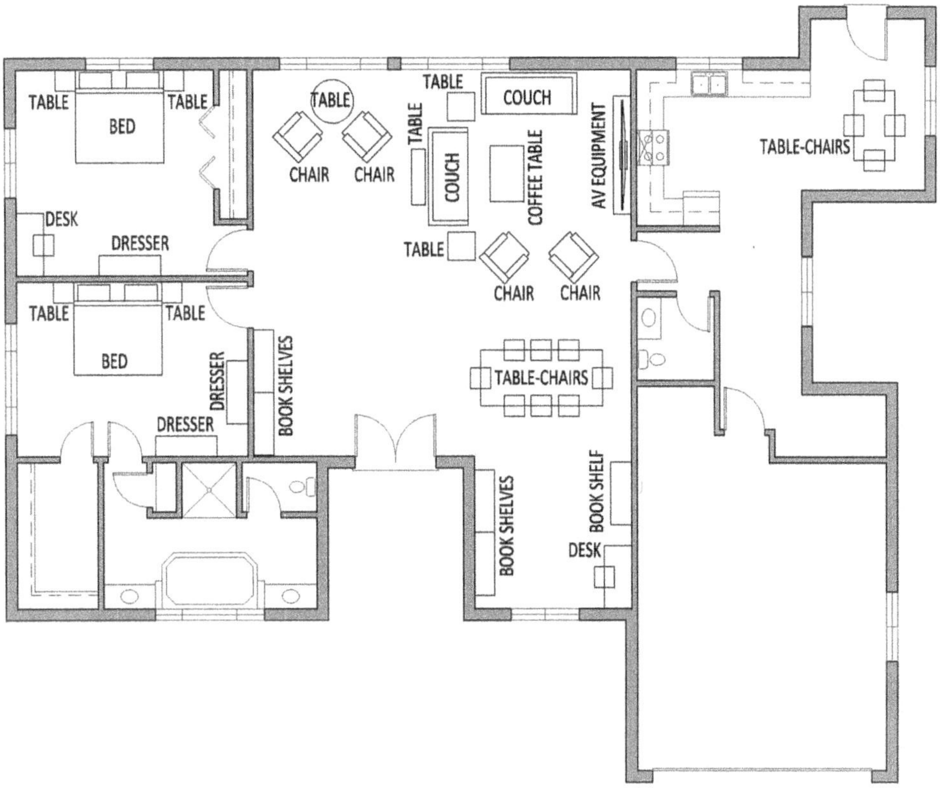

FIGURE 13.24 Furniture and appliances in floor plan.

- Architectural tick setting; know functions of all others
- Text tab
- Text font and size settings; know functions of all others
- Fit tab
- Fit setting; know functions of all others
- Primary Units tab
- Units setting
- Precision setting
- Prefix setting
- Suffix setting
- Zero suppression settings; know functions of all others
- Alternate Units tab
- Multiplier settings; know functions of all others
- Tolerances tab
- Tolerance settings (if needed)

- Geometric constraints
  - Perpendicular
  - Parallel
  - Horizontal
  - Vertical
  - Concentric
  - Tangent
  - Coincident

- Dimensional constraints
  - Horizontal
  - Vertical
  - Aligned
  - Radius
  - Diameter
  - Angle

Review questions

Answer the following based on what you learned in this chapter:

1. Describe the important features found under the Lines tab.
2. Describe the important features found under the Symbols and Arrows tab.
3. Describe the important features found under the Text tab.
4. Describe the important features found under the Fit tab.
5. Describe the important features found under the Primary Units tab.
6. Describe the important features found under the Alternate Units tab.
7. Describe the important features found under the Tolerances tab.
8. Describe the purpose of geometric constraints.
9. What seven geometric constraints are discussed?
10. Describe the purpose of dimensional constraints.
11. What six dimensional constraints are discussed?

Exercises

1. Draw the following shape and label it exactly as shown. Be sure to note all the modifications to the dimensions, including arrowheads and the presence of alternate units and tolerances. You may have to get creative with the leader. (Difficulty level: Intermediate; Time to completion: 10–15 minutes.)

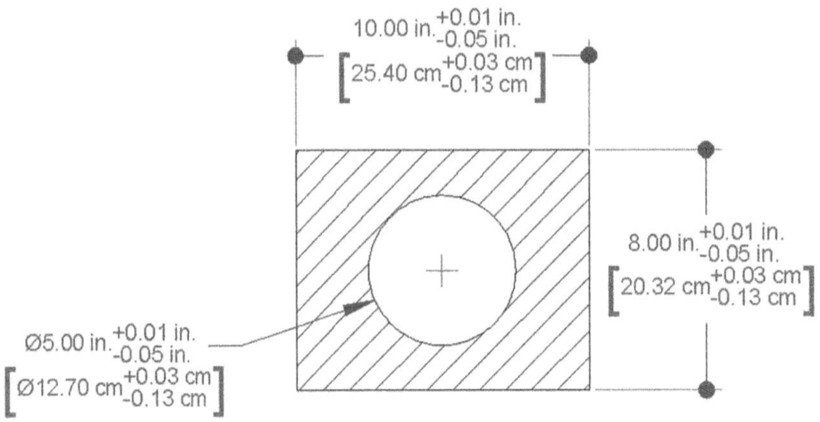

2. Draw a rough parallelogram shape, with one side 20″, as shown on the left. Parallel constrain all four sides, filleting if necessary. Then, carefully, coincident constrain all four corners. Finally, dimension constrain one side, using a value of 30, as seen on the right. If everything is done right, the rectangle should increase in size uniformly. (Difficulty level: Easy; Time to completion: 5 minutes.)

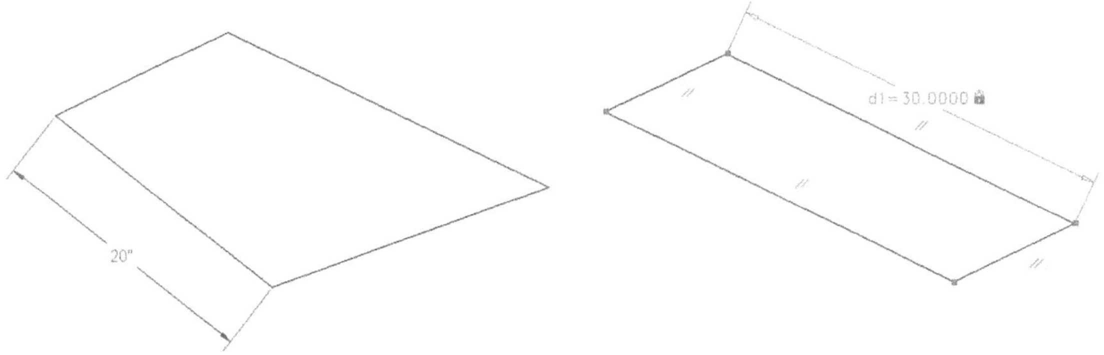

3. Draw the following plate and dimensions, including the tolerances as shown. The tolerance format method used here is Deviation. Carefully place all dimensions in the positions shown for maximum clarity as this is a dimension-dense drawing. (Difficulty level: Easy; Time to completion: 10–15 minutes.)

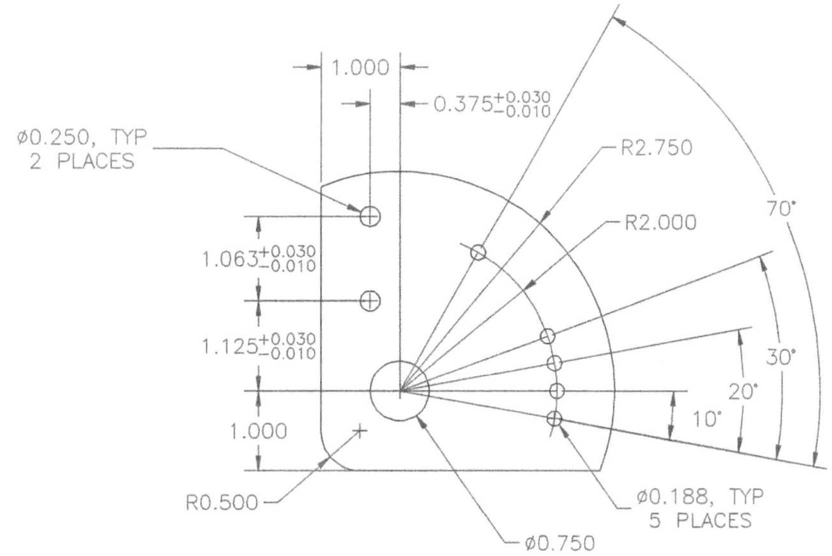

Chapter 14

Options, Shortcuts, CUI, Design Center, and Express Tools

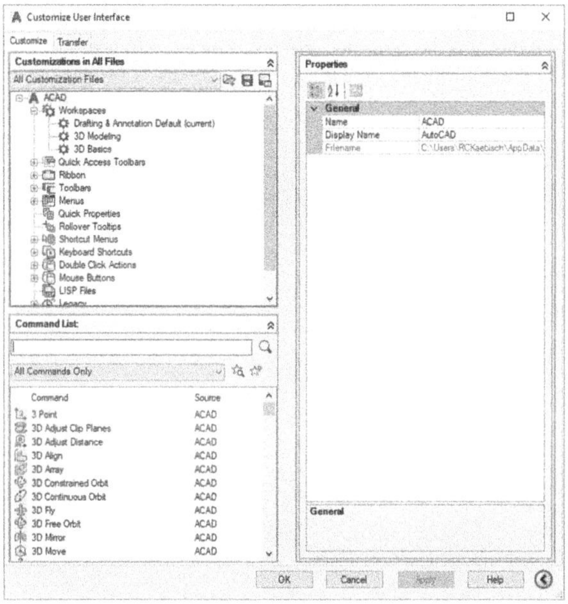

Learning objectives

In this chapter, we introduce a variety of advanced tools and discuss the following:

- Options dialog box
  - Files tab
  - Display tab
  - Open and Save tab
  - Plot and Publish tab
  - System tab
  - User Preferences tab
  - Drafting tab
  - 3D Modeling tab
  - Selection tab
  - Profiles tab
- Shortcuts and the acad.pgp file
- Customize User Interface (CUI)
- Design Center
- Express Tools

By the end of this chapter, you will learn how to significantly customize your environment and speed up your work via importing predrawn blocks and using shortcuts to accelerate command input.

Estimated time for completion of this chapter: 2–3 hours.

Up and Running with AutoCAD 2022. https://doi.org/10.1016/B978-0-323-89923-9.00014-7

14.1 Options

Our topics for this chapter are the Options dialog box, followed by the concept of Shortcuts (the acad.pgp file), and the Customize User Interface (CUI). We then look at the Design Center and the Express Tools. These topics introduce you to basic customization, greatly enhance your efficiency, and will hopefully dispel any remaining frustration with making AutoCAD look and behave how you want it to.

The Options dialog box is your access point for basic and advanced AutoCAD customization menus, and through it, you can change many essential settings and tweak AutoCAD to your personal taste and style. Almost every user finds a few items worthy of changing, and knowing this dialog box is essential. Often, while covering this topic, students remark how they have been annoyed by some feature (or lack of it) and now know how to turn it off or on. This is exactly why we introduce this dialog box or cover it in more detail if you have already seen it.

Note that our goal is not to go over every single item in Options; it is far too big for that, and it is unnecessary anyway. Instead, we outline the most important features and explain why they are necessary. There are instances where you will be advised against changing the default values of one item or another. If you want to know more, press F1 (Help files) while in each of the tabs for a detailed outline of every single feature. Note, however, that you should first have a clear understanding of the selected items this chapter focuses on before going into more detail.

To access the Options dialog box, right-click anywhere in the drawing area, and a menu appears, as shown in Fig. 14.1. Click on the last choice: Options… Alternatively, you can also just type in `options` and press Enter.

A sizable dialog box called *Options* appears. Across the top you will see the following tabs: Files, Display, Open and Save, Plot and Publish, System, User Preferences, Drafting, 3D Modeling, Selection, Profiles, and Online. You will likely see Display as the active default tab. Except for the 3D Modeling tab (which we leave for the online 3D book), we go over each one and introduce all the significant features of each. In some cases, it is a thorough discussion, and in others, just glance at. Go ahead and click on the very first tab (Files) and let us begin.

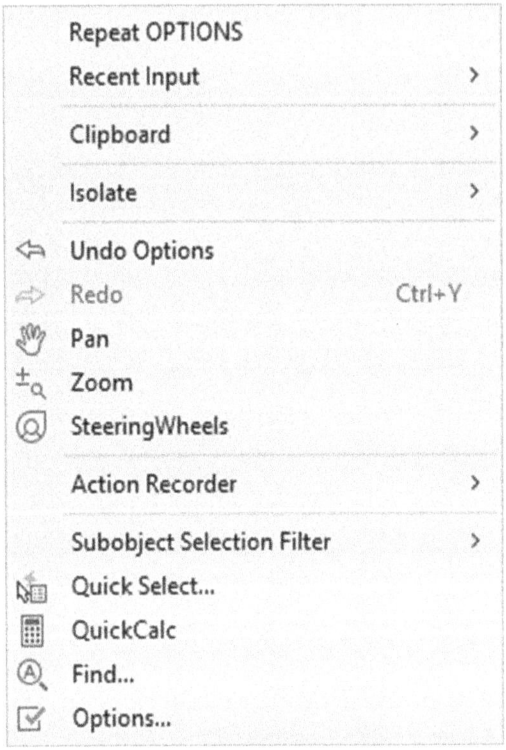

FIGURE 14.1 Accessing Options.

Files tab

This first tab is the easiest to go over for the simple reason that you rarely must change anything here. What are all these listings, then? When AutoCAD, or any software for that matter, installs in your computer, it spreads out all over a certain area of the disk drive (where you indicated it should go on installation) and drops off necessary files as it sees fit.

What you are looking at in Fig. 14.2 is a collection of paths to some of those files as they appear after an installation. They are deemed to be appropriate for a user to access. You can view each path and modify it by clicking on the plus signs. There are quite a few of them, indicating locations of a variety of important AutoCAD functions: Help files, automatic saves, and the like. Some of these you may recognize, some you may not.

Be careful though; default settings are a good thing in this case. If you change the end location of some file, often you (and sometimes AutoCAD itself) may not find it, and difficulties result, such as when you try to move the device driver files. It is strongly advised to leave most of these as they are, as there really is no compelling reason to change the preset locations. There are just a few file paths that *may* need changing, as listed next:

- *Automatic Save File Location*: The default path is C:\users\owner\appdata\local\temp (or something similar). You may want to change that to an easier-to-find locale if you plan on using Autosave.
- *Log File Location*: The default path for these saved text window data logs reside at C:\users\owner\local\autodesk directory (or somewhere near there). If you plan to log certain activity on your project, you may want to relocate these.
- *Action Recorder Settings→Action Recording File Location*: This path should be set to your file's project folder if you plan on using the Action Recorder (a Chapter 15 topic).
- *Plot and Publish Log File Location*: These data logs are for plots, and they also reside in the C:\users\owner\local\autodesk neighborhood. Reset them to some easier-to-find locale if you plan on keeping logs of print jobs.

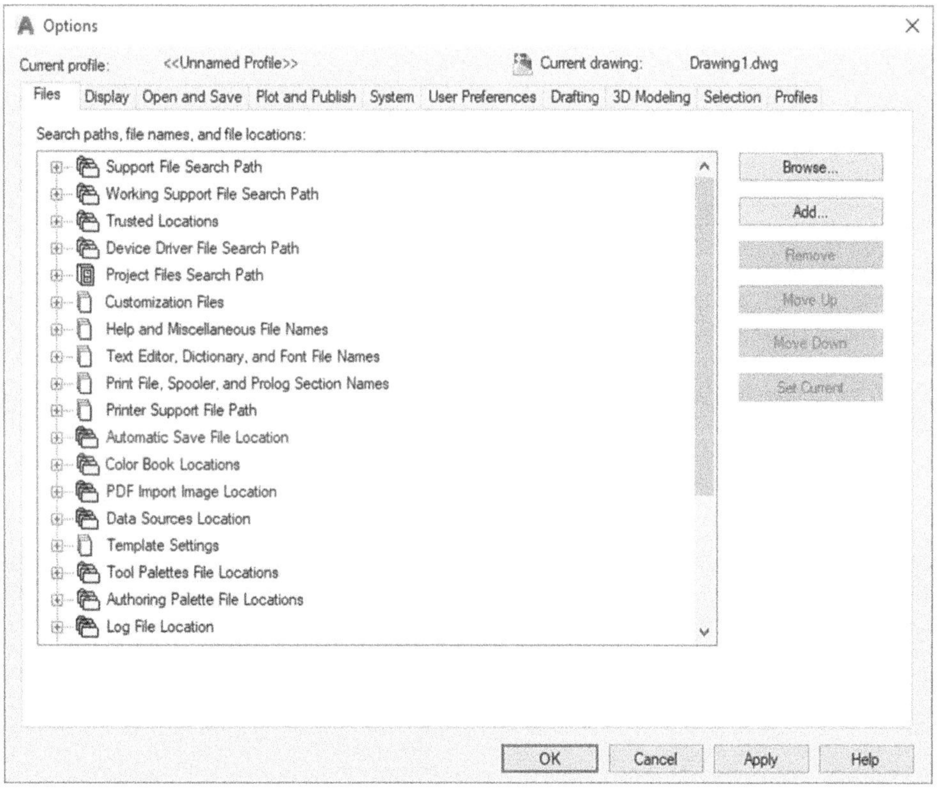

FIGURE 14.2 Files tab.

Other than these four items, the new user will have little to change among these paths, and I can count on just one hand the amount of times I needed to do this in all my years of drafting. These paths would probably change if you are at an office on a networked computer system. If you do wish to learn, via a detailed description, of what each path means, press F1 while you have the Files tab open.

Display tab

From here on forth, we group and discuss everything by category, like how AutoCAD groups everything within these tabs—by a named topic, with frame drawn around it. We list them from top to bottom and left to right. The Display tab (Fig. 14.3) has Window Elements, Layout elements, Display resolution, Display performance, Crosshair size, and Fade control. Among them are quite a few items of interest.

Window Elements

The first thing you will notice is the Light/Dark theme selector. By default, this AutoCAD just like its predecessor, has the Dark theme set. It was changed to Light way back in Chapter 1 for better visibility on a printed page and to save ink. You can certainly set it to Dark if you wish to achieve that somewhat menacing, cutting-edge look. Note that this has nothing to do with the screen's background color (which we discuss shortly), but merely the overall theme of the surrounding Ribbon, border, menus, etc.

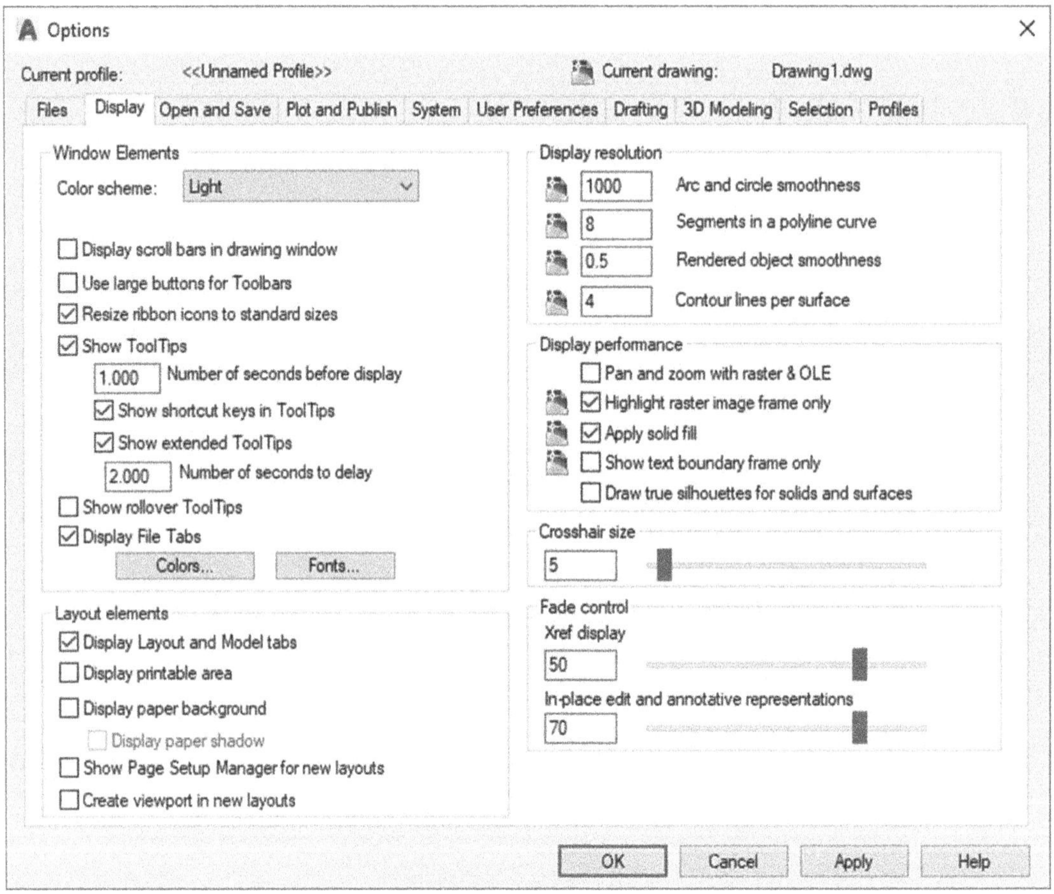

FIGURE 14.3 Display tab.

Working your way downward, you do not really need the first three boxes checked unless you really like the scroll bars (a holdover from AutoCAD's early days) or need to see larger toolbar buttons if you are using toolbars at all; however, those *are* very useful for the visually impaired. Below that is the Ribbon icon resizing, just a minor adjustment, which you can also leave unchecked. The next five boxes feature various ToolTip-related settings, and these are good to leave checked to explain the meaning of items and tools you are exploring. Included in there is the delay time, with 1 or 2 seconds adequate. Leave Display File Tabs checked as well.

The Colors… button just below ToolTips refers to the color of your screen background. The default dark gray background color has an RGB value of 33,40,48. Generally users change this color to black and is preferred by most (but not all) users. Black is a good choice for the simple reason that the dark background throws less light at your eyes from the screen and lessens fatigue (some of this was discussed as far back as Chapter 1). The only catch is that printed paper output is always white, and some people prefer seeing the design as it would appear in real life, on white paper, hence the white background.

Obviously, it is easy to get used to the black background and pretty much disregard the "paper" aspects of the model space, and this is what most users do. In this and most textbooks, however, the screen background color is always white to save ink and blend images smoothly with the white paper.

Other colors are available, of course (all 16 million colors are available as a background), but as we joke in class, if you see someone using a bright neon green background, stay away—this is probably not a sane individual! To change the color scheme for practice, press the Colors… button, and the dialog box shown in Fig. 14.4 appears.

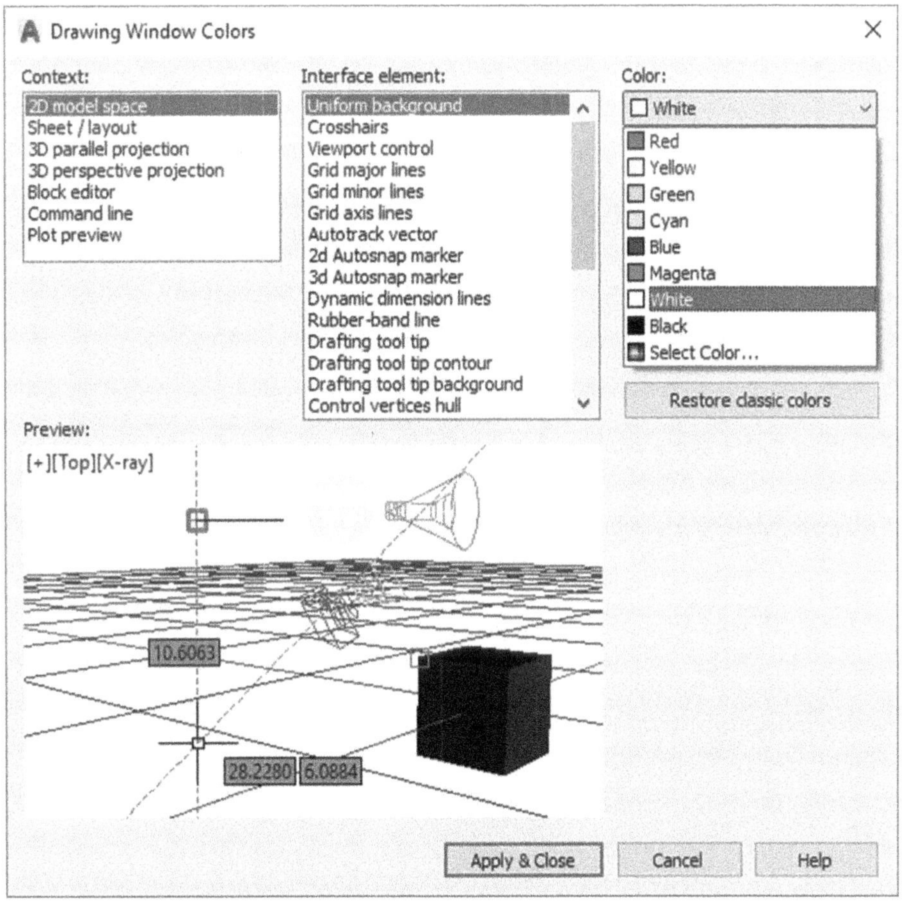

FIGURE 14.4 Drawing Window Colors with White selected.

Here, you can select from the general Context: menu on the left (only the first choice, 2D model space, for now), and select some element in the Interface element: menu on the right (e.g., the Uniform background). Then, pick the color you want this element to have. Fig. 14.4 shows white being chosen. Note the sizable number of choices in both the left and middle columns. AutoCAD allows for the color of just about anything to be changed; browse through all the choices to get familiar. If you change so much that you forget what you did, you can always press the Restore classic colors button on the right of the dialog box. When done, press Apply & Close.

The Fonts... button, to the right of Colors... in the Display tab, refers to the font present on the command line text, not the font in the actual drawing, so unless you have an objection to the Courier New font (the default), do not change anything.

Layout elements

In this category at the bottom left, it is recommended to leave everything its default configuration with all boxes checked except Show Page Setup Manager for new layouts. Generally, all those options (area, background, etc.) make a new user more comfortable in Paper Space, as was discussed in more detail in Chapter 10. As with many such suggestions, it does come down to personal taste. Check out Paper Space with and without these buttons checked to determine how you prefer it.

Display resolution

Moving to the upper right of the Display tab, we have the Display resolution category. Here, it is suggested to leave all but one field as default. That field is the Arc and circle smoothness value, which is 1000 by default. That was set low for historical reasons that have to do with accelerating regenerations on slower, old computers by having arcs and circles be made up of small lines. You can set that to the 20,000 maximum, which is fine for today's fast machines.

Display performance

This category is also a bit of a throwback from the old days of slow computers. These five checked boxes combine to accelerate, or slow down, regeneration of various elements. It is best to leave everything as default, with box 2 and 3 checked and the others unchecked.

Crosshair size

Moving down, we have the Crosshair size setting. Crosshair size is generally a personal preference. Many leave it at a "flyspeck" setting, which is 5, but increasing the size to a full 100 is how AutoCAD used to be set by default. The full screen crosshairs make seeing if you are on the paper or in floating model space easier when you are in a Paper Space viewport, as detailed in Chapter 10, or checking if a drawn line is straight by comparing it to the straight-edge crosshairs. In the end, it is up to you, and experience has shown that newer users prefer the flyspeck, while the veterans stick to the full size.

Fade control

The last category at the bottom right has to do with how dark the xref is when you are editing it in place, a Chapter 17, External References (Xrefs) topic. For now, the default settings are fine.

If you want a more detailed description of what each selection means, press F1 to read the Help files while you have the Display tab open. Alternatively, hover the mouse over each element for a brief description. If you are done, do a final check to see that your settings reflect what is in Fig. 14.3, and let us move on to the next tab.

Open and Save tab

The Open and Save tab (Fig. 14.5) contains the File Save, File Safety Precautions, File Open, Application menu, External References (Xrefs), and the ObjectARX Applications categories. Several important features are worth discussing here, most of them contained inside the two categories on the left-hand side: File Save and File Safety Precautions.

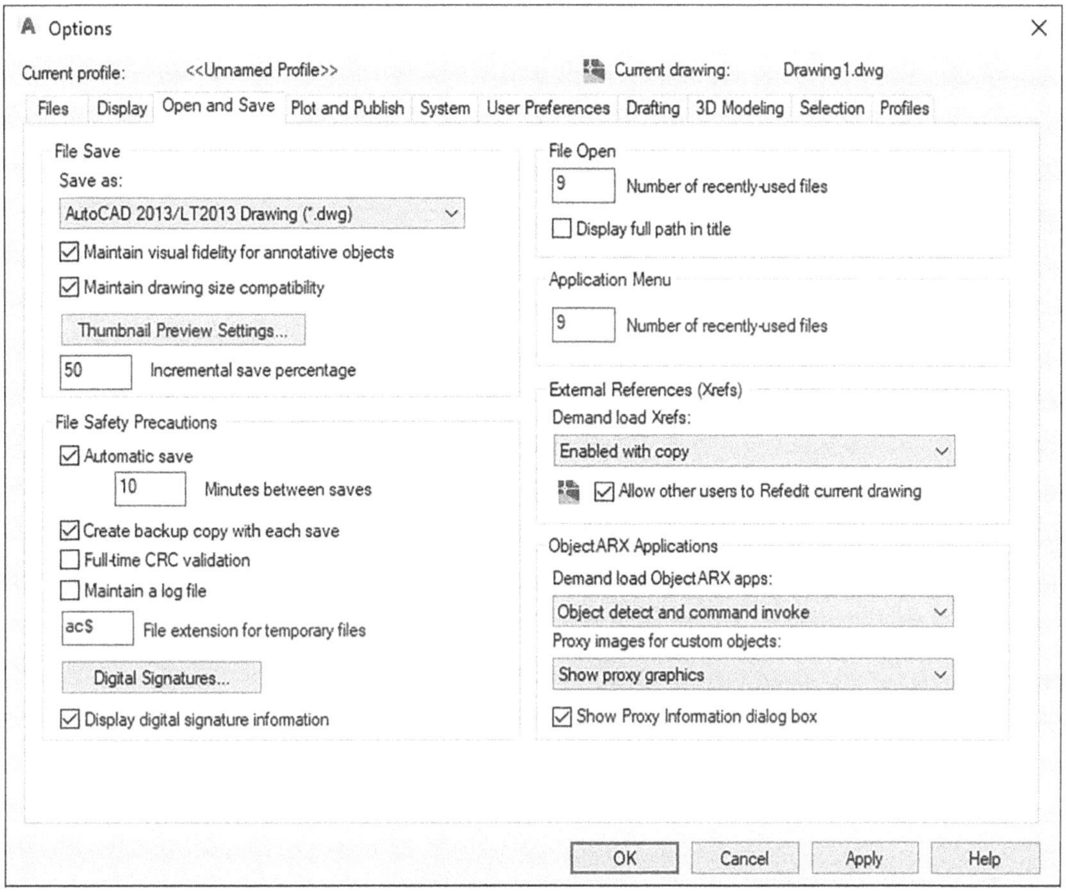

FIGURE 14.5 Open and Save.

File Save

In this category, let us look at the top option, labeled *Save as:* This drop-down menu is very important, as it allows you to set the default to permanently save your files down to an older version of AutoCAD. The significance of this cannot be overstated because you simply cannot open a file created on a more recent version using an older release. AutoCAD is an often updated software, with new releases every year (since 2004), and you can rest assured that not everyone is running the same version. While that may have been an isolated case, AutoCAD, 2015, 2016, and 2017 are still in widespread use, and nothing is more frustrating to someone than getting a copy of your files for collaboration and not being able to open them.

Autodesk of course realizes this and allows you to save files to an older version on a case-by-case basis during the save procedure. AutoCAD even by default saves to the last DWG file format change, which was in 2018. To be safe, if it is not already set this way, select the AutoCAD 2013/LT 2013 Drawing (*.dwg) choice, as seen in Fig. 14.5; that way you "undercut" about five releases and assure that the file can be accessed by just about everyone using version 2013 and higher including the latest release.

If you need to, you can save individual files down even further for collaborators still using versions older than 2013 (all the way back to R14/LT98/LT97 if needed). We can skip over the rest of the options under the File Save category (leave everything as default) and move down to the File Safety Precautions category.

File Safety Precautions

Let us work our way top to bottom in this category. The Automatic save check box is the first thing we encounter, and it is about as close to controversial as something in AutoCAD gets. The suggestion is this: Do not use it. Others may say, why not, it is there. The argument against it is simple: As a computer user, you need to save your work often, no matter what application you use. To rely on automatic saves is to invite trouble when you use a program that does not have an automatic save (or it is not turned on). As an AutoCAD user, **you should be saving your work every 5 to 10 minutes,** something students often do not do in class.

There are also some practical difficulties with Automatic save. For one thing, AutoCAD needs to exit a command to save a drawing. AutoSave does this for you, so even if you are in the middle of something, you are kicked out for AutoSave to run (except while in block editing mode). Other issues are where AutoSave files go and what extension they have. You can set where they go using one of the paths under the File tab, but then all AutoSave files go there, regardless of what actual drawing is open, and overwrite existing files. The extension of AutoSave files is .sv$.

In summary, while AutoSave can be set up, it is not advisable to rely on it; often, you see veteran programmers and computer users ignore this feature in all computer use and save instinctively every few minutes. If you do wish to use AutoSave, simply turn it on by checking the box and setting a time threshold (minutes between saves) just below that.

The Create backup file with each save check box can also be mildly controversial, as it seems like a good idea initially, yet does not stand up under scrutiny. Backup files (if the option is checked) are created every time you save a file. A backup file (.bak) is essentially a regular drawing file (.dwg) that is cloaked with an extension Windows does not recognize, but you do (see Appendix C, File Extensions for more info on all extensions). If the .dwg file is lost, then the .bak file can be renamed with a .dwg extension and, voila, the drawing is there again. So, what is wrong with this? Well, in theory nothing, but in real-life use there are a few issues.

The main problem is that backup files double the size of your project folders; it is like having a copy of every drawing file next to the original. They are just not necessary. The combination of cheap storage media and a heightened sense of risk in the last decade and a half have prompted just about every company to do daily and weekly backups. Also, AutoCAD's drawings rarely "go bad" to the point of being unrecoverable by the Audit and Recover commands. Finally, backup files reside right next to the drawing files themselves, and if you lose the entire folder, they all go. Rarely do you lose just the drawing; even then, going to the previous night's backups is enough to get them back. So, while some would argue that no harm is done by keeping the .bak files, one could also say, why add the extra junk to your company's server? Ultimately, the choice is up to you, although sometimes company policy may dictate the use of backup files, and this is beyond your control.

Moving further down, most users do not need Full-time CRC validation. CRC stands for Cyclic Redundancy Check and is an error checking mechanism for when your drawings seem to crash after you bring something in. This does not happen often, so you generally do not need this feature on.

Maintain a log file makes AutoCAD write the contents of the text window to a log file, also an unnecessary feature for most users. File extensions for temp files can also be left as the default ac$.

Digital Signatures… is a relatively new feature in AutoCAD that addresses the need to secure or authenticate the drawings. This is not often used, as most AutoCAD drawings are not secret in nature, but the idea is intriguing and worth a look. When you click on the button, the box in Fig. 14.6 appears.

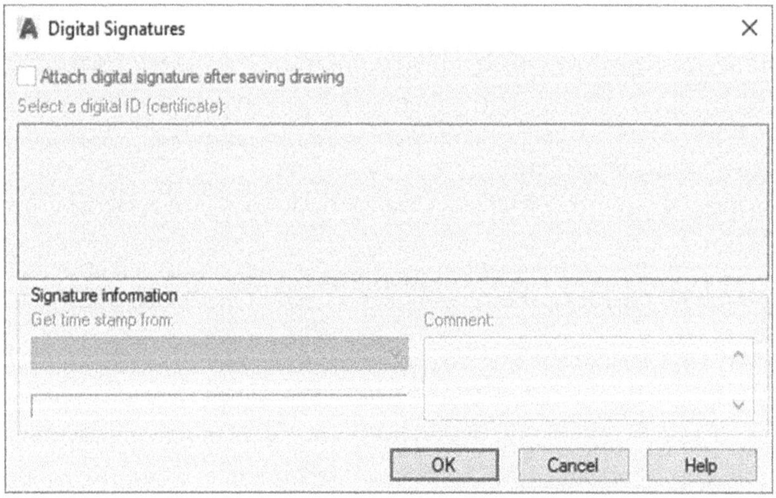

FIGURE 14.6 Security Options.

The first tab allows you to add a password to the drawing. Then, the drawing can be opened only by someone who knows the password; although be careful—if the password is forgotten, the drawing is lost. The second tab has to do with a digital signature. This is a lengthy topic and is sometimes used when collaborating with others. In short, a user installs a certificate from a digital security vendor (VeriSign in this case), which allows for a "digital signature," which in turn prevents changes to the drawing or keeps track of them. This is a very simplified version of the concept, of course, but if your company already uses this, then no further explanation is necessary; if not, you will likely not care about this feature anyway.

File Open

Moving to the top right of the Open and Save tab we have File Open category, which simply shows several recently opened files. Leave as the default max, with Display full path in file unchecked as that may take up a lot of room across.

Applications Menu

This is the same idea as File Open but refers to the application menu's recent documents. Leave it as default.

External References (Xrefs)

These fine points of external reference loading and usage are best left as default.

ObjectARX applications

ObjectARX is a relatively complex topic that requires some programming background to fully understand. It stands for AutoCAD Runtime Extension and is an API for customizing and extending AutoCAD. An API in turn stands for Application Programming Interface, which specifies how software components should interact with each other. ObjectARX Software Development Kits are available to third-party programmers and software engineers to develop plug-in applications for each release of AutoCAD. See Appendix F, AutoLISP Basics and Advanced Customization Tools, for more information on this. As far as this category is concerned, leave everything as default and allow AutoCAD to show and load the associated apps and proxy graphics, which are visual objects that are not "proxied" and have no graphical equivalent.

If you want a more detailed description of what each selection means, press F1 to read the Help files while you have the Open and Save tab open. Alternatively, hover the mouse over each element for a brief description. If you are done, do a final check to see that your settings reflect what is in Fig. 14.5, and let us move on to the next tab.

Plot and Publish tab

The Plot and Publish tab (Fig. 14.7) contains the Default plot settings for new drawings, Plot to file, Background processing options, Plot and publish log file, Auto publish, General plot options, Specify plot offset relative to categories, and two separate buttons at the bottom right: Plot Stamp Settings… and Plot Style Table Settings. Except for a few items, not a tremendous amount is important under this tab for the average user. We run through everything with some general comments and descriptions.

Default plot settings for new drawings

This category selects the default plotter or adds a new one. However, this should not be done from here but rather from the Windows control panel. When new plotters or printers are added, AutoCAD notices them, as well as what the chosen default is. In the end, however, this is of little significance. Recall from Chapter 9 that the printer settings for each drawing are set individually. Therefore, leave everything here as default.

Plot to File

In this category, a drawing can be plotted to a .plt (plot) file and saved in this location. While this is generally an outdated legacy command from the old days of .plt files being sent (or physically carried on a floppy disk) to plotters, it has found some use today when taking drawings to a printing company. The company may not have AutoCAD, or anyone qualified to operate it, so instead they accept the .plt files, which are nothing more than ready-to-print code that can be interpreted by any plotter.

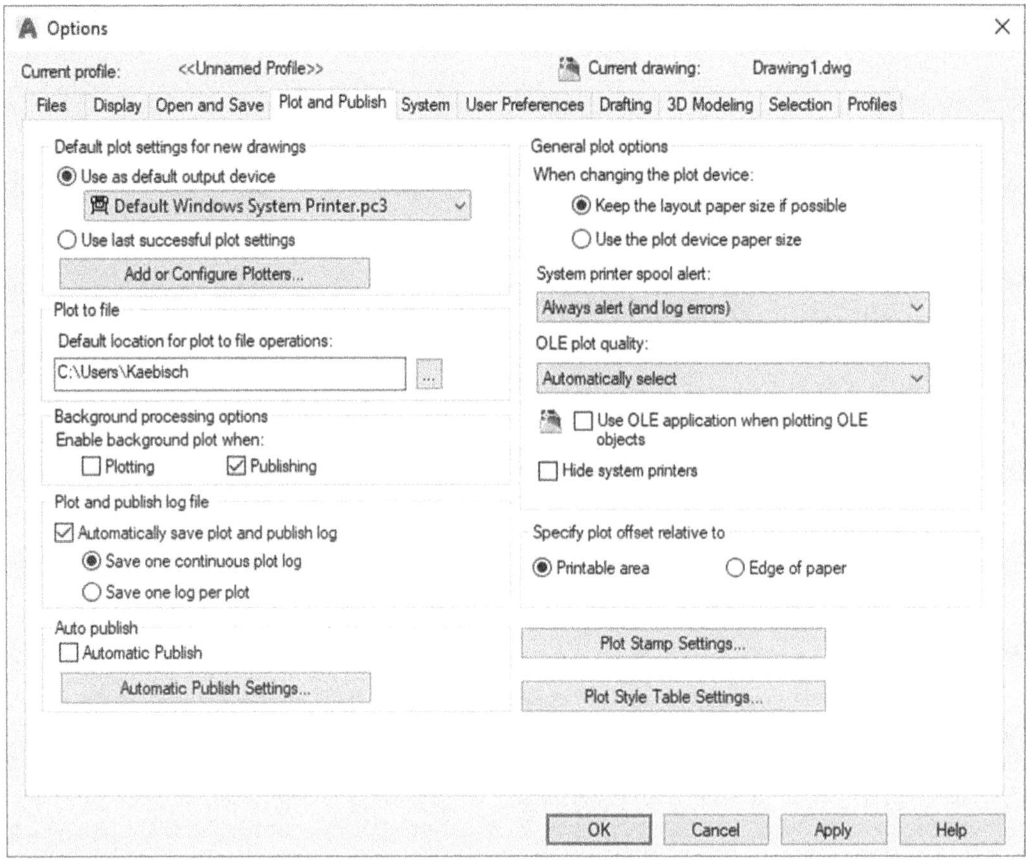

FIGURE 14.7 Plot and Publish.

Background processing options

This category has to do with being able to plot in the background while working on a drawing. With the Plotting option unchecked, you must wait until the plotting is complete before continuing. Publishing may take longer, so you want to allow this by checking it.

Plot and Publish Log File

This category is an option that creates (or turns off) the log feature, which is nothing more than a record of the Job name, Date and time started and completed, Full file path, Selected layout name, Page setup name, Device name, Paper size name, and a few others. It is recommended to disable (uncheck) this feature unless there is a specific reason to create a record of all this information.

Auto Publish

This category is an option that allows you to automatically create design web format (DWF) files. DWF is a topic all unto itself. The basic idea is to generate vector files that can be viewed by others who have the Autodesk Viewer or, if not, then through a browser. You need this box checked only if you are planning on generating these files to give to someone or to put your drawings up on the web.

General Plot Options

Moving to the upper right of the Plot and Publish tab, we have the General plot options category. This has to do with system printers and the output quality of object linked embedded (OLE) items, a topic to be discussed in Chapter 16. The OLE plot quality drop-down choices can be adjusted depending on what is embedded. In this case, leave everything in this category as default, as these are the most reasonable choice for the average user.

Specify Plot Offset Relative To

Leave this category as default, as you generally want to have the printable area as the start point of any plot offsets, not the edge of the paper, to ensure extra room.

Plot Stamp Settings...

This button, at the bottom right, is an item of interest. A plot stamp is simply a text string that appears on all output when activated in the Plot dialog box. This text string has information on a variety of parameters, but the ones of most interest include the drawing name and the date and time. This can be useful for internal check plots (obviously not for final output), to show where the drawing is on the company server or to indicate which paper output is the latest based on the date and time. Bring up the Plot Stamp (Fig. 14.8) by clicking the button and use the Advanced button in the lower left to further refine the plot stamp if need be; these settings mostly have to do with the size and position of the stamp.

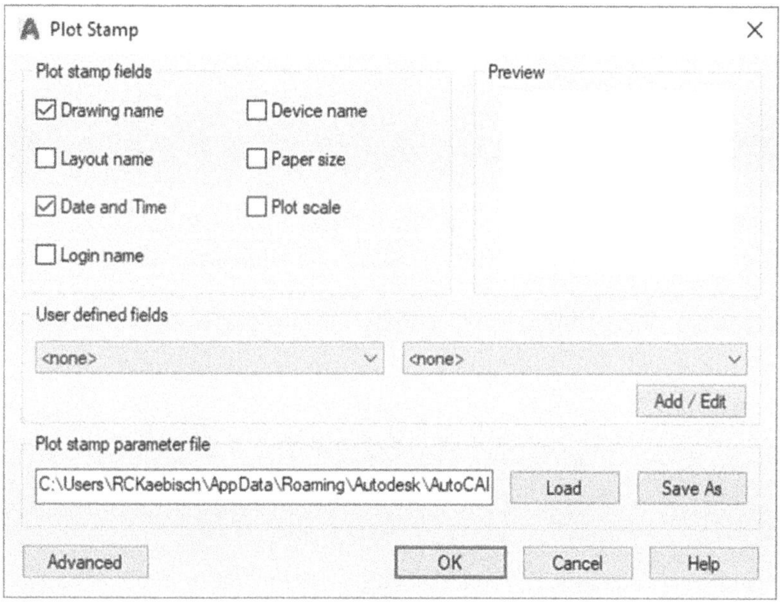

FIGURE 14.8 Plot Stamp.

Plot Style Table Settings

Leave everything under this button as default. These settings are addressed elsewhere, such as the plot dialog box and the plot style manager. See Chapter 19 for more on this.

If you want a more detailed description of what each selection means, press F1 to read the Help files while you have the Plot and Publish tab open. Alternatively, hover the mouse over each element for a brief description. If you are done, do a final check to see that your settings reflect what is in Fig. 14.7, and let us move on to the next tab.

System tab

The System tab (Fig. 14.9) contains the Hardware Acceleration, Current Pointing Device, Touch Experience, Layout Regen Options, General Options, Help, InfoCenter, Security, and the dbConnect Options categories. Much like the previous tab, not a tremendous amount is important under this tab for the average user except for a few items. We run through everything with some general comments and descriptions.

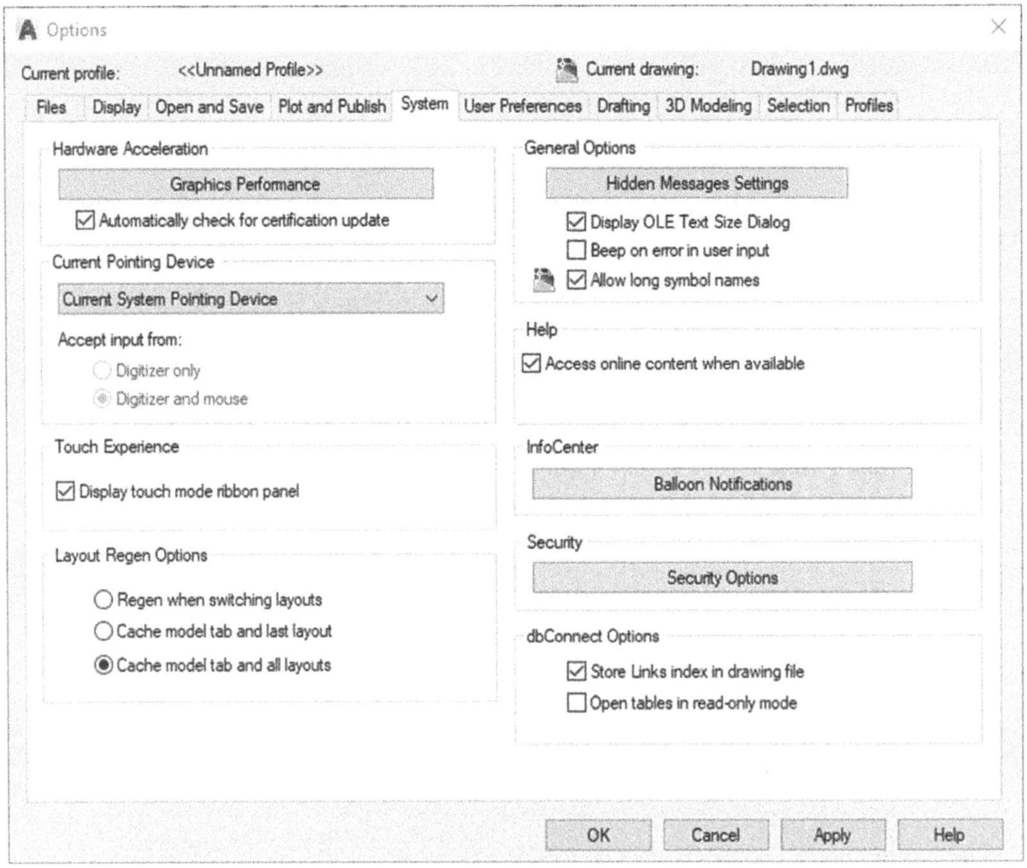

FIGURE 14.9 System tab.

Hardware Acceleration

This category, and the associated Graphics Performance button, is tailored toward 3D performance tweaks, and the category was even called 3D Performance in a previous version of AutoCAD. It details various graphics settings and system descriptions. These settings are already optimized automatically and should be just fine for the average user. Feel free to run through the Tuner Log, which can be seen after a few clicks—lot of interesting information is in there for the tech savvy.

Current Pointing Device

This category allows for setting up digitizer operations, a topic not relevant to most users, who have no digitizer.

Touch Experience

Displays or hides the Touch Mode Ribbon Panel. The on/off switch cancels or enables touch pad operations such as zooming and panning on a laptop touch pad. (HINT: Always use a mouse unless a touchpad is your only option during the zombie apocalypse.)

Layout Regen Options

This category should be left as default; it refers to regeneration of layouts when you switch to them. With today's fast machines an extra regen here and there does not affect productivity.

General Options

Leave everything in this category as default. Here resides perhaps one of the most annoying features of AutoCAD: The "Beep on error in user input" option. This may push a student just starting out and making lots of errors over the edge. You do not want that box checked!

Help

Unless your computer is not connected to the Internet, keep this box checked to access online content.

InfoCenter

The InfoCenter and its Balloon Notifications button refer to the various information balloons that appear with helpful updates or tidbits of information. Although an advanced user may find them annoying, as a student you should keep them around. Briefly explore the button to see what options it allows you regarding these notifications.

Security

The category refers to dealing with *.exe (executable) files and *.lsp (AutoLISP) files. Leave everything as default.

dbConnect options

This category should also be left as default. The *db* refers to database connectivity, which is also not relevant to most users but may be important in some high-end custom applications, where AutoCAD is linked to a database.

If you want a more detailed description of what each selection means, press F1 to read the Help files while you have the System tab open. Alternatively, hover the mouse over each element for a brief description. If you are done, do a final check to see that your settings reflect what is in Fig. 14.9 and let us move on to the next tab.

User Preferences tab

The User Preferences tab (Fig. 14.10) contains the Windows Standard Behavior, Insertion scale, Hyperlink, Fields, Priority for Coordinate Data Entry, Associative Dimensioning, Undo/Redo categories and the Block Editor Settings…, Lineweight Settings…, and Default Scale List… buttons. Some significant items lie under the first category. We describe it in detail and run through everything else with some general comments and descriptions.

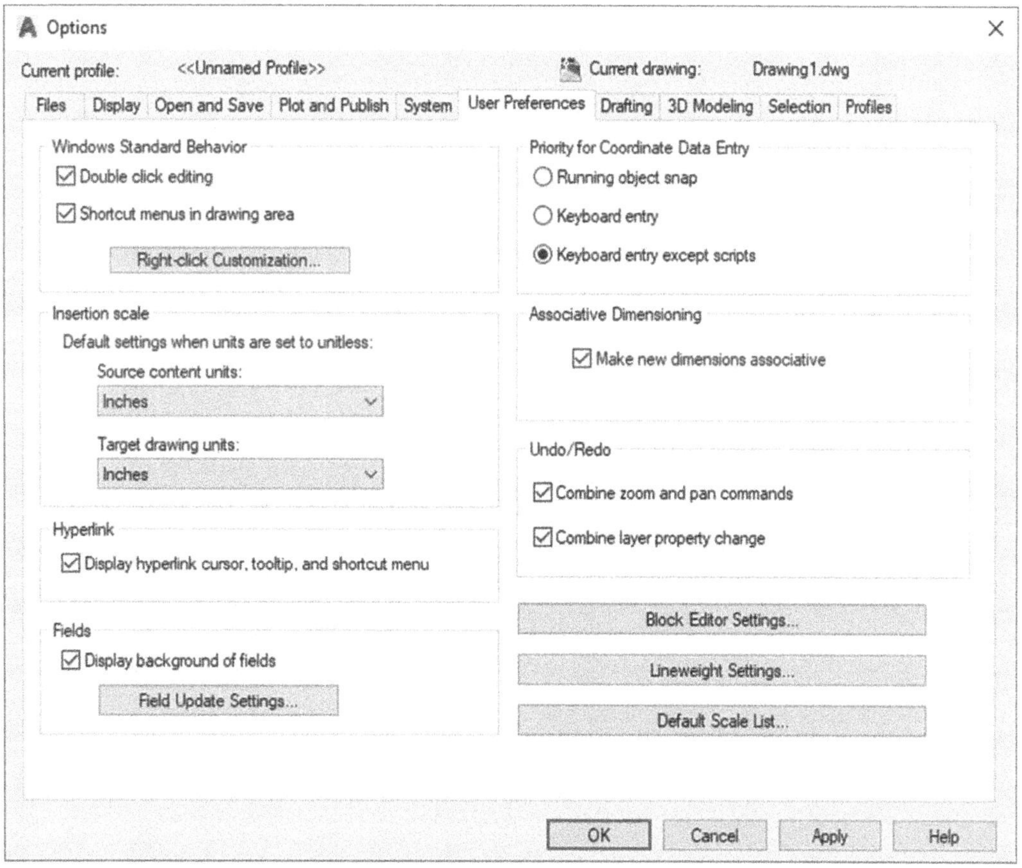

FIGURE 14.10 User Preferences tab.

Windows Standard Behavior

This category concerns speed, which is a topic near and dear to the hearts of most AutoCAD users. Since time equals money, whatever technique accelerates drafting speed is always welcome. One of the more important ones (shortcuts) is discussed later in this chapter. Another important "trick" is to set your mouse buttons to the most likely and needed settings for fast drafting. As such, the Right-click Customization… button is very important. Keeping the boxes above that button checked as a default, click on it, and the dialog box in Fig. 14.11 appears.

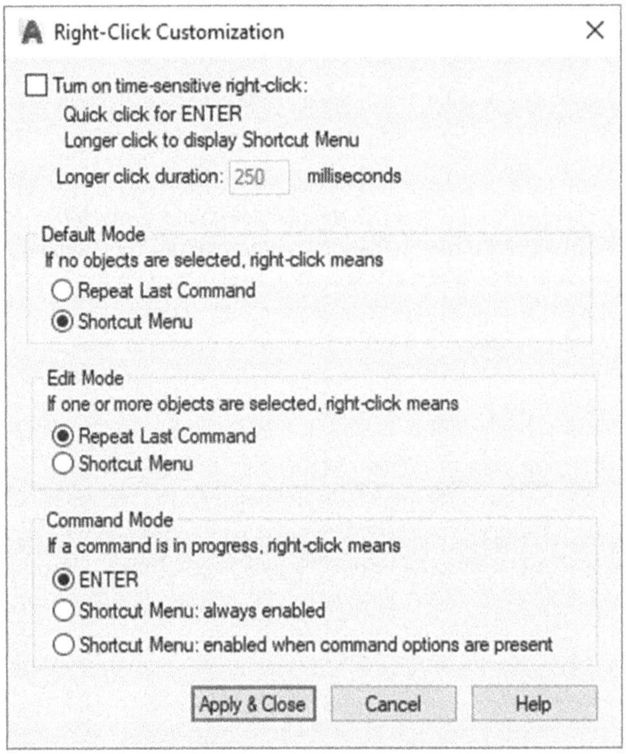

FIGURE 14.11 Right-Click Customization.

Your mouse settings should be as shown in Fig. 14.11, with the last one, Command Mode, set to ENTER. This subtle but powerful setting speeds up your basic drafting by allowing you to instantly finish commands by using the right mouse button instead of having to press Enter on the keyboard or select it from the right-click menu. Also set Edit Mode to Repeat Last Command and Default Mode to Shortcut Menu. The time sensitivity settings at the very top refer to the mouse button "reaction times." The default 250 ms value happens to be a good time for most people, so nothing else needs to be adjusted here. Press Apply and Close.

Insertion scale

This category controls the scale for inserting blocks; leave the default values.

Hyperlink

This category controls visual settings of hyperlinks; leave this box checked as the default is fine. We cover hyperlinks in the next chapter.

Fields

This category sets field preferences, which you should leave as default. A field is text in AutoCAD drawings that automatically updates itself (after a regen command) whenever there is a change in the values to which the field is connected. I have not found them to be widely used.

Priority for Coordinate Data Entry

Moving to the top right of the User Preference tab, this category controls how AutoCAD responds to coordinate data input, also not necessary to change for most users.

Associative Dimensioning

Leave the box under this category checked, since all dimensions should be associative (meaning, if the geometry changes, so do the dimensions associated with it).

Undo/Redo

This category controls undo and redo for zoom and pan; leave both boxes checked to save some time if you must go back a lot of steps. They are then combined and skipped over.

Block Editor Settings…

This button changes the environment of the block editor. You worked with that editor when you first learned dynamic blocks. Clicking on it brings up the Block Editor Settings dialog box, where you can change just about every color, size, and font parameter. For most users, it is best to leave everything as a default.

Lineweight Settings…

This button displays the Lineweight Settings dialog box. These settings can be adjusted when setting up the .ctb files, as described in Chapter 19; so, they are not needed here. Leave everything as default.

Default Scale List…

This button displays the Edit Scale List dialog box. Additional scales are rarely needed but can be added here, if you have some unusual requirements. This may be more of an issue with metric units, as engineers may use more scales than architects.

If you want a more detailed description of what each selection means, press F1 to read the Help files while you have the User Preferences tab open. Alternatively, hover the mouse over each element for a brief description. If you are done, do a final check to see that your settings reflect what is in Fig. 14.10 and let us move on to the next tab.

Drafting tab

The Drafting tab (Fig. 14.12) contains the AutoSnap Settings, AutoSnap Marker Size, Object Snap Options, AutoTrack Settings, Alignment Point Acquisition, and the Aperture Size categories, as well as the Drafting Tooltip Settings…, Lights Glyph Settings…, and Cameras Glyph Settings… buttons. The entire left side of this tab is devoted to snap point markers, those colored geometric objects that represent endpoints, midpoints, and the like. No critical items are here, so we run through everything with some general comments and descriptions.

AutoSnap settings

Leave everything in this category as default, including Colors, unless you prefer something different.

AutoSnap Marker Size

Leave this category as default, unless you have trouble seeing the symbols or find them too big. The sizing of the marker is easily adjusted via the slider.

Object Snap Options

Also leave everything in this category as default. You especially want to make sure that the Ignore hatch objects (the first box) is checked, as that would make the object snap go crazy trying to snap to every piece of a complex hatch pattern.

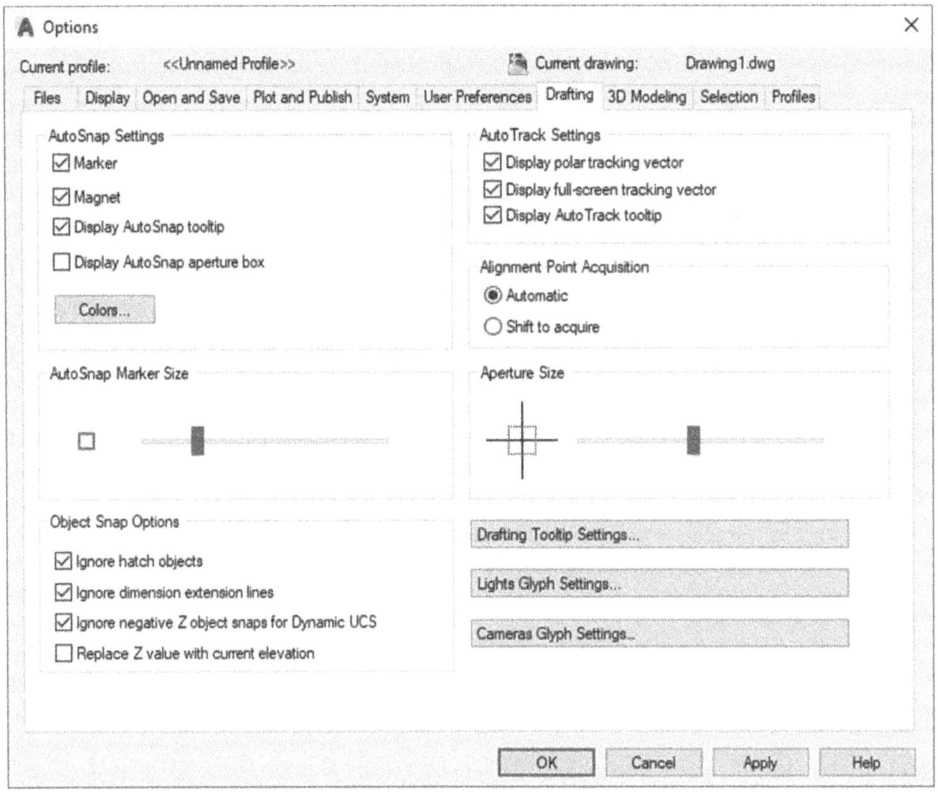

FIGURE 14.12 Drafting tab.

AutoTrack Settings

Moving to the upper right of the Drafting tab, we have the AutoTrack Settings category. It has to do with AutoCAD's OTRACK tracking tool, which is briefly discussed in Chapter 15, along with other miscellaneous topics. Leave these settings as default for proper operation of that tool.

Alignment Point Acquisition

Leave the setting in this category as default.

Aperture Size

Aperture Size is perhaps the one item that may occasionally be adjusted to suit the user's taste, but generally keep it the same size or slightly larger than the AutoSnap marker. If the aperture is too big, it may be distracting or may be confused with small rectangles on the screen.

Drafting Tooltip Settings…

This button features minor nuanced adjustments to drafting tool tips (such as distance or angle displays) and their visibility to the users (the transparency). Leave all them as default.

Lights Glyph Settings… and Cameras Glyph Settings…

These buttons are minor color settings for lighting and cameras and are not relevant at this point; leave them as default.

If you want a more detailed description of what each selection means, press F1 to read the Help files while you have the User Drafting tab open. Alternatively, hover the mouse over each element for a brief description. If you are done, do a final check to see that your settings reflect what is in Fig. 14.12 and let us move on to the next tab.

3D Modeling tab

Settings and adjustments under this tab are not discussed in this first third of the 2D-only chapters of the book.

Selection tab

The Selections tab (Fig. 14.13) contains the Pickbox size, Selection modes, Ribbon options, Grip size, Grips, and Preview categories. Some important items are under the Visual Effects Settings… button under the Preview category, so we cover that in detail, and run through everything else with some general comments and descriptions.

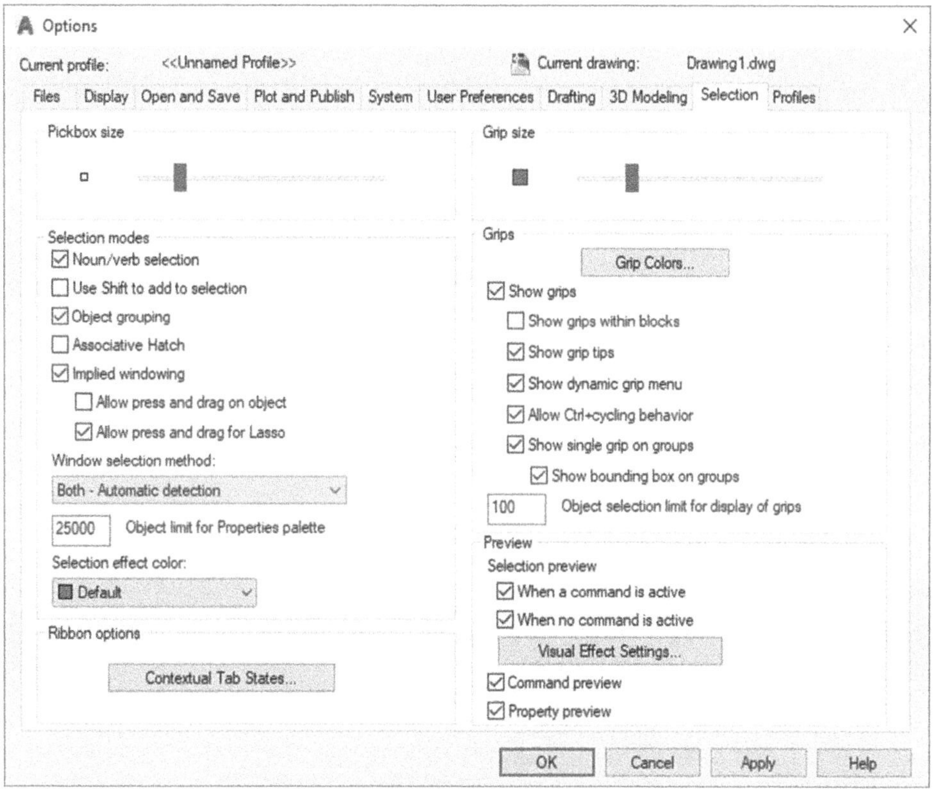

FIGURE 14.13 Selection tab.

Pickbox size

This category is simply what the cursor becomes when you are in the process of picking something. Set it too small and you have trouble selecting the geometry. Set it too big and you select more than you wanted. Generally, leave it at the default size shown in Fig. 14.13.

Selection modes

This category is just more adjustments to object selection. Leave them all as default, as seen in Fig. 14.13, including the Window selection method choice. The Object limit for Properties palette can be left as the ridiculously high default of 25,000. Rarely will you have that many objects selected or even present in your drawing.

Ribbon options

This category provides some options regarding contextual tab states. Leave them as default.

Grip size

Moving to the upper right of the Selection tab, the Grip size category with the slider adjustment should be self-explanatory. Leave the grips as default size unless you prefer bigger or smaller ones.

Grips

This category has yet more grip settings for colors and other adjustments. There is really no reason to change grip color. Blue means cold (unselected), red means hot (selected), and hover can be any color (default Color 11 is just fine). Leave the other settings as default; you want the grips enabled but not inside blocks. Leave the maximum number of grips that can be shown at 100 to not overwhelm the drawing if you select more.

Preview

This category is a relatively new feature in AutoCAD that causes objects to "light up" when your mouse hovers over them. In AutoCAD 2015, this effect was made subtler, eliminated dashing the objects, and only added a gray highlight to them. If you find this feature distracting, you can turn it off by unchecking the first two buttons. To leave it on but modify the settings, click the Visual Effects Settings... button. When you press it, you see what is shown in Fig. 14.14.

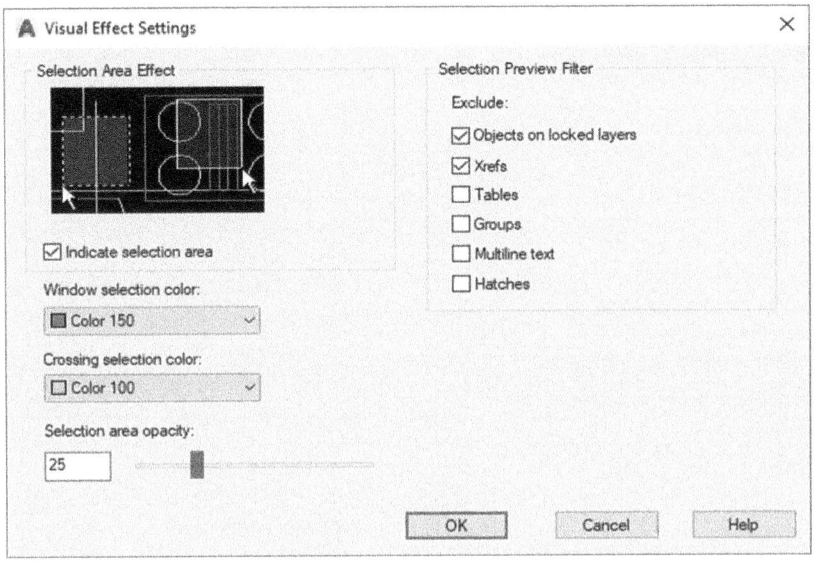

FIGURE 14.14 Visual Effects Settings.

On the left side of this dialog box are various controls for the internal colors of the familiar Window/Crossing as well as the new Lasso tool. You can change those colors as well as their intensities by varying the Selection area opacity slider. You can also eliminate them, as the colors are technically unnecessary; what is important is whether the selection rectangle is dashed or solid. If you prefer minimum fuss with your selection methods, you can remove the colors by setting the opacity at zero or uncheck the Indicate selection area button.

If you want a more detailed description of what each selection means, press F1 to read the Help files while you have the User Selection tab open. Alternatively, hover the mouse over each element for a brief description. If you are done, do a final check to see that your settings reflect what is in Fig. 14.13 and let us move on to the next tab.

Profiles Tab

This tab (Fig. 14.15) is simply there to save your personal settings, so if someone else is using AutoCAD on your computer and prefers other settings (maybe that neon green background color), that person can set his or her own and save them as well. Simply press the Add to List... button and enter a name for the profile, as seen in Fig. 14.16; finish by pressing Apply & Close.

The profile is added to the list (as many as you want can be added) and can be restored by selecting it and pressing Set Current. Profiles can be exported (as .arg files), imported, renamed, and deleted as needed. Profiles were more relevant in the days when PCs were more expensive and each one in a company may have had a few users, but some people still use them to avoid office arguments about exactly how AutoCAD "should look" if one were to "borrow" a PC for the day, then change AutoCAD to suit his or her taste, and forget to reset it to how the original owner or user likes it.

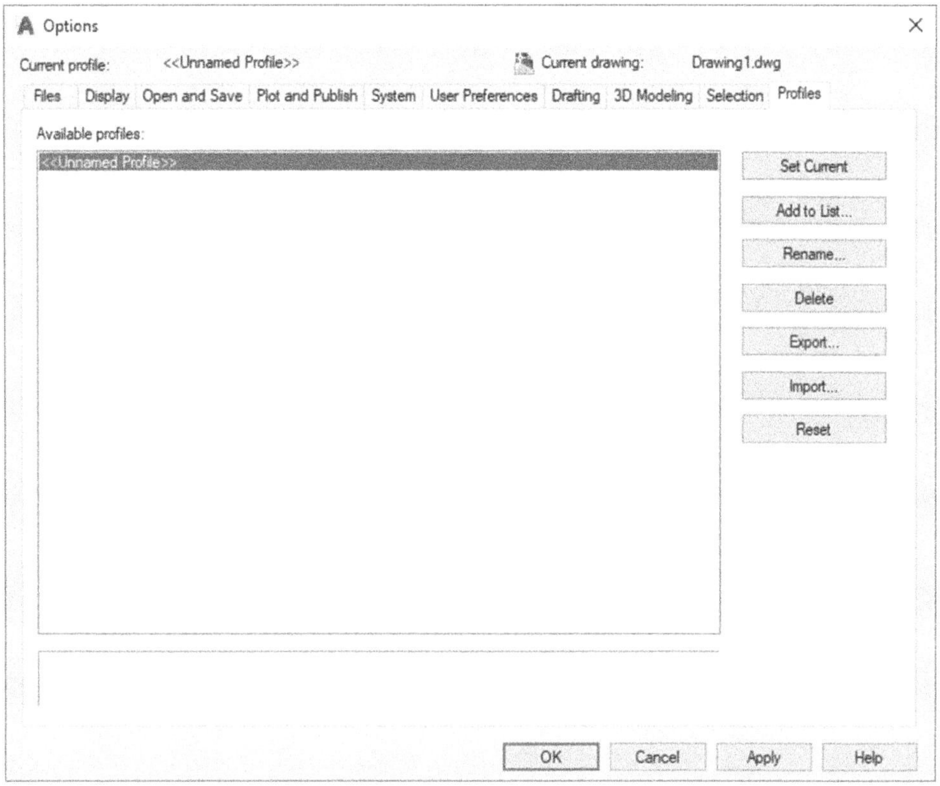

FIGURE 14.15 Profiles tab.

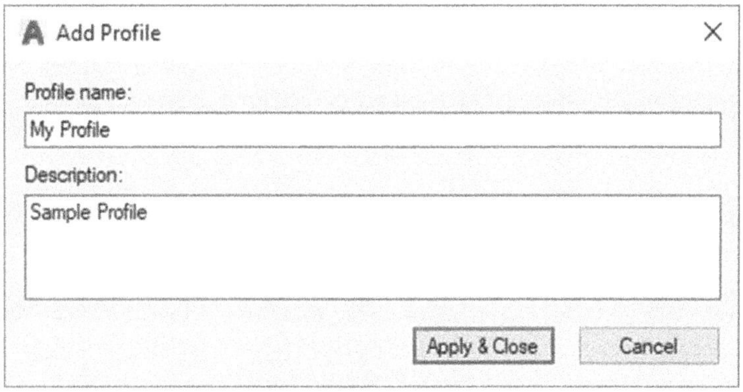

FIGURE 14.16 Add Profile.

14.2 Shortcuts

Recall that it was mentioned in the previous chapters how important speed is to an AutoCAD designer. One of the tools to help you work faster was already mentioned, the mouse Right-Click Settings under the User Preferences tab. Keyboard shortcuts, however, are another important way to accelerate your input. They are useful only if you type. If you are a diehard Ribbon user and abhor the keyboard, then you have no need for them. Many students, however, end up typing quite a lot, even if they also integrate some Ribbon use.

The reasons are simple: Typing is a fast, very direct method to interact with AutoCAD and remains quite popular despite other methods. Less Mouse movement is a good thing when you spend all day drafting. However, to take advantage of this, you need shortcuts, as typing out the full command defeats the speed advantage.

You may have already been using shortcuts all along. Instead of typing in `arc`, you may have just pressed the letter `a`; instead of `move`, the letter `m`. This is exactly what is meant by shortcuts: It is the process of shortening the typed commands so they can be entered faster. What we want to do here is put all this on solid footing and outline exactly how to set your own shortcuts to make best use of them.

PGP File

Here is a basic primer on shortcuts. They have been a part of AutoCAD since the early days, when typing commands was the best (and, in the very beginning, the only) way to interact with the software. Abbreviating the commands was the logical next step to speed up typing, and that capability was integrated into AutoCAD as well.

The abbreviations are stored in a file called acad.pgp, which resides in the Support folder of your AutoCAD installation. AutoCAD comes with a wide range of preinstalled shortcuts, but the file is easily found and meant to be modified to suit the user's taste and habits. The key here is that you need to have your shortcuts memorized (or it makes no sense to use them in the first place), and of course, you need to pick the most efficient abbreviations of the commands. To that end, AutoCAD does not care what you use—it can be names of your relatives if you so choose—but obviously you want to use something that makes sense to you and shortens the typing effort. Ultimately, the first letter, first and second letters, or first and last letters are used for all but a few of the commands.

Here are some suggestions for the Modify commands:

Erase: e
Move: m
Copy: c
Rotate: ro
Scale: s
Trim: t
Extend: ed
Fillet: f
Offset: o
Mirror: mi

The idea here is to abbreviate the most often used commands in the shortest way possible. If, for example, there is a conflict between two commands that start with the same letter (e.g., circle and copy), then you determine which command gets used more often and give that command fewer letters. Copy is generally used more often, so it would be `c` and circle would be `ci`.

Sometimes, doubling up of letters is done, such as with array (aa), because a is taken by arc, and so on. Listed in Fig. 14.17 is a sample acad.pgp file. It is presented here only for illustration purposes and may or may not match what you may want in yours, although it can be a good starting point to creating your own.

A,	*ARC	ED,	*EXTEND	PP	*PURGE
AA,	*ARRAY	EX,	*EXPLODE	Q	*QSAVE
B,	*BHATCH	EL,	*ELLIPSE	R	*REGEN
BR,	*BREAK	F,	*FILLET	R3	*ROTATE3D
BL,	*BLOCK	I,	*INSERT	RO	*ROTATE
C,	*COPY	ISO,	*LAYISO	RE	*RECTANG
CF,	*CHAMFER	L,	*LINE	RR	*RTZOOM
CH,	*CHANGE	LA,	*LAYER	RT	*RTPAN
CI,	*CIRCLE	LF,	*LAYFRZ	S	*SCALE
D,	*DTEXT	LW,	*LAYWALK	ST	*STRETCH
DD,	*DDEDIT	LI,	*LIST	T	*TRIM
MT,	*MTEXT	ISO,	*LAYISO	U	*UNDO
DI,	*DIST	M,	*MOVE	UU	*UCSICON
DD,	*DDEDIT	MM,	*MATCHPROP	W	*WBLOCK
DDC,	*PROPERTIES	MI,	*MIRROR	X	*EXTRUDE
DDO,	*DSETTINGS	MI3,	*MIRROR3D	Z	*ZOOM
DDS,	*DDSTYLE	O,	*OFFSET		
DDU,	*DDUNITS	P,	*PLOT		
DV,	*DVIEW	PE,	*PEDIT		
E,	*ERASE	PL,	*PLINE		

FIGURE 14.17 Sample acad.pgp file.

Altering the PGP File

So, how do you find and alter the pgp file to suit your tastes? The long way is to find it via this path on your "C" drive: C:\Users*Name*\AppData\Roaming\Autodesk\AutoCAD 2022\R2***\enu\Support, but you really do not need to do this, and the shorter way is to access the file through the Edit Aliases command in Customization Panel on the Manage Tab of the Ribbon (See Fig. 14.18). The pgp file then appears on your screen (opened as a Notepad file, which is just a simple ASCII text editor, if you never used it before). Scroll down the file until you see the rather lengthy built-in listing of shortcuts. At this point, you may do one of the three things:

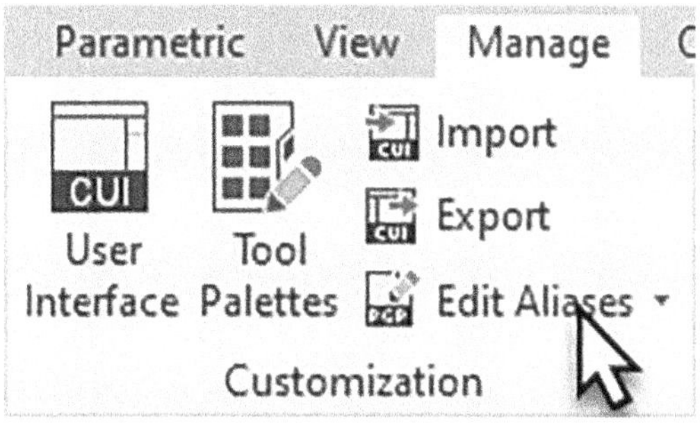

FIGURE 14.18 Edit Aliases command in Ribbon.

1. Leave the entire file as it is and use the shortcuts without modification.
2. Erase or change the shortcuts for individual commands that you either do not use or would like to modify.
3. Erase the entire list and start fresh, entering the shortcuts as you would like them to be, one by one. If this is the case, be sure to follow the format given, where you enter the shortcut, then a comma, then a few spaces, and finally the command preceded by a *, as seen in Fig. 14.17. You need not use all capital letters—that is just a preference.

When done, save the acad.pgp file and close Notepad. At this point, for the changes to take effect, you can either shutdown and restart AutoCAD or (if you are in the middle of a drawing and do not want to do that) type in *reinit* and press Enter. The dialog box in Fig. 14.19 appears.

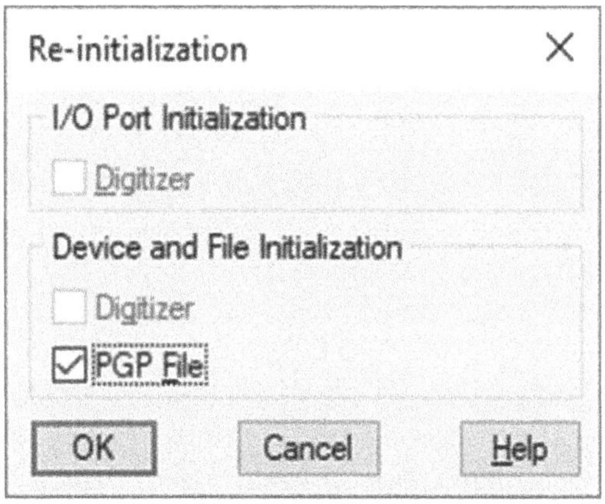

FIGURE 14.19 Re-initialization of the acad.pgp file.

Check off the PGP File box and press OK. The pgp file is ready to go, and you can use it right away. To add or subtract commands, just repeat the steps at the beginning of this section, save the pgp file, and reinitialize.

It is strongly recommended that you at least try to modify and use the pgp file. It is almost always faster to type the shortcuts than to click on icons. You can eventually learn to type quickly and without even glancing at the keyboard (like a good administrative assistant or court reporter), assuming you do not already type fast, as many students do.

14.3 Customize User Interface

The Customize User Interface dialog box (referred to from here on as just the CUI) is the next thing we examine in this chapter. The CUI has to do with changing the look and functionality of menus, toolbars, and other items and, as such, can be of importance to a small set of users who may need to do this for aesthetic or functional reasons.

The CUI is an advanced and somewhat mysterious concept and one not taken advantage of by most users, as either they do not know much about it or do not wish to change their Ribbon or menus. A few though find this customization ability of great interest and spend time customizing the look of AutoCAD to the point of almost having AutoCAD look like another program entirely.

We take the middle ground here and present some CUI basics, demonstrating the creation of a new custom toolbar and leaving the student with some CUI ideas to pursue further if he or she is interested in doing so.

Let us introduce the basic CUI as it opens in default mode. Type in `cui`, press Enter (or alternatively use the Ribbon's Manage tab→User Interface), and after a few seconds, the dialog box in Fig. 14.20 opens. Press the expansion button (small arrow, bottom right) to make it full size if it does not do this automatically.

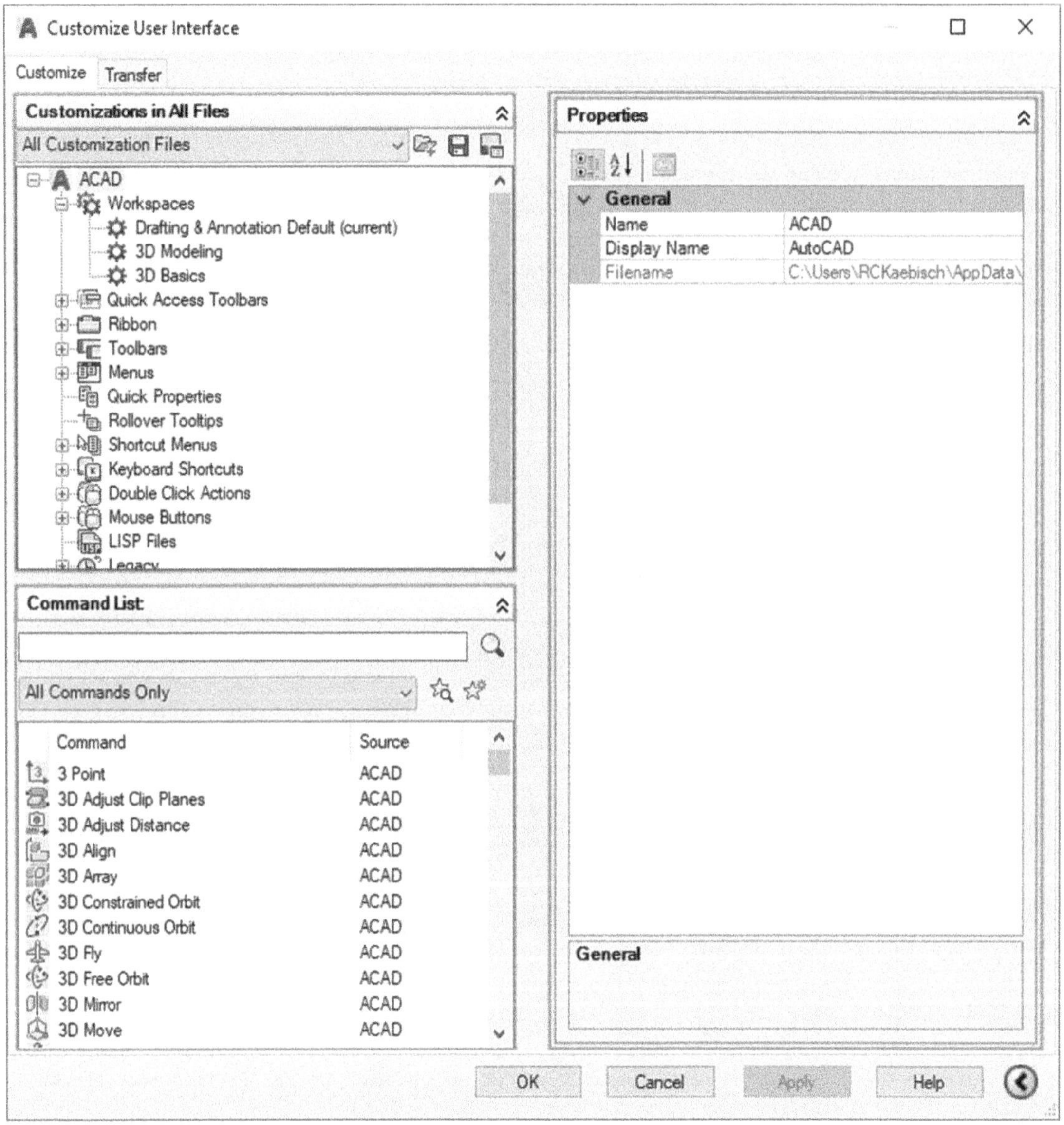

FIGURE 14.20 The Customize User Interface (CUI) dialog box.

In general, the CUI is your window into customizing virtually anything in AutoCAD that you can push, select, or type in. The full list of candidates is in the upper left of the CUI box: Workspaces, Toolbars, Menus, Ribbon Panels, and all the other choices. You may need to minimize the Workspace menu to see all the rest of them. Just to get your feet wet, let us create a custom toolbar that has only the buttons you want and none of the ones you never use.

Step 1. In the upper left-hand corner, find the Toolbars expandable menu and right-click, selecting New Toolbar. Give your new toolbar a name in the name field (`My_New_Toolbar`, in this example). Once it is done, you can click on it and a Toolbar Preview Properties appears on the right of the CUI window, along with an expanded Properties palette.

Step 2. On the bottom left begin selecting the commands you want your new toolbar to have (you can filter the commands to find them more easily) and drag them into the new toolbar name on the upper left. As soon as you begin doing this, a toolbar preview and Button Image category appear on the right-hand side, as seen in Fig. 14.21, when Rectangle, Revcloud, Arc, and Spline were added in.

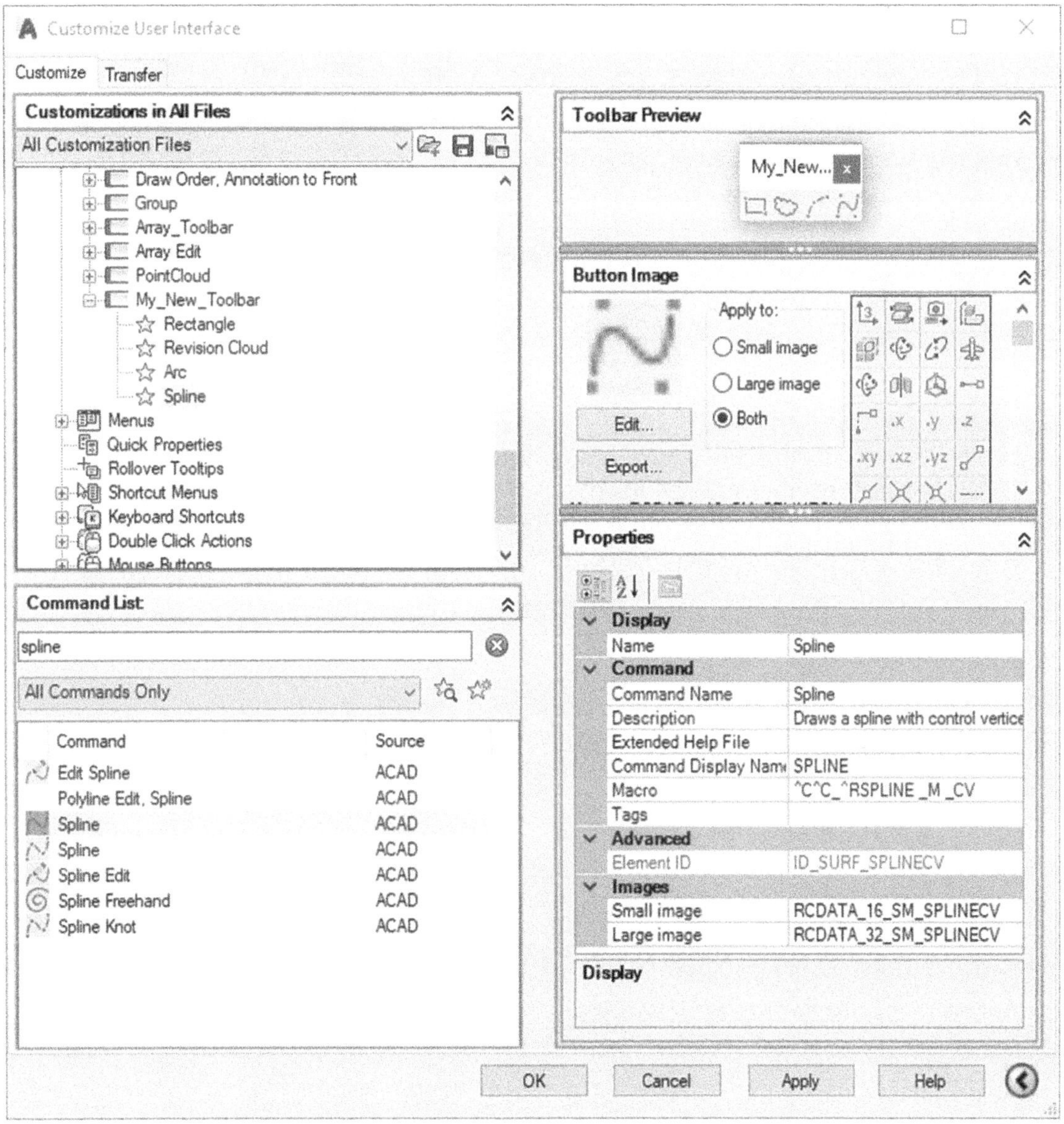

FIGURE 14.21 My_Toolbar—Add Rectangle, Revcloud, Arc, and Spline.

Interestingly, you can even modify the image itself by pressing Edit… in the Button Image category at the upper right, which takes you into the dialog box shown in Fig. 14.22. Here, you can use pixel-by-pixel editing tools to edit the image. You have some additional tools, such as a grid and a Preview window, to see what you have created. Finally, you can save your work and return to the CUI. You can draw some rather creative-looking buttons if you so desire (a few practical jokes of smiley-face icons from years gone by come to mind here). Admittedly, this is not done very often, as most users have better things to do with AutoCAD, or at least one would hope.

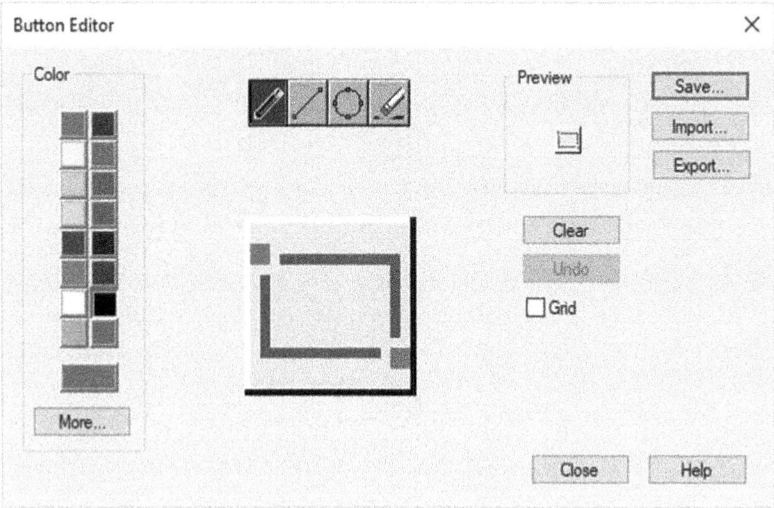

FIGURE 14.22 Button Editor.

Finally, after adding a few more commands (Revcloud, Circle, and Spline) and pressing Apply and OK, the new toolbar is born, as shown in Fig. 14.23; this is certainly a good trick if you like toolbars and want to see only the tools you use on yours.

FIGURE 14.23 The new My_New_Toolbar.

This is just the tip of the iceberg, of course. You can modify all the rest of the choices. Shortcut Keys under the Keyboard Shortcuts are quite useful. This refers to shortcuts such as Ctrl + C for Copy Clip (also known as *accelerator keys*). You can set up your keyboard to accelerate many commands. Just some of the preset ones are shown in Fig. 14.24 at the upper right of the CUI dialog box.

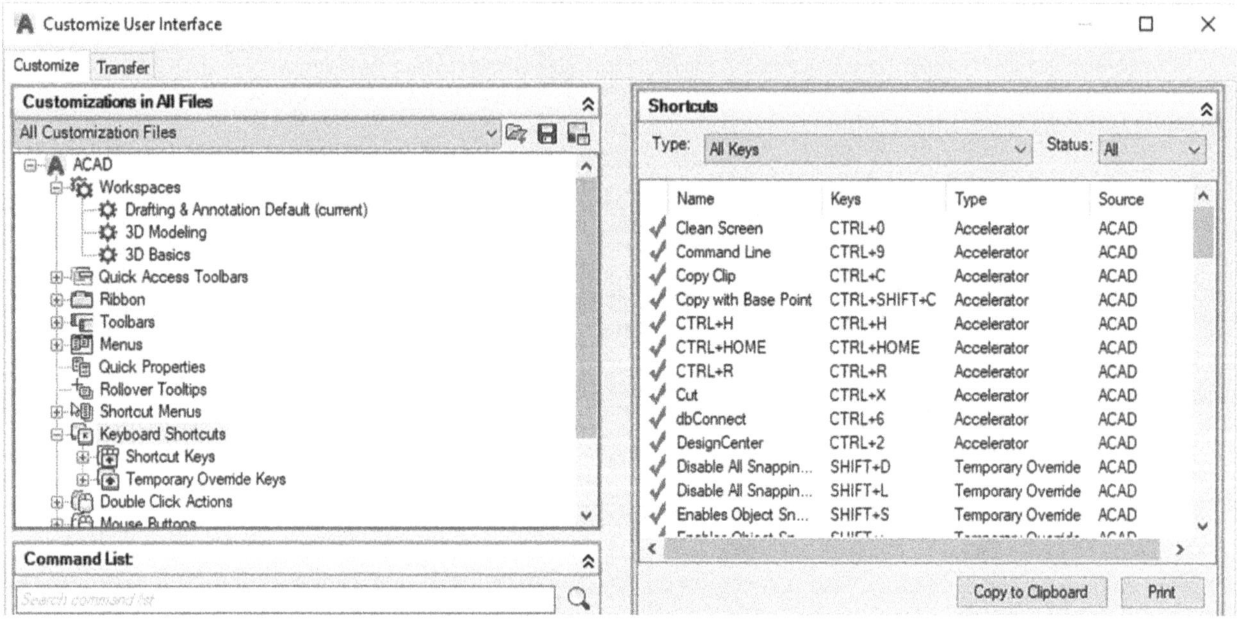

FIGURE 14.24 Accelerator keys.

It may be worth your time to look through the Help files and investigate the CUI further; it really is a bottomless pit of customization, although be careful not to change something just for the sake of change but only if you feel it may help your productivity.

14.4 Design Center

You can probably guess at one of the purposes of the Design Center (DC) as soon as you open it via the Palettes→panel of the View tab of the Ribbon (also via typing in `dc` and pressing Enter, or Ctrl+2). Be sure to adjust all the window panes for good visibility, as shown in Fig. 14.25. The DC can also be docked to any side of the screen.

FIGURE 14.25 AutoCAD Design Center.

What you see in the far left window is basically the Windows Explorer used regularly to view files (such as when you click on My Computer). This is a big clue to the first of two functions of the DC, which are to view the contents of one drawing on the left and, as we soon see, transfer parts of that drawing into the active drawing on the right. Let us give this a try using the process outlined next.

By default, the DC will likely open to the //AutoCAD 20**/sample file folder on the left. Go ahead open an additional few folders under the Sample folder, called *en-us*, followed by *Design Center*. There you see a sample drawing called HVAC—Heating Ventilation Air Conditioning.dwg. You would ideally of course browse for, and choose, an actual drawing you created (that is the whole point, after all), but for our purposes right now, one of these sample drawings is just fine.

Click on the plus sign next to the HVAC file name and expand the folders, revealing a collection of useful data, such as Blocks, Dimstyles, Layers, and Layouts. Now click on a category, such as Blocks. A collection of blocks present in this drawing is revealed in the adjacent window. Click and drag any block from the choices given on the left into the drawing space on the right, as seen in Fig. 14.25 with a block called *Fan and Motor with Belt Guard*. Note that, in Fig. 14.26, the DC is fully docked to the left side of the screen and the window panels are adjusted to give equal spacing to the DC and the drawing area.

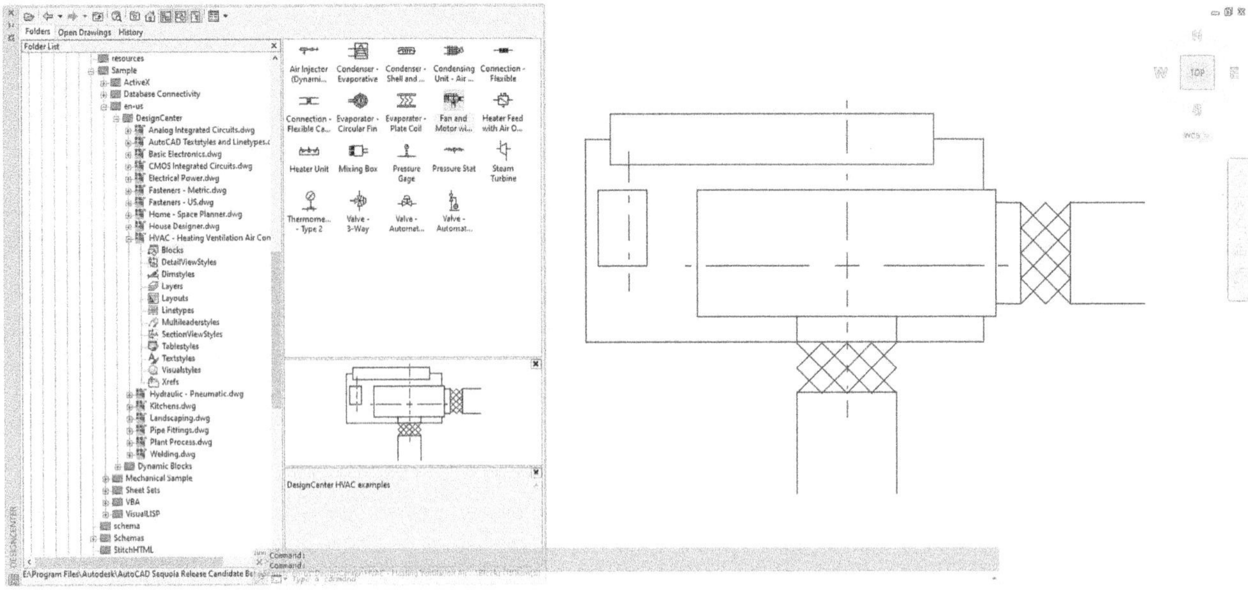

FIGURE 14.26 Design Center insertion of blocks.

In the same manner, you can transfer layers and almost anything else lifted from one drawing to another. This is very useful when you need only a few items, as opposed to everything, and is a great visual alternative to the usual Copy/Paste (Ctrl + C and Ctrl + P).

The next two tabs of the DC are not all that significant. Open Drawings simply looks at the content of the current active drawing, and History is exactly that, the history of all the open drawings thus far in the drawing session.

For the 2019 release, Autodesk had discontinued *Autodesk Seek*. That service and content is now hosted and controlled by BIMobject. Assuming you have Internet access on your computer, you can access their database of manufacturer produced design files, blocks, and 3D models in various file formats. The web address is https://bimobject.com. You can see the home screen in Fig. 14.27.

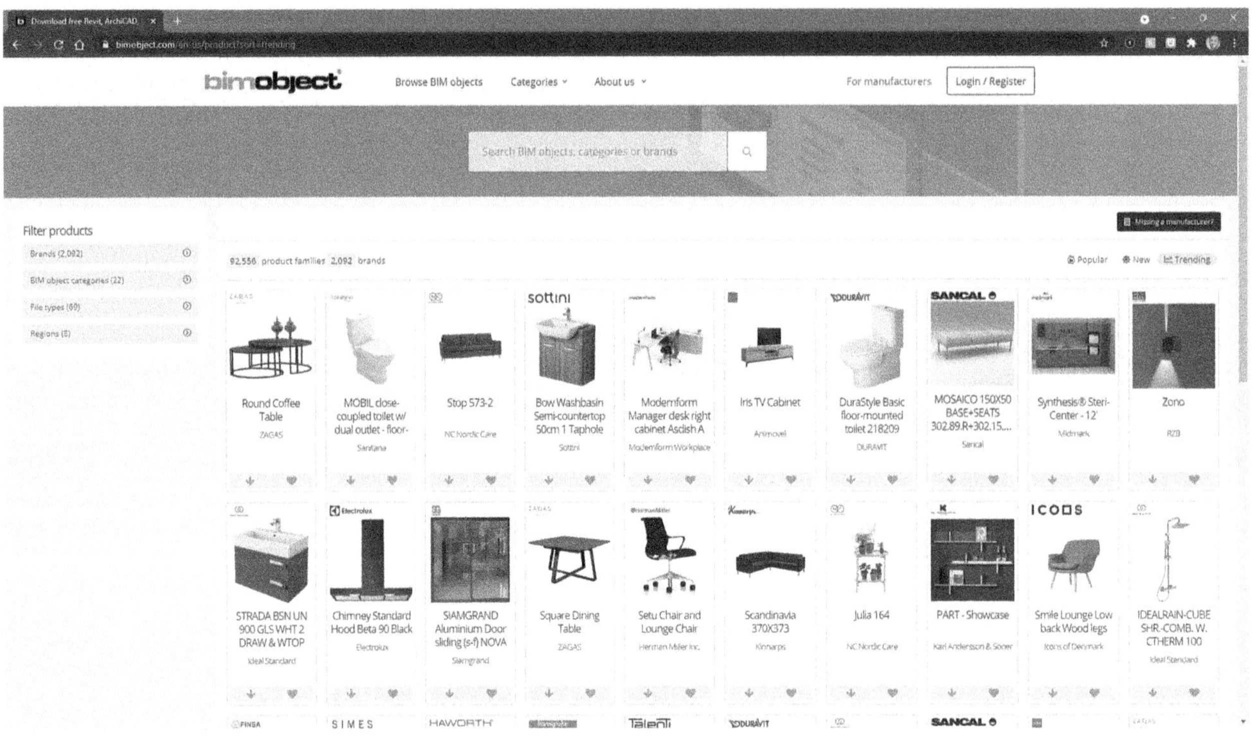

FIGURE 14.27 BIMobject web page.

When you see the web page load, click on the Browse BIM Objects link at the upper left, and you are taken to a splash page that scrolls endlessly and contains numerous BIM (Building Information Modeling) items. To the left are sub-categories and you can choose by Brand, Category, File Type, Region, and Types. Click around to see what is available. For example, if you search for the "60 CM GAS HOB" , select the stainless steel version and you see what is in Fig. 14.28. If you then click the Download button, you get several file choices to save the file/block as a Revit, ArchiCAD, AutoCAD, Sketchup, and other file types (Fig. 14.29). Explore this extensive website further; it is truly a great resource with over 350,000 BIM objects advertised as available as of this writing. NOTE: *You must create an account to access the downloadable content.*

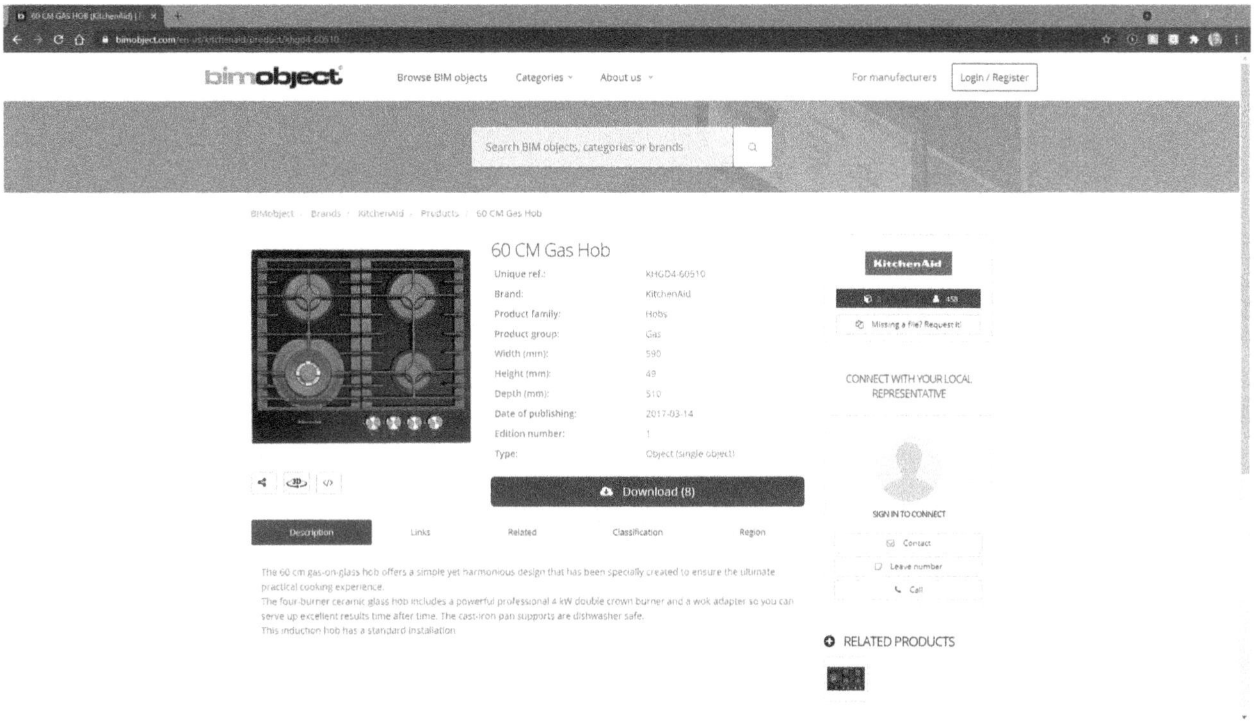

FIGURE 14.28 Example product page.

FIGURE 14.29 Example download file(s) page.

14.5 Express tools

Express Tools are a set of commands grouped together under the Express drop-down menus. These commands add extra functionality to AutoCAD, and while some of them are redundant or not that useful, others became so indispensable that they were later incorporated into regular AutoCAD menus and command sets. It really is a mixed bag, and our goal here is to highlight only the most useful ones, leaving the rest for you to explore if you so choose.

Express Tools, originally called *Bonus Tools*, got their start around Release 14 in the mid-1990s as a collection of useful routines and small programs. Where they originally came from is a matter of debate, but Autodesk likely acquired them from other third-party developers and possibly regular programming-savvy users who felt that AutoCAD just did not have that perfect command and developed their own as a "plug-in."

However, it happened, these tools found their way to Autodesk, and the company began to offer them as an "extra" to AutoCAD (not technically part of the core software). Sometimes, these tools were offered for sale; sometimes, they were free to VIP subscribers; and eventually they became free to all customers, although up until recently they often still needed to be installed in a separate and deliberate step while installing AutoCAD itself.

The first step in learning Express Tools is to check that you have them (they should come standard with the latest AutoCAD), and we proceed from here on with the assumption that you do and we are ready to go over them. Fig. 14.30 shows what the Express Tools Ribbon tab looks like.

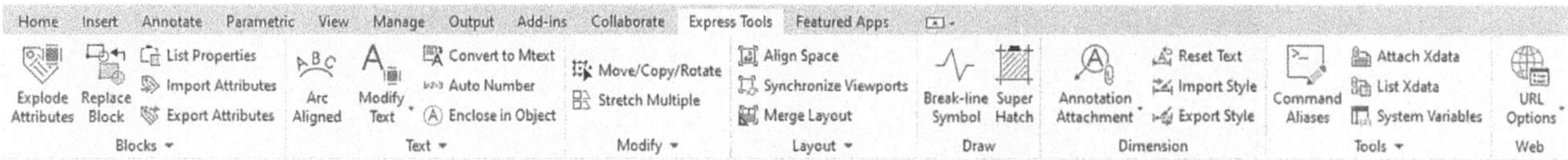

FIGURE 14.30 AutoCAD Express Tools Ribbon access.

Next is a rundown of the most useful Express Tools. If you would like info on those commands that are not specifically covered, refer to the Express Tools Help files for detailed description of each one. The Help files for the Express Tools are the same as the regular AutoCAD Help files (first seen in Fig. 2.26). They are accessed via the usual F1 key, and then just do a search for "Express Tools Reference." The Express Tools Help menu is shown in Fig. 14.31. Sometimes, you may find the functionality of an express tool already a part of AutoCAD in some other way, shape, or form. Occasionally, typed commands call up these tools as well. These are discussed as they come up.

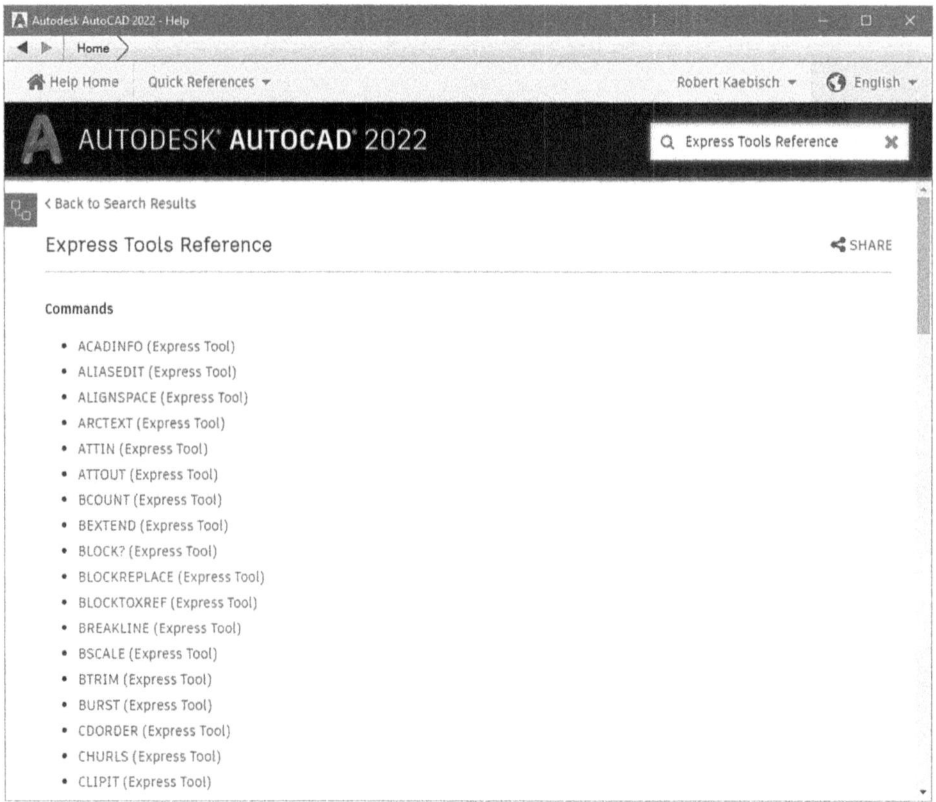

FIGURE 14.31 Express Tools Reference in the Help files.

Blocks Tools

The first panel Blocks (Fig. 14.32). Here you find menu choices, all having to do with blocks, nested objects, shapes, and attributes. Do explore all of them but pay special attention to the ones described in more detail next.

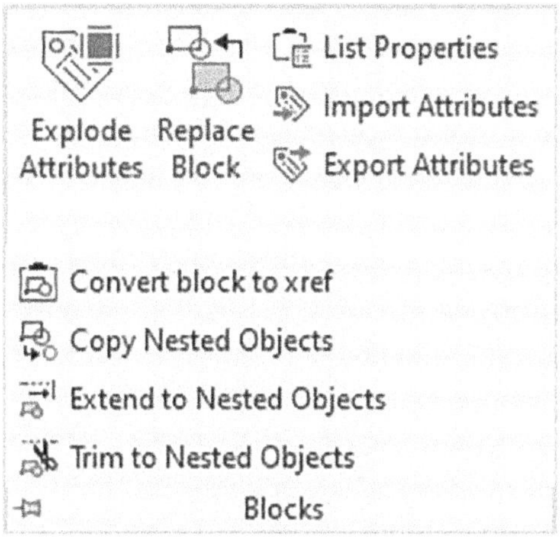

FIGURE 14.32 Express Tools (Blocks).

- *Explode Attributes to Text*: When you explode attributes (Chapter 19 topic), they do not turn into regular text; instead they take the shape of attribute definitions. This useful tool forces them to revert to ordinary text so you need not retype anything.
- *Copy Nested Objects*: This command is found in the Blocks panel dropdown menu and is useful for plucking out nested items, such as those embedded in xrefs (Chapter 17 topic) or blocks. These objects are otherwise stuck inside and not easily accessible for copying. The command line equivalent for this is ncopy and does the same thing. Additional commands below this one allow you to trim and extend to these nested objects as well.

Text Tools

The second panel is Text (Fig. 14.33). Here, you find menu choices, all having to do with text and mtext. Do explore all of them but pay special attention to the ones described in more detail next.

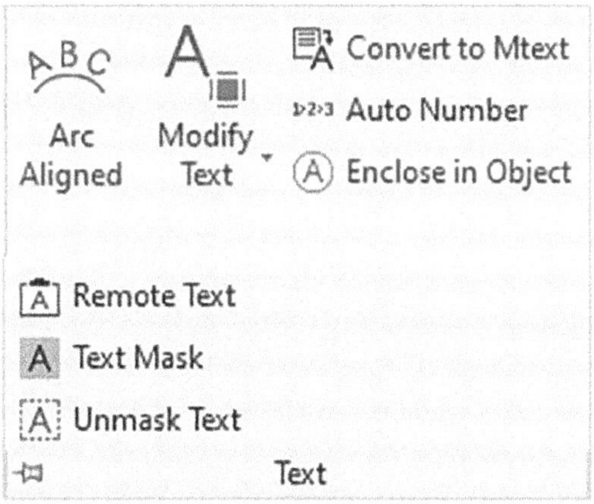

FIGURE 14.33 Express Tools (Text).

- *Convert Text to Mtext*: This command pretty much does what it advertises: It converts lines of text to a paragraph of mtext. It is faster and more efficient than copying the text one line at a time or redoing mtext completely. Simply activate the command, pick the lines of text you want to convert, and press Enter.
- *ArcAligned Text*: This command is used to tackle a problem that has never been addressed by AutoCAD prior to the Express Tools existence: How to create text that stretches around an arc. The procedure is to draw an arc and select this tool (or just type in arctext), then select the arc. The dialog box seen in Fig. 14.34 appears. Enter the actual text and

adjust parameters such as fonts, size, and anything else you want (they all have explanatory tool tips if you float your mouse over them); finally, press OK when done. Your text should look like that shown in Fig. 14.35. If the text is too squeezed or spread out, adjust the sizing. Note that the text can be over or under the arc. Also remember that you need to type in arctext to go back and edit it. Regular ddedit or double-clicking does not work here.

- *Enclose Text with Object*: This command encloses your text with rectangles, slots, or circles. Run through the menu options, setting offsets, shapes, and sizes.

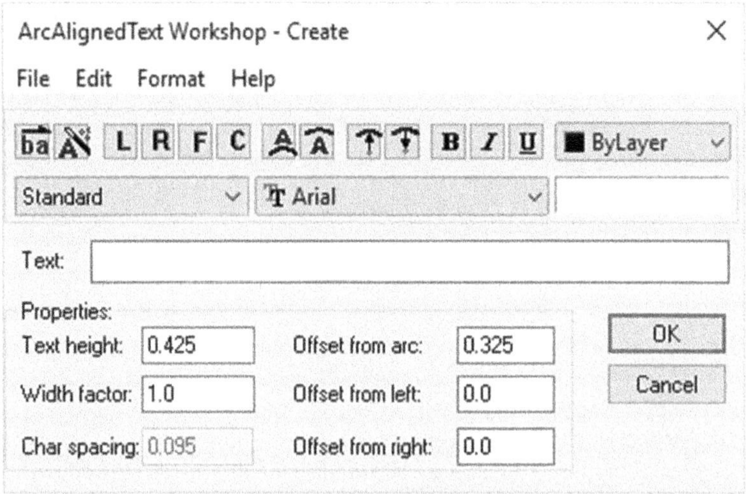

FIGURE 14.34 ArcAligned Text.

FIGURE 14.35 ArcAligned text sample.

Modify tools

Dimension

The third panel is Modify (Fig. 14.36). Here, you find menu choices having to do with various move, copy, rotate, stretch, and other command variations. Several of them duplicate other commands (such as Multiple copy) or are of limited use (such as Move/Copy/Rotate). Do explore all of them, however, but pay special attention to the ones described in more detail next.

- *Extended Offset*: This command has some usefulness when doing many offsets of the same object with a Shift+Pick option.
- *Flatten Objects*: If you go on to study 3D, this is a great tool for flattening (converting) a 3D model into a 2D one.

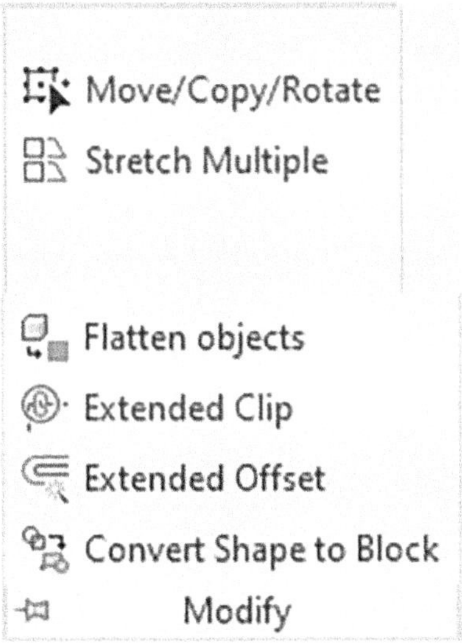

FIGURE 14.36 Express Tools (Modify).

Layout Tools

The panel is Layout tools (Fig. 14.37). Here, you find four menu choices, all having to do with Paper Space viewports. These commands are rarely used or are duplicates of existing commands but do take a few minutes to briefly explore them. We do not go into further detail.

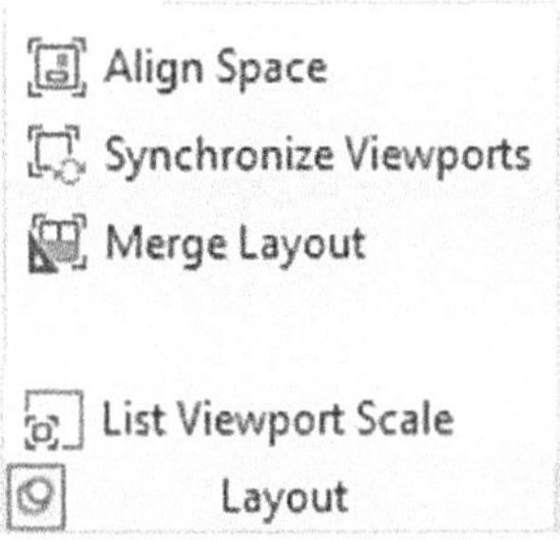

FIGURE 14.37 Express Tools (Layout tools).

Draw Tools

The panel is called Draw (Fig. 14.38). Here, you find just two menu choices, Break-line Symbol and Super Hatch…. We go over both next.

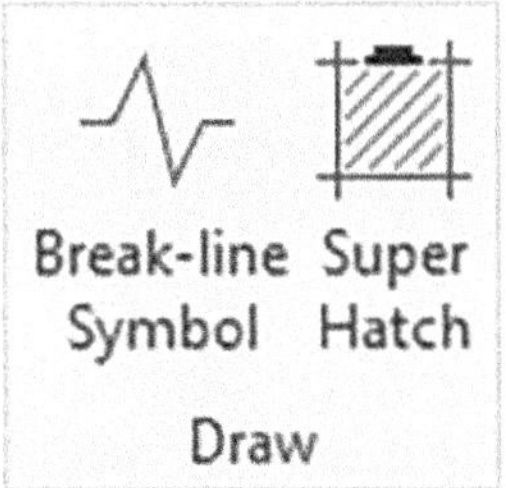

FIGURE 14.38 Express Tools (Draw tools).

- *Break-line Symbol*: Students have found this command one of the most useful among the Express Tools because it seems everyone would like to avoid drawing this symbol from scratch. Here, the process of making one is as easy as clicking two points and indicating a midpoint for the actual break line. Try it by selecting the command and specifying a first and second point along a line; then, using the MIDpoint OSNAP, place the break symbol in the middle. The Size option allows you to alter the break size.
- *Super Hatch…*: This command creates hatch patterns out of ordinary images by basically copying them many times over to fill an area. It was covered in detail toward the end of Chapter 5.

Dimension Tools

The sixth panel is Dimension (Fig. 14.39). Here, you find relatively obscure dimstyle and leader tools. Take a few minutes to explore them, but we do not go into further detail.

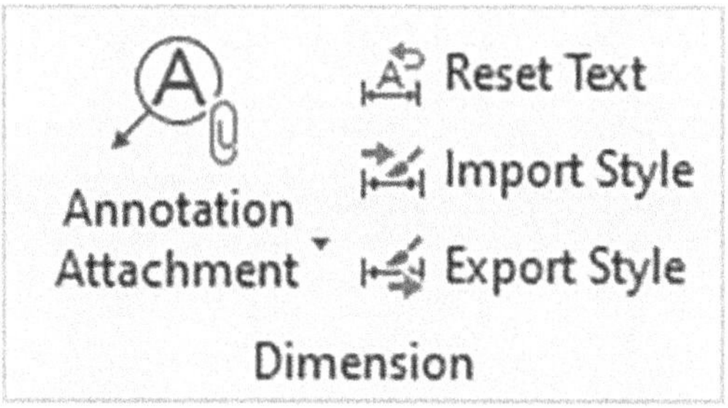

FIGURE 14.39 Express Tools (Dimension tools).

Tools Tools

The seventh panel is Tools (Fig. 14.40). Here, you find nine tools that erve a variety of functions, from editing aliases and variables (the pgp file already discussed and the CUI take care of those) to making shapes and linetypes. Look over the commands on your own, but we do not go into further detail here.

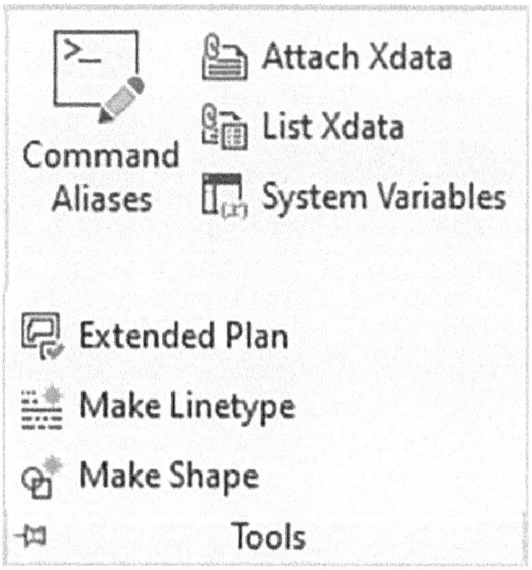

FIGURE 14.40 Express Tools (Tools tools).

Web tools

The ninth category from the top is Web tools (Fig. 14.41). Here, you find three URL/hyperlink-related tools. AutoCAD hyperlinks have not yet been discussed, so we do not go into further detail here. You see a bit more on these tools when hyperlinks are introduced in the next chapter.

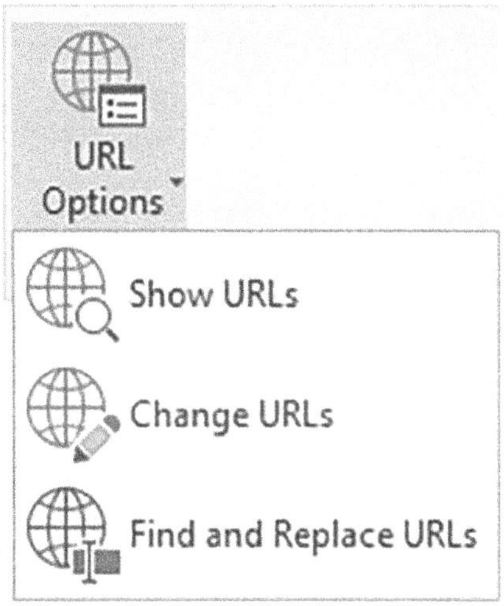

FIGURE 14.41 Express Tools (URL Options tools).

14.6 Level 2 drawing project (4 of 10): Architectural floor plan

For Part 4 of the drawing project, you add in electrical wiring, switches, lights, panels, and outlets. This design is similar in principle, if not in complexity, to a real electrical design.

Step 1. You need the following layers (with suggested colors) for the electrical pieces. The pieces are shown monochrome in Figs. 14.42 and 14.43 for clarity.

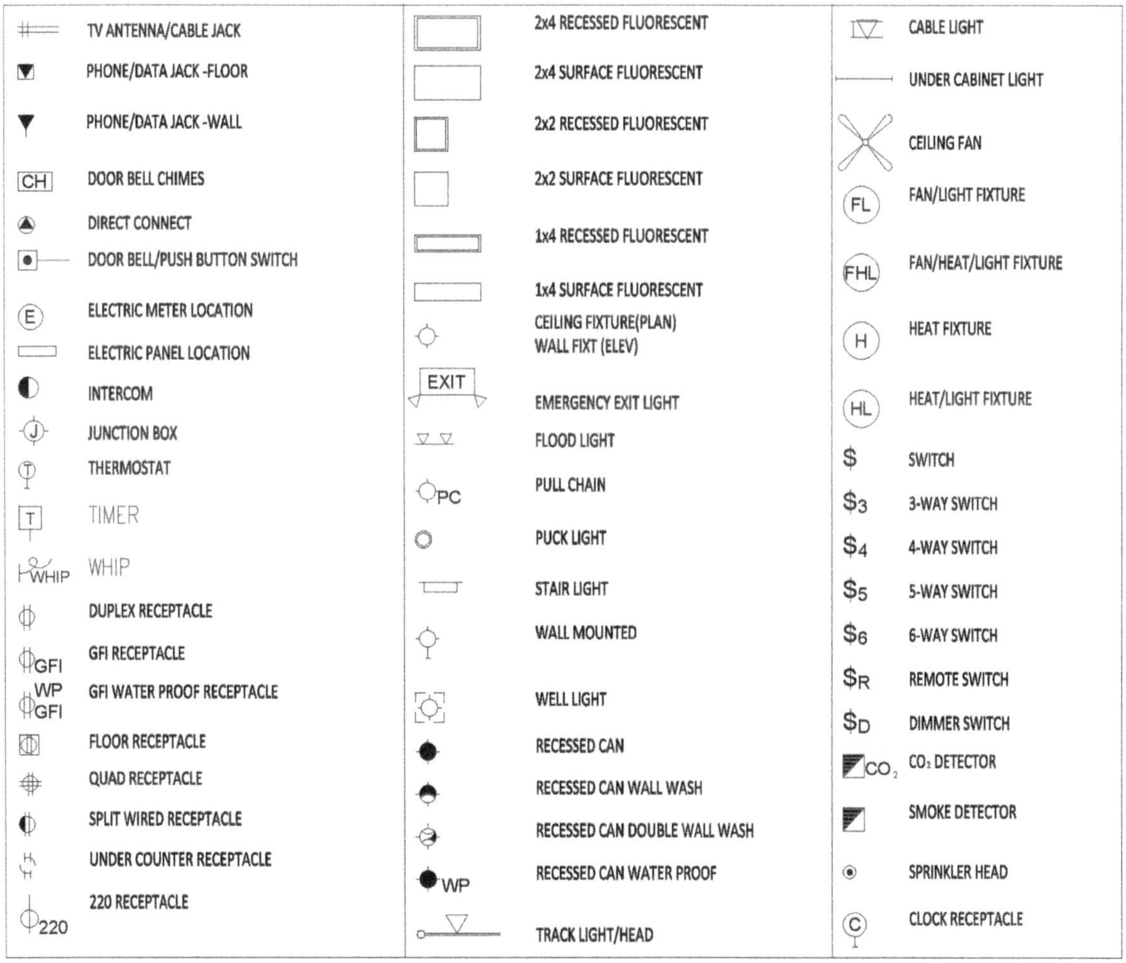

FIGURE 14.42 Various electrical symbols.

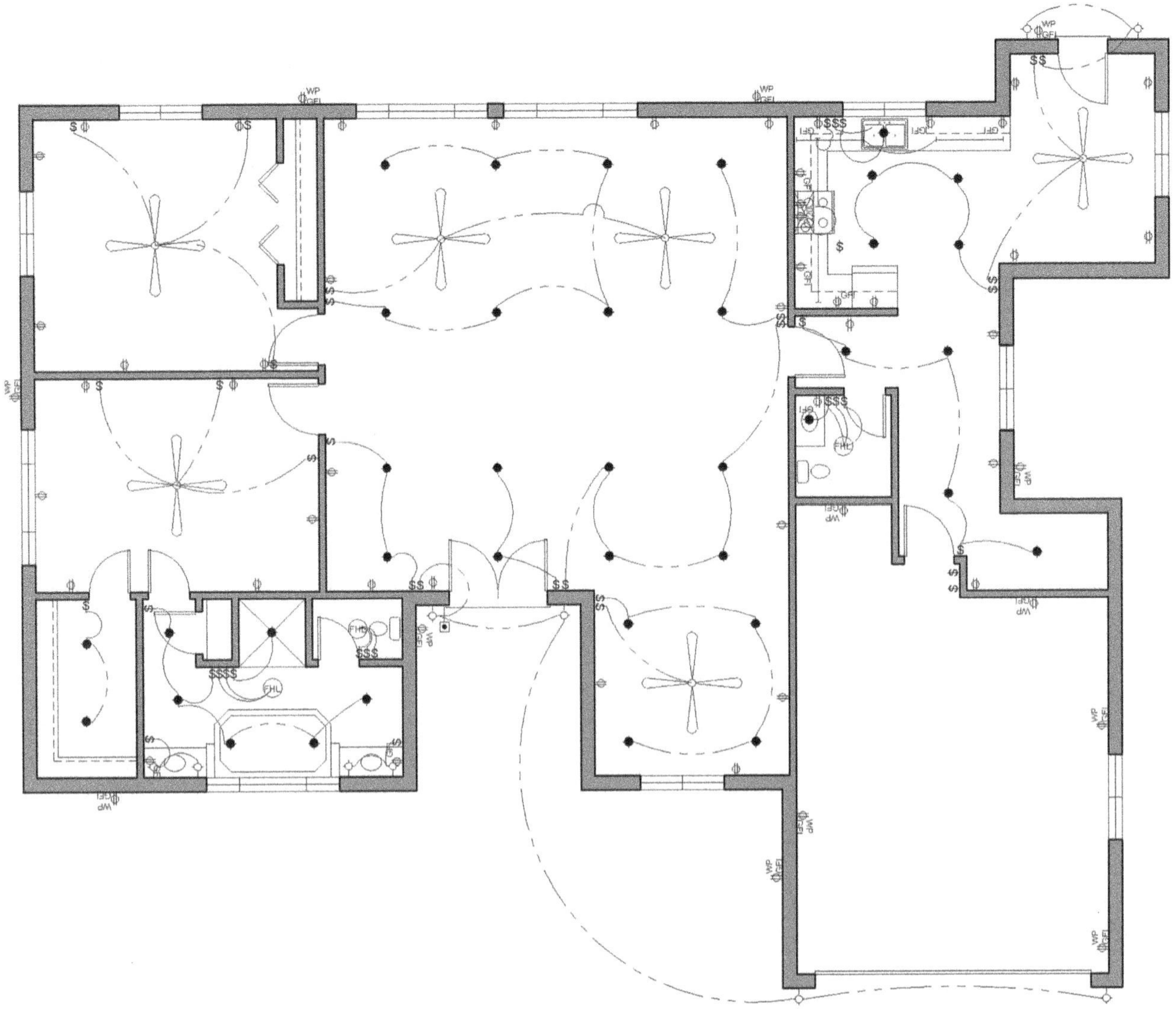

FIGURE 14.43 Electrical layout.

- E-Light (Green).
- E-Outlet (Red).
- E-Panel (White).
- E-Switch (Magenta).
- E-Wiring (Blue, Phantom).

Step 2. Create the electrical symbols as needed from Fig. 14.41 or additional symbols as needed. Make sure the layers and colors are appropriate. Finally, make a block out of each symbol, so you can just copy and rotate them into position. With switches and outlets, there are numerous types (dimmer, three-way, duplex outlet, GFI, etc.), so this only scratches the surface.

Step 3. Lay out the electrical symbols for your house layout and/or as seen in the floor plan in Fig. 14.42. Some of the wiring runs are splines and some are arcs.

Summary

You should understand and know how to use the following concepts and commands before moving on to Chapter 15:

- Files tab

- Display tab
- Open and Save tab
- Plot and Publish tab
- System tab
- User Preferences tab
- Drafting tab
- 3D Modeling tab
- Selection tab
- Profiles tab
- Shortcuts
- pgp file
- CUI
- Design Center
- Express Tools

Review questions

Answer the following based on what you learned in this chapter:

1. What is important under the Files tab?
2. What is important under the Display tab?
3. What is important under the Open and Save tab?
4. What is important under the Plot and Publish tab?
5. What is important under the System tab?
6. What is important under the User Preferences tab?
7. What is important under the Drafting tab?
8. What is important under the Selection tab?
9. What is important under the Profiles tab?
10. What is the pgp file? How do you access it?
11. Describe the overall purpose of the CUI.
12. Describe the purpose of the Design Center.
13. Describe the important Express Tools that are covered.

Exercises

1. Based on what you learned about the Options dialog box, set up your own profile. Be sure to include or at least consider all the variables and settings discussed. Save the profile as Profile_Your_Name. (Difficulty level: Easy; Time to completion: <10 minutes.)
2. Based on what you learned about the pgp file, create your own based on whatever abbreviations you prefer. Be sure to utilize the commands shown in Fig. 14.17 as well as add any of your own that you may want included. (Difficulty level: Easy; Time to completion: 20 minutes.)
3. Based on what you learned about CUI, create your own toolbar that features the following command icons:
   - Line
   - Circle
   - Rectangle
   - Move
   - Copy
   - Rotate
   - Offset

   (Difficulty level: Easy; Time to completion: 15 minutes.)
4. Review the Express Tools. Practice creating another arc-aligned text and break line. Open any multilayer file and review the Layer tools: layfrz, layiso, and laywalk. (Difficulty level: Easy; Time to completion: 20 minutes.)

5. To keep your Level 1 drafting skills sharp, draw the following three-view layout of a nut design. Make use of array as much as possible for the front view. (Difficulty level: Easy; Time to completion: 20–30 minutes.)

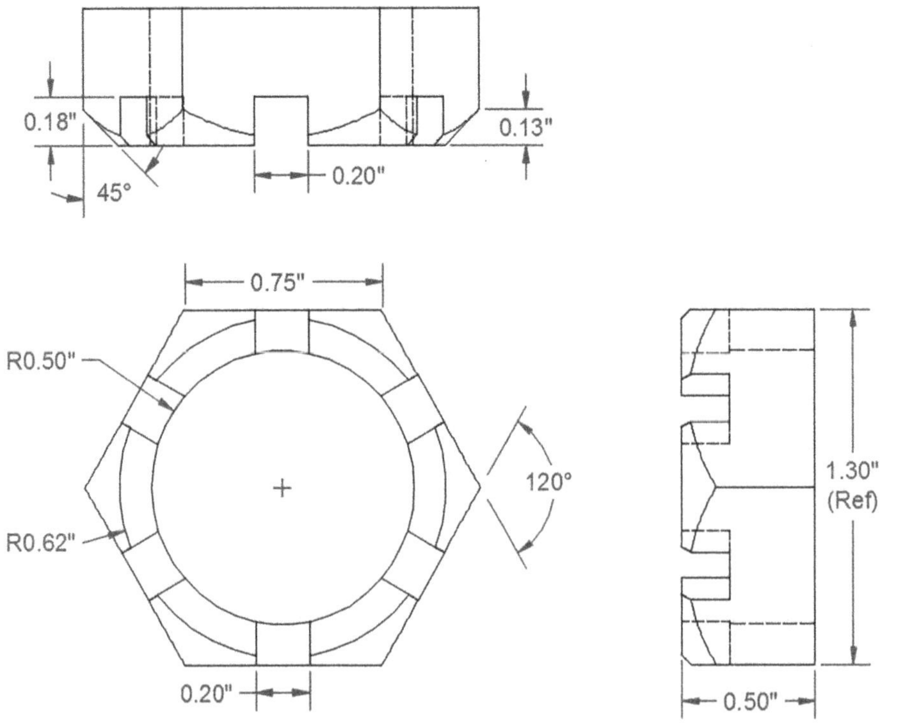

6. To keep your Level 1 drafting skills sharp, draw the following two-view mechanical spindle layout. (Difficulty level: Easy; Time to completion: 20–30 minutes.)

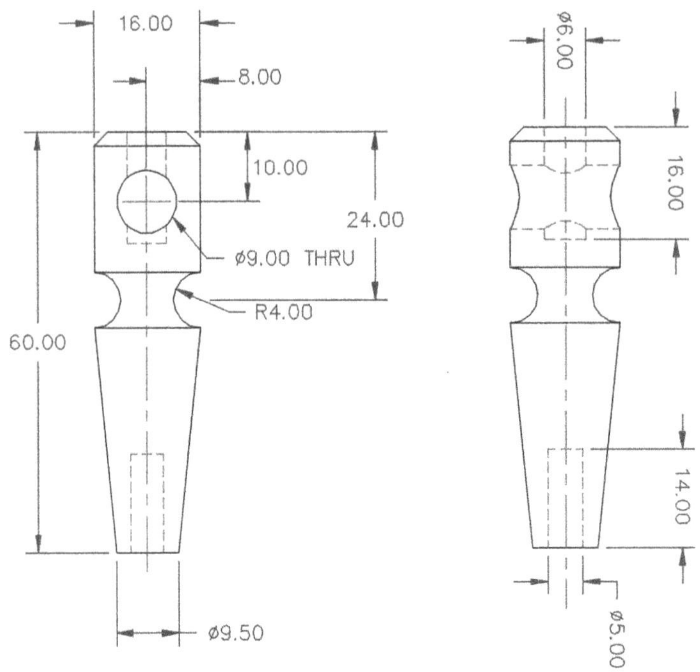

Chapter 15

Advanced design and file management tools

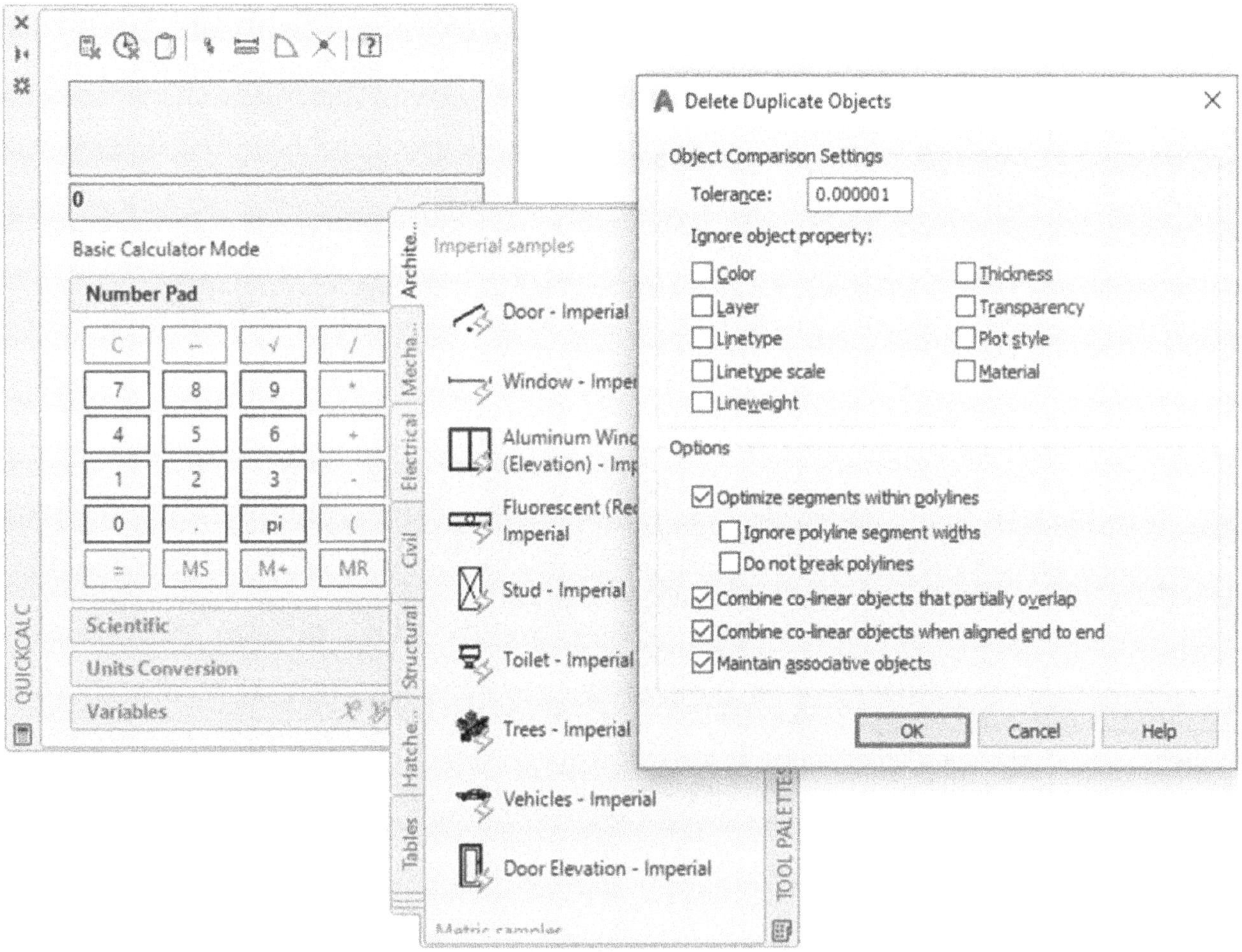

Learning objectives

In this chapter, we continue to introduce a variety of advanced tools and discuss the following:

- Align
- Action recording
- Audit and recover
- Blend
- Break and join
- CAD Standards
- Calculator

Up and Running with AutoCAD 2022. https://doi.org/10.1016/B978-0-323-89923-9.00015-9

- Defpoints
- Divide and point style
- Donut
- Draw order
- eTransmit
- Filter
- Hyperlink
- Lengthen
- Object tracking (OTRACK)
- Overkill
- Point and node
- Publish
- Raster
- Revcloud
- Sheet sets
- Selection methods
- Stretch
- System variables
- Tables
- Tool palette
- UCS and crosshair rotation
- Window tiling
- Wipeout

By the end of this chapter, you will add significantly to your command repertoire.
Estimated time for completion of this chapter: 3 hours.

15.1 Introduction to advanced design and file management tools

This is easily the most diverse and unique chapter you will read in the book. Here, we really explore the "nooks and crannies" of AutoCAD and cover a wide range of very diverse commands. The reason these commands are grouped here is because this text was written from the start to simplify and isolate core topics and essential ideas, without cluttering the main themes with additional AutoCAD commands and concepts. Several more "big ideas" are left, such as xrefs, attributes, and isometric. Those are addressed in upcoming chapters. However, that leaves out a great deal of useful concepts and commands that did not quite fit in anywhere else.

These bite-sized chunks of information form the basis of this chapter. Many of these topics are just random commands, standing alone, not part of any other major concept, while some may be part of a larger family but were not included earlier to avoid getting distracted from a much more important theme. All are useful to one extent or another and are necessary for a well-rounded knowledge of AutoCAD.

Obviously, this is not an exhaustive list; some items have been left out due to space limitations or very limited usefulness, but for the most part, the important ones are here. The list is in alphabetical order. Go through it at your own pace, noting how each concept or command can help you in your design work. Some are whimsical, some are rather odd, and surely you will find one or two that you cannot do without!

15.2 Align

This command does exactly what it advertises, aligns one object to another. In Fig. 15.1, we have a rectangle drawn and rotated to some random angle and below it a straight line. While one can always measure the angle of the rectangle relative to the line and use the rotate command, there is a much easier and accurate way to align the rectangle to the line (or vice versa).

Step 1. Type in `align` and press Enter. You can also go to the Modify panel dropdown menu on the Home tab of the Ribbon.

- AutoCAD says: `Select objects:`

Step 2. Select the object to be moved into alignment (the rectangle). Press Enter.

- AutoCAD says: `Specify first source point:`

Step 3. Pick the lower left corner of the rectangle using the ENDpoint OSNAP (Point 1 in Fig. 15.2).
- AutoCAD says: `Specify first destination point:`

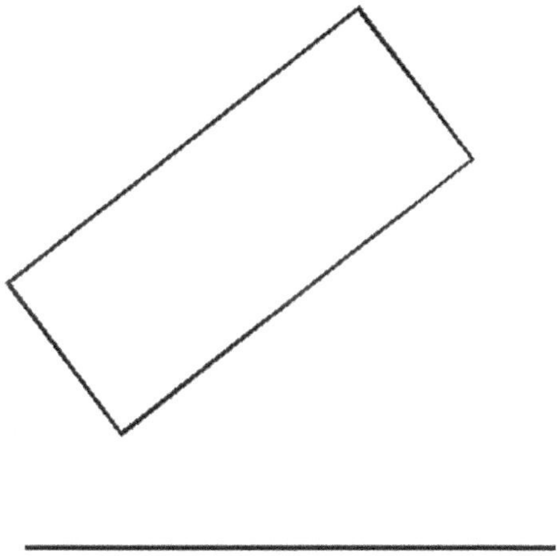

FIGURE 15.1 Rectangle to be aligned with line.

Step 4. Pick the left point of the line using the ENDpoint OSNAP (Point 2 in Fig. 15.2).
- AutoCAD says: `Specify second source point:`

Step 5. Pick the lower right corner of the rectangle using the endpoint OSNAP (Point 3 in Fig. 15.2).
- AutoCAD says: `Specify second destination point:`

Step 6. Pick the right point of the line using the ENDpoint OSNAP (Point 4 in Fig. 15.2).
- AutoCAD says: `Specify third source point or <continue>:`

Step 7. You have no need for a third set of points, so press Enter.
- AutoCAD says: `Scale objects based on alignment points? [Yes/No] <N>:`

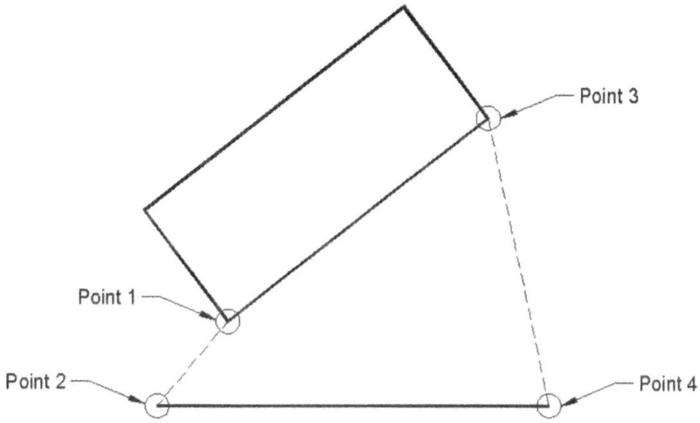

FIGURE 15.2 Alignment points.

You can do something interesting here. Not only can you align the rectangle to the line, you can also scale it up to the line if needed by selecting `Yes`. We do not try this now but go back after learning the basic align command and try this variation of it. For now, just press Enter to accept the default `No` option. The rectangle is aligned to the line, as seen in Fig. 15.3.

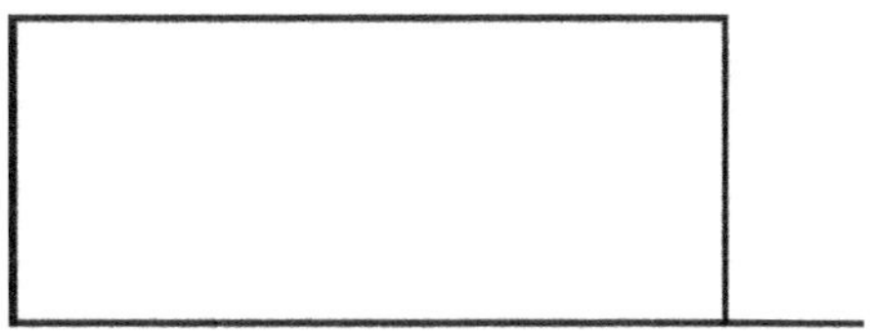

FIGURE 15.3 Alignment, result.

15.3 Action recording

Action recording is a relatively new tool in AutoCAD that is used to build Action Macros and save them as .actm files. *Macros* is short for *macroinstructions*, a computer term that can be defined as a short set of instructions to perform a task. The idea here is to record your on-screen actions and save them for future use. These actions should be some sort of useful steps to build a tedious or repetitive object and not just any simple steps. You can then use this macro to make your life easier later.

To give this a try, open a brand-new file with the Ribbon present. You can use the command without the Ribbon, by simply typing in `actrecord` to start the process and `stop` to end it, but it is easier and more visual with the Ribbon present. Either way, start the command via any of the following methods:

Keyboard: Type in **actrecord** and press Enter
Ribbon: **Manage** tab→**Record**

You then see what is shown in Fig. 15.4. Notice the Ribbon tab's record button became a stop button. The mouse has also acquired a red "Recording" dot. If for any reason the Ribbon did not open to the right tab, just manually select the Manage tab.

Begin drawing anything you like, such as a piece of furniture, as seen in Fig. 15.5. Notice how each action of your mouse, keyboard, menu, Ribbon, and so on is recorded in the log on the upper left. You can even add comments to the macro that will pop up on playback, as well as more advanced scripts and macros via AutoLISP or a more advanced language like Visual Basic.

When done, press the Stop button and the Action Macro dialog box appears. Give the macro a descriptive name (no spaces allowed) and press OK, as seen in Fig. 15.6. Note that, if the Folder Path is darkened out and not editable, it can be easily changed via Options…. Go into the first tab, Files, then find Action Recorder Settings. There you can change the destination folder for your action macros.

The macro has now been completed and stored as a .actm file in a default location (which can be changed via Options… → Files). You can put it to use by erasing the chair and pressing the Play button. The macro executes and the chair is drawn. You then see the message shown in Fig. 15.7.

This is a very simple example to illustrate the basics. You can get quite complex with macros. For example, you can insert user input or a message. Macros recognize only a few dialog boxes (like layers), so if you need to use one, then make it a command line dialog box via the - (hyphen) prior to typing it in. So, hatch would be `-hatch`, and so on.

Experiment with this on your own; there are many possibilities.

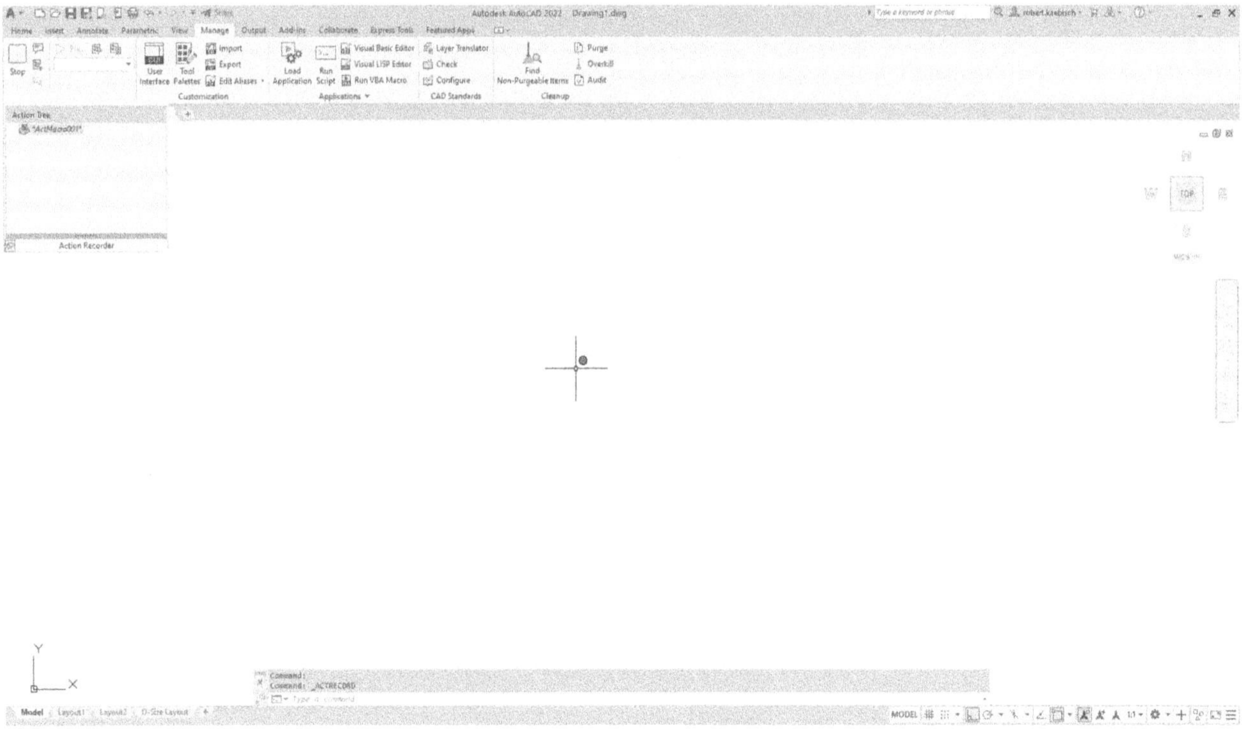

FIGURE 15.4 Action recorder.

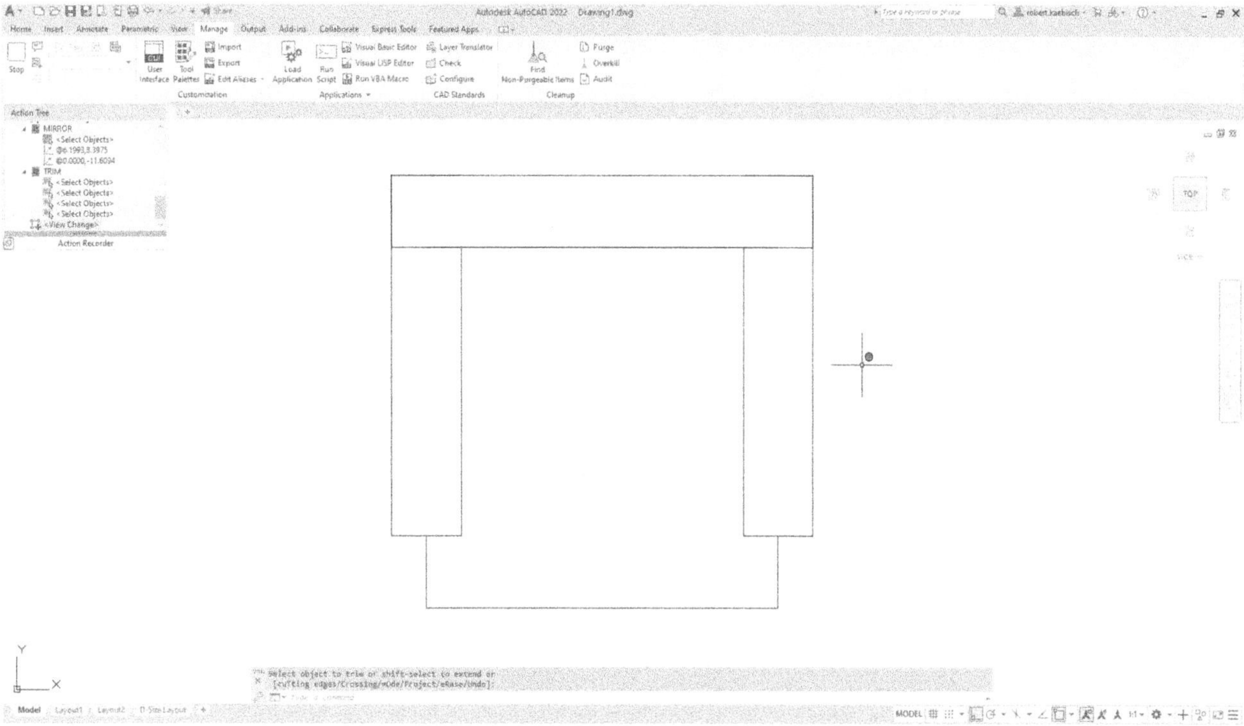

FIGURE 15.5 Recording log.

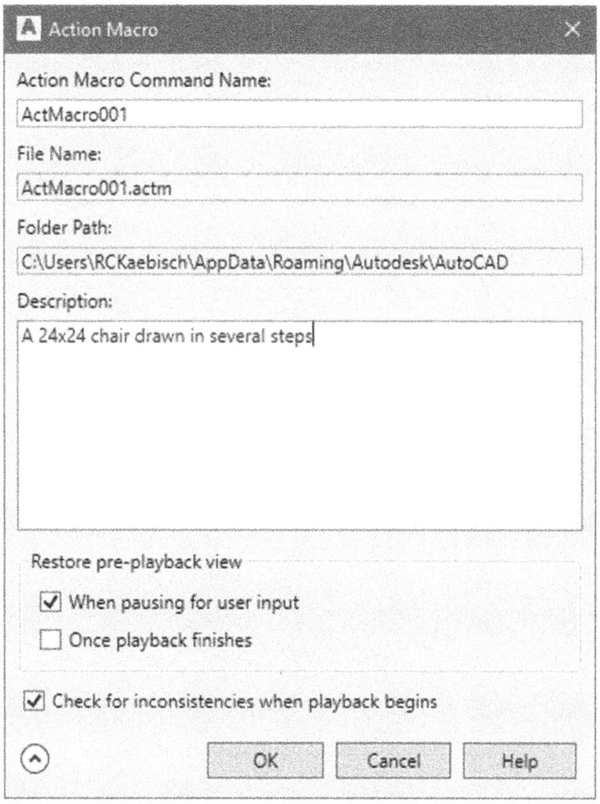

FIGURE 15.6 Action Macro.

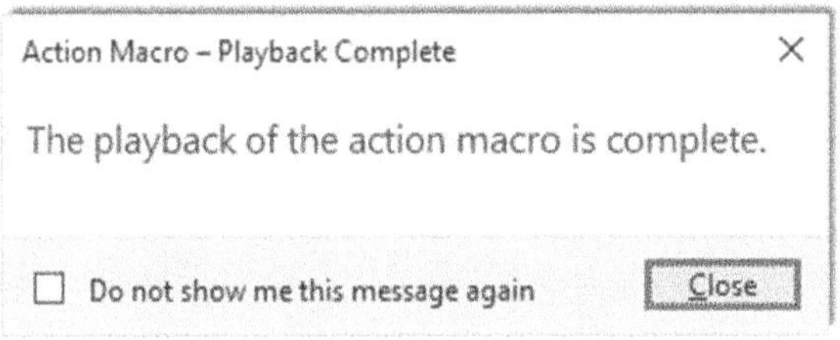

FIGURE 15.7 Action Macro – Playback Complete.

15.4 Audit and recover

Although rare, AutoCAD does sometimes lock up and crash (usually a PC hardware or memory issue but it can also be software related). In the event of a crash, the drawing is automatically saved but may be corrupted when you try to open it up again. Sometimes, drawings just cause AutoCAD to behave oddly for no specific reason (i.e., commands do not work as expected or just fail). In either case, the drawing needs to be audited or recovered.

Auditing is the lower-level check. AutoCAD evaluates the integrity of the drawing and corrects some errors. You simply type in `audit` and press Enter; alternatively, use Application Menu (Big A) → Drawing Utilities tools. AutoCAD asks about fixing any detected errors (say `Yes`), cycles through drawing entities, and in a few seconds generates a log like what you see next (assuming nothing wrong was found or fixed, as seen next):

```
Auditing Header
Auditing Tables
Auditing Entities Pass 1
Pass 1 100 objects audited
Auditing Entities Pass 2
Pass 2 100 objects audited
Auditing Blocks
1 Blocks audited
Total errors found 0 fixed 0
Erased 0 objects
```

Of course, if something is found, fixes are reported. Sometimes, however, the audit is not enough and you must recover the drawing. This is not as drastic as it sounds. The recover command is powerful and almost always brings the drawing back as a matter of routine. I can count on one hand the number of drawings that were lost forever due to corruption in 18 years of CAD design. And, even in those cases, going to the previous night's server backup files restored the drawing, minus a few hours of lost work.

To use this command, open a blank drawing file, type in `recover`, and press Enter. Alternatively, you go to the Application Menu (Big A) Drawing tools. You are prompted to browse and find the drawing. Once you select it and press Open, recover runs and generates its own log, as seen next. This example was also generated with a clean file. If there were items to fix, recover would have shown the actions taken.

```
Drawing recovery.
Drawing recovery log.
Scanning completed.
Validating objects in the handle table.
Valid objects 755 Invalid objects 0
Validating objects completed.
Salvaged database from drawing.
Auditing Header
Auditing Tables
Auditing Entities Pass 1
Pass 1 900 objects audited
Auditing Entities Pass 2
Pass 2 900 objects audited
```

```
Auditing Blocks
19 Blocks audited
Total errors found 0 fixed 0
Erased 0 objects
```

Following the log, you see the message box shown in Fig. 15.8.

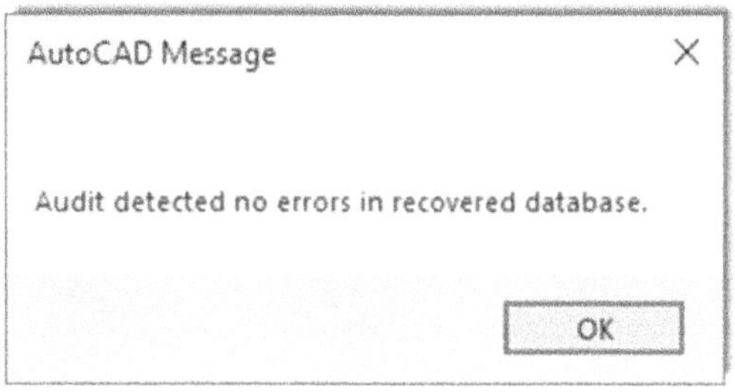

FIGURE 15.8 Recover message.

If something were indeed wrong (and then fixed), the drawing would have just opened as normal. Immediately save it.

15.5 Blend

Blend allows you to join, or blend together, linework that is otherwise separate. This linework typically is straight lines or curves, such as arcs or splines. To a lesser extent, some shapes (such as rectangles) can also be blended, although that does not really yield useful results. In all cases, what the blend command does is create a spline between these shapes, curves, and lines. The spline shape can then be modified if needed. To try it out, create two arcs as seen in the top part of Fig. 15.9, and then follow the steps shown next.

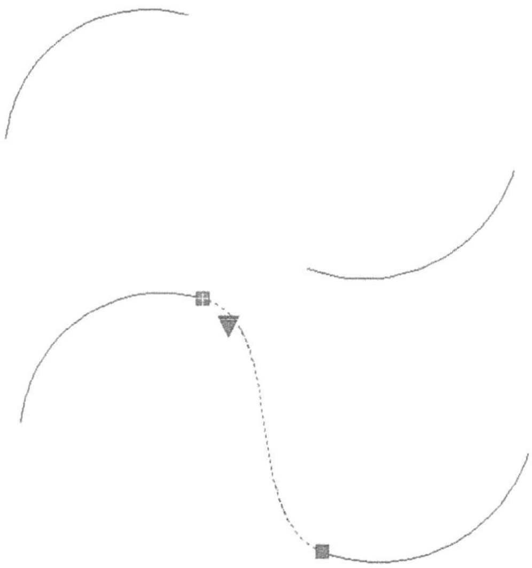

FIGURE 15.9 Blend.

Keyboard: Type in **blend** and press Enter
Ribbon: Home tab→Modify

Step 1. Start up the blend command via any of the preceding methods.

- AutoCAD says: `Continuity = Tangent`
  `Select first object or [CONtinuity]:`

Step 2. Pick the first arc, close to where the blend will start.

- AutoCAD says: `Select second object:`

Step 3. Pick the second arc, close to where the blend will end. You see a live preview of the new spline just as you are about to click. A new spline then is created to connect the arcs.

The bottom part of Fig. 15.9 shows the results. The new blend has been selected, and the grip points and a down arrow are visible. The down arrow reveals a menu if clicked on. This menu lets you choose between showing control vertices or fit points, which presents you with different methods for altering the blend. Experiment with both to see the effects.

15.6 Break and join

The break command takes a geometric object and puts a break in it (removes a section, in other words) or, if applied at an intersection with another object, breaks that object at the intersection without removing any material. The join command is the exact opposite. It adds material to close the break but works only in certain circumstances. Let us examine each command closer.

Break, Method 1

To try out the break command, open a new file and draw a line of any size or direction on your screen.

Keyboard: Type in **break** and press Enter
Ribbon: **Home** tab→**Modify**

Step 1. Start up the break command via any of the preceding methods. Be sure to turn off OSNAP (or use a nearest OSNAP).

- AutoCAD says: `Select object:`

Step 2. Go ahead and click to select the line at some random spot. It turns dashed.

- AutoCAD says: `Specify second break point or [First point]:`

Step 3. Pick another nearby point on the line, and a section between these two points is removed (Fig. 15.10). Note that you can preview the result of the break command as you execute it, like many other modify commands.

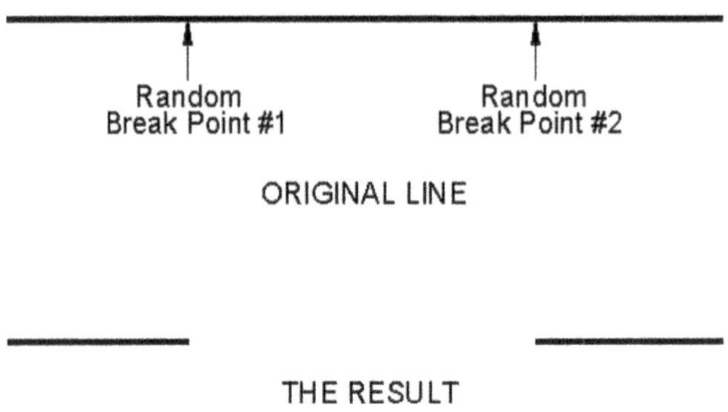

FIGURE 15.10 Break command, Method 1.

In this method, the first point of the break is somewhat random unless you used an osnap. When you selected the line, the first point was automatically chosen. However, as shown in the next method, you can select the line to be broken, then, using the first point option, specify exactly where to start the break. We use this to our advantage to create four lines out of two, as seen next.

Single Break, Method 2

Draw a horizontal line of any size, followed by a perpendicular vertical one that is of equal size. They should cross perfectly in the middle. You have a plus sign. One good trick to do this is to first draw a circle and just connect the opposing quadrant points with two lines, as seen in Fig. 15.11.

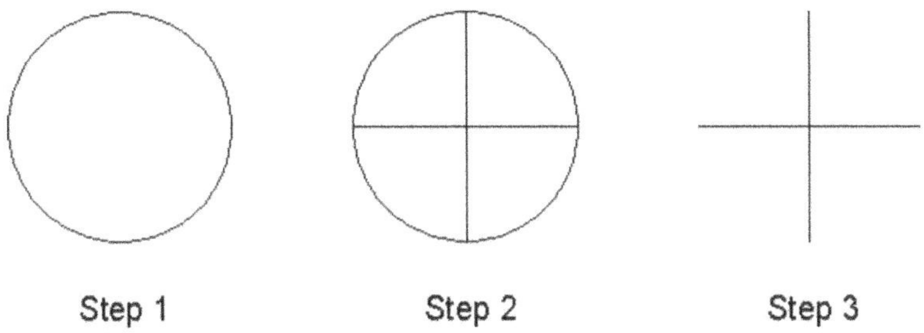

FIGURE 15.11 Setup for break command, Method 2.

Keyboard: Type in **breakatpoint** and press Enter
Ribbon: **Home** tab→**Modify**

The horizontal line is now two equal pieces. Repeat these steps with the vertical line. The result is four lines where there were only two just moments earlier, as seen in Fig. 15.12, when the grips are revealed. All four are broken at their intersection.

Step 1. Start the single break command via any of the previous methods.
- AutoCAD says: `Select object:`

Step 2. Go ahead and click to select the horizontal line at some random spot. It highlights.
- AutoCAD says: `Specify break:`

Step 3. A midpoint or Intersection OSNAP, select the middle of the horizontal line.

Step 1. Start the join command via any of the preceding methods.
- AutoCAD says: `Select source object:`

Step 2. Select the first line.
- AutoCAD says: `Select lines to join to source:`

Step 3. Select the second line.
- AutoCAD says: `Select lines to join to source:`

Step 4. You are done with the two segments, so press Enter.
- AutoCAD says: `1line joined to source`

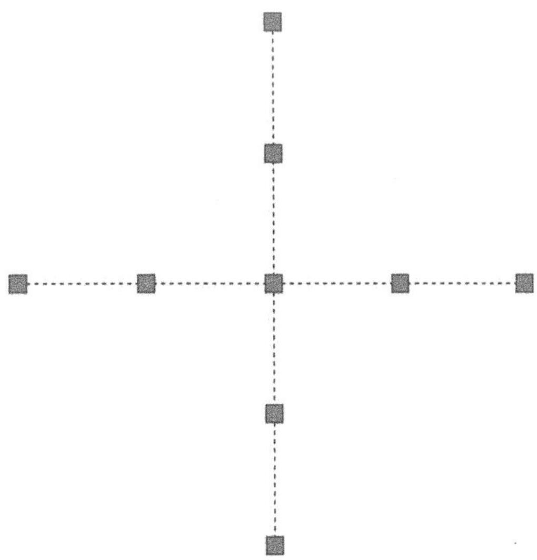

FIGURE 15.12 Break command, Method 2.

Join

The join command is included in this discussion because it is the exact opposite of break. It closes the gap between colinear lines or creates a polyline out of lines with shared endpoints. Note the important disclaimer, here: The join command does not close a gap between lines that are not colinear. This means that, if they are not pointing in the exact same direction and are not actually in line, then it does not work. To try this out this command, draw two coplanar lines (or use break to put a gap in), as seen in Fig. 15.13. Note that join works on an unlimited number of coplanar lines; we are using only two in this example for simplicity.

Keyboard: Type in **join** and press Enter
Ribbon: **Home** tab→**Modify**

FIGURE 15.13 Setup for join command.

A solid line takes the place of the two segments.

15.7 CAD standards

This rather obscure procedure may be valuable to a few organizations, and I have on occasion recommended it to prime contractors who often find their design files bouncing around from subcontractor to subcontractor. After all the respective additions to the drawings are finished, they are sent back to the main contractor. There, it is found that the standards have deteriorated, as everyone did his or her own thing regarding layers and text or dim styles. A way to identify and clean up the mess would be useful, hence, the existence of this tool.

The CAD Standards procedures can be a bit confusing, and the process is not effortless nor does it guarantee perfect results. Depending on the damage, you may still need to do manual cleanup of layer properties, errant fonts, and the like. The best defense is to hold all others who may work on your drawings to a high standard from the start and outline what is acceptable and what is not. Obviously, this can be challenging in just one office, never mind implementing it for designers at other, remote locations.

Here is the basic idea. Come up with one file that is in great shape. This can be a regular job that is singled out for its perfect or near perfect use of company standards and is of the highest quality and integrity. Then, do a Save As for this job under a .dws (drawing standards) extension. This is your standards file. All others are compared to it as they come in from subcontractors or just other designers.

To implement the standards, open a job that just came in and is of questionable CAD quality. Go to the CAD Standards panel of the Manage tab of the Ribbon. Select the Configure command The Configure Standards dialog box appears, as seen in Fig. 15.14.

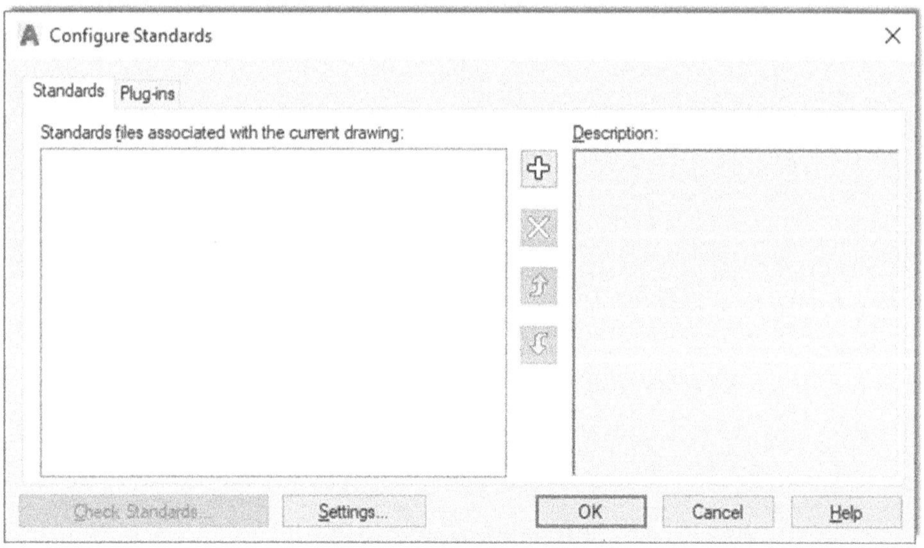

FIGURE 15.14 Configure Standards.

Next use the plus (+) sign to open the saved .dws file. The file appears in the window on the left with some descriptions on the right. The Check Standards button (lower left) also lights up, at which point you press it to execute the check. The Check Standards dialog box appears, as seen in Fig. 15.15.

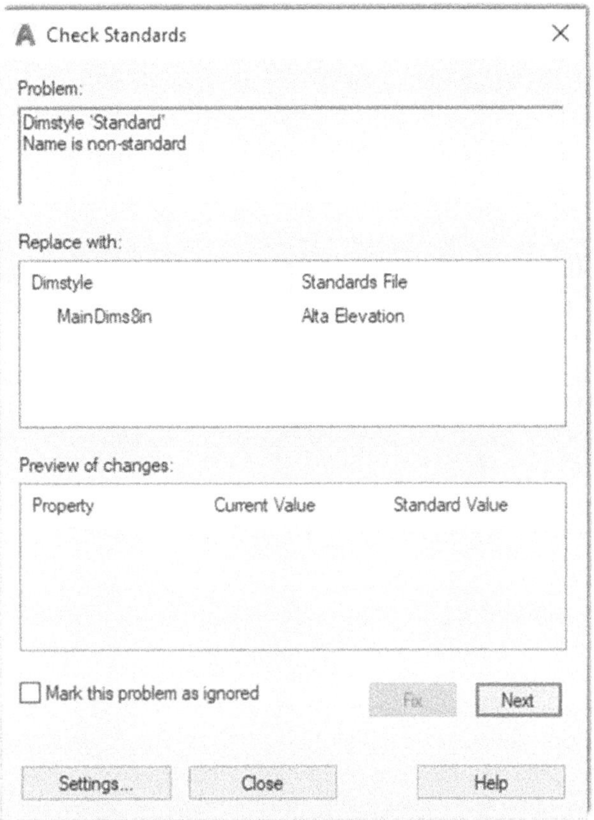

FIGURE 15.15 Check Standards.

Next, you need to run through the fixes, item by item. Continue pressing the Fix button as AutoCAD runs through the fixes and matches up text, layers, and other items. When done, the Check Complete notice shown in Fig. 15.16 displays. In this example, three problems were found but none fixed automatically.

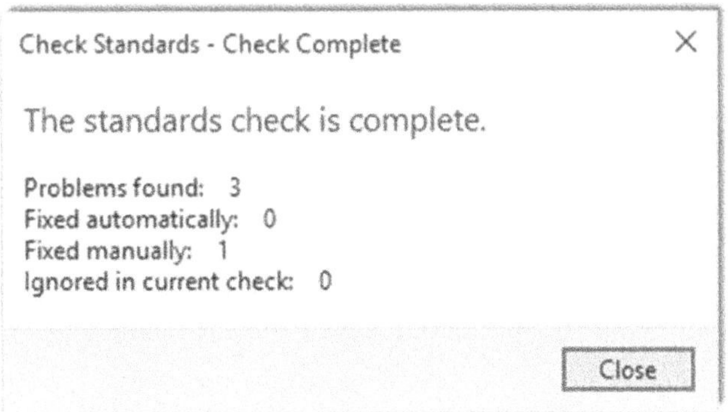

FIGURE 15.16 Check completed.

15.8 Calculator

The calculator is a surprisingly powerful and versatile tool within AutoCAD, and one that is often overlooked by users. The function's power and versatility should come as no surprise, as AutoCAD itself is highly mathematical software, a calculator, performing complex operations "under the hood" as you draw, copy, and rotate objects.

The calculator function can perform all regular and scientific calculations you would expect from a handheld device; in addition, it can assist you in the construction of geometry and even allows you to use AutoLISP, a programming language. A full description of the calculator command can take up dozens of pages, so we explore only the basics here. However, we look at one example of construction assistance using the transparent line command cal and one example of the actual graphic calculator tool (QuickCalc) to do unit conversions. You are encouraged to read the Help files for a wealth of additional information.

First, a brief overview of the basic cal command. This was the original version of the calculator. It is a command line tool enabled by typing in `cal` and pressing Enter. AutoCAD then says: `>>Expression:` at which point you can enter some math function using the familiar characters such as +, –, *, and /. Pressing Enter again yields an answer.

Not too long ago the limit of the calculator was +32,767 and –32,768 as the largest displayable final integer, but that restriction was lifted somewhere around the AutoCAD 2005 release, and you can now work with integer numbers between +2,147,483,647 and –2,147,483,648 (that is 2 to the 15th and 31st powers, respectively, if you really wanted to know where those seemingly random numbers came from).

Let us use the previously described cal command to assist in design because crunching numbers alone can be done by your handheld or Windows calculator. Suppose you need to offset a line one half of some number, but you do not know what that value is (say, one half of 17.625). Instead of reaching for that calculator, let AutoCAD do both: calculate one half of 17.625 as well as offset the line, all in one command. To do this,

1. Draw a line of any reasonable size.
2. Start the offset command via any method.
3. You do not know the offset distance, so let us use the calculator.
4. Type in `'cal` (the apostrophe makes cal, or any command, temporarily transparent, meaning you can enter it without exiting out of another, in-progress command, which in this case is offset). Press Enter.
5. The `>> Expression:` prompt asks for the expression. Enter `17.625/2` and press Enter.
6. Your answer (`8.8125`) appears, and you may resume the offset command.
7. Continuing as usual, select the line and pick a direction to complete the offset and Esc to exit.

This was a very simple case of how the calculator assists in design tasks. It can do a lot more in that regard, and you are encouraged to examine the Help files for additional examples.

The calculator function is not just a typed command line but also now comes in a graphic version. If you use this graphic version, you are using what is referred to as a *QuickCalc*. It is almost a whole new command all to itself and has quite a bit more functionality.

Keyboard: Type in **quickcalc** and press Enter.
Ribbon: **View** tab→**Palettes**

Start up the QuickCalc command via any of the preceding methods and you see Fig. 15.17. The Number Pad and Scientific keypads are self-explanatory. If you ever used a handheld calculator, you should be right at home with this one. Variables, at the bottom, are not covered, but Units Conversion deserves a special mention. It is a very useful tool for converting between a wide variety of Length, Area, Volume, and Angular values. Simply select from the fields on the right (drop-down arrows appear) and enter the values to convert just below that. A converted value appears according to what you requested.

15.9 Defpoints

Defpoints is not a command, but rather a concept that appears often and bears a short mention. Defpoints stands for definition points, and they appear in the drawing (and mysteriously in the Layers dialog box, where most users first notice them) as soon as you add any dimension to your drawing. What are they exactly? They are the tiny points where a

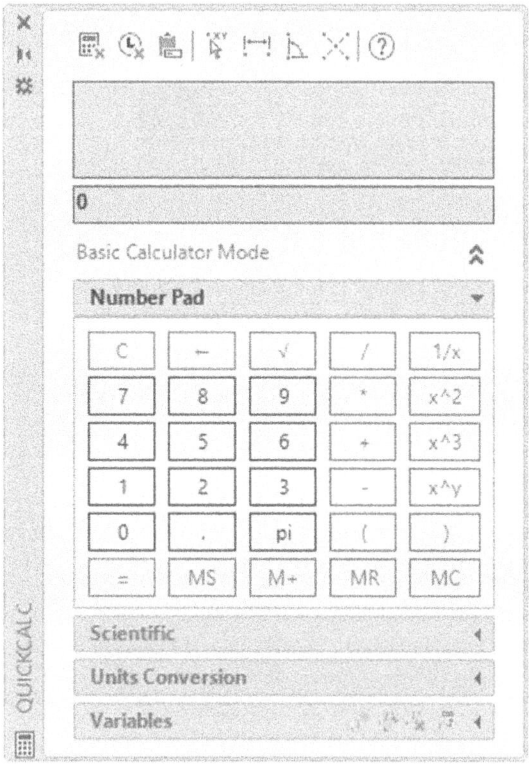

FIGURE 15.17 QuickCalc.

dimension attaches itself to the object being dimensioned. The Defpoints layer does not go away easily and usually prompts questions from students as to its nature and intent. The layer fortunately does not plot, and you can pretty much disregard its existence if you choose. One benefit of it not plotting is that you can put any objects that you want to see in AutoCAD but do not want to see physically printed on the Defpoints layer (such as viewports, as is discussed in Chapter 10). In fact, that is exactly what many users do, though a dedicated VP layer is recommended instead. In conclusion, do not worry too much about Defpoints; just understand what they are and continue.

15.10 Divide and point style

Divide does exactly what it advertises; it divides an object into equally spaced intervals. Note, however, that the object is not actually broken into those segments (that is for a break command to do) but rather just divided by equally spaced points. Then, you can use these points as guides for further work. Let us give this a try.

Step 1. Draw a horizontal line of any size.
Step 2. Type in `divide` and press Enter. Alternatively, you can use the Draw panel dropdown menu n the Home tab of the Ribbon.
- AutoCAD says: `Select object to divide:`

Step 3. Select the line.
- AutoCAD says: `Enter the number of segments or [Block]:`

Step 4. Enter some value, `10`, for example, and press Enter.

The line looks the same, with no apparent divisions. It appears the command has not worked. It worked just fine, and the line is divided into the 10 segments using nine points, as expected. The problem is that you cannot see the points; they are obscured by the line, and we need to change the point style to something more visible. Simply type Ptype or go to the Utilities panel dropdown of the Home tab of the Ribbon. You then see the dialog box shown in Fig. 15.18.

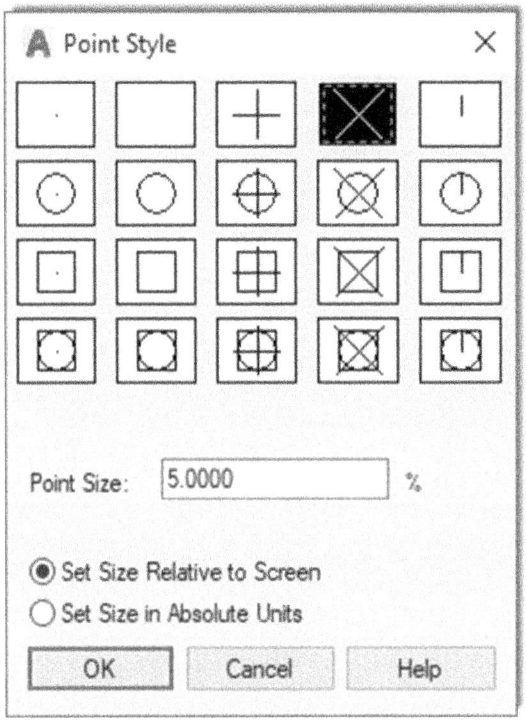

FIGURE 15.18 Point Style.

Select a point style that is not too complex; the X format in the top row (second from right) is a popular one. Press OK, and the divided line appears as you were expecting it, divided into the 10 visible segments (Fig. 15.19).

FIGURE 15.19 Divided line.

So, how exactly can you use this to your benefit? Well, the Node OSNAP, which is discussed later in this chapter, allows you to connect to points. If you activate it, you can attach new geometry, perhaps new lines, to those Xs. This is quite useful in basic construction and layout. Note an interesting quirk. If you zoom in or out, the points change size, but revert to "standard" size upon typing in regen.

15.11 Donut

Although this command sees only occasional use, it is still worth a mention. It creates exactly what it advertises: a donut of varying size. Broken down into its components, a donut is two circles of different size, one inside the other, and a solid hatch fill. You can technically create this shape with existing techniques, but the donut command just makes it easier and much faster. You have control over the sizes of both circles and, as a result, the size of the exterior and interior of the donut. The inner fill between them is always a solid hatch—and if you specify an inner circle diameter of zero, you will create a solid filled circle with just one command. Let us give it a try.

Keyboard: Type in **donut** and press Enter
Ribbon: **Home** tab→**Draw**

Step 1. Start the donut command via any of the preceding methods.

- AutoCAD says: `Specify inside diameter of donut <0.5000>:`

Step 2. You can accept this by pressing Enter or type in another value (we accept the 0.5 in this example).

- AutoCAD says: `Specify outside diameter of donut <1.0000>:`

Step 3. Once again, you can accept this value or type in your own (we accept the 1.0 in this example).

- AutoCAD says: `Specify center of donut or <exit>:`

Step 4. Click anywhere on the screen and the donut appears, as seen in Fig. 15.20.

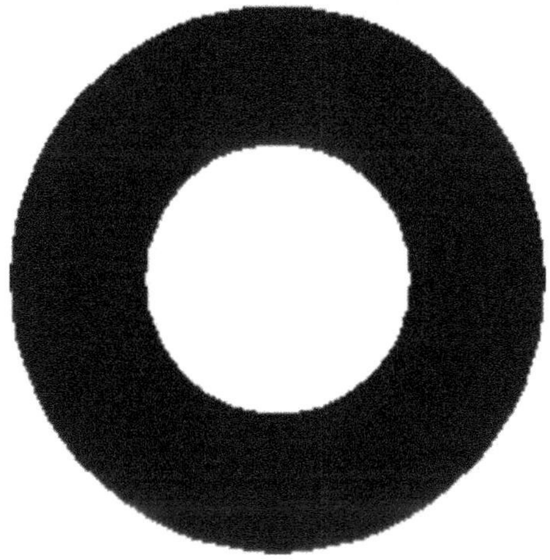

FIGURE 15.20 Donut.

15.12 Draw order

Keyboard: type in **draworder** and press Enter
Ribbon: **Home** tab→**Modify**

Draw Order refers to the stacking order of objects positioned one on top of another. This feature is commonly used in most graphic design programs and has an occasional use in AutoCAD as well. The stacking order is usually not of major concern in day-to-day basic drafting. If items overlap, the one drawn most recently is displayed on top; and in cases of two intersecting lines, the intersection point is 1 pixel in size, hardly worth a second thought.

Sometimes, however, such as when two or more hatch patterns are stacked, the order is critical. We use this feature later in this chapter to great effect when we discuss Wipeout. For now, just know where the Draw Order command is. It is found in the Modify panel dropdown of the Home tab of the Ribbon or simply type draworder.

Bring to Front and Send to Back objects strictly to and from the top and bottom of the order, while Bring Above Objects and Bring Below Objects are more selective, moving items one step at a time. With these four tools, you can arrange any number of objects in any order.

15.13 eTransmit

The purpose of this command is to assist with emailing drawings. The problem is that, when drawings are sent out, items can be inadvertently omitted. These omissions may include something major, like associated xrefs, or something secondary, such as fonts and styles. As a project grows more complex, it is easy to forget to send xrefs and even easier to forget about the unique fonts, styles, and .ctb files that have been generated over time. The recipient of the email may have only some of these, and what they download and reconstructs on their end may not be quite what you may have intended to be seen.

The eTransmit command is a go-getter of sorts. It sniffs out every associated file and everything else that may be relevant to recreate the file(s) in a new location as originally intended. Then, it packages everything together in a folder and sends it out. While veteran users have remarked that an experienced and aware designer needs to always have a clear idea of what to send (eTransmit is a relatively recent addition to AutoCAD after all), it is still a valuable tool for novices and experts alike.

So, what is (and what is not) included in the eTransmit transmittal? Here is a partial list of what is included. You should know of most of these files. If not, refer to Appendix C.

***.dwg**: Root drawing file and any attached external references.
***.dwf**: Design web format files that are attached externally to the root drawing or xrefs.
***.xls**: Microsoft Excel spreadsheet files that are linked to data extraction tables.
***.ctb**: Color-dependent plot style files used to control the appearance of the objects in the drawings of the transmittal set when plotting.
***.dwt**: Drawing template file associated with a sheet set.
***.dst**: Sheet set file.
***.dws**: Drawing standards file associated to a drawing for standards checking.

Here is a partial list of what is not included. Do not worry if you do not recognize all these formats. You will know most of them if you read the Appendices in addition to completing Level 2.

***.arx, *.dbx, *.lsp, *.vlx, *.dvb, *.dll**: Custom application files (ObjectARX, ObjectDBX, AutoLISP, Visual LISP, VBA, and .NET).
***.pat**: Hatch pattern files.
***.lin**: Linetype definition files.
***.mln**: Multiline definition files.

To use eTransmit, save your work but keep the file open. Type in `etransmit` and press Enter. Alternatively, Application Menu (Big A) publish options. You are asked to save the file if it has not been saved yet. Then, the dialog box in Fig. 15.21 appears.

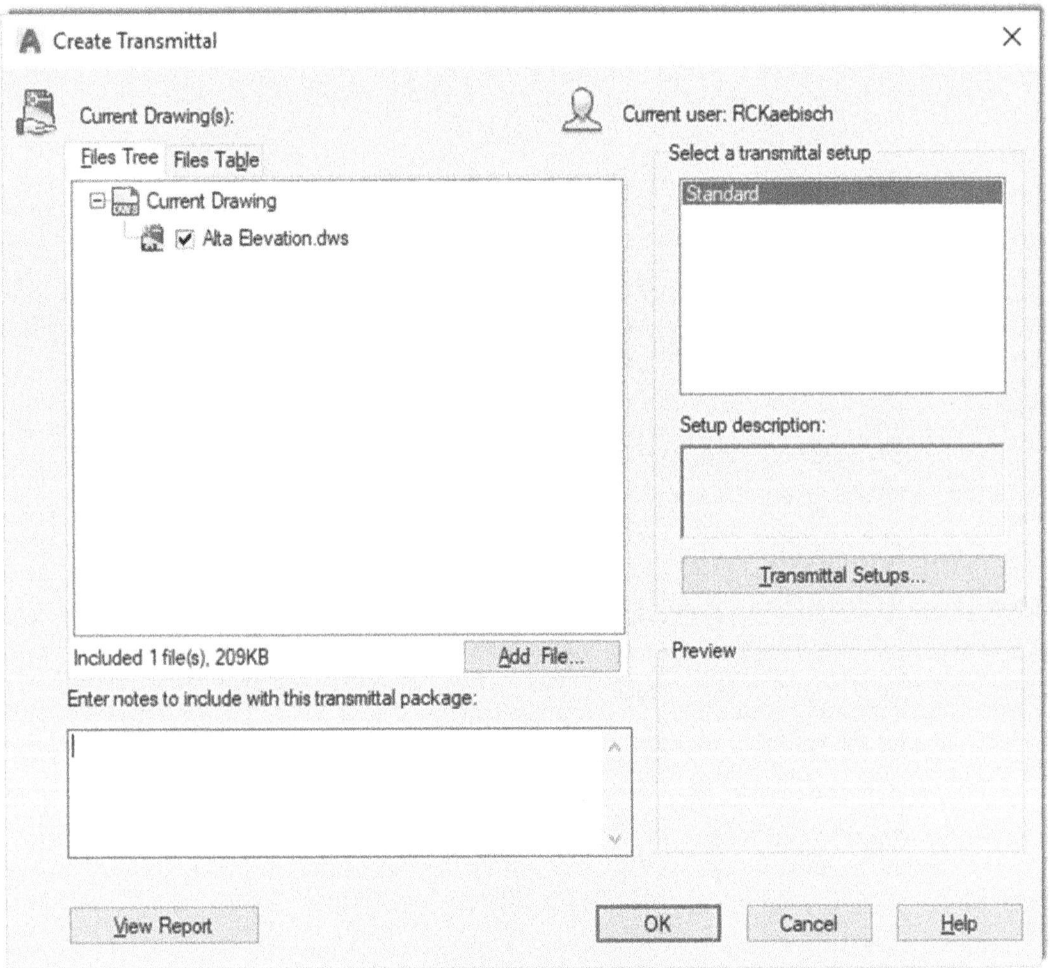

FIGURE 15.21 Create transmittal.

The procedure is highly automated, but you do have to select a few parameters. The current drawing that is open is the one that is packaged together. Some of the included items are shown in the file tree. Expanding the + signs shows specific files (which you can deselect if desired). You also can add more drawings to the transmittal using the Add File… button. If you are fine with just the current drawing (and what is included in it), then press the Transmittal Setups… button. You then see the dialog box shown in Fig. 15.22.

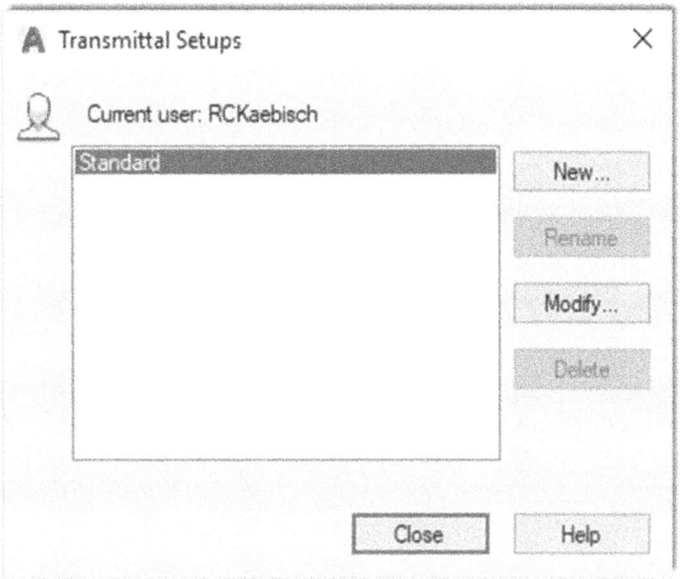

FIGURE 15.22 Transmittal Setups.

Let us create a new setup as an exercise. Press the New… button, and the dialog box in Fig. 15.23 appears. Give the transmittal a name and press Continue. The dialog box in Fig. 15.24 then appears.

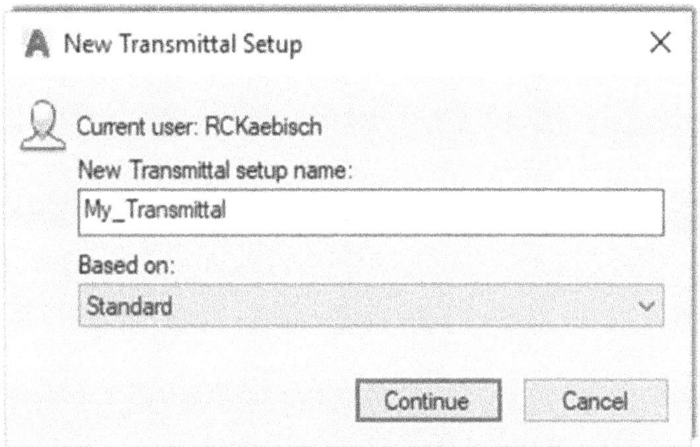

FIGURE 15.23 New Transmittal Setups.

All you really must do here is visually check all the settings. Looking at the fields from the left, top to bottom, we see the Zip option for sending the files (recommended), followed by a variety of information related to formats, names, and locations. Unless you have a specific reason to change something, leave everything as default. On the right-hand side, top to bottom, leave everything under the Actions category as default (unchecked) as well as everything under the Include options category (middle three boxes checked).

When done press OK, then Close in the next dialog box (Fig. 15.22), and finally OK in the main setup box (Fig. 15.21). You are asked where to put the Zip file and for any last-minute name changes. Pressing Save drops the Zip file into its place, and it is ready for emailing.

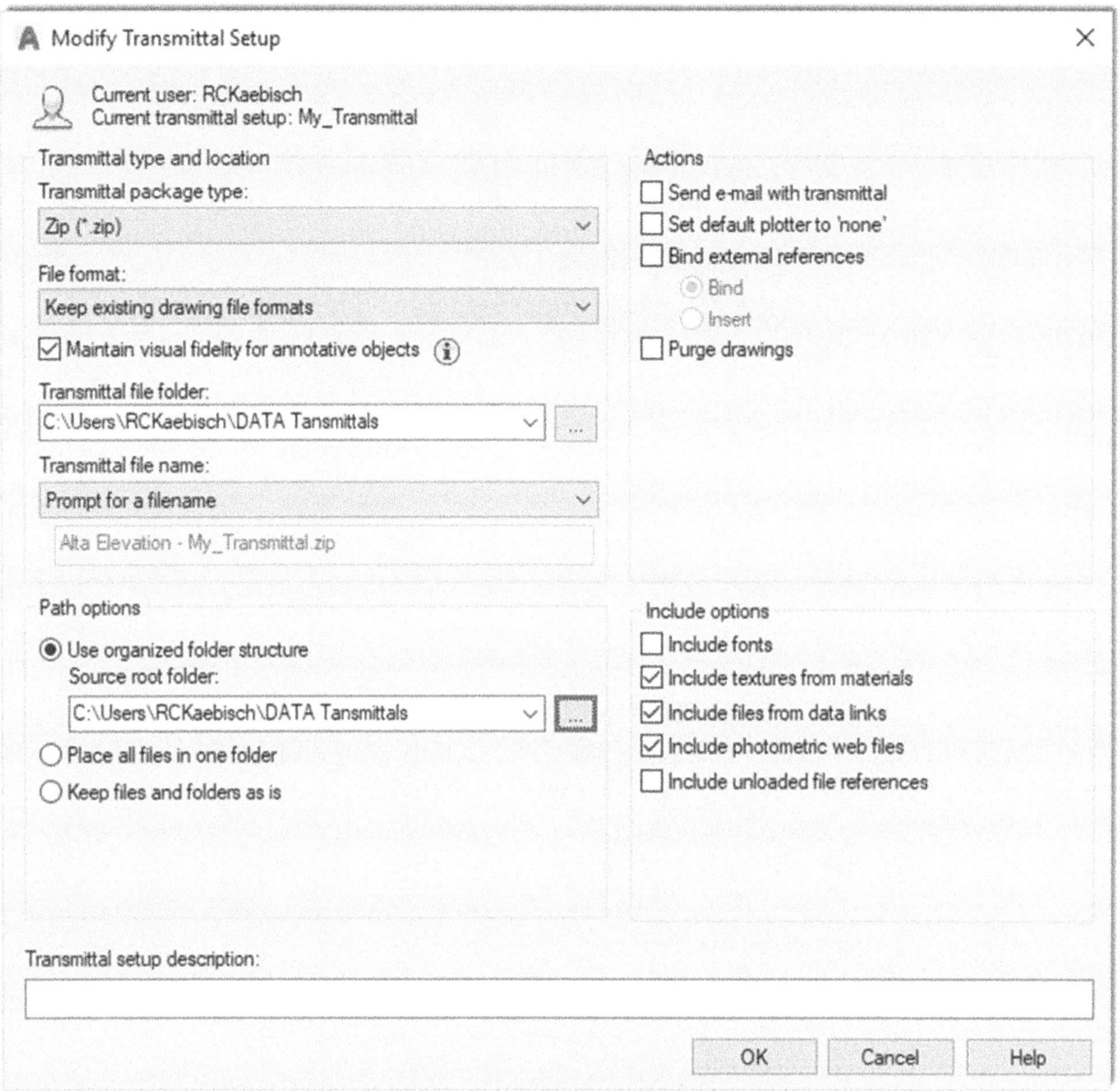

FIGURE 15.24 Modify Transmittal Setup.

15.14 Filter

Filters were briefly discussed in the context of advanced layers (see Chapter 12). While the ideas introduced here are roughly similar, the application is slightly different, so let us review them from the beginning.

Filters are nothing more than a method to isolate and select objects using their features or properties. So, e.g., you can isolate and select all the circles by virtue of them being circles and find green objects because they all have the color green in common. You can also add together properties and create requests like "Find me circles that are red only" and so on. This is a good tool for quickly finding stuff in a complex drawing. Once you find these objects, you can change their properties, delete them, or do whatever else is needed.

To experiment with filters, let us first create a simple drawing that is just a collection of shapes. Draw the set of circles, rectangles, and arcs seen in Fig. 15.25. Then, change the shapes' colors to blue or red.

FIGURE 15.25 Basic shapes for filters.

Type in `filter` and press Enter. There is no Ribbon equivalent. Object Selection Filters appears, as shown in Fig. 15.26. The procedure here is to first select the filter, then add it to the list, and finally apply it.

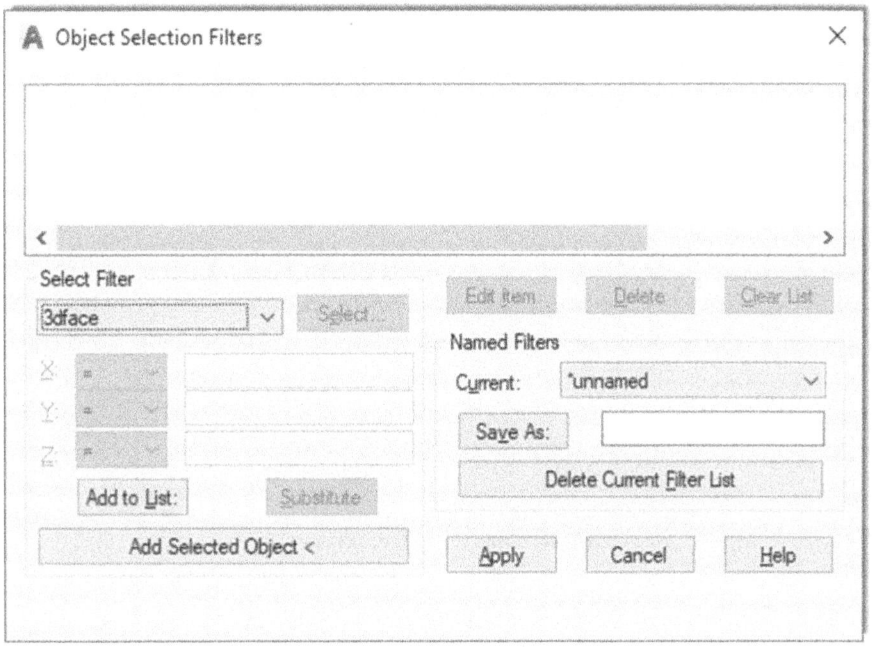

FIGURE 15.26 Object Selection Filters.

Select Arc from the Select Filter drop-down menu on the left. Then, press the Add to List: button just below that. For good measure, let us also add the color red. Select the color from the same Select Filter menu. Notice, at that point, the Select button lights up. Go ahead and select the color red. Then, press Add to List: Your filter should look like Fig. 15.27.

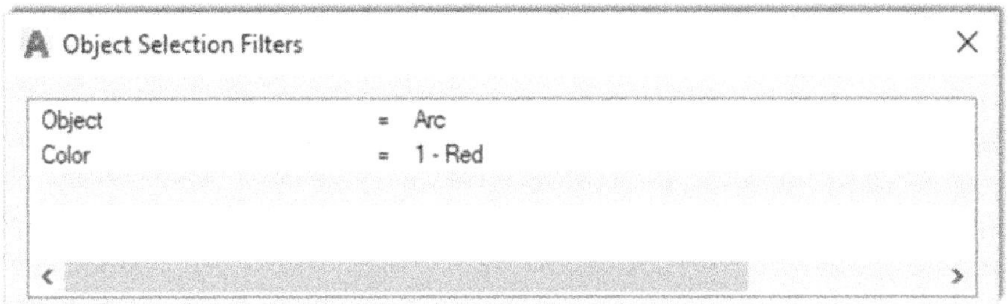

FIGURE 15.27 Objects and colors selected.

Now, press the Apply button. AutoCAD asks you to select objects. Select everything by typing in all. Every red arc in the drawing becomes dashed after the first Enter. Press Enter again and the grips come on, as seen in Fig. 15.28. Now type in `erase` and press Enter. Because they are selected, all the red arcs disappear. Filters can be saved for future use via the Save As: button, and the listing can be modified or deleted via the Edit Item, Delete, and Clear List buttons.

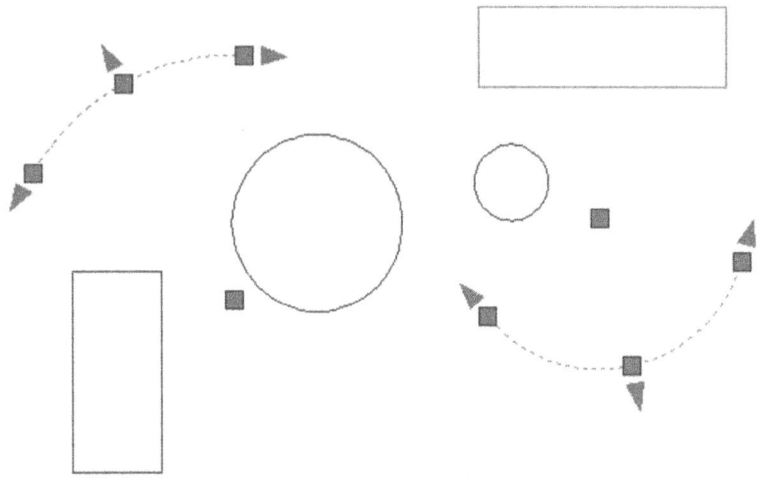

FIGURE 15.28 Segmented arcs selected by filter.

15.15 Hyperlink

The hyperlink command attaches a hyperlink to either an object or text. During this process, you are asked to specify what this hyperlink opens when activated. It can be a website (usually) or a document (sometimes). In either case, it is a great command to quickly access additional useful information. Someone, e.g., can see a drawing of a product and follow the link to a website of the manufacturer or call up an Excel file that lists all the specs of a design. To try out the hyperlink features, draw a symbol for a chair (make a block) and a text website address, as seen in Fig. 15.29.

FIGURE 15.29 Chair and web address.

We would like to link the chair (an *object*) to a furniture website (IKEA, for example) and link the Google text (a text *string*) to the Google website.

Type in `hyperlink` and press Enter. Alternatively, you can use the Hyperlink.... tool in the Data panel of the Insert tab of the Ribbon. AutoCAD asks you to select objects. Go ahead and pick the chair block. The dialog box in Fig. 15.30 then appears.

Enter the name of the desired website in the appropriate field. That is where it says: Type the file or Web page name, not Text to display. Not much else needs to be done, so press OK. Repeat the process for the Google text.

Hover your mouse over the chair then the Google text, and notice there is something new. A hyperlink symbol appears, with text prompting you to hold down the Ctrl key and click to follow the link (Fig. 15.31). If your computer is currently connected to the Internet, go ahead and do that in both cases (chair and Google website) to bring up the respective websites.

There is an interesting variation to this, as mentioned earlier. You need not attach a website to the link. You can also attach a regular document, so it is called up when the hyperlink is clicked on, as before. Pretty much anything can be attached if it can be found and the computer can run the application: Word, Excel, and PowerPoint documents; pictures; videos—you name it.

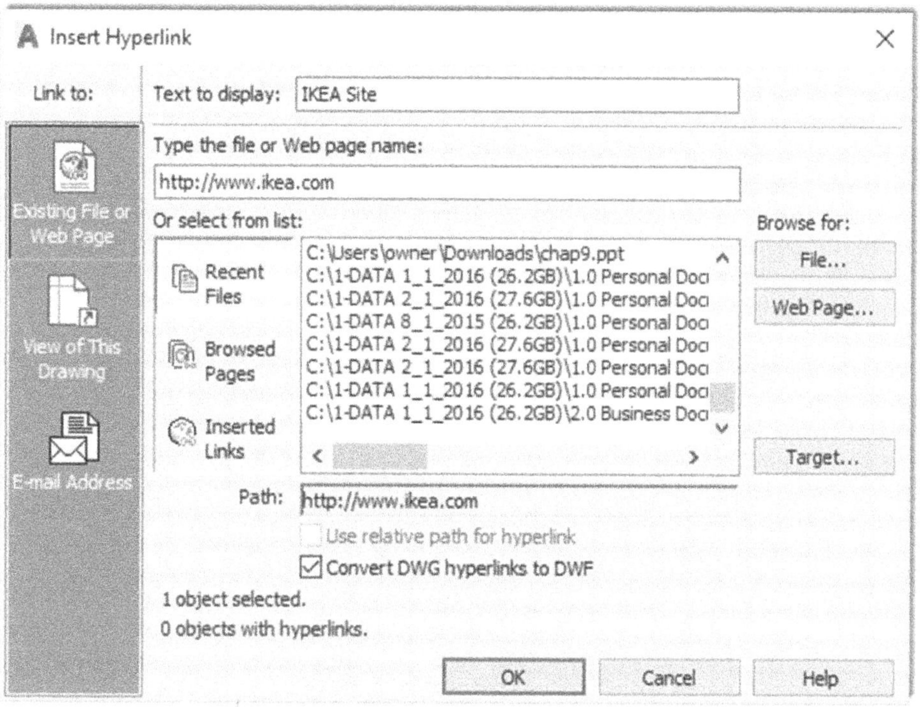

FIGURE 15.30 Insert Hyperlink.

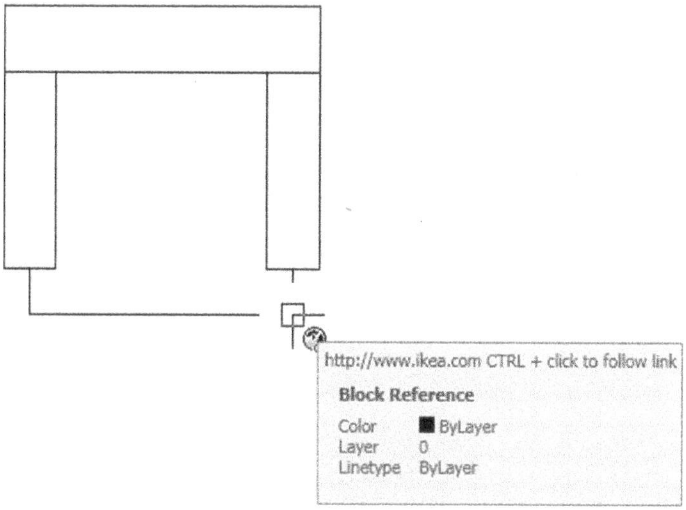

FIGURE 15.31 Hyperlink.

To do this, follow the same steps as before, but this time select the File… button under the Browse for: header on the right side of the dialog box. Find the file of interest, double-click on it, and OK the Hyperlink dialog box. That is it; the document is now attached to your object or text and supersedes any other links you may have.

As a final note, recall that Express Tools has a category called Web Tools. We skipped discussing them in Chapter 14, because the hyperlink command was not yet introduced. Take a few minutes now to look over what is available. These express commands include tools to Show, Change, and Find and Replace URLs. These are useful when you have multiple URLs in your drawing and need to manage them. The three tools are very straightforward to use, and we do not spend any further time on them.

15.16 Lengthen

The lengthen command is quite a bit like the extend command in its effect but goes about doing it differently and is worth adding to your AutoCAD vocabulary. The command works by lengthening nonclosed objects, such as lines or arcs (circles and rectangles do not qualify). You can lengthen the objects by change in size (`Delta`), by percentage added or subtracted (`Percent`), by total final size (`Total`), or dynamically "on the fly" (`Dynamic`). All this applies to angles (of arcs) as well. To try it out, draw an arc and a line (Fig. 15.32).

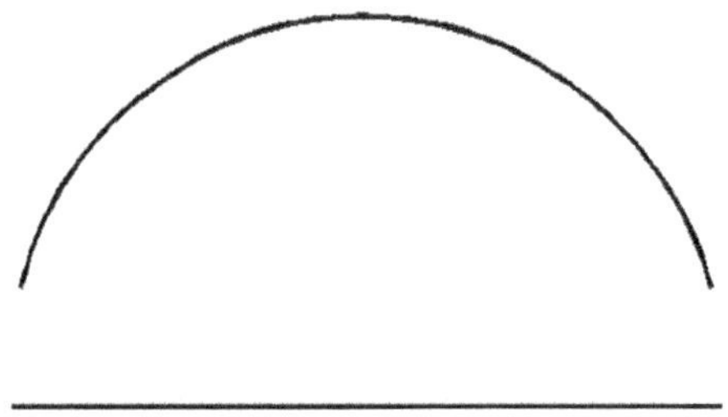

FIGURE 15.32 Arc and line for lengthen.

The process is essentially the same with the arc and other lengthening options. Repeat the command several times and run through each option.

Step 1. Type in `lengthen` and press Enter. Alternatively, use the Modify panel dropdown menu on the Home tab of the Ribbon.

- AutoCAD says: `Select an object or [DElta/Percent/Total/DYnamic]:`

Step 2. Pick the `Percent` choice by typing in `p` and pressing Enter.

- AutoCAD says: `Enter percentage length <100.0000>:`

Step 3. Enter a value such as 150 and press Enter.

- AutoCAD says: `Select an object to change or [Undo]:`

Step 4. Pick the line and it lengthens 150%. Press Esc to exit the command.

15.17 Object snap tracking

Object snap tracking is a drawing aid that allows you to find various OSNAP points of an object and connect to them without drawing anything that touches that object, a sort of remote OSNAP. This function can be activated when the OTRACK button is pressed (turns blue) at the bottom of the AutoCAD screen and is deactivated when the button is pressed again (turns gray). Alternatively, you can also press F11 on the keyboard. The best way to see the purpose and function of this command is to try it. Draw a series of rectangles or squares spaced some distance apart, as shown in Fig. 15.33. You need not number them; that is only for identification purposes.

What we want to do is connect a series of lines from (and to) the midpoints of each of these shapes via right angles (not directly). An example of this may be a cable diagram or an electrical schematic. One way would be to start a line from the midpoint of the right side of shape 1, stop somewhere randomly, and repeat the process from the top of shape 2. Then, you can use trim, extend, or fillet to connect the lines. This, however, is cumbersome and slow.

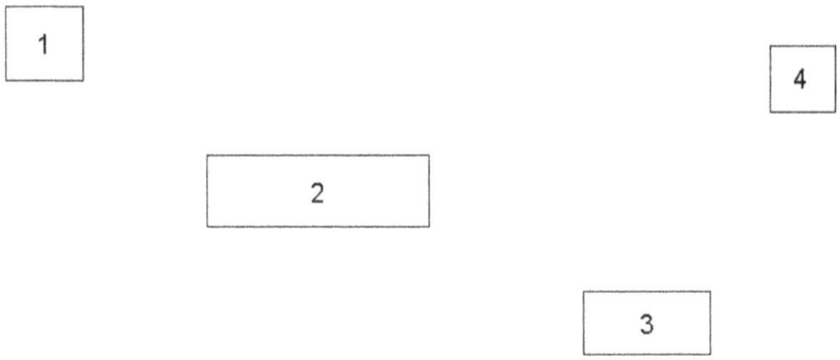

FIGURE 15.33 Rectangles for OTRACK.

A better way is to begin the line from the right side midpoint of shape 1 as before (turn on ORTHO also) but then activate OTRACK and, without clicking, position the mouse over the midpoint of the top of shape 2. A straight dashed vertical line appears. Follow that dashed line back until the mouse meets the straight horizontal dashed line. Then and only then, click once and continue to draw the line to shape 2. This is shown in Fig. 15.34. Notice what OTRACK does. It senses where that midpoint of shape 2 is and gives you a position to confirm and click before proceeding. Complete similar lines from the bottom of shape 2 to the top of shape 3 and then 4.

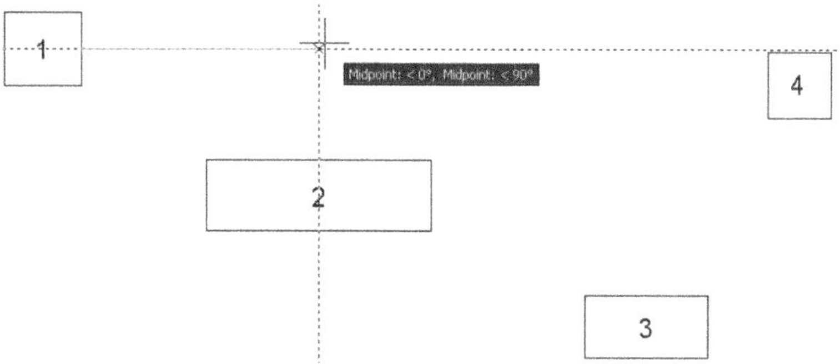

FIGURE 15.34 OTRACK in progress.

15.18 Overkill

Keyboard: type in **overkill** and press Enter
Ribbon: **Home** tab→**Modify**

This interesting command was once under the Express Tools but now is part of the main AutoCAD commands, found n two places on the Ribbon, one in the Modify panel dropdown on the Home tab and once in the Cleanup panel of the Manage tab. (You can also still type its more dramatic original name: `overkill`). What this command does is quite useful. It deletes duplicate objects that overlap each other. In the 2D flat world of AutoCAD, it can be quite challenging and time-consuming to find these overlaps, especially if similar shapes and colors are involved. With overkill, you simply select the offending objects, and AutoCAD figures out which to keep and which to delete, effectively merging them all into one.

To try it out, draw five lines, one on top of another. Although it may not be obvious to a novice, an experienced user notices these lines are a bit darker and "fuzzy," indicating the presence of an overlap; it is subtle but there. Now type in `overkill` and press Enter. AutoCAD asks you to select objects. Select all five lines using a Window or Crossing (not one at a time) and press Enter when done. The dialog box in Fig. 15.35 appears. Note that, if you have seen this command in older versions of AutoCAD, it has been redesigned somewhat.

You can leave all the defaults as they are, pressing OK to have all the lines merge into one line. This is probably the simplest application of the overkill command. However, what if the lines are on different layers?

In that case, you must check off Ignore Layers, and the command merges the layers into one. Numeric fuzz works by comparing the distances between nearly overlapping objects and acts on them if the distance falls inside the fuzz value. Overkill has other useful features as well, so go through the command, referring to the Express Tools Help files for more information.

15.19 Point and node

You have already seen and used points when we discussed the **divide** command earlier in this chapter. In that example, points were created to divide an object. To see them clearly, you needed to modify the form of the point, from a dot to something more visible, using Format → Point Style…

The point command can also be used to create points from scratch, as described next.

Keyboard: Type in **point** and press Enter
Ribbon: **Home** tab→**Draw**

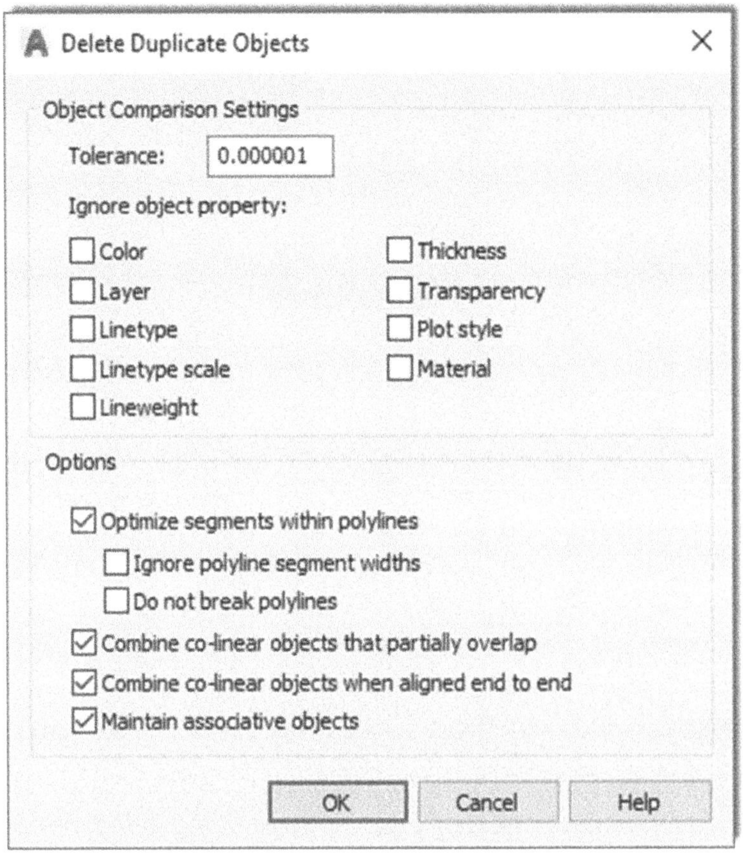

FIGURE 15.35 Delete Duplicate Objects (Overkill).

Step 1. Start the point command via any of the preceding methods.

- AutoCAD says: `Current point modes: PDMODE = 3 PDSIZE = 0.0000Specify a point:`

Step 2. Click anywhere and a point appears.

As you may imagine, points sitting there just by themselves are not too useful but change them to an X (for visibility) and they can be used to create "anchors" to which you can attach geometry. For this, you need an OSNAP point that can do that. Called *Node*, it is one of the 14 OSNAPs. The symbol is a circle with an X superimposed on it. Go ahead and activate Node and practice drawing some lines that originate from the points you just created.

15.20 Publish

The Publish command is a convenient way to print multiple sheets (Layouts) all at once, also referred to as Batch Plotting. This command depends on knowing the basics of Paper Space, so you may want to review Chapter 10 if needed. Assuming, however, that you have done so, the idea here is quite simple. When a file with multiple layouts is opened, the Publish command sees them and allows you to print them all at once, as opposed to clicking on each tab one at a time and using the plot command. For this example, a file called Alta Elevation is opened. It features a design spread out over two layouts: a Plan View and an Elevation View. The Publish command is then started via any of the following methods.

Keyboard: Type in **publish** and press Enter
Ribbon: **Output** tab→**Batch Plot**

The dialog box in Fig. 15.36 then appears. Note that it can be expanded, if it is not, by pressing the Show details button on the lower left.

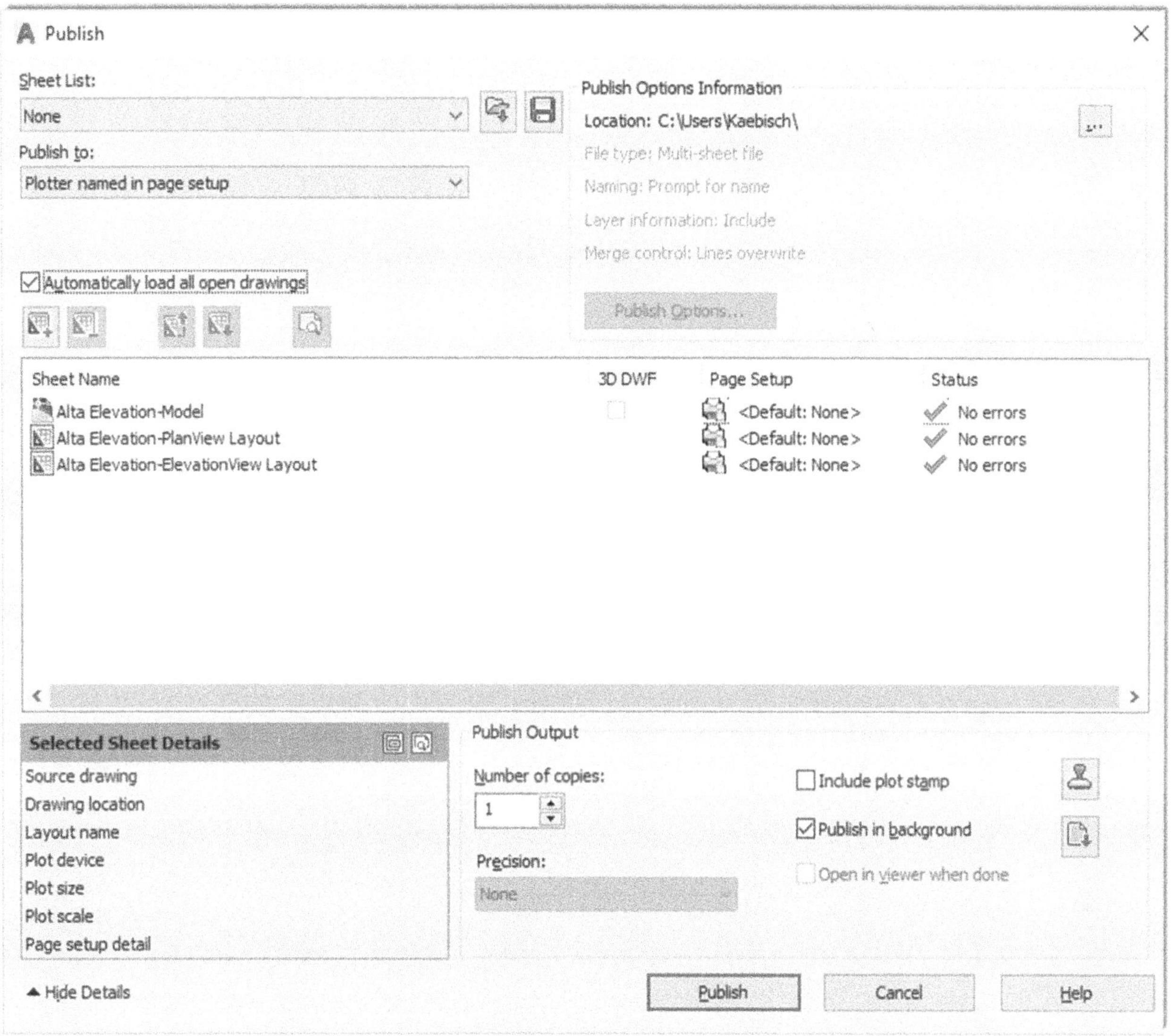

FIGURE 15.36 Publish dialog box.

The dialog box features some options to fine-tune publishing, such as adding additional sheets, plot stamp, deciding how many copies, and a few others. All you must do at this point is just press the Publish button on the bottom, and AutoCAD prints out one copy of every single layout the file contains. It is quick and easy and saves time. Go ahead and explore this very useful command on your own for additional features.

15.21 Raster

Raster is a term you may hear in AutoCAD circles, and it is not so much a command as a concept in graphic design. A raster graphic is nothing more than a bitmap, which in turn is a grid of pixels viewable on any display medium, such as paper or a computer monitor. This method of presenting graphics stands in marked contrast to vector graphics, which are based on underlying mathematical formulas. Vectors can be scaled up or down with no loss of clarity, while bitmaps cannot. So, raster images are basically pictures. These can easily be inserted into AutoCAD, as covered in a previous chapter.

Raster graphics hold a unique place in AutoCAD history. They were needed to implement a massive worldwide conversion from paper-based, hand-drawn designs to CAD files. When it became obvious that computer-aided design was the future, millions upon millions of hand-drawn documents, spanning decades of design work, were scanned into the computer, became backgrounds, and were patiently redrawn. Government and private industries around the industrial world collectively spent large sums of money and many work hours to transform these archives into viable CAD files for future use. Most of this work was done during a 10-year span (1988–98), although some of it was started prior and some work is still ongoing. I recall putting in some significant screen time doing these conversions early in my career, and for a new AutoCAD user, this was excellent, if boring, practice: scan, draw, print, check, and repeat.

Raster scanning software can be quite sophisticated. At the time, we used regular large sheet scanners that simply pulled up the graphic on the screen. We then tugged at it until it was the right size (by comparing a drawn line to a scanned known dimension). Later, we acquired more sophisticated software that recognized shapes and attempted recreating these pieces of geometry with somewhat mixed results. Any smudges, imperfections, or odd shapes could confuse the system. Noise level (sensitivity) settings were used to control how much was picked up, and it was usually better to set it lower, so you could add what was missing rather than constantly erase accidently picked up junk. As an AutoCAD user today, you may never need these tools, but it is good to know what they are just in case.

15.22 Revcloud

This is a command that architects everywhere were screaming for. Revision clouds, or *revclouds* for short, are used to indicate changes on a drawing. While you can use regular ellipses, rectangles, or circles just as well, these shapes blend too easily with actual design work, so the unmistakable cloud shape is a better choice to highlight changes on a print.

Revclouds are made up of small arcs, and that is exactly how they were once drawn, one arc at a time, making it an extremely tedious process. Finally, some program-savvy designers wrote a routine to automate the process. Autodesk then acquired the routine and upgraded it to semiofficial Express Tools status. Even there, it was much too useful to be optional, and it eventually became its own native AutoCAD tool, and the modern revcloud was born.

Some significant new enhancements to the command were introduced several releases ago, and it is easier than ever to use. In the past, you either freehanded the cloud or turned a predrawn geometric object (usually a rectangle or circle) into a revcloud. You can still do this, but now you can create instant rectangular and multipoint clouds as well. In all cases, you can set the size of the arcs that make up the revcloud and tell them which way to face, in or out.

Take some time to explore all the possible options—what follows are just some basic instructions to get you going; fully describing this command would take a few pages.

Keyboard: Type in **revcloud** and press Enter
Ribbon: Home tab→Draw

Step 1. Start up the revcloud command via any of the preceding methods.

- AutoCAD says: `Minimum arc length: 1.0000 Maximum arc length: 1.0000`
  `Style: Normal Type: Rectangular`
  `Specify first corner point or [Arc length/Object/Rectangular/Polygonal/Freehand/Style/Modify] <Object>:`

Step 2. Click anywhere once.

- AutoCAD says: `Specify opposite corner:`

Step 3. Draw a revcloud the same way you would draw a rectangle. When you drag your mouse across the screen, a rectangular revcloud appears. When you click a second time the revcloud is complete.

Your result should look like Fig. 15.37 (you may have to zoom in or out to see the revcloud clearly).

FIGURE 15.37 Rectangular revcloud.

As mentioned before, there is a lot more to this command. When you first started revcloud, you noticed the menu consisting of `Arc length`, `Object`, `Rectangular`, `Polygonal`, `Freehand`, `Style`, and `Modify`. Let us briefly mention each, and then go ahead try all of them one by one. Many of these options were recently added in AutoCAD 2016, so you may not see them if you end up using an older release at work.

- *Arc length*: This option sets the size of the arcs that make up the revcloud. You can enter a value or click through several distances.
- *Object*: This option makes a revcloud out of a geometric shape, such as a rectangle or a circle.
- *Rectangular*: This is now the default method for creating revclouds and was the method used in the multistep example previously.
- *Polygonal*: This option is the same basic idea as the Rectangular option, but you now are free to click around to create as many sides to the revcloud as you want. Useful for creating a complex cloud with many twists and turns, as befitting real-life design changes. This process is reminiscent of creating a Lasso Window or Crossing.
- *Freehand*: This option is the "classic" way of making revclouds, by freehand sketching them around the parts of your drawing that changed. This approach requires a steady hand on the mouse, regarding not just direction but speed also. The size of the arcs changes if you accelerate the mouse across the screen.
- *Style*: This option allows you to change the style of the revcloud from the regular one seen in Fig. 15.37 to one that looks like it was drawn with a calligraphy pen. This is done via a special routine that uses pline arcs of decreasing thickness and is a throwback to the really "old days," when pen and pencil hand drafters took great pains in making their revclouds look fancy, using (you guessed it) calligraphy pens.
- *Modify*: This option was a badly needed one; it allows you to easily modify already drawn revclouds. You select the revcloud you wish to modify, draw the revised section, and following the given prompts, erase what is not needed.

15.23 Sheet sets

Like many new features in AutoCAD, these are not revolutions but rather evolutions of previous ideas. This tool, introduced in AutoCAD 2005, was meant to combine some aspects of Paper Space layouts and the Publish command with a few new twists, to create an "all under one roof" method of setting up the presentation and printing of a project. The idea is to set up sheet sets (1 Layout = 1 Sheet) that define exactly what is needed to plot the entire job correctly. These sheets can come from different drawings, even different jobs, and are grouped together by their respective categories. The idea is that they are now easier to handle and organize.

The industry has shown some reluctance to use a tool that somewhat duplicates existing functionality (even if new features are added), but a few organizations, not surprisingly ones that produce jobs that span dozens of sheets, took to this sheet set idea. We provide a brief overview, and you can decide if you want to pursue this further.

To create a sheet set, you need to set it up. Go to File → New Sheet Set... and you see what is shown in Fig. 15.38. Select the first choice, an example sheet set, and press Next >. What is shown in Fig. 15.39 appears.

Select the appropriate sheet set, Architectural Imperial, e.g., and press Next >. Fig. 15.40 appears. Here, you can give the set a name and location. Press Next >, and what is shown in Fig. 15.41 appears. The various sheet sets, sorted by category, are visible. Press Finish, and the Sheet Set Manager appears, as seen in Fig. 15.42.

This was all just to set up the basic sheets (placeholders). They are still blank and need drawings associated with them. Right-click on the Architectural folder and select Import Layouts as Sheet.... What is shown in Fig. 15.43 appears.

You now must browse for the right drawing associated with the Architectural layout. Continue down the categories until all are filled. This is just the basic idea of sheets sets. It is an extensive tool that can span many more pages of descriptions, and you should explore some of it on your own.

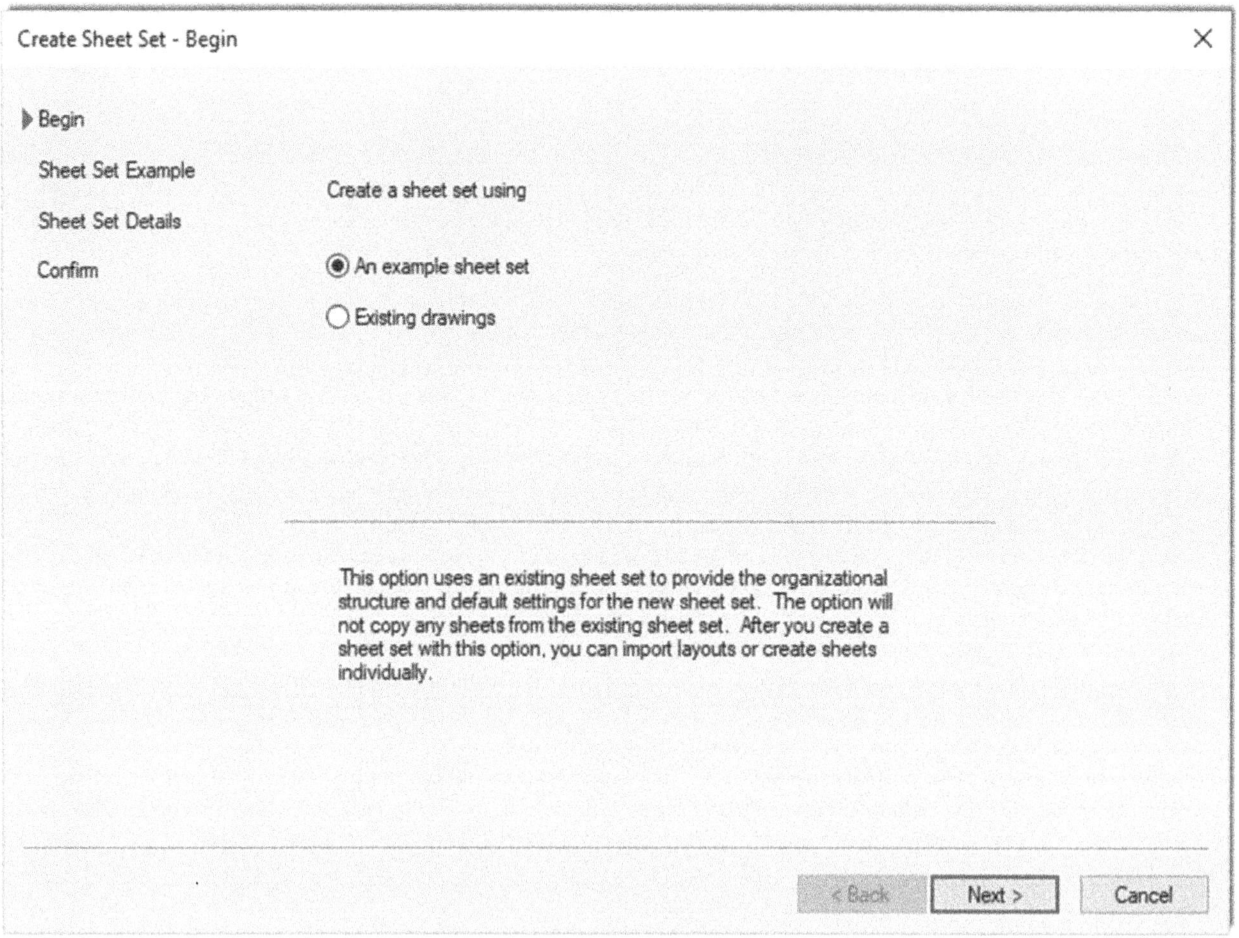

FIGURE 15.38 Create Sheet Set - Begin.

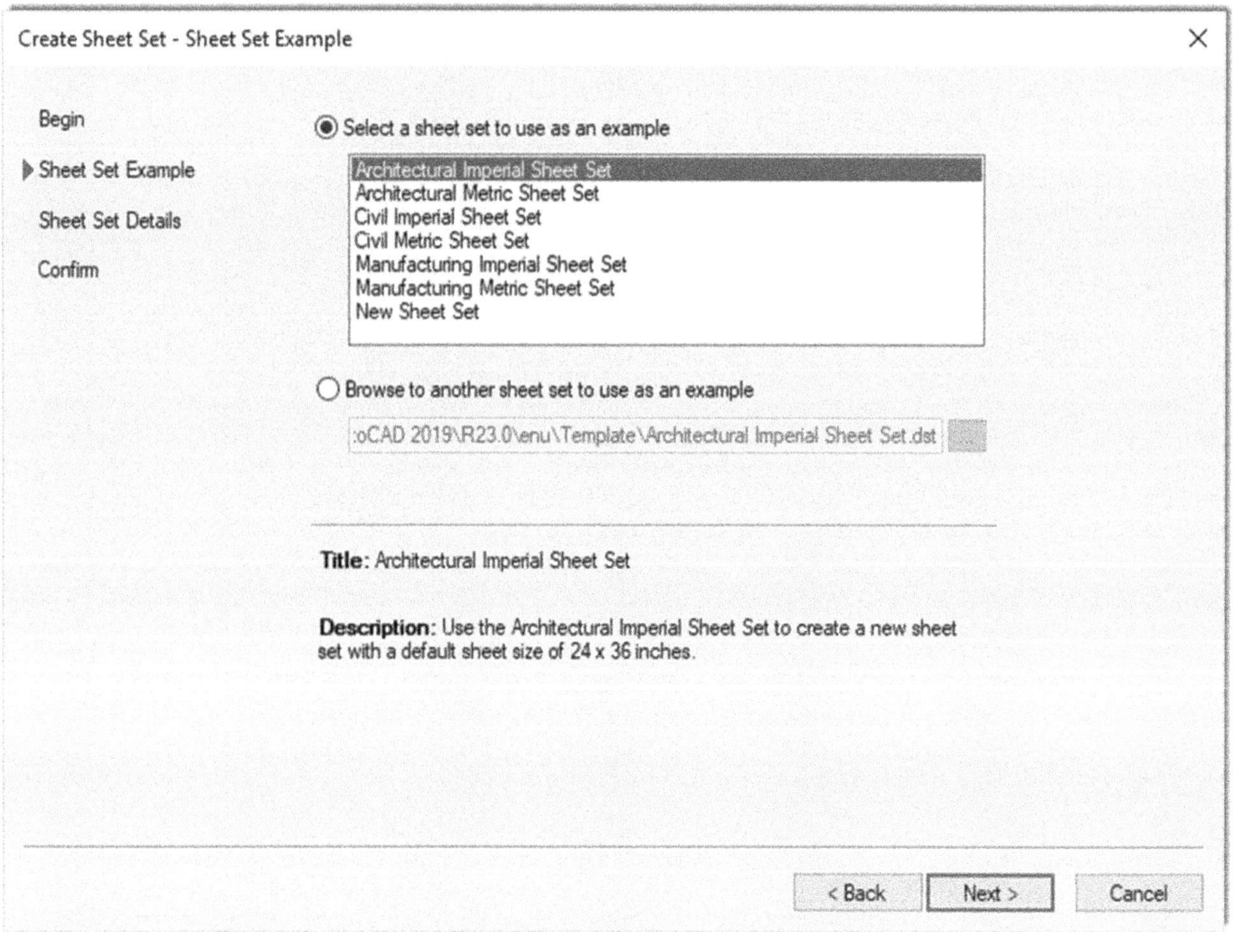

FIGURE 15.39 Create Sheet Set - Sheet Set Example.

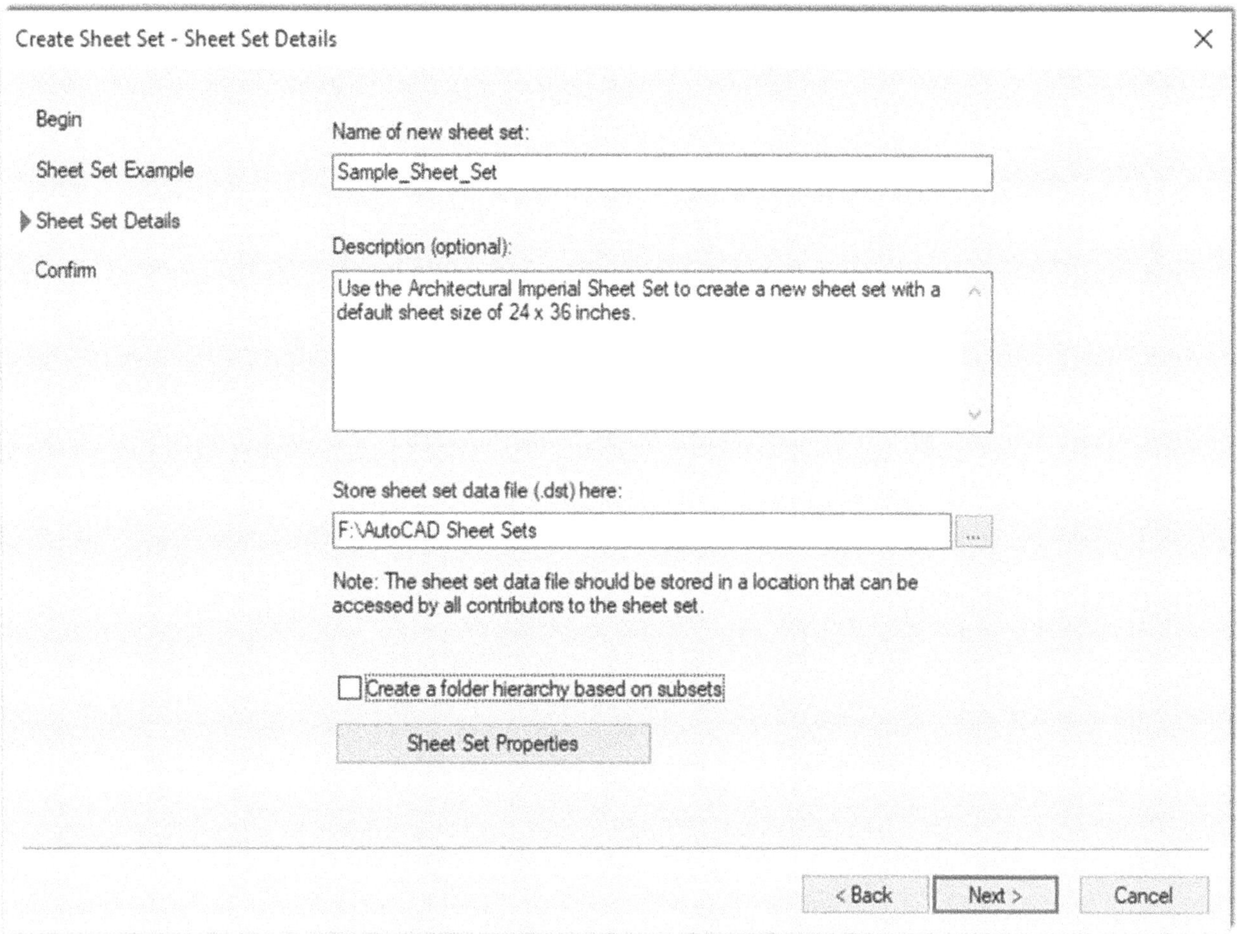

FIGURE 15.40 Create Sheet Set - Sheet Set Details.

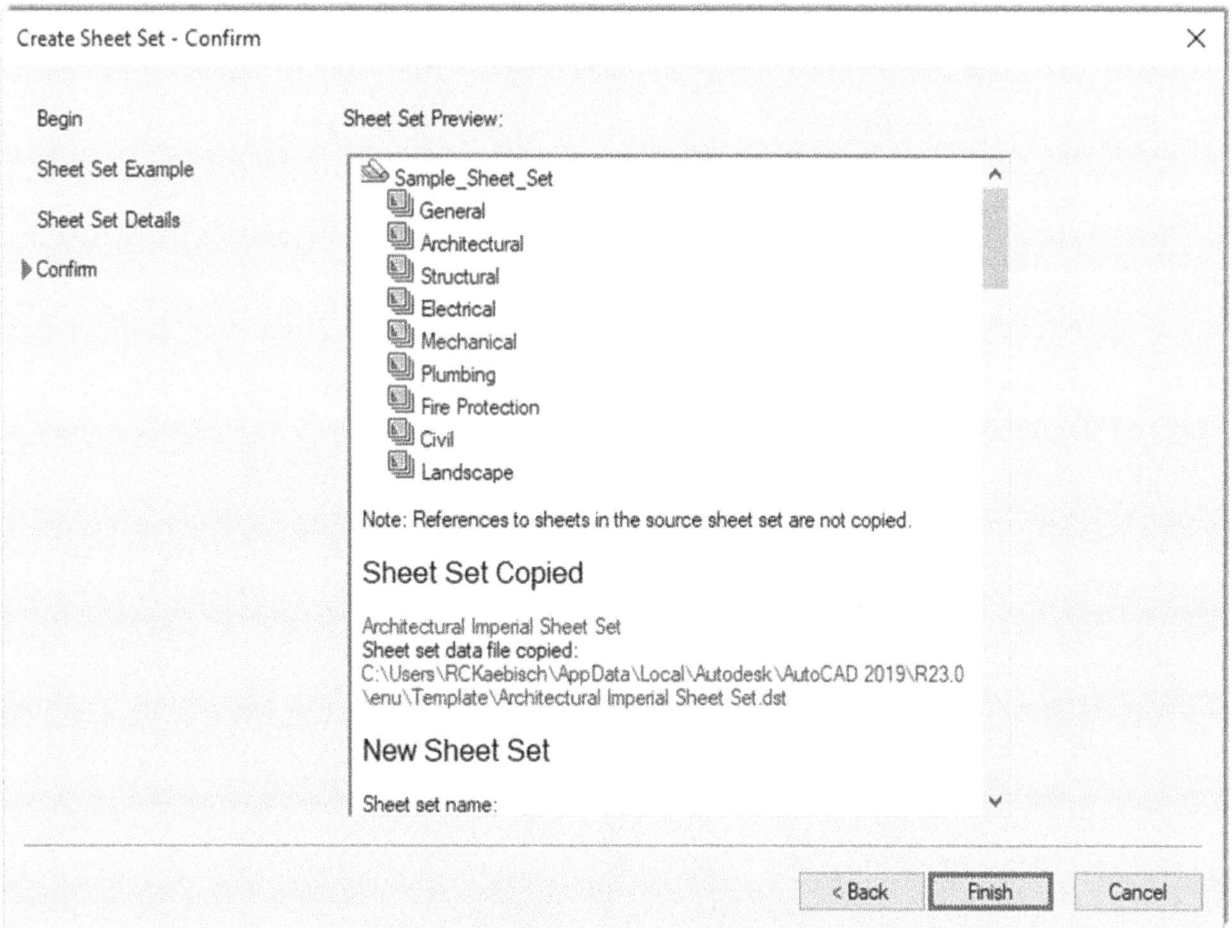

FIGURE 15.41 Create Sheet Set - Confirm.

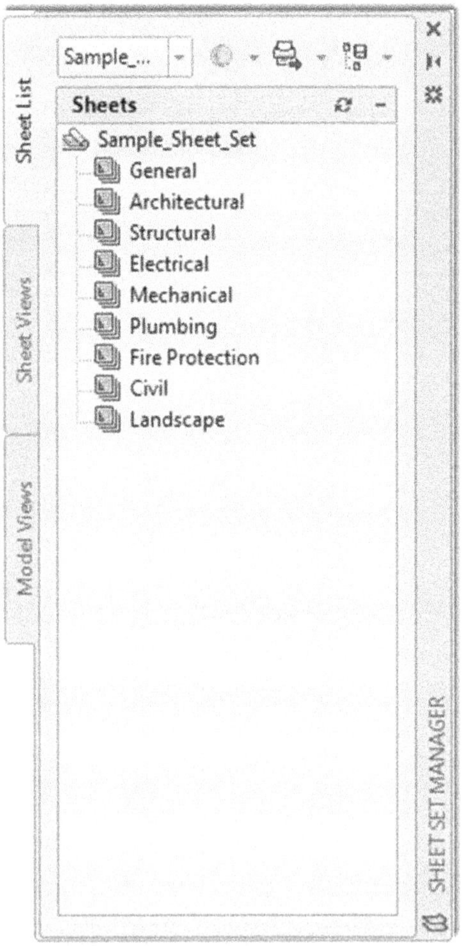

FIGURE 15.42 Sheet Set Manager.

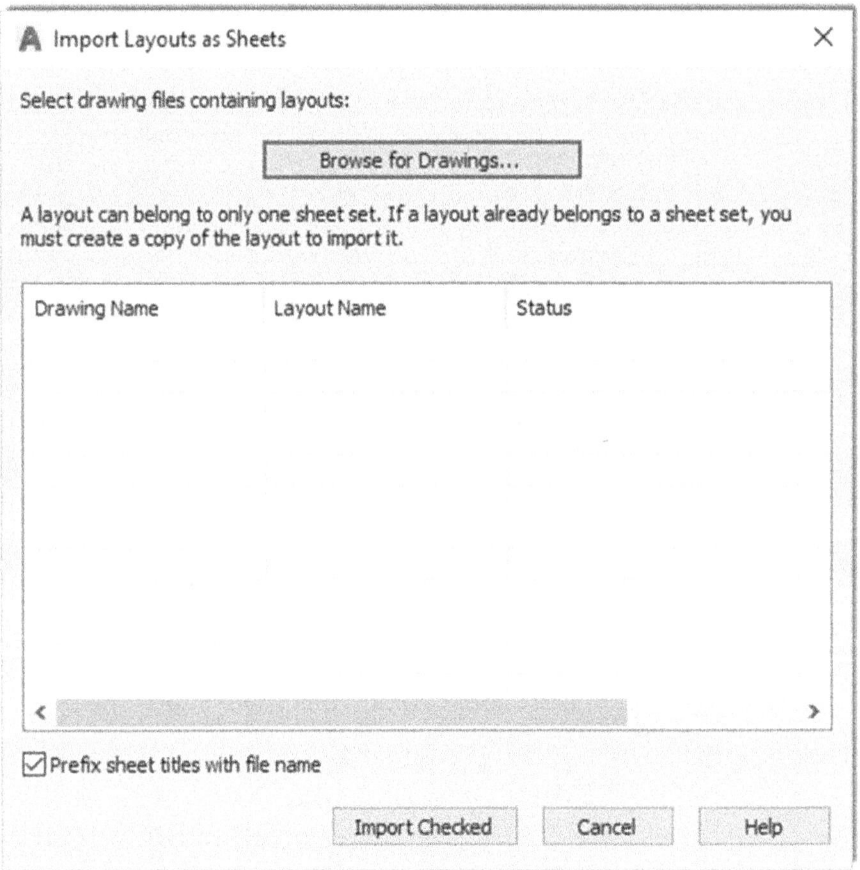

FIGURE 15.43 Import Layouts as Sheets.

15.24 Selection methods

What we outline here are a few other methods of selecting items that you may have not yet seen: the Window Polygon (WP), Crossing Polygon (CP), and Fence (F). All these are applicable when selecting objects for several operations such as erase, move, copy, and trim. These options would be secondary to the relatively new Lasso tool, and are similar in many ways, but they are still good to know. To try out the Window polygon,

Step 1. Draw a set of circles, as seen in Fig. 15.44.

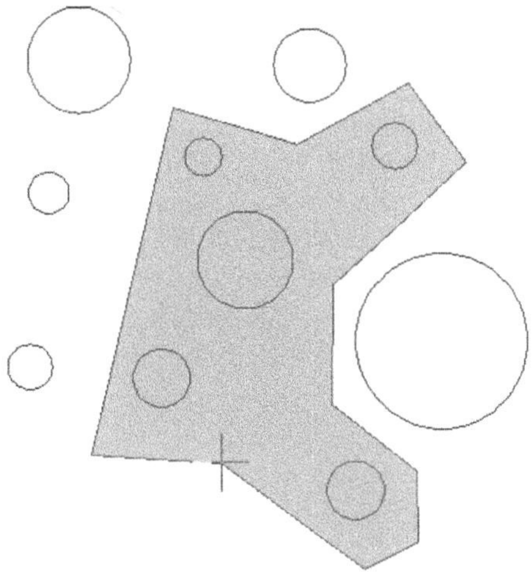

FIGURE 15.44 Window polygon.

Step 2. Begin the erase command via any method you prefer.
Step 3. When it is time to select objects, type in `wp` for `Window polygon` and press Enter.
- AutoCAD says: `First polygon point:`

Step 4. Click anywhere.
- AutoCAD says: `Specify endpoint of line or [Undo]:`

Step 5. Continue to "stitch" your way around the circles you wish to select, and right-click when done.

Notice the difference here between the standard Window selection tool (strictly rectangular in nature) and the flexible "any shape you want it" Window polygon (Fig. 15.44). Its action, of course, is similar; everything it encloses is selected.

To try out the Crossing polygon,

Step 1. Draw another set of circles, as seen in Fig. 15.45, or just undo your previous steps.

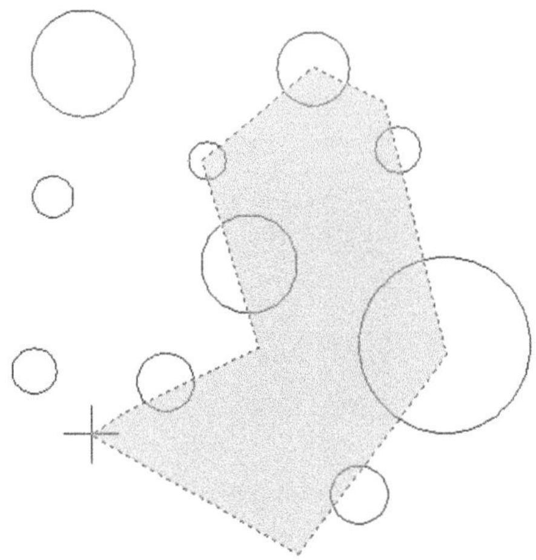

FIGURE 15.45 Crossing polygon.

Step 2. Begin the erase command via any method you prefer.

Step 3. When it is time to select objects, type in `cp` for `Crossing polygon` and press Enter.

- AutoCAD says: `First polygon point:`

Step 4. Click anywhere.

- AutoCAD says: `Specify endpoint of line or [Undo]:`

Step 5. Continue to "stitch" your way through the circles you wish to select, and right-click when done.

As you can see, the Crossing polygon (Fig. 15.45) works the same exact way as the Window polygon, except anything it touches gets selected, much like how the regular Crossing operates.

The Fence option works by having a drawn line by the cutting tool during a trim command and selects multiple lines to be trimmed. Draw a bunch of vertical lines, and one horizontal line, as seen in Fig. 15.46. Then execute the trim command, but when it is time to select the lines to be trimmed, type in f for Fence, and click twice to draw a line across the lines to be trimmed, finally pressing Enter to make them disappear.

In case you discovered it on your own, yes, you can accomplish the exact same thing using a regular Crossing to eliminate the lines once a cutting edge is selected. However, the Fence command comes in handy when going around corners, as the flexible line-based approach can bend around any shape, while the rigid rectangular-shaped Crossing cannot. Try out both methods, as well as using Fence around corners.

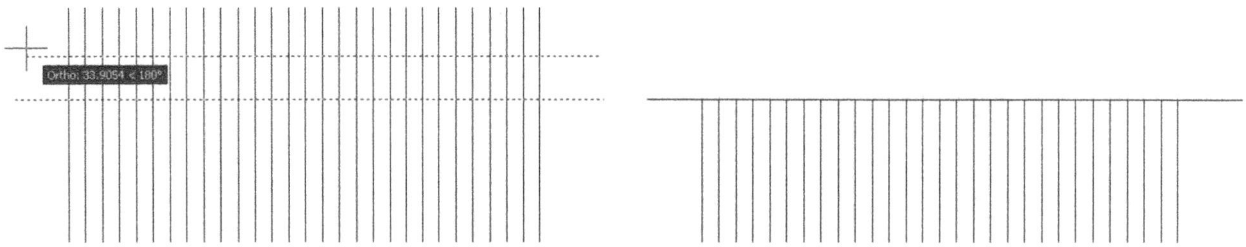

FIGURE 15.46 Fence command and the results.

15.25 Stretch

This is an easy command to learn and use, and it can work wonders for quick shape modifications. The stretch command uses the Crossing tool to literally stretch shapes. Create two rectangles, as seen in Fig. 15.47.

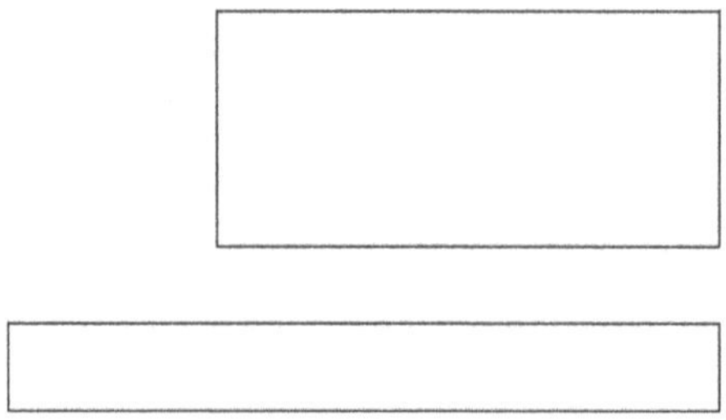

FIGURE 15.47 Rectangles for stretch command.

We want to stretch the right sides of both rectangles some distance.

Keyboard: Type in **stretch** and press Enter
Ribbon: Home tab→Modify

Start the stretch command via any of the preceding methods.

Using a *Crossing* (this does not work with the Window selector), select the first quarter or so of the rectangles, as seen in Fig. 15.48.

FIGURE 15.48 Starting the stretch command.

They become dashed. Press Enter and click anywhere to begin stretching them, as seen in Fig. 15.49. You need to have Ortho on and all OSNAPs off.

FIGURE 15.49 Stretch command in progress.

Finally, click elsewhere and the rectangles have a new length. You did not have to explode and redraw them or use grips. You can use the `Displacement` option to key in specific X, Y, or Z distances or stretch to another OSNAP point. Note that stretch does not work on partially selected circles and it moves, not stretches, fully selected circles and other objects.

15.26 System variables

There is not much to say about system variables except to make you aware of what they are, so if they come up in reading other texts or the AutoCAD Help files, you know with what you are dealing. System variables are Yes/No, On/Off, or 0/1 types of commands used to indicate a *preference* or a *setting*. AutoCAD has over 850 of them for every conceivable variable (something that changes) in the software, and listing them all is futile, although if you wish to see them all, type in `setvar`, press Enter, then type `?`, and press Enter twice.

Many system variables are set graphically via a dialog box, and users may not even be aware or know they are changing a system variable. Perusing the Help files and AutoCAD literature, especially on an advanced level, you find many references to them. Knowing system variables is important for those wishing to learn advanced customization (such as AutoLISP). Here is a useful example that is typed in on the command line. Suppose you are mirroring text and the new set is backwards, as seen in Fig. 15.50.

FIGURE 15.50 Mirrtext.

That is not quite right, as you want text to be readable even after mirroring, so some sort of system variable (SV) needs to be set. The responsible SV is called *mirrtext*, and it is of the 0/1 variety, with the two possible states represented by a 0 and a 1. Somehow, it got set to 1, and needs to be 0. Type in `mirrtext`, press Enter, type in the new value of `0`, and press Enter again. Try the mirror command again—it should work fine now.

As mentioned before, this is just one example; there are hundreds of others. Keep an eye out for them. To help you do that, AutoCAD now features a very handy tool in this latest release called the *System Variable Monitor*. Type in `sysvarmonitor` and press Enter. You can also double-click the relevant icon all the way on the far right of the Drawing Aids strip (it looks like a sine wave with an orange triangle). The new System Variable Monitor dialog box appears and lists several system variables and their currently set values, along with their preferred ones. You can then manually change them to the preferred values as (and if) needed, or if you really messed up something, press Reset all.

15.27 Tables

If you have not yet used tables, here is a quick primer on them. They are pretty much what you might expect, a bunch of rows and columns, with a header on top, ready for you to enter information. You can specify the number of rows and columns and add or take them away later as well. Tables are sort of like low-tech Excel spreadsheets and can be useful in certain situations (like when you need data displayed in a hurry). We cover some basics here and let you explore more on your own.

Keyboard: Type in **table** and press Enter
Ribbon: Home tab→Annotation

Start the table command via any of the preceding methods. You see Fig. 15.51.

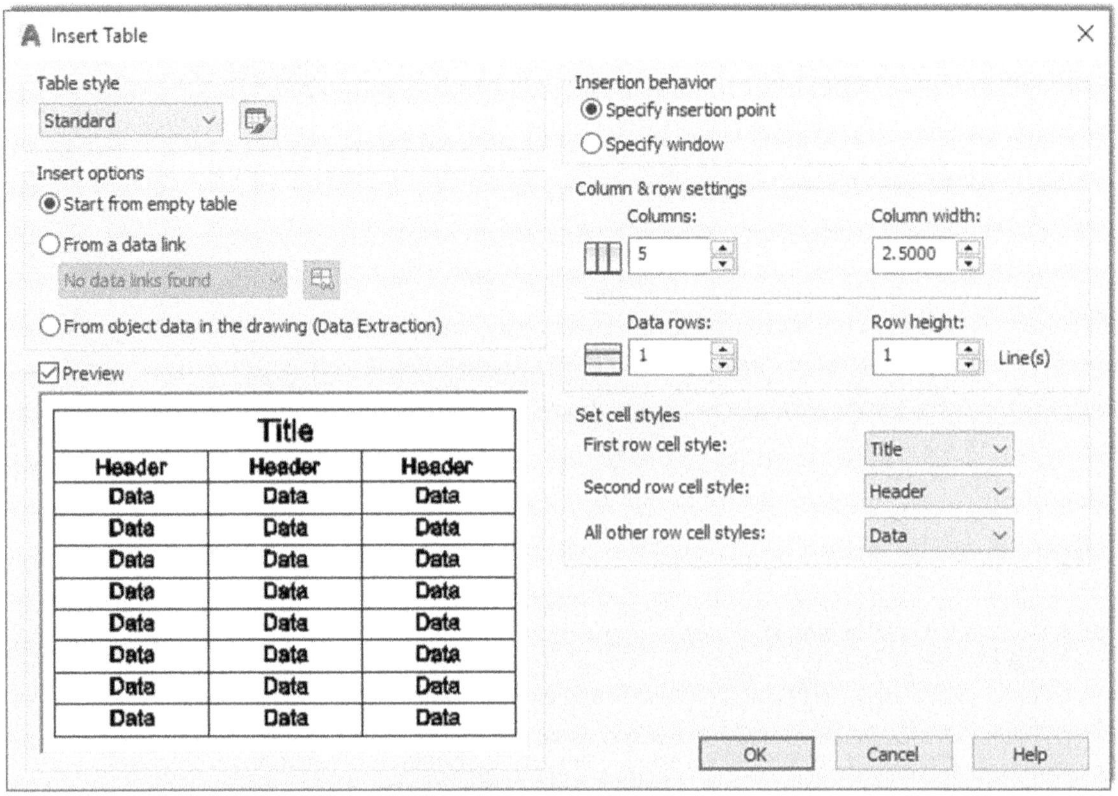

FIGURE 15.51 Insert Table.

Under the Column and row settings category, set your columns and rows as well as their respective widths and heights (we do a 10 ×10 here). When done, press OK, and you are asked to click where you want the table to go. When you position it, you see what is shown in Fig. 15.52.

Here, you are asked to set up the header. Type in "Sample Data Table" and press OK. You see the completed table in Fig. 15.53. You can now fill in the table with data by clicking in each cell and typing in values, as seen in Fig. 15.54.

There is more to this command, such as sizing and modification of the table, as well as some ability to link up with Excel spreadsheets, so go ahead and explore it further. You can also set up styles of tables by first using the tablestyle command, which gives you the dialog box for "designing" a table (Fig. 15.55).

Press New… and give your style a name. You then get the New Table Style dialog box (Fig. 15.56). Here you can work with the general properties, text, and borders of the new table, like an Excel spreadsheet, and design a custom style that fits your needs.

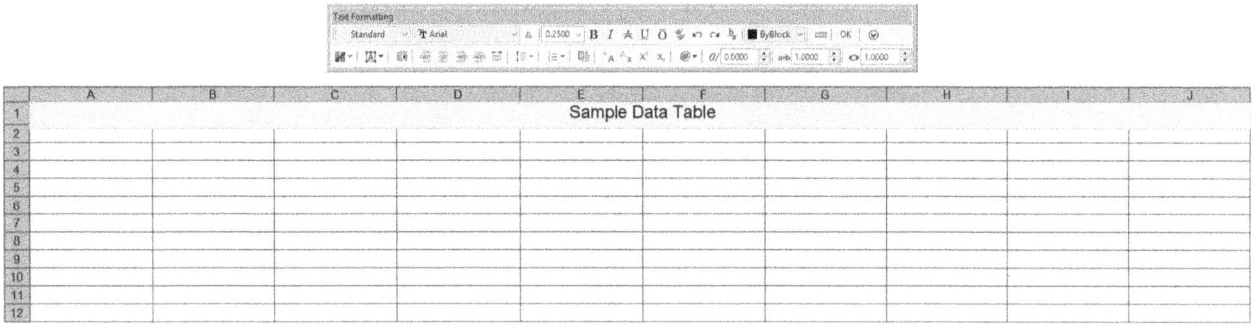

FIGURE 15.52 Insert Table.

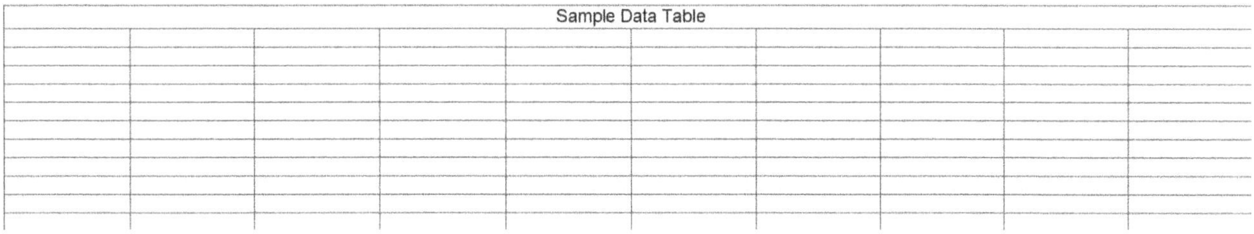

FIGURE 15.53 Fill in header.

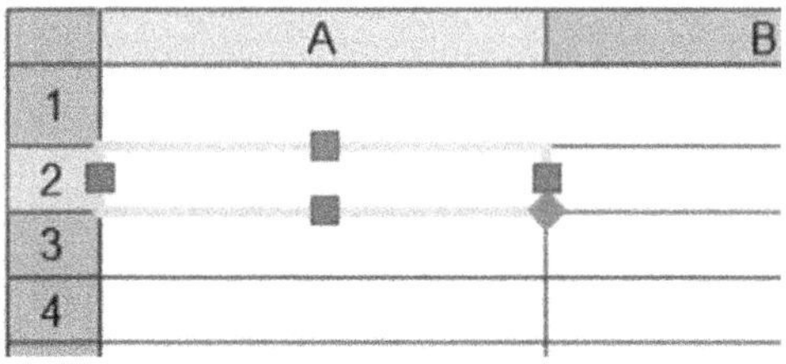

FIGURE 15.54 Fill in individual cells.

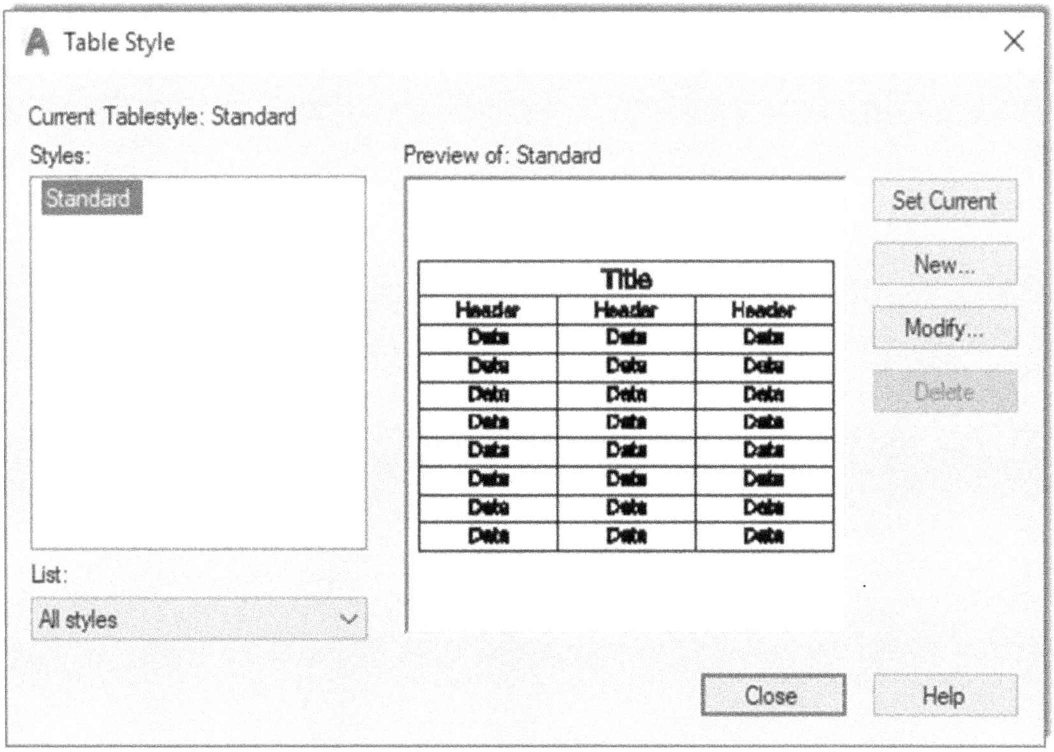

FIGURE 15.55 Table Style.

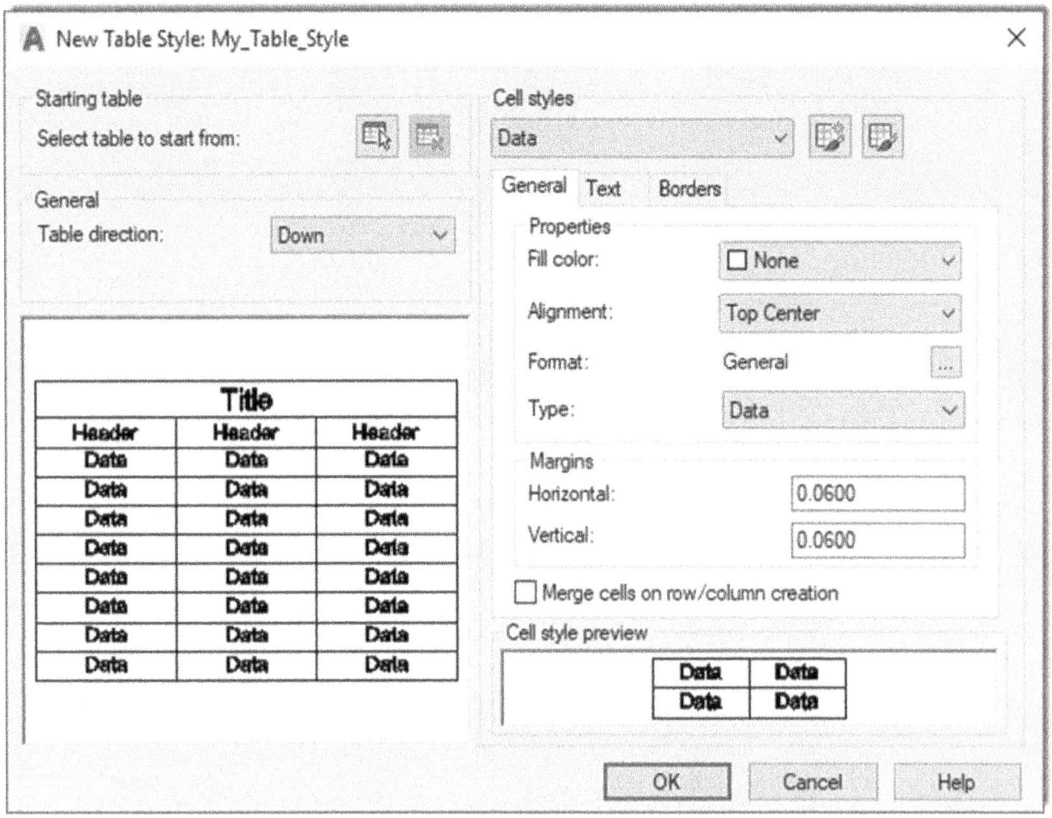

FIGURE 15.56 New Table Style.

15.28 Tool palette

This useful palette, shown in Fig. 15.57, can be accessed by typing in `toolpalettes`, via Palettes panel on the View tab of the Ribbon, or by pressing Ctrl + 3. It may also be on your screen already but hidden away off to the side. To see it, you must hover the mouse over the palette's spine for it to reveal itself. This "autohide" feature can be disabled via a right-click (on the spine) menu, seen in Fig. 15.58.

The tool palette represents a collection of commonly used commands and predefined blocks. What you see when the palette comes up may be different from Fig. 15.57, as it depends completely on what menu is pulled up or what tab is clicked. In all cases, to use this palette, just click on a command (or click an object then move the mouse to the drawing area), and it is executed as though you typed or selected a Ribbon command. To access more menus, simply right-click on the gray "spine." As mentioned before, the menu in Fig. 15.58 appears.

Even more tabs are available in this palette. Go to the bottom left of the tabs (where it looks like several tabs are bunched together) and left-click. Another extensive menu appears, as shown in Fig. 15.59. Take the time to go through each of the menu choices. The number of available blocks is quite extensive.

Although we do not go into this further, you may also customize what tabs you have available by creating your own and adding custom commands and blocks to it. This is like creating new toolbars using the CUI, as covered in Chapter 14. You can also modify existing commands in the tabs; finally, the entire palette can be hidden, docked, and even made transparent, so it is always there but you can see what you are doing underneath it, a neat, if distracting, trick. Explore all the options; most are easy to figure out and self-explanatory.

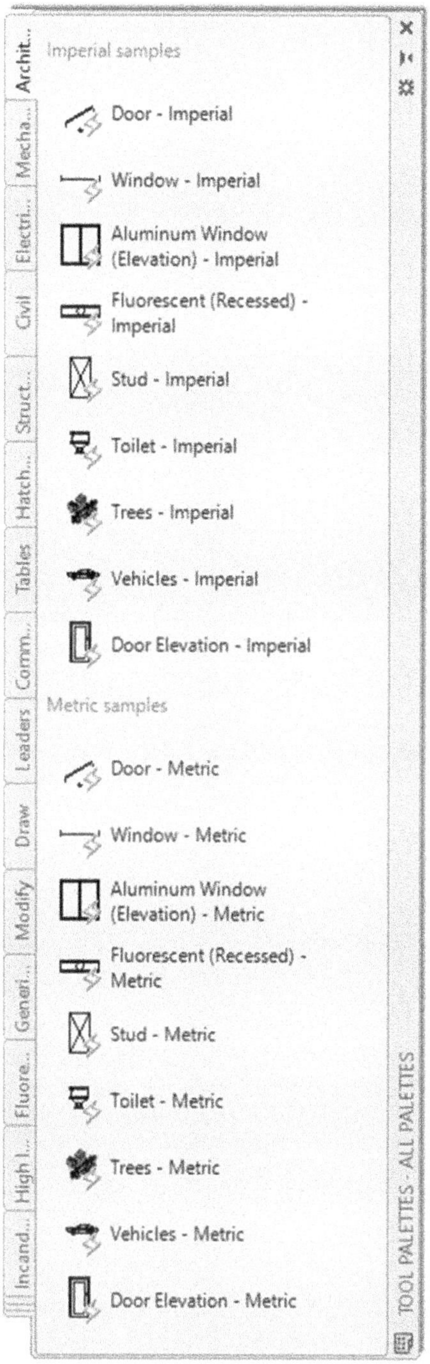

FIGURE 15.57 Tool Palettes, All Palettes—Architecture tab.

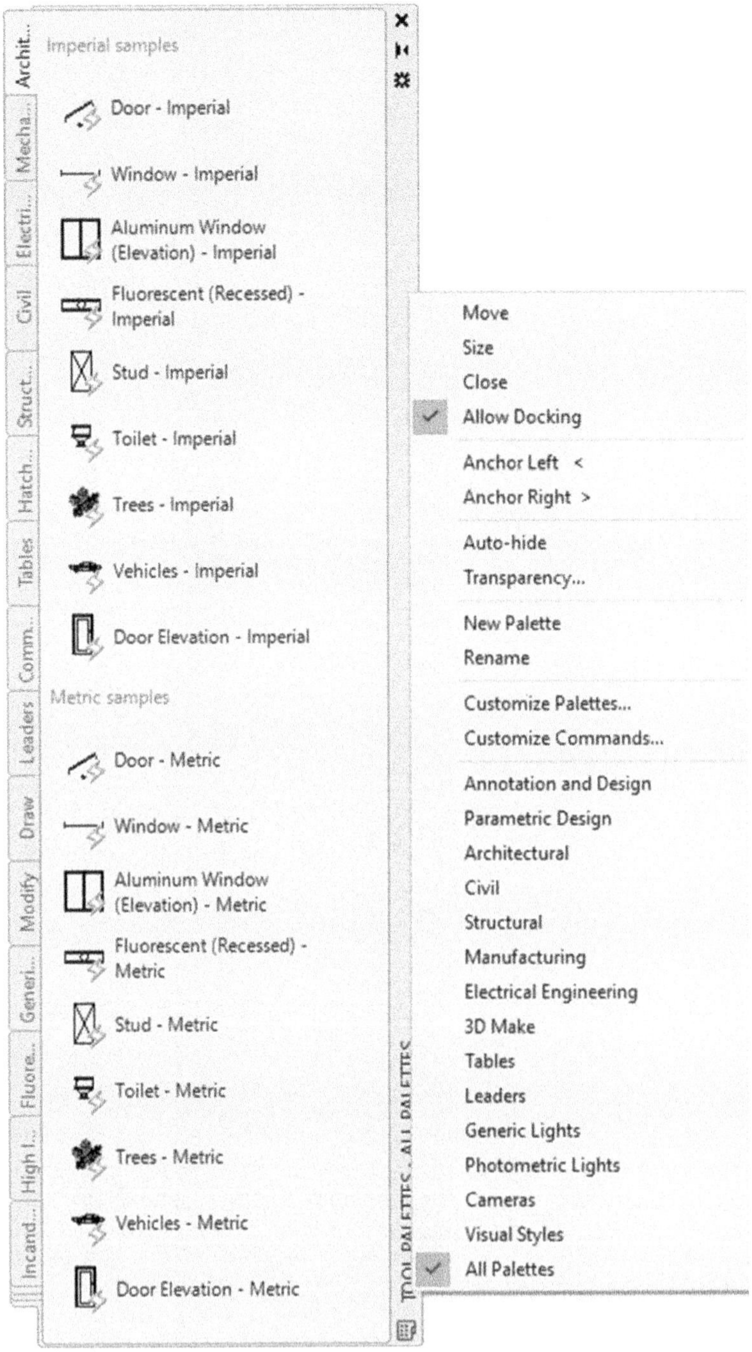

FIGURE 15.58 Spine menu.

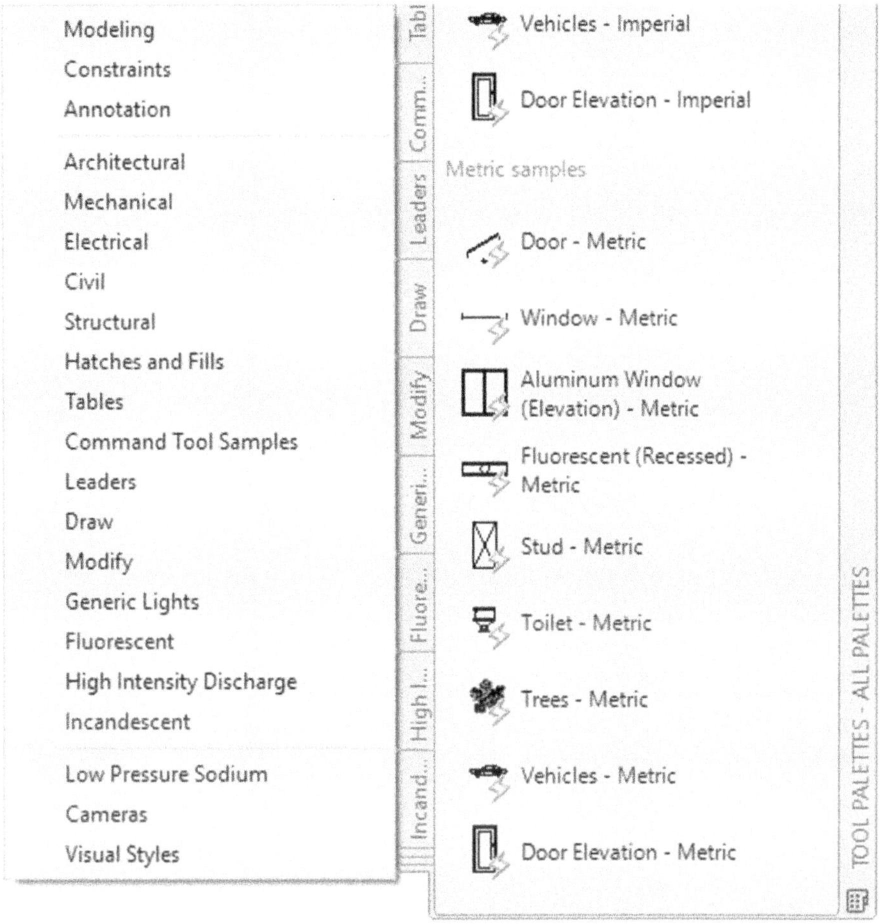

FIGURE 15.59 Tabs menu.

15.29 UCS and crosshair rotation

Rotating the crosshairs or the UCS icon is a very useful trick for working with tilted shapes. Not everything you work on will be perfectly horizontal and vertical. Some designs, or parts thereof, will be rotated at some angle; and it would be nice to align the crosshairs to them. This allows for easier drawing and better orientation for the designer. Commands such as Ortho can now be implemented at the new tilted angle.

Rotating the UCS icon is technically not the same as rotating just the crosshairs, but we do not split hairs (so to speak) over the conceptual differences, and we treat both methods as equivalent for now. There are two ways to rotate the Crosshairs/UCS icon: the legacy method prior to AutoCAD 2012 (using snap or the UCS command) and the new method via the UCS icon grips menu, which we cover first.

Method 1

In AutoCAD 2012, the UCS icon gained some new functionality in the form of grips. It seems like an odd place to add grips—the icon is not a construction object after all—but it works well. To demonstrate, draw a rectangle tilted at a 45 degrees angle and click on the UCS icon. You see three grips appear. The outer ones can be clicked again to manually rotate the icon to any desired angle. The inner one can be used to move the icon around. Each of the grips reveals a menu as well. We ignore the inner grip for now and focus on the outer ones. Select the Rotate Around Z Axis choice (Fig. 15.60), and AutoCAD prompts you to enter a rotation angle (also 45 degrees in this case).

Once you enter the value, the crosshairs rotate and turn a colorful green and red. See the rotation of the crosshairs in Fig. 15.61. If you have Ortho on, you can now draw lines that are aligned to your rectangle.

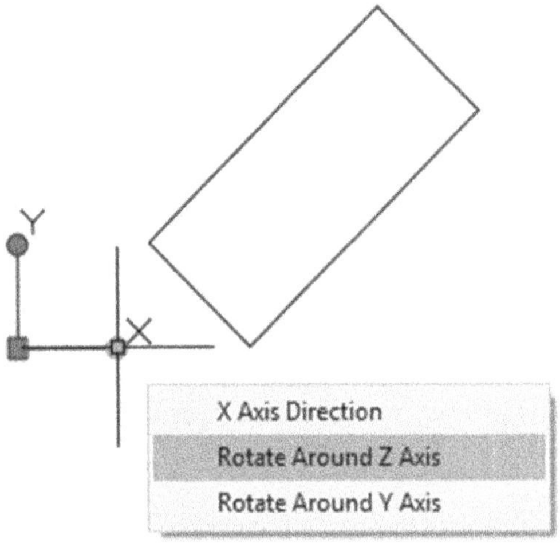

FIGURE 15.60 UCS grip menu.

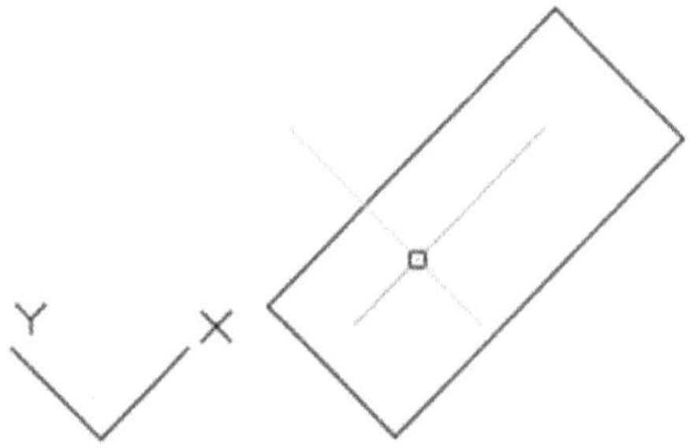

FIGURE 15.61 Rotated crosshairs.

Note that you can achieve the same result by the following steps:

Step 1. Type `ucs` and press Enter.

- AutoCAD says: `Specify origin of UCS or [Face/NAmed/OBject/Previous/View/World/X/Y/Z/ZAxis] <World>:`

Step 2. Type in `z` and press Enter. You always want to rotate around the Z axis in 2D.

- AutoCAD says: `Specify rotation angle about Z axis <90>`

Step 3. Type in 45 degrees for this example. The UCS icon is rotated along with the crosshairs.

To return to the usual orientation of the UCS/crosshairs, you can type `ucs`, press Enter, and type in w for World, or select World from the center grip menu, as seen in Fig. 15.62.

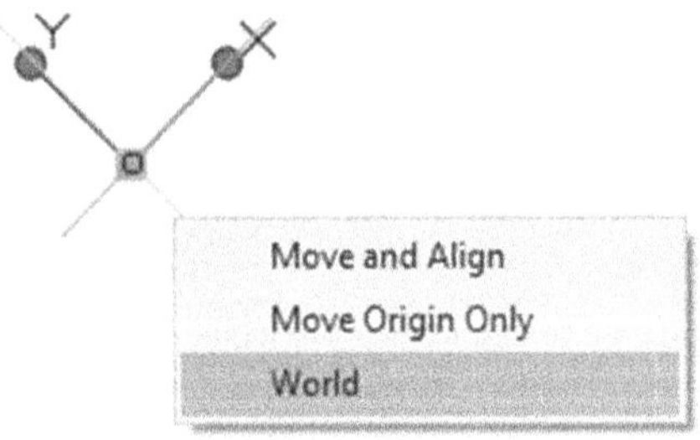

FIGURE 15.62 UCS World.

Method 2

This legacy method does something different: It rotates the crosshairs but not the UCS icon. It is a minor variation on the same idea but one of which you should be aware. In 2D AutoCAD, this makes no practical difference, as stated earlier. In 3D, there is a more significant consequence, and the UCS method is used.

Step 1. Type in `snap` and press Enter.
Step 2. Type in `ro` and press Enter.
Step 3. For the base point, type in `0,0`.
Step 4. For angle of rotation type in per for perpendicular.
Step 5. Select an angled line.
Step 6. Turn off the SNAP command.

To get back to normal, repeat all six steps, except type in `0` at step 4. Finally, to close out this section, an honorable mention needs to go to yet another way to rotate the *view* to match the current UCS. Type in `plan` and press Enter twice.

15.30 Window tiling

By now you should be quite familiar with the multiple drawing tabs that can be found on top of the AutoCAD drawing area if you have several files open. At a minimum you will see the Start Tab with the Learn and Create screens, and then whatever files you may have open. All this was first discussed in Chapter 1, and these relatively new techniques represent a welcome jump forward in opening and managing files as well as easily accessing other design tools.

However, with this approach, each file remains under its own tab and therefore on its own separate screen. What if you wanted to see the files side by side? A method, called *Window Tiling*, addresses this. This is just a simple tool for allowing the viewing of two (or more) drawings side by side in a convenient manner. While each file still is managed under its own tab, they can now be located together in your viewing area. You can stack these vertically or horizontally, with AutoCAD spacing the drawings out evenly.

Open two drawings in one session of AutoCAD and press Restore Down (middle button) at the upper right for one of the drawings, so the two drawings are tiled one on top of the other. You see something like Fig. 15.63, with the Start menu in the background. First, click and minimize that Start menu (or it will be tiled with your drawing files), then View tab of the Ribbon, on the Interface Panel, choose Tile Vertically from the drop-down menu, as seen in Fig. 15.64. After the button is clicked on, you see what is shown in Fig. 15.65. The active drawing is highlighted.

Both drawings are lined up neatly on the screen. You can now click in and out of each and use copy and paste commands to transfer items if necessary. The same thing, of course, can be done for a horizontal tiling.

You can also employ Floating Model Space. If you have more than one monitor you can now click and drag a drawing to another monitor as well. Now you can look at two or more larger views of your drawings across many screens. This is a new feature for AutoCAD as of the 2022 You can also employ Floating Model Space. If you have more than one monitor you can now click and drag a drawing to another monitor as well. Now you can look at two or more larger views of your drawings across many screens. This is a new feature for AutoCAD as of the 2022 version.

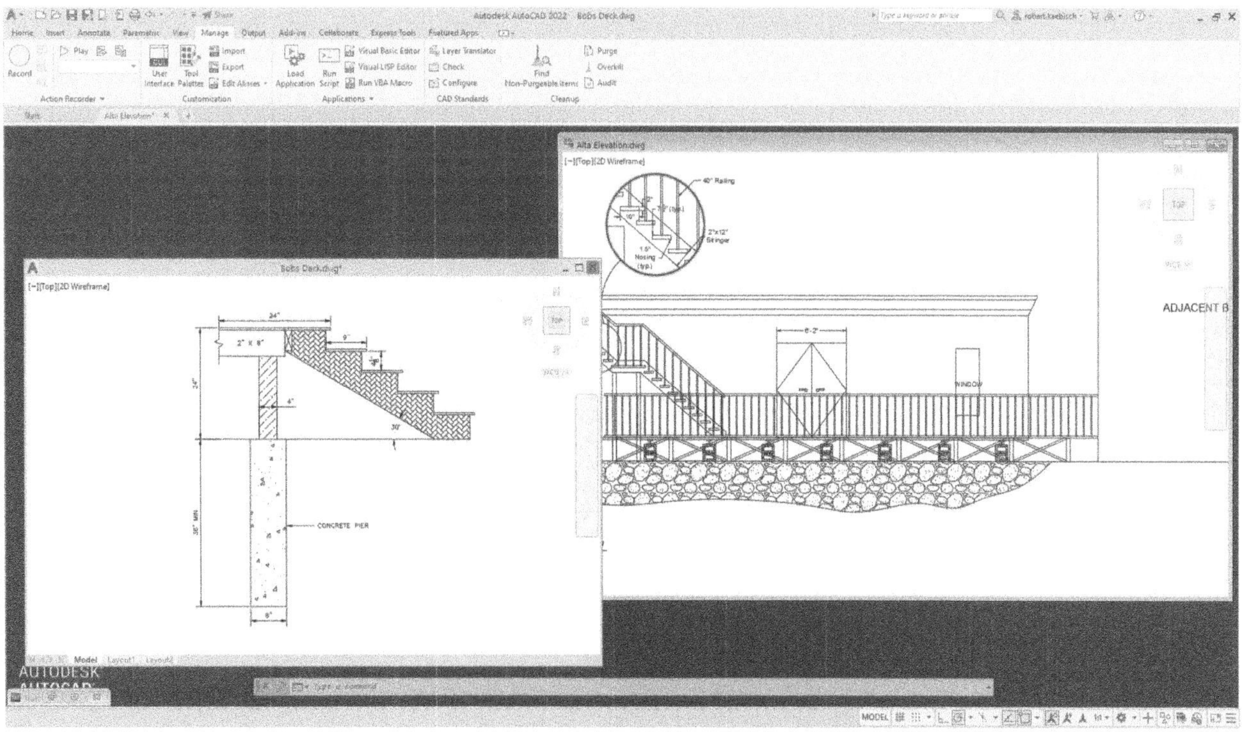

FIGURE 15.63 Window tiling, start.

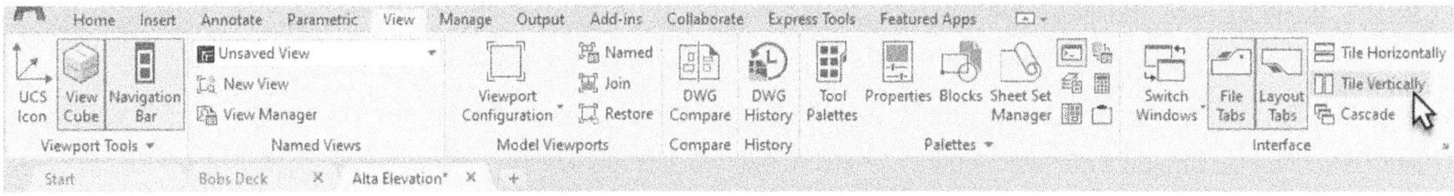

FIGURE 15.64 Drawing Windows, Tile Vertically.

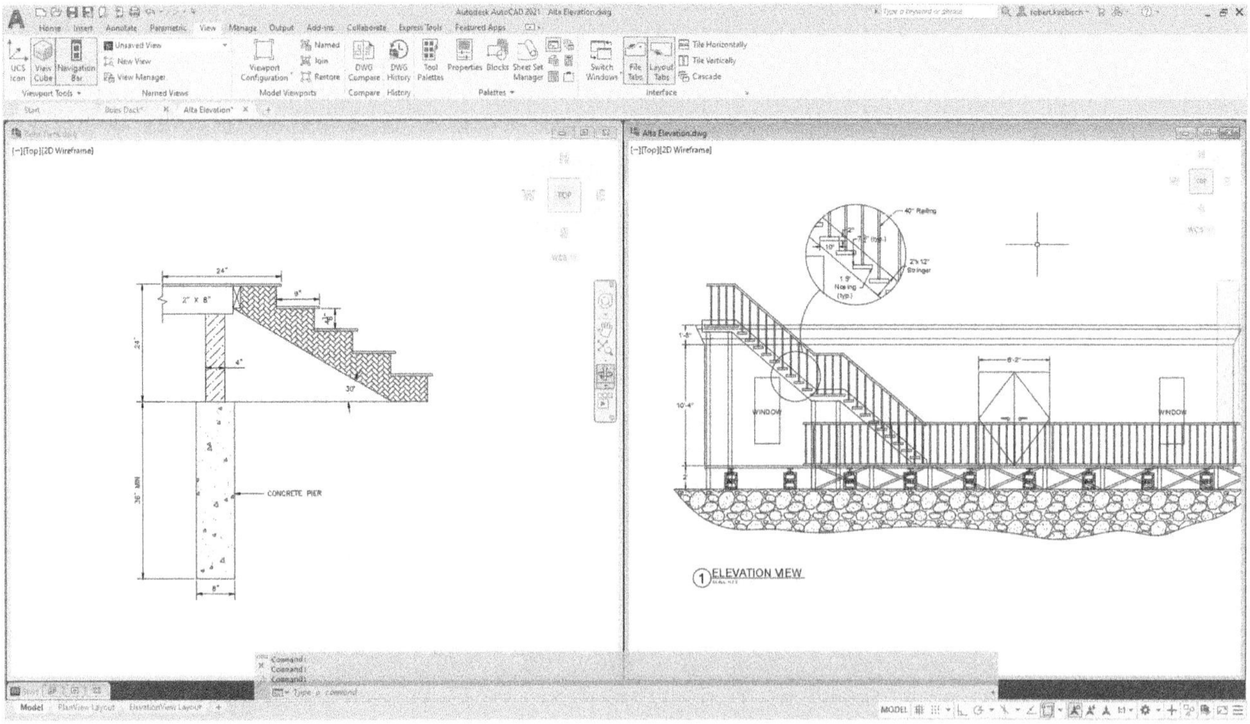

FIGURE 15.65 Drawing Window Tiling, result.

15.31 Wipeout

For reasons that may not be immediately apparent but are explained in detail, this command is one of the most important in 2D, to produce drawings that not only possess extraordinary clarity but are easy to work with and modify. When used properly, wipeout is truly an "insiders'" secret, not known by beginner students, and one of the reasons why drawings done by AutoCAD experts have that intangible "something" that sets them apart from those done by less-experienced designers.

First, let us define what a wipeout is. It is simply a tool that creates a mask (or screen) to block out the viewer's ability to see other drawn geometry covered by the wipeout. To use it, draw a shape of some sort. Then type in `wipeout`, press Enter, and click your way around all or part of the shape. When you press Enter or right-click, the shape disappears behind the wipeout, as seen in Fig. 15.66.

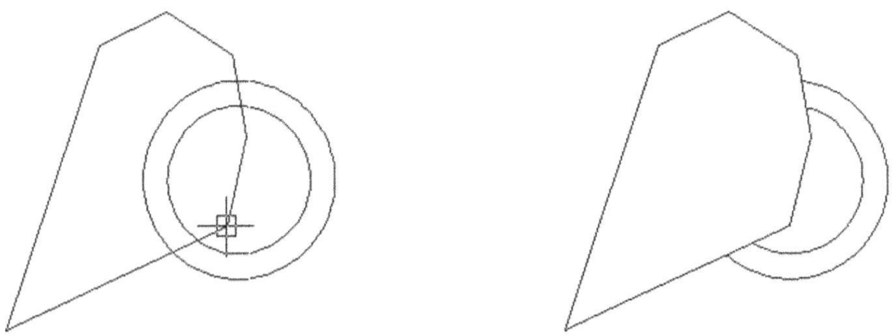

FIGURE 15.66 Creating a basic wipeout, and the result.

Notice the wipeout itself is still visible (something to be addressed later), but the double circle shape has been partially blocked, as expected. Practice using this command a few times just to become familiar with it.

Remember that a wipeout itself is an object. It can be moved, erased, copied, rotated, scaled, and just about anything else you can do to a regular geometric object. You can also click on the wipeout (click on the lines, not the blank center) to activate its grips. Then, the grips can be used to reshape it into any shape whatsoever, as seen in Fig. 15.67.

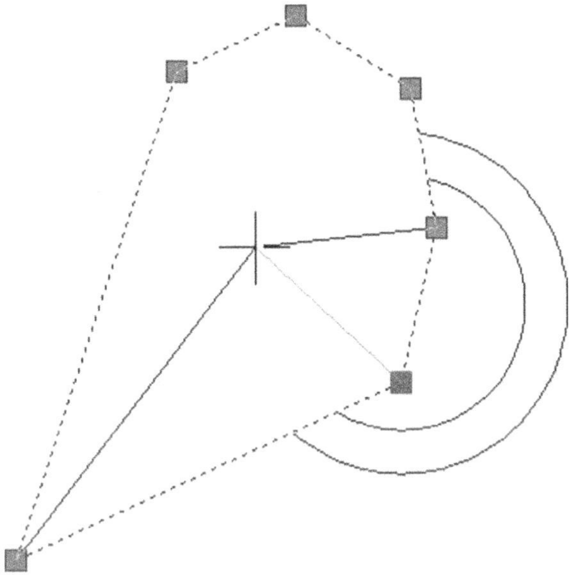

FIGURE 15.67 Wipeout grips.

So, what is so important about wipeout? Well, this command allows us to do something very special in 2D work. We can give our designs *depth.*

When you look around you in the real world, objects are generally not transparent. So, if you are looking at a bus in front of a building, part of the building is obscured by the bus. If another car were to park between you and the bus, it would obscure part of the bus, and so forth. In 3D, this is no problem at all. All objects are solid, and while they can be viewed in transparent wireframe form, a click of the button gives you rendered models that create an instant illusion of depth. This is important, as it matches our everyday experience, and we quickly identify what we are looking at. In the image in Fig. 15.68, the chair and the computer monitor both block a part of the computer desk. It is easy to tell what is in front of what in the rendered image on the left, but not as easy with the wireframe image on the right.

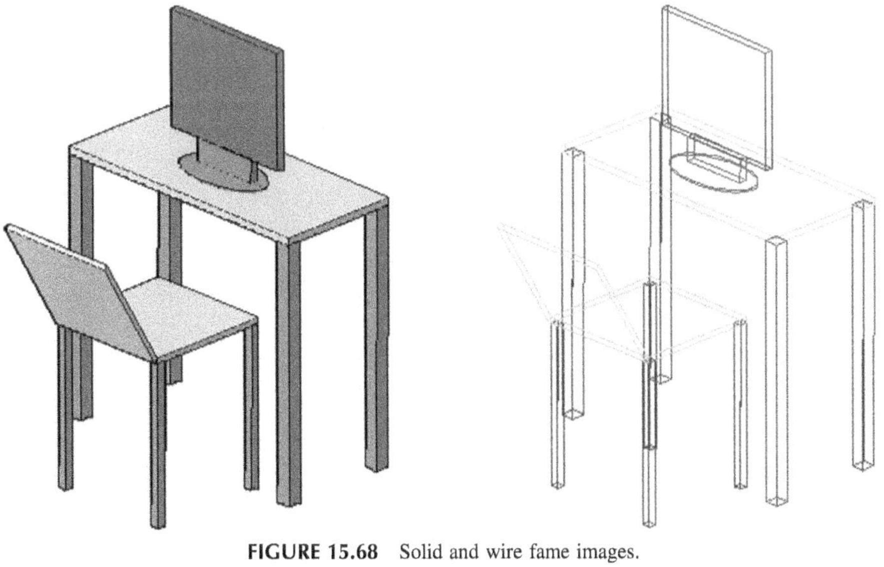

FIGURE 15.68 Solid and wire fame images.

Let us try the same thing with the bus and car, as seen in Fig. 15.69. Which is in front of which? You really cannot tell without shading the image, as in Fig. 15.70.

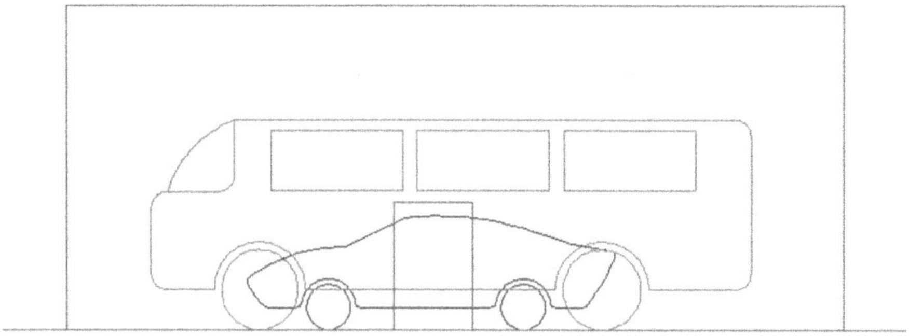

FIGURE 15.69 Bus, building, and car—in wireframe.

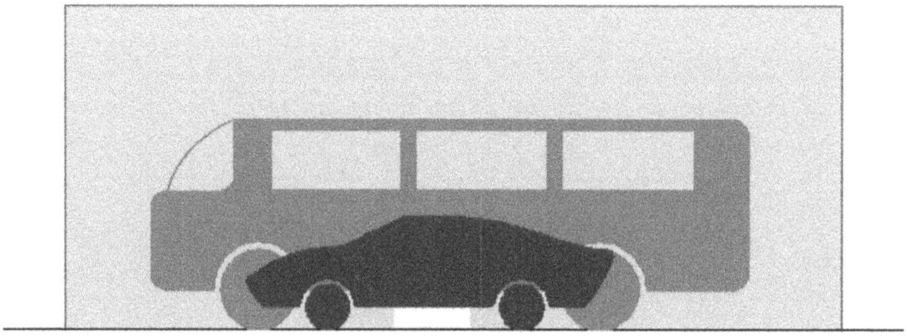

FIGURE 15.70 Bus, building, and car—solid.

In 2D, though, we have a problem: We cannot shade in the conventional sense. Everything is drawn using lines and nothing is solid, even though it may be in real life. In a plan view of an architectural floor plan, this is not an issue, as we are looking straight down, but what about in an elevation view, such as the previous images? A section through a room in an elevation view may feature furniture obscuring walls and walls obscuring other items behind them. Or, in electrical engineering, we have equipment on a rack obscuring the mounting holes. To provide realism and a sense of depth, we need to somehow obscure our view of items not seen. Of course, you can trim linework behind the object, but what if the object moves? You now have a gaping hole. We need a way to temporarily obscure and hide geometry that is tied directly to the object doing the hiding, so if it moves, the linework reappears. Wipeout is the secret to that trick.

In Fig. 15.71, notice how the equipment bolted to the rack obscures part of the rack. Now, if the equipment is moved, such as in Fig. 15.72 (notice the numbers to the left change from 20U–22U to 8U–10U), the wipeout moves with it and it acts like a solid object. The move is successfully done with no trimming necessary.

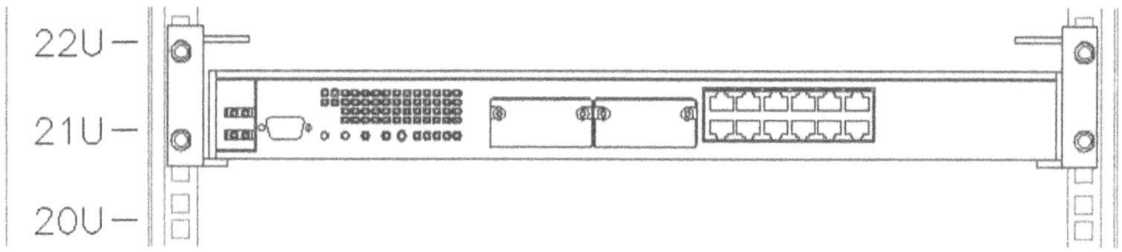

FIGURE 15.71 Electrical gear on a rack with a wipeout.

What if you did not have a wipeout? Such is the case in Fig. 15.73. Notice that the rack is visible underneath the equipment, giving no indication of depth. You would have to trim lines to achieve this.

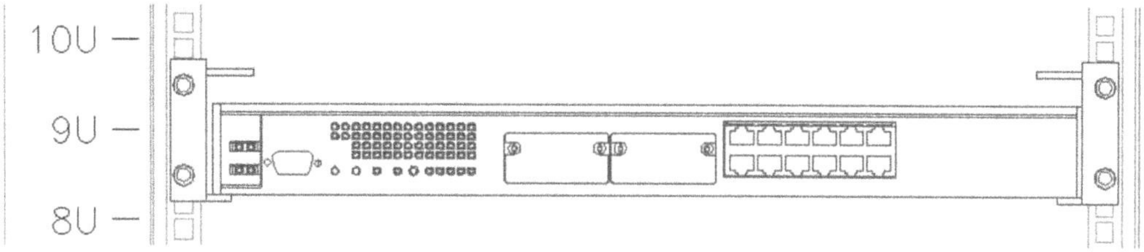

FIGURE 15.72 Electrical gear relocated.

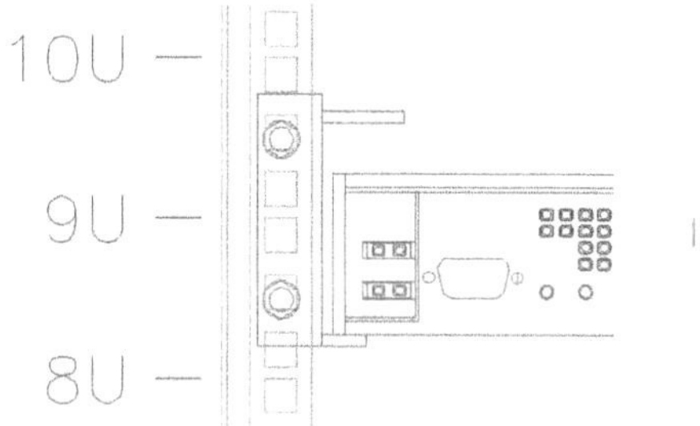

FIGURE 15.73 Same electrical gear with no wipeout.

In summary, when a wipeout is attached to an object, it prevents anything behind it from being seen, exactly as in real life. So, a bookshelf with a wipeout attached blocks the wall behind it and electrical equipment blocks the mounting holes to which it is attached. Move the object and the wipeout moves, also blocking out whatever you happen to be blocking. This is the technique that the pros use to give drawings a uniquely realistic look in 2D design, and you should learn this as well. Let us go through some details.

The essential steps of making this technique work include the following:

Step 1. Create the drawing of what you are designing, such as the equipment rack in Fig. 15.74.

FIGURE 15.74 Wipeout, Step 1.

Step 2. Using wipeout, click to trace the outline of the drawing point by point (just the four corners in this case).
Step 3. Using Tools → Draw Order → Send to Back, put the wipeout behind the design.
Step 4. Make a block out of the design and wipeout.

The object is now "solidified" and can be placed in front of other objects. If you create more objects in this manner, you can simply use the Draw Order tool to shuffle them around top to bottom as needed (if all of them are blocks). In this manner, you can create a rather impressive sense of depth and clarity to an otherwise busy drawing.

Take a close look at Fig. 15.75. It is a larger section of the same cabinet shown in the previous figures. It could not have been done this way without extensive use of the wipeout command. Notice how it is easy to tell which equipment is in front and which is behind the rack. There are, in fact, several layers of objects: the equipment bolted to the front of the rack, then the rack itself, then the equipment in the back of the rack.

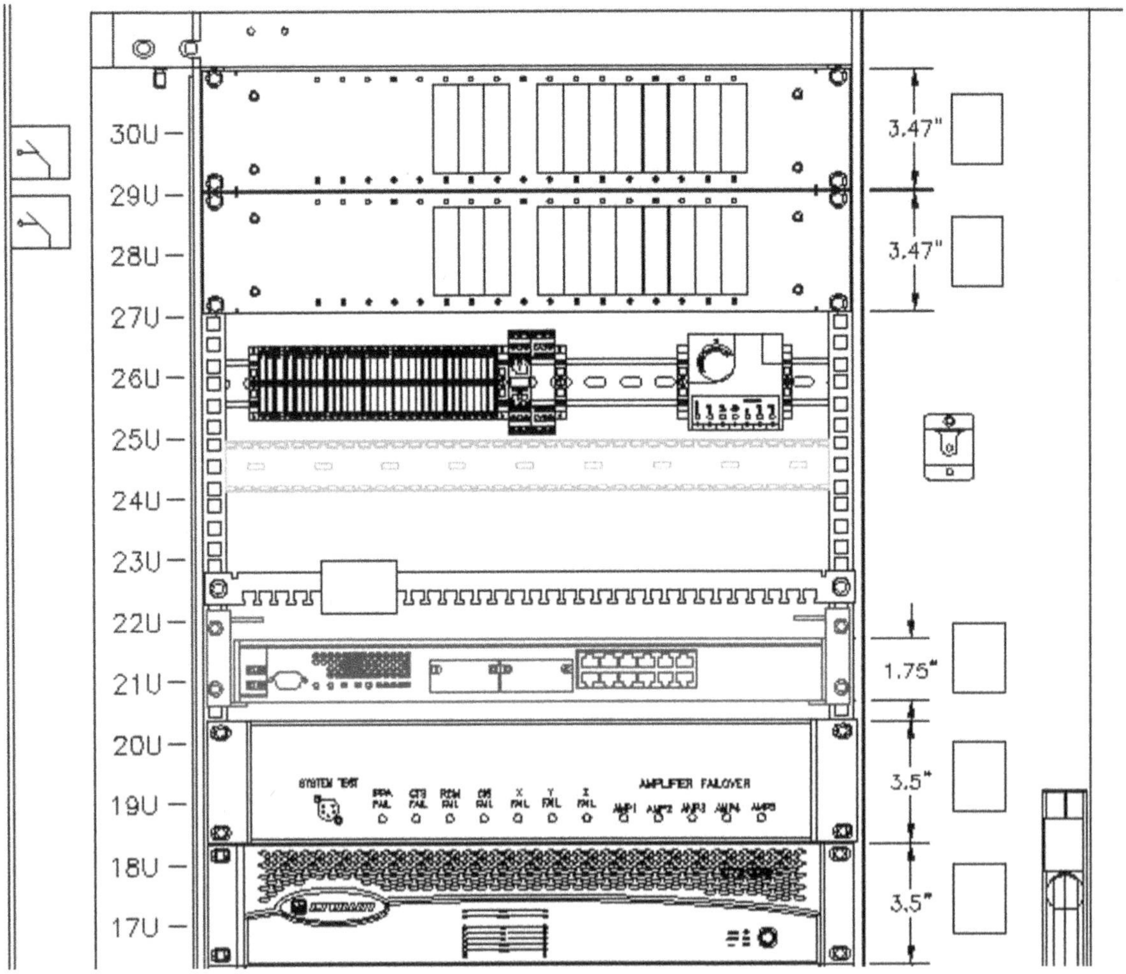

FIGURE 15.75 Full equipment rack with multiple wipeouts.

The final item to mention with wipeouts is that they themselves can be hidden (while their ability to hide objects remains in effect). To do this, simply type in wipeout, press Enter, and select f for Frames in the submenu. AutoCAD then says: `Enter mode [ON/OFF]<ON>:`. If you type in `off`, AutoCAD turns off all wipeouts on the screen, while preserving their effects on other objects.

15.32 Level 2 drawing project (5 of 10): architectural floor plan

For Part 5 of the drawing project, you add carpeting and flooring to your floor plan. This is somewhat arbitrary and may or may not be the type of carpeting, tile, or flooring an architect would select on another design. You may use a different pattern, of course, if you wish. As always, be sure to use the proper layering.

Step 1. At a minimum use the following layer convention:

- A-Floor Covering (Gray_9).

Step 2. Freeze all layers that are not relevant to the carpeting, leaving only walls, wall hatch, appliances, and windows. Draw lines to block off doorways as needed so the hatch does not bleed from one room into another. Finally, create the carpeting/flooring plan shown in Fig. 15.76.

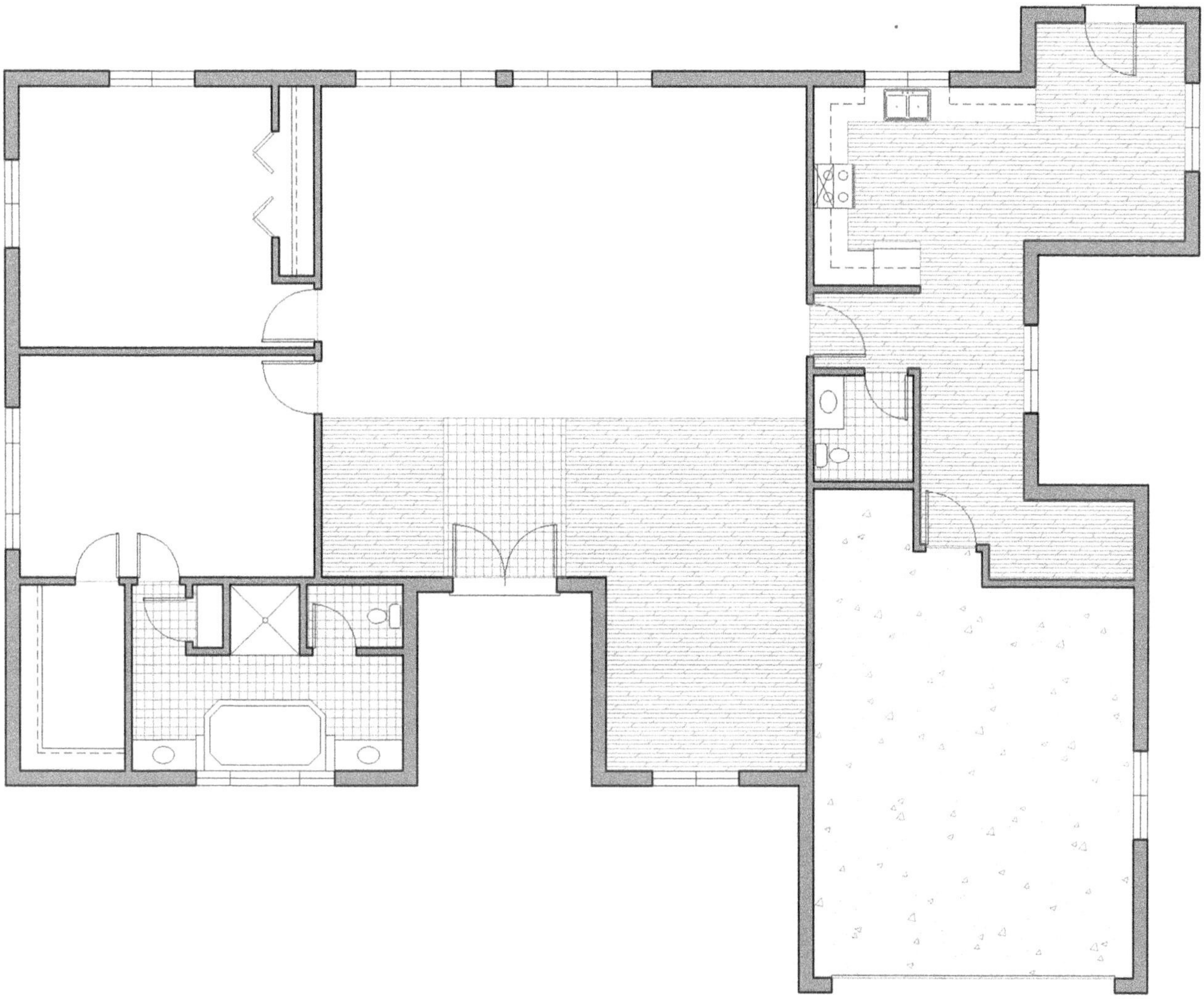

FIGURE 15.76 Carpeting and flooring layout.

Summary

You should understand and know how to use the following concepts and commands before moving on to Chapter 16*:*

- Align
- Action Recording
- Audit and recover
- Blend
- Break and join
- CAD Standards
- Calculator
- Defpoints
- Divide and point style
- Donut
- Draw Order
- eTransmit
- Filter
- Hyperlink
- Lengthen
- Object tracking (OTRACK)
- Overkill
- Point and node
- Publish
- Raster
- Revcloud
- Sheet sets
- Selection methods
- Stretch
- System variables
- Tables
- Tool palette
- UCS and crosshair rotation
- Window tiling
- Wipeout

Review questions

Answer the following based on what you learned in this chapter:

1. What are the purposes of the audit and recover commands? What is the difference?
2. For join to work, lines must be what?
3. The Defpoints layer appears when you create what?
4. Does the divide command cut an object into pieces?
5. What is the idea behind eTransmit?
6. To what can you attach a hyperlink? What can you call up using a hyperlink? How do you activate a hyperlink?
7. What is the idea behind overkill?
8. Can you make a revcloud out of a rectangle? A circle?
9. What three new selection methods are discussed?
10. Do you use a Crossing or a Window with the stretch command?
11. Name a system variable mentioned as an example.
12. Explain the importance of the wipeout command. What does it add to the drawing?

Exercises

1. Draw the following two rectangles according to the dimensions given. Pedit the width to 0.25″ and shade each separately with a solid hatch, color 9. Then, use align to place the 10″ box on top of the 4″ box. (Difficulty level: Easy; Time to completion: 5 minutes.)

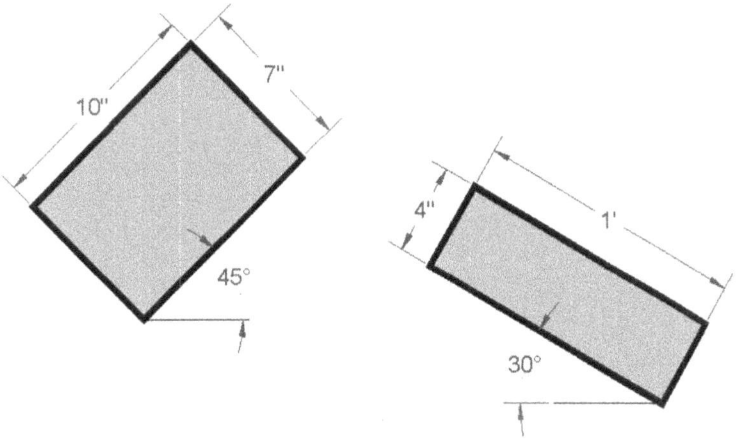

2. Draw the following line and divide it into seven sections using six points of the style shown. Then, pedit the line to a thickness of 0.25″ and break it using the break command (First point option) at each of the points. Finally, offset each alternate section 0.5″ up and 0.5″ down, as shown. (Difficulty level: Easy; Time to completion: 5–7 minutes.)

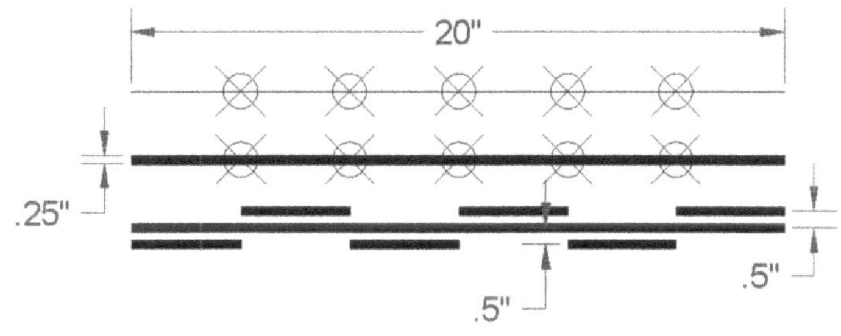

3. Create the following donuts according to the sizing shown. Include the linework for the truck itself. (Difficulty level: Easy; Time to completion: 5–10 minutes.)

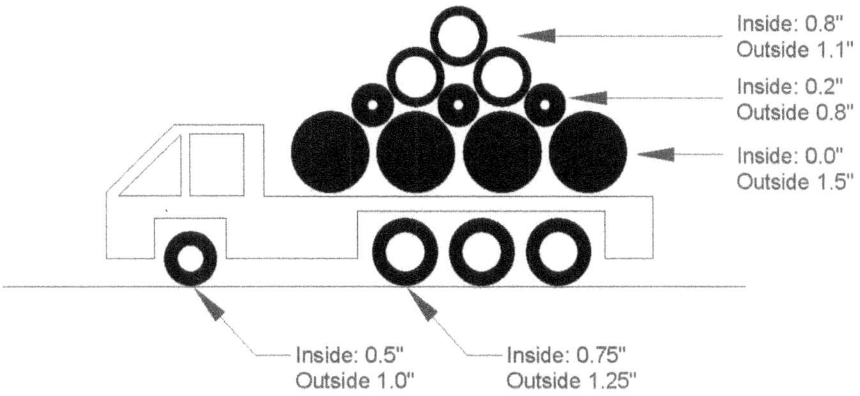

4. Draw the following rectangle, circle, and triangle according to the dimensions shown. Then, use revcloud and the Object option to turn each into a revcloud. Finally, pedit the new shapes to a thickness of 0.25″. (Difficulty level: Easy; Time to completion: <5 minutes.)

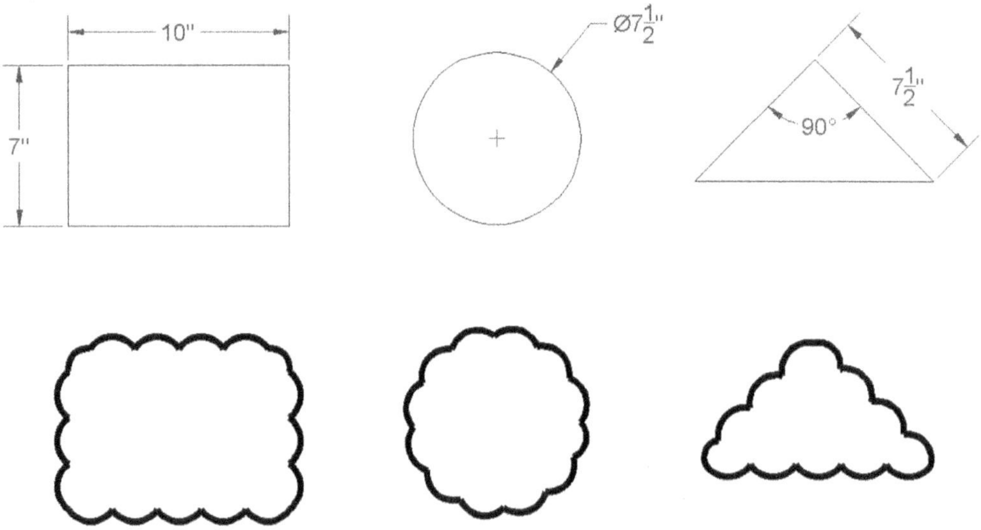

5. Use the OTRACK tool to connect the following shapes as shown. The exact sizing of the shapes is not important but do try to locate them relative to each other as shown in the exercise. (Difficulty level: Easy; Time to completion: 5 minutes.)

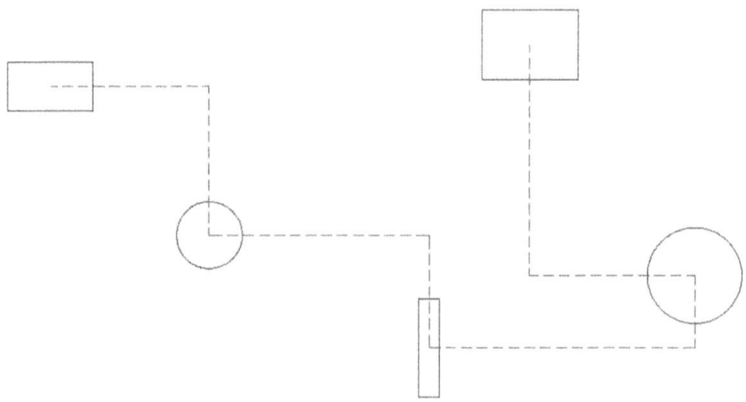

6. Draw a 4 × 4 rectangular array made up of 5″ circles. Then, scale every other one by 50% and shade the smaller circles with a solid hatch, as shown in the image on the left. Next, erase the shaded circles all at once using the erase command in combination with a Window Polygon. Your selection should look somewhat like the image on the right. Repeat again with a Crossing Polygon, then a Fence. (Difficulty level: Easy; Time to completion: 5−10 minutes.)

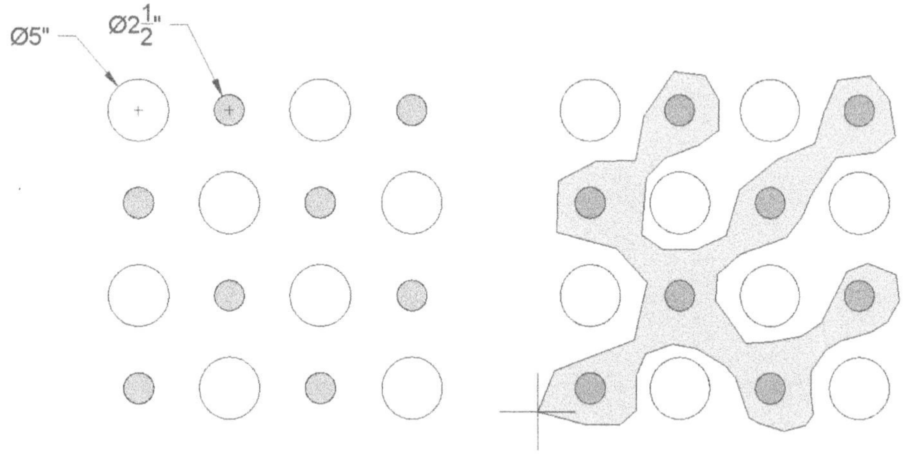

7. Recreate Exercise 1 of Chapter 4 (a table), using the table command as opposed to the basic lines and text approach of the original exercise. The table is reproduced here for your convenience. (Difficulty level: Intermediate; Time to completion: 10−15 minutes.)

LIST OF MAIN PARTS				
ITEM	QTY	DESCRIPTION	SIZE DWG	DWG. #
1	1	SKID ASSEMBLY	A1	83006
2	1	STRAINER ASSEMBLY	A1	83008
3	1	GENERATOR ASSEMBLY	A1	59470-001
4	1	GENERATOR ASSEMBLY	D1	59470-002
5	2	POWER SUPPLY (OIL IMMERSED)	B1	58717-001
6	1	POWER SUPPLY (OIL IMMERSED)	B1	58717-002
7	1	LOCAL CONTROL PANEL ASSEMBLY	D1	83017-001
8	2	LOCAL CONTROL PANEL ASSEMBLY	A1	88017-002
9	1	INSTRUMENT AIR ASSEMBLY	A1	83022
10	1	AIR PUMP	A1	86584

Spotlight on: civil engineering

Civil engineering specializes in the design and construction of bridges, roads, buildings, dams, canals, and other infrastructure. Civil engineering is the oldest of the engineering specialties, and some of its principles were already well established during the building of the pyramids in ancient Egypt and aqueducts in Rome. Civil engineering has several subspecialties, including environmental engineering, geotechnical engineering, structural engineering, transportation engineering, municipal or urban engineering, construction, and water resources among them.

FIGURE 1 *Image Source: www.megacivilengineering.blogspot.com.*

Education for civil engineers starts out like that for the other engineering disciplines. In the United States, all engineers have to attend a 4-year ABET-accredited school for their entry-level degree, a Bachelor of Science. While there, all students go through a somewhat similar program in their first 2 years, regardless of future specialization. Classes taken include extensive math, physics, and some chemistry courses, followed by statics, dynamics, mechanics, thermodynamics, fluid dynamics, and material science. In their final 2 years, engineers specialize by taking courses relevant to their chosen field. For civil engineering, this includes courses in asphalt, concrete, soil, geotechnical, water resources, traffic engineering, seismic, bridge, and advanced structural design.

On graduation, civil engineers can immediately enter the workforce or go on to graduate school. Many civil engineers choose to pursue a Professional Engineer (P.E.) license. The process first involves passing a Fundamentals of Engineering (F.E.) exam, followed by several years of work experience under a registered P.E., and finally sitting for the P.E. exam itself. The P.E. license allows the civil engineer to approve and sign off on project plans, and many engineers offer their consulting services to architects on large building projects, where the expertise of a civil engineer is a major requirement.

Civil engineers can generally expect starting salaries (with a bachelor's degree) in the \$60,000/year range, which is roughly in the middle-to-lower range among engineering specialties. The median pay is about \$82,000. This, of course, depends highly on market demand and location. The highest starting pay of all, \$88,000 and up, is for

FIGURE 2 *Image Source: www.ciprianicharlesdesigns.wordpress.com.*

petroleum engineering graduates. This specialty is a close cousin of civil engineering that specializes in locating and extracting petroleum. A master's degree is highly desirable and required for many management spots. The job outlook for civil engineers is excellent, as the profession is very much needed as the world's infrastructure ages and population growth creates a need for new development.

How do civil engineers use AutoCAD and what can you expect? Industry-wide, AutoCAD enjoys a significant amount of use among civil engineers, as many civil projects are 2D in nature. This is especially true with site plan work and some building design. An example of a civil site plan is shown in Fig. 3, with many of them quite complex and detailed in nature. They use Xrefs for the building footprint(s), plines or splines for the gradients, and other advanced tools. The figure features the architectural layout (footprint), parking spots, landscaping, and site boundaries among other items. Depending on the plan, utility and power lines may be shown, as well as outdoor lighting, gradient lines, and various legal boundaries, such as easements and offsets.

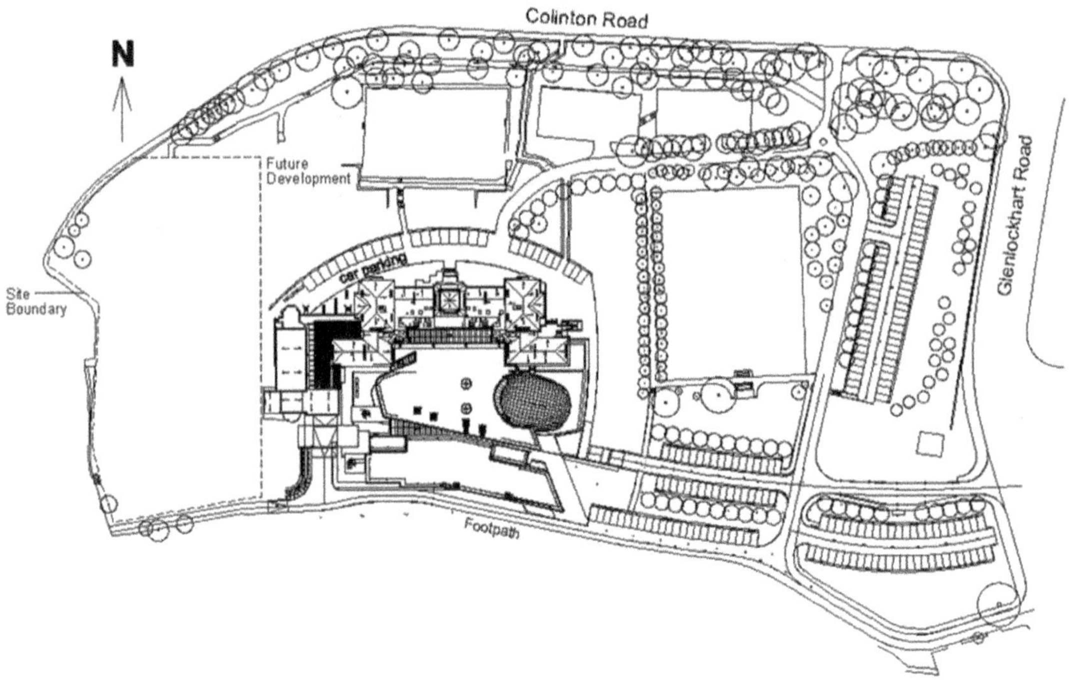

FIGURE 3 An example of a 2D civil engineering site plan. *Image Source: Elliot Gindis.*

As you move more into classic civil engineering projects (bridges, roads, etc.), 2D AutoCAD gives way to dedicated 3D software applications specially geared to this type of work, such as MicroStation, GEOPAK, and InRoads from Bentley. When AutoCAD is used, it is sometimes paired with additional software like Civil 3D, Land Desktop, or various GIS add-ons.

Layering in civil engineering work is highly specialized and depends on the project. Site plan work may include some AIA layers, but mostly they are descriptive and unique to the company; there is no universal standard. Road, Easement Line, Gas Line, and Property Line are all valid layer names.

Chapter 16

Importing and exporting data

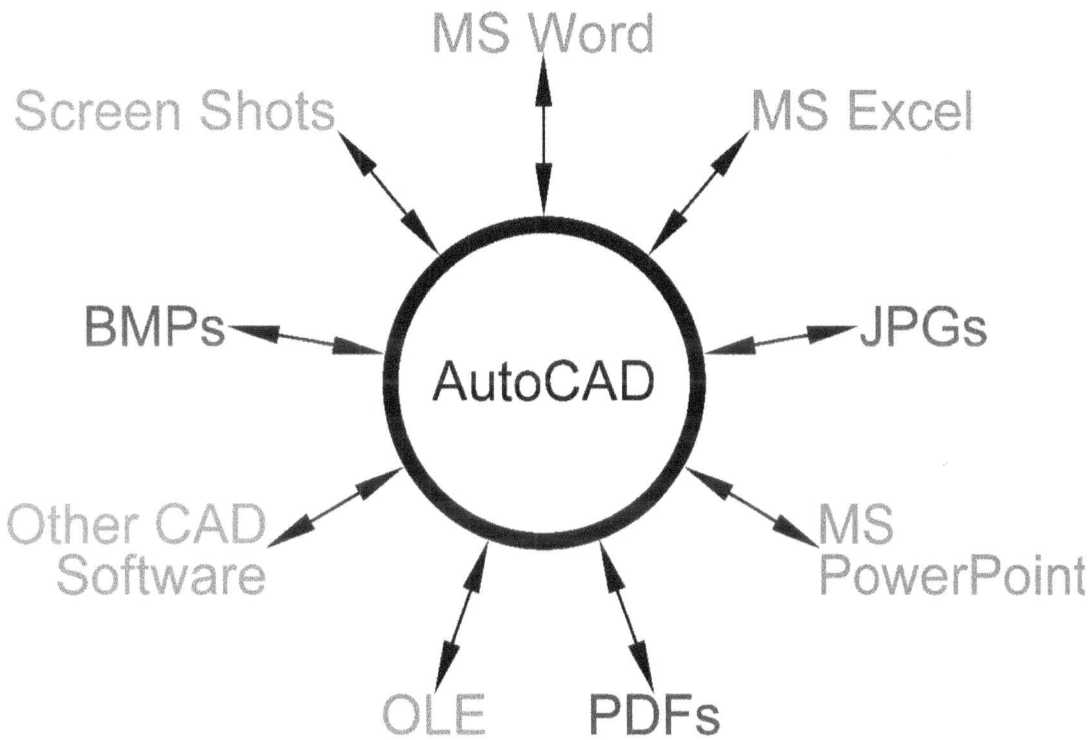

Learning Objectives

In this chapter, we learn how AutoCAD interacts with other software you are likely to use in the course of design work. We cover the following topics:

- AutoCAD and MS Word
- AutoCAD and MS Excel
- AutoCAD and MS PowerPoint
- AutoCAD and JPGs
- AutoCAD, Portable Document Formats (PDFs), and screen shots
- AutoCAD and other CAD software
- Exporting and Save As
- Importing and OLE (object linked embedded)

By the end of this chapter, you will smoothly import and export data among a variety of common office and design applications.

Estimated time for completion of this chapter: 2 hours.

Up and Running with AutoCAD 2022. https://doi.org/10.1016/B978-0-323-89923-9.00016-0

16.1 Introduction to importing and exporting data

AutoCAD and the designers who use it usually do not work in a "software vacuum." They typically interact with not only AutoCAD but also several other applications, some of them generic and used by many other professions, others unique to their own field. As such, there is often a need to either import data into or export data out of AutoCAD. Examples are numerous, such as an architect who wishes to transfer notes typed in MS Word into a project details page in AutoCAD or an engineer who needs to drop in an Excel spreadsheet. Just as often, AutoCAD drawings may need to be inserted into MS Word reports, PowerPoint presentations, and other applications. You may then need to create PDFs of your drawing or insert a PDF or an image into one.

Swapping files among the various software applications has been tricky in the past. Software companies are under no obligation to make such tasks flow smoothly, as it is not their business what else you may have on your PC (which may include programs from a direct competitor). Of course, such thinking scores no points with the customer, so developers do try to minimize compatibility issues whenever possible. Today, such tasks flow much smoother but still require knowledge of procedures that can be unique to each application.

The sheer quantity of software on the market is staggering, so we limit the discussion to

- Software or file types that are used or encountered by virtually everyone who uses a computer (Word, Excel, PowerPoint, JPG, and PDF).
- Software that has special relevance to AutoCAD users (such as other CAD programs).

We focus on two distinct tasks, how to *import* files *from* these applications and how to *export* AutoCAD files *into* them, analyzing each software pairing bidirectionally, and conclude with a brief overview of the export and insert features and OLE.

16.2 Importing and exporting to and from MS office applications

We begin with MS Word, Excel, and PowerPoint, as those are probably the ones most important and relevant to most users. Throughout, unless noted otherwise, we make extensive use of Copy and Paste. These Copy/Paste to clipboard tools are some of the most useful in computing, allowing you to shift around just about anything from one spot to another. The keyboard shortcuts are Ctrl + C for copy and Ctrl + V for paste or you can just right-click and select Copy and Paste.

Be aware that the "copy" is not AutoCAD's standard copy command, so do not use that. Yes, some students mix these up, so let us be clear from the start. Open Word, Excel, and PowerPoint and keep them handy as we go through each one.

Word into AutoCAD

An example of this may be when you would like to bring in extensive typed construction notes onto one of the pages of the AutoCAD layout.

Go ahead and click and drag to highlight the text in Word first and right-click or Ctrl + C to copy to the clipboard. Go to the AutoCAD screen, but do not right-click and paste in the text. If you do this, it appears as an embedded object, with a white background (seen in the top line of Fig. 16.1) and cannot be edited. Instead, create an mtext field of an appropriate size, and paste the copied text into it. The result is easily editable text that can be further formatted, as seen in the bottom line of Fig. 16.1.

FIGURE 16.1 Insertion of MS Word text.

AutoCAD into word

An example of this may be when you are preparing a report and would like to include some of the graphics spread out among the text to enhance clarity or support documentation.

The easiest way is to use the PrtScn (print screen) key, Microsoft's Snipping Tool, or other screen capture software. Make sure you screen background is set to white before doing any screen capture. You can also use the PublishToWeb JPG or PublishToWeb PNG printers that are built in to AutoCAD.

Once inserted, some cropping may be necessary to get rid of excess white space. Select the image and then right-click to bring up a shortcut menu with Wrap, Size, and Format options. A typical view of this procedure is shown in Fig. 16.2 in Microsoft Word 2016.

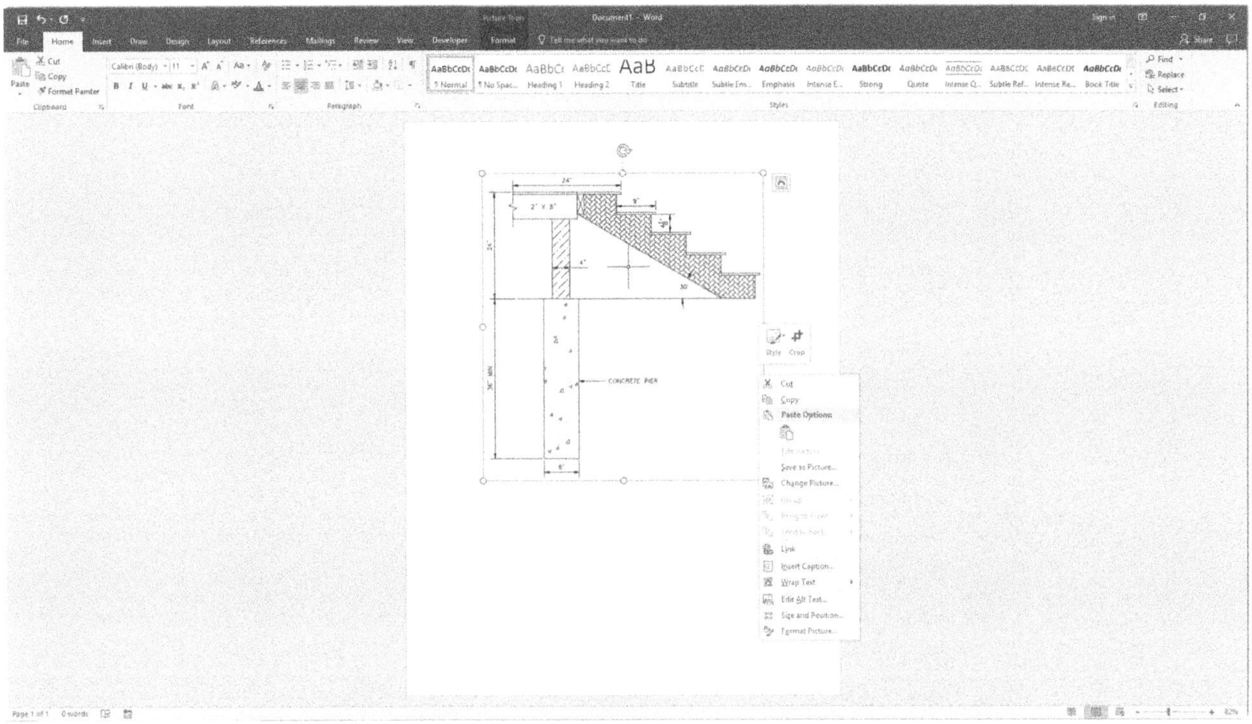

FIGURE 16.2 Insertion of an AutoCAD design into an MS Word document.

Excel into AutoCAD

An example of this may be when you would like to bring a spreadsheet into your electrical engineering drawing that details the type of connections and wiring used in a design.

Once again, simply use Copy/Paste. In Excel, create a small group of cells, number them (adding borders if desired), then highlight all 10 cells, right-click Copy, go into AutoCAD, and right-click for the pop-up menu and choose Clipboard > Paste (or Ctrl+V). As soon as you do this, pick an insertion and the spread sheet appears as shown in Fig. 16.3.

1	2
3	4
5	6
7	8
9	10

FIGURE 16.3 Copy/Paste Excel into AutoCAD.

If you carefully double-click the border of the inserted Excel sheet, the Properties palette will appear as shown in Fig. 16.4.

This is your basic Excel insertion. The group of cells are very much like a block, and you can erase, copy, and move them as needed. You can also scale them by clicking once and moving around the resulting grip points. You cannot mirror or rotate the group of cells, however.

A major question usually arises at this point. Are the inserted Excel cells linked to the actual Excel file from which they came, allowing for changes and updates in AutoCAD and/or Excel? They are not in the true sense of the word. What you can do is double-click inside the border of these inserted cells and Excel is called up temporarily for editing purposes.

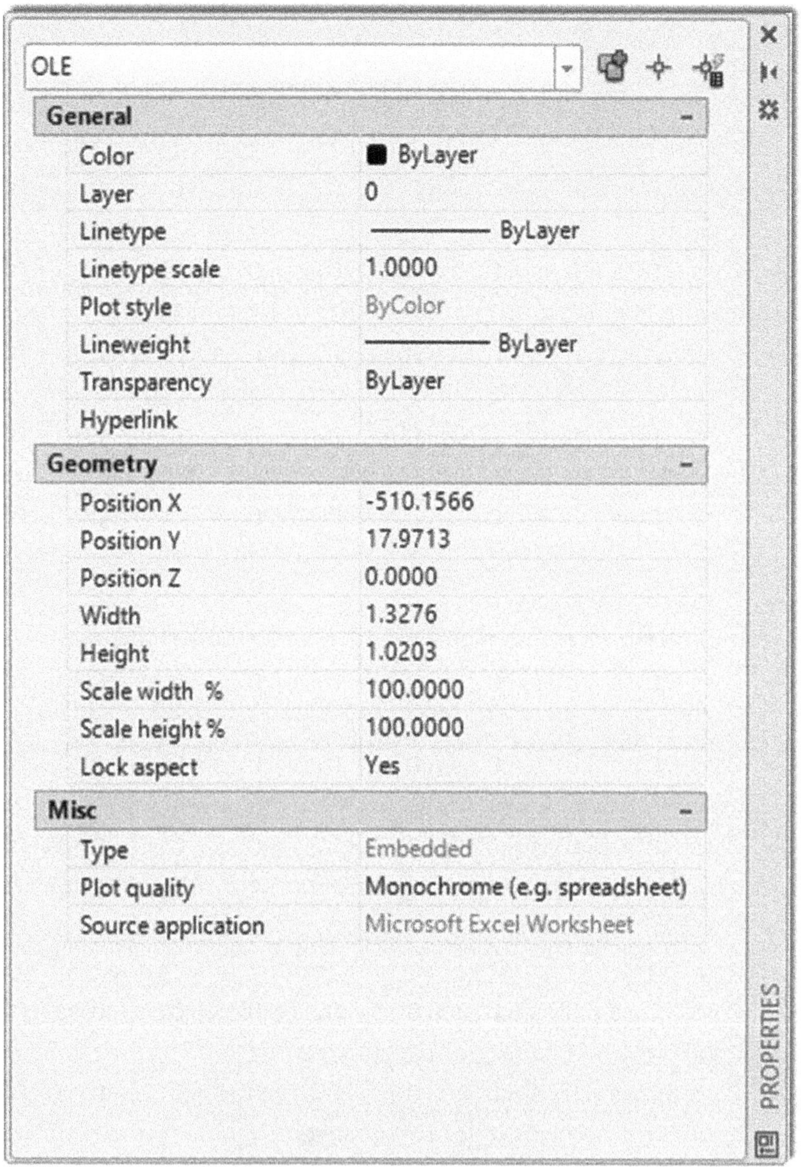

FIGURE 16.4 OLE Properties Panel.

Changes made in this temporary Excel session are reflected in AutoCAD if AutoCAD is open. If you close AutoCAD and work in the Excel sheet by itself, you would have to re-insert the cells again to see the latest changes. So, there is no true link. It is like AutoCAD is using Excel as a temporary table editor and that is it.

Having truly linked files is a very useful idea. There are some fancy Excel/AutoCAD interfaces in the industry, where extensive Excel data are used to generate corresponding AutoCAD drawings, but this trick requires additional programming and extensive customization and is beyond the scope of this book.

AutoCAD into Excel

This action basically pastes an image into Excel with the same Copy/Paste procedure as for Word. This, however, is rarely done, as Excel is primarily for spreadsheet data; and unless you need a supporting drawing, there is just no overwhelming reason to do this.

PowerPoint into AutoCAD

This action is not very common, as PowerPoint is almost always the destination for text and graphics, not an intermediate step to be inserted into something else. However, if it is needed, you can insert slides into AutoCAD via regular Copy/Paste.

AutoCAD into PowerPoint

This makes a lot more sense, as you may want to insert a drawing into a PowerPoint presentation. Fortunately, PowerPoint behaves very much like Word in this case, and one screen capture followed by some cropping is all it takes to insert an AutoCAD design, and no new techniques need to be learned. PowerPoint does not play well with .dwg files.

16.3 Screen shots

Screen shots or screen captures refer to capturing a snapshot of everything on your screen as seen by you, the user, and sent to memory, to be inserted somewhere. This is done by simply pressing the Print Screen button, which on most computers is called *PrtScn*, *SysRq*, or sometimes combined with *F12* on a laptop. You can capture what is on your screen this way and insert it into Word or PowerPoint, then crop out what you do not need to see. While many computer users may know of the Print Screen function to not actually print anything, they may overlook this useful trick for inserting data into other applications. If nothing seems to happen after pressing Print Screen, open a graphic editing program like Paint. Go straight to the paste command and an image of your latest Print Screen should appear. (Using the snipping tool or something similar is much easier.)

16.4 JPGs

This file compression format (used most often with photographs) can be inserted directly into AutoCAD and is a great idea if you would like a picture of what you drafted next to the drawing. Examples include an aerial photo of a site plan for architectural and civil applications and a photo of an engineering design after manufacture for as-built or record drawings. Inserting JPGs is a straightforward process.

Step 1. In Windows file manager, click on the name of a file you want to insert, right-click for the pop-up menu and select Copy.
Step 2. Then in AutoCAD right-click for a pop-up menu, select Clipboard > Paste to bring it into AutoCAD.

- AutoCAD asks for an insertion point: `Specify insertion point <0,0>:`

Step 3. Click anywhere you like.

- AutoCAD then asks for the image size: `Base image size: Width: 0.00, Height: 0.00 Inches`
  `Specify scale factor or [Unit] <1>:`

Step 4. Scale the JPG by moving the mouse to any size you wish (or type in a value or simply press Enter).

- AutoCAD then asks for the rotation angle: `Specify rotation angle <0>:`

Step 5. If you want the image rotated, enter a degree value; otherwise, press Enter.

The result of a JPG insertion is shown in Fig. 16.5. The edge has then been selected (one click), revealing the Ribbon's Image Editor. The image can be erased, copied, moved, rotated, mirrored, and just about anything else you can do to an element. You can also do some very basic image adjusting via the Ribbon or by double-clicking on the image's edge. The dialog box shown in Fig. 16.6 appears, allowing you to adjust Brightness, Contrast, and Fade, as well as reset everything. Any changes apply only to that JPG, not copies of it (if other copies exist).

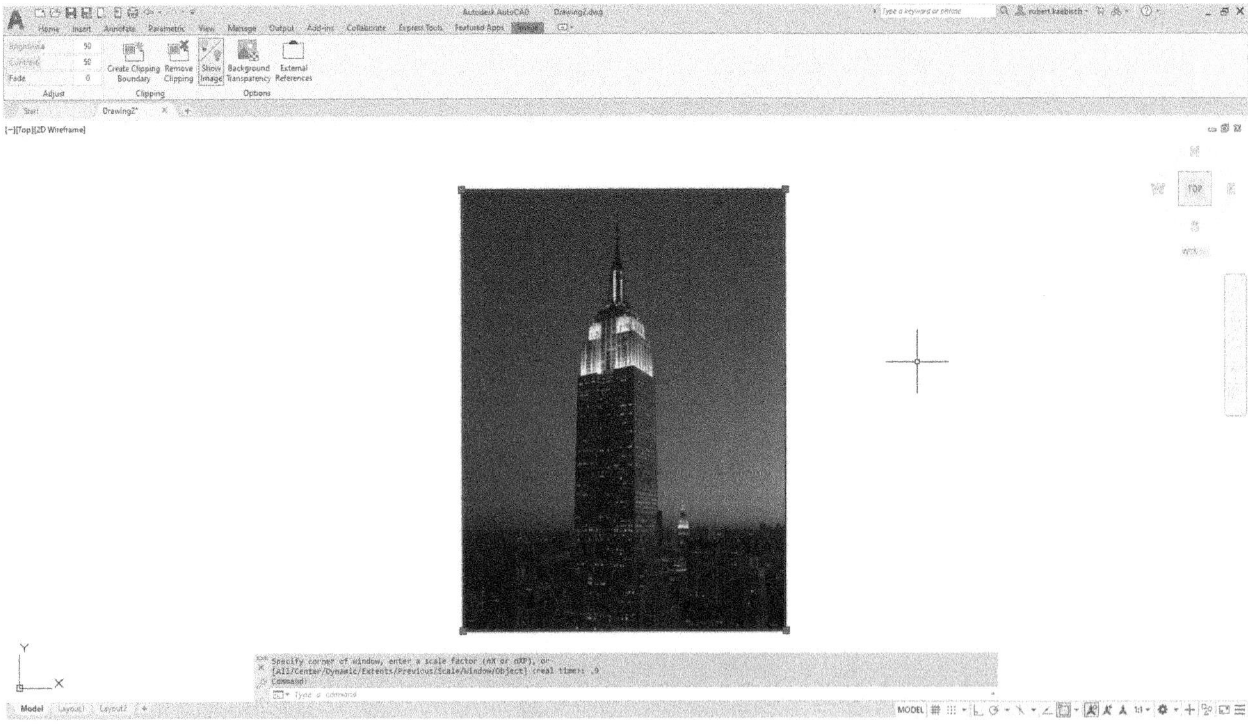

FIGURE 16.5 Embedded JPG.

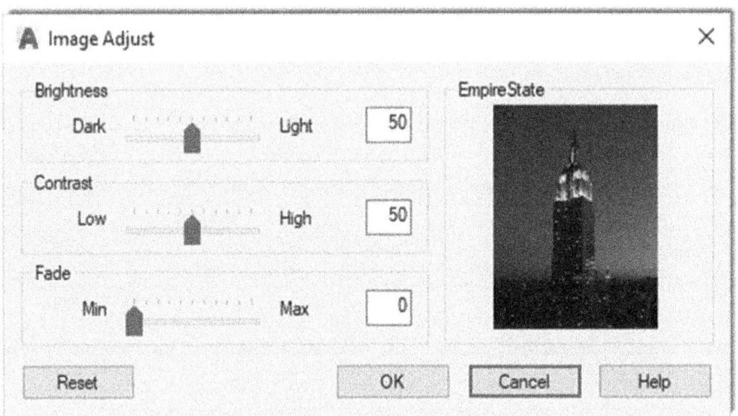

FIGURE 16.6 Image Adjust.

16.5 PDFs

The PDF is, of course, Adobe's popular software for document exchange. AutoCAD always had a good relationship with PDF, allowing users to easily convert drawings by simply printing to PDF—in other words, selecting a PDF "virtual printer" built into AutoCAD, installed with Windows, other PDF printers or, of course, Adobe Acrobat. The Print function would then execute, and AutoCAD asks you where you want the PDF file saved. Once the destination was selected, the PDF file was created. This process was much the same with Word and other software, and continues, with even more enhancements first introduced with AutoCAD 2018.

Another way to generate a PDF, especially if you need to customize the output (or you do not have Acrobat installed), is via the Ribbon's Output tab> Export approach. If you select the PDF option, a "Save As PDF" dialog box appears and allows some tinkering with options and output to finesse the results, as seen in Fig. 16.7.

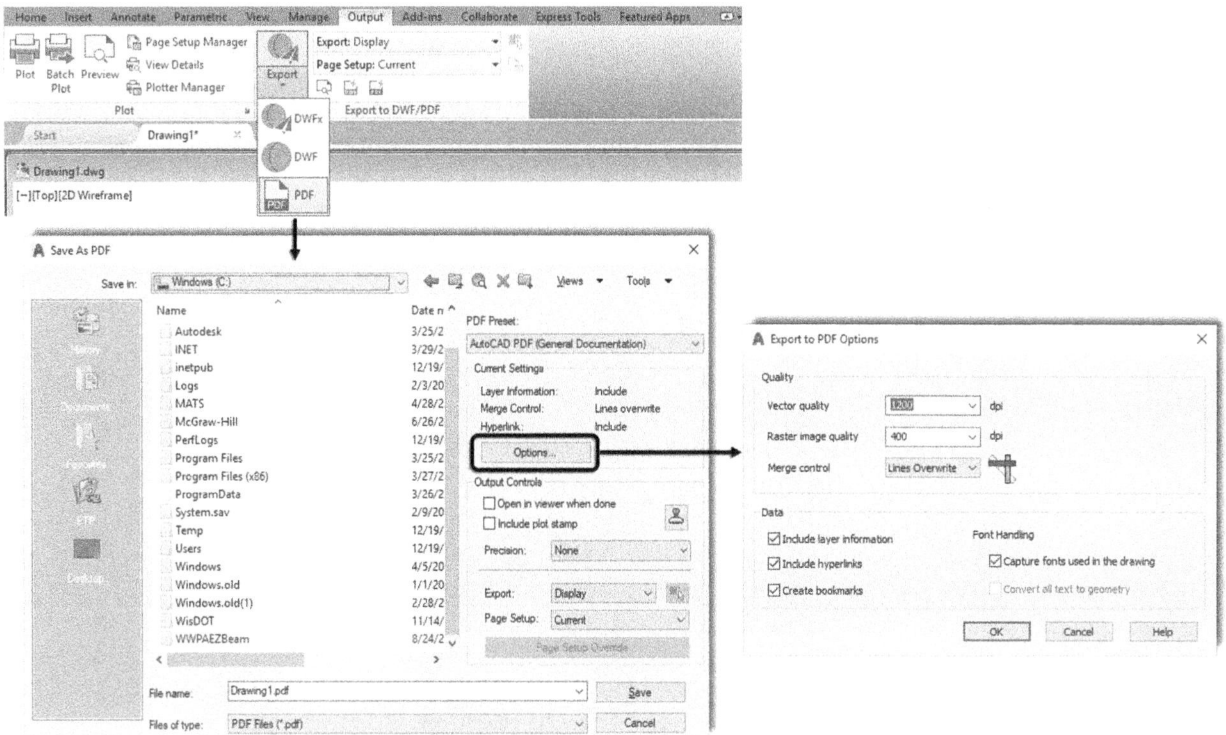

FIGURE 16.7 Save As PDF with options.

To try it out and bring in a PDF, use the PDF Import tool on the Import panel of the Insert tab of the Ribbon You then are prompted to look for the file you want to insert (that dialog box is not shown), and once you select the PDF, you will see the dialog box in Fig. 16.8.

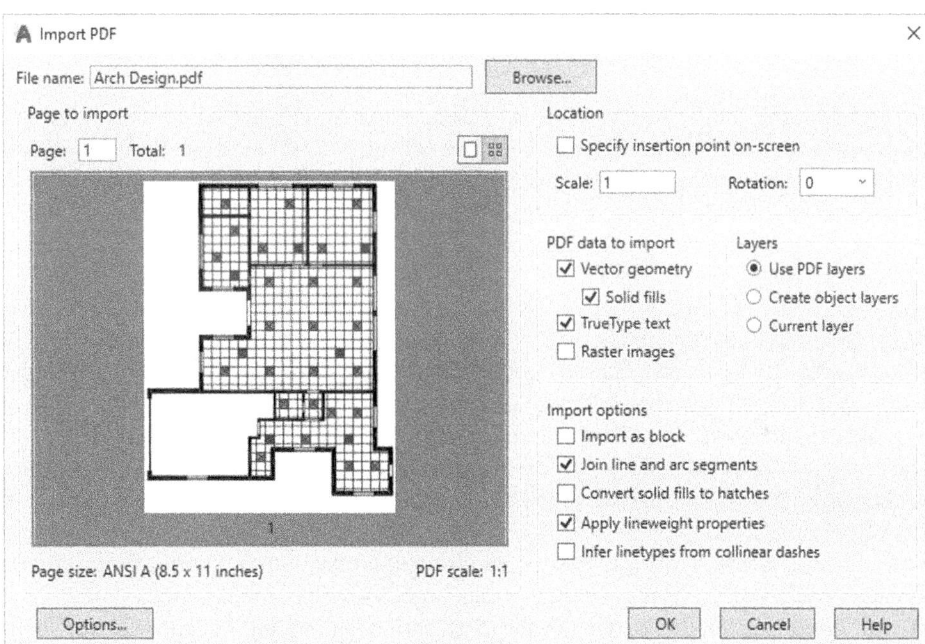

FIGURE 16.8 Import PDF.

Several things may happen depending on the chosen PDF to import. AutoCAD will attempt to turn any and all geometry continued in the PDF into useable AutoCAD objects, including line geometry and even graphics. If there is nothing to change into a useable object the PDF will behave much like the JPG files discussed earlier, with limited editing options in the Ribbon as shown in Fig. 16.5 above. If AutoCAD can recognize text and vector geometry in the PDF you will get useable AutoCAD objects in your drawing that you can manipulate and use as if you created them yourself as shown in Fig. 16.9. This is very powerful and useful tool when exchanging drawings with others. If it is an older PDF, does not contain useable geometry or was not created with AutoCAD, you may get an error or only get some items to import. If the PDF does not contain anything that AutoCAD can use, bring the file in as an external reference (or xref as the cool kids say) as described in the next chapter.

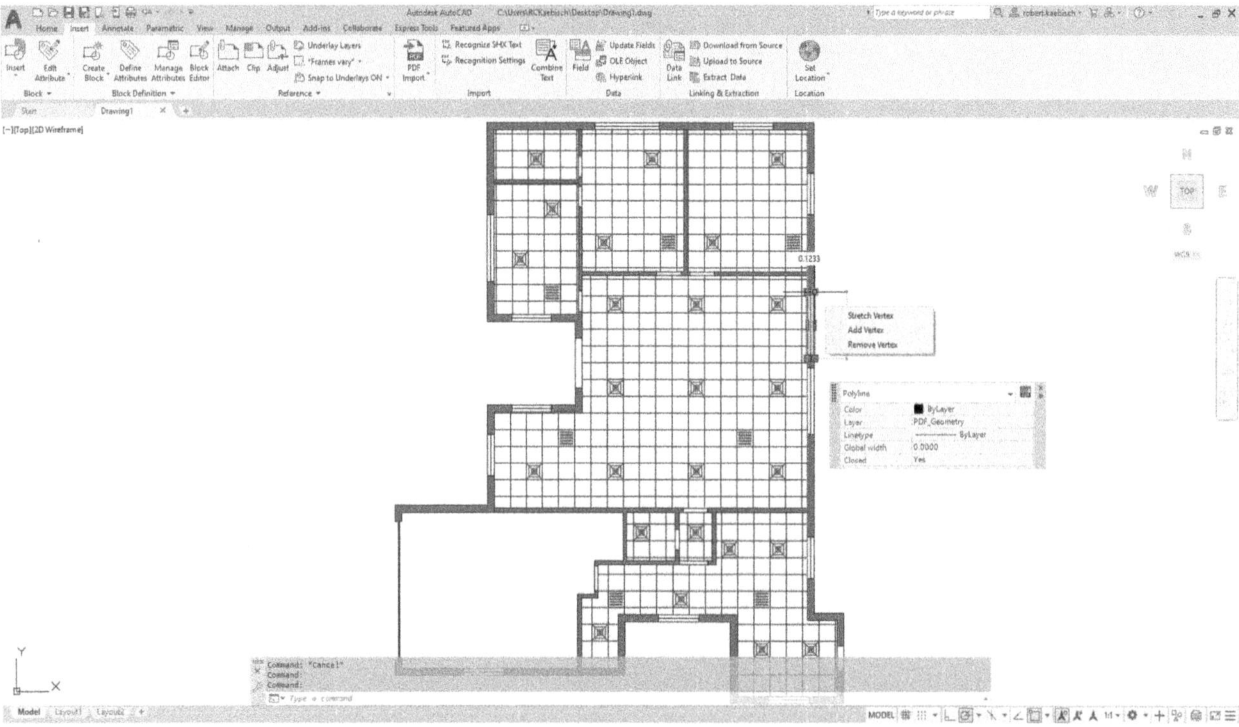

FIGURE 16.9 PDF after Import with useable geometry.

There is considerable need for this in the industry. Some drawings exist only in PDF format with the original dwg files lost or unavailable, and you may need to convert a PDF to a useable drawing form. If the PDF was a created from a CAD file and/or contains vector geometry, things are much easier, and you can get close to a 100% correct conversion. Be sure to verify some measurements on the PDF import to confirm its accuracy and scale.

16.6 Other CAD software

Initially, as AutoCAD gained market dominance, it did not really attempt, nor had any incentive, to "play nice" with other CAD software on the market. Being the 2D industry leader allowed AutoCAD to get away with not accepting any other files in their native format. It was the company's way or the highway, but that slowly changed. One major concession regarding importing and exporting was with MicroStation, AutoCAD's nemesis and only serious competitor. MicroStation always opened and generated AutoCAD files easily but not the other way around.

Finally, in recent releases, AutoCAD listed the .dgn file format as something you can import and export. You were then able to bring in MicroStation files, more or less intact, as well as create ones from AutoCAD drawings. In AutoCAD 2009, the settings box was expanded and that was carried over into the current AutoCAD version. You can now compare layers and other features side by side and make adjustments, so the target file is like the original.

To access this, use the Import DGN function in the Application Menu (Big A). Browse for the dgn file you are interested in importing. Another way around all of this is just to have the MicroStation user generate the AutoCAD file. You can also export .dwg files directly to .dgn as well by using the cascading Application Menu Export function selecting .dgn as the target file. After selecting a save location, you then see the Settings dialog box in Fig. 16.10. Of course, MicroStation accepts the AutoCAD file with no problem, so this may not be necessary. If you would like to learn a bit more about MicroStation in general, see Appendix B.

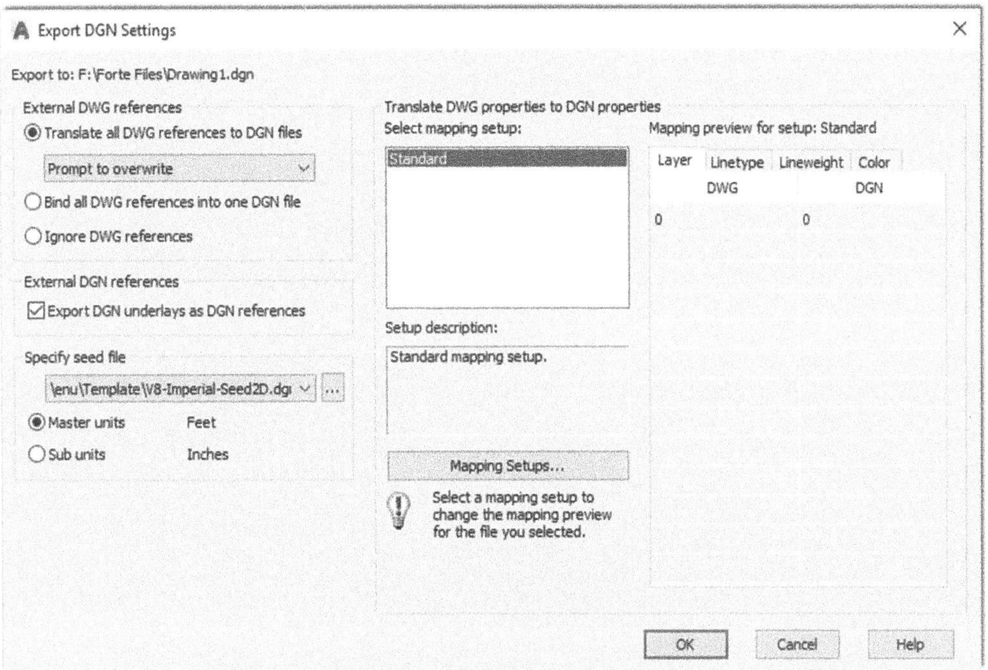

FIGURE 16.10 Export DGN settings.

Let us expand the discussion to other CAD software. What if someone is using another CAD program and needs to exchange files with an AutoCAD user? The .dxf file is the solution, as it is a format agreed on by most CAD vendors as the go-between that can be understood by all applications for easy file sharing. AutoCAD easily opens a .dxf or generates its own .dxf files for any drawing. This is discussed momentarily.

A few observations on sharing files: Among architects, the issue is not much of a problem, as AutoCAD is the main application in this profession, and competitors such as ArchiCAD easily open AutoCAD's files and send a .dxf right back if needed. In engineering, however, AutoCAD has a lesser presence. While some electrical, mechanical, and civil engineers use AutoCAD, many others, especially industrial, rail, aerospace, automotive, and naval engineers, use 3D modeling and analysis software instead, such as CATIA, NX, Pro/Engineer, or SolidWorks.

Until recently, these applications did not interact well with AutoCAD. Their respective "kernels" (the software's core architecture) are significantly different in design and intent from AutoCAD's ACIS 3D kernel. Their native and export file types (such as IGES and STEP) may also be unfamiliar to AutoCAD users. Some of the previously mentioned software can generate .dxf files, but those files are "flattened" and what you get is a 2D snapshot of what used to be a 3D model. AutoCAD was also able to import images from CATIA and SolidWorks using the OLE command (to be discussed soon), but once again, they were just flattened images.

All this can be somewhat confusing. Because so many different CAD packages are on the market, AutoCAD designers need to be aware and up to date on what others in their industry are using and think on their feet when it comes time to exchange files, as there usually is a way to access most file types. Let us now look at the specifics of exporting, importing, and the OLE command to explore the available options.

16.7 Exporting and the save as feature

As previously with the MicroStation example, exporting data can be done by selecting Export from the Application Menu (Big A). What you see is the dialog box in Fig. 16.11.

Your drawing can be exported to the following formats:

- *3D DWF* or *DWFx*: This is to convert the drawing to web format for use on viewers.
- *FBX*: FBX stands for Filmbox and is a proprietary file format developed by Kaydara and now owned by Autodesk. It is used to provide interoperability between digital content creation applications.
- *Metafile*: This is to convert the drawing to a Windows Meta File, a Microsoft graphics format for use with both vectors and bitmaps.
- *ACIS*: ACIS is the 3D solid modeling kernel of AutoCAD as well as a format to which a drawing can be exported. Few other software currently use the ACIS kernel (SolidEdge comes to mind), as Parasolid is the industry standard, so this is a rarely used export format.

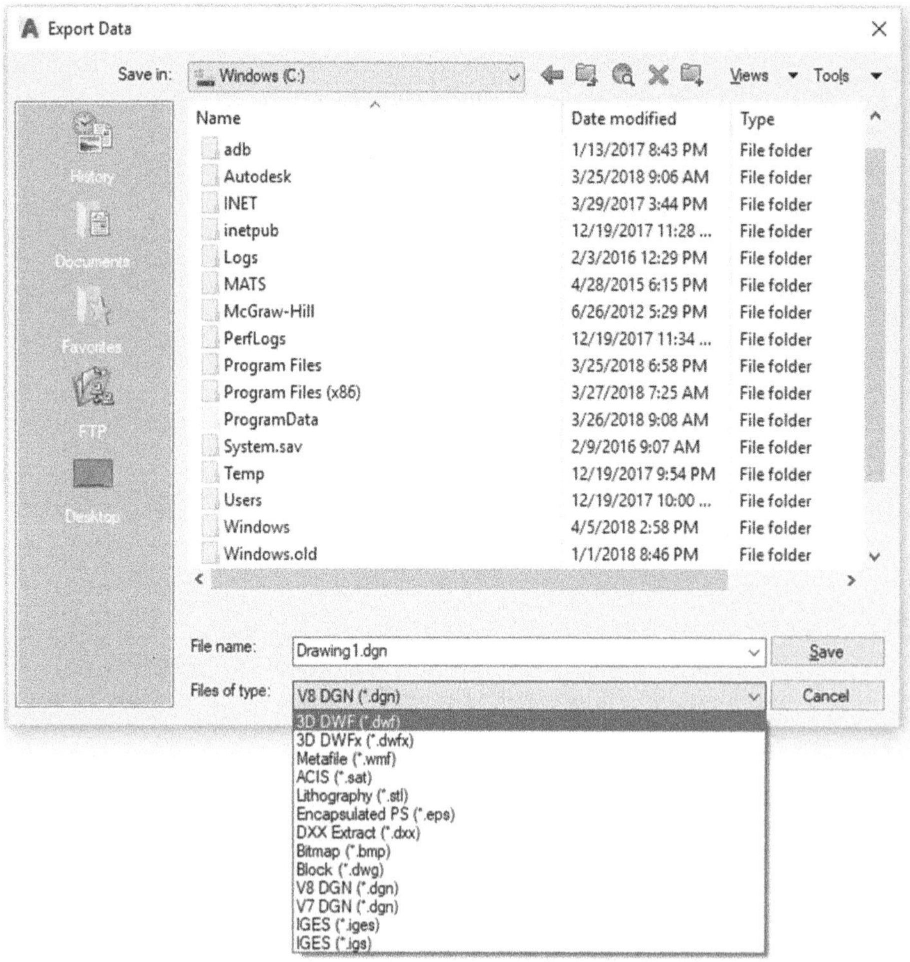

FIGURE 16.11 Export Data dialog box.

- *Lithography*: Stereo lithography files (*.stl) are the industry standard for rapid prototyping and can be exported from most 3D CAD applications, including AutoCAD. Basically, it is a file that uses a mesh of triangles to form the shell of your solid object, where each triangle shares common sides and vertices.
- *Encapsulated PS*: The eps is a standard format for importing and exporting PostScript language files in all environments, allowing a drawing to be embedded as an illustration.
- *DXX Extract*: DXX stands for Drawing Interchange Attribute. This is a .dxf file with only block and attributes information and is not relevant to most users.
- *Bitmap*: This converts the drawing from vector form to bitmap form, with the resulting pixilation of the linework; it is not recommended.
- *Block*: This is another way to create the familiar block covered in Level 1.
- *V8 DGN*: MicroStation Version 8, as described previously.
- *V7 DGN*: MicroStation Version 7, as described previously.
- *IGES*: IGES stands for Initial Graphics Exchange Specification and is a vendor-neutral format for 3D CAD solid modeling files. The files can end in either .iges or .igs.

The Save As dialog box can be accessed anytime you select Save or Save As … from the Appliaction Menu (Big A) or Quick Access toolbar or simply by typing. In either case, you get the dialog box shown in Fig. 16.12.

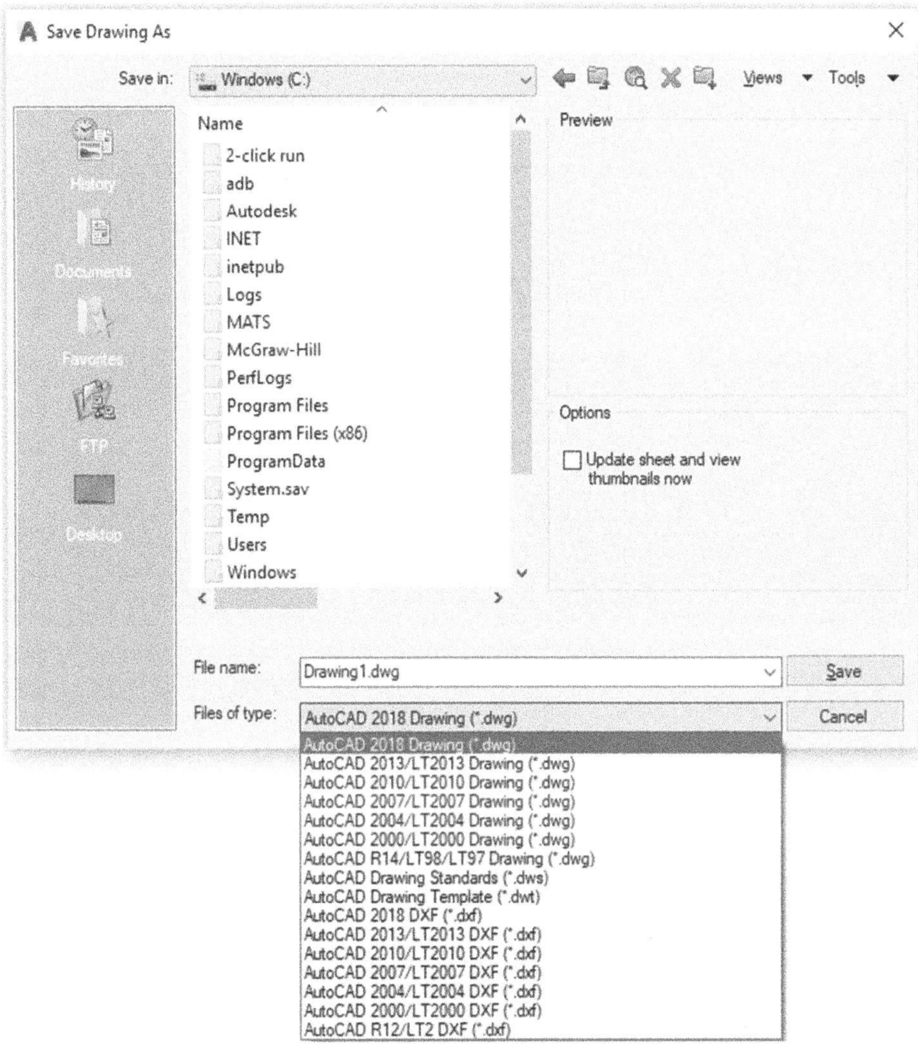

FIGURE 16.12 Save Drawing As dialog box.

The extensions presented here are described in detail in Appendix C, File Extensions, but in summary, you can save your drawing as an older version of AutoCAD (an important step that is automated via a setting in Options, as covered in chapter: Options, Shortcuts, CUI, Design Center, and Express Tools) or as a .dxf file, as described earlier. Note that the. dxf files can also be created going back in time, all the way to Release 12 (that was a mostly DOS-based AutoCAD). The other extensions, .dws and .dwt, are used less often and are also mentioned in Appendix C.

16.8 Inserting and OLE

The final discussion of this chapter features the Insert menu option in general and the OLE Object (type: insertobj) command in specific. The Data panel of the Insert tab of the Ribbon is shown in Fig. 16.13.

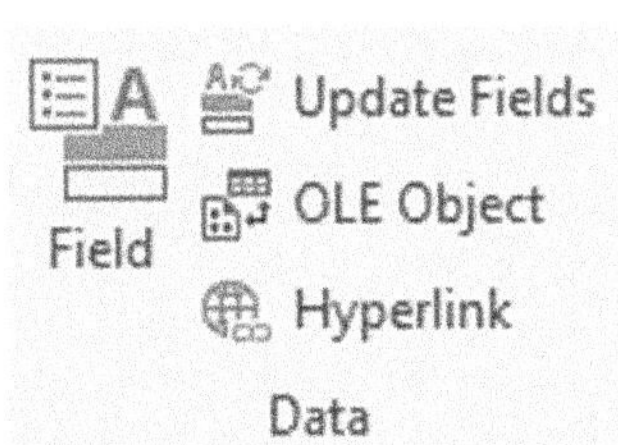

FIGURE 16.13 Data panel of insert tab of ribbon.

Let us then focus on OLE (object linked embedded). OLE is a way to embed (insert) nonnative files into AutoCAD on a provisional basis, meaning these files are not part of AutoCAD (they cannot be anyway, as they are of non-AutoCAD formats), but they are merely dropped in and are visually present. In principle, you can drop in just about anything, with just about any results. Some files sit nicely, others do not appear, and still others may even cause a crash. Let us look at the dialog box.

Select OLE Object … from the Insert menu, and the dialog box in Fig. 16.14 appears.

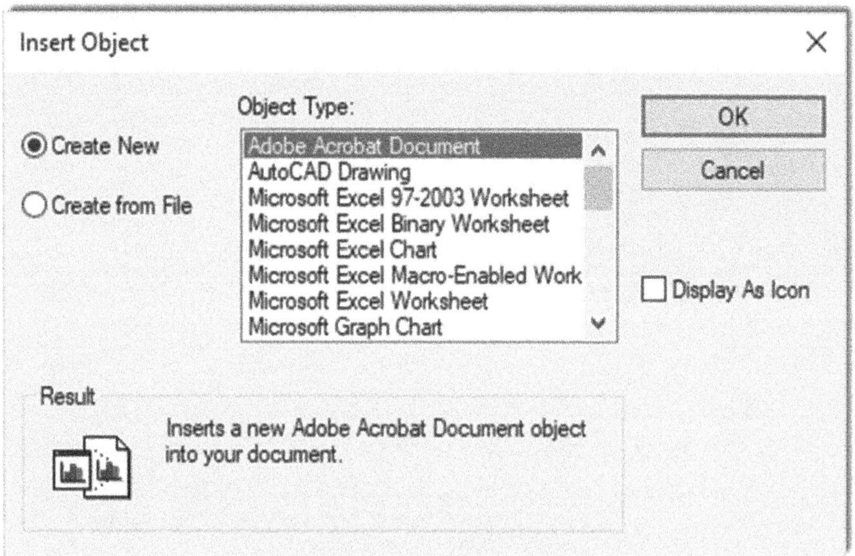

FIGURE 16.14 Insert object.

These, in theory, are the various software file items that can be inserted into AutoCAD. Looking through it, top to bottom, you can see that a wide variety of Adobe, Microsoft, and even Dassault (they own CATIA and SolidWorks) files can be inserted. You can also, in principle, insert video and audio clips (one can envision music files playing on the opening of an AutoCAD drawing) and even Flash animations. Simply pick what you want to insert and select Create from File. Some of the more useful OLE insertions are related to the Microsoft products listed, but much of that topic was already covered in this chapter. To insert a file, select Create from File, and browse for the file in which you are interested.

We do not cover Fields in this text. They are a very powerful tool for automating text entry by getting data from AutoCAD objects and should be explored once you have mastered the first 20 chapters of this book.

16.9 Level 2 drawing project (6 of 10): architectural floor plan

For Part 6 of the drawing project, you complete the internal design of the house by adding a reflected ceiling plan (RCP) and a heating, ventilating, air conditioning (HVAC) system that is made up of diffusers and a few return air ducts. Note that a residential home is not likely to have an RCP, which is so named because it looks exactly as if you held a mirror in your hand and looked down at it to see the ceiling. RCPs are mostly part of commercial space design and include what you generally see in a typical office: white ceiling tiles, lights, and HVAC diffusers. Here, for educational purposes, we add this ceiling into the residential design.

Step 1. Create some appropriate layers, such as

- M-RCP (Gray_8).
- M-Diffuser (Yellow).
- M-Ret_Air (Cyan).

Step 2. Create some HVAC symbols, as shown in Fig. 16.15. Make blocks out of them for easier handling. Be sure to be on that object's layer before making a block.

Step 3. Create the RCP plan by drawing arbitrary vertical and horizontal "starter" lines in the middle of each room and offsetting 2 feet in either direction until you reach the walls. Then, trim as necessary.

Step 4. Add in the diffuser and return air duct symbols as shown in the plan (Fig. 16.16).

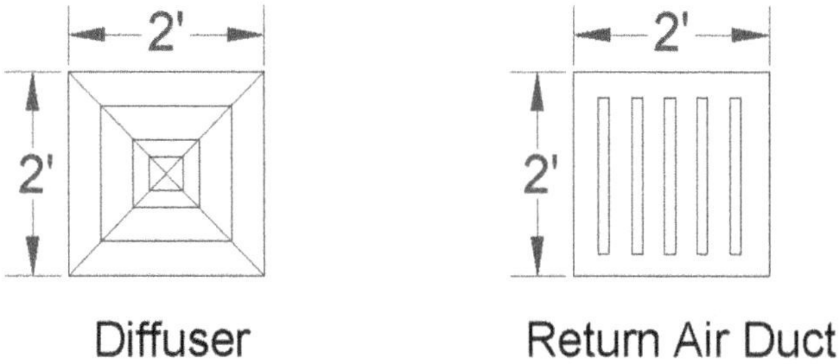

FIGURE 16.15 Diffuser and return air duct symbols.

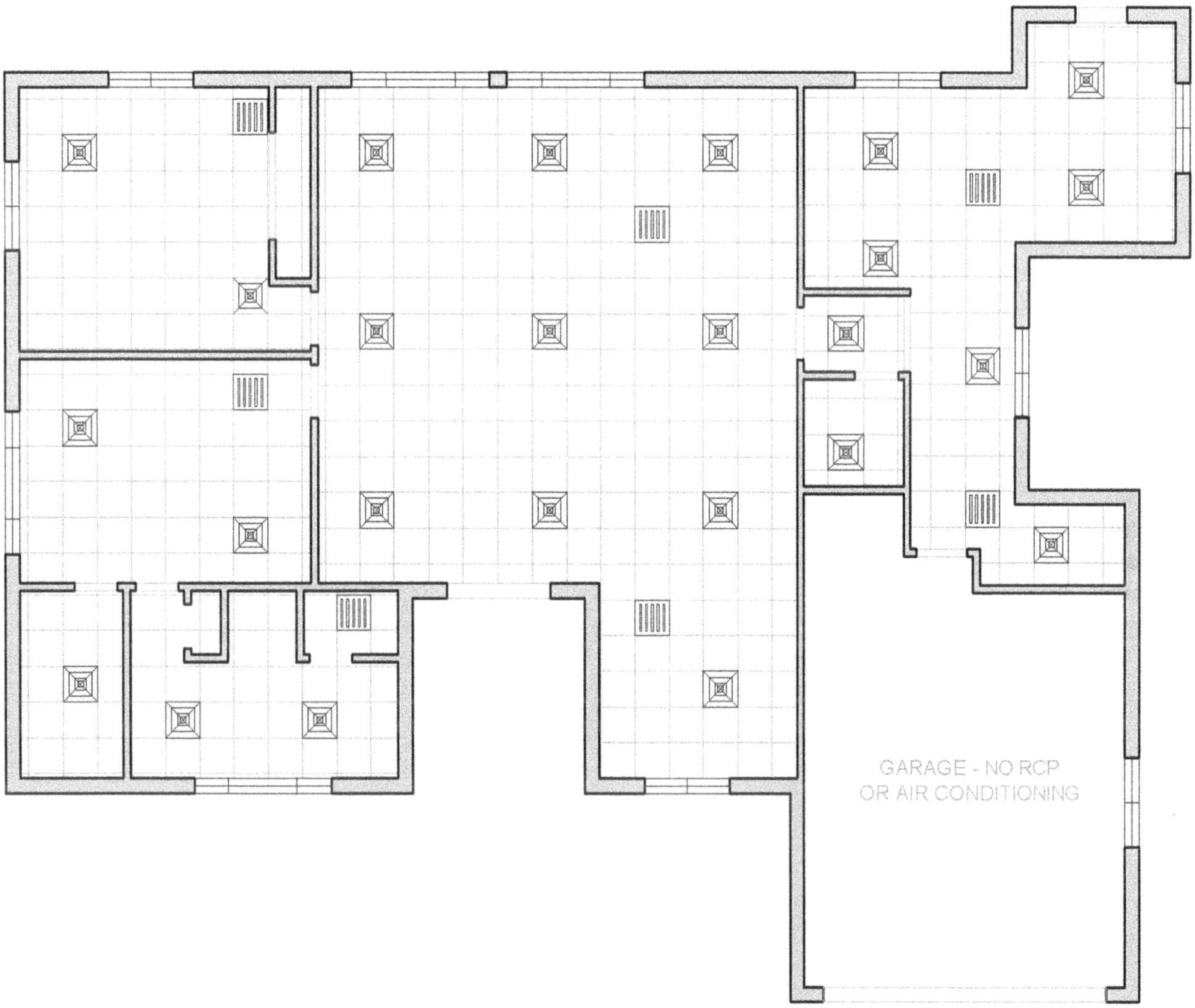

FIGURE 16.16 RCP and HVAC layout.

Summary

You should understand and know how to use the following concepts and commands before moving on to Chapter 17:

- Data transfer from Word to AutoCAD
- Data transfer from AutoCAD to Word

- Data transfer from Excel to AutoCAD
- Data transfer from AutoCAD to PowerPoint
- Screen shots, JPGs, and PDFs
- Interaction with other CAD software
- Exporting and Save As features
- Inserting and OLE

Review questions

Answer the following based on what you learned in this chapter:

1. What is the correct way to bring Word text into AutoCAD?
2. What is the procedure for bringing AutoCAD drawings into Word?
3. What is the correct way to bring Excel data into AutoCAD?
4. What is the procedure for bringing AutoCAD drawings into PowerPoint?
5. What is the procedure for importing JPGs and generating PDFs?
6. Describe the basic ideas of interacting with other CAD software.
7. What are Exporting and Save As?
8. What is OLE?

Exercise

1. At your own pace, review everything presented in this chapter and do the following:
   - Type any short paragraph in MS Word and import it into AutoCAD correctly, using Copy and Paste.
   - Type in several columns of data in MS Excel, and then import them into AutoCAD. Adjust them by scaling up and down.
   - Import any image from AutoCAD into MS Word using:
     - The Print Screen function.
     - Copy and Paste.
   - Export any drawing to a .dxf format.
   - Bring in any JPG (picture) using Copy and Paste. Adjust it in AutoCAD.

(Difficulty level: Easy; Time to completion: 15 minutes.)

Chapter 17

External references (xrefs)

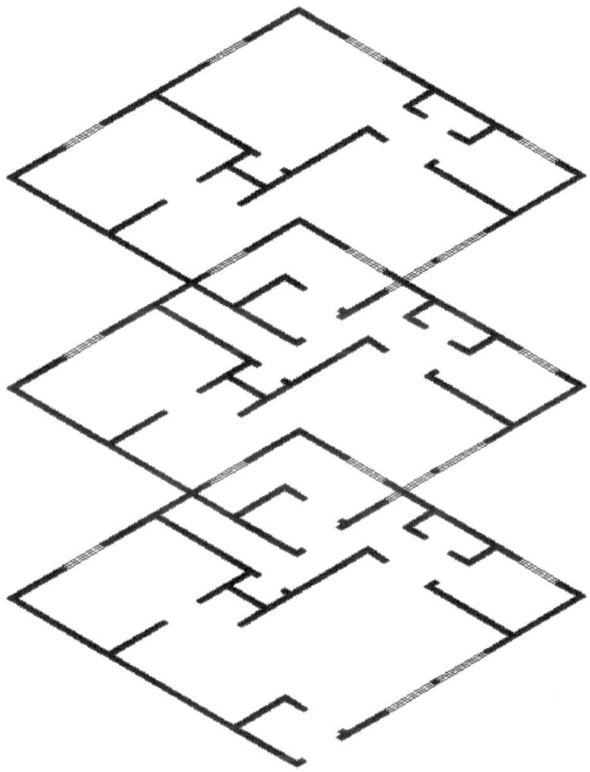

Learning Objectives

In this chapter, we introduce and thoroughly cover the concept of external references (xrefs). We specifically introduce the following topics:

- The primary reasons for using xrefs
- Loading xrefs
- Unloading xrefs
- Binding xrefs
- Updating and editing xrefs
- Layers in xrefs
- Multiple xrefs

By the end of this chapter, you will be well versed in applying this critical concept to your design work.
Estimated time for completion of this chapter: 1–2 hours.

Up and Running with AutoCAD 2022. https://doi.org/10.1016/B978-0-323-89923-9.00017-2

17.1 Introduction to xrefs

External references, *xrefs* for short, are a critically important topic for architectural AutoCAD users and very useful knowledge for all others. Along with Paper Space (see Chapter 10), xrefs form the core of advanced AutoCAD knowledge. Like many advanced topics, you need to understand fundamentally what it is and when to use it. The "button pushing," the mechanics of how to make it work is the easy part. In this introduction, we will discuss what an xref is and why there is a need for this rather clever concept, which has existed since AutoCAD's early days and is a necessary tool with any drafting software.

What is an xref?

As defined, an xref is a file that is electronically attached (referenced) to another, separate, file. It then appears in that new file as a fully visible background, against which you can position new design work. This xref is not really part of the file, but it is merely "electronically paper clipped" to it and cannot be modified in a traditional way; it serves as only a background. The word *xref* is also used as an action word or verb, as in "You need to xref that file in."

Why do we need an xref? What is the benefit?

The preceding paragraph was just a strict definition and may not have yet illuminated the real reason for xrefs. To see why we have this concept in AutoCAD, let us propose the following scenario.

Imagine you are designing a tall office building. It is boxy in shape (like the Aon Center, Chicago), and as such the exterior walls, stair towers and elevator shafts do not change from floor to floor. The interior, however, does. The first floor is a lobby, the second is a restaurant, the third is storage, and floors 4 through 100 are office space. In short, for a designer, once the exterior is done, the rest is all interior-space work. Therefore, going from floor to floor, the exterior design file can be xrefed in and just "hangs out," providing a background for placement of interior walls and other items. This dramatically reduces the size of each file, as a copy of the exterior is merely referenced and never part of the interior files. Got 100 floors of interior work to do? No problem; one xref file of the exterior walls, stair shafts and elevator shafts is all that is needed. It can be xref's in to one drawing for one person, or hundreds of drawings for hundreds of people. To summarize: *Xref is used to reduce file size by attaching a core drawing to multiple files.*

If this were all there was, xref would be a neat trick and would have surely fallen out of favor as computer speed, power, and disk storage space increased over the years. However, the preceding was just a warm-up. The main reason for the xref concept is design changes. If you need to change the shape of the exterior walls or vertical shafts, you need to do it only once, and the change propagates through all files using that xref as soon as it is saved. Think about the implications for a moment. In the old hand-drafting days, each floor had to be drawn separately. Imagine now a last-minute change to the exterior design. Hundreds of sheets would have been updated by hand. Score one for AutoCAD. To summarize: *Xref is used to automate design change updates to core drawings.*

Need another benefit? How about security? The xref file can be placed in a folder that is accessible only to the structural engineer and lead architect, not the interior designer or electrical engineer. Granted, this "need to know" basis for access may not be necessary, but it is good to have a way of keeping someone from shifting structural walls to accommodate the furniture. To summarize: *Xrefs are used to enhance critical file security.*

Ideally, you are now sold on the benefits of an xref. So, how is this all done in practice? The xref is typically not just the exterior walls but also columns and the interior core (elevators and stairs), as those also tend to not change from floor to floor. The designer creates these files, naming them something descriptive, such as Exterior_Walls_Ref or Columns_Core. Next, the xrefs are attached together and then to the first floor file or to the first floor file individually.

The first floor file is the active design file of course, and is named Floor_One_Arch. At first, it is blank, showing just the xref. Then, the interior design is completed. The same designer (or another designer) opens another blank file, attaches the same xref (or set of xrefs), and names this file Floor_Two_Arch, and so on. Xrefs can be attached, detached, refreshed, and bound to the main drawing file. We cover all this shortly. We also cover some xref-related topics, such as layering and methods to edit xrefs in place if absolutely needed.

17.2 Using xrefs

To demonstrate the features of the xref command, we first need to have a building exterior file with which to work. You can select one of your own or one assigned by your instructor, if taking a class. Alternatively (and for good practice), draw the floor plan in Exercise 1 at the end of this chapter. We use that floor plan for the rest of the xref discussion. In Fig. 17.1, the plan is rotated horizontally and the left corner moved to point 0,0. All the dimensions are frozen, and the file is saved as Exterior Walls.dwg.

As of now, nothing is special about this file. We still need to xref it to the working file. Open a blank file and save it as Floor_One_Layout.dwg. Then, type in `xref` and press Enter. (Alternatively, you can use the Attach command from the References panel of the Insert tab of the Ribbon). This will bring up the External References palette as seen in Fig. 17.2.

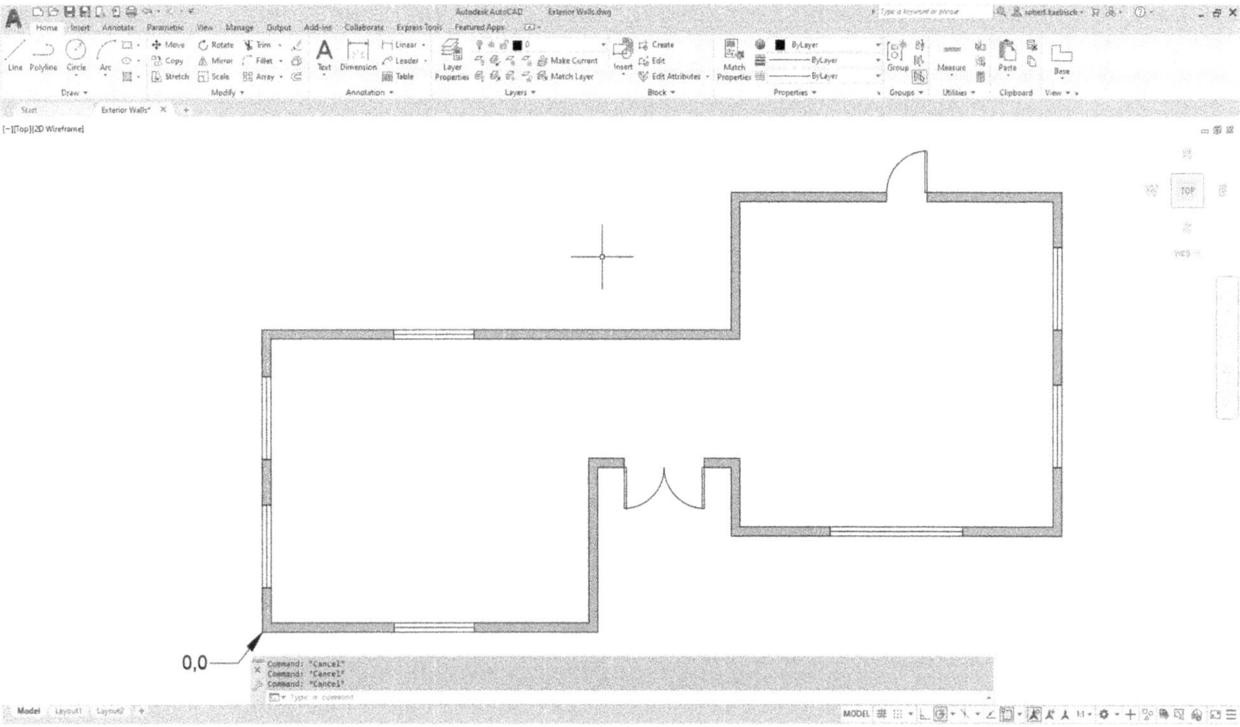

FIGURE 17.1 The file to be used as an xref file.

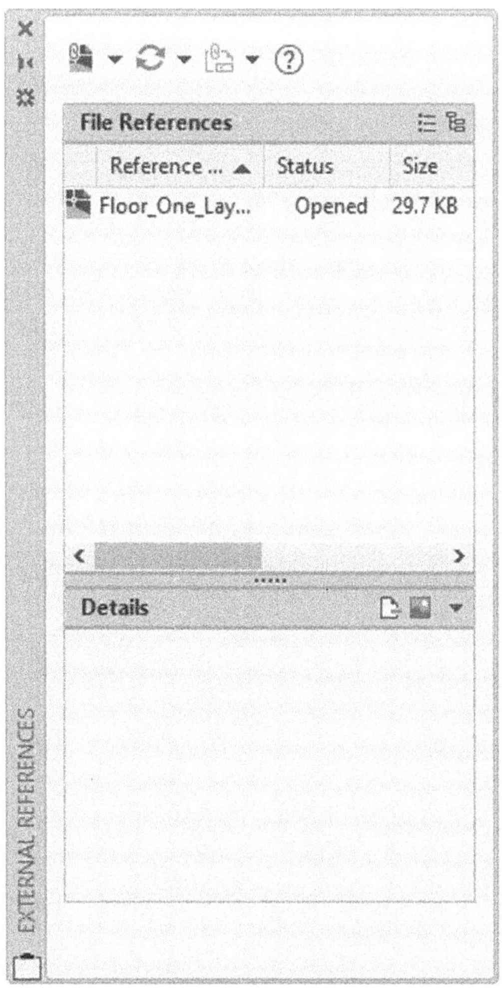

FIGURE 17.2 The xref palette.

Click the drawing file icon with the paperclip at the top left of the palette. You can click on the down arrow to reveal further attachment options (Image, dwf, dgn and PDF) but we do not need those in this case. Simply click on the DWG icon and the Select Reference File browsing window opens. Browse to find your file to xref (Exterior Walls.dwg) and click on Open. The External Reference insertion box appears, as shown in Fig. 17.3.

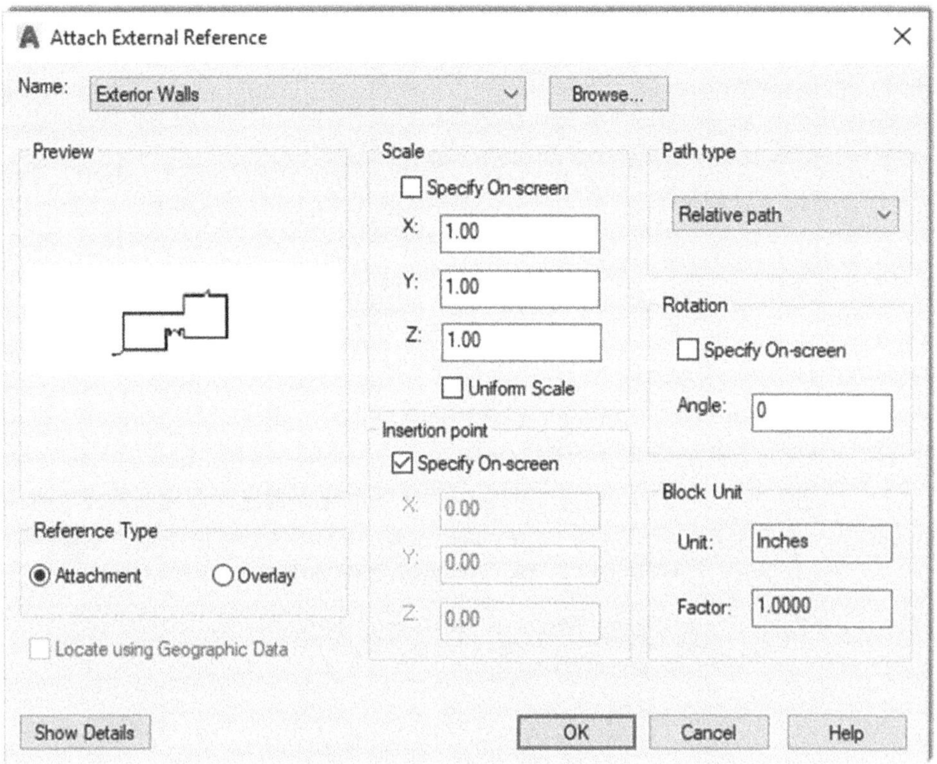

FIGURE 17.3 Attach External Reference.

Examine what is featured here, most of which should be familiar to you (such as Insertion Point, Scale, and Rotation). Uncheck Insertion Point; the fields turn from gray to white. We want the xref to insert at 0,0,0. Finally, press OK.

The xref appears with its lower left corner at 0,0,0. Zoom out to see the entire drawing if not apparently visible. Notice the new addition to the xref palette, as seen in Fig. 17.4. The Exterior Walls file is visible and indicated as Loaded with the appropriate date, time, and path.

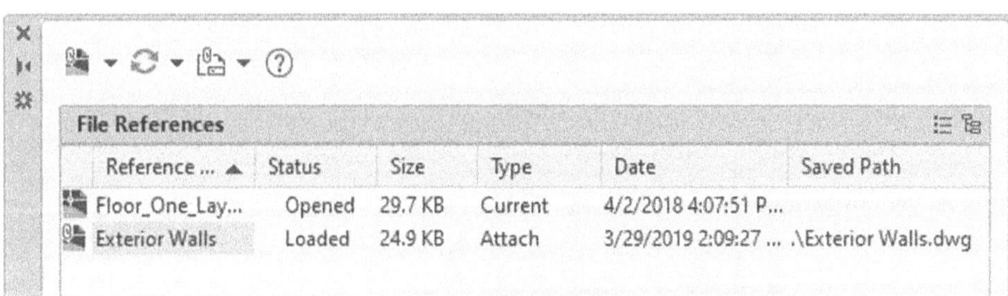

FIGURE 17.4 Xref loaded (palette expanded for clarity).

By default the file comes in faded to indicate it is an xref of a drawing. Don't worry, it will print just fine. The fading amount is controllable in the drop down menu of the Reference panel on the Insert tab of the Ribbon. You can now proceed to add new design work to "over" the file, as the basic xref procedure is complete. Let us skip ahead and explore some of the other available options, now that the xref is loaded. By right-clicking on the Exterior Walls icon, you get access to the important menu in Fig. 17.5.

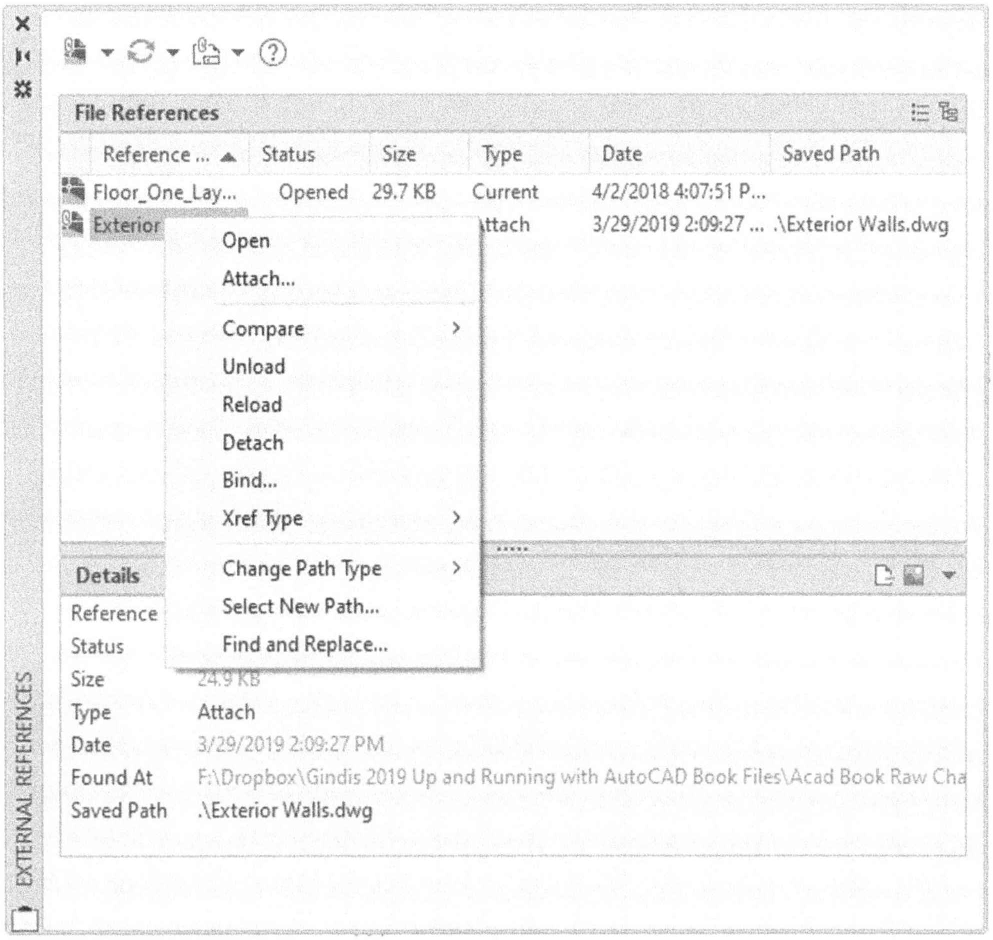

FIGURE 17.5 Xref menu.

The Xref menu includes the following:

- *Open*: This option allows you to open the original xref if you wish to make changes.
- *Attach...*: This option allows you to attach another xref to your working file, using the same procedure as outlined earlier in this chapter.
- *Unload*: This option removes the xref from your working file but preserves the last known path, so if you wish to bring back the xref, you need not browse for it again but just click on Reload.
- *Reload*: The option just referred to for bringing back an unloaded xref.
- *Detach*: This option is more permanent and removes the xref from the working file, requiring you to browse and find it if reattachment is called for again.
- *Bind...*: This option allows you to attach a permanent copy of the xref to your working file and severs all ties with the original. The xref then becomes a block; and if it is exploded, it completely fuses together with the working file. This action is potentially dangerous and is rarely used, as you are now stuck with the xref permanently and changes to the original external file will not be shown here. Two possible uses include permanent drawing, archiving, and emailing the drawing set (so as to not forget the xrefs), though eTransmit can take care of that.

17.3 Layers in xrefs

A very common problem may appear when using xrefs. The xref file itself has layers, often many of them. There may also be several xrefs in one active drawing. And, of course, the drawing itself may have numerous layers of its own. It is almost guaranteed that, on a job of even medium complexity, layer names can and will be duplicated. This is not allowed in AutoCAD, and some method is needed to separate layers in the xref from mixing and being confused with layers in other xrefs or the main design drawing. The solution is quite simple. In the same manner that people have first and last names to tell them apart (especially helpful if the first names are the same), so do layers when xrefs are involved.

All layers that are part of a certain xref have that xref's file name preceding the layer name with a small vertical bar (and no spaces) between them. All layers that are native to the working file have no such prefix, of course.

This simple method ensures no direct duplication by having the xref file name act as the "first name" and the layer name as the "last name." An example is shown in Fig. 17.6. New layers have been added in the host file. Notice the small vertical bar in the xref layer name and the inclusion of the xref file name.

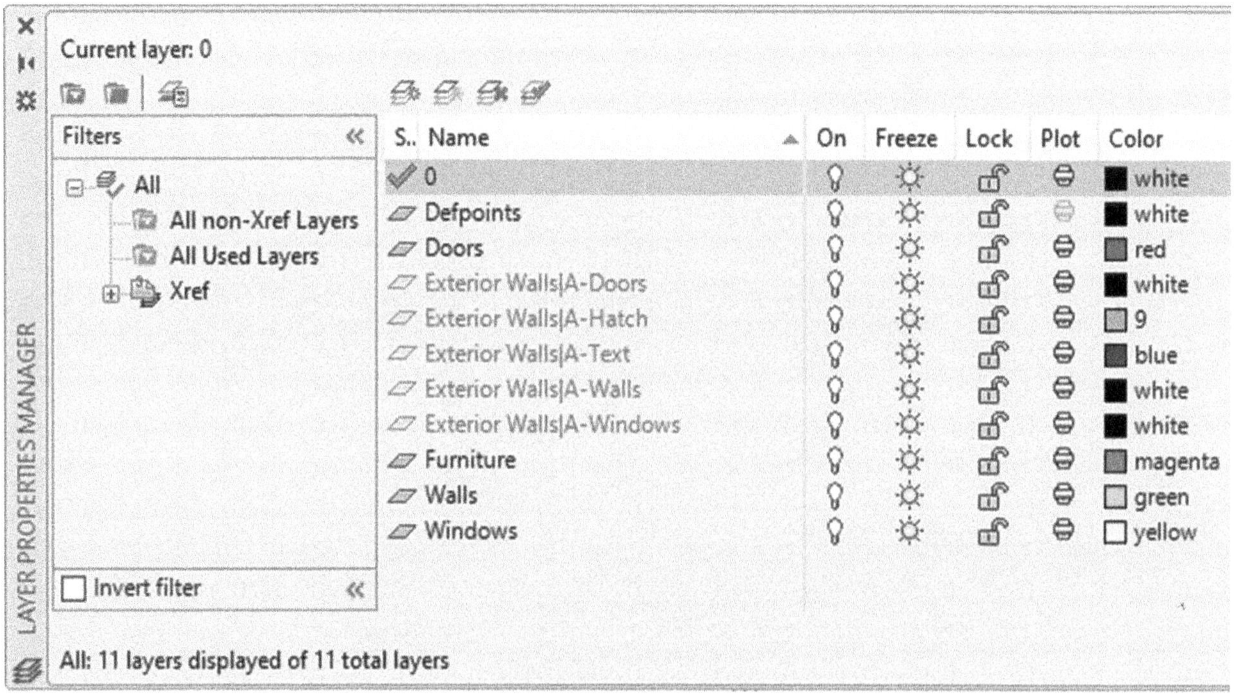

FIGURE 17.6 Xref layering.

When an xref is bound to the drawing (as a *bind not an insert*), the layer's vertical bar changes to a \$0\$ set of symbols, indicating they are now bound and part of the working drawing, as seen in Fig. 17.7. You should be familiar with this if you must examine a bound set.

17.4 Editing and reloading xrefs

At one time, xrefs could not be changed in place. This means you had to open the original xref file to edit it and could not do this from the main working file. This made sense, as you generally want to limit access to the original xref, so it cannot be casually edited.

All things (and software) change, and several releases ago, AutoCAD started allowing xrefs to be edited in place using a technique like editing blocks (somewhat diminishing the security benefit of xrefs, as mentioned earlier). Let us give it a try. Double-click on any of the lines making up the xref while in the main file. The Reference Edit box appears, as seen in Fig. 17.8, with the Ribbon also changing to the (nonfunctional at this point) External Reference tab.

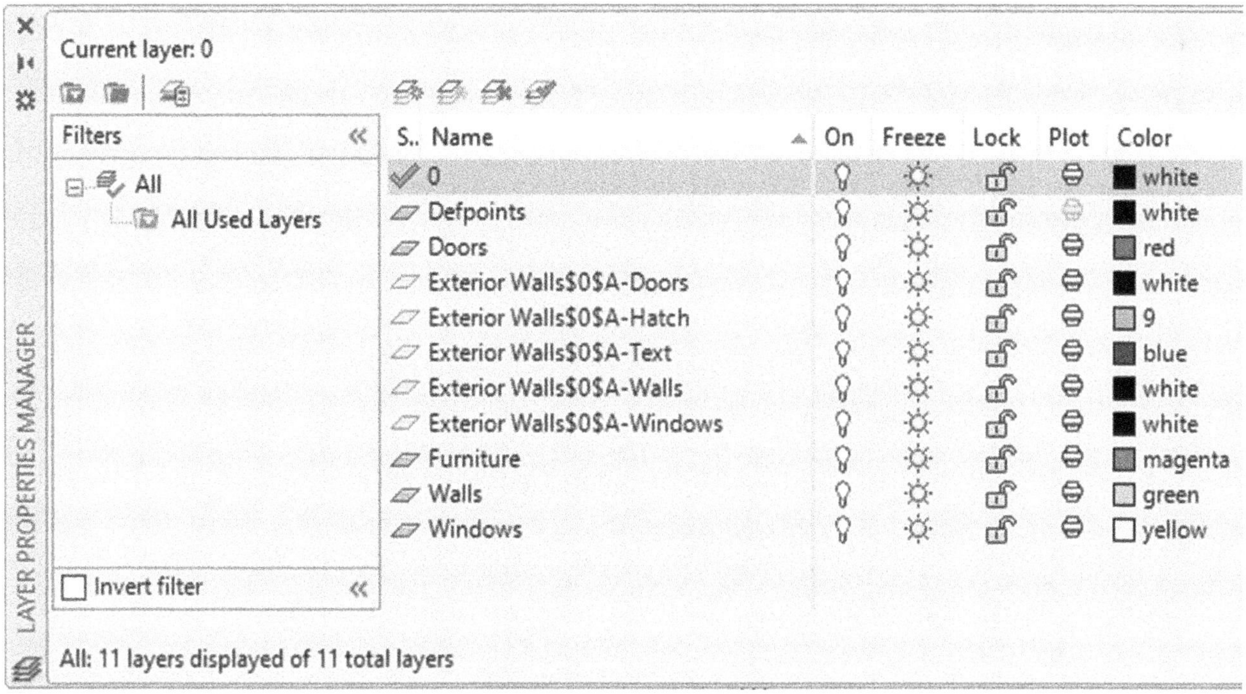

FIGURE 17.7 Xref bound layering.

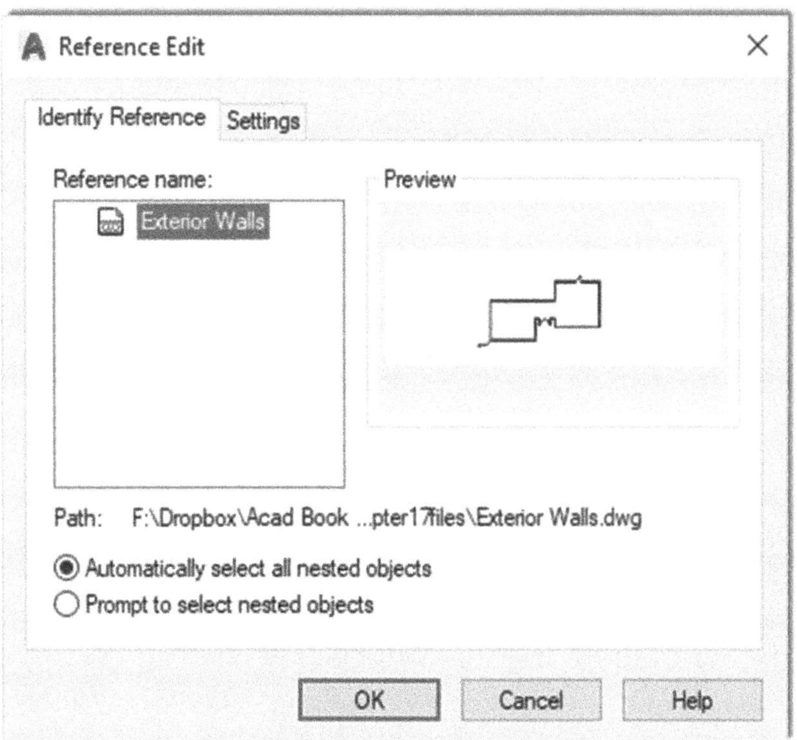

FIGURE 17.8 Reference Edit box.

The xref you clicked on is listed under Reference name. Press OK and you are taken into Reference Edit mode. You may notice a few new features or items of interest:

- The xref has remained brightly lit up while everything that is on the main drawing file has faded down. This is a feature of Refedit. You can now clearly tell apart the xref geometry from the rest. The intensity of the fade can be adjusted as described above.
- The xref can now be edited in place; go ahead and change a few items around.
- The Edit Reference panel pops up at the far right of the Home tab of the Ribbon as shown in Fig. 17.9.

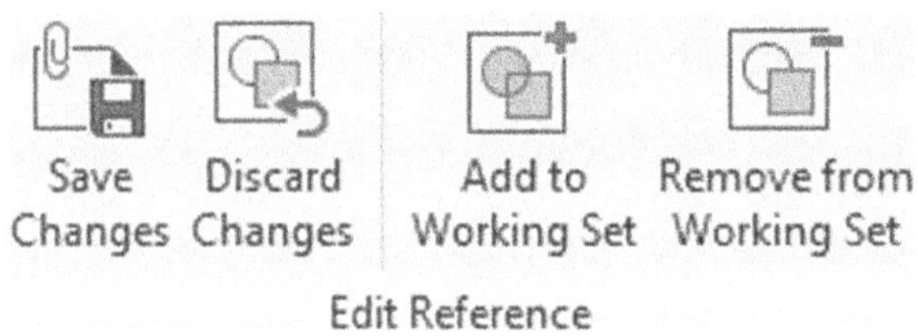

FIGURE 17.9 Edit Reference panel.

When you finish editing the xref, press the Save Changes button in either location. A warning icon appears; press OK, and you are done. The xref has been edited in place.

Just as a reminder, this procedure is only for editing the xref while you have the main working file open, and what you just did can also be done in a far simpler manner by just opening the xref as its own file (although with Refedit you have the advantage of seeing the internal design while editing). In fact, when changes are extensive, the designer may just open the xref itself and work on it all day, without resorting to Refedit. Every time the designer saves his or her work, the changes become permanent and accessible to any other designers who may be working on a job that uses that xref.

A good question may come up at this point. How do other designers know that the xref is changed so they can adjust their interior work accordingly? Many releases ago, the only way to know was for the person working on the xref file to simply tell others via a visit to their cubicle or office, phone, or email. The other users would then press Reload All References, as seen in Fig. 17.10.

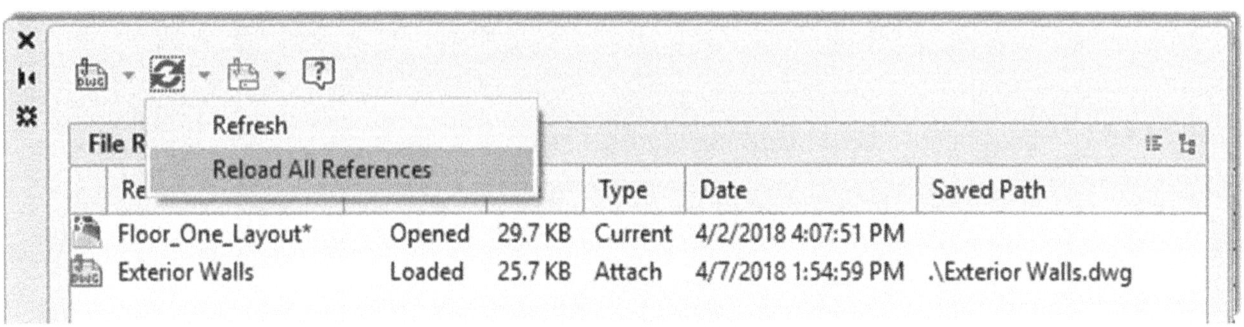

FIGURE 17.10 Reload xref.

Of course, you can still do this, and designers often refresh every few minutes on their own if they know someone is actively working on the xref. However, a new tool appeared recently. It is simply a "blurb" announcement that appears in the lower right corner of AutoCAD's screen, as shown in Fig. 17.11. Press the hyperlink and the xref reloads and refreshes automatically.

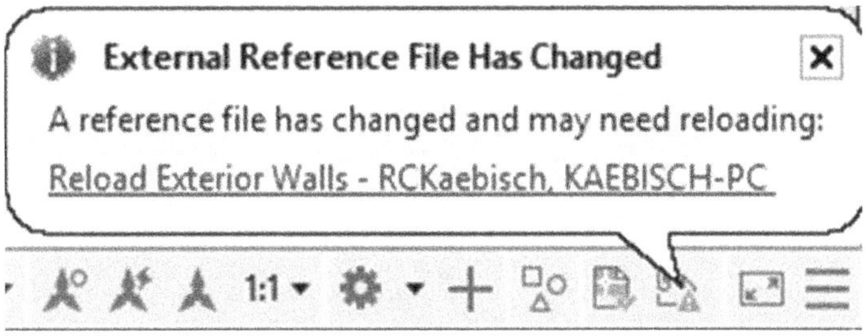

FIGURE 17.11 Reload xref announcement.

Another brief topic in this chapter concerns how to chain xrefs together if there is more than one of them. Some methods are suggested next.

17.5 Multiple xrefs

As mentioned earlier in this chapter, many complex drawings use more than one xref. Typically, you may find the exterior walls, columns, and the core as three separate xrefs. The ceiling grid (RCP) can also be xrefed in for extensive ceiling work, as can the demolition plan and many other combinations.

There are essentially two options. Let us say you work with a total of five xrefs. You can attach them one to another then to a master working file (daisy chain) or one at a time, as shown in Figs. 17.12 and 17.13.

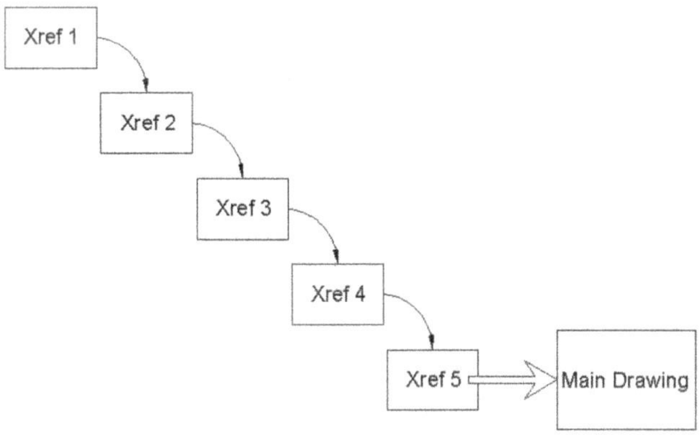

FIGURE 17.12 Multiple xref attach, Method 1.

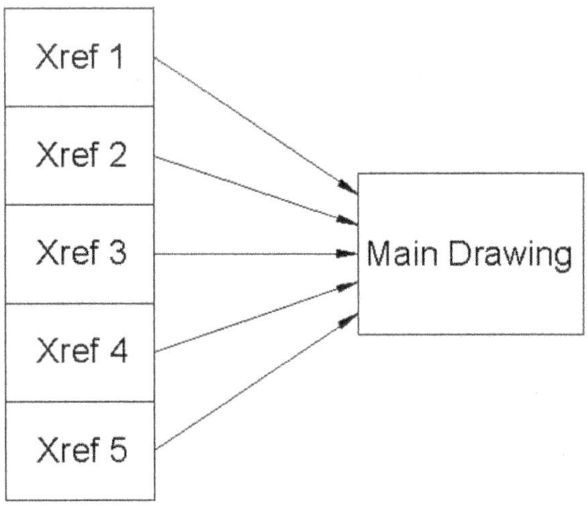

FIGURE 17.13 Multiple xref attach, Method 2.

Which method you use is up to you (there may be company policy or a convention in use at your job). Each method has minor pros and cons, and you should expect both to be used in the industry.

One final word on when not to xref, as this feature gets abused occasionally. Xrefing of furniture, fixtures, and text is not what xref is for. It is intended for multilayered, complex drawings of buildings, systems, or infrastructure and NOT for individual items normally inserted as blocks (furniture and plumbing fixtures) and text. The logic behind xref in the first place is to attach objects that are not likely to change and can serve as the backdrop, so use the tool wisely.

17.6 Ribbon and xrefs

Some xref functions can be accessed via the Ribbon, and we discuss them separately in this section. You can find them under the Insert tab, Reference panel, as seen in Fig. 17.14.

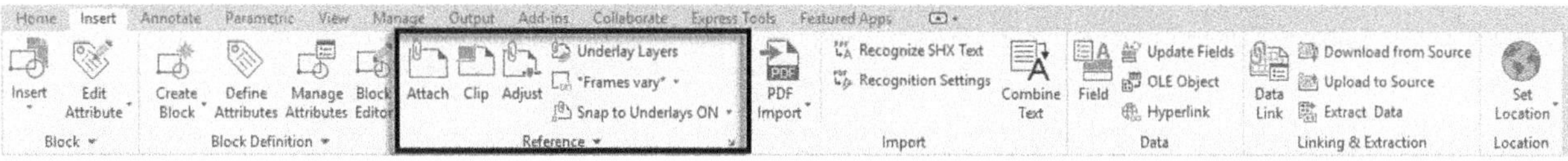

FIGURE 17.14 Xref command via Ribbon, 1.

One important control is the xref fading slider discussed earlier that you find in the Reference panel dropdown menu. This controls the darkness of the xref and was mentioned as also being under the right-click, Options…, Display tab. Some users prefer the xref to be as bold as surrounding design, while some prefer it very light, so they clearly know what is an xref and what is the current design.

Finally, the Ribbon switches to a dedicated xref tab if you click on the xref, as seen in Fig. 17.15, although there is much duplication of the previous tab's commands.

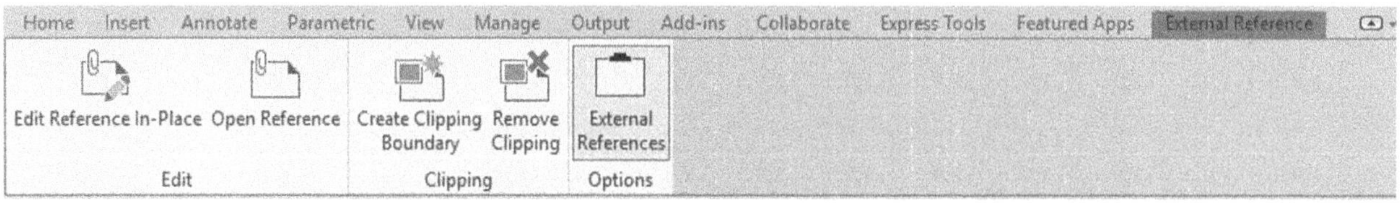

FIGURE 17.15 Xref contextual tab Ribbon, 2.

17.7 Level 2 drawing project (7 of 10): architectural floor plan

For Part 7 of the drawing project, we shift to the outside of the building and create some landscaping in the form of trees, shrubs, and drive/walkways. You need some Chapter 11 tools for the greenery.

Step 1. Create some appropriate layers as shown next. Freeze all layers that are not relevant to the task at hand.

- L-Driveway (Gray_8).
- L-Trees_Shrubs (Color_108).
- L-Walkway (Gray_8).

Step 2. Create the landscaping plan as seen in Fig. 17.16. You need to create plines for the borders of the driveway and the walkways. Then, after adding the hatch patterns (Sand and Brick), erase some or all the plines. For the trees and shrubs, you need to call upon the xline and spline.

Summary

You should understand and know how to use the following concepts and commands before moving on to Chapter 18*:*

- Xref command
- Inserting

- Detaching
- Unloading
- Reloading
- Binding
- Layers in xref
- Nesting xrefs

Review questions

Answer the following based on what you learned in this chapter:

1. What are the several benefits of xref?
2. How is an xref different from a block?
3. How do you attach an xref?
4. How do you detach, reload, and bind an xref?
5. What do layers in an xrefed drawing look like?
6. What do layers look like once you bind the xref?
7. What are some nesting options for xref?

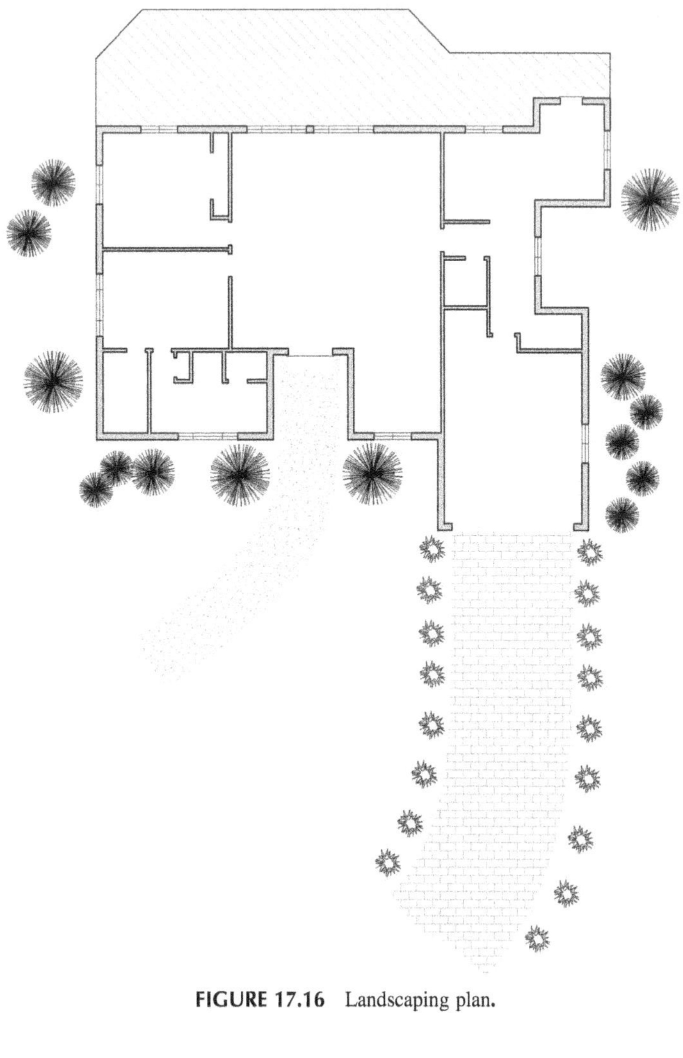

FIGURE 17.16 Landscaping plan.

Exercise

1. Draw the following floor plan to practice the xref command. (Difficulty level: Easy/Moderate; Time to completion: 45–60 minutes.)

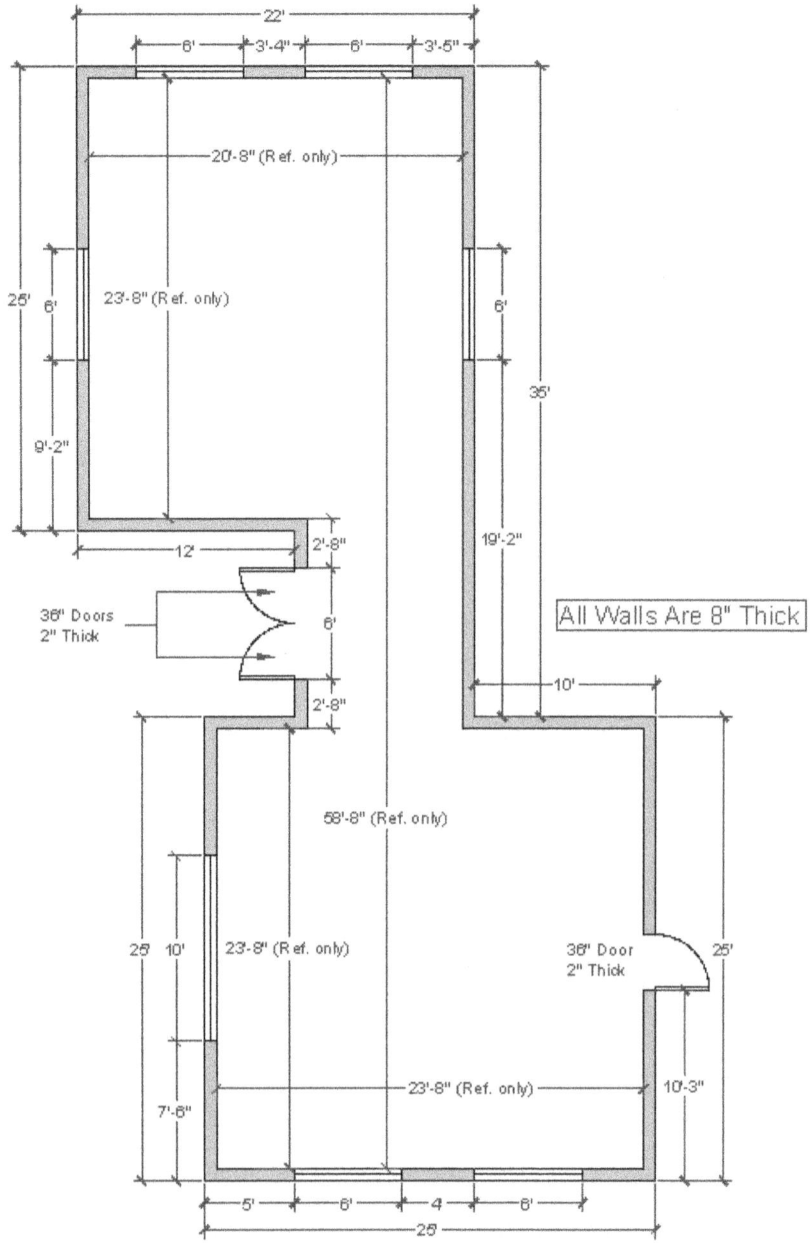

Chapter 18

Attributes

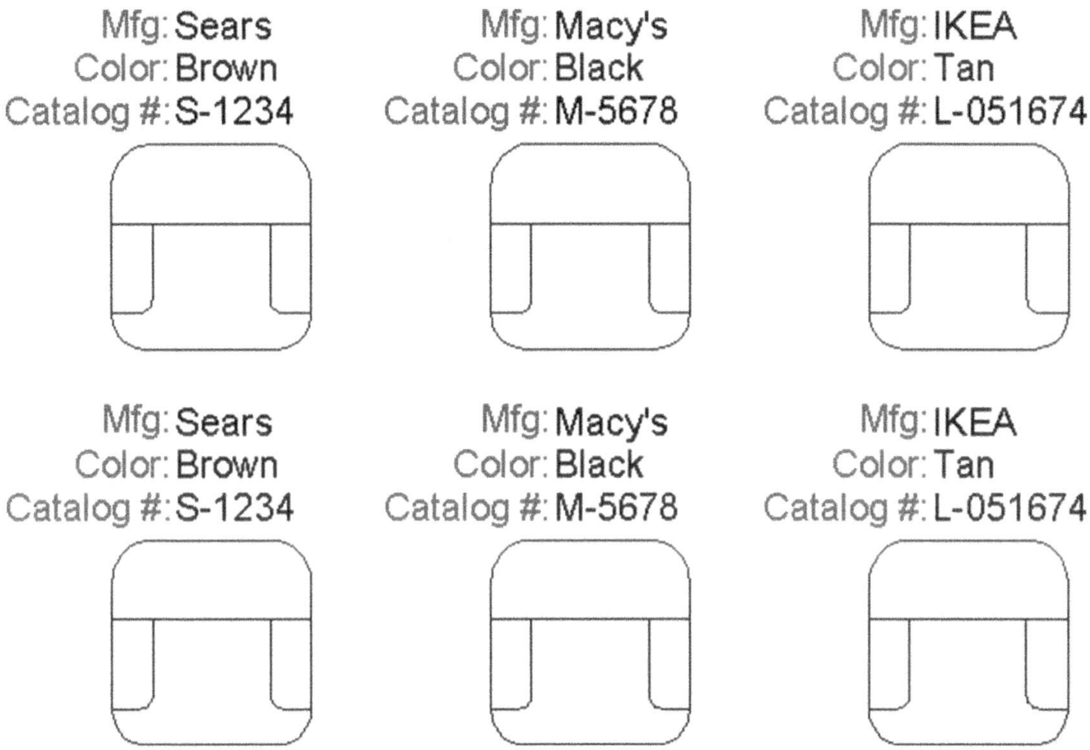

Learning objectives

In this chapter, we introduce and thoroughly cover the concept of attributes. We specifically discuss the following topics:

- The purpose of attributes
- Defining attributes
- Editing attributes and properties
- Extracting data from attributes
- Invisible attributes

By the end of this chapter, you will be well versed in applying this critical concept to your design work.
Estimated time for completion of this chapter: 1–2 hours.

Up and Running with AutoCAD 2022. https://doi.org/10.1016/B978-0-323-89923-9.00018-4

18.1 Introduction to attributes

Attributes is an important advanced topic. An attribute is defined in AutoCAD as information inside a block. The key words here are *information* and *block*. You can have a block without information—you have done this many times before; it is called a *block* or *wblock* but not an attribute. You can also have information in the form of text sitting in your drawing. That is just text or mtext but also not an attribute. However, put the two concepts together and you have what we are now about to explore: Essentially an "intelligent" block, one that has useful, editable, embedded information.

What is the reason for creating attributes in the first place? Why would you want information inside of blocks, and what can you do with this information? The information itself can be anything of value to the designer or client. For an office chair in a corporate floor plan, one may want to know its make, model, color, and catalog number. Then, if there are a variety of chair types for this office, the designer can keep track of all of them. Another example is a list of connections going into an electrical panel. It would be nice to know what and how many. All this information can be entered into the attribute and displayed on the drawing if you wish.

However, this is only half the story, and displaying information is not the main reason for creating attributes. As a matter of fact, this information is often completely hidden, so as not to clutter up the drawing. So, what is the main reason? The key is that this information can be extracted out of the attributes and sent to a spreadsheet (among other destinations). Now you have a truly useful system in place, where extensive information is hidden unobtrusively in the drawing yet can be accessed if needed at the touch of a few buttons and put into an Excel file or other very useful form for cost analysis or just plain old keeping track and organizational purposes.

You may have already heard of these tools in use or envisioned a need for them in architecture with such items as doors and windows schedules. The software keeps track of each door and window in the drawing and links the information to a database that is updated as you add or remove the objects from your design. While basic AutoCAD does not do this automatically, other add-on software does. Underneath it all is the concept of embedded attributes, like those we discuss here.

You should have a good understanding of the preceding few paragraphs. As always, the button pushing is far easier once the concept is crystal clear. We now go ahead and do the following:

- Create a simple design.
- Create a basic attribute definition.
- Make a block out of the simple design and the attribute definition.
- Extract information out of our new attribute.

The information is extracted via a multistep process to two locations, an Excel spreadsheet and a table. In both cases, the information is in an organized tabular form and, while not linked bidirectionally, is an exact reflection of drawing conditions. Let us give it a try.

18.2 Creating the design

We need nothing complicated drawn to learn the basics of attributes. Go ahead and draw a simple chair made up of several rectangles and rounded corners, as shown in Fig. 18.1. Make the chair 30″ × 30″ and approximate the rest of the dimensions relative to those.

We would like to embed the following information in it—manufacturer, color, and catalog number—and put these three items just above the chair. Let us first add some text, a header of sorts, indicating what the three categories are, as seen in Fig. 18.2. Use regular text of any appropriate size, font, and color.

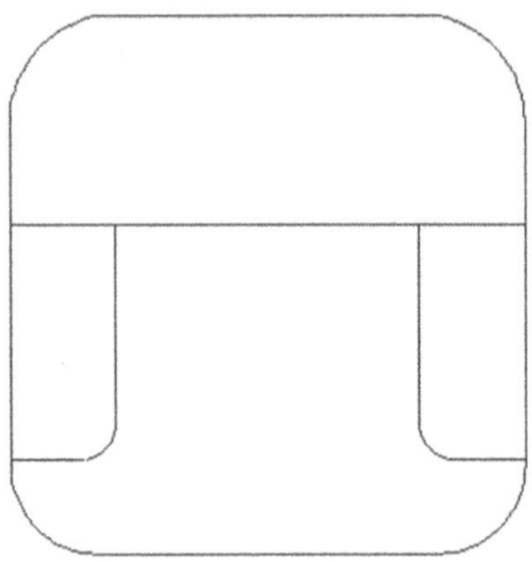

FIGURE 18.1 Basic chair design.

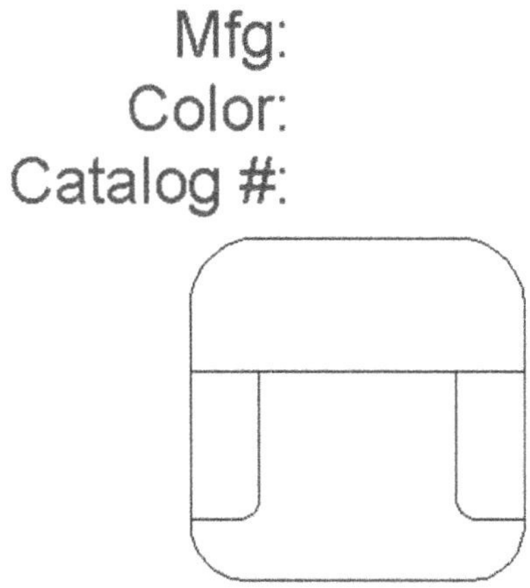

FIGURE 18.2 Basic chair design text.

This is not yet the attribute definition, merely the name of the categories, so we know what we are looking at. You do not have to do this, but it helps clarify things initially, as the attributes are just plain Xs. You now have a simple design and the start of some useful information. We are now ready to make an attribute definition.

18.3 Creating the attribute definitions

Here, we define the new attributes that contain the actual information. The idea for now is to simply assign them all a value of X and change the X to meaningful data later in the process, so we are creating generic definitions, which allow us to create multiple sets of data from one template.

Keyboard: Type in **attdef** and press Enter.
Ribbon: **Insert tab→**

Start the attribute definition via any of the preceding methods. The dialog box in Fig. 18.3 appears.

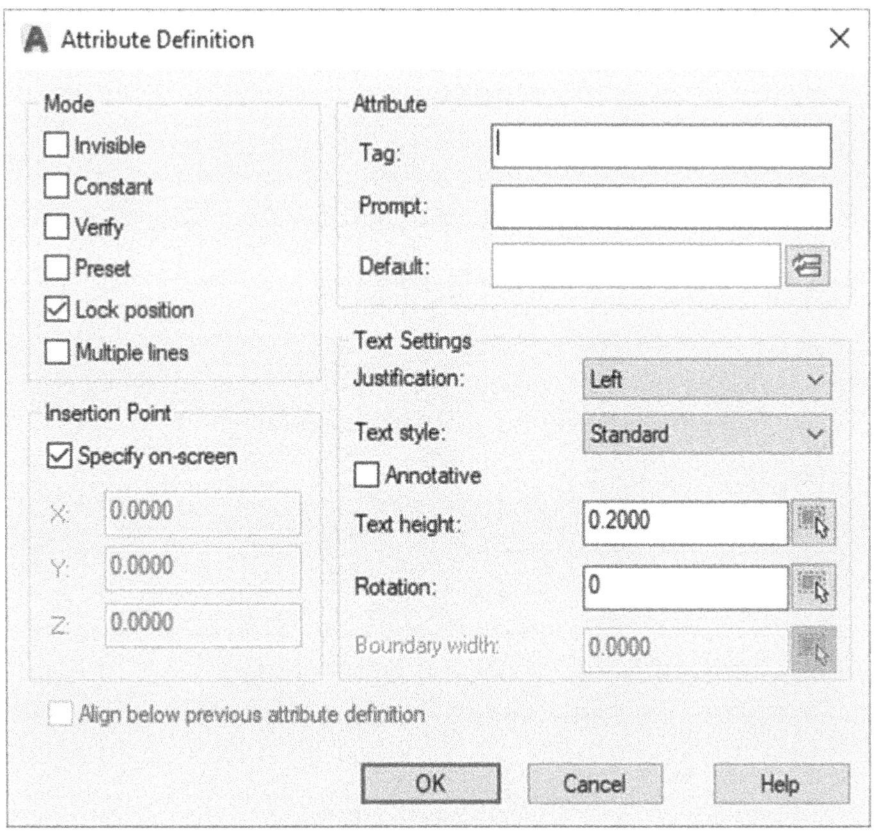

FIGURE 18.3 Attribute Definition.

We are interested in only the upper right of the dialog box for now. Under the Attribute category, let us focus on Tag:, Prompt:, and Default:

- For Tag, type in Mfg for Manufacturer (this is the actual category of the attribute definition).
- For Prompt, type in Please enter manufacturer (this is just a reminder, or question to answer, for the user in a dialog box).
- For Default, type in X (this is a generic value that can be changed by the user later), as seen in Fig. 18.4.

The rest of the information in the dialog box can be left as is for now, although we discuss a feature under Mode (upper left) later in this chapter. One thing you can adjust, however, is the text height, matching the attribute size to the size of the text over the chair. The text is 4″ in height, so you can enter that value into the field.

Press OK, and your attribute definition appears attached to your crosshairs. Go ahead and place it next to the Mfg: text. Note that it is in capital letters by default. Go ahead and go through the same procedure for the remaining two categories, changing the Tag value and slightly modifying the Prompt. The default remains as X. Note also that you cannot have blank spaces in the tag, hence the CATALOG_#. The end results are shown in Fig. 18.5.

We are almost done creating the attribute definitions. Note the differences between the text and the attribute. Double-click on the original text, and you get the standard text editing field. Double-click on the attribute definition, and you get a different, Edit Attribute Definition field, as shown in Fig. 18.6.

At this point, it may not be obvious why we have two identical sets of categories next to each other or why we entered the X. After the next step, what we are doing should become much clearer.

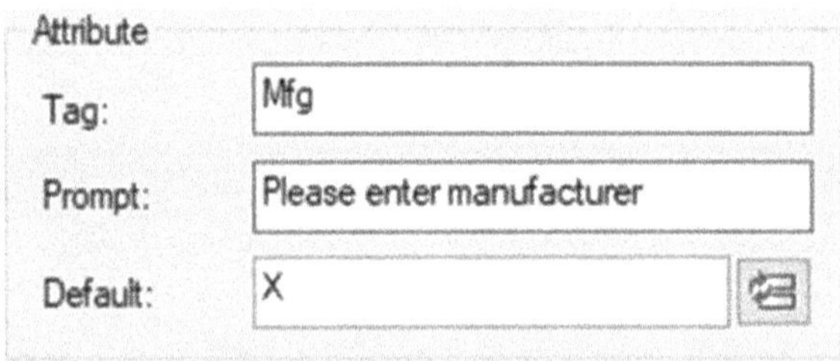

FIGURE 18.4 Attribute fields.

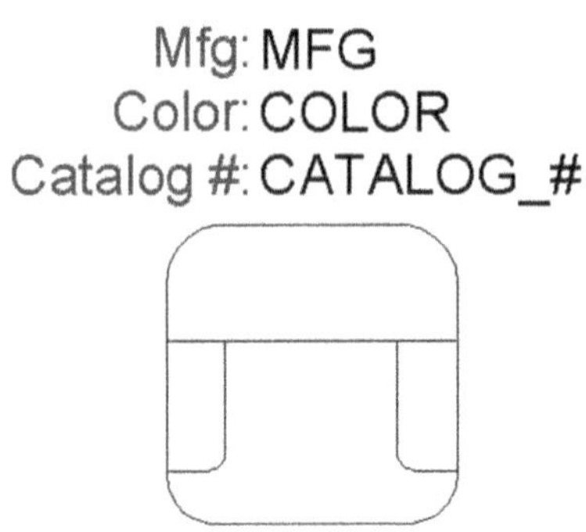

FIGURE 18.5 Attribute fields completed.

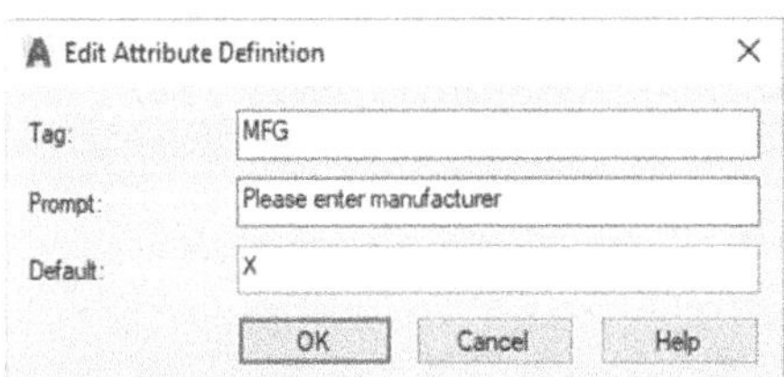

FIGURE 18.6 Edit Attribute Definition.

18.4 Creating the attribute block

This next step should be familiar: Create a regular block named Sample_Chair via any method you prefer. Give it a name, select the objects (both the chair and all the text), and pick an insertion point. Finally, click on OK, and the dialog box shown in Fig. 18.7 appears. This dialog box was upgraded last year to allow for more attributes to be listed without having to flip to another page—the previous version if AutoCAD had just 8 lines total; this one now has 15. In the interest of time and clarity, we stick to just the three attributes shown here, but keep in mind you can enter many more if needed.

The Edit Attributes dialog box allows you to enter specific values for those placeholders (marked as X), but we do not do that at this point. Simply press OK and your new attribute is officially born (Fig. 18.8). Notice all the previous

FIGURE 18.7 Edit Attributes.

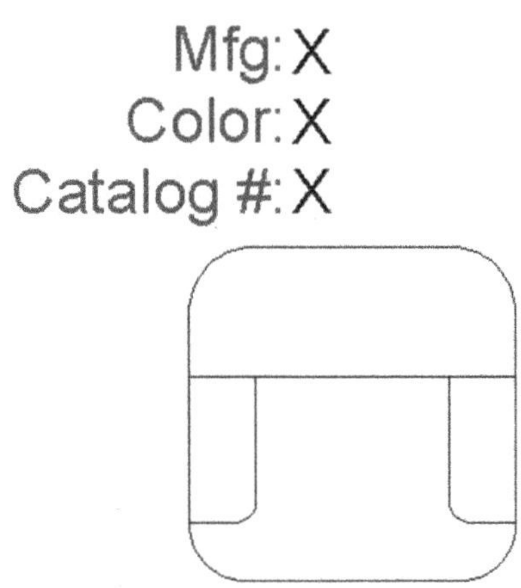

FIGURE 18.8 Edit Attribute.

categories are now just X and ready for customization. This is the process for creating a generic attribute. In the example of the office chairs, you create several copies (the distinct types of chairs) and customize the information for each. Then, copy and distribute the types as needed on the floor plan, as we do next.

18.5 Attribute properties and editing

Now that you have a block with attributes, you need to learn how to edit its fields to add or delete information from an attribute as well as change its color, font, and just about any other property. Some of these techniques overlap each other and duplicate effort, so we need to clearly state what is used for what.

First, if you want to change only the content of an attribute (in other words, what it says), you need to get back to the dialog box shown in Fig. 18.7. It is basic and simple but does not change fonts and colors. Also, the only way to get to it is by typing. The command, attedit, takes care of the content. Indeed, this is the way you should "populate," or fill in, the fields once we start copying the chair. Try it right now and edit the chair to say Sears for the manufacturer, Brown for color, and S-1234 for catalog number.

Now, what if you wanted to change an attribute's properties, as discussed earlier? Well, the method here changes slightly, as you need to deal with the Enhanced Attribute Editor. Although a full command matrix is shown next, the easiest way to get to it is just to double-click on the attribute.

Keyboard: Type in **eattedit** and press Enter.
Ribbon: Home tab→Block→

What you get after either double-clicking or using the previous methods is Fig. 18.9. Here, you can do quite a lot. This Editor not only allows you to fill in the fields but also a variety of properties (colors, layers, fonts, etc.). Click through the three tabs, Attribute, Text Options, and Properties, to explore all the options. Here are some additional pointers on attributes.

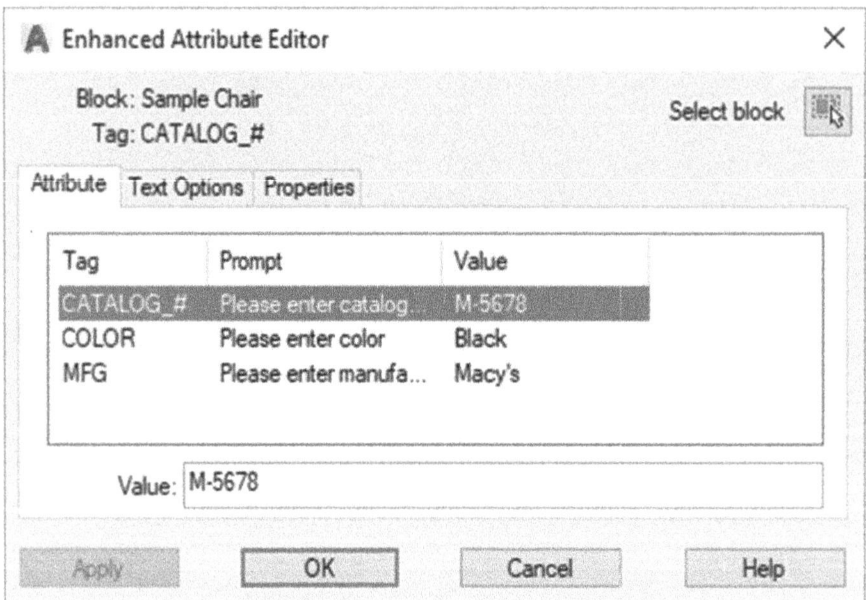

FIGURE 18.9 Enhanced Attribute Editor.

Exploding attributes

Attributes can be easily exploded, but this action destroys the blocks and turns them into what you see in Fig. 18.8, where you have just generic placeholders unattached to a physical design. You can change a few fields and reassemble the block, which is a perfectly valid technique, especially if the attribute is giving you a tough time. Be careful though; exploding complex attributes can suddenly fill your screen with an enormous amount of data, so keep one finger on that Undo button.

Inserting attributes

As you learned in this chapter, attributes are blocks; therefore, you can insert them as needed, as an alternative to copying them. When you insert a block that contains an attribute, you are prompted to fill in the X placeholders at the command line. Indeed, this is how title blocks are often filled out when a generic template block is inserted into a specific job file.

Ideally, everything makes sense now that you see the big picture of what was done. Review this chapter up to now if not. Our next step is to learn how to extract the useful information. For this, you need to copy the block to simulate having several chairs in a real drawing. Then, go ahead and change some of the attribute values. The result is shown in Fig. 18.10.

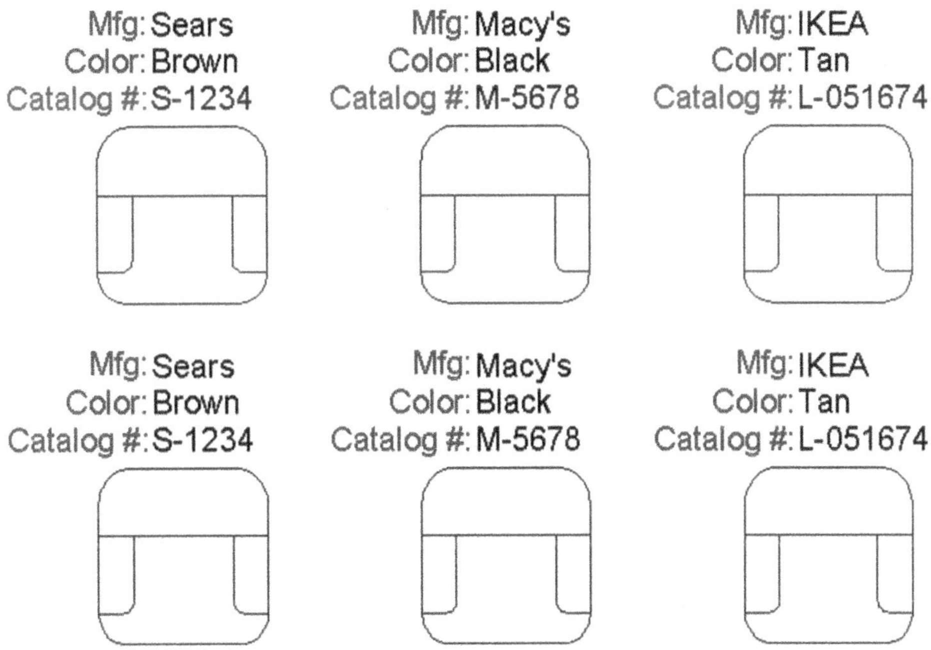

FIGURE 18.10 Various attributes.

18.6 Attribute extraction

Extracting the information is a multistep process, necessarily so because of the number of options offered along the way. Our target extraction is both an Excel spreadsheet and a table embedded in AutoCAD. Remember, of course, that this is a basic generic extraction and more complex spreadsheets and tables can be created to hold such data, which is left to you to explore.

Keyboard: Type in **eattext** and press Enter.
Ribbon: Insert tab→**Linking & Extraction**→

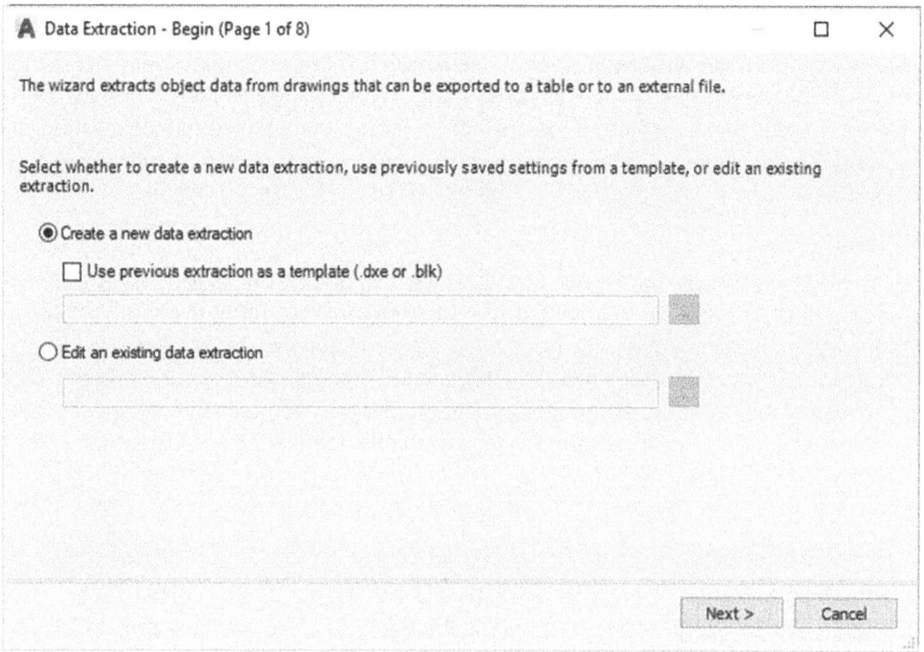

FIGURE 18.11 Data Extraction – Begin (Page 1 of 8).

Step (Page) 1. Begin
Start up the data extraction process via any of the methods shown in the command matrix. The Data Extraction - Begin (Page 1 of 8) dialog box shown in Fig. 18.11 appears. You are going to create a new data extraction, so leave the default top choice but uncheck the Use previous… box. Then, press Next > and you are taken to the Save Data Extraction As dialog box. Save your file as Sample.dxe anywhere you will find it easily on your drive. You are then taken to the next page.

Step (Page) 2. Define Data Source (Fig. 18.12)
Here, we are interested in the data source, or location, of the file to be looked at. Leave the default choice with the Include current drawing box checked off. Press Next>.

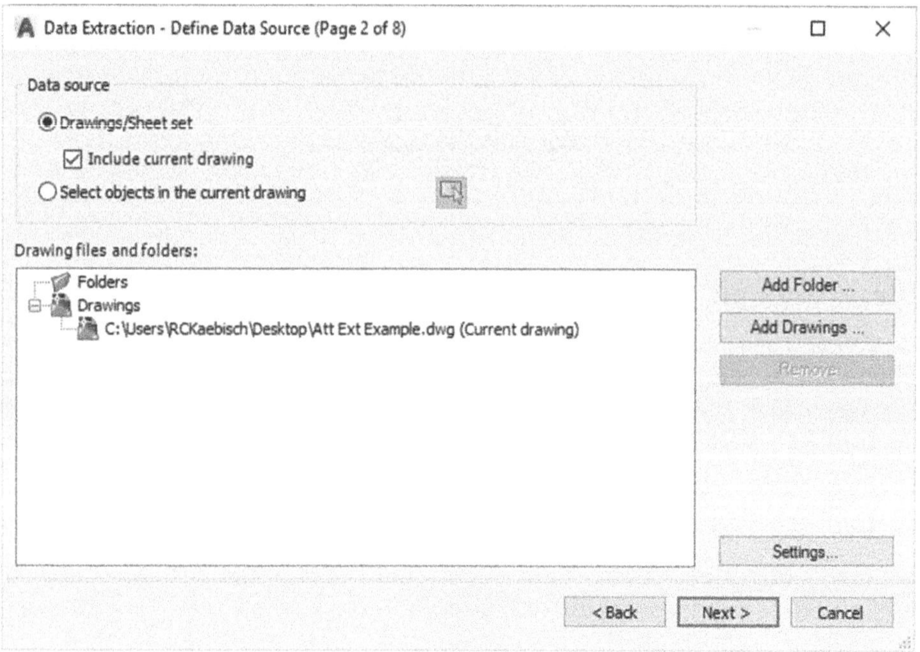

FIGURE 18.12 Data Extraction – Define Data Source (Page 2 of 8).

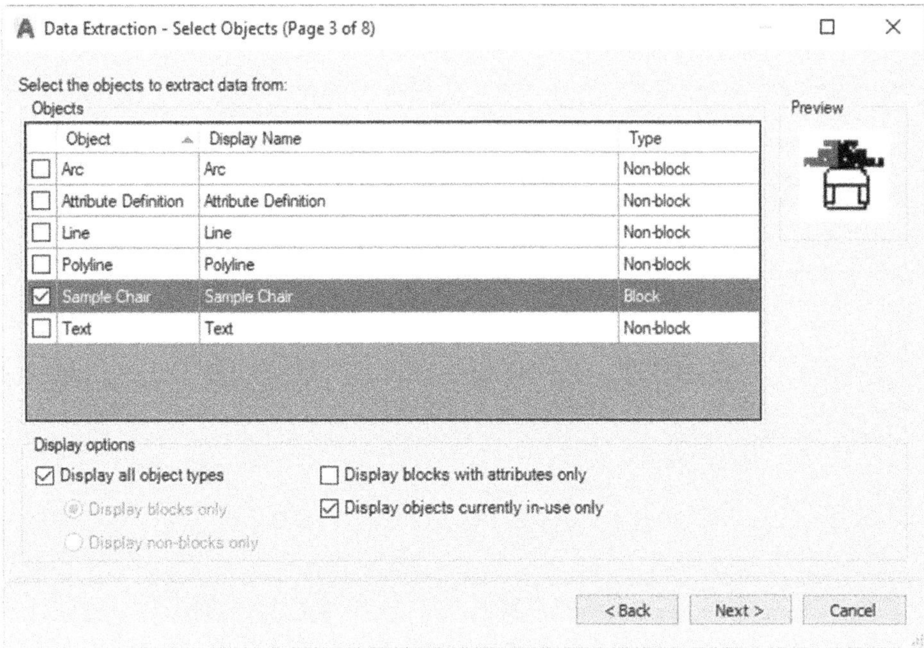

FIGURE 18.13 Data Extraction – Select Objects (Page 3 of 8).

Step (Page) 3. Select Objects (Fig. 18.13)
Here, we are interested in selecting from what we want to extract the data. In our case, there is only one choice, the sample chair, as it is the only available block (as seen under the type column). You can also click on Display blocks only, under the Display options, to isolate the chair. In either case, select it, then press Next>.
Step (Page) 4. Select Properties (Fig. 18.14)

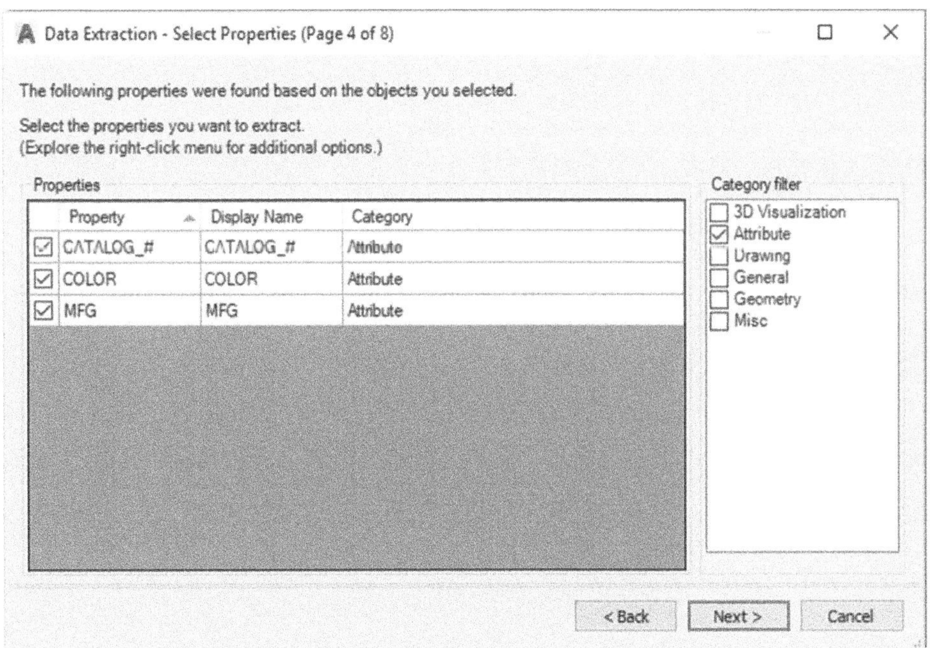

FIGURE 18.14 Data Extraction – Select Properties (Page 4 of 8).

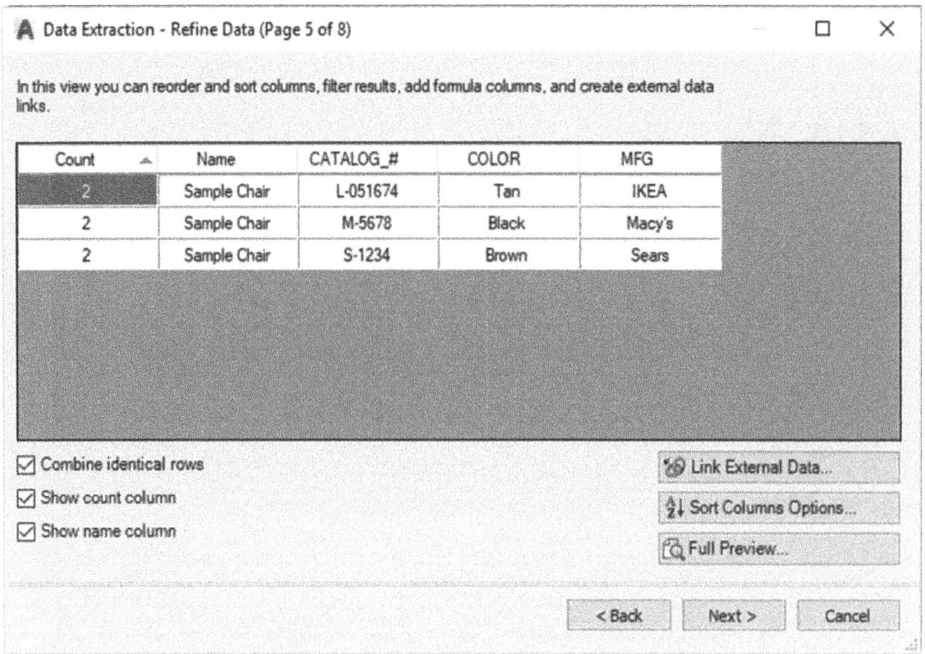

FIGURE 18.15 Data Extraction – Refine Data (Page 5 of 8).

Here, you initially see many property choices. Use the category filter on the right to uncheck all the choices except Attribute. You then easily see the three properties on the left that we created earlier. Check off all three and press Next>.

Step (Page) 5. Refine Data (Fig. 18.15)

This step is for you to refine the data, if needed. It contains some interesting features, such as Link External Data… (to tie in this output to another Excel file) as well as some tools for adjusting and tweaking the columns. Much of this and more can be done in Excel itself after extraction is completed, but it can be useful to adjust the look and content of the table extractions, as tables are less flexible. If you want to change nothing, press Next > for the final few steps.

Step (Page) 6. Choose Output (Fig. 18.16)

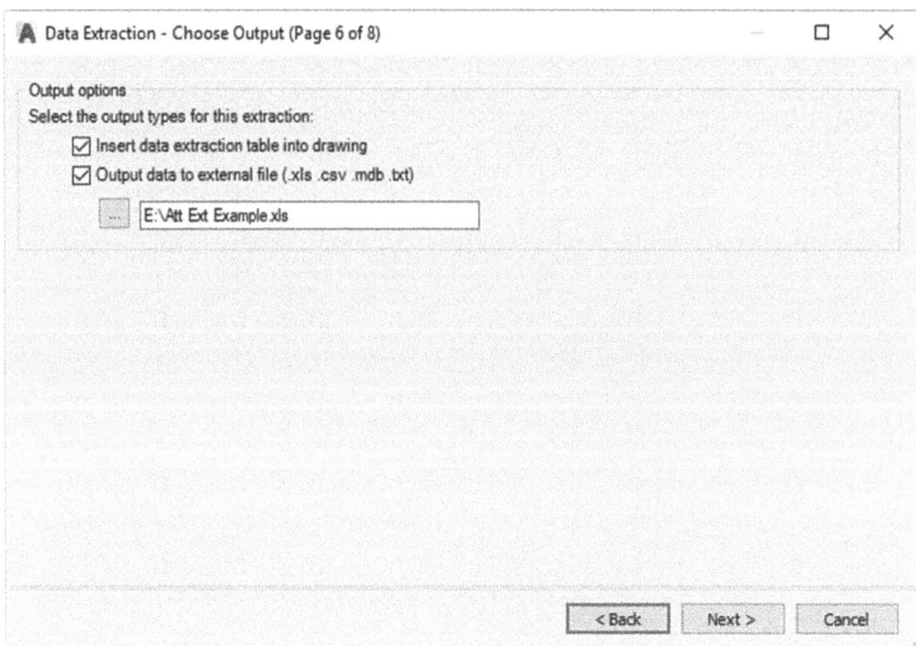

FIGURE 18.16 Data Extraction – Choose Output (Page 6 of 8).

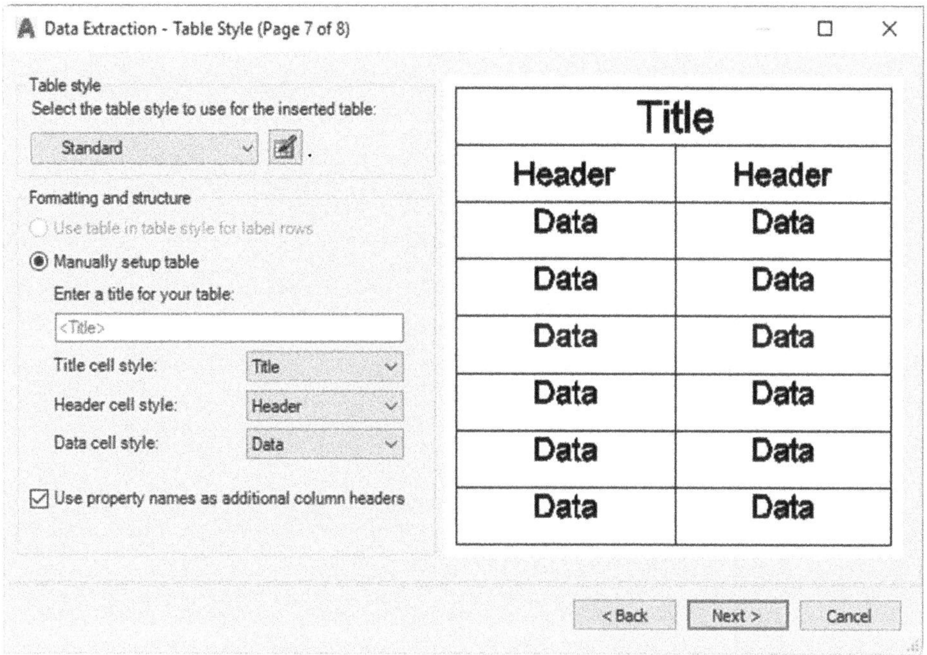

FIGURE 18.17 Data Extraction – Table Style (Page 7 of 8).

Here, we are interested in our destination output. Although the Excel spreadsheet gets all the attention, you can also extract to a table that will be inserted into the drawing itself. Tables are first introduced in Chapter 15, Advanced Design and File Management Tools, so review that section, if necessary. Check off both boxes if you have not done so yet, and finally select the location of the Excel file by using the three dots browse button. When ready, press Next >.

Step (Page) 7. Table Style (Fig. 18.17)

We do not do anything to the table, leaving it as is, but you should review the options available and press Next>when done.

Step (Page) 8. Finish (Fig. 18.18)

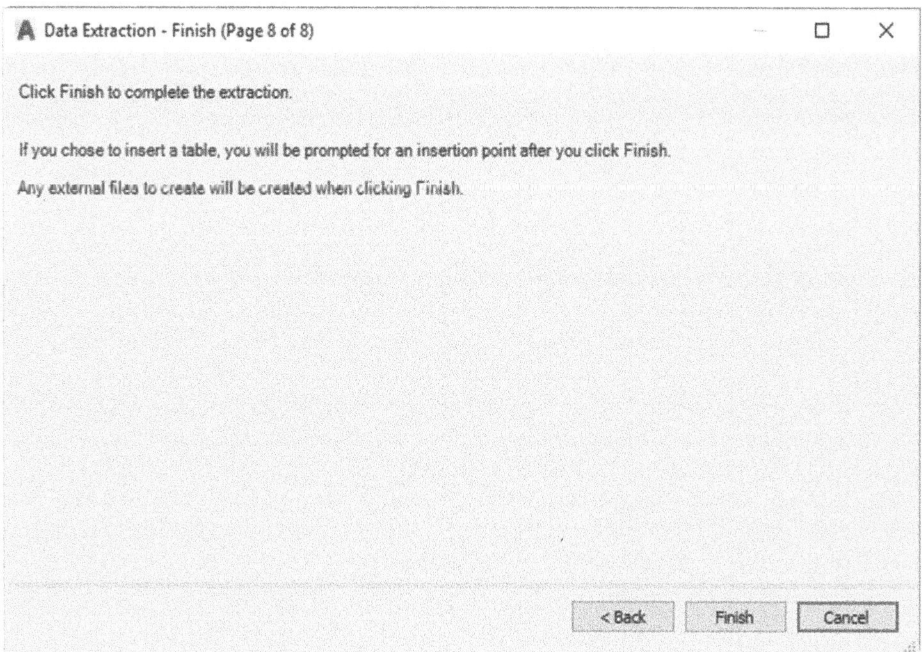

FIGURE 18.18 Data Extraction – Finish (Page 8 of 8).

Count	Name	CATALOG_#	COLOR	MFG
2	Sample Chair	L-051674	Tan	IKEA
2	Sample Chair	M-5678	Black	Macy's
2	Sample Chair	S-1234	Brown	Sears

FIGURE 18.19 Table insertion (AutoCAD).

Count	Name	CATALOG #	COLOR	MFG
2	Sample Chair	L-051674	Tan	IKEA
2	Sample Chair	M-5678	Black	Macy's
2	Sample Chair	S-1234	Brown	Sears

FIGURE 18.20 Table insertion (Excel).

On the final page, there is little to do except read what AutoCAD is telling you and press Finish. You are asked for an insertion point and notice the table attached to the mouse. You may have to zoom in to see it better. Finally, when you can see it, click to place the table. The table is inserted into your drawing, as shown in Fig. 18.19.

It is not fancy, but it gets the information across. The main output, however, is the Excel file. Look for it wherever you may have dropped it off and double-click on it to start Excel. Then, after some cleanup (cell borders, column width, fonts, colors, etc.), select the cells and Copy/Paste them into AutoCAD, as you did in Chapter 16. It looks like Fig. 18.20.

This is attribute data extraction. You should recognize the value in what you did and the potential it holds for easy data access and accounting.

There are other uses for attributes. One common use is formatting a full title sheet and block with generic attributes and saving it as a company standard. In this case, the text is the fixed permanent categories, such as Date, Sheet, Drawn by, and other fields; and the attribute definitions are the variables, such as the actual date, sheet number, and the initials of whoever created the drawing. Then, everything is blocked out and saved for future use, with the Xs representing to-be-filled-in variables. As the need comes up for a new title block, it is inserted as a block into the design's Paper Space with the variables filled out on the spot. We conclude the attributes chapter by briefly going over an important option in the attdef command dialog box.

18.7 Invisible attributes

This option appears as the first check-off box in the upper left of the attdef dialog box and is the most important of all the ones in that column. It is used to make attributes invisible on creation.

Now, why would you do this? As mentioned at the beginning of this chapter, this is a relatively common way of creating attributes for the simple reason that you do not need to see the text and the attribute information. If you know what you created, then there is no need to clutter up the drawing with that list repeatedly. This is especially true if there is a great deal of information (electrical termination panels come to mind) and revealing all would create a dense blanket of text on the drawing, obscuring everything else.

To use the invisible option, simply start the attribute creation process with attdef, check off the Invisible option, then proceed as outlined previously. The attribute, of course, remains visible until the block is created and disappears afterward. To see what you have, there is no problem: Simply double-click on the block or type in attedit, as outlined previously.

As you can see, being invisible for an attribute is no problem at all, as you can easily access it for editing if needed, and it is unobtrusively out of the way until then, the best of both worlds. The only downside, of course, is that you do not know that the attribute is there if you were not the one who set it up. It is always a good habit to double-click on blocks just to see what pops up on an unfamiliar drawing; and indeed, this is a standard investigative procedure in these cases. Fig. 18.21 shows the Invisible option checked off in the attdef dialog box.

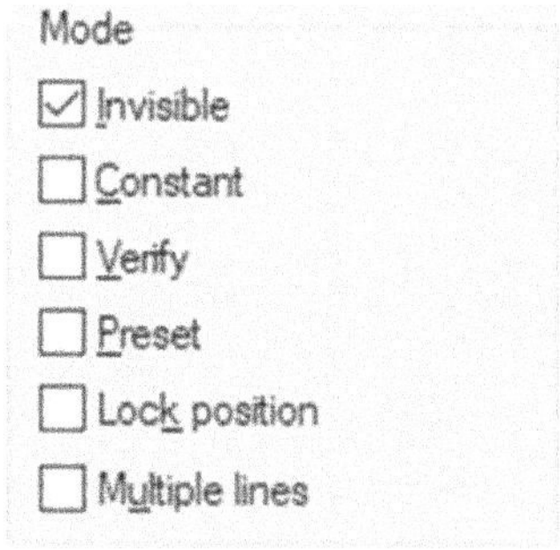

FIGURE 18.21 Invisible option.

18.8 Level 2 drawing project (8 of 10): architectural floor plan

For Part 8 of the drawing project, we finish up the design by creating an exterior elevation. Elevations are the "head on" views of a building. They are projected from overhead or plan views and are meant to give the client and builder a visual of the architect's design. These elevations feature walls, doors, windows, any other exterior design features (such as columns or fireplace chimneys), and of course the roof design.

Step 1. Start a new drawing with an architectural template or set to architectural units. Xref in your floor plan, freezing all layers except the walls, doors, and windows. Then draw x-lines vertically down from each wall corner and opening on the wall you are creating in elevation, as seen in Fig. 18.22.

Step 2. The height of the elevation elements should be based on a section view(s) also xrefed in (Chapter 5, Exercise 10 and Chapter 6, Exercise 9 can be used or created as needed). Then draw x-lines horizontally over from each section feature (footing, foundation, grade line, floor height, window height, roof fascia) in the wall, as seen in Fig. 18.23.

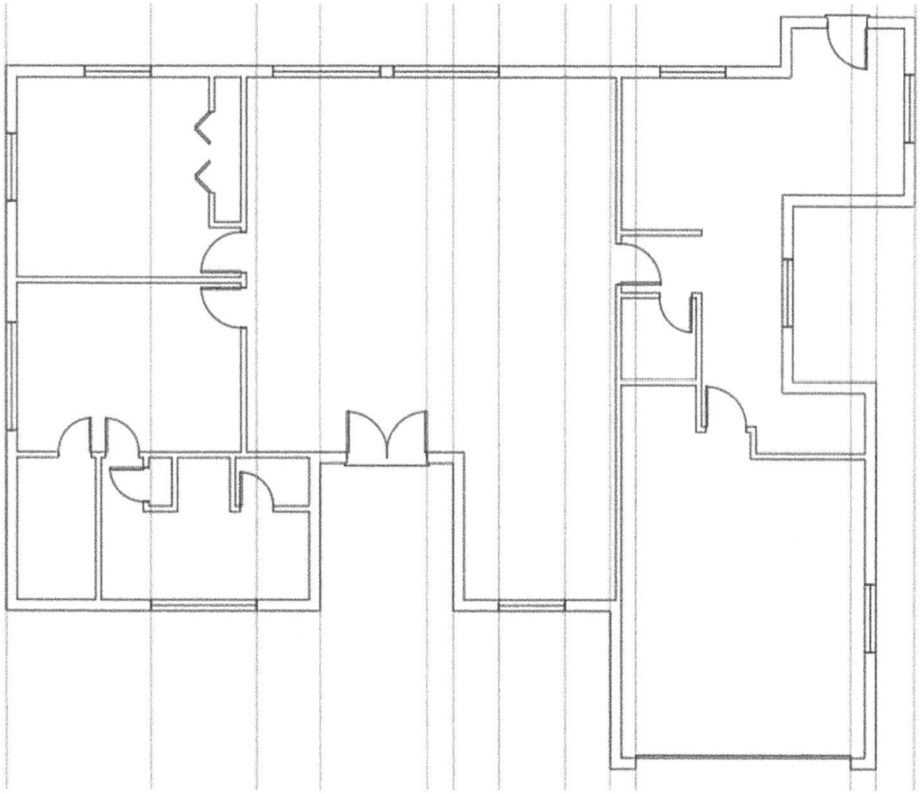

FIGURE 18.22 External Reference of floor plan with x-lines, Step 1.

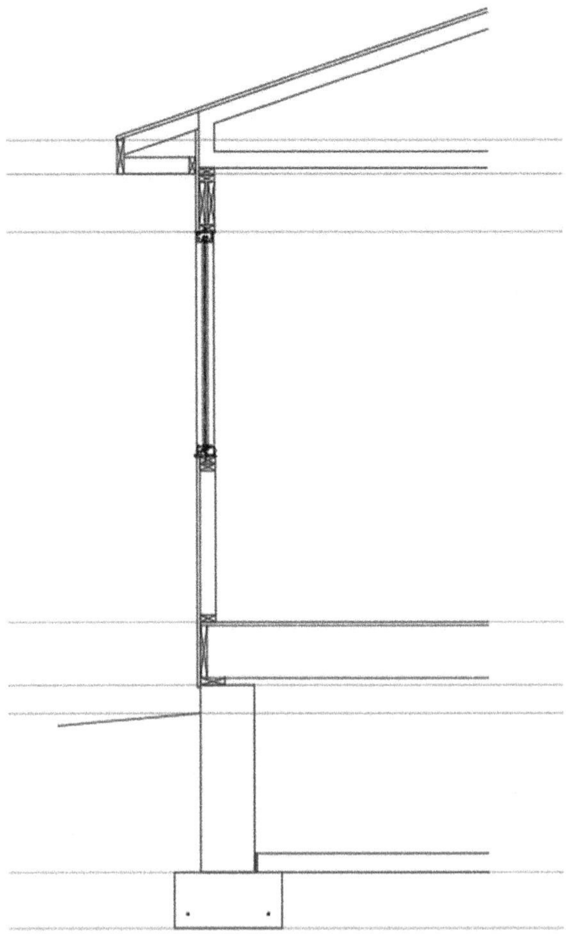

FIGURE 18.23 External Reference of section(s) with x-lines, Step 2.

Step 3. Complete the elevation by "tracing" over the intersecting x-lines from the plan and section as shown in Fig. 18.24. Then add exterior wall hatch and some other finishing touches, such as trim and notes. Fig. 18.25 shows the result. By copying and rotating the xref of the floor plan, you can draw the other elevations quickly as shown in Fig. 18.26. Repeat the same procedure for at least one other side of the house and/or complete all four (Figs. 18.25 and 18.26).

Summary

You should understand and know how to use the following concepts before moving on to Chapter 19*:*

- Attribute definitions
- Attribute blocks
- Editing fields and properties
- Attribute extraction
- Invisible attributes

Review questions

Answer the following based on what you learned in this chapter:

1. What is an attribute and how do you create one?
2. How do you edit attributes?

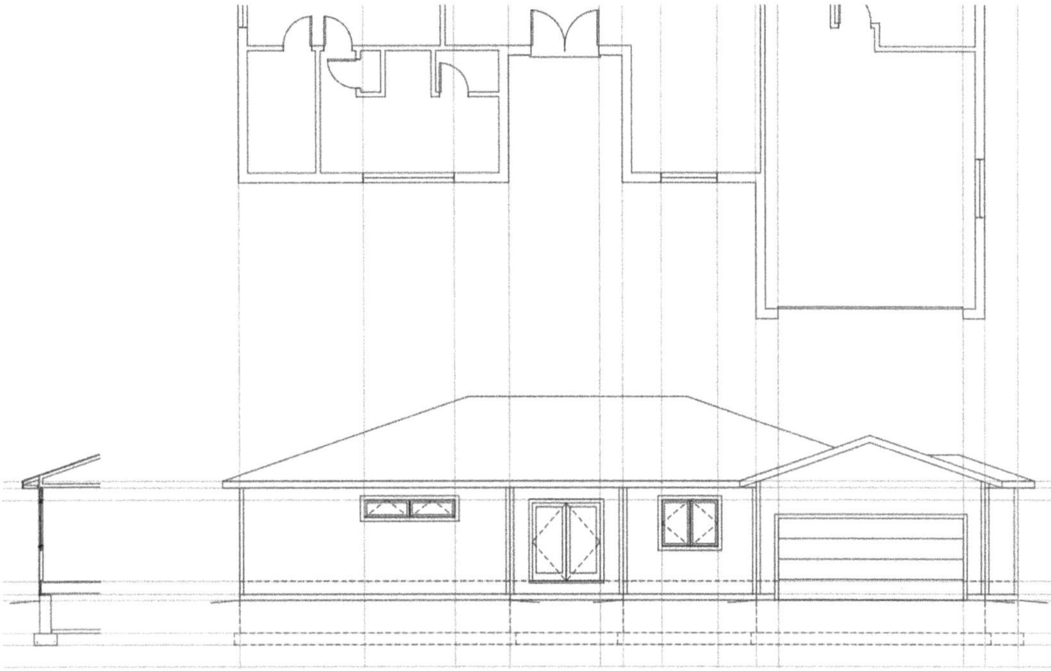

FIGURE 18.24 Elevation created from x-line intersections, Step 3.

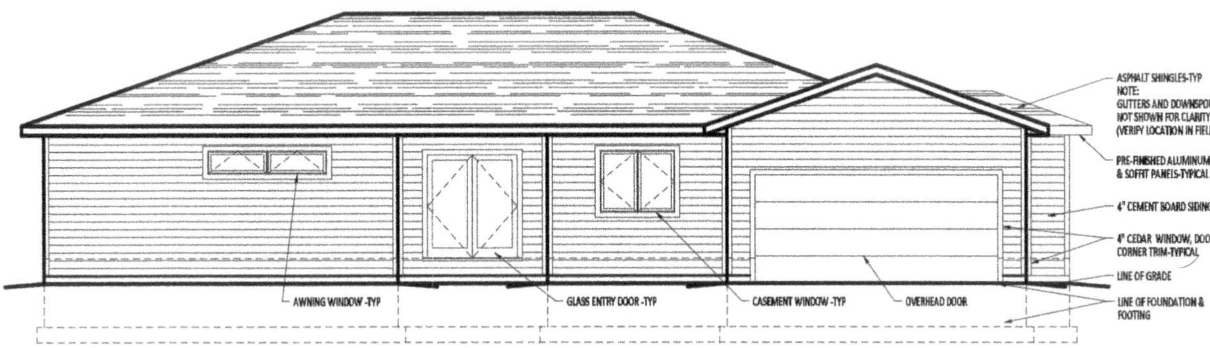

FIGURE 18.25 Completed Front Elevation.

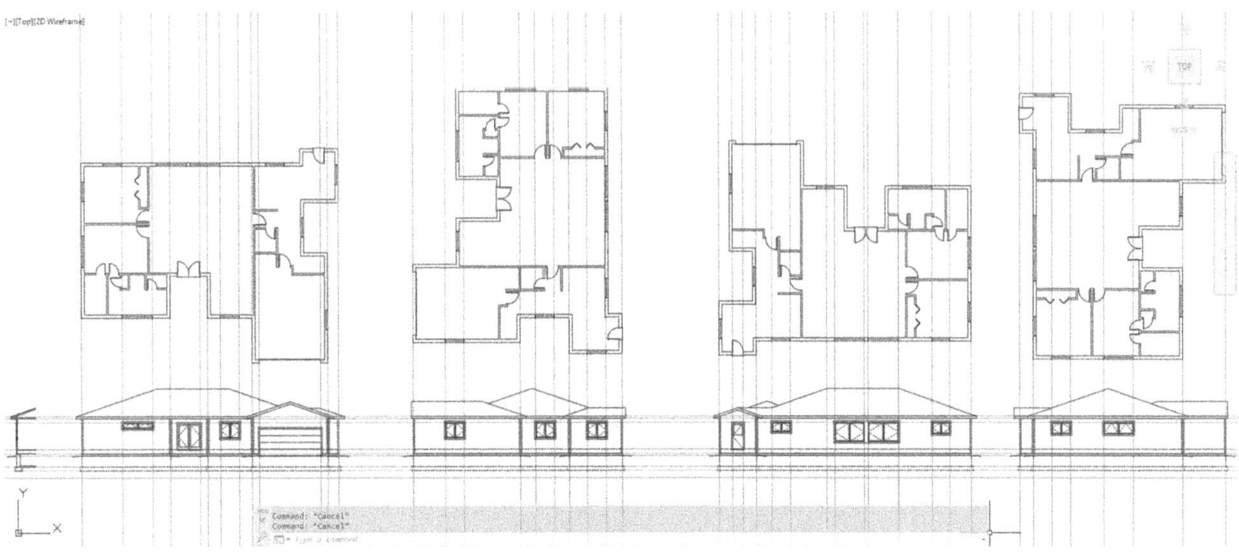

FIGURE 18.26 Elevations being created from copied and rotated External Reference file.

3. How do you extract data from attributes? Into what form?
4. How do you create an invisible attribute? Why can this be important?

Exercise

1A. Draw the following doors:

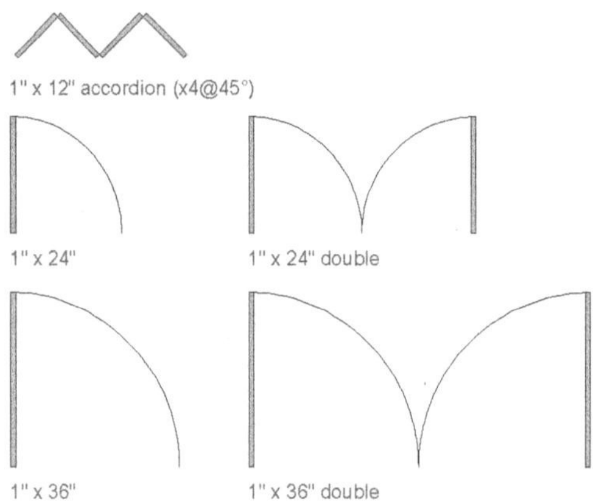

1B. Next, add text attributes to each, as shown by the sample that follows for the 1″ × 12″ accordion:

- *Type:* Accordion double.
- *Size:* 1″ × 12″.
- *Color:* Natural wood.
- *Mfg.:* XYZ Door Company.
- *Stock #:* 654-A-1 × 12.

1C. Create similar attribute data for the remaining four doors. Vary the colors and the stock numbers.
1D. Make a block out of each one and fill in the attribute information.
1E. Copy the entire set once for a total of 10 doors.
1F. Extract the data to Excel.

(Difficulty level: Intermediate; Time to completion: 30 minutes.)

Spotlight on: aerospace engineering

Aerospace engineering is a specialized branch of mechanical engineering that evolved relatively recently in our history, as humankind took to flight, first in the atmosphere then into space. Formally defined, it is the science behind the design and construction of aircraft (aeronautical engineering) and spacecraft (astronautical engineering). These engineers conceptualize, build, and test everything from gliders to missiles, to jet fighters and space rockets. A few also work in naval design, as some of the principles guiding the flow of air (a fluid) around an aircraft readily transfer to the flow of water (also a fluid) around a ship or submarine.

FIGURE 1 Grumman X-29 experimental. *Image Source: Northrop Grumman.*

Aerospace engineers need a mastery of a wide variety of engineering sciences, such as fluid dynamics, structures, control systems, aeroelasticity, thermodynamics, materials, and electrical engineering. It is perhaps one of the more diverse engineering disciplines and specialization is necessary. Three broad areas of specialization include aeronautics, propulsion, and space vehicles.

FIGURE 2 Airbus A380 Airliner. *Image Source: Airbus.*

Education for aerospace engineers starts out similar to that for the other engineering disciplines. In the United States, all engineers typically must attend a 4-year ABET-accredited school for their entry-level degree, a Bachelor of Science. While there, all students go through a somewhat similar program in their first 2 years, regardless of future specialization. Classes taken include extensive math, physics, and some chemistry courses, followed by statics, dynamics, mechanics, thermodynamics, fluid dynamics, and material science. In their final 2 years, engineers specialize by taking courses relevant to their chosen field. For aerospace engineering, this includes aerodynamics, structures, controls, propulsion, and orbital mechanics, among others.

On graduation, aerospace engineers can immediately enter the workforce or go on to graduate school. Although not required and not common in the aerospace industry, some engineers choose to pursue a Professional Engineer (P.E.) license. This would typically be in related mechanical engineering, as there is no aerospace P.E. exam. The process first involves passing a Fundamentals of Engineering (F.E.) exam, followed by several years of work experience, and finally the P.E. exam.

Aerospace engineers can generally expect somewhat higher starting salaries than civil, mechanical, and industrial engineering graduates, on par with electrical and lower than petroleum engineering and some computer science degrees. A master's degree is usually the path to higher starting pay, which with a bachelor's, averaged around $70,000 nationwide. Your specific starting salary, of course, depends highly on locality, current market demand, and even GPA on graduation. The median pay for all aerospace engineers was about $110,000 and the national average pay for all aerospace engineers is about $91,000.

Aerospace engineers often work for government or defense contractors, and as such need to get and maintain a security clearance (Secret, Top Secret, TS/SCI, or TS/SAP), a somewhat unique requirement that may not be necessary for other fields. Keep in mind that the security checks will go as far back as your 18th birthday, so this may not be the field to get into for those who "experimented" extensively with illegal substances or activities while in college.

So how do aerospace engineers use AutoCAD and what can you expect? Industry-wide, AutoCAD use is not widespread due to the nature of the profession, which relies on more sophisticated software to model parts directly in 3D. The industry standard is CATIA, though NX and Pro/Engineer (Creo) are also used. Aerospace engineering also relies heavily on software for initial testing or design validation. Examples include finite element analysis (FEA) and software such as NASTRAN or ANSYS to model loads, deflections, and heat propagation. MATLAB can be used to model vibrations and stresses, while computational fluid dynamics (CFD) and software such as Fluent are used to model the flow of air around the design at various speeds and temperatures (also applicable to the flow of water). See Appendix B if you would like to learn more about these techniques.

AutoCAD, however, can still be found for small part design, aerospace electrical schematics, and overall system layouts. Smaller companies that have little to no modeling, testing, and manufacturing needs also may use AutoCAD due to the associated cost savings. Many home kit airplane designers or builders turned to AutoCAD to produce drawings or just to document ideas. The general trend is to move away from AutoCAD as design requirements become more sophisticated and testing and manufacturing are involved.

AutoCAD layering is far simpler in aerospace engineering than in architecture. There is no AIA standard to follow, only an internal company standard, which may not be as detailed. As always, the layer names should be clear as to what they contain. Fig. 3 shows an aerospace application of AutoCAD for a set of composite wings for a small unmanned aerial vehicle (UAV) design.

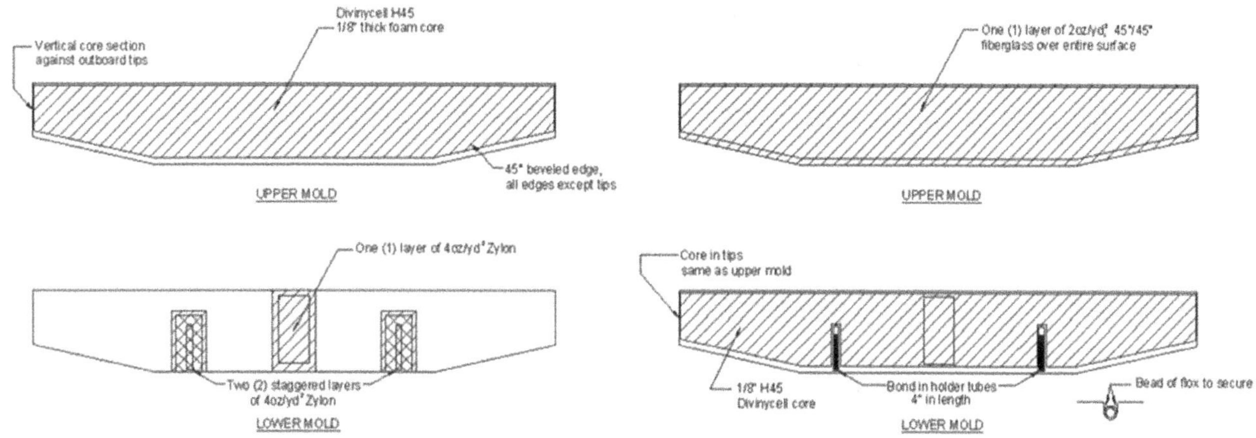

FIGURE 3 Wings for a small unmanned aerial vehicle drawn with AutoCAD. *Images source: Elliot Gindis.*

Fig. 4 shows a typical design of a small general aviation (GA) aircraft wing, with overall dimensions and specs.

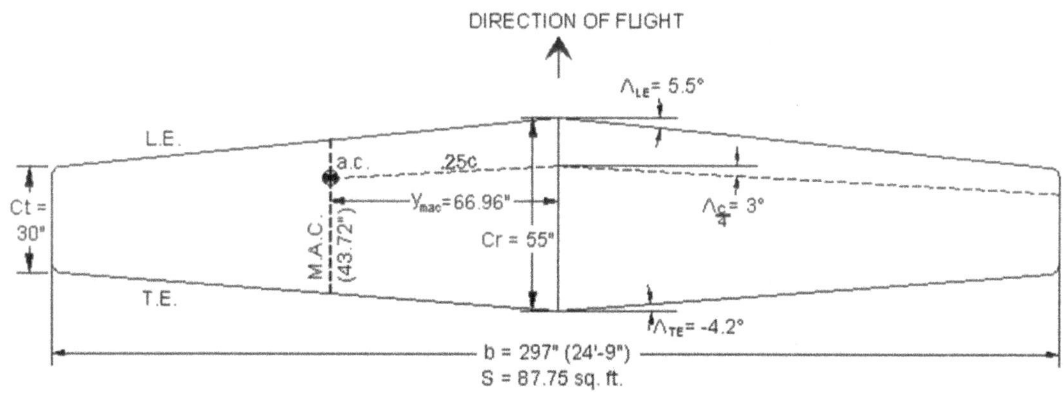

FIGURE 4 Wing specifications for a small, general aviation aircraft drawn with AutoCAD. *Image Source: Elliot Gindis.*

Finally, Fig. 5 shows a side profile of a small private plane (a Lancair Legacy) drawn in AutoCAD for a student project. This side profile study was used for sizing while learning design; the vertical and horizontal numbers are inch position from the tip of the propeller cone and the ground. Note the use of a thicker polyline in the outline to accentuate the profile.

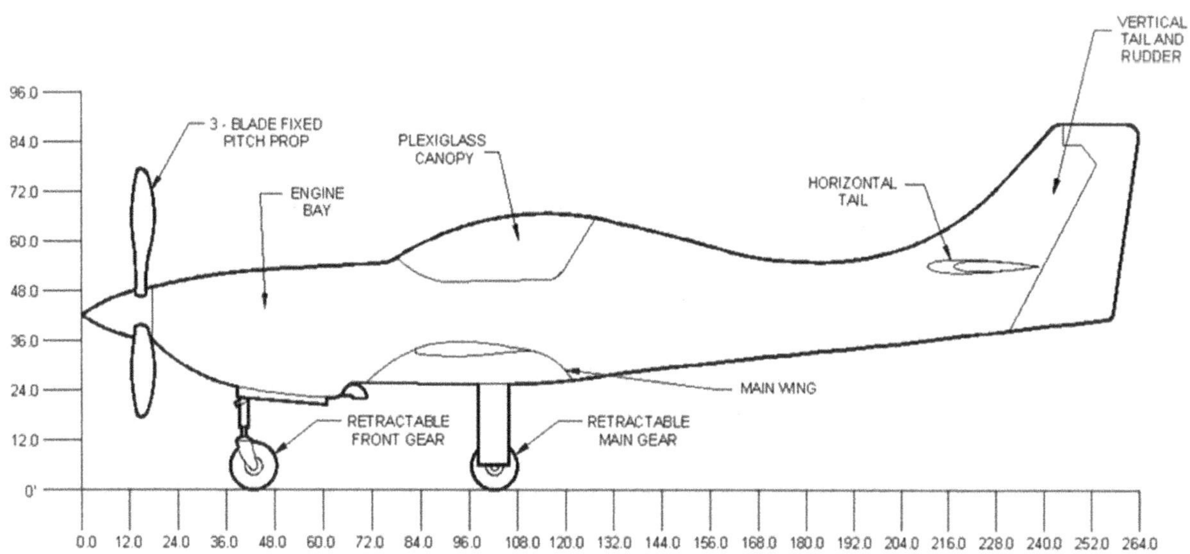

FIGURE 5 A small, general aviation aircraft (Lancair Legacy) drawn with AutoCAD. *Image Source: Elliot Gindis.*

Chapter 19

Advanced output and pen settings

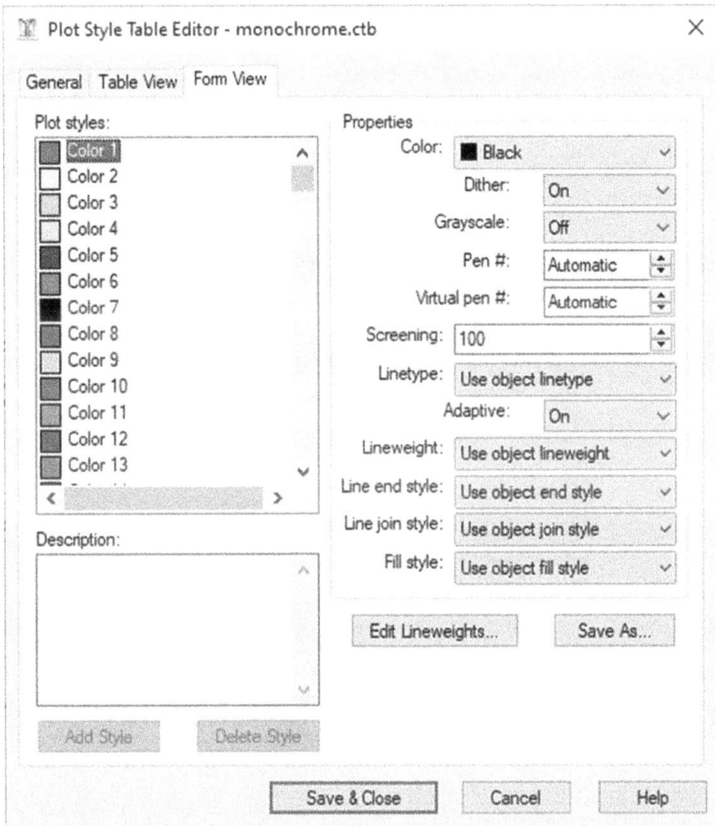

Learning objectives

In this chapter, we introduce and thoroughly cover the concept of advanced output, which includes the concept of the CTB (Color-Dependent Plot Style Table) pen settings file. We specifically discuss the following topics:

- CTB files and their purpose
- Editing lineweights
- Screening
- Lineweight Settings

By the end of this chapter, you will be well versed in applying this critical concept to your design work. Estimated time for completion of this chapter: 1 hour.

Up and Running with AutoCAD 2022. https://doi.org/10.1016/B978-0-323-89923-9.00019-6

19.1 Introduction to advanced output and pen settings

This chapter is the last of the three that talk about some aspect of printing and output. Although general and advanced printing was already explored in Chapters 9 and 10, a crucial piece of information was omitted. This information concerns pen settings and other advanced output techniques, essential knowledge for professional-looking output. The reasons we are doing this are outlined next.

Most designs are rather complex when everything has been added to the drawing. Lines cross other lines, and text is everywhere. The eye needs to be able to quickly and easily focus on what is essential, e.g., the outlines of the walls of a floor plan or the outline of the mechanical part among the many other elements in the design.

This is analogous to a mix of a rock song. Not every instrument can be the loudest; something must give, and a balance needs to be found so the nuances and color of the music can stand out. Usually, when the singer sings, other instruments must be turned down a bit, but when a guitar solo is played, its volume gets turned up, and so on. Otherwise, it would be a loud, obnoxious mess with all sounds competing for attention.

Such is the situation with a pencil and paper drawing. All lines cannot be the same thickness and/or darkness. Such a monochrome mess would be confusing to look at. Draftspersons have known this for as long as the profession has been around and, as a result, have numerous pencils at their disposal, ranging from lightest (hard lead) to darkest (soft lead) and everything in between. They use the thickest/darkest lines for the primary design elements and the thin/lighter leads for all else.

AutoCAD has a similar philosophy. Lines must have differences in thickness for some to stand out more than others. But, how do we show these differences now that we no longer use pencil leads? The answer is to use color. Colors are then assigned line thicknesses, and assuming of course a consistent and intelligent pattern of use on the part of the designer, differences in line thickness can easily be incorporated into the design by assigning certain colors to the primary geometry. As you can see, colors play a dual role. They allow you to tell geometries apart not only on the screen but also on the paper output. The challenge then is to set up a table that assigns colors to certain line thicknesses and, most important, stick to using these colors properly and consistently.

19.2 Setting standards

The thing you need to determine before anything gets touched in AutoCAD is what colors will be used to represent the most important geometry. You did this back in Chapter 3 when layers were introduced. This, of course, is something done once by the CAD manager and usually is applied to all drawings. All we are really doing is establishing a "pecking order," so to speak, for colors. This order corresponds from the most important (hence, thickest) to the least important (the thinnest) pen setting. There are no hard rules here as far as what colors to use; it is arbitrary, although convention favors some colors over others.

Let us say (based on common convention and Chapter 3) that green is the A-Walls layer color in our architecture plan and is assigned the thickest lineweight. The A-Windows layer is set to red and the A-Doors layer yellow. Both those objects are almost as important as the walls but can be set a bit thinner. The A-Furniture and A-Appliances layer is magenta and thinner still. A-Text and A-Dims layers are cyan and thinner yet. Finally, in our example, the thinnest is hatch. Those dense patterns really need to be light, so as not to overwhelm the drawing, so we assign them a gray color and the thinnest lineweight of all.

Although this is just an arbitrary example, it is also exactly the thought process you need to go through to have a rough idea of how things will look. Once this is generally documented, you can move on to the actual settings.

19.3 The CTB file

The CTB (Color-Dependent Plot Style Table) file is the file that needs to be modified to set the pen settings. It is where you assign pen thickness to a color and where we spend our time for the remainder of this chapter. We open the CTB file, give it a name, modify settings, and save. Then, as you set up plots, you assign the CTB file to those plots, which is a permanent, "do-once" action.

There are two ways to set up the CTB file: either by accessing it in the Page Setup Manager (upper right corner) or as a separate action, which is what we do. In the Application menu (Big A), go to Print, and then scroll down to Manage Plot

Styles. You will be taken to the file location where AutoCAD stores its standard, out-of-box plot styles. Click the Add-A-Plot Style Wizard shortcut. The dialog box shown in Fig. 19.1 appears.

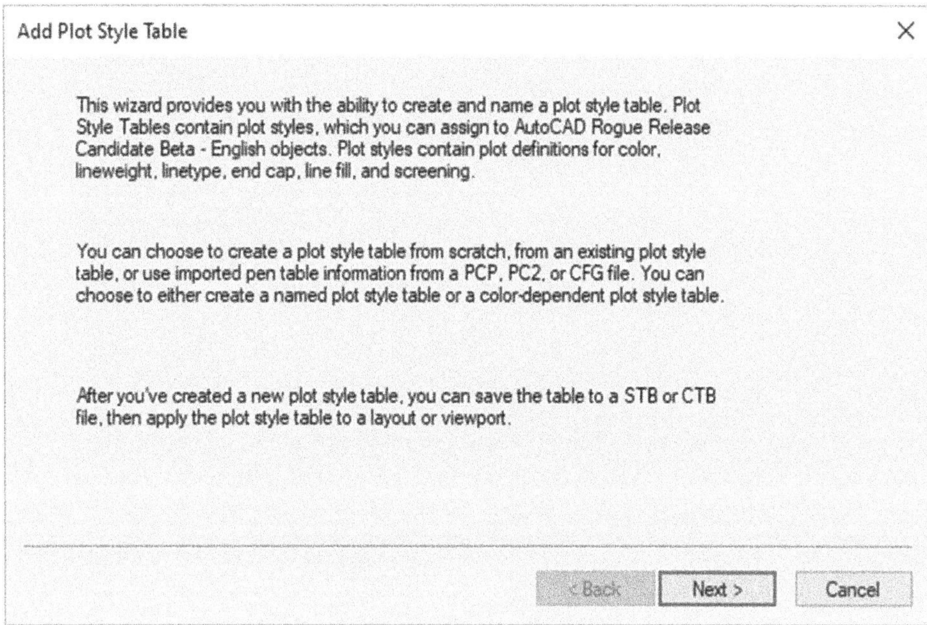

FIGURE 19.1 Add Plot Style Table.

Press Next> and you are taken to the next page of this dialog box (Fig. 19.2), where you select the first option, Start from scratch. This begins a new line setting procedure from the default base values.

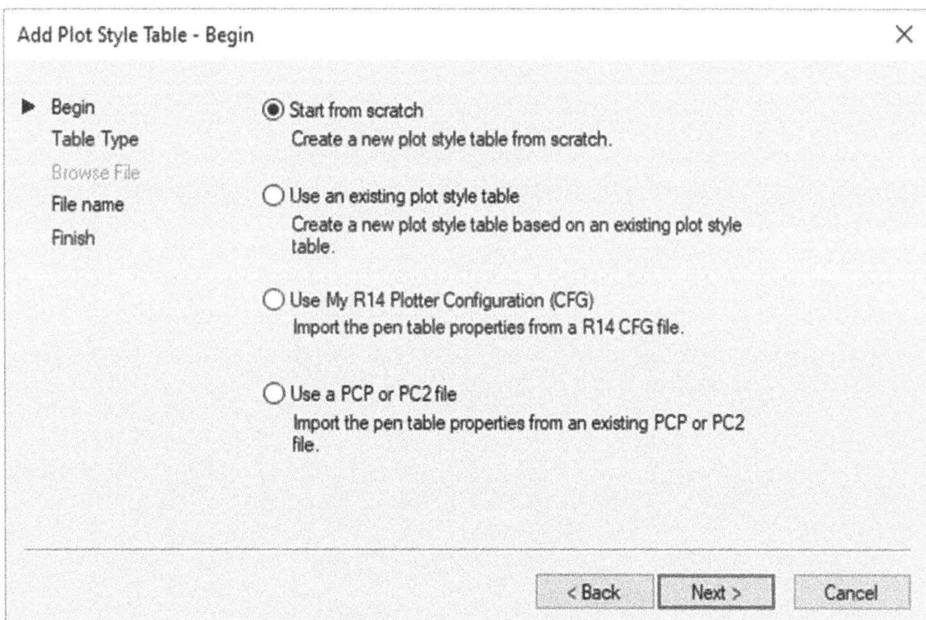

FIGURE 19.2 Add Plot Style Table - Begin.

Press Next> and you are taken to the next page of this dialog box (Fig. 19.3); select the Color-Dependent Plot Style Table choice. This is the only choice we look at; the named plot styles are not covered and are rarely used.

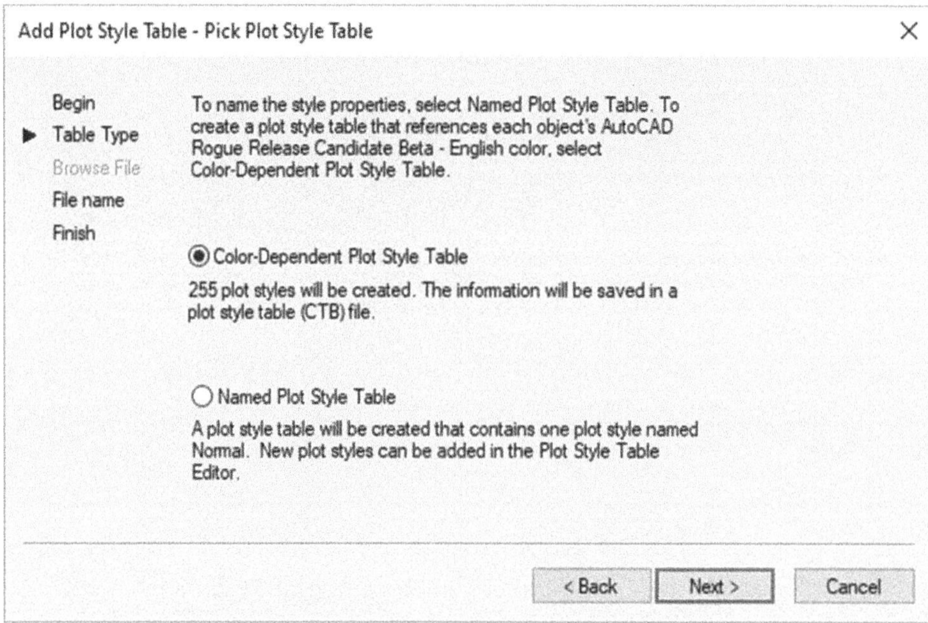

FIGURE 19.3 Add Plot Style Table – Pick Plot Style Table.

Press Next> and you are taken to the next page of this dialog box (Fig. 19.4). Give the new CTB file a descriptive name, like StandardCTB, ColorCTB, or anything else that will be recognized. Here, it is Sample_CTB_File.

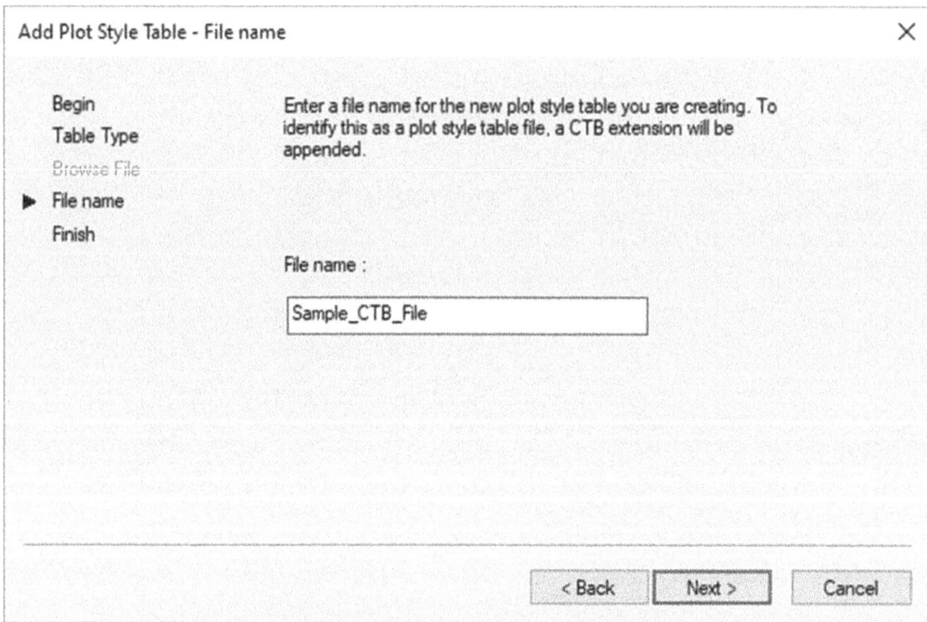

FIGURE 19.4 Add Plot Style Table – File name.

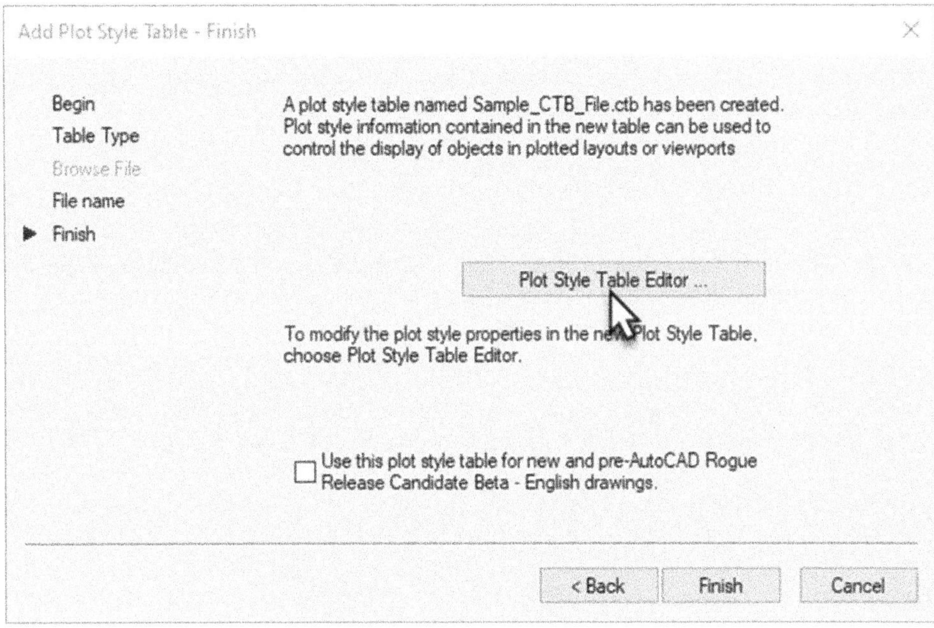

FIGURE 19.5 Add Plot Style Table – Finish.

Press Next> for the final step (Fig. 19.5). Here, we are interested in the button Plot Style Table Editor… to set the lineweights. Press that button and you are taken to our destination, the actual Plot Style Table Editor, shown in Fig. 19.6.

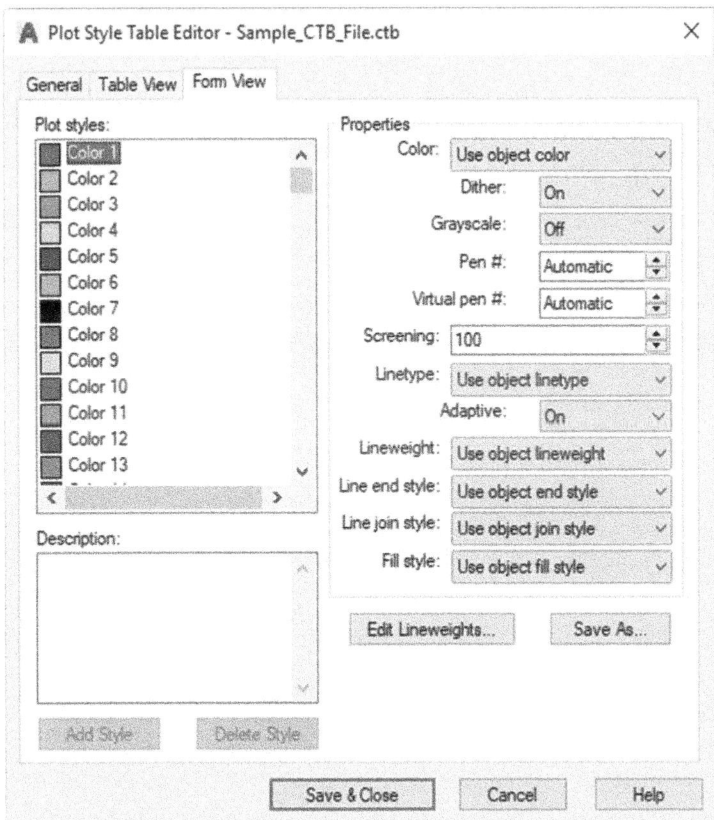

FIGURE 19.6 Plot Style Table Editor for Sample_CTB_File.

Here is where we set the colors, weights, and a few other tricks. Stay in the default Form View (referring to the tabs on top), as we do not need the information in any other format. Familiarize yourself with this editor. On the left is the palette of the available 255 colors, and on the right is a variety of effects and settings you can impart on whatever color or colors you select from the palette. All the tools are there; you just need to proceed carefully.

Step 1

The first thing you need to do is to set all the colors in the color palette to black. As the default settings are now, each color is interpreted literally, which means that if you are sending your design to a color plotter, all the linework will come out in color, exactly as seen on the screen. This is a very desirable outcome in certain situations but not in most (we talk about this more later). For now, we need to make all colors monochrome make all colors so we can move on to lineweight.

- Select all the colors (1–255) in the right side, *Plot styles:* area of the palette by first selecting Color 1, then use the scroll bar to go all way down to see Color 255. Holding down the Shift key and select Color 255 to select all the index colors.
- Then, as all the colors are highlighted blue, select the very first drop-down choice in left, Properties area, called *Color:*, and select Black (Fig. 19.7). All colors are assigned black ink.

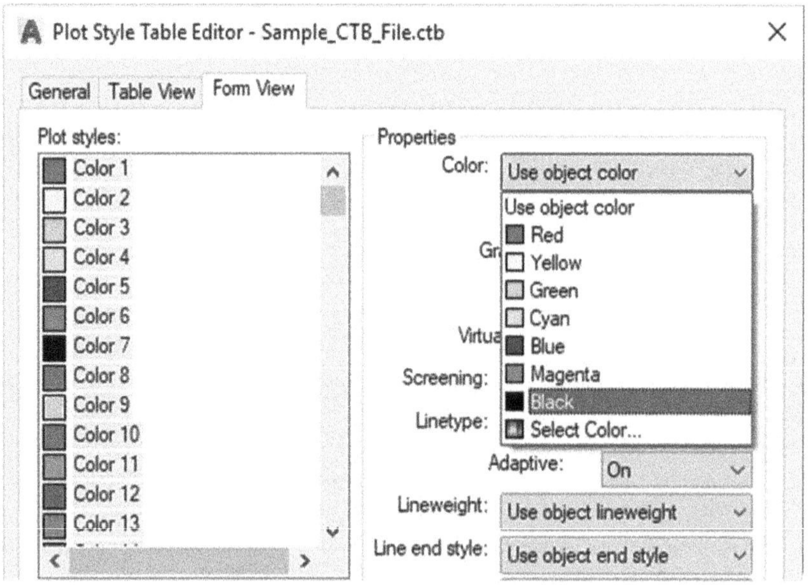

FIGURE 19.7 Selecting black for all colors.

- Clear the palette choices by clicking randomly anywhere in the white empty area of the left side of the palette and test out your changes by clicking random colors. They should all report as black for their *Color:* setting.

Step 2

Next, we need to set the actual thicknesses. This is done in the ***Lineweight:*** drop-down menu, about halfway down the list on the right. Currently, it says ***Use object lineweight***, which simply follow the ayer (or specific object settings in the case of wide polylines) with the default value being 0.250 mm.

Here, we need to discuss a major point you must understand in working with this feature. The actual line sizes are of secondary importance to the relative differences between the sizes. This means that you need not worry too much about how thick a 0.250 mm or 0.050 mm line is What is more important is that there is a large enough difference between the choices (0.20 vs. 0.150 mm as opposed to 0.200 vs. 0.180 mm), so the eye can notice this difference. You can just as easily choose 0.25 or 0.30 mm, but using both near each other would not result in enough contrast. It is the contrasting thickness between lineweights that matters first. As you get up in the higher numbers, all lines can end up being overly thick, so it is best to stick to the range of 0.300 mm on the upper end and 0.050 mm on the lower end. AutoCAD does show a 0.000 mm value as well which just tells a printer to print as thin as it can. Some printers will not accept this value so it is best not to use it. Let us put this all to use and set some values:

- Click on and select color 3 (green); set Lineweight to 0.250 mm.
- Click on and select color 1 (red); set Lineweight to 0.150 mm.
- Click on and select color 2 (yellow); set Lineweight to 0.150 mm.
- Click on and select color 4 (cyan); set Lineweight to 0.090 mm.
- Click on and select color 6 (magenta); set Lineweight to 0.090 mm.
- Click on and select color 9 (gray); set Lineweight to 0.090 mm.

Now, as an applying this theory to the Level 1 apartment drawing. The A-Walls layer is colored green and would appear darkest (0.300 mm), as required. The A-Doors and A-Windows layers were yellow and red, respectively, and are slightly lighter (0.1500 mm), followed by the A-Text and the A-Appliances (or A-Furniture) layers, colored cyan and magenta, respectively, lighter still (0.1500 mm). Last are the gray-colored hatch patterns coming in at (0.1000 mm).

So, will these lineweight choices look good? The drawing will certainly look more professional, and features will stand apart from each other. The exact values and colors are all determined by the architect, engineer, or designer, as this person inspects the results and tweaks the lineweights as needed. This is an extensive trial-and-error procedure, and not all individuals select the same settings; however, in any professional office, these techniques must be used regardless of exactly what specific settings are set. Nothing screams amateur more than a monochrome drawing with equal lineweights.

Step 3

To complete the procedure, press Save & Close and Finish in the Add Plot Style Table - Finish dialog box. Then, open a drawing, such as the Level 1 floor plan, if available. Type in plot and set the drawing up for plotting or printing as outlined in Chapter 10, being sure this time to select CTB_File.Sample.ctb instead of monochrome.ctb in the appropriate field.

19.4 Additional CTB file features

We briefly cover two other useful features about which you should know. The first one was already alluded to earlier. When you selected all the colors and turned them to black, you could have left out a few. These colors would then have been plotted in color. So, this action is certainly not an "all color" or "all black" deal. You can mix and match; and typically, in a piping or electrical plan (for example), the entire floor or site plan is monochrome, but the pipes or cables and wiring are left in vivid red, blue, or green (yellow and cyan do not show up well on white paper). The intent of the design is much clearer. It is a great effect, although it can obviously be derailed by not having a color-capable printer or plotter.

The second feature has to do with the screening effect. This lightens any color of your choosing so less ink is put on the paper during printing. The color is literally "faded out," and it is a very useful trick, as we see with an example. The color most often chosen to be screened are the shades of gray. Color 9 (gray) is natural choice for this effect and already looks faded, although any color can be used.

To screen a color, select it in the editor as just outlined, and go to the Screening field (as seen in Fig. 19.8). Use the arrows or type in a value. Typically, the screening effect is most useful around 30%–40%. Then, proceed as before with saving and plotting.

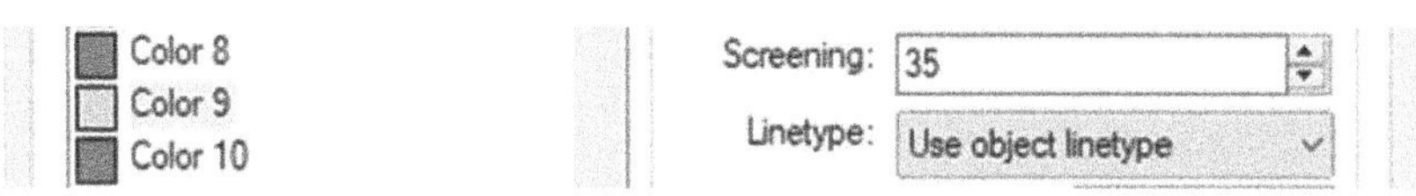

FIGURE 19.8 35% screening of Color 9.

So, what is the effect like? Fig. 19.9 shows two solid fills on the same solid white color: one with no screening (left), meaning it is 100%, and one with the 35% screening (right) as viewed through the plot preview window.

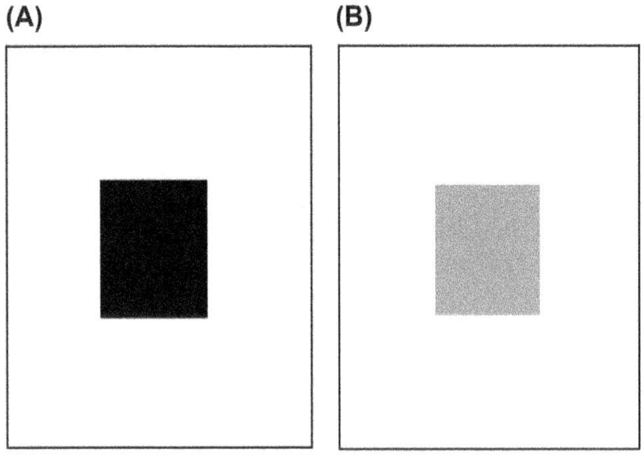

FIGURE 19.9 Plot preview without and with 35% screening.

It is a great trick to use. Here are some ideas where you can do this:

- In a key plan, where the area of work is grayed out and screened, yet the underlying architecture is still visible.
- In an area of work on a demolition plan, where the area to be demolished for a redesign is grayed out, with all the features still visible.
- As a rendering tool, to create contrasting shade and shadow.
- As a rendering tool, to give a sense of depth or solidness to exterior elevations.

19.5 Lineweight VISIBILITY Settings

An additional Lineweight Settings option is available to you. It is found all the way on the right among the bottom screen menu choices, as shown in Fig. 19.10. If it is not there already, you of course must first bring it up via the main menu at the

FIGURE 19.10 Lineweight icon.

far right (the three horizontal bars). If the icon is blue, as shown in Fig. 19.10, the Lineweight feature (called LWT in older versions of AutoCAD) is active. If it is grayed out, then it is not.

Right-clicking on this icon brings up the Lineweight Settings... dialog box (Fig. 19.11).

This feature is not the same as covered earlier in this chapter. The CTB file and setting the lineweights in that manner gets you only see lineweights in the print preview or physically printed results. This means that, on screen, all the lines remain the same size 1 pixel wide, regardless of the CTB settings. The results of the CTB settings can be seen only on paper or in the print previe.

In contrast, the Lineweight option allows you to set line thickness that is visible on screen in the Layer settings for visibility while in Model Space and in Paper Space. It is a rather interesting effect. While in model space, a layer with a lineweight set from 0.00mm to 0.25mm is displayed as one pixel wide, and thicker lineweights use a proportional pixel width greater than 1 pixel wide. Lineweight display in model space does not change with the zoom factor so lineweights

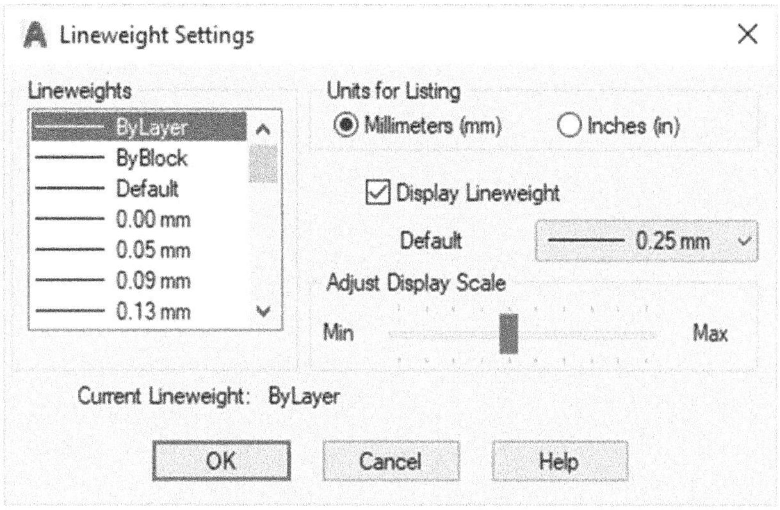

FIGURE 19.11 Lineweight Settings.

can get muddy as you zoom out, away from your design. In a paper space layout, lineweights are displayed as they would be printed, and lineweight display changes with the zoom factor. Once set, the lines remain at that thickness regardless of how much you zoom in or out. Opinions on this effect vary among AutoCAD users. In Model Space mos see it as a distraction and are perfectly fine remembering what colors are set to what line thickness via the CTB method. It can be useful in Paper Space, showing a quick print preview of what would be printed without going into the Plot dialog, Page Setup Manager or Ribbon's Output tab for an actual print preview. It is up to you to experiment and see where you stand.

If the Default lineweight of 0.25 mm (0.010 inch) is too thin for your taste, simply type lineweight and set a different global default in the dialog box shown in Fig. 19.11. Alternatively, you can set lineweights layer by layer by going in Layer Properties Manager, creating a new layer (usually with a color), and setting the value under the Lineweight header, as seen in Fig. 19.12 (with 0.50 mm used for clarity). The result is seen in Fig. 19.13.

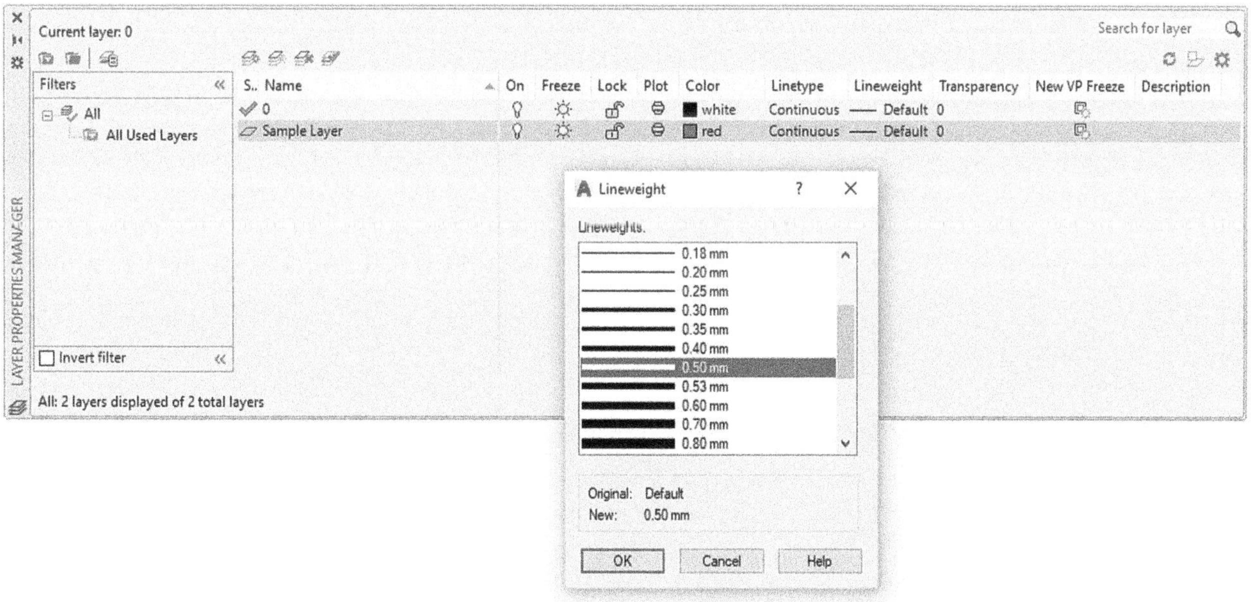

FIGURE 19.12 Lineweight Settings.

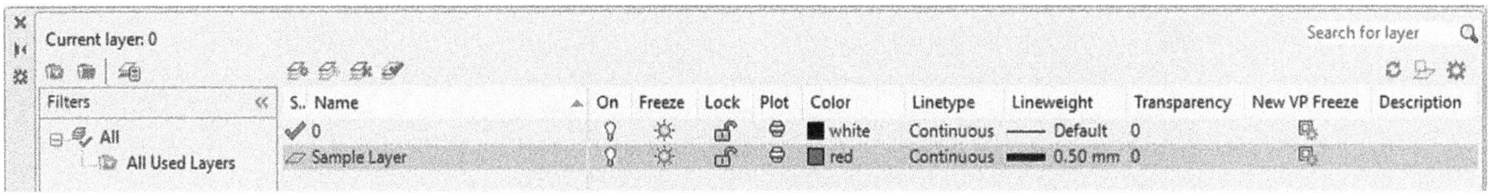

FIGURE 19.13 Lineweight set to 0.50 mm.

After setting this layer as current, draw something. Notice that the line is thin, as if nothing happened. Now press in the Show/Hide Lineweight button to activate it; the line suddenly turns thicker, as seen in Fig. 19.14. This is exactly how it will print.

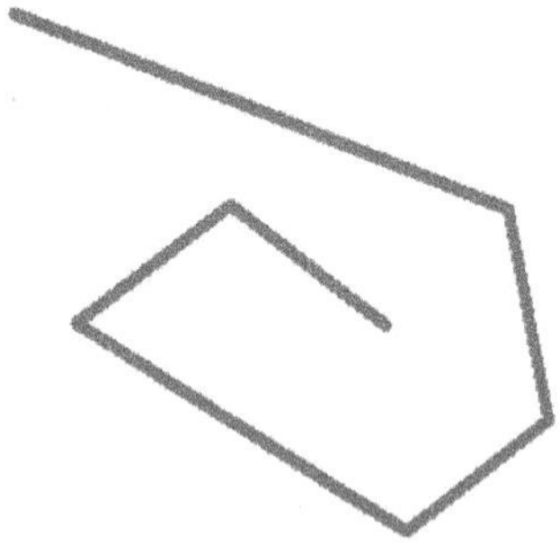

FIGURE 19.14 Line drawn with the 0.50 mm Lineweight Setting.

Note, that CTB settings supersede a layers Lineweight settings. Ideally your CTB file and your layer lineweight setting would match so there is no confusion between the on-screen display and the final output. In the professional world, where shared CAD standard files are housed on a network, the CTB method alone seems to be the most prevalent and has been well established over time. If you are in an educational environment you may not have the ability to keep your lineweight settings outside of your drawing file. In this case it is easy to set your lineweights in each drawing by setting them in the layer properties manager. When you are ready to print, choose the monochrome.ctb file for your output. AutoCAD will print all layers as black but will also follow the lineweights of the layers as you assigned them in the drawing.

If you are in an educational environment and need grayscale or screening, like most things in AutoCAD, there are several options. You can add transparency to your layers in the Layer Properties Manager to lighten them up. This takes some experimenting and can yield interesting results where lines overlap. A final approach is to set your layers with the desired grayscale work to the color 9. Then when printing, set your plot style table to monochrome.ctb and click the button to Edit the Plot Style Table. Next, scroll down and change the Color setting for color 9 to Use object color. Your layers set to color 9 will print as gray following the linetype and lineweight as specified for those layers. Every other color will still print black following their respective layer's linetypes and lineweights. There are lengthy debates online about which way is best and you or your CAD manager must decide. What is more important is to know your options and how to troubleshoot printed output problems if they arise.

19.6 Level 2 drawing project (9 of 10): architectural floor plan

For Part 9 of the drawing project, you do two things. You first set up a master title block that contains attributes. It is reused numerous times when we set up full Paper Space layout tabs in Part 10, in the next chapter. Then, use what you learned in this chapter to create a useable CTB file.

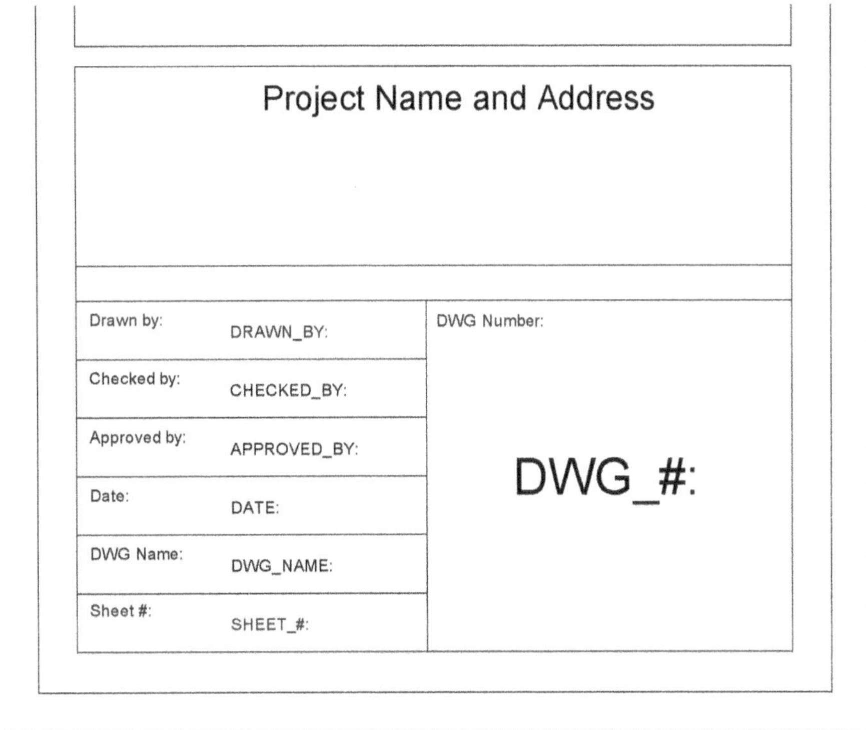

FIGURE 19.15 Title block with attributes.

Step 1. Apply attributes to create an intelligent title block. You can use a modified version of the title block first created in Chapter 10. Start out with that basic template, insert it into a Paper Space tab, and add some extra lines to the lower right part of the title block (see Fig. 19.15). Then, following what you learned in Chapter 18, add text for the following categories, followed by the attributes themselves. Make sure you are drawing on a title block layer (color: white)

- Drawn by:
- Checked by:
- Approved by:
- Date:
- DWG Name:
- Sheet #:
- DWG Number:

Step 2. When you have filled everything in, make a block of the entire title block and either fill in the Xs right away or leave them as variables until you set up the rest of the Paper Space layout in the next chapter. Fig. 19.16 shows the title block filled in with information after it was made into a block.

Step 3. The next step is to assign proper lineweights to the CTB file of the architectural design. Assume the standard we have been following and make green the darkest line, followed by yellow and red, then all the others. The exact values are up to you.

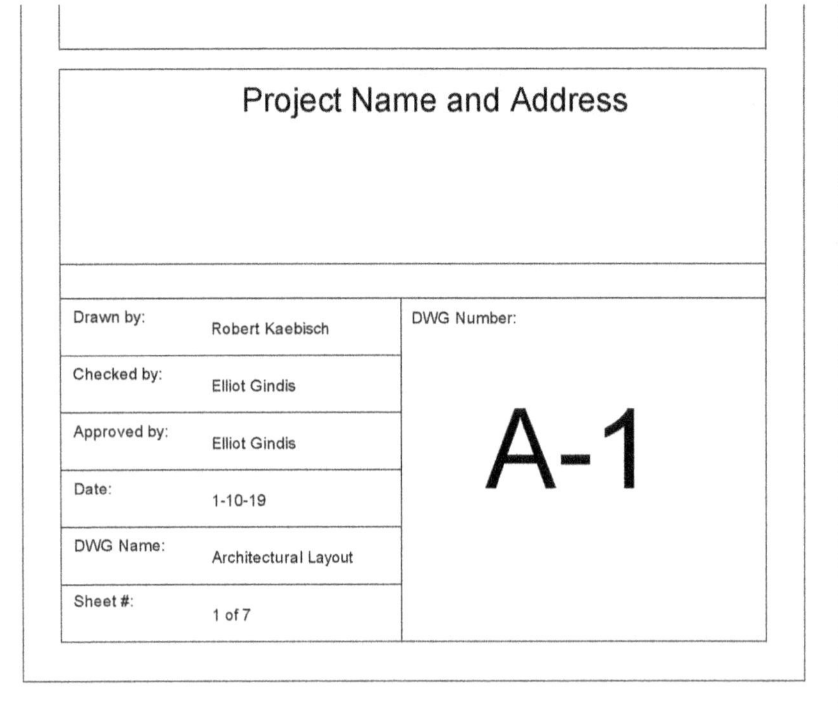

FIGURE 19.16 Title block completed.

Summary

You should understand and know how to use the following concepts before moving on to Chapter 20*:*

- CTB file
- Editing lineweights
- Screening
- Lineweight Settings

Review questions

Answer the following based on what you learned in this chapter:

1. What is a CTB file?
2. Describe the process of setting up a CTB file.
3. What is screening?
4. What is the effect of Lineweight Settings?

Exercise

1. For practice, create a brand-new CTB file. Call it Ch_19_Sample_File. Monochrome all colors and assign the thicknesses to the listed colors as seen next:
   - Color 1, red, 0.2500
   - Color 2, yellow, 0.1900
   - Color 3, green, 0.1500
   - Color 4, cyan, 0.3000
   - Color 5, blue, use object lineweight
   - Color 6, magenta, 0.0900
   - Color 7, white, 0.1000
   - Color 9, gray, use object lineweight
   - Screen Color 9 to 30%.

   (Difficulty level: Easy; Time to completion: <10 minutes.)

Chapter 20

Isometric drawing

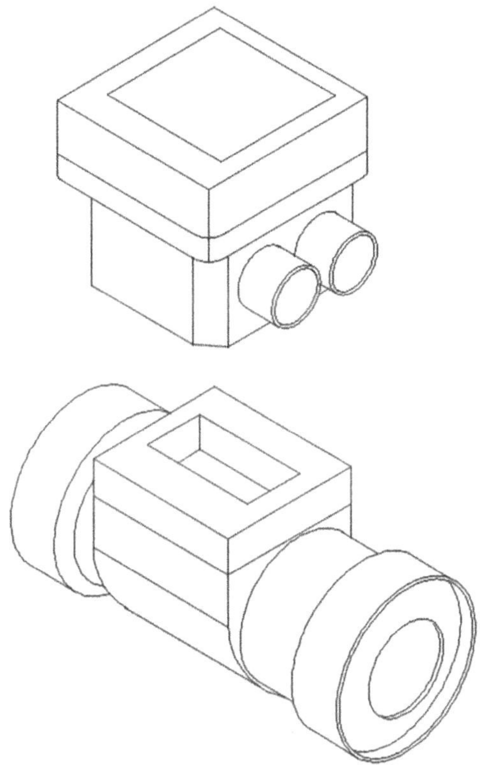

Learning objectives

In this chapter, we introduce isometric drawing. We cover

- General types of graphical projection
- Isometric projection
- When to use it
- When not to use it
- Setting isometric projection
- Changing planes (F5)
- Ellipses in isometric
- Text in isometric

During this chapter, you will create several isometric designs, including a computer desk and mechanical devices. Estimated time for completion of this chapter: 1 hour.

Up and Running with AutoCAD 2022. https://doi.org/10.1016/B978-0-323-89923-9.00020-2

20.1 Introduction to graphical projection

You really cannot talk about isometric projection without first discussing the general (and broader) topic of *graphical projection*. What we are trying to accomplish here is to project a 3D object onto a 2D plane. The various approaches for this are collectively referred to as graphical projection techniques. Isometric projection is just one of those techniques.

None of this is unique to AutoCAD or a new idea in any way. Technical drawing has been around since at least the industrial revolution, and engineers, designers, and architects have been seeking ways to show their designs in three dimensions on a piece of paper for generations. So, in essence (and as it applies to AutoCAD), it means we are drawing a 3D object without resorting to the complexities of 3D space; sort of "faking it" in 2D you could say. This creates the *illusion* of 3D and is often good enough to get across a design idea (or intent) to your customer or client.

Types of projections

There are essentially two broad categories of graphical projections—*parallel* and *perspective*. Each has numerous subtypes as described next:

- **Parallel Projection**—In this type of projection, the lines of sight from the object to the projection plane are parallel to each other. What this means in plain English is that the lines do not *converge* into the distance. This is quite the opposite of what your eyes see in reality; however, this approach is preferred for technical drawing. Examples of parallel projection include *orthographic*, *oblique*, and *axonometric*. Axonometric projection, which can reveal multiple sides, also includes several subtypes including *isometric*, *diametric*, and *trimetric*. The most widely used of these is the isometric projection, which we will discuss in more detail.
- **Perspective Projection**—In this type of projection we are trying to approximate on a flat surface what is seen by the eye. This means usage of concepts such as *vanishing point convergence* and the appearance of *foreshortening*, which means that an object's dimensions along the line of sight are shorter than its dimensions across the line of sight. Perspective projection comes in several "flavors." A *one-point* perspective contains only one vanishing point. A typical example is a set of train tracks. If you stand on them and look into the distance, they converge to a single "vanishing point." A *two-point* perspective has two vanishing points, and a *three-point* has three vanishing points. Both of those perspectives can be easily seen with sharp-angled buildings, and these perspectives are most often represented in outdoor scenes. There is also a four-point and zero-point perspective, but we will not discuss those. AutoCAD does not make use of perspective projection in 2D but does in 3D via the DVIEW command. See the 3D chapters for more information.

20.2 Isometric projection

Using automated tools in 2D AutoCAD, you can quickly and easily draw in isometric projection. Using some additional manual methods, you can also draw in a type of *oblique* projection known as "cavalier", and even diametric and trimetric, but those last two cannot be automatically set, so they would be tedious to create.

So, after all the above definitions and descriptions, what do the main projections that we will be using look like in practice? Let us dive right in with Fig. 20.1, which represents isometric projection, created using the basic T-square and triangle. Note the various angles shown for this type:

Next, look at Fig. 20.2, which represents the cavalier projection, a member of the oblique family. Note the various angles shown for this type:

There is nothing inherently special about these two projections regarding drawing them in AutoCAD. You can easily create both types manually by some simple application of the line and the rotate commands. In other words, draw a few lines at 30 degrees (or rotate some horizontal ones), and then draw a few vertical ones and connect everything together neatly; and you will have the 3D-looking "box" shown in Fig. 20.1. Really, the same thing is in Fig. 20.2, just with more horizontal lines instead of rotated ones.

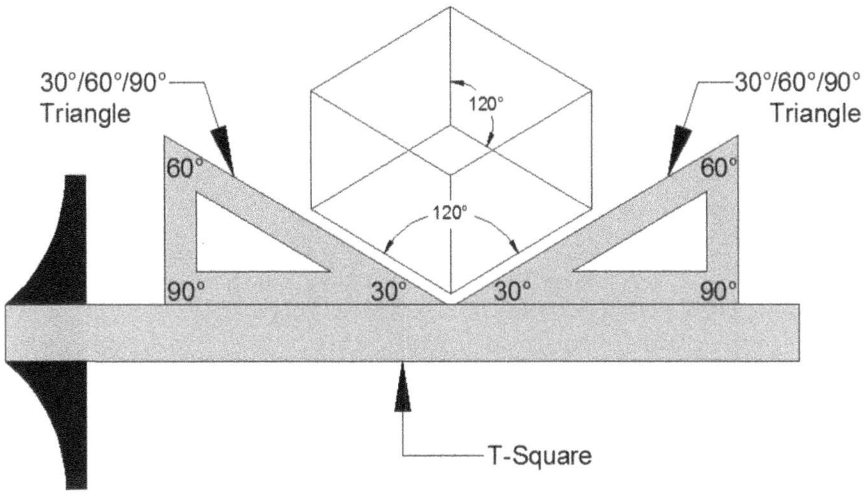

FIGURE 20.1 Axonometric (isometric) projection.

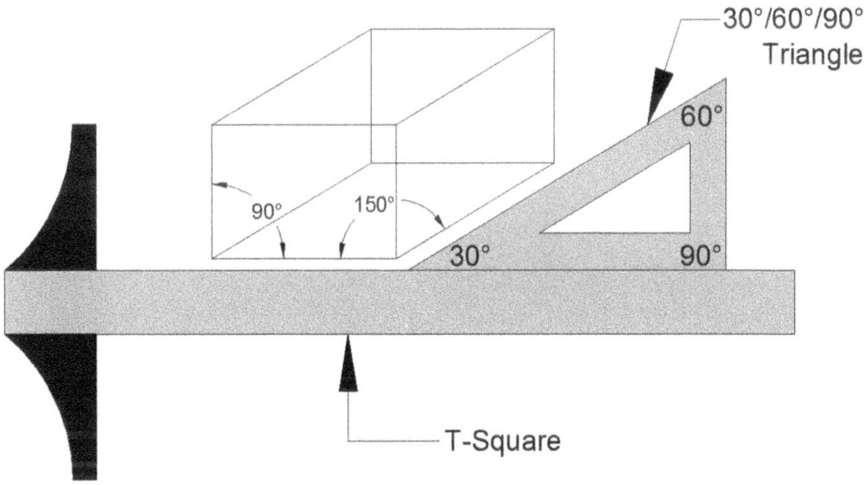

FIGURE 20.2 Oblique (cavalier) projection.

What we are going to do in AutoCAD is automate this process and create Fig. 20.1 faster and easier. Before we do though, one last brief discussion regarding *when* you should use this technique, and when you should not.

Why use isometric projection instead of 3D?

There are a few reasons. One is that it is much easier to learn isometric drawing than to develop proficiency in real 3D. So, for someone who needs only a quick 3D sketch occasionally, just to accent an otherwise excellent 2D presentation, this perfectly fits the bill. Another reason is that 3D takes up more time and computing resources, something that is not always available, and there is also some difficulty in combining 2D and 3D drawings on one sheet. Many students do not go on to take AutoCAD 3D right away (nor should they), and isometric fills that temporary void quite well. For those who do go on, isometric drawing serves as an excellent introduction to 3D and should be reviewed before starting the 3D instruction.

When *not* to use isometric perspective

Isometric drawing is inappropriate when precision 3D drawings are required for design, testing, and manufacturing. Remember, isometric projection is only for visualization. These are *not* true 3D objects and have no real depth, only the illusion of depth. As a result, true measurements and the use of the offset command are not possible in the traditional sense.

20.3 Basic technique

As just mentioned, nothing is inherently special about drawing an isometric design. A basic manual procedure was described earlier in just a few sentences. It is not hard, but it *is* tedious. There is a way to have AutoCAD preset the crosshairs so you are always drawing in isometric 30 degrees mode. To explain it, we need to introduce the idea of planes, so here is a brief introduction to the plane concept.

In isometric drawing, you will always draw on one of three available planes: top, right, or left (see Fig. 20.3). Because you are using a 2D pointing device (a mouse), which exists only in a 2D world, you need to be able to easily move from one plane to the other to be able to draw on it. To toggle between planes, press the F5 key. Your crosshairs line up accordingly, and now you just draw using the line command. Here is an important rule: You almost always have Ortho on while in isometric mode; otherwise, the lines are not straight and you are no longer on any plane (though this *is* useful in rare cases). Let us put it all together and draw an isometric box. After you can do this, more advanced isometric drawings are surprisingly easy. One final piece of advice—set your crosshairs to 100% of the screen; it will help to visualize what you are doing better.

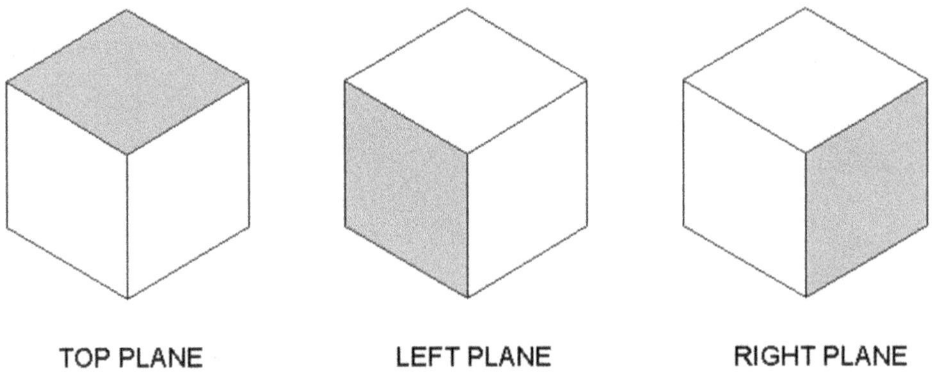

FIGURE 20.3 Isometric planes.

Step 1. Open a blank file.
Step 2. From the Drawing and Construction Aids menu at the lower right, click the Arrow for the Snap settings popup menu and select Snap Settings ..., and the dialog box seen in Fig. 20.4 appears.
Step 3. Pick the Snap and Grid tab, go to the Snap type category (lower left), pick Isometric snap, and press OK. Fig. 20.4 is a screen shot of the Drafting Settings dialog box with Isometric snap selected.
Step 4. Make sure Ortho is activated, and the crosshairs are set at 100% for easier visualization of planes. Notice also how the crosshairs are positioned now that you are in isometric mode.
Step 5. Draw straight lines (as they should be with Ortho) of any size and continue drawing until you have the front face of a box. Very important: Do not try to connect the final line to the first line; you will never get it exactly right. Instead overshoot and click after passing it by, as shown in Fig. 20.5.
Step 6. Now, use the fillet command to fillet the line between Clicks 4 and 5 and Clicks 1 and 2. Fig. 20.6 shows the result.

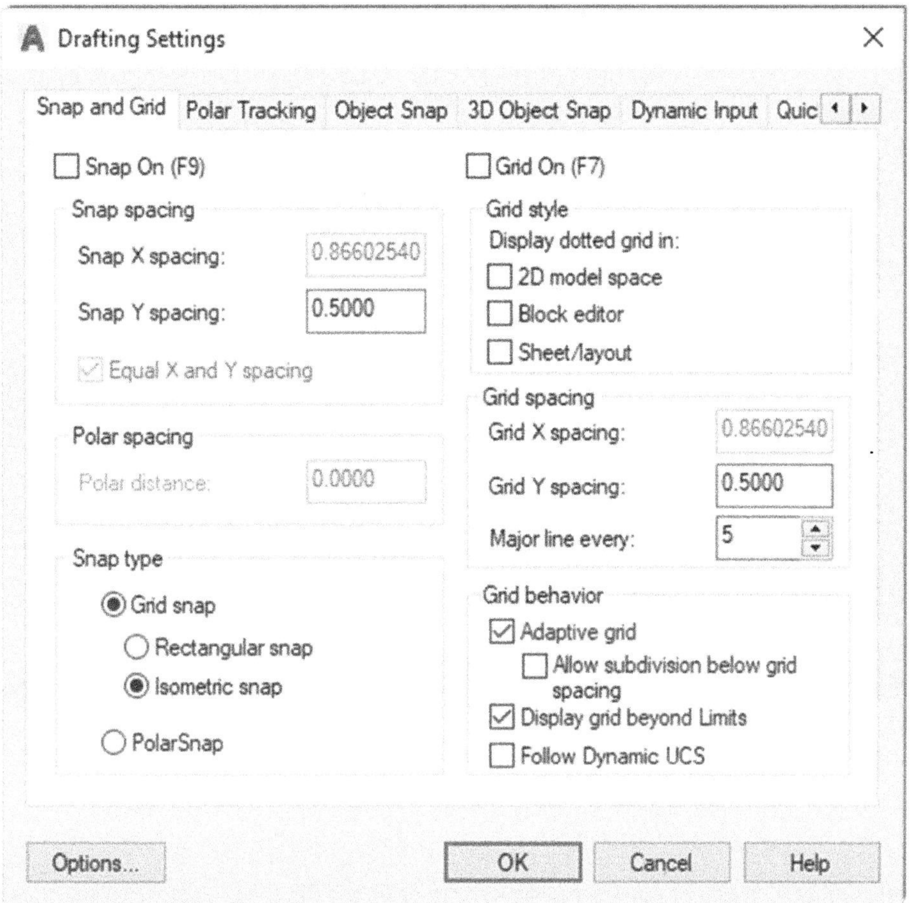

FIGURE 20.4 Drafting Settings with Isometric snap selected.

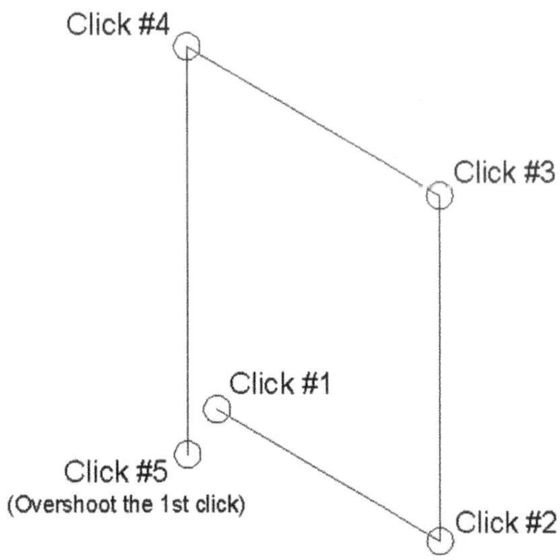

FIGURE 20.5 Isometric drawing, Step 5.

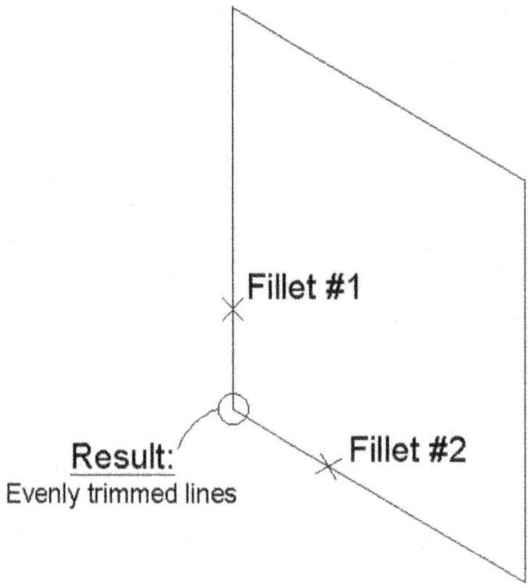

FIGURE 20.6 Isometric drawing, Step 6.

Step 7. Now, press F5. The crosshairs shift to another plane and you can draw another surface.
Step 8. You can press F5 again for the final surface or use the copy command. Be sure to use the ENDpoint OSNAP to cleanly connect lines. Fig. 20.7 shows what you get after a few lines.

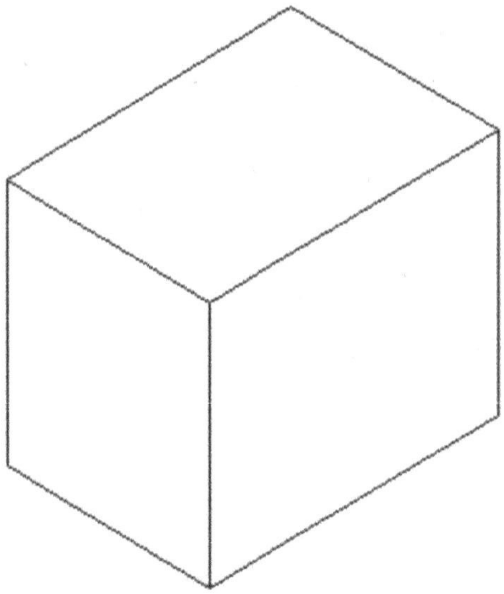

FIGURE 20.7 Isometric drawing, Step 8.

Step 9. And the result, after a few more lines, is seen in Fig. 20.8.

What makes isometric drawing easy is that the techniques do not change for more complex objects. There may be more lines, but it is the same thing over and over.

Here are some additional pointers:

- To draw circles in isometric perspective use ellipses. This command is introduced in Chapter 2 and used to make the bathtub in the apartment drawing at the end of Chapter 4. We say a few more words about using ellipses in isometric design next.

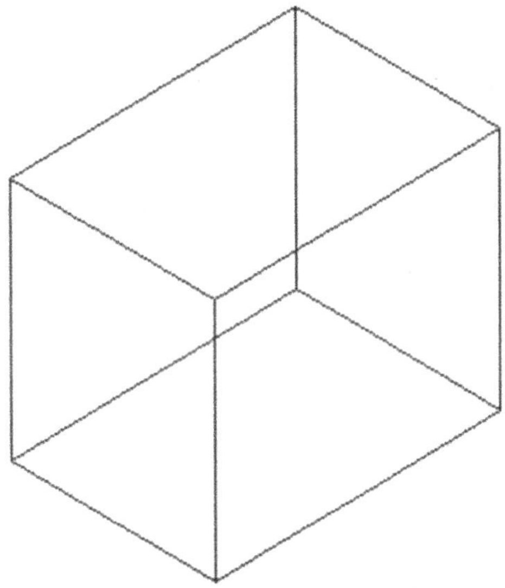

FIGURE 20.8 Isometric drawing, Step 9.

- Always stay in Ortho mode unless you need to deliberately angle a line at some odd angle (see Exercise 3 at the end of the chapter for an example of this.)
- Text can be angled up or down by 30 degrees to align with the Isoplanes. We discuss this shortly.
- Have fun with this. Isometric drawing can be accurate and/or a semi-artistic endeavor and not always as rigid as orthogonal drafting. Eyeball distances but try to make the drawing look good. Do not be afraid to throw in some hatch patterns, as is done in some of the practice exercises in this chapter. Remember though, if you need more precision, you must learn 3D in Chapters 21–30, the 3D sections of the book.

20.4 Ellipses in isometric drawing

A circle angled away from you is an ellipse. These shapes find wide application in isometric drawing, and you will need to use them often in the end-of-chapter exercises.

There are two ways to create an ellipse for isometric use—manually and using the Isocircle option of the Ellipse command. It's of course a bit more tedious to do it manually, but it is still a necessary skill if your isometric plane is not perfectly aligned to the expected values of 120 degrees between lines, as seen in Fig. 20.1.

Let us go ahead and first manually create and place an ellipse in an isometric plane. It may require some adjusting to get right and a bit of a creative eye to do so. Reproducing what was first shown in Chapter 2, we have Fig. 20.9.

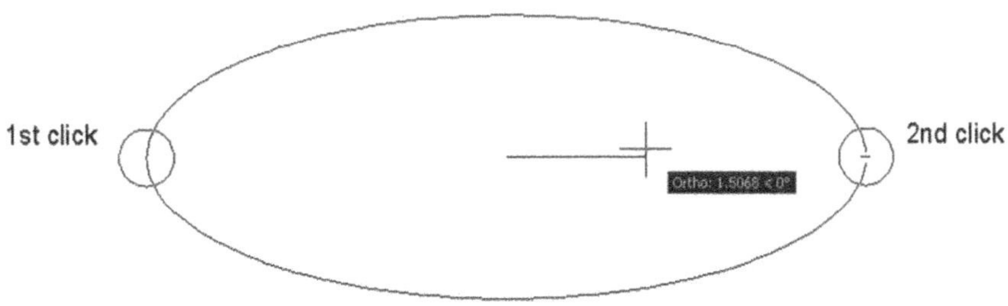

FIGURE 20.9 Basic ellipse.

Once the ellipse is positioned and is the right size, you can rotate it until it looks correct and in the same plane as the rest of the isometric geometry. Fig. 20.10 is an example of some ellipses in an isometric box.

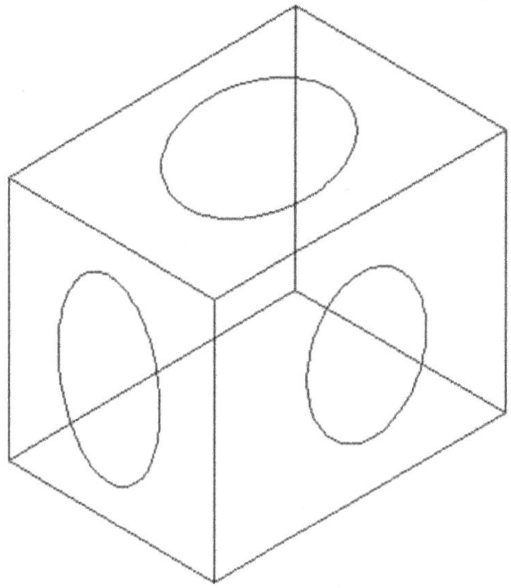

FIGURE 20.10 Isometric box with ellipse.

The more automated Isocircle option is of course easier. For this to work you need to be in one of the isoplanes via the isometric snap option as seen in Fig. 20.4. Cycle through the isoplanes via F5 to land on the one you want to have an ellipse in. Then begin the ellipse command by typing it (the Ribbon tool button will take you to a typical ellipse) and select the `Isocircle` option by pressing `i`. AutoCAD will say: `Specify center of isocircle:`, then as you draw the ellipse you will see: `Specify radius of isocircle or [Diameter]:`. When done, you will see an ellipse properly aligned to the plane. Simply press F5 and repeat for the other planes. You will again get the design shown in Fig. 20.10.

20.5 Text and dimensions in isometric drawing

Text can easily be physically aligned with isoplanes by rotating it plus or minus 30 degrees, or other values depending on the plane, as shown in Fig. 20.11. There is, however, another, arguably better, way to create isometric text for two of those three planes, by applying an oblique angle to it as described next.

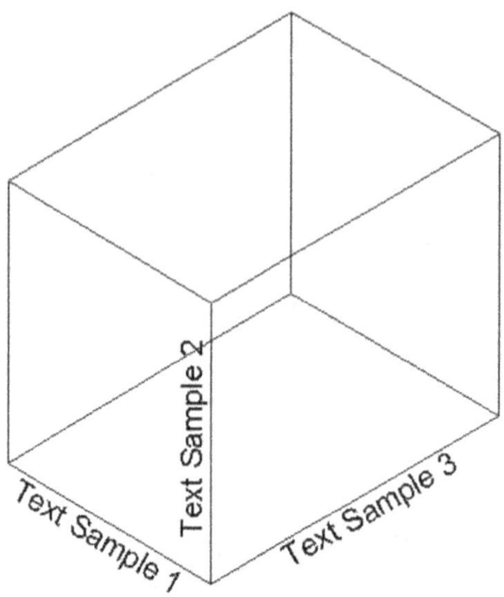

FIGURE 20.11 Text in isometric drawing—rotation.

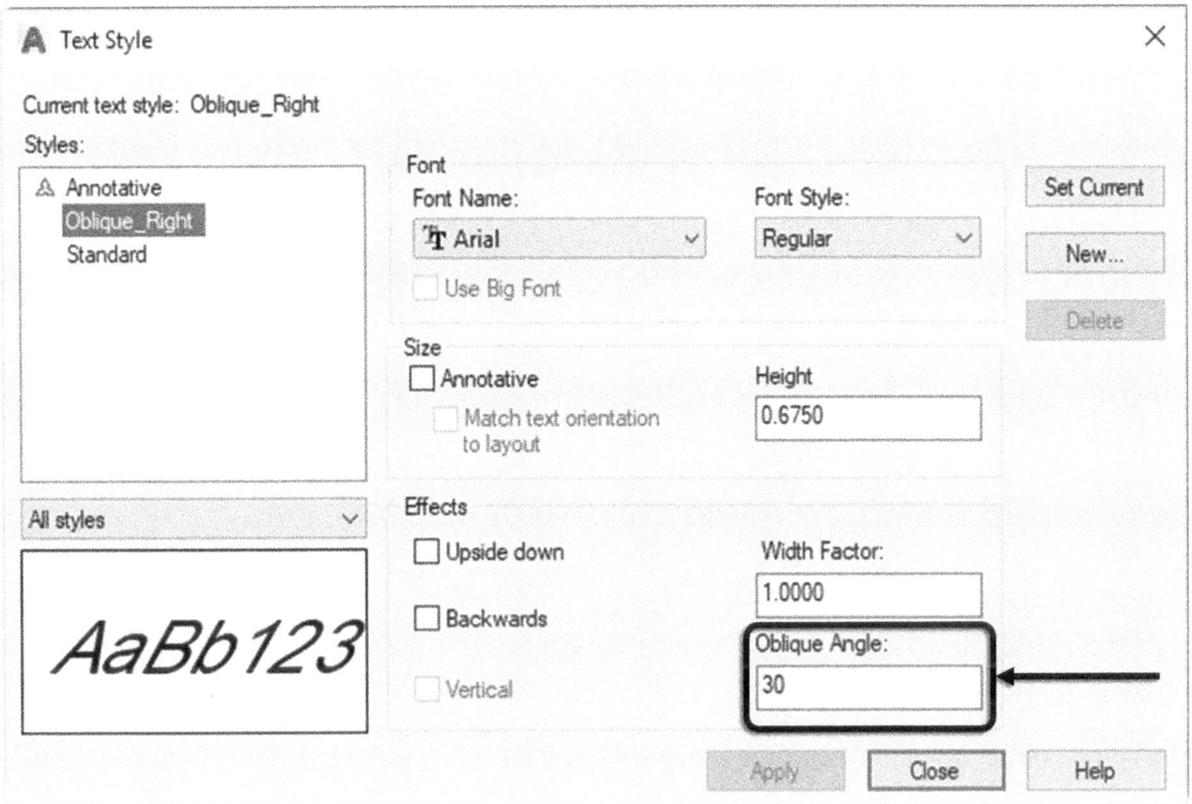

FIGURE 20.12 Setting the Oblique-Right text style.

Go to the Text Style dialog box and create a new style, called Oblique-Right, with an oblique angle of 30 degrees, as seen in Fig. 20.12. Size your text height as needed for your particular isometric design. When done, press Apply, Set Current, and Close.

Go ahead now and rewrite the same text at the bottom of the isometric box as seen in Fig. 20.13. Just as before, make sure it's rotated 30 degrees. Notice now how the new oblique text leans "in plane" with the box. A lot of designers prefer this approach as opposed to just pure rotation.

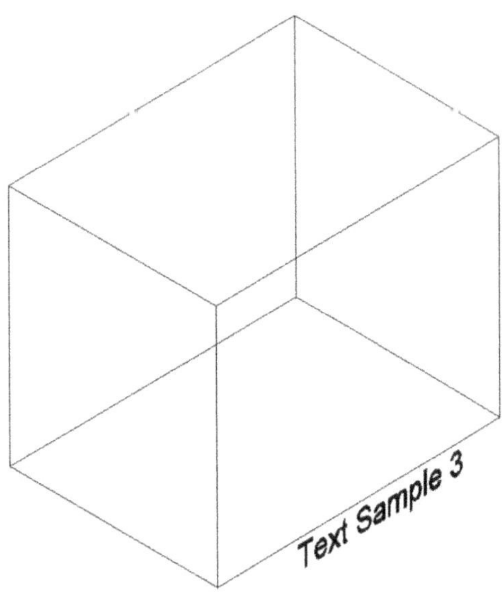

FIGURE 20.13 Oblique-Right text style.

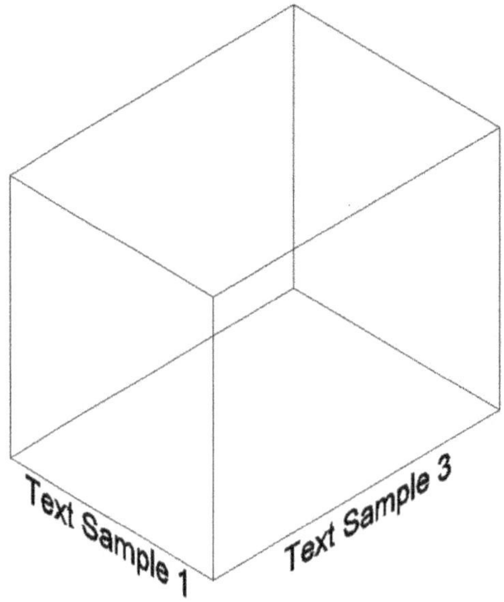

FIGURE 20.14 Oblique-Left text style added.

You can also do this with the opposite plane by creating an Oblique-Left text font and leaning it negative 30 degrees. The result is seen in Fig. 20.14. For the straight vertical text (Text Sample 2 in Fig. 20.11), no obliqueness is needed of course, as it is pointing straight up and can be left as is.

You can also easily dimension an isometric drawing using Aligned and Vertical dimensions, as seen in Fig. 20.15. However, just like with text, you can improve on this by using oblique angles. For dimensions this is a two-step process. First, if needed, set the dimension text to oblique (+30 degrees or −30 degrees) using the techniques just described for regular text. Next, set the dimensions themselves to an oblique angle. If everything is done correctly, the text and dimension's extension lines will match up, as well as match the isoplane you are on.

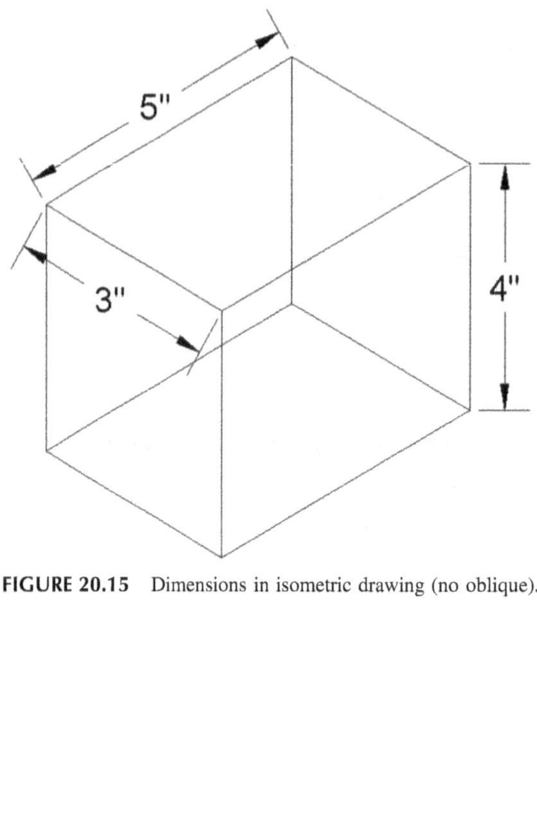

FIGURE 20.15 Dimensions in isometric drawing (no oblique).

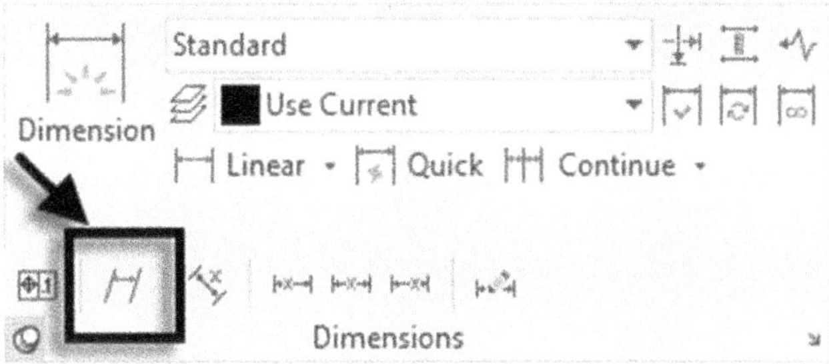

FIGURE 20.16 Dimedit for oblique dimensions.

To set the dimensions to oblique, first create and place them in position via Aligned or another approach, then type in `dimedit` or press the DIMEDIT button on the Ribbon's Annotate tab under the Dimensions drop down (see Fig. 20.16).

AutoCAD will ask you to select the dimension and then`Enter obliquing angle (press ENTER for none):`

Go ahead and enter +30 degrees, −30 degrees or whatever value is appropriate, and the dimension will become oblique. If it does not look right, just undo and repeat with other values. Fig. 20.15 should now look like Fig. 20.17. Note that in this particular example there was no need to oblique the dimension's text, only the dimension itself.

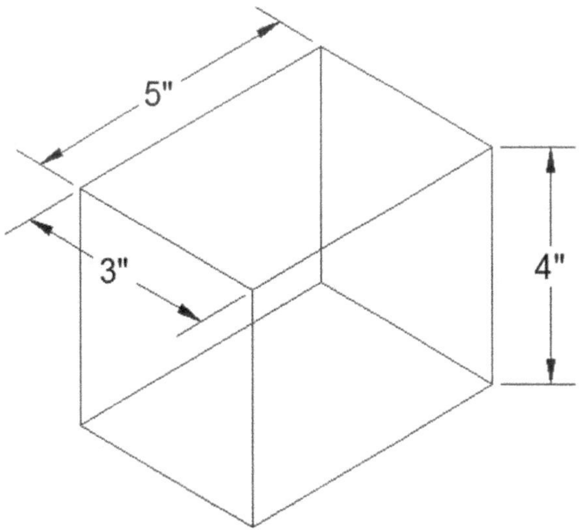

FIGURE 20.17 Dimensions in isometric drawing (oblique).

20.6 Level 2 drawing project (10 of 10): architectural floor plan

For Part 10 of the drawing project, you finish the design by adding the Paper Space layouts. You need to put together a lot of techniques you learned in Levels 1 and 2.

Step 1. In Paper Space, create the following seven layouts:

- Architectural layout
- Furniture layout
- Carpeting layout
- Electrical layout
- RCP_HVAC layout
- Landscaping layout
- Elevations

Step 2. Insert a title block into each Paper Space layout. Label the title block as appropriate for that layout. Make sure all the attribute fields are filled in and page numbers are incremented.
Step 3. Create an mview window for each layout (VP layer, color: white), centering and zooming each layout to the proper scale using the techniques you learned in Chapter 10. The scale will be either ¼″, ⅜″, or ⅛″. Freeze the VP layer or set it to not print when you are done with each of the layouts.
Step 4. In each layout, use advanced layer settings to freeze all layers that are not necessary for that particular view. You will likely need to create as many layer settings as there are Paper Space layouts, as each one shows something different.
Step 5. Add labels and any other title block information you may need.
Step 6. If your computer is connected to a printer or plotter, plot out all the layouts on D-sized paper. Make sure all the following are addressed:

- Paper Space layouts are correct.
- Proper scale is used in each mview window.
- Proper layers show in each mview window.
- The title block is correct for each layout.
- The CTB file is correct and there are varying lineweights.

All the layout plans are shown in Figs. 20.18–20.24.

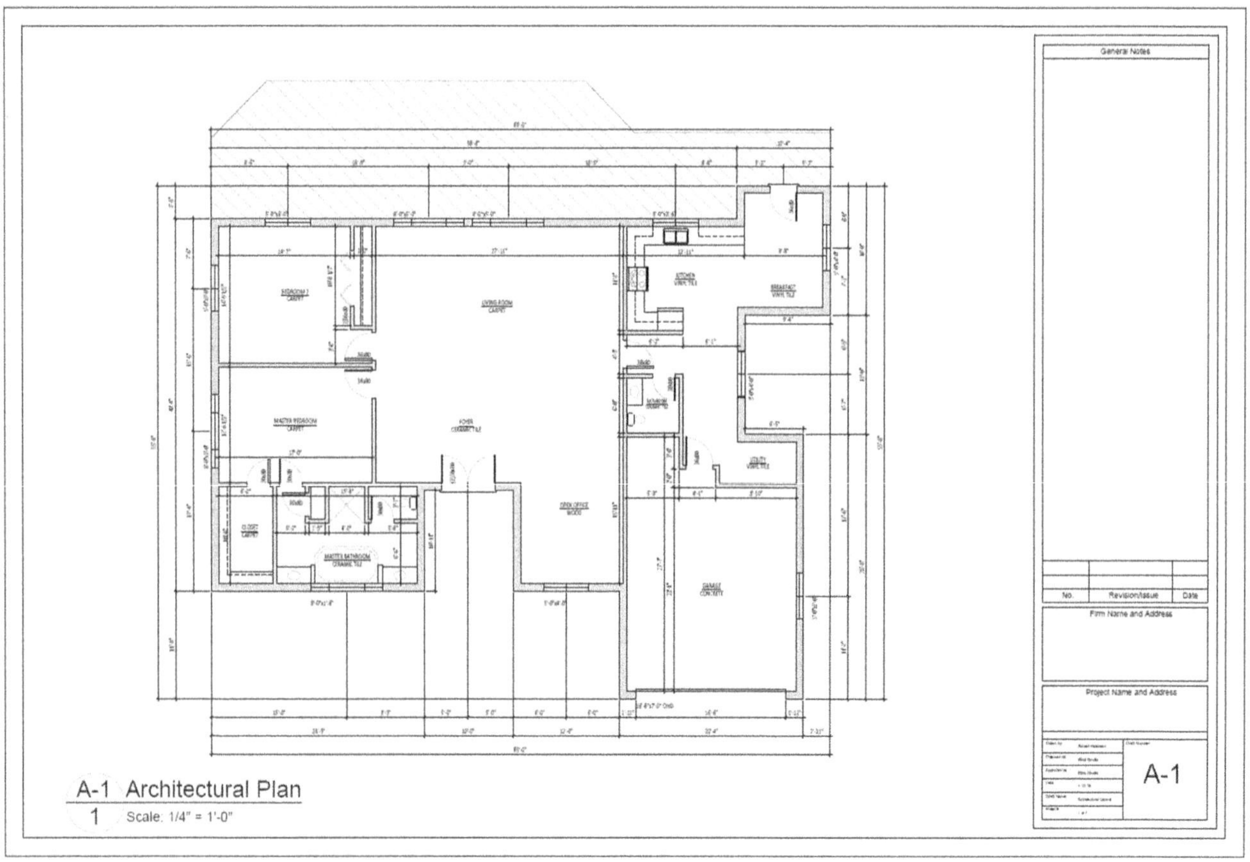

FIGURE 20.18 Architectural Plan.

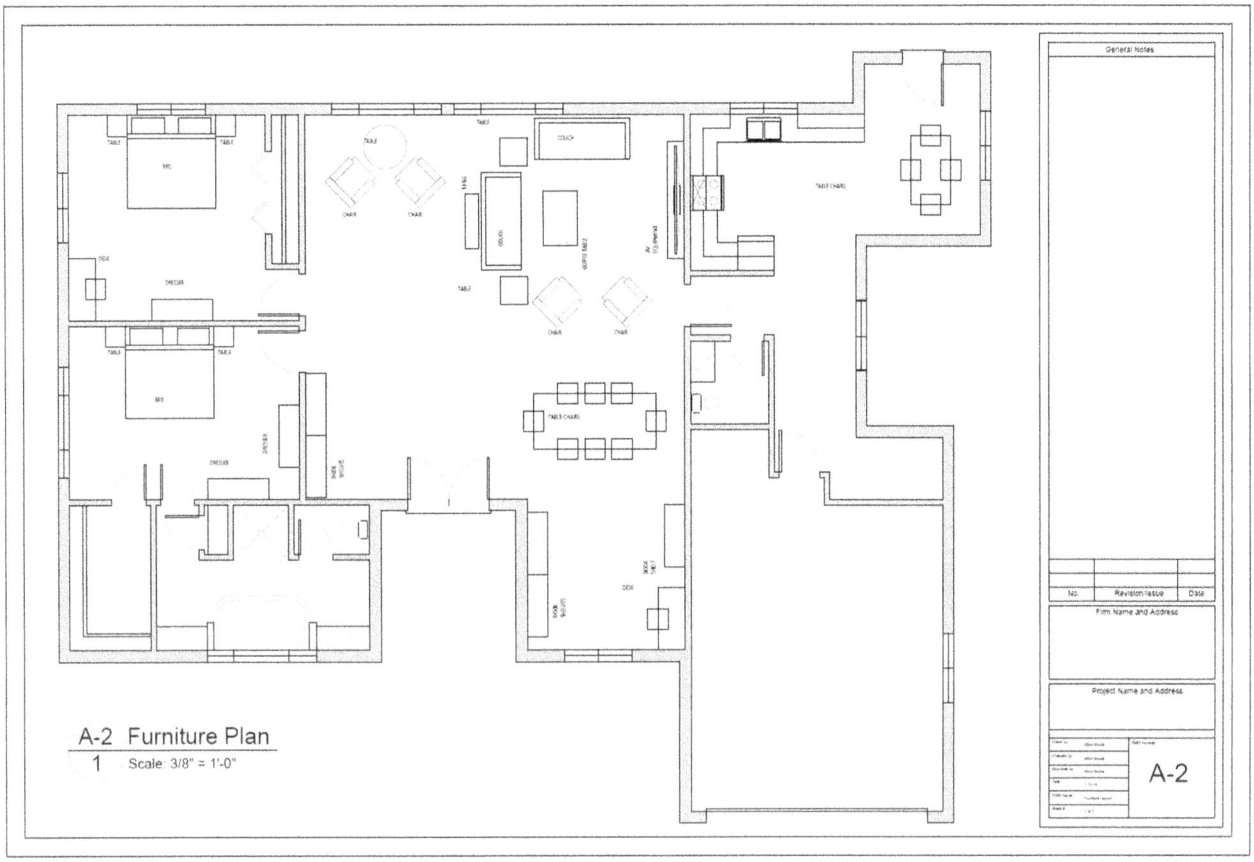

FIGURE 20.19 Furniture Plan.

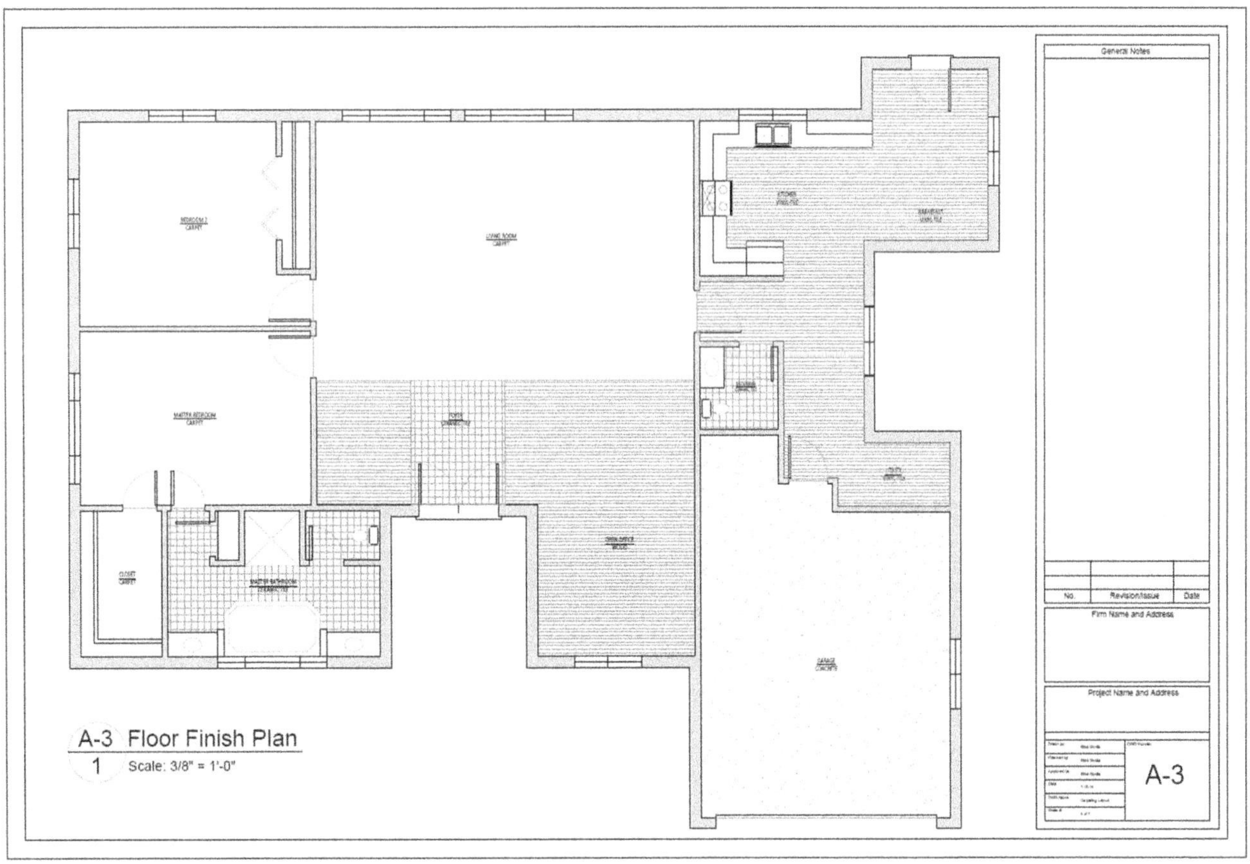

FIGURE 20.20 Floor finish Plan.

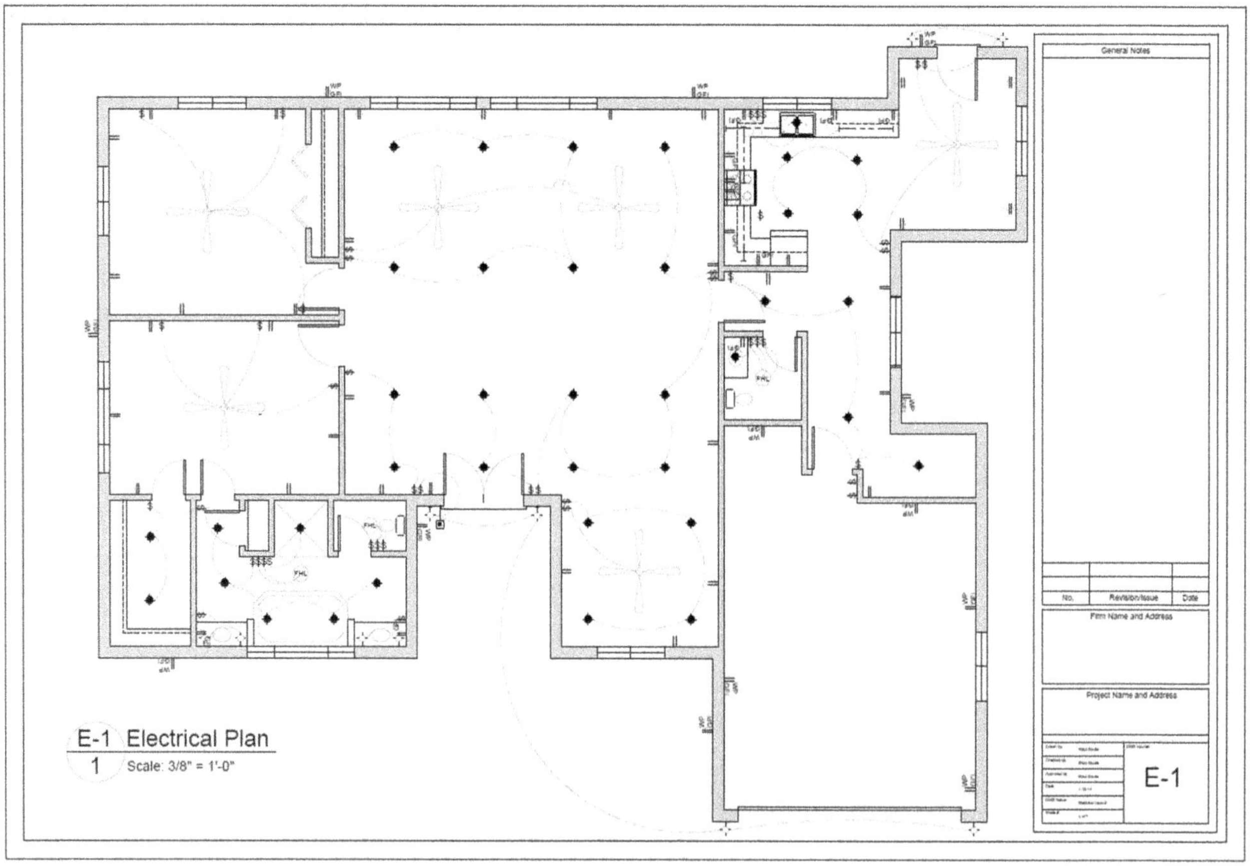

FIGURE 20.21 Electrical Plan.

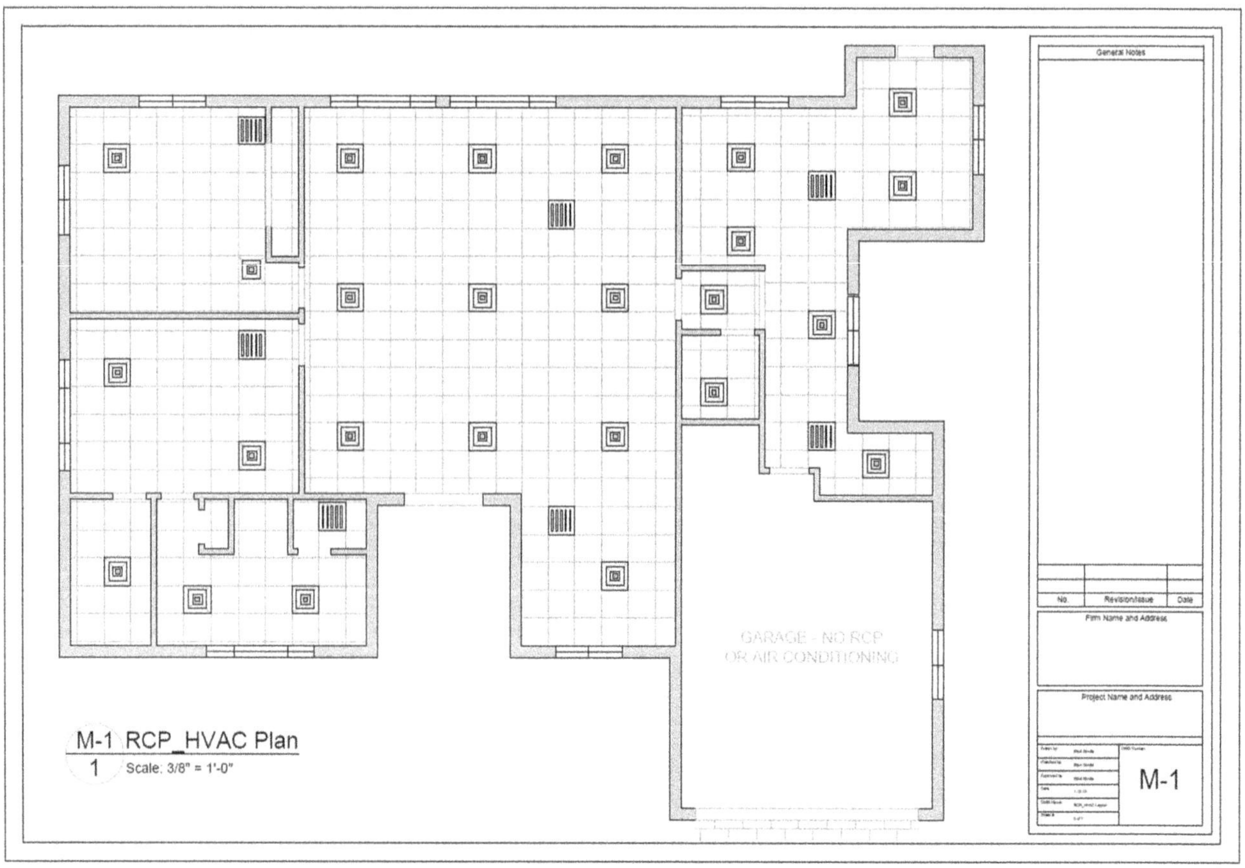

FIGURE 20.22 RCP_HVAC Plan.

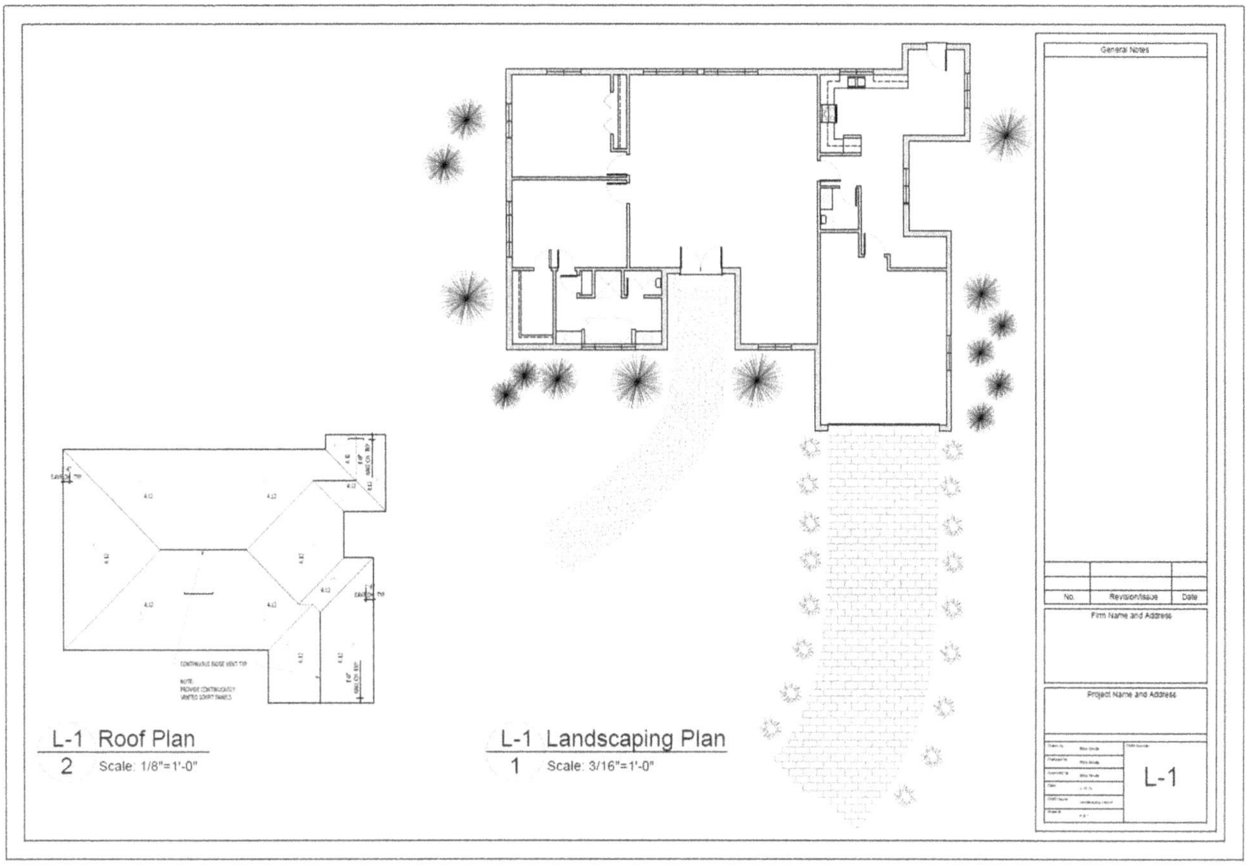

FIGURE 20.23 Landscaping Plan and Roof Plan.

FIGURE 20.24 Elevations.

Summary

You should understand and know how to use the following concepts at the end of this chapter:

- General types of graphical projection
- Isometric projection
  - When to use it
  - When not to use it
- Setting isometric projection
- Planes and changing planes (F5)
- Ellipses in isometric drawing
- Text and dimensions in isometric drawing

Review questions

Answer the following based on what you learned in this chapter:

1. List some instances when isometric drawing is appropriate.
2. List some instances where 3D would be better than isometric drawing.
3. Name the three planes that exist in isometric perspective.
4. In what dialog box is Isometric snap found?
5. What F-key switches the cursor from plane to plane?
6. What is almost always on when drawing in isometric?
7. What is the equivalent of a circle in isometric drawing?

Exercises

1. Draw the following architectural detail. All sizes are approximate and may be estimated. You will need to cycle through all three Isoplanes for the main structure and use ellipses for the fasteners. (Difficulty level: Easy; Time to completion: 10–15 minutes.)

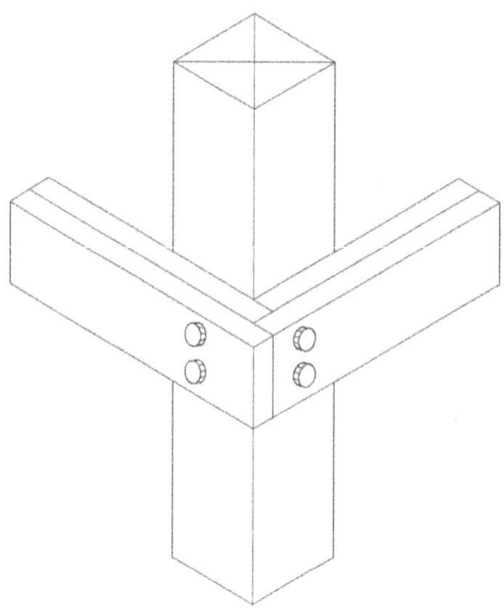

2. Draw the following architectural detail. All sizes are approximate and may be estimated. You will need to cycle through all three Isoplanes for the main structure, use ellipses for the fasteners and a spline for the cutaway. (Difficulty level: Easy; Time to completion: 10–15 minutes.)

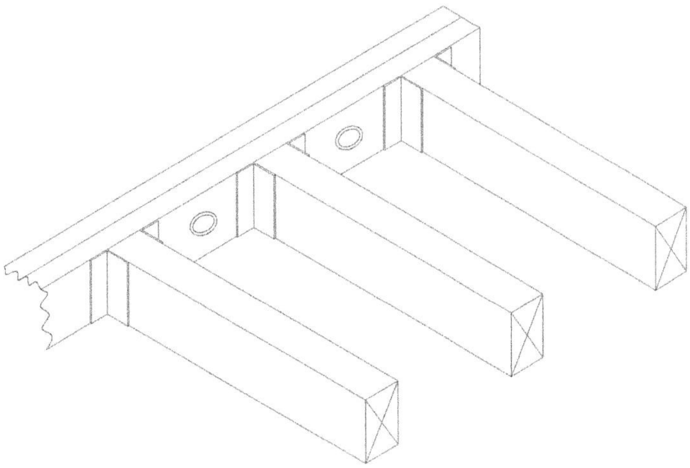

3. Draw the following detail of a concrete barrier. All sizes are approximate and may be estimated. You will need to cycle through all three Isoplanes for the main structure and use one ellipse. Be aware also that, for certain lines, you will break from the Isoplanes and draw freehand. As a last step, add solid hatching as shown. (Difficulty level: Easy; Time to completion: 10–15 minutes.)

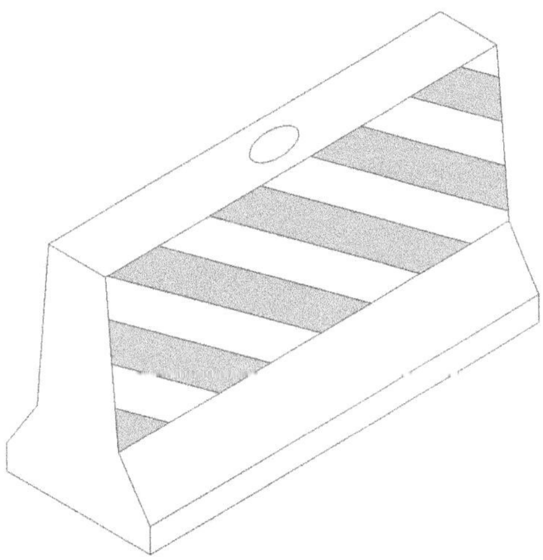

4. Draw the following detail of a nut and bolt. All sizes are approximate and may be estimated. There is more emphasis here on freehand drawing (including using arcs) after some basic Isoplanes linework is established. There is also more usage of ellipses and partial ellipses, which need to be trimmed as necessary. (Difficulty level: Easy; Time to completion: 10–15 minutes.)

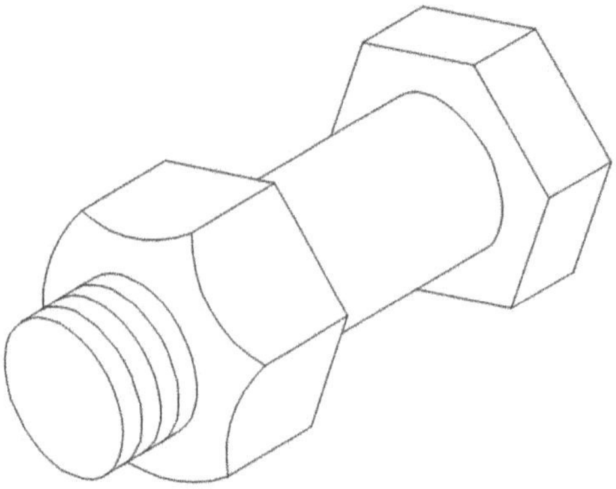

5. Draw the following valve/piping detail. All sizes are approximate and may be estimated. There is more usage of ellipses and partial ellipses, which need to be trimmed as necessary. (Difficulty level: Intermediate; Time to completion: 15–20 minutes.)

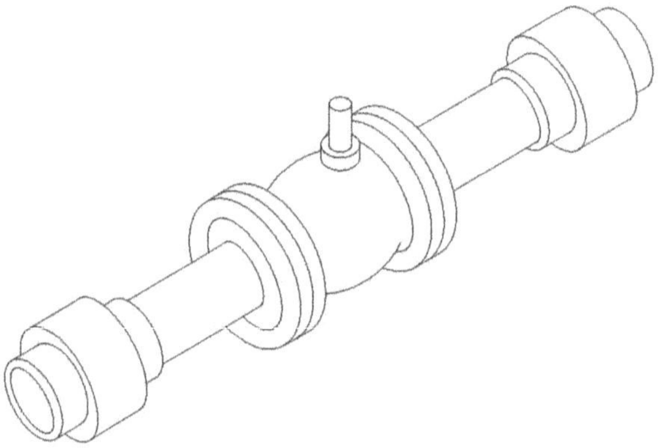

6. Draw the following piping detail. All sizes are approximate and may be estimated. This exercise is like the previous two in its usage of ellipses and partial ellipses, which need to be trimmed as necessary. (Difficulty level: Intermediate; Time to completion: 15–20 minutes.)

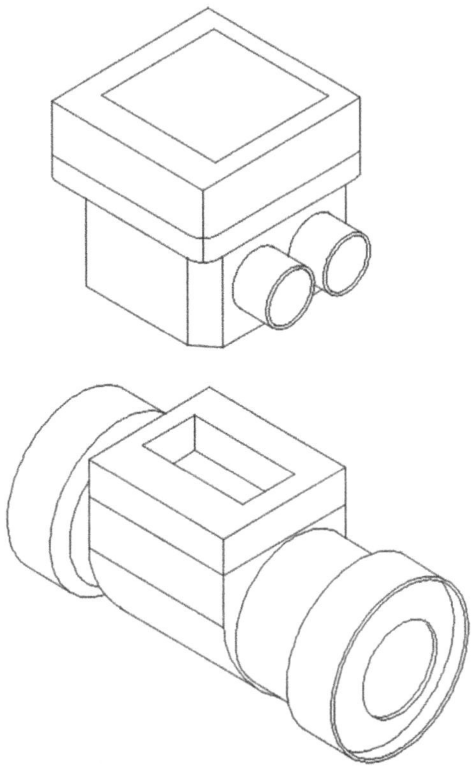

7. Draw the following computer desk. All sizes are approximate and may be estimated. You will need to cycle through all three Isoplanes for the main structure and use ellipses for a few elements. Add text and hatching as shown. (Difficulty level: Moderate; Time to completion: 20–25 minutes.)

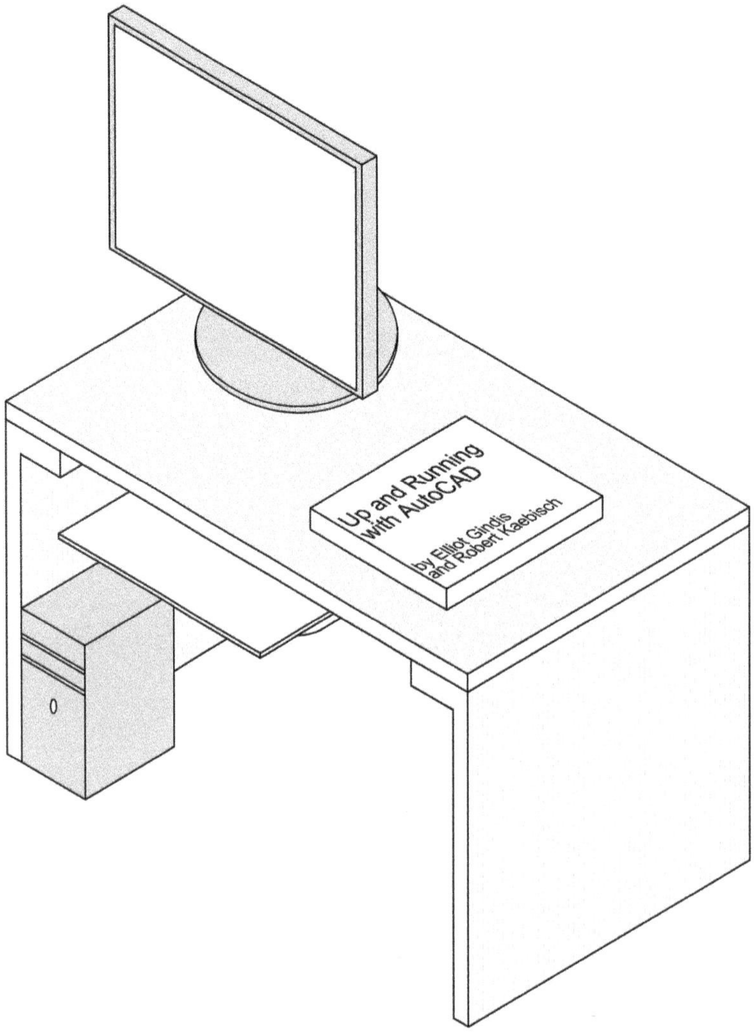

8. Draw the following architectural room detail. This exercise features not only isometric but also a perspective view, where the design converges to a "vanishing point" toward the back. Here, there is little Isoplanes work and quite a bit of improvising and freehand drawing, as isometric does not allow for perspective views. These types of layouts are not very common but still can be encountered in design work. They offer a very dynamic and interesting way to present an idea. A certain amount of artistic license is allowed, so do improvise. Note the various nuances in the design, such as slight rotation of the steel reinforcing bars in the side walls. Add in hatch and text as shown for the last step. (Difficulty level: Advanced; Time to completion: 60–90 minutes.)

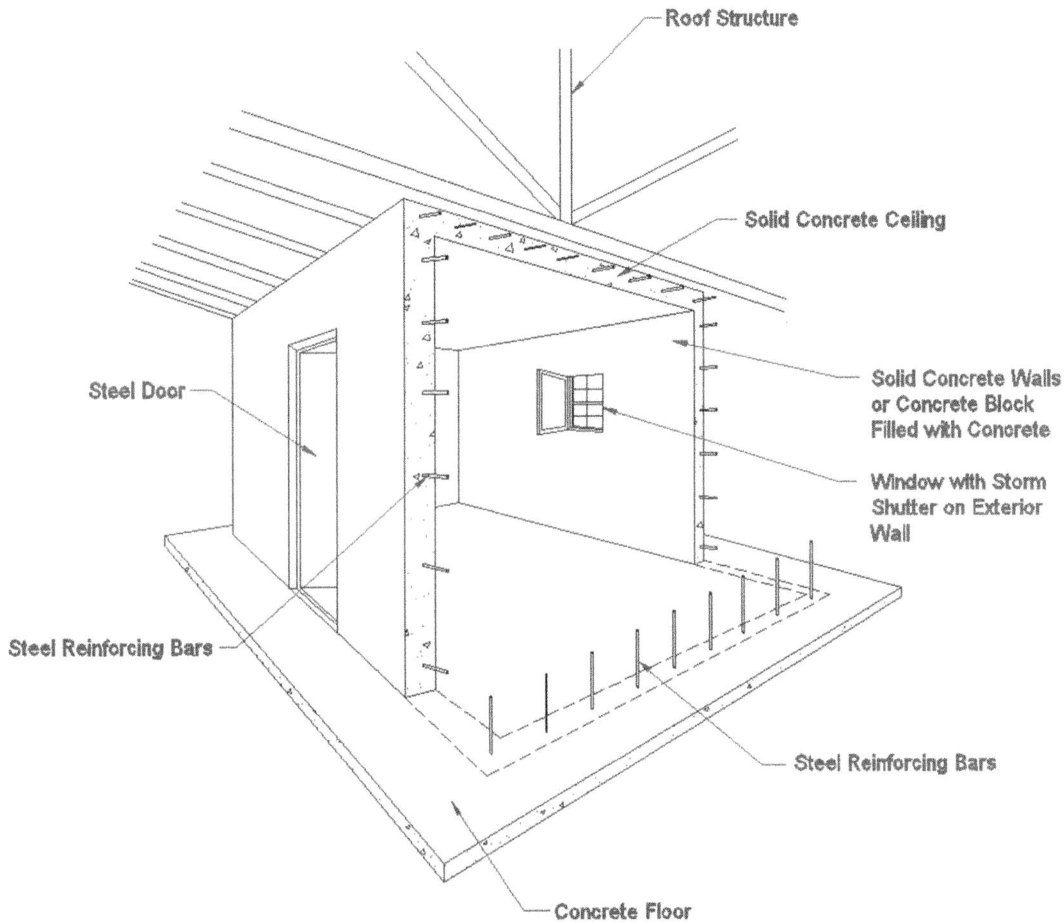

Level 2

Answers to review questions

Chapter 11: Review questions

1. What are the two major properties a pline has and a line does not?
   Property 1: A pline's segments are joined together. Property 2: A pline can have thickness.
2. What is the key command to add thickness to a pline?
   Pedit.
3. What is the key command to create a pline out of regular lines?
   Pedit.
4. What option gives the pline curvature? What option undoes the curvature?
   Spline option. Decurve option.
5. What two major things happen when you explode a pline?
   It loses its thickness and is split into segments.
6. What other options for pline were mentioned?
   Making a pline with variable thickness, adding an arc, and various other vertex options.
7. What is the bpoly command?
   Bpoly creates a pline border on top of a group of ordinary lines forming a closed area. You need to select a point in the enclosed area to accomplish this.
8. Define an xline and specify a unique property it has.
   An xline is a continuous line that has no beginning or end.
9. How is a ray the same and how is it different from an xline?
   A ray has no end, just like an xline, but it does have a beginning, unlike an xline.
10. Describe a spline and list its advantages.
    A spline is a NURBS curve and is used to create organic curvature, something that cannot be done any other way.
11. List some uses for a spline in a drawing situation.
    Gradient lines, cables, electrical wires, pathways, landscaping, for example.
12. What is an mline? List some uses.
    An mline is a multiline. It can be used for everything from architectural walls to highway design.
13. What are some of the items likely to be changed using mlstyle?
    Number of lines making up the mline, distances between those lines, linetypes making up the lines, colors of the lines and fills, end caps, and other details.
14. What are some of the uses of mledit?
    Mainly to edit intersections that are made using mlines; also to create vertexes and make cuts in mlines.
15. Describe the sketch command. What is a record increment? What size is best?
    The sketch command allows you to draw freehand. A record increment is the smallest size piece of line that makes up the sketch pathway; .01 is an optimal size.
16. Describe the basic function of a Geometric Center OSNAP point
    The Geometric Center OSNAP allows you to snap from/to the center (centroid) of complex closed shapes.

Chapter 12: Review questions

1. Describe what is meant by layer management. Why is it needed?
   Layer management describes a set of tools and concepts that, taken together, manage a large amount of layers, so they can be grouped together under some common theme and controlled as one unit (frozen, locked, etc.).
2. Describe the basic premise of a script.
   A script automates a series of input commands and saves the process under a name, so it can be recalled and put to use automating a repetitive task.
3. What can a script file do for layer management?
   A script file can freeze and thaw a large amount of layers by simply executing the script. This was the original way to do layer management.
4. Describe what Layer State Manager does.
   It saves layer settings under a descriptive name, so this setting can be recalled to easily freeze and thaw (among other actions) large amounts of layers all at once.
5. What are the fundamental steps in using Layer State Manager?
   Once the layers are set up as desired, the setting is saved in the LSM.
6. What is layer filtering? What are the two types?
   Layer filtering sorts layers based on common traits (such as color or names), so they can all be worked with at once. Property filter and Group filter.

Chapter 13: Review questions

1. Describe the important features found under the Lines tab.
   Nothing critical is found here, but the colors of the dimension and extension lines are changed.
2. Describe the important features found under the Symbols and Arrows tab.
   Here, the arrowheads are changed to an architectural tick.
3. Describe the important features found under the Text tab.
   Here, the text style and color are changed, with a box added around the text.
4. Describe the important features found under the Fit tab.
   Here, the most important feature is the setting of the overall scale.
5. Describe the important features found under the Primary Units tab.
   Here, everything under Linear dimensions, Zero suppression, and Angular dimensions is discussed.
6. Describe the important features found under the Alternate Units tab.
   Here, the approach to setting alternate units is discussed, including Zero suppression and placement.
7. Describe the important features found under the Tolerances tab.
   The types of tolerances are covered.
8. Describe the purpose of geometric constraints.
   To constrain the positioning of drawn geometry.
9. What seven geometric constraints are discussed?
   Perpendicular, parallel, horizontal, vertical, concentric, tangent, and coincident.
10. Describe the purpose of dimensional constraints.
    To constrain the sizing of drawn dimensions and the geometry they represent.
11. What six dimensional constraints are discussed?
    Horizontal, vertical, aligned, radius, diameter, and angle.

Chapter 14: Review questions

1. What is important under the Files tab?
   It is recommended that nothing be changed under this tab.
2. What is important under the Display tab?
   Here, you can turn off some unnecessary screen tools like scroll bars; change the background colors of the screen, command line, and other features; clean up the Paper Space layouts; and change the crosshair size.
3. What is important under the Open and Save tab?
   Here, you can save a drawing down to an older release of AutoCAD, delete Automatic save and backup files, and set security options.
4. What is important under the Plot and Publish tab?
   Here, you can set and modify plot stamp settings.

5. What is important under the System tab?
   It is recommended that nothing be changed under this tab.
6. What is important under the User Preferences tab?
   Here, you can set right-click customization.
7. What is important under the Drafting tab?
   Here, you can modify the look of the AutoSnap and Aperture boxes.
8. What is important under the Selection tab?
   Here, you can modify the pickbox and grip size as well as set the visual effects settings.
9. What is important under the Profiles tab?
   Here, you can save your settings and preferences under a name for future recall.
10. What is the pgp file? How do you access it?
    The pgp file holds AutoCAD shortcuts. It can be accessed via Tools→Customize→Edit Program Parameters (acad.pgp) in the drop-down menus.
11. Describe the overall purpose of the CUI.
    The Customize User Interface is used to modify the AutoCAD environment with changes to toolbars, palettes, shortcuts, macros, and much more.
12. Describe the purpose of the Design Center.
    The Design Center is useful for bringing entities such as layers and blocks into a drawing from another location as well as downloading symbol library blocks from the Autodesk Seek website.
13. Describe the important Express Tools that are covered.
    Some of the more useful Express Tools include Copy Nested Objects, Explode Attributes to Text, Convert Text to Mtext, Arc-Aligned Text, Enclose Text with Object, Delete Duplicate Objects, Flatten Objects, Break Line Symbol, Super Hatch, Convert PLT to DWG, and Layer Express Tools such as layfrz, layiso, and laywalk.

Chapter 15: Review questions

1. What are the purposes of the audit and recover commands? What is the difference?
   Both commands are used to check up on the integrity of an AutoCAD drawing in cases of errors or a drawing not opening. Audit is the lower-level check for a drawing that is already opened but not running right. Recover is a higher-level check and may be used to try to open a corrupted drawing.
2. For join to work, lines have to be what?
   They have to be coplanar.
3. The Defpoints layer appears when you create what?
   Dimensions.
4. Does the divide command actually cut an object into pieces?
   No, it only inserts points to create sections, but the object is still intact.
5. What is the idea behind eTransmit?
   The eTransmit command allows you to send a drawing via email along with all associated fonts, linetypes, xrefs, and so on.
6. To what can you attach a hyperlink? What can you call up using a hyperlink? How do you activate a hyperlink?
   To an object or text. Either a website or a document. Hold down Ctrl and click.
7. What is the idea behind overkill?
   Overkill allows you to remove multiple stacked lines.
8. Can you make a revcloud out of a rectangle? A circle?
   Yes, via the Object option.
9. What three new selection methods are discussed?
   Window polygon, Crossing polygon, and Fence.
10. Do you use a Crossing or a Window with the stretch command?
    A Crossing.
11. Name a system variable mentioned as an example.
    Mirrtext.
12. Explain the importance of the wipeout command. What does it add to the drawing?
    The wipeout command hides objects by covering them up. It can be used to add depth to a drawing.

Chapter 16: Review questions

1. What is the correct way to bring Word text into AutoCAD?
   Paste copied text from Word into mtext.
2. What is the procedure for bringing AutoCAD drawings into Word?
   For most applications, a Copy/Paste suffices. PrtScn can be used as well.
3. What is the correct way to bring Excel data into AutoCAD?
   Copy/Paste into an OLE.
4. What is the procedure for bringing AutoCAD drawings into PowerPoint?
   Copy/Paste or PrtScn works.
5. What is the procedure for importing JPGs and generating PDFs?
   The JPGs can be Copied/Pasted in, then adjusted. To generate a PDF, select it as the output printer or via the Ribbons output tab.

Chapter 17: Review questions

1. What are several benefits of xref?
   Reduce file size by attaching a core drawing to multiple files, critical file security, and automated design change updates to core drawings.
2. How is an xref different from a block?
   An xref is attached to a drawing and is never truly a part of it. A block is inserted into a drawing and becomes part of it.
3. How do you attach an xref?
   Xref command, browse for the xref file, specify insertion point, and then OK.
4. How do you detach, reload, and bind an xref?
   By opening up a menu via a right-click on the drawing name in the xref palette.
5. What do layers in an xrefed drawing look like?
   They are preceded by the name of the xref itself.
6. What do layers look like once you bind the xref?
   They contain a 0 in the name.
7. What are some nesting options for xref?
   One xref attached to another then to another (etc.) and then to the main drawing or all attached to one drawing simultaneously.

Chapter 18: Review questions

1. What is an attribute and how do you create one?
   An attribute is information inside of a block. To create one, you use the attdef command and fill in the required fields.
2. How do you edit attributes?
   Double-click on the attribute.
3. How do you extract data from attributes? Into what form?
   Data extraction is via the eattext command. It can be extracted into an Excel spreadsheet or an AutoCAD table.
4. How do you create an invisible attribute? Why can this be important?
   By checking off the invisible button during the initial creation of the attribute. This is important so that the information is not visible in the drawing but can still be used.

Chapter 19: Review questions

1. What is a CTB file?
   This is the file that stores information on line thickness and other settings.
2. Describe the process of setting up a CTB file.
   You do this through a multipage Add Plot Style dialog box.
3. What is screening?
   Screening fades the intensity of any color.
4. What is the effect of Lineweight Settings?
   All lines that are made thicker via this option will appear as such on the screen.

Chapter 20: Review questions

1. List some instances when isometric drawing is appropriate.
   If you do not yet have 3D skills or need only a quick and simple 3D-looking model. Also, if computing resources are limited or you do not want to spend time combining 2D and 3D.
2. List some instances where 3D would be better than isometric drawing.
   In any instances where a true 3D model is needed to convey the design intent, including for purposes of shading and rendering. Also when more accuracy is needed.
3. Name the three planes that exist in isometric perspective.
   Top, left, and right.
4. In what dialog box is Isometric snap found?
   ddosnap, Snap, and Grid tab.
5. What F key switches the cursor from plane to plane?
   F5.
6. What is almost always on when drawing in isometric?
   Ortho.
7. What is the equivalent of a circle in isometric drawing?
   Ellipse.

Level 3

Chapters 21–30

Welcome to the fascinating world of 3D computer-aided design. Ideally, you have completely mastered everything in Levels 1 and 2 and are ready to learn a new set of skills. Although 3D is a different world in AutoCAD, it builds on what you learned in 2D; really what we are doing is just adding another dimension (literally).

The world of 3D has come a long way in AutoCAD. From humble beginnings in the 1980s (which left users joking that it is not really 3D but 2.5D), through steady improvements in the 1990s, to a major retooling with release 2007, AutoCAD's 3D capabilities have grown exponentially, and it is now quite capable 3D software.

Note however that AutoCAD still does not do engineering modeling and analysis and cannot compete with CATIA, SolidWorks, SolidEdge, Pro/Engineer (Creo), NX, Inventor, and others, nor was it ever designed to. At its heart AutoCAD is still *drafting* software first and foremost and is *not* meant for complex modeling and simulation. Autodesk has other products, such as Inventor and Revit, which venture into the world of parametric modeling, while AutoCAD is firmly in the world of drafting and visualization. If you need to test a design for structural failure or send the geometry to a CNC mill for manufacturing, AutoCAD is not the software for this.

This, of course, is just fine for thousands of architects and engineers worldwide who use AutoCAD's 3D capabilities for visualization and create superb 3D designs. Some typical uses include 3D architectural models of residential and commercial interiors and exteriors, furniture design, small part design, electrical riser equipment diagrams, and other applications that require a 3D view of the object(s) to visualize, promote, display, and sell the design or concept. Often, these models are exported to specialized rendering software, such as 3ds Max, Form-Z, or Rhino, to obtain photorealistic images. Creating these underlying models is what we focus on here.

Our study of 3D is loosely broken into three major sections. The first part, which starts with Chapter 21, is an introduction to the basics. We cover all the fundamental tools needed to create what can be referred to as *flat design*, meaning objects without major curvature. This broadly refers to architectural models, such as simple 3D floor plans that feature walls, door, windows, and furniture. For the most part, these are elements that feature essentially straight edges. We learn basic tools like Extrusion, 3D Rotate, 3D Mirror, and Booleans.

The second part is an extension of this knowledge into what can be thought of as *curved design*, meaning objects with curvature. This broadly refers to mechanical models, such as gears, shafts, bearings, springs, and shape-formed piping, to name a few examples. There is a great deal of overlap between architecture and mechanical design in terms of what is required of AutoCAD, of course, and you really need to learn everything regardless of what field you are in. Here, we cover tools such as Revolve, Loft, Shell, and Taper.

The third part is all about advanced tools, such as Sections, Vports, Meshes, Dynamic Views, Animations, Lighting, and Rendering, which round out your study of 3D and add realism in the form of lighting and shading as well as some rendering and backgrounds to your design.

A word of caution: Be sure to have truly mastered 2D before proceeding. All the 3D chapters are written with the assumption that you are an experienced 2D user, and few explanations are forthcoming when we use 2D tools to begin 3D projects, especially past Chapter 21. Also, review isometric drawing (see Chapter 20). That seemingly unrelated topic is actually a precursor to 3D, as you will soon see, and you should go back and review it.

Chapter 21

3D Basics

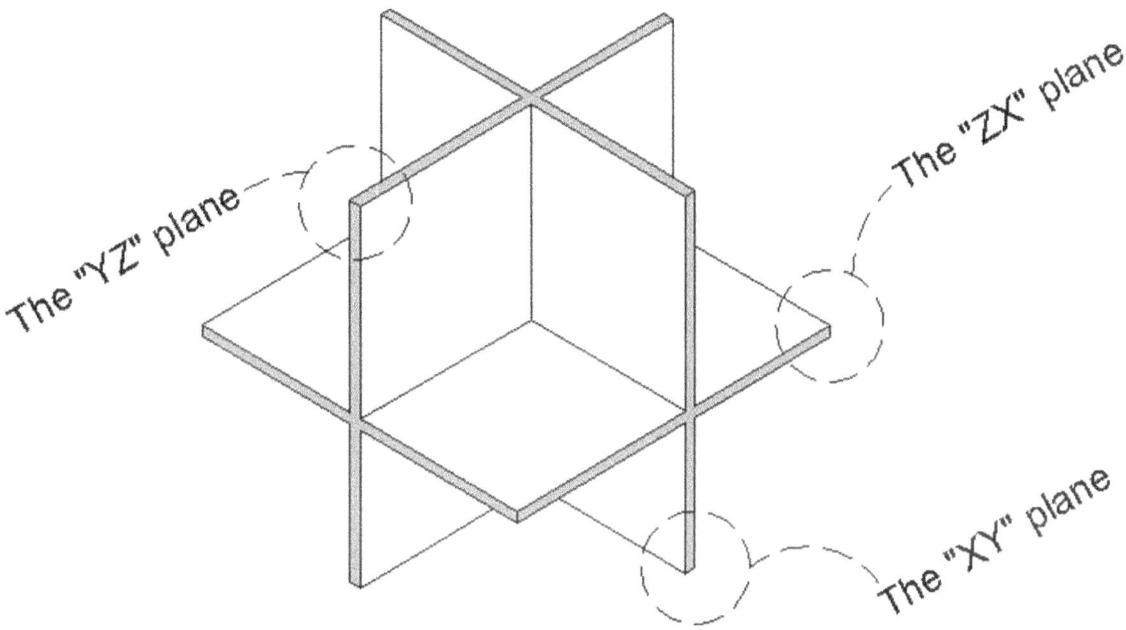

Learning Objectives

In this chapter, we introduce the basics of 3D theory and operating in 3D space. Specific topics include the following:

- Axes, planes, and faces
- 3D workspace - Ribbon, toolbars, and options
- Entering and exiting 3D
- Projecting into 3D
- 3D dynamic views
- Extrude
- Visual styles - Hide, Realistic Shade, and Conceptual Shade
- ViewCube and Navigation Bar

By the end of this chapter, you will understand basic 3D theory and be able to navigate in and out of 3D, navigate inside 3D, as well as extrude, hide, and shade your designs.

Estimated time for completion of this chapter: 1.5 hours.

Up and Running with AutoCAD 2022. https://doi.org/10.1016/B978-0-323-89923-9.00021-4

21.1 Axes, Planes, and Faces

So, what exactly does *drawing in 3D* mean? It is the ability to give depth to objects or to expand them into the "third dimension" from a flat plane. This concept should be intuitively obvious - after all, we live in a 3D world. Everything has not just a length and width but also a depth (or height).

It turns out that, as you were learning AutoCAD in Levels 1 and 2, you were in 3D all along. It is just that you did not see or use this vertical third dimension; therefore, everything appeared flat, similar to not seeing the height of a tall building if you are flying directly above it. All this changes as we discuss this hidden dimension and learn how to project into it. For this, we need to start at the very beginning and learn about axes, planes, and faces, because by definition, using the third dimension involves using a third, and previously ignored, Z axis.

Concept 1. There exists in the Cartesian coordinate system a total of three axes: X, Y, and Z. These axes intersect each other at the 0,0,0 point, as seen in Fig. 21.1 (left), and by definition can be positive or negative, as represented by solid and dashed lines in Fig. 21.1 (right).

Concept 2. A plane is defined as an intersection of two axes. Therefore, the X, Y, and Z axes can define three unique planes: the XY, YZ, and ZX planes, as seen in Fig. 21.2.

Concept 3. A total of six faces of an imaginary cube can then be formed using the three planes. This can be easily seen if we move the planes out and connect them edge to edge, as in Fig. 21.3. For our purposes right now, planes and faces are really the same thing, and we refer only to planes from here on (though the term *faces* does get some use in related concepts).

The preceding discussion on axes and planes is very important; make sure you completely understand everything thus far. We use both axes and planes shortly.

21.2 3D Workspaces, Ribbon, Toolbars, and 3D Options

Before we can start to work in 3D, we need to set up the appropriate tools, so as to not be distracted with trying to find them later on when we need them. Just as in 2D, you can work in 3D via the Ribbon, toolbars, and the occasional typing. Also, quite a few 3D palettes locate many tools in one spot. You often use a combination of all these tools depending on the situation.

AutoCAD simplifies your work in 3D by allowing you to switch to a 3D workspace. Here you find the Ribbon altered, with many of the needed 3D tools in one place. You have two choices, 3D Basics and 3D Modeling, as seen in Fig. 21.4. If you have older versions of AutoCAD installed on your computer, you may see other 3D choices but ignore them and choose 3D Modeling.

You will see your AutoCAD screen change to 3D mode. Note the difference in the Ribbon. It now has a full complement of 3D tools available under multiple tabs (Solid, Surface, Mesh, Render, etc.) while retaining some essential 2D tools. Also be sure to restore the menu bar at the top of the screen (see Chapter 1 if you do not recall how). Finally, we bring up some new 3D toolbars, as described next.

It is very useful to have 3D toolbars on your screen in addition to the Ribbon while you are learning 3D. Despite the overlapping redundancy between the Ribbon and toolbars, a few commands are still found only in the toolbars, and they

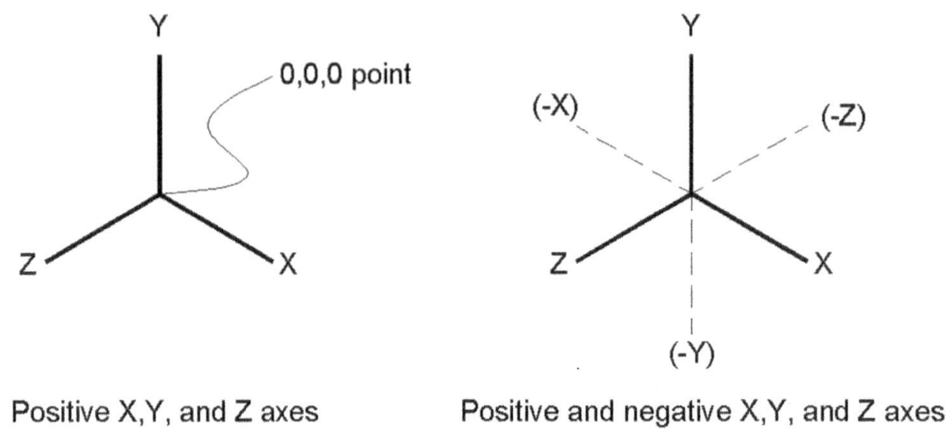

FIGURE 21.1 The X, Y, and Z axes.

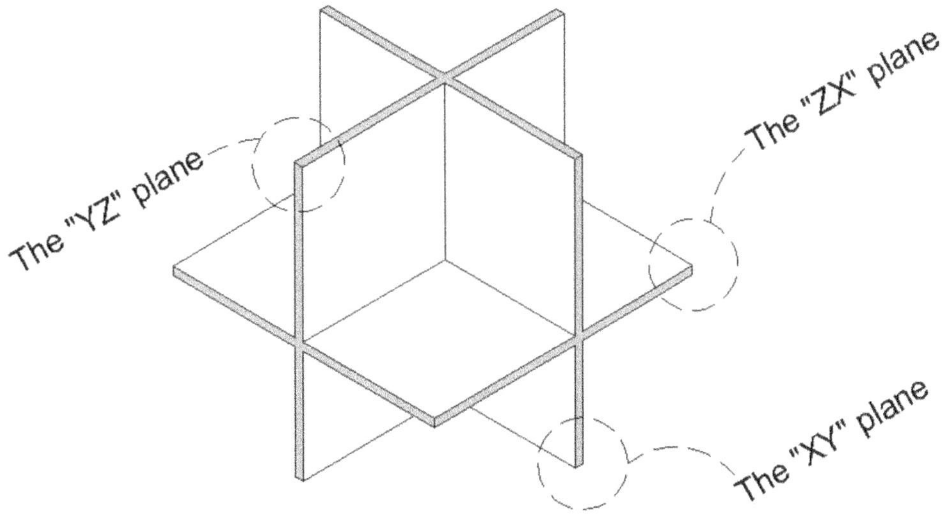

FIGURE 21.2 The XY, YZ, and ZX axes.

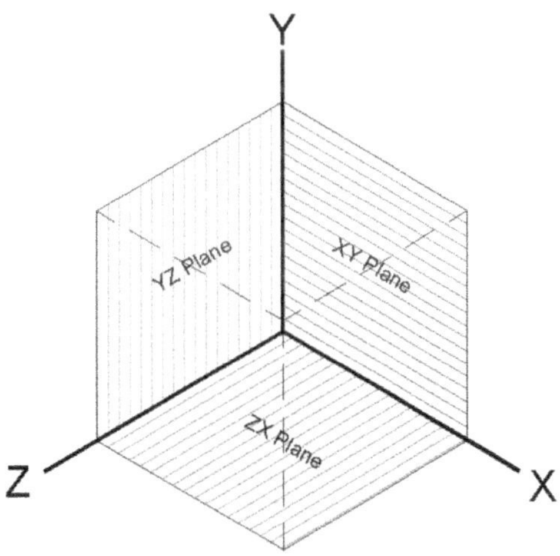

FIGURE 21.3 The XY, YZ, and ZX faces (planes) of a cube.

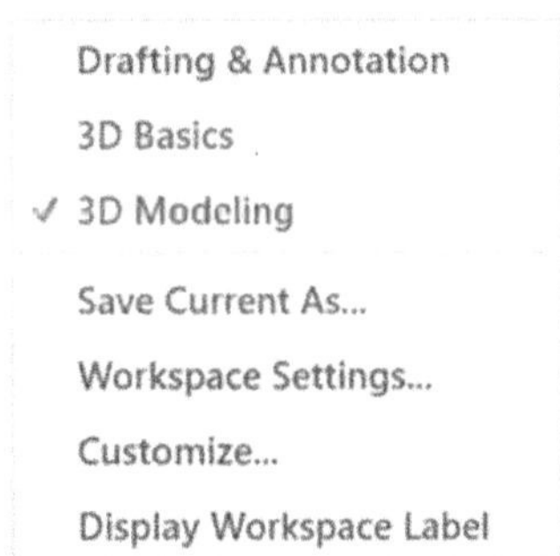

FIGURE 21.4 3D Basics and 3D Modeling workspaces.

are often easier to access, so we build a collection of them as we go. If you prefer the Ribbon to all else, you can start eliminating the toolbars after getting acquainted with them as they are introduced.

Using Tools→Toolbars→AutoCAD in the cascading menus, bring up the toolbars seen in Fig. 21.5. They are View, 3D Navigation, and Visual Styles. Arrange them around the screen area. Your screen should look close to, if not exactly like, Fig. 21.6.

It is worth noting that, although you have less and less typing to do in 3D, a few commands are still available only via this method. Also, do not forget, you still have the cascading menus with full 3D tools as an alternative to all of these, in

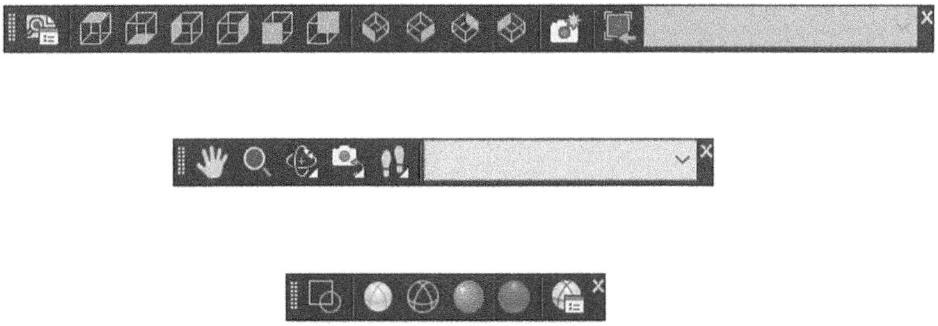

FIGURE 21.5 Essential 3D toolbars.

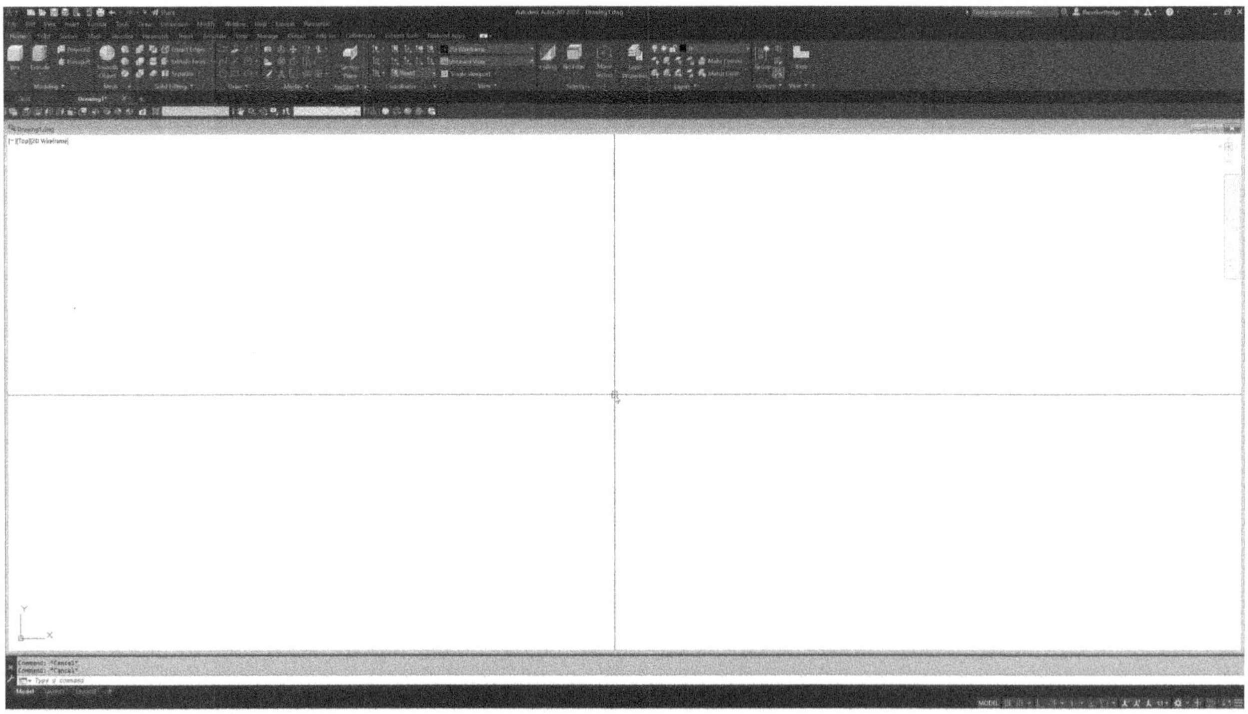

FIGURE 21.6 3D Modeling workspace.

many cases. This covers all four primary methods of interacting with AutoCAD in 3D. As you can see, except for more of an emphasis on graphical inputs (Ribbon and toolbars), the methods remain the same as in 2D.

A few words about 3D and the Options dialog box, first examined in Chapter 14. It is highly recommended that your crosshairs are size 100 (all the way across) in 3D, as this helps a lot with visualizing what you are doing and where in 3D space you are. As a reminder, you can change this under the Display tab. Just as important, you need your UCS icon. In 2D, you may have been tempted to turn it off (it was one of the tips early in Level 1), but here you need it to know where you are in 3D space, and you really cannot work without it. Finally, let us take a look at the 3D Modeling tab under the Options… command that we skipped over in Chapter 14. It is shown in Fig. 21.7.

For now, the only suggestion is to make sure the "Show Z axis in crosshairs" option at the top left is unchecked (it may already be). The 3D concepts are generally much clearer when only two axes are present in the crosshairs. We return to this Options dialog box later in our 3D studies to tweak a few more settings; mostly those at the bottom right (Walk and Fly, ViewCube, etc.). Click on OK and let us jump right into 3D in the next section.

21.3 Entering and Exiting 3D

The key to starting out learning 3D is to get into 3D mode and reveal the third axis. The gateway to entering 3D is any of the isometric views, although we use the SW Isometric View most often. The way to exit 3D is any of the flat views (front, back, left, right, etc.), although we use the Top View most often. Let us give this a try using the View toolbar, as shown in Fig. 21.8.

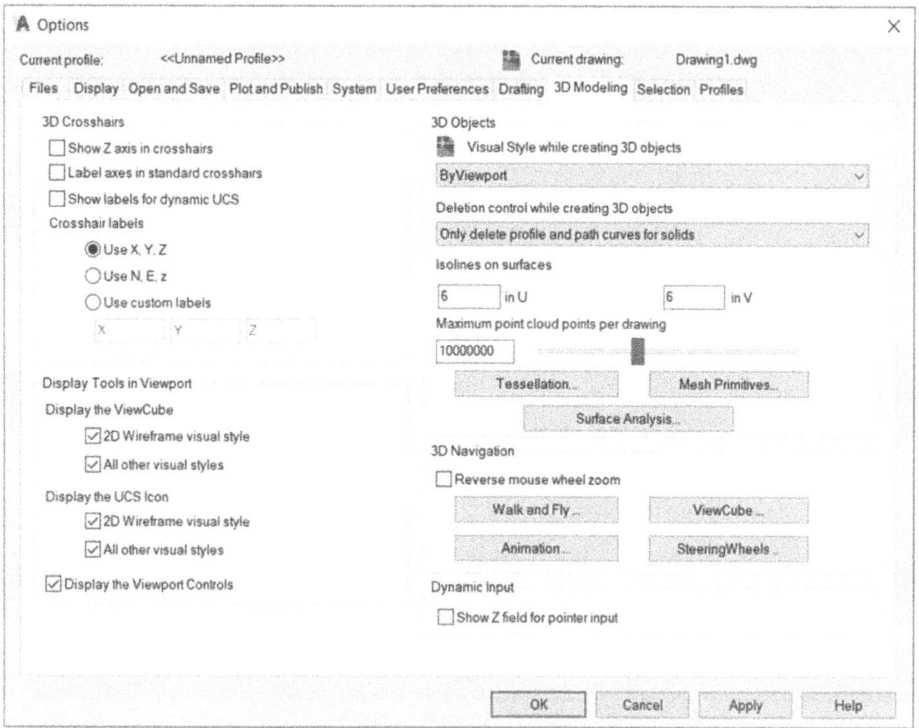

FIGURE 21.7 Options, 3D Modeling tab.

Press the SW Isometric View icon and the screen switches to what you see in Fig. 21.9. Examine what happened after going into SW Isometric. The Z axis, previously unseen, is revealed by rotation of the UCS icon and the associated planes. Note how this is reflected in the new position, or shape, of the crosshairs. Note that your UCS icon may be in the center of the screen; we discuss how to get it into the lower left corner in just a moment.

Go ahead and press the Top View icon and go back to 2D. Now that you understand the two essential views, go ahead and press all the other icons to see what view they give you. Icons 2 - 7 from the left are all 2D views, while icons 8 - 11 are all 3D. When done, return to the Top 2D view.

Now, try the exact same thing via the Ribbon under the Home tab, View, as seen in Fig. 21.10. You have the same basic symbols, so you should know how to proceed.

Finally, let us try the same thing via the cascading menu, as seen in Fig. 21.11. The path is View→3D Views→SW Isometric (or Top). Once again, the symbols are the same, so you should know how to proceed.

As mentioned just before, when you switch to the SW Isometric View, your UCS icon may stay in the middle of the screen. You generally want it tucked away in the lower left corner, as it is in Fig. 21.9. To do this, type in `ucsicon` and press Enter.

- AutoCAD says: `Enter an option [ON/OFF/All/Noorigin/ORigin/Selectable/Properties]<ON>:`

Type in `n` for `Noorigin`, press Enter, and the icon returns to the corner and does not interfere with viewing design work. In later chapters, we ask it to move out of the corner when we need the icon to be aligned with objects we are working on, but for now this is not necessary, and it just gets in the way.

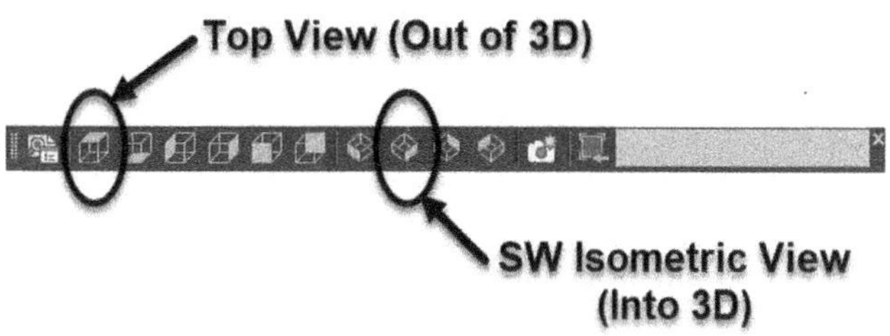

FIGURE 21.8 Top and SW Isometric Views.

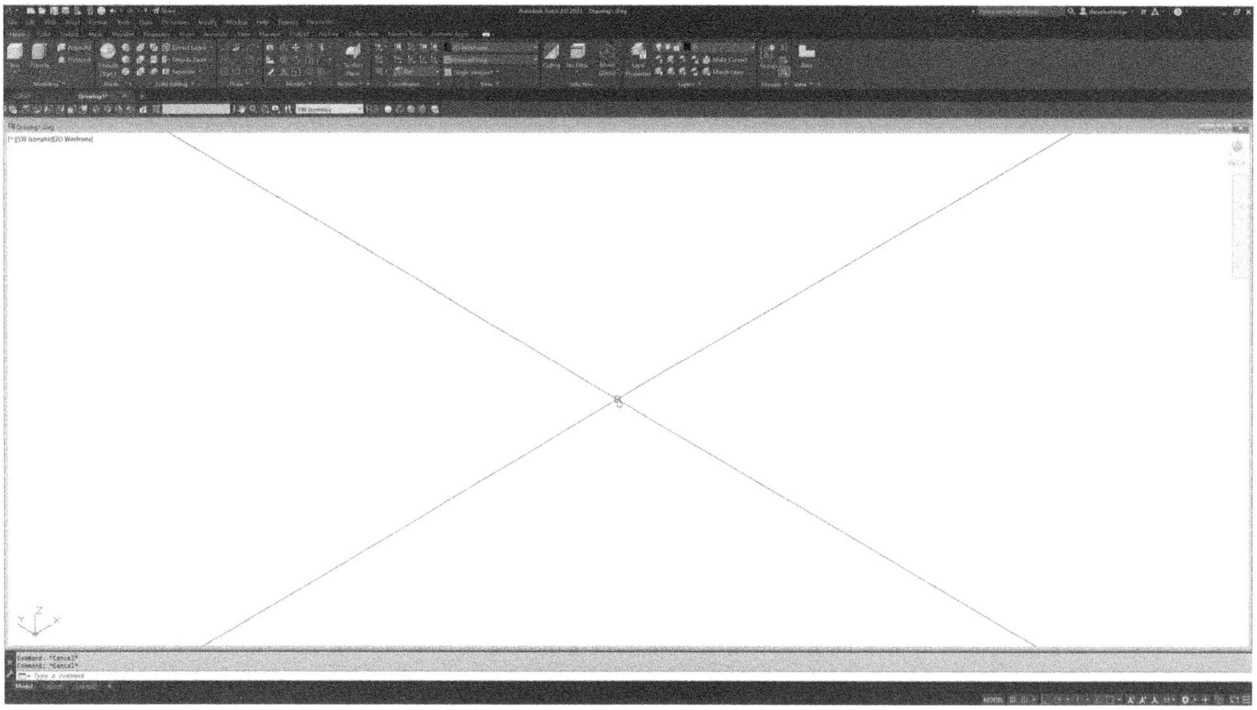

FIGURE 21.9 3D SW Isometric View.

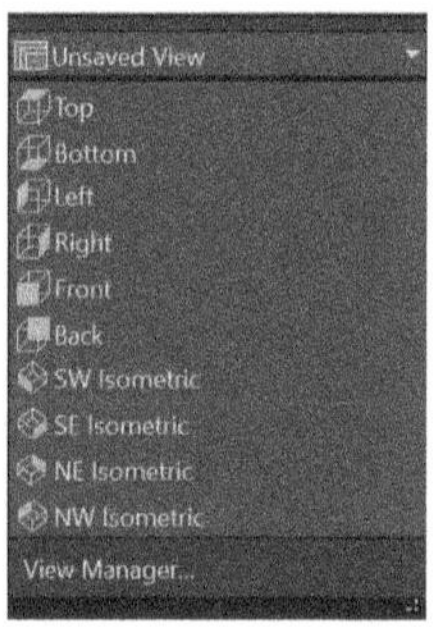

FIGURE 21.10 3D SW Isometric View (Ribbon).

You now know how to get into 3D (SW View) and back to 2D (Top View). So, in 3D, go ahead and draw a rectangle that is 10″×6″. You can do this using the rectangle command or individual lines; at this point it does not matter. The result is shown in Fig. 21.12.

21.4 Projecting into 3D

Here, we come to a roadblock of sorts. Say you would like to draw four sides and a top to create a full 3D box. How do you do this? That is the subject of this section. Six releases ago, in AutoCAD 2015, a significant change was made in how you can project linework into 3D from the World UCS (the XY) plane. We call this Method 1 and describe the older method afterward as Method 2. It is still needed for creating new (closed-shape) geometry on different planes but does not work for lines.

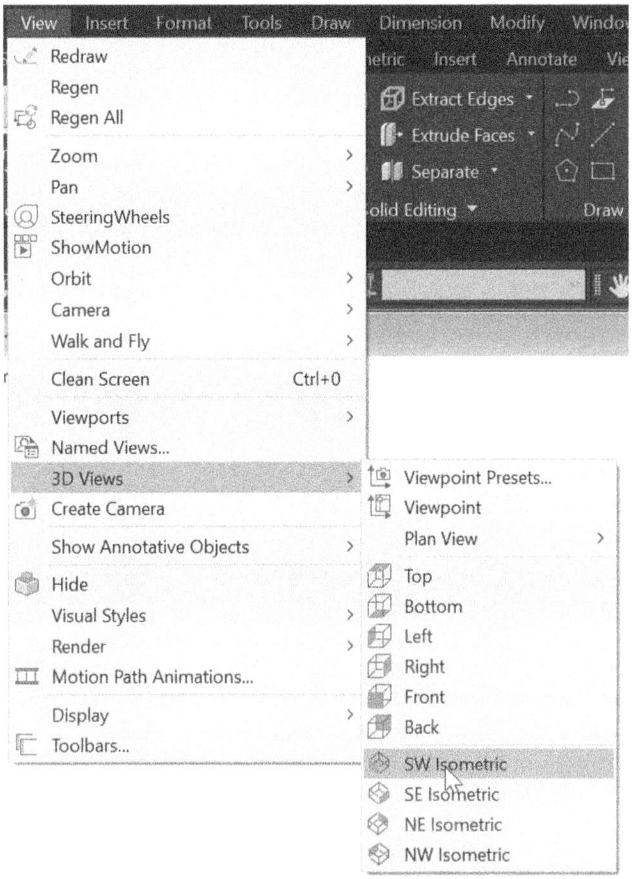

FIGURE 21.11 3D SW Isometric View (cascading menus).

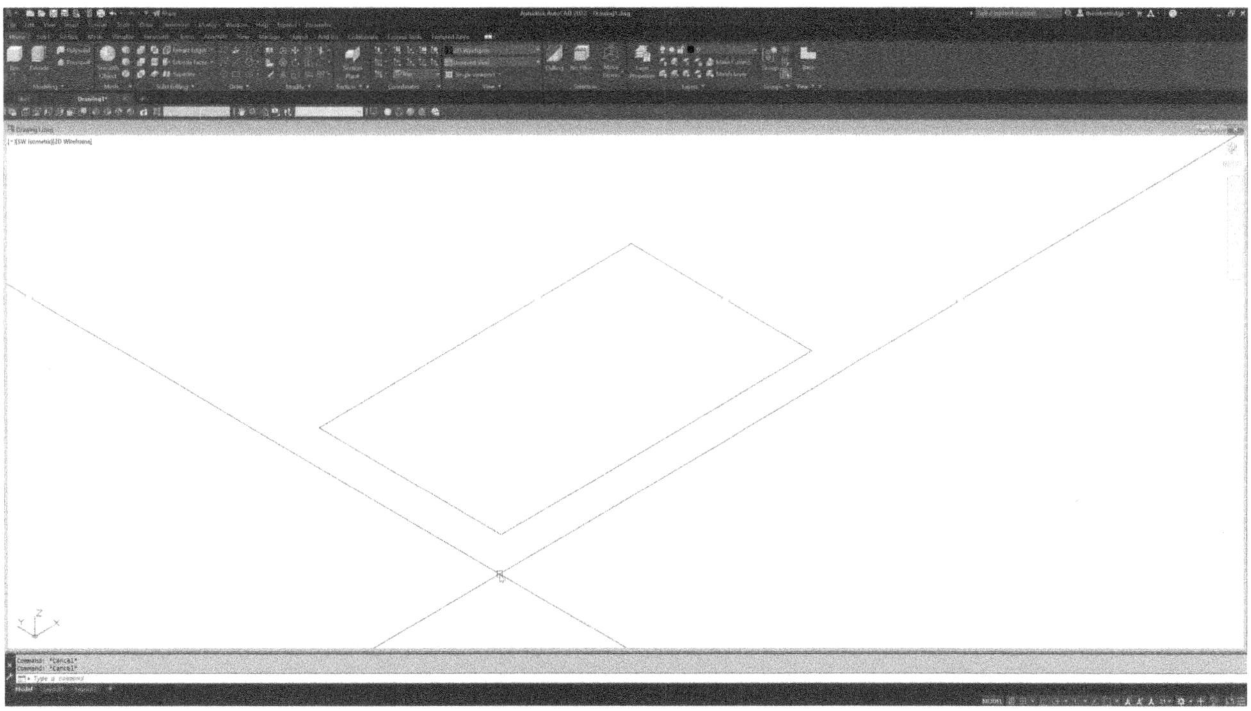

FIGURE 21.12 Rectangle in 3D.

Method 1 (New)

With this new method, you can now project lines, and only lines, into 3D from the XY plane of the SW Isometric view (the one you are in now) without rotating planes, the significance of which is described shortly. To create your 3D box, do the following.

Step 1. Begin the line command, with Ortho on, from any endpoint corner of your new rectangle. The line will appear and go straight up as seen in Fig. 21.13. You can make the line any random length as was done in this example.

Step 2. Either copy the line to each of the other three corners of the rectangle, attaching it precisely or draw three new ones of the same length.

Step 3. Copy the rectangle from the bottom to the top, or draw a new one from precise endpoints of the lines. The result is seen in Fig. 21.14.

Note that you could not have done this with AutoCAD 2014 or any other older version. The line(s) would not have appeared and gone "up" into 3D without you first switching planes. You still need to do this to create new geometry in another plane, so we discuss the classic method next.

Method 2 (Classic)

The following discussion applies to the previous steps if you are working on any pre-AutoCAD 2015 version. Even if you are not, study this discussion, as it will be needed for other 3D construction scenarios. Before we start, go ahead and erase all but the original bottom rectangle of your 3D box.

Your input device (universally a mouse these days) is by its nature a 2D device. It works by going forward-back and left-right on your desk or mouse pad but not straight up. This is obviously a bit of a problem for projecting and drawing into the third dimension. Industry software designers and researchers noticed this a long time ago and came up with a simple and elegant solution that was integrated into numerous software packages over the years, including AutoCAD.

When you need to go into 3D, instead of raising your mouse into the air, you simply switch from the "flat" plane you are on to a vertical plane. The effect is immediate; you can now draw "up" relative to you, the observer. To go back to flat, you switch the planes right back. This is shown graphically in Fig. 21.15.

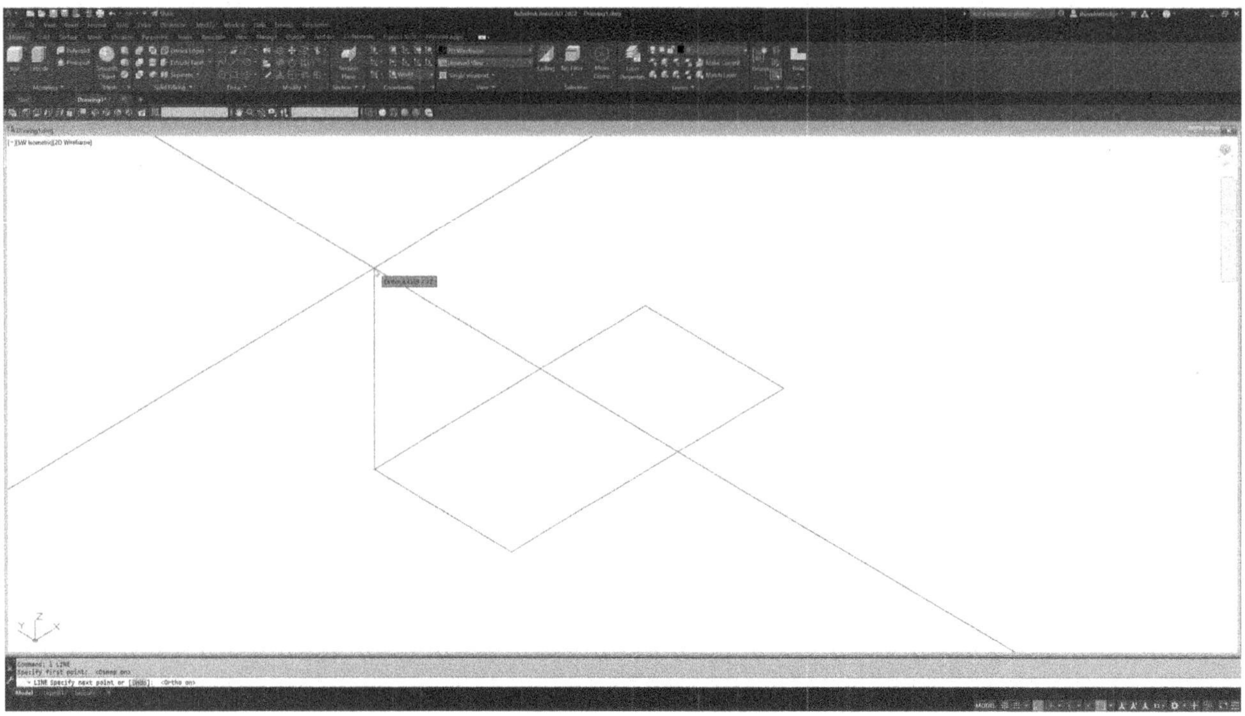

FIGURE 21.13 Method 1, Step 1.

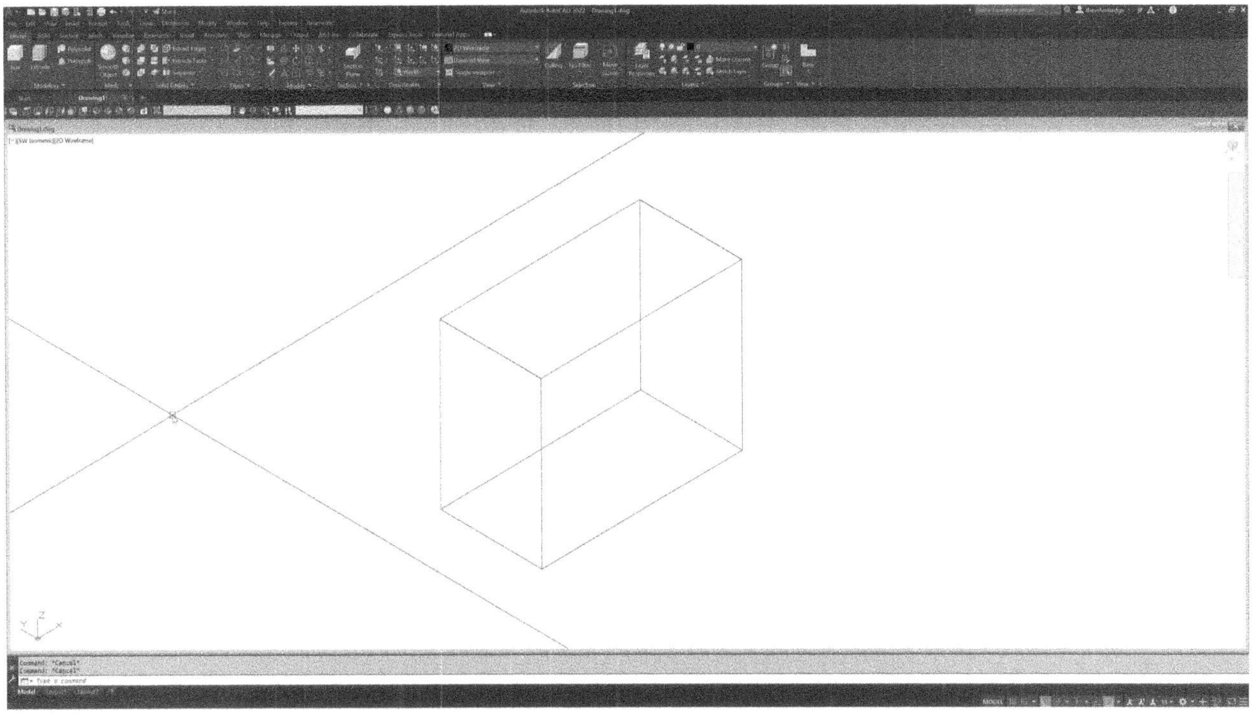

FIGURE 21.14 Method 1, Step 3. Completed 3D box.

You already did this when learning isometric drawing by pressing F5 and cycling through a total of three planes: top, left, and right. We do something similar here in 3D. This is why isometric drawing was important to review. In a way, you already knew some 3D concepts after finishing that chapter.

So, how do you actually rotate these planes? Rotating planes is equivalent to rotating the UCS icon. We introduce the various ways you can rotate the UCS icon via the familiar command matrix. You can do this via typing, cascading menus, toolbars, or the Ribbon. If you wish to continue using the toolbars, you need to add the UCS toolbar to your collection (Fig. 21.16). We discuss most of this toolbar in Chapter 28, but, for now, need only three of the icons, as shown next.

Keyboard: Type in **ucs** and press Enter
Cascading menus: **Tools→New UCS→X** or **Y** or **Z**
Toolbar icon: **UCS** toolbar or or
Ribbon: **View** tab→ or or

To perform the UCS rotation, use any of the methods shown in the preceding command matrix. However, there is a slight difference in procedure, depending on what method you choose. Before you start, read all the way through the steps to get a full understanding!

Step 1(a). If you *typed* in `ucs` and pressed Enter, then,

- AutoCAD says: `Specify origin of UCS or [Face/NAmed/OBject/Previous/View/World /X/Y/Z/ZAxis] <World>:`

Note that the UCS icon is attached to your mouse. This is not relevant to what we are doing at the moment, and you can ignore it, moving on to choosing an axis in Step 2, then proceeding to Step 3. However,

Step 1 (b). If you used the *icons*, *cascading menus*, or the *Ribbon*, then the process of choosing an axis is part of the command right away (as you picked either an X, Y, or Z), and you can skip Step 2.

Step 2. Pick the axis around which you want to rotate: X, Y, or Z.

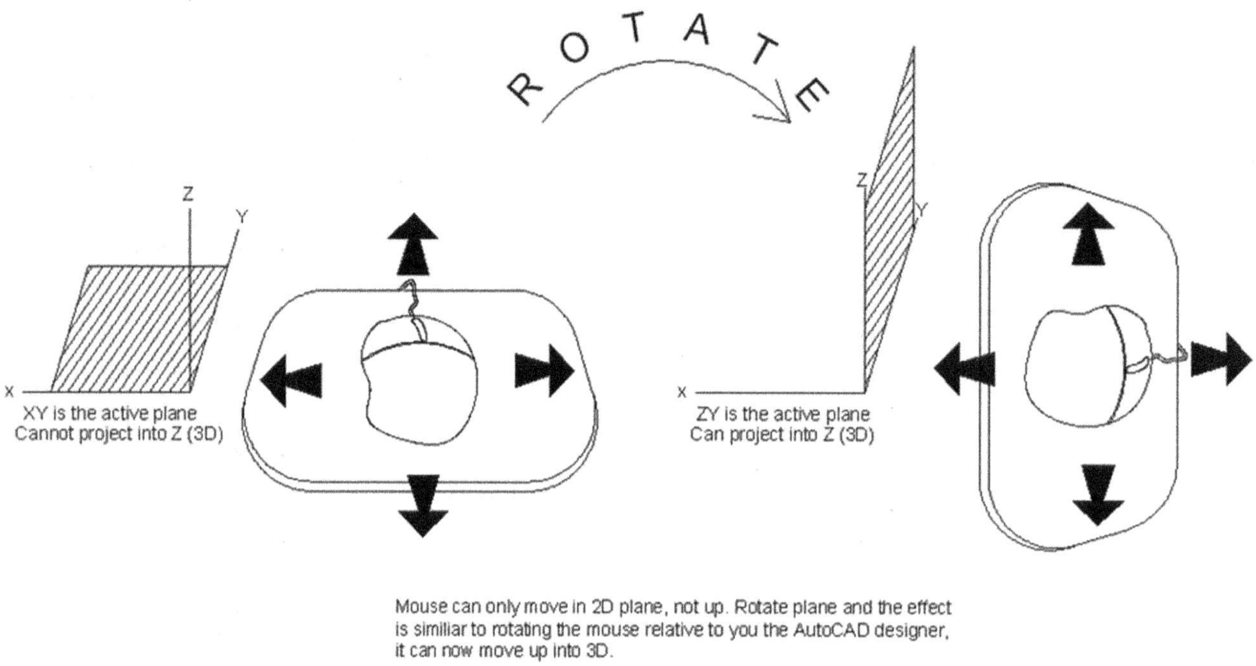

FIGURE 21.15 Rotation of planes.

FIGURE 21.16 UCS toolbar.

A very important skill needs to be learned right at the point of Step 1 or Step 2. Around which axis do we need to rotate the UCS icon to be able to draw "up" into the third dimension: X, Y, or Z? It is something you need to be able to quickly identify. Fortunately, in this case, you have a choice of two out of the three axes: either the X or the Y. Rotating around the Z axis does nothing useful (in this example) and just spins the UCS axes around like the blades of a helicopter. Do you see why? So, type in either choice X or Y and press Enter, or if using the other methods, choose X or Y right away. This example proceeds with the X axis being chosen.

Step 3. After the X axis is selected,

- AutoCAD says: Specify rotation angle about X axis <90>:

Step 4. You can just press Enter, as 90° (the default) is exactly what we need.

You should see what is shown in Fig. 21.17. The X axis is the "hinge" around which we pivoted and the crosshairs now point in the direction we need to go, which is straight up, relative to you.

To bring your UCS icon back to the previous familiar state (Fig. 21.18), you can do any of the following:

Keyboard: Type in **ucs** and press Enter twice.
Cascading menus: **Tools→New UCS→World**
Toolbar icon: **UCS** toolbar
Ribbon: **View** tab→

This is called *restoring the icon to World View*. Practice rotating the UCS icon around all three axes. Notice how the axes act as "hinges" for rotations. You need to recognize which hinge to rotate around later on as we do actual object rotations.

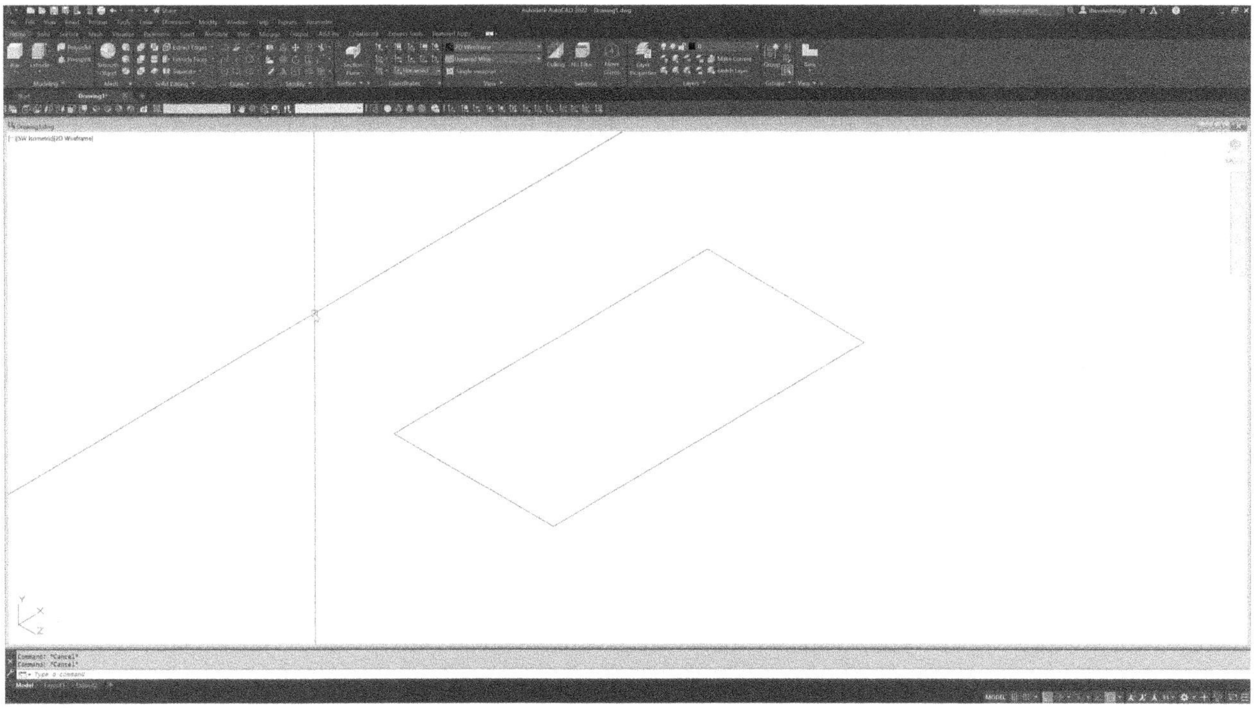

FIGURE 21.17 UCS rotation around the X axis.

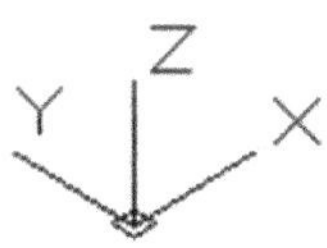

FIGURE 21.18 3D UCS icon reset to World View.

If you were using AutoCAD 2014 or earlier version, you would now complete the box using standard drafting techniques and draw a straight line starting from one of the rectangle's corners and continuing up 8″ (using Ortho and OSNAP points, of course). You would then copy it to the other three corners and copy the bottom four lines (or rectangle) to the top, thereby creating a box. You would have arrived at the same as Fig. 21.14.

However, starting with AutoCAD 2015, and continuing with 2021, you cannot really draw lines along your new plane once you rotated the UCS icon. The software actively discourages you by creating them off in a distance. Go ahead and try it! And, really, there was no need to do this; you can just go "up" without rotating anything. So why did we learn this procedure, aside from knowing how to do it for older AutoCAD releases? Well, now, you can create new geometry such as rectangles and circles in this new plane, as seen in Fig. 21.19. We need this to create extrusions, a very important 3D design method, which we discuss right after we introduce Dynamic Views and Orbit.

To summarize these changes,

- You can draw basic linework straight "up" from the XY plane while in the SW Isometric or other views (using Ortho). There is no need to switch planes.
- You cannot easily draw linework from other planes, but you can draw closed shapes, such as rectangles and circles.

While this relatively new change in AutoCAD's 3D design approach will give textbook authors and instructors everywhere a bit of a headache when trying to describe it (since all the previous versions of AutoCAD did it differently), the change makes sense. Since most 3D design is done via extrude anyway - as we describe soon - this change does not affect much of your day-to-day work, but for the occasional line that needs to be drawn into 3D, AutoCAD has made it easier, removing the UCS rotation step.

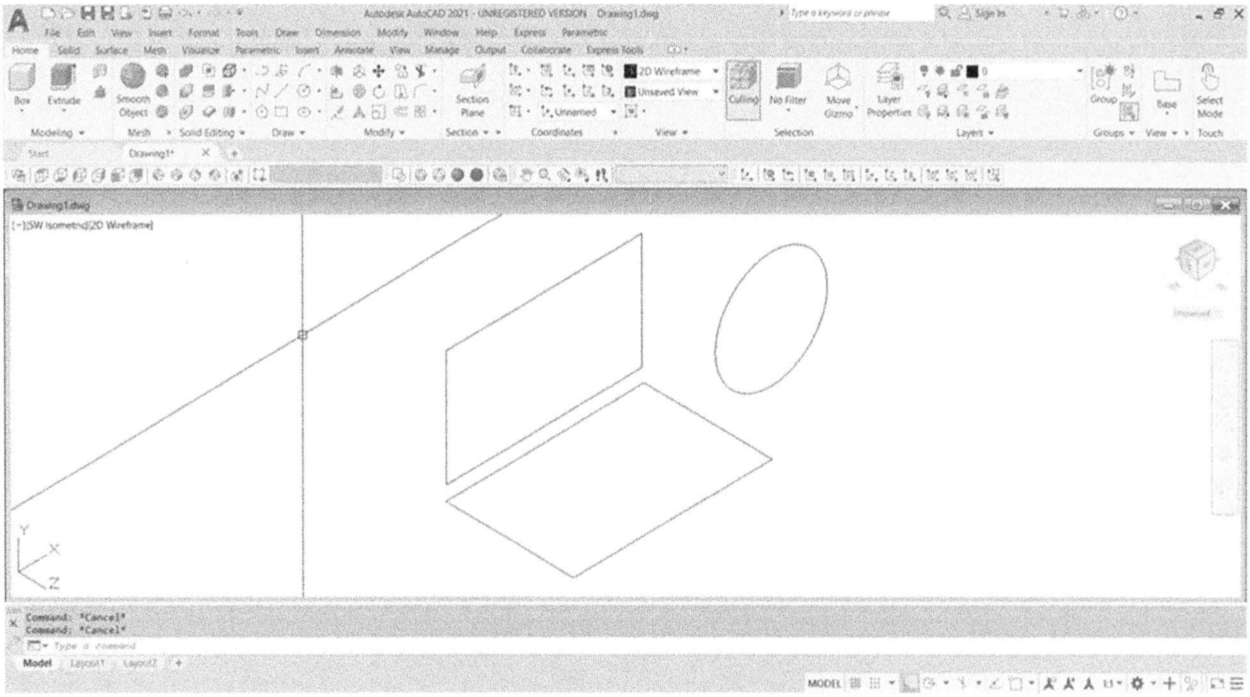

FIGURE 21.19 Drawing new geometry in XY plane.

21.5 3D Dynamic Views

In 3D, you need to constantly spin the object around to get a good look at it from all sides. AutoCAD has powerful 3D Orbit tools to easily look at your design in real time as opposed to just preset views. Remember that, even though we say that we are rotating the object, really we are not but rather rotating our view of it.

The 3D orbit command comes in several flavors. The "3dorbit" version is constrained to vertical and horizontal movements, while the "3dforbit" (the *f* stands for "free"), allows for unrestricted movement and is the version we focus on. There is also the "3dcorbit" (the *c* stands for "continuous"), which rotates the view nonstop until you exit the command. This is more entertaining than useful but does look impressive in a presentation. You can find all of these cascading down from the 3D Navigation toolbar, so feel free to explore them on your own. We next further describe 3D free orbit. To begin the 3D free orbit, use any of the following methods:

Keyboard: Type in **orbit** and press Enter.
Cascading menus: **View→Orbit→Free Orbit**
Toolbar icon: **3D Navigation** toolbar→
Ribbon: **View** tab→**Navigate 2D**

As soon as you click on the Free Orbit icon, a green circle appears, and you are able to do the orbit by clicking down, holding the left mouse button, and moving the mouse around, as seen in Fig. 21.20. When done just press Esc, and the new view is permanent. To restore the familiar SW view, just press that icon.

What we have done thus far is create a wireframe model. This is a common term in computer-aided design (not just AutoCAD) and simply means the model is not shaded or rendered and resembles a wire that has been shaped into something (a frame), hence, wireframe.

Although we introduced a lot of useful concepts thus far, such as axes, planes, 3D views, 3D orbit, and UCS rotation, the wireframes we created cannot be shaded, as they are essentially "hollow" pieces and contain no surfaces. Creating real solids is the subject of the next section, so go ahead and erase the box; we will create it again.

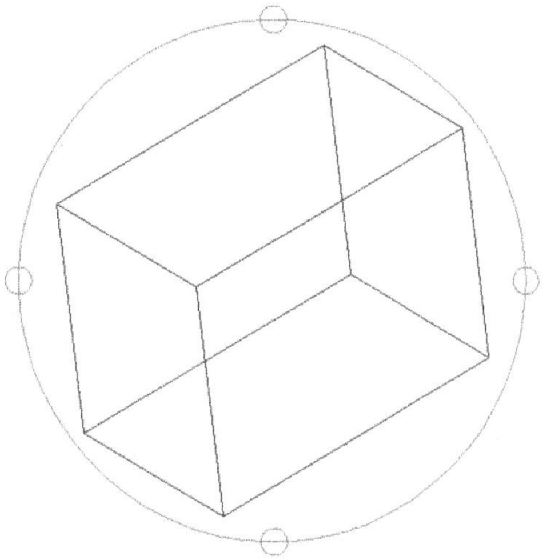

FIGURE 21.20 3D Free Orbit in use.

21.6 Extrude

Extrusion is the method most often used to quickly and easily create solid objects and, as such, is perhaps one of the most important commands in 3D. If you prefer toolbars, you need a new toolbar, Modeling (Fig. 21.21), to access the extrude command via that method.

We begin the same way as before, by drawing a 10″×6″ rectangle but this time use only the rectangle command, not individual lines, then proceed as described next:

Keyboard: Type in **extrude** and press Enter.
Cascading menus: **Draw→Modeling→Extrude**
Toolbar icon: **Modeling** toolbar→ 
Ribbon: **Solid** tab→ 

Step 1. Start the extrude command via any of the preceding methods.

- AutoCAD says:

```
Current wire frame density:ISOLINES=4, Closed profiles creation mode = Solid
Select objects to extrude or [MOde]: _MO Closed profiles creation mode [SOlid/SUrface] <Solid>: _SO
Select objects to extrude or [MOde]:
```

Step 2. Pick the rectangle (AutoCAD will tell you `1 found`) and press Enter.

- AutoCAD says: Specify height of extrusion or [Direction/Path/Taper angle/Expression] <3.7154>:

Step 3. You can move your mouse up and down and create a thickness in real time, but it is more practical to type in a value, so go ahead and type in `4` for the thickness. You should see what is shown in Fig. 21.22.

This new, extruded box may at first seem quite similar to the previous one you tried doing. That is because wireframe is not just a method of construction but also a method of presentation. There are some significant differences, however. This box is a real solid model, not just a collection of "wires" spliced together. We can now hide and shade it as described in the next section.

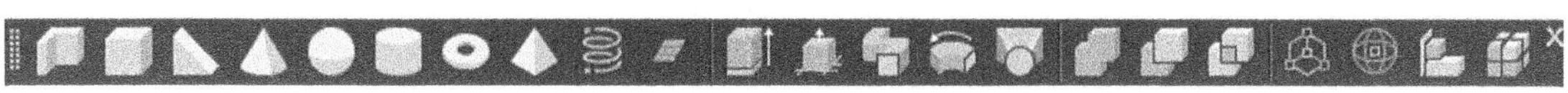

FIGURE 21.21 Modeling toolbar.

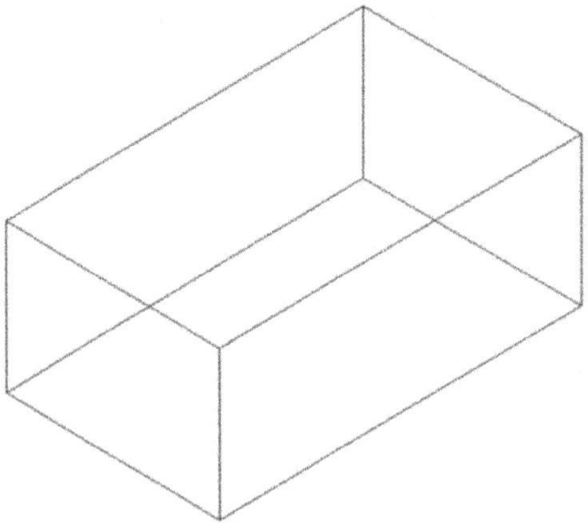

FIGURE 21.22 Extruded box.

21.7 Visual Styles: Hide and Shade

Visual Styles is an important set of tools that allows you to view your design in a variety of useful ways. We explore two variations of Hide and two distinct versions of Shade: realistic and conceptual. We also look at a few other settings, such as Sketchy and X-ray.

The hide command simply hides wireframe linework that you would not see with a solid object. There are two slightly different variations of it. The first type of hide is temporary while you are viewing a stationary design. The hidden view reverts to wireframe if you try to rotate it or regenerate it. This can be done by simply typing in `hide` and pressing Enter, via the Hide icon on the Render toolbar (see Fig. 21.23), or finally via the cascading menu View→Hide.

The other variety of hide is more permanent and the design can be rotated in 3D while staying fully hidden. That is accomplished via the 3D Hidden Visual Style icon on the Visual Styles toolbar (first shown in Fig. 21.5). To return to wireframe, you can no longer just regenerate but must press the 2D Wireframe icon on the same Visual Styles toolbar. All of the previous is summarized in the command matrix shown next. Go ahead and try all the ways to hide your 3D box before moving on.

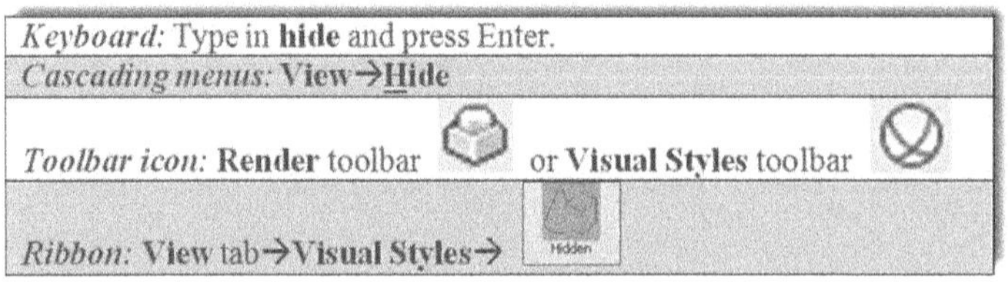

Keyboard: Type in **hide** and press Enter.
Cascading menus: **View→Hide**
Toolbar icon: **Render** toolbar or **Visual Styles** toolbar
Ribbon: **View** tab→**Visual Styles→** Hidden

The result of hiding via the keyboard hide entry, cascading menus, and the Render toolbar's Hide icon is shown in Fig. 21.24. Note how you can no longer zoom and pan. You can still do a Free Orbit, but the hidden lines reappear, so this is useful to just view the design and maybe run off a print.

The result of using the 3D Hidden Visual Style icon as well as the Hidden choice in the Ribbon is shown in Fig. 21.25. Note how the background color of the screen changes (which can be changed in the Options…dialog box, Display tab). Also note how you can now zoom and pan. This version of hide is good for continued work on the design while in hidden mode.

FIGURE 21.23 Render toolbar.

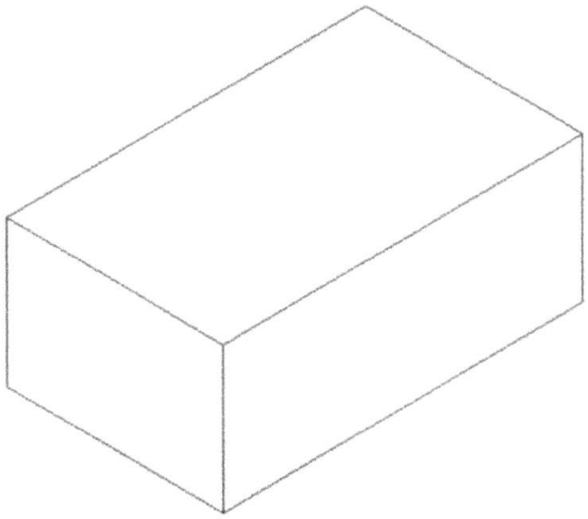

FIGURE 21.24 Hidden box, Method 1.

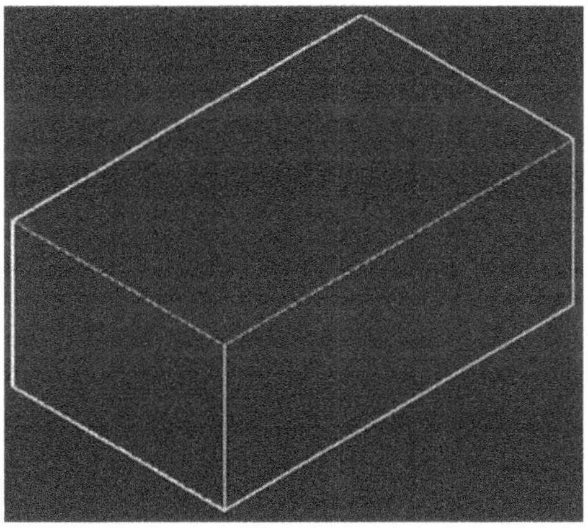

FIGURE 21.25 Hidden box, Method 2.

Let us now give shading a try. As already mentioned, it comes in two versions, conceptual and realistic. Which you use depends on a variety of shape and background color settings. First, we focus on the default shade, which is the Realistic Visual Style. Start shade via any of the following methods:

Keyboard: Type in **shade** and press Enter.
Cascading menus: **View→Visual Styles→Realistic**
Toolbar icon: **Visual Styles** toolbar→ 
Ribbon: **View** tab→**Visual Styles**→ 

The result of this shading is shown in Fig. 21.26.

The other style of shade is set as follows. Note that typing in shade is no longer applicable, as that leads to the realistic version of shade being applied by default.

FIGURE 21.26 Realistic shading box.

Keyboard: None
Cascading menus: View→Visual Styles→Conceptual
Toolbar icon: **Visual Styles** toolbar→
Ribbon: View tab→Visual Styles→

The result of this shading is shown in Fig. 21.27.

You probably did not notice too much of a difference between the two types of shading. That is because the boxes are not really colored themselves; also, they appeared against the default background, altering the appearance of their own coloring. To appreciate the differences between the two types of shading, observe the two sets of colored and shaded shapes in Figs. 21.28 and 21.29.

These shapes are set against a white then a black background with the realistic and conceptual shading on and similar colors added to all the shapes. Notice how the softer shading on the right gives a more pleasing picture of the model. For this reason, the Conceptual Visual Style shading is used more often in presenting designs, both on screen and on paper. It is also highly recommended that you set your background to white or black in 3D for a more accurate representation of colors. To return to the unshaded image, press the 2D Wireframe option or icon.

Two more types of presentation views are quickly mentioned: Sketchy and X-ray. Both can be accessed through the Ribbon's View tab Visual Styles or through the cascading menu's View→Visual Styles. Try them both out on your box.

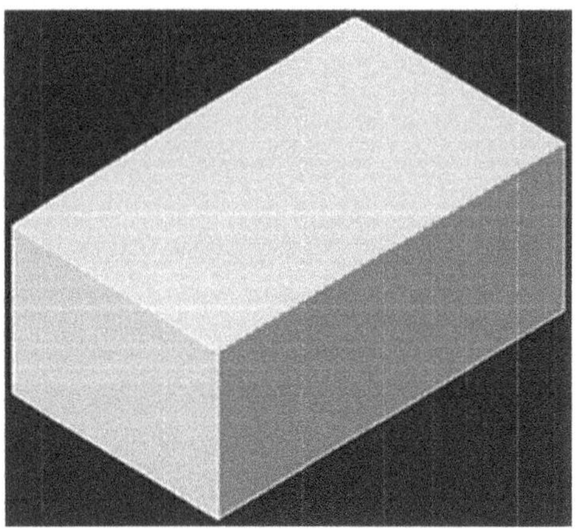

FIGURE 21.27 Conceptual shading box.

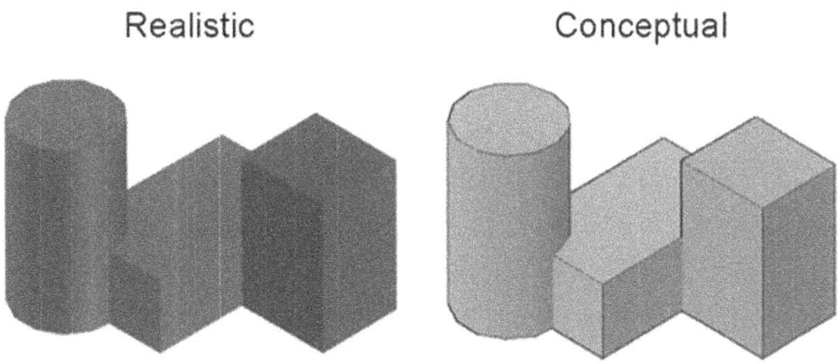

FIGURE 21.28 Realistic vs. conceptual shading (white background).

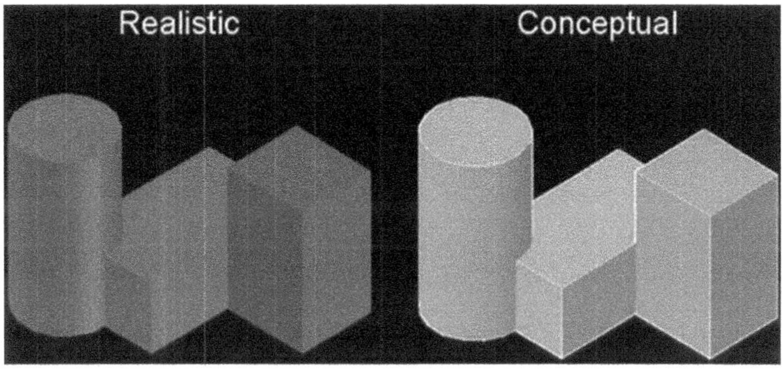

FIGURE 21.29 Realistic vs. conceptual shading (black background).

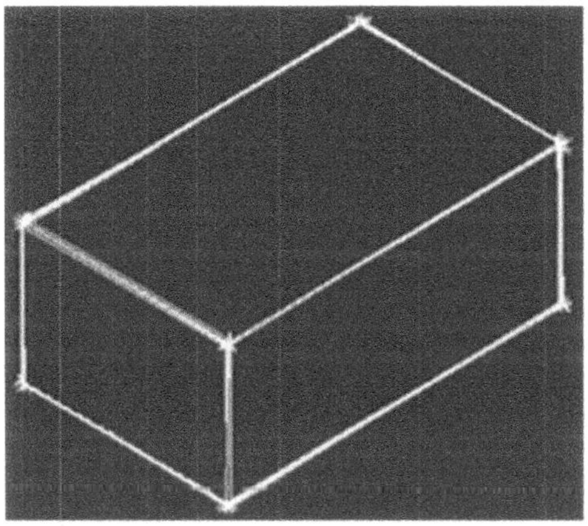

FIGURE 21.30 Sketchy visual style.

Fig. 21.30 shows the sketchy visual style. It presents a "hand-drawn" rendering of your design and may find some use in a more artistic representation. X-ray makes the shaded shape transparent and is seen in Fig. 21.31.

All the views and settings thus far described can be accessed via a set of drop-down menus at the very upper left of the drawing area. The first category from the left, [-], is the Viewport Controls, seen in Fig. 21.32. They present various choices for viewports and whether or not to show the ViewCube, SteeringWheels, and Navigation Bar (to be discussed in the next section). Just to the right of that, under [SW Isometric], are the View Controls, seen in Fig. 21.33. They allow yet another way to set and manage the various views. Finally, the [2D Wireframe] menu contains the Visual Style Controls, as seen in Fig. 21.34. It allows you to set the visual styles without the need for toolbars or the Ribbon.

After the final discussion regarding the ViewCube, SteeringWheels, and the Navigation Bar in the next section, you should be familiar with almost everything in these menus.

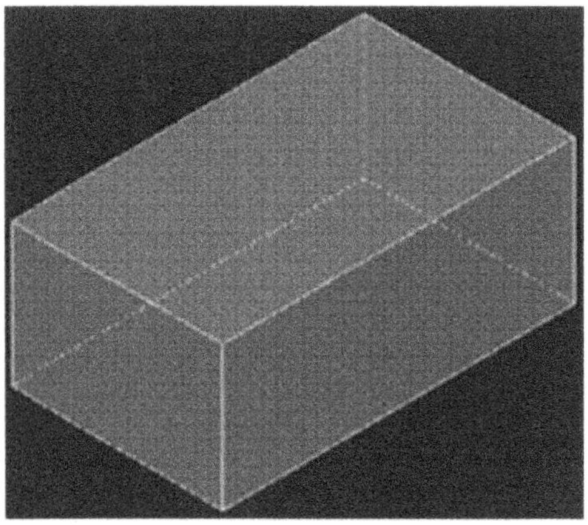

FIGURE 21.31 X-ray visual style.

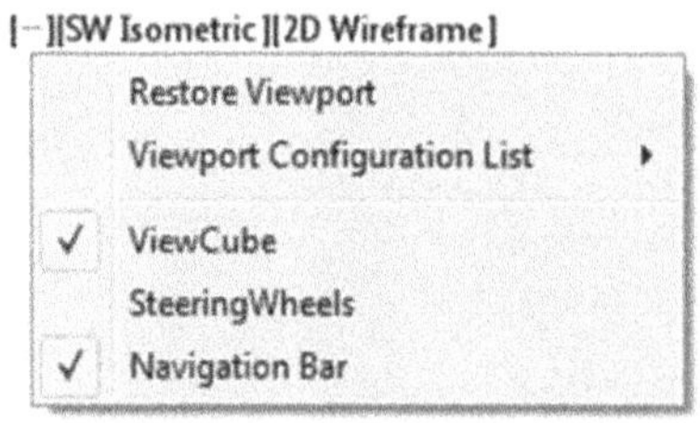

FIGURE 21.32 Viewport controls.

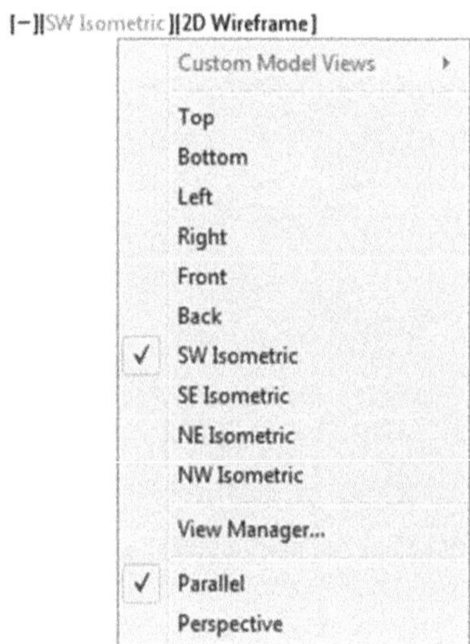

FIGURE 21.33 View controls.

21.8 Viewcube and Navigation Bar

Unless they were turned off on purpose, you almost certainly have been looking at the ViewCube and Navigation Bar since Chapter 1. The top view of this cube and bar reside in the upper right corner of your screen, as seen in the 2D top view in Fig. 21.35. SteeringWheels (an advanced tool not discussed here) is also seen in Fig. 21.35 and is brought up via the first button on the Navigation Bar.

With your box shaded, switch to SW Isometric view. Now that you are in 3D, the bar stays the same, but the cube looks like the one in Fig. 21.36.

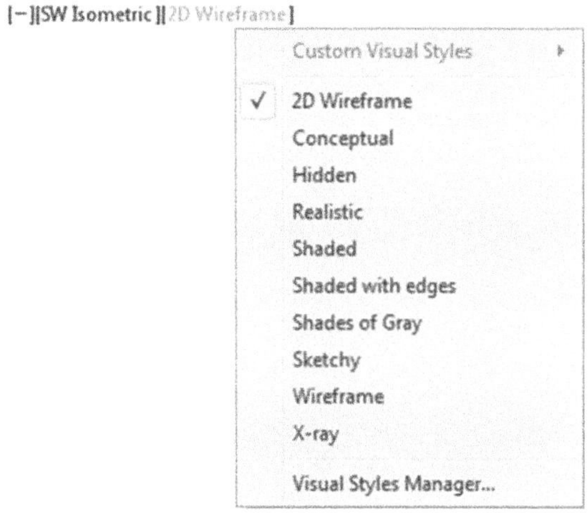

FIGURE 21.34 Visual Style controls.

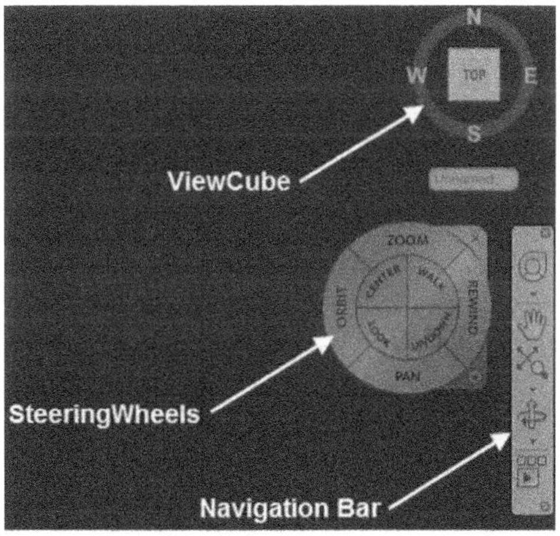

FIGURE 21.35 ViewCube, SteeringWheels, and Navigation Bar.

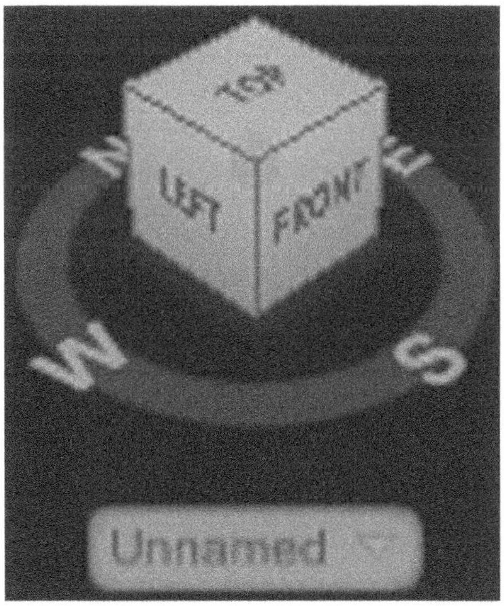

FIGURE 21.36 ViewCube in 3D.

This cube is useful primarily in 3D navigation, and that is why there was no mention of it in the previous 2D chapters. It first appeared in AutoCAD 2009 and is used to dynamically rotate to the graphically shown views, such as top, left, or front. You can click on the faces of the cube, its edges, or its corners; and of course, you can just manually rotate it. Whatever you click on "lights up," and your design moves accordingly. Additionally, you can press the N, S, E, and W buttons. Go ahead and try it out, returning to the SW Isometric View afterward.

The ViewCube presents no completely new ideas but rather combines a host of various tools in one convenient location. Much of it is equivalent to the toolbar view icons, although you now have the ability to also view a design from a "corner" or an "edge."

The buttons on the Navigation Bar should also be generally familiar at this point, except for the top button, Full Navigation Wheel, and the bottom button, ShowMotion (also an advanced topic, not covered in this chapter). The rest of the buttons from the top going down are Pan, Zoom Extents, and the various Orbits. Review the functionality of each button.

One last item to mention with the navigation cube is its ability to put you into perspective view. We cover perspective view in later chapters, but you can try it out briefly now. Hover your mouse around the top left area of the cube and a tiny blue house appears, as seen in Fig. 21.37.

Click on the house icon and your shaded cube switches to a perspective view (Fig. 21.38).

FIGURE 21.37 Additional menu button.

FIGURE 21.38 Perspective view.

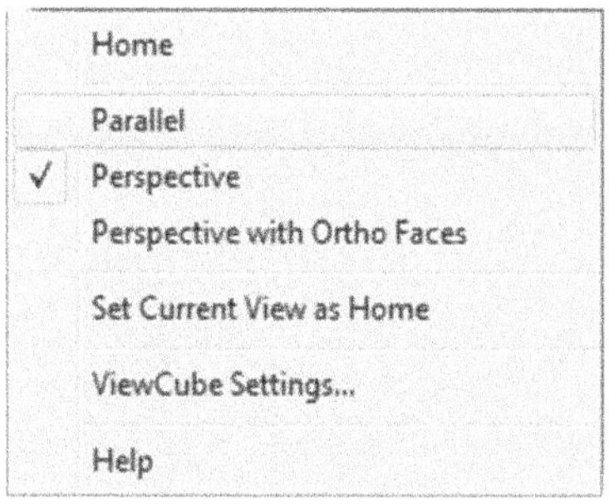

FIGURE 21.39 Perspective view.

To exit out of this view, right-click on the house and you see the menu in Fig. 21.39. Select and click on Parallel and you exit the perspective view.

This is it for the basics. It seems we introduced an endless collection of new concepts, but really it was just six "big ideas," as listed next:

- The concepts of axes and planes.
- How to get in and out of 3D.
- How to rotate planes.
- How to orbit around your design.
- How to extrude the design.
- How to shade the design.

Review everything carefully; your success in the rest of 3D depends on understanding these basic concepts.

Summary

You should understand and know how to use the following concepts and commands before moving on to Chapter 22:

- Axes: X, Y, and Z
- Planes: XY, ZX, and YZ
- Faces
- 3D toolbars: View, 3D Navigation, and Visual Styles
- SW Isometric View
- Top view
- UCS rotation
- Free Orbit
- Extrude
- Hide
- Shade: realistic and conceptual views
- ViewCube and Navigation Bar

Review Questions

Answer the following based on what you learned in this chapter:

1. What three axes are discussed?
2. What three planes are discussed?
3. What 3D toolbars are introduced in this chapter?
4. What command is used most often to create 3D objects?
5. How do you rotate the UCS icon?
6. What does hide do?

7. What does shade do?
8. What are the two types of shade?
9. What is the purpose of the Navigation Bar? ViewCube?

Exercises

1. In 3D SW Isometric Conceptual Visual Style (CVS), set your units to Architectural and draw a 10′×5″ rectangle. Extrude this rectangle to an 8′ height. This is one of several ways to draw a wall. We need this in the coming chapters. Zoom out to extents if necessary and alter the color if you wish. The result is shown next. (Difficulty level: Easy; Time to completion: <3 minutes.)

2. In the 3D SW Isometric CVS, set your units to Architectural and draw a 4″×3″ box. Extrude it to 6″. Then rotate your UCS around the Y axis by 30° and around the X axis by 30°. Finally, draw the same rectangle with the same extrusion. Change the colors if you wish. This is what you should get. (Difficulty level: Easy; Time to completion: <5 minutes.)

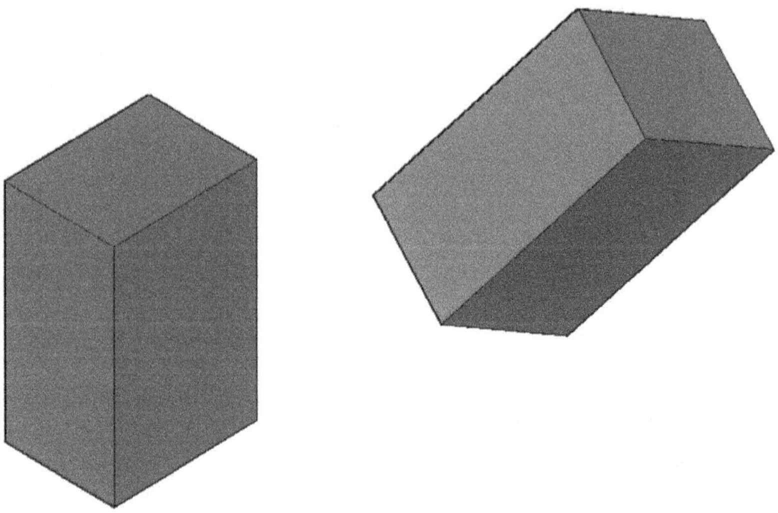

3. In the 3D SW Isometric CVS, set your units to Architectural and draw a 5″ circle, extruding it to 25″. Now rotate your UCS around the X axis by 90° and draw the same extruded shape. Then, restore the UCS position, and rotate it around the Y axis by 90°. Draw the same shape again. You should have the three cylinders shown in the left image. Move two of them center to center with the third one as shown in the right image. (Difficulty level: Easy, Time to completion: 5–7 minutes.)

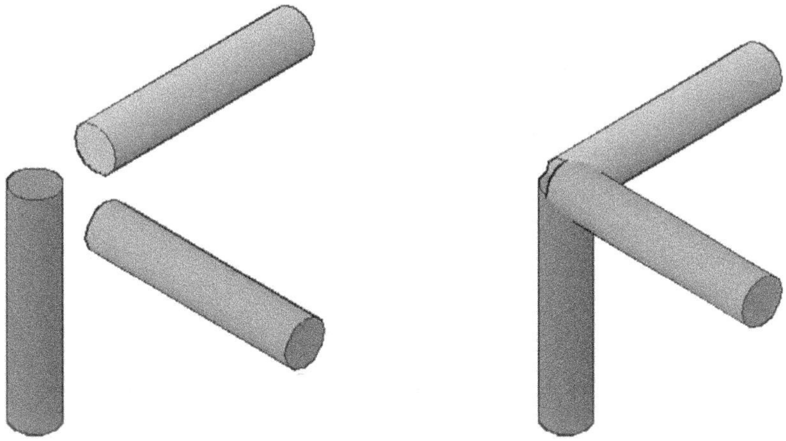

4. In 2D, draw a pline profile of a simple ship, as shown next. No exact dimensions are given, and your profile does not have to look exactly like the one shown but somewhat similar. Then, go into 3D SW Isometric and extrude the shape to the approximate depth shown. Color and 3D Orbit to see the shape in its given orientation. (Difficulty level: Easy, Time to completion: 5 minutes.)

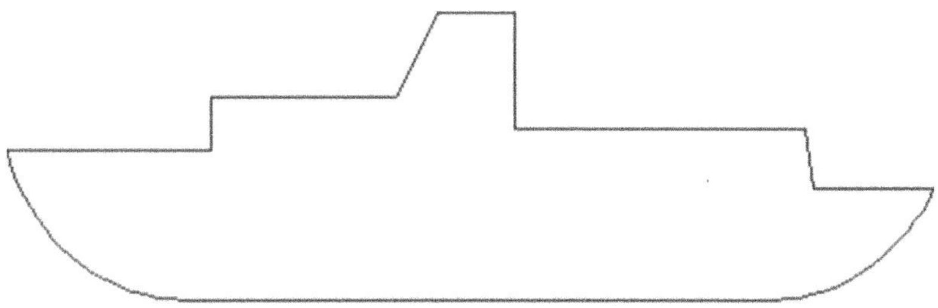

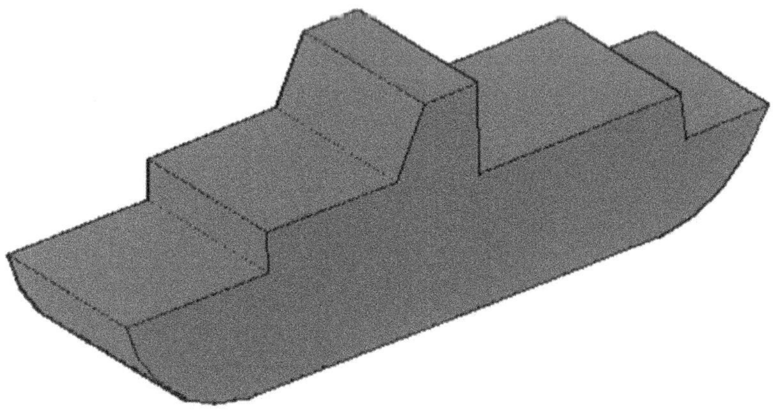

5. In 2D, draw a pline profile of a simple truck cab/frame and wheels as shown next. No exact dimensions are given, and your profile does not have to look exactly like the one shown but somewhat similar. Then, go into 3D SW Isometric and extrude the profiles to the approximate depths shown. Color and 3D Orbit to see the shape in its given orientation. (Difficulty level: Easy, Time to completion: 5–7 minutes.)

6. In 2D, draw a circle profile of a wheel and axle as shown. The full sizing and extrusion of the wheels are given, as well as the sizing and extrusion of the axle between the pairs of wheels. The connecting axle is the same radius but can be any length. Go into 3D SW Isometric and Color and 3D Orbit to see the model from all angles. (Difficulty level: Easy, Time to completion: 7–10 minutes.)

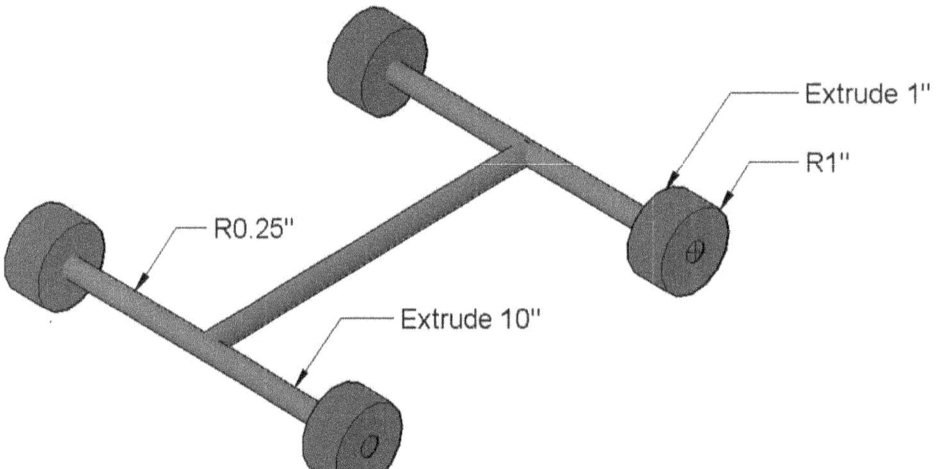

Chapter 22

Object Manipulation

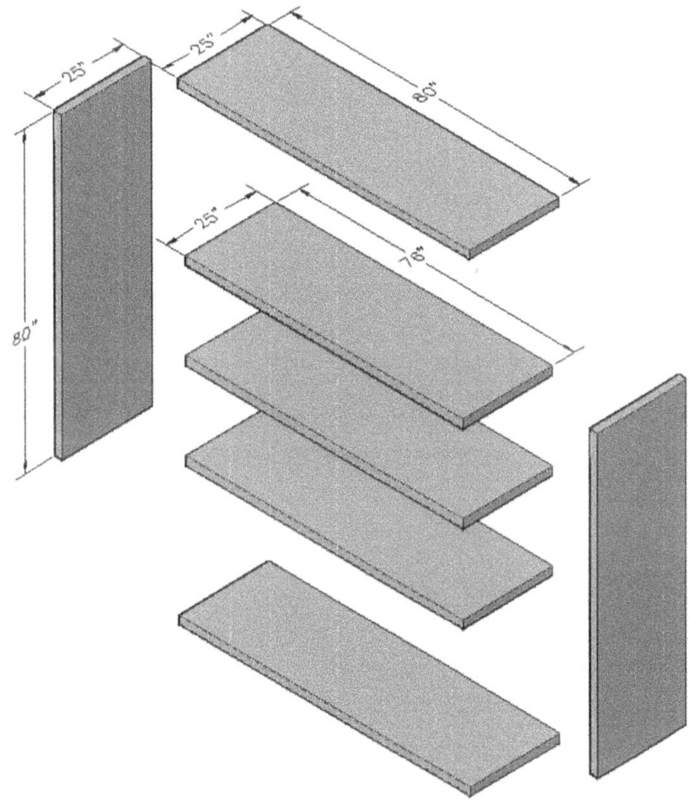

Learning Objectives

In this chapter, we introduce the tools necessary to manipulate objects in 3D space to position them as required by the design. We specifically cover

- Rotate3D
- 3Drotate (gizmo)
- Mirror3D
- 3D array: polar and rectangular
- 3Dscale
- 3Dmove
- Fillets and chamfers in 3D

By the end of this chapter, you will be able to create flat designs of moderate complexity. We introduce multiple small projects throughout this and upcoming chapters to put each new concept to use right away.

Estimated time for completion of this chapter: 2 hours.

Up and Running with AutoCAD 2022. https://doi.org/10.1016/B978-0-323-89923-9.00022-6

22.1 Introduction to Object Manipulation

It was already mentioned that extrude is perhaps the single most important command in 3D AutoCAD. Using it, you can create a wide array of objects that are flat in nature, such as building walls, doors, and some furniture, just to name a few items. However, unless you are designing a perfect box of a house, you need to rotate and mirror these objects into position as well as array objects and add fillets or chamfers for more realistic edges. All these tools and more are covered in this chapter. We go through each one and do a few sample exercises. These tools, together with Boolean operations (introduced in the following chapter), are virtually all you need to model a basic architectural design of a building in 3D.

Rotate3D

Although you may not yet have tried it, you can use the regular rotate command in 3D. You can rotate the object on whichever plane you are on. The rotational axis is perpendicular to the plane and the net result is the object "spinning in place." However, you may often want to rotate the object around another axis, for example, make a flat plate "stand up." Recall from the first chapter that you can use the UCS command to simply rotate your active plane and draw new objects that way, but for existing ones you need to use the 3D version of rotate to bring them into position. Rotate3D gives you this ability and is an important design tool.

This command comes in two versions: rotate3D and 3Drotate. The difference is that the former is command driven and the latter uses a graphical "gizmo" to facilitate rotation. We try both, starting with rotate3D. Draw a flat 10″ × 10″ plate and extrude it to 1″ thickness, as seen in Fig. 22.1, with the Conceptual Visual Style shading on and color added.

Notice the colorful UCS icon at the bottom of the figure. It is very prominent on purpose, because it is the focus of our attention while rotating in 3D. What we want to do right now is rotate the plate so it stands up on its side like a vertical wall. What axis do we pick as a "hinge" around which the plate rotates? Fortunately, we have a choice of two: the X and the Y axes. Rotating around the Z does nothing for us in this case. Notice the similarities of this discussion to the one on UCS rotation in Chapter 21, it is essentially the same idea.

So, now that we know what we want to do, let us give the command a try. There is no Ribbon, cascading menu, or toolbar equivalent, so do the following:

Step 1. Type in `rotate3D` and press Enter.

- AutoCAD says: `Current positive angle: ANGDIR = counterclockwise ANGBASE = 0`
  `Select objects:`

Step 2. Select the plate and press Enter.

- AutoCAD says: `Specify first point on axis or define axis by`
  `[Object/Last/View/Xaxis/Yaxis/Zaxis/2points]:`

Step 3. Here, you need to choose which axis to rotate around. Let us go with X. Type that in and press Enter.

- AutoCAD says: `Specify a point on the X axis <0,0,0>:`

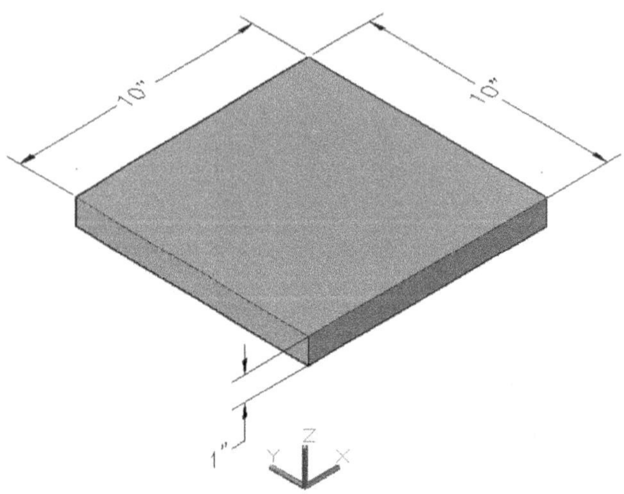

FIGURE 22.1 Flat plate.

FIGURE 22.2 Plate 3D rotation around the X axis.

Step 4. Here, you need to pick the point on the axis of rotation, which should be on the object itself. Using OSNAPs pick the lower left corner point of the plate.

- AutoCAD says: `Specify rotation angle or [Reference]:`

Step 5. Here, you can enter any numerical value, but let us stay with 90° for now. Type in `90` and press Enter. The plate rotates 90° around the X axis and stands on its end, as in Fig. 22.2.

In the same exact manner, practice rotating the plate around the other axes. From where it is right now in Fig. 22.2, a rotation around the Z axis makes it face the lower left side of the page. A rotation around the X axis makes it lay down flat again, and a rotation around the Y axis produces no visible results, aside from a shift in position, as the plate is a perfect 10″ × 10″ square.

3Drotate (Gizmo)

An alternative approach to rotating in 3D is a relatively new tool called a *gizmo*. The net result is exactly the same as with the typed command just covered, but the approach is more graphical in nature. Also, toolbar, Ribbon, and cascading menu approaches are available, as seen here in the command matrix.

Keyboard: Type in **3drotate** and press Enter
Cascading menus: **Modify→3D Operations→3D Rotate**
Toolbar icon: **Modeling** toolbar
Ribbon: **Home** tab→**Modify**

When you start up the command via any of the preceding methods, you are asked to select the object in question and you see the gizmo shown in Fig. 22.3.

The purpose behind this tool is to be able to easily visualize rotational motion. Following the command line directions, pick a base point, then the circular band that best represents the rotational axis of motion you want. Finally, enter an angle, and the shape is rotated. This all closely matches the previous method and detailed steps are not given, as they are essentially the same. A snapshot of the gizmo in action is shown in Fig. 22.4. Practice both approaches and stick with the one you prefer.

Be sure to understand these procedures thoroughly. Next, we put everything together and draw our first real 3D project, a chair, then move on to mirror3D.

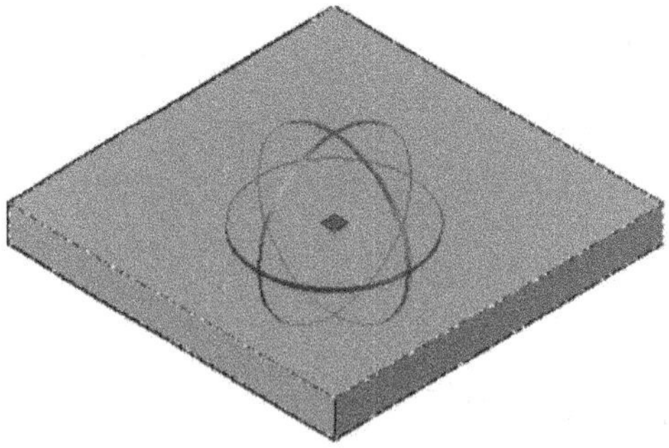

FIGURE 22.3 Gizmo tool.

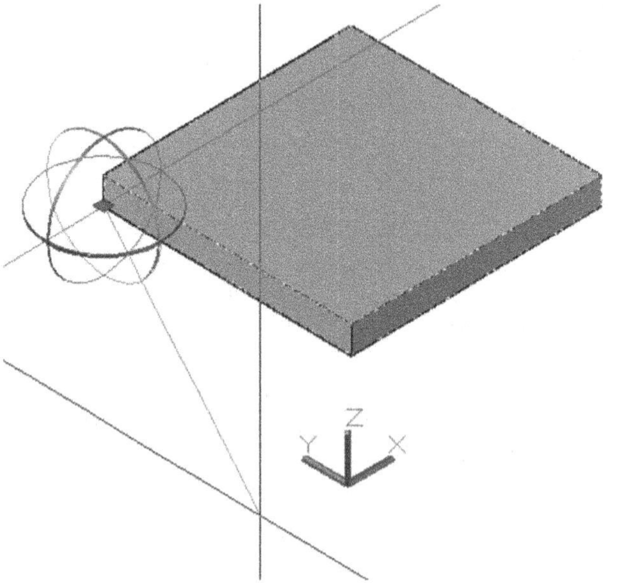

FIGURE 22.4 Gizmo tool rotating an object.

3D Modeling Exercise: Chair

Create the model of a chair based on the dimensions given in Fig. 22.5. The full assembly, at a slightly different angle, is shown in Fig. 22.6. Specifically, you need to

- Create four 2″×2″ rectangles extruded 30″ for the legs.
- Create two 24″×24″ rectangles extruded 2″ for the seat and backrest.
- Rotate3D one of the rectangles 30° from vertical for the reclining backrest.
- Make sure everything fits together perfectly, adding shading and color of your choice.

Mirror3D

Mirror3D is similar in approach to rotate3D. Much like before, you can use the regular 2D mirror command, but this results in mirroring in the current plane only. What we need is the ability to mirror between any planes. As you may have guessed, the key here is selecting the correct plane (as opposed to axis) to mirror over. Just as a reminder, we have three axes (X, Y, and Z) and three unique planes (XY, ZX, and YZ). Unlike the axes, you do not see the planes; it is something you must envision in your mind. Think of them as a thin sheet of glass that slices through parts. Whatever is on the other side is the mirror image. Erase your chair and let us start with a clean screen. Draw the previous 10″×10″ flat plate extruded to 1″ (Fig. 22.7).

We would like to mirror it so there is another plate next to the darker shaded right side. Before proceeding, identify the correct plane. Remember a plane is not an axis; rather it is formed by two axes put together, so your choices once again are XY, ZX, and YZ. The correct answer is ZX. Do you see why? It is important to be able to quickly recognize these planes.

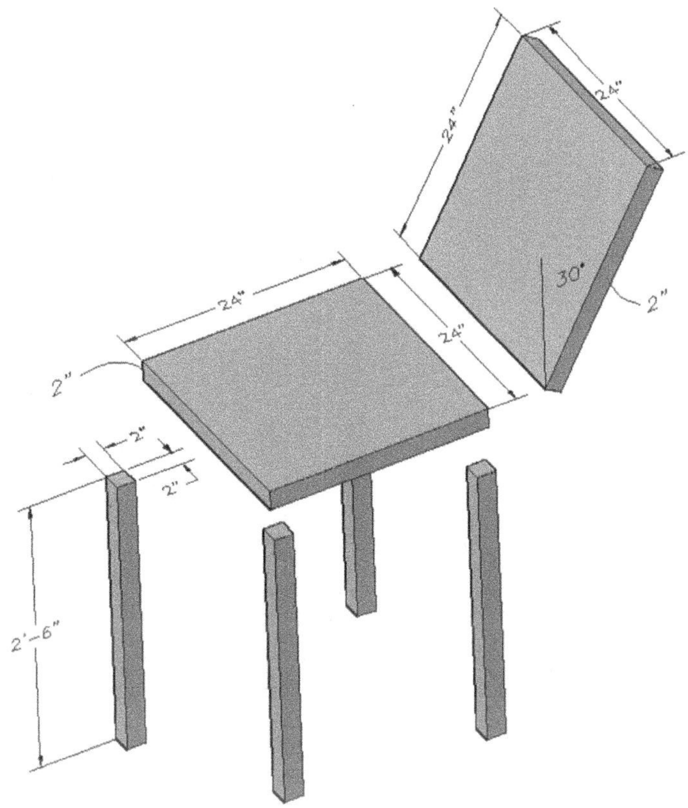

FIGURE 22.5 Drawing project, 3D chair.

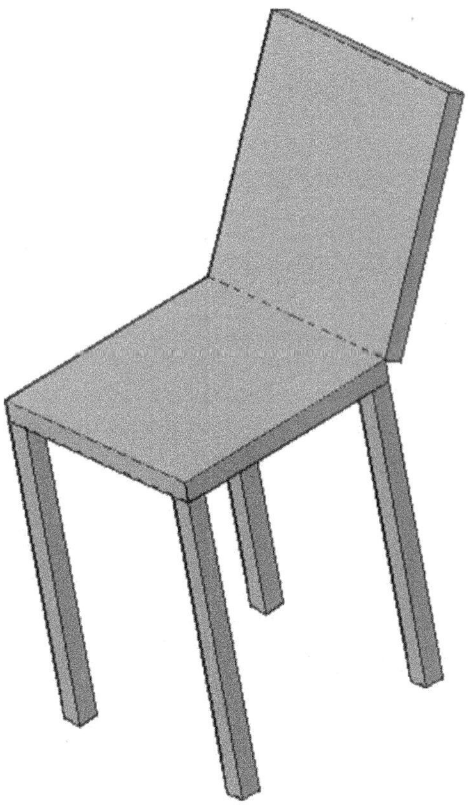

FIGURE 22.6 3D chair, full assembly.

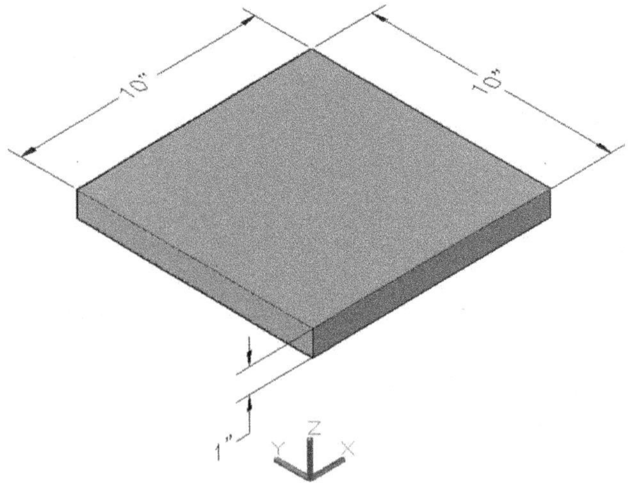

FIGURE 22.7 Flat plate for Mirror 3D.

Let us now run through the full mirroring procedure. Mirror3D can also be written as 3Dmirror, with no effect on the command. No gizmo is associated with it, but there is a Ribbon icon, as seen here in the command matrix.

Keyboard: Type in **mirror3d** and press Enter
Cascading menus: **Modify→3D Operations→3D Mirror**
Toolbar icon: none
Ribbon: **Home** tab→**Modify**

Step 1. Start up the command via any of the preceding methods.

- AutoCAD says: `Select objects:`

Step 2. Select the plate and press Enter.

- AutoCAD says: `Specify first point of mirror plane (3 points) or[Object/Last/Zaxis/ View/XY/YZ/ZX/ 3points] <3points>:`

Step 3. It is here that you need to select the correct plane. As discussed before, type in ZX and press Enter.

- AutoCAD says: `Specify point on ZX plane <0,0,0>:`

Step 4. Pick a point somewhere on the mirroring plane using OSNAPs.

- AutoCAD says: `Delete source objects? [Yes/No] <N>:`

Step 5. This is similar to the 2D version of the mirror command. Press Enter for the default. The mirrored image appears as shown in Fig. 22.8.

Go ahead and mirror the two plates over the remaining planes. The procedure is exactly the same, except you type in YZ for the first mirror plane and XY for the second. The expected results are shown in Figs. 22.9 and 22.10, respectively.

3D Modeling Exercise: Bookshelf

Create the model of a bookshelf based on the dimensions given in Fig. 22.11. The full assembly is shown in Fig. 22.12. Specifically you need to

- Create four 25″×80″ rectangles extruded 2″ for the top, bottom, and side shelves.
- Create three 25″×76″ rectangles extruded 2″ for the inner shelves.
- Rotate3D any parts that need to be rotated according to design requirements.
- Keep in mind that the distances between inner shelves can be a random value.
- Make sure everything fits together perfectly, adding shading and color of your choice.

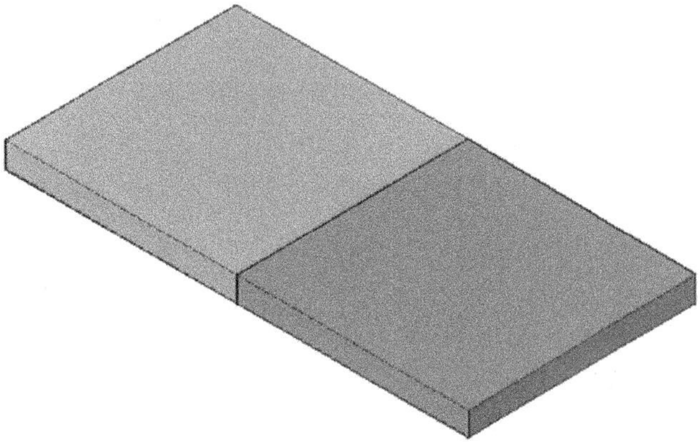

FIGURE 22.8 Flat plate mirrored over the ZX plane.

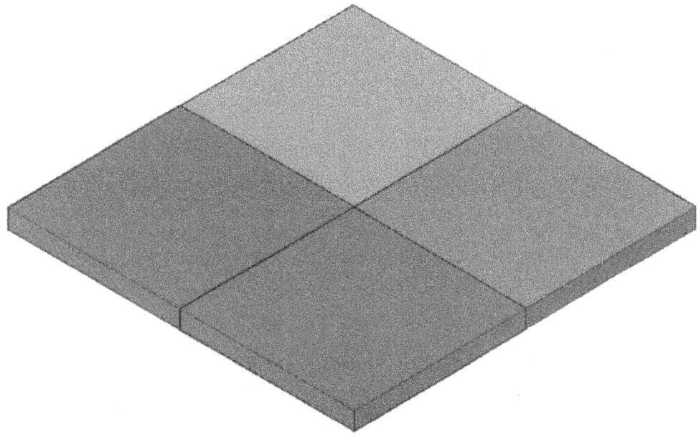

FIGURE 22.9 Flat plates mirrored over the YZ plane.

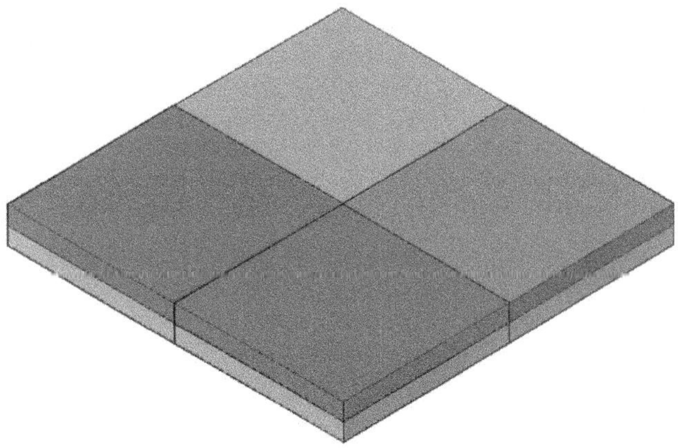

FIGURE 22.10 Flat plates mirrored over the XY plane.

3Darray

The 3Darray is quite similar in principle to the regular 2D one. There is, for starters, both a rectangular and polar form, just as in 2D. With the rectangular array, you just add a 3D component (a level) to the familiar column and row. For the polar array, the one big difference is that two points define the axis of rotation instead of just one point for the center. Another difference is the curious fact that 3Darray is all command-line driven; there is no dialog box.

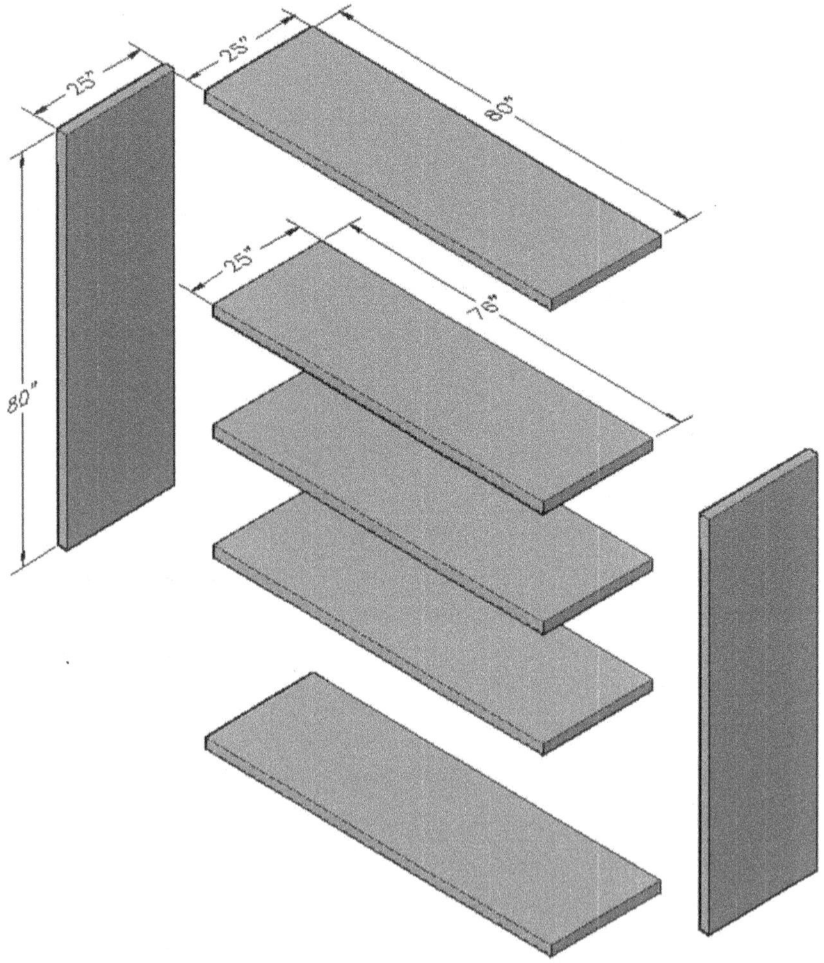

FIGURE 22.11 Drawing project, 3D bookshelf.

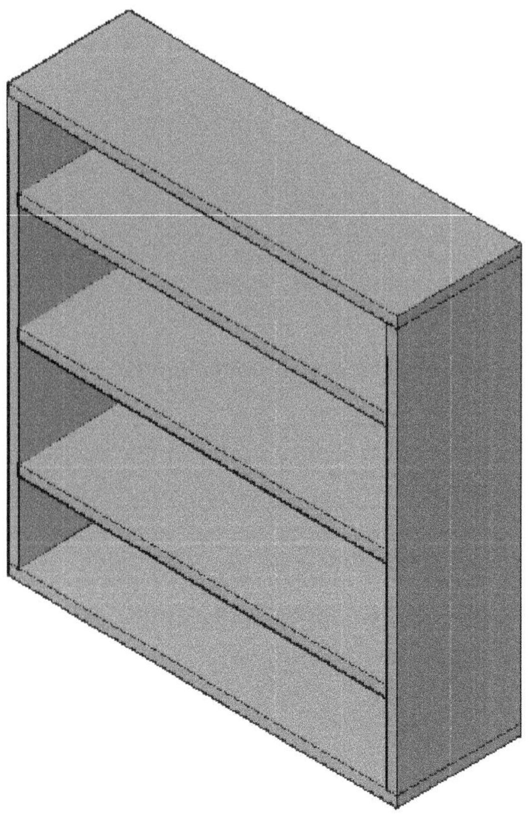

FIGURE 22.12 3D bookshelf, full assembly.

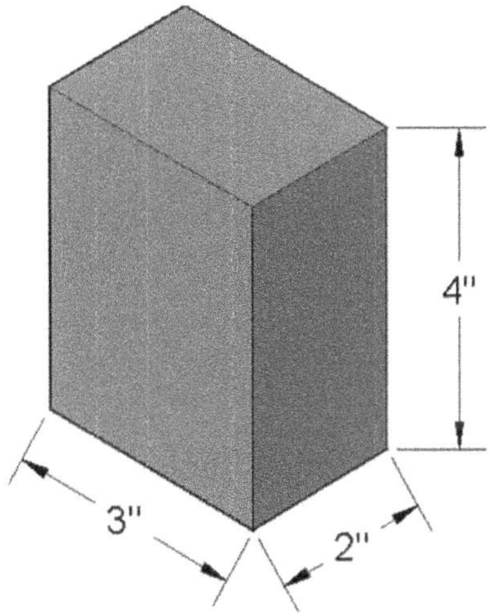

FIGURE 22.13 Extruded block for 3D rectangular array.

Let us go ahead and try out the command, starting with the rectangular array. Clear your screen and draw a 2″×3″ rectangle extruded to 4″, as seen in Fig. 22.13. Add shading and a color of your choice.

Keyboard: Type in **3darray** and press Enter
Cascading menus: **Modify→3D Operations→3D Array**
Toolbar icon: **Modeling** toolbar
Ribbon: **Home** tab→**Modify**

Step 1. Startup 3D array via any of the preceding methods.

- AutoCAD says: Select objects:

Step 2. Select the cube and press Enter.

- AutoCAD says: Enter the type of array [Rectangular/Polar]<R>:

Step 3. Press Enter to accept the default Rectangular array.

- AutoCAD says: Enter the number of rows (—) <1>:

Step 4. Enter a small value, such as 5, and press Enter.

- AutoCAD says: Enter the number of columns (|||) <1>:

Step 5. Enter a small value, such as 5, and press Enter.

- AutoCAD says: Enter the number of levels (...) <1>:

Step 6. Enter a small value, such as 5, and press Enter.

- AutoCAD says: Specify the distance between rows (—):

Step 7. Here, be sure to enter a number large enough to allow for space between the rows, 10 in this case.

- AutoCAD says: Specify the distance between columns (|||):

Step 8. Be sure to enter a number large enough to allow for space between the columns, 10 in this case.

- AutoCAD says: Specify the distance between levels (...):

Step 9. Be sure to enter a number large enough to allow for space between the levels, 10 in this case.

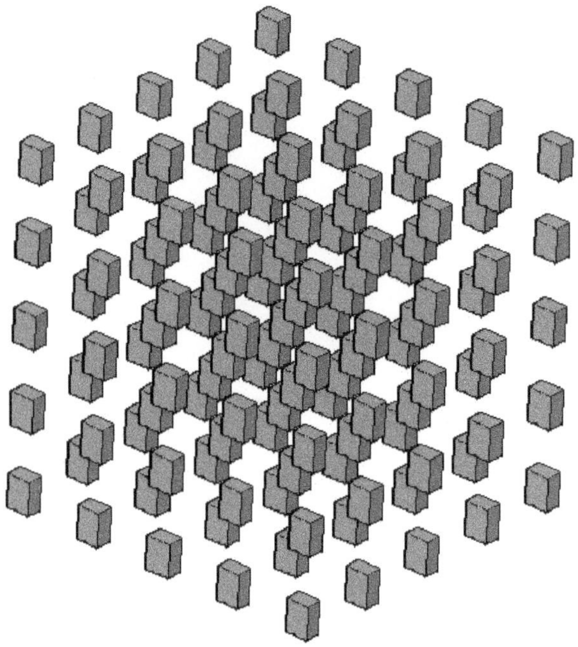

FIGURE 22.14 3D rectangular array.

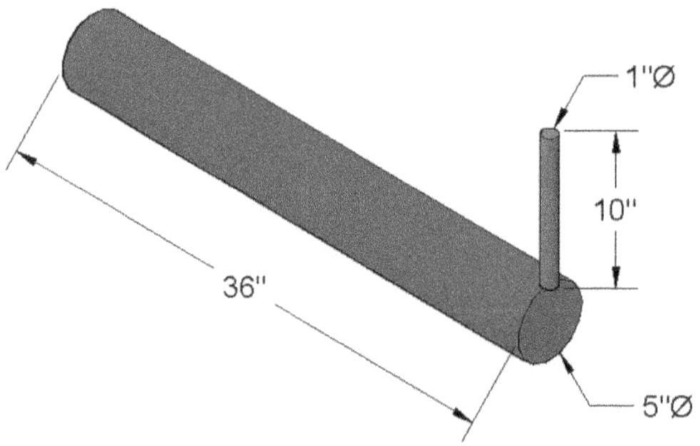

FIGURE 22.15 Axle and spoke for 3D polar array.

The result of the array with the 125 elements is shown in Fig. 22.14. This exact type of array may not be used that often but is still valuable to know. Next, we go over the vastly more useful polar version.

To demonstrate the polar array, we draw an axle with spokes. To create an axle, draw a circle 5″ in diameter and extrude it to 36″. Then, draw a small spoke, say a 1″ circle extruded to 10″, and position it carefully in a perpendicular manner on the tip of the axle. You likely have to rotate3D the spoke into position and attach it using quadrant or center points. After adding shading and color, the result is as shown in Fig. 22.15.

Step 1. Startup 3D array via any of the same methods shown earlier in the matrix for the rectangular array.

- AutoCAD says: `Select objects:`

Step 2. Select the spoke and press Enter.

- AutoCAD says: `Enter the type of array [Rectangular/Polar]<R>:`

Step 3. Press p to accept the polar array.

- AutoCAD says: `Enter the number of items in the array:`

Step 4. You can enter any value, but let us stick to 10 in this case.

- AutoCAD says: `Specify the angle to fill (+=ccw, -=cw) <360>:`

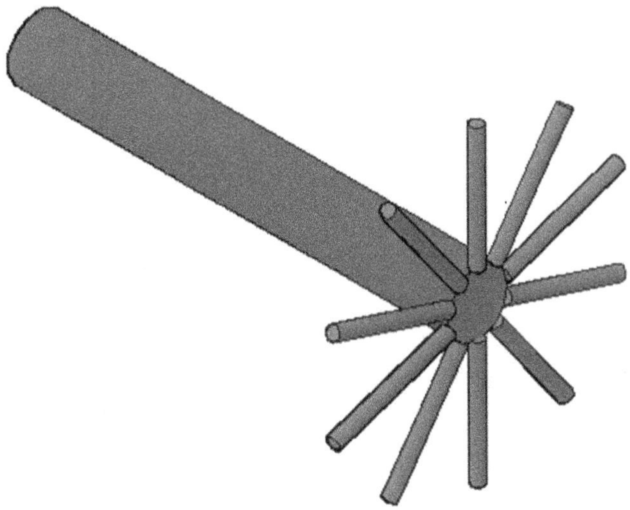

FIGURE 22.16 Completed 3D polar array.

Step 5. Press Enter to accept the default 360°, as we would like to go all the way around.

- AutoCAD says: Rotate arrayed objects? [Yes/No] <Y>:

Step 6. Press Enter for Yes, as we want the spokes to rotate.

- AutoCAD says: Specify center point of array:

Step 7. Here, you need to specify the **center** of rotation. Pick the center of the axle using the Center OSNAP.

- AutoCAD says: Specify second point on axis of rotation:

Step 8. Here, you need to pick the second center point using OSNAPs on the other side of the axle. These two points together indicate the axis of rotation for the spokes. The result is shown in Fig. 22.16.

3Dscale

This 3D version of the scale command is not much different from the regular scale (which you can also use in 3D). There is no icon or cascading menu for this command; you have to type it in or use the Ribbon. Draw a basic 3D shaded cube as you did just before for 3Darray. It is reproduced again with a different color (Fig. 22.17).

Keyboard: Type in **3dscale** and press Enter
Cascading menus: none
Toolbar icon: none
Ribbon: **Solid** tab→**Selection**

Step 1. Start up the 3Dscale command via one of the two preceding methods.

- AutoCAD says: Select objects:

Step 2. Select the cube and press Enter.

- AutoCAD says: Specify base point:

Step 3. You also see a scaling gizmo appear, which snaps to whatever base point you select (Fig. 22.18).

- AutoCAD then says: Pick a scale axis or plane:

FIGURE 22.17 Extruded block for 3D scale.

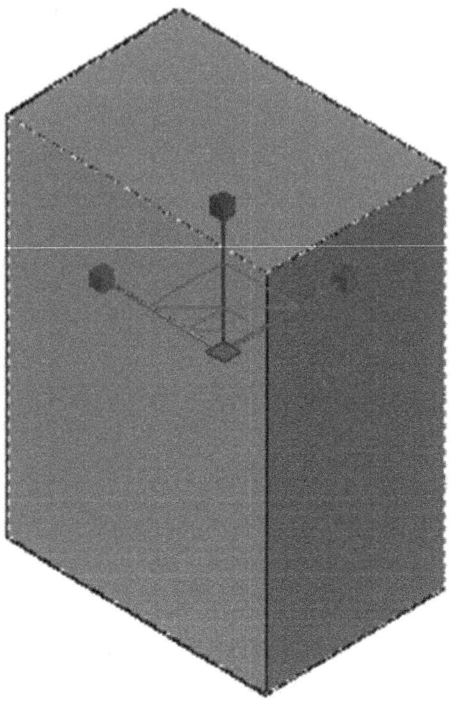

FIGURE 22.18 3Dscale gizmo.

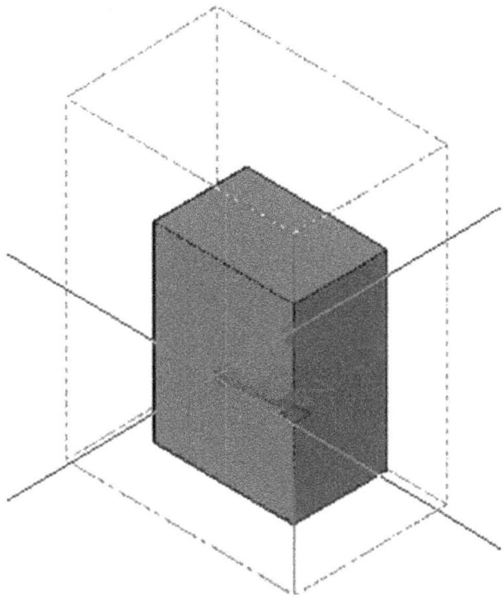

FIGURE 22.19 3Dscale in progress.

Step 4. Pick the XY plane. You can then move the mouse to scale the object or type in a value, as seen in Fig. 22.19.

3Dmove

Just as 3Dscale was somewhat of a duplication of the regular scale command, so is 3Dmove. However, there is a difference, as this command allows for constrained movement of an object along any selected axis or plane, regardless of your UCS orientation or whether or not Ortho is on. In that regard, it is quite useful, because just randomly moving an object in 3D space (with no constraint, such as Ortho) places the object literally anywhere, and usually nowhere near where you think it is in relation to other objects. The 3D space is quite deceptive. For proof of this, try moving something then using 3D Orbit to glance at the objects from a different perspective. You may be surprised where it really is. Let us try the 3Dmove command on the same random cube you just used.

Keyboard: Type in **3dmove** and press Enter
Cascading menus: **Modify→3D Operations→3D Move**
Toolbar icon: **Modeling** toolbar→
Ribbon: **Solid tab→Selection**

Step 1. Start up the 3Dmove command via any of the preceding methods.

- AutoCAD says: `Select objects:`

Step 2. Select the cube and press Enter.

- AutoCAD says: `Specify base point or [Displacement] <Displacement>:`

Step 3. A 3Dmove gizmo appears, but do *not* pick a base point. Instead, use the mouse to carefully select an axis. As soon as you do, the axis turns gold in color. The cube is constrained (even if Ortho is not on), and you can now slide the cube up and down that axis, as seen in Fig. 22.20.

The procedure is the same for the Y and Z axes. If you select a plane, however, you do need to have Ortho on to be constrained to the chosen plane.

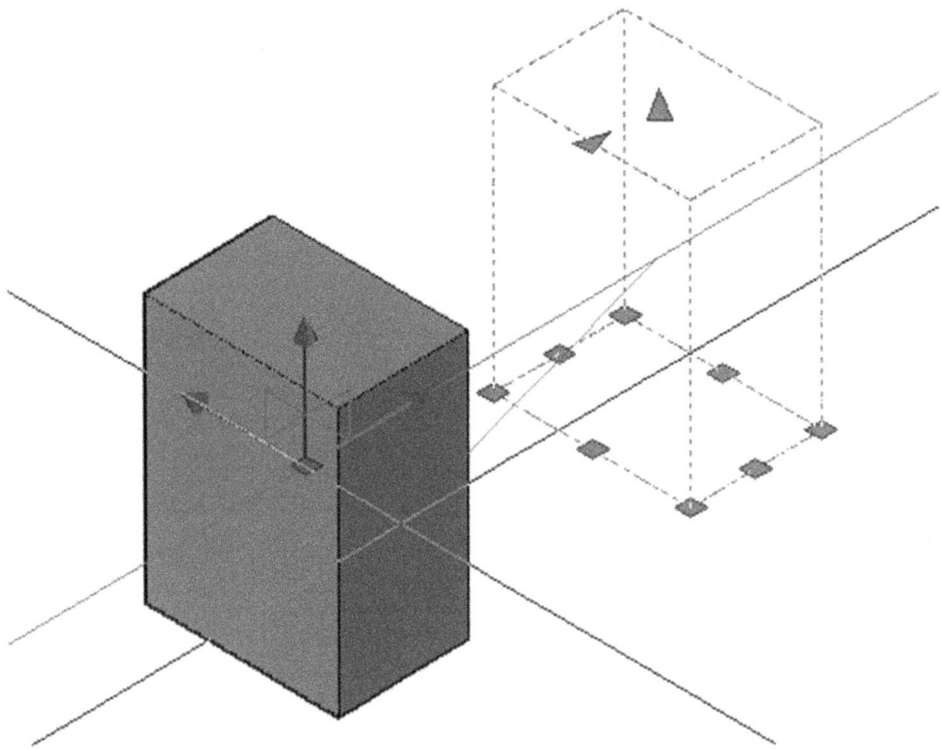

FIGURE 22.20 3Dmove in progress (X axis).

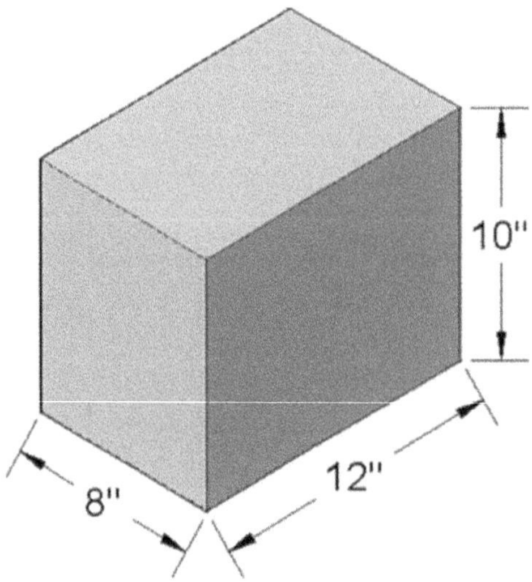

FIGURE 22.21 Extruded block for fillet.

Fillets and Chamfers in 3D

Recall the operation of the fillet command (with a radius of zero) in 2D AutoCAD. You could use the command to trim two intersecting lines or close a gap between those lines. None of that makes any sense when working with 3D models, which are, by definition, already closed shapes. However, recall also that, in 2D, we gave fillets radiuses. We use this feature in 3D by adding radiuses or rounded corners on solid objects. In a way we now use the original and more correct form of the fillet command, which by definition has always been a rounded corner.

There are two ways to work with the fillet command in 3D. One way is by using the 2D approach, that is, typing, toolbar, cascading menu, or Ribbon. The other is via the dedicated Fillet tool available as part of the Solid Editing toolbar. The result is the same but the approach is different. Let us try the 2D version first.

FIGURE 22.22 3D fillet.

Draw a 12″×8″ box and extrude it to 10″, as seen in Fig. 22.21.

Step 1. As first described in Chapter 1, start up fillet via any method you choose.

- AutoCAD says: `Current settings: Mode = TRIM, Radius = 0.0000`
  `Select first object or [Undo/Polyline/Radius/Trim/Multiple]:`

Step 2. Press `r` for `Radius`.

- AutoCAD says: `Specify fillet radius <0.0000>:`

Step 3. Enter a reasonable value such as `1"` and press Enter.

- AutoCAD says: `Select first object or [Undo/Polyline/Radius/Trim/Multiple]:`

Step 4. Select one of the cube edges facing you.

- AutoCAD says: `Enter fillet radius <1.0000>:`

Step 5. Confirm this by pressing Enter and continue selecting the remaining two edges. Each time,

- AutoCAD says: `Select an edge or [Chain/Radius]:`

Step 6. After all three edges are selected, you see them dashed. Finally press Enter.

- AutoCAD says: `3 edge(s) selected for fillet.`

The result is shown in Fig. 22.22.

The color of the fillets may be different from the rest of the object, as seen in Fig. 22.22. This depends on what layer is active as you perform the fillet. In this case it was 0, and the color defaults to the standard gray (shown in dark gray, here), although the block itself is colored (shown here in medium and light gray).

The 3D version of fillet is a bit more dynamic in nature. Create another 12″×10″×8″ box and use the commands seen in the following matrix.

Keyboard: Type in **filletedge** and press Enter
Cascading menus: **Modify→Solid Editing→Fillet edges**
Toolbar icon: **Solid Editing** toolbar→
Ribbon: **Solid** tab→**Solid Editing**

- AutoCAD says: `Radius=1.0000 Select an edge or [Chain/Radius]:`

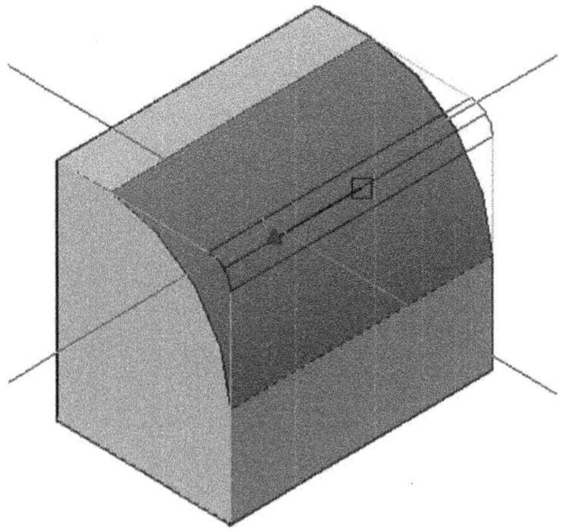

FIGURE 22.23 3D fillet.

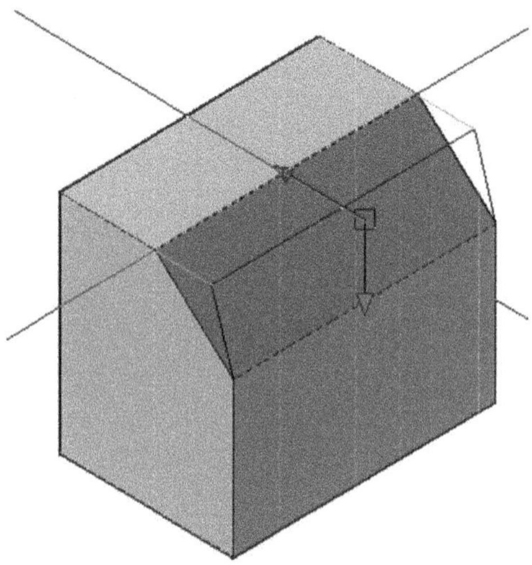

FIGURE 22.24 Dynamic chamfering in 3D.

Select the edge to be filleted and it becomes shaded. You may change the radius or press Enter to accept. You then see a red triangle (a grip). Click on it and drag your mouse back and forth, noting how the fillet changes dynamically, as seen in Fig. 22.23.

Click again and press Enter to confirm the final fillet. While this looks impressive, in most cases, you will likely want to specify an exact fillet size, so the dynamic filleting may not find much use in real design work.

Chamfering is quite similar, so we do not run through the specific steps, but you can also do it through the standard 2D approach or via the 3D dynamic chamfer, as shown next with the command matrix.

Keyboard: Type in **chamferedge** and press Enter
Cascading menus: **Modify→Solid Editing→Chamfer edges**
Toolbar icon: **Solid Editing** toolbar→
Ribbon: **Solid** tab→**Solid Editing**

Once again the edge to be chamfered is shaded and you must enter sizes when prompted or press Enter to accept. You then see two triangular grips this time (Fig. 22.24). Each can be moved to dynamically expand the sides of the chamfer. As with the fillet, you may want to just type in the value.

Review everything covered in this chapter. In the next chapter, we cover Boolean operations, the final key to flat design.

Summary

You should understand and know how to use the following concepts and commands before moving on to Chapter 23:

- Rotate3D
- 3Drotate (gizmo)
- Mirror3D
- 3Darray
- Polar
- Rectangular
- 3Dscale
- 3Dmove
- Fillets in 3D
- Chamfers in 3D

Review Questions

Answer the following based on what you learned in this chapter:

1. Describe the rotate3D procedure. Do you need axes or planes for this command?
2. How is 3Drotate different from rotate3D?
3. Describe the mirror3D procedure. Do you need axes or planes for this command?
4. Describe the 3Darray procedure, both rectangular and polar.
5. Describe 3Dscale and 3Dmove.
6. Describe fillets and chamfers in 3D.

Exercises

1. Create the following table using the rectangle, circle, extrude, fillet, mirror3D, and a few other commands. The exact size of the table can be anything you want, but if you need some values to get started, the tabletop is 50″×30″ and the bottom is 20″×20″. All fillets and thicknesses are 2″. (Difficulty level: Easy; Time to completion: 10 minutes.)

2. Create the following buggy frame. You will need the circle, extrude, 3Darray, donut, copy, and some of the UCS rotation commands. Once again, sizing is not what is important here, but you can make the two main axles 72″ in length and base everything else on that. (Difficulty level: Moderate; Time to completion: 15 minutes.)

3. Create the following helicopter model using the tools you learned in this chapter. While this design will not win any awards, it will really get you using recently learned commands, such as extrude, fillet, rotate3D, and 3Darray. Start out by creating a basic blade profile (an airfoil) using pline. Then, extrude it and 3Darray the resulting blade around a rotor, similar to the axle and spokes in the previous exercise. Create the body using extrude and fillet. Attach a tail using basic extrude and scale down a copy of the main rotor for the tail rotor, placing everything into position. Finally, shade and color the model. Sizing is up to you. (Difficulty level: Moderate; Time to completion: 30 minutes.)

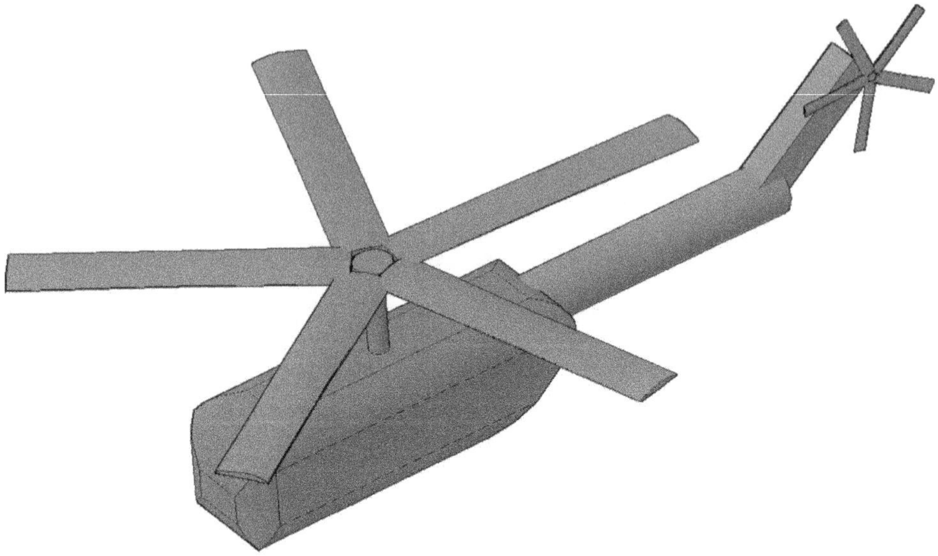

4. Create the following model of a basic couch. It consists of a lower main frame, backside, and left and right sides (which are filleted). Then add in four legs and two pillows. All the shapes, and their sizing and extrusions, are outlined beneath the model. Set up layers and colors as you see fit. (Difficulty level: Moderate; Time to completion: 15–20 minutes.)

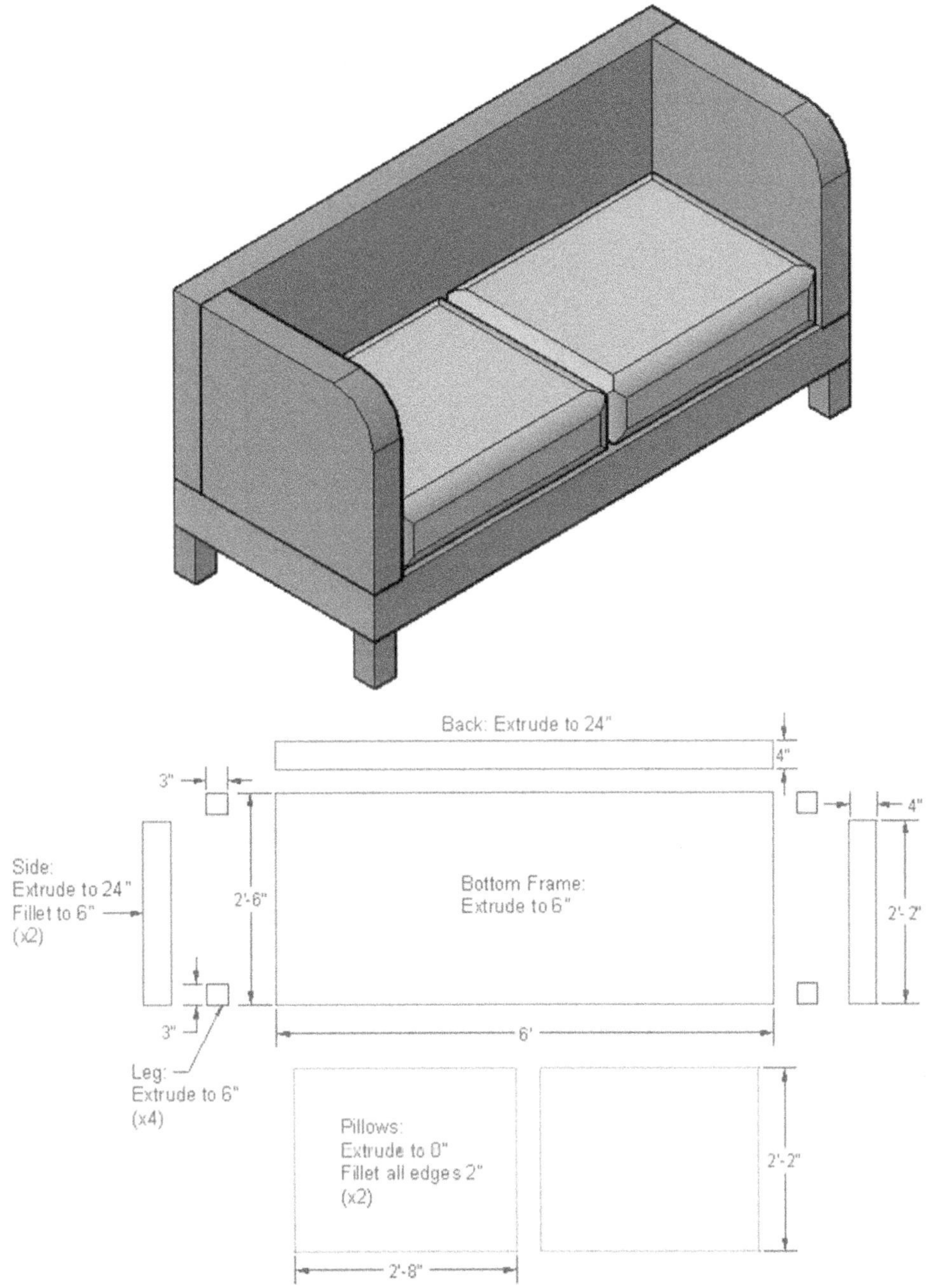

5. Here is one more set of furniture to work on. Create the following table, chair, and floor scene. It is shown from one main angle as well as an additional "exploded" view, and a detail of the wheel. In this exercise, no dimensions are given, so use whatever you think is reasonable - the focus here is on the AutoCAD commands. The floor, legs, and table are all basic extrusions. The chair features extrusions, 3Darray, mirror3D, rotate3D, and some filleting. The "wheels" of the chair are nothing more than an extrusion with a large fillet on one side to form a half-sphere and mirrored. (Difficulty level: Moderate; Time to completion: 30–45 minutes.)

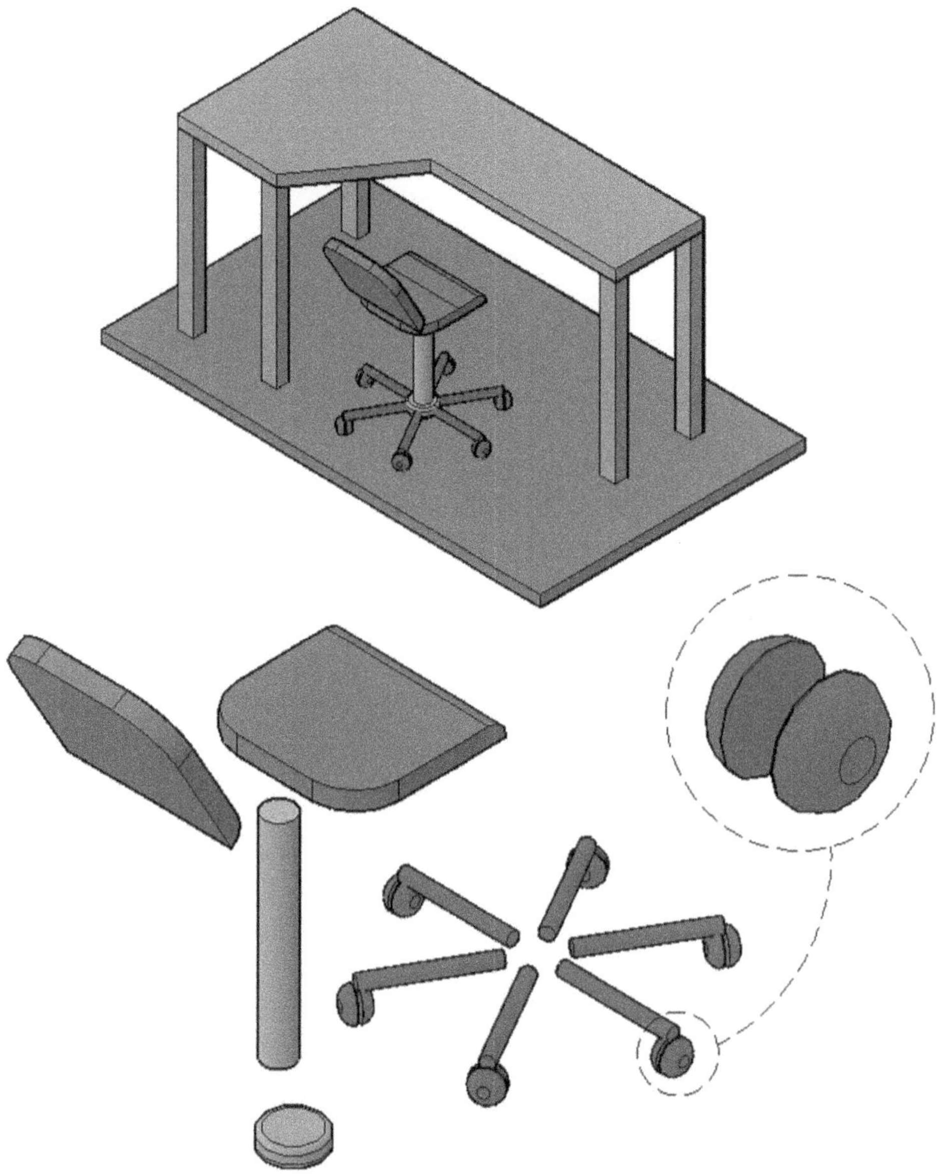

Spotlight on: chemical engineering

Chemical engineering is a specialized branch of engineering that combines chemistry, physics, and sometimes biology to process raw materials (often chemicals) into useful products. Chemical engineers are broadly divided into two categories. The category most people associate with the profession is "process engineering." Here, the focus is on the design and operations of industrial plants. The second category, less known, but just as important, is "product engineering." Here, the engineers are tasked with the development of chemical substances for products such as cleaners, pharmaceuticals, and other items, including food and beverages.

Often students ask how exactly chemical engineering is different from chemistry. The difference is that chemists generally work with basic compositions of materials, on the atomic and molecular level, using this knowledge to synthesize products. Chemical engineers (and some *applied* chemists), however, are more interested in developing higher level "end user" products and industrial processes. These processes utilize scientific principles but do not develop them. In short: chemists come up with our fundamental knowledge; chemical engineers apply it in ways useful to society.

FIGURE 1 *Image Source: www.omanengineering.blogspot.com.*

Education for chemical engineers starts out similar to education for the other engineering disciplines. In the United States, all engineers typically have to attend a 4-year ABET-accredited school for their entry-level degree, a Bachelor of Science. While there, all students go through a somewhat similar program in their first 2 years, regardless of future specialization. Classes for chemical engineers include extensive math, physics, chemistry, and thermodynamics. In their final 2 years, chemical engineers specialize by taking courses in biochemistry, kinetics, transport processes, polymer science, and materials, among others.

FIGURE 2 *Image Source: www.chem-eng.blogspot.com.*

On graduation, chemical engineers can immediately enter the workforce or go on to graduate school. Although not required, some engineers choose to pursue a Professional Engineer (P.E.) license. The process first involves passing a Fundamentals of Engineering (F.E.) exam, followed by several years of work experience under a registered P.E., finally sitting for the Chemical Engineering P.E. exam itself.

Chemical engineers can generally expect starting salaries in the $65,000/year range with a bachelor's degree and $77,000 with a master's degree, which is one of the highest among engineering specialties, just behind petroleum engineering, which is a field that attracts some chemical engineers. The pay with 10 years of experience is over $95,000. This, of course, depends highly on market demand and location. A master's degree is highly desirable and required for many management spots.

So, how do chemical engineers use AutoCAD and what can you expect? One major application of AutoCAD in this profession is P&ID. This is a topic that was mentioned in several Level 1 chapters, with a few corresponding exercises. P&ID stands for "piping and instrumentation diagrams," and these are commonly used in chemical and process engineering with respect to plant design. A basic P&ID diagram is shown in Fig. 3.

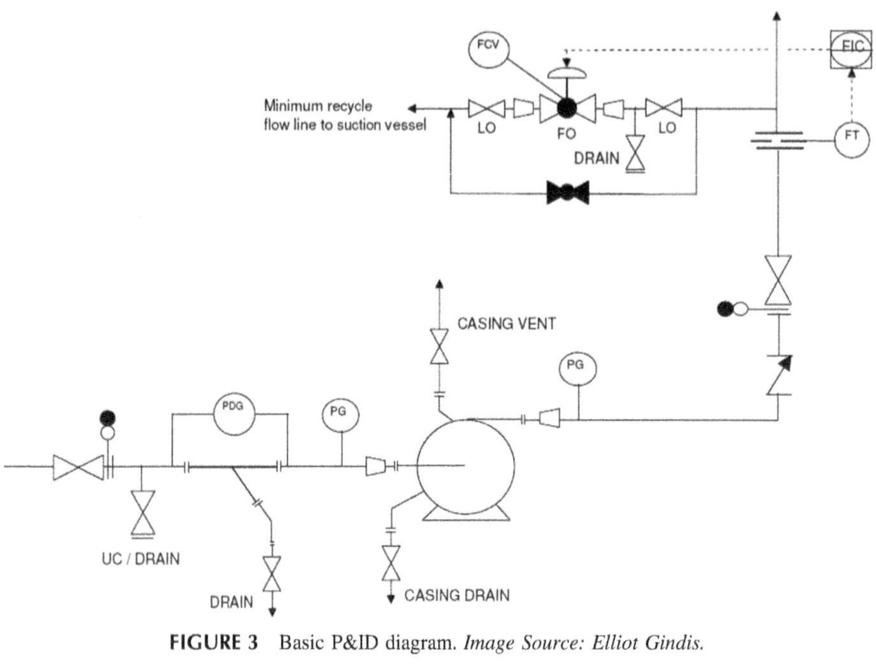

FIGURE 3 Basic P&ID diagram. *Image Source: Elliot Gindis.*

Much like electrical schematics, most P&IDs are 2D in nature (though 3D models are used on occasion) and inherently simple from an AutoCAD perspective, though probably far from simple in the engineering sense. The key here is accuracy and precision! The schematics must be easy to understand and read, and that may be the biggest challenge and responsibility of drafting—to present a complex design in an easy-to-read manner. The layering system is relatively straightforward, with layer names reflecting the functionality of the pieces. Expect to see modular design, symbol libraries, and data-embedded blocks, a.k.a. *attributes*, in this field.

Autodesk has an add-on package to automate and simplify working on P&IDs called, unsurprisingly, *AutoCAD P&ID*. It has tools to assist with error checking, managing data, and symbol libraries. Autodesk also makes Plant 3D, for design and visualization of plant design. There are of course other companies in the market, such as PROCAD, Bentley and its OpenPlant software, Siemens with Cosmos P&ID, and Intergraph's SmartPlant. If your career takes you into this field, you will likely encounter these design and testing packages.

Chapter 23

Boolean operations and primitives

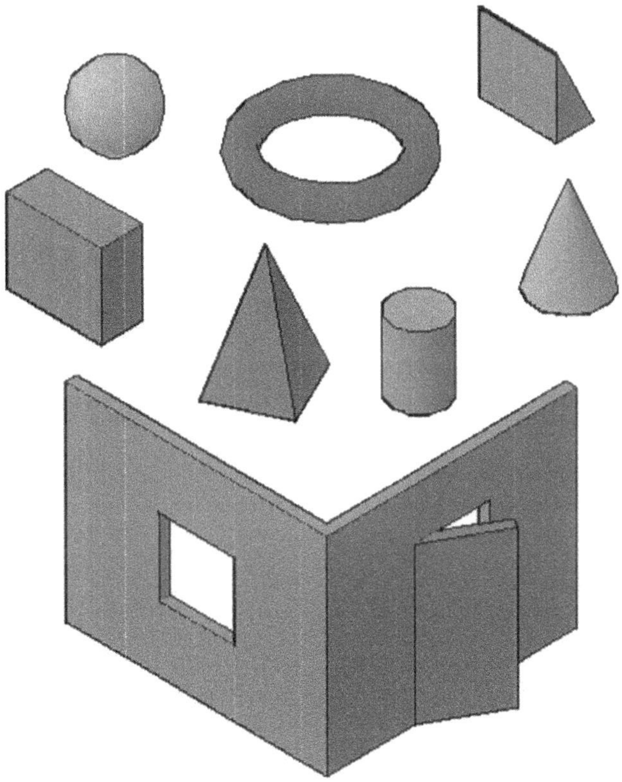

Learning objectives

In this chapter, we introduce a set of tools, called *Boolean operators*, that allow important additive and subtractive interactions between objects. We also introduce the concept of AutoCAD primitive shapes, known as just *primitives*. We specifically look at

- Union
- Subtract
- Intersect
- Box
- Wedge
- Cone
- Sphere
- Cylinder
- Torus
- Pyramid

By the end of this chapter, you will be able to create designs that feature openings and holes as well as complex unions and intersections. Also, you will understand basic primitive shapes and be able to apply them in design work as necessary.

Estimated time for completion of this chapter: 2–3 hours.

Up and Running with AutoCAD 2022. https://doi.org/10.1016/B978-0-323-89923-9.00023-8

23.1 Introduction to boolean operations

We have thus far been able to put together a few fairly simple objects using extrude and the 3D manipulation commands (mirror3D, rotate3D, etc.). What we still miss is the ability for objects to interact with each other, such as being fused together (e.g., two adjoining walls) or subtracted from each other (e.g., a hole in a wall representing a window).

This functionality, referred to as *Boolean operations*, is very important for advanced 3D modeling in both architectural and mechanical design. We cover the three Boolean operations of union, subtract, and intersect. The functionality of each can be represented by two overlapping circles (representing two solids), as shown in Fig. 23.1.

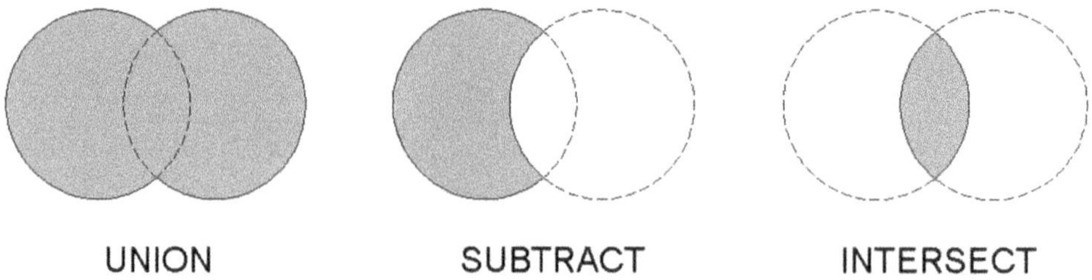

FIGURE 23.1 Boolean operations.

These circles also form the AutoCAD icons for these operations. As you can see from the shaded areas, the solids can be

- Joined together completely.
- Subtracted, leaving only one part of the solid (either one can be chosen).
- Intersected where only the intersecting geometry remains.

All three Boolean operators can be accessed via typing, cascading menus, a toolbar (Modeling toolbar, which you should already have up), and of course the Ribbon.

Union

The union operation fuses solids or regions together into one object. This is an irreversible operation, so be careful before you implement it. Objects do not actually have to be touching to be joined together. Union is not as commonly used as subtract but does find some use when joining together walls or mechanical assemblies.

Let us give union a try by first creating two solid cubes of any size and positioning them to intersect each other, as seen in Fig. 23.2. Notice that their grips have been activated, and you can see that the two cubes are distinct and separate entities, even though they are touching.

Keyboard: Type in **union** and press Enter
Cascading menus: **Modify→Solid Editing→Union**
Toolbar icon: **Modeling** toolbar→
Ribbon: **Solid** tab→**Boolean**

Step 1. Start the union command via any of the preceding methods.
AutoCAD says: `Select objects:`

Step 2. Select both cubes (they turn dashed) and press Enter. Both cubes now are fused, as seen in Fig. 23.3. Notice how the grips reflect one entity, not two. Go ahead and move the objects; they move as one. This becomes even clearer if we switch to wireframe (Fig. 23.4), with a color change for additional clarity.

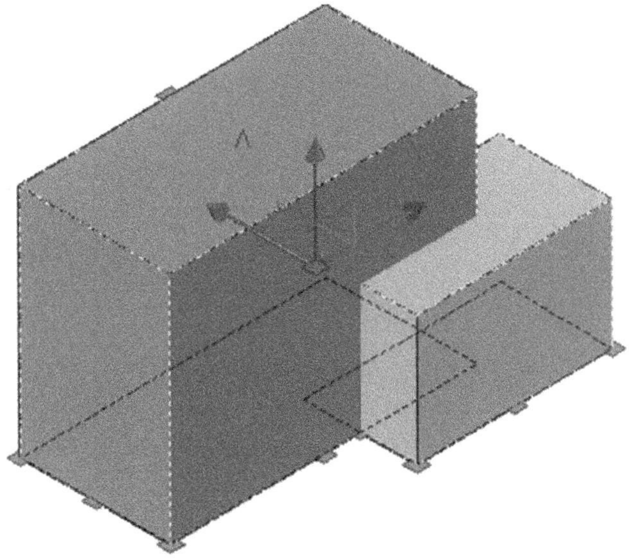

FIGURE 23.2 Two solid cubes before union.

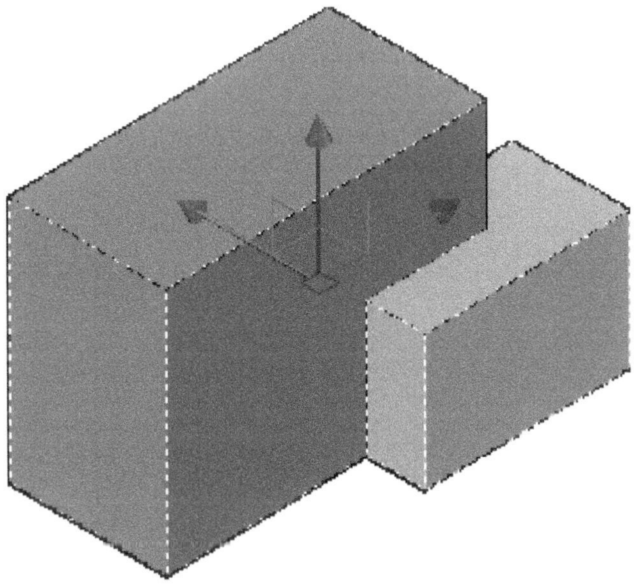

FIGURE 23.3 One solid cube after union (shaded view).

Subtract

This is a frequently used command, and it is quite indispensable in 3D work. As you may have guessed from the name, it involves the process of subtracting one object from another. In AutoCAD, this is the only way to take away material to create an empty space. Examples of this include the creation of doors and windows in 3D architectural floor plans and bolt holes in mechanical design, just to name a few. Let us set up just such a situation with a flat plate and a circle, which are then extruded and subtracted to form a hole.

Staying in wireframe mode (for clarity), draw a 10″×10″ square, extruding it to 2″, as shown in Fig. 23.5.

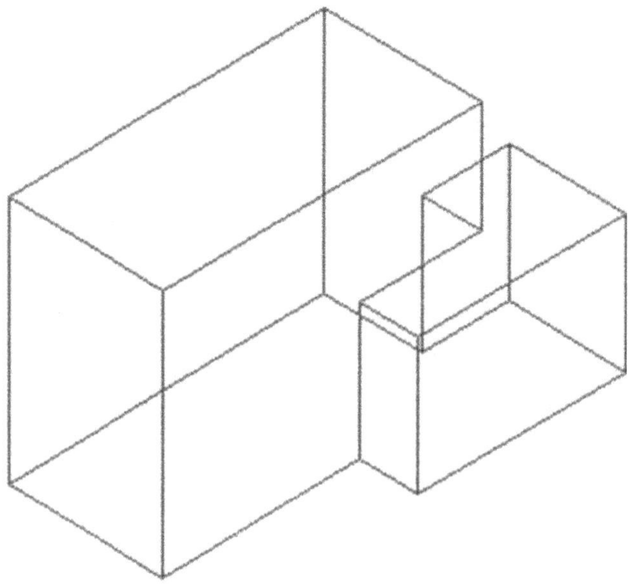

FIGURE 23.4 One solid cube after union (wireframe).

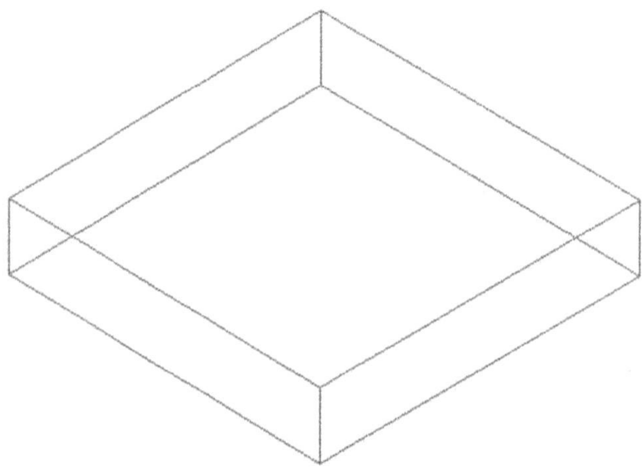

FIGURE 23.5 10" ×10" ×2" block.

Now, draw a guideline from corner to corner (use endpoints) of the top surface of the extruded rectangle and a circle of 2″ radius based on the midpoint of the guideline, as shown in Fig. 23.6. Note that switching to the wireframe view and adding temporary guidelines is a common technique in 3D.

You can erase the guideline and extrude the circle a total of −2″ (yes, that is a negative sign), as seen in Fig. 23.7. Although it seems like we are done, we are not; the "hole" is not yet a hole.

The final step is to subtract the extruded circle from the extruded rectangle, leaving a hole in its place.

Keyboard: Type in **subtract** and press Enter
Cascading menus: **Modify→Solid Editing→Subtract**
Toolbar icon: **Modeling** toolbar→
Ribbon: **Solid** tab→**Boolean**

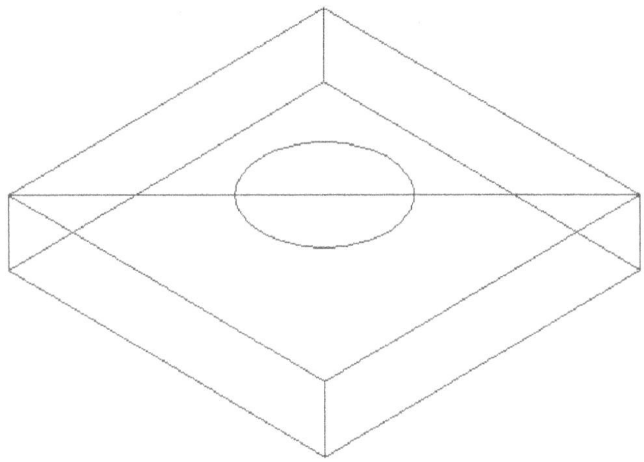

FIGURE 23.6 10" ×10" ×2" block.

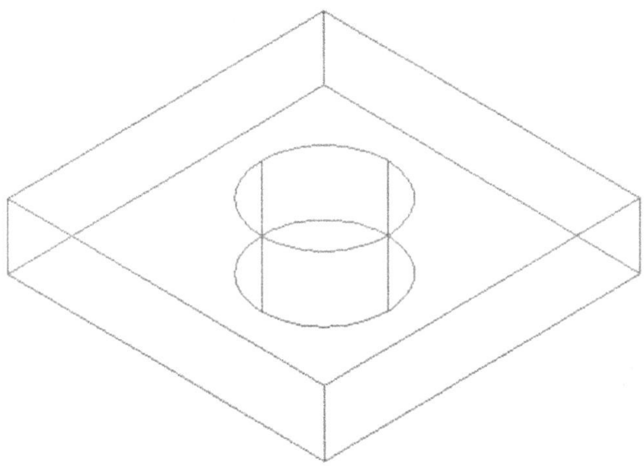

FIGURE 23.7 Circle extruded.

Step 1. Start up the subtract command via any of the preceding methods.

- AutoCAD says: `Select solids, surfaces, and regions to subtract from...`
  `Select objects`

Step 2. Here, you have to be careful. You must first select the material that you want to keep. In this case, it is the extruded square. Go ahead and select it, then press Enter.

- AutoCAD says: `Select solids and regions to subtract...`
  `Select objects:`

Step 3. Select the material you want to get rid of to make the hole. That of course is the extruded circle.

The subtraction operation is performed. Although you may not be able to immediately see what you did, a simple switch to the shaded mode reveals the hole, as seen in Fig. 23.8.

Be sure you thoroughly understand the essence of the subtract command; as mentioned earlier, we use it often. The key is to visualize what material you want removed and where and to create a shape to exactly reflect that. Remember again that the first selection is always for the material you want to keep; the second is for the material you want to remove.

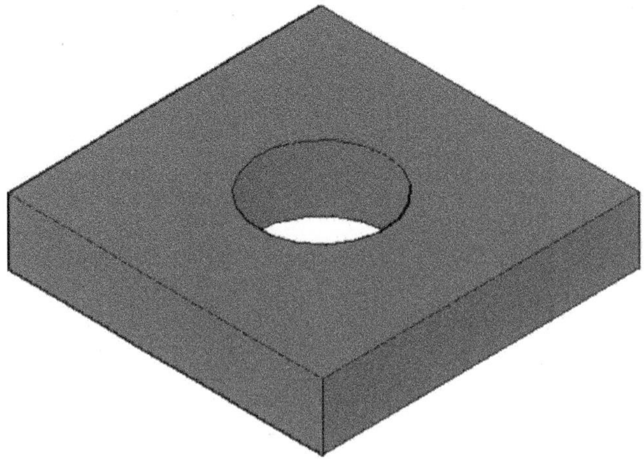

FIGURE 23.8 Completed subtraction operation.

Intersect

The final Boolean command creates a new shape out of the overlapping parts of other shapes and can be used to create some interesting objects. Let us create a simple intersection. Use the previous rectangle with the hole and copy it to the side (with Ortho on) to create two shapes that intersect each other. Fig. 23.9 shows you what you get. Notice the overlap between the shapes. This is what is kept after the intersection command is executed.

Keyboard: Type in **intersect** and press Enter
Cascading menus: **Modify→Solid Editing→Intersect**
Toolbar icon: **Modeling** toolbar→
Ribbon: **Solid** tab→**Boolean**

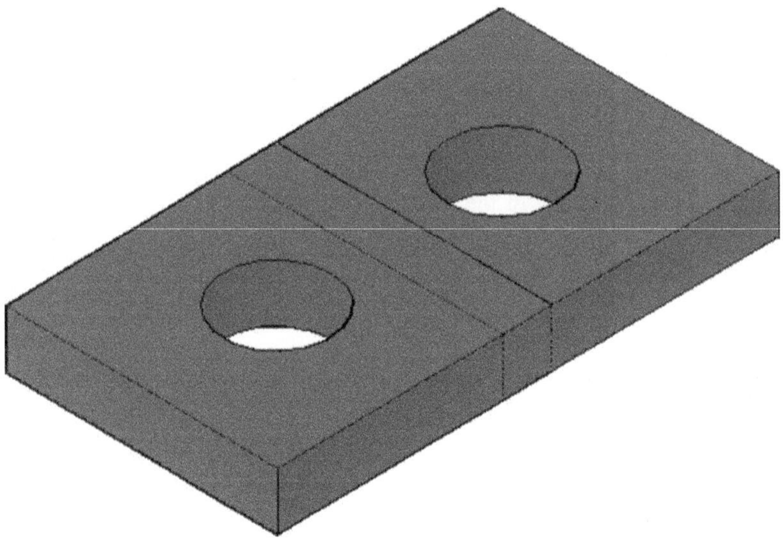

FIGURE 23.9 Intersected shapes.

Step 1. Start intersect via any of the preceding methods.

- AutoCAD says: `Select objects:`

Step 2. Select both pieces and press Enter. The result is shown in Fig. 23.10; only the intersecting part remains.

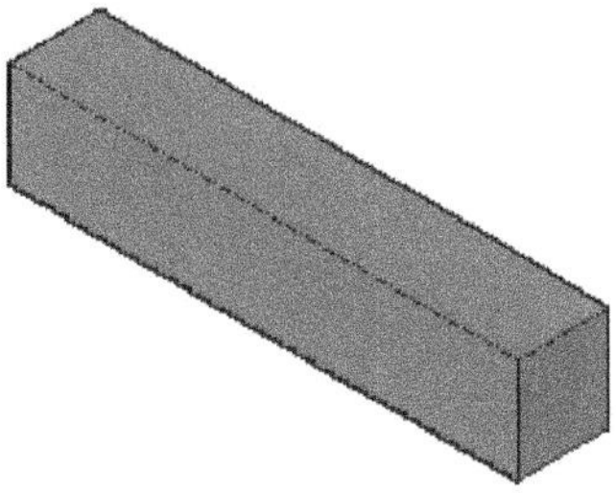

FIGURE 23.10 Completed intersect operation.

3D Modeling exercise: building wall with door and window

This next practice drawing is a very important one. We are creating a basic 3D model of two building walls, complete with a door and a window opening. Although this model is quite simple, it contains many of the important elements needed to create an entire building in 3D, the differences being only the amount of pieces present and some minor refinements. All the essentials are here, however, so follow all the steps closely.

Step 1. Draw two walls that are 8' high and 10' long, with a thickness of 5". One way to do this is to first draw a rectangle that is 5"×10' and extrude it to 8'. Then, copy the structure and reattach the copy at 90° to the original wall. Fig. 23.11 shows what you should see in wireframe mode.

Step 2. Use the union command to fuse the walls together. Fig. 23.12 shows another, slightly rotated, view of the walls in wireframe after the application of union. Notice where they intersect; they are clearly in one piece, not two separate panels.

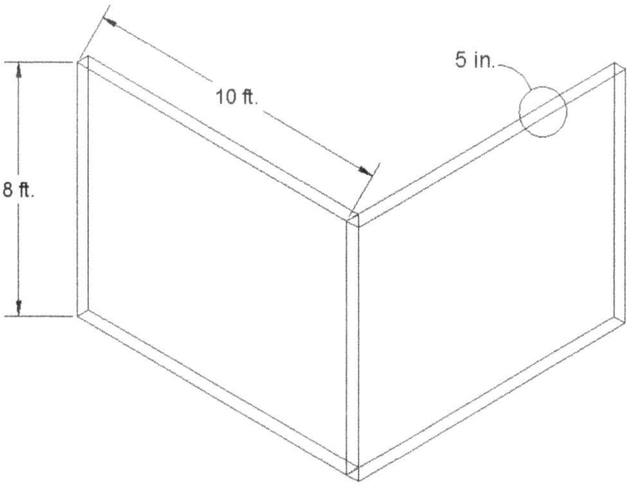

FIGURE 23.11 3D modeling exercise, Step 1.

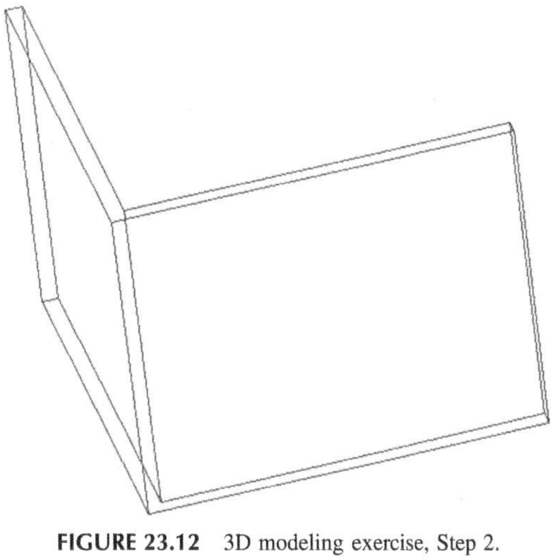

FIGURE 23.12 3D modeling exercise, Step 2.

Step 3. Now create a 36"× 36" window (off to the side) as well as a 36" wide× 72" high door (also off to the side). Both should be 5" thick, as that is the overall thickness of the walls. Rotate3D the objects into position if needed. The result is shown in Fig. 23.13.

Step 4. Let us think carefully about what to do next. The window is actually a hole in the wall, so we need not reuse it. All we need to do is position it in the right place and subtract. The door is a slightly different story. We need one copy of it to make the opening in the first place, and one copy to then install into the new opening (and rotate into an open position). So it is a good idea to make a copy of the door right now, because the process of subtracting deletes the original. Do not forget this, or you will end up redoing the door after using the original to create the opening.

Step 5. After copying the door, insert it carefully into the wall, making sure it is lined up correctly (midpoint of bottom front of door to midpoint of bottom front of wall). Then, insert the window in a similar manner and raise it up (using Ortho) to roughly the halfway point in the wall. You may have your own method of doing this as well, but either way, Fig. 23.14 shows what you should have as a result.

Step 6. Go ahead and subtract out the window and the door. Fig. 23.15 shows what you see in wireframe mode. Fig. 23.16 shows the same thing in shaded mode with color added.

Step 7. Finally, let us put the copy of the door back where it belongs and give it a−30° rotation (negative again). The final result is in Fig. 23.17.

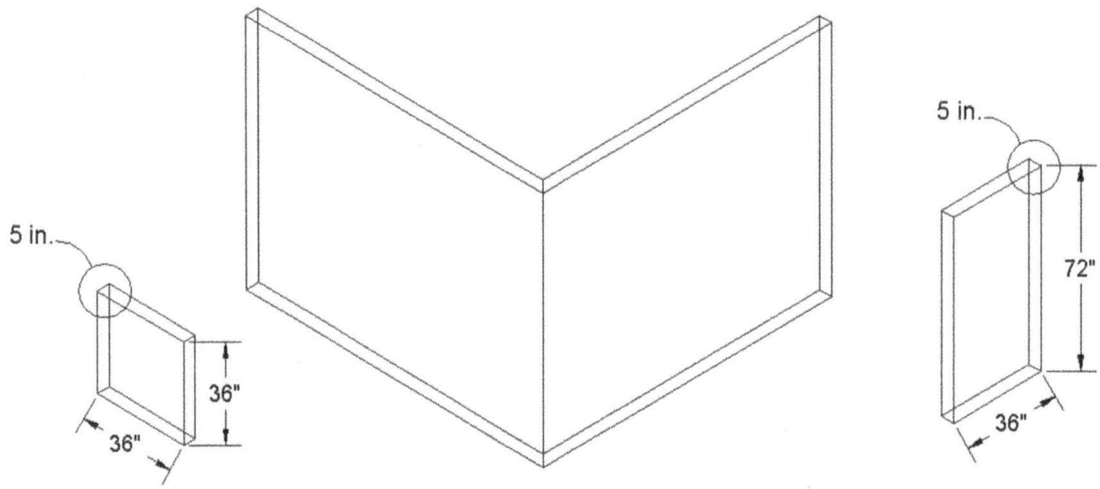

FIGURE 23.13 3D modeling exercise, Step 3.

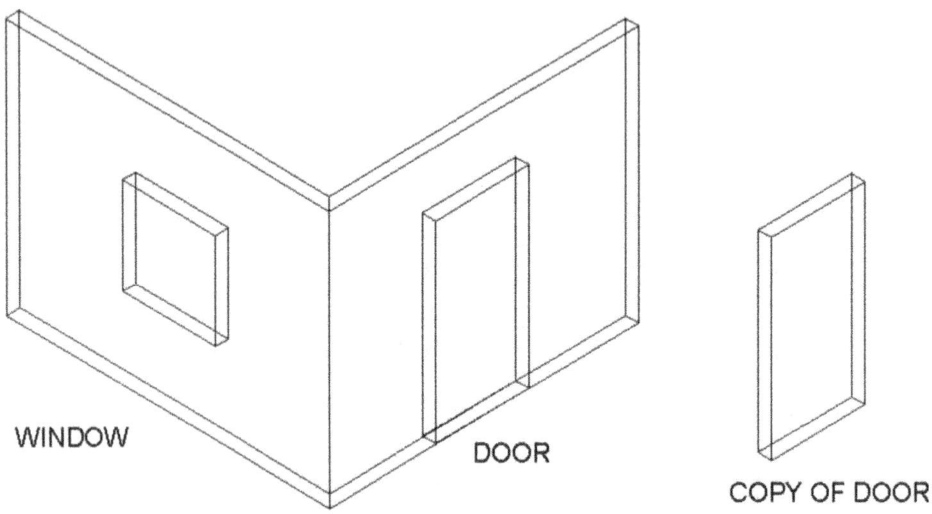

FIGURE 23.14 3D modeling exercise, Steps 4 and 5.

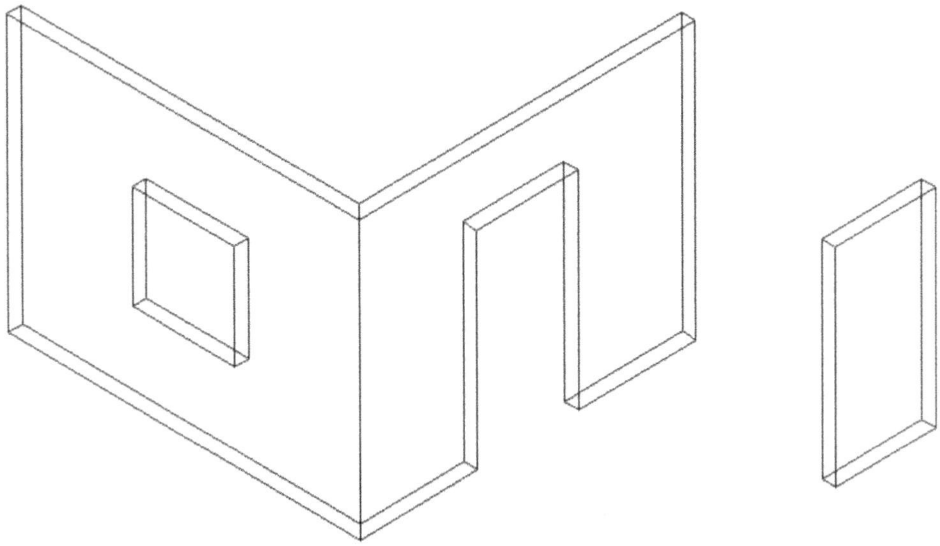

FIGURE 23.15 3D modeling exercise, Step 6 (wireframe).

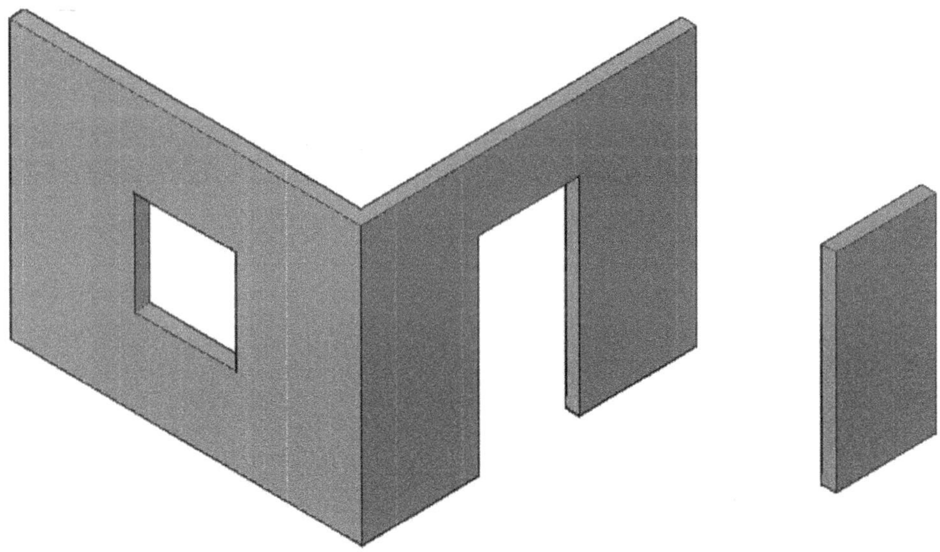

FIGURE 23.16 3D modeling exercise, Step 6 (solid).

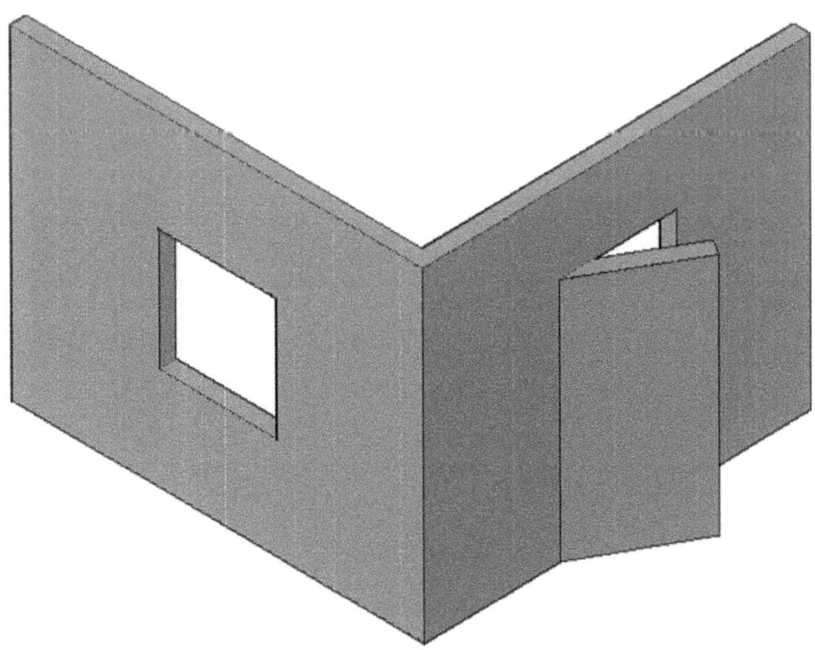

FIGURE 23.17 3D modeling exercise, Step 7.

23.2 Introduction to primitives

We now go over a very basic set of tools that you may find useful in certain situations. Called *primitives*, they are nothing more than ready-to-go shapes built into AutoCAD. Think of them as Lego pieces for computer-aided design; all you need to do is specify a few dimensional values and place them in your drawing as needed. The available primitives are

- Box
- Wedge
- Cone
- Sphere
- Cylinder
- Torus
- Pyramid

As with most commands you learn in AutoCAD, there are four ways to access these tools:

1. Typing in the desired primitive and pressing Enter.
2. Cascading menus under Draw→Modeling.
3. Using the Modeling toolbar; icons 2–8 from the left.
4. Using the Ribbon; Solid tab.

Once you select the primitive via one of these methods, simply enter the shape's dimensional values as prompted (such as height, width, angle) or just click or move the mouse for random values. It is best to have the Conceptual Visual Style shading turned on, as the shapes become clearer, as well as staying in the 3D SW Isometric View, to get the full effect. Hold off with trying the other shapes on the Modeling toolbar, as we go over them in advanced chapters. Finally, all primitives can be erased, moved, copied, and rotated in space, just like any other solid shape.

Box

Keyboard: Type in **box** and press Enter
Cascading menus: **Draw→Modeling→Box**
Toolbar icon: **Modeling** toolbar→
Ribbon: **Solid** tab→**Primitive**

Step 1. To try this first primitive, use any of the preceding methods.

- AutoCAD says: `Specify first corner or [Center]:`

Step 2. Click anywhere on the screen.

- AutoCAD says: `Specify other corner or [Cube/Length]:`

Step 3. A 2D solid rectangle is drawn based on the movements of your mouse, or you can enter a value by pressing `L` for `length`. Click randomly or enter a value.

- AutoCAD says: `Specify height or [2Point] <7.2545>:`

Step 4. Drag your mouse up (or enter a value) and the shape is automatically extruded into a cube, as seen in Fig. 23.18.

Wedge

Keyboard: Type in **wedge** and press Enter
Cascading menus: **Draw→Modeling→Wedge**
Toolbar icon: **Modeling** toolbar→
Ribbon: **Solid** tab→ **Primitive**

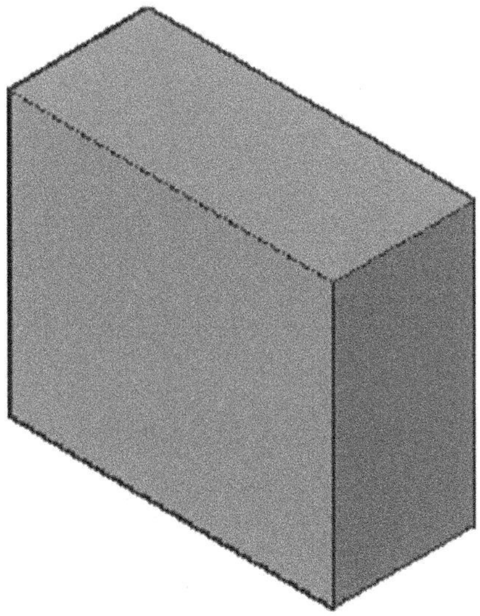

FIGURE 23.18 Box primitive.

Step 1. To try this primitive, use any of the preceding methods.

- AutoCAD says: `Specify first corner or [Center]:`

Step 2. Click anywhere on the screen.

- AutoCAD says: `Specify other corner or [Cube/Length]:`

Step 3. A 2D solid rectangle (the base of the wedge) is drawn based on the movements of your mouse, or you can enter a value by pressing `L` for `length`. Click randomly or enter a value.

- AutoCAD says: `Specify height or [2Point] <2.0000>:`

Step 4. Drag your mouse up (or enter a value) and the wedge shape is automatically extruded, as seen in Fig. 23.19.

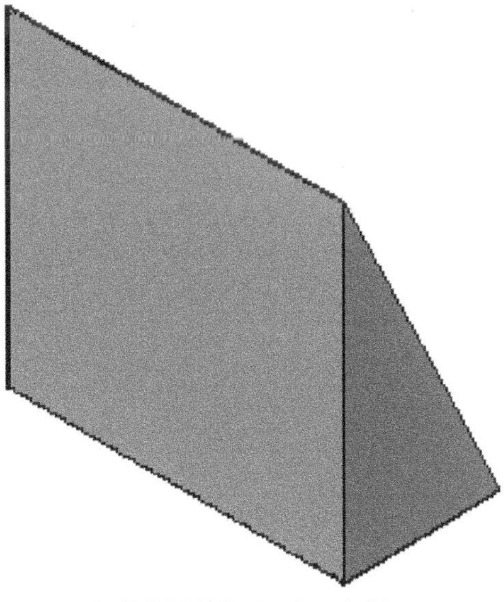

FIGURE 23.19 Wedge primitive.

Cone

Keyboard: Type in **cone** and press Enter
Cascading menus: **Draw→Modeling→Cone**
Toolbar icon: **Modeling** toolbar→
Ribbon: **Solid** tab→ **Primitive**

Step 1. To try this primitive, use any of the preceding methods.

- AutoCAD says: `Specify center point of base or [3P/2P/Ttr/Elliptical]:`

Step 2. Click anywhere on the screen or use the 3P, 2P Ttr features, as in the 2D world.

- AutoCAD says: `Specify base radius or [Diameter] <7.1596>:`

Step 3. A 2D solid circle (the base of the cone) is drawn based on the movements of your mouse, or you can enter a value by typing in a radius or pressing `d` for `diameter`. Click randomly or enter a value.

- AutoCAD says: `Specify height or [2Point/Axis endpoint/Top radius] <6.9580>:`

Step 4. Drag your mouse up (or enter a value) and the cone shape is automatically extruded, as seen in Fig. 23.20. The command has some interesting sub-options, such as `Top radius`, which can chop off the pointy top of the cone. Experiment with these options.

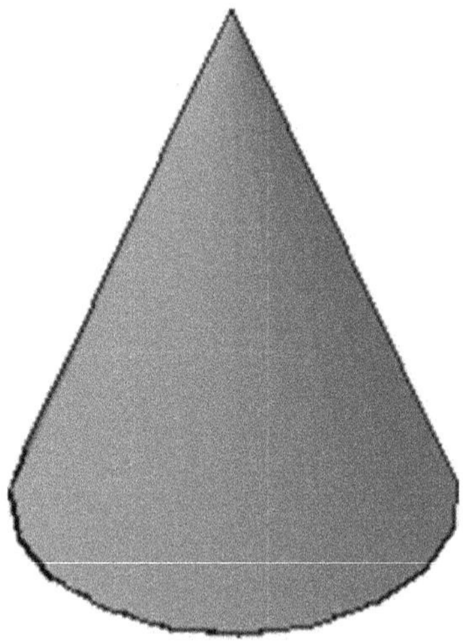

FIGURE 23.20 Cone primitive.

Sphere

Keyboard: Type in **sphere** and press Enter
Cascading menus: **Draw→Modeling→Sphere**
Toolbar icon: **Modeling** toolbar→
Ribbon: **Solid** tab→ **Primitive**

Step 1. To try this primitive, use any of the preceding methods.

- AutoCAD says: `Specify center point or [3P/2P/Ttr]:`

Step 2. Click anywhere on the screen or use the 3P, 2P Ttr features, as in the 2D world.

- AutoCAD says: `Specify radius or [Diameter] <5.7587>:`

Step 3. A solid sphere is drawn based on the movements of your mouse, or you can enter a value by typing in a radius or pressing `d` for `diameter`. The final shape is shown in Fig. 23.21.

FIGURE 23.21 Sphere primitive.

Cylinder

Keyboard: Type in **cylinder** and press Enter
Cascading menus: **Draw→Modeling→Cylinder**
Toolbar icon: **Modeling** toolbar→
Ribbon: **Solid** tab→ **Primitive**

Step 1. To try this primitive, use any of the preceding methods.

- AutoCAD says: `Specify center point of base or [3P/2P/Ttr/Elliptical]:`

Step 2. Click anywhere on the screen or use the 3P, 2P Ttrfeatures, as in the 2D world.

- AutoCAD says: `Specify base radius or [Diameter] <7.9122>:`

Step 3. A solid circle (the base of the cylinder) is drawn based on the movements of your mouse, or you can enter a value by typing in a radius or pressing `d` for `diameter`.

- AutoCAD says: `Specify height or [2Point/Axis endpoint] <22.8198>:`

Step 4. Drag your mouse up (or enter a value) and the cylinder shape is automatically extruded, as seen in Fig. 23.22.

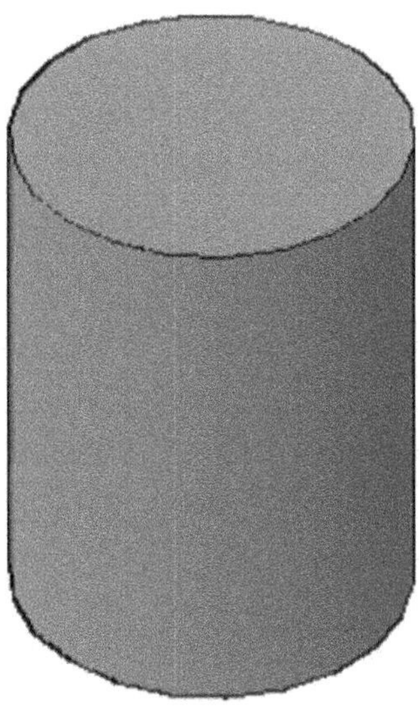

FIGURE 23.22 Cylinder primitive.

Torus

Keyboard: Type in **torus** and press Enter
Cascading menus: **Draw→Modeling→Torus**
Toolbar icon: **Modeling** toolbar→
Ribbon: **Solid** tab→ **Primitive**

Step 1. A torus is basically a donut. To try this primitive, use any of the preceding methods.

- AutoCAD says: `Specify center point or [3P/2P/Ttr]:`

Step 2. Click anywhere on the screen or use the 3P, 2P Ttr features, as in the 2D world.

- AutoCAD says: `Specify radius or [Diameter] <6.3128>:`

Step 3. A hoop is drawn based on the movements of your mouse or you can enter a value by typing in a radius or pressing d for diameter.

- AutoCAD says: `Specify tube radius or [2Point/Diameter] <6.2433>:`

Step 4. Drag your mouse (or enter a value) and the torus shape is automatically extruded, as seen in Fig. 23.23.

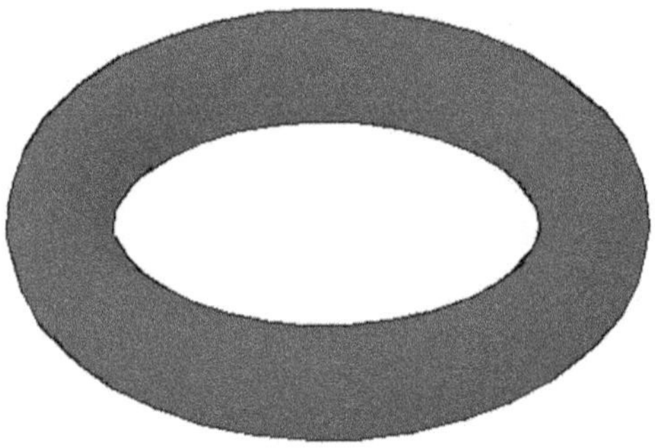

FIGURE 23.23 Torus primitive.

Pyramid

Keyboard: Type in **pyramid** and press Enter
Cascading menus: **Draw→Modeling→Pyramid**
Toolbar icon: **Modeling** toolbar→
Ribbon: **Solid** tab→ **Primitive**

Step 1. To try this primitive, use any of the preceding methods.

- AutoCAD says: `4 sides Circumscribed Specify center point of base or [Edge/Sides]:`

Step 2. You can then click anywhere, or use the Sides option to give the pyramid more than four sides, a useful feature.

- AutoCAD says: `Specify base radius or [Inscribed] <22.2294>:`

Step 3. A pyramid base is drawn based on the movements of your mouse, or you can enter a value. The inscribed feature works in the same manner as it was used with the polygon command.

- AutoCAD says: `Specify height or [2Point/Axis endpoint/Top radius] <38.8857>:`

Step 4. Drag your mouse (or enter a value) and the pyramid shape is automatically extruded, as seen in Fig. 23.24. Just like with the cone, you can vary the top radius of the shape.

23.3 Applying primitives

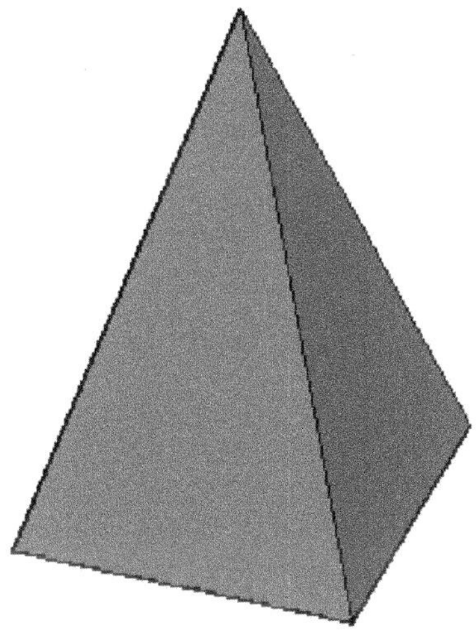

FIGURE 23.24 Pyramid primitive.

So, how useful are these primitives in real-life applications? The best way to describe this is "extremely useful in isolated cases and not used much in most others." For example, if you need a shape not quickly done any other way, such as a sphere or torus (you can easily extrude the other shapes), then primitives are indispensable. But, since most of the 3D objects you will design are not simple shapes, primitives are rarely used. It is still a good skill to have and an easy one to learn. Go over each shape several times, add some color, and be sure to practice 3D orbit to get various views of the objects, as well as switching between wireframe and shaded.

This chapter concludes 3D "flat design." These basic 3D skills go a long way in creating a realistic model of a building or even a simple mechanical device. In the coming chapters, we add additional tools and refinement, starting with curved design. For now, be sure to completely understand everything and practice everything learned thus far.

Summary

You should understand and know how to use the following concepts and commands before moving on to Chapter 24:

- Union
- Subtract
- Intersect
- Box
- Wedge
- Cone
- Sphere

- Cylinder
- Torus
- Pyramid

Review questions

Answer the following based on what you learned in this chapter:

1. What are the three Boolean operations?
2. In what toolbar are they found?
3. Is union a permanent setting?
4. Why is it a good idea to make a copy of something that is being subtracted?
5. Name the seven primitives covered in this chapter.
6. What four methods are used to create primitives?
7. What two primitives cannot be created any other way?

Exercises

1. Using your new Boolean skills, create this washer and bolt assembly. Make the washer 3″ square and the hole (as well as the bolt) 0.75″ in diameter, and size the hex bolt top off of that. You need the circle, extrude, and polygon commands among others. (Difficulty level: Easy; Time to completion: 10 minutes.)

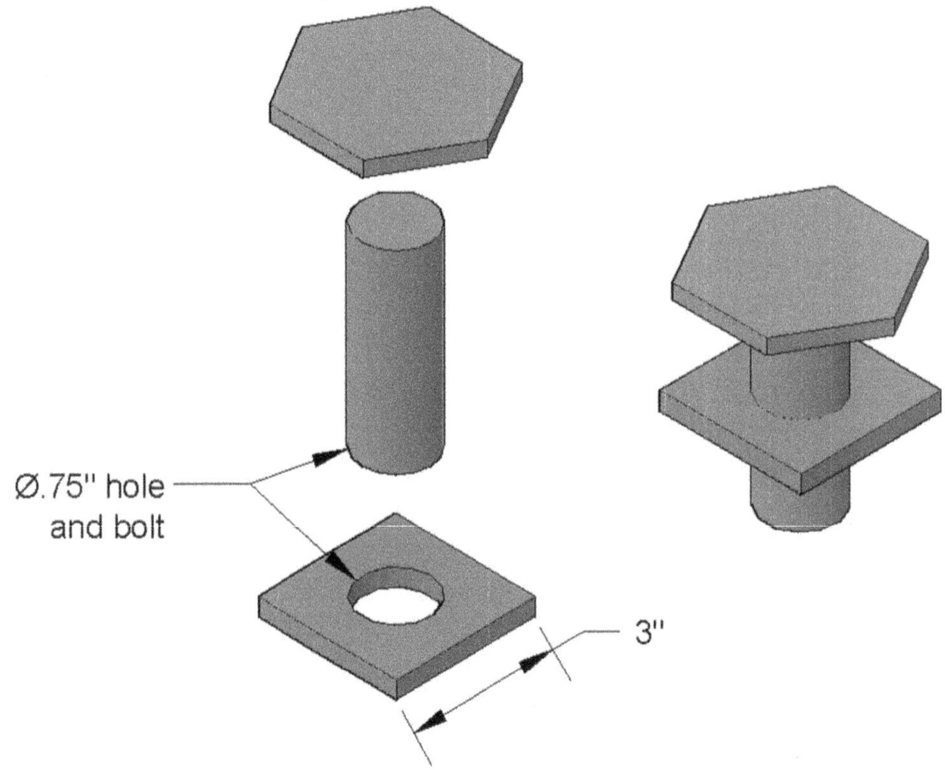

2. Create the following design of a bed. It consists of a thick panel, a mattress, two identical headboards (with cutouts), and two pillows angled at 30°. Sizing is entirely up to you; draw a reasonable-looking base panel and size the rest from that. You need the rectangle, line, arc, pedit, extrude, fillet, mirror3D, and subtract commands among others, as well as some UCS rotations. (Difficulty level: Moderate; Time to completion: 30 minutes.)

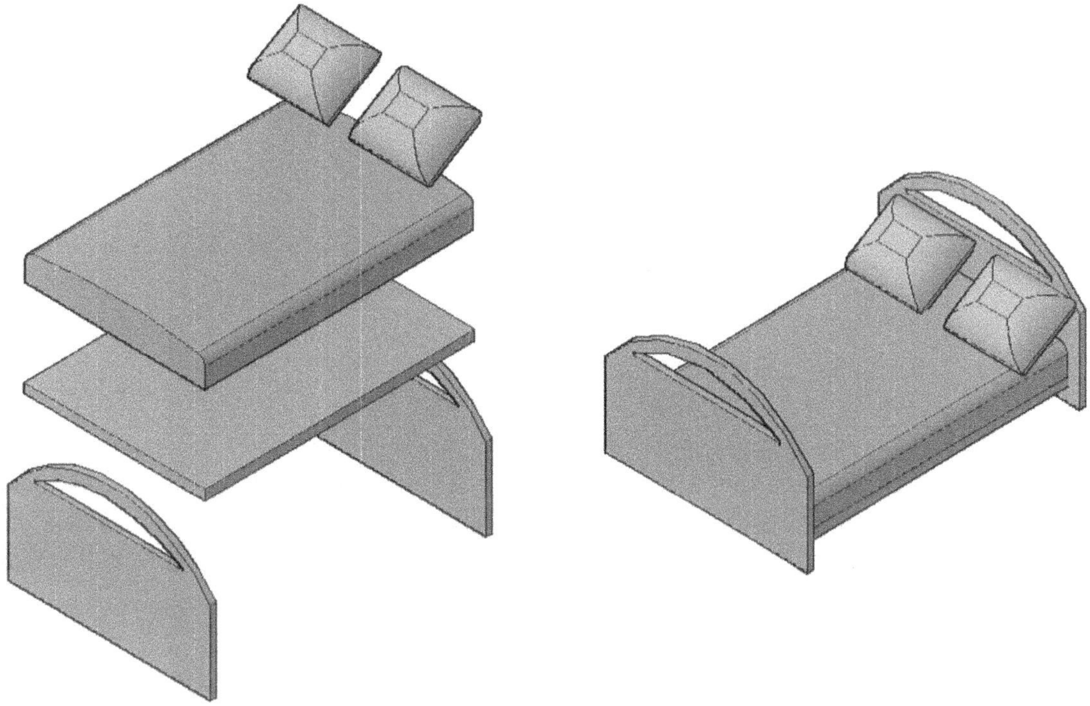

3. Create the following design of a drilled, vented brake rotor. Create a circle of the given size, extrude to 1″, add the center axle cutout, and arrange the drilled holes as shown on the left, arraying them eight times. Extrude and subtract as needed. Shade and color to finish the model as seen on the right. (Difficulty level: Moderate; Time to completion: 15 minutes.)

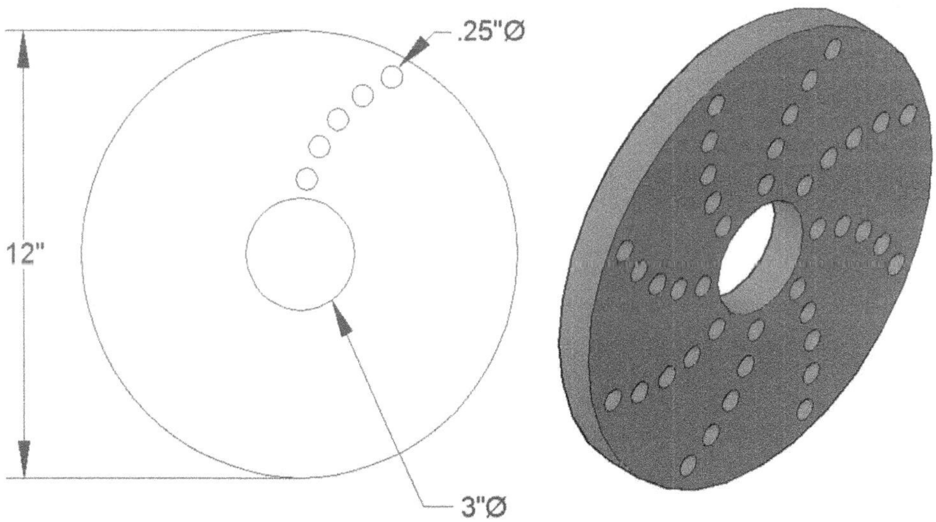

4. Create the following simplified model of construction excavator. Begin all extrusions with basic 2D shapes using some starter dimensions given in the exploded view. Size all others from them. You need to use the subtract command for the wheels/track assembly as well as the cab. In addition, the union command was used to make the cab one piece. Pretty much everything else was done via basic extrusion, 3D rotation, and careful positioning of objects. (Difficulty level: Moderate/Advanced; Time to completion: 45–60 minutes.)

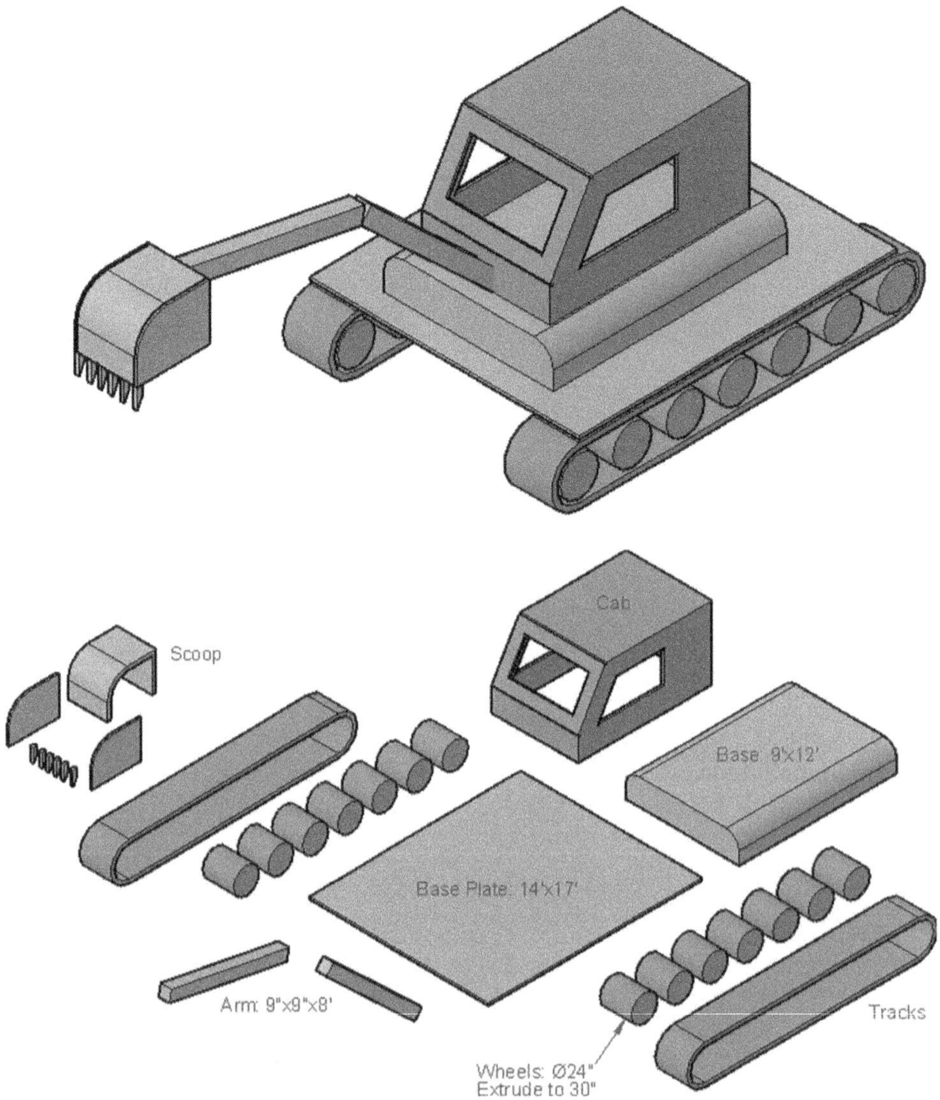

5. Create each primitive shape once via only mouse clicking and once using keyed-in values. (Difficulty level: Easy; Time to completion: 5–10 minutes.)

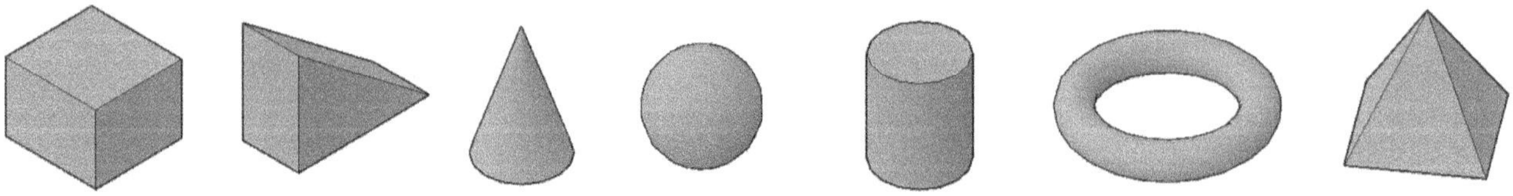

6. Create the following layout using primitives and the rectangle command. All sizes and positioning are approximate. (Difficulty level: Easy; Time to completion: 10 minutes.)

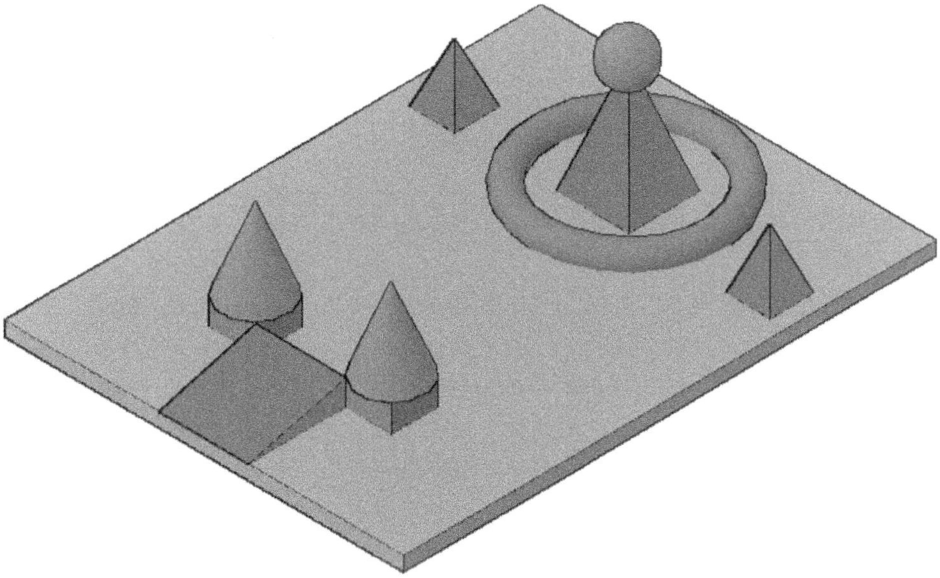

7. Create the following shapes using primitives and Boolean interactions. All sizes and positioning are approximate. (Difficulty level: Easy; Time to completion: 10 minutes.)

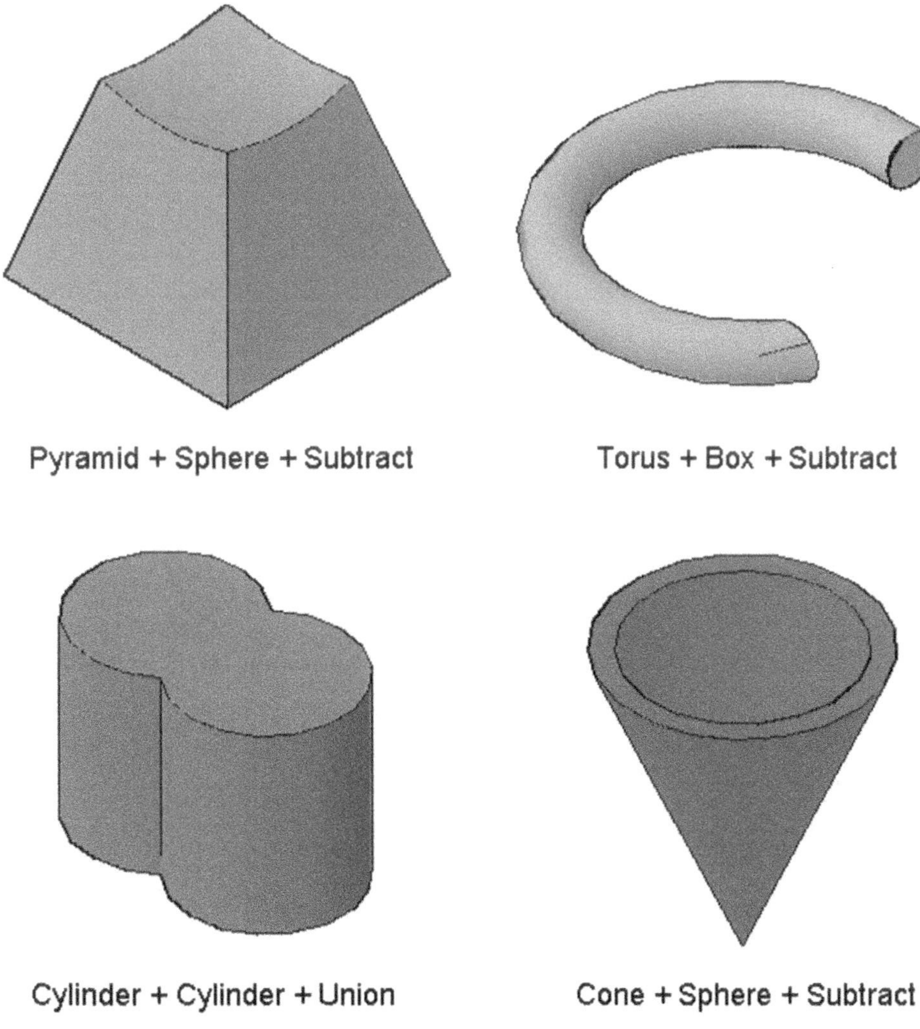

Chapter 24

Solid Modeling

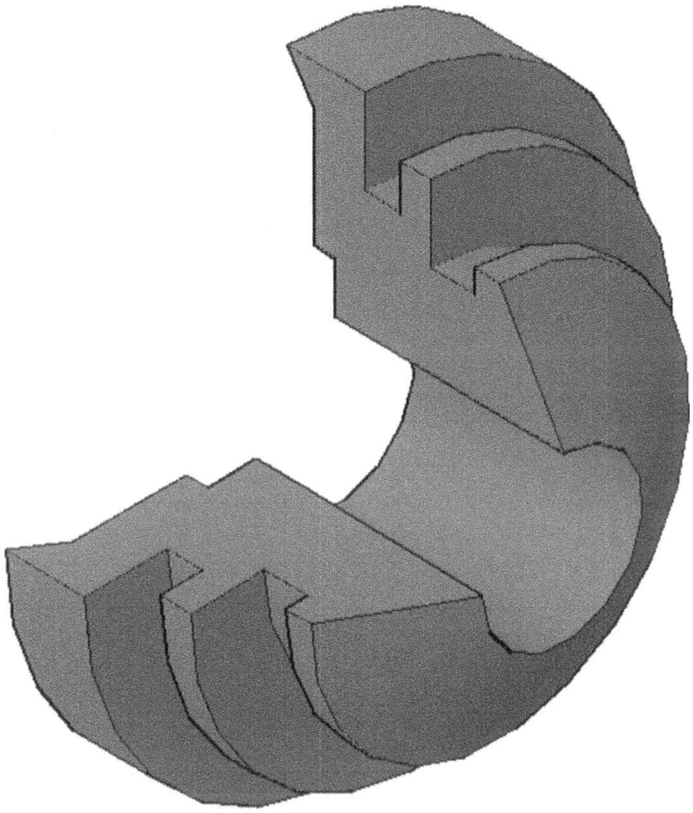

Learning Objectives

In this chapter, we introduce a completely new set of tools in solid modeling, which are used for "curved design." These dramatically expand the range of design work that you can do. We specifically look at

- Revolve
- Shell
- Taper
- Loft
- Path extrusion
- Sweep

By the end of this chapter, you will be able to create designs that were not possible with previous tools, with much of this new knowledge applicable to engineering design.

Estimated time for completion of this chapter: 2–3 hours.

Up and Running with AutoCAD 2022. https://doi.org/10.1016/B978-0-323-89923-9.00024-X

24.1 Introduction to Solid Modeling

With this chapter, we head in a new direction and begin talking about some of the advanced solid modeling tools that make up what can be referred to as *curved design*. While these tools are not necessarily more difficult to master, they do open up a new range of modeling possibilities, including some rather complex 3D designs. We introduce revolve, shell, taper, loft, path extrusion, and sweep. Although some of these commands, such as taper and shell, do not involve curves, they are included with this advanced tools group.

Revolve

As essential as extrude was to flat design, such is the importance of revolve to curved design, and it is the first command we cover. Revolve easily creates complex revolved shapes based on the concept of solid or surface of revolution, and this command has no other equivalent in AutoCAD.

The basic idea is the following. We would like to draw a 2D outline of an object and revolve this outline or "profile" around an axis. If the profile is closed, the result is a solid of revolution; if the profile is open (has gaps) or lines are used instead of polylines, the result is a surface of revolution. This process can create some unique shapes that are impossible to make any other way, but the entire concept rests on the designer's ability to visualize the profile on which the model is built.

In past releases of AutoCAD, the revolve command was a bit picky, and you had to use a closed polyline for the revolution to work, no exceptions. These restrictions have been lifted, and you can create a profile using the regular line command and even leave gaps in it; although, once again, just be aware that this creates surfaces, not solids, as is shown later. Your axis of revolution can be anywhere on or near the object, and you need not revolve all the way around but can use any degree value in between if necessary, such as when you want to show a cross section.

Let us put all this together with some examples. In 2D, create a profile approximately similar to the one shown in Fig. 24.1, using a polyline to ensure a solid. We try it with regular lines later on.

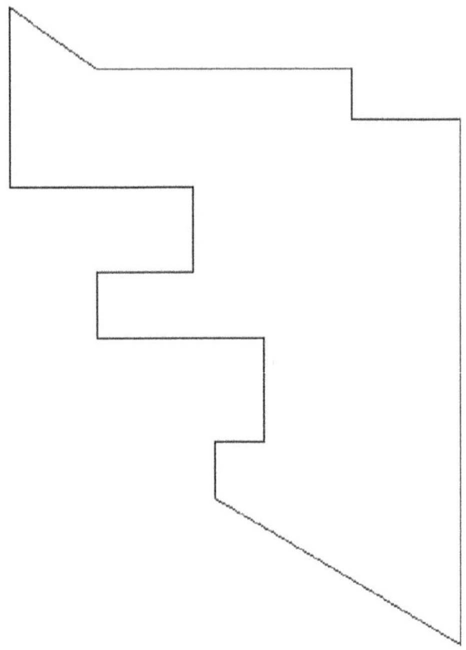

FIGURE 24.1 Polyline profile for revolve.

Keyboard: Type in **revolve** and press Enter
Cascading menus: **Draw→Modeling→Revolve**
Toolbar icon: **Modeling** toolbar→
Ribbon: **Solid** tab→**Solid**

Step 1. Start revolve via any of the preceding methods.

- AutoCAD says: `Current wire frame density:ISOLINES=4,`
  `Closed profiles creation mode = Solid`
  `Select objects to revolve or [MOde]:`

Step 2. Select the profile and press Enter.

- AutoCAD says: `Specify axis start point or define axis by [Object/ X/Y/Z]`

Step 3. Here, AutoCAD is asking for an axis of revolution for the profile. This can be addressed in many ways, and the axis can be on the object or some distance away from it (causing the revolved solid to have a hole about the axis). For this example, we pick two points at the top right and bottom right of the profile, as seen in Fig. 24.2. Using the ENDpoint OSNAP, click on these two locations.

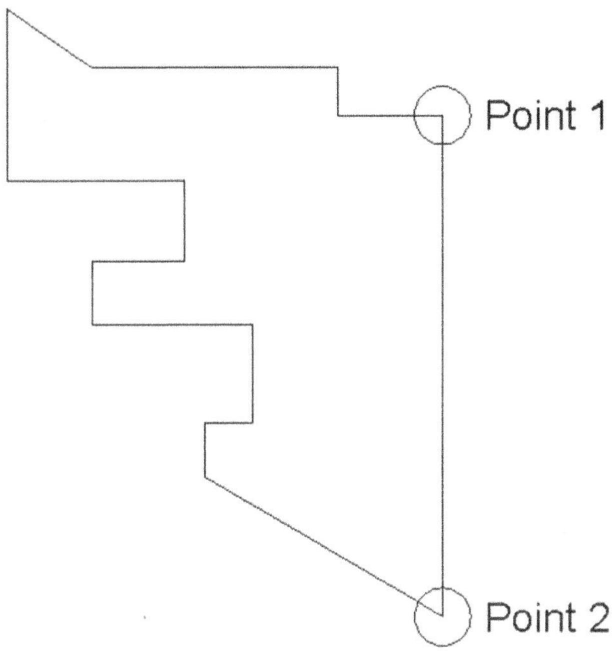

FIGURE 24.2 Selecting rotation axis.

Step 4. After the first click,

- AutoCAD says: `Specify axis endpoint:`

After the second click,

- AutoCAD says: `Specify angle of revolution or [STart angle/Reverse/EXpression] <360>:`

Step 5. Here, you can ask for a full 360° revolution by just pressing Enter, but instead type in 270, so we can see a cutout of the part. If you are in 3D, you can also just trace out the revolution angle with movements of your mouse. For now though, you see a 2D revolved surface, as in Fig. 24.3.

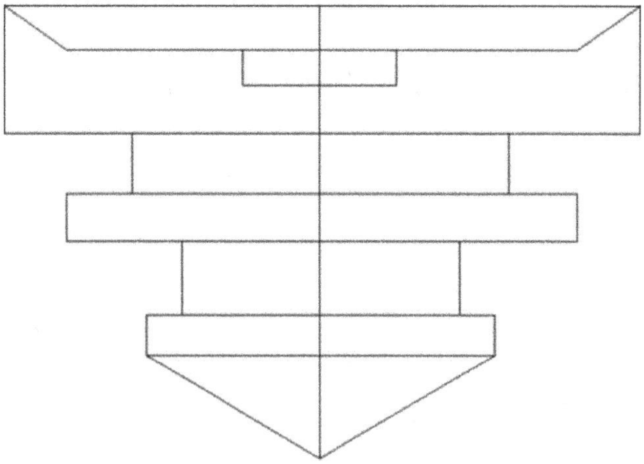

FIGURE 24.3 Revolved profile in 2D wireframe.

To see the full effect of what you have done, switch to the familiar SW Isometric view and also shade and color the solid. The far more dramatic result is seen in Fig. 24.4.

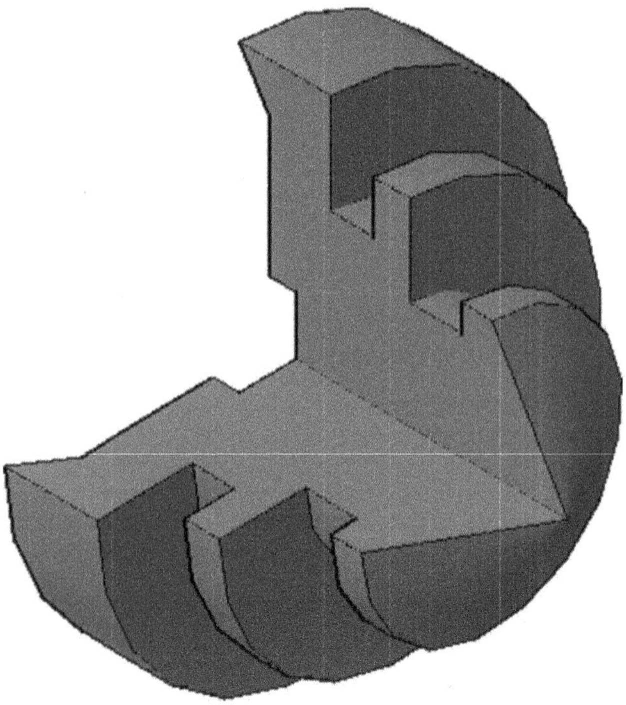

FIGURE 24.4 Completed revolution, shaded and colored.

For some other ideas, undo your last step, and try locating the axis of revolution some distance away from the object (Fig. 24.5). This creates a hole in the center of the object on execution of revolve, as seen in Fig. 24.6.

The axis of revolution can actually be anywhere. In Fig. 24.7, it is above the profile and at an angle, creating a completely different solid of revolution, as seen in Fig. 24.8.

Next, let us try the same profile but with a piece missing (Fig. 24.9). The profile no longer is a closed polyline, and in this case, the revolve command creates a surface, not a solid, a difference you can easily see. The same effect is achieved using plain old lines. Even if they are connected in a loop, AutoCAD still defaults to the surface revolution. Fig. 24.10 shows the result.

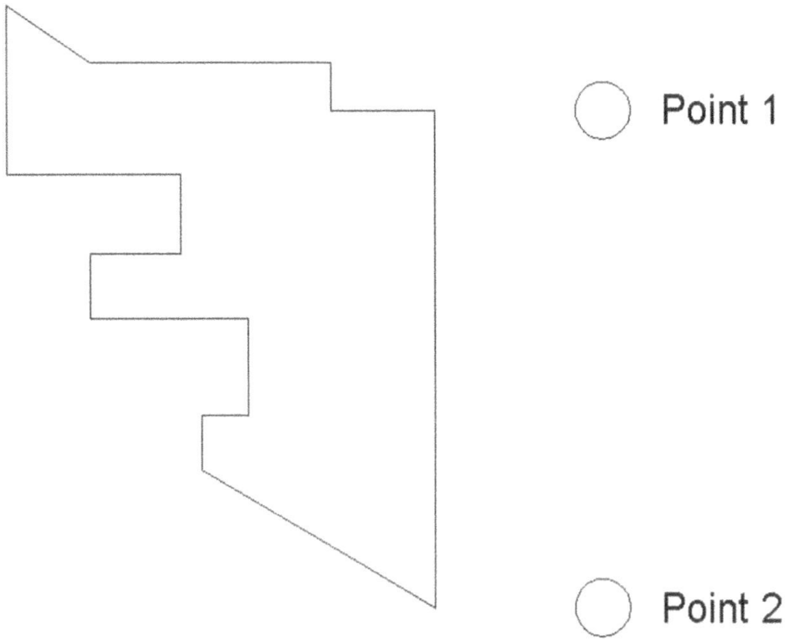

FIGURE 24.5 Offset axis of revolution, Example 1.

FIGURE 24.6 Completed offset revolution, Example 1.

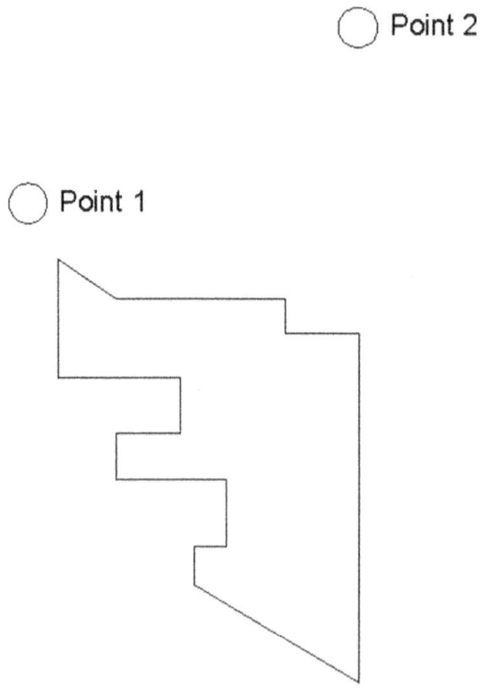

FIGURE 24.7 Offset axis of revolution, Example 2.

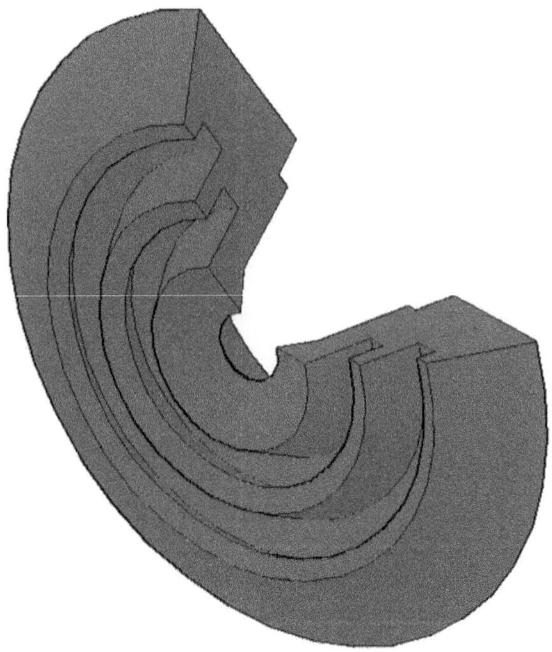

FIGURE 24.8 Completed offset revolution, Example 2.

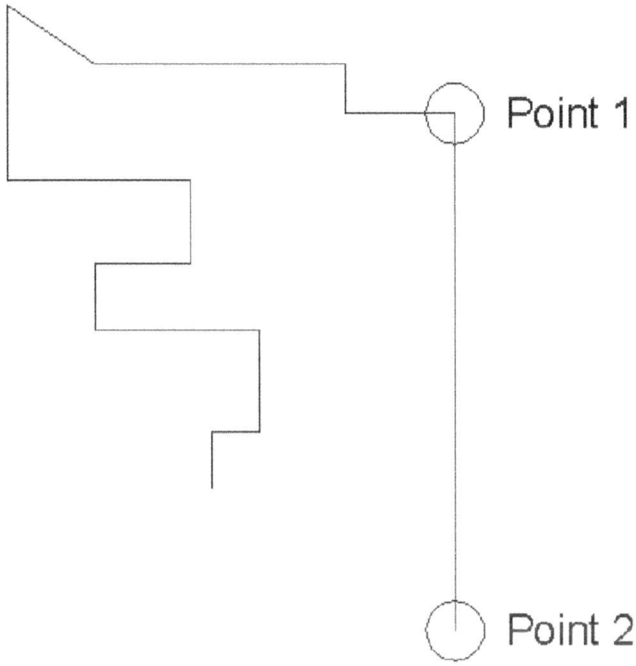

FIGURE 24.9 Surface of revolution profile.

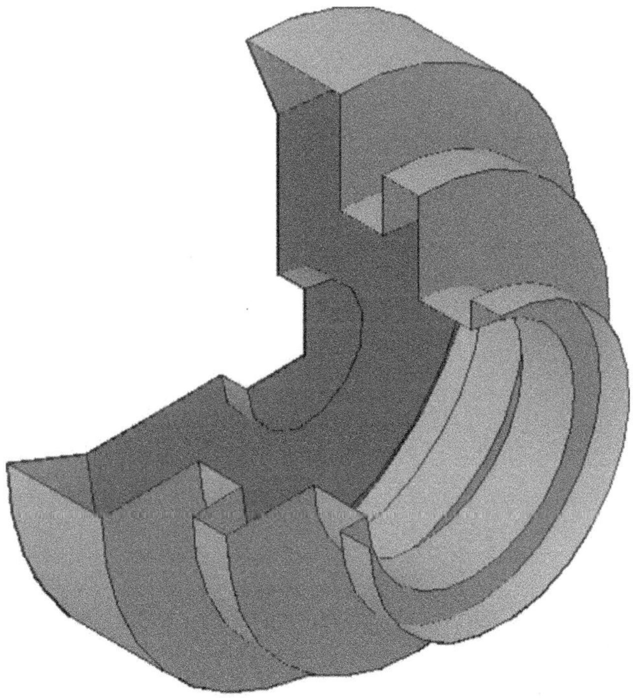

FIGURE 24.10 Completed surface revolution.

Try the following example of revolution, a 3D cup. Draw the profile in Fig. 24.11, pedit, and revolve 360°. Then, switch to 3D, and shade, color, and rotate if needed. You now have your first piece of glassware for a future 3D house, as seen in Fig. 24.12.

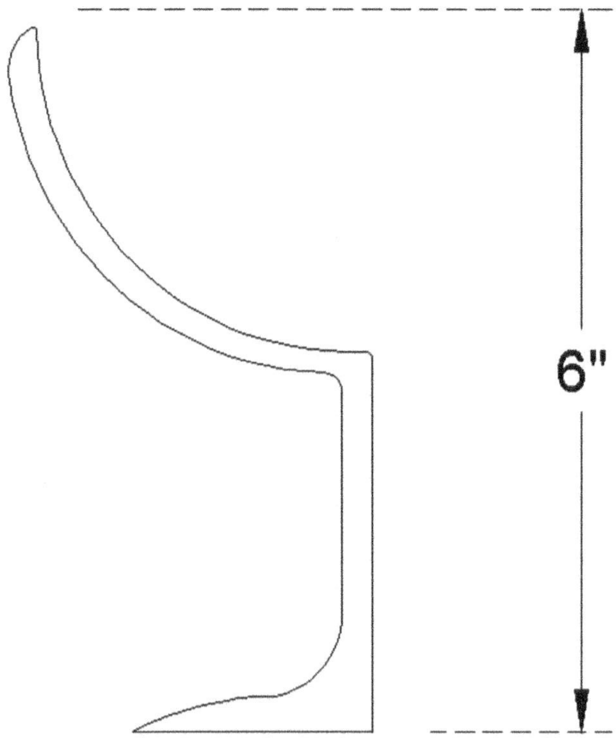

FIGURE 24.11 Cup profile.

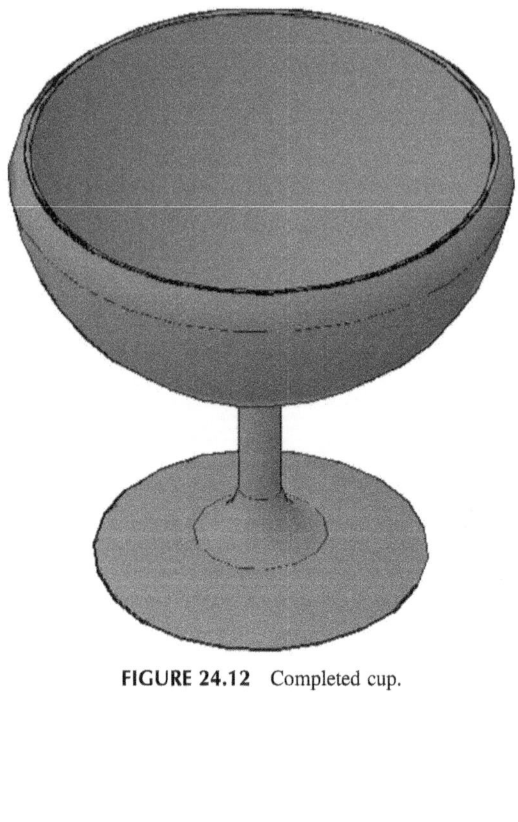

FIGURE 24.12 Completed cup.

As you can see, revolve is a very versatile, powerful command and is really limited only by your imagination and visualization skills. Be sure to go through each example presented thus far. None of the profiles or solids has to look exactly like the ones shown, and indeed, no dimensions are provided (except for the cup height), but be sure to understand the underlying principles and what is being done to produce what you see.

Shell

Shell is one of several AutoCAD 3D commands "borrowed" from engineering solid modeling software due to their usefulness. The command creates a "shell" of a predetermined thickness out of a solid object. The shell is of an even thickness all around and can have as many open surfaces as needed.

Let us demonstrate by example. Draw a 10″×8″ rectangle and extrude it out to 4″, as seen in Fig. 24.13.

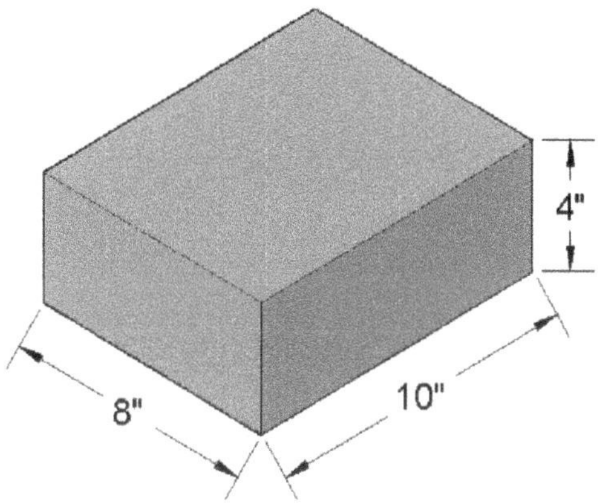

FIGURE 24.13 Basic block for shell.

To activate the command, you cannot type in "shell" (that is an OS command), but you can use other methods, as seen next in the command matrix.

Keyboard: **Solidedit**, Enter. **Body** option, then **Shell**
Cascading menus: **Modify→Solid Editing→Shell**
Toolbar icon: **Solid Editing** toolbar→
Ribbon: **Solid** tab→**Solid Editing**

Step 1. Start the shell command via any of the preceding methods.

- AutoCAD gives you a lengthy response:

```
Solids editing automatic checking: SOLIDCHECK = 1
Enter a solids editing option [Face/Edge/Body/Undo/eXit] <eXit>: _body
Enter a body editing option
[Imprint/seParate solids/Shell/cLean/Check/Undo/eXit] <eXit>: _shell
Select a 3D solid:
```

Step 2. Select the extruded block.

- AutoCAD says: `Remove faces or [Undo/Add/ALL]:`

Step 3. Carefully pick the upper surface of the extruded block. Be careful not to select an edge in this case but an actual flat surface. You may select more than one surface, and it is something you should try the next time you practice the command. For now, just press Enter after selecting the top surface.

- AutoCAD says: `Enter the shell offset distance:`

Step 4. Here, you need to enter a valid shell size (thickness). Some complications can result at this point if your shape has other modifications to it, such as fillets. Then, if the thickness of the walls is too thin, you can "eat through" the material, causing errors or actual holes in the shape. This particular example should proceed smoothly. Enter .5′ for the thickness.

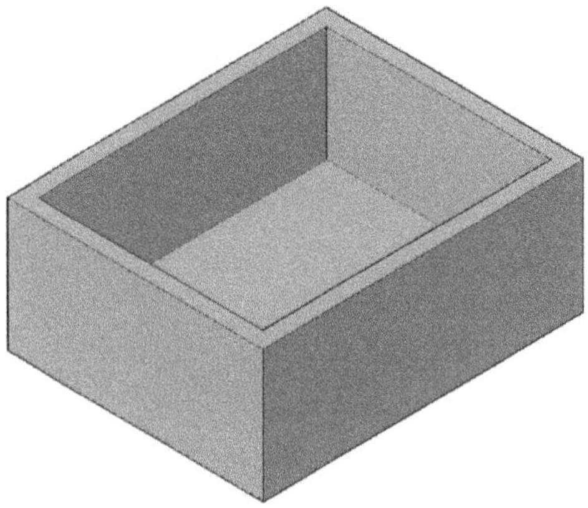

FIGURE 24.14 Completed shell.

- AutoCAD says: `Solid validation started.`
  `Solid validation completed.`
  `Enter a body editing option`

The shell command executes, and the result is as seen in Fig. 24.14.

As suggested earlier, try removing more than one face. The result is an "open" box, as seen in Fig. 24.15. You can remove all four sides if you wish and end up with a flat plate, although that somewhat defeats the whole point of shell.

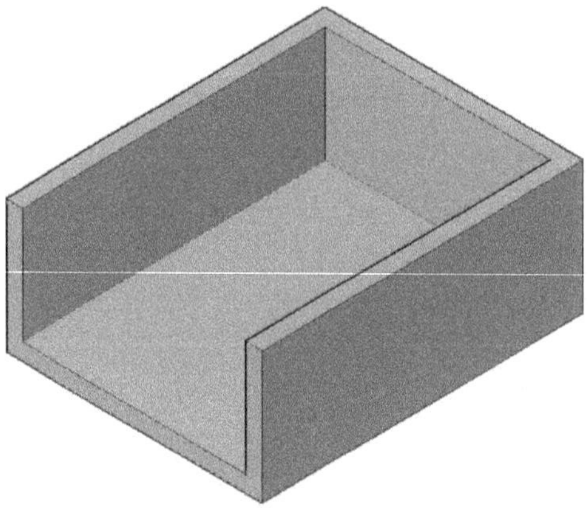

FIGURE 24.15 Completed shell (two faces removed).

Taper

Taper is a useful command that does exactly what it advertises: It tapers (or creates a gradient) on one or more sides of an object. A simple example, and one we will model soon, is a plastic wastebasket commonly found in any office. If you have one nearby, take a look at it. Its base is a slightly smaller rectangle than the top rim. The reason for tapering such items is because of the nature of plastic molds and the manufacturing process. Another useful result is that now these wastebaskets are easily stackable one on top of another.

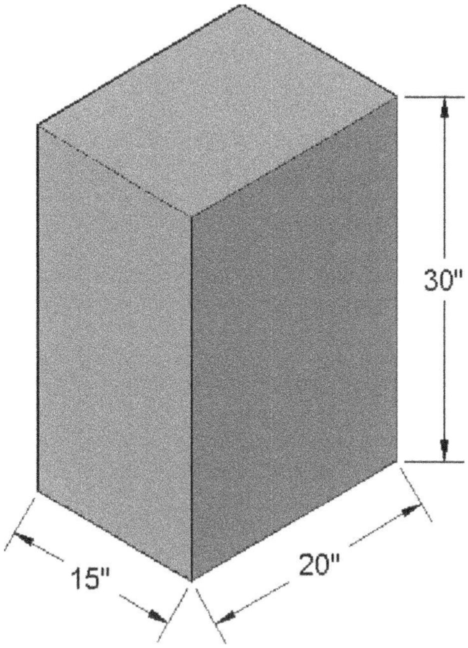

FIGURE 24.16 Box for tapering.

Let us practice this command on a tall rectangular box. Draw a $20'' \times 15''$ rectangle and extrude it to $30''$, as seen in Fig. 24.16.

The taper command is another one that cannot be typed in, and the other input methods must be used.

Keyboard: **Solidedit**, Enter. **Face** option, then **Taper**
Cascading menus: **Modify→Solid Editing→Taper faces**
Toolbar icon: **Solid Editing** toolbar→
Ribbon: **Solid** tab→**Solid Editing**

Step 1. Start the taper command via any of the preceding methods.

- AutoCAD again gives you a lengthy response:

```
_solidedit
Solids editing automatic checking: SOLIDCHECK = 1
Enter a solids editing option [Face/Edge/Body/Undo/eXit] <eXit>: _face
Enter a face editing option
[Extrude/Move/Rotate/Offset/Taper/Delete/Copy/coLor/mAterial/Undo/eXit] <eXit>:
_taper
```

Step 2. Here, the first task is to select all the faces you want to taper. This can be as few as one face or as many as all of them. Be careful though; by *face*, we mean the ones that will taper. Therefore, the top and bottom of the box are not valid choices, only the sides. Also, note that it is easier to select faces when you are in wireframe view. In shaded view, you need to rotate around the object to get all the faces; in wireframe, you can (carefully) pick "through" the shape. When you pick a face it becomes dashed.

- AutoCAD says: `Select faces or [Undo/Remove/ALL]: 1 face found.`

Step 3. When done press Enter.

- AutoCAD `says: Specify the base point:`

Step 4. Here, you need to consider what is meant by *base point*. It is the start of the taper. You would need to pick either the top or the bottom of the block. Since this is a wastebasket, which tapers in as you go down, we pick the midpoint of one of the lines representing the *top face or surface*. If you choose a similar point on the bottom face or surface, then the block tapers up, like a pyramid. Go ahead and select the base point, as discussed.

- AutoCAD says: `Specify another point along the axis of tapering:`

Step 5. This, of course, is the opposite of where the first point was, so pick the midpoint of the *bottom face* or *surface*. In summary, you are giving AutoCAD what amounts to a straight line that tells it in what direction the taper is going.

- AutoCAD says: `Specify the taper angle:`

Step 6. This final request asks for the amount of taper in degrees. Enter a small value, perhaps 4 or 5, and press Enter. The block tapers evenly on all sides. The result in shade mode is shown in Fig. 24.17.

FIGURE 24.17 Completed taper.

3D Modeling Exercise: Wastebasket

Let us put extrude, taper, fillet, and shell together into one exercise.

Step 1. Clear your screen of previous work, and make sure you are in the SW Isometric view. Starting from the beginning, draw a 22" × 18" rectangle and extrude it to 34". The result is a block of material similar to what we just created when we practiced the taper command. It is shown shaded with a different color, in Fig. 24.18.

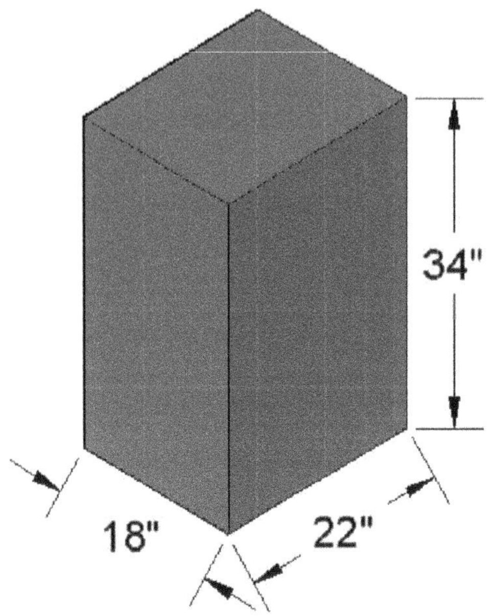

FIGURE 24.18 Wastebasket, Step 1.

Step 2. Taper the block on all four sides to 4°, as seen in Fig. 24.19.

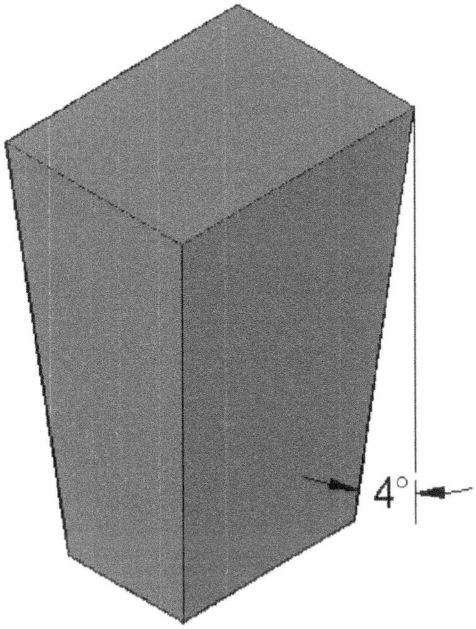

FIGURE 24.19 Wastebasket, Step 2.

Step 3. Fillet the side edges with a 3" radius, and the bottom with a 1.5" radius, as seen in Fig. 24.20.

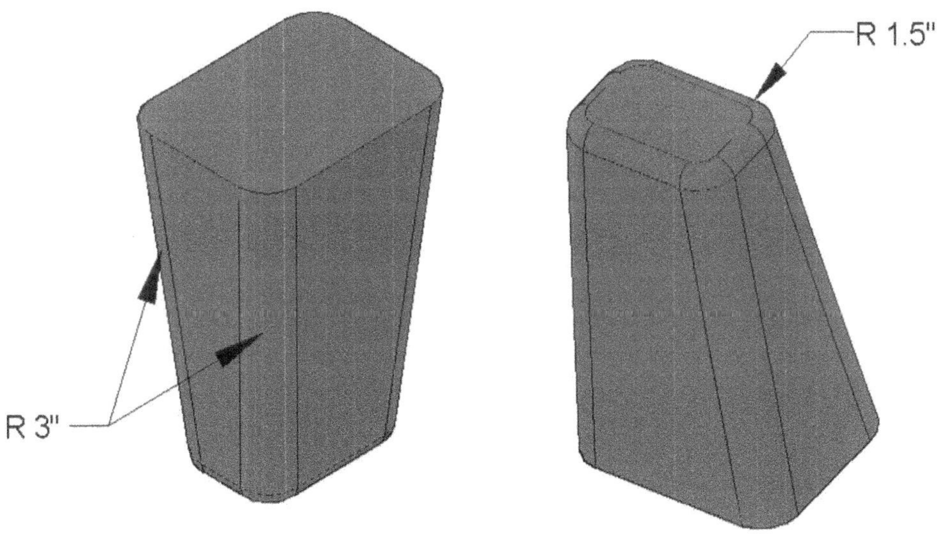

FIGURE 24.20 Wastebasket, Step 3.

Step 4. Shell the wastebasket to a 0.35" thickness, as seen in Fig. 24.21.

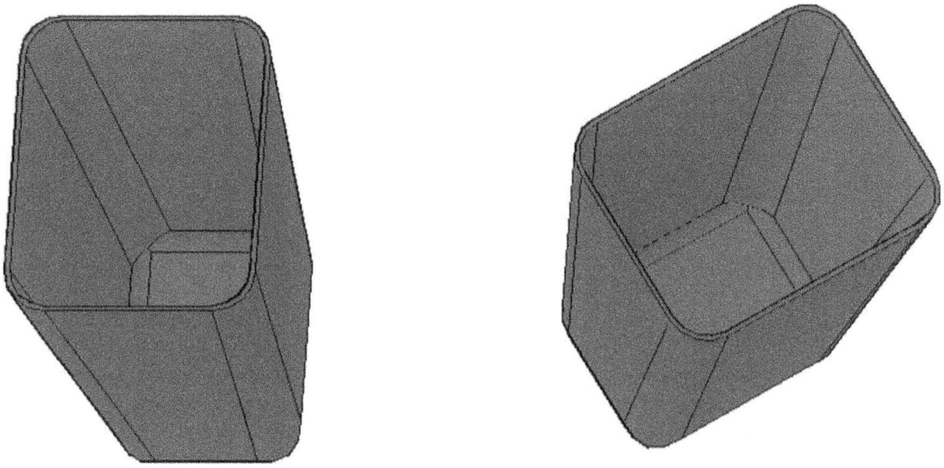

FIGURE 24.21 Wastebasket, Step 4.

Loft

Loft was added along with many other 3D enhancements in a major revision of AutoCAD's 3D capabilities starting with AutoCAD 2007. It is another command "borrowed" from the engineering solid modeling world and is one of the more useful and versatile.

Despite some rather complex surface and part modeling programming "under the hood" with loft, as far as the AutoCAD end user is concerned, the command works on a simple principle. It connects together open or closed profiles. All you have to do is pick those profiles in the order you want them lofted and AutoCAD "connects the dots," so to speak, and creates either a surface (if the profile is open) or a solid (if the profile is closed). Take the very simple example of a classic wooden barrel, as shown in Fig. 24.22.

FIGURE 24.22 Barrel.

While, in this particular case, you can use revolve to model the object, this is only because it is symmetric about the center axis. We often do not have such a convenience, so we need to default to loft instead. The barrel's lengthwise profile can be modeled as three circles, one after another, with the first and third representing the top and bottom, and the second (slightly larger one) representing the fatter middle part, as seen in Fig. 24.23.

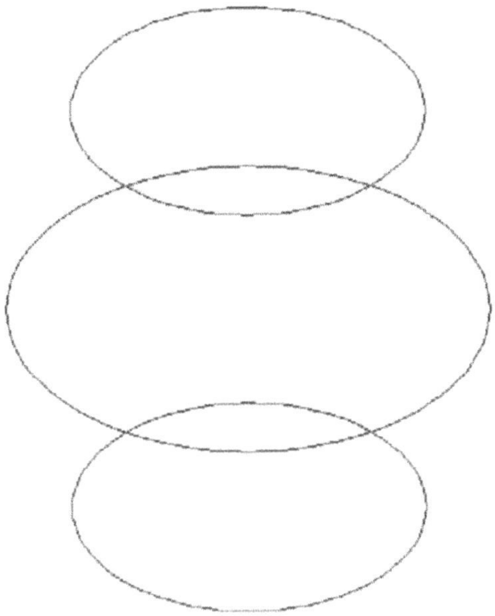

FIGURE 24.23 Three-ring profile of a barrel.

These rings are drawn to a random size, but both outer rings are an equal distance from the larger center one, ensuring our barrel is not lopsided. You can easily set up these 3D profiles by drawing two sizes of circles in 2D Top View, with each having the same center point. Then, switch to a side view and "lift" the smaller circle upward (or the larger one downward). Finally, a quick mirroring completes the profile. You should be able to do this type of 3D work with little effort at this point in the course.

Let us now run the loft command and get a barrel.

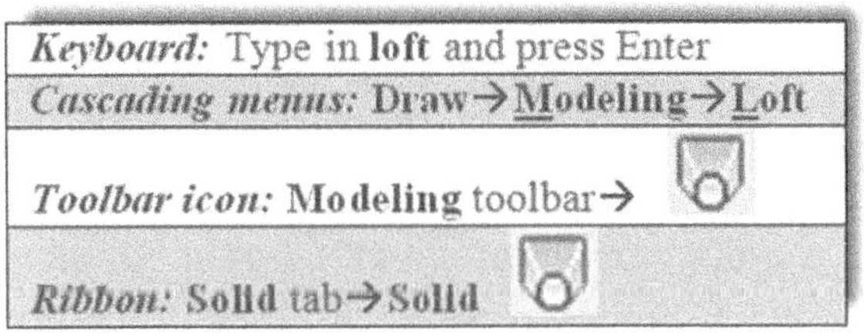

Keyboard: Type in **loft** and press Enter
Cascading menus: Draw→Modeling→Loft
Toolbar icon: **Modeling** toolbar→
Ribbon: **Solid** tab→**Solid**

Step 1. Start the command via any of the preceding methods.

- AutoCAD says:

```
Current wire frame density:ISOLINES=4,
Closed profiles creation mode = Solid
Select cross sections in lofting order or [POint/Join multiple edges/MOde]: _MO
Closed profiles creation mode [SOlid/SUrface] <Solid>: _SO
Select cross sections in lofting order or [POint/Join multiple edges/MOde]:
```

Step 2. Select the hoops, one after another, from top to bottom (or vice versa). Each one becomes dashed as you select it. Selecting the hoops out of order does not necessarily stop the command but gives you unexpected results.

- AutoCAD says:`1 found, 3 total`

Step 3. Press Enter when all three are selected.

- AutoCAD says: Enter an option [Guides/Path/Cross sections only/Settings]<Cross sections only>:

Step 4. Press Enter. The loft is performed and the barrel is visible if you are in the shaded mode. You also see a prominent down arrow, which when pressed, reveals a menu to fine-tune the loft (Fig. 24.24). Experiment with the various

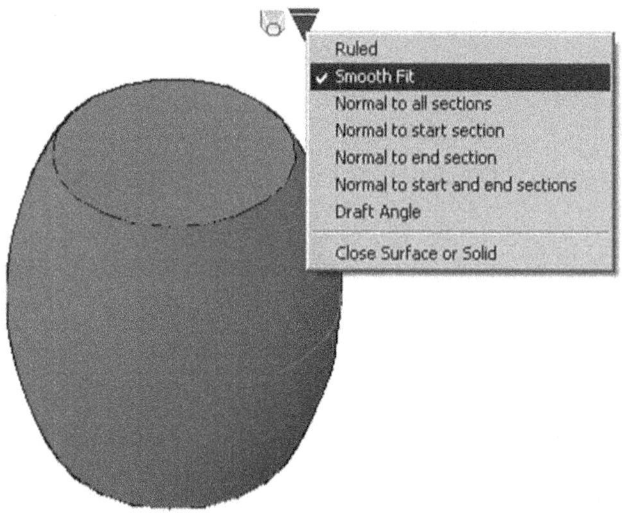

FIGURE 24.24 Loft settings menu.

choices to see the effect on the overall shape. In almost all cases, we want a smooth fit (the default), so after experimenting, restore that choice.

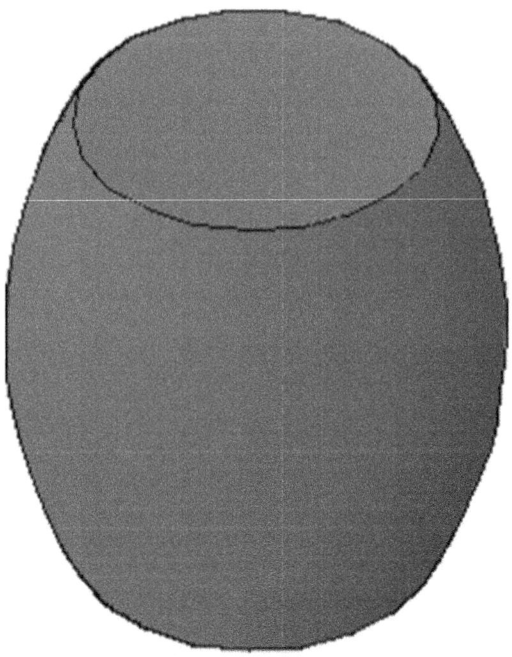

FIGURE 24.25 Loft completed.

The final result is shown again in Fig. 24.25. The down arrow is gone but can be brought back by a single mouse-click on the shape.

While this was a very simple example, the essentials of loft are all there. Because the circles by definition are closed profiles, we get a solid out of this. If, for example, we had three arcs instead (in the same positions), loft would yield a

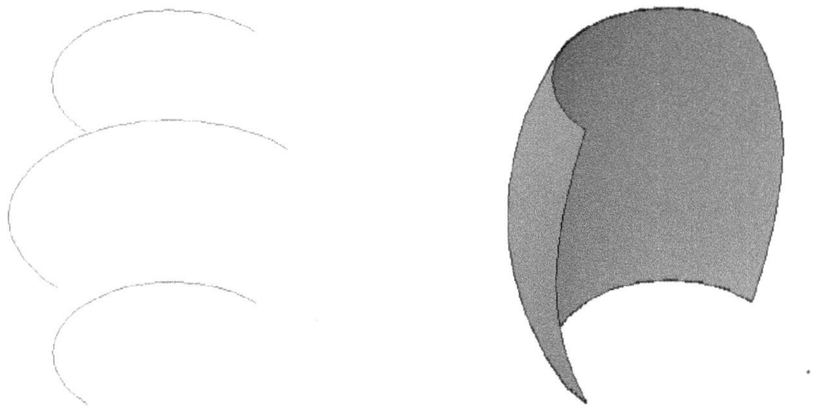

FIGURE 24.26 Loft using arcs.

surface, as seen in Fig. 24.26. Go ahead and try this by trimming the circles. We use this surfacing ability later on to generate some rather complex surfaces quickly and easily.

The true power of loft shows when you need to smoothly merge profiles of varying shapes, sizes, and locations. AutoCAD's Help files have an extensive list of items that qualify as cross sections, so be sure to take a look. Sometimes, you may hit a snag in performing loft due to an unusual combination of elements, but as long as they are not on the same plane (two plates on a dinner table is an example of objects on the same plane), you should be able to complete the

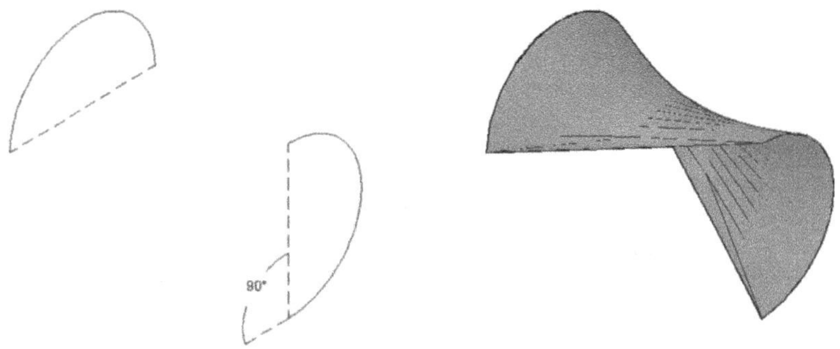

FIGURE 24.28 Loft using two arcs.

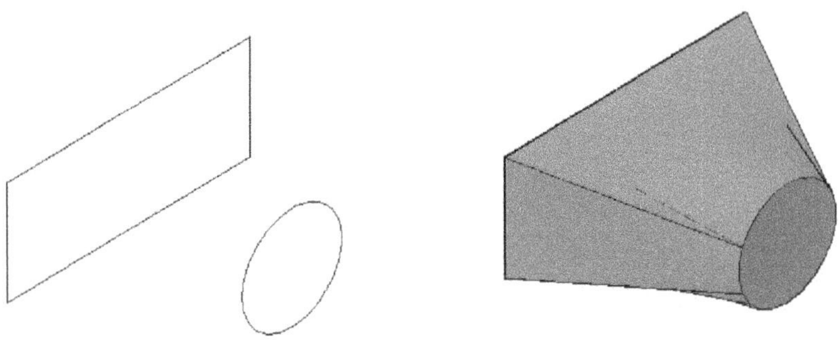

FIGURE 24.27 Loft using a rectangle and a circle.

operation. Some examples shown next include lofting a solid between a rectangle and a circle some distance away (Fig. 24.27) and lofting a surface between two arcs some distance apart (Fig. 24.28).

3D Modeling Exercise: Mechanical Drill Bit

This is a good example of a loft application. We create three profiles, position them, and loft between them, creating a complex shape.

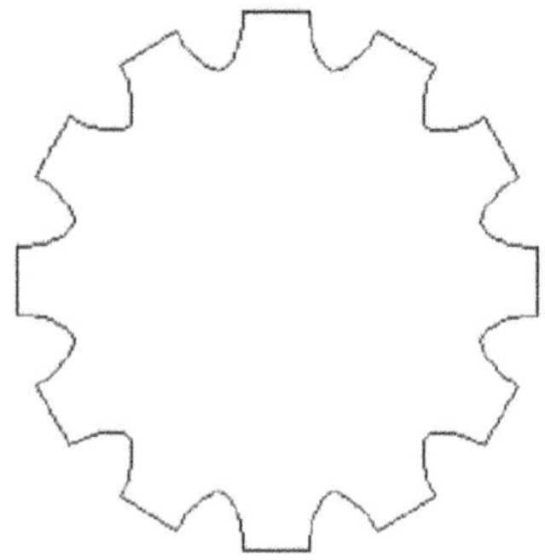

FIGURE 24.29 Drill bit, Step 1.

Step 1. Draw a circle and one tooth, similar to what is found on a gear. Then, array it 12 times around the circle. Finally, do a little bit of cleanup trimming and pedit the entire structure. This is your basic profile, as seen in Fig. 24.29.

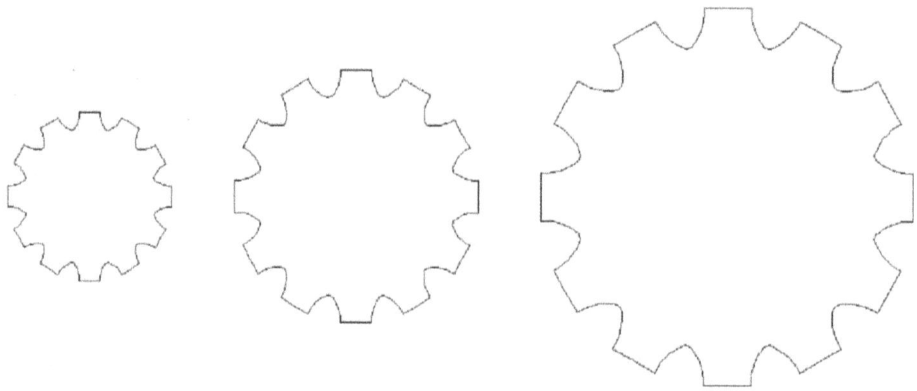

FIGURE 24.30 Drill bit, Step 2.

You need not draw the exact same thing; something close is just fine.

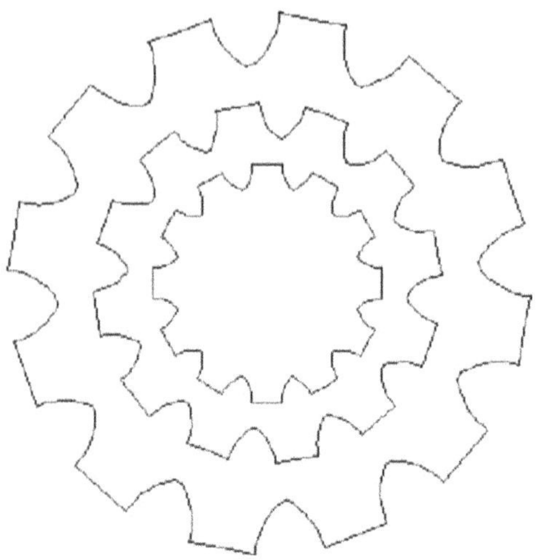

FIGURE 24.31 Drill bit, Step 3.

Step 2. Now copy the profile for a total of three items. Make each new copy 1.5 times larger than the previous one using scale, as seen in Fig. 24.30.

Step 3. Carefully place the three profiles one on top of another using their center points and rotate profiles 2 and 3 (the medium and the largest ones) 10° and 20°, respectively, as seen in Fig. 24.31.

FIGURE 24.32 Drill bit, Step 4.

Step 4. Now, switch to the 3D SW Isometric view and separate the three profiles evenly in size order, as seen in Fig. 24.32. Be careful, you may have to 3Drotate all the profiles to get them to "stand upright." Also, be sure to check the side profiles to make sure they are spaced evenly. Use Ortho, and when done, 3Dorbit around the profile to check that they are located just right.

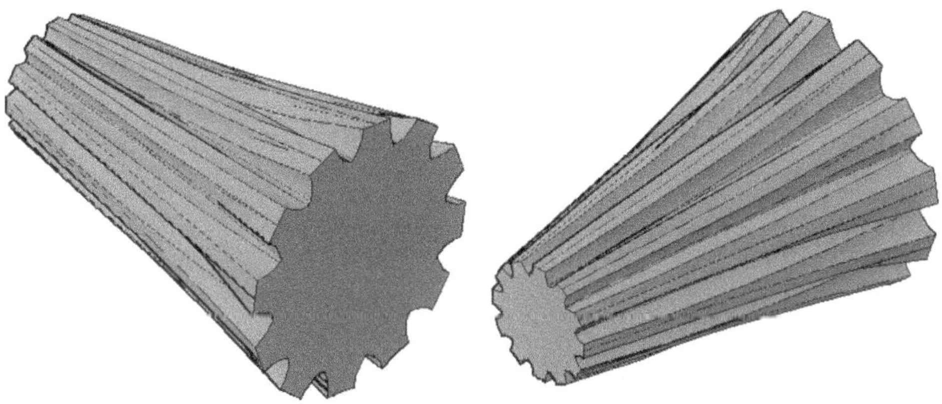

FIGURE 24.33 *Drill bit, Step* 5.

Step 5. Finally, you are ready to execute loft. Fig. 24.33 shows the result.

As you can see from this and the previous examples, loft is a very versatile and powerful command. You can represent virtually any object as long as you are able to visualize and correctly draw a suitable profile.

Path Extrusion

Path extrusion and sweep, which follow next, are commands with a similar mission but different approaches. Path extrusion is the older command and still quite useful in certain situations, while the newer sweep just expands and simplifies the same idea.

The central idea here is to allow extrusion along a given path, which can be curved and otherwise nonlinear (recall that

FIGURE 24.34 Basic path.

the basic extrude command was linear in nature). Path extrusion is technically not a new command, as the option to select a path while extruding was actually always there in the command's submenus, so all we are doing here is finally trying it out.

Draw a circle of any diameter or radius. Then, draw an L-shaped set of lines, as shown in Fig. 24.34, and add a small fillet to the intersection of the lines; finally pedit the lines and arc.

FIGURE 24.35 Basic path, circle rotated.

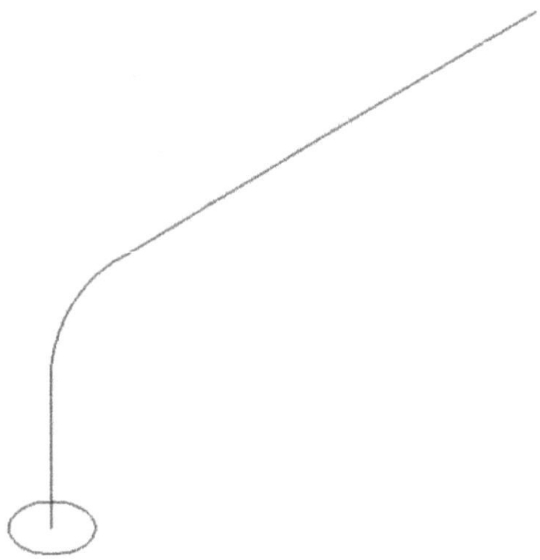

FIGURE 24.36 Basic path in 3D.

This line is the path, and the circle is the profile we would like to extrude. There is one more step before you can do the path extrusion, however. This command requires that the profile (circle) be turned perpendicular to the path (line). Use 3Drotate to rotate the circle 90°. It will look like the one shown in Fig. 24.35. Now switch to 3D. You may need to 3Drotate the pieces into an upright position, as seen in Fig. 24.36.

We can now do the path extrusion. There is no separate path extrusion command, rather it is a submenu of the regular extrude.

Step 1. Startup extrude via any preferred method.

- AutoCAD says: `Current wire frame density: ISOLINES = 4 Select objects to extrude:`

Step 2. Select the circle and press Enter.

- AutoCAD says: `Specify height of extrusion or [Direction/Path/Taper angle]`

Step 3. Select `p` for `Path` and press Enter.

- AutoCAD says: `Select extrusion path or [Taper angle]`

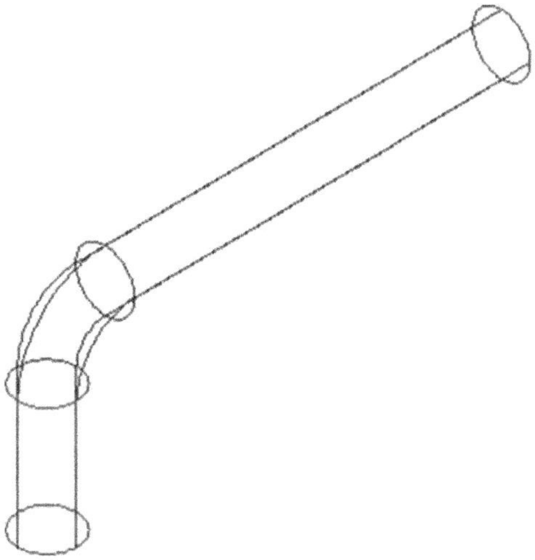

FIGURE 24.37 Path extrusion of the pipe.

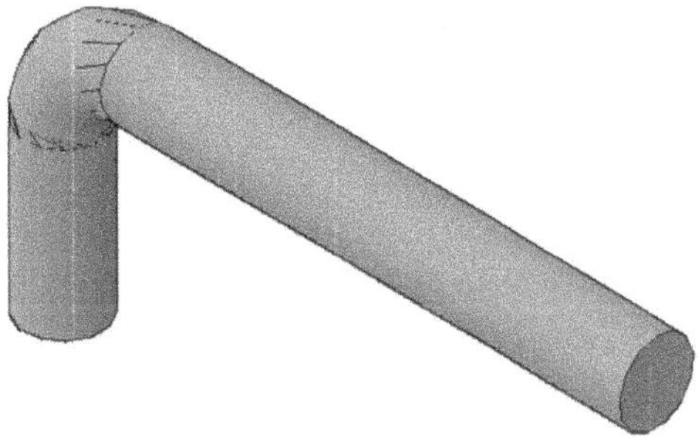

FIGURE 24.38 Path extruded pipe, shaded and colored.

Step 4. Click to select the path itself. You see what is shown in Fig. 24.37 in wireframe.

The extrusion is then shaded and colored to give the final result, as seen in Fig. 24.38. This view is also switched to SE Isometric to show the other side of the pipe.

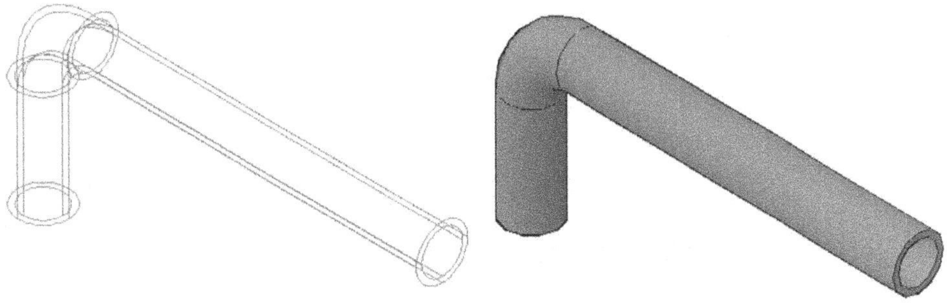

FIGURE 24.39 Path extruded pipe with hollow center.

Fig. 24.39 shows the same path extruded pipe with a hollow center, to actually make it a pipe. This was done by going back to the original profile and simply adding another (smaller) circle profile and extruding it along the same path. Then, a subtraction was done, subtracting one extrusion from the other. As an exercise, go ahead and recreate this yourself.

Sweep

As already mentioned, sweep is a variation on the same theme but with some enhancements. Once again, we have a profile and a path, but instead of carefully lining up the circle to the path, we can select almost any profile, anywhere on the screen,

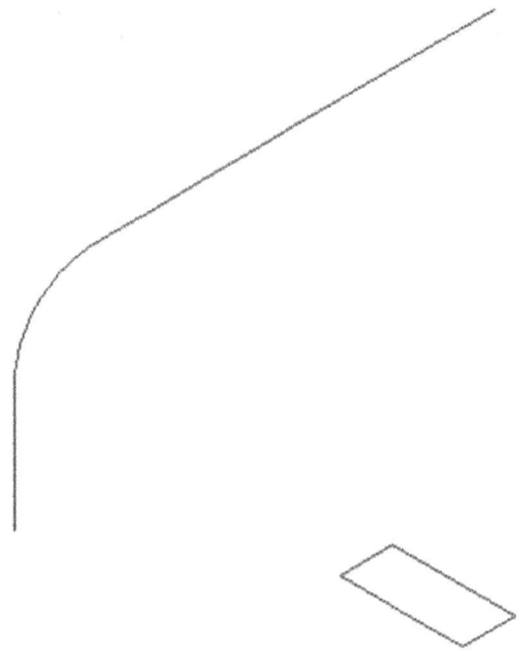

FIGURE 24.40 Rectangle and sweep path.

and click on almost any path to generate a sweep. Therefore, it is a much more flexible command. Fig. 24.40 shows an example with a rectangle used to create a sweep along essentially the same type of path. Notice the profile is nowhere near the path. Go ahead and set this up via the following steps:

Keyboard: Type in **sweep** and press Enter.
Cascading menus: **Draw→Modeling→Sweep**
Toolbar icon: **Modeling** toolbar→
Ribbon: **Solid** tab→**Solid**

Step 1. Start the sweep command via any of the preceding methods.

- AutoCAD says:

```
Current wire frame density: ISOLINES=4, Closed profiles creation mode = Solid
Select objects to sweep or [MOde]: _MO
Closed profiles creation mode [SOlid/SUrface] <Solid>: _SO
Select objects to sweep or [MOde]:
```

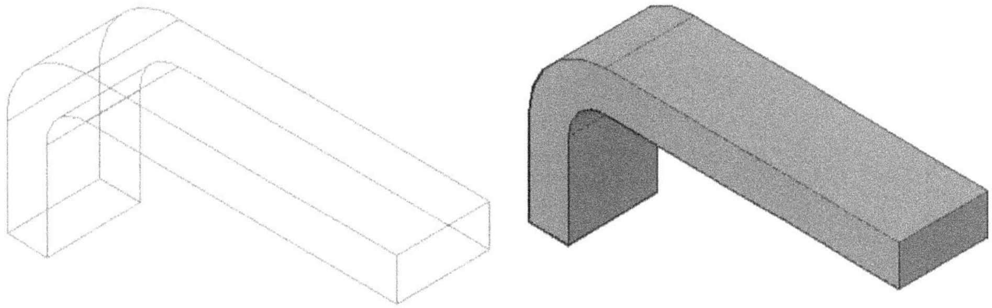

FIGURE 24.41 Rectangle sweep.

Step 2. Pick the rectangle and press Enter.

- AutoCAD says: `Select sweep path or [Alignment/Base point/Scale/Twist]:`

Step 3. Select the path and the sweep is produced, as seen shaded and colored in the SE Isometric view in Fig. 24.41.

There are some limits to this, of course, and those have to do mostly with the size of the profile or path. If the profile does not physically fit the path, AutoCAD tells you that it is unable to produce the sweep. Note that the sweeping action destroys the original profile, so if you think you may need it again, keep a copy of it. This is an important point in creating multiple sweeps later on.

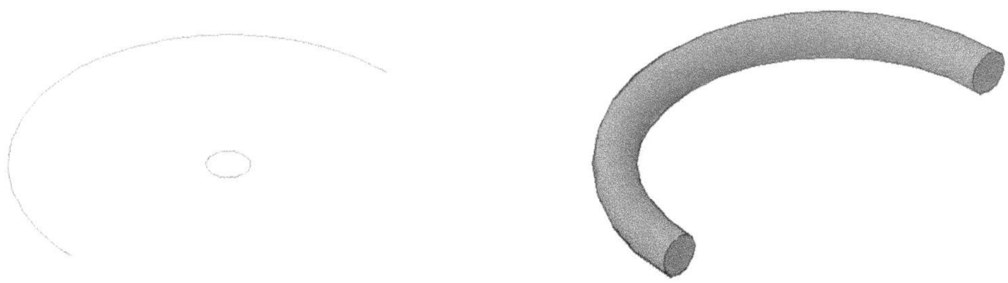

FIGURE 24.42 Circle sweep.

A major advantage of this command over path extrusion is that a sweep can be produced even if the profile is on the same plane as the path, as seen with an arc and a circle in Fig. 24.42.

The sweep command is not bothered by the similar plane. It just takes whatever profile is selected and applies it perpendicularly to whatever path you choose.

This concludes the advanced 3D tools. Practice and learn to apply everything thus far. We have an additional chapter with a few more tools to complete your 3D construction knowledge, but the ones presented in this chapter are the primary ones.

Drawing Challenge: Helical Coil

Using what you know so far, draw the helical coil shown in Fig. 24.43. You have all the tools at your disposal at this point. Think carefully, and do not worry about size or how many coils to do, only the shape. The answer is featured at the end of the next chapter, but do not look it up right away; give it your best shot first. Also, do not be tempted to use AutoCAD's helix tool, which is covered in the next chapter. The results may not necessarily look like the example shown here anyway.

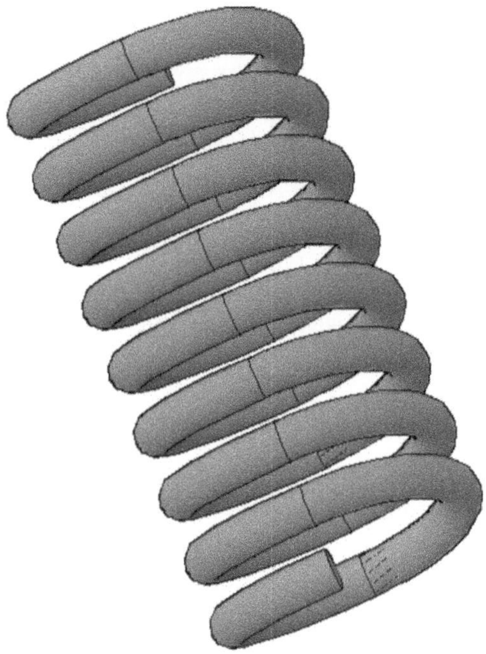

FIGURE 24.43 Helical coil drawing challenge.

There is a reason why you should try to figure this out, and it is mentioned along with the answer in the next chapter. Good luck!

Summary

You should understand and know how to use the following concepts and commands before moving on to Chapter 25:

- Revolve
- Shell
- Taper
- Loft
- Path extrusion
- Sweep

Review Questions

Answer the following based on what you learned in this chapter:

1. Describe the revolve command.
2. Does the profile need to be open or closed? Describe the effect of each.
3. Describe the shell command.
4. How many surfaces can you select with shell?
5. Describe the taper command.
6. Describe the loft command.
7. Describe the path extrusion command.

8. Describe the sweep command.
9. Why is sweep better than path extrusion?

Exercises

1. Use the following profile and axis of rotation (the line below the profile) to revolve it 360° to create the model of a basic tire and rim. Sizing and exact profile shape is up to you, and you need various drawing and editing commands, including pedit, prior to revolve. (Difficulty level: Moderate; Time to completion: 25 minutes.)

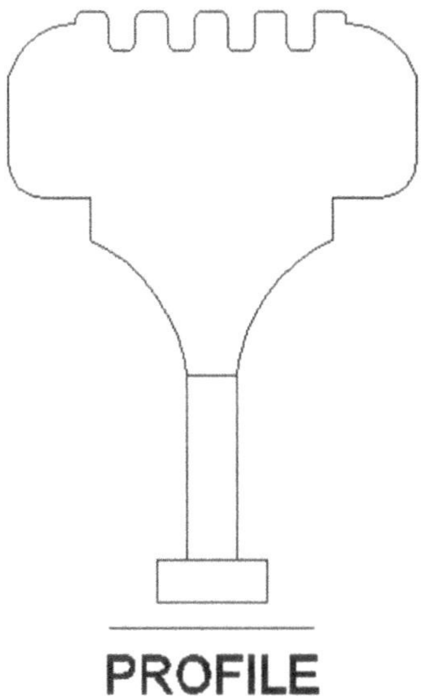

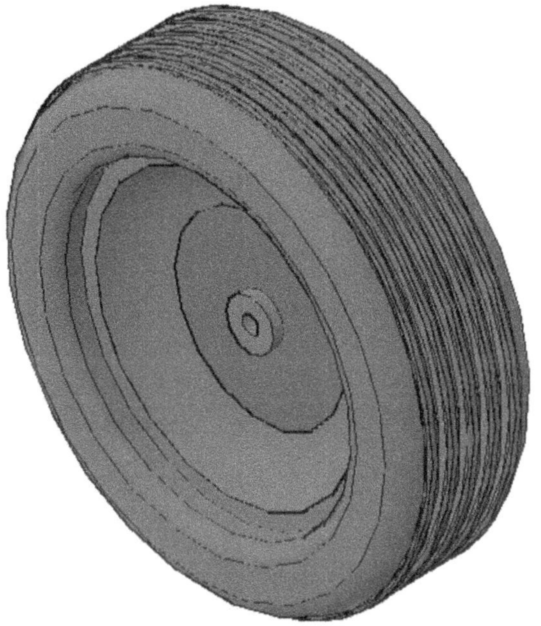

2. Create four 4.5″×6″ rectangles, then explode, trim, and pedit them together into an outline as shown in the profile on the left. Next, extrude them to a height of 5.5″ and shade and color them. Finally, use the shell command (0.2″ thickness) to remove the faces, as shown in the profile on the right. (Difficulty level: Moderate; Time to completion: 10 minutes.)

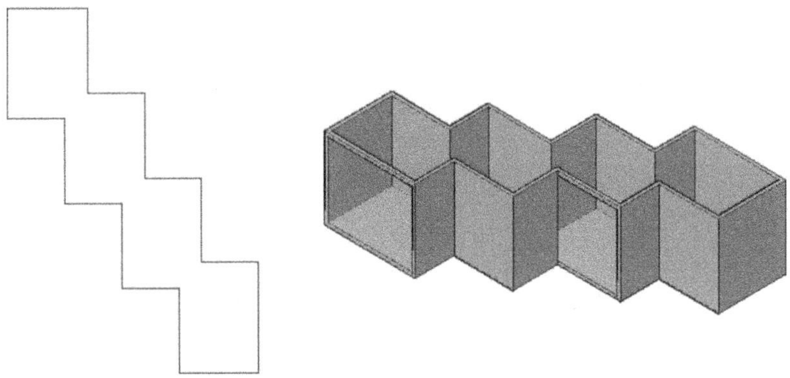

3. Create the following basket using the rectangle (20″ × 10″), extrude (10″), taper (10°), fillet (1.5″), and shell (0.25″) commands. Then, add the handle using arc and sweep (1″ diameter circle). (Difficulty level: Moderate; Time to completion: 15–20 minutes.)

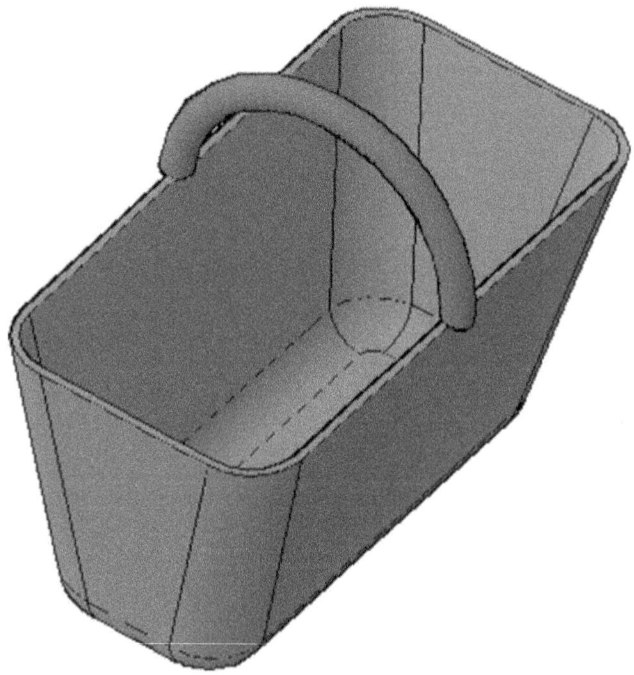

4. Create the following bracket and pole model. An exploded view as well as a breakdown of some individual parts are shown for clarity. A pair of overall dimensions is given, but the rest of the sizing is up to you. (Difficulty level: Moderate; Time to completion: 30–40 minutes.)

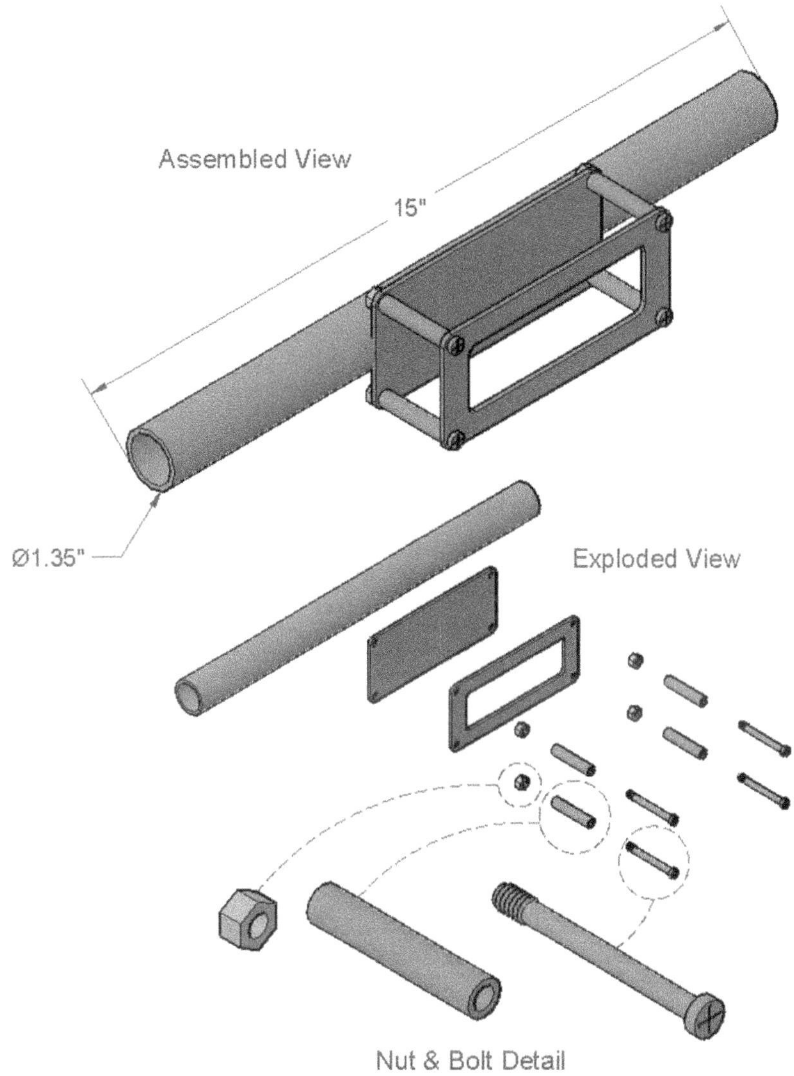

5. Create the following simplified fire extinguisher model. The steps involved, including the overall sizing, are shown for clarity, but you have to improvise the rest of the details. You will need pline, pedit, revolve, shade, subtract, fillet, sweep, and other commands. (Difficulty level: Moderate/Advanced; Time to completion: 60 minutes.)

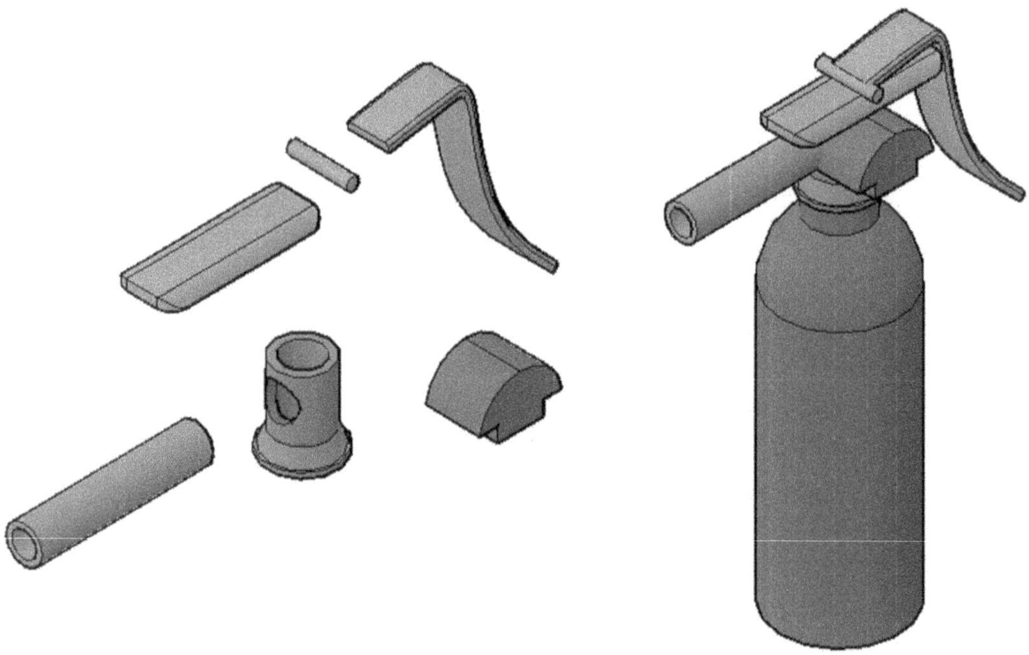

Chapter 25

Advanced Solids, Faces, and Edges

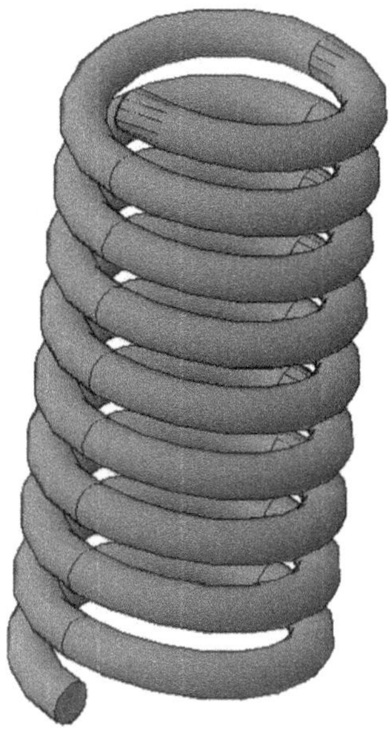

Learning Objectives

In this chapter, we continue to introduce additional advanced tools, including

- Polysolid
- Helix
- 3D path array
- 3Dalign
- Faces
- Move
- Offset
- Delete
- Rotate
- Copy
- Color
- Edges

By the end of this chapter, you will have the complete 3D object design tool set in your hands. Estimated time for completion of this chapter: 2 hours.

Up and Running with AutoCAD 2022. https://doi.org/10.1016/B978-0-323-89923-9.00025-1

25.1 Introduction To Advanced Solids, Faces, And Edges

With this chapter, we continue to introduce advanced 3D tools. While they are not the primary tools you use in day-to-day 3D work, they are still needed to round out your knowledge. Specifically, we add polysolid, helix, 3D path array, and 3Dalign to the list and focus on working with faces (move, offset, delete, rotate, copy, and color) as well as edges (copy and color).

Polysolid

After trying out polysolid, you may never create a wall via the extrude command again. A polysolid is analogous to a polyline. What you get is continuous sections of extruded rectangles similar to continuous sections of a pline in 2D. It is a great tool for quickly constructing 3D walls of a building. Once you begin creating the polysolid, you can continue to "stitch" forward one section at a time by just pointing the mouse in the direction you want to go and entering a distance (usually with Ortho on).

Keyboard: Type in **polysolid** and press Enter
Cascading menus: **Draw→Modeling→Polysolid**
Toolbar icon: **Modeling** toolbar→ 
Ribbon: **Solid** tab→ 

Step 1. Start polysolid via any of the preceding methods.

- AutoCAD says:_Polysolid Height = 4.0000, Width = 0.2500, Justification = Center Specify start point or [Object/Height/Width/Justify] <Object>:

Step 2. These are the default values for the height and width, and they are OK for now. Go into the SW Isometric view and activate Ortho. Then, click anywhere on the screen.

- AutoCAD says:Specify next point or [Arc/Undo]:

Step 3. A connected polysolid is drawn every time you click, so do several clicks and press Enter when done, then shade and color the polysolid. You should have something similar to what is shown in Fig. 25.1 on your screen.

To create a section of a wall similar to what you did in Chapter 24, you need to change the height and width of the polysolid by selecting h for height or w for width in the suboptions.

For *height*,

- AutoCAD says: Specify height <4.0000>:

Enter 72 for the height.

For *width*,

- AutoCAD says: Specify width <0.2500>:

Enter 5 for the width.

Polysolid has two more useful features. The first occurs if you select the Arc suboption while clicking around; you can create an arc section, as seen in Fig. 25.2.

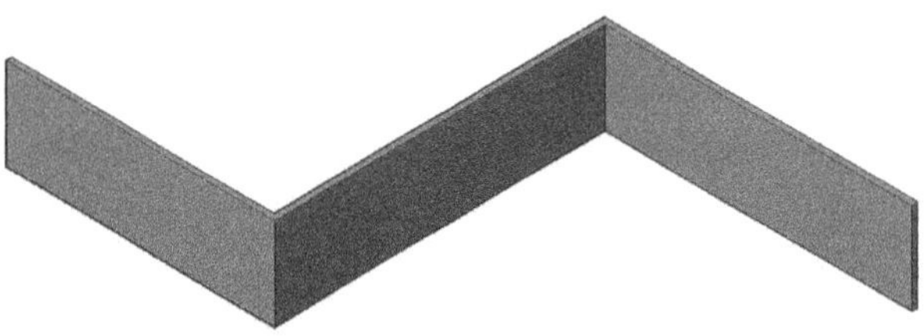

FIGURE 25.1 Basic polysolid.

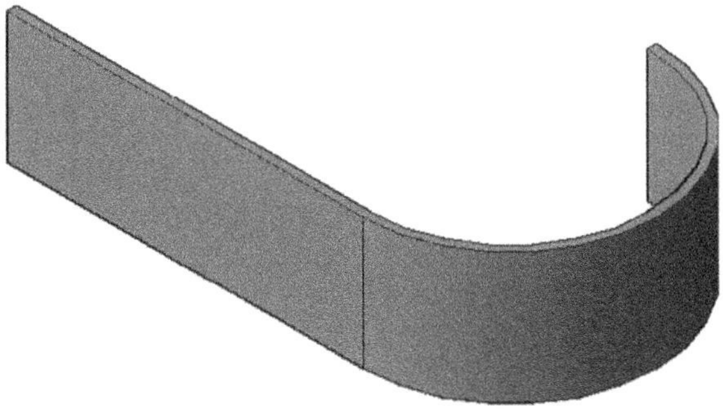

FIGURE 25.2 Basic polysolid with arc.

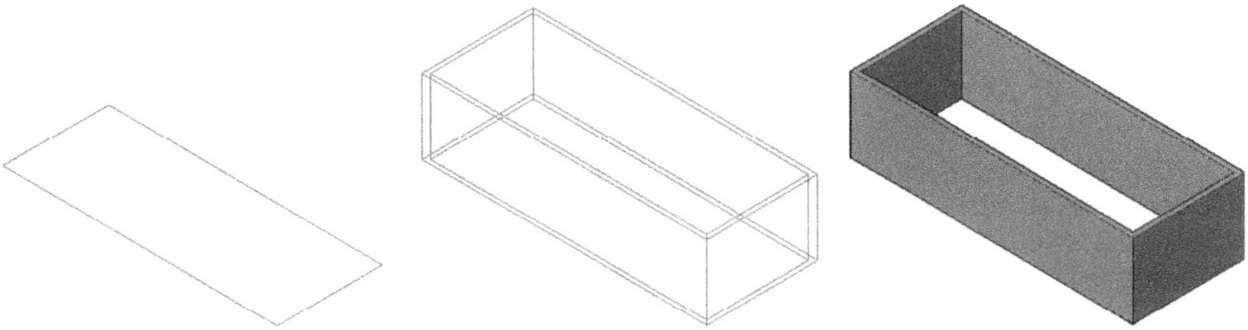

FIGURE 25.3 Rectangle to polysolid.

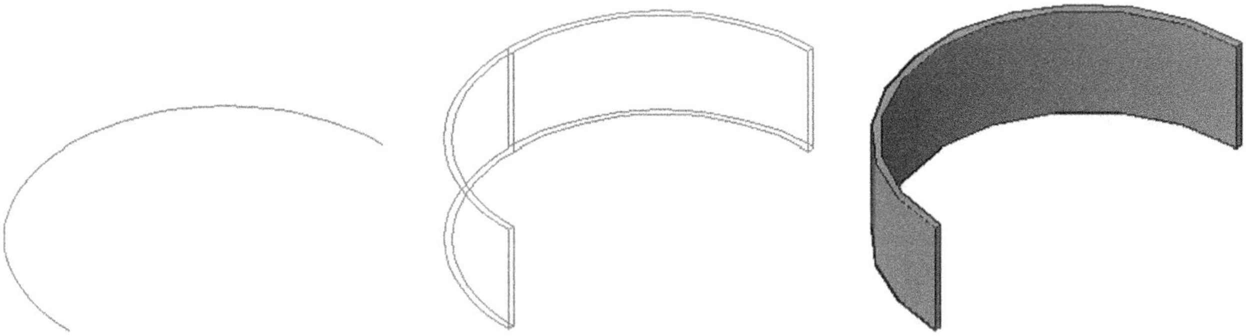

FIGURE 25.4 Arc to polysolid.

The second very useful feature is the ability of the polysolid command to convert regular 2D geometric objects into polysolids using the `Object` suboption. These objects can be rectangles, arcs, circles, and others, as seen in Figs. 25.3 and 25.4. Simply begin the polysolid command, press `o` for `Object`, and select the shape. The height is dictated by the Height setting, which can be easily changed.

Helix

After apparently receiving many requests, called *wish list entries*, on this topic, Autodesk finally included the helix command. It pretty much does what it advertises, creating a basic 3D helix coil, as seen in Fig. 25.5.

Keyboard: Type in **helix** and press Enter
Cascading menus: **Draw→Helix**
Toolbar icon: **Modeling** toolbar→
Ribbon: **Home** tab→**Draw**

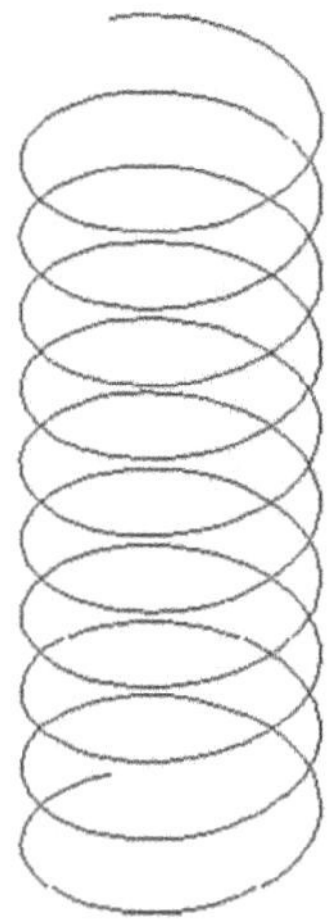

FIGURE 25.5 Basic helix coil.

Step 1. Start helix via any of the preceding methods.

- AutoCAD says: `Number of turns = 3.0000 Twist = CCW`
  `Specify center point of base:`

Step 2. You are asked here to specify where you want the helix coil to be. Generally, you click on any random point to just create the helix. Afterward, you can move it into a specific position according to design needs.

- AutoCAD says: `Specify base radius or [Diameter] <1.0000>:`

Step 3. This is the overall radius or diameter of the bottom part of the coil. Enter any value; 10 is used in this example.

- AutoCAD says: `Specify top radius or [Diameter] <10.0000>:`

Step 4. The top radius or diameter of the coil does not necessarily have to be the same as the bottom; this way your coil can taper in or expand if needed. For this example, however, we choose the top and bottom to be the same size (10).

- AutoCAD says: `Specify helix height or [Axis endpoint/Turns/turn Height/tWist] <1.0000>:`

Step 5. Although you can manually pull the helix into shape at this point, a better approach is to indicate the number of turns and helix height. Type in `t` for `Turns` and press Enter.

- AutoCAD says: `Enter number of turns <3.0000>:`

Step 6. Enter any reasonable value; 10 is used for this example.

- AutoCAD says: `Specify helix height or [Axis endpoint/Turns/turn Height/tWist] <1.0000>:`

Step 7. Now type in the height you wish the helix to have; 40 is used in this example.

The final result is seen on the left-hand side of Fig. 25.6. Now, how would you give the helix a thickness, like a real spring? We recently covered the answer to that: the sweep command. Create a small circle profile positioned near the helix. Using the helix as a path, create a sweep of the circle profile, as seen in the rest of Fig. 25.6.

While this is an easy way to model the coil as presented in the challenge drawing of the previous chapter, resist the temptation to do so and search for other ways.

3D Path Array

The array command was first introduced in Chapter 8. Having gone through a bit of a redesign, array is now Ribbon or command-line driven (no more dialog boxes) and features a new, third variation, the path array. This allows you to select an object and a path. Multiple instances of that object are then created along the path. The same exact thing can be done in 3D.To try this out, we create a spiral staircase. Draw a small step consisting of a 20″×8″ rectangle extruded to a thickness of 1″, as seen in Fig. 25.7. Do not forget to add color and shade.

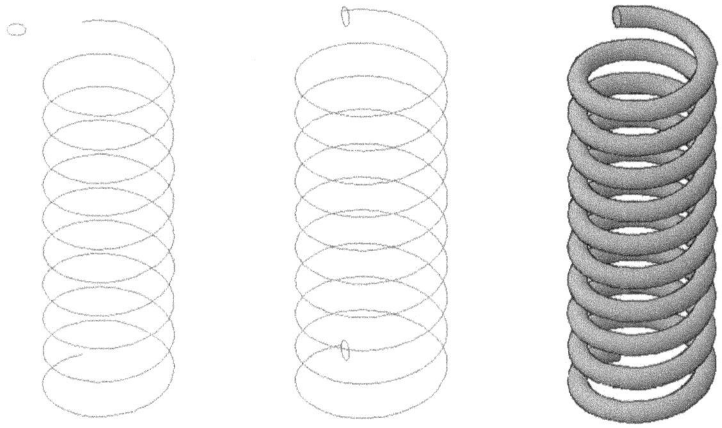

FIGURE 25.6 Completed helix coil with sweep.

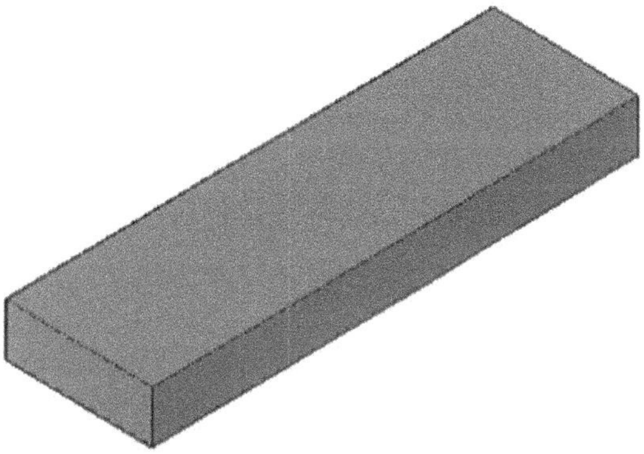

FIGURE 25.7 20" ×8" ×1" step for path array.

Then, using your new helix skills, create a helix of any reasonable size. The one used in this example is 15 ×15, with a height of 50. The step is rotated 90° to align with the bottom of the helix, as seen in Fig. 25.8.

Finally, run through the path array command (it is the same for 3D as for 2D). Just as a reminder, you are asked to select the object and indicate this is a path array. Then, type in `50` when asked how many objects. Finally, press Enter twice to complete the 3D path array command. Your results should look similar to Fig. 25.9, which shows the new staircase from two views. Note that the path is still visible but can be erased if desired, of course.

While this was just a "quick and dirty" example, you can easily see how some time spent crafting a proper set of stairs, handrails, and a frame yields a graceful and impressive design of a spiral staircase with comparatively modest effort. You will try this out in one of the end-of-chapter exercises.

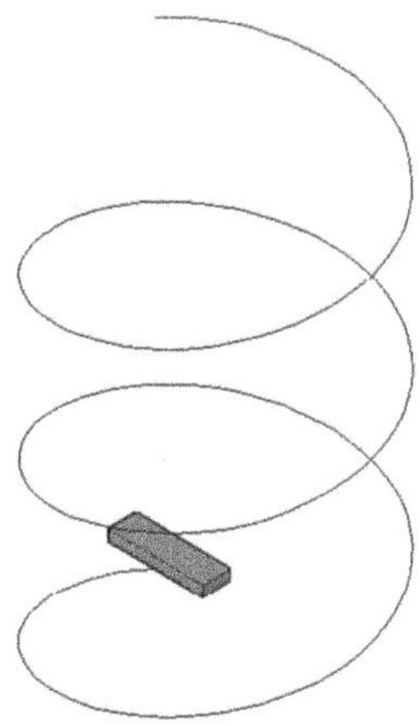

FIGURE 25.8 Helix for path array.

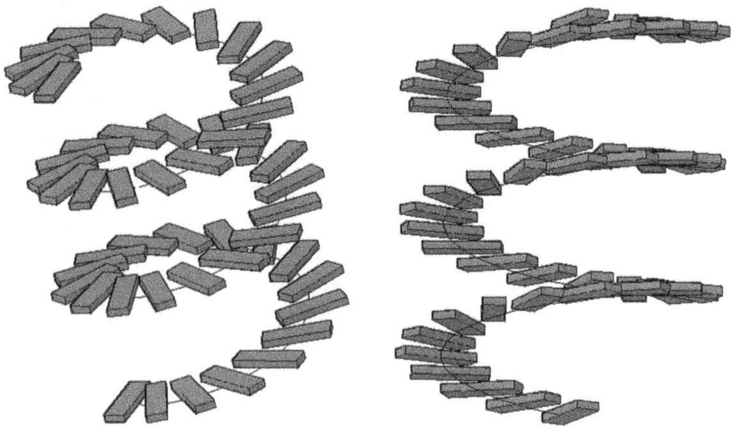

FIGURE 25.9 Completed spiral staircase (two views).

3Dalign

As you may recall from Level 2, aligning objects in 2D can be somewhat tricky, and the align command is quite useful in quickly and easily mating up geometric pieces. In 3D, aligning objects can be even more of a challenge, especially if they are out of alignment in all three axes. Therefore, 3Dalign is almost mandatory to accomplish this. Go ahead, create and position the two blocks as described next.

In Fig. 25.10, on the left (Block 1) is a 25″×15″ block, extruded to 15″. It is even with the X, Y, and Z axes, as seen in the standard SW Isometric view, while a similar, copied Block 2 on the right has been rotated 15° along the X and Y axes and 30° along the Z axis. We need to align Face 1 with Face 2 by moving Block 2 (since it is the rotated one) and bringing it around to have its face match up with the face of the correctly aligned Block 1.

Keyboard: Type in **3dalign** and press Enter
Cascading menus: **Modify→3D Operations→3D Align**
Toolbar icon: **Modeling** toolbar→
Ribbon: **Home** tab→**Modify**

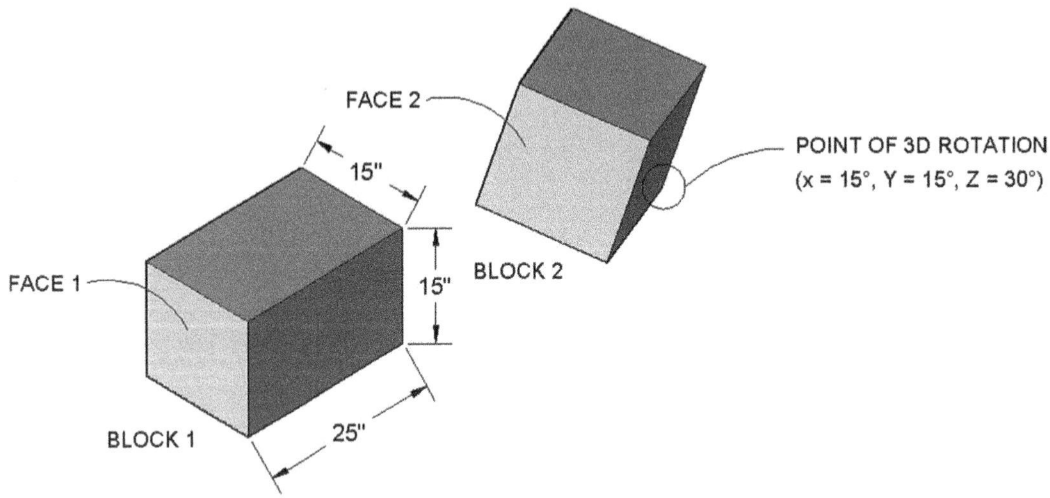

FIGURE 25.10 Blocks for 3Dalign.

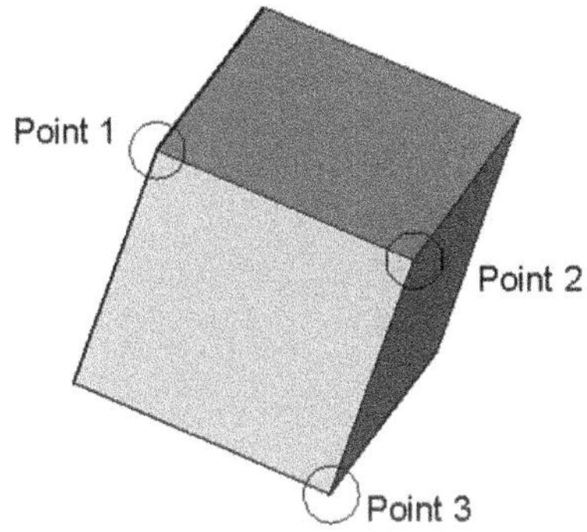

FIGURE 25.11 Selecting points, Step 3.

Step 1. Start the 3Dalign command via any of the preceding methods.

- AutoCAD says: `Select objects:`

Step 2. Select just the misaligned Block 2 (you could, in theory, pick Block 1 as well) and press Enter.

- AutoCAD says: `Specify source plane and orientation ...Specify base point or [Copy]:`

Step 3. Here, you need to click on and choose three consecutive source points on Block 2's face that you want used as points of attachment to Block 1. Several combinations are possible, so let us arbitrarily choose the points shown in Fig. 26.11. As you select these three points,

- AutoCAD says: `Specify second point or [Continue] <C>: Specify third point or [Continue] <C>: Specify destination plane and orientation ...Specify first destination point:`

Step 4. Now, you are ready to specify the destination points. Here you have to be careful and think about what those points actually are, as random choices produce random results. Observe that, in this example, Block 2 needs to be rotated 180° for its lighter face to match up to Block 1's lighter face. Therefore, the destination points need to be mirror images of the source points, not the exact same ones.

The order we picked the source points was Top Left→Top Right→Bottom Right. As we swing Block 2 around, the corresponding points on Block 1 are Top Right→Top Left→Bottom Left. Do you see why? Visualize this on a nearby object, one that you can pick up, locate the points on, and turn around: perhaps a cell phone or a book. Let us now select those points, as seen in Fig. 25.12. As you are doing this,

- AutoCAD says: `Specify first destination point: Specify second destination point or [eXit]<X>: Specify third destination point or [eXit]<X>:`

Step 5. Select the points carefully; Block 2 may be in the way, so you need to trust your OSNAPs (or just stay in wireframe mode to see the points). The final result is shown in Fig. 25.13.

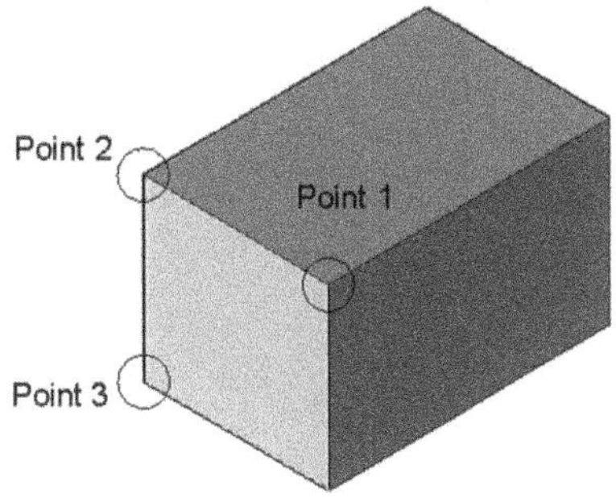

FIGURE 25.12 Selecting points, Step 4.

FIGURE 25.13 3Dalign completed.

25.2 Working With Faces

Suppose you created a hole in a block via extrusion and subtraction, as seen in Fig. 25.14. How would you delete that hole, if it is no longer needed, without redrawing or otherwise damaging the original block? What if you need to move the hole from its center position to another area of the block? Or, what if you just need to change its diameter? All this and more are addressed by the various faces operations, such as move face, copy face, offset face, delete face, rotate face, and even color face.

Being able to easily modify the position, geometry, and sizing of 3D features is a fundamental requirement of all solid modeling programs. The parametric modeling software on the market approaches this in various ways, some better than others. The Spatial Technologies ACIS kernel that powers AutoCAD's 3D modeler is not quite on the level of Parasolid, Granite, or CAA, which power the high-end NX, Creo (Pro/Engineer), and CATIA software, respectively (see Appendix H for more information on kernels), but it does a decent job working with embedded 3D features using the various faces tools. Let us cover them all, one at a time, as they are very important in effective 3D design.

Delete Faces

Everything you create in AutoCAD 3D has faces. They may be the faces of an object, such as the extruded hole or actual surfaces of larger objects. When AutoCAD refers to faces, it is those objects or surfaces that can be moved or copied. In some cases, you can interpret this as moving or copying an entire feature (such as the extruded hole in the upcoming examples) or, in other cases, an actual surface of an object.

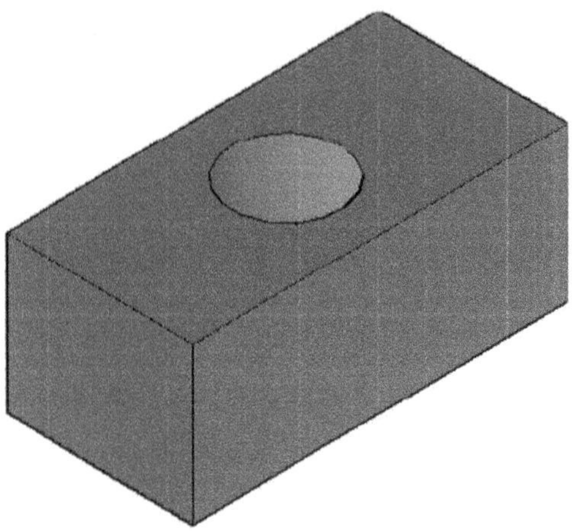

FIGURE 25.14 Block with a hole.

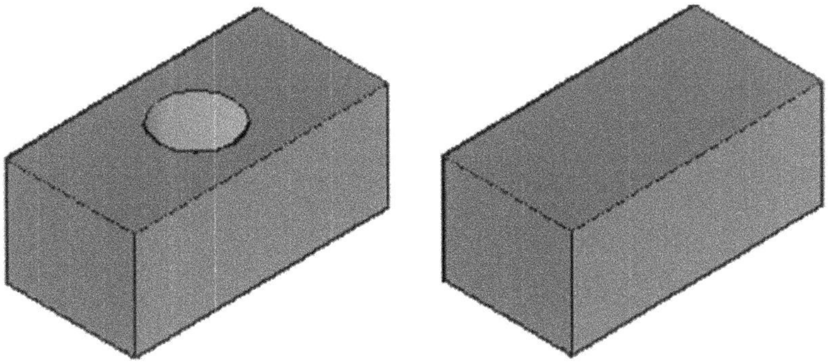

FIGURE 25.15 Delete faces (hole).

Delete faces is probably the easiest of the face commands to start with. It is meant to be used to delete full features, such as an extruded cylinder (hole), not surfaces. To try it out, draw a random size block with a hole in it, as seen in Fig. 25.15.

Keyboard: Type in **solidedit** and press Enter. Then **face**, Enter and **delete**
Cascading menus: **Modify→Solid Editing→Delete faces**
Toolbar icon: **Solid Editing** toolbar→ 
Ribbon: **Home** tab→**Solid Editing**→ 

Step 1. Startup delete face (a solid editing command) via any of the preceding methods.

- AutoCAD brings up a lengthy solid editing menu (unless typing):

```
_solidedit
Solids editing automatic checking: SOLIDCHECK = 1
Enter a solids editing option [Face/Edge/Body/Undo/eXit] <eXit>: _face
Enter a face editing option Extrude/Move/Rotate/Offset/Taper/Delete/Copy/ coLor/mAterial/Undo/
eXit] <eXit>:_color
Select faces or [Undo/Remove]:
```

Step 2. Click to select the hole. Be careful to pick only the hole not surrounding geometry.

- AutoCAD says:`1 face found.`

Step 3. Press Enter.

- AutoCAD says:

```
Solid validation started.
Solid validation completed.
Enter a face editing option.
```

Step 4. The hole is removed, as seen in Fig. 25.15. Press Esc to return to the command line.

The delete faces command can also be used to delete unwanted fillets in 3D. Simply run the command, select the fillet, and it reverts to a 0 radius (sharp corner), as seen in the sequence of Fig. 25.16.

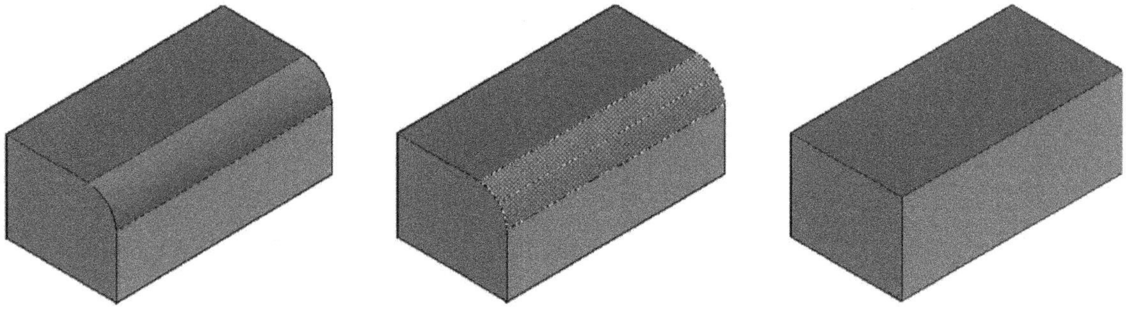

FIGURE 25.16 Delete faces (fillet).

Move Faces

The move faces command moves the hole to another location. Undo the face deletion you just practiced and let us try to move the hole around.

Keyboard: Type in **solidedit** and press Enter. Then **face**, Enter and **move**
Cascading menus: **Modify→Solid Editing→Move faces**
Toolbar icon: **Solid Editing** toolbar→
Ribbon: **Home** tab→**Solid Editing**→

Step 1. Startup move face via any of the preceding methods.

Step 2. AutoCAD shows the same lengthy solid editing menu (not reproduced here) and asks you to select the faces (the hole). From there, the command is exactly like a regular move command. It asks for a base point to move from and a second point to move to. Be sure to have Ortho on when doing this. On completion, the hole is moved, as seen in Fig. 25.17.

Copy Faces

Copy faces copies any element you select in the same exact manner that move faces works. It is more useful for a slightly different application, involving actual faces as opposed to elements like a hole, however. To try out both approaches, leave your hole in its new position at the edge of the block and do the following:

Keyboard: Type in **solidedit** and press Enter. Then **face**, Enter and **copy**
Cascading menus: **Modify→Solid Editing→Copy faces**
Toolbar icon: **Solid Editing** toolbar→
Ribbon: **Home** tab→**Solid Editing**→

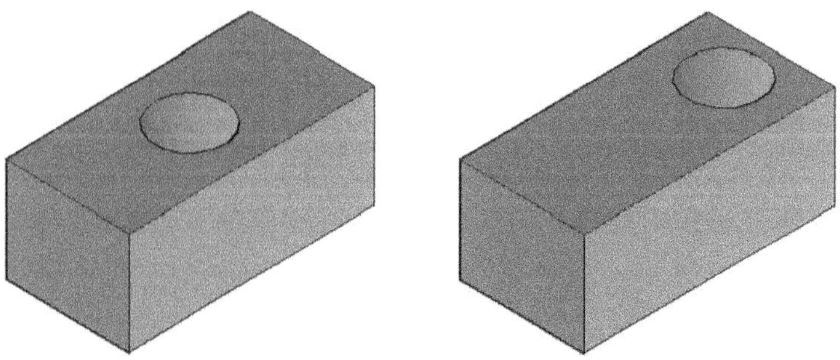

FIGURE 25.17 Move faces, before and after.

Step 1. Startup copy face via any of the preceding methods.

Step 2. You are prompted for a base point to copy *from* and a second point to copy *to*. However, you have to be careful in how you apply this command. If you copy the extruded hole, the new hole is not a solid and cannot be subtracted. What you get is shown in Fig. 25.18.

What happened was that just the edge was copied, and we discuss how to deal with this properly when we discuss edges. Meanwhile, what exactly is copy faces good for? Well, remember that you can move and copy not just features but also surfaces. We copy a surface off to the side and extrude it into a shape, as shown next in Figs. 25.19 and 25.20. This can be useful because you need not create new profiles from scratch but can reuse complex geometry.

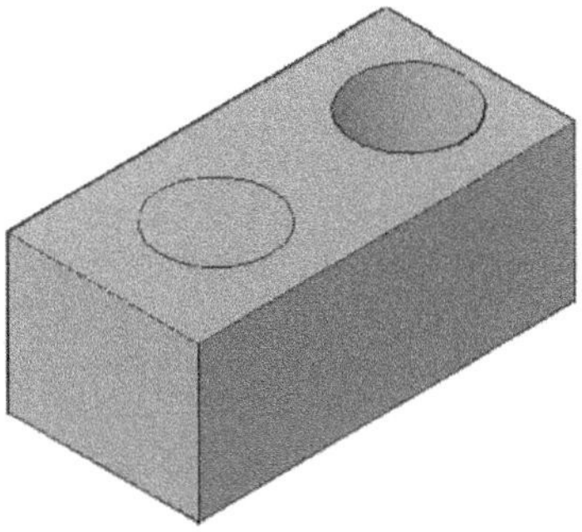

FIGURE 25.18 Copy faces applied to hole.

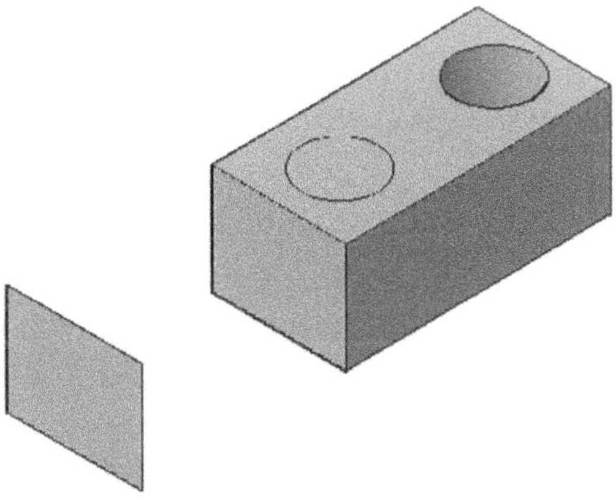

FIGURE 25.19 Copy faces applied to surface.

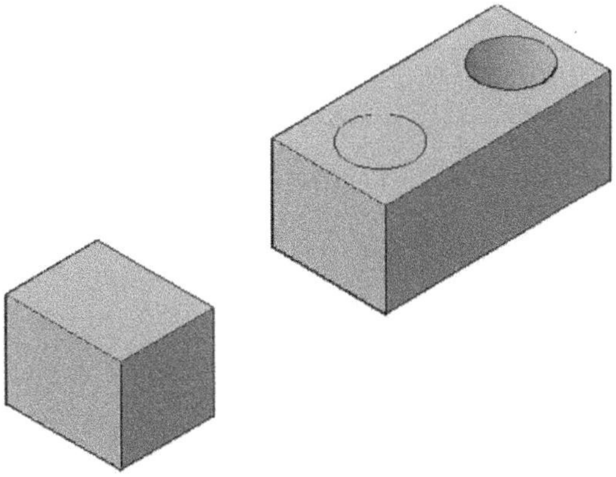

FIGURE 25.20 Copied face extruded.

Offset Faces

Offset faces can accomplish essentially the same thing as extrude. While you are used to extruding 2D objects into the third dimension, extrude and offset faces can also be used to lengthen or shorten what is already a 3D object by extruding or offsetting the face you pick, and the solid material stretches or shrinks to meet the new position of this face. The approach is essentially the same, so we focus on offset faces.

Keyboard: Type in **solidedit** and press Enter. Then **face**, Enter and **offset**
Cascading menus: **Modify→Solid Editing→Offset faces**
Toolbar icon: **Solid Editing** toolbar→
Ribbon: **Home** tab→**Solid Editing**→

Step 1. Startup offset face via any of the preceding methods.

Step 2. AutoCAD once again runs through the lengthy solid editing menu and asks you to select a face. Pick the face you want to offset and press Enter. AutoCAD then asks you to specify the offset distance. Enter some value, and you see the result, similar to what is shown in Fig. 25.21.

Offset faces can also come in handy when attempting to change the size of an embedded feature, such as the hole's diameter in this case. Applying offset faces to the hole, we get the result seen in Fig. 25.22, when adding an offset of 3 to it.

FIGURE 25.21 Offset face 1.

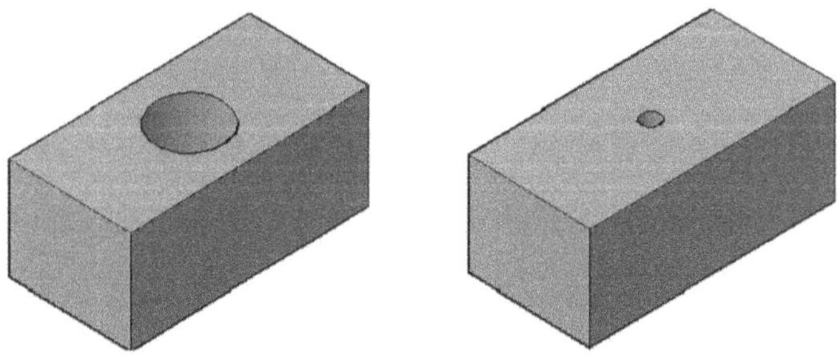

FIGURE 25.22 Offset face 2.

Rotate Faces

Rotate faces is primarily for rotating embedded 3D features. Since it is hard to see the rotation of a circle, let us draw a new part, consisting of a random size plate with a cutout, as seen in Fig. 25.23. Next to the cutout is a small extrusion for reference purposes. We would like to rotate this cutout 90°.

Keyboard: Type in **solidedit** and press Enter. Then **face**, Enter and **rotate**
Cascading menus: **Modify→Solid Editing→Rotate faces**
Toolbar icon: **Solid Editing** toolbar→ 
Ribbon: **Home** tab→**Solid Editing**→ 

Step 1. Startup rotate face via any of the preceding methods.
Step 2. AutoCAD once again gives you a lengthy solid editing menu (not reproduced here) and asks you to pick the faces for rotation. Carefully select both straight and curved faces and press Enter.

- AutoCAD says: `Specify an axis point or [Axis by object/View/Xaxis/Yaxis/Zaxis] <2points>:`

Step 3. You are free to choose any axis for the rotation or use two points to specify a new one. For this example, the Z axis is used.

- AutoCAD says: `Specify the origin of the rotation <0,0,0>:`

Step 4. The origin of the rotation is up to you to choose. For this example though, OTRACK is used to select the exact center of the cutout.

- AutoCAD says: `Specify a rotation angle or [Reference]:`

Step 5. Once again, this is up to you, but 90° is used for this example.

- AutoCAD rotates the cutout and says: Solid validation started. Solid validation completed. Enter a face editing option.

The result is seen in Fig. 25.24.

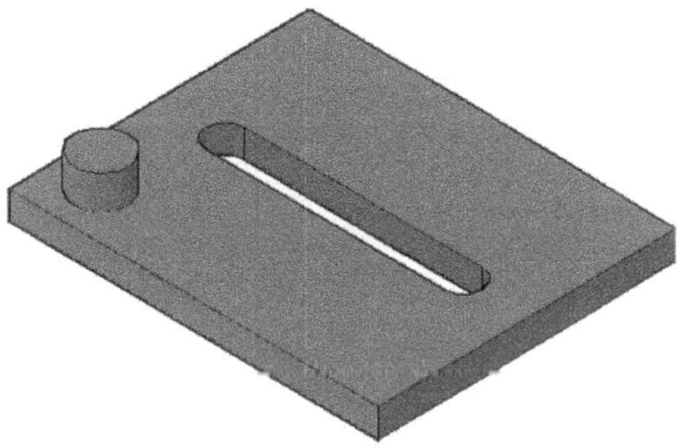

FIGURE 25.23 Plate with groove and extrusion.

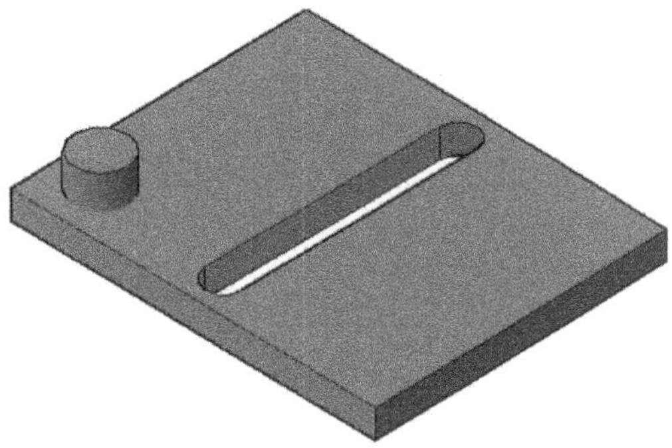

FIGURE 25.24 Groove rotated via rotate face.

Color Faces

AutoCAD gives you the ability to color any of the faces. This is easy to do and is a useful trick to make the faces stand out (we do this with edges as well later on).

Keyboard: Type in **solidedit** and press Enter. Then **face**, Enter and **color**
Cascading menus: **Modify→Solid Editing→Color faces**
Toolbar icon: **Solid Editing** toolbar→
Ribbon: **Home** tab→**Solid Editing**→

Step 1. Startup color face via any of the preceding methods.
Step 2. Select a face, and you see the familiar color dialog box from Chapter 3, Layers, Colors, Linetypes, and Properties.
Step 3. Select a color and the face changes.

25.3 Working With Edges

Working with edges is a short topic, as there are only two items to cover: color edges and copy edges.

Color Edges

Keyboard: Type in **solidedit** and press Enter. Then **edge**, Enter and **color**
Cascading menus: **Modify→Solid Editing→Color edges**
Toolbar icon: **Solid Editing** toolbar→
Ribbon: **Home** tab→**Solid Editing**→

This command is very similar to color faces. You may not see the new colored edge as easily, since it is just that, an edge, basically a thin line.
Step 1. Startup color edge via any of the preceding methods.
Step 2. Select an edge, and you see the familiar color dialog box from Chapter 3.
Step 3. Select a color and the edge changes.

Copy Edges

Keyboard: Type in **solidedit** and press Enter. Then **edge**, Enter and **copy**
Cascading menus: **Modify→Solid Editing→Copy edges**
Toolbar icon: **Solid Editing** toolbar→
Ribbon: **Home** tab→**Solid Editing**→

Recall that, when copy faces was attempted on an extruded hole, there was a problem, as we were unable to subtract the copied hole. Copy edges gets around this problem by copying only the original circle, which can now be extruded. With a simple shape, such as a circle, you could also just redraw it and extrude; but with more complex shapes, this copy edge technique is quite useful. An example follows graphically via Fig. 25.25. Go ahead and try this command out on your own.

Practice all the face and edge editing commands we just covered. We conclude this chapter with an explanation on how the helix was created by hand.

Helical Coil Explained

So how was helical coil drawn in the previous chapter? Have you made an earnest attempt to create it without using the helix tool? Here is the answer. You have to visualize what a helix really is: a bunch of circles connected together. All we need to do is figure out a way to connect them and the rest is easy. Here is one possible way. Draw a circle and "split it in two" by drawing a line across it quadrant to quadrant, trimming off one half and mirroring the half right back, as seen in Fig. 25.26 (you can also use the break command).

Now go into 3D and use rotate3D to raise one half of the circle 15° up (or down) around the Y axis, as seen in Fig. 25.27.

At this point, it is just a matter of connecting a bunch of these segments, as seen in Fig. 25.28. The final step is to use the sweep command to create the thickness for the coil, as seen in Fig. 25.29 (remember to use many circles, as one gets used up for each section). Union the entire coil and assign color and shading. The final coil is shown in Fig. 25.30.

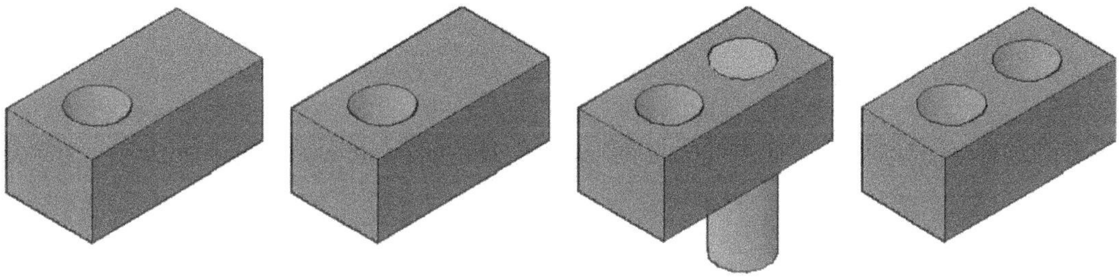

FIGURE 25.25 Edge copied, extruded and subtracted.

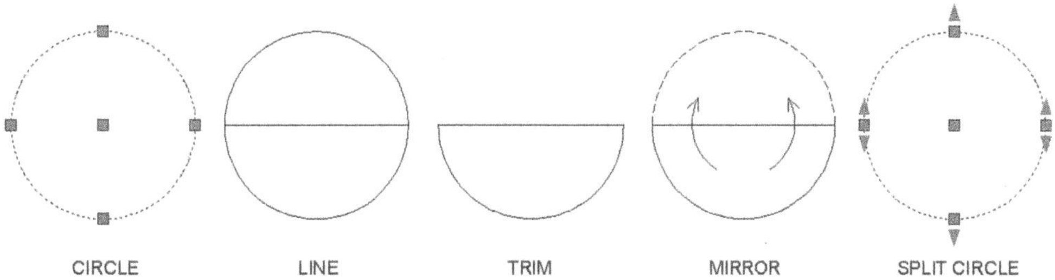

FIGURE 25.26 Creating the helical coil, Part 1.

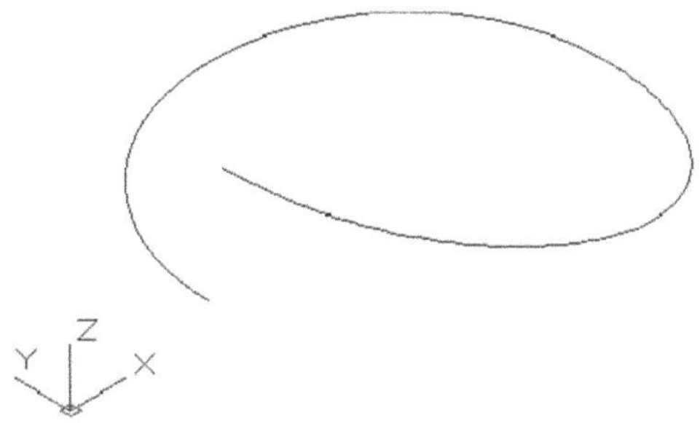

FIGURE 25.27 Creating the helical coil, Part 2.

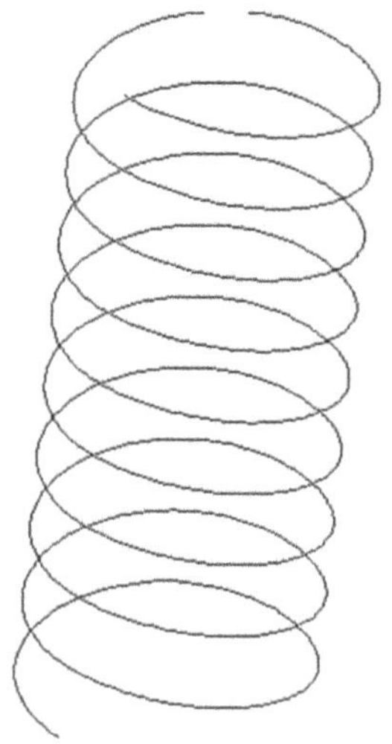

FIGURE 25.28 Creating the helical coil, Part 3.

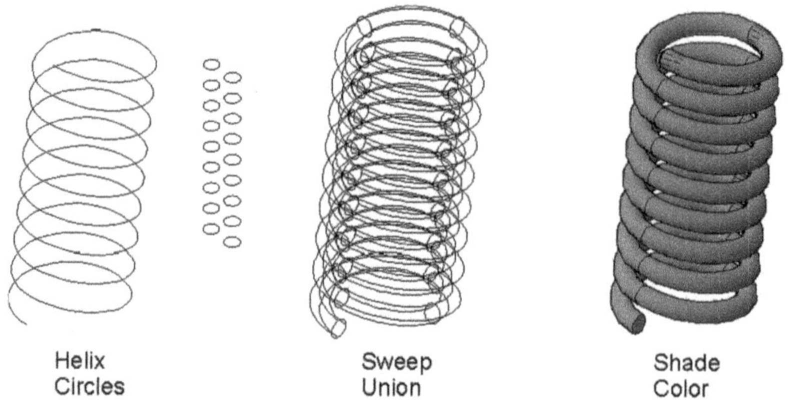

FIGURE 25.29 Creating the helical coil, Part 4.

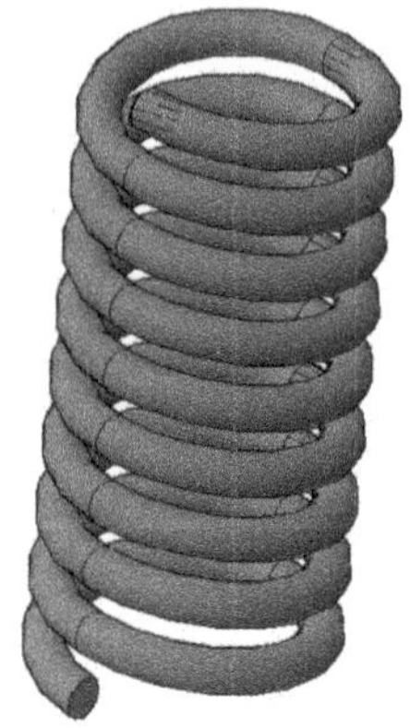

FIGURE 25.30 Completed helical coil.

Why were you asked to create this helix when AutoCAD has an automated command for this? Well, mainly because it is a good exercise in critical thinking in a situation that has no easy answer (at least prior to the introduction of the helix command in AutoCAD 2007). Very often, situations arise in which you have to depict something complex and no ready-to-use tools are available; after all, every possible design scenario cannot be accounted for by AutoCAD. You have to stop, think of what you do have available, and see how you can use these tools creatively to construct something new.

Everything used in the construction of this helix is familiar to you, but the combination of commands is novel. That is the key, finding ways to accomplish a task using the tools you already have. It is a virtual certainty you will run into drawing situations where you need to think outside the box. Ideally, you will remember this exercise and remember to think in terms of fundamental tools.

Summary

You should understand and know how to use the following concepts and commands before moving on to Chapter 26:

- Polysolid
- Helix
- 3D path array
- 3Dalign
- Faces
  - Move
  - Offset
  - Delete
  - Rotate
  - Copy
  - Color
- Edges
  - Copy
  - Color

Review Questions

Answer the following based on what you learned in this chapter:

1. Describe the polysolid command. Can you turn regular objects into polysolids?
2. Describe the helix command.
3. Describe the 3D path array command.
4. Describe the 3Dalign command.
5. What are the six faces commands?
6. What are the two edges commands?

Exercises

1. Create the following wall enclosure using only the polysolid command. All wall heights are 7′. All wall thicknesses are 6″. The rest of the dimensioning is shown. Use the basic "point with mouse and enter dimension" technique, first shown with a line in Chapter 2, to create the polysolid walls. The "mid" dimensions refer to the fact that the given line is positioned at the midpoint of the line to which it is attached. (Difficulty level: Easy; Time to completion: 15–20 minutes.)

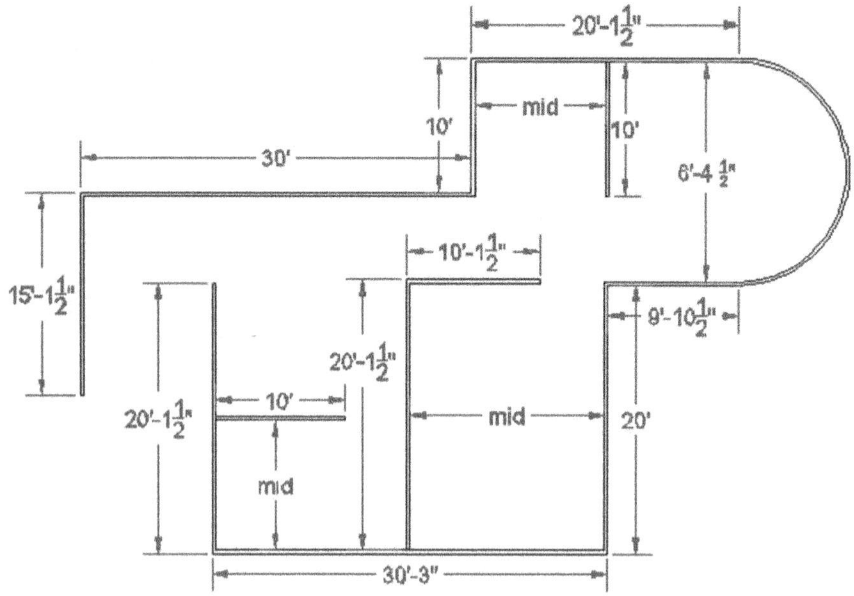

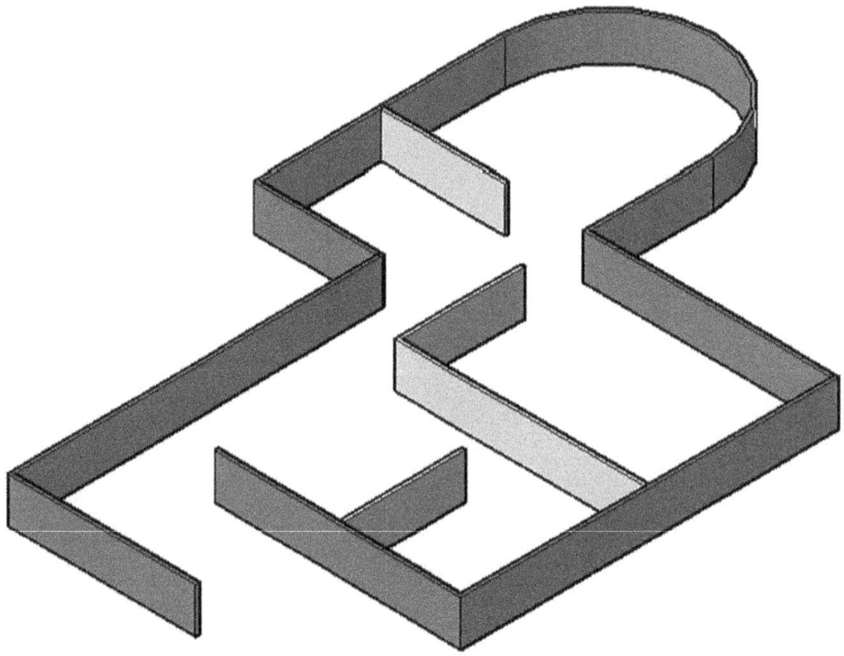

2. Create the following three helix designs based on the data given, including the sweep. The expected result is shown. (Difficulty level: Easy; Time to completion: 10–15 minutes.)
   a. Diameter: 10, top radius: 4, Turns: 12, Twist: CCW, Height: 50. Sweep with circle, d =2.
   b. Diameter: 5, top radius: 10, Turns: 3, Twist: CW, Height: 10. Sweep with rectangle, L =2, W =3.
   c. Diameter: 4, top radius: 4, Turns: 20, Twist: CCW, Height: 40. Sweep with circle, d =2.

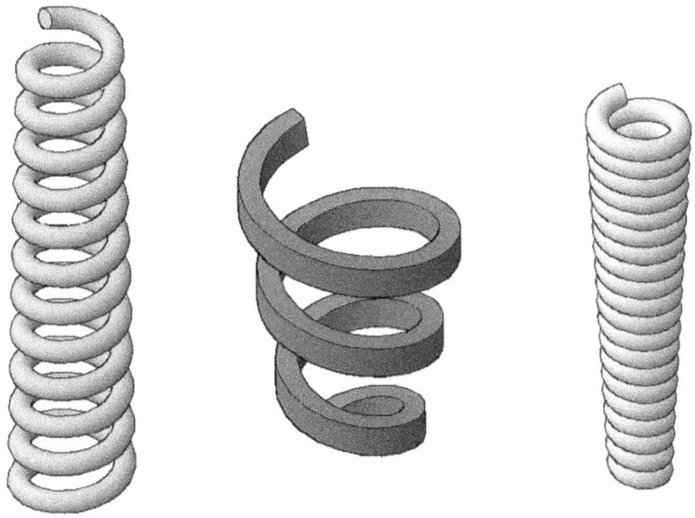

3. Create the following spiral staircase with hand rails. You first need to create a sample stair with the dimensions shown, extruding it to 75. Next, create a set of hand rail poles of any appropriate size and height, positioned as shown, followed by the floor (any size) and the center pole (height of 120). Then, create a helix, centered on the pole, with the following basic features: height: 120, turns: 1.5, base and top radius: 30, twist: CCW. Finally, position the sample stair at the start of the helix, and perform a path array along that helix, similar to the example in this chapter. You can add the top of the handrail using a sweep of a basic profile—the path used was a spline that hopped from midpoint to midpoint of the top of the rail poles. These are the essential steps, but you have to improvise and experiment to get everything in sync and looking similar to the exercise. Color and shade as needed. (Difficulty level: Difficult; Time to completion: 60−90 minutes.)

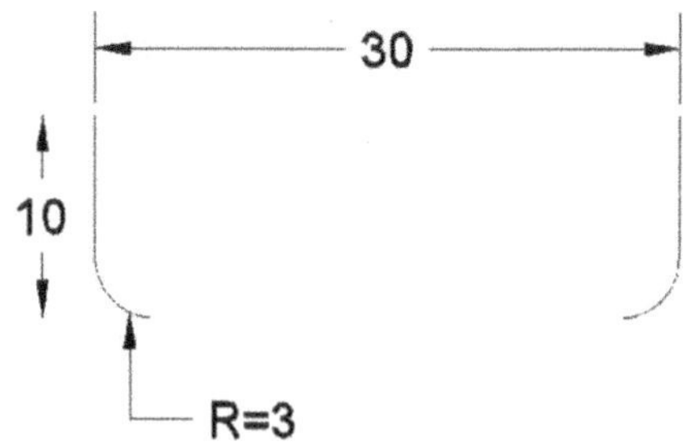

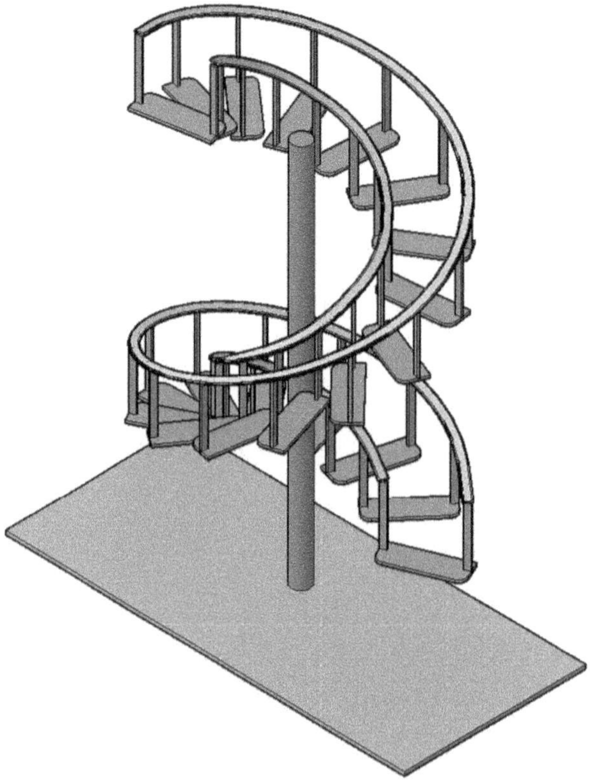

4. Color the faces of an extruded box as follows: top: yellow, left: green, right: purple. Create a hole in the middle as shown. (Difficulty level: Easy; Time to completion: 1–2 minutes.)

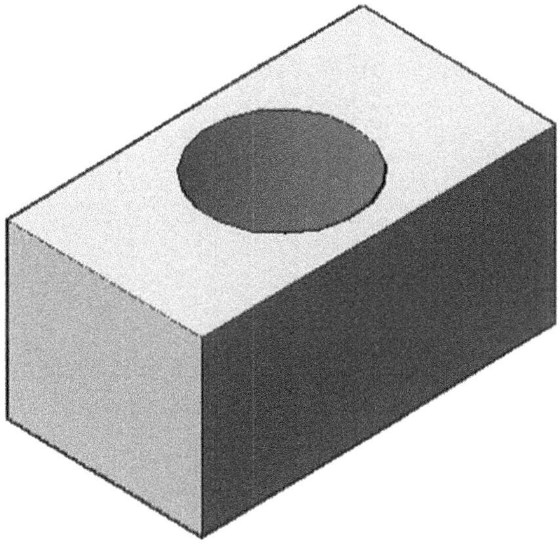

5. Color the faces of the following extruded box with any color of your choosing. Offset the hole to a smaller size as shown. (Difficulty level: Easy; Time to completion: 1−2 minutes.)

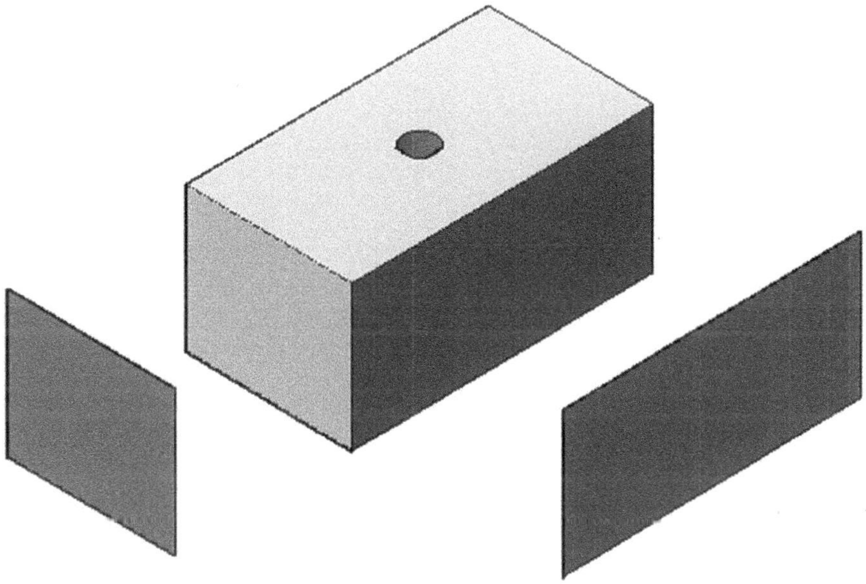

Spotlight on: biomedical engineering

Biomedical engineering is the merging of engineering and medicine (or biology). It is the newest of the engineering disciplines and has just recently become a field in its own right. Engineers who entered this profession were usually mechanical, electrical, or chemical engineers. They collaborated side by side with medical researchers and doctors or attended medical school themselves, becoming a physician/engineer. Today, several schools offer specialized biomedical engineering degrees that combine all these fields.

FIGURE 1 *Image Source: www.osoti.net.*

Biomedical engineers often work in research, and depending on their field of interest, they may specialize in bionics, genetic engineering, tissue engineering (artificial organs), or pharmaceutical engineering. Probably, the one product most often associated with biomedical engineering is artificial body parts, such as the artificial heart or hip and knee replacements. Although old TV shows such as the *Six Million Dollar Man* create an image of an exotic half man, half machine with superhuman strength and speed, most of such applications are quite down to earth and commonplace. You may even know a grandparent or other relative with an artificial implant. Biomedical engineers help design and manufacture these parts.

FIGURE 2 *Image Source: Associated Press.*

Another major field for these types of engineers is medical devices and tools. From heart valves to stents to surgical equipment to scanners of all types, biomedical engineers create devices that interact with the human body. It is a very interesting and diverse field and an active area of research and innovation.

Biomedical engineering is not an easy career to enter and be educated in. You are merging two already challenging fields—engineering and medicine—into one profession. There are two approaches here. In the classic approach, you carry a dual major in engineering (usually mechanical or electrical) and biology or chemistry. You then go on to graduate school for an advanced degree in engineering or to medical school to become a physician, sometimes both. A more recently available (and somewhat easier) approach is to major in biomedical engineering from the start. More and more universities now offer this degree, and you will find yourself taking courses such as anatomy, biomechanics, materials and imaging, chemistry, and biomedical instrumentation. It is truly a hybrid of engineering and medicine. Additional courses include robotics, circuit analysis, and industrial systems, with additional medical classes such as molecular biology.

On graduation, biomedical engineers almost always need to go to graduate or medical school. In fact, some programs discourage students from majoring in this field unless they plan, at the very least, on getting a master's degree, as there is just not enough taught at the undergraduate level for such a complex field. As of 2015, there was no PE licensing for a biomedical engineer. In the past, many biomedical engineers were only engineers and picked up the biological or medical knowledge on the job. As the field sees more growth and specialization, however, the need to have dually trained professionals—part engineer, part medical doctor, especially for research—is greater than ever.

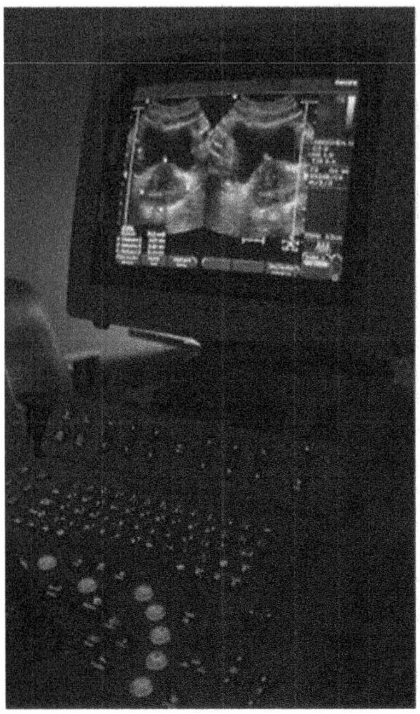

FIGURE 3 *Image Source: Associated Press.*

Biomedical engineers (with just a 4-year biomedical degree) can generally expect starting salaries in the $50,000/year range, which is toward the low end among engineering specialties. The median pay nationwide was $64,000 in 2017. However, as stated earlier, many go on to graduate school or even further, as required for research; and the median salary for 2014 was a very respectable $94,000 for those with advanced degrees.

So, how do biomedical engineers use AutoCAD and what can you expect? Probably the top area AutoCAD is to be found is biomedical instrumentation. The "engineering" of biomedical engineering is often of the electrical kind, and you will find plenty of schematics and wiring for scanners and imaging devices designed in AutoCAD. Things get a bit more complex as you move into implant design. Here, the "engineering" tends to lean more toward the mechanical side, and the implants themselves feature smooth organic curvature and complex material requirements. Often the appropriate software is a 3D parametric solid modeling package, not AutoCAD. One well-known company in that exact specialty, Stryker Howmedica Osteonics of New Jersey, uses Pro/Engineer (Creo) for its design work.

Chapter 26

Surfaces and meshes

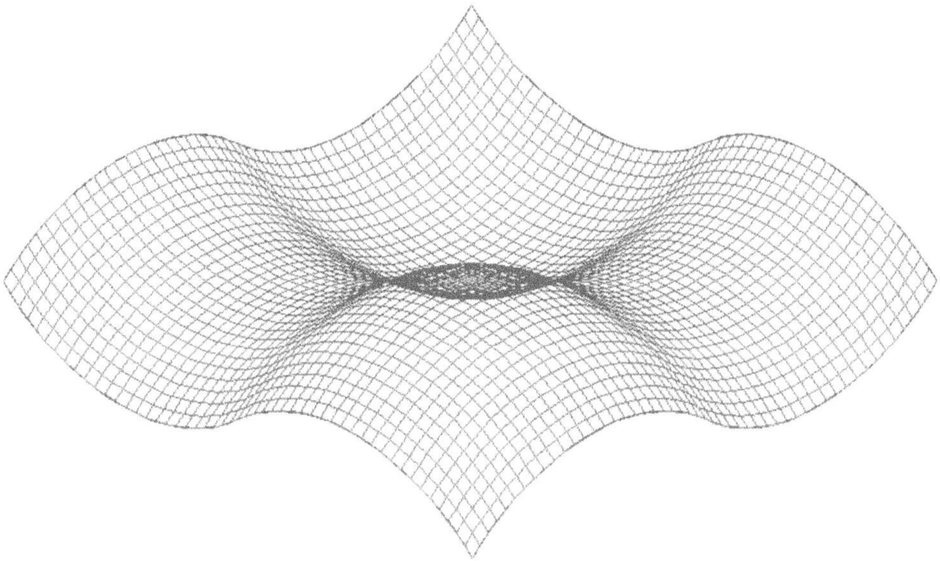

Learning Objectives

In this chapter, we focus on surfaces instead of solid objects. These objects have no depth the way solids do but possess many desirable properties for design work. Specifically, we mention

- Planar surface
- Region
- Rulesurf
- Tabsurf
- Revsurf
- Edgesurf
- 3Dface
- 3Dmesh
- Smooth Mesh Primitives
- Mesh smoothing
- Face move and rotate
- Collapse face or edge
- Convert to solid or surface

By the end of this chapter, you will be able to create a variety of surfaces and meshes.
Estimated time for completion of this chapter: 2 hours.

Up and Running with AutoCAD 2022. https://doi.org/10.1016/B978-0-323-89923-9.00026-3

26.1 Introduction to Surfaces and Meshes

In this chapter, we explore some of AutoCAD's surfacing and meshing capabilities. They underwent some significant enhancements in the AutoCAD 2011 release. A surface and a mesh are often the same, since in all but a few cases, a surface, by definition, is made up of a mesh. A mesh can then be described as an interconnecting weave of lines representing a surface (as opposed to a solid).

AutoCAD has several general approaches to creating, and working with, these objects. The most recent approach parallels what has been done in advanced 3D software like CATIA for some time now. You create a rough "blob" representing a general shape, then you adjust it one mesh intersection or corner at a time, pushing, pulling, and sculpting it like a piece of clay.

Another method involves creating a "framework" for the surface, a sort of skeleton, if you will. This framework is generally made using splines, which can be shaped easily via control points (grips). You can then create surfaces on top of this framework.

Finally, surfaces can be created via ordinary design methods already covered. Recall how you revolved an "open" profile to create a surface model in Chapter 24. In the same chapter, you also used the loft command to create surfaces between two arcs. Loft is actually quite a useful command for basic surfacing of noncomplex curvature.

We start from the beginning and cover "legacy" meshing capabilities. These tools, such as planar surface, region, and the "surf" family (rulesurf, tabsurf, revsurf, and edgesurf), have been around for quite some time and are useful for basic surfacing and meshing tasks. We then move on to some advanced legacy tools, such as 3D faces and meshes, and discuss the new family of tools introduced (or enhanced) in AutoCAD 2011. Remember, more than ever, these tools depend on your imagination and creativity to shine. They work best when combined by a skilled user to create some very advanced shapes.

26.2 Surfacing Commands

Planar Surface

Perhaps the simplest thing you can surface is a rectangle, and this is where we start. Here, you just have a basic closed shape and want to cover it with a surface, and planar surface or just *planesurf*, is the best way to do this without using extrude or making it a true solid. To try it out, switch to 3D and create a rectangle of any size, as seen in Fig. 26.1.

Keyboard: Type in **planesurf** and press Enter
Cascading menus: **Draw→Modeling→Surfaces→Planar**
Toolbar icon: **Modeling** toolbar→
Ribbon: **Surface** tab→**Create**

Step 1. Start the planesurf command via any of the preceding methods.

- AutoCAD says: `Specify first corner or [Object] <Object>:`

Step 2. Using OSNAPs, select one of the corners of the rectangle.

- AutoCAD says: `Specify other corner:`

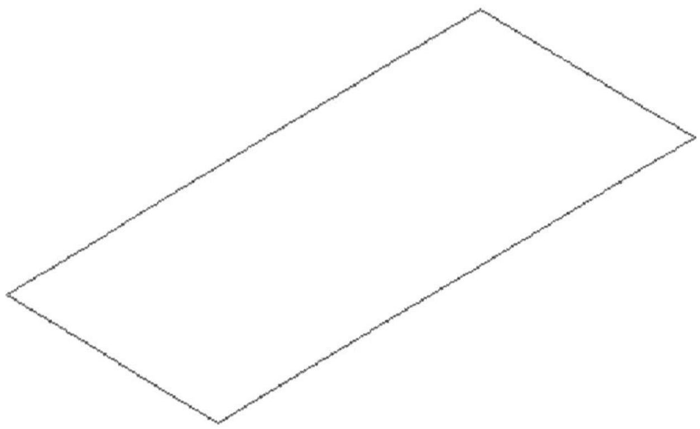

FIGURE 26.1 Basic rectangle in 3D.

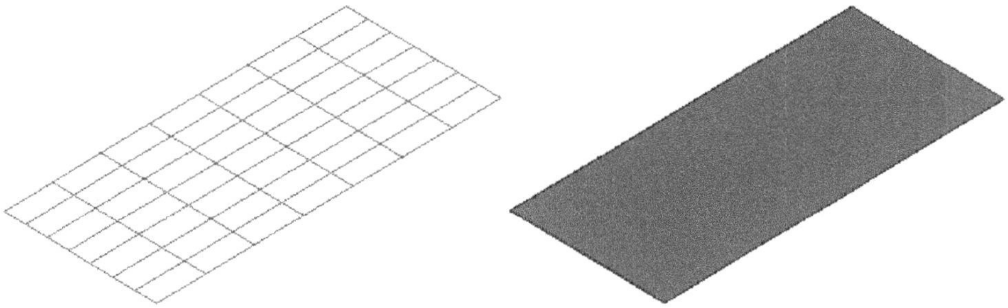

FIGURE 26.2 Rectangle with planesurf.

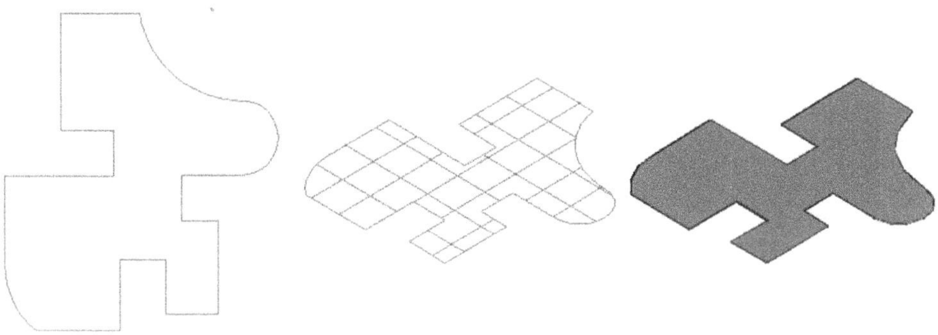

FIGURE 26.3 Shape with planesurf (Object).

Step 3. Select the diagonally *opposite* corner of the rectangle. A surface is formed, as seen in the wireframe and shade modes in Fig. 26.2.

You can also use planesurf to create surfaces out of objects by typing in `o` for `Object`. You can then just select the object and turn it into a surface; no clicking of points is needed. This is useful when a curvature, such as an arc, is involved and point-by-point clicking is not practical. Fig. 26.3 shows a closed border drawn with a pline, with planesurf then applied in 3D.

Region

Region gets a brief mention here as an alternate method to create a simple surface. The difference here is that a region is a surface but not a mesh, a rare instance when this is the case. You are not able to edit it in any way, and that may be just fine for simple shapes. To try out region, create another rectangle or circle in 3D with shading on.

Keyboard: Type in **region** and press Enter
Cascading menus: **Draw→Region**
Toolbar icon: **Draw toolbar→**
Ribbon: **Home** tab→ **Draw→**

Step 1. Start the region command via any of the preceding methods.

- AutoCAD says: `Select objects:`

Step 2. Pick your shape and press Enter.

- AutoCAD says: `1 loop extracted.`

  `Region created.`

A region is created.

Rulesurf

We now go over the family of "surfs." These are all legacy commands that perform different varieties of surfacing, depending on the given starting conditions.

Rulesurf stands for "ruled surface" and is a handy tool to create surfaces between various straight lines or curves (without cross sections) that are not necessarily straight or parallel to each other or even closed. It fills a niche somewhere between the loft and planar surface options. To try it out, create two lines at a random angle to each other, as seen in Fig. 26.4.

Keyboard: Type in **rulesurf** and press Enter
Cascading menus: **Draw→Modeling→Meshes→Ruled Mesh**
Toolbar icon: none
Ribbon: **Mesh** tab→**Primitives**

Step 1. Start the rulesurf command via any of the preceding methods.

- AutoCAD says:

```
_rulesurf
Current wire frame density: SURFTAB1 = 6
Select first defining curve:
```

Step 2. Select the first line, as seen in Fig. 26.5.

Step 3. Select the second line at roughly the same upper location (if you select a bottom location the surface self-intersects). The shaded results are shown in Fig. 26.6.

As with many of these surfacing techniques, at first, the results may not seem very useful, but with some imagination and creativity, you can use them to fulfill rather complex surfacing needs. For example, in Fig. 26.7, we create two arcs, rotate one of them 90°, and connect them end to end.

You can then apply rulesurf to these arcs and get a unique result, as seen in Fig. 26.8. The arcs become one surface. It is then shaded and rotated for a different view. The density of the mesh can be set by the SURFTAB1 and SURFTAB2 system variables by simply inputting a higher value.

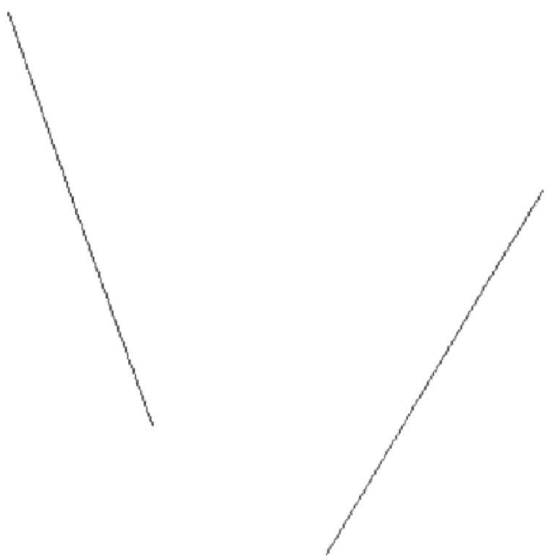

FIGURE 26.4 Random lines for rulesurf.

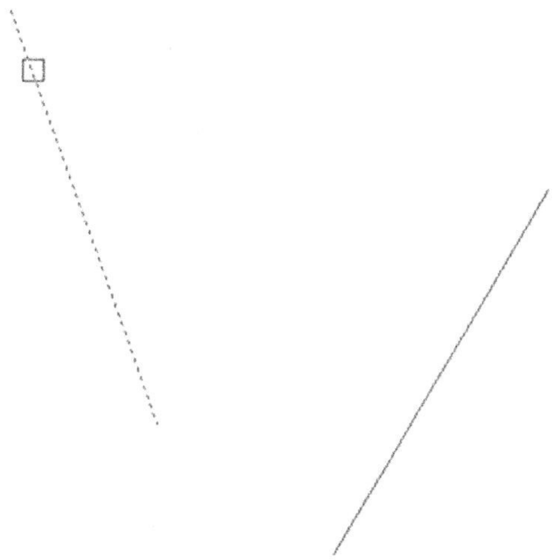

FIGURE 26.5 Select first line for rulesurf.

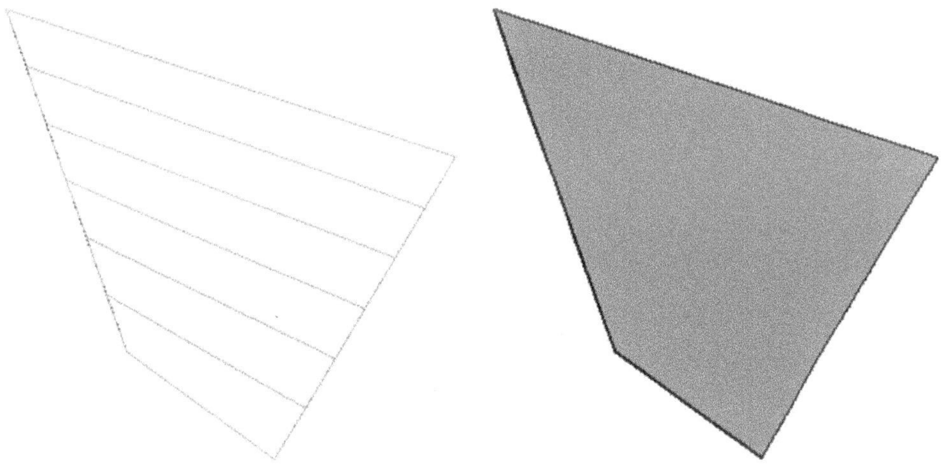

FIGURE 26.6 Final result for rulesurf.

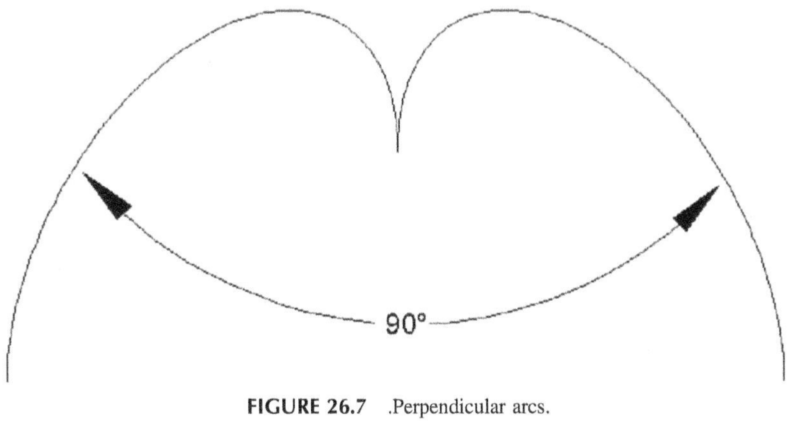

FIGURE 26.7 .Perpendicular arcs.

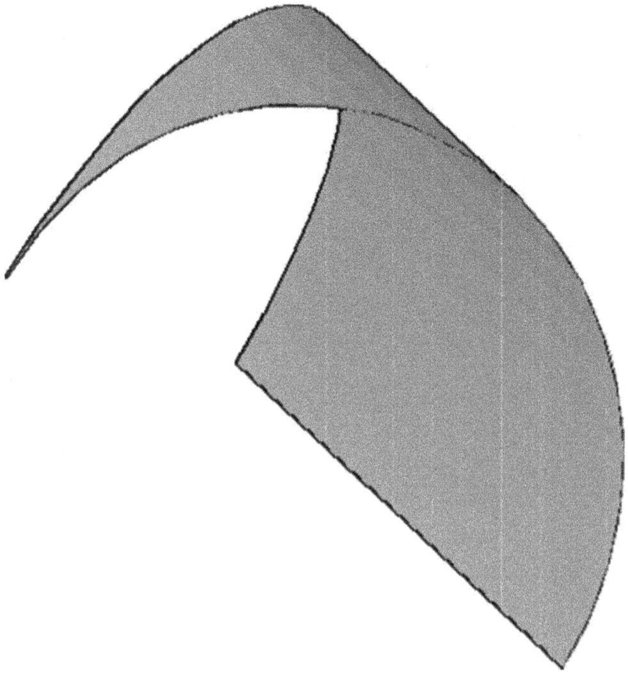

FIGURE 26.8 Final result for rulesurf of arcs.

Tabsurf

Tabsurf stands for "tabulated surface" and is a surfaces version of the extrude command, with some elements of path extrusion mixed in. The command requires a path curve and a direction vector. To try it out, create a circle with a perpendicular line, as seen in Fig. 26.9.

Keyboard: Type in **tabsurf** and press Enter
Cascading menus: **Draw→Modeling→Meshes→Tabulated Mesh**
Toolbar icon: none
Ribbon: **Mesh** tab→**Primitives**

Step 1. Start the tabsurf command via any of the preceding methods.

- AutoCAD says: `Current wire frame density: SURFTAB1 = 50Select object for path curve:`

Step 2. Select the circle.

- AutoCAD says: `Select object for direction vector:`

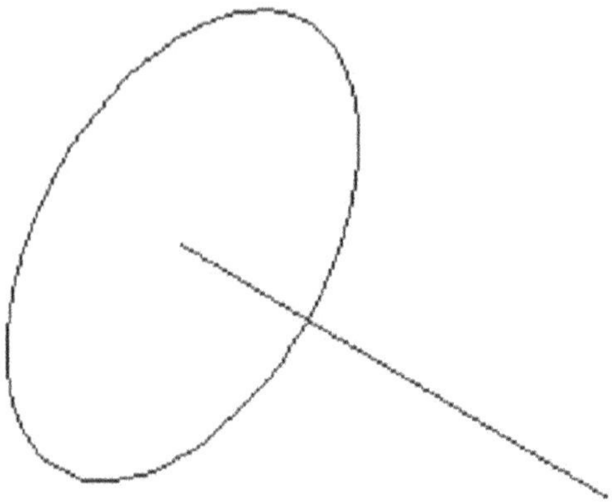

FIGURE 26.9 Circle and line for tabsurf.

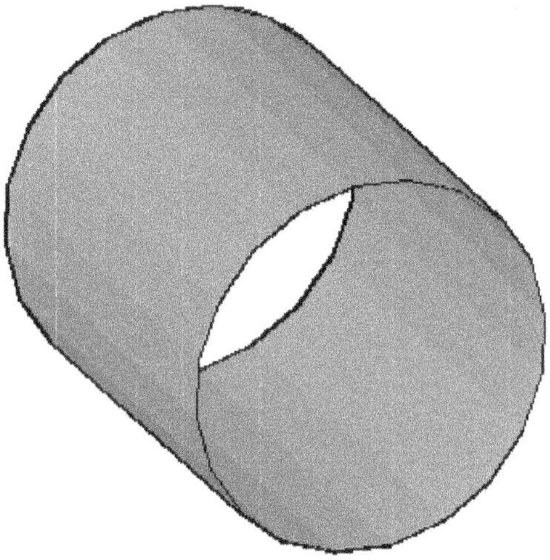

FIGURE 26.10 Final result for tabsurf.

Step 3. Select the line.

The result, shown in Fig. 26.10, is a surface projected along the direction vector, not unlike path extrusion or regular extrude, with the difference, however, that the object is not a solid but a surface (be sure to set the SURFTAB1 and SURFTAB2 system variables to a high value for smoothness). A similar example with an I-beam is shown in Fig. 26.11.

The final tabsurf example is slightly more complex. In 2D, you need to draw a string of connected arcs, all pedited together, as seen in Fig. 26.12. Go into 3D and rotate the string of arcs into a straight up position if need be. Then, draw a perpendicular direction vector, attached to one end of the arc string, as seen in Fig. 26.13, finally tabsurf the shape, as seen in Fig. 26.14. These exercises should generate some ideas as far as applying the techniques to actual design work.

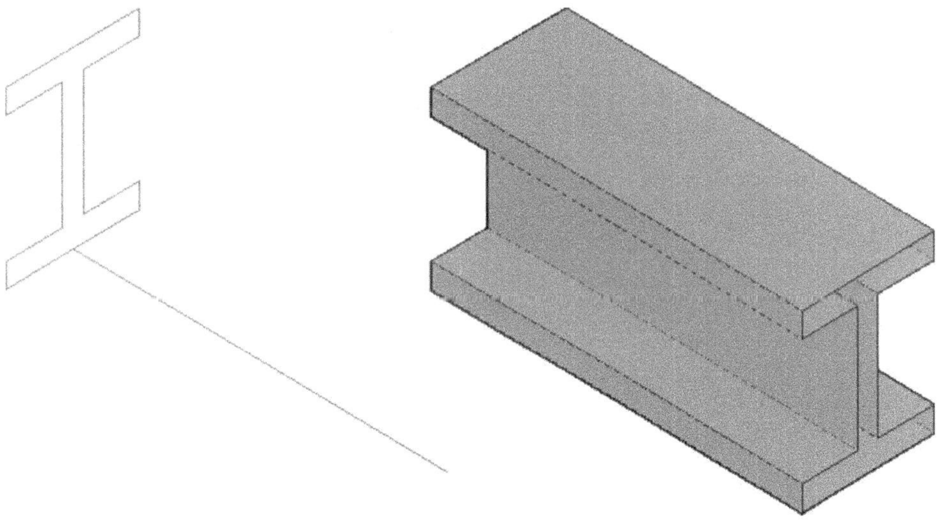

FIGURE 26.11 Tabsurf example for I-beam.

FIGURE 26.12 Arcs in 2D.

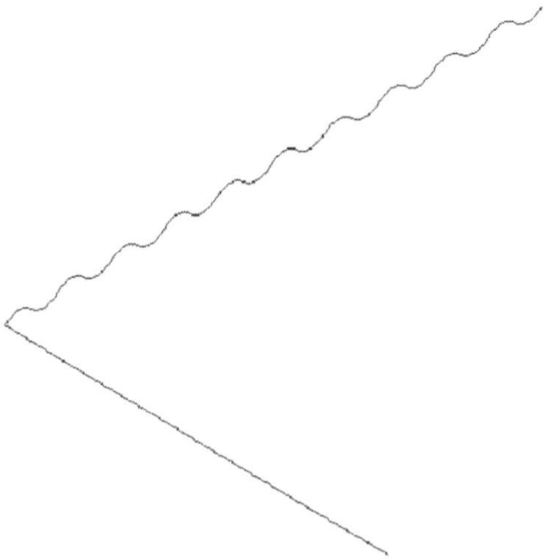

FIGURE 26.13 Arcs with a direction vector in 3D.

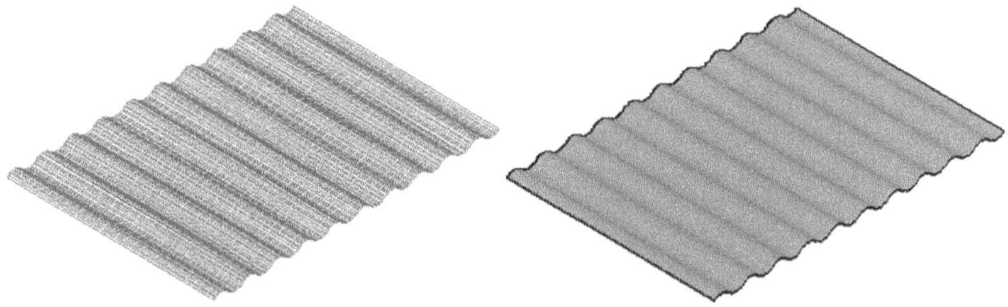

FIGURE 26.14 Tabsurf example completed.

Revsurf

Revsurf stands for "revolved surface" and is a close relative of the revolve command covered in earlier 3D chapters, with the only difference in many cases being that revsurf produces somewhat smoother curves, if the SURFTAB1 and SURFTAB2 system variables are set high enough (around 50 is a good starting point). Generally, you can treat this command as an alternative to revolve. To illustrate it, create the profile seen in Fig. 26.15; it is similar to the cup profile used with revolve. Be sure to pedit the profile.

Keyboard: Type in **revsurf** and press Enter
Cascading menus: **Draw→Modeling→Meshes→Revolved Mesh**
Toolbar icon: none
Ribbon: **Mesh** tab→**Primitives** 

Step 1. Start the revsurf command via any of the preceding methods.

- AutoCAD says: `Current wire frame density: SURFTAB1 = 50 SURFTAB2 = 6`

  `Select object to revolve:`

Step 2. Select the profile.

- AutoCAD says: `Select object that defines the axis of revolution:`

Step 3. Select the rotation axis (the vertical line).

- AutoCAD says: `Specify start angle <0>:`

Step 4. Press Enter.

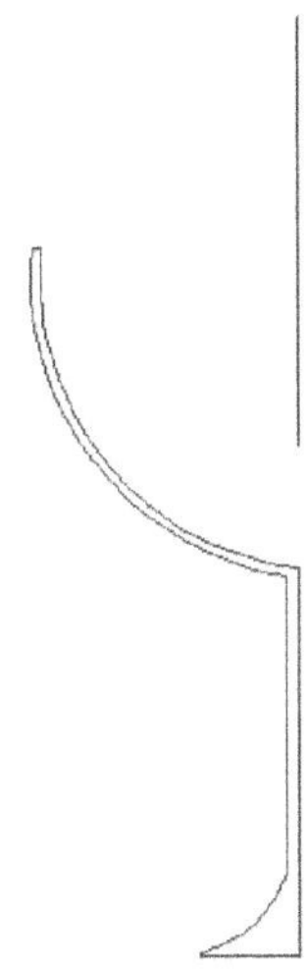

FIGURE 26.15 Profile for revsurf.

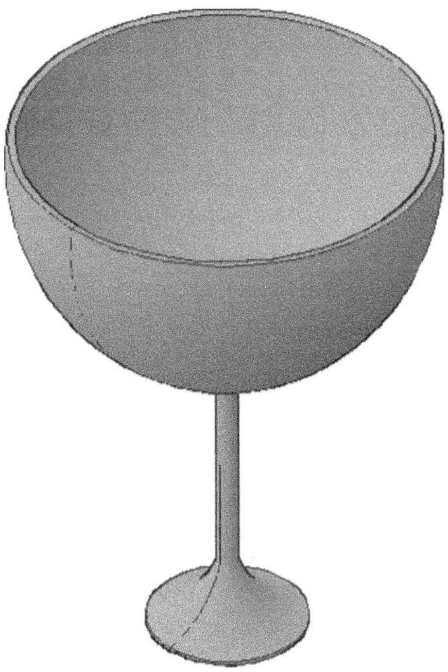

FIGURE 26.16 Revsurf example completed.

- AutoCAD says: `Specify included angle (+=ccw, -=cw) <360>:`

Step 5. Key in any value or press Enter for the full 360°. The result, shaded, is shown in Fig. 26.16.

Edgesurf

The edgesurf command creates a Coons surface patch mesh or a mesh in the M and N directions between a set of edges that are attached to each other with endpoints in a closed loop. The hard part is of course properly defining these edges. Once that is done correctly, AutoCAD fills in the rest and a sophisticated surface is created.

For a good mesh rendering, you need to make it dense, using the SURFTAB1 and SURFTAB2 system variables set to 50 or more (too much, though, slows down your system's performance). Now, let us try to create a basic mesh as an example. In 2D, draw two arcs and pedit them as seen in Fig. 26.17.

FIGURE 26.17 Two connected arcs for edgesurf.

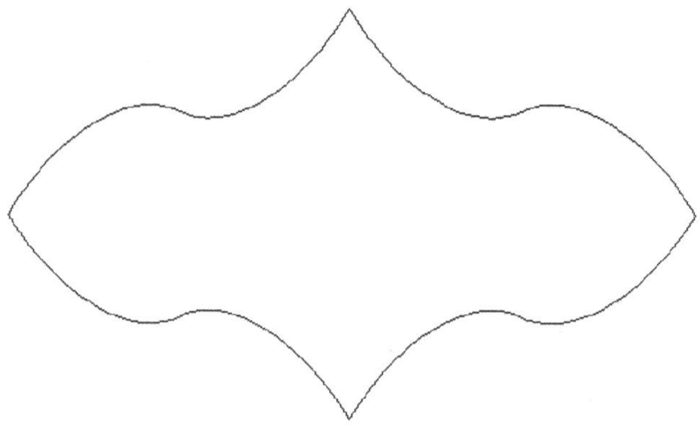

FIGURE 26.18 Two connected arcs for edgesurf.

Now switch to 3D, and using the copy and rotate3D commands, arrange them in a closed loop, with each side 90° perpendicular to its neighbor, as seen in Fig. 26.18.

Keyboard: Type in **edgesurf** and press Enter
Cascading menus: **Draw→Modeling→Meshes→Edge Mesh**
Toolbar icon: none
Ribbon: **Mesh** tab→**Primitives**

Step 1. Start the edgesurf command via any of the preceding methods.

- AutoCAD says: `Current wire frame density: SURFTAB1 = 50SURFTAB2=50`

  `Select object 1 for surface edge:`

Step 2. Select the four edges, one after the other.

- AutoCAD says: `Select object 2 for surface edge:`
  `Select object 3 for surface edge:`
  `Select object 4 for surface edge:`

Step 3. After the final edge is selected, you see the mesh image shown on the left in Fig. 26.19. On the right, it is shaded and slightly rotated for a better view.

Creating these meshes can be time consuming and occasionally troublesome, but if you follow the basic rules of pediting all the curves (edges) and connecting them together, you should be fine. The sophistication and versatility of this command is rather impressive, and almost any curve can be represented, if you can find a clever way to create the necessary edges. The only limitation is that you are limited to four edges at a time, but by connecting separate meshed objects together, you can get around this.

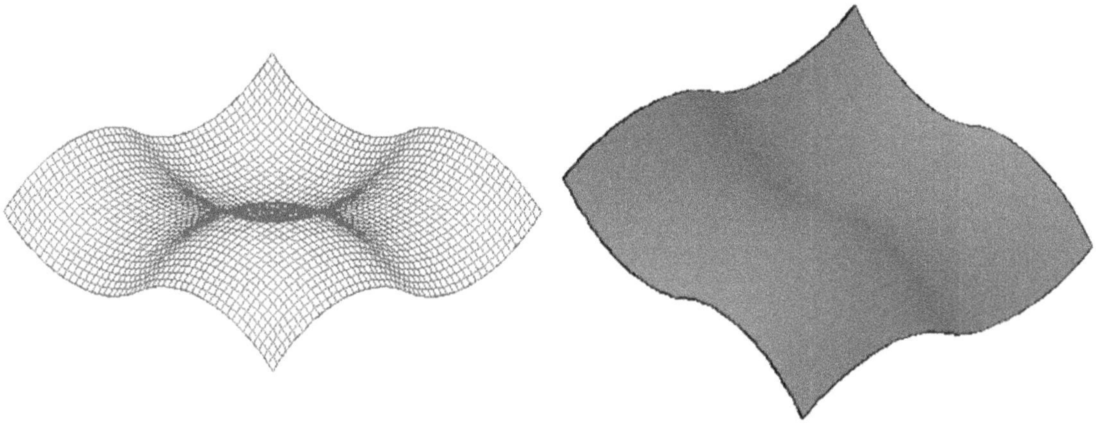

FIGURE 26.19 Final result for edgesurf.

3Dface and 3Dmesh

Our final brief discussion on legacy mesh commands focuses on free-form meshes, ones that require no cross-section profile or framework. However, just because the framework is not required, does not mean it is not necessary to a meaningful design. The 3Dface command is actually the one referred to in this chapter introduction as a way to dress up a "framework." The idea is to simply click your way around points in the design, and when AutoCAD senses a closed path, it installs a surface.

Let us give this a try on a random set of surfaces. The design shown in Fig. 26.20 can be created by carefully rotating the UCS icon while drawing straight lines to establish a "skeleton" shape. Then, using grips, you can move the intersections of lines together to other intersections. Finally, you can add new lines by just drawing from endpoint to endpoint. Try to create a number of faces that can be filled in. Your shape need not look exactly like this one.

The command is activated via typing or a cascading menu.

Step 1. In 3D mode, with shading and Ortho on, type in `3dface` and press Enter, or use the cascading menu Draw → Modeling → Meshes → 3D Face.

- AutoCAD says: `3dface Specify first point or [Invisible]:`

Step 2. Using OSNAP endpoints, click around each facet of the shape. After each click,

- AutoCAD says: `Specify second point or [Invisible]:`
  `Specify third point or [Invisible]`

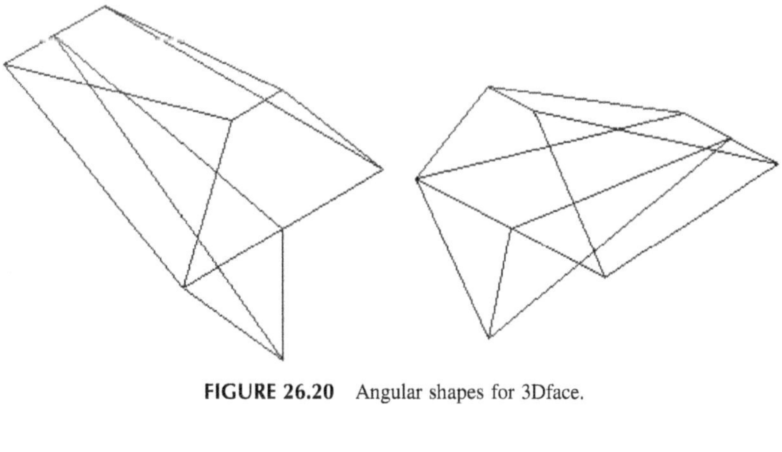

FIGURE 26.20 Angular shapes for 3Dface.

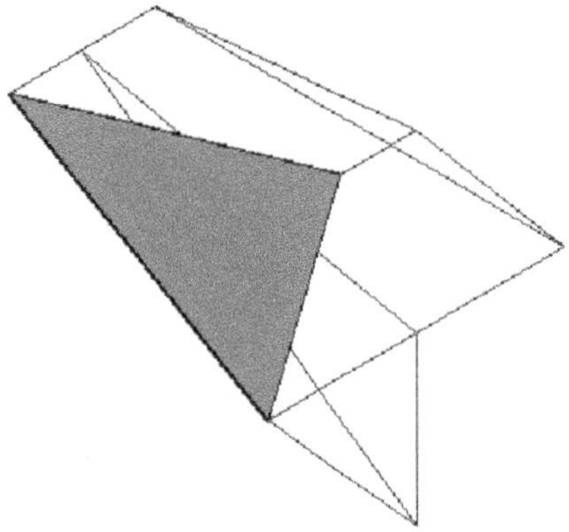

FIGURE 26.21 3Dface.

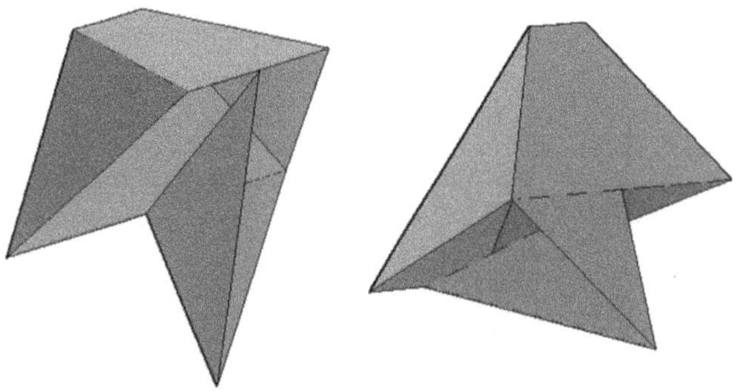

FIGURE 26.22 3Dface completed.

Step 3. After three or four clicks (depending on which facet you chose to do first), you should be able to get what is seen in Fig. 26.21. Continue covering the other faces, and add color, to get what is shown in Fig. 26.22 (viewed from several angles).

Little is said of 3Dmesh. This command is a good subject for programming and automation. The 3Dmesh command asks you for the size of the mesh in the M and N directions and location (vertex) points. These can be entered manually or as points on a framework. Using AutoLISP, a programmer can automate meshing of complex surfaces, but this is beyond the scope of the book.

The 3Dface and 3Dmesh commands conclude a basic overview of legacy commands, ones that have been available for years. A great many new mesh tools were introduced back in AutoCAD 2011, and we look at some here. These tools involve "sculpting" a shape, which is a popular way to create meshes in advanced 3D solid modeling software, as mentioned at the start of this chapter.

26.3 Smooth Mesh Primitives

The way to start "sculpting" a shape is to first create one that has built-in mesh surfaces. This can be easily done via smooth mesh primitives. Primitives, as introduced in Chapter 23, are simple solid shapes, such as a box, sphere, and cone. You again create these shapes, but now they have mesh surfaces that can be manipulated into complex shapes. To create mesh primitives, you can use the Ribbon, the Smooth Mesh Primitives toolbar (Fig. 26.23), or cascading menus, as seen in the command matrix.

FIGURE 26.23 Smooth Mesh Primitives toolbar.

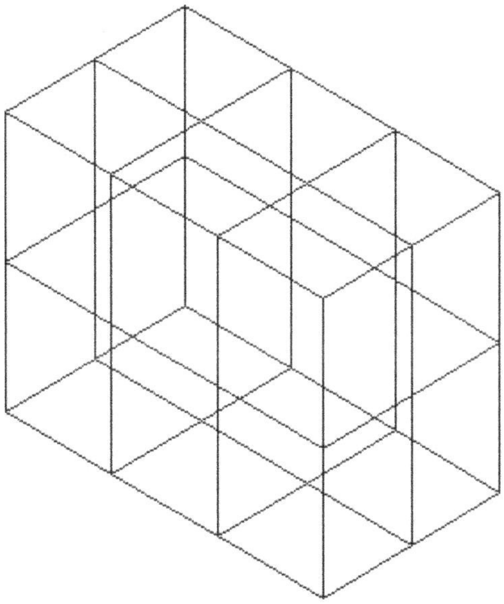

FIGURE 26.24 Smooth Mesh Primitives box.

Keyboard: Type in **mesh** and press Enter, then **b** for **Box**
Cascading menus: **Draw→Modeling→Meshes→Primitives→Box**
Toolbar icon: **Smooth Face Primitives** toolbar
Ribbon: **Mesh** tab→**Primitives**

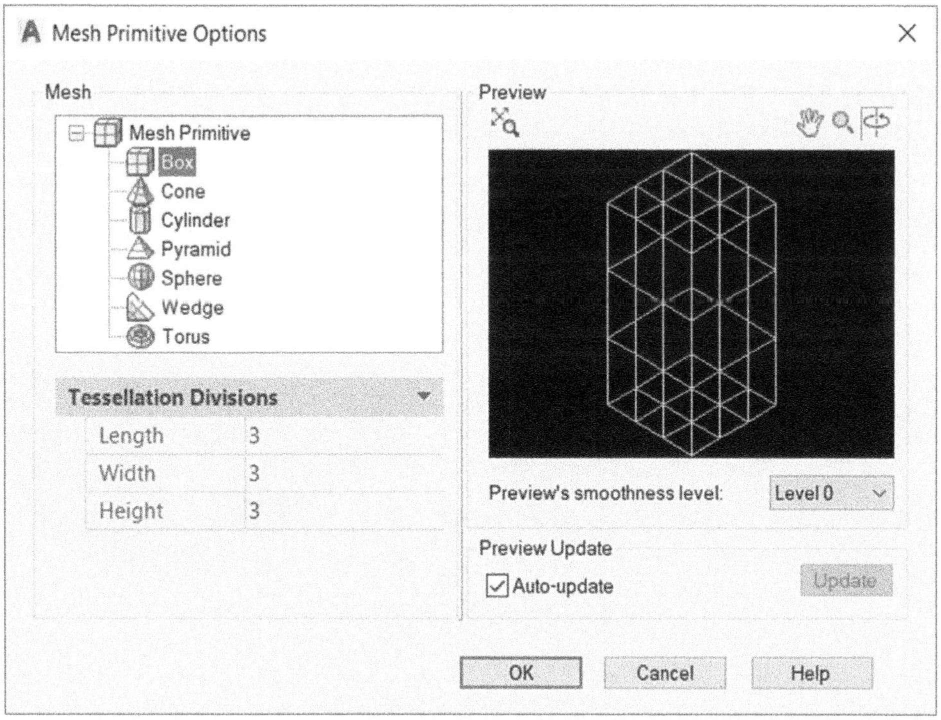

FIGURE 26.25 Mesh Primitives Options.

The steps to create the box are similar to the regular primitives box (review chapter: Boolean Operations and Primitives, if needed), but the results, viewed in wireframe, are definitely not the same, as seen in Fig. 26.24.

The primitive is now composed of multiple faces that can be shaped and morphed into the desired final shape. You can alter the number of faces (technically called *tessellation divisions*) of the primitive via the Mesh Primitive Options dialog box (Fig. 26.25). It can be accessed via the small arrow at the bottom right of the Ribbon's Mesh, Primitives tab, which will first bring up the Mesh Tessellation Options. Then just click the Mesh Primitives... button. You will see what is in Fig. 26.25.

Select the shape with which you want to work and change the tessellation values, that is, length, width, or height, thereby creating more faces. Having more working surfaces on a shape yields smoother shapes and more "control points" with which to alter the meshes. It also allows for more accuracy in shaping the part. The trade-off is more complexity and possible introduction of errors, as well as an extra load on computing resources if the meshes are extremely dense.

How dense to make the mesh grip is a common engineering dilemma in finite element analysis (FEA) for stresses and loads and computational fluid dynamics (CFD) for fluid flow modeling. With modern computers, taxing the resources is less of an issue but can still be a problem if you add one too many digits to the density value. Since you may be working with individual mesh facets in this case, make the overall grid only as dense as you, the user, not the computer, can comfortably handle. This can also be automated via the mesh smoothing command, as described next.

Mesh Modification 1. Smoothness

You can smooth out the primitive shapes automatically via the mesh smoothing command, best accessed via the Ribbon's Mesh tab. Your choices are Smooth More (increases the number of facets in the mesh), Smooth Less (decreases the number of facets in the mesh), or Refine Mesh (same as Smooth More but works only after that command is employed). Take a few minutes to try out the results on the box. After one Smooth More, followed by a Refine Mesh, you get the modified version of the shape shown in Fig. 26.26, as seen in wireframe and realistic visual style mode side by side. Notice the marked increase in facets.

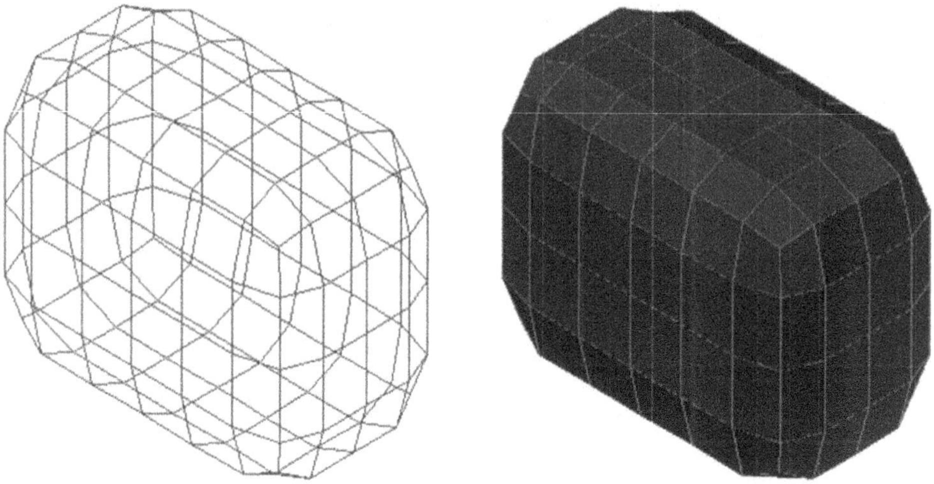

FIGURE 26.26 Smooth and Refined Mesh.

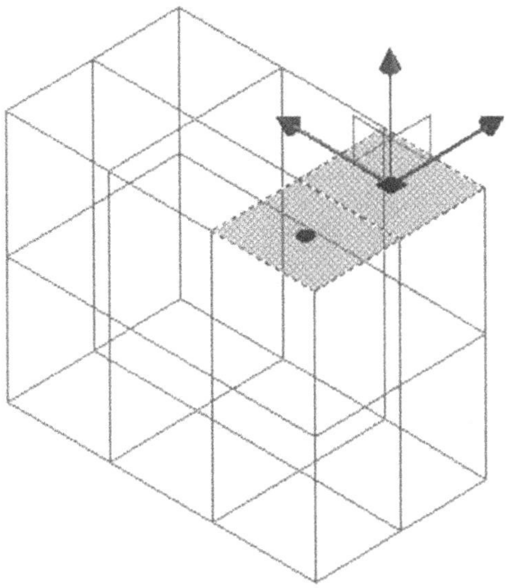

FIGURE 26.27 Facets selected.

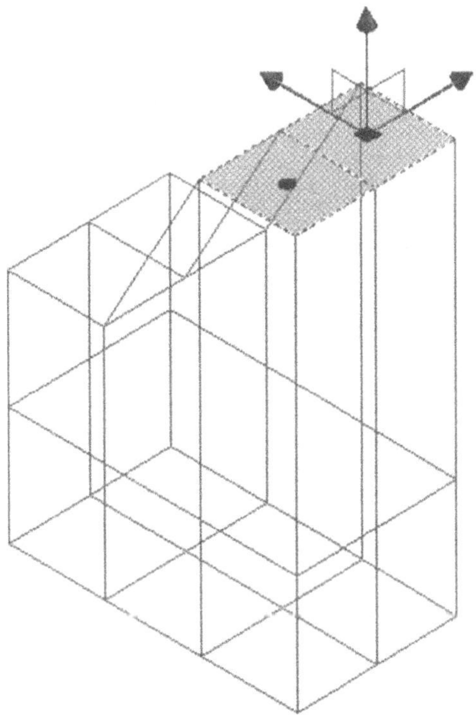

FIGURE 26.28 Facets removed.

Mesh Modification 2. Filters and Gizmos

Return your shape to the original box via the undo command. Next, we explore the face filter and the move and rotate gizmo (first seen in chapter: Object Manipulation). From Mesh tab, Subobject, select Move Gizmo followed by Face Filter. Then, click on the top right two facets, as seen in Fig. 26.27.

The mesh facets are highlighted and the move gizmo appears. With the UCS icon aligned vertically and Ortho on, move the facets up, noticing how the entire shape is "morphed," as seen in Fig. 26.28.

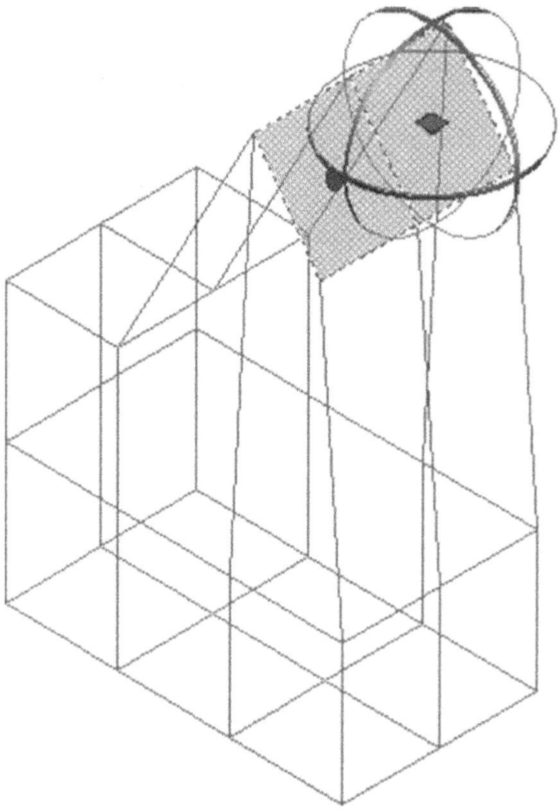

FIGURE 26.29 facets rotated.

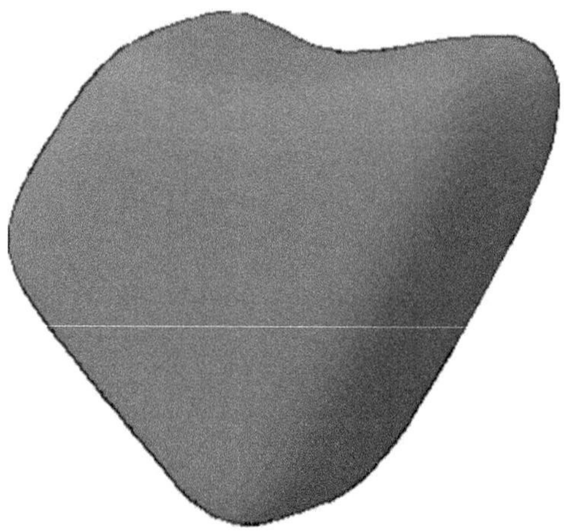

FIGURE 26.30 Final mesh shape.

In a similar manner, select the rotate gizmo and rotate the facets to a different angle, as seen in Fig. 26.29. After a few rounds of adding smoothness and another shading change, the shape is presented in Conceptual Visual Style (Fig. 26.30).

Mesh Modification 3. Additional Tools

Using smoothness, filters, and gizmos, you can already do quite a bit of shape morphing. Here are some additional tools to consider:

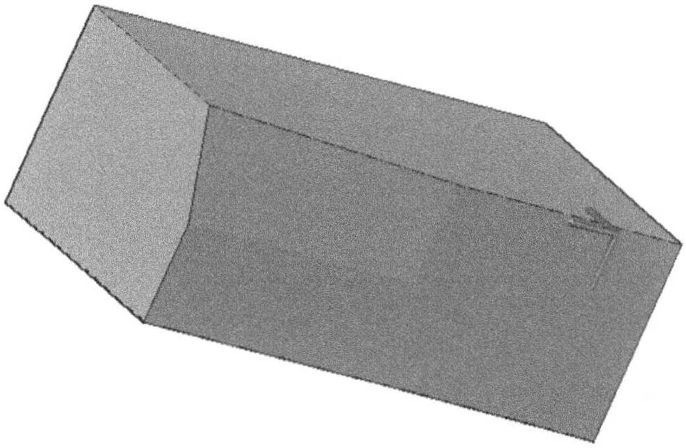

FIGURE 26.31 Collapsed face.

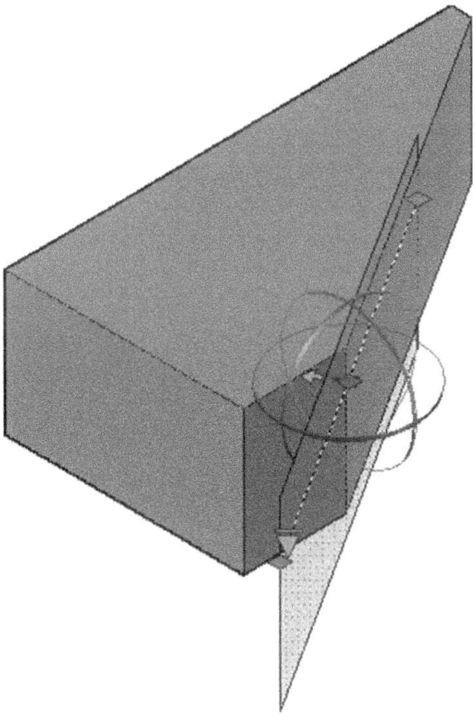

FIGURE 26.32 Section Plane.

- *Collapse Face or Edge*: This tool, found under the Ribbon's Mesh, Mesh Edit, makes a selected face or edge disappear on a shape, with the remaining ones closing ranks around the void, as seen in Fig. 26.31 with a mesh box.
- *Convert to Solid or Surface*: This tool, found under the Ribbon's Mesh, Convert Mesh, converts mesh objects to solids or surfaces. You do not see a tremendous difference in appearance, but the shape is editable in a different way from an organic mesh. For example, individual facets can no longer be selected in a solid.
- *Section Plane*: This tool, found under the Ribbon's Mesh, Section, allows you to easily section a mesh shape to view the interior or just a piece of it. This tool can be used on regular solids as well. Simply press Section Plane and select a method of locating the plane. The result of sectioning a simple box is shown in Fig. 26.32. The section itself can be rotated via the gizmo.

Summary

You should understand and know how to use the following concepts and commands before moving on to Chapter 27:

- Planar surface
- Region
- Rulesurf
- Tabsurf
- Revsurf
- Edgesurf
- 3Dface
- Smooth mesh primitives
- Mesh smoothing
- Face move and rotate
- Collapse face or edge
- Convert to solid or surface

Review Questions

Answer the following based on what you learned in this chapter:

1. Describe the planar surface command.
2. Describe the region command.
3. Describe rulesurf and its applications.
4. Describe tabsurf and its applications.
5. Describe revsurf and its applications.
6. Describe edgesurf and its applications.
7. Describe 3Dface.
8. What are the Smooth Mesh Primitives?

Exercises

1. Create the following structure using several rectangles. The base and top are both 100″ × 100″ and the walls are 150″ high (or you can use your own sizing if you wish). Then, create an arbitrary window in the same plane as one of the walls. Create a region out of the two walls and the window, subtracting the window out. Then use the planesurf command to mesh the floor and ceiling, as seen on the left. Finally add color and shading. Note that, when in shaded mode, you cannot tell which surface is a region and which is a planesurf. For extra practice, set both views side-by-side using viewports and the mview command, as seen in the exercise image. (Difficulty level: Easy; Time to completion: 10 minutes.)

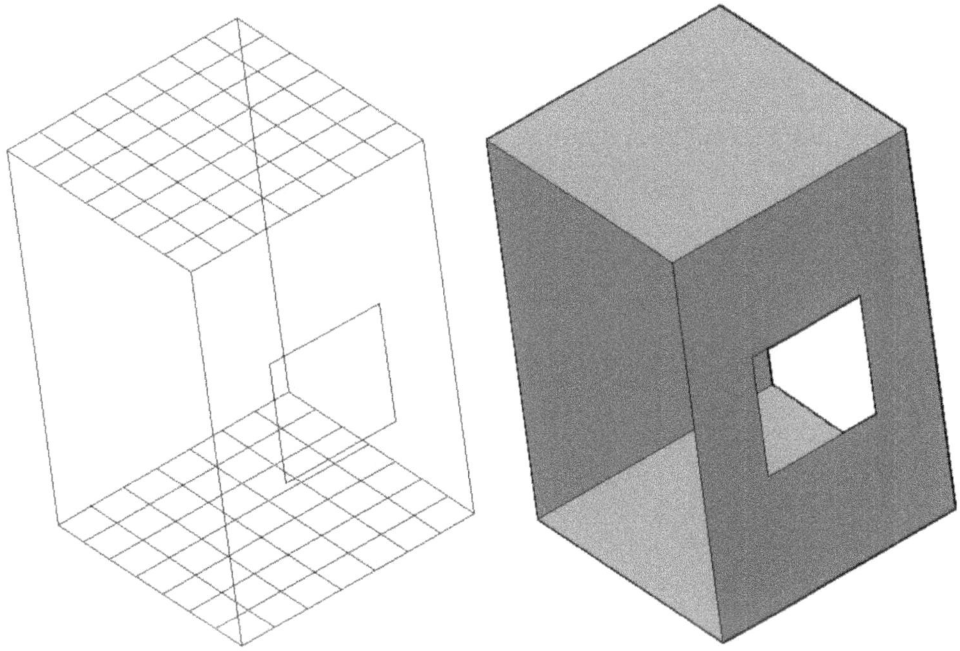

2. Create the following structure using several lines and arcs. Sizing is entirely arbitrary in this case, but be sure to replicate the basic shape. Use grip points as an aid to maneuver linework into position. After the basic frame is laid out, as seen in the upper image, add surfaces via the rulesurf command. Remember to select points that are roughly across from each other on the various lines. The one time you do not do that is for the larger end of the shape (on the right), where you pick opposite sides, creating the unique "bow-tie" effect. Finally, add color and shading. For extra practice, set both views one on top of the other using viewports and the mview command, as seen in the exercise image. (Difficulty level: Easy; Time to completion: 10 minutes.)

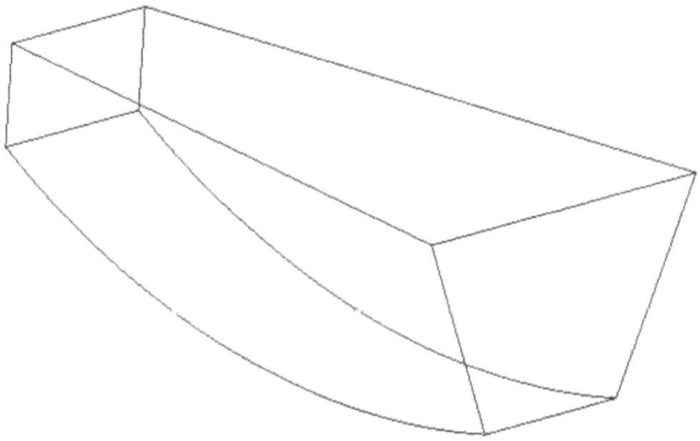

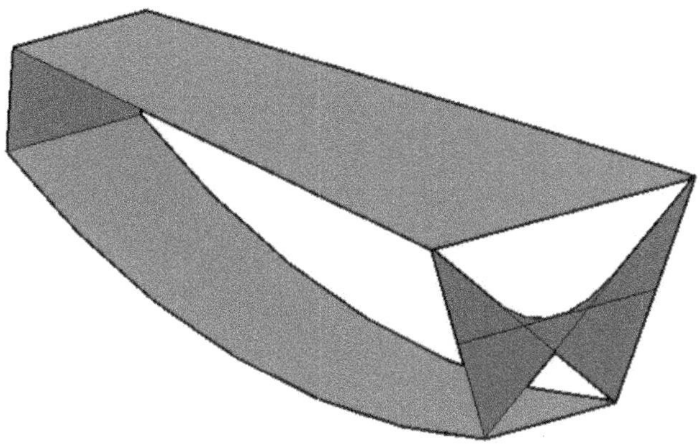

3. Create the following floor and column structure using linework, rectangles, extrusion, and the tabsurf command. The column is a pline profile, similar to Fig. 26.11, with dimensions given. Use tabsurf to extrude it out to the given height. The rest of the exercise is basic extruding and positioning. Sizing of the floor and beams as well as exact positioning is up to you. For extra practice, set up the profile, shaded 3D column, and the final layout side by side and on top of each other using viewports and the mview command, as seen in the exercise image. (Difficulty level: Easy; Time to completion: 15–20 minutes.)

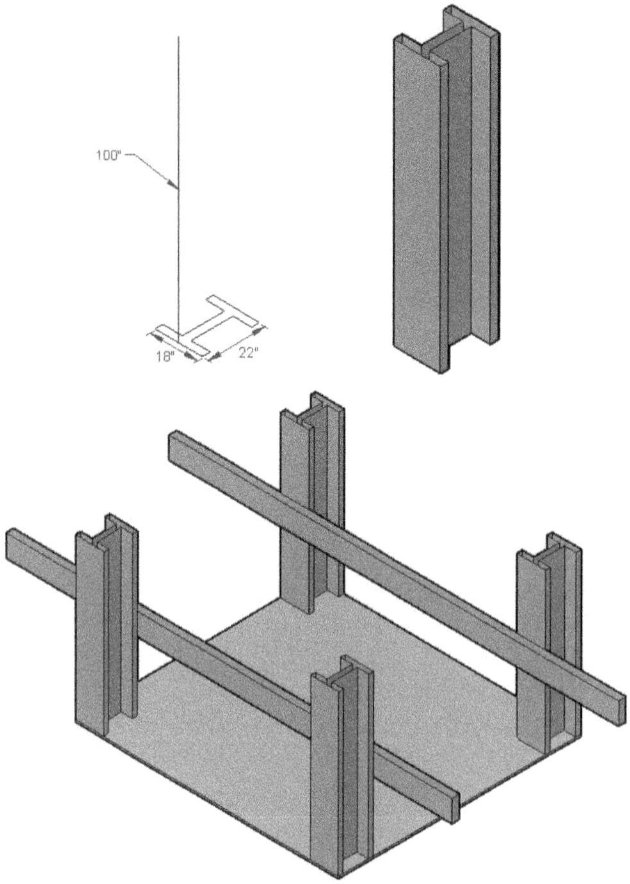

4. Create the following profile and axis of rotation. Then, revolve the profile using the revsurf command, with the 360° degree option selected. For extra practice, set up the profile, revolved shape, and the final shaded 3D view side by side and on top of each other using viewports and the mview command, as seen in the exercise image. (Difficulty level: Easy; Time to completion: 10–15 minutes.)

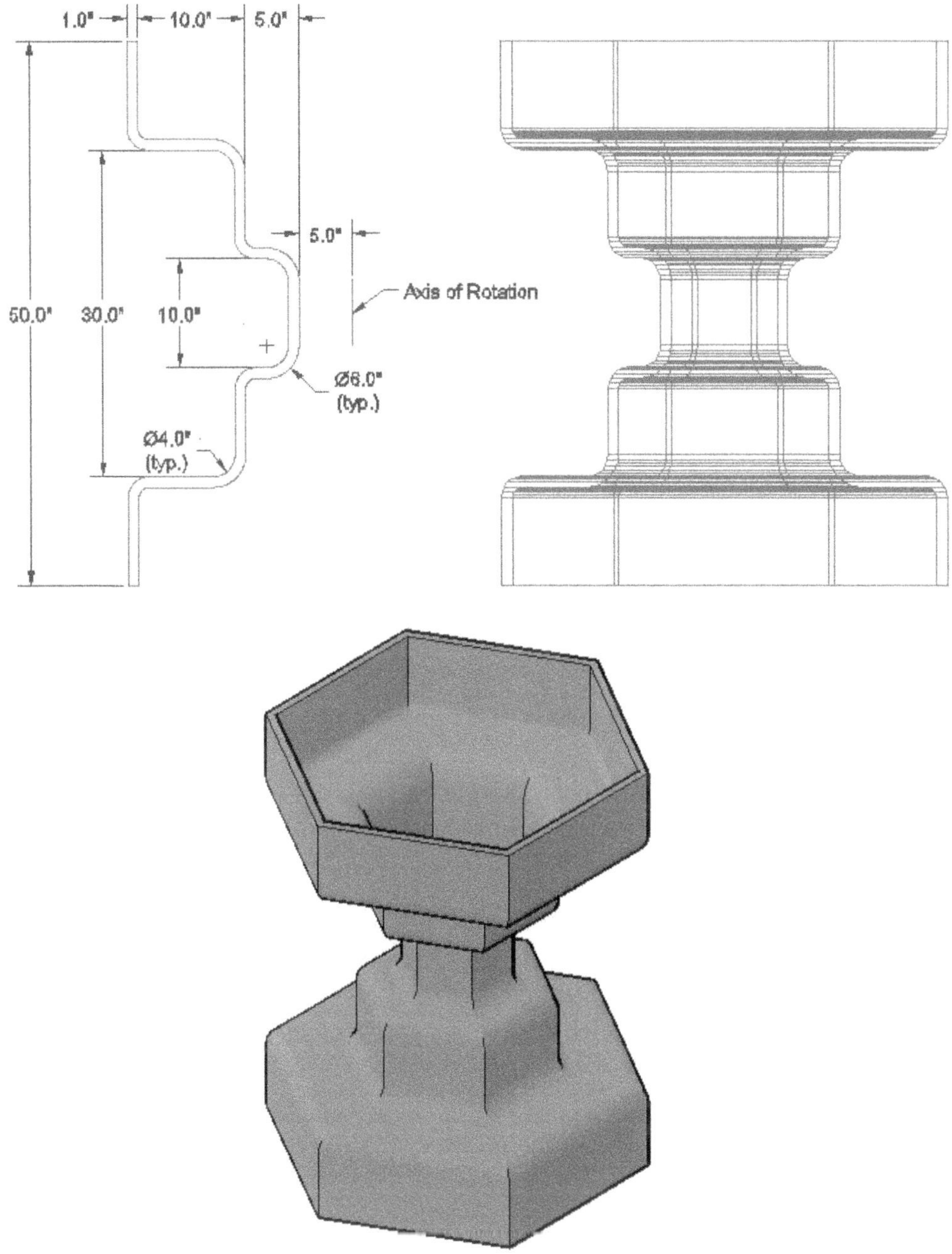

5. Create the following simplified hood, fenders, and windscreen design of a jeep-like vehicle. All dimensioning is arbitrary, but you need to use the region and edgesurf command to create the surfaces as shown in the top image. Carefully create profiles and edges for all the pieces; once a few critical ones are done, the rest is just copying. For extra practice, set up the profiles and the shaded final assembly one on top of the other using viewports and the mview command, as seen in the exercise image (Difficulty level: Easy; Time to completion: 10–15 minutes.)

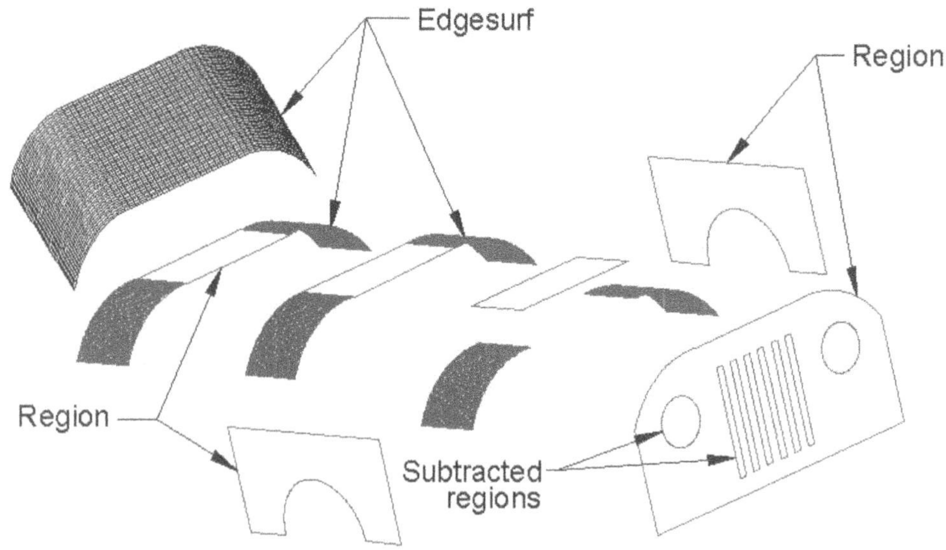

6. Based on 3D design principles and the edgesurf command, create the following mesh chair. You need to create a frame in 3D space, attach arcs in strategic locations, and finally execute the edgesurf command for the mesh. A visual step-by-step breakdown is shown for clarity, but you still need to figure out the steps involved. One possible approach is to draw a wireframe 3D box (not a solid) and use that as a reference point to attach the frame. All sizing is arbitrary, and is up to you. (Difficulty level: Intermediate/Advanced; Time to completion: 30 minutes.)

Chapter 27

Slicing, sectioning, layouts, and vports

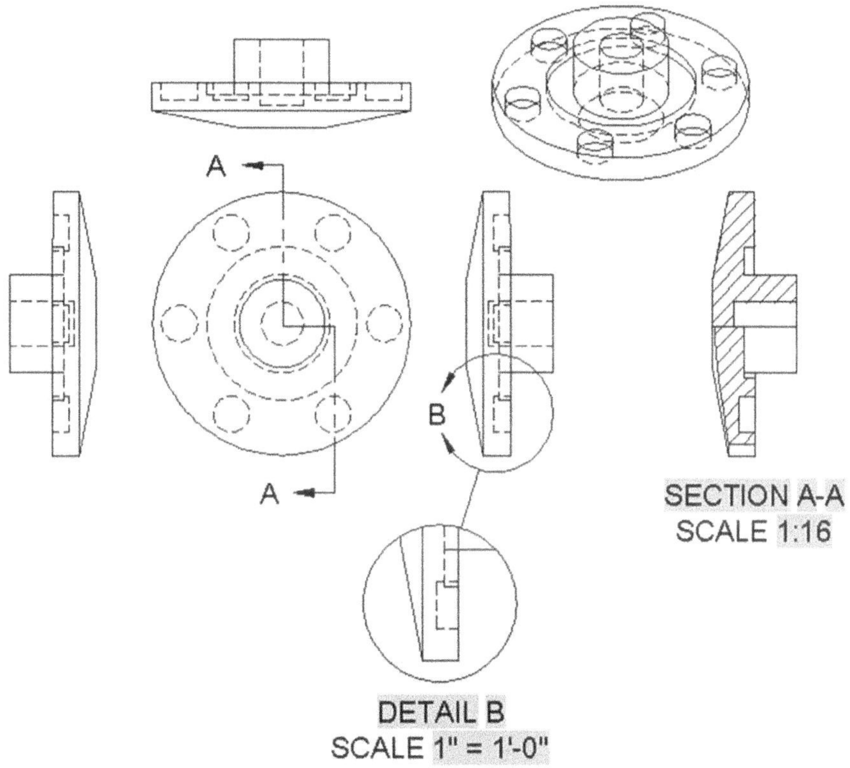

Learning objectives

In this chapter, we introduce and thoroughly cover the concepts of sectioning and presenting views. We specifically introduce the following topics:

- Slice
- Section
- Base view
- Projected view
- Section view
- Detail view
- Additional layout options
- Flatten and Flatshot
- Vports

By the end of this chapter, you will be able to slice and section 3D models, create complete 2D layouts of the 3D models, and present your model in multiple viewports.

Estimated time for completion of this chapter: 2 hours.

Up and Running with AutoCAD 2022. https://doi.org/10.1016/B978-0-323-89923-9.00027-5

27.1 Introduction to slicing, sectioning, layouts, and vports

This is the first of three chapters that deals with the various ways you can detail, present, view, clarify, and interact with your 3D design. This is a broad topic, which covers slicing, sectioning, layouts, and viewports in this chapter, with more discussion on dynamic views, path animation, and many other tools in Chapter 28 and Chapter 29. Eight releases ago, AutoCAD 2013, introduced significant enhancements to create 2D layouts from 3D models (similar to advanced 3D parametric design software), and we cover this in detail after the basics of slicing and sectioning.

27.2 Slicing and sectioning

You already know how to rotate and view your design from various angles, but what if you wanted to get inside of it? The only approach thus far was to revolve a shape less than 360°, revealing the inside. What about a nonrevolved design or being able to look inside a fully revolved view? Here, sectioning comes in handy. Section views are used in both engineering and architecture to show the inner workings of mechanical parts and architectural elevations. We focus here on simple mechanical devices, but you can apply the techniques to any design.

To illustrate slicing and sectioning, we need to create a basic shape to use as an example. Create the profile in Fig. 27.1 using a pline, and revolve it 360°. Color and rotate it bottom side up, as seen in the figure. Then, add a circle of 6″ diameter and extrude it 3″ as shown, polar arraying it six times. Explode the array and subtract the pieces out to create the cavities, as seen in Fig. 27.2.

Slice

Slice is more basic than sectioning (though technically the same idea), so we start here. The slice command chops off a piece of the part so you can see inside. You can either keep the chopped-off side or get rid of it. We demonstrate both approaches. The most common way to indicate where to slice is via planes (XY/YZ/ZX), but you can also indicate via three points or even use an object as the "slicer."

Keyboard: Type in **slice** and press Enter
Cascading menus: **Modify→3D Operations→Slice**
Toolbar icon: none
Ribbon: **Home** tab→**Solid Editing**

Step 1. Begin the slice command via any of the preceding methods.

- AutoCAD says: `Select objects to slice:`

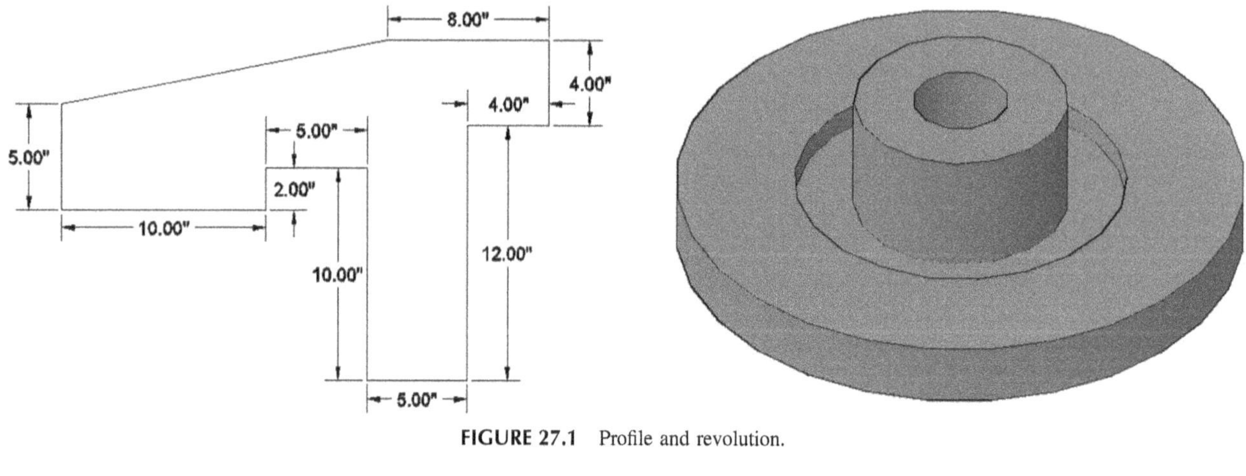

FIGURE 27.1 Profile and revolution.

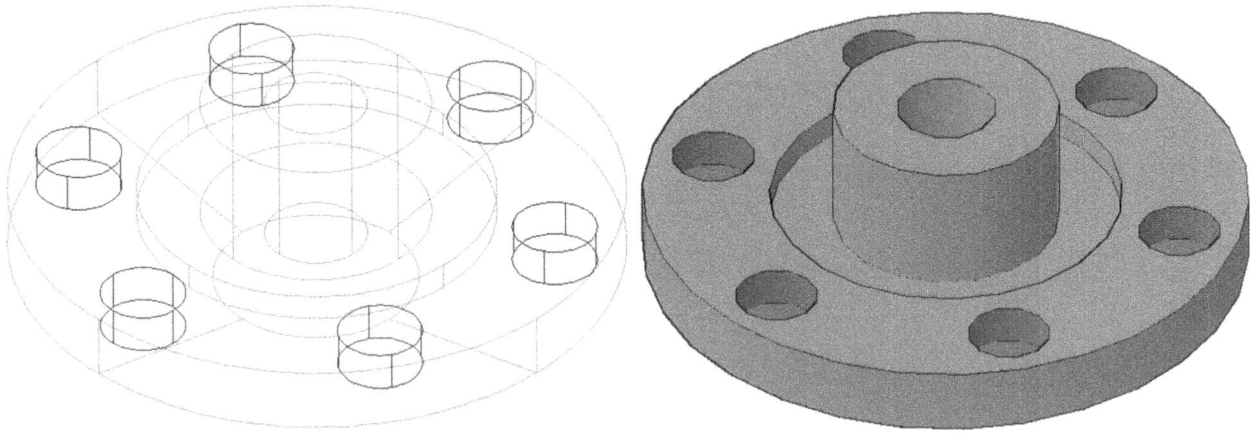

FIGURE 27.2 Add cavity, array and subtract

Step 2. Select the revolved shape and press Enter.

- AutoCAD says: `Specify start point of slicing plane or [planar Object/Surface/Zaxis/ View/XY/YZ/ZX/ 3points] <3points>:`

Step 3. We use the planes as a slicer. If your UCS is in the standard (World) configuration (as seen in Fig. 27.3), then select ZX as the plane.

- AutoCAD says: `Specify a point on the ZX-plane <0,0,0>:`

Step 4. Here, you just need to pick a point where you want the plane to start. It is infinite in both planar directions, so you are just locating an anchor point. Using OSNAPs, select the lower left quadrant (also seen in Fig. 27.3).

- AutoCAD says: `Specify a point on desired side or [keep Both sides] <Both>:`

Step 5. Here AutoCAD is asking which side of the slice you want to keep. Pick the one at the top of the screen by clicking on it, or in its general area, and you will see one half of the part disappear (Fig. 27.4)

If you want to keep both sides and have the slice act literally like a knife and no more, then select `B` for `Both sides` in Step 4. You can then just move one of the pieces off to the side and reveal the inside of both, as seen in Fig. 27.5. Go ahead and undo what you just did and run through the slice command again, keeping both sides this time.

Finally, you can also use an object as a slicer, as seen here with a randomly sized rectangle (turned into a region for clarity) standing in for a plane (Fig. 27.6). Select the `O` for `Object` option in Step 2, and delete one of the sliced sides.

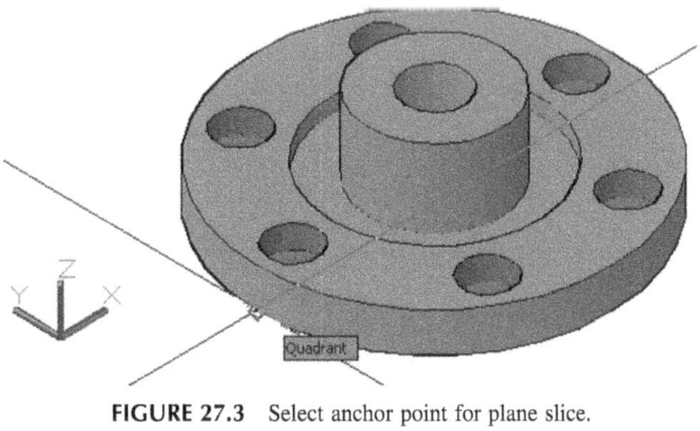

FIGURE 27.3 Select anchor point for plane slice.

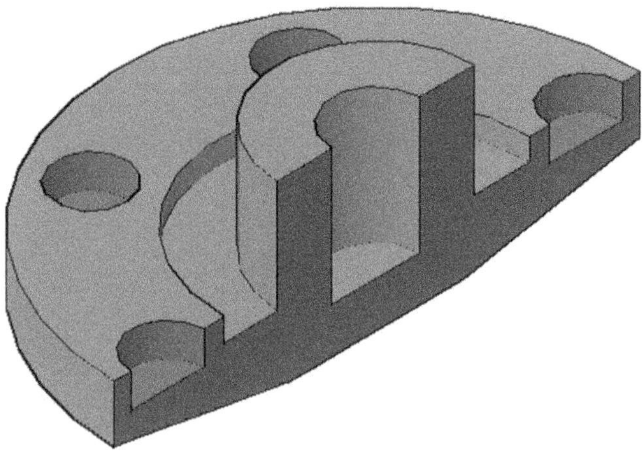

FIGURE 27.4 Completed slice.

Section

A section is related to the slice. Now, instead of cutting off a piece of the 3D model, we are just creating a cross-section plane. This plane can then be moved to another view and worked on as if it is a model all its own. So the critical difference here is that we are not directly modifying the original part but using it as the basis for something new.

Let us go back to the original unmodified model of Fig. 27.2. We want to create a section in the exact same plane (and using the same basic theory) as with the slice example. We then move the newly created section out of the model and place it somewhere else.

Step 1. This command is available only via typing, so go ahead type in `section` and press Enter.

- AutoCAD says: `Select objects:`

Step 2. Select the 3D model.

- AutoCAD says: `Specify first point on Section plane by [Object/Zaxis/ View/XY/YZ/ZX/3points] <3points>:`

Step 3. Select the ZX plane as in the previous example with slice.

- AutoCAD says: `Specify a point on the ZX-plane <0,0,0>:`

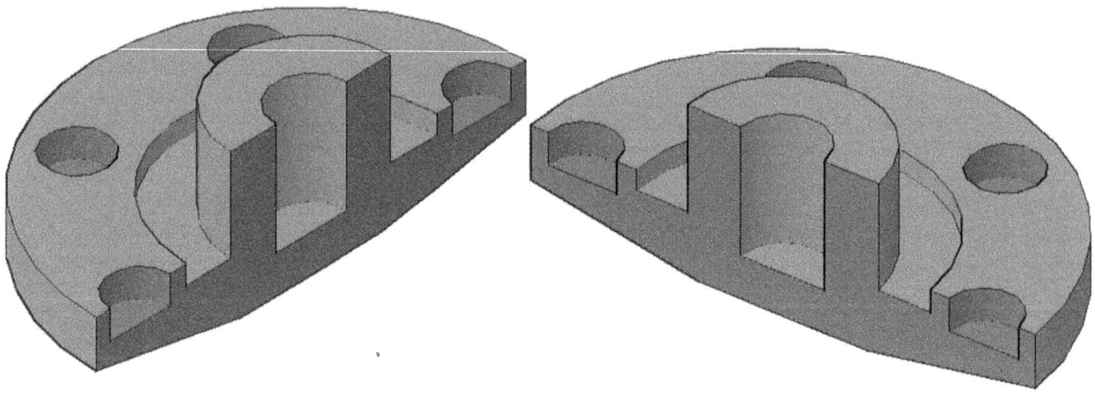

FIGURE 27.5 Completed slice - Both sides option.

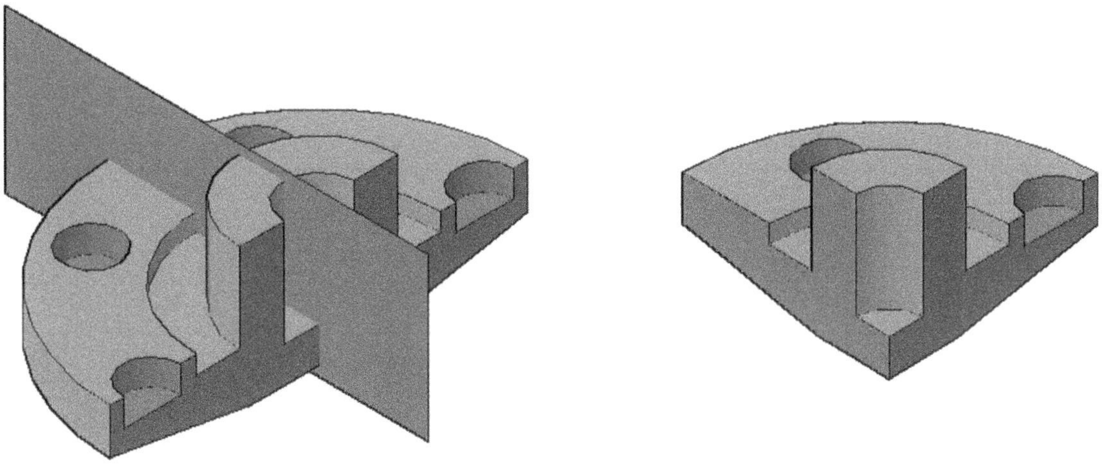

FIGURE 27.6 Completed slice - Object option.

Step 4. Select the lower quadrant, also similar to the slice example. The section is created. You may have to switch to wireframe to see it, as shown on the left-hand side of Fig. 27.7. On the right-hand side the section is moved outside the 3D part, and both are shaded.

The section itself can now be extruded if needed and of course labeled with dimensions, which is usually the reason behind creating it in the first place: to annotate it for clarity.

27.3 Layouts and details

Although you have worked with layouts and Paper Space before, this was limited to 2D views of 2D designs. But, what if you have a 3D model and want to quickly create 2D views of it automatically, including sections and enlarged details? This is something that was available for years in high-end design software like CATIA, NX, SolidWorks, and others. AutoCAD finally incorporated this feature back in the 2013 release. You can now create instant views, sections, and details after just a few clicks.

This process is relatively simple, and no viewports are involved, but it does follow strict drafting standards and depends on you knowing what these standards and conventions are in order to create a meaningful layout. Use of parent views, projections, sections, and details flows directly from hand drafting. Here is a quick review.

- *Parent View*: This is the view of the model that generates all other views. If it changes in any way, you have to update all the other associated views. AutoCAD, of course, does this for you.

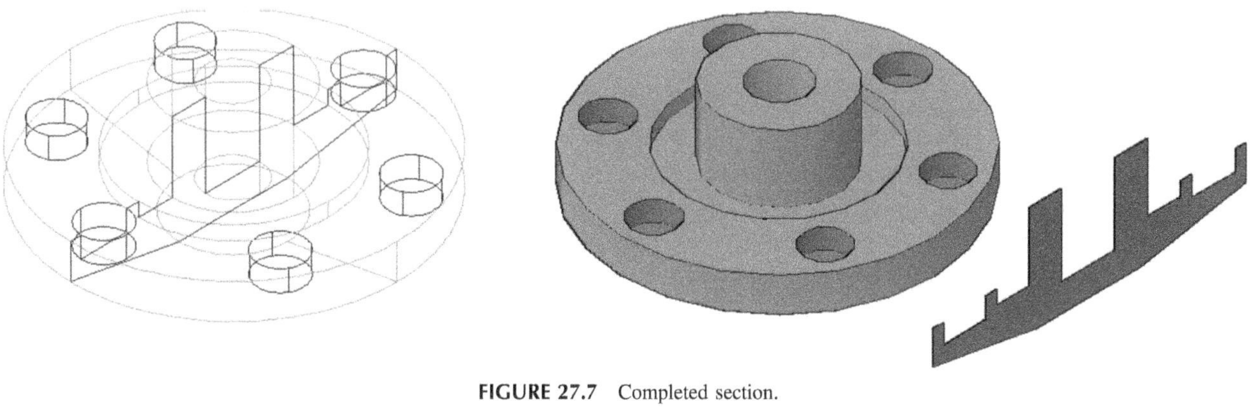

FIGURE 27.7 Completed section.

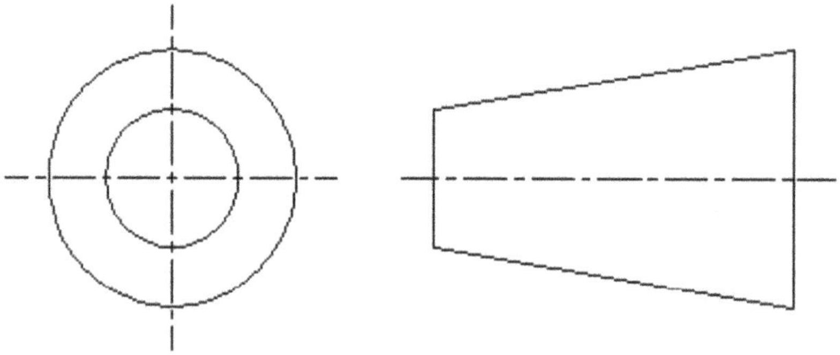

FIGURE 27.8 Third Angle view projection.

- *Projections*: These can be front, back, top, bottom, side, and isometric in nature. Be aware of how projected views are oriented from the parent view and where they are located. In the commonly used Third Angle projection, a head-on view turns to the right and faces the parent view (relative to you, the observer). Fig. 27.8 demonstrates the convention used. There is also a First Angle projection, where the side view is the main one, not the front.
- *Sections*: These can be generally classified as straight sections, which run from one end of the part to another, or jogged sections that turn at various angles as they make their way through the part.
- *Details*: These are blowups or enlargements of a part of the model for clarification purposes. A detail can be attached to the view that generated it, or it can be free standing.

With these in mind, we explore the available options and create a full layout of the mechanical part we have been using.

Creating these layouts is done primarily via the Ribbon. There are no toolbars or cascading menus to use. Typed equivalents exist, and they will be shown, but the Ribbon is the simplest way to proceed. Go ahead and select a shaded top view of the same 3D model we have been using. Be sure you are in the 3D Modeling workspace and select the Ribbon's Home tab. You should have what is shown in Fig. 27.9, with the Base function circled at the very upper right of the screen.

NOTE: *If you are using an older version of AutoCAD, specifically Release 2020, 2019, 2018 and 2017, etc.,* the following commands (VIEWBASE, VIEWPROJ, VIEWSECTION, and VIEWDETAIL) are only available on 64-bit computer systems and will not appear on 32-bit machines. AutoCAD 2016 allowed the*se commands in both 64 and 32 bit.* Therefore, if you are trying to use these commands and they are not appearing, check your AutoCAD release, and operating system. None of this is an issue on AutoCAD 2021 as the software does not support 32-bit systems in the first place.

VIEWBASE Establishes the Base (Parent) View

Step 1. In the Ribbon's Home tab, View panel, select the Base option. Base View From Model Space.

- AutoCAD says: `Specify model source [Model space/File] <Model space>: _M`

  `Select objects or [Entire model] <Entire model>:`

Step 2. Select the entire model.

- AutoCAD says: `Enter new or existing layout name to make current or [?] <Layout 1>:`

Step 3. Type in Sample_Layout or any other name you wish. Press Enter. AutoCAD switches to Paper Space and a new tab with the name you gave it appears. A view of the part is attached to the movements of your mouse.

- AutoCAD says: `Restoring cached viewports - Regenerating layout.`
  `Type = Base and Projected Hidden Lines = Visible and hidden lines Scale = 1:16`
  `Specify location of base view or [Type/sElect/Orientation/Hidden lines/Scale/Visibility] <Type>:`

Step 5. Click to place the view anywhere on the screen.

- AutoCAD says: `Select option [sElect/Orientation/Hidden lines/Scale/Visibility/Move/eXit] <eXit>:`

Step 6. Press Enter, then click again to place another view below the first one

- AutoCAD says: `Specify location of projected view or <eXit>:`

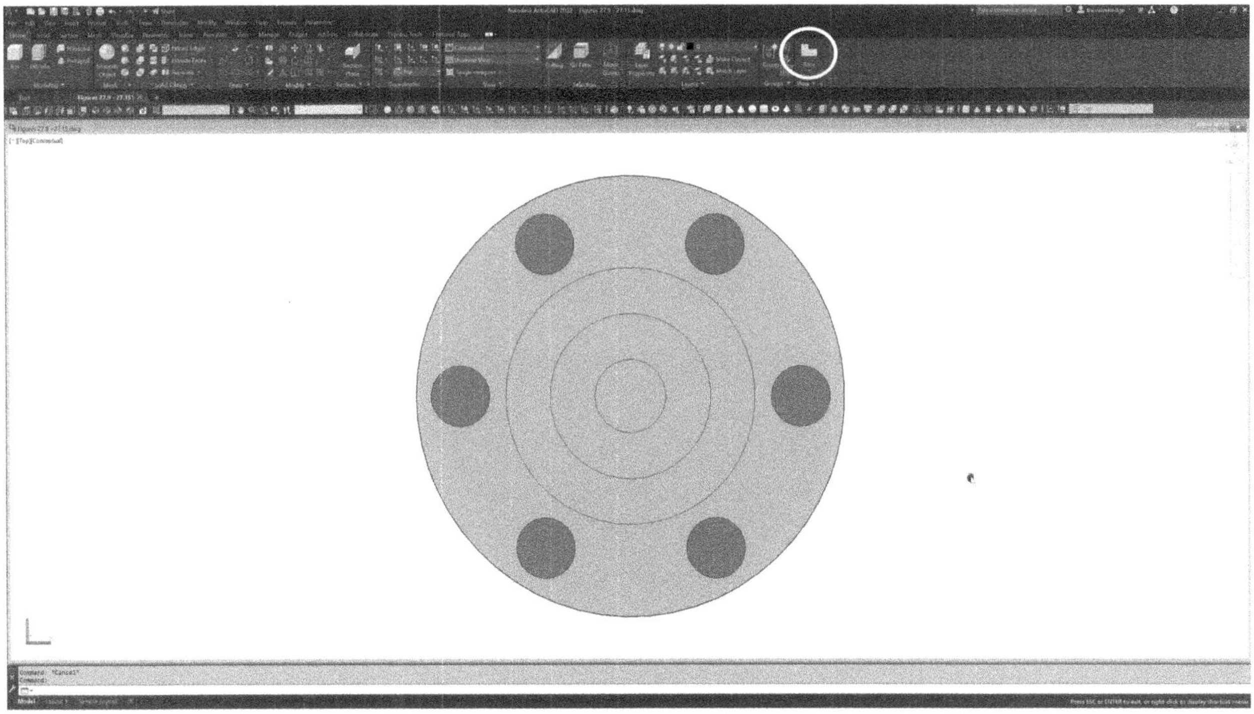

FIGURE 27.9 Setup for Layouts.

Step 7. Press Enter one final time to get two views as seen in Fig. 27.10.

- AutoCAD says: `Base and 1 projected view(s) created successfully.`

Note that you need not stop at just two views; you could continue to add additional views to the right, left, top, bottom, and isometric, as desired. As practice, undo what you just did and repeat all the steps, this time adding an additional isometric view, as seen in Fig. 27.11. Adding additional projections (or removing them) can easily be done at the next step, so just two is adequate in the initial setup, but we keep the isometric projection.

Next, we would like to add (or delete) additional projections. How many and what type need to be shown on a drawing is really up to the engineer or designer and company standards and needs. We establish just a few more projections, leaving enough room for sections and details. These new views inherit the same scale and other parameters of the parent view that you choose.

VIEWPROJ Establishes Additional Projections

Step 1. Select, by clicking once, whichever of the existing projections you would like to use as the "parent" projection in order to generate some new ones. In this example, the front view in the middle is chosen. Once it is clicked on, the Ribbon changes to the Drawing View tab. Select the Projected icon, as seen in Fig. 27.12.

- AutoCAD says: `Specify location of projected view or<eXit>:`

Step 2. The new views are attached to the movements of your mouse. Note how it can be located only in certain planes; left right, top, and bottom. Click to place the projections one time to the left and one time to the right of the main view, pressing Enter when done. The result is shown in Fig. 27.13.

Note that, within the proper plane, you can delete or move any of these projected views. The scale of each view can also be adjusted, if needed, by clicking on the view once to reveal a set of grips. The grip that looks like an arrow shows a set of scale values if you click on it again. We need not do this at this time, but it is necessary to know how to, just in case.

Now that you have all the views set up, you may need to add sections and details. The process for adding section views is described next.

VIEWSECTION Creates Section Views

Step 1. Select, by clicking once, whichever view you want to section—the main view in the center in this example. The Ribbon once again changes to the Drawing View tab seen in Fig. 27.12. Select the Section icon, just to the right of Projected. As you drop down the menu, take note of the five types of sections available: Full, Half, Offset, Aligned, and

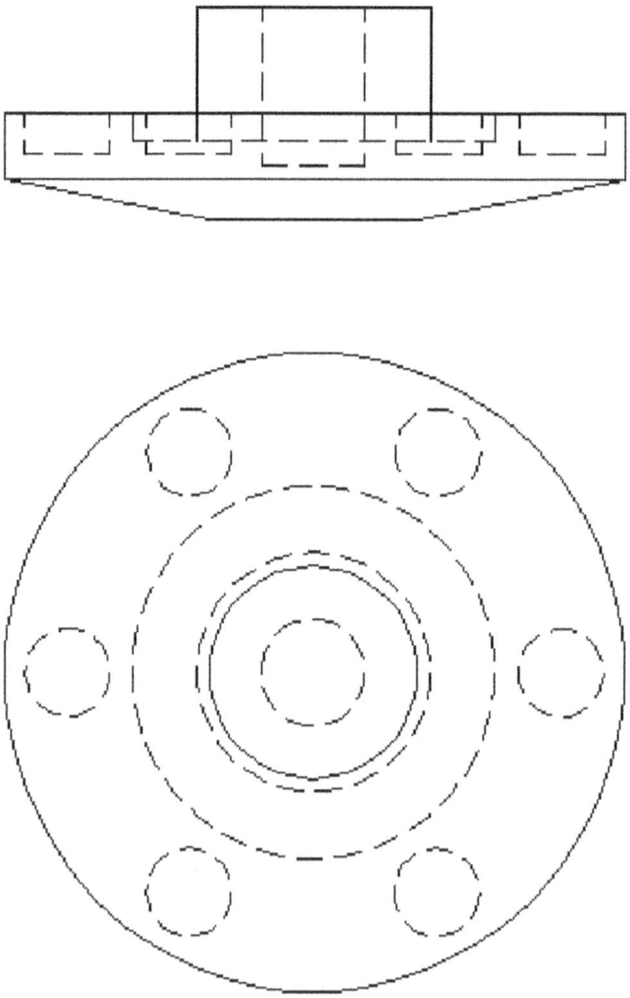

FIGURE 27.10 Base and projected views.

From Object. We try only the Offset one, leaving the rest for you to practice. They all work on the same principle. Click the Offset option and the Ribbon changes to the Section View Creation tab.

- AutoCAD says: `Hidden Lines = Visible lines Scale = 1:16 (From parent)`
  `Specify start point or [Type/Hidden lines/Scale/Visibility/Annotation/hatCh]`
  `<Type>: _t`
  `Select type [Full/Half/OFfset/Aligned/OBject/eXit] <eXit>: _of`
  `Specify start point:`

Step 2. An "A" and an arrow, the start of a section designator, are attached to your mouse movements. Click your way down the main top view, as seen in Fig. 27.14. Note how the section lines pass through several of the cutouts in the model. Seeing those cutouts in the section is usually of interest to the designer, so be sure to trace a path that reveals useful information, not just a random cut.

Step 3. When done press Enter and a section view appears, attached to the motions of your mouse, in plane.

- AutoCAD says: `Specify location of section view or:`

Step 4. Move your mouse off to the right, some distance from the main view, and click to place the section view.

- AutoCAD says: `Select option [Hidden lines/Scale/Visibility/Projection/Depth/Annotation/hatCh/Move/eXit] <eXit>:`

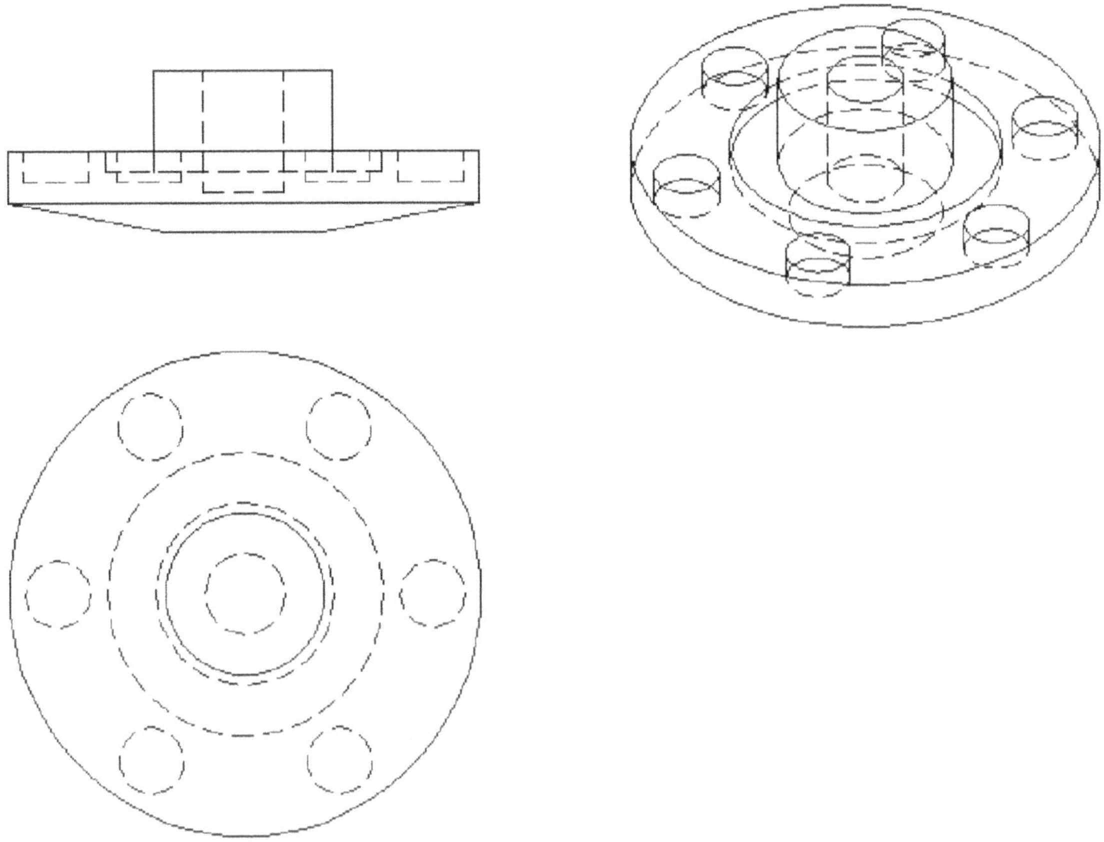

FIGURE 27.11 Isometric view added.

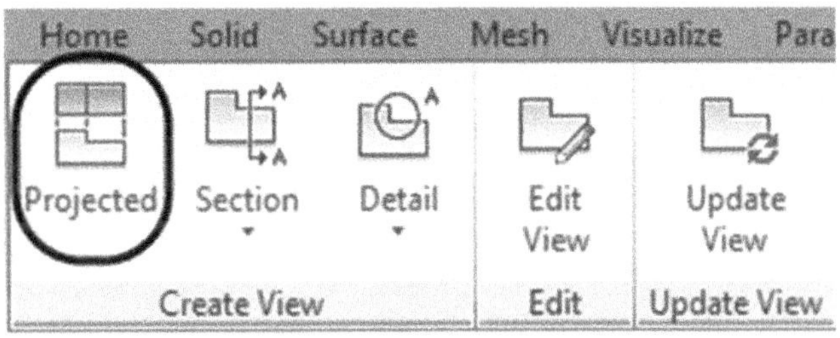

FIGURE 27.12 Drawing View tab, Projected.

These are additional options, some of which we discuss shortly, but for now, just press Enter to complete the command. Your view loses its shading and retains hatching, as is proper for a section view. Below it, editable text appears, indicating the name of the section and its scale (same as the parent view). The final result is shown in Fig. 27.15.

The final major step involves creating details. As mentioned, these are nothing more than larger views of existing geometry and are primarily for clarification purposes of hard-to-see parts of the design.

VIEWDETAIL Creates Detail Views

Step 1. Select, by clicking once, whichever view you want to section - the bottom of one of the side projections on the right is chosen in this example. The Ribbon once again changes to the Drawing View tab seen in Fig. 27.12. Select the Detail icon, just to the right of Section. As you drop down the menu, take note of the two types of details available: Circular and Rectangular. We try only the Circular one, leaving the Rectangular one for you to practice. They all work on the same principle. Click on the Circular option and the Ribbon changes to the Detail View Creation tab.

- AutoCAD says:

```
Boundary = Circular Model edge = Smooth Scale = 1" = 1'-0"
Specify center point or [Hidden lines/Scale/Visibility/Boundary/model Edge/Annotation] <Boundary>: _b
```

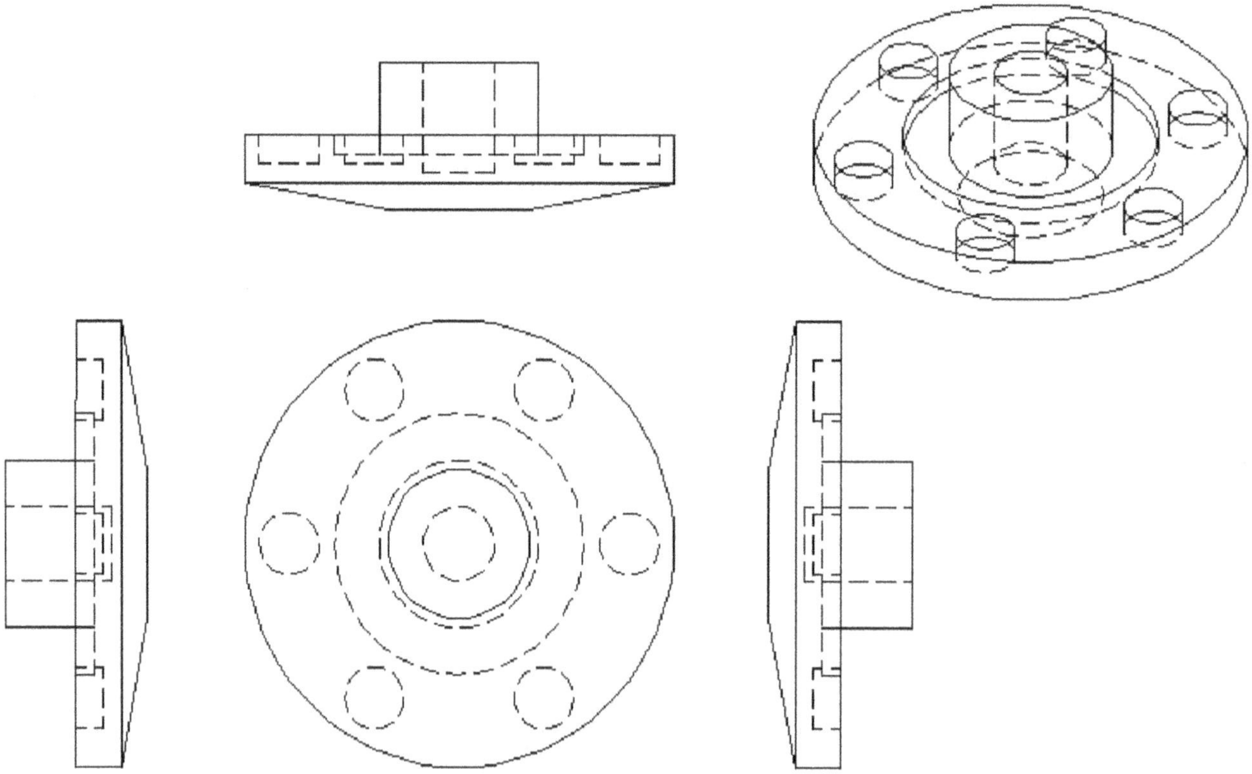

FIGURE 27.13 Additional projections.

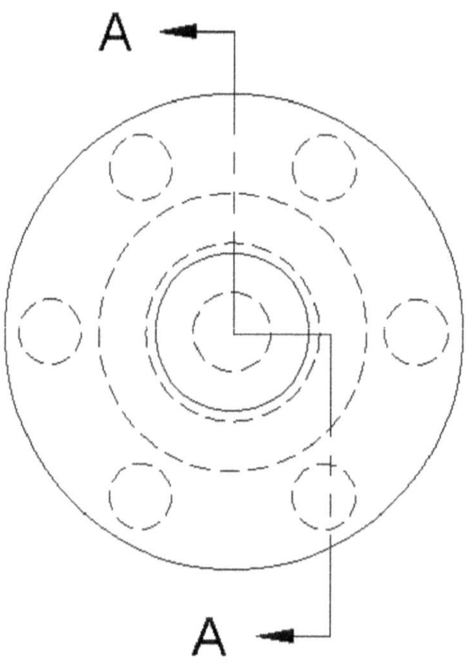

FIGURE 27.14 Offset section.

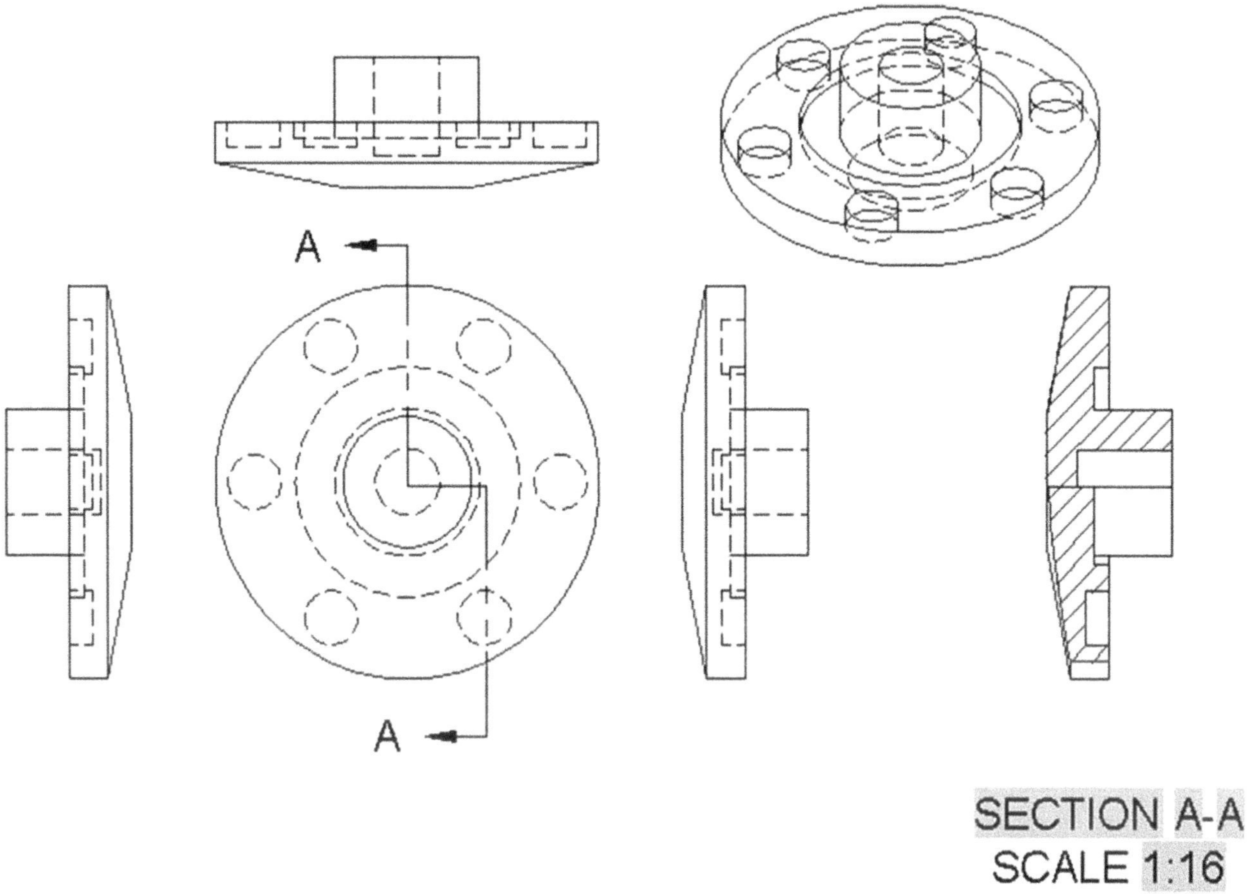

FIGURE 27.15 Offset section.

```
Select boundary type [Circular/Rectangular/eXit] <Circular>: _c
Specify center point or [Hidden lines/Scale/Visibility/Boundary/model Edge/Annotation] <Boundary>:
```

Step 3. Click somewhere in the lower region of the projection. You see a circular detail marker with the letter B.

- AutoCAD says: `Specify size of boundary or [Rectangular/Undo]:`

Step 4. Move the mouse out slowly until the circle is big enough to cover the area you want to detail.

- AutoCAD says: `Specify location of detail view:`

Step 5. As you are positioning the detail, be sure to select the Smooth with connection line under Model Edge of the Ribbon's Detail View Creation tab. Click somewhere below the view - or anywhere there is room.

- AutoCAD says: `Select option [Hidden lines/Scale/Visibility/Boundary/model Edge/Annotation/Move/eXit] <eXit>:`

Step 6. Press Enter, and the detail loses its shading, appearing with some editable text.

- AutoCAD says: `Detail view created successfully.`

The final result is shown in Fig. 27.16.

Additional layout options

What we just covered are the essentials of creating 2D layouts from a 3D model. Note, of course, that this should all take place within the confines of a title block; something that was omitted for clarity in the preceding discussion. You first want to create or import the title block into Paper Space, then use the available room to properly position all the views, sections, and details. This can sometimes take considerable effort to position well for maximum clarity.

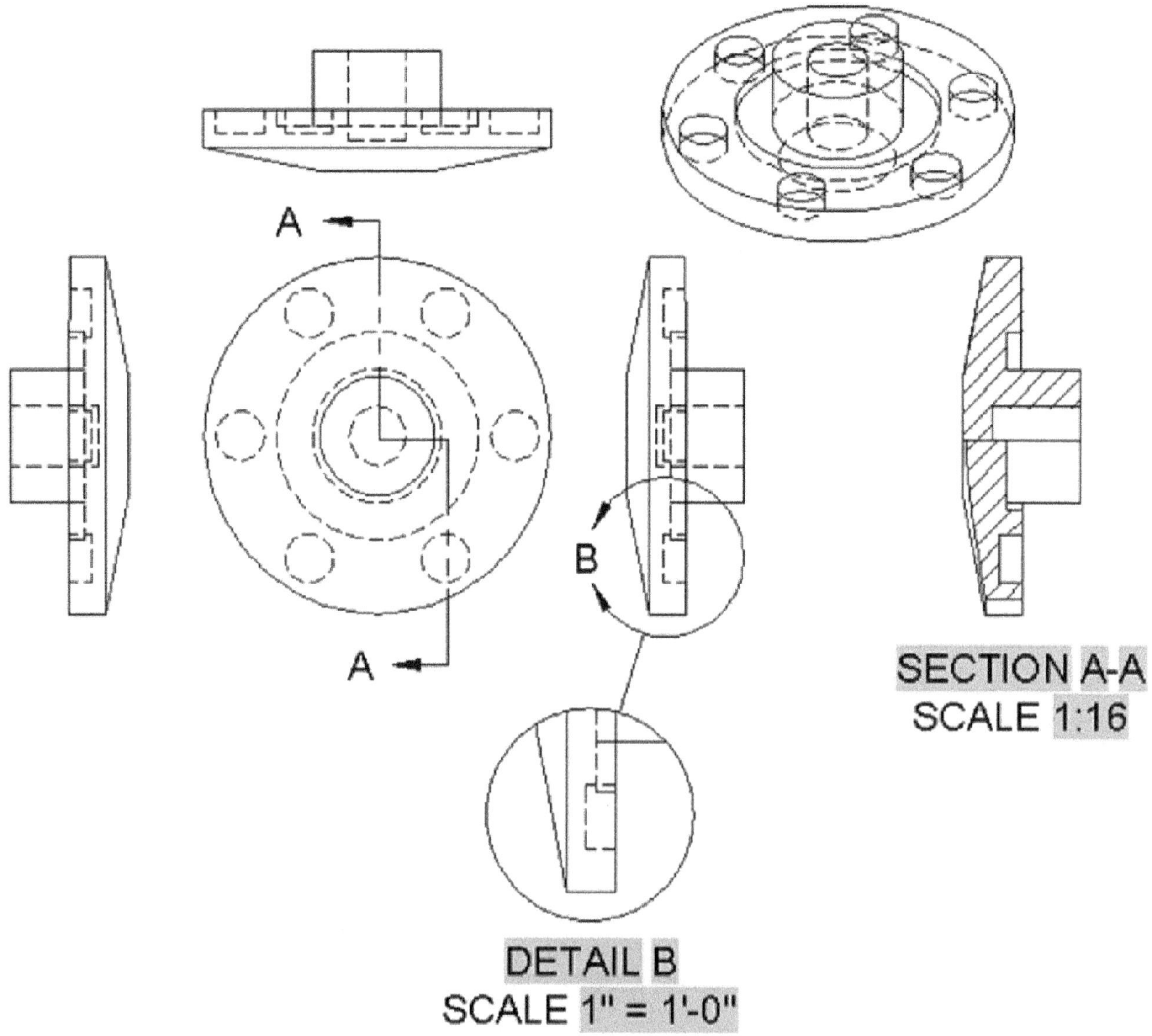

FIGURE 27.16 Full set of projections, sections, and details.

Of course, we glossed over a few other options. These mainly have to do with appearances of the views. So far, they have appeared as basic linework, with hidden and visible lines. You can easily change the appearance to include shading, hidden lines, and visible lines. This is done by selecting, by clicking once, any of the views. The Ribbon then changes to the Drawing View tab. Select the Edit View icon under the Edit panel. The Ribbon then changes to the Drawing View Editor, and you can select from several hidden, visible, and shaded choices, such as seen in Fig. 27.17. When done press OK, or Cancel if it is not what you wanted.

The result is shown in Fig. 27.18. Take some time to look over all the options, including ones for modifying the text that appears with the new sections and details.

27.4 Flatten and flatshot

The ability to create 2D flat views from 3D models was the central theme of the previous discussion, and we now add another set of tools to accomplish this. Flatten and Flatshot are two sides of the same coin. The commands exist to create 2D views, for illustrative purposes, out of 3D models. Note that these two commands do not create intelligent layouts, as

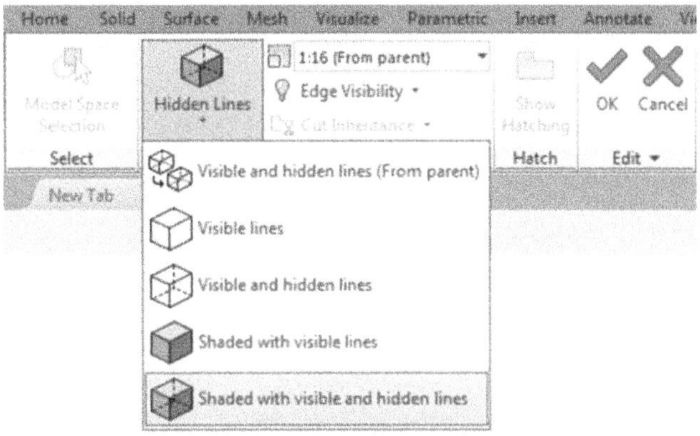

FIGURE 27.17 Shaded with visible and hidden lines.

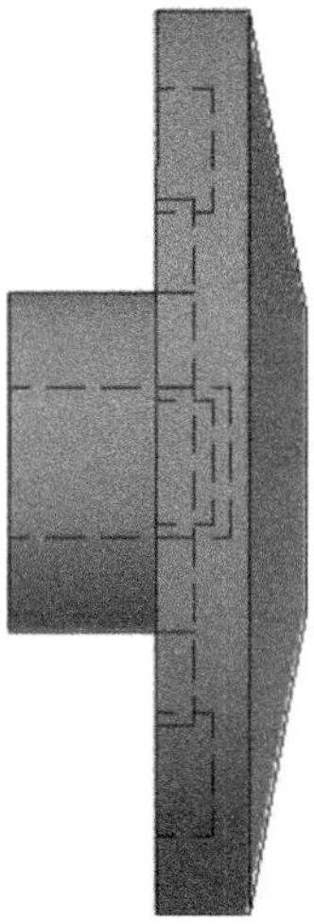

FIGURE 27.18 Result of modifying the view.

just discussed, rather they are used for a simple 2D "snapshots" of the model. What you see is what you get with these. The differences between Flatten and Flatshot are as follows:

- *Flatten*: The simpler of the two commands, it creates a 2D snapshot by just converting the current 3D view and model to 2D. You have to be careful using this one, as the original model is destroyed on conversion to 2D.
- *Flatshot*: A slightly more involved command, it gives you additional options in regard to scaling, positioning, and exporting the new 2D view. Most important, it does not destroy the original model but makes a 2D copy of it.

Both Flatten and Flatshot allow you to select whether or not to show hidden lines. Let us give them both a try with one of the side views we generated in the previous exercise.

Flatten

Step 1. There is no toolbar, cascading menu or Ribbon alternative, so type in flatten and press Enter.

- AutoCAD says: `Select objects to convert to 2d...`

  `Select objects:`

  Step 2. Select the side view of the 3D model and press Enter.

- AutoCAD says: `Remove hidden lines? <No>:`

Step 3. You can pick yes or no. In this example, we keep the hidden lines. Press Enter. The flattened 2D model appears as seen in Fig. 27.19.

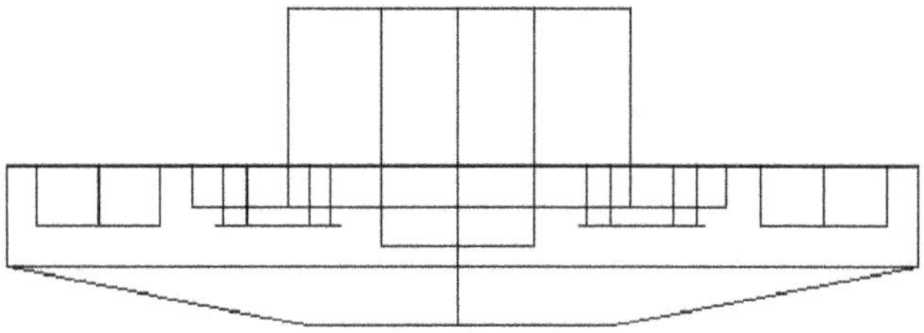

FIGURE 27.19 Flatten results.

Flatshot

Step 1. There is no toolbar or cascading menu alternative, but you can use the Ribbon's Home tab, Section panel, or type in Flatshot and press Enter. The dialog box shown in Fig. 27.20 appears.

Step 2. Press the Create button on the dialog box, and the 2D model appears on the screen attached to the movements of your mouse.

- AutoCAD says: `Units: Inches Conversion: 1.0000`
  `Specify insertion point or [Basepoint/Scale/X/Y/Z/Rotate]:`

Step 3. Click anywhere on the screen to place the model.

- AutoCAD says: `Enter X scale factor, specify opposite corner, or [Corner/XYZ] <1>:`

Step 4. Press Enter to accept the 1 to 1 X scale factor (unless you want to change it).

- AutoCAD says: `Enter Y scale factor <use X scale factor>:`

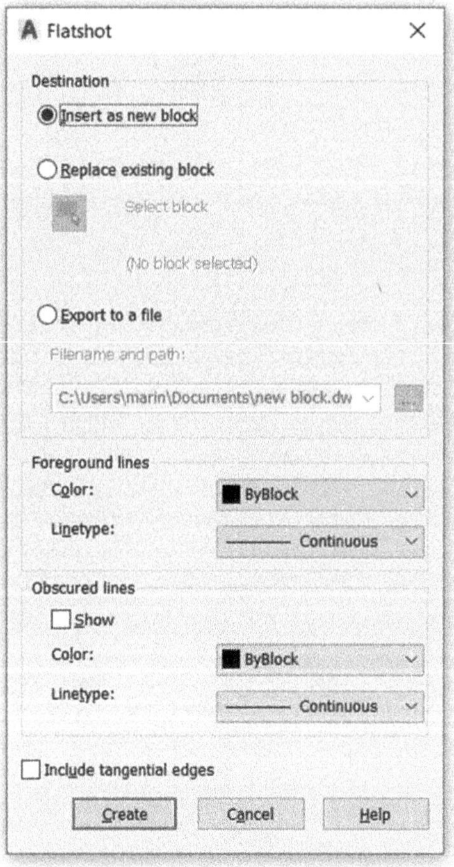

FIGURE 27.20 Flatshot dialog box.

Step 5. Press Enter to accept the 1 to 1 Y scale factor (unless you want to change it).

- AutoCAD says: `Specify rotation angle <0>:`

Step 6. Press Enter to accept the 0 angle of rotation (unless you want to rotate the view). You are done; the 3D model looks like what is shown in Fig. 27.21.

Make a note of some of the options in the Flatshot dialog box, such as exporting the file or changing the colors of the hidden lines, if you wish to do so.

27.5 Vports

Vports, which is short for *viewports*, is a unique tool in AutoCAD that allows you to view the same model from multiple angles all at once. The screen can be split in one of the several available ways to best fit the situation. Your design then appears in all screens, and you can assign each screen its own view and properties, such as shade or wireframe mode. Back in AutoCAD 2015, some enhancements were added to this feature. To access vports, let us first bring up a simple 3D model, such as the wastebasket created in Chapter 24, and shown again in Fig. 27.22.

Once you have this file opened or created the basket again, you can access vports via any of the following methods.

Keyboard: Type in **vports** and press Enter
Cascading menus: **View→Viewports→New Viewports…**
Toolbar icon: **Layouts** toolbar→
Ribbon: **Vizualize** tab→**Model viewports**

This brings up the dialog box seen in Fig. 27.23. Note that, if you use the Ribbon, you just choose one of these configurations right away, and the dialog box is bypassed.

As you can see from the configuration choices on the left, you have a few ways to arrange the screen. The default view is always Single, and that is the choice you will select later on, when you want to get back to the familiar single screen. For now, to try this out, choose the Four: Equal configuration and press OK. You should see what is shown in Fig. 27.24.

Notice that the view of the wastepaper basket has been duplicated four times and has not undergone any further changes. Keep in mind that only the view has been duplicated, not the actual 3D model. You may have to zoom to extents in each viewport to get the model to take up all the available area.

The current viewport is highlighted with a blue frame, and you can also see the crosshairs and the ViewCube/Navigation Bar combination in the upper right corner. You can now easily jump from screen to screen by just clicking once in it. Go ahead and visit each one and modify the views to a combination of top, side, and isometric, as seen in Fig. 27.25.

Now, try working in one viewport. Notice that anything you do in one is reflected in the others, as it should be, since it is just one model with which you are working. Another popular configuration, Four: Left, is shown in Fig. 27.26. Here, you have one primary large view and three supplemental views to see other parts of the model.

How you use these vports (or whether you use them at all) is up to you and your style of working in 3D. In some cases, they are very useful; in others, it is best to just rotate the model as you need and not maintain a constant view.

Another use for these vports is in plotting out the design for presentations, and it is here that this command gets a chance to really shine. While printing all the vports in Model Space is not possible, if you switch to Paper Space and repeat the vports Four: Equal command, you get what is shown in Fig. 27.27. Note that you have to actually draw the area where you want the vports to go before they appear and that the Ribbon's vports commands do not work in Paper Space. Zoom to extents in each one, and add some other modifications to arrive at what is shown in the figure.

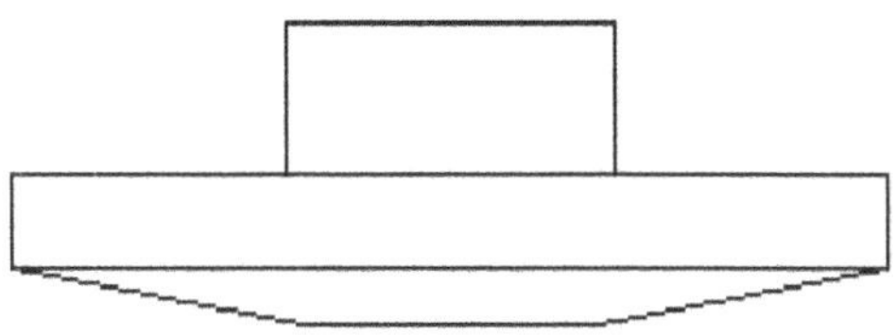

FIGURE 27.21 Flatshot results.

FIGURE 27.22 Wastebasket model.

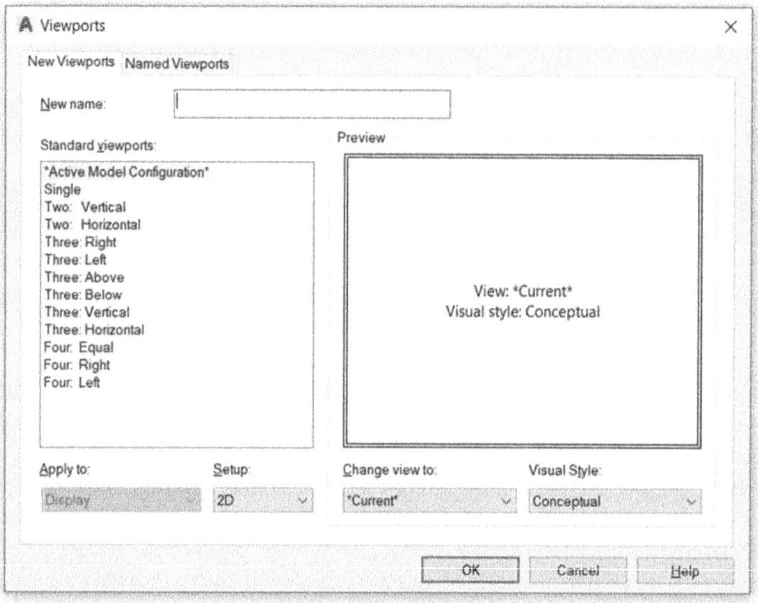

FIGURE 27.23 Viewports.

Each view is still active and can be modified, the same as in Model Space, but now also the overall scene with all the viewports can be plotted or printed for an impressive display.

AutoCAD now has the ability to dynamically resize viewports and add/subtract new ones. This essentially frees you from the premade configurations available and allows you to position and size the viewports in virtually any way desired. To try this out, first start with the Four: Equal Split configuration seen in Fig. 27.24, then look for a tiny + and – seen in the frame of each viewport (to the right of the arrow in Fig. 27.28).

Press the plus or minus sign and drag a new viewport frame to either side until you have it where you want it. The frame is green while you are dragging it (in this case to the left), as seen in Fig. 27.29 by the black arrows. When you stop, a new

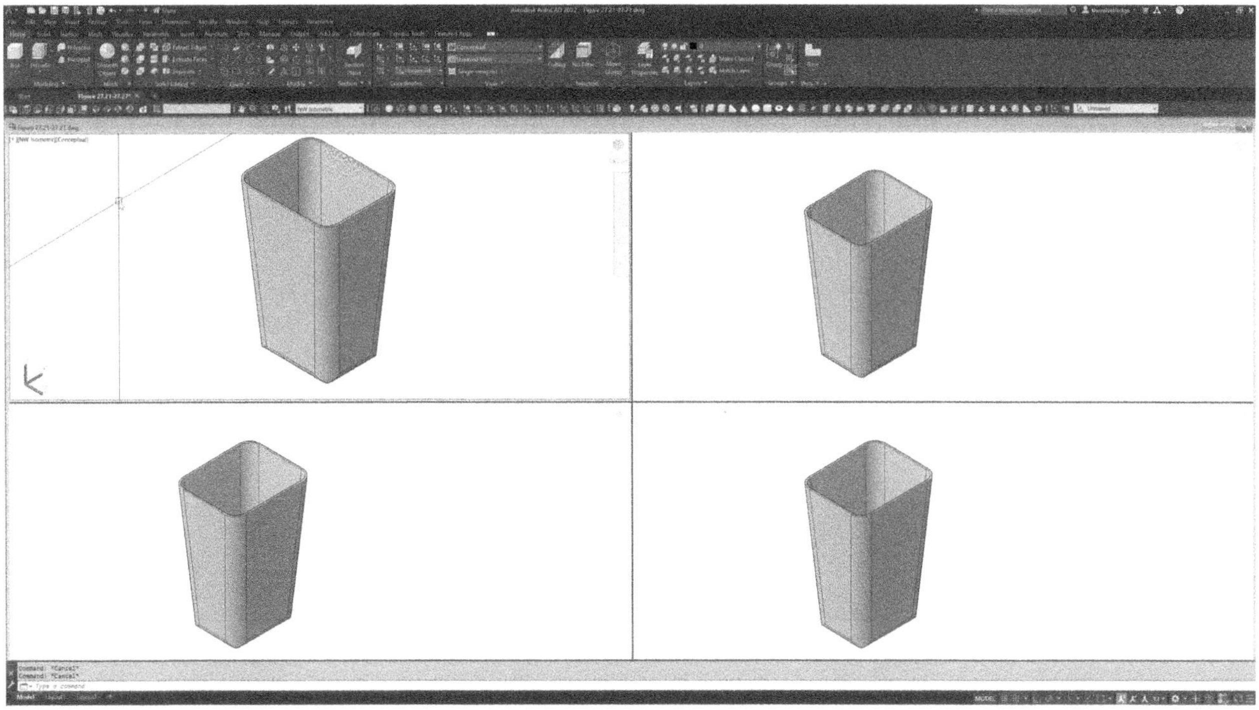

FIGURE 27.24 Four: Equal split.

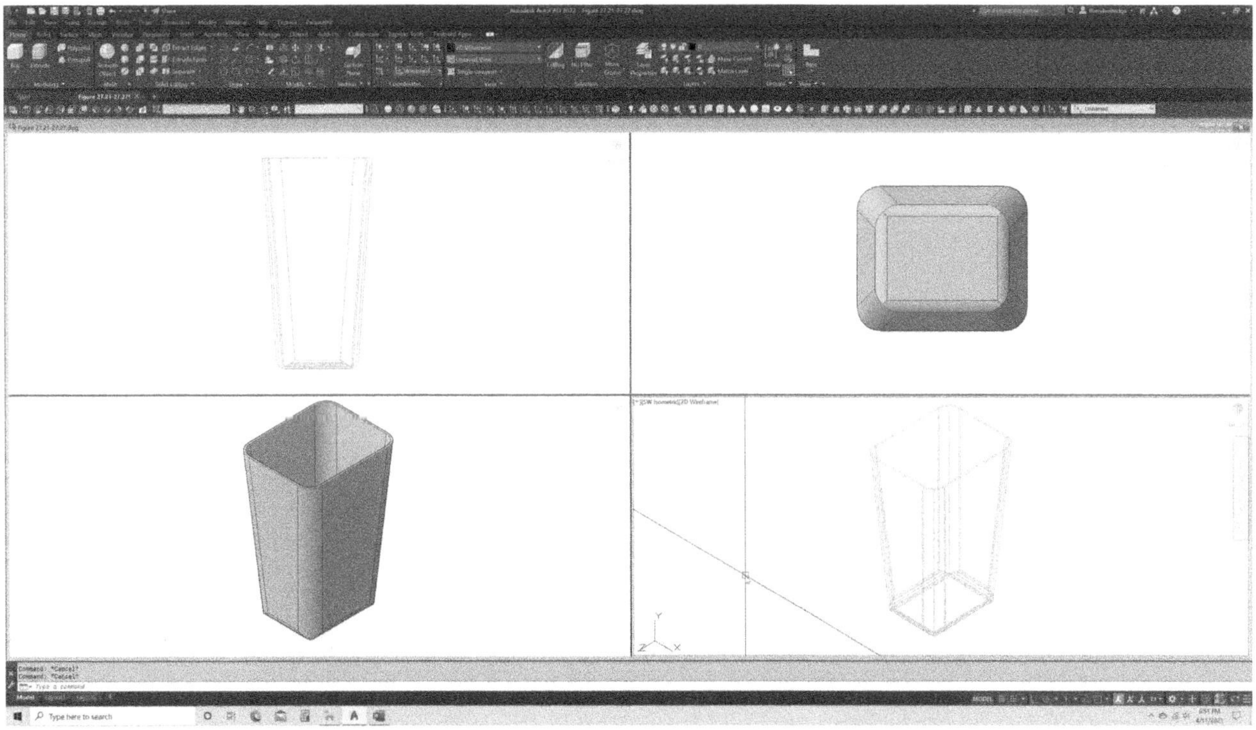

FIGURE 27.25 Four: Equal split with view modifications.

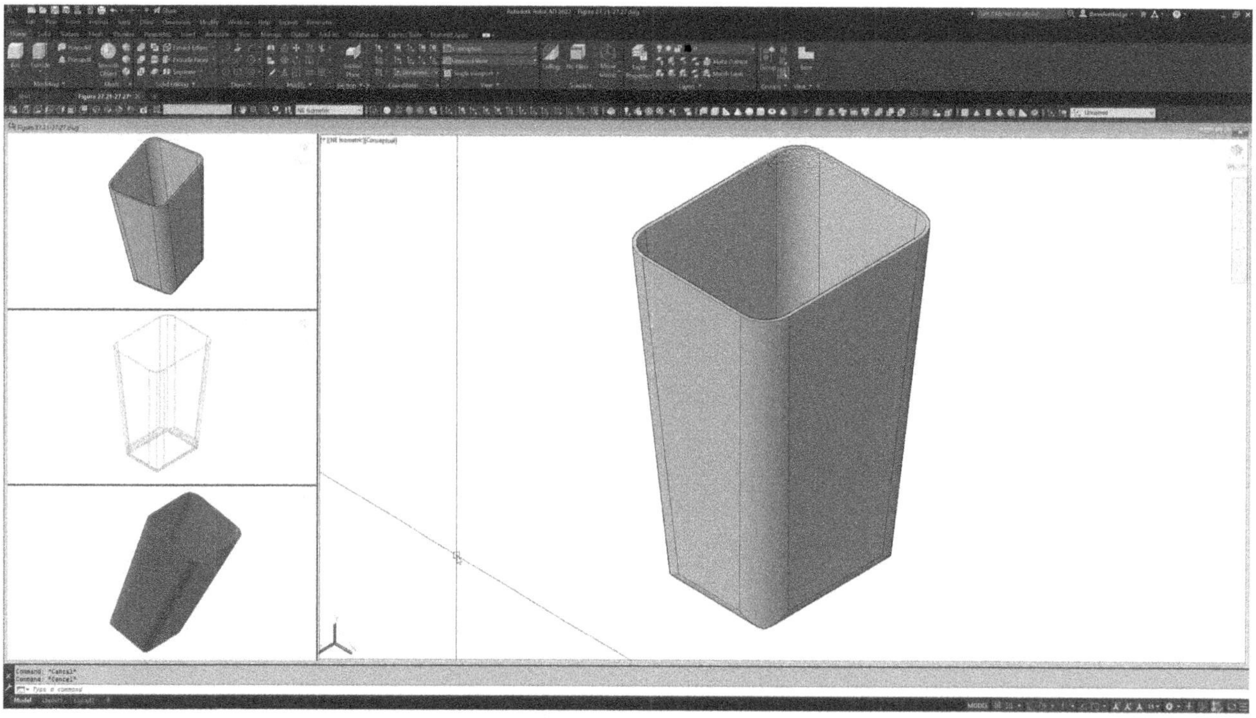

FIGURE 27.26 Four: Left split.

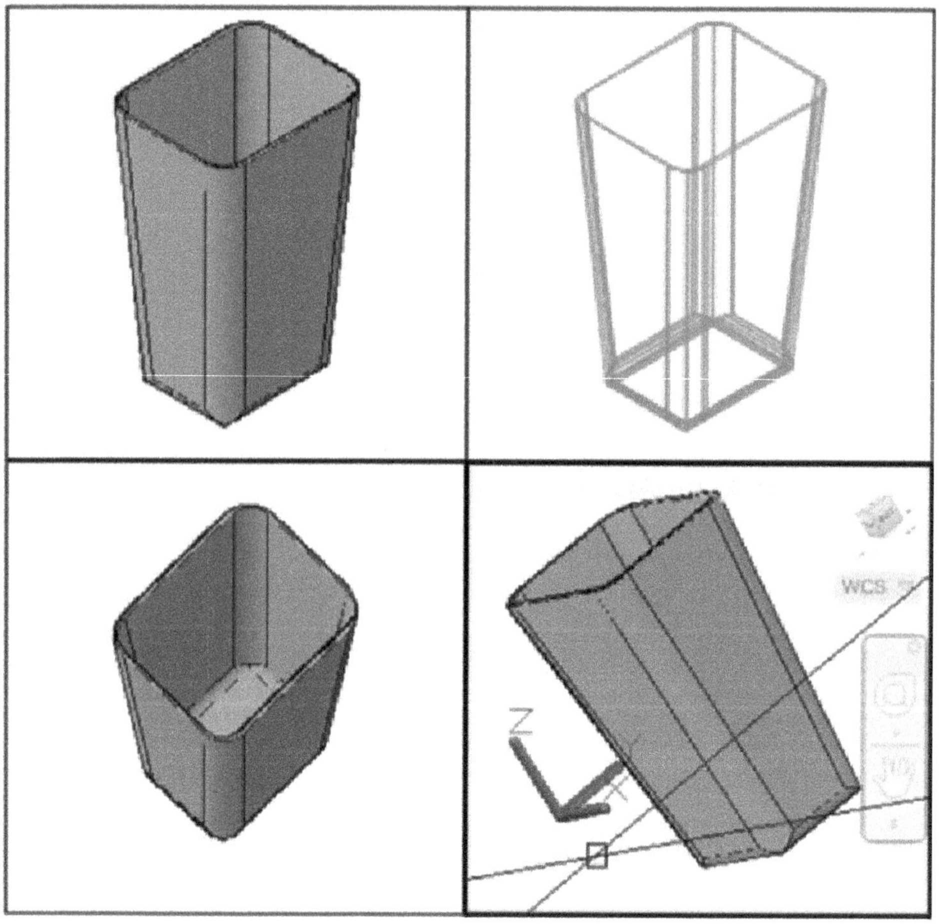

FIGURE 27.27 Four: Equal in Paper Space.

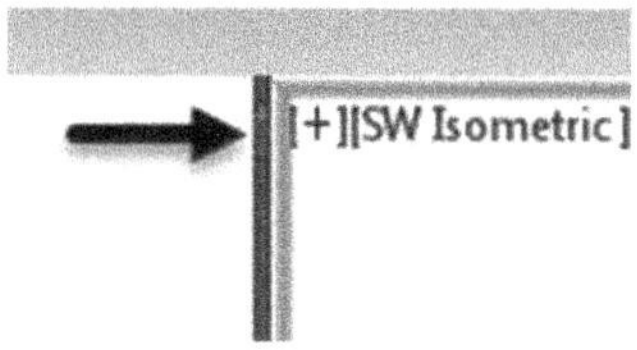

FIGURE 27.28 Viewport controls.

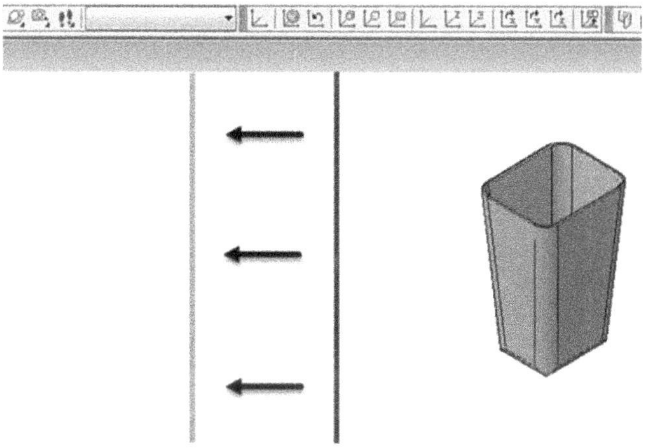

FIGURE 27.29 Creating new viewport.

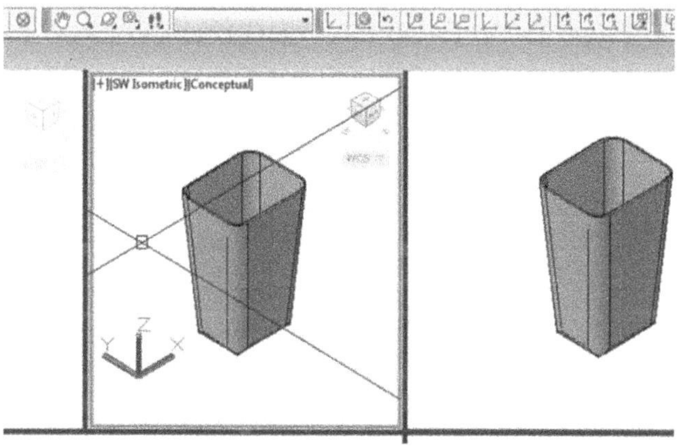

FIGURE 27.30 New viewport completed.

viewport is created, as seen in Fig. 27.30. Existing viewports can also be adjusted dynamically by clicking and dragging the fames in either the vertical or horizontal direction.

Finally (and this is purely for entertainment purposes), you can split the screen into 16 viewports (the maximum is 64), as seen in Fig. 27.31. The standard vports dialog box does not let you do this, but type in -vports (a hyphen means it is "command line" only) and press Enter.

- AutoCAD says: `Enter an option [Save/Restore/Delete/Join/SIngle/?/2/3/4]<3>:`

Enter the value 4 to split the screen into four views. Then, click into each new view and split it again in the exact same manner. In case you are wondering, no, this does not serve any real purpose aside from somewhat testing out your computer's memory.

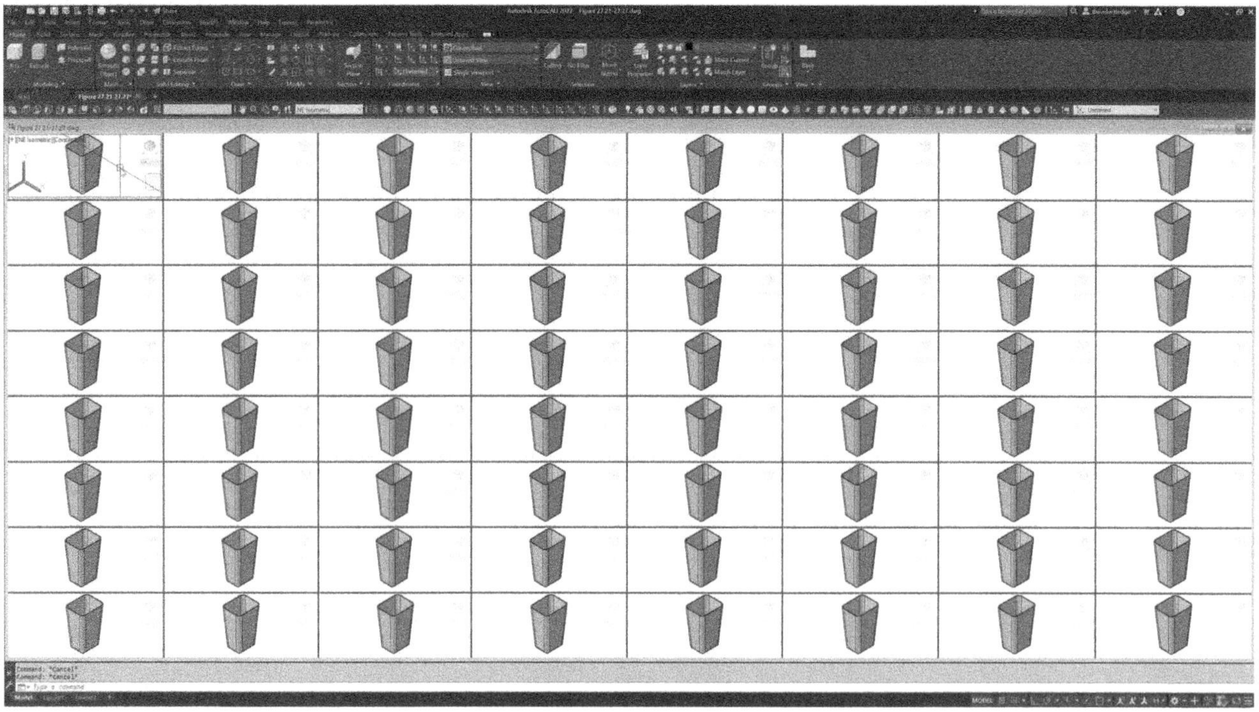

FIGURE 27.31 16 vports.

Summary

You should understand and know how to use the following concepts and commands before moving on to Chapter 28:

- Slice
- Section
- Base view
- Projected view
- Section view
- Detail view
- Additional layout options
- Flatten
- Flatshot
- Vports

Review questions

Answer the following based on what you learned in this chapter:

1. Describe the slice command.
2. Describe the section command.
3. What is the major difference between slice and section?
4. What does the VIEWBASE command accomplish?
5. What does the VIEWPROJ command accomplish?
6. What does the VIEWSECTION command accomplish?
7. What does the VIEWDETAIL command accomplish?
8. What is the critical difference between flatten and flatshot?
9. What are vports useful for? List two functions.

Exercise

1. To practice the concepts you learned in this chapter, let us create another shape (somewhat similar to one of the Level 1 projects) and use the slice, section, and layout tools to present it. You then create vports to view it from multiple angles. Although this is one exercise, it is broken down into multiple parts. (Difficulty level: Advanced; Time to completion: 60−90 minutes.)

   Step 1. In 2D, create the following shape. It is a half-cylinder with a piston inside and a spring. The helix for the spring is based on the circle created via the TTR option.

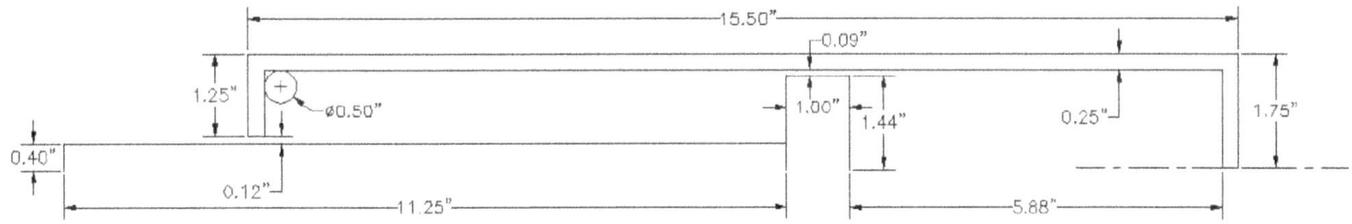

   Step 2. Revolve the inner parts around the centerline, seen in Step 1, to create the piston and use the helix command to create the spring. Adjust the diameter and number of turns of the helix to fit the appropriate area, as shown. Revolve the outer shell to create the cylinder. Shown next are the individual parts, unshaded and shaded, as well as the full assembly. Of course, you do not see the inside structure once you complete Step 2.

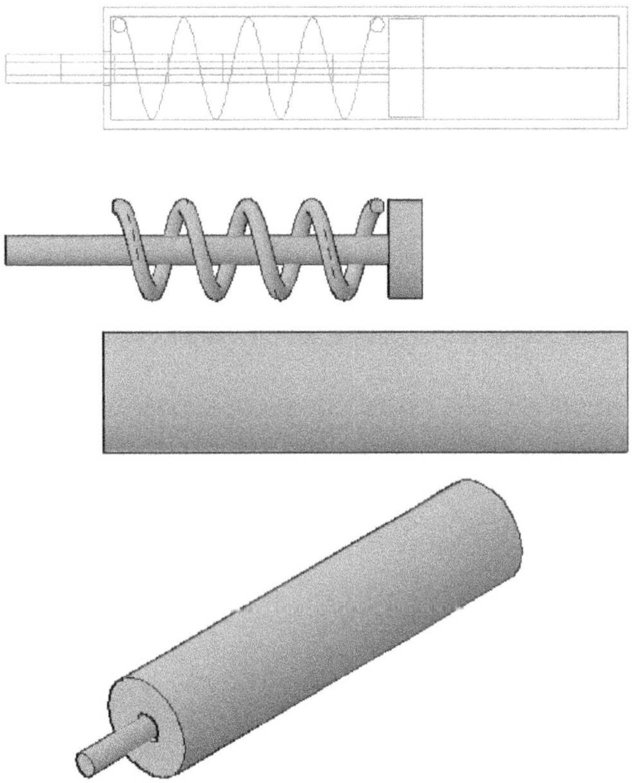

   Step 3. Slice the view lengthwise as seen next. Use the YZ axis as the cutting plane.

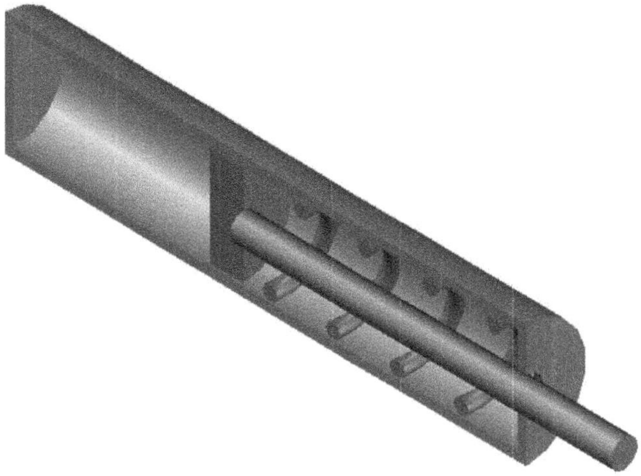

Step 4. Section the view lengthwise, as seen next. Use the YZ axis as the cutting plane.

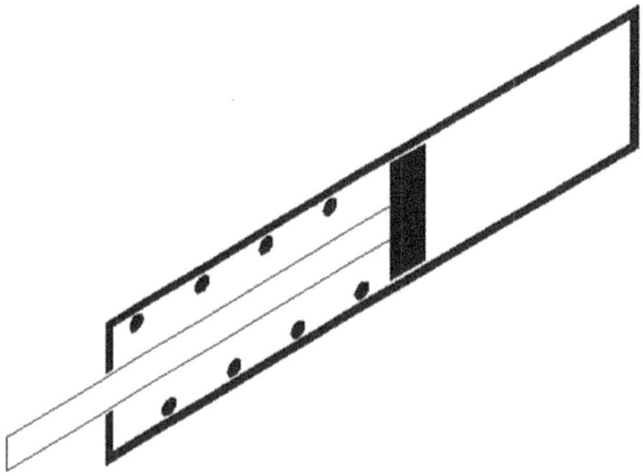

Step 5. Create a layout according to the steps of Section 27.3. A suggested layout is shown next, but you can improvise.

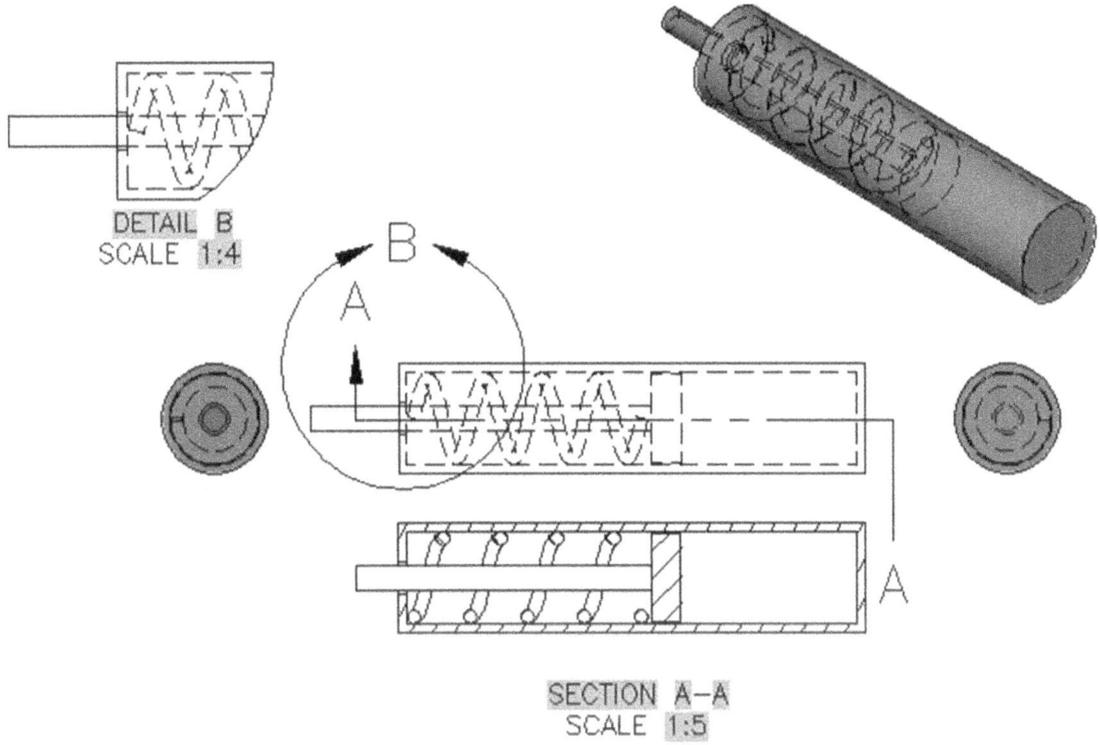

Step 6. As a final step, create multiple view of the design using the Vports command, as seen next.

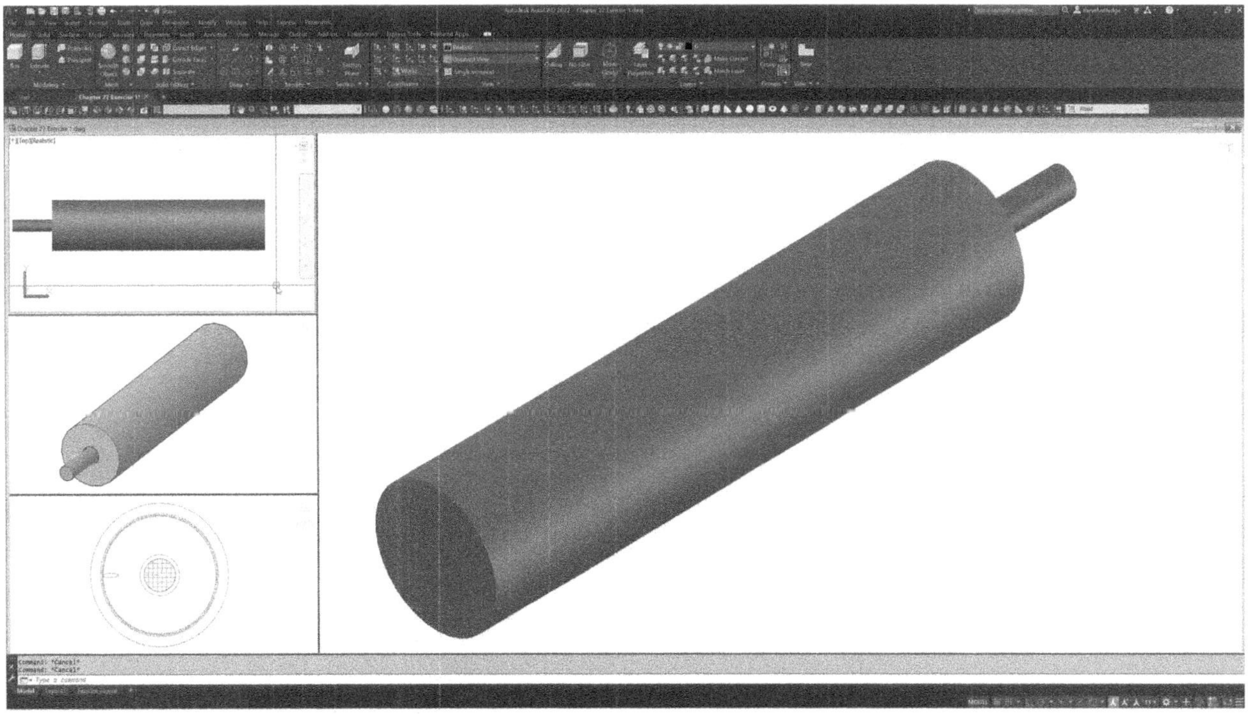

Chapter 28

Advanced UCS, Views, Text, and Dimensions in 3D

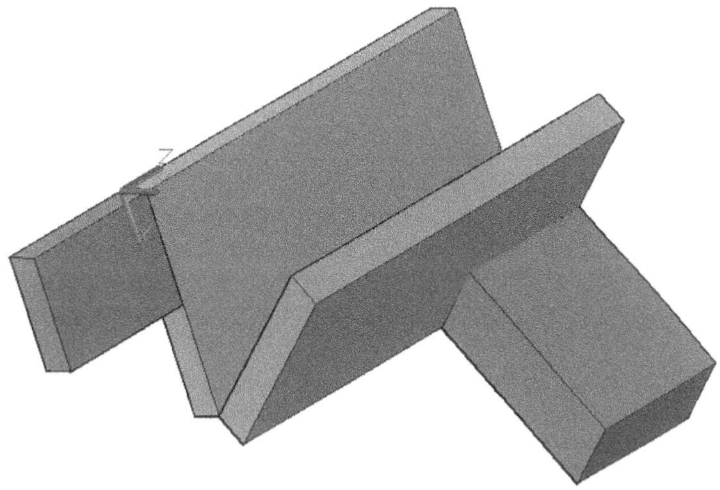

Learning Objectives

In this chapter, we complete the design toolbox and expand on your ability to use surfaces of objects to create new shapes. We cover

- Advanced UCS
- UCS
- World
- UCS previous
- Face UCS
- Object
- View
- Origin
- Z-axis vector
- 3 Point
- X, Y, Z
- Apply
- Named UCS
- Views
- Text and dimensioning in 3D

By the end of this chapter, you will be able to create designs regardless of the surface of orientation.
Estimated time for completion of this chapter: 1 hour.

Up and Running with AutoCAD 2022. https://doi.org/10.1016/B978-0-323-89923-9.00028-7

28.1 Introduction To Advanced Ucs, Views, Text, And Dimensions In 3d

This chapter begins with some additional tools needed to effectively work in 3D space. Advanced UCS refers to additional functionality with the Universal Coordinate Symbol. This symbol is not static in regard to the surfaces with which you work. You can relocate it to a more convenient plane as you see fit to create new design features more effectively.

Once we cover this new functionality, we do a quick introduction to saving views so you can come back and reuse them. This is important in complex designs, where you do not want to continually "reinvent the wheel" after you find new views to work more effectively. Finally, we discuss using text and dimensions in 3D space. Although this is not necessarily different from 2D space, the orientation may throw you off, and some tips and tricks are needed.

28.2 Advanced Ucs

So far the Universal Coordinate System (UCS) icon has been sitting in the background, acting as a reference point when we need to rotate something about an axis or mirror something about a plane. The icon itself is also rotated occasionally, when we need to draw into another plane. This is all good, but we can make the icon work even harder for us and add some more usefulness to its repertoire. To investigate it further, bring up UCS's two toolbars: UCS and UCS II, as seen in Fig. 28.1. We cover these additional icons (and of course their typed, cascading menu, and Ribbon counterparts).

Most of the new commands have something to do with aligning the UCS icon with chosen geometry. This is the central theme in learning UCS's new features. Why is such alignment important? Because this gives you a new "platform" off which to work. Suppose you have an object that is leaning in some direction and you need to draw something else that sits on it. You can draw this new object and rotate it into position, but a better approach is to right away align yourself with the tilted surface and use that as a base from which to draw. All this should tie in with the fundamental theory presented way back in Chapter 21 concerning drawing planes and their rotation. Let us jump right in, and you will clearly see what we are talking about very soon.

First of all, though, we need something on which to practice. The various UCS functions are best understood by just trying them out, so we use the shape in Fig. 28.2 for all UCS demonstrations. Draw a rectangle of any size and extrude it, then rotate3D it some value (perhaps 20°) and mirror3D it for symmetry. The result in the SW Isometric view is shown in Fig. 28.2.

UCS

This first tool presents many options but is primarily used for aligning the UCS via several clicks of the mouse.

Keyboard: Type in ucs and press Enter
Cascading menus: Tools→New UCS→Origin
Toolbar icon: UCS toolbar→
Ribbon: Home tab→Coordinates

Step 1. Start the UCS command via any of the preceding methods.

- AutoCAD says: `Specify origin of UCS or [Face/NAmed/OBject/Previous/View/World/X/Y/Z/ZAxis] <World>:`

FIGURE 28.1 UCS and UCS II toolbars.

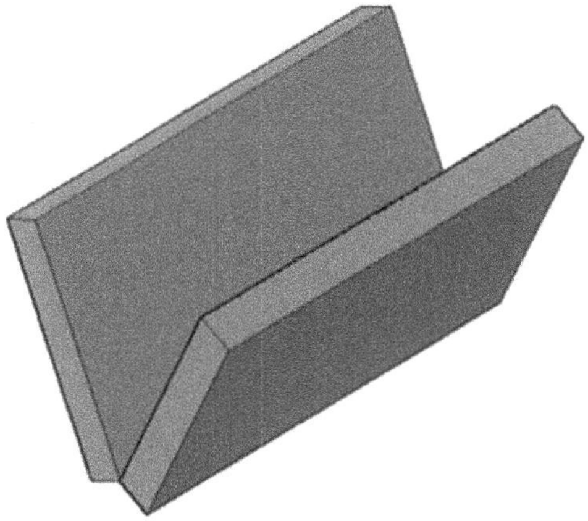

FIGURE 28.2 Basic shapes for UCS practice.

Step 2. Click somewhere on the object to locate the origin (always using OSNAPs).

- AutoCAD says: `Specify point on X-axis or <Accept>:`

Step 3. Here, you can just press Enter, but if you click on another point, the positive X axis of the UCS icon passes through that point.

- AutoCAD says: `Specify point on the XY plane or <Accept>:`

Step 4. Here, you can again press Enter, but if you click on a third and final point somewhere on the object the Y axis passes through it.

Notice how the UCS is now aligned to the points you indicated. We revisit this a bit later with the `Object` and `3 Point` options. The basic UCS command can also be used to rotate the UCS around any of the axes, although there are actually dedicated icons, as we explore soon.

World

This next tool should be a familiar one, as it "resets" the UCS icon and was mentioned in earlier chapters. The result is that the UCS is returned to the World view from whatever configuration it may have been in previously. Try it out via the following methods.

Keyboard: Type in **ucs**, press Enter, then Enter again
Cascading menus: Tools→New UCS→World
Toolbar icon: UCS toolbar→
Ribbon: Home tab→Coordinates

UCS Previous

This next tool simply returns you to whatever UCS configuration you had previously. It is quite useful if you are using just two settings and need to toggle quickly back and forth. Try it out via the following methods.

Keyboard: Type in **ucs**, press Enter, then **p** and Enter
Cascading menus: Tools→New UCS→Previous
Toolbar icon: UCS toolbar→
Ribbon: Home tab→Coordinates

Face UCS

This next tool is one of two very important alignment tools (the other being `3 Point`) that are the core of this section on advanced UCS. Here, you can align your UCS to any face on the screen as long as it is flat. The major advantage here, of course, is that you do not need to know the angle of this face.

Keyboard: Type in **ucs**, press Enter, then **f** and Enter
Cascading menus: Tools→New UCS→Face
Toolbar icon: UCS toolbar→
Ribbon: Home tab→Coordinates

Step 1. Start the command via any of the preceding methods.

- AutoCAD says:

```
Current ucs name: *NO NAME*
Specify origin of UCS or [Face/NAmed/OBject/Previous/View/World/X/Y/Z/ZAxis]<World>: _fa
Select face of solid object:
```

Step 2. Click on any face to which you need to align, and the face is highlighted.

- AutoCAD says: `Enter an option [Next/Xflip/Yflip] <accept>:`

You can easily change the direction of the UCS icon while keeping the 0,0,0 origin in one spot by using the Xflip, Yflip, or Next options before pressing Enter to accept.

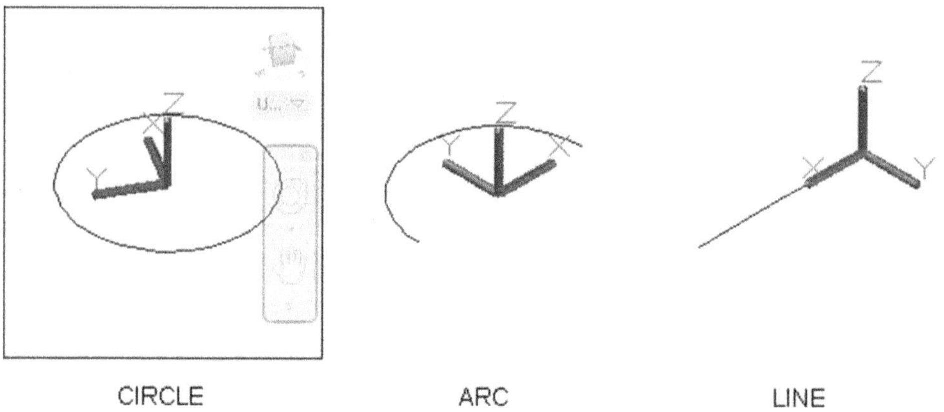

FIGURE 28.3 Object UCS alignment.

Object

This next tool is an interesting variation on Face UCS. Here, you can select almost any object and the UCS icon aligns itself to that object in various preset ways. With an arc or circle, the alignment is to the center; with a line, it is a point closest to where you clicked; and so forth.

Keyboard: Type in **ucs**, press Enter, then **ob** and Enter
Cascading menus: **Tools→New UCS→Object**
Toolbar icon: UCS toolbar→
Ribbon: **Home tab→Coordinates**

A full listing of how it reacts is in the Help files. In Fig. 28.3, we see the UCS icon aligned to a few of these objects.

View

This next UCS tool establishes a new coordinate system with the XY plane perpendicular to your viewing direction (parallel to your screen).

Keyboard: Type in **ucs**, press Enter, then **v** and Enter
Cascading menus: **Tools→New UCS→View**
Toolbar icon: UCS toolbar→
Ribbon: **Home tab→Coordinates**

This is equivalent to going to the Top View while still in 3D, as seen in the before and after screen shots of Fig. 28.4. This technique probably will not be used too often but is the only way to accomplish this if needed.

Origin

This next tool establishes a new origin for the UCS icon.

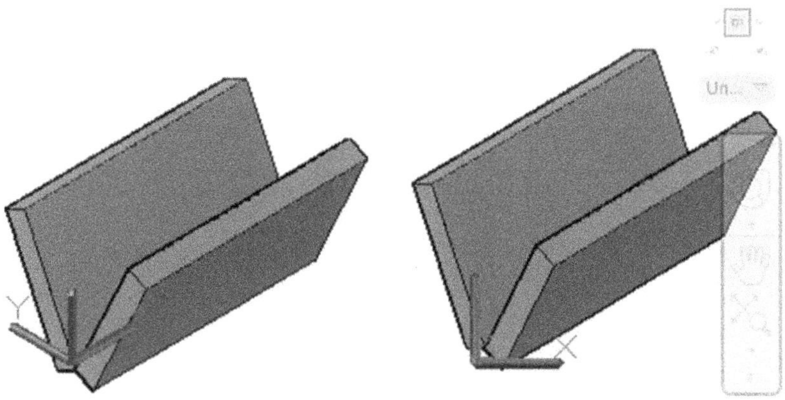

FIGURE 28.4 View UCS.

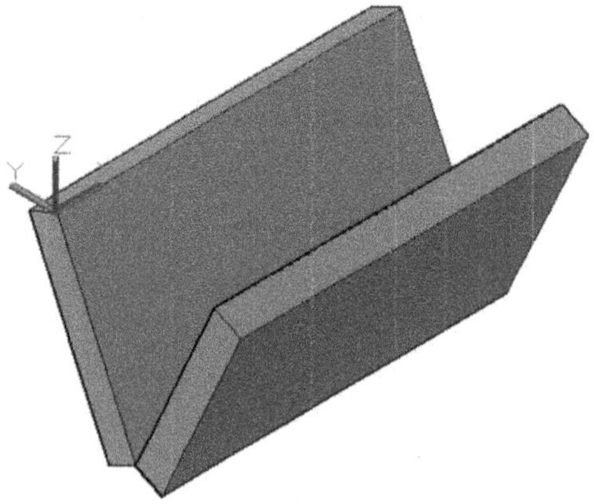

FIGURE 28.5 Origin UCS.

Keyboard: Type in **ucs**, press Enter, then **o** and Enter
Cascading menus: Tools→New UCS→Origin
Toolbar icon: UCS toolbar→
Ribbon: Home tab→Coordinates

Simply start the command and click anywhere on or near the object; the UCS moves its origin there, as seen in Fig. 28.5. Be sure that the UCS icon Origin option is on, or the icon does not move anywhere.

Z-Axis Vector

The next tool simply points the Z axis in a specified direction and is just another way to orient the UCS icon.

Keyboard: Type in **ucs**, press Enter, then **za** and Enter
Cascading menus: Tools→New UCS→Z Axis Vector
Toolbar icon: UCS toolbar→
Ribbon: Home tab→Coordinates

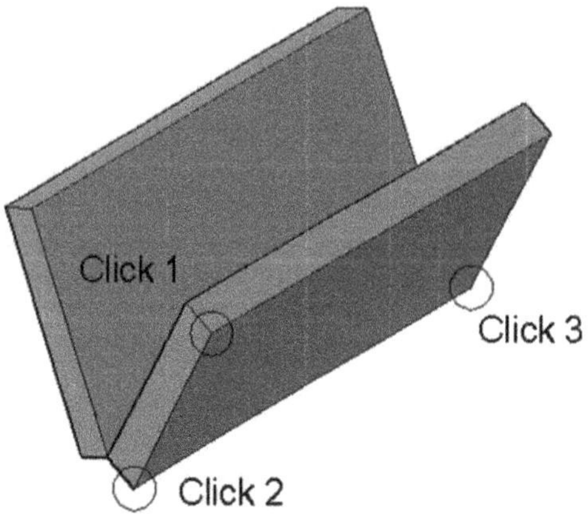

FIGURE 28.6 Refining 3-Point UCS.

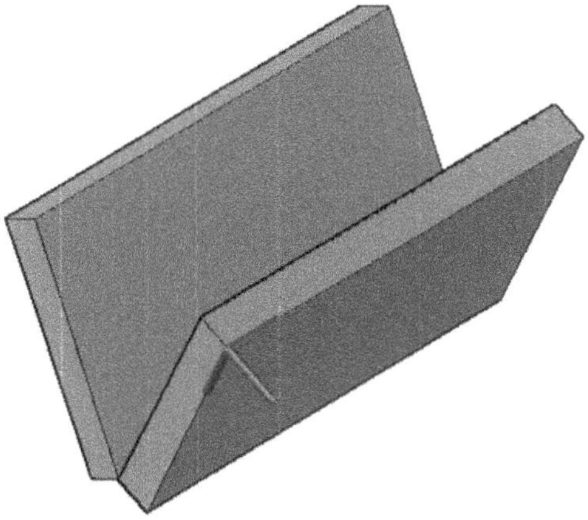

FIGURE 28.7 3-Point UCS.

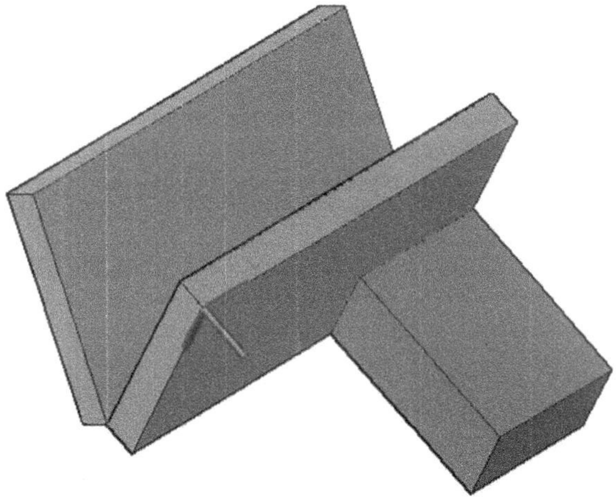

FIGURE 28.8 3-Point UCS, extrusion 1.

3 Point

This tool is the other extremely useful orientation tool (along with Face UCS). It works on the concept of three points being used to indicate a plane. We present a full demonstration of its use. This demonstration also is relevant to the Face UCS method, as it also orients the UCS to a plane, although in a different way.

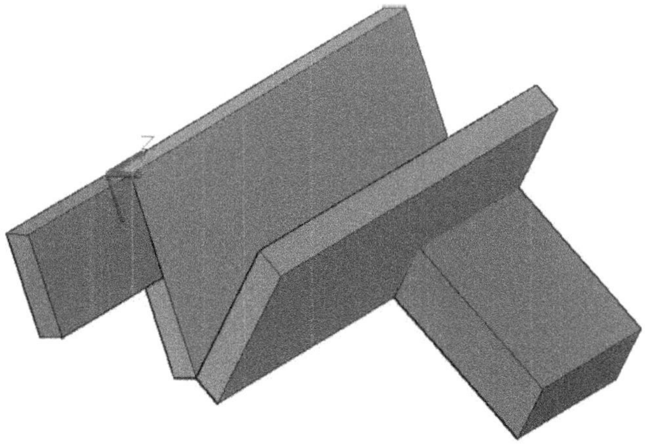

FIGURE 28.9 3-Point UCS, extrusion 2.

Keyboard: Type in **ucs**, press Enter, then **3p** and Enter
Cascading menus: **Tools→New UCS→3 Point**
Toolbar icon: **UCS** toolbar→
Ribbon: **Home** tab→**Coordinates**

Step 1. Start the command via any of the preceding methods.

- AutoCAD says:
  `Specify origin of UCS or`
  `[Face/NAmed/OBject/Previous/View/World/X/Y/Z/ZAxis] <World>: _3`
  `Specify new origin point <0,0,0>:`

Step 2. Select the three points in consecutive order to define a plane, as seen in Fig. 28.6.

- With each click, AutoCAD says:
  `Specify point on positive portion of X-axis <1.00,0.00,0.00>:`
  `Specify point on positive-Y portion of the UCS XY plane 0.00,1.00,0.00>`

When done, the crosshairs and the UCS icon align themselves with the face defined by the three clicks, as seen in Fig. 28.7.

You can now easily draw something based on that surface, as seen in Fig. 28.8. Performing the same 3-Point alignment and extrusion on the front surface, we have what is shown in Fig. 28.9.

Note that, as mentioned before, Face UCS works the same way, except that you are just selecting the entire face, not defining it as with 3 Point. But, once you establish the new UCS orientation, proceed in the same manner to create new geometry.

X, Y, and Z

These tools are simply icons for the familiar UCS rotation you learned about in the first few chapters of Level 3. Simply press the icons instead of typing in the UCS planes, then enter the angle for the rotation. That is all you need to do, and the UCS rotates accordingly.

Keyboard: Type in **ucs**, press Enter, then **x**, **y** or **z** and Enter
Cascading menus: **Tools→New UCS→X, Y or Z**
Toolbar icon: **UCS** toolbar→
Ribbon: **Home** tab→**Coordinates**

Taking a look next at the UCS II toolbar, we examine two icons, the Named UCS and the drop-down window next to it. The first icon (UCS) is the same as the one on the previous main UCS toolbar, and we do not go over it again.

Named UCS

If you are going to spend some time with a UCS configured in a certain useful way, you may want to save it in case you need it again. AutoCAD allows you to save as many custom UCS presets as you want. To save a view, do the following:

Keyboard: Type in **ucsman** and press Enter
Cascading menus: **Tools→Named UCS...**
Toolbar icon: **UCS II** toolbar→
Ribbon: **Home** tab→**Coordinates**

Step 1. Type in UCS and press Enter.

- AutoCAD says: `Specify origin of UCS or [Face/NAmed/OBject/Previous/View/World/X/Y/Z/ZAxis]<World>:`

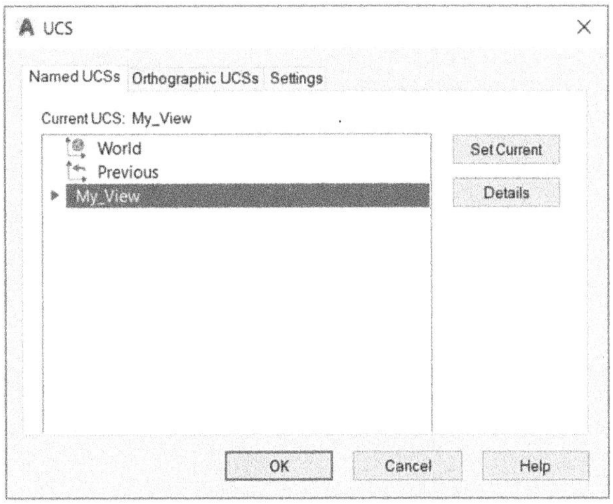

FIGURE 28.10 Named UCS.

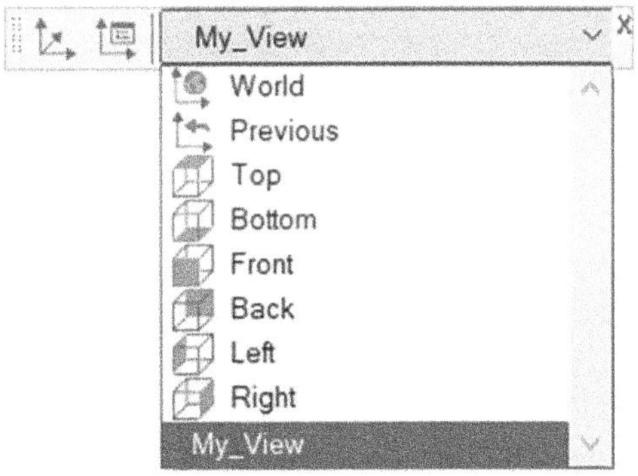

FIGURE 28.11 Named UCS toolbar drop-down menu.

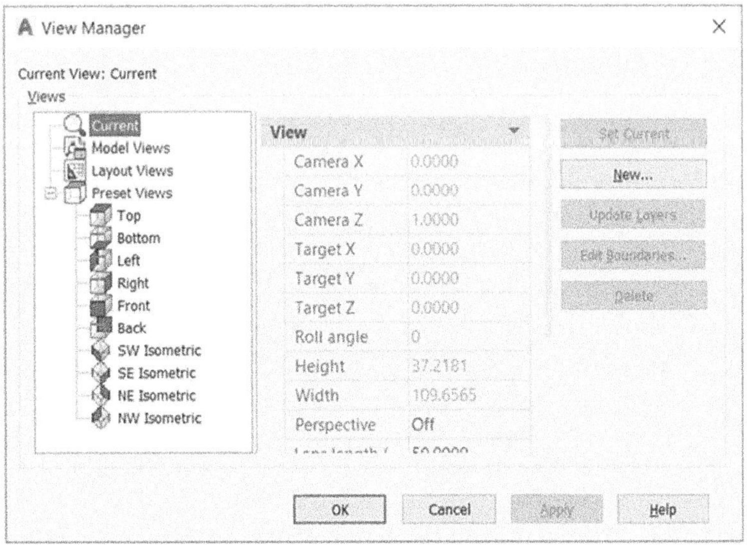

FIGURE 28.12 View Manager.

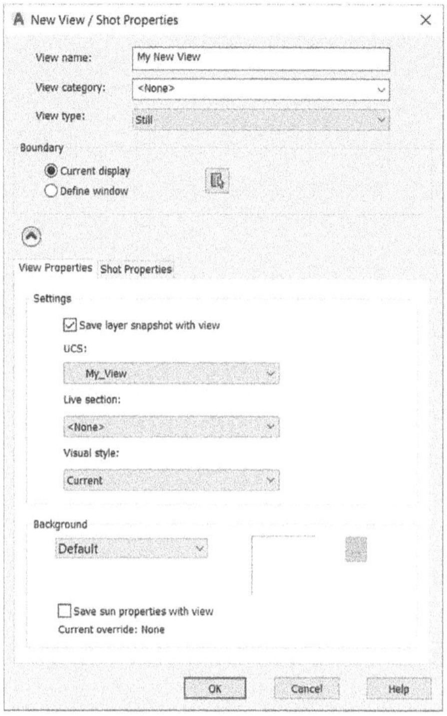

FIGURE 28.13 New View/Shot Properties.

Step 2. Type in `save` and press Enter. The Save option does not appear in the preceding menu, but it works.

- AutoCAD says: `Enter name to save current UCS or [?]`

Step 3. Enter a name for your view, which in this case is just a generic MyView, and press Enter.

The new view is now in AutoCAD's memory, saved for future use. It can be found in the UCS dialog box, which is described next.

The dialog box (Fig. 28.10) can be called up via any of the preceding methods. Notice the presence of the highlighted My_View UCS setting. You can go to other settings and always go back to it later.

The UCS setting also appears in the toolbar drop-down menu, as seen in Fig. 28.11. This drop-down menu is a convenient way to quickly access not only your custom UCS settings but also the preset ones, including World and Previous.

28.3 Views And View Manager

In a closely related topic to saving preset UCS views, you can also save general views of your design. This is a very convenient tool if you find a good view with which to work and may want to come back to later. The views are saved and managed through the View Manager dialog box. Access the command through any of the following methods:

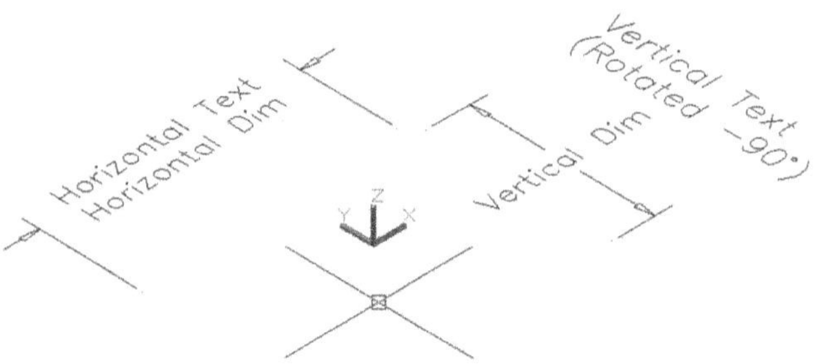

FIGURE 28.14 3D dimensions and text, Example 1.

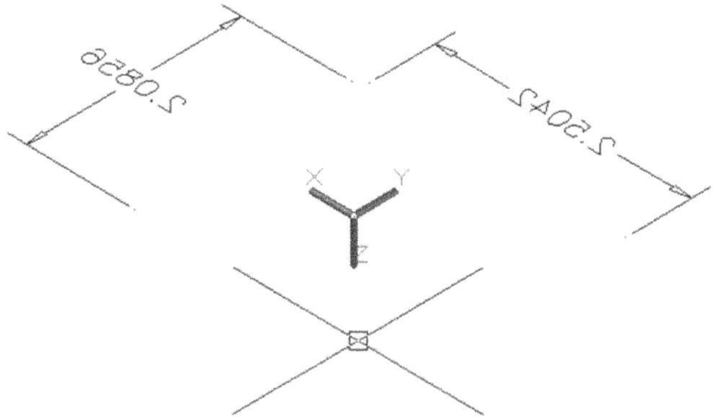

FIGURE 28.15 3D dimensions and text, Example 2.

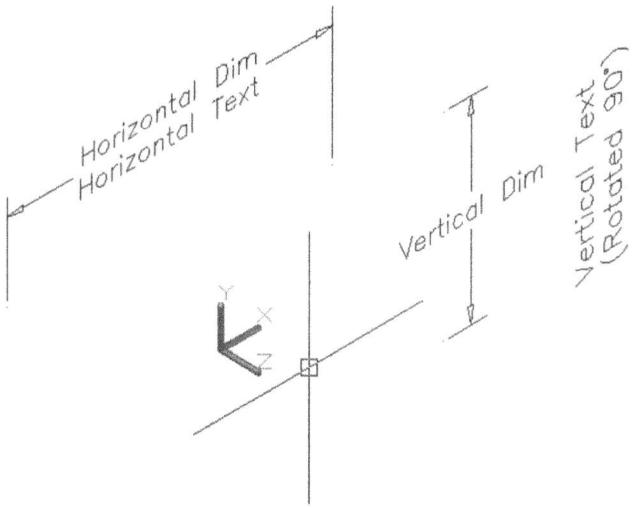

FIGURE 28.16 3D dimensions and text, Example 3.

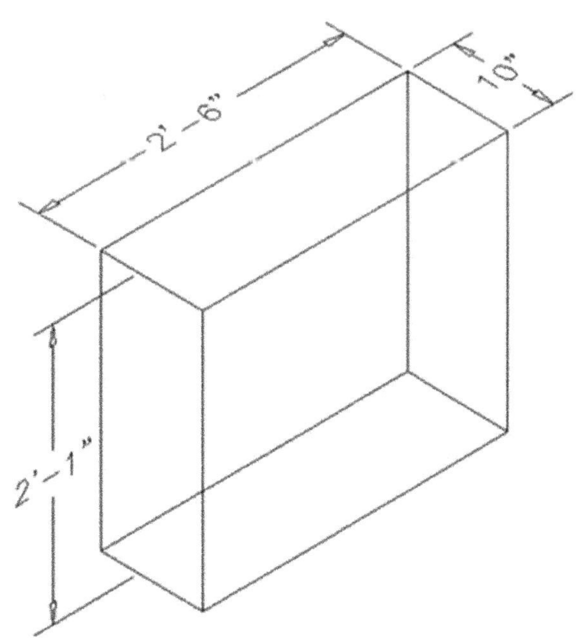

FIGURE 28.17 3D dimensioned cube.

Keyboard: Type in view and press Enter
Cascading menus: View→Named Views...
Toolbar icon: View toolbar→
Ribbon: View tab→Views

You then see the dialog box in Fig. 28.12. Press the New... button and the New View/Shot Properties dialog box appears, as seen in Fig. 28.13. We are not concerned with the Shot Properties tab for now, so just focus on the View Properties tab. Enter a descriptive view name in the first field and press OK. The name then appears in the View Manager. To retrieve the view, just bring up the View Manager again and double-click on the view you want retrieved.

28.4 Text And Dimensions In 3d

Creating text and dimensions in 3D is not necessarily any different than in 2D; you just use the standard text, mtext, or ddim commands. However, placing them correctly takes some practice because, in 3D, what is up or down is not always clear. You have to pay very close attention to what plane you are on and which way is "up"; otherwise, the dimensions and text appear in the wrong place or the text is inverted. Let us run through a few examples. In each of these cases, pay close attention to everything shown, such as the position of the UCS icon (and the associated crosshairs), as well as the positioning of the dimension and text.

The easiest case is to use the standard World UCS view (SW Isometric) that you are accustomed to in this course. If the Z axis is pointed up, then the dimensions and text come out as shown in Fig. 28.14.

Now, what would happen if we reversed the orientation of the Z axis and it faced down instead? As Fig. 28.15 shows, if we attempt to then add dimensions and text (horizontal or vertical), we are faced with a complete reversal and mirror imaging of the text and dims, which is not the desired outcome.

As this example shows, you have to be very careful about flipping the UCS icon around when dimensioning, as the results may be unpredictable (or at least seem that way). What happens in this example is that you are writing on the other side of the face. So, how do we add dimensions along the Z axis? The next example addresses this.

Let us bring the Z axis back to World view and rotate the UCS icon 90° along the X axis, as seen in Fig. 28.16. Notice the effect this has on horizontal and vertical dimensions.

With these two UCS settings you should be able to dimension just about anything, as the dimensioned cube in Fig. 28.17 indicates, although do not be surprised to run into some difficulties if you are dealing with complex rotated surfaces.

Summary

You should understand and know how to use the following concepts and commands before moving on to Chapter 29:

- Advanced UCS
- UCS
- World
- UCS previous
- Face UCS
- Object
- View
- Origin
- Z-axis vector
- 3 point
- X, Y, Z
- Apply
- Named UCS
- View

- Text and dimensioning in 3D

Review Questions

Answer the following based on what you learned in this chapter:

1. List all 13 advanced UCS functions and what they do.
2. What are views useful for?

Exercise

1. Create the following fictional 3D suspension and tire model. A breakdown of the parts is shown for clarity, along with hints, but you still need to know the details of creating each part. Some sizing is given, but the focus is on the commands. Use layers. (Difficulty level: Advanced; Time to completion: 1–2 hours.)
   Step 1. Create the tire and rim via plines using the basic sizing as shown. Revolve both around the given centerline 360° for the final shapes shown.

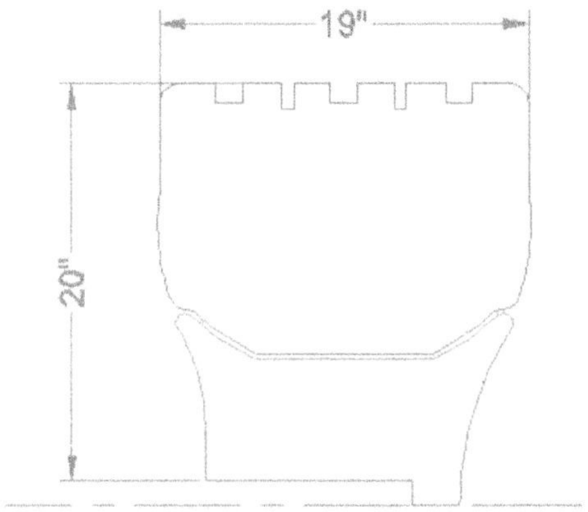

   Step 2. The strut is two sets of extruded circles. The coil spring is a helix and was demonstrated in previous chapters. You can choose which method you want to use. The axle is a 50" long, 3" diameter extruded cylinder.

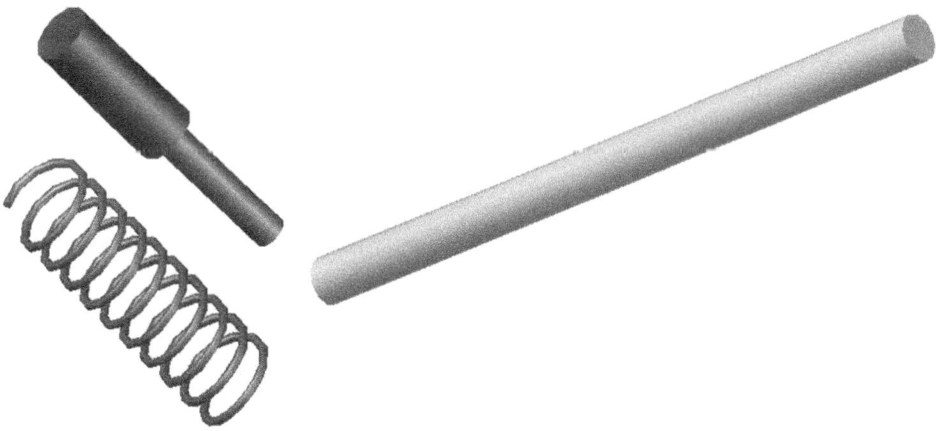

   Step 3. The control arms are sweep or path extrusions, your choice. The cylinders are extrusions. For the axle boot, you need to use revolve. The back panel is a 28"×13"×2" extrusion.

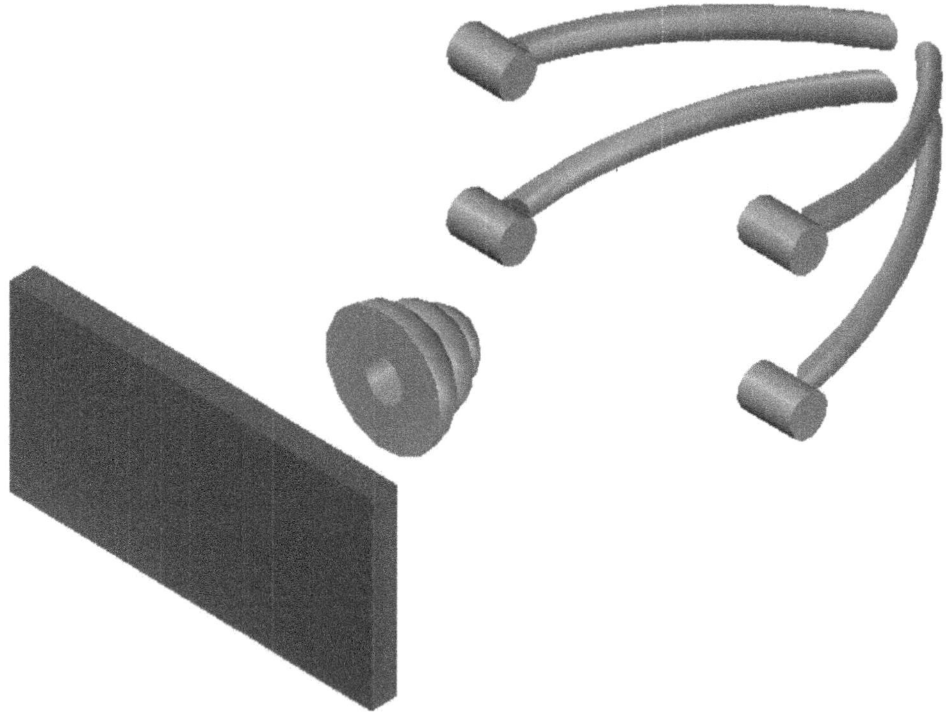

Step 4. The disk brake requires extrusion, subtraction, 3Darray, and other commands. It has a 10" diameter. Size the rest from that. The panel is about 12"×8"×1".

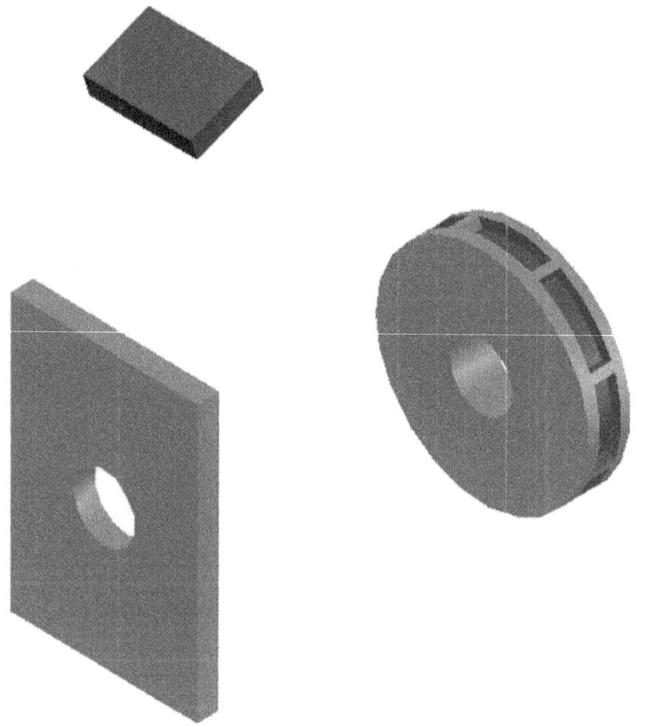

Step 5. The final assembly is shown next. Note that you need to carefully 3Drotate the strut and spring into position, as they are tilted off axis.

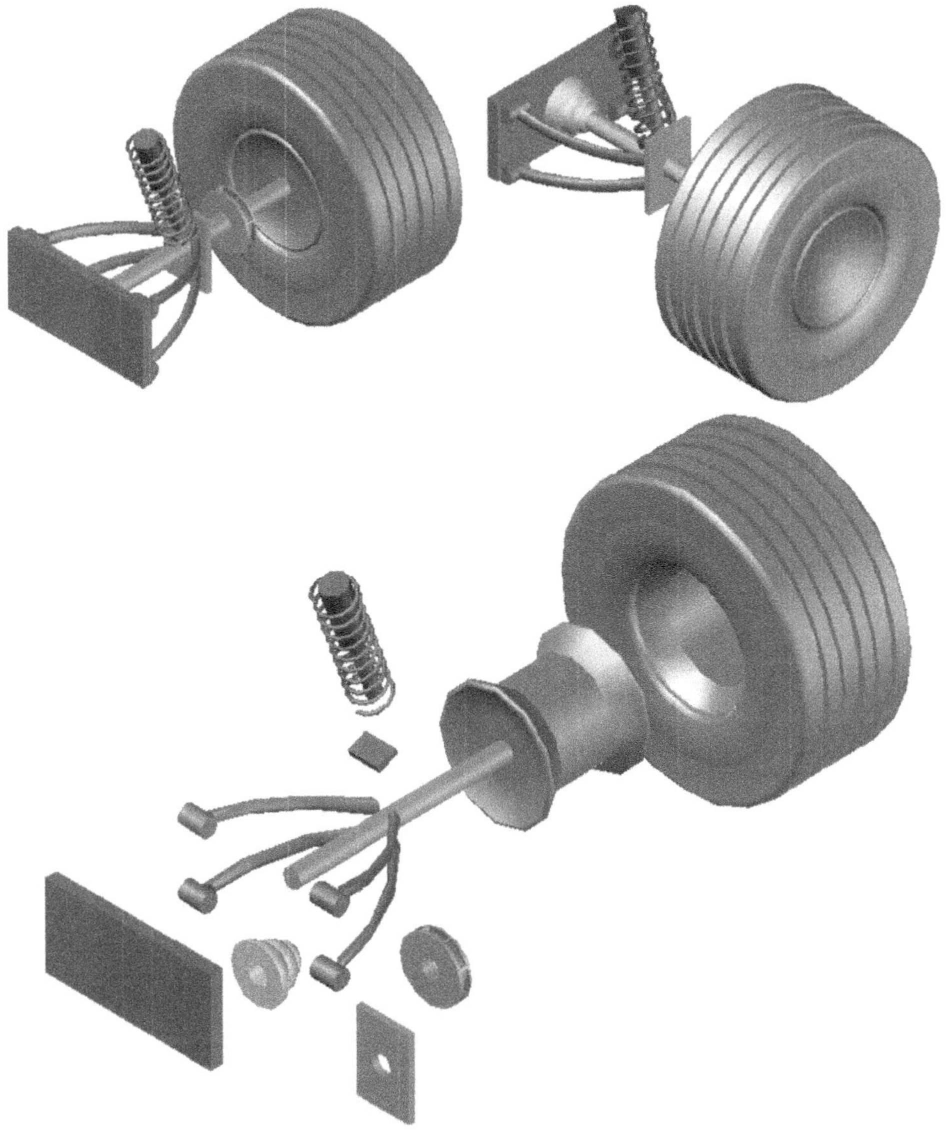

Spotlight on: drafting, CAD management, teaching, and consulting

Overview

Our final Spotlight On focuses on drafting, CAD management, teaching, and consulting. The first two of these topics may seem like rather odd choices; after all, haven't you been studying them throughout the book? While it is true that you have been learning how to draft and may have even read the appendix dedicated to CAD management (see Appendix E), we really have not discussed what it is like to actually have a career working as a drafter or CAD manager. While all the engineering, architecture, and design professions we profiled earlier are excellent career paths, many students using this textbook began learning AutoCAD to be a drafter, and this is what we focus on. We also look beyond doing drafting to teaching drafting, either in a classroom or consultant capacity.

All these topics are very personal to me. While I am working as an engineer right now, I started off my career in the mid-1990s as an AutoCAD drafter and kept at it for 10 years full time (actually double time, you could say, considering the long days I worked). It was not until later in life I returned to engineering and away from pure drafting. Although today I still use AutoCAD regularly and still teach, write, and consult, I will never forget the decade of AutoCAD work, in the trenches of New York City. Drafting, CAD management, consulting, teaching, and corporate training was my life—14 hours per day, 6 days per week. Although one could say that the "Golden Age" of AutoCAD has somewhat passed, with enough qualified candidates finally filling the unending industry demand, it still can be a rewarding and well-paid career.

FIGURE 1

AutoCAD training requirements

The educational requirements for an AutoCAD drafting career are more rigorous now than at the peak of demand when I got started. Although I had minimal experience back then, I was hired anyway, and the company trained me in-house over a grueling 6 month period; AutoCAD boot camp you could say. Now, 20 years later, this is much less common. You need to attend one of the many community college or tech school programs to get either a certificate or a full diploma in drafting. This can take anywhere from a few months (basic certificate) all the way up to 2 years (for an associate degree). Either way, you graduate "ready to go" and can be productive at a company right away.

If necessary, take both exams offered by Autodesk: User and Professional (see Appendix K) for a good "foot in the door" to a company. However, I consider these exams inadequate in measuring the overall knowledge of an individual. They just are not comprehensive enough. The proof of your skill remains in the ability to draft quickly and efficiently, and many companies will set you in front of a CAD station to draft a given sample. I was on both sides of that fence, doing a few of those and later on devising tests for new hires. Being fast and efficient is still the best way to stand out from the pack. I cannot say it enough—practice your basics!

Architectural drafting

So, what is it really like once you get hired as a draftsperson? This, of course, depends on where you go. You can draw a line between the world of architecture and engineering. They are different and have their own pros and cons. On your first day on the job, in either profession, you will be shown company standards and given some time to customize your CAD station. Be sure to get AutoCAD looking like you want it. Gone are the days when you shared workstations. You will be on your own computer, so be sure to customize anything that boosts your productivity (pgp file, toolbars, etc.).

Be sure to take note of the printer/plotter configurations, and what is available. Be sure to also identify the lead CAD designer or the one most knowledgeable in AutoCAD. Not only are you guaranteed to get stuck on some

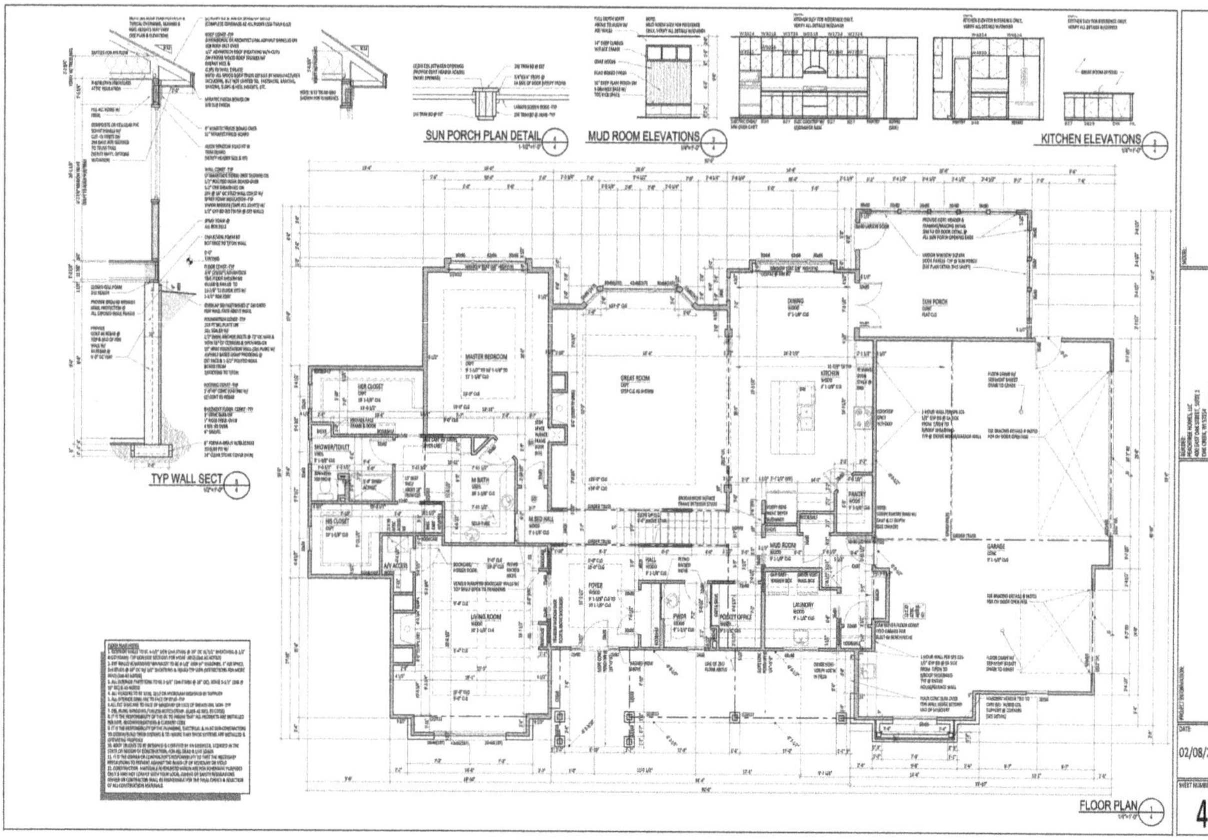

FIGURE 2 *Image source: Robert C. Kaebisch.*

features of the software but also you need to know exactly what standards are to be followed—not all of them are enforced, and there may be a reason why. You also need to learn all the network paths and where to retrieve and save drawings.

Day-to-day work at an architecture office revolves around whichever project is assigned to you. In my experience, it is more stressful in architecture, given the deadlines inherent in that profession. Engineering offices in larger companies (such as Siemens and IBM, two companies with which I spent significant time) operate on larger budgets and longer timeframes, giving you a chance to develop additional skills for the job. Architects bill per hour and you need to be on your game from day one. Expect to see Paper Space and xrefs as a matter of routine. Expect also (in some offices anyway) an emphasis on 3D. More and more architects create computer renderings for clients, often using additional software such as Rhino, SketchUp, and Revit. By and large, however, regular 2D floor plans will be your bread and butter. Most architects I worked with used redline markups, basically red ink on a paper printout; and you will quickly learn the meaning of the phrase "they bled on it." You will take those changes and execute the actual drafting, then print a new sheet for more markups or a final check.

As you gain experience, the architect may allow you to lay out the design or changes to the design, especially if they are routine, then just approve what you do. This is a great way to get to know the profession in case you do want to become an architect. Although I was never interested in an architecture career, somewhere out there exist entire floors in buildings that I designed. Expect to also be out in the field to do surveys or "as-builts." You show up with a camera and pen and paper and take down information, measurement, and pictures to bring back to the office. Sometimes you can bring AutoCAD with you on a laptop and do the drafting right on the spot. These surveys have gone high tech in recent years with widespread usage of hand-held laser measuring tools.

The money you can earn while drafting for architects is good but not great (except of course at the very top). Drafting time is submitted as part of the overall bid, but generally, the allowed rate is not high by the time everything is negotiated. I left architecture after a few years to move to larger engineering companies, with a resulting boost in my pay. To that end, while I was working in architecture offices, I noticed the high rate of turnover in their drafting staff, with exceptions of course. One huge benefit of this type of drafting is the experience you get on AutoCAD. A sizable architecture project, especially with a 3D component, uses virtually all of AutoCAD's capability. If you have a choice in the matter, start your career in architecture, and then move into engineering drafting. You will gain critical knowledge that will come in very handy if you want to be a CAD manager and for teaching or corporate training.

Engineering drafting

Although this work is not fundamentally different from architecture—you are still using AutoCAD after all—as mentioned, some aspects of the job do change. In my experience, the projects are larger, more complex, and in many ways more diverse. At Siemens, where we worked on the New York City subway's signaling system, the work spanned the mechanical, electrical, and even structural fields. Although my use of AutoCAD was limited to mostly intermediate-level drafting and there was little in the way of 3D modeling, I did have to interact with varying CAD systems and other software. Most of Chapter 16 was written based on that experience. The money also was far better, as you would expect from a multimillion-dollar contract. Keep in mind of course that this was a major firm and engineering work done at a small office is not going to be the same.

Engineering drafting involves the same level of detail as architecture. You may encounter a lot of electrical schematics, cross sections, and use of attributes to hold information. A lot of our layouts of complex communication equipment were simple from a drafting perspective, same lines, circles, and rectangles repeatedly, but complex in its intricacy. When you have thousands of pieces coming together, you must approach things differently. Use of blocks, nested blocks, and symbol libraries are routine. We had to be very careful in how we presented complex information, which brings us to another major point.

Especially in the engineering world, your job is not just drafting but designing a good layout. It is the next step up from just doing redline markups. You must constantly take a step back from your drawing and ask if this will be easy to read by someone who is seeing it for the first time. Use wipeouts to hide what need not be shown (see chapter: Advanced Design and File Management Tools), and use darker outlines for major pieces, reserving lighter lines for the secondary items. Be creative in positioning design elements on a crowded drawing and always ask yourself: Does this make sense? The purpose of drawing is to convey information. If it fails in that regard, you have not done your job as a draftsperson.

CAD management

What is meant by CAD management is simply that you are the top draftsperson in the organization, and it is your job is to not only draft but also keep an eye on standards, resolve any AutoCAD questions and concerns, and perhaps supervise other draftspersons. No special training is needed, only experience. I spent several years as a CAD manager at Siemens, hiring and overseeing several individuals in that time, one of whom became my business partner in my company, Vertical Technologies, a few years later.

The key here, of course, is knowing AutoCAD inside and out and being able to explain to nontechnical managers what the issues are. You also must be able to give reasonable estimates of how long something will take and any technical hurdles on which you need support. My ability to teach AutoCAD clearly to students was partially developed through my experience explaining issues to managers. You would be amazed at how complex and mysterious this software and its abilities can seem to some. As a CAD manager, you will find yourself attending far more meetings than if you were just a draftsperson. Expect to also find yourself teaching others how to do things right.

The final piece of advice for CAD managers is this: Be careful hiring. As described in Appendix E, the best person to hire is the one that is fastest and most accurate on AutoCAD, as demonstrated by having the person do a simple drawing, such as the floor plan from Chapter 3. Experience has shown me that if someone does not know an advanced concept, which can be taught in a few minutes. But only extensive time spent drafting can instill speed and accuracy.

Teaching AutoCAD

Not everyone enjoys or wants to teach. Teaching is a calling more than a job, though teaching a small group of adult professionals is a far cry from a room full of fourth graders, so we will not pretend it is the same thing. However, it is still very rewarding and a lot of fun. You learn even more about AutoCAD through teaching because you really must dissect it for students and troubleshoot some unique problems students get themselves into. I stepped into teaching by accident when, in 1999, a coworker at Liz Claiborne (where we did store planning and design) wanted to go to Europe for the summer after a bad breakup and asked me to fill in for her class at the prestigious Pratt Institute of Design in midtown Manhattan. I do not think she ever came back to the United States, and my temporary teaching gig stretched into many years. I also taught at the New York Institute of Technology and other, private, training centers all over the New York City area.

If you think you may want to get into teaching, the best thing to do is to get as much diverse experience as possible. An instructor who has limited experience, even if he or she knows AutoCAD well, just is not able to speak to all the ways AutoCAD is used in the industry. If, however, you have worked across a wide array of specializations and have the patience to teach, this may be for you. You also must know AutoCAD backward and forward. You must display confidence in your knowledge and confidence while doing public speaking. You will be in front of anywhere from 5 to 20 students at a time.

To get a job teaching AutoCAD you really must just watch for an opening and drop off your resume at potential schools. Most schools and educational centers want job experience, so getting a foot in the door is not easy. It also matters a great deal what type of program you teach. A community college may offer an entire degree in drafting. Then, you have to be a true "educator," (such as Robert Kaebisch, the coauthor of this textbook), possibly with a degree in education, as you would also be responsible for teaching math, science, and maybe technical writing. I always taught in continuing education programs. This is far easier in terms of non-AutoCAD credentials. My students were all adults, and I had to teach only AutoCAD, usually for 8 weeks at a time. I preferred this approach, as I did not really wish to teach math and science. This may be the way to go for you as well, if you want to get into AutoCAD instruction, by seeking out these types of programs. They may be part of a university or college, such as Pratt or NYIT, or private companies such as Netcom, CDM, or RoboTECH, where I taught on and off.

One additional benefit is that the hourly pay is higher in these schools than in a full-time teaching position at a college. This is because you work on "contract" and do not get benefits, resulting in significantly higher pay. On the flip side, you may not work as many hours as a "full timer" would, and because of this may make less in the long run. Because I already had a full-time job in the daytime and purchased my own benefits, this was perfect for me; and I taught in the evenings and often on weekends for many years.

AutoCAD consulting

The final discussion here focuses on AutoCAD consulting. What is meant by *consulting* is any short- or long-term, and usually highly paid, specialized service that you offer a client. With AutoCAD that can be drafting, but generally, you will rarely do that, as drafters can be hired cheaper through placement agencies. Most, if not all, of the consulting that I did refers to corporate training. These were short-term (2- to 5-day) training seminars, also referred to as *boot camps*. They are held at the company's offices and are intense 8- to 9-hour per day classes, meant to educate the engineers or architects in the proper use of AutoCAD, as well as problem resolution and optimizing AutoCAD to their specific needs.

This type of consulting is the pinnacle of AutoCAD use and AutoCAD training. You fly or drive to the corporate offices and immerse yourself in an intense, high-pressure environment, teaching a very demanding group of individuals, who paid a great deal of money to have you there, while at the same time giving them ideas on how to best optimize AutoCAD to their business and resolving thorny issues with their current approach. It is a tough environment, and it took me over 7 years to truly establish myself in this field, with the opening of my consulting firm in 2003. Prior to that, I was sent to these jobs through several private training centers with which I collaborated.

If you wish to get into this type of work, you need to overcome a few hurdles. You must have absolute mastery of every aspect of AutoCAD, followed by a wide-ranging knowledge of many industries. You could be teaching structural engineers 1 week and electrical engineers the following week. I once even had the interesting experience of training a squad of FBI special agents assigned to counterterrorism. What did they need with AutoCAD, you ask? They were designing special nonfragmentation concrete barriers to secure a political convention and needed to learn 3D AutoCAD in a hurry. Why didn't they hire a company to create the barriers for them? The work was too sensitive in nature and they needed the product fast; too fast apparently for the lengthy bidding process to find a vendor to design these. The agents designed the barriers themselves (with input from structural engineers) and just had a company manufacture them.

You will also need a solid knowledge of computers, operating systems, and IT networks, as well as have experience in all versions of AutoCAD and LT. You just never know what you are going to get. You must be willing to travel for a week at a time, often on short notice. The best way to get started is to first partner up with an established training center, then branch out on your own, when you get enough experience. A website advertising your services, of course, helps a great deal, as does reputation and performance. Many of my clients were repeat business.

Consulting work like this was very well paid for a long time. Typical fees for 1 day of work ranged from $1200 to $2200 or more. I typically charged on the lower end of that range (if I did not have to fly to the client) because I had no middleman to pay and wanted to be very competitive. Over the years, as more and more AutoCAD experts matured in the market, the prices I could charge dropped significantly. It is now a more a matter of convenience, the companies may have in-house experts, but just don't have the time to pull them off projects to teach newbies or expand the skill sets of intermediate-level coworkers, hence the hiring of outside instructors. It is still a good business to pursue and thought I do not do these gigs as often anymore, I still enjoy teaching, which almost makes up for the lower revenue per job. Well, almost ...

Best of luck to you, in whatever you decide to do!

Chapter 29

Dview, camera, walk and fly, path animation

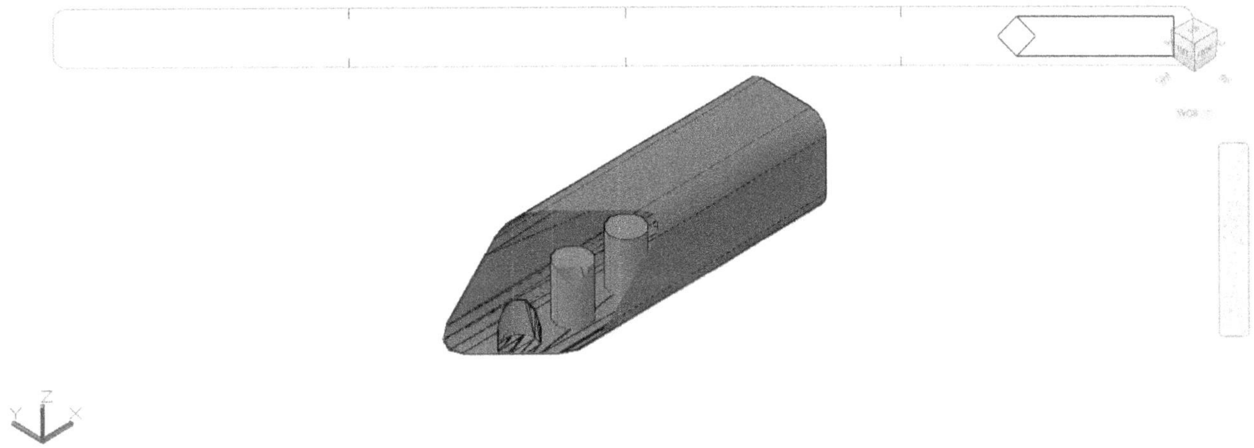

Learning objectives

In this chapter, we look at advanced presentation views and various animation options. The specific topics are as follows:

- Dview
- CAmera
- TArget
- Distance
- POints
- PAn
- Zoom
- TWist
- CLip
- Hide
- Off
- Undo
- Cameras
- Walk and Fly
- Path animation

By the end of this chapter, you will be able to present your model in new and unique ways as well as build basic walk-through animation.

Estimated time for completion of this chapter: 2 hours.

Up and Running with AutoCAD 2022. https://doi.org/10.1016/B978-0-323-89923-9.00029-9

29.1 Dynamic view

Dynamic view, or dview, is just one command but it has many options. Some of the options are less useful or duplicate existing commands, while a few others are very important and worth a closer look. The full list of dview commands includes `CAmera`, `TArget`, `Distance`, `POints`, `PAn`, `Zoom`, `TWist`, `CLip`, `Hide`, `Off`, and `Undo`.

Out of these options, two are in high demand by AutoCAD designers and worth covering in detail. The first is the `Distance` option, which allows you to set perspective; the other is the `Clip` option, which allows you to set clipping planes and strip away parts of the design to look inside. We, of course, cover all of the mentioned options but with a special focus on these two.

Before we try any of these options, you need to get used to something unique to the dview command. In a blank file, go to SW Isometric and turn on shading. Then type in `dview` and press Enter (there is no Ribbon, toolbar, or cascading menu equivalent).

- AutoCAD says: `Select objects or<use DVIEWBLOCK>:`

Here, you are being asked to what you would like to apply dynamic view options. We do not yet have a 3D design, so just press Enter; you see the house in Fig. 29.1.

This little house appears as a point of reference only if there is nothing else on your screen and is perhaps one of AutoCAD's oddest quirks, prompting many a student to exclaim "what the heck is that" in the middle of class. You can see the effect of whatever you do in dview on this house, but immediately after running the option, the house disappears. In the interest of keeping something permanent on the screen, draw a simple 20″×5″ rectangle and extrude it to 5″. Then, add a 1″ fillet to its sides, as seen in Fig. 29.2. Now, we can proceed with learning dview.

Camera

This command works on the principle of a camera rotating around an object. You are asked to enter the camera angle from the XY plane as well as the angle in the XY plane from the X axis. You are also given the option to use the mouse to just click on a suitable position, which is what most users end up doing.

To try it out, type in `dview` and press Enter. Select the block when prompted, and type in `ca` for `CAmera`. Spin your mouse around and observe the results.

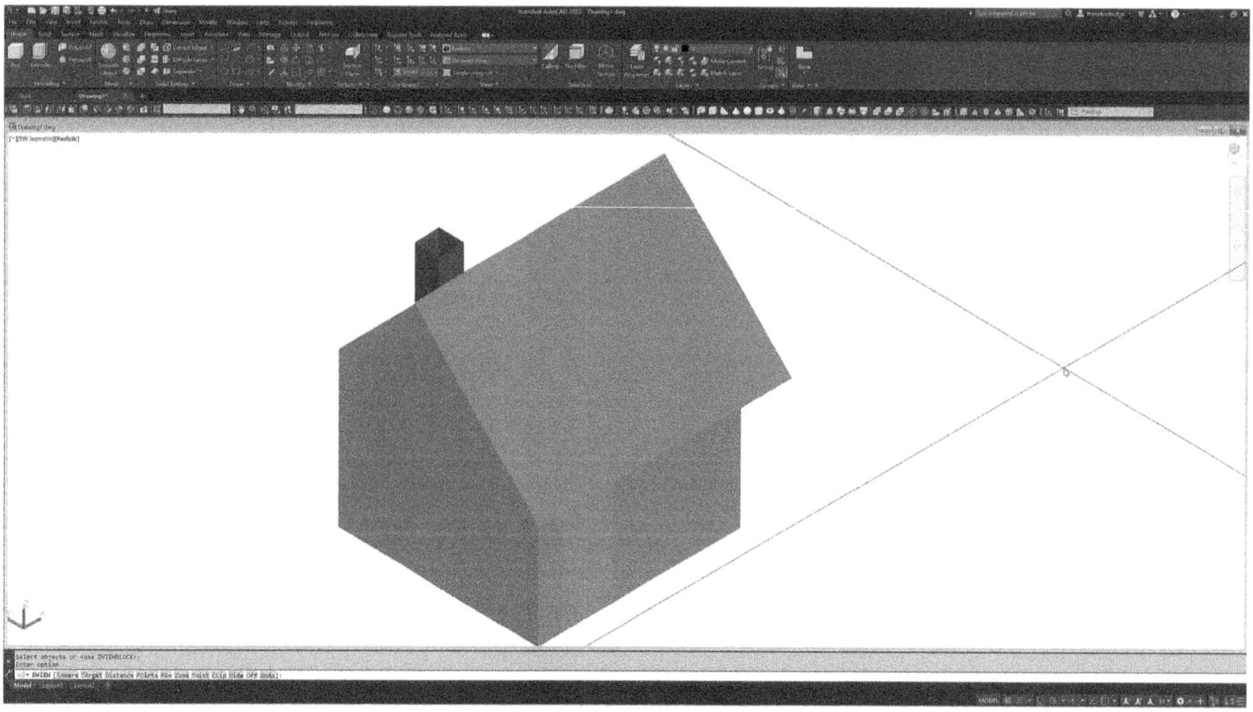

FIGURE 29.1 Dview "house."

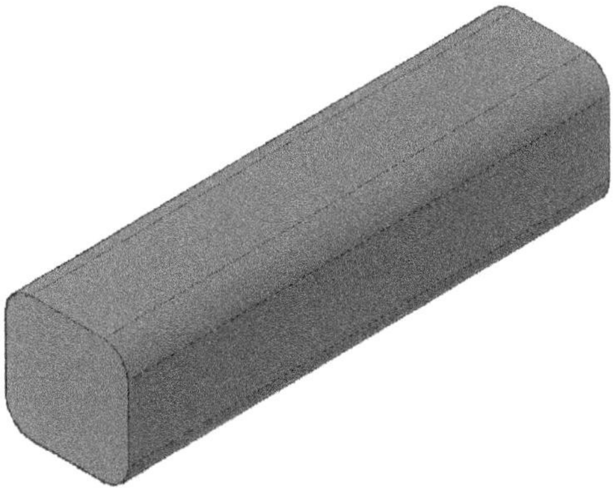

FIGURE 29.2 Extruded shape for dview.

Target

This command is similar to `CAmera`, but works on the principle of the target rotating around the camera. It has the same basic options of entering angle values or mouse clicking. To try it out, type in `dview` and press Enter. Select the block when prompted, and type in `ta` for `TArget`. Spin your mouse around and observe the results.

Distance

We explore this option in more detail. Distance is the method used to set a perspective view. These views of objects are more realistic and feature foreshortened geometry with lines that tend toward a vanishing point, as in reality. For more information on perspective views with detailed descriptions of 0-point, 1-point, 2-point, and 3-point perspectives (AutoCAD uses 3-point), see any graphic design or art textbook. Distance can be applied to any object, but just as in real life, the effect is more pronounced with larger objects.

Also, be aware that this view is only for presentations; it cannot be used during construction of the design, as perspective changes dimension values by its very nature. To try out this option, type in `dview` and press Enter as before. Select the object and press Enter. Then, type in `d` for the `Distance` option. A slider horizontal bar appears at the top of the screen, as seen in Fig. 29.3.

This yellow and red slider is the camera to target distance meter, with 1″ being full size, and increasing to the right (decreasing to the left) of the 1x. Slide over to the right until you get to 9x and click. The object remains in perspective mode along with the UCS icon. Press Enter to accept the settings and get out of dview, being careful not to press Esc at this point. You can also set this view as described in an earlier chapter via the menu of the little house icon at the top right of the screen, as seen in Fig. 29.4.

This perspective view, while not of much use in small part engineering drawings, is quite useful for 3D architectural and civil engineering design presentation. Once in Perspective view, print the drawing and, to get out of this view, change to wireframe.

Points

This option allows you to select a specific view based on the locations of three points. The first is the current camera position, and you need to select the second (target point) and third (camera point) points. This tool has some interesting applications when you need to set a specific view, such as looking down a hallway in a building. You can then set your points in strategic locations, such as one end of the hallway for one point and the top of an opposing door for another point. For the most part, however, designers prefer to just use 3D Orbit and zoom or pan to get the right view.

Pan

This is just a variation on the basic pan command that shifts the image without changing the level of magnification. Give it a try.

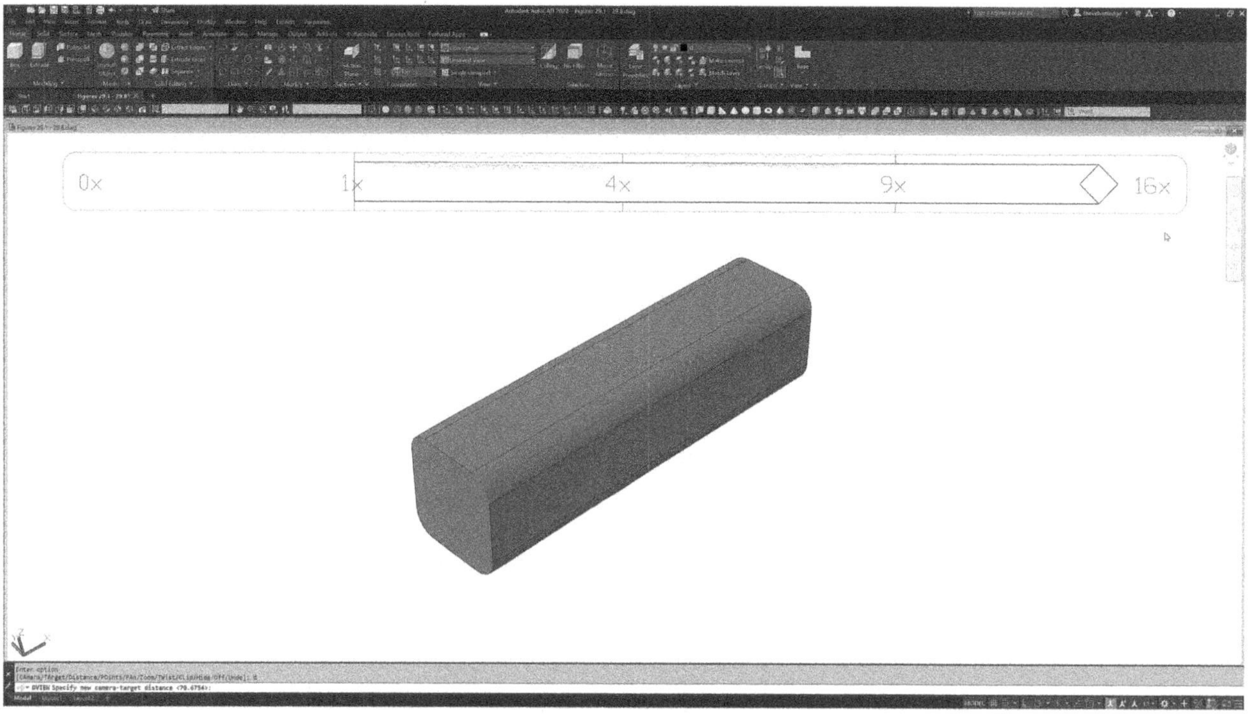

FIGURE 29.3 DVIEW, Distance option.

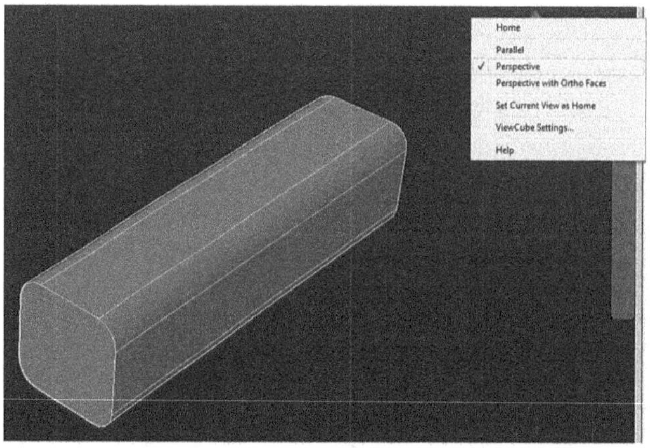

FIGURE 29.4 Perspective view, alternate settings.

Zoom

This is also just a variation on the zoom command that utilizes the same slider you first saw with the distance option. Give it a try.

Twist

This is another variation of the 3D Orbit command, where the object is rotated or "twisted" around. Give it a try.

Clip

This is the second of two dview commands that deserve a longer discussion. Clip introduces the concept of "clipping planes," which (in various forms) is a common tool in 3D computer-aided design. The idea is to create a flat plane that slices open a design to reveal the internal workings. The plane itself is invisible, only its effect is seen, and can be applied to the front or back of the object. When you reveal as much as you need to, you set the view and print the design. This tool, like most other dview options, is not meant to be permanent and is not for use during design work, only for final presentation.

To illustrate the `Clip` tool, add some geometry in the form of five 2″ diameter cylinders to the inside of the previously used block, as seen in wireframe mode in Fig. 29.5. In full shade mode, the inside geometry cannot be seen, as shown in Fig. 29.6.

Let us apply a clipping plane to the front and strip away some of the outside material to show the insides. Type in `dview`, press Enter, and select the object, pressing Enter again. Type in `cl` for `CLip` and press Enter.

- AutoCAD says: `Enter clipping option [Back/Front/Off]<Off>:`

Clipping planes can be thought of as flat plates, with one in the front and one in the back. As the object is moved back and forth, the clipping plane cuts in, or clips, the object either from the front or the back. Typically, you use the front clipping plane, so type in `f` and press Enter.

- AutoCAD says: `Specify distance from target or [set to Eye(camera)/ON/OFF]<451.8572>:`

You also see a yellow slider (similar to the one seen when learning distance), which can be moved back and forth to establish the position of the clipping plane, as seen in Fig. 29.7.

As you can see, the internals of the part can be revealed easily. Fig. 29.8 shows a view with the back clipping plane set in a similar manner, though the results are not as useful. You can actually 3D Orbit through a clipping plane and get a full view of the internal workings of a design from any angle. To remove the planes, you have the off command in the clipping menu.

Hide

This command is essentially the same as the standard hide command learned in Chapter 21.

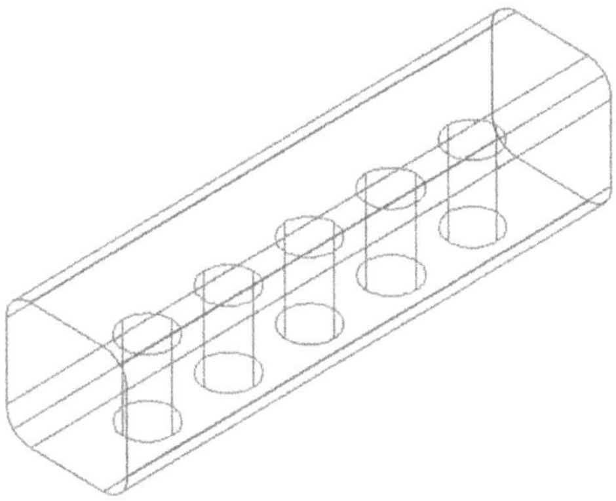

FIGURE 29.5 Internal geometry.

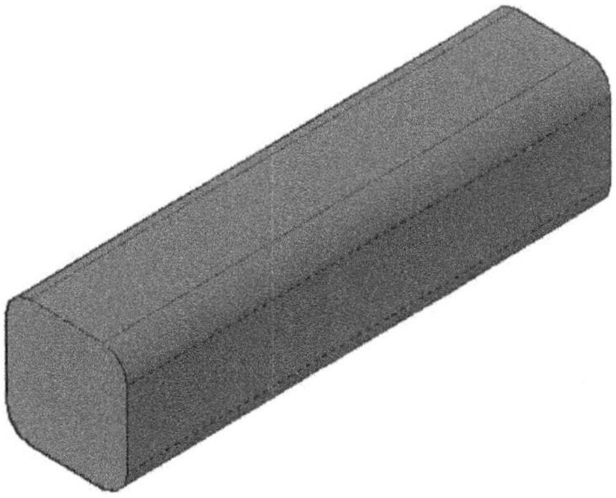

FIGURE 29.6 Shade mode.

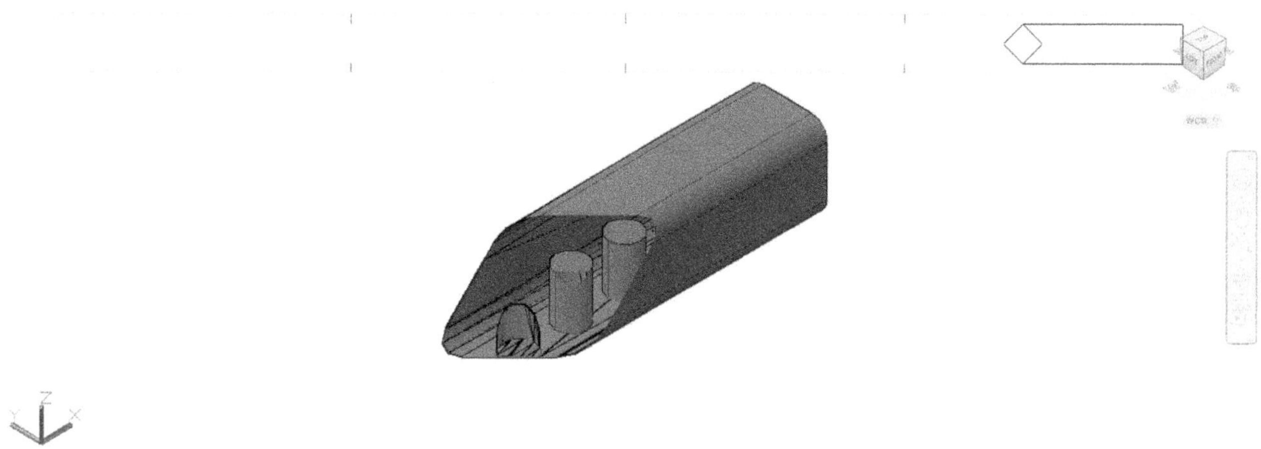

FIGURE 29.7 Front clipping plane.

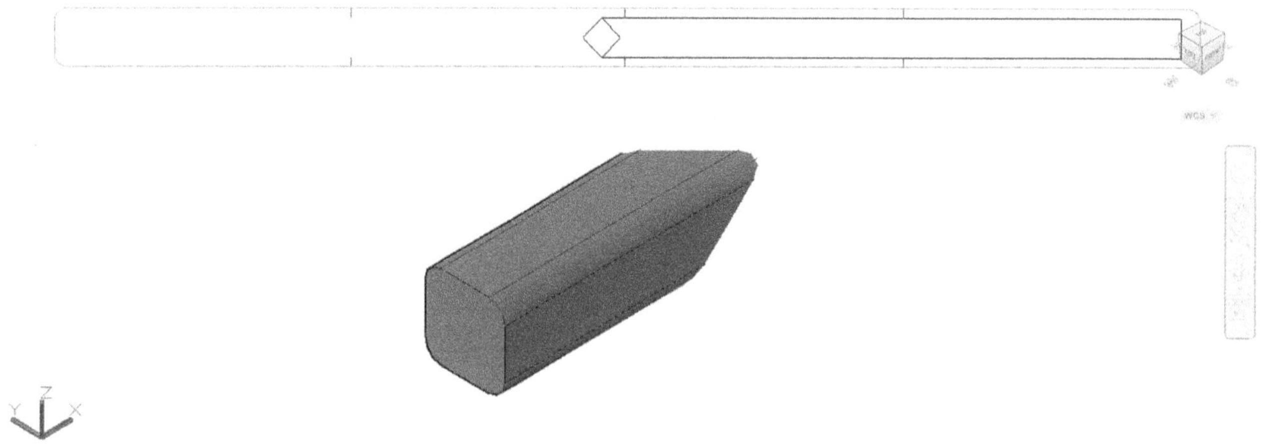

FIGURE 29.8 Front clipping plane.

Off

This command turns off perspective viewing, although you can do the same thing by switching to wireframe mode.

Undo

Undo basically undoes the last command in dview and can be used multiple times, just like the regular undo command in 2D AutoCAD.

29.2 Cameras

The camera tool provides an interesting twist on setting views. The idea is pretty much what it sounds like: You set up a camera at some strategic location and a view is created that looks out through that camera in perspective. This view is saved in the View Manager, and you can double-click on it to go into it. You generally stay in that view but are able to move around (rotate, pan, zoom, etc.). These views are used for presenting and plotting a design, and you can easily switch out of them at will. Let us take a look at some of the details.

Step 1. First of all, create a few random blocks of any size, and then shade and color them while in SW Isometric, as seen in Fig. 29.9.

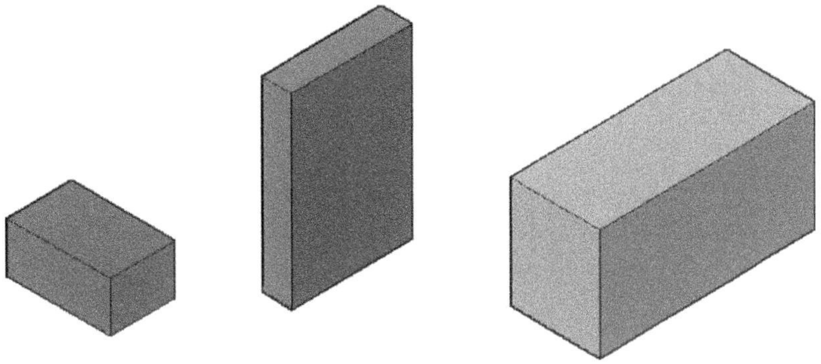

FIGURE 29.9 Blocks for cameras.

Keyboard: Type in **camera** and press Enter
Cascading menus: View→Create Camera
Toolbar icon: View toolbar→
Ribbon: none

Step 2. Start the camera command via any of the preceding methods.

- AutoCAD says: `Current camera settings: Height=4.0000 Lens Length=100.0000 mm`

`Specify camera location:`

Your values may of course be different, but they all can be adjusted. You also see a small camera symbol attached to the mouse, as shown in Fig. 29.10.

Step 3. Position the camera in a location that is some distance away from the objects and at a good angle to view the blocks. As soon as you click to position the camera, you see the "target view cone" shown in Fig. 29.11. This is the camera's field of view.

- AutoCAD also says: `Specify target location:`

Step 4. Drag out the view cone to cover the blocks and click again.

- AutoCAD says: `Enter an option [?/Name/LOcation/Height/Target/LEns/Clipping/View/eXit]`

`<eXit>:`

Here, you can change some of the camera values, but for now just press Enter. The little camera symbol sits at the location you specified, looking in the direction you pointed. If you check the View Manager, you find Camera 1 there.

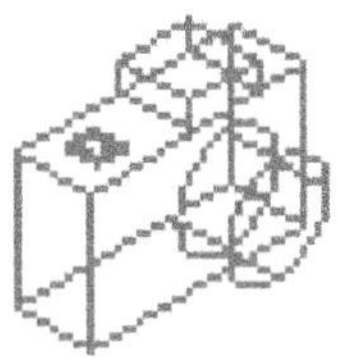

FIGURE 29.10 Camera symbol.

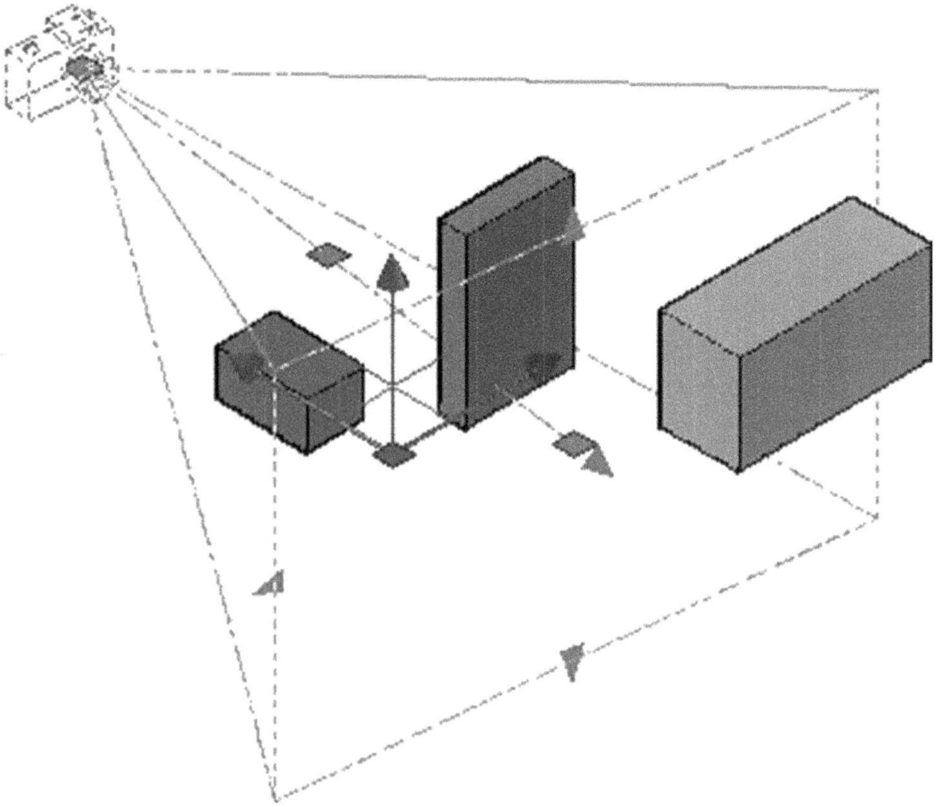

FIGURE 29.11 Camera field of view.

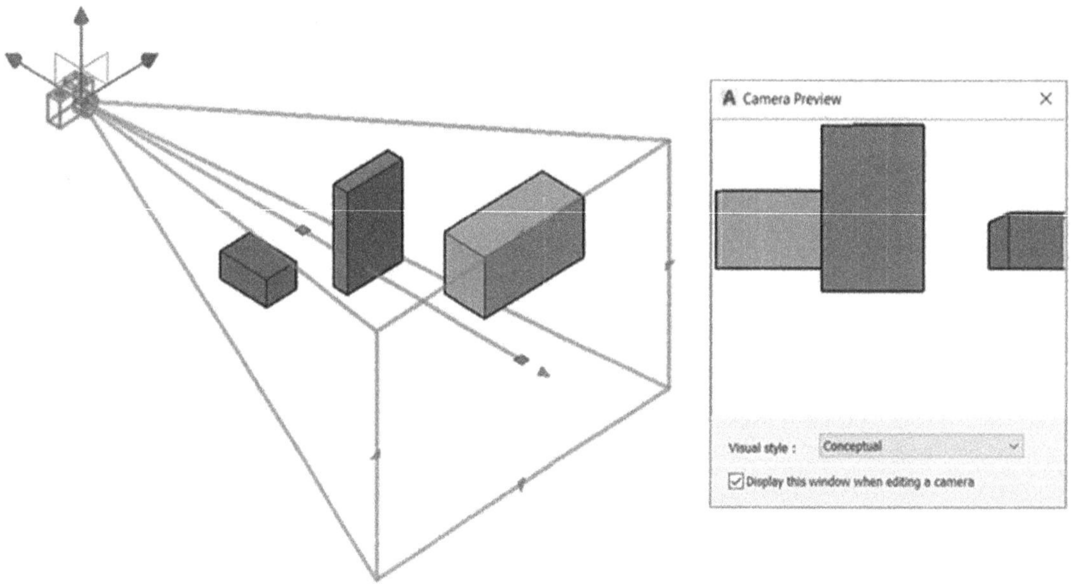

FIGURE 29.12 Camera Preview.

Before you go into that view, however, you may want to first check what it looks like by clicking on the camera. You see what is shown in Fig. 29.12.

You can now adjust the camera settings via the blue grips (e.g., widen the field of view), and these actions are reflected in the preview window on the right. When done, go to the View Manager (via the view command) and select the Camera 1 view, as seen in Fig. 29.13. Press Set Current, then OK.

You then see the Camera Perspective view (Fig. 29.14), which in this case is a floor level view of the blocks. Notice also how the screen is split between a "ground" and a "sky." This view can be rotated and zoomed or panned. Go ahead and try all of those. To get out of it, go to the little "house" introduced earlier (Fig. 29.15) and select Parallel.

Camera placement is an extensive topic that allows for creativity in where you put the camera and the resulting generation of some unique views. You should spend some time exploring this further.

29.3 Walk and fly

Walk and Fly sound fancier than they really are. All we are talking about here is the ability to move around and through a 3D design using keyboard input (typically the arrow keys). Walking involves forward/reverse and left/right movements and you remain in the XY plane. Flying is the same movements but unconstrained by the XY plane, so you have the appearance of flying through the design as you move forward and in reverse.

The best analogy to this is the experience of playing a first person action video game, where you see what the character sees as you move through hallways and rooms while shooting at the bad guys - that last part of course is missing in AutoCAD, though a few well-placed petitions to Autodesk may get designers thinking!

All the Walk and Fly movements can be saved and played back in an animation, and this is usually how things would be done, as a client would typically not want to sit around as you stumble your way through a 3D house model. It takes practice to smoothly toggle between Walk and Fly modes; we are not talking precise joystick control here. During your strolls, you can keep track of where you are in the big picture via a Position Locator palette, which also allows you to fine-tune some aspects of Walk and Fly. Walk and Fly are always conducted in perspective view, and AutoCAD sets that view automatically (after a prompt) before anything begins.

To begin exploring this, we need a small architectural floor plan with which to work. Our apartment, first started in Level 1 and worked on throughout some of the chapters, fits the bill perfectly. If not, you can use any 3D model. If you still have the apartment model saved, bring it up in 2D and clean it up a bit by removing everything but the walls, as seen in Fig. 29.16.

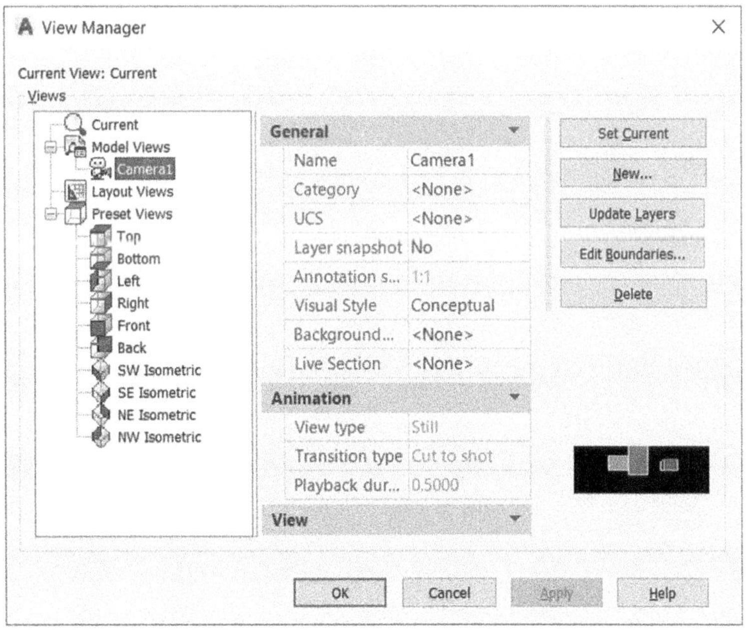

FIGURE 29.13 View Manager with Camera 1.

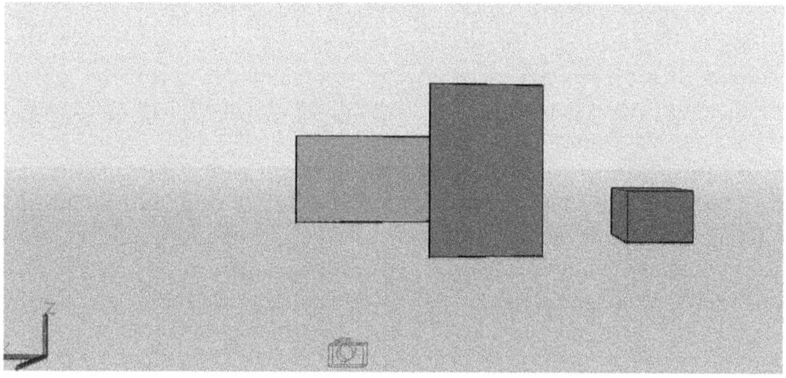

FIGURE 29.14 Camera 1 view in perspective.

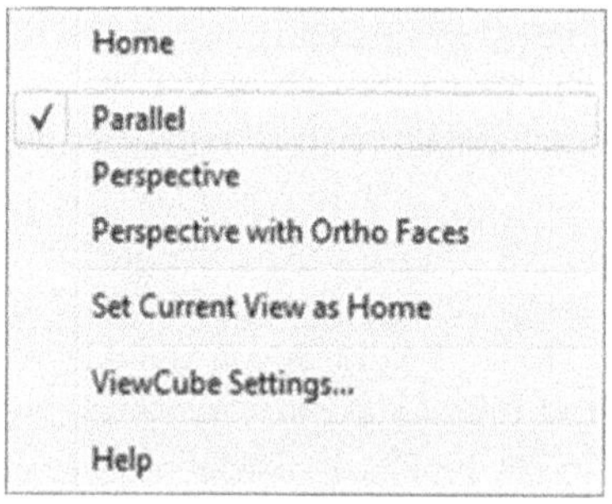

FIGURE 29.15 Exiting Camera 1 view.

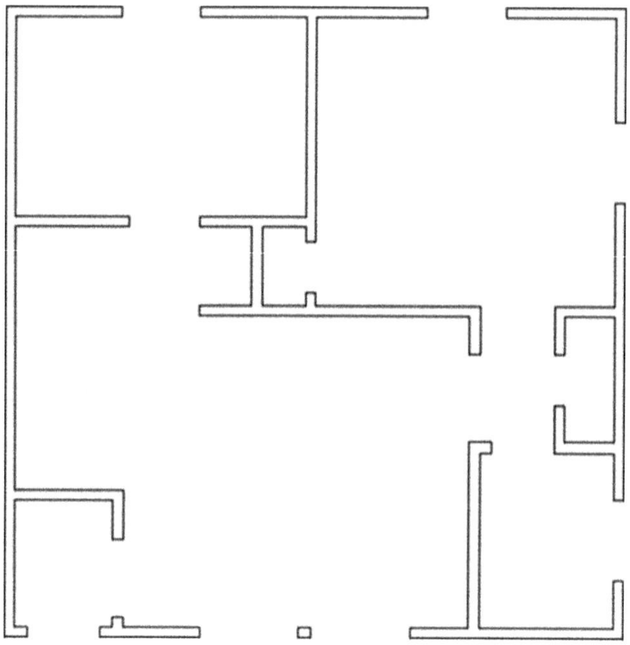

FIGURE 29.16 2D apartment.

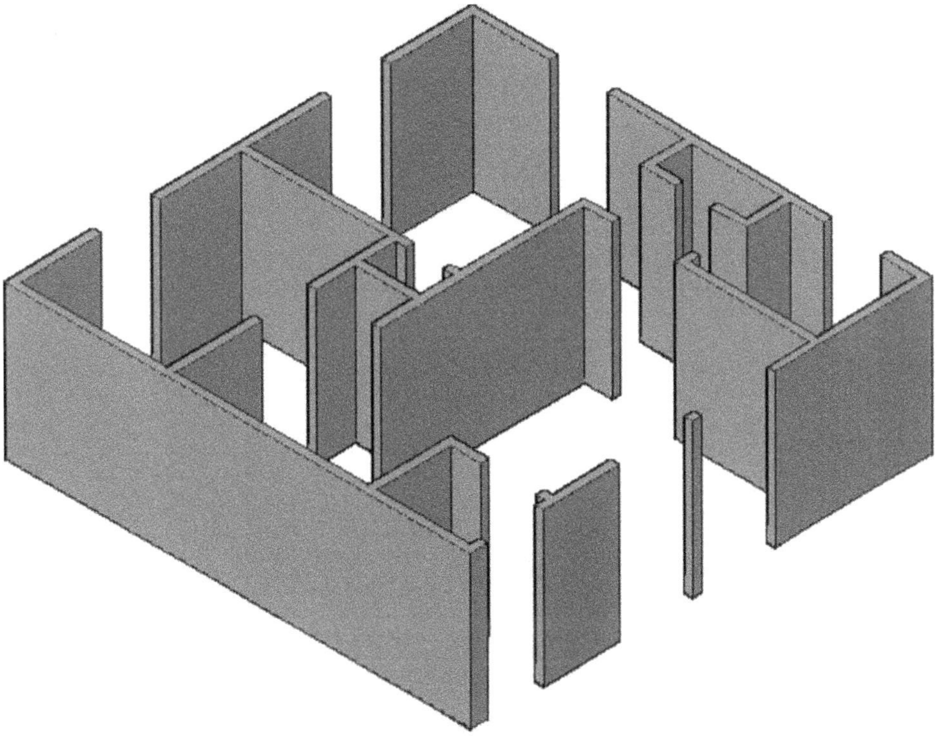

FIGURE 29.17 3D apartment: extruded, colored and shaded.

Now, pedit all the visible linework. Then, extrude all the walls out to a height of 8′ (96″). You should have what is shown in Fig. 29.17 after shading, coloring, and switching to the SW Isometric view.

To finish the design, let us add a floor and fill in the gaps in the wall where there needs to be windows. For the floor, simply draw an extruded rectangle from corner to corner of the apartment, making sure that it is located at the bottom (no roof at this time), of course. Then, draw rectangles above and below the window openings, extruding 24″ down and 36″ up. That makes for 36″ windows.

Do the same for the door openings, extruding 12″ down, thereby making the doors 96″ - 12″ = 84″, or 7′ tall. Finally, add the doors themselves where needed, according to the original floor plan (see Level 1 images). The result is shown in Fig. 29.18. Be sure to union the wall sections together and add some color for clarity.

After you are done, bring up the Walk and Fly toolbar, as shown in Fig. 29.19. The toolbar consists of Walk (the feet), Fly (the airplane), and a third icon for settings, which we explore as well. Let us try the Walk tool. When you first attempt to use it, you likely get the prompt shown in Fig. 29.20. After clicking on Change, the design enters Perspective view and the Position Locator palette appear on the screen (Fig. 29.21).

We look at this palette in detail later on, but for now set it aside and let us focus on navigating the design. You can navigate it (walk) via the tools shown in Fig. 29.22.

The information needed to use Walk and Fly is spelled out in the list. While you can certainly use the letter keys (W, S, A, and D), it is probably easier to use the keyboard arrows (Up, Down, Left, and Right). To toggle between Walk and Fly, press the F key. Take some time to practice walking and flying via the four keyboard arrows, with the letter F key as the toggle between the two. Once you have the basic idea of Walk and Fly, let us take a look at the settings icon. When you press that, you see the box shown in Fig. 29.23.

The main settings have to do with the Display Instruction and Display Position window and palette. You can get rid of both of them if needed; although they are helpful, they are not essential to Walk and Fly operation. Just as important, however, is what is found at the bottom of the dialog box, the drawing units and Steps per second. You can increase both to Walk and Fly through the design faster.

Do not forget also to use the mouse itself to rotate and spin the design around if needed. Walk and Fly are just part of the equation. Let us take another look at the Position Locator. In the top window, you see the design's position relative to you, the observer, as seen in Fig. 29.21. Although you are not likely to get lost while traveling through this simple one-

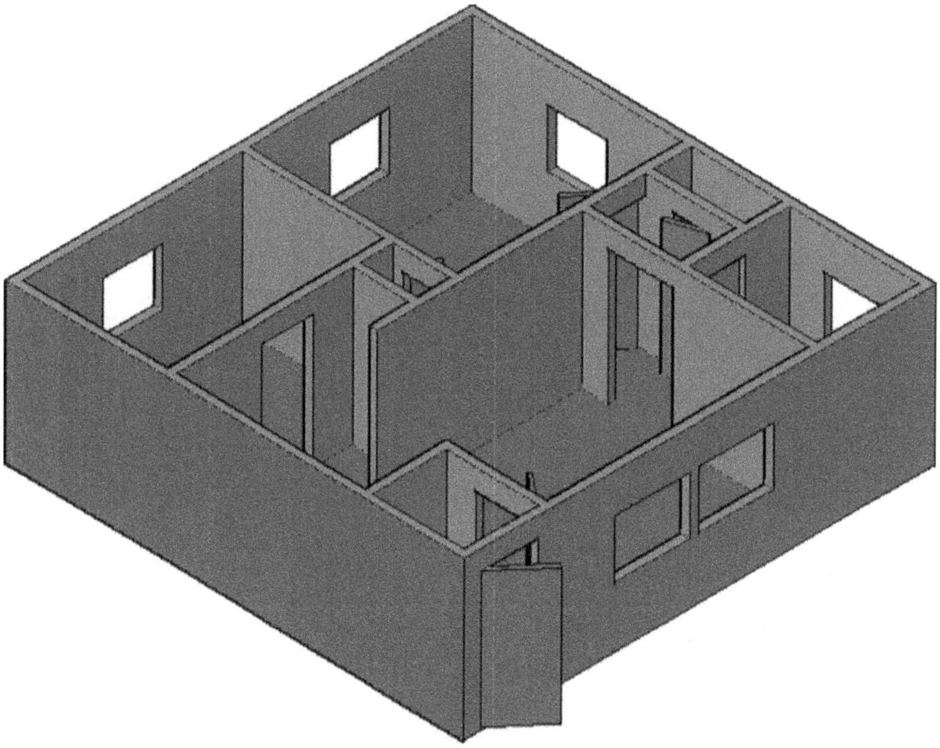

FIGURE 29.18 3D apartment with windows, doors and floor.

FIGURE 29.19 Walk and Fly toolbar.

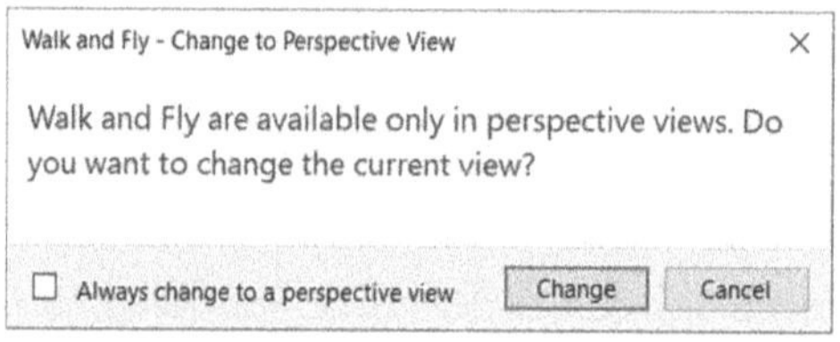

FIGURE 29.20 Walk and Fly warning prompt.

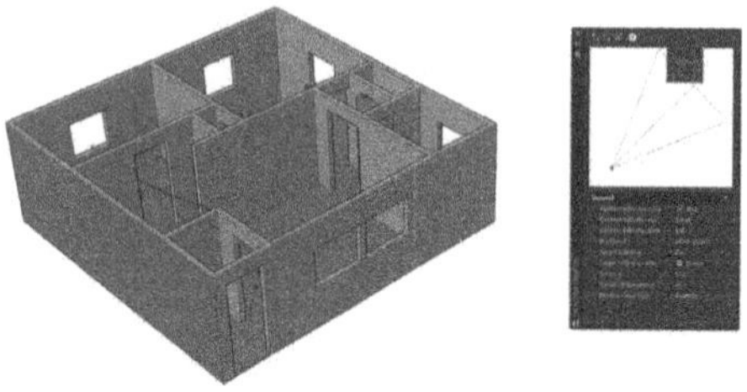

FIGURE 29.21 Position Locator palette.

Up arrow / W key	Move forward
Down arrow / S key	Move backward
Left arrow / A key	Move left
Right arrow / D key	Move right
Drag mouse	Look around & turn
F key	Toggle Fly mode

FIGURE 29.22 Walk and Fly navigation tools.

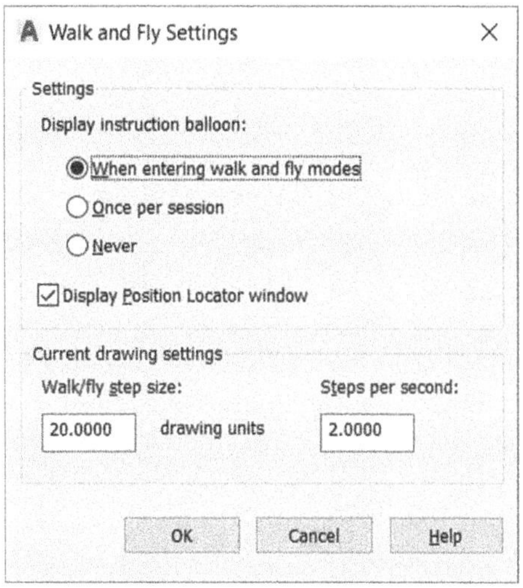

FIGURE 29.23 Walk and Fly settings.

bedroom design, you may in a more sophisticated walk-through, and this "bird's-eye view" of where you are can be helpful. The rest of the general settings below that preview window are related to how the preview window looks.

29.4 Path animation

Path animation has to do with creating animations of designs. These animations are not the classic type, where we have objects moving and interacting (e.g., AutoCAD cannot animate the motions of pistons inside cylinders of a car engine model). Rather, what we refer to is an animated walk-through, a sort of frame-by-frame set of scenes strung together into a smooth motion.

To begin animation, type in anipath and press Enter. Alternatively, you can use the drop-down menu: View→Motion Paths Animation…. The dialog box shown in Fig. 29.24 appears.

This is the combined settings and animation dialog box. Here, you can set your preferences, preview the animation, and export it to the desired format. Some of the options are listed in more detail:

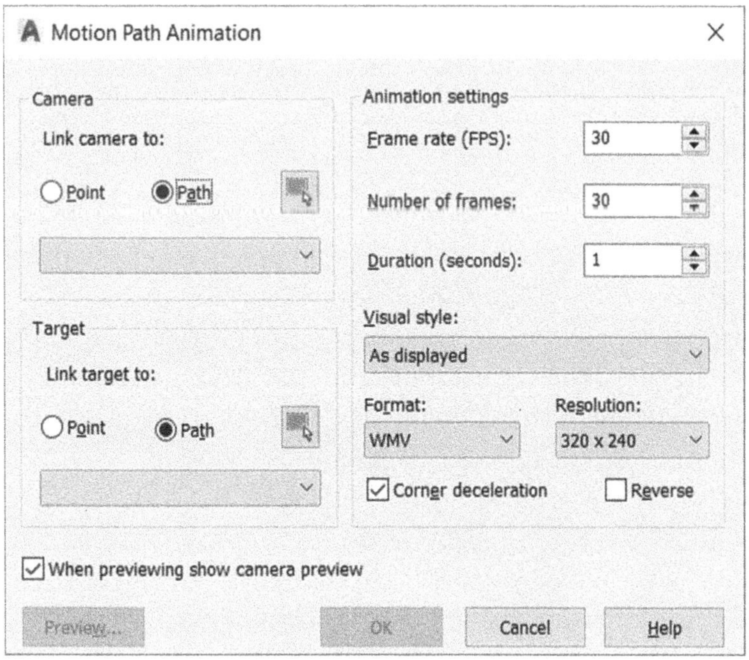

FIGURE 29.24 Motion Path Animation.

- *Frame rates*: This is the number of frames per second that play at normal playback speeds: the more frames, the larger the output file. The maximum is 60, and 40 is usually just fine. The number of frames divided by the duration yields the Frame rate (FPS).
- *Resolution*: This determines how sharp the image is: The larger is the resolution, the larger is the output file. The range is from 160×120 to 1024×768. The default, 320×240, is usually adequate.
- *Format*: The available exporting formats are avi, mov, mpg, and wmv.
- *Camera or Target*: You can specify either one; we use camera.

As the name implies, motion path animation depends on a path. Once the path is defined, along with a number of animation-related parameters, AutoCAD sends a camera down the path and plays back what it sees as it travels. It is smooth and continuous and a good way to present a tour of an entire building.

To try this out, cancel the dialog box and create a path in the building using pline. This is by far the hardest and most time-consuming task, and it is also the most important. A good path ensures a clear, smooth view of the interior. Here are some tips for path placement:

- Place the path in the middle of doorways using temporary guidelines. If you use the edges of the door, the camera literally slams into the door frame as it travels, and the view is briefly obscured.
- Add fillets to round out the turns instead of right angles for smoother motion.
- Create a realistic and steady path, like one that a real person would walk. Do not jump from floor to ceiling without a good reason.

In Fig. 29.25, several doors are removed for clarity, and a sample path through our 3D apartment is carefully drawn in using pline. The path comes in through the front door and walks through the living room and into the kitchen, following a typical path a person would take.

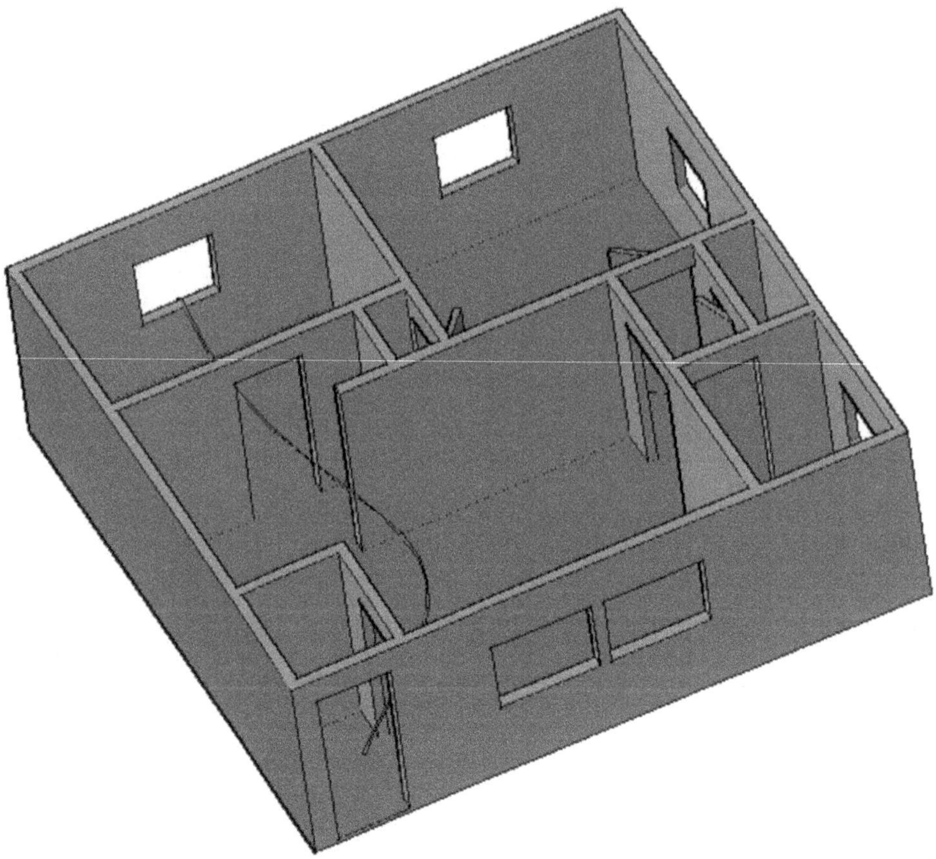

FIGURE 29.25 Motion pathway.

This path may or may not be ideal and may have to be fixed later, but it is fine for now. When done, select View→Motion Paths Animation… again, and follow these steps:

Step 1. The first step is to link the camera to a point or a path. Typically, the camera is linked to the path, so in the upper left-hand area, use the button to select the path and confirm its name as Path 1 in the field.

Step 2. Next, you need to link the target to a point or a path. If you choose Path, then the Target and Camera settings are the same, meaning the camera follows the path and turns when the path turns. If you choose Point (and select a distant point), then the camera is always fixed on that point as it follows its path.

Step 3. Now, you need to run through the animation settings. Here, everything should be more or less intuitive, as all three fields are linked. So, if you select a frame rate of 10 and specify 120 frames, then 120/10=12, so there will be 12 seconds of "footage" or animation. Select values that are appropriate to the task at hand. If you have a large walk-through, you may need to run the animation for a minute or more. Just be aware that the files generated can get big very quickly, and overall performance and quality also depend on your computer (RAM, video card, etc.).

Step 4. Finally, select the Visual Style, Format, and Resolution and press Preview…. A preview window appears and runs through the animation, as seen in Fig. 29.26.

When done, simply close the Animation Preview via the X in the upper right corner; you are taken back to the Motion Path Animation dialog box. Press OK, and you are taken to the Save As dialog box and asked where you would like to save

FIGURE 29.26 Animation Preview.

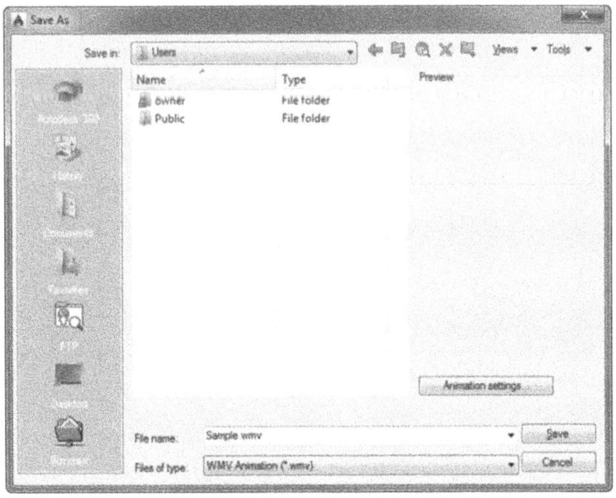

FIGURE 29.27 Saving animation.

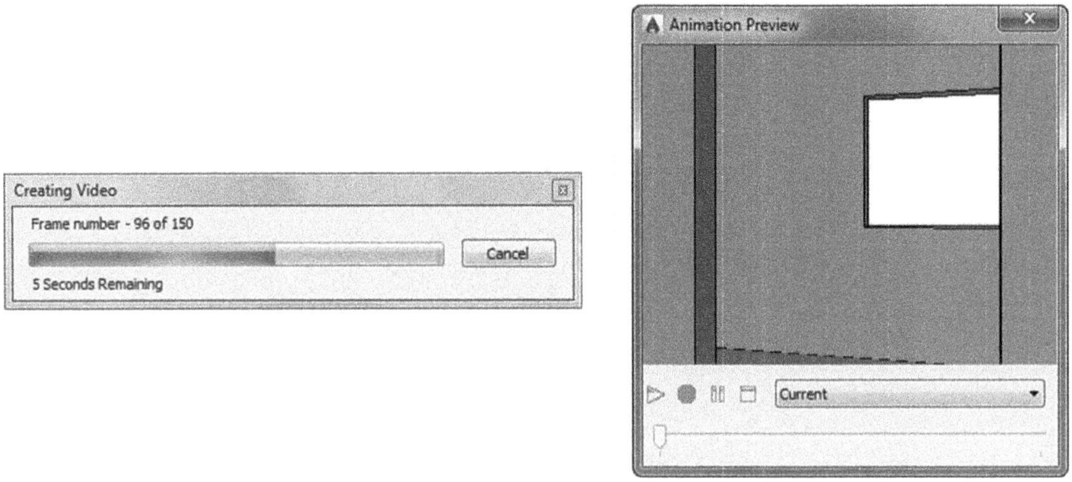

FIGURE 29.28 Animation generation.

the file, as well as to confirm the output format (Fig. 29.27). Select a location and the generation sequence commences, as seen in Fig. 29.28. The file is generated and you can watch it with whatever player you have on your computer.

Summary

You should understand and know how to use the following concepts and commands before moving on to Chapter 30:

- Dview
  - CAmera
  - TArget
  - Distance
  - POints
  - PAn
  - Zoom
  - TWist
  - CLip
  - Hide
  - Off
  - Undo
- Cameras
- Walk and Fly
- Path animation

Review questions

Answer the following based on what you learned in this chapter:

1. List all 11 dview functions and what they do.
2. List any additional views that are covered.
3. What is the purpose of Walk and Fly?
4. How are the Walk and Fly functions controlled?
5. How do you create path animations?

Exercises

1. Create the following fictional concrete barrier with steel reinforcing. All dimensions are given. First, model the interior panels and circular cross members. Then, enclose everything in the outer structure, aligning the elements carefully. Finally, after coloring and shading, use the dview command with the Clip option to Front clip the design to the approximate depth illustrated. For extra practice, set up a series of viewports via the mview command to display the wireframe, shaded, and the clipped design as shown. (Difficulty level: Intermediate; Time to completion: 30 - 45 minutes.)

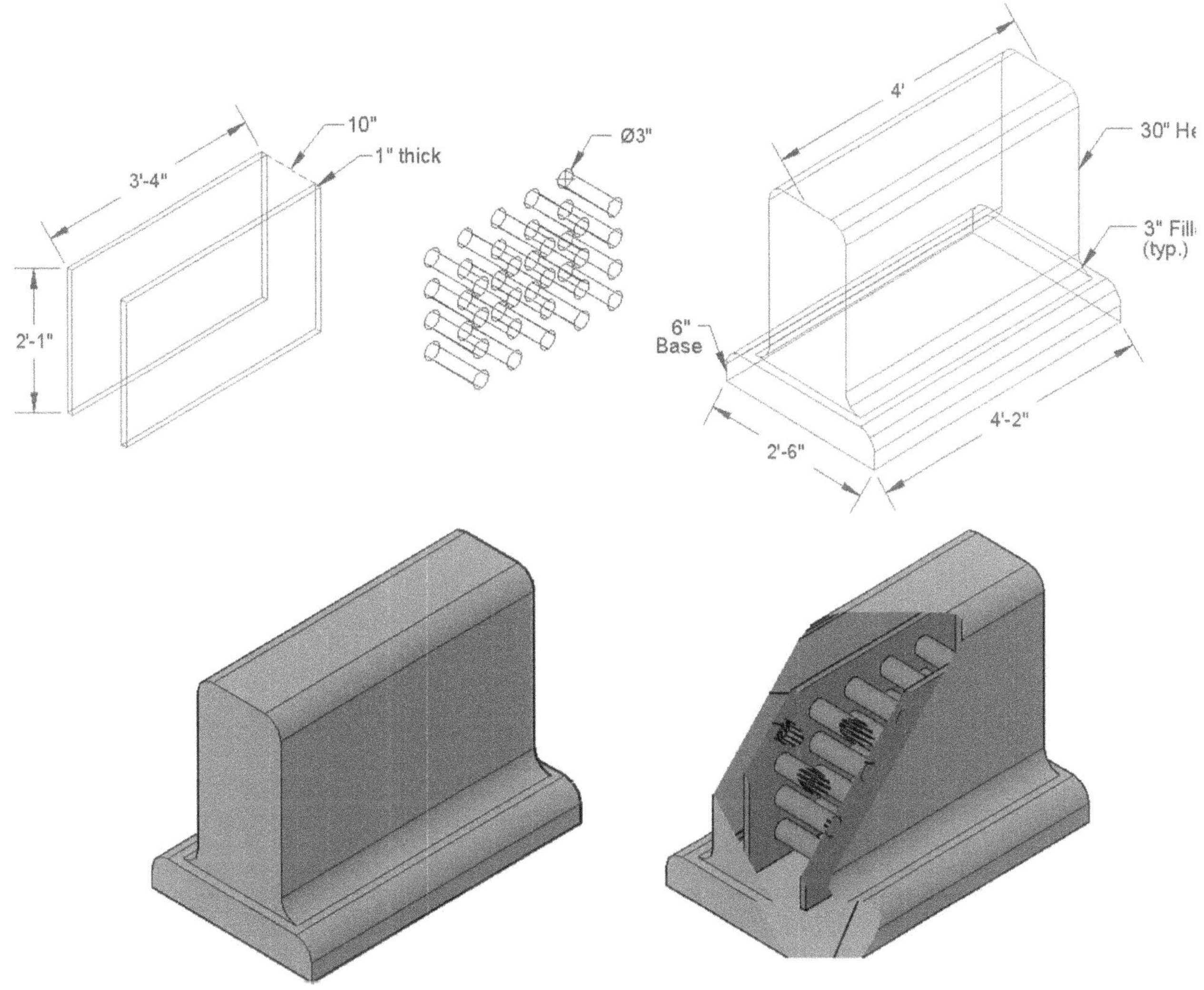

2. The following exercise is more in-depth and requires some set up. When you are done, you will get a chance to practice the camera, Walk and Fly, and the path animation commands, all using the same model. (Difficulty level: Advanced; Time to completion: 30 - 45 minutes for setup, 60 - 90 minutes for exercise.)
   Step 1. In 2D, create a basic floor plan using the given dimensions. The dimensions themselves are only for your reference and are not needed in the drawing.
   Step 2. Trim out all the necessary linework to form four independent "cross-shaped" sections as shown. Everything in the design is symmetrical. Pedit all the wall sections and double-check for stray lines and any other errors.
   Step 3. Extrude all four sections to the height shown. Add a floor to the design, extruded to the depth shown.

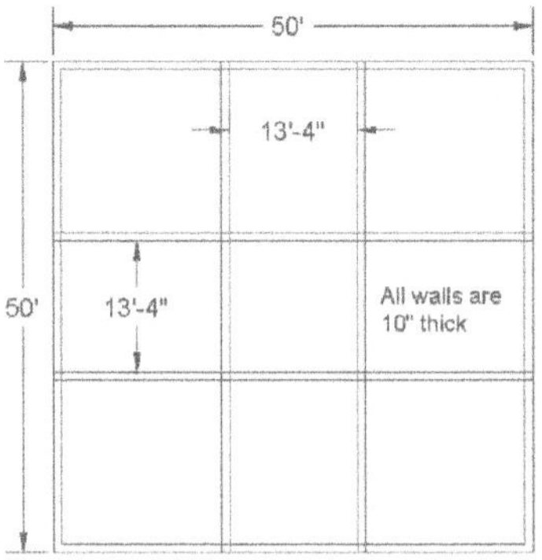

Step 1

Step 2

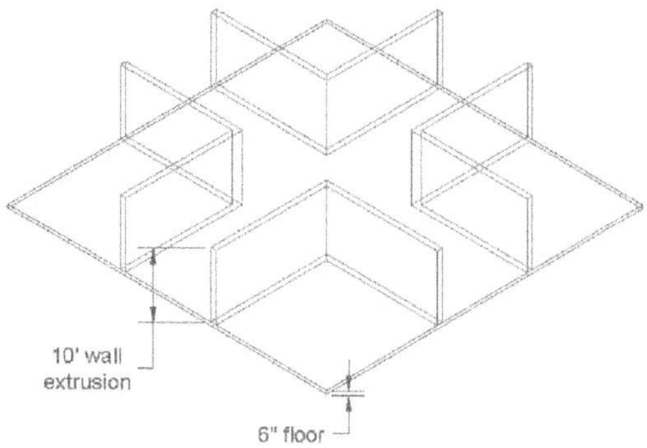

Step 3

Step 4. Add four walls with windows to the four openings of the floor plan design as seen in the diagrams. These can be of any thickness, and the windows can be of any size, but make sure they are centered and all walls match up without gaps or overlaps. Color and shade your model.

Step 5. Add in furniture to three of the four "hallways" of the design. The furniture can be what is shown (a detail is provided) or something of your own design. Be sure to position it evenly and center it on the floor space. Add a set of swinging doors to the one end that has no furniture. Door sizing is up to you.

Step 6. As a final step, add a pole with a 2' diameter and 10' height to the geometric center of the design as shown. This step concludes the basic design. Check it over before proceeding to the first part of the actual exercise; the camera command.

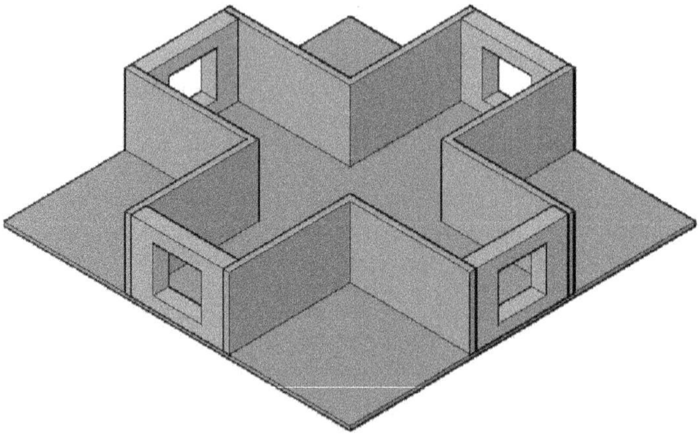

Step 4

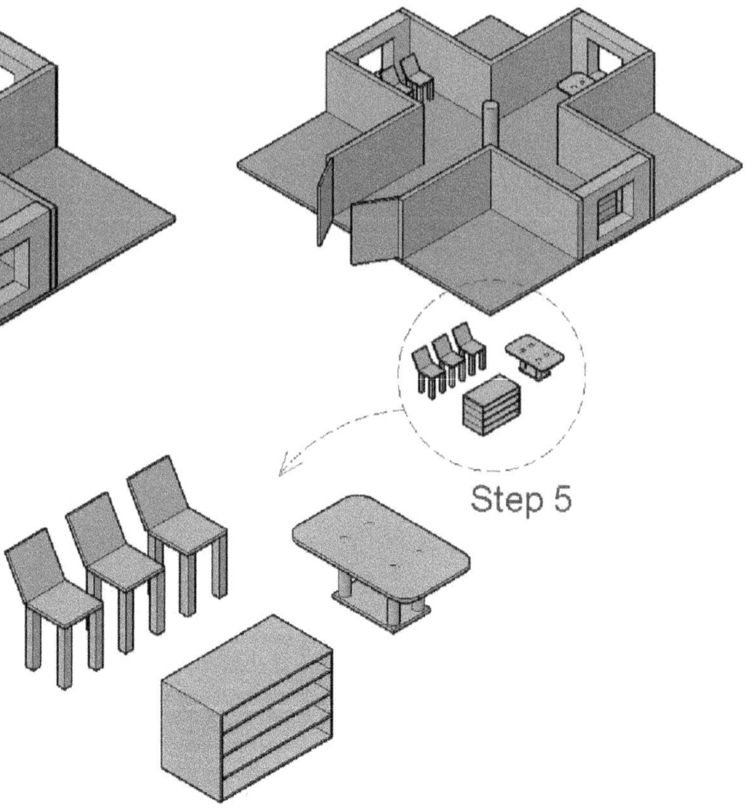

Step 5

Camera: To practice the camera command, first set up a few cameras at various strategic locations in your new 3D floor plan. Recall from earlier in this chapter that you need to use OSNAPs to position where the camera is located with the first click, then where it is pointing with the second click. Every time you position the camera, check to see what exactly it is seeing by clicking on it and bringing up the Camera Preview. When done, use the View Manager to view each of these views. A sample positioning of five cameras is shown with arrows giving you a rough idea of what they are looking at. Finally, three sample previews are shown, followed by one full view. Be sure to pick varied and interesting camera locations on your 3D model to recreate something similar to what is shown, or an entirely new set of vantage points of your choosing.

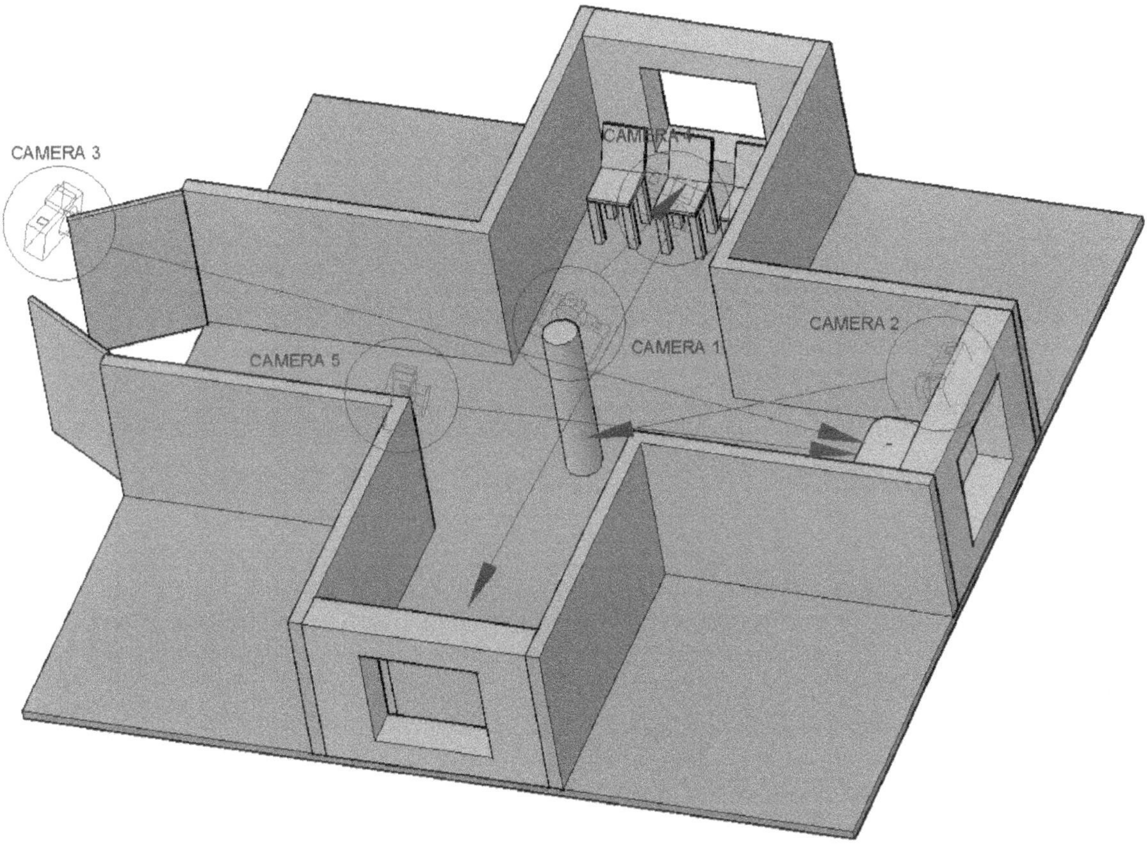

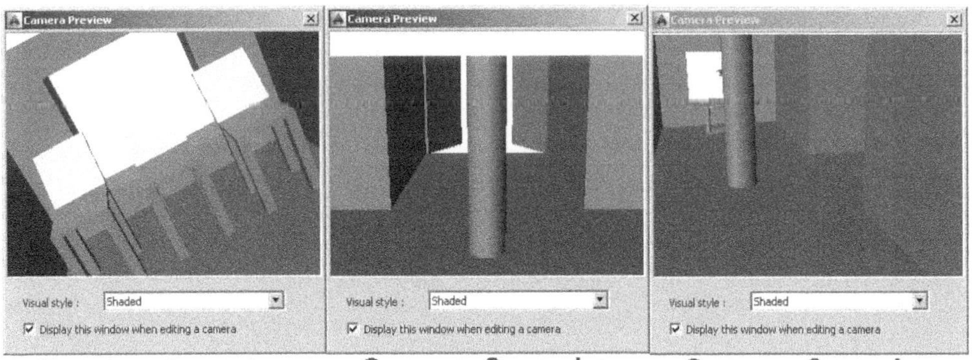

Camera 1 preview Camera 2 preview Camera 3 preview

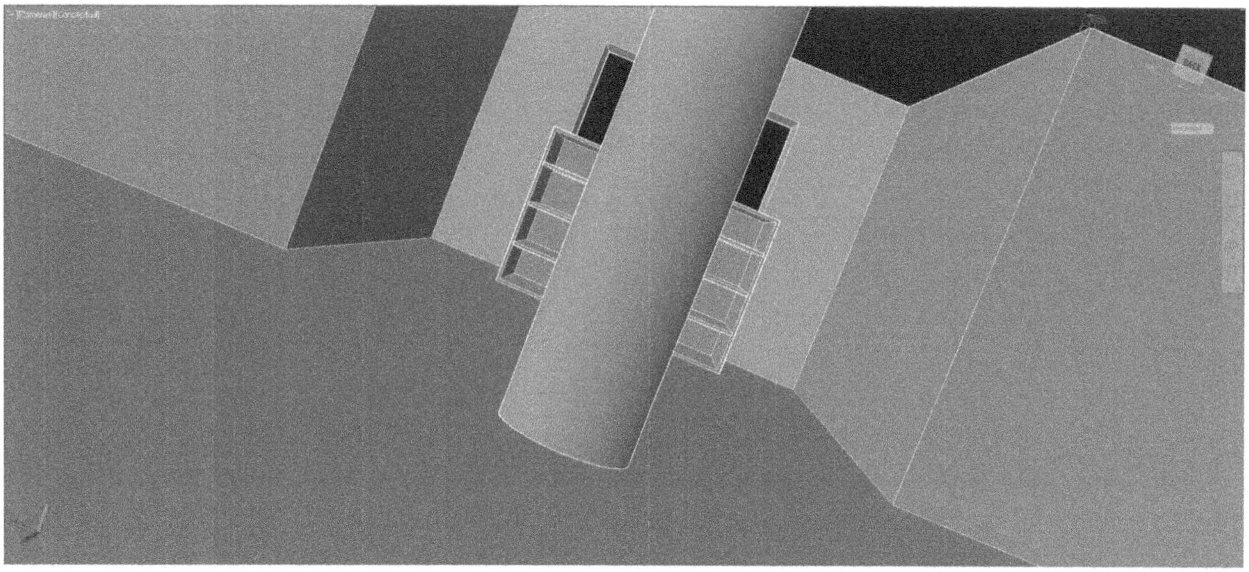

Walk and Fly: To practice walk and fly simply begin at the front entrance and navigate your way through any of the hallways of the 3D model. There is no need for any images to illustrate this, but do try to walk into all the far corners of the model.

Path Animation: Begin the path animation command by setting up a reasonable path to follow around all corners of the 3D model, similar to the Walk and Fly "trip" that you just did. Keep the path at eye level (5′ off the ground) and create a smooth pline or spline path. One way to do this is to carefully position ½ size copies of the main center pole at strategic locations all around the inside of the model as shown next (with several layers frozen for clarity). Note how the spline path is overlaid on the top surfaces of the smaller poles. When these poles are deleted you will have a well-positioned path for the animation camera to follow.

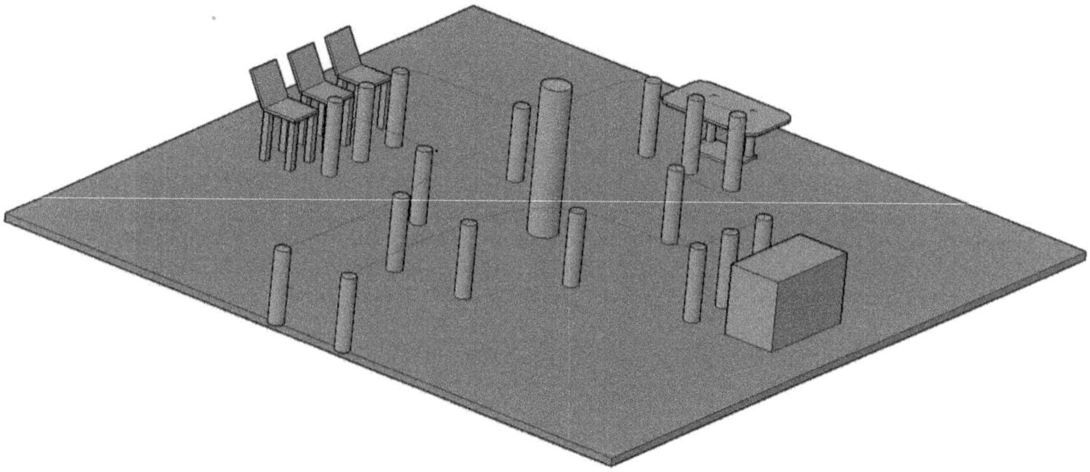

Here, the path is cleared of all poles and Path1 and Path2 are chosen. Set the duration to 30 seconds, and leave the default FPS and Number of frames.

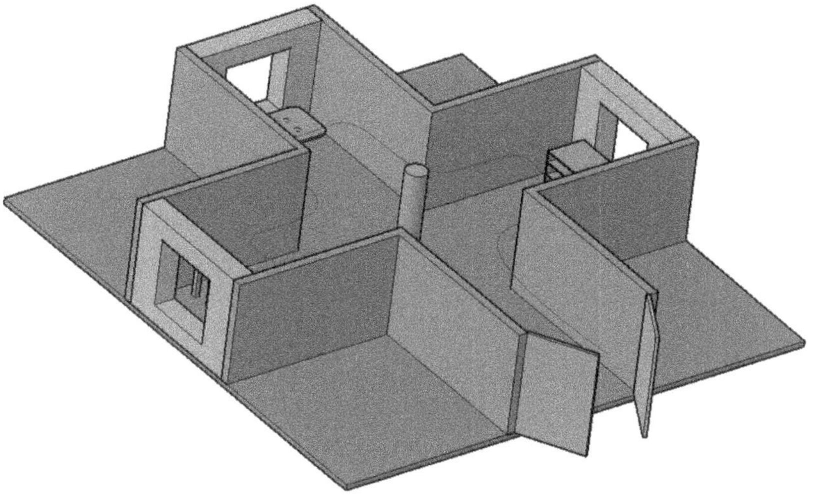

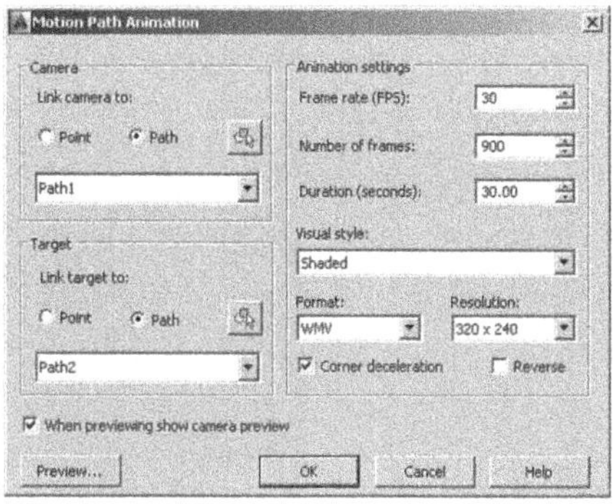

A shot of the animation preview is shown next. Set up several varying paths inside and outside your mode for additional practice.

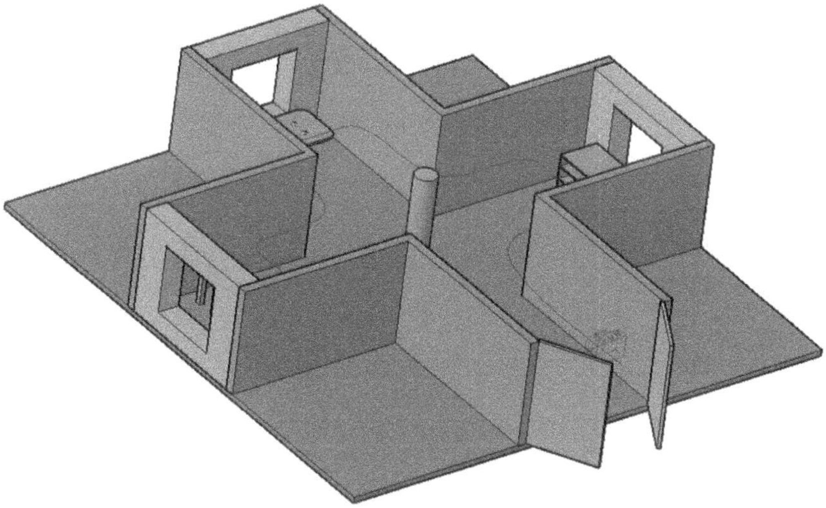

3. The third and final exercise of this chapter has nothing to do with the tools introduced here but rather serves as a "cleanup" of our tire and suspension model presentation. Using the Chapter 28, Exercise 1 design, set up a title block and viewports as shown in the following illustration. (Difficulty level: Advanced; Time to completion: 20 minutes.)

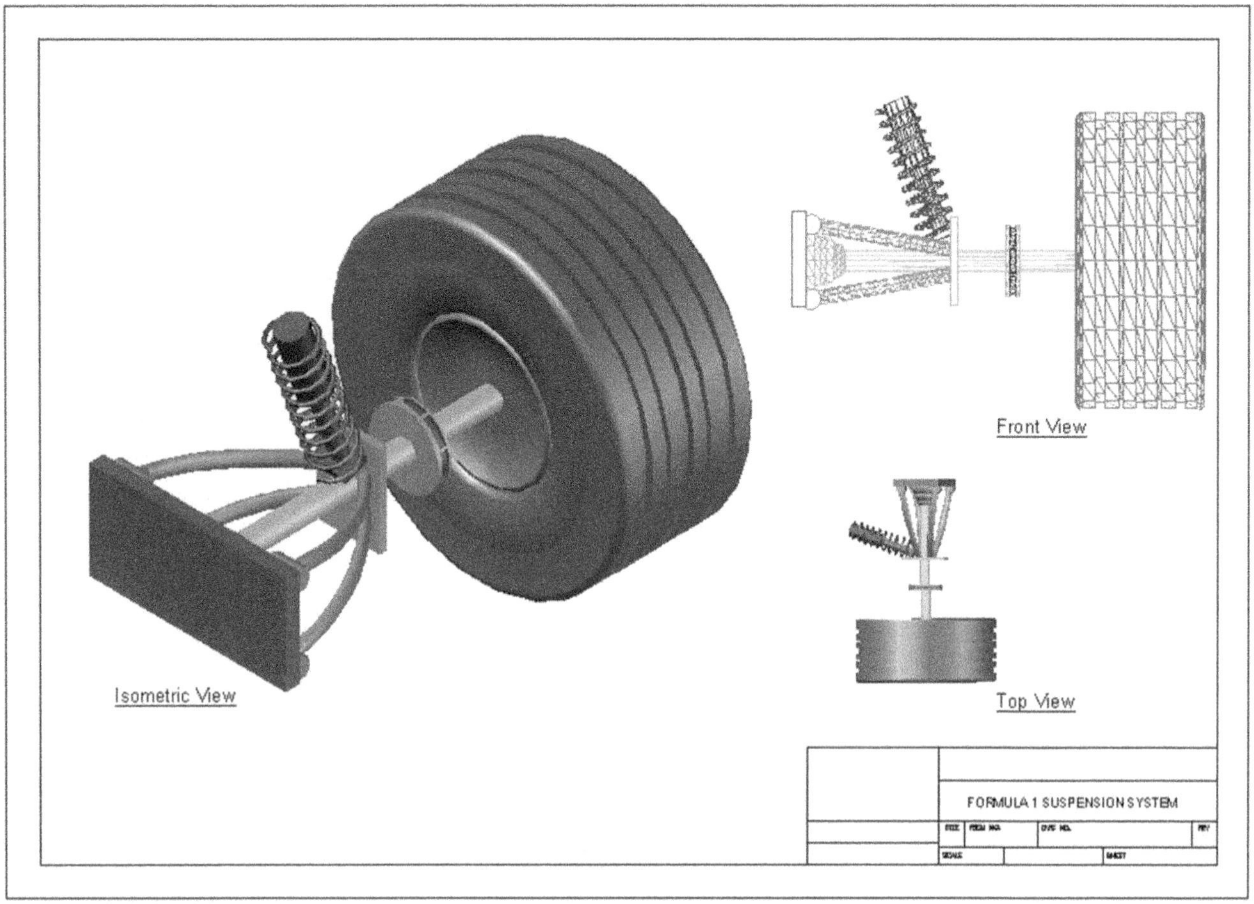
Front View
Isometric View
Top View
FORMULA 1 SUSPENSION SYSTEM

Chapter 30

Lighting and rendering

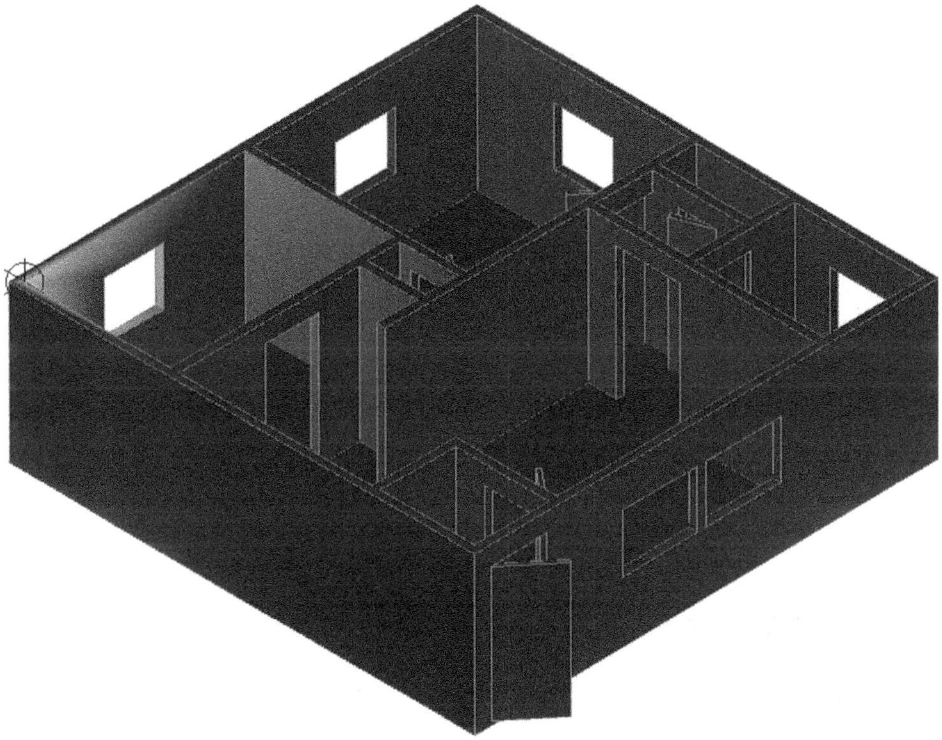

Learning objectives

In this final chapter, we explore lighting and rendering and discuss

- Point light
- Spotlight
- Distant light
- Shadows
- Sun and Sky
- Materials
- Rendering

By the end of this chapter, you will be able to add various lighting effects and basic materials to your design. Estimated time for completion of this chapter: 2 hours.

Up and Running with AutoCAD 2022. https://doi.org/10.1016/B978-0-323-89923-9.00030-5

30.1 Introduction to lighting and rendering

This final chapter of this textbook is dedicated to "finishing" a design. This means the application of lighting, shadows, environmental effects, backgrounds, materials, and rendering. It is something that is done last, after a design is completed, as these effects do not stand alone but are applied to an existing model. All finishing effects are loosely grouped under two major categories: *lighting* and *rendering*.

It should be stated up front that, despite major improvements in AutoCAD's lighting and rendering ability (including a new rendering engine for AutoCAD 2016), it is still not quite on par with available dedicated rendering software. Packages such as Form Z, Rhino, and the industry standard 3ds Max (3D Studio Max) can create stunning photorealistic images and, in some cases, advanced animation. Autodesk has had a bit of a dilemma with AutoCAD. Should it pursue advanced rendering development and risk taking sales away from its own 3ds Max or add to AutoCAD to please those few customers who may need some rendering ability? After all, the vast majority of AutoCAD designs are still plain old 2D schematics or floor plans, and it makes little sense to load AutoCAD with rendering power that customers rarely need.

The answer is somewhere in the middle. AutoCAD always had limited lighting and rendering ability, going back to the Advanced Modeling Extension of the early releases, and slowly new features were added. These were just enough to satisfy casual 3D users but never so much that they would not consider the expensive dedicated software. One way around this, of course, is to purchase reasonably priced add-ons, like AccuRender (see Appendix I). This software adds dramatically to the existing capabilities without expensive investment.

In the meantime, AutoCAD's development team (always in need of new features to advertise as a new version rolls out) slowly but surely kept improving the built-in lighting and rendering. Today, with the latest version, Release 2021, we have quite a bit to discuss. It is still not 3ds Max, but if you have never worked with advanced dedicated effects software, you may be impressed. It is also a great introduction into that world. Let us take a look first at lighting and related topics (such as Shadows and Sun and Sky), followed by materials and rendering.

30.2 Lighting

Lighting in digital art and CAD software is based on several established techniques and principles, and AutoCAD is no exception. A model can be lit in one of several ways, which involve variations on *ambient*, *directional*, or *spotlight* light sources. This, of course, mimics our everyday experiences, where we have a variety of lighting conditions, from diffused sunlight to the intense headlights of a car and everything in between.

Most of the options that have to do with lighting are either on the Lights toolbar or the Visualize tab of the Ribbon. Both are shown in Fig. 30.1, along with shadows and the sun options that we discuss later.

The three main options for lighting are

- *Point light*: This light, similar to a light bulb, originates from one point and radiates in all directions.
- *Spotlight*: This light, similar to a flashlight, originates from one point, but radiates in only one direction.
- *Distant light*: This light, similar to the sun, originates from a distant source and does not fall off in intensity.

FIGURE 30.1 Ribbon Visualize tab (with relevant panels circled) and the Lights toolbar.

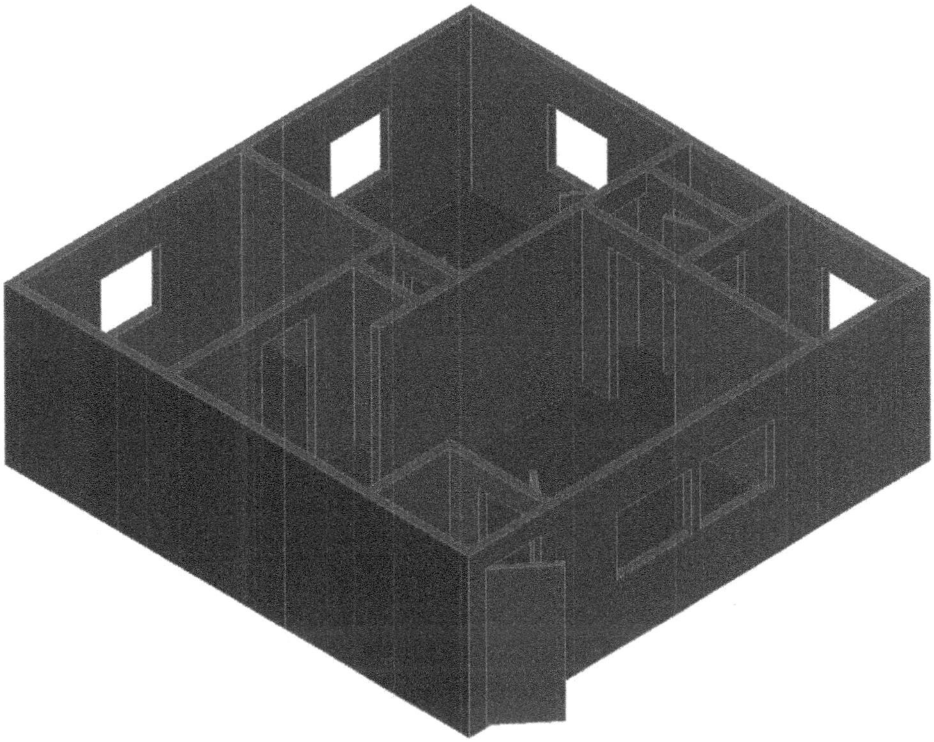

FIGURE 30.2 RVS shading.

Let us give these various lights a try. After discussing the basics of each, we list the options available with each light type, which are more or less similar from one lighting scenario to another. Bring up the same 3D apartment model used in the previous chapter and switch to the Realistic Visual Style (RVS), as seen in Fig. 30.2.

Point light

The first lighting condition to try is the most often used one, the New Point Light option.

Keyboard: Type in **light** and press Enter
Cascading menus: View→Render→Light→New Point Light
Toolbar icon: Lights toolbar→
Ribbon: Visualize tab→Lights panel→

Step 1. Begin the New Point Light via any of the preceding methods. You may see a warning dialog box, asking if you want to turn off default lighting (not shown here). Go ahead and agree to that.
Step 2a. If you type in `light`,

- AutoCAD says: `Enter light type [Point/Spot/Web/Targetpoint/Freespot/freeweB/Distant] <Point>:`

Go ahead and select `P` for `Point` and press Enter. Go on to Step 2b.
Step 2b. If you used any of the other methods or just completed Step 2a,

- AutoCAD says: `Specify source location <0,0,0>:`

Step 3. Select the extreme upper left corner of the building.

- AutoCAD says: `Enter an option to change [Name/Intensity factor/Status/Photometry/shadoW/Attenuation/filterColor/eXit] <eXit>:`

Step 4. We go over some of these options in detail momentarily, but for now press i for Intensity.

- AutoCAD says: `Enter intensity (0.00 –max float) <1.0000>:`

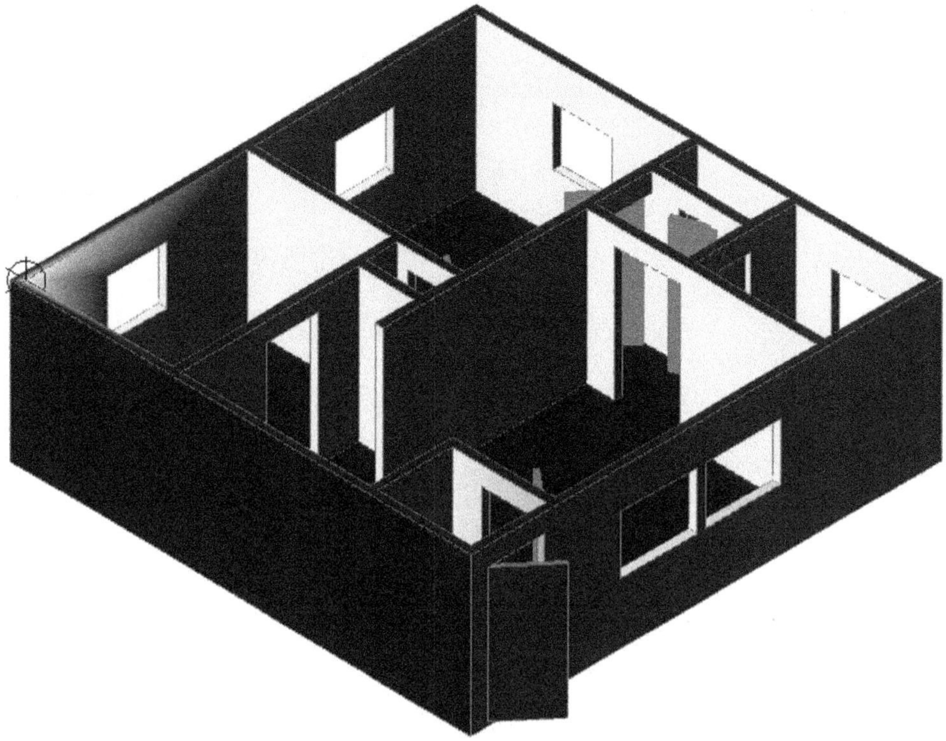

FIGURE 30.3 Point Light (LIGHTINGUNITS 0).

Step 5. Enter the value 20. Then, press `c` for `Color`.

- AutoCAD says: `Enter true color (R,G,B) or enter an option [Index color/Hsl/colorBook] <255,255,255>:`

Step 6. Press `i` for `Index color` and enter color number 30 (it is one of the yellows).

The result is a spotlight located at an upper left corner of the 3D model (notice the circular marker), shining yellow light on the entire structure (Fig. 30.3). Note that the light goes through the walls, illuminating everything equally, as if the walls were not even there. We change this later with the LIGHTINGUNITS system variable. You can locate as many point lights as you want; we do only one example of this.

Spotlight

The New Spotlight option is similar in theory to the New Point Light option just covered.

Keyboard: Type in **light** and press Enter
Cascading menus: View→Render→Light→New Spotlight
Toolbar icon: Lights toolbar→ 
Ribbon: Visualize tab→Lights panel→ 

Step 1. Begin the Spotlight via any of the preceding methods. You may see a warning dialog box, asking if you want to turn off default lighting (not shown here). Go ahead and agree to that.
Step 2a. If you type in `light`,

- AutoCAD says: `Enter light type [Point/Spot/Web/Targetpoint/Freespot/freeweB/Distant] <Point>:`

Go ahead and select `S` for `Spot` and press Enter. Go on to Step 2b.
Step 2b. If you used any of the other methods or just completed Step 2a,

- AutoCAD says: `Specify source location <0,0,0>:`

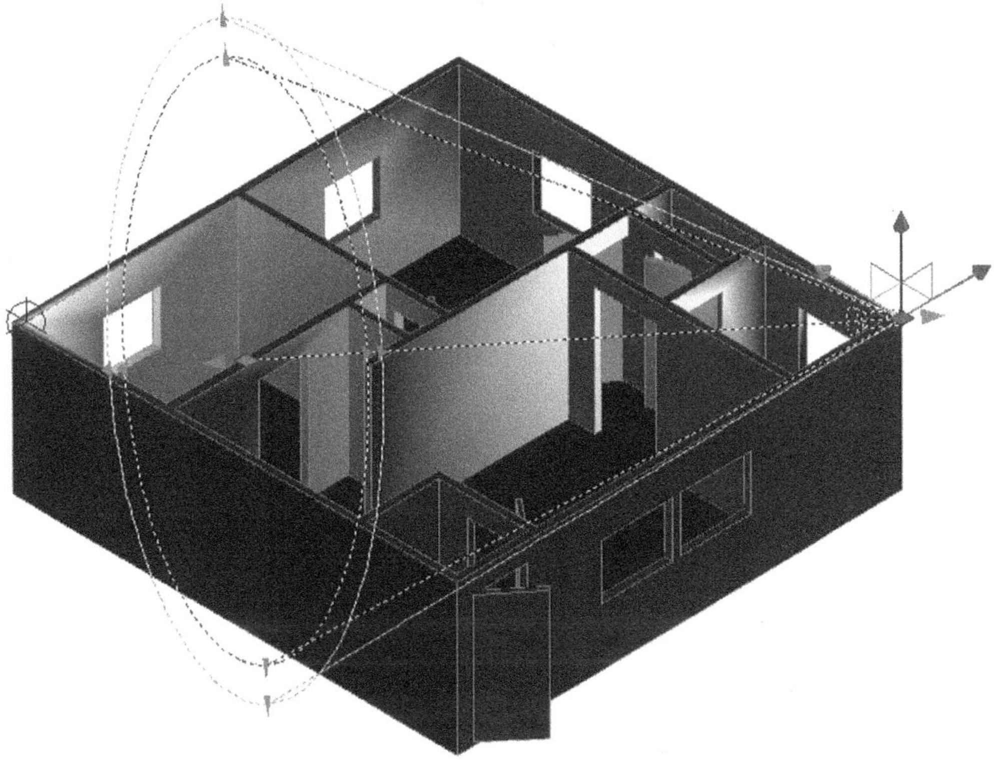

FIGURE 30.4 Spotlight.

Step 3. Select a corner of the building (a flashlight is attached to the crosshairs).

- AutoCAD says: `Specify target location <0,0,-10>:`

Step 4. Select a point in such a way as to indicate a direction for the spotlight to shine. A large menu appears that has essentially the same choices seen previously.

- AutoCAD says: `Enter an option to change [Name/Intensity factor/Status/Photometry/Hotspot/Falloff/shadoW/Attenuation/filterColor/eXit] <eXit>:`

Step 5. Using values of 100 for the Intensity factor and 255,127,0 for the filterColor, go ahead and adjust those options.

Fig. 30.4 shows the results. The spotlight is attached to the right corner of the design. It is clicked on to show its light cone (called a *glyph*) and grips. These glyphs can be adjusted via the grips to vary the "sweep" of the light source and the area it affects (illuminates).

Distant light

The final light type we try out is New Distant Light. This is more of a floodlight that saturates everything and is not used as often.

Keyboard: Type in **light** and press Enter
Cascading menus: View→**Render**→**Light**→**New Distant Light**
Toolbar icon: Lights toolbar→ 
Ribbon: **Visualize** tab→**Lights** panel→ 

Step 1. Begin the Distant Light via any of the preceding methods. You may see a warning dialog box asking if you want to disable distant lights when the lighting unit is photometric (not shown here). Allow it in this case!
Step 2a. If you type in `light`,

- AutoCAD says: `Enter light type [Point/Spot/Web/Targetpoint/Freespot/freeweB/Distant] <Point>:`

Go ahead and select d for Distant and press Enter. Go on to Step 2b.

Step 2b. If you used any of the other methods or just completed Step 2a,

- AutoCAD says: Specify light direction FROM <0,0,0> or [Vector]:

Step 3. Pick a direction from which the light comes.

- AutoCAD says: Specify light direction TO <1,1,1>:

Step 4. Pick a direction into which the light goes. The familiar menu appears (not shown this time), and you can set some options to complete the distant light setting.

Lighting options

You probably noticed that each of these lighting approaches featured (with slight variations) several lengthy menus of options. We touched on a choice here and there as we set up sample light scenarios. Let us go ahead and describe in detail what these options mean. After you read through all of them, go back and retry all the Point, Spot, and Distant lights, varying the options and comparing the results.

- *Name*: You can give the light setting a name so you know to what you are referring later on. Use logical names like Bedroom_Spotlight_1.
- *Intensity factor*: This is the brightness of the light and you can enter almost any value. The exact number is dictated by design needs and often arrived at by trial and error.
- *Status*: This simply means *on* or *off*. You can position a light then turn it off, as with a regular light switch, if not needed in a scenario.
- *Photometry*: This option depends on the LIGHTINGUNITS system variable that was mentioned earlier, to be set to 1 or 2 (International or American with Photometry enabled), not 0, which is Generic lighting, as seen in Fig. 30.3. Once that is set, photometry refers to the *perceived power emitted by a light source*. The result of setting that system variable is that a more realistic light gradient is shown, where the intensity fades as you move farther from the lamp source, as seen in Fig. 30.5.There are three additional options under the Photometry menu choice:

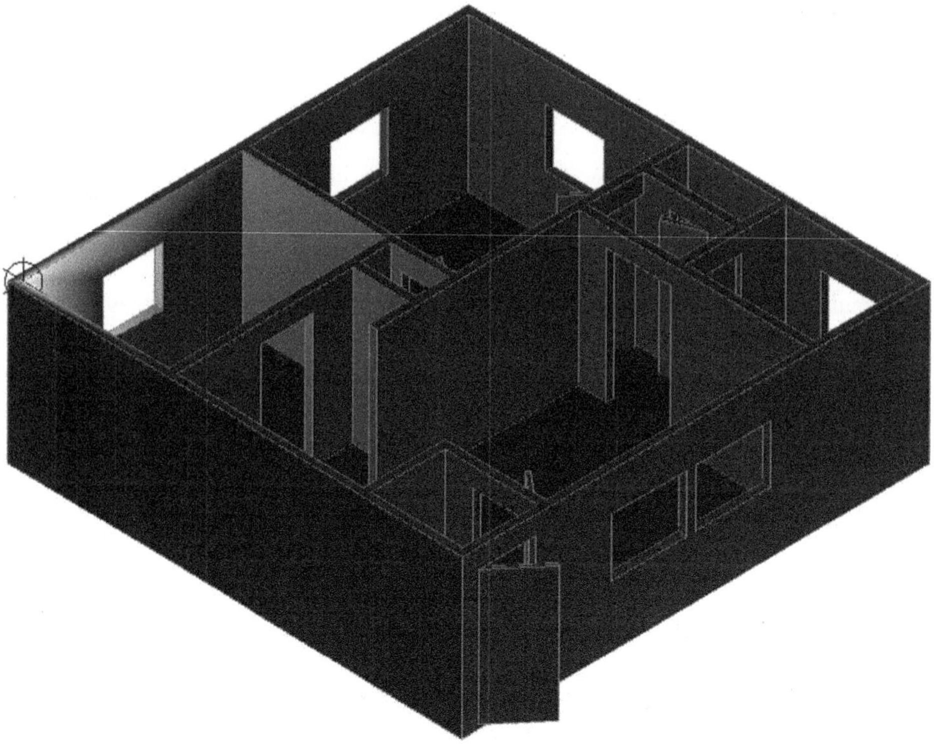

FIGURE 30.5 Point Light (LIGHTINGUNITS 1).

- *Intensity*: This is similar to the previous generic intensity setting, but allows for more options. The default values are in Candelas (Cd), the SI unit of luminous intensity. You can also press f for the Flux option (Lm) or press i for Illuminate, in which case you can use Lux (Lx) or Foot Candle (Fc) values. As a suboption, you can even use Distance. These settings are familiar to any architect or designer who works with lighting. Photometry is a science unto itself; experiment with the various options, and if you have never worked with lighting before, seek out information online or in a library on this interesting topic.
- *Color*: This option assigns color to the light (otherwise, it is the default white) but works differently from the color filters in the main menu. Here, color represents a type of lamp or a Kelvin temperature setting. If you type in a question mark and press Enter twice you get this list of standard colors or lamps:
  - "D65"
  - "Fluorescent"
  - "Coolwhite"
  - "Whitefluorescent"
  - "daylightfluorescenT"
  - "Incandescent"
  - "Xenon"
  - "Halogen"
  - "Quartz"
  - "Metalhalide"
  - "mErcury"
  - "Phosphormercury"
  - "highpressureSodium"
  - "Lowpressuresodium"

You can choose a lamp type and it is shown. Another interesting option is the ability to use temperature as a light setting. Go back one step and type in `k` for `Kelvin` and enter a temperature value. As mentioned before, it takes experience and knowledge to use these tools properly and there is a science of proper lighting. For those who work in this field, AutoCAD gives some basic tools with which to work; and this is definitely an invitation to learn more.

- *eXit*: This option exits the Photometry command.

shadow: This command controls the appearances of shadows when your light source encounters an obstruction (such as a piece of furniture). The options under that menu choice include

- *Off*: This option turns off shadows.
- *Sharp*: This displays shadows with sharp edges (less realistic, but better computer performance).
- *soFtmapped*: This entire option has to do with soft shadows and how much computer resources (map size and memory) to allocate to the shadow. As with all such settings, there is a trade-off between quality and system performance.
- *softsAmpled*: This is another set of settings for realistic shadows.

Attenuation: Attenuation is the gradual loss in intensity of any kind of energy source through a medium, which in this case is the decay of our light source as distance from that source increases. This is another interesting topic all unto itself. A variety of settings are available here, such as

- *attenuation Type*: The two choices are Inverse Linear and Inverse Squared, which have to do with the strength of the light (whether the value is squared or linear) as the distance increases.
- *Use limits*: This option turns limits on and off (see next two option descriptions).
- *attenuation start Limit*: This option specifies where the light starts.
- *attenuation End limit*: This option specifies where the light ends.
- *eXit*: This option exits the attenuation command.

filterColor: This is simply the color of the light emitted with no regard to temperature or type of lamp, as seen with Photometry settings. You can enter RGB values, Index colors, or use any of the other color choices AutoCAD provides.

Hotspot: This command, seen in the Spotlight option only, specifies the angle that defines the brightest cone of light, which is known as the beam angle. This value can range from 0° to 160°.

FIGURE 30.6 Properties palette.

Falloff: This command, also seen in the Spotlight option only, specifies the angle that defines the full cone of light, which is also known as the *field angle*. This value can also range from 0° to 160°.

To complete our tour of AutoCAD's lighting features, there are a few more items to cover. Perhaps the most important one is the Properties palette as it applies to lighting. Double-click on any light to call it up, as seen with the spotlight in Fig. 30.6.Using the Properties palette you can change virtually any aspect of the newly created light, with the added benefit that you can see the effect right away.

To keep track of all your lights, you can activate the Lights in Model palette (Fig. 30.7) by clicking on the Lights List icon on the toolbar or the arrow flyout on the Ribbon's Render tab. This palette shows every light created, which is why it is important to give them descriptive names. In our example, we did not, and they are simply referred to as Pointlight4 and Spotlight5.

Shadows, sun, and sky

Shadows are an integral part of the sun and sky settings, which are nothing more than shadows customized to a specific location and time. You can get a simple shadow right away, however. Draw and position some simple 3D shapes, make sure they are lifted off the ground (above zero in the XY plane), and in the Ribbon's Visualize tab, find the shadow settings and select Ground Shadows (Fig. 30.8). The result is seen in Fig. 30.9.

Sun and Sky, as mentioned already, refer to a background light setting where you can set a particular city location anywhere in the world and its particular latitude and longitude are factored into the lighting conditions. To set the time zone, you need access to Live Map Data through AutoCAD Mobile App (formerly called Autodesk 360). If you have this,

FIGURE 30.7 Lights in Model palette.

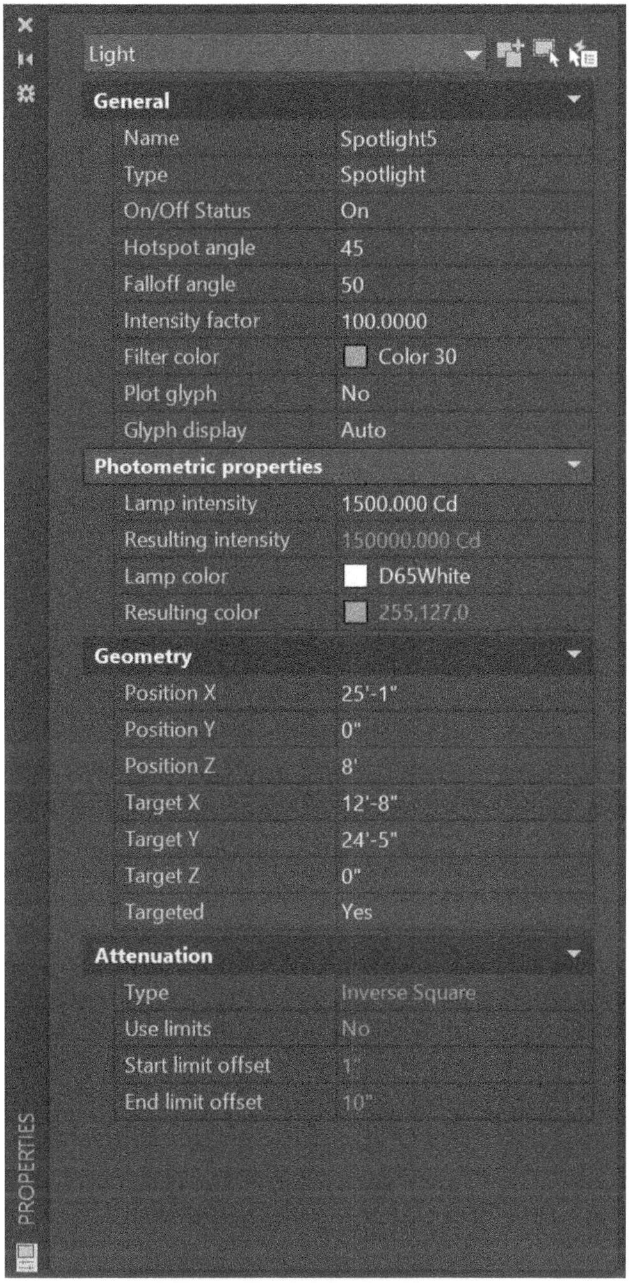

FIGURE 30.8 Ribbons Ground Shadows.

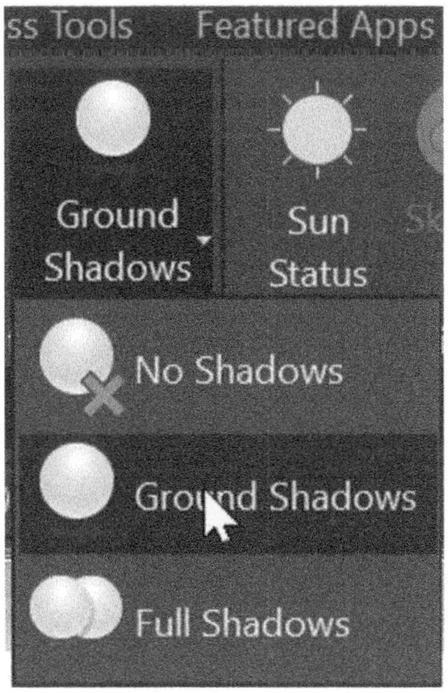

FIGURE 30.9 Basic ground shadows.

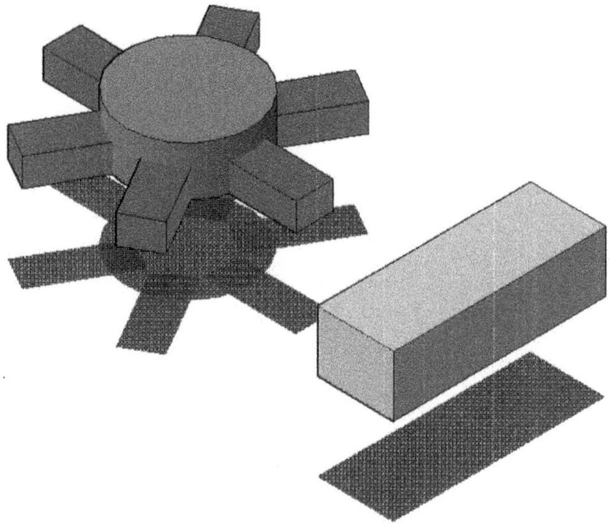

FIGURE 30.10 Geolocation – Live Map Data.

then press the Set Location drop-down icon in the Render, Sun & Location tab (it looks like a globe) and select From Map. A pop-up box (Fig. 30.10) then asks you if you want to use Live Map Data. If you say Yes, you are prompted to sign into Autodesk 360. Note that you can also import maps via the *.kml or *.kmz files if you have used those before.

Note that this is new for AutoCAD 2014 thru 2021. Prior to that, in AutoCAD 2013 and older versions, you could have just picked to enter the location values and seen the dialog box shown in Fig. 30.11. It is reproduced here in case you end up working on an older release. The Geographic Location dialog box allows you to set your particular region, country, or city via drop-down menus. You can use the provided time zones as an alternative. You can also click anywhere on the map or enter values manually, although not too many people have that kind of information handy. When done, click on OK.

Finally, let us bring up sun properties via the final icon on the Lights toolbar. The Sun Properties palette appears, as seen in Fig. 30.12.

Using this palette and a bit of trial and error (at least until you understand what each property does), you can set the ambient sunlight properties. The exact description of each category is found in the Help files, reproduced here for convenience and in edited form for clarity.

General: This is where you set the general properties of the sun.

- *Status*: This option turns the sun on and off. If lighting is not enabled in the drawing, this setting has no effect.
- *Intensity Factor*: This option sets the intensity or brightness of the sun. The range is from 0 (no light) to maximum. The higher is the number, the brighter the light.

FIGURE 30.11 Geographic Location, from AutoCAD 2013.

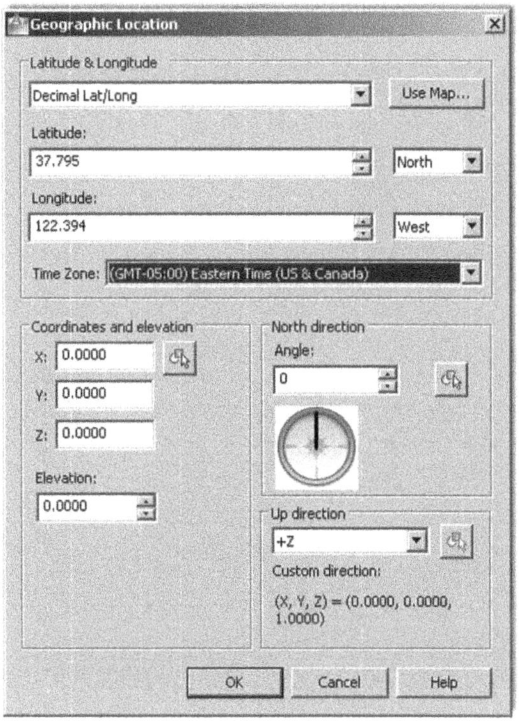

FIGURE 30.12 Sun Properties palette.

- *Color*: This option controls the color of the light. Enter a color name or number or click on Select Color to open the Select Color dialog box.
- *Shadows*: This option turns on and off the display and calculation of shadows for the sun. Turning shadows off increases performance.

Sky Properties: This is where you set the general properties of the sky.

- *Status*: This option determines if the sky illumination is computed at render time. This has no impact on the viewport illumination or the background. It simply makes the sky available as a gathered light source for rendering. Note: This does not control the background. Its values are Sky Off, Sky Background, and Illumination.
- *Intensity Factor*: This option provides a way to magnify the effect of the sky's light. Its values are 0.0 to MAX.
- *Haze*: This option determines the magnitude of scattering effects in the atmosphere. Its values are 0.0 to 15.

Horizon: This category of properties pertains to the appearance and location of the ground plane.

- *Height*: This option determines the absolute position of the ground plane relative to world zero. This parameter represents a world-space length and should be formatted in the current length unit. Its values are +to MAX.
- *Blur*: This option determines the amount of blurring between ground plane and sky. Its values are 0 to 10.
- *Ground Color*: This option determines the color of the ground plane by selecting a color from the drop-down list. You can also select the Select Color dialog box to make a color choice.

Advanced: This category of properties adds various artistic effects.

- *Night Color*: This option specifies the color of the night sky by selecting a color from the drop-down list. You can also select the Select Color dialog box to make a color choice.
- *Aerial Perspective*: This option specifies if aerial perspective is applied. Its values are On or Off.
- *Visibility Distance*: This option specifies the distance at which 10% haze occlusion results. Its values are 0.0 to MAX.

Sun Disk Appearance: This category of properties pertains to the background only, and they control the appearance of the sun disk.

- *Disk Scale*: This option specifies the scale of the sun disk (1.0=correct size).
- *Glow Intensity*: This option specifies the intensity of the sun's glow. Its values are 0.0 to 25.0.
- *Disk Intensity*: This option specifies the intensity of the sun disk. Its values are 0.0 to 25.0.

Sun Angle Calculator: This category of properties sets the angle of the sun.

- *Date*: This option displays the current date setting.
- *Time*: This option displays the current time setting.
- *Daylight Saving*: This option displays the current setting for daylight savings time.
- *Azimuth*: This option displays the azimuth, the angle of the sun along the horizon clockwise from due north. This setting is read-only.
- *Altitude*: This option displays the altitude, the angle of the sun vertically from the horizon. The maximum is 90°, or directly overhead. This setting is read-only.
- *Source Vector*: This option displays the coordinates of the source vector, the direction of the sun. This setting is read-only.

Rendered Shadow Details: This category of properties specifies the properties of the shadows.

- *Type*: This option displays the setting for shadow type. This setting is read-only when the display of shadows is turned off. The selections are Sharp and Soft (mapped), which display the Map size option, and Soft (area), which displays the Samples option. Soft (area) is the only option for the sun in photometric workflow (LIGHTINGUNITS set to 1 or 2).
- *Map size* (Standard lighting workflow only): This option displays the size of the shadow map. This setting is read-only when the display of shadows is turned off. Its values are 0 to 1000.
- *Samples*: This option specifies the number of samples to take on the solar disk. This setting is read-only when the display of shadows is turned off. Its values are 0 to 1000.
- *Softness*: This option displays the setting for the appearance of the edges of shadows. This setting is read-only when the display of shadows is turned off. Its values are 0 to 50.0.

Geographic Location: This category of properties displays the current geographic location settings. This information is read-only. When a city is not stored with latitude and longitude, the city does not appear in the list. Use the Edit Geographic Location button to open the Geographic Location dialog box.

30.3 Materials and rendering

Rendering in AutoCAD is a two-part story. The first part involves materials and simply adding them to designs. AutoCAD has a sizable built-in library of materials, with most residing in the tool palettes; and adding them is simply a matter of click and drag onto the design. These materials are pictures, usually bitmaps, and they simply cover the design similar to wallpaper. You can also create and modify your own materials, within limits.

The other side to rendering is actual rendering, where more advanced settings are applied and the software makes passes over the design, rendering the surfaces as desired. In regard to both material application and rendering, other dedicated software packages have far more capabilities than AutoCAD, as explained earlier in this chapter, but this is still a good introduction to this fascinating world. And, remember that quality of output is often a result of knowledge. An expert using limited software may still achieve better results than a novice on superior software. If you wish to pursue this type of work further, consult Appendix I, Lighting, Rendering, Effects, and Animation for additional information on this topic.

Materials

Let us jump right into materials by introducing the two relevant palettes (materials is very much a palette-driven feature): the Materials Browser and the Materials Editor. Both are shown in Fig. 30.13 and can be accessed via the cascading menus: Tools→Palettes→Materials Browser (or Materials Editor).

Let us first take a look at the Materials Browser. It has expanded significantly over the last several releases of AutoCAD and now features over 1200 materials (a small sample is seen in Fig. 30.14). These materials are bitmap graphics designed in-house at Autodesk (or acquired) and are superimposed onto geometry in a simple manner, akin to wrapping paper.

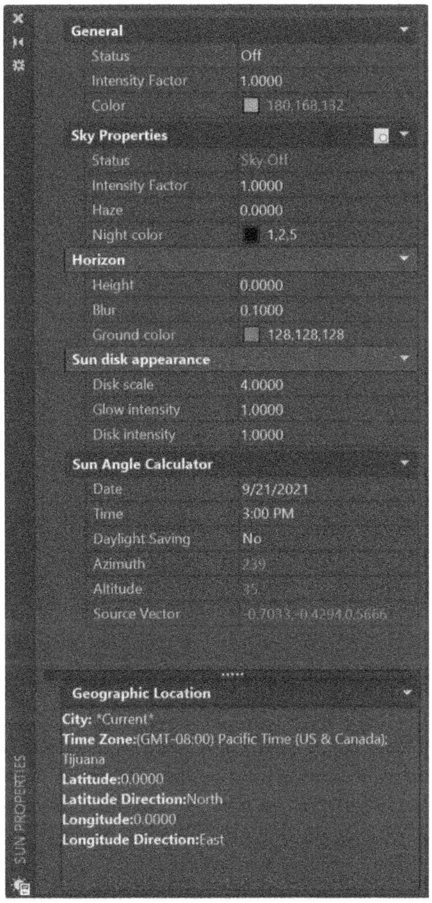

FIGURE 30.13 Materials Editor and Materials Browser.

They are not "procedural materials" and do not become one with the design (see Appendix I for more on this). The Materials Browser allows you to visually search the materials database and create a library of materials relevant to your design intent.

Applying these materials without further modification is simply a matter of click and drag. Draw a cube of any size, as seen in Fig. 30.15, and select a random material. Then, click and drag the material onto the cube, which acquires those properties instantly. Some additional examples are shown in Fig. 30.16.If you right-click on any of the materials in the selection list, you can add it to the Material Browser's My Materials collection for future reuse. These materials can be deleted and renamed, if necessary. The main Autodesk library, however, is locked, as indicated by a padlock. You can sort My Materials (or the main library) according to Name, Category, Type, or Color as well as filter what is shown by All, Applied, Selected, or Unused. Spend some time getting acquainted with the Materials Browser, so you have a good idea of what is available and what you can do with these materials.

The Materials Editor is exactly that: It edits and modifies existing materials. To enter something into the editor, double-click on it in the browser - there are also a pair of small icons for this at the bottom of the material, when viewed as a thumbnail. The material appears in the editor's preview window, with associated menus below that. An example is shown with the previously used Ashlar material (Fig. 30.17). Some of the more useful settings that can be modified include Reflectivity, Transparency, and Self Illumination. Explore all these settings to see what the effects are. When done, the material appears in the Materials Browser. Be sure to rename it before using.

Rendering

Our final discussion involves rendering. Because of AutoCAD's improved materials and excellent visualization and coloring options, rendering in general is not something that most users will require on a regular basis. AutoCAD's rendering ability and the available options are not extensive, though admittedly much improved in the last decade.

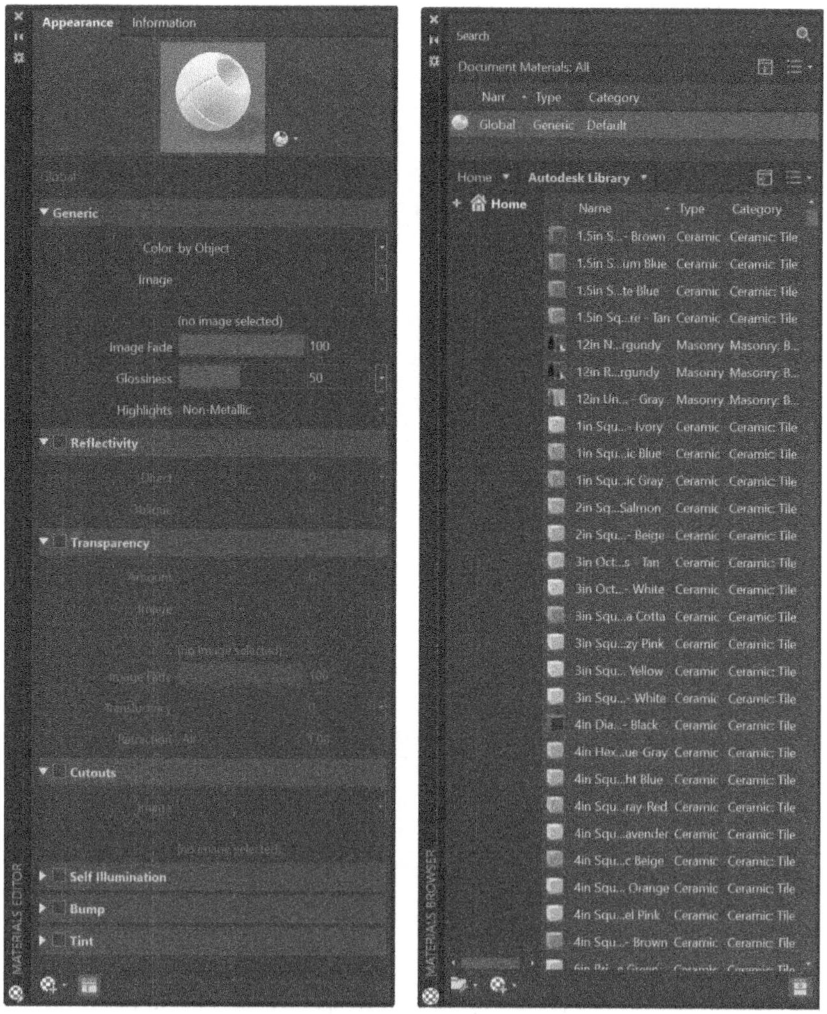

FIGURE 30.14 Expanded Materials Browser.

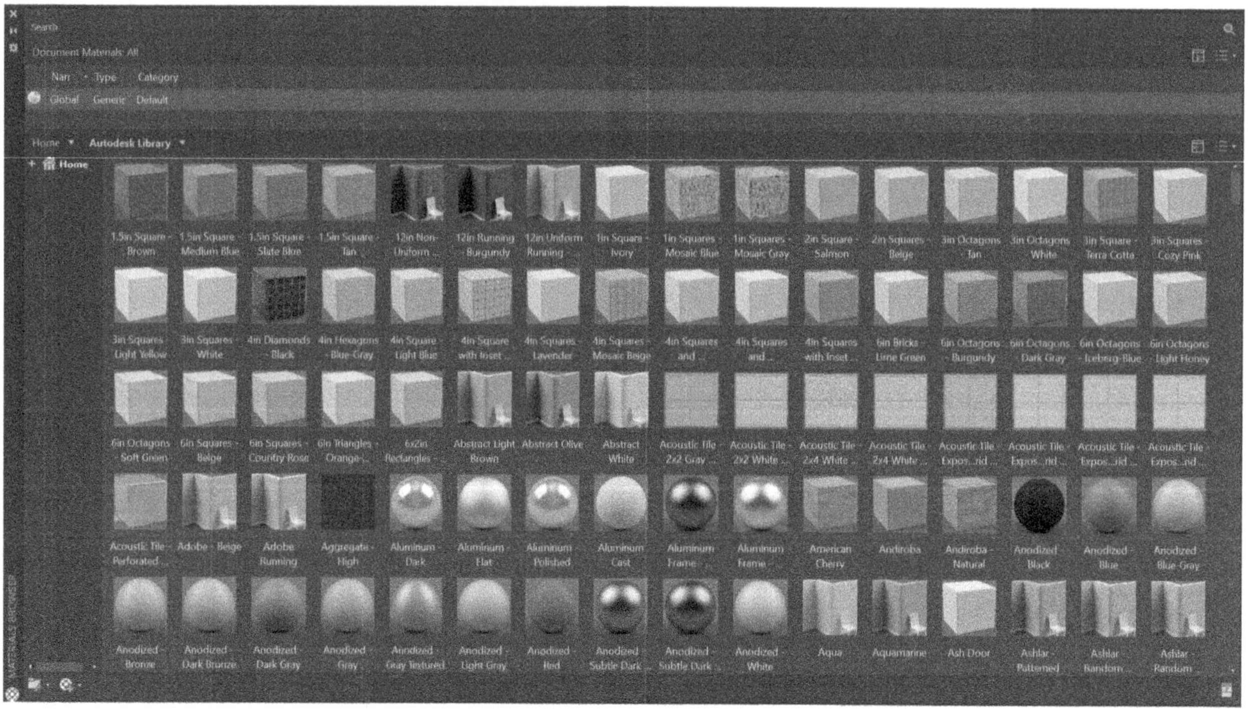

FIGURE 30.15 Material application.

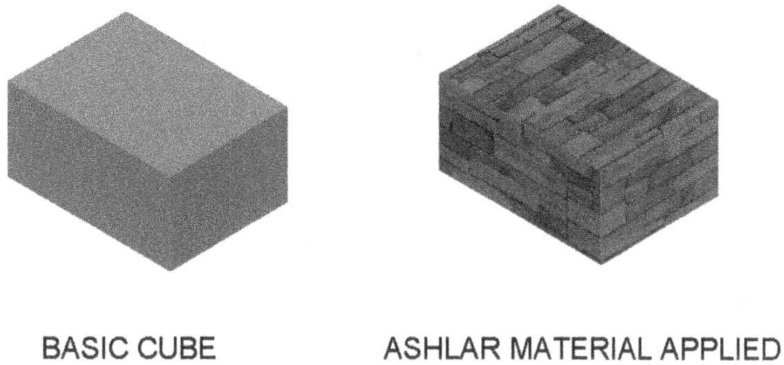

FIGURE 30.16 Material application.

FIGURE 30.17 Materials Editor with a material loaded.

The latest version features a new rendering engine and a simplified process to get what you want. Before we go any further though, let us first describe exactly what rendering is.

In advanced modeling software, *rendering* is a general catch-all term that refers to advanced colorizing, materials, and lighting effects applied to a design. In AutoCAD, these procedures are separate from rendering, and we have already covered them in this chapter or elsewhere. What rendering means in *our* discussion is simply a process to generate output that can be saved to a file. Any addition of colors, materials, and lighting is done prior to this. Then, the process of rendering (of a ready-to-go model) produces a tangible output in the form of a JPG, or another format such as a BMP (bitmap), TIF, PNG, and so forth, that can then be printed on a LaserJet or a plotter. These renderings can also be inserted into a presentation or just given to a client. The bottom line is this: You are getting a pretty picture of your model to use as you please.

To try out rendering, bring up the model of our familiar apartment (Fig. 30.18) and, if you wish, add a new material for some variety, as shown in the figure. The gateway to rendering is the Ribbon's Visualize tab, Render panel and to a lesser extent the Render toolbar (they both have similar options), and both are shown in Fig. 30.19.

Most of the settings you deal with prior to starting a rendering session have to do with the type and quality of the output, so we take a look at this first. Starting the actual rendering process involves just one push of a button. The new (back in AutoCAD 2016) rendering interface and palettes make these settings easier, more intuitive, and even humorous.

The first item to take a look at is the size of the desired output via the Render to Size drop-down menu in the Render panel, as seen in Fig. 30.20. Here, you can choose your desired level from relatively basic SVGA to Full HDTV to a much denser 300 dpi on an A3 sheet (a very sharp, high-dpi setting for this sized sheet). Needless to say, each upgrade in pixel resolution comes with a penalty in rendering time and computer resource usage.

Additional output settings can be accessed via the appropriately titled More Output Settings... menu choice at the bottom, which brings up a small dialog box (Fig. 30.21) that allows you to make custom Width, Height, and Resolution settings. Also, here, you select other output file formats by checking the box under Save Image and clicking Browse....

This action brings up the Render Output File dialog box (Fig. 30.22), where you can drop down the Files of type: field at the bottom. As Fig. 30.22 shows, your available output formats are BMP, PCX, TGA, TIF, JPEG (or JPG), and PNG, with the JPG photograph format being the most familiar to most people. Just as important, you can specify the location of the rendering output, which can be right next to the drawing file itself or in a separate output folder.

There is another area of the Render panel, just to the right of the teapot, where you can set rendering by number of levels (or passes) the rendering engine makes or also by time, as seen in Fig. 30.23. The low (1 level), Medium (5 levels), and High (10 levels), of course, make your rendering sharper as you go up the scale, at the usual cost in time and resources.

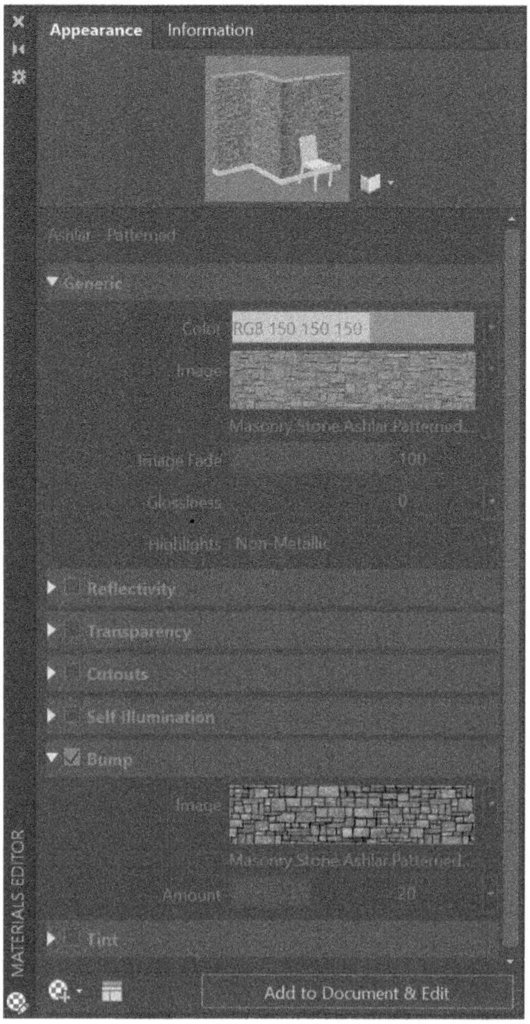

FIGURE 30.18 Rendering model.

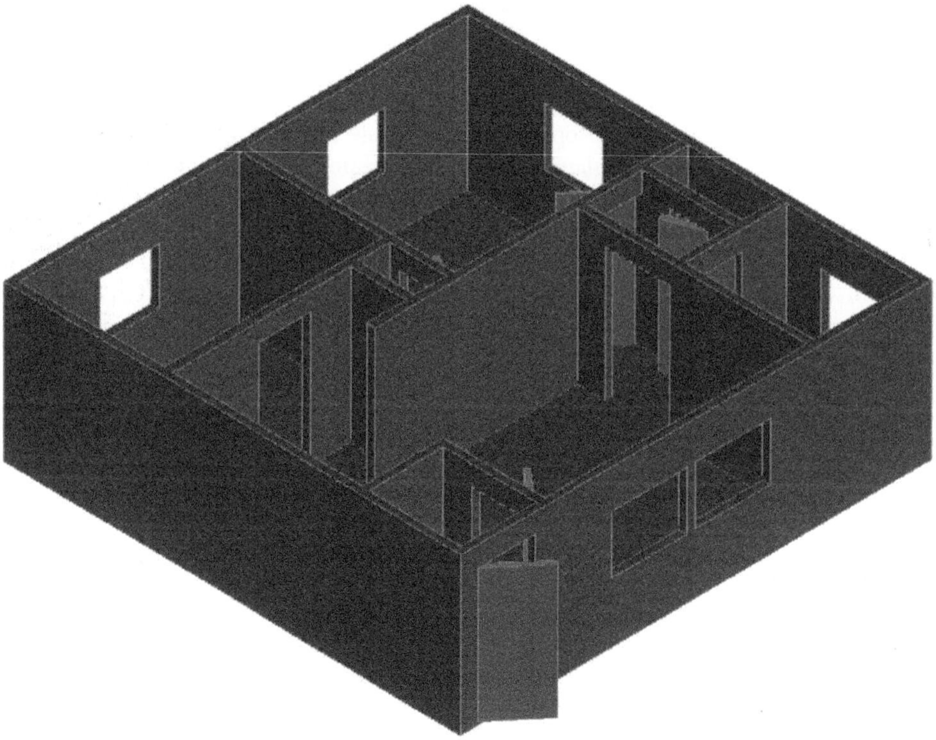

FIGURE 30.19 Ribbon's Render panel and the Render toolbar.

FIGURE 30.20

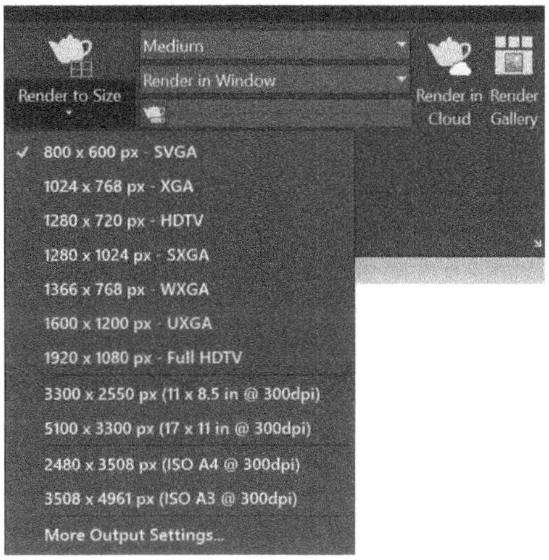

FIGURE 30.21 Render to Size Output Settings.

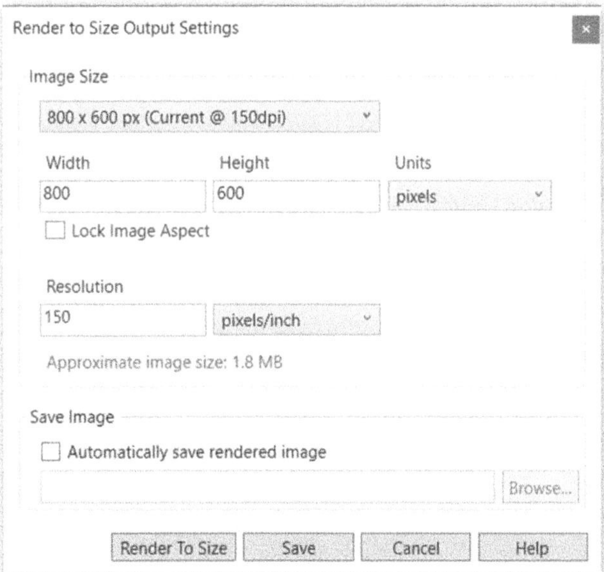

FIGURE 30.22 JPEG Image Option in the Render Output File dialog box.

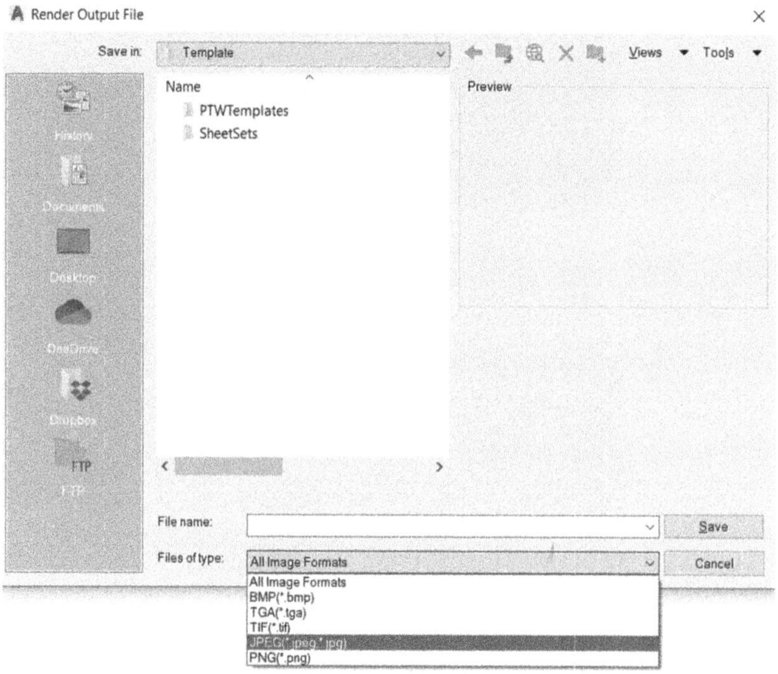

FIGURE 30.23 Rendering levels and times.

Speaking of time, the next three presets render until a time limit is reached, regardless of how many levels are covered. They are, in a bit of rare Autodesk humor, tied to office events. Coffee-Break Quality is 10 minutes of rendering time (as if anyone limits their coffee break to a mere 10 minutes!), Lunch Quality is one hour, and Overnight Quality is 12 hours, assuming of course you are not an entry-level attorney or Wall Street firm employee, working 16-hour days.

At the bottom of the panel seen in Fig. 30.23 is access to the Render Presets Manager via Manage Render Presets… (one of a few ways to get to it). Shown in Fig. 30.24, it allows for some additional settings for your output, with a few features duplicated from elsewhere. Here, levels and times can be set to any intermediate values as opposed to preset ones.

Finally we have one more palette, the Render Environment and Exposure palette (Fig. 30.25). It can be accessed via the drop-down menu on the Render panel (also seen in Fig. 30.25). Here, you can apply some advanced settings with background and lighting and adjust the exposure and white balance of the rendering.

After all these "presettings" are done, you are ready to render your model. Press the teapot icon on the Render toolbar; the rendering window appears and AutoCAD begins making passes over your model. Fig. 30.26 is a screen shot of this process about 28% complete on a very low-resolution rendering of the 3D apartment model. When it is done, the image is saved and ready for use in the format and location you requested in the Render Output File dialog box.

FIGURE 30.24 Advanced Render Settings.

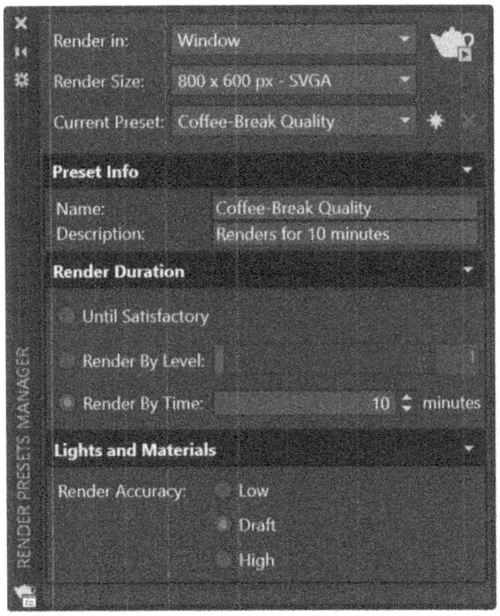

FIGURE 30.25 Render Environment and Exposure palette.

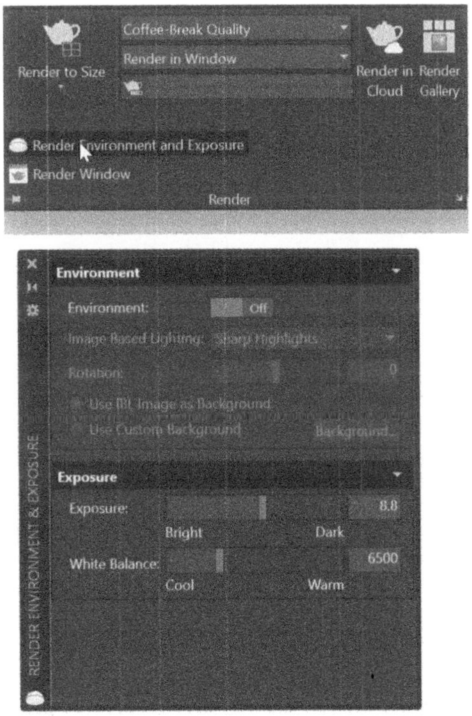

FIGURE 30.26 Render Environment and Exposure palette.

Summary

You should understand *and know how to use the following concepts and commands at the conclusion of this chapter:*

- Point light
- Spotlight
- Distant light
- Shadows
- Sun and Sky
- Materials
- Rendering

Review questions

Answer the following based on what you learned in this chapter:

1. Describe point light.
2. Describe spotlight.
3. Describe distant light.
4. Describe sun and sky.
5. Describe how to bring up materials.
6. Describe basic rendering procedures.

Exercise

Create the following 20″×50″ box, extruded to 10″ and shelled out to 1″. Attach a point light to both corners. Set one point light to an intensity of 100 and a blue color. Set the other to an intensity of 75 and a red color. Then, render the figure to a JPG file. (Difficulty level: Easy; Time to completion: 5–10 minutes.)

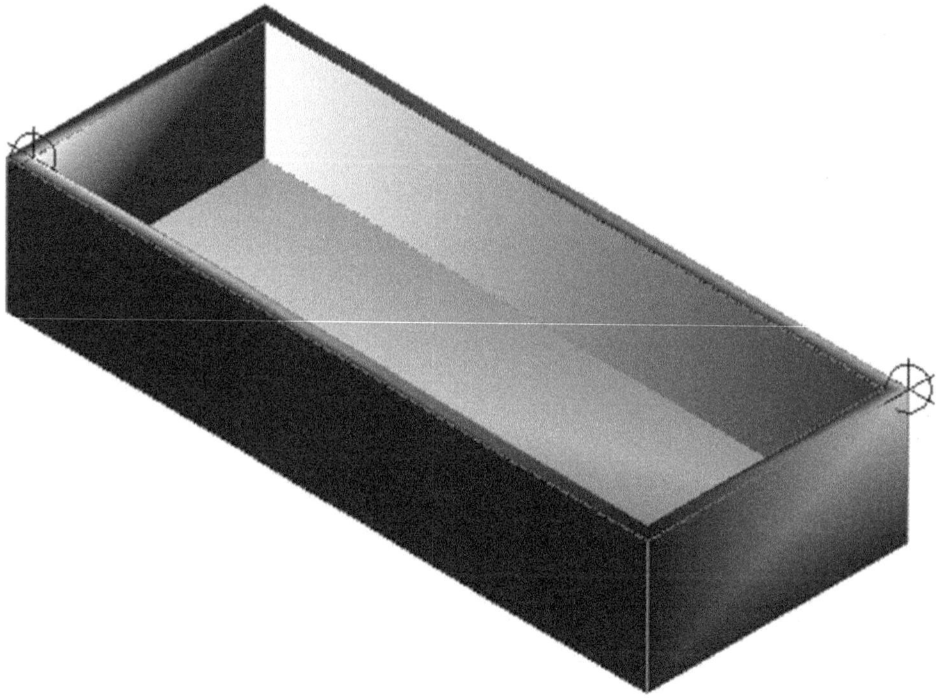

Level 3

Answers to review questions

Chapter 21 Review questions

1. What three axes are discussed?
   X, Y, and Z.
2. What three planes are discussed?
   XY, ZX, and YZ.
3. What 3D toolbars are introduced in this chapter?
   View, 3D Navigation, Visual Styles, UCS, Modeling, and Render.
4. What command is used most often to create 3D objects?
   Extrude.
5. How do you rotate the UCS icon?
   Via the UCS command or UCS toolbar icons.
6. What does hide do?
   Hide turns off lines that would not be visible if the shape were a solid.
7. What does shade do?
   Shade adds colored shading, with light effects, to your design.
8. What are the two types of shade?
   The two types are conceptual and realistic.
9. What is the purpose of the Navigation bar and ViewCube?
   Both present additional 3D navigation and viewing options.

Chapter 22 Review questions

1. Describe the rotate3D procedure. Do you need axes or planes for this command?
   Rotate3D rotates objects around axes. You need axes.
2. How is 3Drotate different from Rotate3D?
   3Drotate features a graphical gizmo to assist in the rotation process. The net effect is the same as rotate3D.
3. Describe the mirror3D procedure. Do you need axes or planes for this command?
   The mirror3D command mirrors objects around planes. You need planes.
4. Describe the 3Darray procedure, both rectangular and polar.
   Rectangular 3Darray is command-line driven and asks for rows, columns, and levels as well as the distances between all of these. Polar 3Darray is also command-line driven and asks for number of items, angle to fill, and two points that define the axis of rotation for the array.
5. Describe 3Dscale and 3Dmove.
   Both commands are similar to their 2D versions; however, 3Dmove features movements constrained by axes.
6. Describe fillets and chamfers in 3D.
   A 3D fillet must have a radius and can be performed instantly on any edge, as well as dynamically in real time. A 3D chamfer is similar.

Chapter 23 Review questions

1. What are the three Boolean operations?
   Union, subtract, and intersect.
2. In what toolbar are they found?
   Modeling toolbar.
3. Is union a permanent setting?
   Yes
4. Why is it a good idea to make a copy of something that is being subtracted?
   Because you may need that part for later design work and the subtract operation deletes it.
5. Name the seven primitives covered in this chapter.
   Box, wedge, cone, sphere, cylinder, torus, and pyramid.
6. What four methods are used to create primitives?
   Typing, cascading menus, Modeling toolbar, and the Ribbon's Solid tab.
7. What two primitives cannot be created any other way?
   Sphere and torus.

Chapter 24 Review questions

1. Describe the revolve command.
   This command creates a revolved shape around a chosen axis of revolution.
2. Does the profile need to be open or closed? Describe the effect of each.
   Either way works; however, a closed profile yields a solid, and an open one creates a surface of revolution.
3. Describe the shell command.
   This command creates a hollow cavity inside a solid based on a given wall thickness.
4. How many surfaces can you select with shell?
   You can select a minimum of one and a maximum of all of them. Typically, only one or two are selected.
5. Describe the taper command.
   This command tapers (adds slope to) a solid.
6. Describe the loft command.
   This command creates a solid out of a sequence of profiles.
7. Describe the path extrusion command.
   This command creates a solid extrusion along an indicated pathway.
8. Describe the sweep command.
   The sweep command is similar to a path extrusion but more general in nature and creates a solid out of virtually any profile and along virtually any pathway.
9. Why is sweep better than path extrusion?
   It is more tolerant of geometric variations and inconsistencies.

Chapter 25 Review questions

1. Describe the polysolid command. Can you turn regular objects into polysolids?
   This command creates a solid shape out of linework or basic geometric objects. Yes.
2. Describe the helix command.
   This command creates a wire helix based on indicated radiuses of the top and bottom along with the number and spacing of the coils.
3. Describe the 3D path array command.
   The 3D path array works in a similar manner to the 2D path array. A solid is drawn, a path is selected (in our example, a helix), and the solid is duplicated along the path.
4. Describe the 3Dalign command.
   This command aligns objects in 3D space via a three-point tangency.
5. What are the six faces commands?
   Move face, offset face, delete face, rotate face, copy face, and color face.
6. What are the two edges commands?
   Copy edge and color edge.

Chapter 26 Review questions

1. Describe the planar surface command.
   A planar surface is a very simple mesh surface shading tool for basic geometry and surfaces.
2. Describe the region command.
   A region is similar to a planar surface but is not a mesh.
3. Describe rulesurf and its applications.
   Rulesurf (ruled surface) is used to create surfaces between straight lines or curves.
4. Describe tabsurf and its applications.
   Tabsurf (tabulated surface) is used to create surfaces along a path.
5. Describe revsurf and its applications.
   Revsurf (revolved surface) is similar to the revolve command but produces surface revolutions.
6. Describe edgesurf and its applications.
   The edgesurf command is used to create mesh (Coon) surfaces.
7. Describe 3Dface.
   3Dface is a free form mesh creating tool, usually used to cover surfaces.
8. What are the Smooth Mesh Primitives?
   These are basic primitive shapes that are made up of mesh faces, which in turn can be worked with to "sculpt" an organic form.

Chapter 27 Review questions

1. Describe the slice command.
   The slice command creates a section view through a part by deleting the surrounding material.
2. Describe the section command.
   The section command creates a section plane, which can now be worked on.
3. What is the major difference between slice and section?
   The slice command slices then allows you to delete material along a section plane. The section command does not allow you to delete material, rather it creates the section plane itself.
4. What does the VIEWBASE command accomplish?
   It establishes the base (parent) view.
5. What does the VIEWPROJ command accomplish?
   It establishes additional projections.
6. What does the VIEWSECTION command accomplish?
   It creates section views in the projections.
7. What does the VIEWDETAIL command accomplish?
   It creates a detail view from a projection.
8. What is the critical difference between flatten and flatshot?
   Flatten creates a 2D model from a 3D model via conversion and destruction of the original. Flatshot creates a 2D copy of the original 3D model and does not destroy it.
9. What are vports useful for? List two functions.
   Vports allow you to view and print a model from multiple angles all at once.

Chapter 28 Review questions

1. List all 13 advanced UCS functions and what they do.
   UCS, World, UCS previous, face UCS, object, view, origin, Z axis vector, 3 Point, X, Y, Z, and apply. See Chapter 28, Advanced UCS, Views, Text, and Dimensions in 3D for detailed descriptions of each.
2. What are views useful for?
   Views come in handy when you want to save a particular viewpoint of your model that is convenient or useful.

Chapter 29 Review questions

1. List all 11 dview functions and what they do.
   Camera, Target, Distance, Points, Pan, Zoom, Twist, Clip, Hide, Off, and Undo. See Chapter 29, Dview, Camera, Walk and Fly, Path Animation for detailed descriptions of each.
2. List any additional views that are covered.
   The camera tool provides for additional viewing tools.
3. What is the purpose of Walk and Fly?
   To create a walking tour through a 3D model.
4. How are Walk and Fly functions controlled?
   Via the keyboard arrows or hot keys.
5. How do you create path animations?
   You first create a pathway, followed by setting up a camera and fine-tuning the animation by specifying the frame rate, resolution, and format.

Chapter 30 Review questions

1. Describe point light.
   The point light is a direct lighting effect and the most often used of the lighting options. It shines evenly from a point in space indicated by the user.
2. Describe spotlight.
   The spotlight is somewhat similar to the point light but more concentrated, akin to a flashlight, with less reach through walls, casting softer shadows.
3. Describe distant light.
   The distant light is similar to a floodlight, saturating everything in its cone.
4. Describe sun and sky.
   Sun and sky refer to the background lighting and the ability to set a particular worldwide location, which in turn accounts for latitude and longitude.
5. Describe how to bring up materials.
   Materials are accessed via a palette called the Materials Browser (also called the Materials Editor) and can be clicked and dragged onto shapes.
6. Describe basic rendering procedures.
   Rendering involves the render command and saving the output to a file such as in JPG format.

Appendix A

Additional information on AutoCAD

Appendix A outlines additional information on AutoCAD, such as who makes it, what AutoCAD LT is, how AutoCAD is purchased and how much it costs, differences between releases, AutoCAD for the Mac, a history of AutoCAD, AutoCAD releases, other major Autodesk products, and AutoCAD-related websites.

A.1 WHO MAKES AUTOCAD?

As you may already know, AutoCAD is designed and marketed by Autodesk, Inc. (NASDAQ: ADSK), a software giant headquartered in San Rafael, California. Its website is www.autodesk.com. The company was founded in 1982 by programmers John Walker and Dan Drake along with several other associates and went public in 1985 (see the history). Autodesk today is headed by president and CEO Andrew Anagnost, has over 10,000 employees worldwide, and posted revenues of over $3.27 billion in FY 2020. Although the company markets about 100 software applications, AutoCAD and AutoCAD based products, remain their flagship products and the biggest single source of revenue (estimated to be over 45% of the total).

It is difficult to pin down the exact size of AutoCAD's worldwide "installed base," meaning the number of copies of the software that have been purchased and used over the years. Some estimates from 2008 show this to be 2.8 million units of AutoCAD and 3.6 million units of AutoCAD LT. An Autodesk press release from the same year claimed 4.16 million seats but did not specify if that was only AutoCAD or LT as well. Industry analysts have estimated that Autodesk currently sells about 40,000 copies of AutoCAD per quarter or 160,000 seats per year. Due to piracy, the actual installed base is probably significantly higher. Also, the number of seats can vary depending on how you count educational versions and those "add-on" vertical software purchases that need AutoCAD to run but are counted separately.

Any way you look at this, though, these numbers indicate that AutoCAD is indisputably the world's number one drafting software, with a market share often estimated to be somewhere in the 55%–60% range in the 2D world (some analysts say 85%) and 30%–35% among all CAD software (both 2D and 3D) worldwide. Even though AutoCAD has over a dozen major competitors and hundreds of smaller ones, this sort of market dominance is likely to assure AutoCAD's relevance for years to come.

A.2 WHAT IS AUTOCAD LT?

Contrary to popular belief, LT does not stand for "lite" (or as we joke in class, "AutoCAD with fewer calories"). It also does not mean "less trouble" or "learning and training" or even "limited," although Autodesk likes to throw that last term around during conferences. Instead, it is generally taken to mean "laptop" and was originally meant for slower and weaker early laptop computers, a sort of portable AutoCAD, if you will. Today, laptops are just as powerful as desktops and LT exists to compete in the lower-end CAD market with similarly priced software. It is essentially a stripped-down AutoCAD. What is missing is detailed exhaustively on the Autodesk website, but it essentially boils down to no 3D ability, less advanced customization, and no network licensing. In addition to these, some other minor missing features are scattered throughout the software (like lack of multi-lines), but the differences are not that great; and aside from the lack of 3D, it takes a trained eye to spot LT. It really looks and acts like AutoCAD and with good reason. It was created by taking regular AutoCAD and "commenting out" (thereby deactivating) portions of the C++ code that Autodesk wanted to reserve for its pricier main product.

Having said this, however, AutoCAD LT is an excellent product (its estimated sales are said to outnumber AutoCAD) and is recommended for companies that do not have ambitious CAD requirements. The truth and something Autodesk does not always promote is that some users never need the full power of AutoCAD and do just fine with LT while paying a fraction of the cost of full AutoCAD. Electrical engineers especially need nothing more than LT for basic linework and diagrams, as do architects and mechanical engineers who deal with only 2D work. The size of the files you anticipate working on, by the way, is of no consequence. LT lacks certain features but not the ability to work with multistory skyscraper layouts, if needed.

A.3 HOW IS AUTOCAD PURCHASED AND HOW MUCH DOES IT COST?

There was a major change to Autodesk's policy regarding AutoCAD pricing and purchase options. It was hinted at for some time throughout 2015 and was finally implemented on January 31, 2016. You can no longer outright purchase AutoCAD for a one-time price. Instead, there exists a "subscription" model, where you pay as you go (monthly or annually) for usage and licensing of the software. This is a major departure from the last 30+ years and follows an overall trend among some software vendors to "rent" out their products as opposed to a one-time sale. The specifics of this new policy are outlined next, along with the older approach presented for historical reasons and to put things into perspective.

For the most part, the way AutoCAD was sold, up until this change, was via a perpetual license or the "buy-it-once" method. It is the way we approach most purchases—you contact a seller, pay for the product, and it is yours to use forever or until you want a newer version of that product or something else entirely.

To purchase a perpetual license of AutoCAD in the 1980s and 1990s, you would have contacted an authorized dealer (or reseller). The sale was then usually arranged by phone, paid for with a credit card, and AutoCAD was shipped on floppy disks (then CDs and later DVDs) via registered mail. In more recent times the process moved online, and since 2000 Autodesk's website was configured for direct purchase and the reseller was often passed over—although resellers still often had access to better pricing and offered tech support. Over the years, printed command references and manuals were phased out and even physical media was hard to find as customers were just downloading the software purchase and associated authorization codes. The full AutoCAD was never sold in physical or online stores, but its sibling AutoCAD LT was in fact found in such retailers as Staples and Office Depot as well as Amazon.com.

So how much did it all cost? Here are some recent historical data. In January 2015, AutoCAD 2015 was shipping or downloading for $4195. An upgrade from 2014 or 2013 was $2095. For its part, AutoCAD LT was a far more modest $1200. A maintenance subscription was also available as an option. For 2015, it was an extra $545 per license, bringing the cost of AutoCAD to $4740. For the extra cost you would get several layers of support and other benefits such as cloud access and discounts on future purchases, and of course free upgrades every year.

So where are we now with this new policy? Well, if you purchased AutoCAD before January 31, 2016, you were grandfathered into the perpetual license and could use the software for as long as you wished. All purchases since then, including the latest AutoCAD version (to be released in late March), have only the subscription "rental" option. The cost for one license, as outlined on the Autodesk website, is either $200 per month or $1610 per year. There is also a 3 year ($4345) subscription available. For AutoCAD LT the prices are $50 monthly or $400 per year. LT also has a 3 year ($1080) subscription available.

So, is this new pricing a good thing for AutoCAD customers? That depends on who you are and how you use and manage the software. It is certainly a win for Autodesk when it comes to small firms, as these often do not upgrade for many years and now must. However, it should benefit some of the larger enterprise users who do upgrade annually and now can spread out their software expenses over a full year cycle. The monthly rental plan is also good for when you hire some drafting help for a few months and will not need the license once the overflow work is done. The software used to be counted as an asset for companies as well, but with the subscription model it is counted as a liability.

With all of that stated, recent trends point to the fact that AutoCAD tools and functionality will continue to remain very similar to previous versions. Minor improvements, but few game-changing "must-have" tools and updates work their way into the software overtime. It is possible that many vendors and customers will resent the forced upgrades going forward.

A.4 ARE THERE SIGNIFICANT DIFFERENCES BETWEEN AUTOCAD RELEASES?

This is a common question at the start of training sessions, and the answer is hinted at in the last paragraph of the preceding software-pricing discussion. Many schools are authorized training centers and, as such, have the latest version of AutoCAD installed. That may or may not be the case with the students and their respective companies. No issue seems to cause as much confusion and anxiety as this one. To put your fears to rest, if you are learning AutoCAD with this book and have

AutoCAD 2020, 2019, 2018, 2017, or even earlier versions at home or at work do not worry. The differences, while real, are for the most part relatively minor and cosmetic (except perhaps for the introduction of the Ribbon in 2009). While AutoCAD has certainly been modernized, it has not fundamentally changed since it moved to Windows and away from DOS over 20 years ago and finally stabilized with R14 in 1997. If you were to open that release, you would feel right at home with 90% of the features.

While Autodesk's marketing department may insist that the latest version of AutoCAD has some critically important new features or enhancement, the fact remains that the vast majority of what is truly important has already been added to AutoCAD and new features are incremental improvements. Most of what you learn in one version is applicable to several versions backward and forward, especially at a beginner's level.

A.5 IS THERE AN AUTOCAD FOR THE MAC?

The short answer is "Yes." After a nearly 20-year absence, Autodesk announced in August 2010 that the Macintosh platform is once again supported, and AutoCAD for the Mac was shipped in November of that year. Why was AutoCAD offered for the Mac up until 1992, and then removed from the market, only to be reintroduced in 2010? The reasons are purely financial ones. In the late 1980s and early 1990s, for those old enough to remember, the PC was not yet the dominant force it is today. Mac and other manufacturers were all strong competitors and there were many choices on the market. Autodesk decided to go with the leading two, PC and Mac, and invested in both. As the PC became the de facto world computer standard and Apple went into a decline, resources had to be focused on the PC, as it takes a lot of effort to reprogram and support a major piece of software for another configuration.

We all know what happened to Apple Inc., of course. Under leadership from the late Steve Jobs, it roared back strong with a line of high-demand consumer products, such as the iPod, iPhone, and iPad. Its computers also regained some of its former market share with superb desktop and laptop offerings.

Autodesk took notice of these developments. Such is the nature of the software business: Your fortunes are closely tied in with the ups and downs of hardware, operating systems, and chip makers. The Mac once again began to look like an attractive platform, and Autodesk invested in reprogramming AutoCAD to support it. When exactly this effort started is not known outside the company, but it was likely authorized a few years before the official 2010 launch, as it takes a lot of lead time to rework code of such complexity. This move should net Autodesk quite a few new sales of AutoCAD and spells major trouble for other Mac-based CAD software such as VectorWorks, which became popular precisely *because* AutoCAD was not available for the Mac and catered to those designers who refused to switch to a PC. Fig. A.1 is a screenshot of AutoCAD for the Mac from Autodesk's promotional literature.

It remains to be seen how successful AutoCAD for the Mac will be in the long run, of course. The program has been rewritten from the ground up as a native Mac application, not simply a port from Windows, so the software itself certainly seems to have a solid foundation. However, the original late-2010 release seemed, at the time, to be a bit of a work in progress, as evidenced by the fact that some standard AutoCAD features were not yet supported. Things have gotten better in the last few years, of course. The full (and shrinking) list of what is still not available can be found on the Autodesk website, but it includes the following. If you completed Levels 1, 2, and the 3D chapters of this textbook, you should recognize all these tools. Here is a partial list of AutoCAD for the Mac missing features (as of this writing):

- Quick Properties Palette
- Layer State Manager
- Tool Palettes
- Navigation Bar
- AutoCorrect Command Entry
- Hatch Preview
- Express Tools
- Material Mapping
- Advanced Rendering
- Camera Creation
- Walkthroughs
- DGN Underlay
- Hyperlinks
- dB Connect
- Visual LISP

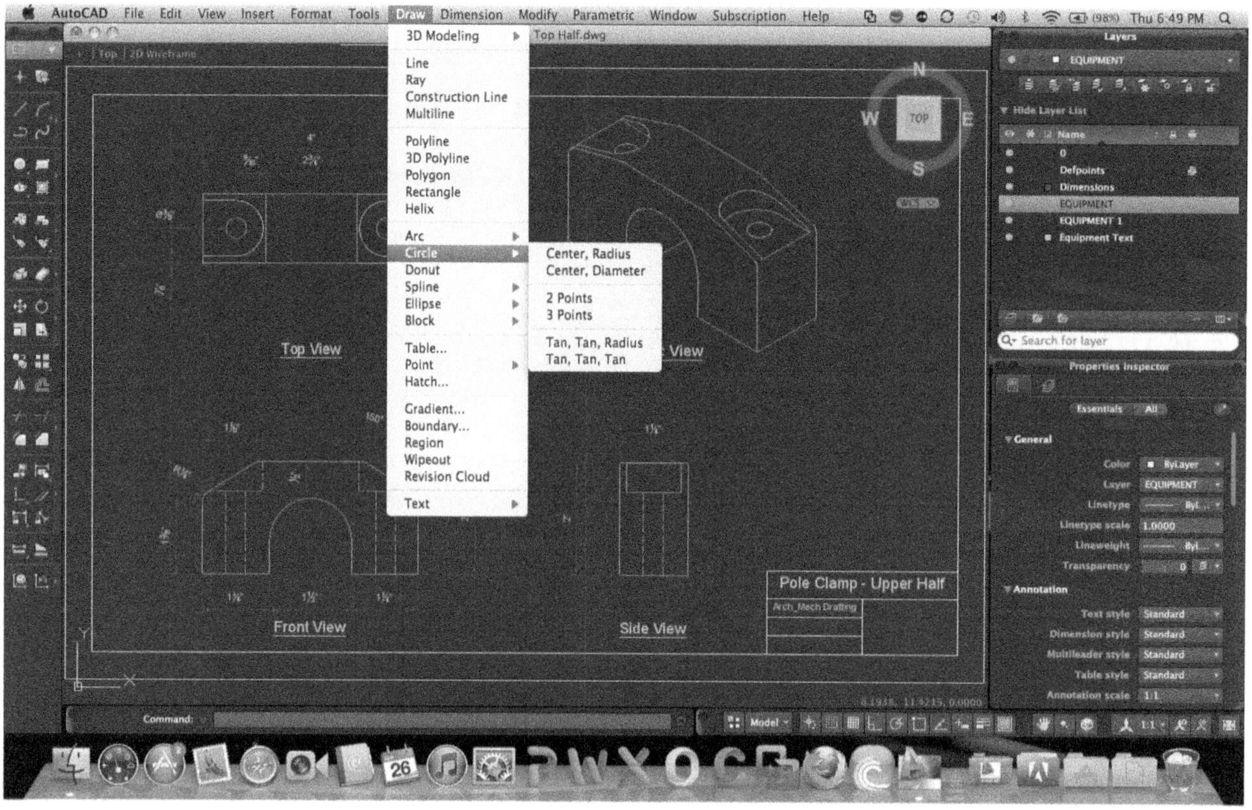

FIGURE A.1 AutoCAD for the Mac screenshot.

- Action Recorder
- Drawing Password Protect
- CUI Import/Export

Some of these are reasonably important, some are more obscure, and it is likely many of these issues will be resolved in the coming releases, or even by the time this textbook goes to press. Other than that, AutoCAD for the Mac is not entirely different from the PC version; after all, it still must behave like AutoCAD. Although the Ribbon is removed, the cascading menus and toolbars are still present, and the software can accept files from its PC-based cousins. AutoCAD for the Mac was priced the same as AutoCAD for the PC. Its subscription fees are currently the same as its PC-based cousin. There is an LT version for the Mac as well. It also mirrors the PC version in past and present pricing.

A.6 A BRIEF HISTORY OF AUTODESK AND AUTOCAD

The father of Autodesk and AutoCAD is John Walker. As a programmer by training, he formed Marinchip Systems in 1977 and, by 1982, seeing an opportunity with the emergence of the personal computer, decided to create a new "software only" company with business partner Dan Drake. The plan was to market a variety of software, including a simple drafting application they acquired earlier. Based on code written in 1981 by Mike Riddle and called *MicroCAD* (later changed briefly to *Interact*), the software would sell for under $1000 and would run on an IBM PC. It was just what they were looking for to add to their other products.

Curiously, the still unnamed company did not see anything special in the future of only MicroCAD and focused its energy on developing and improving other products as well. Among them were Optical (a spreadsheet), Window (a screen editor), and another application called *Autodesk* or *Automated Desktop*, which was an office automation and filing system meant to compete with a popular, at the time, application called *Visidex*. This Automated Desktop was seen as the most promising product; and when the time came to finally incorporate and select a name, Walker and company (after rejecting such gems as Coders of the Lost Spark) settled on *Autodesk Inc.* as the official and legal name.

Written originally in SPL, MicroCAD was ported by Walker onto an IBM machine and recoded in C. As it was practically a new product now, it was renamed *AutoCAD-80*, which stood for Automated Computer Aided Design. AutoCAD-80 made its formal debut in November 1982 at the Las Vegas COMDEX show. As it was new and had little competition at the time, the new application drew much attention. The development team knew a potential hit was on their hands, and a major corner was turned in AutoCAD's future. AutoCAD began to be shipped to customers, slowly at first, and then increased dramatically. A wish list of features given to the development team by customers was actively incorporated into each succeeding version—they were not yet called *releases*.

In June 1985, Autodesk went public and was valued at $70 million; in 1989 it was at $500 million and in 1991 at $1.4 billion. With each release, milestones, large and small, such as the introduction of 3D in Version 2.1 and porting to Windows with Releases 12 and 13, were achieved. In a story not unlike Microsoft, Autodesk, through persistence, innovation, firm belief in the future of personal computers, and a little bit of luck, came to dominate the CAD industry.

Fig. A.2 shows perhaps the most famous early AutoCAD drawing of all, a fire hose nozzle drawn in 1984 by customer Don Strimbu on a very early AutoCAD version. This was the most complex drawing on AutoCAD up to that point, used by the development team as a timing test for machines, and featured on the covers of many publications, including *Scientific American* in September 1984.

FIGURE A.2 A nozzle drawn by Don Strimbu in 1984 on a very early version of AutoCAD.

A.7 AUTOCAD RELEASES

Autodesk first released AutoCAD in December 1982, with Version 1.0 (Release 1) on DOS, and continued a trend of releasing a new version sporadically every 1–3 years until R2004, when it switched to a more regular, once a year schedule. A major change occurred with R12 and R13 when AutoCAD was ported to Windows and support for UNIX, DOS, and Macintosh was dropped. R12 and R13 were transitional versions (R13 is better left forgotten), with both DOS and Windows platforms supported. Not until R14 did AutoCAD regain its footing, as sales dramatically increased. Fig. A.3 is AutoCAD's release timeline. A release history follows:

Version 1.0 (Release 1), December 1982.
Version 1.2 (Release 2), April 1983.
Version 1.3 (Release 3), August 1983.
Version 1.4 (Release 4), October 1983.
Version 2.0 (Release 5), October 1984.
Version 2.1 (Release 6), May 1985.
Version 2.5 (Release 7), June 1986.

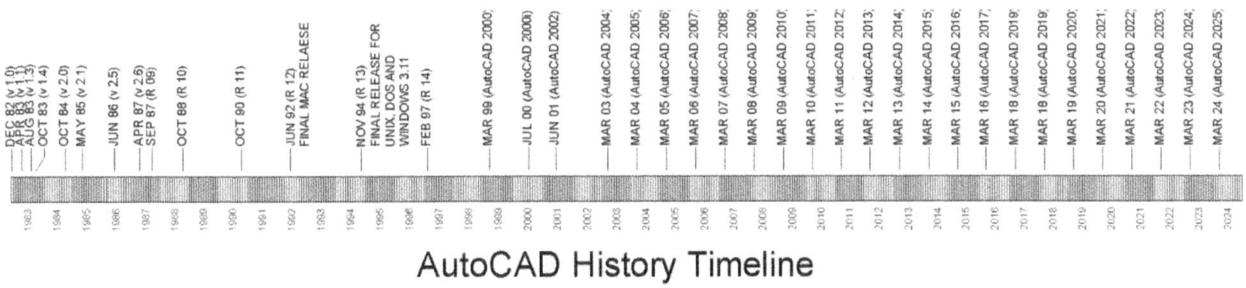

FIGURE A.3 AutoCAD timeline, 1982 through 2024 (projected).

Version 2.6 (Release 8), April 1987.
Release 9, September 1987.
Release 10, October 1988.
Release 11, October 1990.
Release 12, June 1992 (last release for Apple Macintosh until 2011).
Release 13, November 1994 (last release for UNIX, MS-DOS, and Windows 3.11).
Release 14, February 1997.
AutoCAD 2000 (R15.0), March 1999.
AutoCAD 2000i (R15.1), July 2000.
AutoCAD 2002 (R15.6), June 2001.
AutoCAD 2004 (R16.0), March 2003.
AutoCAD 2005 (R16.1), March 2004.
AutoCAD 2006 (R16.2), March 2005.
AutoCAD 2007 (R17.0), March 2006.
AutoCAD 2008 (R17.1), March 2007.
AutoCAD 2009 (R17.2), March 2008.
AutoCAD 2010 (R18.0), March 2009.
AutoCAD 2011 (R18.1), March 2010.
AutoCAD 2012 (R18.2), March 2011.
AutoCAD 2013 (R19.0), March 2012.
AutoCAD 2014 (R19.1), March 2013.
AutoCAD 2015 (R20.0), March 2014.
AutoCAD 2016 (R20.1), March 2015
AutoCAD 2017 (R21.0), March 2016.
AutoCAD 2018 (R22.0), March 2017.
AutoCAD 2019 (R23.0), March 2018.
AutoCAD 2020 (R23.1), March 2019.
AutoCAD 2021 (R24.0), March 2020.
AutoCAD 2022 (R24.1).
AutoCAD 2023 (R24.2), March 2022 (anticipated).

A.8 MAJOR AUTODESK PRODUCTS

Autodesk currently markets about 100 software products, some major, and some obscure. Only one application sold today is the company's original product, which is AutoCAD. The rest have been developed in-house from scratch over the years, purchased outright from the original developers, or as is often the case, acquired in takeovers and mergers. Many products are designed to "hitch a ride" on the back of AutoCAD and extend its power to serve a niche market with additional functionality and menus not found in the base product. Those, often homegrown applications, are referred to as *verticals*. In other cases, Autodesk markets radically different software, such as one used for high-end animations and cartoons. Those have been acquired to extend the company's reach into new markets.

The following lists Autodesk's flagship products that you should know about, as you may come across them in the industry; also listed are the popular add-on products purchased with AutoCAD.

Flagship products:

- *AutoCAD*: Autodesk's flagship product for 2D and 3D computer-aided drafting and design.
- *AutoCAD LT*: A limited functionality version of AutoCAD for an entry-level 2D only market.
- *Alias*: Advanced software for creating complex surfaces for automotive and other industrial design.
- *3ds Max*: A powerful high-end rendering and animation software product.
- *Inventor*: A mid- to high-end parametric solid modeling and design package for engineering.
- *Revit*: A mid- to high-end parametric software for architectural and building design.
- *Maya*: A very high-end animation, effects, and rendering software.

AutoCAD Verticals (Software built on the AutoCAD base platform):

- *Civil 3D*: BIM functionality for civil engineering applications.
- *AutoCAD Architecture*: Additional functionality for architecture design applications.

- *AutoCAD Electrical*: Additional functionality for electrical engineering applications.
- *AutoCAD MEP*: Additional functionality for building system design (mechanical, electrical, plumbing).
- *AutoCAD Mechanical*: Additional functionality for mechanical engineering.
- *AutoCAD Plant 3D*: Additional functionality for plant (factory) designs including creating piping and instrumentation (P&ID) diagrams and designs.

These are just some of the major software applications in Autodesk's inventory. A significant number of products, including all the less known or obscure ones, have been omitted. A full and comprehensive list of all of them can be found on the Autodesk website (www.autodesk.com/products). You are encouraged to look them over to stay ahead of the curve and know what other design professionals may be using.

A special mention should be made regarding AutoCAD WS (recently renamed *AutoCAD mobile app*). WS was Autodesk's first mobile app for AutoCAD. It allows you to view cloud-based drawings and edit them in a limited manner on a mobile device, such as a phone or tablet. AutoCAD WS was free at its basic level but did feature several low-cost paid plans that in turn gave you greater access to tools, storage, and other goodies. You can download this app from Google Play, Amazon, and other sources. It is a very convenient approach for managers and designers with limited drafting needs, who are often in the field or just need to quickly look at a drawing and approve it while on the go.

A.9 AUTOCAD-RELATED WEBSITES

Next is a listing of important and valuable websites for AutoCAD- and general CAD-related information. Online content has grown from occasionally useful to indispensable, and experienced AutoCAD users must have at least the following at their fingertips when the need arises. Note that link fidelity is not guaranteed, and except for the few major ones at the start of the list, the sites may and do change.

- *Autodesk Inc* (www.autodesk.com): Mentioned previously, it is home base for the makers of AutoCAD. Here, you find locations of AutoCAD dealers and training centers as well as endless information on AutoCAD and other software products and services. It is well worth visiting occasionally.
- *Autodesk User Group International* (www.augi.com): The Holy Grail for all things related to AutoCAD. This site is the officially sanctioned user community for Autodesk and its products. AUGI puts out numerous publications and runs conferences and training seminars all over the United States and worldwide. Its website is packed full of tips, tricks, articles, and forums where you can ask the community your hardest AutoCAD questions. Claiming over 100,000 members, AUGI has the distinction of being the only organization that can officially submit "wish lists" directly to Autodesk, meaning that, if you are a member and have an idea for a new feature or how to improve AutoCAD, submit it via the AUGI community, and Autodesk may just listen to you.
- *Cadalyst magazine* (www.cadalyst.com): A great magazine dedicated to all things CAD, including solid modeling and rendering but with a large amount of coverage dedicated specifically to AutoCAD. A must read for AutoCAD managers, the magazine features informative articles on all aspects of the job, including programming, industry trends, and advanced topics. The subscription was free up until 2010. You can still access free archived issues from as far back as 2004, but the magazine now charges a fee for the new issues.
- *AutoCAD Lessons on the Web* (www.cadtutor.net): A well-organized and extensive site for learning AutoCAD that includes some tutorials on 3ds Max, VIZ, Photoshop, and web design.
- *The Business of CAD, Part I* (www.upfrontezine.com): The website describes itself thus: "Launched in 1985, the upFront.eZine e-newsletter is your prime independent source of weekly business news and opinion for the computer-aided design (CAD) industry." The site is a no-nonsense, cut-through-the-hype source for CAD (mostly AutoCAD but others, too) information, latest releases, and analysis of Autodesk business decisions, a great resource for a CAD manager.
- *The Business of CAD, Part II* (http://worldcadaccess.typepad.com): Edited by blogger Ralph Grabowski, this site is a spin-off blog from the preceding upFront.eZine site. It also addresses news in the media about all things related to CAD and how they will affect you, as well as general computer topics; it is recommended for the serious CAD manager or user though it has not been updated since 2017.
- *CADLearning The Blast* (https://www.cadlearning.com/theblast/): This site provides a free online magazine with some of the best tips and tricks for users of all levels showcasing many types of CAD programs. Their goal is to provide

industry expertise that you can incorporate into your everyday workflows. They offer additional paid subscription consulting as well but it is not required to sign up for that to get their online magazine.

As with many topics in this brief appendix, what is listed here only scratches the surface. Remember that you are not the only one out there using AutoCAD, so feel free to widen your view and see what else is out there if you have questions or would just like to learn more. At a minimum, it is recommended to join AUGI, signup for The Blast and get a subscription to *Cadalyst*.

Appendix B

Other CAD software, design and analysis tools, and concepts

While AutoCAD may be the market leader for drafting software and is often used as a generic term for all CAD software, it certainly does not operate in a vacuum and is not the only game in town. The list that follows summarizes other major players in the drafting field. The reason you should know about them is that you may run into them in use and may have to exchange files if collaborating. Saying “I’ve never heard of that” is usually not a good way to start a technical conversation in your area of expertise, so take note of the following products.

B.1 MicroStation

(www.Bentley.com)

A major, and really the only serious, competitor to AutoCAD, MicroStation is produced and marketed by Bentley Systems. For historical reasons, the software is widely used in government agencies and the civil engineering community. It comes in full and PowerDraft variations (akin to AutoCAD and LT). The current version is MicroStation CONNECT v10, and its file extension is .dgn. Relatively recently, AutoCAD began to open these files via the import command. MicroStation, on the other hand, has always been able to open AutoCAD’s .dwg files. Of course, many techniques and even third-party software are available to convert one file format quickly and easily to another quickly as well.

MicroStation was designed from a different perspective than AutoCAD, and this shows mainly in the way the user interacts with this software. It is more icon- and menu-driven than AutoCAD; it allows little typing and had no Ribbon Interface until the most recent release. However, MicroStation executes the same exact commands, and you can use it to design anything that you can do in AutoCAD. One aspect that users may find frustrating when switching from AutoCAD is that MicroStation commands are for the most part named differently, so you need to learn a new vocabulary. Blocks, e.g., are called *cells*, layers are called *levels*, snap points are referred to as *tentative points*, and so on. Bentley designers have done their best to differentiate their product from AutoCAD, but the result can be like “using a different part of the brain” when working on it, as described by one former AutoCAD user in an online blog.

That said, however, MicroStation is very capable software and, in some areas, superior to AutoCAD, such as its ability to handle multiple Xrefs, 3D rendering, design simulation, and review, change and design management, and digital file security, among other advanced features. Its main disadvantage (interface opinions aside) is that it is not an industry standard outside some civil engineering communities, which creates occasional collaboration and file sharing issues as well as difficulty in finding qualified users and training classes. Bentley’s pricing is also a bit cumbersome with various a-la-carte modules that can be purchased separately. Overall, the software costs about the same as AutoCAD.

MicroStation and AutoCAD users have historically divided themselves into somewhat fanatical partisan groups, each extolling the virtues of its respective software over the other. However, an objective view of both applications yields an opinion that, like any software, both have their pros and cons, and those who disparage one over the other usually do not know what they are talking about.

B.2 ArchiCAD

(www.graphisoft.com)

ArchiCAD is a very popular application among some architects. It is geared toward the Macintosh but is also available for Windows. Developed by a Hungarian firm, Graphisoft at around the same time as AutoCAD, ArchiCAD took a different approach, allowing for a parametric relationship between objects as well as easier transitions to 3D, as all objects are created inherently with depth. These features, common to BIM software, can also be found in Revit and other applications and is the direction the industry seems to be headed as well (see Appendix N), making ArchiCAD a bit ahead of its time. The software has a loyal installed user base of about 100,000 architects worldwide and can interoperate fully with AutoCAD files.

B.3 TurboCAD

(www.turbocad.com)

TurboCAD is a mid-range 2D/3D drafting software, first developed and marketed in South Africa in 1985 and brought to the United States the following year. It was marketed as an entry-level basic drafting tool (it once sold for as low as $99!), but with constant upgrading and development, TurboCAD has grown over the years to a $1600 software product for the architectural and mechanical engineering community. It is currently owned by IMSI/Design LLC and operates on both Windows and Macintosh. The current release is TurboCAD Pro Platinum (for Windows) or Mac Pro v10 (for Mac) and a mid-range 2D/3D version TurboCAD Expert for around $500. There is also a downgraded 2D-only TurboCAD Deluxe for around $250 similar to AutoCAD LT. Some versions of the software have a subscription sales model as well, and many have specialized add-ons (Civil, Mechanical, etc.)

B.4 Other design software

Some other notable CAD applications include RealCAD, ViaCAD, SharkCAD, AllyCAD, PowerCADD, CorelCAD, DataCAD, VariCAD, VectorWorks (formerly MiniCAD), IntelliCAD, Trimble's Sketch-Up, Form Z, and Rhino, although the last two are often used only for modeling and visualization. This list is not exhaustive by any means, as hundreds of other CAD programs are available.

Many of my students are interested in other aspects of design, and questions often come up as to how AutoCAD fits into the "big picture" of a much larger engineering or design world. To address some of these questions, the following information, culled from a chapter written in 2007 for an undergraduate machine design theory workbook, is included. AutoCAD students, especially engineers, may find this informative. The subject matter is extensive and this brief overview only scratches the surface. Ideally, if the reader was not aware of these concepts and the engineering design and analysis software described, this serves as a good introduction and a jumping off point to further inquiry.

- **CAD (computer-aided design):** This is a broad generic term for using a computer to design something. It can also be used in reference to a family of software that allows you to design, engineer, and test a product on the computer prior to manufacturing it. CAD software can be used for primarily 2D drafting (such as with AutoCAD) or full 3D parametric solid modeling. The latter is the software that aerospace, mechanical, automotive, and naval engineers commonly use. Examples include Computer-Aided Three-dimensional Interactive Application (CATIA), NX, Pro-Engineer (Creo), IronCAD, and SolidWorks, among others. Included under the broad category of CAD are testing and analysis software.
- **FEA (finite element analysis):** This is used for structural and stress analysis, testing thermal properties, and much more as related to frames, beams, columns, and other structural components. Examples of FEA software include NASTRAN, ALGOR, and ANSYS.
- **CFD (computational fluid dynamics):** This is software used to model fluid flow around and through objects to predict aerodynamic and hydrodynamic performance. Examples of CFD software include Fluent, ANSYS-CFX, and Flow 3-D.
- **CAM (computer-aided manufacturing):** The basic idea is this: We designed and optimized our product, now what? How do we get it from the computer to the machine shop to build a prototype? The solution is to use a CAM software package that takes your design (or parts that make up your design) and converts all geometry to information useable by the manufacturing machines, so the drills, lathes, and routing tools can create the part out of a chunk of metal. Examples of dedicated CAM software include Mastercam and FastCAM.

- **CAE (computer-aided engineering):** Strictly defined, CAE is the "use of information technology for supporting engineers in tasks such as analysis, simulation, design, manufacture, planning, diagnosis, and repair." We use this term to define an ability to do this all from one platform, as explained next.
- **CNC/G-Code (computer numerical code):** This is the language (referred to as *G-Code*) created by the CAM software and read by the CNC machines. G-Code existed for decades. It was originally written by skilled machinists based on the geometry of the component to be manufactured (and sound manufacturing practices). This code can still be written by hand. It is not hard to understand. Most lines start with G, hence its name; and they all describe some function that the CNC machine does as related to the process. After some practice, the lines start to make sense and you can roughly visualize what is being manufactured, just by reading them.

As you read the CAD and CAM descriptions you may be wondering why CAD companies have not provided a complete solution to design, test, and manufacture a product using one software package. They have but to varying degrees of success. For a number of years, an engineering team designing an aircraft, e.g., had to typically create the design (and subsystems) in a 3D program like CATIA, then import the airframe for FEA testing into NASTRAN, followed by airflow testing in CFD software (and a wind tunnel), and finally send each component through CAM software like Mastercam to manufacture it (a greatly simplified version of events, of course).

Obviously, data could be lost or corrupted in the export/import process, and a lot of effort was expended in making sure the design transitioned smoothly from one software application to another. While the final product was technically a result of CAE, software companies had something better in mind: "one-stop shopping." As a result, CAE today means one source for most design and manufacturing needs. CATIA, e.g., has "workbenches" for FEA analysis and CAM. The code generated by CATIA is imported into the CNC machines that (typically after some corrections and editing) produce virtually anything you can design, with some limitations, of course.

As you may imagine, not everything is rosy with this picture. As the specialized companies got better and better at their respective specialties (CAM, FEA, or CFD), it became harder and harder for the CAD companies to catch up and offer equally good solutions embedded in their respective products. Mastercam is still the undisputed leader in CAM worldwide, and NASTRAN is still the king of FEA. CATIA's version of these has gotten better but is still a notch behind. The situation is the same with Pro/E (Creo), NX, and other CAD software.

These high-end 3D packages are also referred to as *product lifecycle management* (PLM) software, a term you will hear a lot of in the engineering community. As described already, these software tools aim to follow a product from design and conception all the way through testing and manufacturing in one integrated package.

Here are some more specific details on the CAD software mentioned in the preceding section. You may run into many of these throughout your engineering career. It is well worth being familiar with these products. The information is up to date as of this writing but can change at any time.

B.4.1 CATIA

(www.3ds.com)

CATIA is the world's leading CAD/CAM/CAE software. Developed in France by Dassault for its Mirage fighter jet project (although originally based on Lockheed's CADAM software), it is now a well-known staple at Boeing and many automotive, aerospace, and naval design companies. It has even been famously adopted by architect Frank Geary for certain architectural projects. CATIA has grown into a full PLM solution and is considered the industry standard, although complex and with a not always user-friendly reputation. Besides CATIA, Dassault Systems also markets other CAD software such as the more affordable and much lauded SolidWorks (discussed separately) and even an AutoCAD LT clone called DraftSight.

B.4.2 NX

(www.plm.automation.siemens.com)

Another major CAD/CAM/CAE PLM software solution that was developed by a company called UGS and was originally known as Unigraphics. NX is essentially a blend of this Unigraphics software and SDRC's popular I-DEAS software, which UGS purchased and took over. A few years later, Siemens PLM Software took over UGS and renamed this blended product to NX. It is used by GM, Rolls Royce, Pratt and Whitney, and many others. The current version runs on Windows and, surprisingly, the Mac OS. Similar to Dassault Systems, Siemens PLM Software makes an entire suite of CAD solutions, including a lower-end NX called SolidEdge (discussed separately).

B.4.3 PTC Creo elements/Pro (formerly Pro/ENGINEER)

(www.ptc.com)

The third major high-end CAD/CAM/CAE PLM software package, originally called Pro/ENGINEER, is historically significant. It caused a major change in the 3D CAD industry when first released in 1987 by introducing the concept of *parametric modeling*. Rather than models being constructed like a mound of clay with pieces added or removed to make changes, the user constructs the model as a list of features (or parameters), which are stored by the program and can be used to change the model by modifying, reordering, or removing them. The software also pioneered the use of feature trees and dimension-driven design in general. CATIA, NX SolidWorks, and many others all now work this way. The current software is no longer called Pro/ENGINEER; the name was changed to Creo Elements/Pro in late 2010 (it was also called Wildfire), and then to just Creo Design Suite. The suite features 3D modeling software (Creo Parametric, Direct, Options Modeler) and 2D drafting software (Creo Sketch, Layout, Schematics) and others. The flagship product and the direct descendant of Pro/ENGINEER is the Parametric package.

B.4.4 SolidWorks

(www.solidworks.com)

SolidWorks is a mid-range CAD/CAM/CAE software product, first introduced in 1995 by a Massachusetts company of the same name. Dassault Systems, the makers of CATIA, acquired the company in 1997. The SolidWorks installed base is estimated at over 2 million users, making the product a major player in the 3D CAD field. SolidWorks is marketed as a lower-cost competitor to major CAD packages, but the software still has extensive capabilities and caught on quickly in smaller design companies. It comes in three flavors: Premium, Professional, and Standard, along with many modules and options. The software has an enviable reputation of being powerful, yet easy-to-use and many universities and colleges teach it.

B.4.5 Inventor

(www.autodesk.com)

Inventor is another mid-range product, from the makers of AutoCAD, for solid modeling and design. Considered a low- to mid-range product when first introduced, Inventor matured to become a major competitor to SolidWorks, SolidEdge, and Creo Parametric. Like AutoCAD, Inventor moved to a subscription model and the associated costs are \$240 monthly and \$1950 for 1 year, rising to \$5805 for 3 years.

B.4.6 IronCAD

(www.ironcad.com)

IronCAD is another 3D CAD software package in the mid-range market. Originally developed by Visionary Design Systems, the software gained ground in recent years due to a user-friendly reputation and a more intuitive approach to 3D design. It retails for about \$1200. A downgraded lower-cost IronCAD Draft also is available for about \$600. The software recently added a new feature called *TraceParts*, which is a web-based content center (a library) with free 2D and 3D standard parts available from numerous manufacturers, making mechanical design easier.

B.4.7 SolidEdge

(www.plm.automation.siemens.com)

Solid Edge is a mid-range 3D CAD product from Siemens PLM Software, the same company that makes the high-end NX package. It is comparable to and competes with, SolidWorks and Inventor in both cost and functionality. It comes in two flavors: Classic and Premium, and has many specialized modules for Design, Simulation, Manufacturing, etc. There is also a 2D version called SolidEdge 2D Drafting.

Listed next are some of the major software products in FEA and CFD, as mentioned earlier.

B.4.8 NASTRAN

Originally developed for NASA as open-source code for the aerospace community to perform structural analysis, NAsaSTRucturalANalysis, or NASTRAN was acquired by several corporations and marketed in numerous versions.

NEiNastran is just one, commonly used flavor of it. NASTRAN in its pure form is a solver for finite element analysis and cannot create its own models or meshes. Developers, however, have added pre- and postprocessors to allow for this. The current release for NEiNastran is V9, though other companies have different designations for their products. NEiNastran was acquired by Autodesk in 2014 and renamed Autodesk NASTRAN. As of March 21, 2018, Autodesk no longer sells Nastran as a stand-alone product to new customers. Autodesk Nastran In-CAD software now includes the Autodesk Nastran Solver and is available only in the Product Design and Manufacturing Collection, which include simulation.

B.4.9 ANSYS

Ansys, Inc., a Pennsylvania-based company, is a major player in the engineering software industry and makes a range of products for FEA and CFD. One of its flagship offerings is ANSYS Fluent, the industry standard for CFD software, holding about 40% of the market. It imports geometry, creates meshes (using GAMBIT software) and boundary conditions, and solves for a variety of fluid flows, using the Navier–Stokes equations as theoretical underpinnings. The current release is Fluent 6.3. The entire suite of products is marketed as ANSYS 16.0 and includes not just Fluent but packages for structures, electronics, and systems.

Appendix C

File extensions

Appendix C outlines a variety of AutoCAD and other commonly encountered software file extensions for student recognition and familiarity. File extensions are the bane of the computer user's existence. They number in the hundreds (AutoCAD alone claims nearly 200), and even if you pare the list down to a useful few, it is still quite a collection. Here, we focus on those extensions that are important to an AutoCAD user, as chances are you will encounter most, if not all, of them while learning the software top to bottom. They are listed alphabetically (with explanations and descriptions) but grouped in categories to bring some order to the chaos and set priorities for what you really need to know and what you can just glance over. Following AutoCAD's list is a further listing of extensions from other popular, often-used software, which you may find useful if you are not already familiar with most of them.

C.1 AutoCAD primary extensions

- **.bak** (backup file): A .bak is an AutoCAD backup file created and updated every time you save a drawing, provided that setting is turned on in the Options dialog box. The .bak file has the same name as the main .dwg file, generally sits right next to it, and can be easily renamed to a valid .dwg if the original .dwg is lost.
- **.dwf** (design web format file): The .dwf is a format for viewing drawings online or with a .dwf viewer. The idea here is to share files with others who do not have AutoCAD in a way that enables them to look but not modify. If this sounds like Adobe .pdf, you are correct; .dwf is a competing format.
- **.dwg** (drawing file format): The .dwg is AutoCAD's famous file extension for all drawing files. This format can be read by many of Autodesk's other software products as well as by some competitors. The format itself changes somewhat from release to release, but these are internal programming changes and invisible to most users.
- **.dwt** (drawing template): The .dwt is AutoCAD's template format and is essentially a .dwg file. The idea is to not repeat setup steps from project to project but use this template. It is equivalent to just saving a completed project as a new job and erasing the contents.
- **.dxf** (drawing exchange format): The .dxf is a universal file format developed to allow for smooth data exchange between competing CAD packages. If an AutoCAD drawing is saved to a .dxf file, it can be read by most other design software (in theory anyway). This, of course, is not always the case, and some CAD packages open .dwg files anyway without this step, but the .dxf remains an important tool for collaboration. The .dxf is similar in principle and intent to IGES and STEP for those readers who may have worked with solid modeling software and understand what those acronyms mean.

C.2 AutoCAD secondary extensions

- **.ac$:** A temporary file generated if AutoCAD crashes; it can be deleted if AutoCAD is not running.
- **.ctb:** This color table file is created when pen settings and thicknesses are set (see chapter: Advanced Output and Pen Settings topic).
- **.cui:** This customizable user interface file is for changes to menus and toolbars (see chapter: Options, Shortcuts, CUI, Design Center, and Express Tools topic).
- **.err:** This error file is generated on an AutoCAD crash. You can delete it.

- **.las:** This layer states file can be imported and exported (see chapter: Advanced Layers topic).
- **.lin:** This linetype definition file can be modified and expanded (see Appendix: Custom Linetypes and Hatch Patterns).
- **.lsp:** This LISP file is a relatively simple language used to modify, customize, and automate AutoCAD to effectively expand its abilities.
- **.pat:** This hatch definition file can be modified and expanded (see Appendix: Custom Linetypes and Hatch Patterns).
- **.pgp:** This program parameters file contains the customizable command shortcuts (see chapter: Options, Shortcuts, CUI, Design Center, and Express Tools topic).
- **.scr:** This script file is used for automation (mentioned in the chapter: Advanced Layers).
- **.sv$:** This is another temporary file for the AutoSave feature. It appears if AutoSave is on and remains in the event of a crash. It can be then renamed to .dwg or deleted.

C.3 Miscellaneous software extensions

These extensions are for software and formats likely to be encountered daily or at least occasionally by AutoCAD users and are always good to know. Some extensions can be found in AutoCAD as export/import choices.

- **.ai:** Adobe Illustrator files.
- **.bmp:** Bitmap, a type of image file format that, like jpg, uses pixels.
- **.dgn:** The MicroStation file extension. This competitor to AutoCAD can open native .dwg files. Appendix B, Other CAD Software, Design and Analysis Tools, and Concepts discusses this software.
- **.doc and .docx:** MS Word files.
- **.eps:** Encapsulated PostScript file.
- **.jpg:** The .jpeg (joint photographic experts group) is a commonly used method (algorithm) of compression for photographic images. It is also a file format (discussed in the chapter: Importing and Exporting Data).
- **.pdf:** The famous Adobe Acrobat file extension. You can print any AutoCAD drawing to a .pdf with printer drivers installed with the AutoCAD program. It is one of the few extensions to become a verb, as in, "Can you PDF that invoice?"
- **.png:** Portable Network Graphics is a raster-graphics file format that supports lossless data compression. PNG was developed as an improved, non-patented replacement for Graphics Interchange Format (GIF).
- **.ppt and .pptx:** MS PowerPoint file extension.
- **.psd:** The famous Photoshop file extension. The software itself also became a verb, as in, "Can I Photoshop my ex-boyfriend/girlfriend out of this photo?"
- **.rfa:** Revit family files, similar yet very different than AutoCAD blocks)
- **.rvt:** Revit project files
- **.skp:** file extension used by SketchUP
- **.vsd:** MS Visio drawing file extension. Visio is a simple and popular "drag and drop style" CAD package for electrical engineering schematics and flowcharts, among other applications. It accepts AutoCAD files.
- **.xls and .xlsx:** MS Excel file extension.

This list is by no means complete, but it covers many of the applications discussed in Chapter 16, Importing and Exporting Data.

Appendix D

Custom linetypes and hatch patterns

This appendix outlines the basic procedures for creating custom linetypes and hatch patterns if you need something that is not found in AutoCAD itself or with any third-party vendor. All linetypes and hatch patterns were originally "coded" by hand, and there is nothing inherently special or magical about making them; they just take some time and patience to create. Though it is admittedly rare for a designer to need something that is not already available on the web, creating these entities is a good skill to acquire. We just present the basics here, and you can certainly do additional research and come up with some fancy designs and patterns.

D.1 Linetype definitions (basic)

All linetypes in AutoCAD (about 38 standard ones, to be precise) reside in the acad.lin or the acadiso.lin file. These are linetype definition files that AutoCAD accesses when you tell it to load linetypes into a drawing. In AutoCAD 2020, these files can be found in the Autodesk\AutoCAD_2020_English_Win_32 bit_dlm\x86\en-us\acad\Acad\Program Files\Root\UserData Cache\en-us\Support folder. Anything you do to these files, including adding to them, immediately shows up the next time you use linetypes. Our goal here is to open the acad.lin file, analyze how linetypes are defined, and then make our own.

Locate and open (usually with Notepad) the acad.lin file and look at it closely. Here is a reproduction of the first six linetype definitions: Border (Standard, 2, and X2) and Center (Standard, 2, and X2).

```
*BORDER, Border____ . ____ . ____ . ____ . ____ .
A,.5,2.25,.5,2.25,0,2.25
*BORDER2, Border (.5x)__ . __ . __ . __ . __ . __ . __ . __ .
A,.25,-.125,.25,2.125,0,2.125
*BORDERX2, Border (2x)________ . ________ . ___
A,1.0,2.5,1.0,2.5,0,2.5
*CENTER, Center______________________________
A,1.25,-.25,.25,-.25
*CENTER2,Center (.5x)________________________
A,.75,2.125,.125,2.125
*CENTERX2,Center (2x)_________________________
A,2.5,2.5,.5,2.5
```

Let us take just one linetype and look at it closer. Here is the standard size center line:

```
*CENTER, Center______________________________
A,1.25,2.25,.25,2.25
```

The two lines you see are the header line and the pattern line; both are needed to properly define a linetype. The header line consists of an asterisk, followed by the name of the linetype (CENTER), then a comma, and finally the description of the linetype (Center). In this case, they are the same.

The pattern line is where the linetype is actually described, and the cryptic looking "numerical code" (A,1.25,2.25,.25,2.25) is the method used to describe it and needs to be discussed for it to make sense.

The A is the alignment field specification and is of little concern (AutoCAD accepts only this type so you always see A). Next are the linetype specification numbers, and they describe the linetype using dashes, dots, and spaces. Let us go over them, as really this is the key concept.

The basic elements are.

- Dash (pen down, positive length specified, such as 0.5).
- Dot (pen down, length of 0).
- Space (pen up, negative length specified, such as –0.25).

Now, you just have to mix and match and arrange these in the order you want them to be to define a linetype you envisioned.

What would the following look like?

```
(A,.75,-.25,0,-.25,.75)
```

Well, you know that dashes are positive, so everywhere you see .75, there will be a dash of that length. Spaces are negative, so everywhere you see –2.25, there will be a space of that length, and finally a 0 is a dot. This is the result:

—— · ——

These are the essentials of creating basic linetypes. AutoCAD, however, also allows for two other types to be created, and they are part of the complex linetype family: string complex and shape complex.

D.2 Linetypes (string complex and shape complex)

The problem with the previously mentioned linetypes is obvious; you have only dashes, dots, and spaces with which to work. While this is enough for many applications, you may occasionally (especially in civil engineering and architecture) need to draw boundary or utility lines that feature text (Fence line, Gas line, etc.) or you may want to put small symbols between the dashes. It is here that strings and shapes come in. String complex linetypes allow you to insert text into the line. Here is a gas line example:

```
*GAS_LINE,Gas line —GAS—GAS—GAS—GAS—GAS—GAS—
A,.5,-.2,["GAS",STANDARD,S=.1,R=0.0,X=20.1,Y=2.05],2.25
```

The header line is as discussed before, and the pattern line still starts with an A. After the A, we see a dash of length .5 followed by a space of length .2; so far it is familiar. Next is information inside a set of brackets, defined as follows:

```
["String", Text Style, Text Height (S), Rotation (R), X-Offset (X), Y-Offset (Y)].
```

Finally, there is a .25 space and the string repeats. The meaning of the information inside the brackets is as follows:

- String: This is the actual text, in quotation marks (GAS, in our example).
- Text Style: The predefined text style (STANDARD, in our example).
- Text Height: The actual text height (.1, in our example).
- Rotation: The rotation of the text relative to the line (0.0, in our example).
- X-Offset: The distance between the end of the dash and the beginning of the text as well as the end of the text and the beginning of the dash (a space of .1, in our example).
- Y-Offset: The distance between the bottom of the text and the line itself. This function centers the text on the line.

With these tools you can create any text string linetype. Shape complex linetypes are basically similar, aside from one difference: Instead of a text string, we have a shape definition inserted. Shape definitions are a separate topic and are not covered in detail in this textbook. Briefly, however, shape files are descriptions of geometric shapes (such as rectangles or circles but can be more complex) and are written using a Text Editor and compiled using the compile command. They are saved under the *.shp extension. AutoCAD accesses these when it reads the linetype definitions. If you are familiar with shapes, then you can proceed in defining shape complex linetypes as with string complex ones.

D.3 Hatch pattern definitions (basic)

Knowing the basics of linetype definitions puts you on familiar ground when learning hatch pattern definitions. Even though they are more complex, the basics and styles of description are similar.

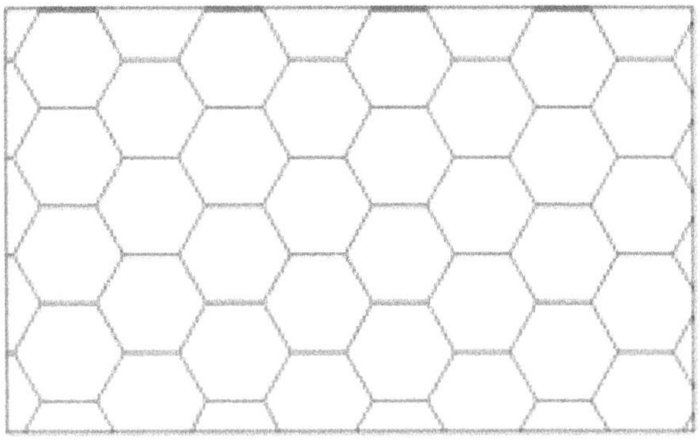

FIGURE D.1 Honeycomb pattern.

Hatch pattern definitions are found in the acad.pat and acadiso.pat files, and they reside in the same folder as the .lin files. Open the acad.pat file and look at a few definitions. Some of them can get long and involved due to the complexity of the pattern (Gravel, for example), but Fig. D.1 shows a simpler one to use as a first example: Honeycomb.

The pattern definition file that describes it looks like this:

```
*HONEY, Honeycomb pattern
0, 0,0, .1875,.108253175, .125,2.25
120, 0,0, .1875,.108253175, .125,2.25
60, 0,0, .1875,.108253175, 2.25,.125
```

The first line (header line) is as discussed previously with linetypes. The next three lines are the hatch descriptors, and they have somewhat different definitions as shown next:

Angle, X-Origin, Y-Origin, D1, D2, Dash Length, where (in this example)

- Angle: Angle of the pattern's first hatch line from the X axis (0°, in our example).
- X-Origin: X coordinate of the hatch line (0,0, in our example).
- Y-Origin: Y coordinate of the hatch line (.1875, in our example).
- D1: The displacement of the second line (.1083, in our example).
- D2: The distance between adjoining lines (.125, in our example).
- Length: Length of dashes or spaces.

The pattern is repeated two more times to form the overall shape and repeated many more times in the overall hatch pattern.

Appendix E

Principles of CAD management

So, you have been hired on as a CAD manager or maybe forced into this due to your substantial AutoCAD skills—or was it because no one else wanted the job? Either way, you are now the top dog, head honcho, and must take the bull by the horns, well many bulls, (i.e., coworkers) who run around and draft whatever and how they please just to get the job out. Then again, maybe you are simply a student finishing your AutoCAD education. The bottom line is that you could use some ideas and a short discussion on this elusive science of CAD management. Well, you came to the right appendix.

All jokes aside, this is an important topic. You must take everything you learned in Levels 1 and 2 and apply it, while keeping others in line. You must be the keeper and the enforcer of standards and the technical expert who can resolve any issue. A true CAD manager who does nothing else is not all that common. You may have to multitask and do your own work while doing the management, especially in smaller companies.

Here are a few rules, ideas, and thoughts gathered from 20 years of experience as a CAD consultant, educator, and CAD manager. They outline some of the challenges and issues and provide some possible answers, to those in charge of the software used to design a lot of what is around us.

E.1 Part 1: Know the eight golden rules of AutoCAD

1. Create a solid template as a starting point for all projects.
2. Always draw one-to-one (1:1).
3. Never draw the same item twice.
4. Avoid drawing standard parts.
5. Use layers correctly.
6. Do not explode anything.
7. Use accuracy: Ortho and OSNAPs.
8. Save often.

Notice that most of those rules were discussed in some way in the first few chapters of Level 1. It is the basic stuff that people screw up, not some fancy new feature. Let us briefly go over these one by one; they bear repeating.

1. *Create a template from which all projects are started from:* At a minimum, everyone in a company should start new drawing files from the same template file (even a company of one person). The template should have the company standard layers (see below), dimension styles, text styles, leader styles, table styles, layout sheet(s) with title blocks, units, and limits set. This also makes work more productive by giving someone a kit of tools to start with. If you or your users must start with the blank, content-less acad.dwt, starting every project from scratch, it will lead to a standardless environment and wasting time setting up every job from scratch.
2. *Always draw one-to-one (1:1)*: This is the most important one. Always draw items to real life sizes. Endless problems arise if users decide to scale items as they draw them. Leave scales and scaling to the pencil-and-paper folks. A drawing must pass a simple test. Measure anything that is known to be a certain value, for example, a door opening that is 3 feet wide, by using the list command. AutoCAD should tell you its value as 36 inches or 3 feet and nothing else. You must enforce this as a CAD manager.

3. *Never draw the same item twice*: In hand drafting, just about everything needed to be drawn one item at a time. Some inexperienced CAD users do the same in AutoCAD. That is not correct. AutoCAD and computer drafting in general require a different mindset, philosophy, and approach. If you have already drawn something once, then copy it, do not create it again. While this one may be obvious, another overlooked tool is the block or wblock command. If you anticipate using the new item again (even if there is a small chance), then make a block out of it. As a CAD manager, you need to sometimes remind users of this. Remember: Time = $$.
4. *Avoid drawing standard parts*: This, like the preceding rule, aims to prevent duplicate effort. Before starting a project, as a CAD manager, you need to see to it that any standard parts (meaning ones that are not unique but constantly reused in the industry) are available for use and are not drawn and redrawn. Chances are good that you either already have these parts or can get the "cads," as they say, from the part supplier. Another resource is the web. Numerous sites have pre-drawn blocks available for use. Buy them, borrow them, or download them, but do not redraw items that someone already has done before. Again, Time = $$.
5. *Use layers correctly*: Layers are there to make life easier, and they are your friends. While overdoing it is always a danger (500 + layers is pushing it), you need to use layers and place items on the appropriate ones. As a CAD manager, you need to develop a layering standard (more on that later) and enforce it. Users in a hurry do not always appreciate what having all items organized in layers does ("it all comes out the same on paper!" they may say), so you must stand firm and enforce the standard. And do not use layer 0. It cannot be renamed with a descriptive name, so why use it?
6. *Do not explode anything*: How do you destroy an AutoCAD drawing and render weeks of work and thousands of dollars in work hours useless? Why, explode everything, of course. Short of deleting the files, this is the best way to ruin them. While this is generally a far-fetched scenario, remind users to not ruin a drawing in small ways, like occasionally exploding mtext, dimension, or hatch pattern. They are liable to do so, if these items do not behave as they are supposed to and users explode them to manually force them into submission. Not a good idea; lack of skill is not an excuse. Remind them to ask you for help if a dimension just does not look right. There is always a way to fix something. Leave the exploding for the occasional block or polygon.
7. *Use accuracy: Ortho and OSNAPs*: Drawing without accuracy is not drafting, it is doodling. Most users learn the basics, such as how to draw straight lines and connect them, but you may still see the occasional project manager (who gets to draft on average only once a month) fix a designer's linework and forget to turn on OSNAPs. When zoomed out, the lines look OK and print just fine (reason enough to throw accuracy out the window for the occasional user), but a closer inspection reveals they are not joined together, causing a world of problems for a variety of functions, such as hatch, area, distance, and other commands. Besides, it is just plain sloppy. How accurate is AutoCAD? Enough to zoom from a scale model of the earth to a grapefruit (more on that later), so be sure to use all the available accuracy.
8. *Save often*: Let us state the obvious. There is no crash-proof software or operating system. Both AutoCAD and Windows can lock up and crash at some point. Remind users to save often and do daily and weekly file backups. But you already knew this, right?

E.2 Part 2: Know the capabilities and limitations of AutoCAD

As a CAD manager, you must know what you must work with. All computer-aided design software, including AutoCAD, has an operating range for which the software is designed and optimized. Some tasks it excels at, you need to leverage that advantage, and some tasks are outside the bounds of expectation. AutoCAD cannot be all things to all people (although you would be hard-pressed to tell based on Autodesk's marketing). What follows is a short discussion on the capabilities and limitations and how you need to best take advantage of what you have.

Capabilities

Often, as a CAD manager, you are asked to provide an estimate on how long it will take to complete the drafting portion of a job or just recreate something in CAD. Sometimes, these requests come from individuals (senior management) who have had extensive experience in hand drafting, and they expect an answer in the form of a significantly smaller time frame. AutoCAD has an aura of magic about it, as if a few buttons just need to be pushed and out pops the design. Where *is* that "Finish Project" icon?

The surprising truth is that a professional hand drafter can give an average AutoCAD designer a run for the money if the request is a simple floor plan. After all, the hand drafter is not burdened with setting up layers, styles, dims, and other CAD prep work. But, this is not where AutoCAD's power lies. Once you begin to repeat patterns and objects, AutoCAD takes a commanding lead. A layout that took 1 hour to draft by hand or via computer can be copied in seconds on the computer but adds another hour to the hand drafter's task.

Do not be surprised by the seeming obviousness of this. I have met many users and students who do not always see where exactly CAD excels; they think work is always faster on a computer. It may not be and may lead to erroneous job estimates. But, in all cases with patterns, copying, and use of stored libraries, CAD is much quicker. So, when estimating a job, look for

- Instances where the offset command can be used.
- Instances where there are identical objects.
- Minimal cases of unique, uneven, and nonrepetitive linework.
- Minimal cases of abstract curves.
- An abundance of standard items.

Knowing what to look for enables you to make more accurate estimates. Remember, the more repetition, the greater is the CAD advantage.

This is just one example, of course. Other advantages include accuracy (discussed next) and ease of collaboration. Another interesting advantage noticed by me is that non-artistic people or ones who would never have been able to draft with pencil, paper, and T-square often excel at CAD, and skills such as neatness and a steady hand become less critical.

In support of Golden Rule 7, a brief discussion on accuracy follows. This is one often brought up in class. Search the web for a drawing called SOLAR.DWG. Download and open the file. It is a drawing of the solar system (Fig. E.1) created in the early days of AutoCAD by founder John Walker and has become quite legendary over the years. You see, the drawing is to scale. This means that the orbits of all the planets are drawn to full size as well as correct relative to each other. Zoom in to find the Earth. Then, zoom in to see the moon around the Earth. Zoom in further and find the lunar landing module on the moon. Zoom in again and find a small plaque on one of the legs of the landing module. On it is a statement from humankind to anyone else who may read it. What does it say? And what is the date?

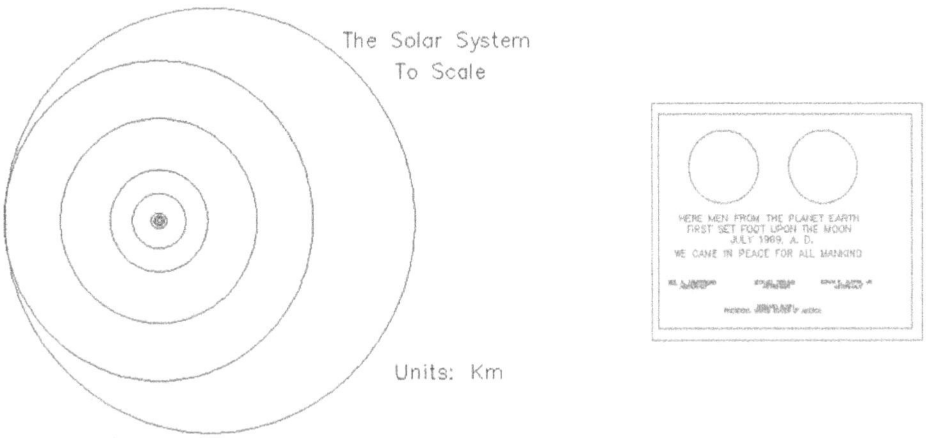

FIGURE E.1 The solar system and plaque to scale. *Source: John Walker.*

Think about what you just witnessed. You have zoomed in from the orbit of Pluto to look at a 5 × 4″ plaque. That is quite an amazing magnification. To put things into perspective, this is the equivalent of looking at the Earth then at a grapefruit or looking at a grapefruit then an atom. Can you design microchips in AutoCAD? Safe to say, yes. Can you connect walls together in a floor plan? Absolutely.

Limitations

One of the chief limitations of AutoCAD is its 3D capabilities. Not that they are shabby, but sometimes they are pushed to do what they were not created for. The essence of the observation is that AutoCAD is not modeling and analysis software, rather it is visualization software when it comes to 3D. It was not designed to compete with Revit, CATIA, NX, Inventor, or SolidWorks. Parametric design, dimension-driven design, FEA, and CFD analysis are not what AutoCAD is about.

These packages are dedicated engineering tools. AutoCAD was intended to be powerful 2D software with tools for presenting a design in 3D when necessary. As such, it is good at pretty pictures, to sell or present an idea.

Another limitation of the software, an obvious one about which a CAD manager needs to be aware, is that AutoCAD does not do the thinking for you. It contains little to no error checking, interference checking, or design fidelity tools, so garbage in gives you garbage out. Be aware as a CAD manager that AutoCAD lets users do almost anything, and the word *anything* can swing in either a positive or a terribly negative direction. Finally, understand the learning curve of this software. It will take a few months for a novice user to get up to speed. Keep an eye on that person until then.

E.3 Part 3: Maintain an office CAD standard

A standard is a necessity for a smooth-running office, especially for one with more than a few AutoCAD users. A standard ensures uniformity and a professional look to the company's output, minimizes CAD errors, and makes it easy for new users to get up to speed. A CAD manual needs to be created, published, and given to all users who may come in contact with AutoCAD. This manual should have, at a minimum,

- The exact location of all key files and the overall office computer file structure, indicating the naming convention for all jobs and the locations of libraries, templates, and standard symbol folders.
- A listing of the office layering convention. This may follow the AIA standard in an architecture office or just an accepted internal company standard.
- A listing of acceptable fonts, dimension styles, and hatch patterns.
- A listing of print and plot procedures, including *.ctb files.
- A listing of proper xref naming procedures.
- A description of how Paper Space is implemented. List title blocks, logos, and necessary title block information.
- A policy on purging files, deleting backup files, and AutoSave.

Keep the manual as short as possible while including all the necessary information. Do not attempt to cover every scenario that could possibly be encountered, as the longer the manual, the less likely it will be looked at cover to cover.

E.4 Part 4: Be an effective teacher and hiring manager

Part of being a CAD manager is sometimes explaining the finer points of AutoCAD to new hires or updating the occasional user as well as promotion of better techniques and habits. If you studied Levels 1 and 2 cover to cover, you may have noticed a concerted effort to simplify as much as possible. You should do the same with any potential students. While this is not meant to be a teaching guide, some basic points apply. If you are tasked with training new hires, then try to observe the following:

- Cover basic theory first—creating, editing, and viewing objects.
- Cover accuracy next, such as Ortho and OSNAP.
- Move on to the fundamental topics in the order of layers, text, hatch, blocks, arrays, dimensions, and print and plot, similar to Level 1.
- If advanced training is needed, cover advanced linework, followed by Xref, attributes, and Paper Space, in that order. The rest of Level 2 can wait, as some of it is geared mainly to CAD managers anyway.
- Stress to users the secrets to speed: the pgp file, right-click customization, and intense practice of the basics; remember the 95/5 rule from the second paragraph of Chapter 1.
- Accurately assess the skill level of users and do not assign advanced tasks to those not yet ready, unless the drawings are not critical for a particular project. I have personally seen the subtle damage to a drawing that can be caused by an inexperienced drafter.

Also, be ready to test potential hires with an AutoCAD test. This test does not have to be long; 45 min to an hour should be the absolute maximum, although truthfully, an experienced user can tell if another person knows AutoCAD within 3 minutes of watching him or her draft. The trick is to maximize every minute and squeeze the most amount of demonstrated skill out of a potential recruit, which means do not have that person engage in lengthy drawing of basic geometry but rather perform broad tasks.

For example, the floor plan from Level 1, Chapter 3 is a good test for an architectural drafting recruit (the drawing can be simplified further if necessary, by eliminating one of the closets or even a room). It should take an experienced user about 15 minutes total to set up a brand-new drawing (fonts, layers, etc.) and draft the basic floor layout (most students can do it in an hour after graduating Level 1—not too bad). Then, it should take another 15 minutes to add text, dims, and maybe some furniture, followed by another 5–10 minutes of hatch and touchups, followed by printing. So, it is roughly a 45-minute test and runs the recruit through just about every needed drafting skill. When looking over the drawing, note the layers chosen, accuracy demonstrated, and how much was completed in what time frame.

Note the following. It is not critical if a new hire does not know xrefs or Paper Space. These can be explained in a few minutes and learned to perfection "on the job." However, nothing but hard time spent drafting teaches someone how to draft quickly and efficiently with minimal errors. This is a requirement, and do not hire anyone who does not have that, as it will take time climbing that learning curve. Exceptions should be granted, of course, for individuals hired for other skills, with AutoCAD being of only secondary importance.

The bottom line (in my hiring experience) is that a fast, accurate drafter is far more desirable than a more knowledgeable one with a sloppy, slow manner of drafting. Realize also that extensive knowledge is not always a good substitute for experience, and while many drafters are fine workers and can come in and be productive on an entry level; be sure you understand what you are getting. I have for a long time encouraged constant practice in class, even at the cost of occasionally not covering an advanced topic or two, knowing full well what these students will face in a saturated market full of experienced AutoCAD professionals.

As a final word, be sure the AutoCAD test is fair but challenging. Leave units in decimal form as the student needs to demonstrate a grasp of nuances and situational awareness and the ability to change units to architectural if the foot/inch input does not work. Include in the test.

- Basic linework to be drawn.
- A handful of layers to set (colors, linetypes, etc.).
- Some text to write (set fonts, sizes, etc.).
- Some hatches to create.
- Some arrays to create.
- Some blocks to create.
- Some dimensions to put in.
- Plot of output.

E.5 Part 5: Stay current and competent

While we are talking about technical competency, you, as a recent advanced class graduate or CAD manager, are not off the hook for continuing to add to your skill set. It is much too easy to settle into a comfortable routine of drafting or designing and not learn anything outside the comfort zone (and not just with AutoCAD either). Some thoughts on this topic follow.

First, realize that it is virtually impossible for one individual to know everything about AutoCAD, although some claim to come close. It is truly an outrageous piece of software, developed and updated constantly by hundreds of software programmers and designers. This statement should make you realize that there is always more to learn (do not throw your hands up in despair, however), and you will not run out of new, easier, and creative ways to do a task. The main trait shared by all successful AutoCAD experts is that they are genuinely interested in the software and in the concept of computer-aided design in general.

To maintain your skills or advance further, look through the following list and determine what applies to you, then take the steps needed to acquire that knowledge. Some of this you will learn as part of studying all levels of the book; others you have to initiate on your own:

- *Learn the 3D features.* Many users do not know 3D well, and it is a good way to stand out from the crowd. It is fun to learn and is an important side of AutoCAD.
- *Learn basic AutoLISP.* This relatively simple programming language allows you to customize AutoCAD and write automation routines, among other uses. While its use dropped off somewhat in recent years, it is still a valuable skill.
- *Explore AutoCAD's advanced features that are sometimes overlooked*, such as sheet sets, dynamic blocks, eTransmit, security features, and CUI. Keep notes of key features and effects (a "cookbook" in software lingo).
- *Read up on the latest and greatest from AutoCAD*, new tips, and tricks, and what the industry is talking about. Cadalyst.com and Augi.com are two good sites for this. You should also get a *Cadalyst* subscription. It is no longer free but still worth the cost. At the very least, check up on the latest in the AutoCAD world on all the relevant websites.
- *Do not work in isolation*; talk to others in your field and see how they do things. Also, be sure to take an AutoCAD update class if upgrading to a new release.

- *Be curious about other design software*, including AutoCAD add-on programs (verticals) such as AutoCAD Mechanical, Electrical, Civil, and MEP. Be knowledgeable about as much software as possible. In the world of CAD, Revit and SolidWorks are two software packages to keep an eye on in the future.
- *Enjoy what you are doing*. If you are not, you should seek other work duties in your profession. AutoCAD is very miserable to deal with if you hate it. Occasionally, you may think it is just one big software virus out to get you; just do not let this be a daily thought.

Appendix F

AutoLISP basics and advanced customization tools

You may have heard of AutoLISP, Visual LISP, VBA (Visual Basic for Applications), .NET Framework, Active X, and ObjectARX. What exactly are they? These are all programming languages, environments, and tools used, among other things, to customize and automate AutoCAD beyond what can be done by simpler methods such as shortcuts, macros, and the CUI, all explored earlier in this book.

To properly discuss all these methods would take up quite a few volumes of text, and indeed much has been written concerning all of them. Our goal here is to introduce the very basics of AutoLISP and only gloss over the rest, so at least you have an idea of what they are. If advanced customization or software development work interests you, you can pursue it much further with other widely available publications.

F.1 Overview 1: AutoLISP

This is really the very beginning of the discussion. AutoLISP is a built-in programming language that comes with AutoCAD and is itself a dialect of a family of historic programming languages, collectively referred to as LISP (list processing language). LISP was invented by an MIT student in 1958 and found wide acceptance in artificial intelligence among other research circles. The language and its developer, J. McCarthy, pioneered many expressions and concepts now found in modern programming.

AutoLISP is a somewhat scaled-back version of LISP uniquely geared toward AutoCAD. It allows for interaction between the user and the software, and indeed one of the primary uses of AutoLISP is to automate routines or complex processes. Once the code is written, it is saved and recalled as needed. The code is then executed as a script (no compiling needed) and performs its operation.

One advantage of AutoLISP over other methods, discussed shortly, is that it requires no special training, except some familiarity with syntax, and is thus quite well suited for nonprogrammers. While it is not a compiled, object-oriented language such as C++ or Java and therefore not suitable for writing actual software applications, it also does not need to be for its intended use.

F.2 Overview 2: Visual LISP

Over the years, AutoLISP has morphed into a significantly more enhanced version, called VLISP. More specifically, Visual LISP was purchased by Autodesk from another developer and became part of AutoCAD since its release in 2000. It is a major improvement and features a graphical interface and a debugger. Visual LISP also has Active X functionality (more on that later). However, Visual LISP still reads and uses regular AutoLISP, and that is still where a beginner needs to start before moving up. We briefly touch on Visual LISP and its interface (the vlide command) later. The subject is an entire book all in itself.

F.3 Overview 3: VBA, .NET, Active X, and ObjectARX

VBA is a language based on Visual BASIC, a proprietary Microsoft product. VBA allows you to write routines that can run in a host application (AutoCAD) but not as stand-alone programs. It has some additional power and functionality over Visual LISP.

.NET Framework is another Microsoft software technology that was intended as a replacement for VBA. It features precoded solutions to common programming problems and a virtual machine to execute the programs.

Active X, also developed by Microsoft, is used to create software components that perform specific functions, many of which are found in Windows applications. Active X controls are small program-building blocks and are often used on the internet, such as for animations and unfortunately also for some viruses and spyware applications.

ObjectARX is the highest level of customization and programming you can do with AutoCAD, short of working on developing AutoCAD itself. Unlike the scripting-type languages and methods discussed so far, ObjectARX (ARX stands for AutoCAD Runtime eXtension) is an application programming interface. Using C++, programmers can create Windows DLLs (Data Link Libraries) that interact directly with AutoCAD itself.

Everything needed to do this is part of the ObjectARX Software Development Kit, freely distributed by Autodesk. This is not easy to master, as you would expect from professional-level software coding tools. Typical users are third party developers who need a way to write complex software that interacts seamlessly with AutoCAD. ObjectARX is specific to each release of AutoCAD and even requires the use of the same compiler Autodesk uses for AutoCAD itself.

Hopefully this discussion gives you some insight into what is out there for AutoCAD customization. This is only a cursory overview, of course, and you can pursue these topics much further. Any journey to learn advanced AutoCAD customization, however, begins with AutoLISP; and this is where we go next to introduce some basics of this programming language.

F.4 AutoLISP fundamentals

As mentioned before, despite the proliferation of new tools, AutoLISP is still very much in use, thousands of chunks of code are still being written annually, and many are easily found online for just about any desired function. The main reason for AutoLISP's past and present (to some extent) popularity remains the fact that AutoCAD designers are not programmers and AutoLISP is accessible enough for almost anyone to jump in and learn. The value it adds to your skill set is substantial. Let us cover some basics and get you writing some code. AutoLISP code can be implemented in two distinct ways:

- You can type and execute a few lines of code right on the command line of AutoCAD. This is good for a simple routine that is useful only in that drawing.
- You can load saved AutoLISP code and execute it, usually the case with more complex and often-used routines.

We generally use the second method of writing and saving our routines. At this stage, it is helpful to know how (and where) to save AutoLISP code and, by extension, how to load what is written.

All AutoLISP code is saved under the *.lsp extension. Once the code is written, it can theoretically be stored just about anywhere (AutoCAD Express folder contains a sizable number of built-in routines). To load the created or downloaded code, just type in `uppload` and the dialog box in Fig. F.1 appears.

Then, you just browse around, find the AutoLISP code you are interested in, click on it, and press Load. Once loaded, you can close the dialog box; the code is ready to execute with a command input. There is a way to automatically load all existing code on starting AutoCAD by adding it to the Startup Suite (lower right corner, the briefcase icon).

Now that you know how to save new code and where to find it, we should try writing a simple example that illustrates some AutoLISP basics. Here are two critical points that need to be understood before you begin:

- AutoLISP is closely tied to AutoCAD commands, so to write something useful, you need to know the commands well. If you do not know how to do it in AutoCAD manually, AutoLISP is of no use.
- Approach AutoLISP for an answer to a specific problem. State what you want to solve or automate, then write out the steps to do it manually. Only then, using proper syntax, write and test the AutoLISP code.

Following this philosophy, let us state a problem we would like to solve. We would like to modify the zoom to extents command by having it zoom to only 90% of the screen, not 100%. That way we can see everything and nothing gets lost at the edges of the screen. But, how do we do this by manually first?

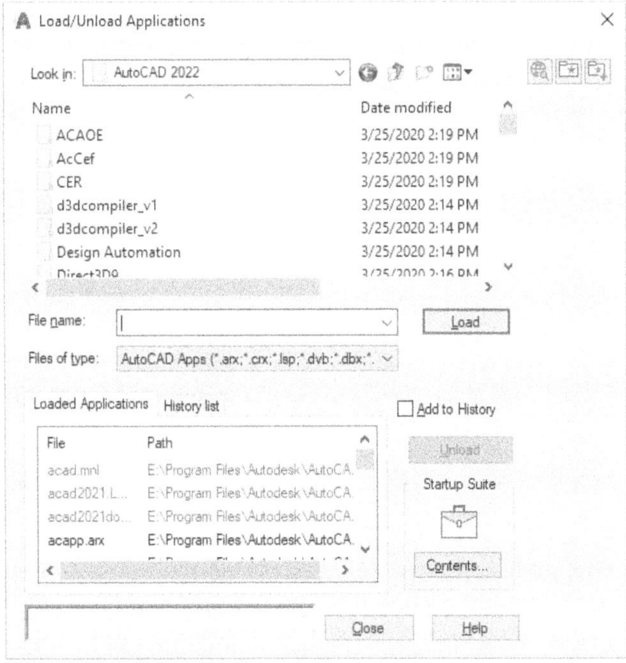

FIGURE F.1 Load/Unload Applications.

You, of course, type in `zoom`, press Enter, select `e` for Extents, and press Enter again. The drawing is zoomed all the way. To bring it back a bit, you repeat the zoom command but type in `.90x` instead of `e`. The drawing then zooms in a bit to fill only 90% of the screen.

Let us now automate these steps so all you must do is type in `zz` (a random choice for a command name; you could pick anything) instead of the two-step process just described.

We need to:

- Open the Notepad editor (we use the built-in Visual LISP editor later).
- Save the blank file as ZoomSample.lsp in any folder you wish.
- Define the function zz by typing in exactly the following: `(defun C:zz ()`.
- List the relevant commands in order by typing in exactly the following:`(command "zoom" "e" "zoom" ".90x")`.
- Finally add `(princ)` and close the expression with another).

So, here is the result:

```
(defun C:zz ()
(command "zoom" "e" "zoom" ".90x")
(princ))
```

Save the file again and let us run it. Open any drawing, load the AutoLISP routine using appload or the drop-down menu and run it by typing in `zz` and pressing Enter. The drawing should zoom out to 90% as expected.

This simple routine has some essential ingredients of all AutoLISP code:

- It starts out with a parenthesis, indicating that a function follows.
- The defun or define function (an AutoLISP expression) is next followed by C, indicating it is an AutoCAD command, and finally the actual command, zz.
- The set of opening and closing parentheses can hold arguments or variables.
- The next line features command, an AutoLISP expression meaning AutoCAD commands (in quotation marks, meaning it is a string) follow.
- The (princ) function is added so the routine does not constantly return a "nil" value.
- Finally, another, matching parenthesis closes out the program.

Here are two simple routines that can be used to rotate the UCS crosshairs to align them to some geometry (this general task was covered in chapter: Advanced Design and File Management Tools).

The first routine does the rotation and is called *perp*:

```
(defun C:perp ()
(command "snap" "ro""" "per"pause "snap" "off")
(princ))
```

The second one brings the crosshairs back to normal and is called *flat*:

```
(defun C:flat ()
(command "snap" "ro""" "0"pause "snap" "off")
(princ))
```

The perp and flat AutoLISP routines feature two new concepts. The first one is the pause function, which allows for user input. In this case, it is for the user to select the line to which the crosshairs are to be aligned. Then, the function resumes and turns off snap. The other concept is the two quotation marks back to back with nothing between them. They simply force an Enter.

Study these three examples closely. They are simple; many AutoLISP routines are, as that is all that is needed to get the job done.

Here is what we learned so far:

- *String*: Anything that appears in quotes (often an AutoCAD command).
- *Expression*: AutoLISP built-in commands, such as defun.
- *Functions*: These can be user defined (perp, flat) or built-in (pause, princ, command).
- *Arguments*: Functions are sometimes followed by arguments in parentheses.

F.5 Variables and comments

Let us move on to other concepts. To take a step up in sophistication from what we did so far, we need to introduce program variables. Variables can be numbers, letters, or just about anything else that can change. The reason they are needed is because they allow for user interaction. Because the user input is not known when the code is first written, a variable is used. When that information is acquired, the program proceeds accordingly. Variables are a fundamental concept in all programming.

Variables in AutoLISP can be global (defined once for the entire routine) or local (defined as needed for one-time use). Local is usually how variables get defined in AutoLISP. The function that creates variables is *setq*. To set a variable equal to whatever the user inputs (such as the radius of a circle), you write setqRadCircle. The next logical step with variables is to pick up on whatever the user inputs. For this we need the getstring function, which returns anything the user types in.

Here is how these concepts look in practice, with the \n simply meaning that a new line is started.

```
(setqRadCircle (getstring "\nCircle Radius: "))
```

In addition to getstring are many other get functions, such as getfiled (asks user to select a file using a dialog box), getangle, and getdist.

As in any programming language, you can comment on your work, so you or someone else can understand your intent when examining the code at a later time. In AutoLISP, comments are preceded by a semicolon. You can add more than one semicolon to indicate sections or headers:

```
;This is a comment
;;So is this
```

F.6 Advanced features and the vlide command

We have only scratched the surface of AutoLISP, and if you enjoyed writing and running these simple routines, then consider taking a full class on this subject or reading specialized books. There is much more to learn. For example,

- *The IF → THEN statements*: This is a common feature in all of programming. If certain conditions are met, then the program does something; if not, it does something else.
- *The WHILE function*: This is another important concept in programming. This function continues to do something until a condition is met.

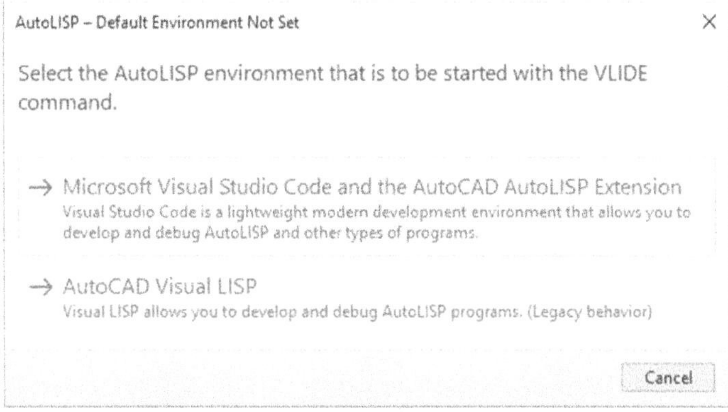

FIGURE F.2 Vlisp default environment dialog box.

Finally, we come to the VLISP command and the Visual LISP Editor (See Fig. F.3. Everything you typed earlier into Notepad can be typed into the Visual LISP Editor (also referred to in programming as an Integrated Developers Environment), which has many additional useful features, such as debugging, error tracing, and checks for closed quotations and parentheses (formatting tools). Everything you type is also color coded for easy identification. Simply type in vlisp and press Enter to bring up the editor if it is installed. If it is not installed, you will get a warning box (see Fig F.2) that directs you to choose the default behavior for the vlisp or vlide commands. You must choose between Microsoft Visual Studio Code (https://code.visualstudio.com/) or AutoCAD Visual LISP. Please note that Autodesk is retiring the current editor and will remove it in a future release. They recommend using the AutoCAD AutoLISP Extension for Visual Studio (VS) Code to create new (and updating all existing) AutoLISP programs.

In-depth coverage of Visual LISP and Microsoft Visual Studio Code is beyond the scope of the book, but you are encouraged to pick up a specialized textbook and try to progress with this invaluable coding skill.

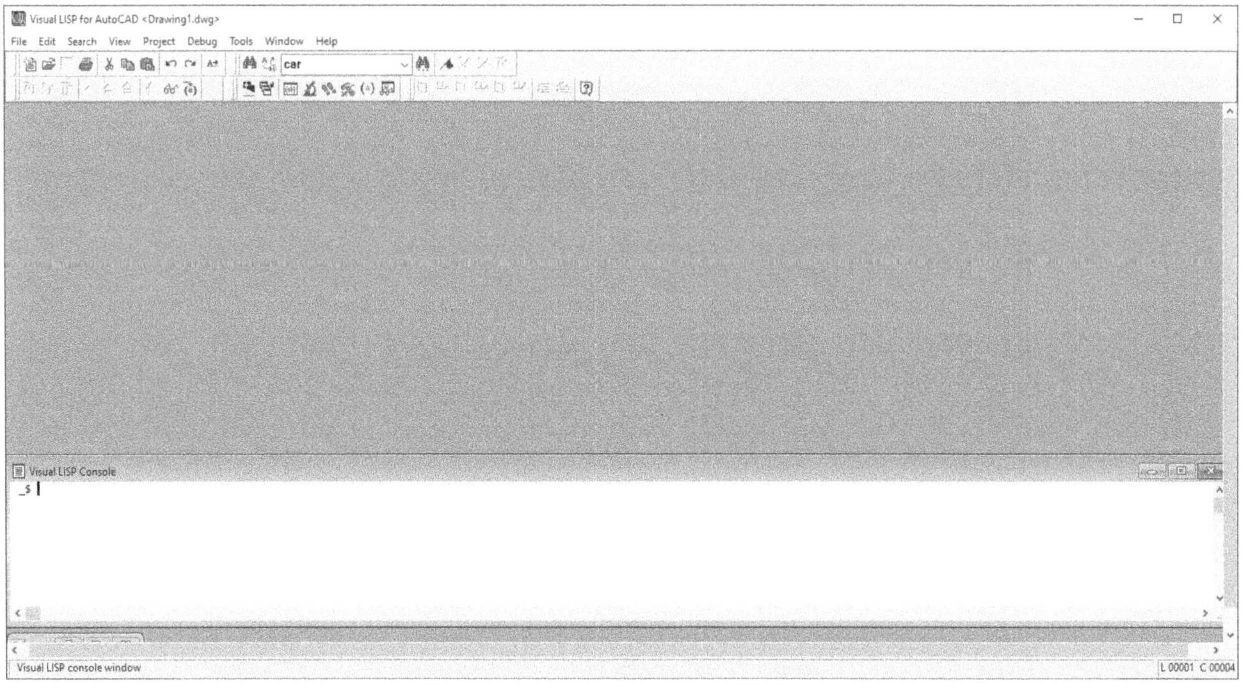

FIGURE F.3 Visual LISP Editor.

Appendix G

PC hardware, printers/plotters, networks, and the cloud

This collection of topics can easily span volumes, and quite a bit has been written about all of them. We try to focus on hardware and networks from an AutoCAD user's point of view exclusively.

Software generally goes hand in hand with the hardware. In recent years, computers have become a lot like cars - very reliable and are usually packed full of standard features that were considered high-end not too long ago. With both you will likely get a solid product that will be right for most ordinary tasks. Remember though that CAD in general, and AutoCAD specifically, is not ordinary software. It is a high-end, powerful application that places heavy demands on the associated hardware and does not run at its peak on a clunky PC. Much like the case where the owner of a sports car is more likely to be interested in (and look under the hood of) the car than the owner of a commuter economy car, so should AutoCAD designers express interest in what is "under the hood" of their computer. After all, they are manipulating and creating multi-megabyte sophisticated drawings or 3D models, not merely typing up a memo on Word; and it is in their best interest to use a PC that meets or exceeds AutoCAD's demands.

Fortunately, as previously mentioned, it is easier than ever to have the right PC. Processor power and the downward spiraling costs of RAM and graphic cards have outpaced AutoCAD's requirements, and most folks, if they spend enough for a mid-range system, will likely get everything they need. Furthermore, anyone who engages in gaming or the visual arts (like digital film editing) will already have the right PC for the job - or Mac, if working with AutoCAD for the Mac. What is good for the goose (games and video editing) is also good for the gander (AutoCAD).

G.1 PC Hardware

By its nature, AutoCAD is not as fussy as some other CAD software out there (a few 3D solid modeling applications come to mind), but it does need a mid-range PC to operate optimally. Here is a rundown of all the computer components:

- *Processor*: The "brain" of the computer, also referred to as the CPU, it is where all the calculations and operations are performed. Intel and AMD are the major players in the PC market. Which is better is a moot point; while Intel was once the only game in town, these days the two companies constantly one-up each other with performance benchmarks. At this point, both manufacturers produce high-quality products, and the Intel versus AMD comparison is essentially meaningless to the average user. Intel processors are organized by families such as i3, i5, i7, i9 or Xeon. The numbers following that are the CPU's power and its intended placement (PC, mobile, laptop, etc.). If a letter follows that, it is used for additional classification such as K, meaning "unlocked for overclocking," or a power saving chip, and so forth. As of this writing the top performing Intel processor is the Xeon W-3275M. It has 28 cores, with a base frequency of 2.5 GHz, which can be boosted up to 4.4 GHz. It has a list price of $7,400. You can get nearly identical performance of the out of the i9-10980XE for only $950. As you can see processors are not cheap. On the AMD side, the Ryzen Threadripper Pro 3995WX is by far the king. In bench tests it beats the best performing Intel by more than 2x, with 64 cores, a base frequency of 2.4 GHz, which can be boosted up to 4.2 GHz. Like the Intel, its performance comes with a high cost at $5,500. A great alternative to both is the AMD Ryzen 9 5900X which performs nearly identically to the Intel but costs only $550. All these CPUs feature multiple cores and something called *hyperthreading technology*.

Multiple cores are independent processing units embedded on the main chip, and hyperthreading allows for multiple tasks to be done at once. To take advantage of this, your software must be written for it. As of this writing, AutoCAD is not optimized for these technologies.

- *Motherboard*: The motherboard is the main mounting and control unit for the CPU (and associated cooling hardware), the RAM, and the video card. The motherboard is also where you will find the connections for the hard drive, power, CD-ROM, and just about everything else. Most of the interconnection wiring also originates or terminates here. Motherboards are obviously very critical and expensive parts of the overall system. When selecting a motherboard, you must match it to the CPU via the stated socket type and number.
- *RAM*: Random access memory is the next consideration, and the more the merrier. RAM has a direct effect on performance, as it is there that software resides while in operation. More available RAM allows for more functions to execute faster. Autodesk recommends 8 GB minimum but go as high as your motherboard can support. Older motherboards accept at least 4 GB and some of the newer ones handle up to 64 GB of memory. Some well-known RAM manufacturers are Corsair and Kingston.
- *Hard drive*: Also known as the storage disk, main drive, and the like, it needs to be at least 1.8 GB to install AutoCAD. Fortunately, this is rarely a problem, as most machines have drives on the order of 100–250 GB. It is not unusual to find a terabyte-sized (1TB) and even larger drives. Hard drives are classified by the "old style" disk drive, which involve spinning magnetized platters. The newer type, solid state (SSD), features much faster data access, but for a slightly higher cost and currently smaller storage size than its spinning disk counterpart. A good combination is to use two drives, an SSD as the primary drive for blazing fast startups of Windows and various applications, with a secondary regular drive for data storage. Hard drives are hooked up to the motherboard via the older IDE hookup or the newer SATA. Some users opt for several drives stacked next to each other in what is known as a RAID configuration. This adds security in case of failure but, of course, increases the cost. Some well-known brands include Seagate, Toshiba, and Western Digital. Prices can range from $50 to $250 and up.
- *Video graphics card*: Perhaps the most overlooked part of the PC, the video or graphics card is like a processor but is specially designed to control the screen images or graphics. It mounts on the motherboard and needs to be of at least mid-range level if you are planning on doing a lot of 3D work, including rendering. Autodesk recommends only a "workstation-class" graphics card but get the best one possible under your allowed budget. High-quality graphic cards are not cheap; a high-end card can easily top out at over $1000 (and up). One in the $300-500 range should be just fine for AutoCAD applications. Some well-known video card manufacturers include ATI, NVidia, AMD, and EVGA.
- *OS*: Until late 2010, operating systems that AutoCAD could run on numbered in the single digits (precisely one... Windows). In November of that year, after a 20-year break, Autodesk announced AutoCAD would once again be available for the Mac. AutoCAD still does not run on UNIX or Linux. Windows 7 SP1, Windows 8.1 and Windows 10 are currently the main Microsoft operating systems AutoCAD is certified compatible for. Windows 8.1 was released in October of 2013. It has proved problematic in the computing world at large but seem to be just fine for AutoCAD. Windows 10, released in 2015, will work just fine. As far as the older XP operating system, note the following warning from Autodesk: "AutoCAD 2014 is the final version of AutoCAD software that will be supported on the Microsoft Windows XP operating system." Keep this in mind if you are planning on installing AutoCAD 2015+ and still have XP running.
- *Sound (audio) card*: This add-on is not technically required for AutoCAD, the way it may be for video editing and gaming, but it is good to have for a complete PC. Sound cards are not terribly expensive, and top of the line ones can be had for well under $200. Some major manufacturers include ASUS and Creative Sounds. The sound card enhances the audio output over the base motherboard and provides some outputs for speakers and microphones.
- *Optical drive (CD/DVD)*: An optical drive is needed if you plan to read from or burn to any disks such as CDs or DVDs. Most modern drives can handle all types of CDs, DVDs, and Blue Ray, both reading from and burning to. The current speeds are about 48x write, 24x rewrite, and 48x read, with prices in the $30 to $60 range. Some machines come without CD drives now, relying on USB, internet downloads, and other methods for acquiring data, but I would keep the CD drive around for some time still - it is cheap and still very useful.
- *Monitor*: Get the best and largest flat screen monitor you can afford; your eyes will thank you many times over. Better yet, get two and put them side by side. Once you have that sort of setup, there is no going back. You can also go to three separate monitors if your video card would support it. Other people are going to ultra-wide monitors that can simulate one, two or three monitors on one super wide screen. Prices on monitors have also dropped in recent years and you should be able to find an excellent 24″ or larger monitor easily for under $200.
- *Mouse and keyboard*: Get a laser precision mouse (forget roller ball, if you can even find one like that) and the best soft-touch full size keyboard with number pad that you can find. Do not try to cut corners here, as you will be in physical contact and using these devices constantly while working on the PC, so get the good stuff. There are some ergonomic mice around that are formed for better hand positioning.

Whether building your own PC or ordering an assembled one, take the PC purchase seriously and try to set realistic budget expectations. CAD workstations are in the same family as gaming stations and you need to budget $1,500 (or more) for a serious machine, not including the monitor. Some well-known vendors include Dell, but if you are inclined to assemble your own, check out online stores such as www.Newegg.com or www.PCPartsPicker.com and shop sales. You will get phenomenal performance for a fraction of the cost, if you know what you are doing when assembling the machine.

AutoCAD Requirements (Windows)

Finally, as a conclusion to this section, here is what Autodesk recommends for the last AutoCAD version, as posted on its website.

- WINDOWS
  - OS:
    - 64-bit Microsoft® Windows® 8.1 and Windows 10 OR
  - Processor:
    - Basic: 2.5–2.9 GHz processor
    - Recommended: 3+ GHz processor
    - Multiple processors: Supported by the application8 GB RAM (16 GB recommended).
  - Memory
    - Basic: 8 GB
    - Recommended: 16 GB
  - Display
    - Conventional Displays:
    - 1920 x 1080 with True Color
    - High Resolution & 4K Displays:
    - Resolutions up to 3840 x 2160 supported on Windows 10, 64-bit systems (with capable display card)
  - Graphics Card
    - Basic: 1 GB GPU with 29 GB/s Bandwidth and DirectX 11 compliant
    - Recommended: 4 GB GPU with 106 GB/s Bandwidth and DirectX 11 compliant
  - Disk Space
    - 7.0 GB
  - Internet access is required for licensing and installation
- MAC
    - Operating System
      - Apple® macOS® Big Sur v11 (requires Update 2021.1)
      - Apple® macOS® Catalina v10.15
      - Apple macOS Mojave v10.14
      - Apple macOS High Sierra v10.13
    - Model
      - Basic: Apple Mac Pro® 4.1, MacBook Pro® 5.1, iMac® 8.1, Mac mini® 3.1, MacBook Air®, MacBook® 5.1
      - Recommended: Apple Mac® models supporting Metal Graphics Engine
      - Apple Mac models with M series chip are supported under Rosetta 2 mode.
    - CPU Type
      - 64-bit Intel CPU
      - Recommended: Intel Core i7 or higher
    - Memory
      - Basic: 4GB
      - Recommended: 8GB or higher
    - Display Resolution
      - Basic: 1280 x 800 display
      - High Resolution: 2880 x 1800 with Retina Display
    - Disk Space
      - 3 GB free disk space for download and installation
    - Display Card
      - Recommended: Mac native installed graphics cards

As you can see these requirements are not astronomical by any means. They are the minimums, so aim to build or configure a machine that is above this.

G.2 Printers and Plotters

There are essentially two ways to output a paper drawing from AutoCAD, a printer or a plotter. Printers are generally Inkjet or LaserJet and black and white or color capable. Prices can range from a $49 desktop Inkjet to a "sky's the limit" corporate high-speed color LaserJet. Which to get depends entirely on budget and desired level of quality for output. Once AutoCAD is installed, it automatically senses any printer attached to the PC (assuming it works and has drivers in the first place). Be aware that most small desktop printers cannot handle 11″×17″ paper. The 11″×17″ format is very important in AutoCAD production, as many jobs are handed in on this size paper or printed as half size check plots, so be sure your printer can handle it.

Plotters (Figure G.1) are a familiar sight in architecture or engineering offices. Anything sized above an 11″×17″ sheet of paper needs to be routed to a plotter, all the way up to a 48″×36″, or larger in some cases. Plotters as a rule are more expensive than printers and buying one should be a carefully researched decision. Most plotters are color capable and accept almost any type of paper, although bond or vellum is the norm. Hewlett-Packard (HP) is the premier manufacturer of plotters, with some models on sale well over $20,000, though their T120, T520, and T730 series plotters are much more affordable. Other companies, such as OCE, Canon, and Encad, round out the field. The typical average cost of a professional-grade plotter hovers around $5000–$6000. Do your research carefully when selecting one and be aware of recurring costs such as ink and paper.

G.3 Networks

This topic is an entire profession all to itself, of course. A computer network is defined as a group of interconnected computers. Networks can range from a local area network (LAN) to a city, country, or worldwide (WAN, or wide area network) one. Networks allow for collaboration and sharing of information from computer to computer and have their own set of hardware familiar to any network engineer, such as routers, hubs, switches, repeaters, and other gear. Networks are operated via a protocol or set of instructions, such as the IEEE 802.3 protocol running on Ethernet technology for a typical LAN. The medium of data transfer between computers is usually a standard Cat5e cable but can be wireless or fiber optic for larger networks.

As an AutoCAD designer, you will likely encounter a local area network consisting of a server (host) with some other computers (clients) connected to it at your company. All the data are stored on the server, and once you log on to the network, your computer can access what it needs.

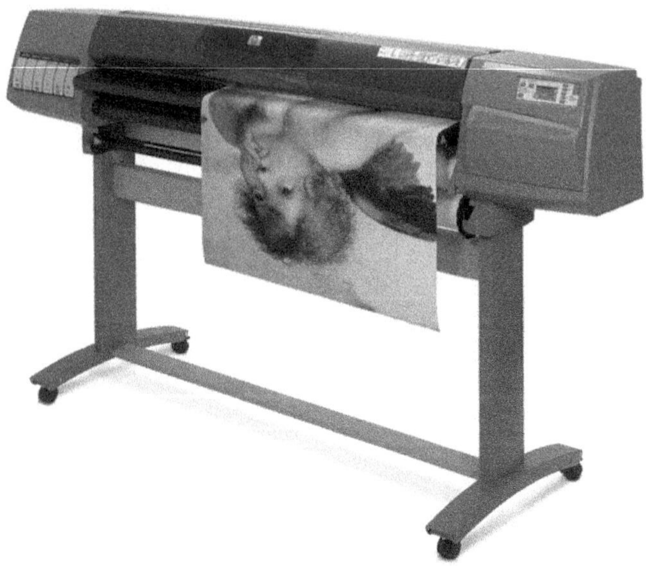

FIGURE G.1 A typical professional plotter. *Source: Hewlett-Packard*

In regard to networks, AutoCAD can be installed on a per-seat or network license method. Per seat is cost effective to a certain limit (usually ten seats). If your company requires more, a site license is purchased, which is cheaper and allows more computers to run the software, which now resides on the server only.

G.4 The Cloud

The *cloud* refers to two things here: Cloud computing and cloud storage. You are probably familiar with both, even if you did not specifically refer to them as such. Cloud computing is a general term for use of computing resources (hardware and software) that are delivered as a service over a network. You, the user, are the front end. The service is the back end, and you communicate with the service via the internet. A simple example is Gmail. You do not store the email messages and all the attachments on whatever computer you may be using, it is all on Google's servers. You simply access it remotely as needed. Cloud storage is a related concept. You store data, personal or company, on a remote server and upload and download as needed.

These explanations are, of course, highly simplified, and a great deal of technology and innovation lie "behind the scenes" of this broad high-tech industry. Cloud computing offers a great many services, such as software, platform, a test environment, and data. Many companies and now individuals use this technology. It is not really new, with underlying principles dating back to the 1950s, but was held back mostly due to low-speed point-to-point connections up until the 1990s. After all, moving data and information is the cornerstone of cloud computing and storage, and until fast internet and other data transfer methods became the norm, a lot of computing and storage was local in nature. The term *cloud* and the cloud symbol first appeared in the early 1990s as well and was a popular metaphor to represent the complex (and unseen) internet computer architecture in network diagrams.

So what does this all have to do with AutoCAD? Well, AutoCAD's output - the drawings - are data. Such data need to be stored and sometimes shared as needed. Just like storing music in the cloud (such as with iTunes), you can store these drawings in a similar manner. This also allows for safe, remote archiving of a company's designs and easy collaboration between individuals that are far away.

Autodesk jumped in the cloud game a few years ago by offering something called *Autodesk 360*. This is a service that, according to Autodesk itself, "provides a broad set of features, cloud services and products that can help you improve the way you design, visualize, simulate, and share your work with others anytime, anywhere." The goal here is to create a collaborative work environment that is hosted completely by Autodesk on its servers. This is a new direction for the company, and no doubt a lucrative one. Users create an account, once they subscribe, and log in via secure, password protected means. They can then share designs and collaborate in a "digital workspace."

It is not known outside Autodesk how many companies signed on to the cloud service, but it is clear that this idea in general is the way things will be done in the future, not just with AutoCAD but in other areas of design. By outsourcing the physical and even the software component of a design firm to a cloud-based service, huge cost savings can potentially be realized. Keep an eye on this in the future, and if interested in learning more, research the topic online.

Appendix H

What are kernels?

If you have taken an active interest in 3D CAD software and are learning what is "under the hood," then you may have read or heard about kernels. What are they and what do farming and planting terms have to do with software? Well, as you may have already deduced, the word *kernel* is a metaphor for describing the heart of the software, a starting point from which everything else grows, the same as a kernel of corn. This analogy is quite accurate: A 3D modeling kernel is the computer code that is the "brains" of any 3D software package, such as CATIA, NX, SolidWorks, Inventor, and of course AutoCAD.

As it turns out, kernels are quite rare. This is because they are extremely difficult, expensive, and labor-intensive to develop from scratch. As a result, while hundreds of 3D-capable CAD software applications are available, there are only a small handful of kernels. If you wanted to start a new CAD company and develop a competitor to any of the established products, then chances are good that you would simply purchase a kernel license and design your software around the kernel, not try to create your own. In many cases, this is indeed what happens in the industry.

Why would kernel developers license their kernels, as opposed to using them for just their own CAD programs? Well, for starters, they do use them in such a way, and most kernel developers have a popular 3D CAD program for sale. But, the licensing fee to "rent out" kernel code is substantial, so they do this as well to increase the revenue stream.

What customers get when they purchase the kernel is a large chunk of code made up of what are called *classes* and *components*—essentially mathematical functions that perform specific modeling tasks. The CAD company then writes its own code to interface with the kernel via numerous APIs (application programming interfaces) and essentially creates the new software around the kernel.

This new CAD software is unique, according to the design philosophy and talent of the CAD company and may be different in how it operates compared to another company's software that uses the same kernel. In essence, however, most modelers are essentially the same, and one can switch relatively easily from, say, CATIA to NX or Creo Elements (Pro/ENGINEER). All three of these software applications feature parametric modeling, part trees, dimension-driven design, and other hallmarks of solid modeling. The differences are chiefly in the interfaces, ease of use, extra features offered with each package, and of course price.

One notable exception is AutoCAD. The kernel it uses, ACIS, is made by Spatial Technologies. Its modeling methods are somewhat different than the other kernels on the market, giving AutoCAD a unique (and not necessarily the best) approach to 3D. The essential difference has to do with the purpose and goal of the ACIS kernel, which is to simply represent objects in 3D, not to analyze and design with them. Models created by ACIS lean more toward "pretty pictures" than engineering tools. This is not surprising because that is exactly what is needed by AutoCAD design professionals—a representation of a 3D model to bring across an idea of a design. There is simply no need for highly sophisticated 3D operations found in the high-end solid modeling software and would be a waste of money to add it to AutoCAD when most customers would never use it.

These other, more high-end, kernels include Parasolid (used by NX, SolidWorks, and SolidEdge), Granite (used by Creo Elements [Pro/ENGINEER]), and CAA (used by CATIA). Autodesk set out to develop its own kernel in recent years, and the fruit of that labor is in its Inventor software.

Appendix I

Lighting, rendering, effects, and animation

The world of digital effects is a very expansive one, and under that umbrella falls everything from simple shaded models (extensively used in the 3D chapters this book) all the way up to full-blown Hollywood-style effects. Architects and engineers use some of these digital effects and techniques, which are roughly in the middle of the pack as far as sophistication goes.

What architects look for includes visualization renderings that present a building or an interior scene in the most realistic way possible, so the clients see what they are getting and can offer input. They may also want some basic animation, such as a walk-through. Some high-end requests may even include photorealistic scenes that have weather effects (wind, precipitation, etc.) or realistic images of people interacting with the design.

Engineers have the same basic goals, to present a design to a client, although their renderings must often account for metals, plastics, and other materials, including material properties. Engineering animation needs are also more sophisticated. An architectural animation of a walk-through is essentially a collection of snapshots played back in succession. With engineering animations, complex part interaction must be accounted for and a full motion path must be shown. The entire animation must also be looped for constant playback. An example of this is a cutaway of an engine, with all the components (pistons, connecting rods, crankshaft, etc.) working in unison to demonstrate the design in operation.

As far as engineering applications go, AutoCAD plays a relatively minor role. From the beginning, the intent of AutoCAD was to be a superb 2D drafting tool, not a 3D solid modeling and testing software. Other software has evolved over the years (with some of these applications predating AutoCAD) for use in such cases. Examples, quoted elsewhere in this book, include CATIA, Creo Elements (Pro/ENGINEER), NX, SolidWorks, SolidEdge, IronCAD, Autodesk's own Inventor, Revit, and many others. These software applications often take care of the drafting, modeling, rendering, testing, presentation, animation, and manufacturing in one complete package. Therefore, AutoCAD rarely finds its way into truly high-end situations; it is simply not built for that.

In the world of architecture and related fields, however, AutoCAD 3D models often find themselves in the middle of the rendering world. These models are imported into a variety of software and are the "skeletons" on which advanced rendering and even animation techniques are applied. Studio VIZ was the affordable choice for many professionals until it was discontinued a few years back. Its replacement was the top of the line, more expensive 3ds Max Design. This software can create stunning photorealistic renderings that feature advanced use of lights, shadows, and materials. 3ds Max software can also do sophisticated animations, even for full feature films, but it is not commonly used for that, as even more advanced software is available for that purpose. There are many other software solutions for creating renderings, of course. Two popular ones include Rhino and Form Z. Both software packages can generate their own shapes and solids but are often used as pure rendering engines with imported models.

The world of animation is tied in with rendering, as the animator simply animates an already complete rendering and adjusts for changes in lights and shadows. A major player in that field is Maya, formerly of Alias Wavefront but now owned by Autodesk. Maya has extensive customizing capabilities, and studios take its core (the kernel) and write custom code around it, tuned to their specific production needs. Maya in various forms was one of the software applications behind movies and TV shows such as *Avatar*, *Frozen*, and *Game of Thrones*, although the amount of customization needed for such high-end productions practically turned Maya into whole new software.

The software listed thus far are just some of the high-end market products. Many other companies in the field produce software in many other price ranges. If you are interested in rendering and animation, it is a great field to get into, and many such students passed through my classroom, as they wanted to start with being able to model 3D designs (in plain old AutoCAD) and move on from there.

Appendix J

3D printing

J.1 Overview and history

3D printing is not really "printing" as most people use and understand that word; a more descriptive name is *additive manufacturing*. It is a process where material is precisely ejected from a specialized computer-controlled machine and added together, layer-by-layer, until a 3D object is produced. The data source used to create this object is a 3D CAD model, or sometimes a 3D scan of an existing object; and the material used in the manufacture can be almost anything, as long as it is suitable for projecting out of a nozzle or something similar (a liquid, gel, or powder would work). It must then, of course, be able to solidify into a stable physical object that you can handle and use. That, in simplified terms, is the essence of 3D printing.

This technology has received a lot of publicity in recent years, but it is not as new as it seems. Evolving from the science of *stereolithography* (pioneered by Chuck Hall in 1984), early versions of additive manufacturing machines used soft polymers (plastics) as the source material. These polymers would be melted down before passing through a nozzle and solidify (or were cured) rapidly after being built up into layers, eventually forming an object. Polymers remain the most popular source material, but other materials (even edible food items) were used in the 1990s and 2000s, as the technology matured and dropped drastically in price. There are now home-based, relatively affordable 3D printers for hobbyists and entrepreneurs. Uses for 3D printing vary widely, with primary applications in product development, prototyping, biotech and dental devices, art, and education among others. It is a rapidly growing field, and one to keep an eye on.

There are four distinct parts to a 3D printing process, as described next.

J.2 Printing process—part 1 (creating the model)

The first step in creating a 3D printed model is to create a viable model of what you want to make. You can go about it in several different ways. The most direct is to design something yourself from scratch via any of the popular 3D solid modeling software packages, such as CATIA, SolidWorks (or AutoCAD). Alternatively, you can go on one of several 3D printing "marketplace" websites from companies like Thingiverse, Shapeways, or Threeding and find, trade, or buy a premade model of something you want. Finally, if you want a 3D model of something you already have (in other words, you just want to replicate it), then you can also do a 3D scan of the object via a 3D scanner. These types of advanced scanners use lasers or soft touch probes to map the surface shape of an object, creating a resultant point cloud, and then use a reconstruction process to build up a 3D model of it.

J.3 Printing process—part 2 (exporting and preparing the model)

Once the model is created it needs to be exported to an STL (STereoLithography) file, which is the native language format for 3D printers and generally most computer-aided manufacturing and rapid prototyping software. Most solid modeling and drafting software can export out to STL, including AutoCAD.

There are a few ways to export your model. The most basic way is to just select File→Export, then choose the *.stl format from the drop-down menu at the bottom of the dialog box. You can accomplish the exact same thing by typing in `3dprintservice` and press Enter or select Publish→Send to 3D print Service on the Application menu (the AutoCAD symbol at the upper left).

However, if you use this approach, then exporting is not the end of the model preparation; you need to now "slice" your model. This converts it to a series of thin layers and generates the code needed to by the 3D printer. Several commercial and open-source slicers are available, such as Skeinforge, Cura, and Simplify3D.

Starting with AutoCAD 2017 there is another option—but only if you are running a 64-bit system; you are out of luck on a 32-bit one. This new approach is a separate application called *Print Studio*. It shipped with AutoCAD 2017 and should also come with the latest version but may need to be installed separately. If it is there, you will see the symbol for it in the Output tab of the Ribbon (in the 3D Modeling Workspace) or you can use the Application menu, Print → 3D Print. If it is not installed, you will be given the opportunity to install it. This application preps up your model and sends it to a 3D printer, if one is hooked up to your computer. If your AutoCAD has it ready to go, try it out!

J.4 Printing process—part 3 (printing the model)

The actual printing itself is almost fully automated and is the relatively easy part of the process, where after hours of designing and testing a model (if you did that vs. just purchasing one), you finally just get to watch it being built. There are a number of popular printing approaches of varying complexity and cost. Many of them continue to use historically proven polymer, powder, gel, or small beads as source material, heated up for easy passage through an applicator nozzle then rapidly cooled to form the solid object. The source material may be coiled up and unfurled into the nozzle mechanism (or be gravity fed from a hopper) at a steady pace.

Extrusion deposition uses this exact approach with polymers, which can be styrene, polycarbonate, polyethylene, or polystyrene. Other methods use liquids or, in the case of *bioprinting* (an active research field), even living tissue.

Regardless of the source material, the actual printing process is very similar. The movable nozzle passes over the base plate (on which the model rests), ejects material, and builds up the model one pass at a time, using the slices as a guide (Fig. J.1). The printer size varies widely, from smaller "desktop" units to larger commercial machines, including a very large unit called the *Big Area Additive Manufacturing* (BAAM) *machine*, used in several industries such as aerospace.

J.5 Printing process—part 4 (finishing model)

Once the model is completed, depending on the material, it may need to sit on the base plate for a while to cure fully. It then needs to be "finished," which can range from just a quick polish all the way to additional material removal. In latter case, material is removed because the model is printed slightly oversized. This is because a greater degree of accuracy and resolution can be achieved in the final product by removing some material in a secondary step, as opposed to trying to achieve this in the original manufacturing process. Finally, these 3D models can be coated, painted, or otherwise "post processed" for use or display (Fig. J.2).

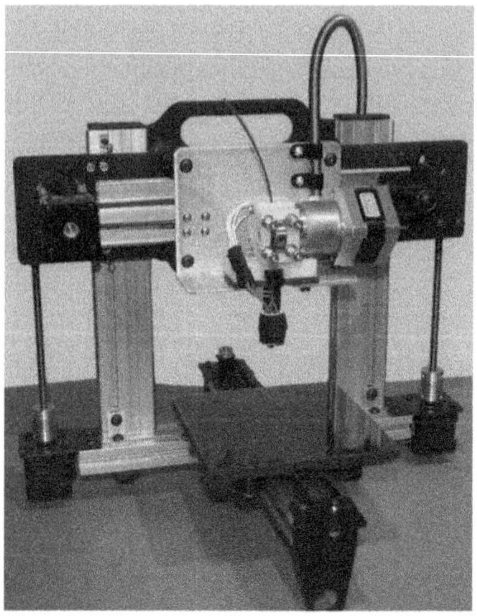

FIGURE J.1 A basic 3D printer. *Source: ORDbot Quantum.*

FIGURE J.2 A 3D-printed architectural model. *Source: Rietveld Architects NY.*

J.6 Applications, related issues, and the future of 3D printing

Needless to say, all of this just scratches the surface of this emerging technology. 3D printing has already found use in the manufacture of consumer apparel, automotive body panels, construction blocks, hand tools, medical implants, and of course extensive use in art, design, and for educational purposes (Fig. J.3).

The overall speed with which the model can be built is not all that fast, thereby limiting the usefulness of 3D printing in mass manufacturing. This, however, is likely to be overcome in the future; and already machines with multiple applicator heads are making use of faster curing materials. Other machines use multiple types of materials at one time.

The technology has had its share of controversy as well. With lower cost do-it-yourself 3D printers becoming widely available, issues of intellectual property, trademarks, and copyrights have arisen. There has been some concern as well of 3D home manufacturing of plastic guns that can evade detection at airports. Social commentators have written about a new approach to manufacturing, where almost anyone can be a producer of an in-demand product and a centralized factory no longer is employed.

As stated earlier—and it bears repeating—this is a rapidly expanding field and one to keep an eye on. This appendix will be updated from edition to edition as this technology matures and grows.

FIGURE J.3 A 3D-printed mechanical model. *Source: Stratasys.com.*

Appendix K

AutoCAD certification exams

It seems like there is an exam or certification or badge for just about every profession or occupation. Although AutoCAD and drafting in general are not really part of the greater information technology world, where certification exams are the benchmarks of competence and almost as valuable as college degrees, there still exists a series of exams for AutoCAD designers to take if they wish to "formalize" their knowledge, hang up a certificate in their office and share a digital badge from Acclaim in their online profiles.

These exams have changed significantly over the years and varied in the number of questions, difficulty level, and the amount of time given to complete them. The typical method of delivery today is online at a test center, with users logging in and taking the exam in a timed fashion, although, in years past, there have been exams with no time limit. You are NOT able to use the Help files, manuals or texts in the exam.

Are these exams important? This question does not have an easy answer. The belief among some established AutoCAD professionals is that the exams are quite limited and do not accurately assess an individual's true mastery of the software, which can really be tested only by having the person draw a complex design. It is partially Autodesk's fault for not developing these exams into the powerhouse institutions that Microsoft, Cisco, Novell, and others have done with theirs. There is however reason to believe that, at the very least, if you can answer the random set of questions on these exams, then you have a good grasp of a multitude of AutoCAD topics.

Do employers ask for these certifications? Historically, there was little demand for them in the past. As AutoCAD was exploding in popularity, anyone who was reasonably good was hired and allowed to mature on the job. In recent years, as the supply of knowledgeable AutoCAD designers finally caught up with the demand, employers wanted ready-made experts and needed a way to separate the "wheat from the chaff." This would explain the significant growth of testing for AutoCAD competence and the newfound interest from students as to how to study for and pass these exams. So, yes, these exams may be important to get a foot in the door with certain companies but having passed the exam does not necessarily guarantee a knowledgeable AutoCAD expert and should not be the sole reason to hire someone.

Let us go over the exams as they exist now for AutoCAD; it is anticipated that the latest exams will be similar, as they are meant to be release-independent. Two levels (or tiers) are available:

- AutoCAD Certified User
- AutoCAD Certified Professional
- Other Autodesk Software Exams
  - Fusion 360 Certified User
  - Revit Architecture Certified User
  - Revit Architecture Certified Professional
  - AutoCAD Civil 3D Certified Professional
  - 3ds Max Certified User
  - 3ds Max Certified Professional
  - Inventor Certified User
  - Inventor Certified Professional
  - Maya Certified User
  - Maya Certified Professional

The *AutoCAD Certified User exam* (ACU) verifies entry-level skills in key Autodesk products and is designed for students and instructors who wish to demonstrate basic proficiency. It is recommended that students have passed an AutoCAD course and have at least 50 hours of hands on experience with the software. The exam includes multiple-choice and performance-based questions. The following are some examples of topics covered in the exam. Note that these are subject to change without notice; use this listing as a sample guide only:

- Apply basic drawing skills
- Draw objects
- Draw with accuracy
- Modify objects
- Use Additional drawing techniques
- Organize objects
- Reuse existing content
- Annotate drawings
- Layouts and printing

The *AutoCAD Certified Professional* (ACP) validates more advanced skills, including complex workflow and design challenges and is designed for students seeking a competitive advantage in a specific product area. It is recommended that students have passed an AutoCAD course and have at least 400 hours of hands on experience with the software. The exam includes multiple-choice and performance-based questions. The following are some examples of topics covered in the exam. Note that these are subject to change without notice; use this listing as a sample guide only:

- Draw objects
  - Draw lines and rectangles
  - Draw circles, arcs, and polygons
- Draw with accuracy
  - Use object-snap tracking
  - Use Coordinate Systems
  - Make isometric drawings
- Modify objects
  - Move and copy objects
  - Rotate and scale objects
  - Create and use arrays
  - Trim and extend objects
  - Offset and mirror objects
  - Use grip editing
  - Fillet and chamfer objects
- Use additional drawing techniques
  - Draw and edit polylines
  - Blend between objects with splines
  - Apply hatches and gradients
- Organize objects
  - Change object properties
  - Alter layer assignments for objects
  - Control layer visibility
  - Assign properties by object or layer
  - Manage layer properties
- Reuse existing content
  - Work with blocks
  - Manage block attributes
  - Reference external drawings and images
- Annotate drawings
  - Add and modify text
  - Use dimensions
  - Add and modify multileaders
  - Create and assign annotative styles
  - Use tables

- Layouts and printing
  - Create layouts
  - Use viewports
  - Set printing and plotting options

Both exams are timed and you must earn at least 80% to pass. Practice sample tests are available, and it is highly recommended to go over these at least to see what types of questions to expect. The cost for taking these exams is around $75–$100 per exam, but that is subject to change, of course, so check the latest information on the Certiport website and with your local authorized testing center. Once you have passed the ACP, it is recommended that you maintain your certification by retaking the ACP exam once every 3 years.

Finally there is the *Autodesk Certified Instructor* (ACI) certification for those wishing to teach AutoCAD and have a credential to show those training centers that ask for it. As of this writing the basic requirements are listed here:

- Must be an Affiliated Instructor with an authorized, active Autodesk Learning Partner.
- Complete Autodesk Global Instructor Authorization Form and Approved Instructor requirements.
- Obtain an Autodesk Instructor ID Number from affiliated Learning Partner.
- Create an Autodesk Learning Central User Profile; log in to ALC; review site's navigation.
- Complete Autodesk Learning Central Course ID: CRS-16 -11138.
  - Course Subject Matter: continuing professional development training, improving instructional skills, and instructional methodologies
- Active Use of Autodesk Training Evaluation System (TES).
  - Instructor's active use must be within 3-months of Approved Instructor authorization.
  - Approved Instructor authorization is noted in Instructor's TES Profile.
  - Must enroll all Autodesk Software Product courses taught in TES.
  - Must provide course surveys to all training participants via TES.

Information on purchasing and scheduling your exam can be found at https://www.autodesk.com/certification. The exams themselves are administered by Certiport at https://certiport.pearsonvue.com/Certifications/Autodesk/Certifications/Overview.

Appendix L

AutoCAD employment

AutoCAD students vary widely in their backgrounds. Some are already in mid-career and are registered architects or engineers looking to pick up new skills. Others are recent college graduates entering the job market with architecture or engineering degrees. These students need AutoCAD as a small part of their job, or maybe even the main part of their job, but are ultimately hired for their engineering or design knowledge and may draft in only the first few years of their career as they learn the ropes.

Another very large group of students, however, are attracted to CAD drafting in and of itself and attend one of the many programs at local community colleges or tech/vocational schools that ultimately lead to a certificate or even a 2-year associate degree in architectural or engineering drafting. These students intend to enter the workforce as professional AutoCAD drafters, a career track. This can lead to CAD management positions as well as serve as a springboard for a career in engineering or architecture, if the students decide to return to school to complete the bachelor's or master's degree. Even if they do not, a professional draftsperson is a well-paid professional who is an asset and, as CAD software increases in sophistication, is needed more than ever. So the question is then: how do you get your "foot in the door" and begin and AutoCAD career?

Once you graduate and are ready to look for work, there are several things to consider. First, decide if you would like to concentrate on architectural, mechanical, electrical, or some other drafting field. Although I moved around a lot in my early (mid 1990s) days of consulting work, this approach does not work in today's saturated market. It is easier to succeed if you focus your interests instead of jumping around from company to company and field to field. What you ultimately decide to pursue should be something you are interested in (houses vs. mechanical devices, for example) as well as something you are good at. Generally, while learning AutoCAD in school, you are drawn to one type of work over another; go with your gut feeling. Then consider your options to find that type of work. Two major sets of resources are presented next:

- *Staffing firms*: This is one of the best ways to find a CAD job, though sometimes staffing firms prefer experienced CAD designers, as those command higher salaries and therefore higher commissions and fees for the staffing agencies. These staffing firms can either place you outright with a client (collecting a onetime finder's fee) or place you on an hourly rate as one of their own workers "on loan" to a client (collecting a premium over your pay). It may or may not matter which route you go, although being hired on directly with benefits may be more desirable, even if the initial salary is lower. Search for staffing firms (also informally called *job shops*) on the internet in your area. Some well-known, large, nationwide firms include Kelly Staffing, Manpower Technical, and Aerotek.
- *Job boards*: These include LinkedIn, Indeed, Glassdoor, and ZipRecruiter, among others. Often these sites are populated with postings from staffing firms (see the preceding), but independent companies can be found as well. Simply enter AutoCAD and your city or town into the search engine and listings appear. A unique resource is Craigslist.org. You may have already used it for everything from apartment hunting to selling something, and its job boards are excellent and very informal and direct. The only downside is that Craigslist.org is most utilized by residents and companies in large metropolitan centers, and if you live in a small town far from major cities, you may have slim pickings (even if the actual job scene is pretty good).

Other places to look include your school's employment office and of course social media. Sites such as Facebook and especially LinkedIn are a great way to get the word out that you are looking via a resume posting. And finally, never forget the value of word of mouth and personal contacts. Let everyone know you are looking for AutoCAD work, and just maybe someone's cousin's friend knows of someone looking to hire.

Appendix M

The future of AutoCAD

It is a bit of a gamble to try to nail down future trends in technology, and the world of computer-aided drafting and design is no exception. One must only look back a few years to find plenty of "experts" who were off in their predictions. New trends and innovations, appearing (and sometimes disappearing) rapidly, ensure that charting the long-term future is more guesswork than a hard science. Who would have predicted just a decade ago that AutoCAD would be again run on a Mac, that we would be storing our data and even collaborating "in the cloud," and that you would be able to draft on a hand-held tablet or easily see your designs in virtual reality or augmented reality?

Still, however, it is possible to discuss where AutoCAD, as well as the CAD industry in general, may be going in the coming years by sticking to the relatively near future. In this compressed timeframe of a few years we can look at technologies and trends that are already being implemented or will soon be here. Although the computer-aided drafting design industry can be considered relatively conservative, taking its time to implement new ideas (as opposed to the lightning fast telecom and entertainment industries), trends, and directions do emerge within it, and some fundamental changes are on the way. Let us take a closer look by first revisiting a bit of history.

M.1 2D design, AutoCAD, and verticals

Over its long history, AutoCAD has matured and grown into a stable, efficient, effective, and very popular 2D drafting application. By design, it was meant to be everything to everyone, and for many years it was. Spanning just about every design and manufacturing industry in existence, in its peak days (arguably 1994–2006), AutoCAD was found in the offices of nearly every small- to medium-size engineering and architectural firm. This was its greatest strength and, surprisingly, its greatest weakness. By being so "generic," it found wide application in many industries, but on the flip side, it was not specialized software and could not be made such without reworking its fundamental core.

Autodesk, of course, did not want to reinvent AutoCAD and risk losing its main audience, but AutoCAD was clearly not specialized enough for many uses. Autodesk remedied this by adding "verticals" of all kinds. This was add-on software that ran on top of AutoCAD and extended its functionality for the mechanical, architectural, civil, electrical, and other markets. With Autodesk's marketing muscle, these verticals proved relatively successful, and an entire generation of CAD designers learned Mechanical Desktop, Architectural Desktop (ADT for short), and many others, while competitors closed in with specialized software of their own. A pattern emerged where just about any engineering or architectural subfield had a choice between AutoCAD plus a vertical add-on or another competitor or two to choose from.

M.2 3D design and solid modeling

Meanwhile, a parallel world was evolving and thriving on its own, quite apart from AutoCAD and most types of architecture. This was the high-end and somewhat exclusive world of 3D solid modeling. For a select group of companies with the need (and the resources), using some off-the-shelf 2D application, no matter how good, was never an option. These industries included the aerospace, automotive, naval, and rail design, among others. The products these companies were designing were simply too complex to invest in 2D drafting. When they left pencil and paper behind in the 1960s and 1970s, they moved right into the world of 3D by commissioning internal development (or purchase) of high-end solid modeling software. Though true 3D software did not appear until the early 1980s, such was the beginning of applications such as Unigraphics, CATIA, and their predecessor, the famous CADAM, a Lockheed staple.

These 3D modeling software applications, including later products such as Pro/Engineer (later renamed to Creo), NX (a merger of Unigraphics and I-Deas), SolidWorks, and others, had features and capabilities that were specialized to high-end engineering design, and they introduced new concepts along the way. For example, when designing a product with hundreds of parts (most of them interacting with each other), techniques such as part trees, parent/child relationships, and parametric design had to be invented to keep track of everything, including interferences, bills of materials, and much more. It made no sense to do anything in 2D, as parts had to be visualized and virtual models had to be built, analyzed, and tested. It was only at the very end of the process that 2D views of individual parts were produced and given to machinists to manufacture.

With the advent of CNC (computer numerical code) even that was no longer necessary, and information could be transmitted directly from a model to the manufacturing machine, with a machinist overseeing setup and quality control. All these techniques developed slowly and deliberately in response to a clear need in the industry. These software applications were complex, expensive, and virtually unrecognizable to an AutoCAD user. They also often required specialized training, and to this day, you generally do not sit down to learn CATIA or NX unless you are a working (or student) engineer. There really was no industry of pure drafters as there was with AutoCAD. You had to know what you were doing in your specialty and how to interpret the results of your design or analysis.

M.3 Merging of industries

If you have studied the 3D chapters of this textbook, you may already see where this narrative is going. It was inevitable that the folks from the 2D architecture, engineering, and construction world would look at their peers in the 3D world and ask, "Can we also do this?" The fundamental premise they saw was very simple: We all live in a 3D world, and this is the way all design should be done. Nowhere was this more evident than in the world of architecture. As early as the late 1980s, visionaries such as Frank Geary commissioned Dassault, the makers of CATIA, to modify their product for architecture/building design. Architects all over the world still designs everything in 2D (some still via pencil and paper), and 3D modeling was the rare exception, not the rule. Because AutoCAD was the undisputed software of choice among architects and interest was building in a 3D approach, the ultimate clash of industries was bound to happen.

Autodesk's response was to borrow several techniques and incorporate them into the 2D and 3D capabilities of AutoCAD, a fact mentioned in a few chapters of this text. Dimension-driven design (see , Chapter 13) and 3D-to-2D layouts (see Chapter 27 of the 3D sections) are just some of the features that filtered their way down from solid modeling into plain AutoCAD. Clearly, however, AutoCAD itself was not fundamentally changing; and another approach was needed to fill the need of customers who wanted to take things into the 3D world, both in engineering and architecture. As they were doing this, competitors wasted no time in pushing their own solutions. In the world of solid modeling, numerous mid- to low-priced products for engineers emerged, such as IronCAD, IntelliCAD, SolidWorks, and Solid Edge. In the world of architecture and design, ArchiCAD and even Rhino touted their 3D capabilities. Specialized software was also available from MicroStation, and these companies captured a significant portion of the civil engineering market. Autodesk came back strong with Inventor, a successful solid modeling software of its own. But the biggest jump toward merging 2D and 3D design was yet to come.

M.4 Revit and BIM

If migrating to 3D was all there was to architecture and building design that would have been done a long time ago. Something else was missing from AutoCAD and similar 2D programs, a set of features that had been present in 3D solid modeling software for quite some time. The architecture world wanted those features as much, if not more, than the 3D capabilities alone. These features are best described as "relational databases" and "parametric change engines."

Simply put, solid modeling software had far more intelligence than drafting software. Objects in solid modeling were all part of a database that tracked their positions and attributes. Everything created was part of something bigger, and everything was linked together. If you created a bracket in 3D, the 2D views were already part of that bracket, the part was tracked in a bill of materials, and its physical properties were also cataloged. If you wanted a section view, e.g., it was just a click away. The bracket was not just a collection of lines, it was a dynamic entity.

To bring these features to the architecture world was the premise behind Revit, a "solid-modeling-for-architects" software acquired by Autodesk in 2002 (originally developed by Revit Technology Corporation). Technically, Revit was what is referred to as building information modeling (BIM) software. This type of software goes one huge step beyond 2D line drafting of a facility by digitally representing all its physical and functional characteristics in a powerful relational database. This type of software is useful not just during the design phase but throughout the entire life cycle of the facility, including its management and operation. Revit (which is short for REVise InsTantly) is, of course, not the only such software on the market; and MicroStation, ArchiCAD, Vectorworks, ARCHIBUS, and CATIA's Digital Project are all viable competitors. Revit, however, is the leading software among its peers and is credited with bringing BIM to a new generation of users.

Revit, and other available BIM software, has revolutionized building and facility design. Solid building models are first constructed immediately in 3D. Then, section, details, and a bill of materials can easily be generated by "slicing" the model up in to see the 2D views. All views are part of the one model, and any changes are instantly updated throughout the entire project. An architect/designer can now keep track of all entities (doors, windows, etc.) in the design, and link cost and materials to them. Square footages and even energy usages can now be tracked in real time as the design progresses. Time and cost are also factored in (the fourth and fifth dimensions in BIM). Finally, Revit has advanced rendering tools and can produce excellent rendered images without third party software.

All this was already incorporated into mechanical design for years, but only relatively recently was it widely available to architects. Considering the advantages, such as waste minimization, shorter turnaround times, and better error detection, it is surprising this was not available sooner. There are, of course, some disadvantages to BIM software such as Revit, with the chief among them being the significant retraining and upfront investment needed. It may also be not the right fit for very small projects, where basic AutoCAD is more than enough.

M.5 The future of AutoCAD

So, after all this discussion of solid modeling, Revit, and BIM, what is the future of AutoCAD? It is probably better to split this into several questions: What is the future of AutoCAD itself, and what is the future of computer-aided design in general, considering this overall trend toward 3D? Will AutoCAD stay the way it is? Will it morph into a different software altogether?

The consensus is a resounding "No!" AutoCAD enjoys a very special place in the overall scheme of things. It is a very well-established and successful 2D drafting package. There will always be a need for this software in its present form. Too many clients in the design world require nothing more than what AutoCAD already offers. By being so good at the simple and basic task of drafting, AutoCAD will remain relevant for the foreseeable future. It is unlikely Autodesk will significantly tinker with its formula for success and its main cash cow. Companies have invested 10, 20, or even 30+ years in their AutoCAD systems, standards, and libraries, and it is difficult to just give that up to go to a new system if your current system is efficient and profitable.

However, Autodesk is changing AutoCAD in small ways. It is, e.g., incorporating as much advanced 3D as possible without fundamentally changing the identity of the software. Autodesk is also doing its best to keep AutoCAD looking modern and cutting edge by incorporating such design techniques as "heads up" display for data entry. It is also trying its best to eliminate the archaic-looking command line—but for the sake of the legacy users, the command line is expected to remain available, if needed. AutoCAD has also followed basic software modernization trends over the years, such as incorporating the Microsoft Office–like Ribbon Interface design style and being offered for sale via download as opposed to physical DVDs or CDs.

The industry itself, however, is moving ever slowly, but surely, toward 3D solid modeling and BIM. A wave is spreading across the architectural profession to incorporate BIM into basic design from the ground up and not treat this approach as something exotic but rather as a new fundamental approach. Autodesk clearly recognizes this trend, and its Revit software is well positioned in the market. Expect to see this trend continue, and if you are inclined toward architectural drafting, you may be well advised to also learn Revit or another BIM-oriented software, such as ArchiCAD, Digital Project Designer, or Bentley Architecture.

In summary, it is doubtful AutoCAD itself will ever morph into another software under pressure from the industry trends; it is simply too profitable, and it will remain relevant "as is" for years to come. Autodesk will continue its yearly update and "sprucing up" of the interface, improving the basic commands, and adding more and more 3D features up to the limit of its ACIS modeling kernel, but its core will remain the same. AutoCAD's market share however will likely shrink in the coming years, with some of that going to Revit and other BIM software. Some mechanical engineers will also switch to 3D, but since that type of modeling software was already available for a long time, the main drift away from AutoCAD will be in the architectural world, not engineering.

As with any predictions, it remains to be seen how things will turn out, but the consensus is that the just-discussed scenarios are the most likely for the immediate future.

Index

Note: 'Page numbers followed by "*f*" indicate figures and "*t*" indicate tables.'

B

C

D

E

I

J

K

L

Q

R